Lecture Notes in Computer Science   1450
Edited by G. Goos, J. Hartmanis and J. van Leeuwen

Springer
*Berlin
Heidelberg
New York
Barcelona
Budapest
Hong Kong
London
Milan
Paris
Singapore
Tokyo*

Luboš Brim Jozef Gruska
Jiří Zlatuška (Eds.)

Mathematical Foundations of Computer Science 1998

23rd International Symposium, MFCS'98
Brno, Czech Republic, August 24-28, 1998
Proceedings

Springer

Series Editors

Gerhard Goos, Karlsruhe University, Germany
Juris Hartmanis, Cornell University, NY, USA
Jan van Leeuwen, Utrecht University, The Netherlands

Volume Editors

Luboš Brim
Jozef Gruska
Jiří Zlatuška
Masaryk University, Faculty of Informatics
Botanická 68a, 602 00 Brno, Czech Republic
E-mail: {brim, gruska, zlatuska}@fi.muni.cz

Cataloging-in-Publication data applied for

Die Deutsche Bibliothek - CIP-Einheitsaufnahme

Mathematical foundations of computer science 1998 : 23rd
international symposium ; proceedings / MFCS '98, Brno, Czech
Republic, August 24 - 28, 1998. Luboš Brim ... (ed.). - Berlin ;
Heidelberg ; New York ; Barcelona ; Budapest ; Hong Kong ;
London ; Milan ; Paris ; Singapore ; Tokyo : Springer, 1998
   (Lecture notes in computer science ; Vol. 1450)
   ISBN 3-540-64827-5

CR Subject Classification (1991): F, C.2, G.2

ISSN 0302-9743
ISBN 3-540-64827-5 Springer-Verlag Berlin Heidelberg New York

This work is subject to copyright. All rights are reserved, whether the whole or part of the material is concerned, specifically the rights of translation, reprinting, re-use of illustrations, recitation, broadcasting, reproduction on microfilms or in any other way, and storage in data banks. Duplication of this publication or parts thereof is permitted only under the provisions of the German Copyright Law of September 9, 1965, in its current version, and permission for use must always be obtained from Springer-Verlag. Violations are liable for prosecution under the German Copyright Law.

© Springer-Verlag Berlin Heidelberg 1998
Printed in Germany

Typesetting: Camera-ready by author
SPIN 10638164   06/3142 – 5 4 3 2 1 0   Printed on acid-free paper

Foreword

The 23rd International Symposium on the Mathematical Foundations of Computer Science (MFCS'98) was held in Brno, Czech Republic, during August 24–28, 1998.

It was organized at Masaryk University in Brno by the Faculty of Informatics in co-operation with universities in Aachen, Haagen, Linz, Metz, Pisa, Szeged, Vienna, and other institutions. MFCS'98 formed one part of a federated conferences event, the other part being CSL'98, the annual conference of the European Association of Computer Science Logic. This federated conferences event consisted of common plenary sessions, invited talks, several parallel technical programme tracks, a dozen satellite workshops organised in parallel with MFCS'98, and tutorials.

MFCS'98 was the 23rd in a series of conferences organized on a rotating basis between the Czech Republic, Poland, and Slovakia, aiming at bringing together specialists in various fields of theoretical computer science and stimulating mathematical research in relevant areas. Previous meetings of the series took place in Jabłonna, 1972; Štrbské Pleso, 1973; Jadwisin, 1974; Mariánské Lázně, 1975; Gdańsk, 1976, Tatranská Lomnica, 1977; Zakopane, 1978; Olomouc, 1979; Rydzyna, 1980; Štrbské Pleso, 1981; Prague, 1984; Bratislava, 1986; Carlsbad, 1988; Porąbka-Kozubnik, 1989; Banská Bystrica, 1990; Kazimierz Dolny, 1991; Prague, 1992; Gdańsk, 1993; Košice, 1994; Prague, 1995; Kraków, 1996; and Bratislava, 1997.

MFCS'98 marked the 25th anniversary of the first MFCS meeting which took place in Czechoslovakia at Štrbské pleso. MFCS'73 is remembered for taking a very broad, advanced, and stimulating view of the theoretical foundations of computing, and for the high scientific and organizational standard. Preparation of MFCS'98 and its satellite events was undertaken with the intention of continuing in this honorable tradition.

There were 168 submissions sent to the program committee in response to the call for papers, which was distributed primarily electronically to major computer science departments, individual researchers, and electronic mailing lists.

All but one submissions were received electronically over the Internet. Every paper was assigned to three program committee members for review and referee reports were collected electronically using e-mail or WWW-based forms from the individual referees over the period March 25 – May 11, 1998.

Program committee meetings were organized in a distributed way, based on a very efficient combination of electronic, telephone, and physical meetings and break-away discussions devoted to borderline or unclear cases. Clear timelines were maintained for individual steps of the process and these were used both by program committee members meeting physically and by those joining the selection meeting electronically.

Electronic pre-meeting discussions started on May 12, 1998 based on an electronic-mailing-list-enabled discussion of paper evaluations and and on completion of the reviews so that every paper had been reviewed by at least three separate referees. The fully-electronic part of the meeting concluded on May 15. During the weekend of May 15 and 16, five program committee members (denoted by an asterisk in the program committee members listing) met at the Faculty of Informatics in Brno. They were provided with Internet access using several computers, telephone and fax lines, and conducted a very careful selection of 72 papers eventually selected for conference presentation.

The selection process concluded on May 16 at 15:00 after having passed through most stages of the selection procedure using almost entirely electronic contact with nearly all members of the program committee. These were continuously provided with information concerning the actual state of the selection process, and they returned their reactions and opinions using e-mail, telephone, and fax, as the basis for the ultimate decisions.

Based on the information already known at the date of writing this, the Federated CSL/MFCS'98 Conferences event consisted of more than two hundred talks presented within up to eight parallel technical tracks including 34 invited talks for four plenary CSL/MFCS sessions, more than 10 invited talks for MFCS, five for CSL, and still others for the satellite workshops (taking typically 2–3 days each). Out of these, 10 invited and 71 submitted MFCS talks are presented in this volume. Last but not least, four tutorials were organized in the two days preceding and following the symposium: *Abstract state machines* by E. Börger (Pisa) and Yu. Gurevich (Ann Arbor), *The Theorema system: An introduction with demonstrations* by B. Buchberger (RISC–Linz) and T. Jebelan (RISC–Linz), *Approximation algorithms* by P. L. Crescenzi (Florence), J. Díaz (Rome), and A. Marchetti-Spaccamela (Rome), and *Quantum computing and quantum logic* by C. H. Bennett (IBM T. Watson Center, Yorktown Heights) and K. Svozil (Vienna).

The main organizer of the Federated CSL/MFCS'98 Conferences was the Faculty of Informatics of Masaryk University, the very first specialized faculty of its kind established in the Czech Republic four years ago. The Organizing Committee was chaired by Jan Staudek.

Special thanks go to Antonín Kučera as the program committee secretary and to Vladimiro Sassone who supplied the WWW-based system which was used to conduct most of the work in an electronic environment. Without these, the program committee's task of formulating a really outstanding program (given the volume of high-quality submissions) would have been immensely more complicated.

Luboš Brim as co-editor has performed the principal editing work needed in connection with collecting the final versions of the papers and tidying things up for the final appearance of the MFCS'98 proceedings as a Springer LNCS volume using LNCS LaTeX style.

Last but not least, we would like to express our thanks to the invited speakers, the authors of contributed papers, tutorial presenters and also to the workshop speakers and organizers for contributing significantly and setting new bounds for the scope and size of MFCS'98.

Brno, June 1998
                                    Jozef Gruska and Jiří Zlatuška

MFCS'98 Program Committee

S. Abramsky (*Edinburgh*)
J. Diaz (*Barcelona*)
J. Gruska, co-chair (*Brno*)*
T. Henzinger (*Berkeley*)
G. Mirkowska (*Pau*)
U. Montanari (*Pisa*)
M. Paterson (*Warwick*)
J. Sgall (*Prague*)*
J. Tiuryn (*Warsaw*)
P. Vitányi (*Amsterdam*)
M. Wirsing (*Munich*)*

B. Buchberger (*Linz*)
V. Diekert (*Stuttgart*)
I. Guessarian (*Paris*)
R. J. Lipton (*Princeton*)
F. Moller (*Uppsala*)
J. Nešetřil (*Prague*)
G. Păun (*Bucharest*)
W. Thomas (*Kiel*)
U. Vaccaro (*Salerno*)*
P. Voda (*Bratislava*)
J. Zlatuška, co-chair (*Brno*)*

MFCS'98 Organizing Committee

M. Brandejs
L. Brim
T. Dudaško
J. Foukalová
I. Hollanová
D. Janoušková
A. Kučera
L. Motyčková

J. Obdržálek
M. Povolný
P. Smrž
P. Sojka
J. Srba
J. Staudek, *chair*
P. Starý
Z. Walletzká

Referees

S. Abramsky
L. de Alfaro
J-P. Allouche
N. Alon
Th. Altenkirch
R. Alur
C. Alvarez
E. Asarin
V. Auletta
R. Backofen
J. Balcazar
R. Banach
M. Bauderon
B. Bauer
D. Beauquier
B. Berard
J. Berstel
C. Blundo
L. Boasson
M. Bonet
F. Brandenburg
L. Brim
V. Bruyere
B. Buchberger
H. Buhrman
G. Buntrock
D. Caucal
P. Cegielski
P. Cenciarelli
B. Chlebus
C. Choffrut
E. Contejean
B. Courcelle
P. Cousot
K. Čulík
J. Dassow
P. Degano
A. Degtyarev
J. Desel
M. Dezani
J. Diaz

V. Diekert
W. Drabent
M. Droste
B. Durand
S. Edwards
F. Esposito
C. De Felice
H. Fernau
M. Fisher
R. Freivalds
C. Frougny
J. Gabarro
L. Gargano
P. Gastin
R. Gavalda
D. Giammarresi
P. Di Gianantonio
R. Gilleron
S. Gilmore
F. Gire
R. van Glabbeek
V. Glasnák
S. Gnesi
W. Goerigk
P. Goldberg
E. Graedel
S. Grigorieff
J.-F. Groote
D. P. Gruska
J. Gruska
I. Guessarian
D. Guijarro
D. Guller
R. Hennicker
T. Henzinger
U. Hertrampf
J. Hromkovič
P. Indyk
P. Jančar
K. Jansen
M. Jantzen

T. Jebelean
S. Kalvala
J. Karhumäki
J. Kari
B. Kirsig
R. Klasing
E. P. Klement
A. Knapp
I. Korec
P. Kosiuczenko
J. Krajíček
H. Kreowski
M. Křetínský
D. Krizanc
A. Kučera
M. Kudlek
W. Kuich
M. Kunde
O. Kupferman
A. Kurz
Ch. Lueth
K.-J. Lange
E. Laporte
S. Lasota
V. Laurent
J. van Leeuwen
G. Lenzi
S. Leonardi
B. Leoniuk
R. Lipton
M. Lohrey
A. Lopes
A. de Luca
C. Lüth
G. Manzini
G. De Marco
L. Margara
M. Marin
B. Martin
C. Martinez
A. Masini

O. Matz
G. Mauri
J. Mazoyer
P.-A. Mellies
S. Merz
B. Meyer
P. Michel
G. Mirkowska
F. Moller
U. Montanari
A. Muscholl
M. Napoli
Ph. Narbel
J. Nešetřil
R. De Nicola
D. Niwinski
M. Novotný
S.-O. Nyström
P. Olveczky
C.-H. L. Ong
A. Osterloh
L. Pacholski
J. Padberg
D. Pardubská
M. Parente
M. Paterson
D. Pattinson
P. Pau
G. Păun
G. Persiano
H. Petersen
G. Pighizzini
R. Pinzani
M. Plátek
A. Podelski
L. Polák
M. Prasad
R. De Prisco

P. Pudlák
S. K. Rajamani
J. Rehof
K. Reinhardt
A. Restivo
B. Reus
J. M. Robson
H. Rolletschek
L. Rosaz
J. Rosický
M. de Rougemont
P. Rozière
W. Rytter
C. Sahinalp
J. Sakarovitch
A. Salwicki
A. De Santis
P. Savicky
V. Scarano
J. Schicho
I. Schiering
Ph. Schnoebelen
A. Schoenegge
W. Schreiner
A. Schubert
J. Šefránek
S. Seibert
G. Sénizergues
M. Serna
J. Sgall
P. Sgall
R. Silvestri
L. Škarvada
K. Skodinis
P. Sosík
P. Spirakis
B. Sprick
L. Staiger

M. Staněk
J. Steinbach
M. Steinby
P. Stevens
C. Stirling
H. Stoerrle
J. Šturc
W. Thomas
S. Tiga
S. Tison
J. Tiuryn
E. Tomuta
J. Tromp
J. Tyszkiewicz
U. Vaccaro
E. Valkema
D. Vasaru
M. Veanes
B. Victor
P. Vitanyi
P. Voda
H. Vogler
I. Walukiewicz
A. Weiermann
J. Wiedermann
Th. Wilke
M. Wirsing
A. Woods
Th. Worsch
H. Yassine
Sheng Yu
J-B. Yunes
S. Žák
M. Zawadowski
Li Zhang
K. Zikan
J. Zlatuška
A. Zvonkine

Table of Contents

Invited Papers

Hypergraph Traversal Revisited: Cost Measures and Dynamic Algorithms . . 1
  G. Ausiello, G. F. Italiano, and U. Nanni

Defining the Java Virtual Machine as Platform for Provably Correct Java
Compilation 17
  E. Börger, W. Schulte

Towards a Theory of Recursive Structures 36
  D. Harel

Modularization and Abstraction: The Keys to Practical Formal Verification 54
  Y. Kesten, A. Pnueli

On the Role of Time and Space in Neural Computation 72
  W. Maass

From Algorithms to Working Programs: On the Use of Program Checking
in LEDA 84
  K. Mehlhorn, S. Näher

Computationally-Sound Checkers 94
  S. Micali

Reasoning About the Past 117
  M. Nielsen

Satisfiability – Algorithms and Logic 129
  P. Pudlák

The Joys of Bisimulation 142
  C. Stirling

Towards Algorithmic Explanation of Mind Evolution and Functioning 152
  J. Wiedermann

Contributed Papers

Complexity of Hard Problems

Combinatorial Hardness Proofs for Polynomial Evaluation 167
  M. Aldaz, J. Heintz, G. Matera, J. L. Montaña, and L. M. Pardo

Minimum Propositional Proof Length is NP-Hard to Linearly
Approximate 176
    M. Alekhnovich, S. Buss, S. Moran, and T. Pitassi

Reconstructing Polyatomic Structures from Discrete X-Rays:
NP-Completeness Proof for Three Atoms 185
    M. Chrobak, Ch. Dürr

Locally Explicit Construction of Rődl's Asymptotically Good Packings ... 194
    N. Kuzjurin

Logic – Semantics – Automata

Proof Theory of Fuzzy Logics: Urquhart's **C** and Related Logics 203
    M. Baaz, A. Ciabattoni, Ch. Fermüller, and H. Veith

Nonstochastic Languages as Projections of 2-Tape Quasideterministic
Languages 213
    R. Bonner, R. Freivalds, J. Lapiņš, and A. Lukjanska

Flow Logic for Imperative Objects 220
    F. Nielson, H. R. Nielson

Expressive Completeness of Temporal Logic of Action 229
    A. Rabinovich

Rewriting

Reducing AC-Termination to Termination........................... 239
    M. C. F. Ferreira, D. Kesner, and L. Puel

On One-Pass Term Rewriting....................................... 248
    Z. Fülöp, E. Jurvanen, M. Steinby, and S. Vágvölgyi

On the Word, Subsumption, and Complement Problem for Recurrent Term
Schematizations 257
    M. Hermann, G. Salzer

Encoding the Hydra Battle as a Rewrite System 267
    H. Touzet

Automata and Transducers

Computing ϵ-Free NFA from Regular Expressions in $O(n \log^2(n))$ Time ... 277
    Ch. Hagenah, A. Muscholl

Iterated Length-Preserving Rational Transductions..................... 286
    M. Latteux, D. Simplot, and A. Terlutte

The Head Hierarchy for Oblivious Finite Automata with Polynomial
Advice Collapses 296
  H. Petersen

The Equivalence Problem for Deterministic Pushdown Transducers into
Abelian Groups 305
  G. Sénizergues

Typing

The Semi-Full Closure of Pure Type Systems 316
  G. Barthe

Predicative Polymorphic Subtyping 326
  M. Benke

A Computational Interpretation of the $\lambda\mu$-Calculus 336
  G. M. Bierman

Polymorphic Subtyping Without Distributivity 346
  J. Chrząszcz

Concurrency – Semantics – Logic

A (Non-elementary) Modular Decision Procedure for LTrL 356
  P. Gastin, R. Meyer, and A. Petit

Complete Abstract Interpretations Made Constructive 366
  R. Giacobazzi, F. Ranzato, and F. Scozzari

Timed Bisimulation and Open Maps 378
  Th. Hune, M. Nielsen

Deadlocking States in Context-Free Process Algebra.................. 388
  J. Srba

Circuit Complexity

A Superpolynomial Lower Bound for a Circuit Computing the Clique
Function with At Most $(1/6)\log\log n$ Negation Gates.................. 399
  K. Amano, A. Maruoka

On Counting AC^0 Circuits with Negative Constants..................... 409
  A. Ambainis, D. M. Barrington, and H. LêThanh

A Second Step Towards Circuit Complexity-Theoretic Analogs of Rice's
Theorem 418
  L. A. Hemaspaandra, J. Rothe

Programming

Model Checking Real-Time Properties of Symmetric Systems 427
    E. A. Emerson, R. J. Trefler

Locality of Order-Invariant First-Order Formulas 437
    M. Grohe, T. Schwentick

Probabilistic Concurrent Constraint Programming: Towards a Fully
Abstract Model 446
    A. Di Pierro, H. Wiklicky

Lazy Functional Algorithms for Exact Real Functionals 456
    A. K. Simpson

Structural Complexity

Randomness vs. Completeness: On the Diagonalization Strength of
Resource-Bounded Random Sets 465
    K. Ambos-Spies, S. Lempp, and G. Mainhardt

Positive Turing and Truth-Table Completeness for NEXP Are
Incomparable 474
    L. Bentzien

Tally NP Sets and Easy Census Functions 483
    J. Goldsmith, M. Ogihara, and J. Rothe

Average-Case Intractability vs. Worst-Case Intractability 493
    J. Köbler, R. Schuler

Formal Languages

Shuffle on Trajectories: The Schützenberger Product and Related
Operations 503
    T. Harju, A. Mateescu, and A. Salomaa

Gaußian Elimination and a Characterization of Algebraic Power Series 512
    W. Kuich

D0L-Systems and Surface Automorphisms 522
    L.-M. Lopez, P. Narbel

About Synchronization Languages 533
    I. Ryl, Y. Roos, and M. Clerbout

Graphs and Hypergraphs

When Can an Equational Simple Graph Be Generated by Hyperedge
Replacement ? 543
   K. Barthelmann

Spatial and Temporal Refinement of Typed Graph Transformation
Systems 553
   M. Große–Rhode, F. Parisi–Presicce, and M. Simeoni

Approximating Maximum Independent Sets in Uniform Hypergraphs 562
   Th. Hofmeister, H. Lefmann

Representing Hyper-Graphs by Regular Languages 571
   S. La Torre, M. Napoli

Turing Complexity and Logic

Improved Time and Space Hierarchies of One-Tape Off-Line TMs 580
   K. Iwama, Ch. Iwamoto

Tarskian Set Constraints Are in NEXPTIME 589
   P. Mielniczuk, L. Pacholski

$\forall\exists^*$-Equational Theory of Context Unification is Π_1^0-Hard 597
   S. Vorobyov

Speeding–Up Nondeterministic Single–Tape Off-Line Computations by One
Alternation 607
   J. Wiedermann

Binary Decision Diagrams

Facial Circuits of Planar Graphs and Context-Free Languages 616
   B. Courcelle, D. Lapoire

Optimizing OBDDs Is Still Intractable for Monotone Functions 625
   K. Iwama, M. Nozoe, and S. Yajima

Blockwise Variable Orderings for Shared BDDs 636
   H. Preuß, A. Srivastav

On the Composition Problem for OBDDs with Multiple Variable Orders .. 645
   A. Slobodová

Combinatorics on Words

Equations in Transfinite Strings 656
   Ch. Choffrut, S. Horvath

Minimal Forbidden Words and Factor Automata 665
    M. Crochemore, F. Mignosi, and A. Restivo

On Defect Effect of Bi-Infinite Words 674
    J. Karhumäki, J. Maňuch, and W. Plandowski

On Repetition-Free Binary Words of Minimal Density 683
    R. Kolpakov, G. Kucherov, and Y. Tarannikov

Trees and Embeddings

Embedding of Hypercubes into Grids 693
    S. L. Bezrukov, J. D. Chavez, L. H. Harper, M. Röttger,
    and U.-P. Schroeder

Tree Decompositions of Small Diameter 702
    H. L. Bodlaender, T. Hagerup

Degree-Preserving Forests 713
    H. Broersma, A. Huck, T. Kloks, O. Koppius, D. Kratsch, H. Müller,
    and H. Tuinstra

A Parallelization of Dijkstra's Shortest Path Algorithm 722
    A. Crauser, K. Mehlhorn, U. Meyer, and P. Sanders

Picture Languages – Function Systems/Complexity

Comparison Between the Complexity of a Function and the Complexity
of Its Graph 732
    B. Durand, S. Porrot

IFS and Control Languages 740
    H. Fernau, L. Staiger

One Quantifier Will Do in Existential Monadic Second-Order Logic over
Pictures 751
    O. Matz

On Some Recognizable Picture-Languages 760
    K. Reinhardt

Communication – Computable Real Numbers

On the Complexity of Wavelength Converters 771
    V. Auletta, I. Caragiannis, Ch. Kaklamanis, and P. Persiano

On Boolean vs. Modular Arithmetic for Circuits and Communication
Protocols 780
    C. Damm

Communication Complexity and Lower Bounds on Multilective
Computations 789
    J. Hromkovič

A Finite Hierarchy of the Recursively Enumerable Real Numbers 798
    K. Weihrauch, X. Zheng

Cellular Automata

One Guess One–Way Cellular Arrays 807
    Th. Buchholz, A. Klein, and M. Kutrib

Topological Definitions of Chaos Applied to Cellular Automata Dynamics . 816
    G. Cattaneo, L. Margara

Characterization of Sensitive Linear Cellular Automata with Respect to
the Counting Distance 825
    G. Manzini

Additive Cellular Automata over $\mathbb{Z}_p$ and the Bottom of (CA,$\leq$) 834
    J. Mazoyer, I. Rapaport

Author Index 845

Hypergraph Traversal Revisited: Cost Measures and Dynamic Algorithms*

Giorgio Ausiello[1], Giuseppe F. Italiano[2], and Umberto Nanni[1]

[1] Dipartimento di Informatica e Sistemistica
Università di Roma "La Sapienza", Italy
{ausiello,nanni}@dis.uniroma1.it
[2] Dipartimento di Matematica Applicata ed Informatica
Università "Ca' Foscari" di Venezia, Italy
italiano@dsi.unive.it, http://www.dsi.unive.it/~italiano

Abstract. Directed hypergraphs are used in several applications to model different combinatorial structures. A *directed hypergraph* is defined by a set of nodes and a set of *hyperarcs*, each connecting a set of *source* nodes to a single *target* node. A *hyperpath*, similarly to the notion of path in directed graphs, consists of a connection among nodes using hyperarcs. Unlike paths in graphs, however, hyperpaths are suitable of many different definitions of measure, corresponding to different concepts arising in various applications. In this paper we consider the problem of finding optimal hyperpaths according to several measures. We also provide results that may shed some light on the intrinsic complexity of finding optimal hyperpaths.

1 Introduction

A *directed hypergraph* is a generalization of the concept of directed graph. It was first introduced in [2] to represent functional dependencies in relational data base schemata. While directed graphs are normally used for representing *one-to-one* functional relations over finite sets, in several areas of computer science the need for more general types of functional relations arises, such as *many-to-one* relations or even functional relations with variable arity. A directed hypergraph is useful exactly in these scenarios: it consists of a finite set of nodes N and a set of *hyperarcs*; a hyperarc generalizes the notion of a graph edge and is defined by a pair: a nonempty set of nodes $S \subseteq N$ and a node $i \in N$. Intuitively, a hyperarc from $i_1, \ldots, i_k$ to j models a function relating $i_1, \ldots, i_k$ to j. Clearly, a directed graph is a special case of directed hypergraph. Classical examples of combinatorial structures that can be easily represented by hypergraphs include functional dependencies in relational databases [2] (where an hyperarc from $A_1, \ldots, A_k$ to B represents a functional dependency between attributes $A_1, \ldots, A_k$ and attribute B), Horn formulae in propositional calculus [6] (in which a hyperarc represents

---

* Work supported in part by EU ESPRIT Long Term Research Project ALCOM-IT under contract no. 20244.

a Horn clause $p_1, \ldots, p_k \to q$, where $p_1, \ldots, p_k$ and q are propositional symbols), implications in problem solving (where hypergraphs can be used as an alternative to and-or graphs), Datalog [10], Operations Research [9], and Petri Nets [1].

In several applications of directed hypergraphs the notions of path and of traversal are required. Intuitively, there is a hyperpath from a set of nodes S to a single node t if there is a hyperarc from a set Z to t and if there exist hyperpaths from S to any of the nodes in Z. While paths and shortest paths in directed graphs are standard concepts, and efficient algorithms for their computation are well known, for hypergraphs the corresponding notions of hyperpaths and shortest hyperpaths are very subtle, and defining cost measures of hyperpaths turns out to be a more delicate issue. Hyperpaths can be measured and compared according to a much broader range of measures than simple graphs, and indeed several measures have been already proposed in the literature (see e.g., [3,5,11,12,14,15]). As shown in [5], some of these measures lead, not surprisingly, to NP-hard optimization problems; however, if the measure function on hyperpaths matches certain conditions, the problems turn out to be solvable in polynomial time.

In this paper, we provide a uniform framework for metrics in directed hypergraphs. We define a general notion of measure function on hyperpaths, and specialize it to some of the most intuitive measures, some of which have already been considered in the literature. Furthermore, we relate the hardness of the optimization problem to the presence of cycles in optimal hyperpaths: if optimal hyperpaths for the problem considered contain no cycles or very few cycles, then the problem can be solved in polynomial time.

2 Directed Hypergraphs and Their Metrics

In this section we introduce the notion of directed hypergraph, hyperpath, and define some cost measures on hyperpaths.

Definition 1. *A directed hypergraph $\mathcal{H}$ is a pair $\langle N, H \rangle$ where N is a set of nodes and H is a set of hyperarcs. Each hyperarc is an ordered pair $\langle S, t \rangle$ from an arbitrary nonempty set $S \subseteq N$ (source set) to a single node $t \in N$ (target node).*

Note that a directed graph is a special case of directed hypergraph, where all the source sets have cardinality one. There are several parameters which can be taken into account for directed hypergraphs:
- The *number of nodes*: $n = |N|$;
- The *number of hyperarcs*: $h = |H|$;
- The *source area* a, that is the sum of cardinalities of all the source sets: $a = \sum_{S \in \mathcal{S}} |S|$, where $\mathcal{S}$ denotes the set of source sets: $\mathcal{S} = \{S \mid \langle S, t \rangle \in H \text{ for some } t \in N\}$;
- The *nonsingleton source area* a', that is the sum of cardinalities of all the nonsingleton source sets: $a' = \sum_{S \in \mathcal{S}_M} |S|$, where $\mathcal{S}_M$ denotes the set of non-singleton source sets: $\mathcal{S}_M = \{S \mid \langle S, t \rangle \in H \text{ for some } t \in N, \text{ and } |S| > 1\}$;

- The *size* s, that is the overall length of the description of the hypergraph (also denoted as $|\mathcal{H}|$). If we represent a directed hypergraph by means of adjacency lists we have that $|\mathcal{H}| \equiv s = n + a' + h$.

In the special case where a directed hypergraph is a directed graph, the number of vertices is equal to the number of nodes n, and the number of edges is $m = h$. Furthermore, $a = n$, $a' = 0$, and $s = n + m$.

Definition 2. *Let $\mathcal{H} = \langle N, H \rangle$ be a directed hypergraph. A hypergraph $\mathcal{H}' = \langle N', H' \rangle$ such that (a) $N' \subseteq N$, (b) $H' \subseteq H$, and, for each $\langle S, t \rangle \in H'$, $S \subseteq N'$, is called a* subhypergraph *of $\mathcal{H}$. We denote this by $\mathcal{H}' \subseteq \mathcal{H}$. Furthermore, let $H' \subseteq H$ be a set of hyperarcs in $\mathcal{H}$. Let $N' \subseteq N$ be the union of source sets and target nodes of hyperarcs in H'. The hypergraph $\mathcal{H}' = \langle N', H' \rangle$ is said to be the* subhypergraph of $\mathcal{H}$ induced by H'.

Figure 1 shows an example of hypergraph and subhypergraph.

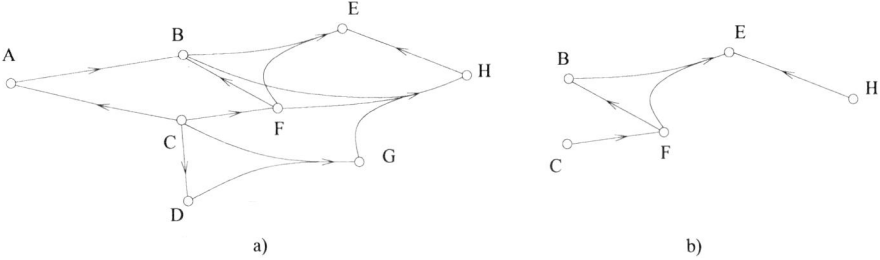

Fig. 1. (a) A directed hypergraph $\mathcal{H}$ and (b) a subhypergraph $\mathcal{H}'$ induced by the set of hyperarcs $\{\langle C, F \rangle, \langle F, B \rangle, \langle BF, E \rangle, \langle H, E \rangle\}$.

We now define the notion of hyperpath in directed hypergraphs. Before, we recall some terminology on directed graphs. A *path* π in a directed graph is a sequence of edges $e_1, e_2, \ldots, e_k$ and vertices $v_0, v_1, \ldots, v_k$, such that $e_i = (v_{i-1}, v_i)$, $1 \leq i \leq k$. A path π is *simple* if no vertex is repeated twice. It is a cycle if $v_0 = v_k$. If $v_i = v_j$ for some $i \neq j$, then the path π contains a cycle as a subpath. Given a path π from x to y, we can describe π in different ways. One simple-minded description of a path is to give the sequence of all the edges in π, as traversed while going from x to y. Notice that this description may contain the same edge more than once and may not even be bounded, since π may contain a cycle which is traversed an unbounded number of times. We refer to this description of a path as *unfolded*. An alternative description may be the subgraph of G containing exactly the edges of π (note that each edge is considered only once). If the graph G is finite, this description is always bounded, and the path is referred to as *folded*. We now turn to directed hypergraphs, and show that there is an even deeper difference between folded and unfolded hyperpaths.

Definition 3. *Let $\mathcal{H} = \langle N, H \rangle$ be a directed hypergraph, $X \subseteq N$ be a non-empty subset of nodes, and y be a node in N. There is a hyperpath from X to y in $\mathcal{H}$ if either (a) $y \in X$ (extended reflexivity); or (b) there is a hyperarc $\langle Z, y \rangle \in H$ and hyperpaths from X to each node $z_i \in Z$ (extended transitivity).*

The recursive definition of a hyperpath is naturally described by a tree labeled on the nodes, referred to as the *hyperpath tree*, and defined as follows.

Definition 4. *Let $\mathcal{H} = \langle N, H \rangle$ be a directed hypergraph, $X \subseteq N$ be a non-empty subset of nodes, and y be a node in N such that there is a hyperpath from X to y. A hyperpath tree from X to y is a tree $t_{X,y}$ defined as follows: (a) if $y \in X$ (extended reflexivity) $t_{X,y}$ is empty; (b) if there is a hyperarc $\langle Z, y \rangle \in H$ and hyperpaths from X to each node $z_i \in Z$ (extended transitivity), then $t_{X,y}$ consists of a root labeled with hyperarc $\langle Z, y \rangle$ having as subtrees the hyperpath trees t_{X,z_i} for each node $z_i \in Z$.*

Note that the hyperpath tree $t_{X,y}$ is such that its root has the target node y in its label. Furthermore, if $\langle S, t \rangle$ is the label of a leaf in the hyperpath tree, the source set S is contained in X. We refer to the hyperpath tree as the *unfolded representation* of a hyperpath. As in the case of a path in a directed graph, this representation describes explicitly the sequence of hyperarcs, as traversed while going from X to y. Once again, the same hyperarc may appear more than once in the hyperpath tree. In what follows, we will use interchangeably the terms unfolded hyperpath and hyperpath tree. As in the case of paths on graphs, also for hyperpaths there is an alternative and more concise description:

Definition 5. *Let $\mathcal{H} = \langle N, H \rangle$ be a directed hypergraph. Let $X \subseteq N$ be a non-empty subset of nodes of $\mathcal{H}$, and $y \in N$ be a node such that there is a hyperpath from X to y in $\mathcal{H}$. A folded hyperpath $h_{X,y}$ from X to y is given by the subhypergraph of $\mathcal{H}$ induced by the hyperarcs in the unfolded hyperpath $t_{X,y}$.*

Throughout this paper, we refer to a folded hyperpath more simply as a hyperpath. As a consequence of Definition 5, either $h_{X,y}$ is the empty hypergraph or there is a hyperarc $\langle Z, y \rangle$ in $h_{X,y}$ and, for each node $z_i \in Z$, there exists a hyperpath h_{X,z_i} from X to z_i which is a subhypergraph of $h_{X,y}$. Figure 2 shows an unfolded and a folded hyperpath of the hypergraph given in Figure 1.

Definition 6. *Let $\mathcal{H} = \langle N, H \rangle$ be a directed hypergraph. Let $X \subseteq N$ be a non-empty subset of nodes of $\mathcal{H}$, and let x_i be a node in X. If there is a nonempty hyperpath h_{X,x_i} in $\mathcal{H}$ consisting of at least one hyperarc, then h_{X,x_i} is a hypercycle. A hyperpath is cyclic if it contains at least one hypercycle as a subhypergraph, and is acyclic otherwise.*

We remark that the unfolded representation of a cyclic hyperpath may use several times the same hyperarc. As in the case of directed graphs, this implies that the unfolded representation of a hyperpath can be much larger than its folded representation. However, differently from the case of directed graphs, this is not the only case in which this can happen: indeed, there may be even acyclic

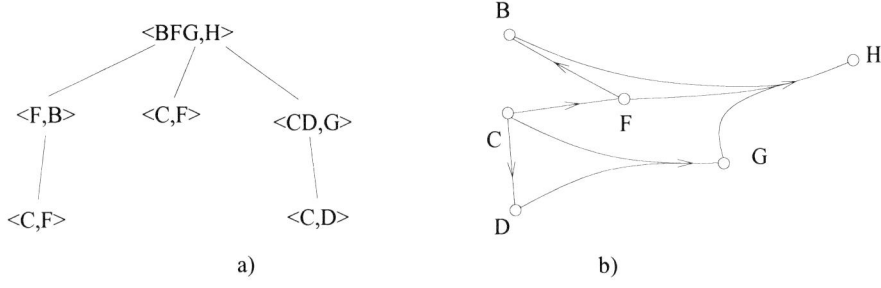

Fig. 2. (a) Unfolded and (b) folded hyperpath from C to H in the hypergraph $\mathcal{H}$ of Figure 1(a).

hyperpaths whose unfolded representation is exponentially larger than the size of their folded representation.

Let G be a directed graph, where each edge has associated a *cost*. Then the cost of a path is simply given by the sum of the costs of all its edges. Differently from the case of paths in directed graphs, hyperpaths in directed hypergraphs have a much more complex structure: this yields different ways of measuring the cost of the same hyperpath. We start with few natural definitions.

Definition 7. *The number of hyperarcs of a hyperpath $h_{X,y} = \langle N_h, H_h \rangle$ is defined to be the cardinality of the set of hyperarcs: $n(h_{X,y}) = |H_h|$.*

Definition 8. *The size of a (folded) hyperpath $h_{X,y} = \langle N_h, H_h \rangle$ is the sum of the number of hyperarcs and source area:*

$$s(h_{X,y}) = |H_h| + \sum_{S_i \in \mathcal{S}(h_{X,y})} |S_i|$$

where $\mathcal{S}(h_{X,y})$ is the set of all the source sets in $h_{X,y}$.

Definition 9. *A weighted hypergraph $\mathcal{H}$ is a triple $\langle N, H, w \rangle$ where $\langle N, H \rangle$ is a directed hypergraph and w is a measure function, which assigns a real weight to all hyperarcs of $\langle N, H \rangle$.*

Definition 10. *The cost $c(h_{X,y})$ of a hyperpath $h_{X,y} = \langle N_h, H_h \rangle$ is the sum of the costs of its hyperarcs: $c(h_{X,y}) = \sum_{\langle S,t \rangle \in H_h} w_{\langle S,t \rangle}$.*

Note that the *size* of a hyperpath gives the overall length of the description of $h_{X,y}$, and that Definition 10 includes as a special case Definition 7 whenever $w_{\langle S,t \rangle} = 1$ for each hyperarc $\langle S, t \rangle$.

Theorem 1. *[11] Let $\mathcal{H} = \langle N, H \rangle$ be a directed hypergraph, x and y be two nodes in N, and k be an integer. The following problems are NP-complete.*

($\mathcal{P}_1$) *Find a hyperpath $h_{x,y}$ with k hyperarcs or less.*
($\mathcal{P}_2$) *Find a hyperpath $h_{x,y}$ of cost k or less.*
($\mathcal{P}_3$) *Find a hyperpath $h_{x,y}$ of size k or less.*

Finding hyperpaths with minimum *source area* or minimum *number of source sets* are easily shown to be NP-hard problems as well.

We now define a general notion of measure function on hyperpaths, and next specialize it to some of the most intuitive measures that give rise to polynomially solvable optimization problems.

Definition 11. *Given a directed hypergraph $\mathcal{H} = \langle N, H \rangle$, the corresponding functional hypergraph $\mathcal{H}_F = \langle N, H; F \rangle$ is defined as follows. Each hyperarc $\langle X, y \rangle \in H$ is associated to a triple $F_{\langle X, y \rangle} = (w_{\langle X, y \rangle}, \psi_{\langle X, y \rangle}, f_{\langle X, y \rangle})$, where:*

- $w_{\langle X, y \rangle} \in \Re$ *is the weight of the hyperarc;*
- $\psi_{\langle X, y \rangle}$ *is a function from $|X|$-tuples of reals to reals: $\psi_{\langle X, y \rangle} : \Re^{|X|} \to \Re$;*
- $f_{\langle X, y \rangle}$ *is a function from a pair of reals to reals: $f_{\langle X, y \rangle} : \Re^2 \to \Re$.*

Let $h_{S,t}$ be a hyperpath from a set of nodes S to a single node t, and let $\langle Z, t \rangle$ be the last hyperarc in $h_{S,t}$, with $Z = \{z_1, z_2, \ldots, z_k\}$: $h_{S,t} = h_{S,z_1} \cup h_{S,z_2} \cup \ldots \cup h_{S,z_k} \cup \{\langle Z, t \rangle\}$. Then we use function $w()$ to take into account the cost of the hyperarc $\langle Z, t \rangle$, and $\psi()$ to take into account the costs of the hyperpaths h_{S,z_i}. We finally use function $f()$ to combine these two costs. This is formalized in the following definition.

Definition 12. *Given a functional directed hypergraph $\mathcal{H}_F = \langle N, H; F \rangle$ μ is a functional measure on hyperpaths (or measure function) if there exists a constant μ_o such that for each hyperpath $h_{S,t}$ the following is true: (a) if $h_{S,t} = \emptyset$ then $\mu(h_{S,t}) = \mu_o$; and (b) if $h_{S,t} = h_{S,z_1} \cup h_{S,z_2} \cup \ldots \cup h_{S,z_k} \cup \{\langle Z, t \rangle\}$, then $\mu(h_{S,t}) = f_{\langle Z,t \rangle}(w_{\langle Z,t \rangle}, \psi_{\langle Z,t \rangle}(\mu(h_{S,z_1}), \mu(h_{S,z_2}), \ldots, \mu(h_{S,z_k})))$.*

In the following we assume that all the weights of hyperarcs are positive, and the value of the measure functions are non-negative. We now define some of the most intuitive cost measures of hyperpaths, and see how they can be obtained by specializing this definition. Consider a functional hypergraph $\mathcal{H}_F = \langle N, H; F \rangle$, and let $h_{X,y}$ be a hyperpath in $\mathcal{H}_F$.

In Definition 10 we defined the cost of a (folded) hyperpath as the sum of the costs of its hyperarcs. In the case of unfolded hyperpaths we define the *traversal cost* as the cost of the root plus the cost of all its subtrees. In other words if a hyperarc is traversed more than once, its cost is repeatedly taken into account.

Definition 13. *The traversal cost $c_t(t_{X,y})$ of an unfolded hyperpath $t_{X,y}$ is inductively defined as follows:*

a) if $t_{X,y}$ is empty (i.e. $y \in X$) then: $c_t(t_{X,y}) = 0$;
b) if the unfolded hyperpath $t_{X,y}$ has root $\langle Z, y \rangle$ with subtrees $t_{X,z_1}, t_{X,z_2}, \ldots, t_{X,z_k}$, then:

$$c_t(t_{X,y}) = w_{\langle Z, y \rangle} + \sum_{z_i \in Z} c_t(t_{X,z_i}).$$

Note that the traversal cost can be obtained with the following choices: $f(x, y) = x + y$ and $\psi(x_1, x_2, \ldots, x_k) = \sum_{i=1}^{k} x_i$.

Definition 14. *The* rank $r(h_{X,y})$ *of an acyclic hyperpath $h_{X,y}$ is recursively defined as follows:*

a) *if $h_{X,y}$ is an empty hypergraph then $r(h_{X,y}) = 0$;*
b) *if $h_{X,y}$ has one hyperarc $\langle Z, y \rangle$ entering y, with $Z = \{z_1, z_2, \ldots, z_k\}$ and $h_{X,z_i} \subset h_{X,y}$, then: $r(h_{X,y}) = w_{\langle Z, y \rangle} + \max_{z_i \in Z}\{r(h_{X,z_i})\}$.*

Note that the rank is obtained with the following choices: $f(x, y) = x + y$ and $\psi(x_1, x_2, \ldots, x_k) = \max_{1 \leq i \leq k}\{x_i\}$. If we consider the unfolded hyperpath $t_{X,y}$ defined in Definition 4, we can recursively define the *rank* $r_t(t_{X,y})$ of the unfolded hyperpath as the sum of the cost of the root plus the maximum *rank* among its children. For acyclic hyperpaths, we have that the two possible definitions of ranks (for folded and unfolded hyperpaths) actually coincide: $r(h_{X,y}) = r_t(t_{X,y})$. We remark that this is not necessarily true, if we try to define a *rank* for cyclic hyperpaths as well. Note that the *rank* corresponds to the maximum cost path from the root to a leaf in the hyperpath tree.

Several other measures are of interest in some applications. For instance, for the choice $f(x, y) = x + y$ and $\psi(x_1, x_2, \ldots, x_k) = \min_{1 \leq i \leq k}\{x_i\}$, we obtain the *gap* of a hyperpath; for $f(x, y) = \min\{x, y\}$ and $\psi(x_1, x_2, \ldots, x_k) = \min_{1 \leq i \leq k}\{x_i\}$ we obtain the *bottleneck* of a hyperpath; and for $f(x,y) = \max\{x, y\}$ and $\psi(x_1, x_2, \ldots, x_k) = \max_{1 \leq i \leq k}\{x_i\}$ we obtain the *threshold* of a hyperpath. Intuitively, the bottleneck of a hyperpath $h_{X,y}$ corresponds to the minimum weight of a hyperarc in $h_{X,y}$, and similarly, the threshold would correspond to the maximum weight of a hyperarc in $h_{X,y}$. In the special case where a directed hypergraph is a directed graph, the cost, traversal cost, rank and gap collapse to the usual definition of length of a (weighted) path. Furthermore, the size is twice the numbers of edges (due to the source area), the bottleneck coincides with the bottleneck capacity of the path (i.e., the edge with minimum capacity of the path), and the threshold is the maximum cost in the path.

3   Classes of Measure Functions

In his definition of grammar problems [12], Knuth introduces the concept of superior function for context-free grammars.

Definition 15. *[12] A function $g(x_1, \ldots, x_k)$ from $(\Re^+)^k$ into $\Re^+$ is a* superior function *(SUP) if it is monotone nondecreasing in each variable and if: $g(x_1, \ldots, x_k) \geq \max(x_1, \ldots, x_k)$.*

Examples of superior functions are $\max_{1 \leq i \leq k}\{x_i\}$, and $\Sigma_{i=1}^{k} x_i$. In a *superior context-free grammar* all the productions are of the form: $Y \longrightarrow g(X_1, \ldots, X_k)$ where the capital letters are nonterminal symbols, and g is a superior function (possibly different for each production). For any nonterminal symbol Y of such a grammar with terminal symbols T, let $L(Y) = \{\alpha \mid \alpha \in T^*, \text{ and } Y \xrightarrow{*} \alpha\}$ be the set of terminal strings derivable from Y. If each terminal symbol is given a constant value in $\Re^+$, it is possible to define a function *val* as a composition of superior functions, that is $val : T^* \to \Re^+$ which, given any string $\alpha \in T^*$, provides a corresponding value $val(\alpha)$. The *Grammar Problem* consists of finding the smallest value that can be associated with a nonterminal symbol Y, namely: $m(Y) = \min_{\alpha \in L(Y)}\{val(\alpha)\}$.

By analogy with superior functions, it is possible to introduce a dual notion.

Definition 16. *A function $g(x_1, \ldots, x_k)$ from $(\Re^+)^k$ into $\Re^+$ is an* inferior function *(INF) if it is monotone nondecreasing in each variable and if: $g(x_1, \ldots, x_k) \leq \min(x_1, \ldots, x_k)$.*

Examples of inferior functions are: $\min_i\{x_i\}$, and the product $\Pi_i\{x_i\}$ restricted to the case $0 \leq x_i \leq 1$, $i = 1, \ldots, k$.

Ramalingam and Reps in [15] introduced a slight generalization of superior functions.

Definition 17. [15] *A function $g(x_1, \ldots, x_k)$ from $(\Re^+)^k$ into $\Re^+$ is a* weakly superior function *(WSUP) if it is monotone nondecreasing in each variable and if, for $1 \leq i \leq k$, $g(x_1, \ldots, x_k) < x_i \Rightarrow g(x_1, \ldots, x_i, \ldots, x_k)$*
$$= g(x_1, \ldots, \infty, \ldots, x_k).$$

If a function is *SUP*, then it is *WSUP*; examples of weakly superior functions that are not superior are $\min_{1 \leq i \leq k}\{x_i\}$, $\min_{1 \leq i \leq k}\{x_i\} * 2$, and any constant function. Again, it is possible to introduce a dual definition.

Definition 18. *A function $g(x_1, \ldots, x_k)$ from $(\Re^+)^k$ into $\Re^+$ is a* weakly inferior function *(WINF) if it is monotone nondecreasing in each variable and if, for $1 \leq i \leq k$, $g(x_1, \ldots, x_k) > x_i \Rightarrow g(x_1, \ldots, x_i, \ldots, x_k) = g(x_1, \ldots, \infty, \ldots, x_k)$.*

Also in this case we have that the class *WINF* contains the class *INF*. Examples of weakly inferior functions that are not inferior functions are: $\max_i\{x_i\}$, $\max_i\{x_i\}/2$, and any constant function. In the following we investigate the relationship among *SUP*, *INF*, *WSUP*, and *WINF*, as well as their compositions. As a straightforward consequence of the above definitions, we have that: *SUP* $\subset$ *WSUP* and *INF* $\subset$ *WINF*.

The following lemma, summarizes the properties holding in the case of a generic composition of functions.

Lemma 1. *If we are given the functions $f, g_1, g_2, \ldots, g_h$ then their composition $f(g_1(\ldots), g_2(\ldots), \ldots, g_h(\ldots))$ is such that if $f, g_1, g_2, \ldots, g_h \in$ SUP (resp. INF, WSUP, WINF), then $f(g_1, g_2, \ldots, g_h) \in$ SUP (resp. INF, WSUP, WINF).*

We now relate these properties to our definition of hypergraphs. Recall that in a functional hypergraph, each hyperarc $\langle X, y \rangle \in H$ is associated to a triple $(w_{\langle X,y \rangle}, \psi_{\langle X,y \rangle}, f_{\langle X,y \rangle})$. The measure μ of any hyperpath $h_{S,y}$ having (X, y) as the last hyperarc is given by

$$\mu(h_{S,y}) = f_{\langle X,y \rangle}(w_{\langle X,y \rangle}, \psi_{\langle X,y \rangle}(\mu(h_{S,x_1}), \mu(h_{S,x_2}), \ldots, \mu(h_{S,x_k}))).$$

If all the functions $f_{\langle X,y \rangle}$, $\psi_{\langle X,y \rangle}$ (for all $\langle X, y \rangle \in H$) of a functional hypergraph $\mathcal{H}_F$ are, say, superior functions, then the overall measure function μ is a superior function as well. Analogous considerations apply to any combination described in Lemma 1. As a special case, we will consider in many cases functional hypergraphs where all $\psi_{\langle X,y \rangle}$, with $\langle X, y \rangle \in H$, are uniform, as well as all $f_{\langle X,y \rangle}$. In many applications (such as those mentioned in Section 1) we are exactly in this situation.

Both Knuth [12] and Ramalingam and Reps [15] considered also the class of *strict* (weakly) superior functions, that are characterized by a strict monotonicity between the value the function at hand and each of its arguments: this leads to the classes *SSUP*, *SWSUP*, and may be further generalized to *SINF*, and *SWINF*. In this cases we have the following results.

Lemma 2. *If we compose a function f with functions $g_1, g_2, \ldots, g_h$, the following properties hold:*
1. *If $f \in SSUP$, and $g_i \in SUP$ for all i (or vice versa) then $f(g_1, g_2, \ldots, g_h) \in SSUP$;*
2. *If $f \in SINF$, and $g_i \in INF$ for all i (or vice versa) then $f(g_1, g_2, \ldots, g_h) \in SINF$;*
3. *If $f \in SWSUP$, and $g_i \in WSUP$ for all i (or vice versa) then $f(g_1, g_2, \ldots, g_h) \in SWSUP$;*
4. *If $f \in SWINF$, and $g_i \in WINF$ for all i (or vice versa) then $f(g_1, g_2, \ldots, g_h) \in SWINF$.*

4 Hyperpath Optimization Problems on Directed Hypergraphs

An optimization problem $\mathcal{P}$ on hyperpaths is characterized by an *optimization criterion* $opt \in \{\min, \max\}$, and a measure function μ on hyperpaths. In the following we will make use of the notion of the unfolded representation of a hyperpath. We recall that this is a tree whose nodes are the hyperarcs that are used to build up the hyperpath, and that may appear themselves several times, if the hyperpath is cyclic.

Definition 19. *An optimization problem $\mathcal{P} = (opt, \mu)$ is k-cycle-convergent (CY-CONV) for some $k \geq 0$ if for each optimal hyperpath $h_{S,t}$ between a source set S and a target node t there exists a subhyperpath $h_{S,t}^k \subseteq h_{S,t}$ whose unfolded representation contains each node at most $(k+1)$ times as a target, and such that*

$\mu(h_{S,t}^k) = \mu(h_{S,t})$. *An optimization problem that is 0-cycle-convergent is said to be* cycle-invariant.

In other words, if we are given a k-cycle-convergent measure function on hyperpaths for $k > 0$, an optimal hyperpath may be cyclic. In the remainder of this paper, we will always consider the case $k = 0, 1$.

Definition 20. *An optimization problem $\mathcal{P} = (opt, \mu)$ is* cycle-unbounded *(CY-UNBOUND) if there exist optimal hyperpaths whose unfolded representation contains the same node as a target an unbounded number of times.*

As in the case of digraphs, finding an optimal (acyclic) solution in the presence of cycles often leads to NP-hard optimization problems. The following theorem states some properties that characterize optimization problems on hyperpaths.

Theorem 2. *Let $\mathcal{P} = (opt, \mu)$ be an optimization problem on hypergraphs. Then the following properties hold:*
a) the minimization of a superior function is cycle-invariant;
b) the maximization of an inferior function is cycle-invariant;
c) the minimization of a weakly superior function is 1-cycle-convergent;
d) the maximization of a weakly inferior function is 1-cycle-convergent.

Proof. To prove (a), consider an optimization problem where μ is a superior function, and let $h_{S,t}$ be a cyclic optimal hyperpath between a source set S and a target node t. Given a hyperpath h, let $T(h)$ be the corresponding unfolded representation. Consider the unfolded hyperpath $T(h_{S,t})$, whose nodes are hyperarcs of the hyperpath: if $h_{S,t}$ is cyclic, there must be a branch in such tree containing two nodes (Y_1, z), and (Y_2, z) with the same target node z. Without loss of generality, suppose that the subtree T_1 rooted at (Y_1, z) contains the subtree T_2 rooted at (Y_2, z). By cutting away from $T(h)$ the whole subtree T_2 we still obtain a folded hyperpath with target node z whose measure can not be larger than $\mu(h_{S,t})$, due to the inequality induced by the concept of superior function between the value of a function and each of its arguments. We can repeatedly delete subtrees of the initial tree $T(h_{S,t})$ until no cycles are in the tree, obtaining an acyclic subhyperpath $h_{S,t}^a$ such that $\mu(h_{S,t}^a) \leq \mu(h_{S,t})$. Analogous considerations can be used to prove (b).

To prove case (c), consider the cyclic hyperpath $h_{S,t} = \{\bigcup_{i=1,\ldots,k} h_{S,z_i}\} \cup \{\langle Z, t \rangle\}$, where $Z = \{z_1, \ldots, z_k\}$. For $1 \leq i \leq k$, define $\mu_i = \mu(h_{S,z_i})$. If the measure function is weakly superior, there is no loss of generality in assuming that we can split the nodes in the set Z into two subsets Z_R and Z_I, respectively called the *relevant* and the *irrelevant* items of Z, defined as follows:
1. If $z_i \in Z_R = \{z_1, \ldots, z_d\}$, then $\mu(h_{S,t}) \geq \mu_i$.
   For each relevant node $z_i \in Z_R$ (a set that might be empty, in case of a constant function), the value of $\mu(h_{S,t})$ is monotone nondecreasing with respect to the values $\mu_i = \mu(h_{S,z_i})$;

2. If $z_j \in Z_I = \{z_{d+1}, \ldots, z_k\}$, then $\mu(h_{S,t}) = \mu(h_{S,t})\big|_{\mu_j=\infty}$

Note that the value of $\mu(h_{S,t})$ does not depend on the values $\mu_j = \mu(h_{S,z_j})$, for $j \in \{d+1, \ldots, k\}$.

Consider the unfolded hyperpath $T(h_{S,t})$ with root $\langle Z, t \rangle$ having children $\langle S, z_i \rangle$, for $i \in \{1, \ldots, d, d+1, \ldots, k\}$. In the given hypotheses, consider the two families of subtrees rooted at the children of $\langle Z, t \rangle$:

1. For the relevant nodes $z_i \in \{z_1, \ldots, z_d\}$, the value of $\mu(h_{S,t})$ does depend on the content of the unfolded hyperpaths $T(h_{S,z_i})$: call *relevant* the edges connecting the root $\langle Z, t \rangle$ and the relevant children $\langle S, z_i \rangle$, with $z_i \in \{z_1, \ldots, z_d\}$;

2. For the irrelevant nodes $z_j \in \{z_{d+1}, \ldots, z_k\}$, $\mu(h_{S,t})$ does not depend on the actual value of $\mu(h_{S,z_j})$: the role of these *irrelevant* subhyperpaths is to propagate the reachability from the root, regardless their measure. Hence we can replace the generic hyperpath h_{S,z_i} with an acyclic hyperpath h_{S,z_i}^a (note that this hyperpath may have to contain node t as a possible target in some intermediate node) without increasing the value of the measure $\mu(h_{S,t})$.

We can recursively proceed on the subtrees $T(h_{S,z_i})$ until all relevant edges have been found, and the irrelevant subtrees have been replaced by acyclic subhyperpaths. It is easy to check that the subtree $T(h_{S,t}^R)$ induced by the relevant edges has the same property that we have exploited in case (*a*): if a target node x appears twice in a branch of $T(h_{S,t}^R)$, say (Y_1, x), and (Y_2, x) with the former node above the latter, we can cut away the whole subtree T_2 rooted at (Y_2, x) and still obtain a folded hyperpath with target node z whose measure can not be larger than $\mu(h_{S,t})$.

Case (*d*) can be proved similarly.

The following theorem can be proved similarly.

Theorem 3. *Consider an optimization problem on hypergraphs. The following properties hold:*
a) The maximization of a strict superior function is cycle-unbounded;
b) The minimization of a strict inferior function is cycle-unbounded.

5 Some Examples of Measure Functions

Using the results developed in the previous sections, we can characterize in a general and unified framework several optimization problems on hypergraphs, as summarized in the table of Figure 3. In this table we define some uniform measure functions μ in terms of the constituent functions $f(w, \psi)$ and $\psi(\mu_1, \ldots, \mu_k)$. The resulting properties of μ and the properties of the optimization problems are derived directly from the arguments given in Sections 3 and 4 (mainly Lemmas 1 and 2 and Theorems 2 and 3).

We also provide a characterization of the *closure* problem as an optimization problem: in this case we can use the min function with uniform hyperarcs

measure function μ	$f(w,\psi)$	$\psi(\mu_1,\ldots,\mu_k)$	resulting properties	MIN problem	MAX problem
rank $w > 0$	+ SSUP	max SUP,WINF	SSUP	CY-INV	CY-UNBOUND
gap $w > 0$	+ SSUP	min WSUP,INF	SWSUP	1-CY-CONV	CY-UNBOUND
bottleneck $w > 0$	min WSUP,INF	min WSUP,INF	WSUP,INF	1-CY-CONV	CY-INV
threshold $w > 0$	max SUP,WINF	max SUP,WINF	SUP,WINF	CY-INV	1-CY-CONV
traversal cost $w > 0$	+ SSUP	$\sum$ SSUP	SSUP	CY-INV	CY-UNBOUND
closure $w = 1$	min WSUP,INF	min WSUP,INF	WSUP,INF	1-CY-CONV	CY-INV

Fig. 3. Characterization of measure functions on hypergraphs.

weights equal to 1. Of course, the *minimum closure* corresponds to the traditional transitive closure over hypergraphs, investigated in [6,7].

We conclude this section by mentioning that the problems of finding a hyperpath of rank, or gap, or traversal cost of k or more are all NP-complete. As shown in the table of Figure 3, all the related optimization problems are cycle-unbounded. In the next sections, we will present efficient algorithms for k-cycle-convergent ($k = 0, 1$, thus including cycle-invariant) optimization problems.

6 Algorithms for Optimization Problems

Several authors have proposed algorithms that can be used to find optimal hyperpaths. With a different formalism, Knuth [12] proposed a generalization of Dijkstra's algorithm to solve the grammar problem described in Section 3. This algorithm can be easily adapted to find the optimal hyperpaths from a single *source node* (the *axiom* of the grammar), to all other nodes in a functional hyperpath $\mathcal{H} = \langle N, H \rangle$ with $n = |N|$ nodes, $h = |H|$ hyperarcs, and a total size $s = |\mathcal{H}|$. Knuth's algorithm requires $O(h \log n + s)$ worst case time. The running time can be reduced to $O(n \log n + s)$ by using Fibonacci heaps [8]. The algorithm is based on Dijkstra's shortest path algorithm, and uses a priority queue. A generic node x is enqueued when the distance of a neighbor from the source has been computed. The priority of x in the queue may decrease if further neighbors provide better connection from the source. When a node is dequeued, its distance from the source is computed. We will refer to this algorithm as *Sort-by-Priority*.

The dynamic maintenance of hyperpaths has been studied by several authors [4,6,7,11,13,15]. Here we only consider the *incremental single source problem*. This problem consists of maintaining the optimal hyperpaths from a single source node to every other node, while performing insertions of hyperarcs (*insert* operations), or weight decreases (*decrease* operations).

Ramalingam and Reps provide in [15] a dynamic solution of this problem. Their solution is more general and applies also to a whole set of hyperarc operations of various kinds. Their algorithm, which we refer to as RR, can be considered as a dynamic version of *Sort-by-Priority*. The main idea behind the algorithm is to use a priority queue: only those nodes whose distance from the root has to be changed are inserted in the priority queue. Let δ be the set of nodes that change their distance from the source. The time complexity of RR is given in terms of *output complexity*, i.e., as a function of a parameter $||\delta||$, which represents the cardinality of the set δ *plus* all the hyperarcs incident to δ. Namely RR requires $O(||\delta|| \log ||\delta||)$ worst-case time to update the optimal hyperpaths from a single source after a hyperarc insertion or a weight decrease. Note that in the worst case $||\delta|| = \Theta(|\mathcal{H}|)$.

With the terminology proposed in this paper, the time bound of RR depends on both the measure function and the type of update. Namely, if the function is *SSUP* or *SINF*, RR processes a hyperarc insertion or a weight decrease with the following costs: (i) each node z in the set δ is enqueued exactly once, and (ii) all the hyperarcs $\langle X, y \rangle$ whose source set X contains z are scanned as soon as node z is dequeued. The same situation holds even in case of a weight decrease operation and when the measure function is *SWSUP* or *SWINF* (such as for *gap*). In case of an *insert* operation and a *SWSUP* or *SWINF* function, instead, it seems possible that each node z in the set δ might be enqueued once for each hyperarc $\langle X, z \rangle \in H$.

Another approach can be used to find optimal hyperpaths, as described in [11]. The algorithm to handle the insertion of a hyperarc $\langle X, y \rangle$, which we will refer to as *Sort-by-Structure*, can be described as follows:

Insert$(X, y; w)$

1. Compute the nodes that become reachable from the source set with the insertion of the new hyperarc. Collect those nodes (together with node y) in a simple queue Q;

2. Extract a node z from Q and compute the optimal path from the source to z. Next, scan all the hyperarcs $\langle S, t \rangle$ whose source set S contains z, possibly inserting node t in queue Q if a better path from the source to node t has been found (where the notion of "better" clearly depends on the optimization criterion);

3. Repeat the previous step until Q becomes empty.

Step 1 consists of computing the set of nodes reachable from S: a dynamic solution of this problem was proposed in [7]. In case of a weight decrease operation this step is skipped, and queue Q is initialized with node y alone. *Sort-by-Structure* can be used with small changes also in case of the "static" problem as an alternative to Knuth's algorithm. Furthermore *Sort-by-Structure* can be easily adapted to handle acyclic hypergraphs in $O(s)$ total time in any incremental sequence of updates.

This solution does not make use of priorities, but the nodes whose optimal from the source is to be improved are enqueued in a simple queue Q. Again, each node may be enqueued several times. If we consider an incremental sequence of operations, consisting in both insertion of new hyperarcs and weight decreases, each node can be enqueued is $O(W)$ times to process the whole sequence of updates, where W denotes the codomain of function μ, i.e., the number of possible values that the measure of a hyperpath may assume. The function $f_{\langle X,z\rangle}(w_{\langle X,z\rangle}, \psi_{\langle X,z\rangle}(\ldots))$ associated to each hyperarc $\langle X,z \rangle$ is (re)computed at most once for each improvement of nodes in the set X. This leads to an overall bound of $O(Ws)$ total time in any incremental sequence of updates, in the hypothesis that the (re)computation of each function associated to hyperarcs requires constant time.

As an example, in artificial intelligence, reasoning with uncertainty leads to the use of *fuzzy logic*. In this application finding optimal hyperpaths corresponds to the highest degree of confidence supporting a given conclusion [4]; this is a natural situation where a small value of W has to be used.

We remark that *Sort-by-Structure* and *Sort-by-Priority* can be combined by using a priority queue in Step 2: this actually provides the best bound for the general case, since each node is enqueued at most once in each call to the update procedure, also in case of the *WSUP* and *WINF* measure functions. We will refer to this version of the algorithm as *Improved Sort-by-Priority*.

The previous considerations lead to the following theorems, whose proofs are omitted here.

Theorem 4. *Let $\mathcal{P} = (opt, \mu)$ be an optimization problem on a functional hypergraphs $\mathcal{H}_F = \langle N, H; F \rangle$, with $n = |N|$, $h = |H|$, and $s = |\mathcal{H}|$. If W is the cardinality of the codomain of function μ, then the following bounds hold:*
a) *if $\mathcal{P}$ is cycle-invariant, then*

- *Sort-by-Priority finds a solution in $O(s + n \log n)$ time;*
- *Sort-by-Structure finds a solution in $O(Ws)$ time;*

b) *if $\mathcal{P}$ is 1-cycle-convergent, then*

- *Sort-by-Priority finds a solution in $O(W(s + n \log n))$ time;*
- *Sort-by-Structure finds a solution in $O(Ws)$ time.*

In the special case that the functional hypergraph is acyclic, then Sort-by-Structure finds a solution in $O(s)$ time.

Note that in case of acyclic hypergraphs the classes of cycle-invariant and k-cycle-convergent measure functions collapse to the same class.

Theorem 5. *Let $\mathcal{P} = (opt, \mu)$ be an optimization problem on a functional hypergraphs $\mathcal{H}_F = \langle N, H; F \rangle$. The problem of maintaining optimal hyperpaths from a source set S to every other node $z \in N$ under a sequence of both hyperarc insertions and weight decreases can be solved with the following time bounds:*
a) *if $\mathcal{P}$ is cycle-invariant, then Sort-by-Structure can process the sequence of updates in $O(Ws)$ total time;*

b) if $\mathcal{P}$ is 1-cycle-convergent, then Improved Sort-by-Priority can process the sequence of updates in $O(W(s + n \log n))$ total time;
c) if the functional hypergraph is acyclic, then Sort-by-Structure can process the sequence of updates in $O(s)$ total time.

For all the algorithms mentioned in this section the space required is $O(s)$ and the solution is computed in explicit form, i.e., queries about the optimal path from the source to any other node in the hypergraph can be answered in constant time. Finally, we mention that in the hypothesis that the τ_F is the maximum cost of (re)computing any of the functions associated to hyperarcs of the functional hypergraph, then all the time bounds have to multiplied by a factor of τ_F.

Acknowledgments

We are grateful to Roberto Giaccio for many discussions throughout this work.

References

1. P. Alimonti, E. Feuerstein, and U. Nanni. Linear time algorithms for liveness and boundedness in conflict-free petri nets. In *1st Latin American Theoretical Informatics*, volume 583, pages 1–14. Lecture Notes in Computer Science, Springer-Verlag, 1992.
2. G. Ausiello, A. D'Atri, and D. Saccà. Graph algorithms for functional dependency manipulation. *Journal of the ACM*, 30:752–766, 1983.
3. G. Ausiello, A. D'Atri, and D. Saccà. Minimal representation of directed hypergraphs. *SIAM Journal on Computing*, 15:418–431, 1986.
4. G. Ausiello, R. Giaccio. On-line algorithms for satisfiability formulae with uncertainty. *Theoretical Computer Science* 171:3–24, 1997.
5. G. Ausiello, R. Giaccio, G. F. Italiano, and U. Nanni. Optimal traversal of directed hypergraphs. Manuscript, 1997.
6. G. Ausiello and G. F. Italiano. Online algorithms for polynomially solvable satisfiability problems. *Journal of Logic Programming*, 10:69–90, 1991.
7. G. Ausiello, G. F. Italiano, and U. Nanni. Dynamic maintenance of directed hypergraphs. *Theoretical Computer Science*, 72(2-3):97–117, 1990.
8. M. L. Fredman and R. E. Tarjan. Fibonacci heaps and their uses in improved network optimization algorithms. *Journal of the ACM*, 34:596–615, 1987.
9. G. Gallo, G. Longo, S. Nguyen, and S. Pallottino. Directed hypergraphs and applications. Discrete Applied Mathematics 42 (1993) 177-201.
10. G. Gallo and G. Rago. A hypergraph approach to logical inference for Datalog formulae. Technical Report 28/90, Dip. di Informatica, Univ. of Pisa, Italy, 1990.
11. G. F. Italiano and U. Nanni. On line maintenance of minimal directed hypergraphs. In *3rd Italian Conf. on Theoretical Computer Science*, pages 335–349 World Scientific Co., 1989.
12. D. E. Knuth. A generalization of Dijkstra's algorithm. *Information Processing Letters*, 6(1):1–5, 1977.
13. P. B. Miltersen. On-line reevaluation of functions. Technical Report DAIMI PB-380, Comp. Sci. Dept., Aarhus University, January 1992.

14. S. Nguyen and S. Pallottino. Hyperpaths and shortest hyperpaths. *Combinatorial Optimization*, 1403:258–271, 1989.
15. G. Ramalingam and T. Reps, An Incremental Algorithm for a Generalization of the Shortest Path Problem, *Journal of Algorithms*, 21:267–305, 1996.

Defining the Java Virtual Machine as Platform for Provably Correct Java Compilation

Egon Börger[1] and Wolfram Schulte[2]

[1] Università di Pisa, Dipartimento di Informatica, I-56125 Pisa, Italy
boerger@di.unipi.it
[2] Universität Ulm, Fakultät für Informatik, D-89069 Ulm, Germany
wolfram@informatik.uni-ulm.de

Abstract. We provide concise abstract code for running the Java Virtual Machine (JVM) to execute compiled Java programs, and define a general compilation scheme of Java programs to JVM code. These definitions, together with the definition of an abstract interpreter of Java programs given in our previous work [3], allow us to prove that any compiler that satisfies the conditions stated in this paper compiles Java code correctly. In addition we have validated our JVM and compiler specification through experimentation.

The modularity of our definitions for Java, the JVM and the compilation scheme exhibit orthogonal language, machine and compiler components, which fit together and provide the basis for a stepwise and provably correct *design–for–reuse*. As a by-product we provide a challenging realistic case study for mechanical verification of a compiler correctness proof.

1 Introduction

Every justification showing that a proposed compiler behaves well is relative to a definition of the semantics of source and target language. In our previous work [3] we have developed a platform independent, rigorous yet easily manageable definition for an interpreter of Java programs, which captures the intuitive understanding Java programmers have of the semantics of their code. In this paper we provide a mathematical (read: rigorous and platform independent) yet practical model of an interpreter for the Java Virtual Machine, which formalizes the concepts presented in the JVM specification [6], as far as they are needed for the compilation of Java programs. We also extract from the JVM specification the definition of a scheme for the compilation of Java to JVM code and prove its correctness.

Main Theorem. *Every compiler that satisfies the conditions listed in this paper compiles Java programs correctly into JVM code.*

We split the JVM and the compilation function into an incremental sequence of four machines and functions—whose structure corresponds to the conservative extension relation among the modular components we exhibited for Java [3]— and define the JVM at two levels of abstraction: a ground model with an abstract class file and abstract instructions, and a refined model where the abstract instructions are implemented by concrete JVM instructions. The structure of our Java machine is carried over mutatis mutandis to the basic structure of the abstract interpreter we are defining here for the JVM as target machine for Java compilation.

In sections 2 to 5 we define the sequence of successively extended JVM machines $JVM_\mathcal{I}$, $JVM_\mathcal{C}$, $JVM_\mathcal{O}$ and $JVM_\mathcal{E}$ for the compilation of programs from the imperative core $Java_\mathcal{I}$ of Java and its extensions $Java_\mathcal{C}$ (by classes, eg. procedures), $Java_\mathcal{O}$ (by object-oriented features, eg. class instances) and $Java_\mathcal{E}$ (by exceptions). We discuss here only the single threaded JVM, although our approach could easily include also multiple threads (see our multi-agent Java model with threads in [3]). We skip those language constructs which can be reduced by standard program transformation techniques to the core constructs dealt with explicitly in our Java models. We still do not consider Java packages, compilation units, visibility of names, strings, arrays, input/output, loading, linking and garbage collection. These features are the object of further refinements of the JVM model presented here. For proof details, the instruction refinement, an extensive bibliography and the discussion of related work we refer the interested reader to an extended version of this paper [1].

2 $JVM_\mathcal{I}$ and the Compilation of $Java_\mathcal{I}$ Programs

For the specification of Java, the JVM and the proof machinery, we use Abstract State Machines (ASMs). ASM specifications have a simple mathematical foundation [5], which justifies their intuitive understanding as "pseudo code" over abstract data. We define the basic JVM, called $JVM_\mathcal{I}$, which is used as the target for compiling Java's statements and expressions over primitive types. We prove that $JVM_\mathcal{I}$ executes the compilation of $Java_\mathcal{I}$ programs correctly.

The following grammars recall the syntax of $Java_\mathcal{I}$ [3] and introduce the corresponding instruction set $JVM_\mathcal{I}$:

$Exp ::= Lit$
$\quad | \ Uop \ Exp$
$\quad | \ Exp \ Bop \ Exp$
$\quad | \ Var$
$\quad | \ Var = Exp$
$\quad | \ Exp\,?\,Exp\,\colon Exp\,\colon$
$Stm ::= ;$
$\quad | \ Exp;$
$\quad | \ Lab : Stm$
$\quad | \ \textbf{break} \ Lab;$
$\quad | \ \textbf{continue} \ Lab;$
$\quad | \ \textbf{if} \ (Exp) \ Stm \ \textbf{else} \ Stm$
$\quad | \ \textbf{while} \ (Exp) \ Stm$
$\quad | \ \{ \ Stm^* \}$

$Instr ::= \textbf{const} \ (Lit)$
$\quad | \ \textbf{uapply} \ (Uop)$
$\quad | \ \textbf{bapply} \ (Bop)$
$\quad | \ \textbf{load} \ (Varnum \times Typ)$
$\quad | \ \textbf{store} \ (Varnum \times Typ)$
$\quad | \ \textbf{dup} \ (Typ)$
$\quad | \ \textbf{pop} \ (Typ)$
$\quad | \ \textbf{ifZero} \ (Lab)$
$\quad | \ \textbf{goto} \ (Lab)$
$\quad | \ \textbf{label} \ (Lab)$

$Varnum == Nat$
$Code \quad == Instr^*$

The JVM$_\mathcal{I}$ instruction set bears a close resemblance to a traditional stack machine like the P-machine. JVM$_\mathcal{I}$ provides instructions to load constants, to apply various unary and binary operators, to load and store a variable, to duplicate and to remove values, and to jump unconditionally or conditionally to a label. Variable locations in the JVM are represented by natural numbers. A JVM$_\mathcal{I}$ program is a sequence of instructions.

The universes *Lit, Uop, Bop, Var, Typ, Lab* contain Java literals, unary and binary operators, variables, primitive types and labels, respectively. With the exception of *Var*, these universes are also used in the JVM$_\mathcal{I}$.

2.1 The Machine JVM$_\mathcal{I}$ for Imperative Code

The JVM is a typed word-oriented stack-machine running the given bytecode *code* : *Code*. As a consequence the central dynamic part of a JVM$_\mathcal{I}$ state consists of a program counter *pc*, a local variable environment *loc* and an operand stack *opd*. The following declarations show their formalization: the first column defines the used types, the second column defines the state, and the third column defines the condition on the initial state. (We consider sequences as isomorphic to functions having an interval of natural numbers starting at 0 as their domain.)

$$
\begin{array}{lll}
Pc \;== Nat & pc \;:\; Pc & pc \;= next_{unlab}(0, code) \\
Loc == Varnum \to Word & loc \;:\; Loc & loc = \emptyset \\
Opd == Word^* & opd : Opd & opd = \epsilon
\end{array}
$$

The close analogy between the abstract and concrete program counters in Java$_\mathcal{I}$ and JVM$_\mathcal{I}$, the memories for local variables and for intermediate values, and their initializations reflects the refinement process, which applied to the machine Java$_\mathcal{I}$ yields JVM$_\mathcal{I}$. This correspondence will guide the justification of the correctness of this first step towards an implementation of Java on the JVM.

Local variables and the operand stack store values of the abstract universe *Word*. *Word*s are supposed to hold at least 32-bit quantities. Java's values, which occupy at most 32-bits, are represented on the level of the JVM as single *Word*s. Java's 64-bit values are mapped to multiple consecutive locations in the local environment and on the operand stack in an implementation dependent way. We define JVM values (*Val*) as sequences of Words, i.e. *Val == Word**. A valid word sequence has length one (32-bit) or two (64-bit). The JVM implements values and operations on Java datatypes as follows. Booleans are represented as integers: 0 is used for *false*, and 1 for *true*. Operations working on boolean, byte, short or char are not supported by the JVM. Instead, upon retrieving the value of a boolean, byte, char or short, it is automatically cast into an int. When writing a value to a boolean, byte, char or short variable, an int is passed and the JVM truncates it to the relevant size.

For the JVM$_\mathcal{I}$ we use two static code traversing functions *next* and *jump*, which yield the next statement to be executed and the next statement after the given labeled statement, respectively. Both functions are defined using an auxiliary function $next_{unlab}$ that skips label instructions. (The expression $\iota\, x \mid p(x)$ denotes the uniquely determined object x that satisfies $p(x)$.)

$next(pc, code)\quad = next_{unlab}(pc + 1, code)$
$jump(l, code)\quad = next_{unlab}(\iota\, pc \mid code(pc) = \text{label}\,(l))$
$next_{unlab}(pc, code) = min\{pc' \mid pc' \geq pc \wedge \forall l \mid code(pc') \neq \text{label}\,(l)\}$

We also use the following JVM$_\mathcal{I}$ macros, where the homonymy with Java$_\mathcal{I}$ macros reflects the refinement relations on which our correctness proof is based.

$proceed\quad\ == pc := next(pc, code)$
$goto(l)\quad\ \ == pc := jump(l, code)$
$pc\ \text{is}\ instr\ == code(pc) = instr$

The following rules define the semantics of the JVM$_\mathcal{I}$ instructions.

if pc is const (lit)
then
    $opd := \widetilde{lit} \cdot opd$
    $proceed$
if pc is uapply $(\odot)\wedge$
    $(v, opd') = split(\mathcal{A}(\odot), opd)$
then
    $opd := \widetilde{\odot}\, v \cdot opd'$
    $proceed$
if pc is bapply $(\otimes)\wedge$
    $(v_2, v_1, opd') = split(\mathcal{A}(\otimes), opd)\wedge$
    $(\otimes \in DivMods) \Rightarrow (v_2 \neq 0)$
then
    $opd := v_1 \widetilde{\otimes} v_2 \cdot opd'$
    $proceed$
if pc is dup $(t)\wedge$
    $(v, opd') = split(t, opd)$
then
    $opd := v \cdot v \cdot opd'$
    $proceed$
if pc is pop $(t)\wedge$
    $(v, opd') = split(t, opd)$
then
    $opd := opd'$
    $proceed$

if pc is load (x, t)
then
    if $sizeof(t) = 1$ **then**
      $opd := loc(x) \cdot opd$
    else if $sizeof(t) = 2$ **then**
      $opd := loc(x) \cdot loc(x+1) \cdot opd$
    $proceed$
if pc is store $(x, t)\wedge$
    $(v, opd') = split(t, opd)$
then
    $opd := opd'$
    if $sizeof(t) = 1$ **then**
      $loc(x) := v(0)$
    else if $sizeof(t) = 2$ **then**
      $loc(x+1) := v(1)$
      $loc(x) := v(0)$
    $proceed$
if pc is goto (l) **then**
    $goto(l)$

if pc is ifZero $(l)\wedge$
    $w \cdot opd' = opd$
then
    $opd := opd'$
    if $w = 0$ **then** $goto(l)$
          else $proceed$

A const instruction pushes the JVM value $\widetilde{lit}$ (one or two words) on the operand stack. An unary (binary) operator changes the value(s) on top of the operand stack. The unary (binary) operators are assumed to have the same meaning as in Java (i.e. $\widetilde{\odot}$ ($\widetilde{\otimes}$)), although they may operate on extended domains. In order to abstract from the different value sizes, we use the function $split : (Typ^*, Opd) \rightarrow (Val^*, Opd)$, which given a sequence of n types and the operand stack, takes the top n values from the operand stack, such that the ith value has the size of the ith type. The function $\mathcal{A}(op)$ returns the argument types of op. The instructions dup and pop duplicate and remove the top stack value, respectively. A load instruction loads the value stored under the location x on top of the stack. If the type of x is a double or long, the next two locations are pushed on top of the stack. A store instruction stores the top (two) word(s) of the operand stack in the local environment at offset x (and $x+1$). A goto instruction causes execution to jump to the next instruction determined by the label. The ifzero instruction is a conditional goto. If the value on top of the operand stack is 0, execution continues at the next instruction determined by the label, otherwise execution proceeds.

The abstract nature of the $JVM_\mathcal{I}$ instructions is reflected in their parameterization by types and operators. It allows us to restrict our attention to a small set of JVM instructions (or better instruction classes) without losing the generality of our model with respect to the JVM specification [6]. The extended version of this paper [1] shows how to refine these parameterized instruction to JVM's real ones.

2.2 Compilation of $Java_\mathcal{I}$ Programs to $JVM_\mathcal{I}$ Code

This section defines the compiling function from $Java_\mathcal{I}$ to $JVM_\mathcal{I}$ code. More efficient compilation schemes can be introduced but we leave optimizations for further refinement steps.

The compilation $\mathcal{E} : Exp \rightarrow Code$ of (occurrences of) $Java_\mathcal{I}$ expressions to $JVM_\mathcal{I}$ instructions is standard. The resulting sequence of instructions has the effect of storing the value of the expression on top of the operand stack.

To improve readability, we use the following conventions for the presentation of the compilation: We suppress the routine machinery for a consistent assignment of (occurrences of) Java variables x to JVM variable numbers $\overline{x}$. Similarly, we suppress the trivial machinery for label generation. Label providing functions lab_i, $i \in Nat$, are defined on occurrences of expressions and statements, are supposed to be injective and to have disjoint ranges. Functions $\mathcal{T}$ defined on occurrences of variables and expressions return their type. We abbreviate: 'Let e be an occurrence of exp in $\mathcal{E}(e) = \ldots$' by '$\mathcal{E}(e \text{ as } exp) = \ldots$'.

$$\begin{aligned}
\mathcal{E}(lit) &= \text{const}\,(lit) \\
\mathcal{E}(\odot\, e) &= \mathcal{E}\,e \cdot \text{uapply}\,(\odot) \\
\mathcal{E}(e_1 \otimes e_1) &= \mathcal{E}\,e_1 \cdot \mathcal{E}\,e_2 \cdot \text{bapply}\,(\otimes) \\
\mathcal{E}(x) &= \text{load}\,(\overline{x}, \mathcal{T}(x)) \\
\mathcal{E}(x\ =\ e) &= \mathcal{E}\,e \cdot \text{dup}\,(\mathcal{T}(e)) \cdot \text{store}\,(\overline{x}, \mathcal{T}(x)) \\
\mathcal{E}(e \text{ as } e_1 ?\ e_2 : e_3 :) &= \mathcal{E}\,e_1 \cdot \text{ifZero}\,(lab_1(e)) \cdot \\
&\quad \mathcal{E}\,e_2 \cdot \text{goto}\,(lab_2(e)) \cdot \text{label}\,(lab_1(e)) \cdot \mathcal{E}\,e_3 \cdot \text{label}\,(lab_2(e))
\end{aligned}$$

Also the compilation $\mathcal{S} : Stm \rightarrow Code$ of $Java_\mathcal{I}$ statements to $JVM_\mathcal{I}$ instructions is standard. The compilation of break lab; and continue lab; uses the auxiliary function $target : Stm \times Lab \rightarrow Stm$. This function provides for occurrences of statements and labels the occurrence of the enclosing labeled statement in the given program.

$$\begin{aligned}
\mathcal{S}(;) &= \epsilon \\
\mathcal{S}(e;) &= \mathcal{E}e \cdot \texttt{pop}\,(\mathcal{T}(e)) \\
\mathcal{S}(\{s_1 \ldots s_m\}) &= \mathcal{S}s_1 \cdot \ldots \cdot \mathcal{S}s_m
\end{aligned}$$

$$\begin{aligned}
\mathcal{S}(s \text{ as if } (e)\, s_1 \text{ else } s_2) &= \mathcal{E}e \cdot \texttt{ifZero}\,(lab_1(s)) \cdot \\
&\quad \mathcal{S}s_1 \cdot \texttt{goto}(lab_2(s)) \cdot \texttt{label}(lab_1(s)) \cdot \mathcal{S}s_2 \cdot \texttt{label}(lab_2(s)) \\
\mathcal{S}(s \text{ as while } (e)\, s_1) &= \texttt{label}\,(lab_1(s)) \cdot \mathcal{E}e \cdot \texttt{ifZero}\,(lab_2(s)) \cdot \\
&\quad \mathcal{S}s_1 \cdot \texttt{goto}\,(lab_1(s)) \cdot \texttt{label}\,(lab_2(s)) \\
\mathcal{S}(s \text{ as } lab : s_1) &= \texttt{label}\,(lab_1(s)) \cdot \mathcal{S}s_1 \cdot \texttt{label}\,(lab_2(s)) \\
\mathcal{S}(s \text{ as continue } lab;) &= \texttt{goto}\,(lab_1(target(s, lab))) \\
\mathcal{S}(s \text{ as break } lab;) &= \texttt{goto}\,(lab_2(target(s, lab)))
\end{aligned}$$

Correctness Theorem for Java$_\mathcal{I}$/JVM$_\mathcal{I}$. Via the refinement relation and under the assumptions stated above, the result of executing any Java$_\mathcal{I}$ program in the machine Java$_\mathcal{I}$ is equivalent to the result of executing the compiled program on the machine JVM$_\mathcal{I}$.

3 JVM$_\mathcal{C}$ and the Compilation of Class Code

In this section we extend the basic JVM$_\mathcal{I}$ machine to the machine JVM$_\mathcal{C}$, which handles class (also called static) fields, class methods and class initializers. JVM$_\mathcal{C}$ thus stands for a machine that supports modules, module-local variables and procedures. We add the clauses for compiling class field access, class field assignment, class method calls and return statements to the definition of the Java$_\mathcal{I}$ compilation function.

The following grammar shows the extension of the syntax of Java$_\mathcal{I}$ to the syntax of Java$_\mathcal{C}$. Furthermore, we define the corresponding JVM$_\mathcal{C}$ instructions:

$$\begin{aligned}
Exp &::= \ldots \\
&\mid FieldSpec \\
&\mid FieldSpec = Exp \\
&\mid MethSpec(Exp^*) \\
Stm &::= \ldots \\
&\mid \texttt{return}; \\
&\mid \texttt{return}\, Exp; \\
Init &::= \texttt{static}\, Stm
\end{aligned}$$

$$\begin{aligned}
Instr &::= \ldots \\
&\mid \texttt{getstatic}\,(FieldSpec \times Typ) \\
&\mid \texttt{putstatic}\,(FieldSpec \times Typ) \\
&\mid \texttt{invokestatic}\,(MethSpec) \\
&\mid \texttt{return}\,(Typ) \\
Fcty &== (Typ^* \times Typ) \\
FieldSpec &== (Class \times Field) \\
MethSpec &== (Class \times Meth \times Fcty)
\end{aligned}$$

JVM$_\mathcal{C}$ provides instructions to load and store class fields, and to call and to return from class methods. Both grammars are based on the same abstract definition of field and method specifications. Field specifications consist of a class and a field name, because Java and the JVM allow fields in different classes to have the same name. Method specifications additionally have a functionality (a sequence of argument types and a result type, which can be void), because Java and the JVM support classes with methods having the same name but taking different parameter types.

Field and method specifications use the abstract universes *Class*, *Field* and *Method*. *Class* is assumed to stand for fully qualified Java class names, *Field* and *Method* for identifiers.

3.1 The Machine JVM_C for Class Code

JVM and Java programs are structured into classes, which establish the program's execution environment. For a general, high-level definition of a provably correct compilation scheme from Java to JVM Code, we can abstract from many data structure specifics of the particular JVM class format. This format is called class file in the JVM specification [6].

Our abstract class file refines in a natural way the class environment of $Java_C$, providing for every class its kind (whether it is a class or an interface), its superclass (if there is any), a list of the interfaces the class implements, and a table for fields and methods. Class files do not include definitions for fields or methods provided by any superclass.

$$
\begin{aligned}
&Env &&== Class \rightarrow ClassDec \\
&ClassKind &&::= AClass \mid AnInterface \\
&ClassDec &&== (\,kind : ClassKind \times super : [Class] \times ifaces : Class^* \times \\
&&&\quad fTab : Field \rightarrow FieldDec \times mTab : (Meth \times Fcty) \rightarrow MethDec)
\end{aligned}
$$

In JVM_C fields and methods can only be static. Fields have a type and optionally a constant value. If a method is implemented in the class, the method body defines its code.

$$
\begin{aligned}
&FieldDec &&== (fKind : MemberKind \times fTyp : Typ \times fConstVal : [Val]) \\
&MethDec &&== (mKind : MemberKind \times mBody : [Code]) \\
&MemberKind &&::= Static
\end{aligned}
$$

In JVM_C we have a fixed environment $env : Env$, defined by the given program. The following functions operate on this environment. The function *mCode* retrieves for a given method specification the method's code to be executed. The function *fInitVal* yields for a given field specification the field's constant value, provided it is available; otherwise, the function returns the default value of the field's type (where $default : Typ \rightarrow Val$).

$$
\begin{aligned}
&mCode(c, m, f) = mBody(mTab(env(c))(m, f)) \\
&fInitVal(c, f) = \textbf{case } fTab(env(c))(f) \textbf{ of } (\_,\_, val) : val \\
&\qquad\qquad\qquad\qquad\qquad\qquad\qquad (\_, fTyp, [\,]) : default(fTyp)
\end{aligned}
$$

The function *supers* calculates the transitive closure of *super*. The function *cfields* returns the set of all fields declared by the class.

$$
\begin{aligned}
&supers : Class \rightarrow Class^* \\
&cfields : Class \rightarrow \mathcal{P}\, FieldSpec
\end{aligned}
$$

For these functions the homonymy to Java$_C$ functions shows the data refinement relation in going from Java$_C$ to JVM$_C$.

Due to the presence of method calls in JVM$_C$ we have to embed the one single JVM$_I$ frame (pc, loc, opd) into the JVM$_C$ frame stack *frames*, enriched by a fourth component which always holds the dynamic chain of method specifications. This embedding defines the refinement relation between JVM$_I$ and JVM$_C$. We refine the static function *code*, so that it always denotes the code stored in the environment under the current method specification *mspec*. The current class, method and functionality are denoted by *cclass*, *cmeth* and *cfcty*, respectively, where $mspec = (cclass, cmeth, cfcty)$.

$$\begin{array}{llll}
pcs & : & Pc^* & \qquad pc == top(pcs) \\
locs & : & Loc^* & \qquad loc == top(locs) \\
opds & : & Opd^* & \qquad opd == top(opds) \\
mspecs & : & MethSpec^* & \qquad mspec == top(mspecs) \\
frames & == & (pcs, locs, opds, mspecs) & \qquad code == mCode(mspec)
\end{array}$$

Before a class can be used its class initializers must be executed. At the JVM level class initializers appear as class methods with the special name <clinit>. Initialization must be done lazily, i.e. when a class is *first used* in Java, and when a reference is *resolved* in the JVM. Resolution is the process of checking symbolic references from the current class to other classes and interfaces. Since Java's notion of class initialization does not correspond to the related class resolution notion of the JVM, we name the initialization related functions and sets differently. A class can be in one of three states. We introduce a dynamic function *res*, which records the current resolution state. A class is *resolved*, if resolution for this class is in progress or done.

$res : Class \to ResolvedState$
$ResolvedState ::= Unresolved \mid Resolved \mid InProgress$
$resolved(state) = state \in \{InProgress, Resolved\}$

The JVM specification [6] uses symbolic references, namely field and method specifications, to support binary compatibility, cf. [4]. As a consequence, the calculation of field offsets and of method offsets is implementation dependent. Therefore, we keep the class field access abstract and define the storage function for class fields to be the same in Java$_C$ and JVM$_C$, namely

$glo : FieldSpec \to Val.$

The runs of JVM$_C$ start with calling the class method *main* of a distinguished class *Main* being part of the environment. However, before *main* is executed, its class *Main* has to be initialized. Therefore, the frame stack initially has two entries: the *main* method at the bottom and the <clinit> method on the top. All classes are initially unresolved and all fields are set to their initial values. This initialization also refines the corresponding conditions imposed on Java$_C$:

$$\begin{array}{ll}
pcs = start(clinit') \cdot start(main') & res = \{(c, Unresolved) \mid c \in dom(env)\} \\
locs = \epsilon \cdot \epsilon & glo = \{(fs, fInitVal(fs)) \mid c \in dom(env), \\
opds = \epsilon \cdot \epsilon & \phantom{glo = \{} fs \in cfields(c)\} \\
mspecs = clinit' \cdot main' &
\end{array}$$

The method specifications $clinit'$ and $main'$ denote the class methods `<clinit>` and $main$ of class $Main$. The macro $start$ returns the first instruction of the code of the given method specification.

$clinit'$ == $proc(Main, $ `<clinit>` $)$ $start(ms)$ == $next_{unlab}(0, mCode(ms))$
$main'$ == $proc(Main, main)$ $proc(c, m)$ == $(c, m, (\epsilon, \text{void}))$

The following rules for JVM$_C$ define the semantics of the new JVM instructions, provided the class of the field or method specification is already resolved. A `getstatic` instruction loads the value (one or two words), stored under the field specification in the global environment, on top of the operand stack. A `putstatic` instruction stores the top (two) word(s) of the operand stack in the global environment at the given field specification. An `invokestatic` instruction pops the arguments from the stack and sets pc to the next instruction. The arguments of the invoked method are placed in the local variables of the new stack frame, and execution continues at the first instruction of the new method. A `return` instruction is 'inverse' to `invokestatic`. It pops a value from the top of the stack and pushes it onto the operand stack of the invoker. All other items in the current stack are discarded. (If the return type is void, $split$ returns the empty sequence as its value.)

if pc is `getstatic` $((c, f), t) \wedge$
  $resolved(res(c))$
then
  $opd := glo(c, f) \cdot opd$
  $proceed$
if pc is `putstatic` $((c, f), t) \wedge$
  $resolved(res(c)) \wedge$
  $(v, opd') = split(t, opd)$
then
  $opd := opd'$
  $glo(c, f) := v$
  $proceed$

if pc is `invokestatic` $(c, m, (ts, t)) \wedge$
  $resolved(res(c)) \wedge (t_1, \ldots, t_n) = ts \wedge$
  $(v_n, \ldots, v_1, opd') = split(t_n, \ldots, t_1, opd)$
then
  $call(next(pc, code), v_1 \cdot \ldots \cdot v_n,$
    $opd', (c, m, (ts, t)))$
if pc is `return` $(t) \wedge$
  $(v, opd') = split(t, opd)$
then
  $return(v)$

The macros $call$ and $return$ update the frames as follows:

$call(pc, loc, opd, mspec)$ ==
  let $pc_0 \cdot pcs' = pcs$
    $opd_0 \cdot opds' = opds$ **in**
  pcs := $start(mspec) \cdot pc \cdot pcs'$
  $locs$:= $loc \cdot locs$
  $opds$:= $\epsilon \cdot opd \cdot opds'$
  $mspecs$:= $mspec \cdot mspecs$

$return(v)$ ==
  if $len(pcs) = 1$ **then**
    $pcs(0) := undef$
  else let $opd_0 \cdot opd_1 \cdot opds' = opds$ **in**
    pcs := $pop(pcs)$
    $locs$:= $pop(locs)$
    $opds$:= $(v \cdot opd_1) \cdot opds'$
    $mspecs$:= $pop(mspecs)$

Execution starts in a state in which no class is resolved. A class is resolved, when it is first referenced. Before a class is resolved, its superclass is resolved (if any). Interfaces are not resolved at this time, although this is not specified in Java's language reference manual [4]. On the level of the JVM resolution leads to three rules. First, resolutions starts, i.e. the class method `<clinit>` is implicitely called, when the class referred to in a `get-`, `put-` or `invokestatic` instruction is not resolved. Second, the class initializer records the fact that class initialization is in progress and calls the superclass initializer recursively. Third, after having executed the class initializer, it is recorded that the class is resolved.

if (pc is **putstatic** $((c,\_),\_) \vee$
    pc is **getstatic** $((c,\_),\_) \vee$
    pc is **invokestatic** $(c,\_,\_)) \wedge$
    $\neg resolved(res(c))$
then
    $call(pc, \emptyset, opd, proc(c, \texttt{<clinit>}))$

if $res(cclass) = Unresolved$
then
    $res(cclass) := InProgress$
    if $supers(cclass) \neq \epsilon \wedge$
        $\neg resolved(res(super(cclass)))$
    then
        $call(pc, \emptyset, opd,$
                $proc(super(cclass),\texttt{<clinit>}))$
if pc is **return** $(t) \wedge cmeth = \texttt{<clinit>}$
then
    $res(cclass) := Resolved$

Firing the second rule depends on the condition that the current class is *Unresolved*—this is the reason why we called the initializer in the first rule. To suppress the simultaneous firing of other rules we strengthen the macro 'is':

pc is $instr == code(pc) = instr \wedge resolved(res(cclass))$

This guarantees that an instruction can only be executed, if the current class is resolved. Opposite to the second rule, the third rule fires simultaneously with the previously presented rule for the **return** instruction.

3.2 Compilation of Java$_C$ Programs to JVM$_C$ Code

The compilation of Java$_\mathcal{I}$ expressions is extended by defining the compilation of class field access, class field assignment, and by the compilation of calls of class methods.

$\mathcal{E}(fspec)$ $= \texttt{getstatic}\,(fspec, \mathcal{T}(fspec))$
$\mathcal{E}(fspec = e)$ $= \mathcal{E}e \cdot \texttt{dup}\,(\mathcal{T}(e)) \cdot \texttt{putstatic}\,(fspec, \mathcal{T}(fspec))$
$\mathcal{E}(mspec(e_1,\ldots,e_n)) = \mathcal{E}e_1 \cdot \ldots \cdot \mathcal{E}e_n \cdot \texttt{invokestatic}\,(mspec)$

We add the clause for **return** statements to the Java$_\mathcal{I}$ compilation.

$\mathcal{S}(\textbf{return } e;) = \mathcal{E}e \cdot \texttt{return}\,(\mathcal{T}(e))$
$\mathcal{S}(\textbf{return};) \;\;\;\;= \texttt{return}\,(\textit{void})$

To compile a class initializer (the *Init* phrase) means to compile its statement as the body of the static <clinit> method.

The extension of Java$_\mathcal{I}$/JVM$_\mathcal{I}$ to Java$_C$/JVM$_C$ is conservative, i.e. purely incremental. For the proof of the *Correctness Theorem for* Java$_C$/JVM$_C$ it therfore suffices to extend the theorem from Java$_\mathcal{I}$/JVM$_\mathcal{I}$ to the new expressions and statements occurring in Java$_C$/JVM$_C$.

4 JVM$_\mathcal{O}$ and the Compilation of Java$_\mathcal{O}$ Programs

In this section we extend the machine JVM$_\mathcal{C}$ to JVM$_\mathcal{O}$. This machine handles the object-oriented features of Java programs, namely instances, instance creation, instance field access, instance method calls with late binding, type casts and null pointers. We add the corresponding new phrases to the definition of the compilation function.

We recall the grammar for the new expressions of Java$_\mathcal{O}$ and define the corresponding JVM$_\mathcal{O}$ instructions:

$Exp ::= \ldots$
| this
| new $ConstrSpec\,(Exp^*)$
| $ConstrSpec\,(Exp^*)$
| $Exp.FieldSpec$
| $Exp.FieldSpec = Exp$
| $Exp.MethSpec\{CallKind\}(Exp^*)$
| Exp instanceof $Class$
| $(Class)\,Exp$
$ConstrSpec == (Class \times Typ^*)$

$Instr ::= \ldots$
| new $(Class)$
| getfield $(FieldSpec \times Typ)$
| putfield $(FieldSpec \times Typ)$
| dup$\_(Typ^*)$
| invokeinstance $(MethSpec \times CallKind)$
| instanceof $(Class)$
| checkcast $(Class)$
$CallKind ::= Constr \mid Nonvirtual \mid Virtual \mid Super$

Java$_\mathcal{O}$ uses constructor specifications to uniquely denote overloaded instance constructors. JVM$_\mathcal{O}$ provides instructions to allocate a new instance, to access or assign its fields, to duplicate values, to invoke instance methods and to check instance types. Java$_\mathcal{O}$ and JVM$_\mathcal{O}$ use the universe $CallKind$, to distinguish the particular way in which instance methods are called.

4.1 The Machine JVM$_\mathcal{O}$ for Object-Oriented Code

JVM$_\mathcal{O}$ uses the same abstract class file as JVM$_\mathcal{C}$. However, instance fields and instance methods—in opposite to class fields and class methods—are not static but dynamic. So we extend the universe $MemberKind$ as follows:

$MemberKind ::= \ldots \mid Dynamic$

The JVM specification [6] fixes the class file. However, the specification does not explain how instances are stored or instance methods are accessed. So we extend the signature of JVM$_\mathcal{C}$ in JVM$_\mathcal{O}$ in the same way as the signature of Java$_\mathcal{C}$ is extended in Java$_\mathcal{O}$. We introduce the following static functions (homonymy with Java$_\mathcal{O}$ functions) that look up information in the global environment:

$dfields \quad : Class \to \mathcal{P}\,FieldSpec$
$dlookup \quad : Class \times MethSpec \to Class$
$compatible : Class \times Class \to Bool$

The function $dfields$ determines the instance fields of a class and of all its superclasses (if any). The function $dlookup$ returns the first (super) class for the given method specification, which implements this method. The expression

$compatible(myType, tarType)$ returns $true$ if $myType$ is assignment compatible with $tarType$ [4]. Note that at the JVM level, there is no special lookup function for constructors. Instead, Java's constructors appear in the JVM as instance initialization methods with the special name <init>.

$JVM_\mathcal{O}$ and $Java_\mathcal{O}$ have the same dynamic functions for memorizing the class and the instance field values of a reference. In both machines they are initially empty. References can be obtained from the abstract universe Ref, which is assumed to be a subset of $Word$. (Likewise, we also assume that $null$ is an element of $Word$.)

$classOf : Ref \rightarrow Class$ $classOf = \emptyset$
$dyn\quad : Ref \times FieldSpec \rightarrow Val$ $dyn\quad = \emptyset$

The following rules define the semantics of the new instructions of $JVM_\mathcal{O}$, provided that the involved class is resolved.

if pc is $\mathbf{new}\,(c) \wedge$
$\quad resolved(res(c))$
then
$\quad$ **extend** Ref **by** r
$\quad\quad classOf(r) := c$
$\quad\quad$ **vary** fs **over** $dfields(c)$
$\quad\quad\quad dyn(r, fs) := fInitVal(fs)$
$\quad\quad opd := r \cdot opd$
$\quad\quad proceed$
if pc is $\mathbf{getfield}\,((c, f), t) \wedge$
$\quad resolved(res(c)) \wedge$
$\quad r \cdot opd' = opd \wedge$
$\quad r \neq null$
then
$\quad opd := dyn(r, (c, f)) \cdot opd'$
$\quad proceed$
if pc is $\mathbf{putfield}\,((c, f), t) \wedge$
$\quad resolved(res(c)) \wedge$
$\quad (v, r, opd') = split(t, c, opd) \wedge$
$\quad r \neq null$
then
$\quad opd := opd'$
$\quad dyn(r, (c, f)) := v$
$\quad proceed$
if pc is $\mathbf{dup\_}(t_1, t_2) \wedge$
$\quad (v_2, v_1, opd') = split(t_2, t_1, opd)$
then
$\quad opd := v_2 \cdot v_1 \cdot v_2 \cdot opd'$
$\quad proceed$

if pc is $\mathbf{invokeinstance}\,((c, m, (ts, t)), k) \wedge$
$\quad resolved(res(c)) \wedge (t_1, \ldots t_n) = ts \wedge$
$\quad (v_n, \ldots, v_1, r, opd') =$
$\quad\quad split(t_n, \ldots, t_1, c), opd) \wedge$
$\quad r \neq null$
then
$\quad call(next(pc, code), r \cdot v_1 \cdot \ldots \cdot v_n,$
$\quad\quad opd', (c', m, (ts, t)))$
where
$\quad c' =$ **case** k **of**
$\quad\quad Constr\quad\quad : c$
$\quad\quad Nonvirtual : cclass$
$\quad\quad Virtual\quad\quad : dlookup(\,classOf(r),$
$\quad\quad\quad\quad\quad\quad\quad\quad\quad m, (ts, t))$
$\quad\quad Super\quad\quad\,\, : dlookup(\,super(cclass),$
$\quad\quad\quad\quad\quad\quad\quad\quad\quad m, (ts, t))$
if pc is $\mathbf{instanceof}\,(c) \wedge$
$\quad resolved(res(c)) \wedge$
$\quad r \cdot opd' = opd$
then
$\quad opd := (r \neq null \wedge$
$\quad\quad\quad\quad compatible(classOf(r), c) \cdot opd'$
$\quad proceed$
if pc is $\mathbf{checkcast}\,(c) \wedge$
$\quad resolved(res(c)) \wedge$
$\quad r \cdot opd' = opd \wedge$
$\quad (r = null \vee compatible(classOf(r), c))$
then
$\quad proceed$

A **new** instruction allocates a fresh reference using the domain extension update of ASMs. The $classOf$ the reference is set to the given class, the class instance fields are set to default values, and the new reference is pushed on the operand stack. A **getfield** instruction pops the target reference from the stack, retrieves the value of the field identified by the given field specification from the dynamic store and pushes one or two words on the operand stack. A putfield instruction pops a value and the target reference from the stack and sets the dynamic store at the point of the target reference and the given field specification to the popped value. A dup_ instruction duplicates the top value and inserts the duplicate below the top value on the stack. An invokeinstance instruction pops the arguments and the target reference (which denotes the instance whose method is being called) from the stack and sets pc to the next instruction. The method's implementing class is being located. If the call kind is

- *Constr*, the method specification denotes a constructor; its code is located in the given class. (The given method m must be <init>.)
- *Nonvirtual*, the method specification denotes a private method; its code is located in the current class. (The given class c must be $cclass$.)
- *Virtual*, the implementing class is looked up dynamically, starting at the class of the target reference.
- *Super*, the method is looked up dynamically, starting at the superclass of the current class. (The given class c must be $super(cclass)$.)

Once a method has been located, invoke calls the method: The arguments for the invoked method are placed in the local variables of the new stack frame, placing the target reference r (denoting this in Java) in $loc(0)$. Execution continues at the first instruction of the new method. An instanceof instruction pops a reference from the operand stack. If the reference is not *null* and assignment compatible with the required class, the integer 1 is pushed on the operand stack, otherwise 0 is pushed. A checkcast instruction checks that the top value on the stack is an instance of the given class.

If the class c of a field or method specification or if the explicitly given class c of a new, an instanceof or a checkcast instruction is not resolved, the JVM first resolves c, i.e. calls c's <clinit> method, before the instruction is executed.

if $(pc$ is new $(c) \vee pc$ is putfield $((c, \_), \_) \vee pc$ is getfield $((c, \_), \_) \vee$
pc is invokeinstance $((c, \_, \_), \_) \vee pc$ is instanceof $(c) \vee pc$ is checkcast $(c)) \wedge$
$\neg resolved(res(c))$
then
$call(pc, \emptyset, opd, proc(c, <\text{clinit}>))$

4.2 Compilation of Java$_\mathcal{O}$ Programs to JVM$_\mathcal{O}$ Code

Since there are no new statements in Java$_\mathcal{O}$, only the compilation of Java$_\mathcal{C}$ expressions has to be extended to the new Java$_\mathcal{O}$ expressions. The reference this is implemented as the distinguished local variable number 0.

$\mathcal{E}(\text{this})$ $= \text{load}\,(0, \mathcal{T}(\text{this}))$
$\mathcal{E}(\text{new}\,(c,\,ts)\,(e_1,\ldots,e_n)) = \text{new}\,(c) \cdot \text{dup}\,(c) \cdot \mathcal{E}\,e_1 \cdot \ldots \cdot \mathcal{E}\,e_n \cdot$
  $\text{invokeinstance}\,((c, \texttt{<init>}, (ts, \texttt{void})), \textit{Constr})$
$\mathcal{E}((c,\,ts)\,(e_1,\ldots,e_n)) = \text{load}\,(0, \mathcal{T}(\text{this})) \cdot \mathcal{E}\,e_1 \cdot \ldots \cdot \mathcal{E}\,e_n \cdot$
  $\text{invokeinstance}\,((c, \texttt{<init>}, (ts, \texttt{void})), \textit{Constr})$
$\mathcal{E}(e.\textit{fspec}) = \mathcal{E}\,e \cdot \text{getfield}\,(\textit{fspec}, \mathcal{T}(\textit{fspec}))$
$\mathcal{E}(e_1.\textit{fspec} = e_2) = \mathcal{E}\,e_1 \cdot \mathcal{E}\,e_2 \cdot$
  $\text{dup}\_(\mathcal{T}(e_1), \mathcal{T}(e_2)) \cdot \text{putfield}\,(\textit{fspec}, \mathcal{T}(\textit{fspec}))$
$\mathcal{E}(e.\textit{mspec}\{k\}(e_1,\ldots,e_n)) = \mathcal{E}\,e \cdot \mathcal{E}\,e_1 \cdot \ldots \cdot \mathcal{E}\,e_n \cdot \text{invokeinstance}\,(\textit{mspec}, k)$
$\mathcal{E}(e \text{ instanceof } c) = \mathcal{E}\,e \cdot \text{instanceof}\,(c)$
$\mathcal{E}((c)\,e) = \mathcal{E}\,e \cdot \text{checkcast}\,(c)$

Due to the conservativity of the extension of $\text{Java}_\mathcal{C}/\text{JVM}_\mathcal{C}$ to $\text{Java}_\mathcal{O}/\text{JVM}_\mathcal{O}$, for the proof of the *Correctness Theorem for* $\text{Java}_\mathcal{O}/\text{JVM}_\mathcal{O}$ it suffices to extend the theorem from $\text{Java}_\mathcal{C}/\text{JVM}_\mathcal{C}$ to the new expressions occurring in $\text{Java}_\mathcal{O}/\text{JVM}_\mathcal{O}$.

The definitions of class initialization for $\text{Java}_\mathcal{O}$ in [4] and resolution for $\text{JVM}_\mathcal{O}$ in [6] do not match because **instanceof** and class cast expressions in Java do not call the initialization of classes. In opposite, the JVM effect is to execute the initialization of the related class if it is not initialized yet. Under the assumption that also in Java these instructions trigger class initialization, these instructions preserve the theorem for $\text{Java}_\mathcal{O}/\text{JVM}_\mathcal{O}$.

5 JVM$_\mathcal{E}$ and the Compilation of Exception Treatment

In this section we extend JVM$_\mathcal{O}$ to JVM$_\mathcal{E}$ that handles exceptions. We add the compilation of the new Java$_\mathcal{E}$ statements and refine the compilation of jump and **return** statements.

The following grammars list the new statements of Java$_\mathcal{E}$ and the new JVM$_\mathcal{E}$ instructions. JVM$_\mathcal{E}$ provides instructions to raise an exception, to jump to and to return from subroutines embedded in methods.

$Stm ::= \ldots$	$Instr ::= \ldots$
$\mid$ **throw** Exp;	$\mid$ **athrow**
$\mid$ **try** Stm **catch** $(Typ, Var, Stm)^*$	$\mid$ **jsr** (Lab)
$\mid$ **try** Stm **finally** Stm	$\mid$ **ret** $(Varnum)$

5.1 The JVM$_\mathcal{E}$ Machine for Executing Exceptions

The JVM supports **try/catch** or **try/finally** by exception tables that list the exceptions of a method. When an exception is raised this table is searched for the handler. Exception tables refine the notion of method body as follows:

$MethDec\ == (mKind : MemberKind \times mBody : [Code \times Exception^*])$
$Exception\ == (from, to, handle : Lab \times catchTyp : [Class])$

The labels *from* and *to* define the range of the protected code; *handle* starts the exception handler for the optional type *catchTyp*. If no *catchTyp* is given (as is the case for `finally` statements), any exception is caught. We refine the function $mCode$ from $JVM_\mathcal{C}$ and introduce a new function $mExcs$, which returns the exceptions of the given method specification.

$$mCode(c, m, f) = fst(mBody(mTab(env(c))(m, f)))$$
$$mExcs(c, m, f) = snd(mBody(mTab(env(c))(m, f)))$$

If a class initializer raised an exception, which is not handled within the method, Java and therefore the JVM require that the method's class must be labeled as erroneous. So we extend the domain of $ResolvedState$ in the same way as we did for Java:

$$ResolvedState ::= \ldots \mid Error$$

If the thrown exception is not an `Error` or one of its subclasses, then $Java_\mathcal{E}$ and $JVM_\mathcal{E}$ throw an `ExceptionInInitializerError`. If a class should be resolved but is marked as erroneous, Java and therefore implicitely the JVM require that a `NoClassDefFoundError` is reported.

We formalize the run-time system search for a handler of an exception by a recursively defined function $catch$. This function first searches the active method using $catch'$. If no handler is found (the exception handler list is empty), the current method frame is discarded, the invoker frame is reinstated and $catch$ is called recursively. A handler is found if the pc is protected by some brackets $from$ and to, and the thrown exception is compatible with the $catchType$. In this case the operand stack is reduced to the exception and execution continues at the address of the exception handler. When $catch'$ returns from a `<clinit>` method, the method has thrown an uncaught exception; according to the strategy presented above the method's class must be labeled as erroneous.

$catch(r, ((pc \cdot pcs, loc \cdot locs, opd \cdot opds, mspec \cdot mspecs), res)) =$
  $catch'(mExcs(mspec))$ **where**
    $catch'(\epsilon) =$
      if $pcs = \epsilon$ **then**
        $((undef \cdot pcs, loc \cdot locs, opd \cdot opds, mspec \cdot mspecs), res)$
      else let $(c, m, \_) = mspec$
        $res' =$ **if** $m =$ `<clinit>` **then** $res \oplus \{(c, Error)\}$ **else** res **in**
        $catch(r, ((pcs, locs, opds, mspecs), res'))$
    $catch'((from, to, handle, catchTyp) \cdot excs) =$
      if $jump(from, mCode(mspec)) \leq pc < jump(to, mCode(mspec)) \land$
        $(catchTyp = [] \lor compatible(classOf(r), catchTyp))$ **then**
        $((jump(handle, mCode(mspec)) \cdot pcs, loc \cdot locs, r \cdot opds, mspec \cdot mspecs), res)$
      else $catch'(excs)$

The following rules define the semantics of $JVM_\mathcal{E}$ instructions. The `athrow` instruction pops a reference from the stack and throws the exception represented by that reference. The `jsr` instruction is used to implement Java's `finally` clause. This instruction pushes the address of the next instruction on the operand stack and jumps to the given label. This requires that the universe Pc (called

ReturnAddress in the JVM specification) is embedded in *Word*. The address, which is put on top of the stack, is used by ret to return from the subroutine, wherefore the return address first has to be stored in a local variable.

if pc is athrow $\land$
$\quad r \cdot opd' = opd \land$
$\quad r \neq null$
then
$\quad (frames, res) := catch(r, (frames, res))$

if pc is ret (x)
then
$\quad pc := loc(x)$

if pc is jsr (lab)
then
$\quad opd := next(pc, code) \cdot opd$
$\quad pc := goto(lab)$

if $res(cclass) = Error$
then
$\quad fail(\text{NoClassDefFoundError})$

If the current class is erroneous, the last rule throws a NoClassDefFoundError using the macro *fail(c)*. This macro replaces the following instruction sequence:

new (c), dup, invokeinstance $((c, <\texttt{init}>, (\epsilon, \texttt{void})), Constr)$, athrow

Whether or not the constructor is called is semantically irrelevant, as long as the constructors only call superclass constructors.

We refine in the obvious way rules that raise run-time exceptions. A typical representative of this rule kind is the refinement of bapply. It throws an ArithmeticException, if the operator is an integer or long division or remainder operator and the right operand is 0.

if pc is bapply $(\otimes) \land (0, v_1, opd') = split(\mathcal{A}(\otimes), opd) \land (\otimes \in DivMods)$
then
$\quad fail(\text{ArithmeticException})$

JVM$_\mathcal{E}$ throws a NullPointerException if the target reference of a getfield, putfield or invokeinstance instruction is *null*, or if the reference of the athrow instruction is *null*. The machine throws a ClassCastException, if the reference on top of stack is neither *null* nor assignment compatible with the required type.

5.2 Compilation of Java$_\mathcal{E}$ Statements to JVM$_\mathcal{E}$ Instructions

Since there are no new expression in Java$_\mathcal{E}$, only the compilation of Java$_\mathcal{O}$ statements has to be extended to the compilation of the new Java$_\mathcal{E}$ statements.

For try/catch statements, the compiled try clause is followed by a jump to the end of the compiled statement. Next the handlers are generated. Each handler stores the exception into the 'catch' parameter, followed by the code of the catch clause and a jump to the end of the compiled statement. For try/finally statements s, the try clause is compiled followed by a call to the embedded subroutine, which is generated for the finally clause. The subroutine first stores the return address into a fresh variable $ret(s)$, and finally calls ret $(ret(s))$. The handler for exceptions that are thrown in the try clause starts at $lab_3(s)$. The handler saves an exception of class Throwable, which is left on the operand stack, into the fresh local variable $exc(s)$, calls the subroutine, and rethrows the exception. Variable providing functions *exc*, *ret* and also *val* that is used below, return for occurences of statements *fresh* variable numbers. This means that any returned variable number must be unused when the exception, return address or return value is stored, and this variable definition must reach its corresponding use.

$\mathcal{S}(\texttt{throw } e;) = \mathcal{E} e \cdot \texttt{athrow}$
$\mathcal{S}(s \texttt{ as try } s_0 \texttt{ catch }(c_1, x_1, s_1) \ldots (c_m, x_m, s_m)) =$
 $\quad \texttt{label}\,(lab_1(s)) \cdot \mathcal{S} s_0 \cdot \texttt{goto}\,(lab_3(s)) \cdot \texttt{label}\,(lab_2(s)) \cdot$
 $\quad \texttt{label}\,(lab_{3+1}) \cdot \texttt{store}\,(\overline{x_1}, c_1) \cdot \mathcal{S} s_1 \cdot \texttt{goto}\,(lab_3(s)) \cdot \ldots \cdot$
 $\quad \texttt{label}\,(lab_{3+m}) \cdot \texttt{store}\,(\overline{x_m}, c_m) \cdot \mathcal{S} s_m \cdot \texttt{goto}\,(lab_3(s)) \cdot$
 $\quad \texttt{label}\,(lab_3(s))$
$\mathcal{S}(s \texttt{ as try } s_1 \texttt{ finally } s_2) =$
 $\quad \texttt{label}\,(lab_1(s)) \cdot \mathcal{S} s_1 \cdot \texttt{jsr}\,(lab_2(s)) \cdot \texttt{goto}\,(lab_4(s)) \cdot$
 $\quad \texttt{label}\,(lab_2(s)) \cdot \texttt{store}\,(ret(s), \texttt{ReturnAddress}) \cdot \mathcal{S} s_2 \cdot \texttt{ret}\,(ret(s)) \cdot$
 $\quad \texttt{label}\,(lab_3(s)) \cdot \texttt{store}\,(exc(s), \texttt{Throwable}) \cdot \texttt{jsr}\,(lab_2(s)) \cdot$
 $\quad \quad \quad \quad \quad \texttt{load}\,(exc(s), \texttt{Throwable}) \cdot \texttt{athrow} \cdot$
 $\quad \texttt{label}\,(lab_4(s))$

If a jump statement is nested inside a `try` clause of a `try/finally` statement and its corresponding target statement contains `try/finally` statements, then all `finally` clauses between the jump statement and the target have to be executed in innermost order. The compilation uses the function *takeFinallyUntilTarget* : $Stm \times Lab \rightarrow Stm^*$, which given an occurrence of a statement and a label, returns in innermost order all occurrences of `try/finally` statements up to the target statement. For `return e` the compiler stores the result of the compiled expression e in a *fresh* temporary variable *val*. The compiler then generates code to jump to all outer `finally` statements in this method using the static function *takeFinally* : $Stm \rightarrow Stm^*$. Thereafter, the local variable *val* is pushed back onto the operand stack and the intended `return` instruction is executed.

$\mathcal{S}(s \texttt{ as break } lab;) = \texttt{let } (s_1, \ldots, s_m) = takeFinallyUntilTarget(s, lab) \texttt{ in}$
 $\quad \texttt{jsr}\,(lab_2(s_1)) \cdot \ldots \cdot \texttt{jsr}\,(lab_2(s_m)) \cdot \texttt{goto}\,(lab_2(target(s, lab)))$
$\mathcal{S}(s \texttt{ as continue } lab;) = \texttt{let } (s_1, \ldots, s_m) = takeFinallyUntilTarget(s, lab) \texttt{ in}$
 $\quad \texttt{jsr}\,(lab_2(s_1)) \cdot \ldots \cdot \texttt{jsr}\,(lab_2(s_m)) \cdot \texttt{goto}\,(lab_1(target(lab, s)))$
$\mathcal{S}(s \texttt{ as return } e;) = \texttt{let } (s_1, \ldots, s_m) = takeFinally(s) \texttt{ in}$
 $\quad \mathcal{E} e \cdot \texttt{store}\,(val(s), \mathcal{T}(e)) \cdot$
 $\quad \texttt{jsr}\,(lab_2(s_1)) \cdot \ldots \cdot \texttt{jsr}\,(lab_2(s_m)) \cdot \texttt{load}\,(val(s), \mathcal{T}(e)) \cdot \texttt{return}\,(\mathcal{T}(e))$
$\mathcal{S}(s \texttt{ as return};) = \texttt{let } (s_1, \ldots, s_m) = takeFinally(s) \texttt{ in}$
 $\quad \texttt{jsr}\,(lab_2(s_1)) \cdot \ldots \cdot \texttt{jsr}\,(lab_2(s_m)) \cdot \texttt{return}\,(\texttt{void})$

In the generation of an exception table inner `try` phrases are concatenated before the outer ones. This guarantees that exceptions are searched in innermost order.

$\mathcal{X}(s \text{ as try } s_0 \text{ catch } (c_1, x_1, s_1), \ldots (c_m, x_m, s_m)) =$
$\quad \mathcal{X} s_0 \cdot (lab_1(s), lab_2(s), lab_{3+1}, c_1) \cdot \mathcal{X} s_1 \cdot \ldots \cdot$
$\quad \quad (lab_1(s), lab_2(s), lab_{3+m}, c_m) \cdot \mathcal{X} s_m$
$\mathcal{X}(s \text{ as try } s_1 \text{ finally } s_2) = \mathcal{X} s_1 \cdot (lab_1(s), lab_2(s), lab_3(s), []) \cdot \mathcal{X} s_2$
$\mathcal{X}(\{ s_1 \ldots s_n \}) \quad \quad \quad = \mathcal{X} s_1 \cdot \ldots \cdot \mathcal{X} s_n$
$\mathcal{X}(\text{if } (e) \, s_1 \text{ else } s_2) \quad = \mathcal{X} s_1 \cdot \mathcal{X} s_2$
$\mathcal{X}(\text{while } (e) \, s) \quad \quad \quad = \mathcal{X} s$
$\mathcal{X}(lab : s) \quad \quad \quad \quad \quad = \mathcal{X} s$
$\mathcal{X}(\_) \quad \quad \quad \quad \quad \quad \quad = \epsilon$

If during execution of a class initializer an exception is thrown and this is not an Error or one of its subclasses, then Java$_\mathcal{E}$ and JVM$_\mathcal{E}$ throw an ExceptionInInitializerError. We refine the compilation of the phrase *Init* as follows:

$\mathcal{S}(\text{static } s) =$
$\quad \mathcal{S}(\text{try } s \text{ catch } (\text{Exception}, x,$
$\quad \quad \quad \text{throw new } (\text{ExceptionInInitializerError}, (\epsilon, \text{void})) \, ();))$

Due to the conservativity of the extension of Java$_\mathcal{O}$/JVM$_\mathcal{O}$ to Java$_\mathcal{E}$/JVM$_\mathcal{E}$, for the proof of the *Correctness Theorem for* Java$_\mathcal{E}$/JVM$_\mathcal{E}$ it suffices to extend the theorem from Java$_\mathcal{O}$/JVM$_\mathcal{O}$ to expression and statement execution in finally and error handling code, and to prove the following

Exception Lemma. The execution of code in Java$_\mathcal{E}$ and the execution of the corresponding compiled code in JVM$_\mathcal{E}$ produce exceptions at corresponding values of the program counters in Java$_\mathcal{E}$ and JVM$_\mathcal{E}$, for the same reasons, with the same failure classes (if any) and trigger the same exception handling.

6 Conclusion

We have presented implementation independent, rigorous yet easy to understand abstract code for the JVM as target machine for compilation of Java programs. Our definition captures faithfully the corresponding explanations of the Java Virtual Machine specification [6] and provides a practical basis for the mathematical analysis and comparison of different implementations of the machine. In particular it allowed us to prove the correctness of a general scheme for compiling Java programs into JVM code. Additionally, we have validated our work by a successful implementation in the functional programming language Haskell. The extended version of this paper [1] includes the proof details, the instruction refinement, an extensive bibliography and the discussion of related work. In an accompanying study [2] we refine the present JVM model to a defensive JVM, where we also isolate the bytecode verifier and the resolution component (including dynamic loading) of the JVM. This JVM can be used to execute compiled Java code as well as any bytecode that is loaded from the net.

Acknowledgment. We thank Ton Vullinghs for comments on this work. The first author thanks the IRIN (Institut de Recherche en Informatique de Nantes, Université de Nantes & École Centrale), in particular the *Équipe Génie logiciel, Méthodes et Spécifications formelles* for the good working environment offered during the last stage of the work on this paper.

References

[1] E. Börger and W. Schulte. Defining the Java Virtual Machine as platform for provably correct Java compilation. Technical report, Universität Ulm, Fakultät für Informatik. Ulm, Germany, 1998.
[2] E. Börger and W. Schulte. A modular design for the Java VM architecture. In E. Börger, editor, *Architecture Design and Validation Methods*. Springer LNCS, to appear, 1998.
[3] E. Börger and W. Schulte. A programmer friendly modular definition of the semantics of Java. In J. Alves-Foss, editor, *Formal Syntax and Semantics of Java(tm)*, Springer LNCS, to appear. 1998.
[4] J. Gosling, B. Joy, and G. Steele. *The Java(tm) Language Specification*. Addison Wesley, 1996.
[5] Y. Gurevich. Evolving algebras 1993: Lipari guide. In E. Börger, editor, *Specification and Validation Methods*. Oxford University Press, 1995.
[6] T. Lindholm and F. Yellin. *The Java(tm) Virtual Machine Specification*. Addison Wesley, 1996.

Towards a Theory of Recursive Structures*

David Harel**

Dept. of Applied Mathematics and Computer Science
The Weizmann Institute of Science, Rehovot, Israel
harel@wisdom.weizmann.ac.il

Abstract. In computer science, one is interested mainly in finite objects. Insofar as infinite objects are of interest, they must be computable, i.e., recursive, thus admitting an effective finite representation. This leads to the notion of a recursive graph, or, more generally, a recursive structure, model or data base. This paper summarizes recent work on recursive structures and data bases, including (i) the high undecidability of many problems on recursive graphs and structures, (ii) a method for deducing results on the descriptive complexity of finitary NP optimization problems from results on the computational complexity (i.e., the degree of undecidability) of their infinitary analogues, (iii) completeness results for query languages on recursive data bases, (iv) correspondences between descriptive and computational complexity over recursive structures, and (v) zero-one laws for recursive structures.

1 Introduction

This paper provides a summary of work — most of it joint with Tirza Hirst — on infinite recursive (i.e., computable) structures and data bases, and attempts to put it in perspective. The work itself is contained in four papers [H,HH1,HH2,HH3], which are summarized, respectively, in Sections 2, 3, 4 and 5.

When computer scientists become interested in an infinite object, they require it to be computable, i.e., recursive, so that it possesses an effective finite representation. Given the prominence of finite graphs in computer science, and the many results and open questions surrounding them, it is very natural to investigate recursive graphs too. Moreover, insight into finite objects can often be gleaned from results about infinite recursive variants thereof. An infinite recursive graph can be thought of simply as a recursive binary relation over the natural numbers. Recursive graphs can be represented by the (finite) algorithms, or Turing machines, that recognize their edge sets, so that it makes sense to investigate the complexity of problems concerning them.

---

* Preliminary versions of this paper appeared in STACS '94, *Proc. 11th Ann. Symp. on Theoretical Aspects of Computer Science*, Lecture Notes in Computer Science, Vol. 775, Springer-Verlag, Berlin, 1994, pp. 633–645, and in *Computer Science Today*, Lecture Notes in Computer Science, Vol. 1000, Springer-Verlag, 1995, pp. 374–391.
** Incumbent of the William Sussman Chair of Mathematics.

Indeed, a significant amount of work has been carried out in recent years regarding the complexity of problems on recursive graphs. Some of the first papers were written in the 1970s by Manaster and Rosenstein [MR] and Bean [B1,B2]. Following that, a variety of problems were considered, including ones that are NP-complete for finite graphs, such as k-colorability and Hamiltonicity [B1,B2,BG2,Bu,GL,MR] and ones that are in P in the finite case, such as Eulerian paths [B2,BG1] In most cases (including the above examples) the problems turned out to be undecidable. This is true even for highly recursive graphs [B1], i.e., ones for which node degree is finite and the set of neighbors of a node is computable. Beigel and Gasarch [BG1] and Gasarch and Lockwood [GL] investigated the precise level of undecidability of many such problems, and showed that they reside on low levels of the arithmetical hierarchy. For example, detecting the existence of an Eulerian path is Π_3^0-complete for recursive graphs and Π_2^0-complete for highly recursive graphs [BG1].

The case of Hamiltonian paths seemed to be more elusive. In 1976, Bean [B2] had shown that the problem is undecidable (even for planar graphs), but the precise characterization was not known. In response to this question, posed by R. Beigel and B. Gasarch, the author was able to show that Hamiltonicity is in fact *highly* undecidable, *viz*, Σ_1^1-complete. The result, proved in [H] and summarized in Section 2, holds even for highly recursive graphs with degree bounded by 3. (It actually holds for planar graphs too.) Hamiltonicity is thus an example of an interesting graph problem that becomes highly undecidable in the infinite case.[1]

The question then arises as to what makes some NP-complete problems highly undecidable in the infinite case, while others (e.g., k-colorability) remain on low levels of the arithmetical hierarchy. This was the starting point of the joint work with T. Hirst. In [HH1], summarized in Section 3, we provide a general definition of infinite recursive versions of NP optimization problems, in such a way that MAX CLIQUE, for example, becomes the question of whether a recursive graph contains an infinite clique. Two main results are proved in [HH1], one enables using knowledge about the infinite case to yield implications to the finite case, and the other enables implications in the other direction. The results establish a connection between the descriptive complexity of (finitary) NP optimization problems, particularly the syntactic class MAX NP, and the computational complexity of their infinite versions, particularly the class Σ_1^1. Taken together, the two results yield many new problems whose infinite versions are highly undecidable and whose finite versions are outside MAX NP. Examples include MAX CLIQUE, MAX INDEPENDENT SET, MAX SUBGRAPH, and MAX TILING.

The next paper, [HH2], summarized in Section 4, puts forward the idea of infinite recursive relational data bases. Such a data base can be defined simply as a finite tuple of recursive relations (not necessarily binary) over some countable domain. We thus obtain a natural generalization of the notion of a finite relational data base. This is not an entirely wild idea: tables of trigonometric

---

[1] Independent work in [AMS] showed that perfect matching is another such problem.

functions, for example, can be viewed as a recursive data base, since we might be interested in the sines or cosines of infinitely many angles. Instead of keeping them all in a table, which is impossible, we keep rules for computing the values from the angles, and vice versa, which is really just to say that we have an effective way of telling whether an edge is present between nodes i and j in an infinite graph, and this is precisely the notion of a recursive graph.

In [HH2], we investigate the class of computable queries over recursive data bases, the motivation being borrowed from [CH1]. Since the set of computable queries on such data bases is not closed under even simple relational operations, one must either make do with a very humble class of queries or considerably restrict the class of allowed data bases. The main parts of [HH2] are concerned with the completeness of two query languages, one for each of these possibilities. The first is quantifier-free first-order logic, which is shown to be complete for the non-restricted case. The second is an appropriately modified version of the complete language QL of [CH1], which is proved complete for the case of "highly symmetric" data bases. These have the property that their set of automorphisms is of finite index for each tuple-width.

While the previous topic involves languages for *computable* queries, our final paper, [HH3], summarized in Section 5, deals with languages that express *non*-computable queries. In the spirit of results for finite structures by Fagin, Immerman and others, we sought to connect the computational complexity of properties of recursive structures with their descriptive complexity, i.e, to capture levels of undecidability syntactically as the properties expressible in various logical formalisms. We consider several formalisms, such as first-order logic, second-order logic and fixpoint logic. One of our results is analogous to that of Fagin [F1]; it states that, for any $k \geq 2$, the properties of recursive structures expressible by Σ_k^1 formulas are exactly the generic properties in the complexity class Σ_k^1 of the analytical hierarchy.

[HH3] also deals with zero-one laws. It is not too difficult to see that many of the classical theorems of logic that hold for general structures (e.g., compactness and completeness) fail not only for finite models but for recursive ones too. Others, such as Ehrenfeucht–Fraisse games, hold for finite and recursive structures too. Zero-one laws, to the effect that certain properties (such as those expressible in first-order logic) are either almost surely true or almost surely false, are considered unique to finite model theory, since they require counting the number of structures of a given finite size. We introduce a way of extending the definition of these laws to recursive structures, and prove that they hold for first-order logic, strict Σ_1^1 and strict Π_1^1. We then use this fact to show non-expressibility of certain properties of recursive structures in these logics.

While recursive structures and models have been investigated quite widely by logicians (see, e.g., [NR]), the kind of issues that computer scientists are interested in have not been addressed prior to the work mentioned above. We feel that that this is a fertile area for research, and raises theoretical and practical questions concerning the computability and complexity of properties of recursive structures, and the theory of queries and update operations over recursive data bases. We hope that the work summarized here will stimulate more research on these topics.

2 Hamiltonicity in Recursive Graphs

A *recursive directed graph* is a pair $G = (V, E)$, where V is recursively isomorphic to the set of natural numbers $\mathcal{N}$, and $E \subset V \times V$ is recursive. G is *undirected* if E is symmetric. A *highly recursive graph* is a recursive graph for which there is a recursive function H from V to finite subsets of V, such that $H(v) = \{u \mid \langle v, u \rangle \in E\}$.

A *one-way* (respectively, *two-way*) *Hamiltonian path* in G is a 1-1 mapping p of $\mathcal{N}$ (respectively, $\mathcal{Z}$) onto V, such that $\langle p(x), p(x+1) \rangle \in E$ for all x.

Bean [B2] showed that determining Hamiltonicity in highly recursive graphs is undecidable. His reduction is from non-well-foundedness of recursive trees with finite degree, which can be viewed simply as the halting problem for (nondeterministic) Turing machines. Given such a tree T, the proof in [B2] constructs a graph G, such that infinite paths in T map to Hamiltonian paths in G. The idea is to make the nodes of G correspond to those of T, but with all nodes that are on the same level being connected in a cyclic fashion. In this way, a Hamiltonian path in G simulates moving down an infinite path in T, but at each level it also cycles through all nodes on that level. A fact that is crucial to this construction is the finiteness of T's degree, so that the proof does not generalize to trees with infinite degree, Thus, Bean's proof only establishes that Hamiltonicity is hard for Π_1^0, or co-r.e.

In [H] we have been able to show that the problem is actually Σ_1^1-complete. Hardness is proved by a reduction (that is elementary but not straightforward) from the non-well-foundedness of recursive trees with possibly *infinite* degree, which is well-known to be a Σ_1^1-complete problem [R]:

Theorem: *Detecting (one-way or two-way) Hamiltonicity in a (directed or undirected) highly recursive graph is Σ_1^1-complete, even for graphs with $H(v) \leq 3$ for all v.*

Proof sketch: In Σ_1^1 is easy: With the $\exists f$ quantifying over total functions from $\mathcal{N}$ to $\mathcal{N}$, we write

$$\exists f \; \forall x \; \forall y \; \exists z \; (\langle f(x), f(x+1) \rangle \in E \wedge (x \neq y \to f(x) \neq f(y)) \wedge f(z) = x).$$

This covers the case of one-way paths. The two-way case is similar.

We now show Σ_1^1-hardness for undirected recursive graphs with one-way paths. (The other cases require more work, especially in removing the infinite branching from the graphs we construct in order to obtain the result for highly recursive graphs. The details can be found in [H].)

Assume a recursive tree T is given, with nodes $\mathcal{N} = 0, 1, 2, 3, \ldots$, and root 0, and whose *parent-of* function is recursive. T can be of infinite degree. We construct an undirected graph G, which has a one-way Hamiltonian path iff T has an infinite path.

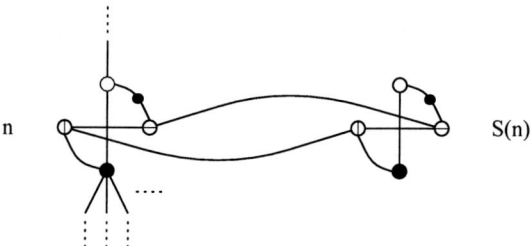

Figure 1

For each element $n \in \mathcal{N}$, G has a cluster of five internal nodes, n^u, n^d, n^r, n^l and n^{ur}, standing, respectively, for *up, down, right, left* and *up-right*. For each such cluster, G has five internal edges:

$$n^l \text{ —— } n^d \text{ —— } n^u \text{ —— } n^{ur} \text{ —— } n^r \text{ —— } n^l$$

For each edge $n \longrightarrow m$ of the tree T, n^d —— m^u is an edge of G. For each node n in T, let $S(n)$ be n's distance from the root in T (its *level*). Since $S(n) \in \mathcal{N}$, we may view $S(n)$ as a node in T. In fact, in G we will think of $S(n)$ as being n's *shadow node*, and the two are connected as follows (see Fig. 1):[2]

$$n^r \text{ —— } S(n)^r \quad \text{and} \quad S(n)^l \text{ —— } n^l$$

To complete the construction, there is one additional root node g in G, with an edge g —— 0^u.

Since T is a recursive tree and S, as a function, is recursive in T, it is easy to see that G is a recursive graph. To complete the proof, we show that T has an infinite path from 0 iff G has a Hamiltonian path.

(*Only-if*) Suppose T has an infinite path p. A Hamiltonian path p' in G starts at the root g, and moves down G's versions of the nodes in p, taking detours to the right to visit n's shadow node $S(n)$ whenever $S(n) \notin p$. The way this is done can be seen in Fig. 2. Since p is infinite, we will eventually reach a node of any desired level in T, so that any $n \notin p$ will eventually show up as a shadow of some node along p and will be visited in due time. It is then easy to see that p' is Hamiltonian.

---

[2] Clearly, given T, the function $S : \mathcal{N} \to \mathcal{N}$ is not necessarily one-one. In fact, Fig. 1 is somewhat misleading, since there may be infinitely many nodes with the same shadow, so that the degree of both up-nodes and down-nodes can be infinite. Moreover, $S(n)$ itself is a node somewhere else in the tree, and hence has its own T-edges, perhaps infinitely many of them.

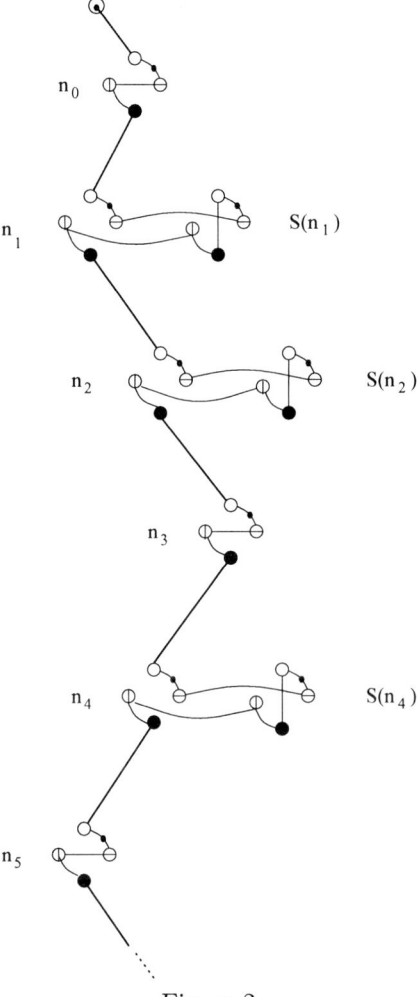

Figure 2

(*If*) Suppose G has a Hamiltonian path p. It helps to view the path p as containing not only the nodes, but also the edges connecting them. Thus, with the exception of the root g, each node in G must contribute to p exactly two incident edges, one incoming and one outgoing.

We now claim that for any n, if p contains the T-edge incident to the up-node n^u, or, when $n = 0$, if it contains the edge between g and 0^u, then it must also contain a T-edge incident to the down node n^d.

To see why this is true, assume p contains the T-edge incident to n^u (this is the edge leading upwards at the top left of Fig. 1). Consider n^{ur} (the small black node in the figure). It has exactly two incident edges, both of which must therefore be in p. But since one of them connects it to n^u, we already have in p the two required edges for n^u, so that the one between n^u and n^d cannot be in p. Now, the only remaining edges incident to n^d are the internal one connecting it to n^l, and its T-edges, if any. However, since p must contain exactly two edges incident to n^d, one of them must be one of the T-edges. △

In fact, Hamiltonicity is Σ_1^1-complete even for planar graphs [HH1].

3 From the Finite to the Infinite and Back

Our approach to optimization problems focuses on their descriptive complexity, an idea that started with Fagin's [F1] characterization of NP in terms of definability in existential second-order logic on finite structures. Fagin's theorem asserts that a collection C of finite structures is NP-computable if and only if there is a quantifier-free formula $\psi(\overline{x}, \overline{y}, S)$, such that for any finite structure A:

$$A \in C \Leftrightarrow A \models (\exists S)(\forall \overline{x})(\exists \overline{y})\psi(\overline{x}, \overline{y}, S).$$

Papadimitriou and Yannakakis [PY] introduced the class MAX NP of maximization problems that can be defined by

$$\max_S |\{\overline{x}:\ A \models (\exists \overline{y})\psi(\overline{x}, \overline{y}, S)\}|,$$

for quantifier-free ψ. MAX SAT is the canonical example of a problem in MAX NP. The authors of [PY] also considered the subclass MAX SNP of MAX NP, consisting of those maximization problems in which the existential quantifier above is not needed. (Actually, the classes MAX NP and MAX SNP of [PY] contain also their closures under L-reductions, which preserve polynomial-time approximation schemes. To avoid confusion, we use the names MAX Σ_0 and MAX Σ_1, introduced in [KT], rather than MAX SNP and MAX NP, for the 'pure' syntactic classes.)

Kolaitis and Thakur [KT] then examined the class of all maximization problems whose optimum is definable using first-order formulas, i.e., by

$$\max_S |\{\overline{w}:\ A \models \psi(\overline{w}, S)\}|,$$

where $\psi(\overline{w}, S)$ is an arbitrary first-order formula. They first showed that this class coincides with the collection of polynomially-bounded NP-maximization problems on finite structures, i.e., those problems whose optimum value is bounded by a polynomial in the input size. They then proved that these problems form a proper hierarchy, with exactly four levels:

$$\text{MAX } \Sigma_0 \subset \text{MAX } \Sigma_1 \subset \text{MAX } \Pi_1 \subset \text{MAX } \Pi_2 = \bigcup_{i \geq 2} \text{MAX } \Pi_i$$

Here, MAX Π_1 is defined just like MAX Σ_1 (i.e., MAX NP), but with a universal quantifier, and MAX Π_2 uses a universal followed by an existential quantifier, and corresponds to Fagin's general result stated above. The three containments are known to be strict. For example, MAX CLIQUE is in MAX Π_1 but not in MAX Σ_1.

We now define a little more precisely the class of optimization problems we deal with[3]:

Definition: (See [PR]) An NPM problem is a tuple $F = (\mathcal{I}_F, S_F, m_F)$, where

- $\mathcal{I}_F$, the set of *input instances*, consists of finite structures over some vocabulary σ, and is recognizable in polynomial time.
- $S_F(I)$ is the space of *feasible solutions* on input $I \in \mathcal{I}_F$. The only requirement on S_F is that there exists a polynomial q and a polynomial time computable predicate p, both depending only on F, such that $\forall I \in \mathcal{I}_F$, $S_F(I) = \{S\colon |S| \leq q(|I|) \land p(I, S)\}$.
- $m_F\colon \mathcal{I}_F \times \Sigma^* \to \mathcal{N}$, the *objective function*, is a polynomial time computable function. $m_F(I, S)$ is defined only when $S \in S_F(I)$.
- The following decision problem is required to be in NP: Given $I \in \mathcal{I}_F$ and an integer k, is there a feasible solution $S \in S_F(I)$, such that $m_F(I, S) \geq k$?

This definition (with an additional technical restriction that we omit here; see [HH1]) is broad enough to encompass most known optimization problems arising in the theory of NP-completeness.

We now define infinitary versions of NPM problems, by evaluating them over infinite recursive structures and asking about the existence of an infinite solution:

Definition: For an NPM problem $F = (\mathcal{I}_F, S_F, m_F)$, let $F^\infty = (\mathcal{I}_F^\infty, S_F^\infty, m_F^\infty)$ be defined as follows:

- $\mathcal{I}_F^\infty$ is the set of *input instances*, which are infinite recursive structures over the vocabulary σ.
- $S_F^\infty(I^\infty)$ is the set of *feasible solutions* on input $I^\infty \in \mathcal{I}_F^\infty$.
- $m_F^\infty\colon \mathcal{I}^\infty \times S_F \to \mathcal{N} \cup \{\infty\}$ is the *objective function*, satisfying

$$\forall I^\infty \in \mathcal{I}_F^\infty, \forall S \in S_F^\infty(I^\infty)\ (m_F^\infty(I^\infty, S) = |\{\overline{x}\colon \psi_F(I^\infty, S, \overline{x})\}|).$$

- The decision problem is: Given $I^\infty \in \mathcal{I}_F^\infty$, does there exist $S \in S_F^\infty(I^\infty)$, such that $m_F^\infty(I^\infty, S) = \infty$? Put another way:

$$F^\infty(I^\infty) = \text{TRUE} \quad \text{iff} \quad \exists S(|\{\overline{x}\colon \psi_F(I^\infty, S, \overline{x})\}| = \infty).$$

Due to the conditions on NPM problems, F^∞ can be shown not to depend on the Π_2-formula representing m_F. This is important, since, if some finite problem F could be defined by two different formulas ψ_1 and ψ_2 that satisfy the condition but yield different infinite problems, we could construct a finite structure for which ψ_1 and ψ_2 determine different solutions.

Here is the first main result of [HH1]:

Theorem: If $F \in \text{Max } \Sigma_1$ then $F^\infty \in \Pi_2^0$.

---

[3] We concentrate here on maximization problems, though the results can be proved for appropriate minimization ones too.

A special case of this is:

Corollary: For any NPM problem F, if F^∞ is Σ_1^1-hard then F is not in MAX Σ_1.

It follows that since the infinite version of Hamiltonicity is Σ_1^1-complete and thus completely outside the arithmetical hierarchy, an appropriately defined finitary version cannot be in MAX Σ_1. Obviously, the corollary is valid not only for such problems but for all problems that are above Π_2^0 in the arithmetical hierarchy. For example, since detecting the existence of an Eulerian path in a recursive graph is Π_3^0-complete [BG1], its finite variant cannot be in MAX Σ_1 either.

In order to be able to state the second main result of [HH1], we define a special kind of *monotonic* reduction between finitary NPM problems, an *M-reduction*:

Definition: Let $\mathcal{A}$ and $\mathcal{B}$ be sets of structures. A function $f: \mathcal{A} \to \mathcal{B}$ is *monotonic* if $\forall A, B \in \mathcal{A}$ $(A \leq B \Rightarrow f(A) \leq f(B))$. (Here, $\leq$ denotes the substructure relation.) Given two NPM problems: $F = (\mathcal{I}_F, S_F, m_F)$ and $G = (\mathcal{I}_G, S_G, m_G)$, an *M-reduction* g from F to G is a tuple $g = (t_1, t_2, t_3)$, such that:

- $t_1: \mathcal{I}_F \to \mathcal{I}_G$, $t_2: \mathcal{I}_F \times S_F \to S_G$, and $t_3: \mathcal{I}_G \times S_G \to S_F$, are all monotonic, polynomial time computable functions..
- m_F and m_G grow monotonically with respect to t_1, t_2 and t_3 (see [HH1] for a more precise formulation).

We denote the existence of an *M*-reduction from F to G by $F \propto_M G$. The second main result of [HH1] shows that *M*-reductions preserve the Σ_1^1-hardness of the corresponding infinitary problems:

Theorem: Let F and G be two NPM problems, with $F \propto_M G$. If F^∞ is Σ_1^1-hard, then G^∞ is Σ_1^1-hard too.

The final part of [HH1] applies these two results to many examples of NPM problems, some of which we now list with their infinitary versions. It is shown in [HH1] that for each of these the infinitary version is Σ_1^1-complete. Mostly, this is done by establishing monotonic reductions on the finite level, and applying the second theorem above. From the first theorem it then follows that the finitary versions must be outside MAX Σ_1.

Here are some of the examples:

1. MAX CLIQUE: I is an undirected graph, $G = (V, E)$.

$$S(G) = \{Y: Y \subseteq V, \forall y, z \in Y \; y \neq z \Rightarrow (y, z) \in E\}$$
$$m(G, Y) = |Y|$$

The maximization version is:

$$\max_{Y \subseteq V} |\{x: x \in Y \land \forall y, z \in Y \; y \neq z \Rightarrow (y, z) \in E\}|$$

MAX CLIQUE$^\infty$: I^∞ is a recursive graph G. Does G contain an infinite clique?

2. MAX IND SET: I is an undirected graph $G = (V, E)$.

$$S(G) = \{Y \colon Y \subseteq V, \ \forall y, z \in Y \ (y, z) \notin E\}$$
$$m(G, Y) = |Y|$$

$$\max_{Y \subseteq V} |\{x \colon x \in Y \ \land \ \forall y, z \in Y \ (y, z) \notin E\}|$$

MAX IND SET$^\infty$: I^∞ is a recursive graph G. Does G contain an infinite independent set?

3. MAX SET PACKING: I is a collection C of finite sets, represented by pairs (i, j), where the set i contains j.

$$S(C) = \{Y \subseteq C \colon \forall A, B \in Y \ A \neq B \Rightarrow A \cap B = \emptyset\}$$
$$m(C, Y) = |Y|$$

MAX SET PACKING$^\infty$: I^∞ is a recursive collection of infinite sets C. Does C contains infinitely many disjoint sets?

4. MAX SUBGRAPH: I is a pair of graphs, $G = (V_1, E_1)$ and $H = (V_2, E_2)$, with $V_2 = \{v_1, \ldots, v_n\}$.

$$S(G, H) = \{Y \colon Y \subseteq V_1 \times V_2, \ \forall (u, v), (x, y) \in Y, \ u \neq x$$
$$\land \ v \neq y \ \land \ (u, x) \in E_1 \Leftrightarrow (v, y) \in E_2\}$$
$$m((G, H), Y) = k \text{ iff } v_1, \ldots, v_k \text{ appear in } Y,$$
$$\text{but } v_{k+1} \text{ does not appear in } Y.$$

MAX SUBGRAPH$^\infty$: I^∞ is a pair of recursive graphs, H and G. Is H a subgraph of G?

5. MAX TILING: I is a grid D of size $n \times n$, and a set of tiles $T = \{t_1, \ldots, t_m\}$. (We assume the reader is familiar with the rules of tiling problems.)

$$S(D, T) = \{Y \colon Y \text{ is a legal tiling of some portion of } D \text{with tiles}$$
$$\text{from } T\}$$
$$m(\{D, T\}, Y) = k \text{ iff } Y \text{ contains a tiling of a full } k \times k \text{ subgrid of } D.$$

MAX TILING$^\infty$: I^∞ is a recursive set of tiles T.
Q: Can T tile the positive quadrant of the infinite integer grid?

We thus establish closely related facts about the level of undecidability of many infinitary problems and the descriptive complexity of their finitary counterparts. More examples appear in [HH1].

Two additional graph problems of interest are mentioned in [HH1], planarity and graph isomorphism. The problem of detecting whether a recursive graph is planar can be shown to be co-r.e. Determining whether two recursive graphs are isomorphic is arithmetical for graphs that have finite degree and contain only finitely many connected components. More precisely, this problem is in Π_1^0 for highly recursive trees; in Π_3^0 for recursive trees with finite degree; in Σ_2^0 for highly recursive graphs; and in Σ_4^0 for recursive graphs with finite degree. As to the isomorphism problem for general recursive graphs, Morozov [Mo] has recently proved, using different techniques, that the problem is Σ_1^1-complete.

4 Completeness for Recursive Data Bases

It is easy to see that recursive relations are not closed under some of the simplest accepted relational operators. For example, if $R(x, y, z)$ means that the yth Turing machine halts on input z after x steps (a primitive-recursive relation), then the projection of R on columns 2 and 3 is the nonrecursive halting predicate. This means that even very simple queries, when applied to general recursive relations, do not preserve computability. Thus, a naive definition of a recursive data base as a finite set of recursive relations will cause many extremely simple queries to be non-computable.

This difficulty can be overcome in essentially two ways (and possibly other intermediate ways that we haven't investigated). The first is to accept the situation as is; that is, to resign ourselves to the fact that on recursive data bases the class of computable queries will necessarily be very humble, and then to try to capture that class in a (correspondingly humble) complete query language. The second is to restrict the data bases, so that the standard kinds of queries *will* preserve computability, and then to try to establish a reasonable completeness result for these restricted inputs. The first case will give rise to a rich class of data bases but a poor class of queries, and the second to a rich class of queries but a poor class of data bases. In both cases, of course, in addition to being Turing computable, the queries will also have to satisfy the consistency criterion of [CH1], more recently termed *genericity*, whereby queries must preserve isomorphisms.

The first result of [HH2] shows that the class of computable queries on recursive data bases is indeed extremely poor. First we need some preparation.

Definition: Let D be a countable set, and let $R_1, \ldots, R_k$, for $k > 0$, be relations, such that for all $1 \leq i \leq k$, $R_i \subseteq D^{a_i}$. $B = (D, R_1, \ldots, R_k)$ is a *recursive relational data base* (or an r-db for short) *of type* $a = (a_1, \ldots a_k)$, if each R_i, considered as a set of tuples, is recursive.

Definition: Let $B_1 = (D_1, R_1, \ldots, R_k)$ and $B_2 = (D_2, R_1', \ldots, R_k')$ be two r-db's of the same type, and let $u \in D_1^n$ and $v \in D_2^n$, for some n. Then (B_1, u) and (B_2, v) are *isomorphic*, written $(B_1, u) \cong (B_2, v)$, if there is an isomorphism between B_1 and B_2 taking u to v. (B_1, u) and (B_2, v) are *locally isomorphic*, written $(B_1, u) \cong_l (B_2, v)$, if the restriction of B_1 to the elements of u and the restriction of B_2 to the elements of v are isomorphic.

Definition: An *r-query* Q (i.e., a partial function yielding, for each r-db B of type a, an output (if any) which is a recursive relation over $D(B)$) is *generic*, if it preserves isomorphisms; i.e. for all B_1, B_2, u, v, if $(B_1, u) \cong (B_2, v)$ then $u \in Q(B_1)$ iff $v \in Q(B_2)$. It is *locally generic* if it preserves local isomorphisms; i.e., for all B_1, B_2, u, v, if $(B_1, u) \cong_l (B_2, v)$ then $u \in Q(B_1)$ iff $v \in Q(B_2)$.

The following is a key lemma in the first result:

Lemma: If Q is a recursive r-query, then Q is generic iff Q is locally generic.

Definition: A query language is *r-complete* if it expresses precisely the class of recursive generic r-queries.

Theorem: The language of first-order logic without quantifiers is r-complete.

We now prepare for the second result of [HH2], which insists on the full set of computable queries of [CH1], but drastically reduces the allowed data bases in order to achieve completeness.

Definition: Let $B = (D, R_1, \ldots, R_k)$ be a fixed r-db. For each $u, v \in D^n$, u and v are *equivalent*, written $u \cong_B v$, if $(B, u) \cong (B, v)$. B is *highly symmetric* if for each $n > 0$, the relation $\cong_B$ induces only a finite number of equivalence classes of rank n.

Highly symmetric graphs consist of a finite or infinite number of connected components, where each component is highly symmetric, and there are only finitely many pairwise non-isomorphic components. In a highly symmetric graph, the finite degrees, the distances between points and the lengths of the induced paths are bounded. A grid or an infinite straight line, for instance, are not highly symmetric, but the full infinite clique is highly symmetric. Fig. 3 shows an example of another highly symmetric graph.

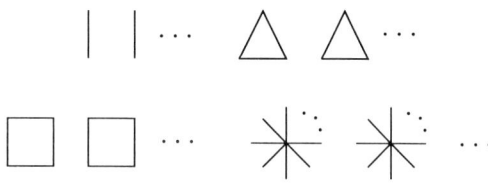

Figure 3

A *characteristic tree* for B is defined as follows. Its root is Λ, and the rest of the vertices are labeled with elements from D, such that the labels along each path from the root form a tuple that is a representative of an equivalence class of $\cong_B$. The whole tree covers representatives of all such classes. No two paths are allowed to form representatives of the same class. We represent a highly symmetric data base B by a tuple

$$C_B = (T_B, \cong_B, C_1, \ldots, C_k),$$

where T_B is some characteristic tree for B, and each C_i is a finite set of representatives of the equivalence classes constituting the relation R_i. We also require that $\cong_B$ be recursive, and that T_B be highly recursive (in the sense of Section 2).

We say that a query Q on a highly symmetric data base is *recursive* if the following version of it, which is applied to the representation C_B rather than to the data base B itself, is partial recursive: whenever $Q(C_B)$ is defined, it yields a finite set of representatives of the equivalence classes representing the relation $Q(B)$.

We now describe the query language QL_s. Its syntax is like that of the QL language of Chandra and Harel [CH1], with the following addition: the test in a **while** loop can be for whether a relation has a single representative, and not only for a relation's emptiness. The semantics of QL_s is the same as the semantics of QL, except for some minor technical adaptations that are omitted here. As in [CH1], the result of applying a program P to C_B is undefined if P does not halt; otherwise it is the contents of some fixed variable, say X_1.

Definition: A query language is *hs-r-complete* if it expresses precisely the class of recursive generic queries over highly symmetric recursive data bases.

Theorem: QL_s is hs-r-complete.

The proof follows four main steps, which are analogous to those given in the completeness proof for QL in [CH1]. The details, however, are more intricate.

In [HH2] a number of additional issues are considered, including the restriction of recursive data bases to finite/co-finite recursive relations, completeness of the generic machines of [AV], and BP-completeness.

5 Expressibility vs. Complexity, and Zero-One Laws

One part of [HH3] proves results that relate the expressive power of various logics over recursive structures to the computational complexity (i.e., the level of undecidability) of the properties expressible therein. We summarize some of these, without providing all of the relevant definitions. In the previous section, we mentioned the result from [HH2] to the effect that the very restricted language of quantifier-free first-order relational calculus is r-complete; i.e., it expresses precisely the recursive and generic r-queries. Here we deal with languages that have stronger expressive power, and hence express also non-recursive queries.

There are many results over *finite* structures that characterize complexity classes in terms of logic. One of the most important of these is Fagin's theorem [F1], mentioned in section 2 above, which establishes that the properties of finite structures expressible by Σ_1^1 formulas are exactly the ones that are in NP. This kind of correspondence also holds between each level of the quantifier hierarchy of second-order logic and the properties computable in the corresponding level of the polynomial-time hierarchy.

In order to talk about recursive structures it is convenient to use the following definition, which we adapt to recursive structures from Vardi [V]

Definition: The *data complexity* of a language L is the level of difficulty of computing the sets $Gr(Q_e) = \{(B, u) | u \in Q(B)\}$ for an expression e in L, where Q_e is the query expressed by e, and B denotes a recursive data base (i.e., structure). A language L is *data-complete* (or *D-complete* for short) for a computational class C if for every expression e in L, $Gr(Q_e)$ is in C, and there is an expression e_0 in L such that $Gr(Q_{e_0})$ is hard for C.

Here we restrict ourselves to the consistent, or generic, queries, which are the ones that preserve isomorphisms. In fact, we require that they preserve the isomorphisms of *all* structures, not only recursive ones, under the assumption that there exist oracles for their relations. That is, Q is considered here to be generic if for all B_1, B_2, if $B_1 \cong B_2$ then $Q(B_1) \cong Q(B_2)$, where $Q(B)$ is the result of applying Q to oracles for the relations in B.

We now provide a very brief description of the main results of this part of [HH3]:

1. First-order logic expresses generic queries from the entire arithmetical hierarchy, but it does not express all of them. For example, the connectivity of recursive graphs is arithmetical, but is not expressible by a first-order formula.

2. The logical formalism E-Σ_1^1, which consists of existential second-order formulas, is D-complete for the complexity class Σ_1^1 of the analytical hierarchy, but there are queries, even arithmetical ones, that are not expressible in E-Σ_1^1. However, over ordered structures (that is, if a built-in total order is added to the vocabulary), all Σ_1^1 properties are expressible in E-Σ_1^1.

3. For $k \geq 2$, a stronger result is proved, analogous to Fagin's result for finite structures: the logical formalism E-Σ_k^1 expresses precisely the generic properties of the complexity class Σ_k^1. This means that every generic query over some vocabulary σ that is expressible by a Σ_k^1 formula over interpreted recursive predicates, is also expressible by an uninterpreted E-Σ_k^1 formula over σ.[4]

4. Monadic E-Σ_1^1, where the second-order quantifiers are restricted to range over unary relations (sets), is D-complete for Σ_1^1, and strict E-Σ_1^1 is D-complete for Σ_2^0.

5. Consider fixpoint logic, which is obtained by adding least fixpoint operators to first-order formulas [CH2, I, Mos]. Denote by FP$_1$ positive fixpoint logic, in which the least fixpoint operator is restricted to positive formulas, and by FP the hierarchy obtained by alternating the least fixpoint operator with the first-order constructs. In finite structures, the FP hierarchy collapses, and a single fixpoint operator suffices [I]. In contrast, for recursive structures FP$_1$ is D-complete for Π_1^1, and hence ¬FP$_1$ (negations of formulas in FP$_1$) is D-complete for Σ_1^1. The data complexity of FP is exactly Δ_2^1, and an example is shown of a query expressible in FP that is hard for both Σ_1^1 and Π_1^1.

---

[4] In the direction going from expressibility in E-Σ_k^1 to computability in Σ_k^1, the second-order quantifiers are used to define a total order and predicates + and ∗, which, in turn, are used to define the needed elementary arithmetic expression. Each subset of elements must contain a minimum in the defined order, which requires for its definition a universal second-order quantifier. This explains why the result requires $k \geq 2$.

The second part of [HH3] deals with 0–1 laws on recursive structures.

If C is a class of finite structures over some vocabulary σ and if P is a property of some structures in C, then the *asymptotic probability* $\mu(P)$ on C is the limit as $n \to \infty$ of the fraction of the structures in C with n elements that satisfy P, provided that the limit exists. Fagin [F2] and Glebskii et al. [GKLT] were the first to discover the connection between logical definability and asymptotic probabilities. They showed that if C is the class of all finite structures over some relational vocabulary, and if P is any property expressible in first-order logic, then $\mu(P)$ exists and is either 0 or 1. This result, known as the *0–1 law for first-order logic*, became the starting point of a series of investigations aimed at discovering the relationship between expressibility in a logic and asymptotic probabilities. Several additional logics, such as fixpoint logic, iterative logic and strict E-Σ_1^1, have been shown by various authors to satisfy the 0–1 law too.

A standard method for establishing 0–1 laws on finite structures, originating in Fagin [F2], is to prove that the following *transfer theorem* holds: there is an infinite structure $\mathbf{A}$ over σ such that for any property P expressible in L:

$$\mathbf{A} \models P \text{ iff } \mu(P) = 1 \text{ on } C.$$

It turns out that there is a single countable structure $\mathbf{A}$ that satisfies this equivalence for all the logics mentioned above. Moreover, $\mathbf{A}$ is characterized by an infinite set of *extension axioms*, which, intuitively, assert that every type can be extended to any other possible type. More specifically, for each finite set X of points, and each possible way that a new point $y \notin X$ could relate to X in terms of atomic formulas over the appropriate vocabulary, there is an extension axiom that asserts that there is indeed such a point. For example, here is an extension axiom over a vocabulary containing one binary relation symbol R:

$$\forall x_1 \forall x_2 \bigg(x_1 \neq x_2 \Rightarrow \exists y \, (y \neq x_1 \wedge y \neq x_2 \wedge$$

$$(y, x_1) \in R \wedge (x_1, y) \notin R \wedge (y, x_2) \notin R \wedge (x_2, y) \in R) \bigg).$$

Fagin realized that the extension axioms are relevant to the study of probabilities on finite structures and proved that on the class C of all finite structures of vocabulary σ, $\mu(\tau) = 1$ for any extension axiom τ. The theory of all extension axioms, denoted T, is known to be ω-categorical (that is, every two countable models are isomorphic), so that $\mathbf{A}$, which is a model for T, is unique up to isomorphism. This unique structure is called the *random countable structure*, since it is generated, with probability 1, by a random process in which each possible tuple appears with probability 1/2, independently of the other tuples. The random graph was studied by Rado [Ra], and is sometimes called the *Rado graph*.

Now, since all countable structures are isomorphic to $\mathbf{A}$ with probability 1, the asymptotic probability of each (generic) property P on countable structures is trivially 0 or 1, since this depends only on whether $\mathbf{A}$ satisfies P or not. Hence, the subject of 0–1 laws over the class of all countable structures is not interesting.

As to recursive structures, which are what we are interested in here, one is faced with the difficulty of defining asymptotic probabilities, since structure size is no longer applicable.

The heart of this part of [HH3] is a proposal for a definition of 0–1 laws for recursive structures.

Definition: Let $\mathcal{F} = \{F_i\}_{i=1}^{\infty}$ be a sequence of recursive structures over some vocabulary, and let P be a property defined over the structures in $\mathcal{F}$. Then the *asymptotic probability* $\mu_{\mathcal{F}}(P)$ is defined to be

$$\mu_{\mathcal{F}}(P) = \lim_{n \to \infty} \frac{|\{F_i|\ 1 \le i \le n,\ F_i \models P\}|}{n}.$$

Definition: Let $\mathcal{F} = \{F_i\}_{i=1}^{\infty}$ be a sequence of recursive structures over some vocabulary σ. We say that $\mathcal{F}$ is a *T-sequence* if $\mu_{\mathcal{F}}(\tau) = 1$ for every extension axiom τ over σ.

As an example, a sequence of graphs that are all isomorphic to the countable random graph **A** is a *T*-sequence. We shall use U to denote one such sequence. Here is another example of a *T*-sequence: take $\mathcal{F} = \{F_n\}_{n=1}^{\infty}$, where each F_n is a graph satisfying all the *n*-extension axioms and is built in stages. First take n distinct and disconnected points. Then, at each stage add a new point z for every set $\{x_1, \ldots, x_n\}$ from previous stages and for every possible extension axiom for it, and connect z accordingly.

Definition: Let P be a property of recursive structures. We say that the *0–1 law holds for* P if for every *T*-sequence $\mathcal{F}$ the limit $\mu_{\mathcal{F}}(P)$ exists and is equal to 0 or 1. The *0–1 law holds for a logic L on recursive structures* if it holds for every property expressible in L.

Here are some of the results proved in [HH3] for this definition of 0–1 laws over recursive structures.

Theorem: The 0–1 law holds for all properties of recursive structures definable in first-order logic, strict E-Σ_1^1 and strict E-Π_1^1. Moreover, if **A** is the countable random structure, P is such a property and $\mathcal{F}$ is a *T*-sequence, then $\mathbf{A} \models P$ iff $\mu_{\mathcal{F}}(P) = 1$.

However, the property of a graph having an infinite clique, for example, is shown not to satisfy the 0–1 law, so that the law does not hold in general for E-Σ_1^1-properties.

As a result of the theorem, a property for which the 0–1 law does not hold is not expressible in first-order logic, strict E-Σ_1^1 or strict E-Π_1^1. In fact, we have the following:

Theorem: Every property on recursive structures that is true in **A**, but does not have probability 1 on some *T*-sequence, is not expressible by an E-Π_1^1 sentence or by a strict E-Σ_1^1 sentence.

In way of applying the techniques, we show in [HH3] that the following properties are not expressible by an E-Π_1^1 sentence or by a strict E-Σ_1^1 sentence:

a recursive graph having an infinite clique, a recursive graph having an infinite independent set, a recursive graph satisfying all the extension axioms, and a pair of recursive graphs being isomorphic.

Acknowledgements: I would like to thank Richard Beigel, who by asking the question addressed in Section 2, introduced me to this area. His work with Bill Gasarch has been a great inspiration. Very special thanks go to Tirza Hirst, without whom this paper couldn't have been written. Apart from Section 2, the results are all joint with her, and form her outstanding PhD thesis.

References

[AV]   S. Abiteboul and V. Vianu, "Generic Computation and Its Complexity", *Proc. 23rd Ann. ACM Symp. on Theory of Computing*, pp. 209–219, ACM Press, New York, 1991.

[AMS] R. Aharoni, M. Magidor and R. A. Shore, "On the Strength of König's Duality Theorem", *J. of Combinatorial Theory (Series B)* **54**:2 (1992), 257–290.

[B1]   D.R. Bean, "Effective Coloration", *J. Sym. Logic* **41** (1976), 469–480.

[B2]   D.R. Bean, "Recursive Euler and Hamiltonian Paths", *Proc. Amer. Math. Soc.* **55** (1976), 385–394.

[BG1] R. Beigel and W. I. Gasarch, unpublished results, 1986-1990.

[BG2] R. Beigel and W. I. Gasarch, "On the Complexity of Finding the Chromatic Number of a Recursive Graph", Parts I & II, *Ann. Pure and Appl. Logic* **45** (1989), 1–38, 227–247.

[Bu]   S. A. Burr, "Some Undecidable Problems Involving the Edge-Coloring and Vertex Coloring of Graphs", *Disc. Math.* **50** (1984), 171–177.

[CH1] A. K. Chandra and D. Harel, "Computable Queries for Relational Data Bases", *J. Comp. Syst. Sci.* **21**, (1980), 156–178.

[CH2] A.K. Chandra and D. Harel, "Structure and Complexity of Relational Queries", *J. Comput. Syst. Sci.* **25** (1982), 99–128.

[F1]   R. Fagin, "Generalized First-Order Spectra and Polynomial-Time Recognizable Sets", In *Complexity of Computations* (R. Karp, ed.), SIAM-AMS Proceedings, Vol. 7, 1974, pp. 43–73.

[F2]   R. Fagin, "Probabilities on Finite Models", *J. of Symbolic Logic,* **41**, (1976), 50 – 58.

[GL]   W. I. Gasarch and M. Lockwood, "The Existence of Matchings for Recursive and Highly Recursive Bipartite Graphs", Technical Report 2029, Univ. of Maryland, May 1988.

[GKLT] Y. V. Glebskii, D. I. Kogan, M. I. Liogonki and V. A. Talanov, "Range and Degree of Realizability of Formulas in the Restricted Predicate Calculus", *Cybernetics* **5**, (1969), 142–154.

[H]   D. Harel, "Hamiltonian Paths in Infinite Graphs", *Israel J. Math.* **76**:3 (1991), 317–336. (Also, *Proc. 23rd Ann. ACM Symp. on Theory of Computing*, New Orleans, pp. 220–229, 1991.)

[HH1] T. Hirst and D. Harel, "Taking it to the Limit: On Infinite Variants of NP-Complete Problems", *J. Comput. Syst. Sci.*, to appear. (Also, *Proc. 8th IEEE Conf. on Structure in Complexity Theory*, IEEE Press, New York, 1993, pp. 292–304.)

[HH2] T. Hirst and D. Harel, "Completeness Results for Recursive Data Bases", *J. Comput. Syst. Sci.*, to appear. (Also, *12th ACM Ann. Symp. on Principles of Database Systems*, ACM Press, New York, 1993, 244–252.)

[HH3] T. Hirst and D. Harel, "More about Recursive Structures: Zero-One Laws and Expressibility vs. Complexity", in preparation.

[I] N. Immerman, "Relational Queries Computable in Polynomial Time", *Inf. and Cont.* **68** (1986), 86–104.

[KT] P. G. Kolaitis and M. N. Thakur, "Logical definability of NP optimization problems", *6th IEEE Conf. on Structure in Complexity Theory*, pp. 353–366, 1991.

[MR] A. Manaster and J. Rosenstein, "Effective Matchmaking (Recursion Theoretic Aspects of a Theorem of Philip Hall)", *Proc. London Math. Soc.* **3** (1972), 615–654.

[Mo] A. S. Morozov, "Functional Trees and Automorphisms of Models", *Algebra and Logic* **32** (1993), 28–38.

[Mos] Y. N. Moschovakis, *Elementary Induction on Abstract Structures*, North Holland, 1974.

[NR] A. Nerode and J. Remmel, "A Survey of Lattices of R. E. Substructures", In *Recursion Theory*, Proc. Symp. in Pure Math. Vol. 42 (A. Nerode and R. A. Shore, eds.), Amer. Math. Soc., Providence, R. I., 1985, pp. 323–375.

[PR] A. Panconesi and D. Ranjan, "Quantifiers and Approximation", *Theor. Comp. Sci.* **107** (1993), 145–163.

[PY] C. H. Papadimitriou and M. Yannakakis, "Optimization, Approximation, and Complexity Classes", *J. Comp. Syst. Sci.* **43**, (1991), 425–440.

[Ra] R. Rado, "Universal Graphs and Universal Functions", *Acta Arith.*, **9**, (1964), 331–340.

[R] H. Rogers, *Theory of Recursive Functions and Effective Computability*, McGraw-Hill, New York, 1967.

[V] M. Y. Vardi, "The Complexity of Relational Query Languages", *Proc. 14th ACM Ann. Symp. on Theory of Computing*, 1982, pp. 137–146.

Modularization and Abstraction: The Keys to Practical Formal Verification[*]

Yonit Kesten[1] and Amir Pnueli[2]

[1] Ben Gurion University, ykesten@bgumail.bgu.ac.il,
[2] Weizmann Institute of Science, amir@wisdom.weizmann.ac.il

Abstract. In spite of the impressive progress in the development of the two main methods for formal verification of reactive systems – Model Checking (in particular symbolic) and Deductive Verification, they are still limited in their ability to handle large systems. It is generally recognized that the only way these methods can ever scale up is by the extensive use of abstraction and modularization, which breaks the task of verifying a large system into several smaller tasks of verifying simpler systems.

In this methodological paper, we review the two main tools of compositionality and abstraction in the framework of linear temporal logic. We illustrate the application of these two methods for the reduction of an infinite-state system into a finite-state system that can then be verified using model checking.

The modest technical contributions contained in this paper are a full formulation of abstraction when applied to a system with both weak and strong fairness requirements and to a general temporal formula, and a presentation of a compositional framework for shared variables and its application for forming *network invariants*.

1 Introduction

In spite of the impressive progress in the development of the two main methods for formal verification of reactive systems – Model Checking (in particular symbolic) and Deductive Verification, they are still limited in their ability to handle large systems. It is generally recognized that the only way these methods can ever scale up to handle industrial-size designs is by the extensive use of abstraction and modularization, which break the task of verifying a large system into several smaller tasks of verifying simpler systems.

In this methodological paper, we review the two main tools of compositionality and abstraction in the framework of linear temporal logic. We illustrate the application of these two methods for the reduction of an infinite-state system into a finite-state system that can then be verified using model checking.

---

[*] This research was supported in part by a gift from Intel, a grant from the U.S.-Israel bi-national science foundation, and an *Infrastructure* grant from the Israeli Ministry of Science and the Arts.

To simplify matters, we have considered two special classes of infinite-state systems for which the combination of compositionality and abstraction can effectively simplify the systems into finite-state ones. The first class is where the unboundedness of the system results from its structure. These are parameterized designs consisting of a parallel composition of finite-state processes, whose number is a varying parameter. For such systems, the source of complexity is the control or the architectural structure. We describe the techniques useful for such systems as *control abstraction*, since it is the control component that we try to simplify. Another source for state complexity is having data variables which range over infinite domains such as the integers. We refer to the techniques appropriate for simplifying such systems as *data abstraction*.

Many methods have been proposed for the uniform verification of parameterized systems, which is the subject of our control abstraction. These include explicit induction ([EN95], [SG92]) network invariants, which can be viewed as implicit induction ([KM95], [WL89], [HLR92], [LHR97]), methods that can be viewed as abstraction and approximation of network invariants ([BCG86], [SG89], [CGJ95]), and other methods that can be viewed as based on abstraction ([ID96], [EN96]). The approach described here is based on the idea of *network invariants* as introduced in [WL89], and elaborated in [KM95] into a working method.

There has been extensive study of the use of data abstraction techniques, mostly based on the notions of *abstract interpretation* ([CC77], [CH78]). Most of the previous work was done in a branching context which complicates the problem if one wishes to preserve both existential and universal properties. On the other hand, if we restrict ourselves to a universal fragment of the logic, e.g. ACTL*, then the conclusions reached are similar to our main result for the restricted case that the property ψ contains negations only within assertions.

The paper [CGL94] obtains a similar result for the fragment ACTL*. However, instead of starting with a concrete property ψ and abstracting it into an appropriate ψ^α, they start with an abstract ACTL* formula Ψ evaluated over the abstract system $\mathcal{K}^\alpha$ and show how to translate (concretize) it into a concrete formula $\psi = \mathcal{C}(\Psi)$. The concretization is such that $M^\alpha_\forall(\psi) = \Psi$.

The survey in [CGL96] considers an even simpler case in which the abstraction does not concern the variables on which the property ψ depends. Consequently, this is the case in which $\psi^\alpha = \psi$.

A more elaborate study in [DGG97] considers a more complex specification language – L_μ, which is a positive version of the μ-calculus.

None of these three articles considers explicitly the question of fairness requirements and how they are affected by the abstraction process.

Approaches based on simulation and studies of the properties they preserve are considered in [BBLS92] and [GL93].

A linear-time application of abstract interpretation is proposed in [BBM95], applying the abstractions directly to the computational model of *fair transition systems* which is very close to the FKS model considered here. However, the method is only applied for the verification of safety properties. Liveness, and therefore fairness, are not considered.

2 A Computational Model: Fair Kripke Structure

As a computational model for reactive systems, we take the model of *fair kripke structure* (FKS) [KPR98], which is a slight variation on the model of *fair transition system* [MP95]. Such a system $\mathcal{K}: \langle V, W, \mathcal{O}, \Theta, \rho, \mathcal{J}, \mathcal{C} \rangle$ consists of the following components.

- $V = \{u_1, ..., u_n\}$: A finite set of typed *system variables*, containing data and control variables. The set of *states* (interpretation) over V is denoted by Σ.
- $W = \{w_1, \ldots, w_n\} \subseteq V$: A finite set of *owned variables*. These are the variables that only the system itself can modify. All other variables can also be modified by the environment. A system is said to be *closed* if $W = V$.
- $\mathcal{O} = \{o_1, \ldots, o_n\} \subseteq V$: A finite set of *observable variables*. These are the variables whose values (and identities) must be preserved in some of the abstractions we will consider. It is required that $V = W \cup \mathcal{O}$, i.e., for every system variable $u \in V$, u is either *owned*, *observable*, or both.
- Θ : The *initial condition* – an *assertion* (first-order state formula) characterizing the initial states.
- ρ : A *transition relation* – an assertion $\rho(V, V')$, relating the values V of the variables in state $s \in \Sigma$ to the values V' in a $\mathcal{K}$-successor state $s' \in \Sigma$.
- $\mathcal{J} = \{J_1, \ldots, J_k\}$: A set of *justice* requirements (also called *weak fairness requirements*). The justice requirement $J \in \mathcal{J}$ is an assertion, intended to guarantee that every computation contains infinitely many J-state (states satisfying J).
- $\mathcal{C} = \{\langle p_1, q_1 \rangle, \ldots \langle p_n, q_n \rangle\}$: A set of *compassion* requirements (also called *strong fairness requirements*). The compassion requirement $\langle p, q \rangle \in \mathcal{C}$ is a pair of assertions, intended to guarantee that every computation containing infinitely many p-states also contains infinitely many q-states.

We require that every state $s \in \Sigma$ has at least one $\mathcal{K}$-successor. This is often ensured by including in ρ the *idling* disjunct $V = V'$ (also called the *stuttering* step). In such cases, every state s is its own $\mathcal{K}$-successor.

Let $\sigma : s_0, s_1, s_2, ...$, be an infinite sequence of states, φ be an assertion, and let $j \geq 0$ be a natural number. We say that j is a φ-position of σ if s_j is a φ-state.

Let $\mathcal{K}$ be an FKS for which the above components have been identified. We define a *computation* of $\mathcal{K}$ to be an infinite sequence of states $\sigma : s_0, s_1, s_2, ...$, satisfying the following requirements:

- *Initiality:* s_0 is initial, i.e., $s_0 \models \Theta$.
- *Consecution:*   For each $j = 0, 1, ...$, the state s_{j+1} is a $\mathcal{K}$-successor of the state s_j.
- *Justice:*   For each $J \in \mathcal{J}$, σ contains infinitely many J-positions

- *Compassion:* For each $\langle p,q \rangle \in \mathcal{C}$, if σ contains infinitely many p-positions, it must also contain infinitely many q-positions.

For an FKS $\mathcal{K}$, we denote by $\mathcal{C}omp(\mathcal{K})$ the set of all computations of $\mathcal{K}$. An FKS $\mathcal{K}$ is called *feasible* if $\mathcal{C}omp(\mathcal{K}) \neq \emptyset$, namely, if $\mathcal{K}$ has at least one computation. The feasibility of an FKS can be checked algorithmically, using symbolic model checking methods, as presented in [KPR98].

All our concrete examples are given in SPL (Simple Programming Language), which is used to represent concurrent programs (e.g., [MP95], [MAB+94]). Every SPL program can be compiled into an FKS in a straightforward manner. In particular, every statement in an SPL program contributes a disjunct to the transition relation. For example, the assignment statement

$$\ell_0 : y := x+1;\ \ell_1 :$$

can be executed when control is at location ℓ_0. When executed, it assigns $x+1$ to y while control moves from ℓ_0 to ℓ_1. This statement contributes to ρ the disjunct

$$\rho_{\ell_0}:\quad at\_\ell_0 \wedge at\_\ell_1' \wedge y' = x+1 \wedge x' = x.$$

The predicates at_ℓ_0 and at_ℓ_1' stand, respectively, for the assertions $\pi_i = 0$ and $\pi_i' = 1$, where π_i is the control variable denoting the current location within the process to which the statement belongs.

3 Operations on FKS's

There are several important operations, one may wish to apply to FKS's.

The first useful set of operations on programs and systems is forming their parallel composition, implying that the two systems execute concurrently. Consider the two fair Kripke structures $\mathcal{K}_1 = \langle V_1, W_1, \mathcal{O}_1, \Theta_1, \rho_1, \mathcal{J}_1, \mathcal{C}_1 \rangle$ and $\mathcal{K}_2 = \langle V_2, W_2, \mathcal{O}_2, \Theta_2, \rho_2, \mathcal{J}_2, \mathcal{C}_2 \rangle$. There are several ways of forming their parallel composition.

3.1 Asynchronous Parallel Composition

The systems $\mathcal{K}_1$ and $\mathcal{K}_2$ are said to be *compatible* if $W_1 \cap W_2 = \emptyset$ and $V_1 \cap V_2 \subseteq \mathcal{O}_1 \cap \mathcal{O}_2$. The first condition requires that a variable can only be owned by one of the systems. The second condition requires that variables known to both systems must be observable in both.

For compatible systems $\mathcal{K}_1$ and $\mathcal{K}_2$, we define their asynchronous parallel composition, denoted by $\mathcal{K}_1 \| \mathcal{K}_2$, to be the system $\mathcal{K} = \langle V, W, \mathcal{O}, \Theta, \rho, \mathcal{J}, \mathcal{C} \rangle$, where

$$\begin{array}{lll} V = V_1 \cup V_2 & W = W_1 \cup W_2 & \mathcal{O} = \mathcal{O}_1 \cup \mathcal{O}_2 \\ \Theta = \Theta_1 \wedge \Theta_2 & \mathcal{J} = \mathcal{J}_1 \cup \mathcal{J}_2 & \mathcal{C} = \mathcal{C}_1 \cup \mathcal{C}_2 \\ \rho = (\rho_1 \wedge pres((V_2 - V_1) \cup W_2)) \vee (\rho_2 \wedge pres((V_1 - V_2) \cup W_1)). \end{array}$$

For a set of variables $U \subseteq V$, the predicate $pres(U)$ stands for the assertion $U' = U$, implying that all the variables in U are preserved by the transition.

Obviously, the basic actions of the composed system $\mathcal{K}$ are chosen from the basic actions of its components, i.e., $\mathcal{K}_1$ and $\mathcal{K}_2$. Thus, we can view the execution of $\mathcal{K}$ as the *interleaved execution* of $\mathcal{K}_1$ and $\mathcal{K}_2$.

As seen from the definition, $\mathcal{K}_1$ and $\mathcal{K}_2$ may have disjoint as well as common system variables, and the variables of $\mathcal{K}$ are the union of all of these variables. The initial condition of $\mathcal{K}$ is the conjunction of the initial conditions of $\mathcal{K}_1$ and $\mathcal{K}_2$. The transition relation of $\mathcal{K}$ states that at any step, we may choose to perform a step of $\mathcal{K}_1$ or a step of $\mathcal{K}_2$. However, when we select one of the two systems, we should also take care to preserve the private variables of the other system. For example, choosing to execute a step of $\mathcal{K}_1$, we should preserve all variables in $V_2 - V_1$ and all the variables owned by $\mathcal{K}_2$.

The justice and compassion sets of $\mathcal{K}$ are formed as the respective unions of the justice and compassion sets of the component systems.

Asynchronous parallel composition corresponds to the SPL parallel operator ∥ constructing a program out of concurrent processes.

3.2 Synchronous Parallel Composition

We define the *synchronous parallel composition* of $\mathcal{K}_1$ and $\mathcal{K}_2$, denotes by $\mathcal{K}_1 |\!|\!| \mathcal{K}_2$, to be the system $\mathcal{K} = \langle V, W, \mathcal{O}, \Theta, \rho, \mathcal{J}, \mathcal{C} \rangle$, where,

$$V = V_1 \cup V_2 \qquad W = W_1 \cup W_2 \qquad \mathcal{O} = \mathcal{O}_1 \cup \mathcal{O}_2$$
$$\Theta = \Theta_1 \wedge \Theta_2 \qquad \mathcal{J} = \mathcal{J}_1 \cup \mathcal{J}_2 \qquad \mathcal{C} = \mathcal{C}_1 \cup \mathcal{C}_2$$
$$\rho = \rho_1 \wedge \rho_2.$$

As implied by the definition, each of the basic actions of system $\mathcal{K}$ consists of the joint execution of an action of $\mathcal{K}_1$ and an action of $\mathcal{K}_2$. Thus, we can view the execution of $\mathcal{K}$ as the *joint execution* of $\mathcal{K}_1$ and $\mathcal{K}_2$.

As will be shown in the next section, the main use of the synchronous parallel composition is for coupling a system with a *tester* which tests for the satisfaction of a temporal formula, and then checking the feasibility of the combined system.

3.3 Modularization of an FKS

Let P be an SPL program and $\mathcal{K}$ its corresponding FKS. The standard compilation of a program into an FKS views the program as a *closed system* which has no interaction with its environment. In the context of compositional verification, we need an *open system* view of an FKS, which takes into account not only actions performed by the system but also actions (in particular, variable changes) performed by the environment.

Let $\mathcal{K} : \langle V, W, \mathcal{O}, \Theta, \rho, \mathcal{J}_K, \mathcal{C}_K \rangle$ be an FKS, such that $s \notin V$. The *modular version* of $\mathcal{K}$, is given by $\mathcal{K}_M : \langle V_M, W_M, \mathcal{O}_M, \Theta_M, \rho_M, \mathcal{J}_M, \mathcal{C}_M \rangle$, where,

$$V_M = V \cup \{s\} \qquad W_M = W \qquad \mathcal{O}_M = \mathcal{O} \cup \{s\}$$
$$\Theta_M = \Theta \qquad \mathcal{J}_M = \mathcal{J} \qquad \mathcal{C}_M = \mathcal{C}$$
$$\rho_M = (\rho \wedge s') \vee (W' = W \wedge \neg s').$$

That is, $\mathcal{K}_M$ the modular version of $\mathcal{K}$ allows as an additional action a transition which preserves the values of all variables owned by $\mathcal{K}$ but allows all other shared variables to change in an arbitrary way. This provides the most general representation of an environment action. The *scheduling variable s* is used to ensure interleaving between the module and its environment. We refer to a system obtained as the modularization of another FKS as a *Fair Kripke Module* (FKM).

We define a *modular computation* of $\mathcal{K}$ to be any computation of $\mathcal{K}_M$. A property φ is said to be *modularly valid* over FKS $\mathcal{K}$, denoted $\mathcal{K} \models_M \varphi$, if φ is $\mathcal{K}_M$-valid.

3.4 Modular Composition

We define the *modular composition* of the compatible FKM's $\mathcal{K}_1$ and $\mathcal{K}_2$, denoted by $\mathcal{K}_1 \|_M \mathcal{K}_2$, to be the FKM $\mathcal{K}_M : \langle V_M, W_M, \mathcal{O}_M, \Theta_M, \rho_M, \mathcal{J}_M, \mathcal{C}_M \rangle$, where,

$$\begin{aligned}
V_M &= V_1 \cup V_2 & W_M &= W_1 \cup W_2 & \mathcal{O}_M &= \mathcal{O}_1 \cup \mathcal{O}_2 \\
\Theta_M &= \Theta_1 \wedge \Theta_2 & \mathcal{J}_M &= \mathcal{J}_1 \cup \mathcal{J}_2 & \mathcal{C}_M &= \mathcal{C}_1 \cup \mathcal{C}_2 \\
\rho_M &= \exists s_1, s_2 : \text{boolean} \cdot (s = s_1 \vee s_2) \wedge \neg(s_1 \wedge s_2) \wedge \rho_1[s \mapsto s_1] \wedge \rho_2[s \mapsto s_2]
\end{aligned}$$

A step in the execution of $\mathcal{K}_M$ is either a step of system $\mathcal{K}_1$ where $s = s_1 = 1$ and $s_2 = 0$, or a step of system $\mathcal{K}_2$ where $s = s_2 = 1$ and $s_1 = 0$, or an environment step where $s = s_1 = s_2 = 0$. A step of $\mathcal{K}_1$ (similarly $\mathcal{K}_2$) is governed by the transition relation $\rho_1[s \mapsto s_1]$, which is the assertion ρ_1 in which all references to s and s' are replaced by references to s_1 and s_1', respectively.

For closed-system $\mathcal{K}_1$ and $\mathcal{K}_2$ which are composed in parallel, we can first modularize each of the systems individually and then form their modular composition. Alternately, we can form first the asynchronous parallel composition of the two systems and then modularize the combined system. It can be seen that both processes yield the same FKM, which is expressible by the equivalence

$$(\mathcal{K}_1)_M \|_M (\mathcal{K}_2)_M \quad \sim \quad (\mathcal{K}_1 \| \mathcal{K}_2)_M.$$

3.5 Sealing Off an Open System

Assume we have an FKM, consisting of a modular composition of several FKM's:

$$\mathcal{K}_M = \mathcal{K}_1 \|_M \ldots \|_M \mathcal{K}_K.$$

System $\mathcal{K}_M$ is still an open system, admitting arbitrary interference by the environment. Once we know that all the processes in the system have been included (possibly including a process that represents the environment) and no further interaction with the external world is expected, way may *seal off* the system, formally excluding any further external communication. This is done by declaring all variables to be owned by the system, and eliminating the scheduling variable s.

Let $\mathcal{K}_M : \langle V, W, \mathcal{O}, \Theta, \rho, \mathcal{J}, \mathcal{C} \rangle$ be an FKM representing an open system. The result of sealing off $\mathcal{K}_M$ is an FKS $\mathcal{K}_c : \langle V_c, W_c, \mathcal{O}_c, \Theta_c, \rho_c, \mathcal{J}_c, \mathcal{C}_c \rangle$, where

$$\begin{aligned}
V_c &= V - \{s\} & W_c &= V & \mathcal{O}_c &= \mathcal{O} - \{s\} \\
\Theta_c &= \Theta & \mathcal{J}_c &= \mathcal{J} & \mathcal{C}_c &= \mathcal{C} \\
\rho_c &= \rho[s \mapsto \text{true}]
\end{aligned}$$

Note that sealing off an FKM is the inverse operation to modularizing an FKS, and therefore taking the asynchronous composition of two closed systems is equivalent to the FKS obtained by sealing off the modular composition of their modular versions, as stated by the following equivalences:

$$([\mathcal{K}]_M)_C \sim \mathcal{K} \qquad ([\mathcal{K}_1]_M \|_M [\mathcal{K}_2]_M)_C \sim \mathcal{K}_1 \| \mathcal{K}_2.$$

4 Requirement Specification Language: Temporal Logic

As a requirement specification language for reactive systems we take *temporal logic* (TL) [MP91]. For simplicity, we consider only the future fragment of TL.

We assume an underlying assertion language $\mathcal{L}$ which contains the predicate calculus and interpreted symbols for expressing the standard operations and relations over some concrete domains. A *temporal formula* is constructed out of state formulas (assertions) to which we apply the boolean operators $\neg$ and $\vee$ (the other boolean operators can be defined from these), and the basic temporal operators $\bigcirc$ (*next*) and $\mathcal{U}$ (*until*).

A *model* for a temporal formula p is an infinite sequence of states $\sigma : s_0, s_1, ...$, where each state s_j provides an interpretation for the variables mentioned in p.

Given a model σ, we present an inductive definition for the notion of a temporal formula p holding at a position $j \geq 0$ in σ, denoted by $(\sigma, j) \models p$.

- For a state formula p, $(\sigma, j) \models p \iff s_j \models p$

  That is, we evaluate p locally, using the interpretation given by s_j.
- $(\sigma, j) \models \neg p \iff (\sigma, j) \not\models p$ • $(\sigma, j) \models p \vee q \iff (\sigma, j) \models p$ or $(\sigma, j) \models q$
- $(\sigma, j) \models \bigcirc p \iff (\sigma, j+1) \models p$
- $(\sigma, j) \models p \mathcal{U} q \iff$ for some $k \geq j, (\sigma, k) \models q$,

  and for every i such that $j \leq i < k, (\sigma, i) \models p$

Additional temporal operators can be defined by $\Diamond p = true \mathcal{U} p$ (*eventually*) and $\Box p = \neg \Diamond \neg p$ (*henceforth*).

For a temporal formula p and a position $j \geq 0$ such that $(\sigma, j) \models p$, we say that j is a p-*position* (in σ). If $(\sigma, 0) \models p$, we say that p holds on σ, and denote it by $\sigma \models p$. A formula p is called *satisfiable* if it holds on some model. A formula p is called *valid*, denoted by $\models p$, if it holds on all models.

Given an FKS $\mathcal{K}$ and a temporal formula p, we say that p is $\mathcal{K}$-*valid*, denoted by $\mathcal{K} \models p$, if p holds on all models which are computations of $\mathcal{K}$.

An algorithm for model checking whether a temporal formula p is valid over a finite-state FKS $\mathcal{K}$ is presented in [KPR98]. The paper presents a version of the algorithm using explicit state enumeration methods as well as a symbolic version. Based on the ideas developed in [LPS81] and [CGH94], the approach calls for the construction of a *tester* for the negation of p. This is an FKS $\mathcal{K}_{\neg p}$ whose computations are all the sequences which satisfy the negated formula p. Then we form the *synchronous parallel composition* $\mathcal{K}_{comb} = \mathcal{K} \| \mathcal{K}_{\neg p}$ and check for feasibility. If $\mathcal{K}_{comb}$ is found to be feasible, this implies that $\mathcal{K}$ has a computation which violates p and therefore p is not valid over $\mathcal{K}$. If $\mathcal{K}_{comb}$ is found to be infeasible, we can conclude that p is $\mathcal{K}$-valid.

5 Control Abstraction

Let $\sigma : s_0, s_1, \ldots$ be an infinite sequence of V-states, and let $U \subseteq V$ be a subset of V. We say that the infinite state sequence $\tilde{\sigma} : \tilde{s}_0, \tilde{s}_1, \ldots$ is a U-*preserving variant* of σ if s_i and $\tilde{s}_i$ agree on the interpretation of the variables in U, for every $i = 0, 1, \ldots$. We define $\sigma : s_0, s_1, \ldots$ to be an *observation* of the FKS $\mathcal{K} = \langle V, W, \mathcal{O}, \Theta, \rho, \mathcal{J}, \mathcal{C} \rangle$ if σ is a $\mathcal{O}$-preserving variant of a computation of $\mathcal{K}$. Let $Obs(\mathcal{K})$ denote the set of all observations of system $\mathcal{K}$.

The FKS $\mathcal{K}_A = \langle V_A, W_A, \mathcal{O}_A, \Theta_A, \rho_A, \mathcal{J}_A, \mathcal{C}_A \rangle$ is defined to be *observation compatible* with $\mathcal{K} = \langle V, W, \mathcal{O}, \Theta, \rho, \mathcal{J}, \mathcal{C} \rangle$ if $\mathcal{O}_A = \mathcal{O}$. The FKS $\mathcal{K}$ is an *abstraction* of the observation compatible $\mathcal{K}$, denoted by $\mathcal{K} \sqsubseteq \mathcal{K}_A$, if $Obs(\mathcal{K}) \subseteq Obs(\mathcal{K}_A)$. We refer to $\mathcal{K}$ and $\mathcal{K}_A$ as the *concrete* and *abstract* systems, respectively.

It can be shown that all the FKS operations defined in Section 3 are monotonic with respect to the abstraction relation. In particular, if $\mathcal{K} \sqsubseteq \mathcal{K}_A$ then $(\mathcal{K} \|_M \mathcal{K}_2) \sqsubseteq (\mathcal{K}_A \|_M \mathcal{K}_2)$ and $\mathcal{K}_M \sqsubseteq \mathcal{K}_A$. Furthermore, if p is a temporal formula whose free variables belong to $\mathcal{O} = \mathcal{O}_A$, then

$$\mathcal{K}_A \models p \quad \text{implies} \quad \mathcal{K} \models p.$$

This indicates how we propose to use abstraction in order to simplify the verification task. Namely, given a property p to be verified over a complex system $\mathcal{K}$, we use abstraction in order to derive a simpler system $\mathcal{K}_A$ and then verify that p is $\mathcal{K}_A$-valid. Note that the implication is still in one direction. Namely, validity over the abstract system implies concrete validity but not, necessarily vice versa. The most striking applications of this strategy are when $\mathcal{K}$ is an infinite-state system, while its abstraction $\mathcal{K}_A$ is finite-state and thus amenable to model checking.

In this section, we concentrate on cases in which the system is a parallel composition $P(n)$: $P_1 \| \cdots \| P_n$, where each P_i is a finite-state system. The unbounded number of states for system $P(n)$ comes from the fact that we consider an infinite *family* of systems, and yet wish to verify uniformly (i.e., for every value of n) that the property p is valid.

The method and one of the examples presented in this section are taken from [KM95]. The main differences between the two presentations are that, while [KM95] considers processes communicating by synchronous message passing we have reformulated the framework to communication by shared variables. D

For simplicity, assume that the property p only refers to the observable variables of P_1 and that processes $P_2, \ldots, P_n$ are identical (up to renaming). The strategy we propose can be summarized as follows:

1. Generate FKM $\mathcal{K}_i$, representing the modular behavior of process P_i, for $i = 1, \ldots, k$.
2. Derive a *network invariant* $\mathcal{I}$, which is an FKM intended to form an abstraction for the modular composition $\mathcal{K}_2 \|_M \cdots \|_M \mathcal{K}_n$ for any value of n.
3. Confirm that $\mathcal{I}$ is indeed a network invariant, by model checking that $P_2 \sqsubseteq \mathcal{I}$ and that $(\mathcal{I} \|_M \mathcal{I}) \sqsubseteq \mathcal{I}$.
4. Model check $\mathcal{K} \models p$, where $\mathcal{K}$ is the closed system $(P_1 \|_M \mathcal{I})_C$.

It can be proven that this strategy is sound [KM95]. Namely, if $\mathcal{K} \models p$ then $P(n) \models p$ for every $n > 0$. Step 2 in this strategy is the only one requiring ingenuity and which cannot be fully mechanized. However, while presenting the examples, we will provide some explanations for the choices we made.

5.1 Mutual Exclusion by Semaphores

As our first running example, we use a program that manages mutual exclusion by semaphores. The program consists of n processes. Each process $P[i]$ cycles

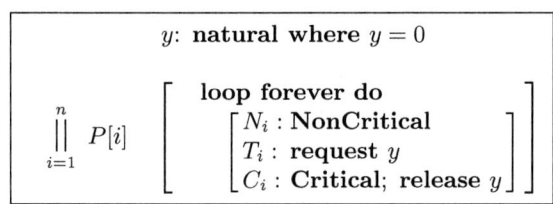

Fig. 1. Program MUX-SEM.

through three possible locations: N_i, T_i, and C_i. Location N_i represents the non-critical activity which the process can perform without coordination with the other processes. Location T_i, is the "trying" location, at which a process decides it needs to access its critical location. At the trying location, the process waits for the semaphore variable y to become 1. On entering the critical section C_i, the process sets y to 0. Finally, C_i is the critical location which should be reachable only exclusively by one process at a time. On exit from the critical section, variable y is reset to 1.

In Fig. 2, we present the FKM corresponding to one of the (identical) processes of program MUX-SEM. In this diagrammatic representation, nodes correspond to sets of states. For example, the node labeled by $(N, 1)$ corresponds to the two states $\langle N, y : 1, s : 1 \rangle$ and $\langle N, y : 1, s : 0 \rangle$. To simplify the presentation, we used a single node to represent these two states which only differ in the interpretation of s. All edges connecting a node to itself have been eliminated. A solid edge in the diagram represents a step of the module itself, while a dotted edge represents a step of the environment. Thus, while only the process can decide to move from control state N to control state C, only the environment can change y from 1 to 0 (and vice versa) while control is still at N. The fairness requirements associated with this FKM are the justice requirement $\neg C$ ensuring that the system will not stay forever at control location C, and the compassion requirement $(T \wedge y = 1, C)$ guaranteeing that if the process is waiting at T and y equals 1 infinitely many times, then control will eventually proceed to C.

A first abstraction we can apply to $\mathcal{K}_1$ is to observe that as far as the sequences of values of the observable variables are concerned, there is no need to distinguish between the control locations N and T. This leads to the FKM $\mathcal{K}_2$ presented in Fig. 3.

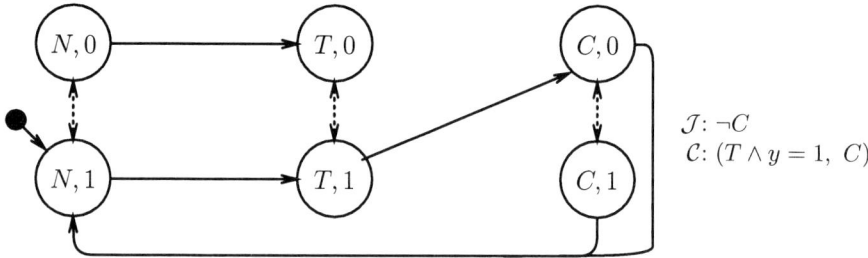

Fig. 2. The FKM $\mathcal{K}_1$ corresponding to process $P[1]$ of program MUX-SEM.

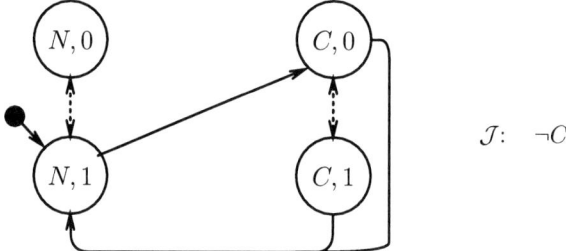

Fig. 3. The FKM $\mathcal{K}_2$ abstracting FKM $\mathcal{K}_1$.

Note that the compassion requirement has been eliminated. This implies that the system can tolerate a behavior in which it never sets y to 0 beyond a certain point. However, such a behavior was allowed also in FKM $\mathcal{K}_1$ by remaining forever at control location N beyond a certain point.

A useful heuristic that often leads to the generation of network invariants is forming the sequence of FKM's $\mathcal{I}_1 = \mathcal{K}_2$, $\mathcal{I}_2 = \mathcal{K}_2\|_M\mathcal{K}_2$, $\mathcal{I}_3 = \mathcal{I}_2\|_M\mathcal{K}_2$, ..., and comparing every successive $\mathcal{I}_i$'s, hoping that the sequence will converge. Trying this approach with the FKM $\mathcal{K}_2$ fails. Comparing $\mathcal{I}_2\colon\mathcal{K}_2\|_M\mathcal{K}_2$ with $\mathcal{I}_1\colon\mathcal{K}_2$, we find that $\mathcal{I}_2$ can generate the observation (displaying the values of y and s)

$$\cdots\ \langle 1,1\rangle,\ \langle 0,1\rangle,\ \langle 1,0\rangle,\ \langle 0,1\rangle,\ \cdots,$$

which cannot be generated by $\mathcal{K}_1$. A step $\langle 1,-\rangle \to \langle 0,1\rangle$ in this behavior corresponds to the module setting y to 0, which corresponds to an entry to the critical section. A step $\langle 0,-\rangle \to \langle 1,0\rangle$ corresponds to a step in which the environment changes y from 0 to 1. Thus, this behavior displays a situation in which $\mathcal{I}_2$ enters twice the critical section before exiting even once, provided the environment raises the value of y, while one of the components of $\mathcal{I}_2$ was still in the critical section. In a similar way, we find that $\mathcal{I}_3$ can enter its critical sections three times in succession, if the environment cooperates, which cannot be done by $\mathcal{I}_2$. This shows that the sequence $\mathcal{I}_1,\mathcal{I}_2,\ldots$ will never converge.

Looking closer at this example, we realize that the factor that differentiates between $\mathcal{I}_1$ and $\mathcal{I}_2$ and between $\mathcal{I}_2$ and $\mathcal{I}_3$ is their response to a behavior of the environment which will never be realized in the closed system, namely raising the semaphore variable to 1 while one of the processes is in its critical section. This leads us to the next abstraction $\mathcal{K}_3$, presented in Fig. 4.

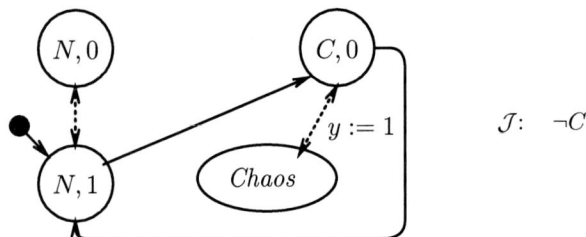

Fig. 4. The FKM $\mathcal{I} = \mathcal{K}_3$ with chaos.

The system $\mathcal{K}_3$ behaves as $\mathcal{K}_2$ as long as the environment behaves properly. However, once it detects that the environment raised the value of y from 0 to 1 while the system was in the critical section, it goes into a *chaos* control state in which "anything goes". That is, all arbitrary sequences of values for the observable variables will be accepted from this point on. It is obvious that $\mathcal{K}_3$ is an abstraction of $\mathcal{K}_2$ because it differs from $\mathcal{K}_2$ in all the additional behaviors it is ready to generate once it reached the *chaos* state.

It is not difficult to verify that $\mathcal{I} = \mathcal{K}_3$ is a network invariant. We model checked that $\mathcal{K}_2 \sqsubseteq \mathcal{K}_3$ and that $(\mathcal{K}_3 \|_M \mathcal{K}_3) \sqsubseteq \mathcal{K}_3$.

It only remains to perform step 4 in the abstraction strategy presented in the beginning of the section. We form the closed-system FKS $\mathcal{K} = (\mathcal{K}_1 \|_M \mathcal{I})_M$ and use model checking to verify the liveness property $\mathcal{K} \models \Box(N_1 \to \Diamond C_1)$. This has been done and established that process $P[1]$ of program MUX-SEM has the property of accessibility for any number of processes.

5.2 The Dining Philosophers Problem

As a more advanced example, we applied the technique described above to the problem of the dining philosophers. As originally described by Dijkstra, n philosophers are seated at a round table. Each philosopher alternates between a thinking phase and a phase in which he becomes hungry and wishes to eat. There are n chop-sticks placed around the table, one chop-stick between every two philosophers. in order to eat, each philosopher needs to acquire the chop-sticks on both sides. A chop-stick can be possessed by only one philosopher at a time.

An solution to the dining philosophers problem, using semaphores, is presented by program DINE-CONTR of Fig. 5.

In this program, philosophers $P[2], \ldots, P[n]$ reach first for the fork on their left (represented by semaphore variable $c[j]$ for philosopher j), and then for their right fork (semaphore $c[j \oplus_n 1]$). Philosopher $P[1]$ behaves differently, reaching first for his right fork ($c[2]$) and only later for his left fork ($c[1]$). We wish to prove the liveness property of accessibility for each of the philosophers.

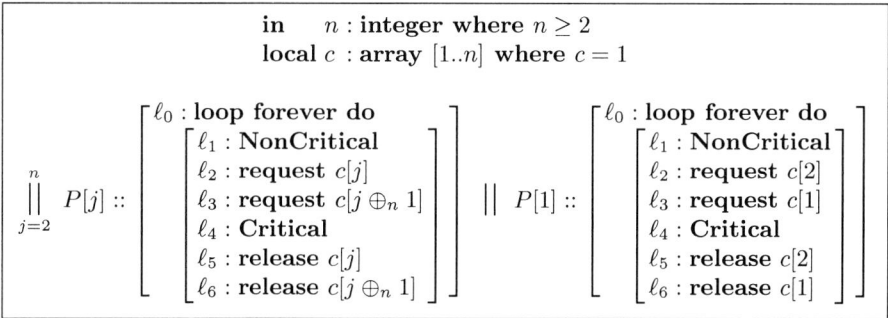

Fig. 5. Program DINE-CONTR: solution with one contrary philosopher.

Proceeding through a sequence of abstraction steps similar to the previous example, we finally wind up with the FKM $\mathcal{I}_{contr}$ presented in Fig. 6.

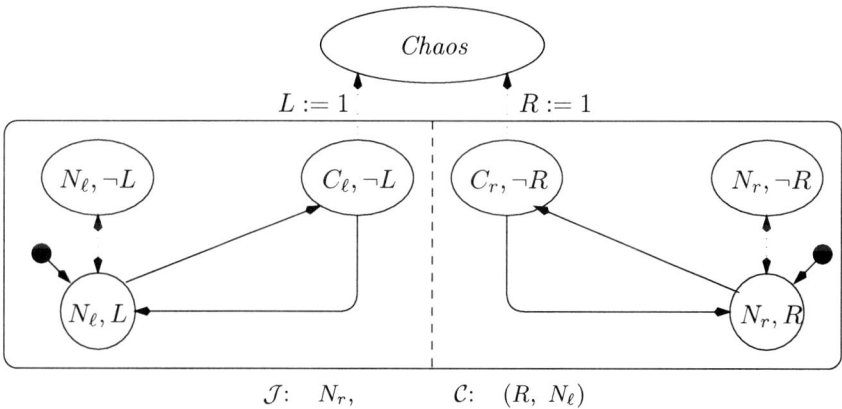

Fig. 6. The FKM $\mathcal{I}_{contr}$, the network invariant for program DINE-CONTR.

The diagram of Fig. 6 consists of two components that operate in parallel, one taking care of the left semaphore L and the other handling the right semaphore R. Whenever an environment fault is detected, i.e. the environment raises a semaphore that has been lowered by the system, both components escape to the *chaos* state after which all behaviors are possible. It is straightforward to verify (using model checking) that $\mathcal{I}_{contr}$ abstracts any of the processes $P[2], \ldots, P[n]$ and that $(\mathcal{I}_{contr} \|_M \mathcal{I}_{contr}) \sqsubseteq \mathcal{I}_{contr}$. It follows that $\mathcal{I}_{contr}$ is a network invariant for any sequence of regular philosophers. We can combine $\mathcal{I}_{contr}$ with $P[1]$ to establish the accessibility properties of the contrary philosopher $P[1]$. We can also verify the accessibility property for all ordinary philosophers.

6 Data Abstraction

In this section, we present a general methodology for *data abstraction*, strongly inspired by the notion of *abstract interpretation* [CC77]. Consider an FKS $\mathcal{K} = \langle V, W, \mathcal{O}, \Theta, \rho, \mathcal{J}, \mathcal{C} \rangle$, and let Σ denote the set of states of $\mathcal{K}$, the *concrete state*. Let $\alpha : \Sigma \mapsto \Sigma_A$ be a mapping of concrete states into *abstract states*. The strategy of *verification by data abstraction* can be summarized as follows:

Strategy 1 (Verification by Data Abstraction)

- *Define an abstraction mapping* α *to abstract the concrete* FKS $\mathcal{K}$ *into an abstract* FKS $\mathcal{K}_A^\alpha$.
- *Abstract the temporal property* ψ *into an abstract property* ψ^α.
- *Verify* $\mathcal{K}_A^\alpha \models \psi^\alpha$.
- *Infer* $\mathcal{K} \models \psi$.

The main question is how to define the abstractions $\mathcal{K}_A^\alpha$ and ψ^α such that $\mathcal{K}_A^\alpha \models \psi^\alpha$ implies $\mathcal{K} \models \psi$.

Example 1. Consider program ANY-Y of Fig. 7, for which we wish to establish the invariance property $\psi: \Box(y \geq 0)$.

$$P_1 :: \begin{bmatrix} \ell_0 : \textbf{while } x = 0 \textbf{ do} \\ \left[\ell_1 : y := y + 1 \right] \\ \ell_2 : \end{bmatrix} \quad \| \quad P_2 :: \begin{bmatrix} m_0 : x := 1 \\ m_1 : \end{bmatrix}$$

x, y: **integer where** $x = y = 0$

Fig. 7. Program ANY-Y: A simple concurrent program.

Program ANY-Y is an infinite-state system since the integer variable y can assume arbitrarily high values. To reduce the complexity of this system, we may consider an abstract variable Y, ranging over the finite (abstract) domain $\{neg, zero, pos\}$. The abstraction function α maps the domain of y into the domain of Y as follows:

$\alpha(y)$: **if** $y < 0$ **then** *neg* **else-if** $y = 0$ **then** *zero* **else** *pos*

With this mapping, we can obtain the abstract version of ANY-Y, called ANY-Y$^\alpha$, and presented in Fig. 8.

A corresponding α-abstraction of the property to be verified is given by $\psi^\alpha: \Box(Y \in \{zero, pos\})$. Since program ANY-Y$^\alpha$ is a finite-state program, we can use model checking in order to verify ANY-Y$^\alpha \models \psi^\alpha$. Following the verification-by-data-abstraction strategy, we can infer ANY-Y $\models \Box(y \geq 0)$.

$$\boxed{\begin{array}{c} x\colon \textbf{integer} \qquad \textbf{where } x = 0 \\ Y\colon \{neg, zero, pos\} \textbf{ where } Y = zero \\ P_1 :: \begin{bmatrix} \ell_0 : \textbf{while } x = 0 \textbf{ do} \\ \left[\ell_1 : Y := \begin{pmatrix} \textbf{if} & Y = neg \\ \textbf{then} & \{neg, zero\} \\ \textbf{else} & pos \end{pmatrix}\right] \\ \ell_2 : \end{bmatrix} \quad \| \quad P_2 :: \begin{bmatrix} m_0 : x := 1 \\ m_1 : \end{bmatrix} \end{array}}$$

Fig. 8. Program ANY-Y$^\alpha$: Abstracted version of ANY-Y.

6.1 Safe Abstraction of Temporal Formulas and Systems

To provide a syntactic representation of the abstraction mapping, we assume a set of *abstract variables* V_A and a set of expressions $\mathcal{E}^\alpha$, such that the equality $V_A = \mathcal{E}^\alpha(V)$ syntactically represents the semantic mapping α. Thus, for Example 1, the expressions $\mathcal{E}^\alpha_{\pi_1}$, $\mathcal{E}^\alpha_{\pi_2}$, $\mathcal{E}^\alpha_x$, and $\mathcal{E}^\alpha_Y$ are given by π_1, π_2, x, and **if** $y < 0$ **then** *neg* **else-if** $y = 0$ **then** *zero* **else** *pos*, respectively.

Let p be an assertion (state formula). We define two abstraction operators over p.

$$M^\alpha_\forall(p)\colon \forall V\ (V_A = \mathcal{E}^\alpha(V) \rightarrow p(V)) \quad \text{and} \quad M^\alpha_\exists(p)\colon \exists V\ (V_A = \mathcal{E}^\alpha(V) \land p(V))$$

The assertion $M^\alpha_\forall(p)$ holds for an abstract state $S \in \Sigma_A$ iff the assertion p holds for all concrete states $s \in \Sigma$ such that $s \in \alpha^{-1}(S)$, i.e., all states s such that $S = \alpha(s)$. This can also be expressed by the inclusion $\alpha^{-1}(\|M^\alpha_\forall(p)\|) \subseteq \|p\|$, where $\|p\|$ and $\|M^\alpha_\forall(p)\|$ represent the sets of states which satisfy the assertions, respectively.

The assertion $M^\alpha_\exists(p)$ holds for an abstract state $S \in \Sigma_A$ iff the assertion p holds for *some* concrete state $s \in \Sigma$ such that $s \in \alpha^{-1}(S)$, i.e., some state s such that $S = \alpha(s)$. This can also be expressed by the inclusion $\|p\| \subseteq \alpha^{-1}(\|M^\alpha_\exists(p)\|)$.

We respectively refer to $M^\alpha_\forall(p)$ and $M^\alpha_\exists(p)$ as the *universal* and *existential* *abstraction* of the formula p.

An assertion p which is a sub-formula of the temporal formula ψ is called a *maximal state sub-formula* of ψ if p is not properly contained in any other state sub-formula of ψ. Sub-formula p is said to have a *positive polarity* in ψ if it is contained under an even number of negations. Otherwise, p is said to have a *negative polarity*.

We define ψ^α, the α-induced abstraction of the temporal formula ψ to be a formula obtained by replacing every p a maximal state sub-formula of positive polarity by $M^\alpha_\forall(p)$ and every q a maximal state sub-formula of negative polarity by $M^\alpha_\exists(p)$

Example 2. Consider, for example, the temporal formula $\psi : (\Box(y \geq 2) \rightarrow \Box(y \geq -1))$. Applying the abstraction α presented in Example 1 yields the formula

$$\psi^\alpha = (\Box(M_\exists^\alpha(y \geq 2)) \rightarrow \Box(M_\forall^\alpha(y \geq -1))) =$$
$$(\Box(Y = pos) \rightarrow \Box(Y \in \{zero, pos\})).$$

Since ψ^α is valid, we can safely conclude that so is ψ. ⌟

Next, we consider the abstraction of an FKS $\mathcal{K}$ into $\mathcal{K}^\alpha$ such that $\mathcal{K}^\alpha \models \psi^\alpha$ implies $\mathcal{K} \models \psi$.
We define the *α-abstracted version* of $\mathcal{K}$ to be the FKS $\mathcal{K}^\alpha = \langle V_A, \Theta^\alpha, \rho^\alpha, \mathcal{J}^\alpha, \mathcal{C}^\alpha \rangle$, where

$$\Theta^\alpha = M_\exists^\alpha(\Theta) \qquad\qquad \rho^\alpha = M_\exists^\alpha(\rho)$$
$$\mathcal{J}^\alpha = \{M_\exists^\alpha(J) \mid J \in \mathcal{J}\} \qquad \mathcal{C}^\alpha = \{(M_\forall^\alpha(p), M_\exists^\alpha(q)) \mid (p,q) \in \mathcal{C}\}$$

Example 3. Let us show how the definition of $\mathcal{K}^\alpha$ leads to the construction of program ANY-Y$^\alpha$, the abstracted version of program ANY-Y, as presented in Example 1.

The initial condition Θ refers to y only through the conjunct $y = 0$. The corresponding abstraction is given by

$$M_\exists^\alpha(y = 0) =$$
$$\exists y.(Y = \text{if } y < 0 \text{ then } \textbf{\textit{neg}} \text{ else-if } y = 0 \text{ then } \textbf{\textit{zero}} \text{ else } \textbf{\textit{pos}}) \wedge (y = 0)$$
$$\sim\quad (Y = zero)$$

The transition relation ρ refers to y only through the part of a disjunct given by $y' = y + 1$. The corresponding abstraction is

$$M_\exists^\alpha(y' = y+1) =$$
$$\exists y, y' \left(\begin{array}{l} Y = \text{if } y < 0 \text{ then } \textbf{\textit{neg}} \text{ else-if } y = 0 \text{ then } \textbf{\textit{zero}} \text{ else } \textbf{\textit{pos}} \\ \wedge\, Y' = \text{if } y' < 0 \text{ then } \textbf{\textit{neg}} \text{ else-if } y' = 0 \text{ then } \textbf{\textit{zero}} \text{ else } \textbf{\textit{pos}} \\ \wedge\, y' = y+1 \end{array} \right)$$
$$\sim Y' = (\text{if } Y = \textbf{\textit{neg}} \text{ then } \{\textbf{\textit{neg}}, \textbf{\textit{zero}}\} \text{ else } \textbf{\textit{pos}}.)$$

Reconstructing a program from the abstracted components Θ^α and ρ^α, we obtain program ANY-Y$^\alpha$ presented in Fig. 8. ⌟

The following claim guarantees the safety of the abstractions jointly applied to the system and the property we wish to verify.

Claim. If the abstracted formula $M_\forall^\alpha(\psi)$ is valid over the abstracted FKS $\mathcal{K}^\alpha$, then ψ is valid over $\mathcal{K}$. That is,

$$\mathcal{K}^\alpha \models M_\forall^\alpha(\psi) \qquad \text{implies} \qquad \mathcal{K} \models \psi.$$

6.2 Determination of the Abstract Domain

The theory presented above assumed that the abstract domain, represented by the abstract variables V_A and their types, and the abstraction mapping α are already given. In this subsection, we consider some recommendations for the choice of an appropriate mapping, given an FKS $\mathcal{K}$ and a temporal property ψ.

Modularization and Abstraction: The Keys to Practical Formal Verification   69

$$\text{local } y_1, y_2 : \textbf{natural where } y_1 = y_2 = 0$$

$$\left[\begin{array}{l} \ell_0 : \textbf{loop forever do} \\ \left[\begin{array}{l} \ell_1 : \textbf{NonCritical} \\ \ell_2 : y_1 := y_2 + 1 \\ \ell_3 : \textbf{await } y_2 = 0 \lor y_1 < y_2 \\ \ell_4 : \textbf{Critical} \\ \ell_5 : y_1 := 0 \end{array}\right] \end{array}\right] \;\|\; \left[\begin{array}{l} m_0 : \textbf{loop forever do} \\ \left[\begin{array}{l} m_1 : \textbf{NonCritical} \\ m_2 : y_2 := y_1 + 1 \\ m_3 : \textbf{await } y_1 = 0 \lor y_2 \le y_1 \\ m_4 : \textbf{Critical} \\ m_5 : y_2 := 0 \end{array}\right] \end{array}\right]$$

$$- P_1 - \qquad\qquad\qquad - P_2 -$$

Fig. 9. Program BAKERY-2: the Bakery algorithm for two processes.

Assuming that the FKS $\mathcal{K}$ is derived from a program P, let $p_1, p_2, \ldots, p_k$ be the set of all atomic formulas referring to the data (non-control) variables appearing within conditions in the program P and within the temporal formula ψ. As a running example, we use program BAKERY-2, presented in Fig. 9.

Program BAKERY-2 is obviously an infinite-state system, since the variables y_1 and y_2 can assume arbitrarily large values.

The temporal properties we wish to establish for program BAKERY-2 are given by

$$\psi_{exc} : \Box \neg (at\_\ell_4 \land at\_m_4) \qquad \psi_{acc} : \Box (at\_\ell_2 \;\rightarrow\; \Diamond at\_\ell_4),$$

The property ψ_{exc} requires *mutual exclusion*, while ψ_{acc} requires *accessibility* for process P_1.

For program BAKERY-2, the atomic data formulas are $y_1 = 0$, $y_2 = 0$, and $y_1 < y_2$. Note that the formula $y_2 \le y_1$ is equivalent to the negation of $y_1 < y_2$ and needs not be included as an independent atomic formula.

Proceeding with the general case, the abstract system variables consist of the concrete control variables, which are left unchanged, and a set of abstract boolean variables $B_{p_1}, B_{p_2}, \ldots, B_{p_k}$, one for each atomic data formula. The abstraction mapping α is defined by

$$\alpha: \quad \{B_{p_1} = p_1, B_{p_2} = p_2, \ldots, B_{p_k} = p_k\}$$

That is, the boolean variable B_{p_i} has the value *true* in the abstract state iff the assertion p_i holds at the corresponding concrete state.

It is straightforward to compute the α-induced abstractions of the initial condition Θ^α and the transition relation ρ^α. In Fig. 10, we present program BAKERY-2(with a capital B), which is obtained by the abstraction as described in the preceding subsections.

Since the properties we wish to verify refer only to the control variables (through the at_ℓ and at_m expressions), they are not affected by the abstraction. Program BAKERY-2 is a finite-state program, and we can apply model checking to verify that it satisfies the two properties of mutual exclusion and accessibility. By Claim 6.1, we can infer that the original program BAKERY-2 also satisfies these two temporal properties.

local $B_{y_1=0}, B_{y_2=0}, B_{y_1<y_2}$: **boolean where** $B_{y_1=0} = B_{y_2=0} = 1, B_{y_1<y_2} = 0$

$$\left[\begin{array}{l} \ell_0 : \textbf{loop forever do} \\ \left[\begin{array}{l} \ell_1 : \textbf{NonCritical} \\ \ell_2 : (B_{y_1=0}, B_{y_1<y_2}) := (0, \neg B_{y_2=0}) \\ \ell_3 : \textbf{await } B_{y_2=0} \vee B_{y_1<y_2} \\ \ell_4 : \textbf{Critical} \\ \ell_5 : B_{y_1=0} := 1 \end{array}\right] \end{array}\right] \quad \| \quad \left[\begin{array}{l} m_0 : \textbf{loop forever do} \\ \left[\begin{array}{l} m_1 : \textbf{NonCritical} \\ m_2 : (B_{y_2=0}, B_{y_1<y_2}) := (0, 0) \\ m_3 : \textbf{await } B_{y_1=0} \vee \neg B_{y_1<y_2} \\ m_4 : \textbf{Critical} \\ m_5 : B_{y_2=0} := 1 \end{array}\right] \end{array}\right]$$

$$- P_1 - \qquad\qquad\qquad - P_2 -$$

Fig. 10. Program BAKERY-2: the Bakery algorithm for two processes.

References

[BBLS92]   S. Bensalem, A. Bouajjani, C. Loiseaux, and J. Sifakis. Properties preserving simulations. *CAV'92*, vol. 663 of *LNCS*, pp 251–263, 1992.

[BBM95]   N. Bjørner, I.A. Browne, and Z. Manna. Automatic generation of invariants and intermediate assertions. 1st *Intl. Conf. on Principles and Practice of Constraint Programming*, vol. 976 of *LNCS*, pp 589–623, 1995.

[BCG86]   M.C. Browne, E.M. Clarke, and O. Grumberg. Reasoning about networks with many finite state processes. *PODC'86*, pp 240–248, 1986.

[CC77]   P. Cousot and R. Cousot. Abstract interpretation: A unified lattice model for static analysis of programs by construction or approximation of fixpoints. *POPL'77*, 1977.

[CGH94]   E.M. Clarke, O. Grumberg, and K. Hamaguchi. Another look at LTL model checking. *CAV'94*, vol. 818 of *LNCS*, pp 415–427, 1994.

[CGJ95]   E.M. Clarke, O. Grumberg, and S. Jha. Verifying parametrized networks using abstraction and regular languages. *CONCUR'95*, pp 395–407, 1995.

[CGL94]   E.M. Clarke, O. Grumberg, and D.E. Long. Model checking and abstraction. *ACM Trans. Prog. Lang. Sys.*, 16(5):1512–1542, 1994.

[CGL96]   E.M. Clarke, O. Grumberg, and D.E. Long. Model checking. *Model Checking, Abstraction and Composition*, vol. 152 of *Nato ASI Series F*, pages 477–498. Springer-Verlag, 1996.

[CH78]   P. Cousot and N. Halbwachs. Automatic discovery of linear restraints among variables of a program. *POPL'78*, pp 84–96, 1978.

[DGG97]   D. Dams, R. Gerth, and O. Grumberg. Abstract interpretation of reactive systems. *ACM Trans. Prog. Lang. Sys.*, 19(2), 1997.

[EN95]   E. A. Emerson and K. S. Namjoshi. Reasoning about rings. *POPL'95*, 1995.

[EN96]   E.A. Emerson and K.S. Namjoshi. Automatic verification of parameterized synchronous systems. *CAV'96*, LNCS, 1996.

[GL93]   S. Graf and C. Loiseaux. A tool for symbolic program verification and abstraction. *CAV'93*, vol. 697 of *LNCS*, pp 71–84, 1993.

[HLR92]   N. Halbwachs, F. Lagnier, and C. Ratel. An experience in proving regular networks of processes by modular model checking. *Acta Informatica*, 29(6/7):523–543, 1992.

[ID96]   C.N. Ip and D. Dill. Verifying systems with replicated components in Murφ. *CAV'96*, *LNCS*, 1996.

[KM95] R.P. Kurshan and K.L. McMillan. A structural induction theorem for processes. *Information and Computation*, 117:1–11, 1995.

[KPR98] Y. Kesten, A. Pnueli, and L. Raviv. Algorithmic verification of linear temporal logic specifications. *ICALP'98*, LNCS, 1998.

[Lam77] L. Lamport. Proving the correctness of multiprocess programs. *IEEE Trans. Software Engin.*, 3:125–143, 1977.

[LHR97] D. Lesens, N. Halbwachs, and P. Raymond. Automatic verification of parameterized linear networks of processes. *POPL'97*, 1997.

[LPS81] D. Lehmann, A. Pnueli, and J. Stavi. Impartiality, justice and fairness: The ethics of concurrent termination. *ICALP'81*, vol. 115 of *LNCS*, pp 264–277, 1981.

[MAB+94] Z. Manna, A. Anuchitanukul, N. Bjørner, A. Browne, E. Chang, M. Colón, L. De Alfaro, H. Devarajan, H. Sipma, and T.E. Uribe. STeP: The Stanford Temporal Prover. Technical Report STAN-CS-TR-94-1518, Dept. of Comp. Sci., Stanford University, Stanford, California, 1994.

[MP91] Z. Manna and A. Pnueli. *The Temporal Logic of Reactive and Concurrent Systems: Specification*. Springer-Verlag, New York, 1991.

[MP95] Z. Manna and A. Pnueli. *Temporal Verification of Reactive Systems: Safety*. Springer-Verlag, New York, 1995.

[SG89] Z. Shtadler and O. Grumberg. Network grammars, communication behaviors and automatic verification. *CAV'89*, vol. 407 of *LNCS*, pp 151–165, 1989.

[SG92] A.P. Sistla and S.M. German. Reasoning about systems with many processes. *J. ACM*, 39:675–735, 1992.

[WL89] P. Wolper and V. Lovinfosse. Verifying properties of large sets of processes with network invariants. *CAV'89*, vol. 407 of *LNCS*, pp 68–80, 1989.

On the Role of Time and Space in Neural Computation

Wolfgang Maass

Institute for Theoretical Computer Science
Technische Universität Graz
Klosterwiesgasse 32/2
A-8010 Graz, Austria
maass@igi.tu-graz.ac.at

Abstract. We discuss structural differences between models for computation in biological neural systems and computational models in theoretical computer science.

1 Introduction

One of the most interesting scientific developments during the next two decades will be the unraveling of the structure of computation in living organisms. Since the information processing capabilities of living organisms are in many aspects superior to those of our current artificial computing machinery, this is likely to have significant consequences for the way in which computers and robots will be designed in the year 2020.

Traditionally theoretical computer science has played the role of a scout that explores novel approaches towards computing well in advance of other sciences. Curiously enough, this did not happen so far in the case of computation in living organisms, and it may be worthwhile to ponder for a moment about the possible reasons for that. One obstacle may result from the fact that theoretical computer science has become to a large extent "technique-driven", i.e., one typically looks for new problems that can be solved by variations and extensions of a body of fascinating mathematical tools that one has come to like, and that form the heart of current theoretical computer science. In contrast, to have a serious impact on theoretical research in neural computation, a theoretical researcher has to be to a larger extent "problem-driven", i.e., he/she has to employ and develop those mathematical concepts and tools that are most adequate for the problem at hand.

On the positive side, I would like to mention a success story regarding an earlier very fruitful interaction between the areas which are nowadays called computational neuroscience and theoretical computer science. McCulloch and Pitts [McCulloch and Pitts, 1943] developed an abstract mathematical model for computation in living organisms: circuits of *"McCulloch-Pitts neurons"* or *threshold gates*, as they are now called in theoretical computer science. Kleene [Kleene, 1956] proved that they were equivalent to his notion of a *finite automaton*. Thus "historically finite automata were first used to model neuron nets"

[Hopcroft and Ullman, 1979]. As we can all see, both models turned out to be a very fruitful for the subsequent development of both fields involved.

Within the scientific discipline of neuroscience a new subfield has emerged during the 90's that is called *computational neuroscience*. However a look at any recent issue of leading journals (e.g. Neural Computation, Network: Computation in Neural Systems, Computational Neuroscience) or conference proceedings in this new area (e.g. of the Annual Conference on Computational Neuroscience) may have a sobering effect on a theoretical computer scientist who is ready to develop adequate computational theories for computational neuroscience: There exists a large amount of interdisciplinary work in computational neuroscience and a fair number of theoretical work has already been done in this area. But so far the theoretical work in computational neuroscience has been dominated by approaches from theoretical physics, information theory, and statistics.[1]

One of the main obstacles for a theoretical computer scientist who is ready to tackle theoretical problems about computing in biological neural systems is the diversity of models for neural computation that are proposed by neuroscientists, and the diversity of opinions among leading neuroscientists regarding the right way to understand computations in the brain. This has the effect that it is hardly possible to identify abstract models and theoretical problems in computational neuroscientists that are of undisputed significance. In fact, it is hard to identify solid empirical facts regarding computation in biological neural systems on which most researchers and laboratories agree.[2] This concerns especially the first questions that a theoretical computer scientist is likely to ask:

- How is information encoded in biological neural systems?
- What are the computational units of biological neural systems, and what functions can they compute?
- How are biological computations organized and programmed?
- Which maps from inputs to outputs are computed by specific neural systems?

We will focus in this short article on one issue on which most neuroscientists seem to be able to agree, and which may point to a fruitful area for future contributions from theoretical computer science to this field: that time and space appear to play a different role for computations in biological neural systems than for computations in currently existing computational models in theoretical computer science. This may provide some "food for thought" for the development of new models that are both of interest from the point of view of theoretical

---

[1] For a theoretical computer scientist from Europe a survey of the current state of computational neuroscience may have an additional sobering effect: most leading journals and conferences are based in the USA.

[2] This is a consequence of the fact that most of the basic questions about computations in living organisms cannot be answered directly through suitable experiments. Hence the available answers are typically based on indirect empirical evidence, that often varies with details of the experimental setup, the specific neural system and species that is studied, and the methods for data-analysis that are employed. Bad tongues say that the answers also depend on the theoretical hypothesis that the researcher wants to support through the experiment.

computer science, and which also provide fruitful new hypotheses regarding the organization of computations in living organisms.

In the remainder of this article I will illustrate in a few examples specific directions into which currently existing computational models from theoretical computer science need to be evolved in order to take into account the different role that time and space play in biological neural systems. In view of the space constraints for this article we cannot give here a survey of relevant results and theories. But we will give in the last section some pointers to up-to-date survey articles and books.

2 On the Role of Time in Neural Computation

The most biology-like computational model that we traditionally consider in theoretical computer science are circuits consisting of threshold gates or sigmoidal gates. These gates compute functions from $\mathbf{R}^n$ into $\mathbf{R}$ of the form

$$\langle x_1,\ldots,x_n \rangle \mapsto \sigma(\sum_{i=1}^{n} w_i x_i + w_0) \tag{1}$$

for some fixed "activation function" $\sigma : \mathbf{R} \to \mathbf{R}$ (e.g., $\sigma(y) = \text{sign}(y)$ in the case of a threshold gate, or $\sigma(y) = 1/(1+e^{-y})$ in the case of a sigmoidal gate) and suitable parameters $w_0, \ldots, w_n \in \mathbf{R}$. A computation in such circuit is defined with the help of some input-dependent schedule, which decides which gates carry out their computational operation (1) at which discrete time step. This computation schedule is particularly obvious in the case of a layered feedforward circuit, where all gates on level ℓ carry out their computational operation at time step ℓ.

In contrast, the *output* of a biological neuron consists of "action potentials" or *"spikes"* (see Fig. 1). In other words: a *biological neuron* does not output any number or bit, instead it simply *marks points in time*. The *input* to a biological

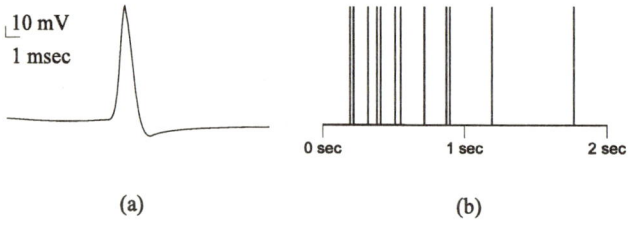

Fig. 1. a) Typical action potential (spike). b) A typical spike train produced by a neuron (each firing time marked by a bar)

neuron v consists of trains of pulses, socalled excitatory postsynaptic potentials (EPSP's) or inhibitory postsynaptic potentials (IPSP's) of a shape as indicated

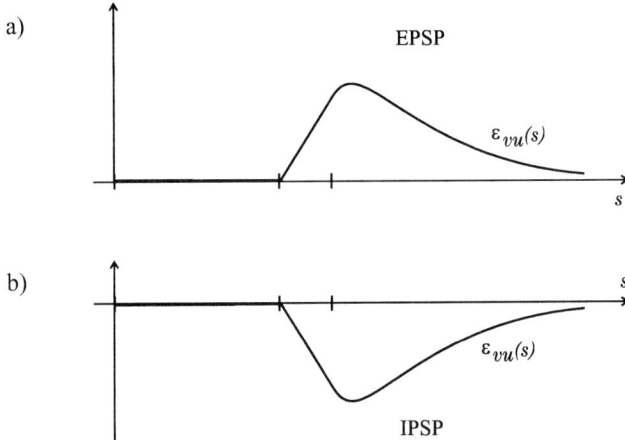

Fig. 2. a) Typical time course of an excitatory postsynaptic potential (EPSP). b) Typical time course of an inhibitory postsynaptic potential (IPSP). The vertical axis indicates the membrane voltage of the neuron v

in Fig. 2. More precisely: About 1000 to 10000 other neurons u are each connected to v by a *synapse*. The synapse from neuron u to neuron v transforms the output spike train of neuron u (which is of a type as illustrated in Fig. 1 b)) into a train of EPSP's or IPSP's in neuron v. One usually assumes that neuron u only causes EPSP's or only causes IPSP's in other neurons v.

According to the *spike response model* (see [Gerstner and van Hemmen, 1994] and [Gerstner, 1998]) one can model the response of the membrane potential of neuron v at time t to a spike train with spikes at times $t_1, t_2, \ldots$ from a presynaptic neuron u by a function of the form

$$\text{response}_{vu}(t) := \sum_i w_{vu}(t) \cdot \varepsilon_{vu}(t - t_i) \ .$$

One assumes in this model that neuron v "*fires*" – and thereby emits a spike – at time t whenever the resulting total membrane potential

$$h_v(t) := \sum \{\text{response}_{vu}(t) \mid u \text{ has a synapse to } v\}$$

at neuron v reaches the *firing threshold* $\vartheta_v(t)$ of neuron v. One refers to this model as a *leaky integrate-and-fire neuron* or *spiking neuron*.

Let us first assume for simplicity that the synaptic "weights" $w_{vu}(t)$ and the firing threshold $\vartheta_v(t)$ do not depend on the time t. Then the spiking neuron v can in principle simulate a threshold gate, provided that all presynaptic neurons u that represent an input variable with value 1 fire at a common time T_{input}, and all presynaptic neurons u that represent an input variable with value 0 do not fire at all during a certain time interval; see Fig. 3. To be precise, one also has to make an assumption about the shapes of the response functions, for example that

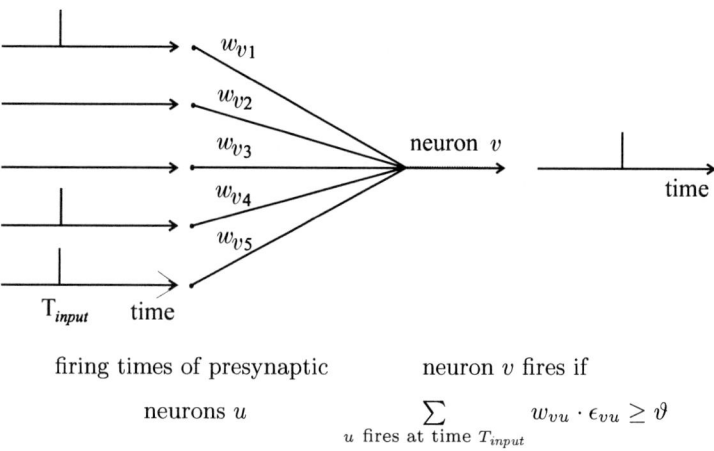

Fig. 3. Simulation of a threshold gate by a spiking neuron.

the response functions $\varepsilon_{vu}(s)$ for different u are identical except for their sign. On the other hand empirical data show that an input of the type indicated in Fig. 4 is more typical. In order to illustrate that in such asynchronous mode a spiking

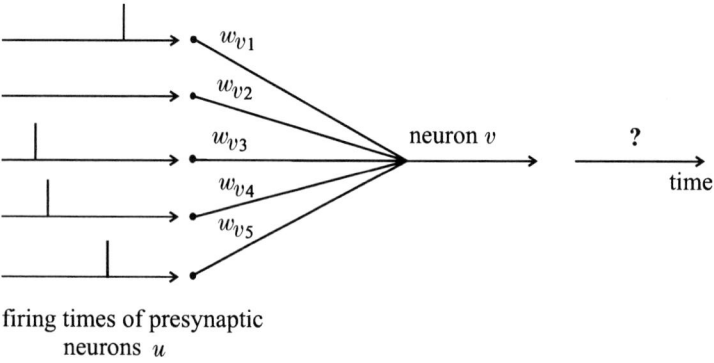

Fig. 4. Typical input for a biological spiking neuron v, where its output cannot be easily described in terms of conventional computational units.

neuron can carry out computational operations that are not at all reflected in the model of a threshold gate or sigmoidal gate (1), we consider for some arbitrary fixed parameters $0 < c_1 < c_2$ the following function $ED_n : \mathbf{R}^n \to \{0, 1\}$:

$$ED_n(x_1, \ldots, x_n) = \begin{cases} 1, \text{ if there are } j \neq j' \text{ so that } |x_j - x_{j'}| \leq c_1 \\ 0, \text{ if } |x_j - x_{j'}| \geq c_2 \text{ for all } j \neq j' \end{cases}.$$

Note that this function $ED_n(x_1, \ldots, x_n)$ (where ED stands for "element distinctness") is in fact a partial function, which may output arbitrary values in case that $c_1 < \min\{|x_j - x_{j'}| : j \neq j' \text{ and } j, j' \in \gamma_i\} < c_2$. Therefore hair-trigger situations can be avoided, and a single spiking neuron can compute this function ED_n even if there is a small amount of noise on its membrane potential $h_v(t)$. We assume here that the inputs $x_1, \ldots, x_n$ to the spiking neuron v are given through the firing times $x_1, \ldots, x_n$ of n presynaptic neurons. On the other hand

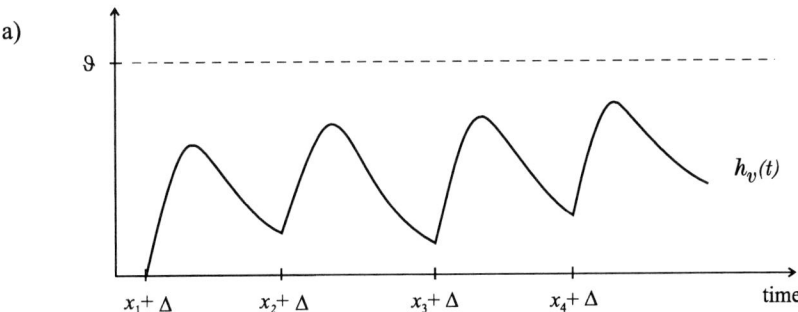

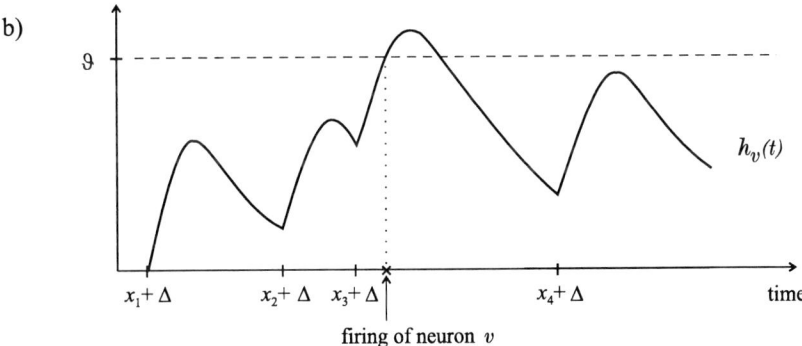

Fig. 5. a) Typical time course of the membrane potential $h_v(t)$ if $ED_4(x_1, x_2, x_3, x_4) = 0$. b) Time course of $h_v(t)$ in the case where $ED_4(x_1, x_2, x_3, x_4) = 1$ because $|x_3 - x_2| \leq c_1$.

the following results show that the same partial function ED_n requires a substantial number of gates if computed by circuits consisting of McCulloch-Pitts neurons (threshold gates) or sigmoidal gates.

Theorem 1. *Any layered threshold circuit that computes ED_n needs to have at least $\log(n!) \geq \frac{n}{2} \cdot \log n$ threshold gates on its first layer.*

The *proof* of Theorem 1 relies on a geometrical argument, see [Maass, 1997].

Theorem 2. *Any feedforward circuit consisting of arbitrary sigmoidal gates needs to have at least $\frac{n-4}{2}$ gates in order to compute ED_n.*

The *proof* of Theorem 2 is more difficult, since sigmoidal gates output *analog numbers* rather than *bits*. Therefore a multilayer circuit consisting of sigmoidal gates can have larger computational power than a circuit consisting of threshold gates (see [Maass et al., 1991,DasGupta and Schnitger, 1996]). The proof procedes in an indirect fashion by showing that any sigmoidal neural net with m gates that computes ED_n can be transformed into another sigmoidal neural net that "shatters" *every* set of $n - 1$ different inputs with the help of $m + 1$ programmable parameters. According to [Sontag, 1997] this implies that $n - 1 \leq 2(m + 1) + 1$. We refer to [Maass, 1997] for further details. ∎

Remark 1. This is the largest lower bound for *any* concrete function in P that has been achieved to date for the size of circuits consisting of sigmoidal gates.

So far we have assumed that the firing threshold $\vartheta_v(t)$ and the synaptic weights $w_{vu}(t)$ are independent of the time t. This is certainly not the case for a biological neuron. In a first approximation one may assume that $\vartheta_v(t)$ shoots up to an extremely high value for a few ms after each firing of v and then returns to a "resting value" ϑ. This threshold dynamics enforces an upper bound on the maximal firing rate of a neuron, which usually is in the range of a few hundred Hz (although *typical* firing rates in the cortex are well below 100 Hz).

The dependence of synaptic weights $w_{vu}(t)$ on the time t is substantially more complex. It has been shown that different synapses exhibit quite heterogeneous dependencies on the preceding firing times of the presynaptic neurons [Dobrunz and Stevens, 1997]. Therefore one has to view synapses as another type of *active computational units* in neural computation. They cannot really be viewed as passive "registers" that store a single parameter – the "synaptic weight" w_{vu} – that remains fixed during a computation. Traditionally one views the "synaptic weights" w_{vu} as parameters that collectively contain the "program" of a computation in a neural circuit. Hence the fact that in biological neurons the values $w_{vu}(t)$ of these parameters are highly dynamic has drastic consequences: It is no larger clear which parameters *store* the program of a neural computation. Obviously that makes it even less clear how *learning algorithms* (i.e., algorithms that adjust the parameters that store the "program" of a neural computation) operate in biological neural systems.

These issues lead us to another significant structural difference between computations in biological neural systems and those computations that are usually studied in theoretical computer science. The inputs and outputs of computations in biological neural systems are typically vectors of *time-series*, rather than vectors of *numbers*. The processing of the t-th input $\underline{x}(t)$ may depend on the preceding inputs $\underline{x}(1), \ldots, \underline{x}(t-1)$. In that respect biological neural computation corresponds to computations carried out by finite state transducers (Mealy- or Moore-machines) or *filters* (as considered in signal processing and systems theory), rather than to computation carried out by feedforward circuits or Turing machines. For analyzing the computational power of neural circuits for computations on time series it is essential for the model that in reality the thresholds $\vartheta_v(t)$ and weights $w_{vu}(t)$ are functions of time that may depend on the preceding history of the computation. The following example [Maass and Zador, 1998b] illustrates that even on the abstract level of threshold circuits one can observe an increase in computational power resulting from history-dependent weights in connection with a sequential input presentation

Consider a threshold gate with n inputs, that receives an input xy of $2n$ bits in two subsequent batches x and y of n bits each. We assume that the n weights $w_1, \ldots, w_n$ of this gate are initially set to 1, and that the threshold of the gate is set to 1. We adopt the following very simple rule for changing these weights between the presentations of the two parts x and y of the input: the value of w_i is changed to 0 during the presentation of the second part y of the input if the i-th component x_i of the first input part x was non-zero. If we consider the output bit of this threshold gate after the presentation of the second part y of the input as the output of the whole computation, this threshold gate with "dynamic synapses" computes the boolean function $F_n : \{0,1\}^{2n} \to \{0,1\}$ defined by $F_n(x,y) = 1 \iff \exists i \in \{1,\ldots,n\}(y_i = 1 \text{ and } x_i = 0)$. One might associate this function F_n with some novelty detection task since it detects whether an input bit has changed from 0 to 1 in the two input batches x and y.

It turns out that this function cannot be computed by a small circuit consisting of threshold gates of the usual type, that receives all $2n$ input bits xy as one batch. In fact, one can prove that any feedforward circuit consisting of the usual type of "static" threshold gates, which may have arbitrary weights, thresholds and connectivity, needs to consist of at least $\frac{n}{\log(n+1)}$ gates in order to compute F_n. This lower bound can easily be derived from the lower bound from [Maass, 1997] for another boolean function $CD_n(x,y)$ from $\{0,1\}^{2n}$ into $\{0,1\}$ which gives output 1 if and only if $x_i + y_i \geq 2$ for some $i \in \{1,\ldots,n\}$, since $CD_n(x,y) = F_n(1-x,y)$.

The article [Maass and Zador, 1998a] surveys empirical data on the temporal dynamics of synapses and theoretical investigations of their possible computational role.

One fundamental open problem for the theory of neural computation is the question how information is encoded in spike trains. There appears to be no unique answer. Rather, the *neural code* seems to vary from system to system, and even the same neural system may apply different neural codes for different computational tasks (see [Rieke et al., 1997] and [Recce, 1998]). It is shown in Fig. 6 that typical interspike intervals are relatively long in comparison with the total computation time of some neural systems. Furthermore interspike intervals tend to be highly irregular. Hence it is rather difficult for a biological neuron to find out within the given computation time the current firing rates of presynaptic neurons (especially in view of the relatively short time duration of most EPSP's). Therefore the popular assumption that information is primarily communicated between neurons through their firing rates is somewhat dubious.

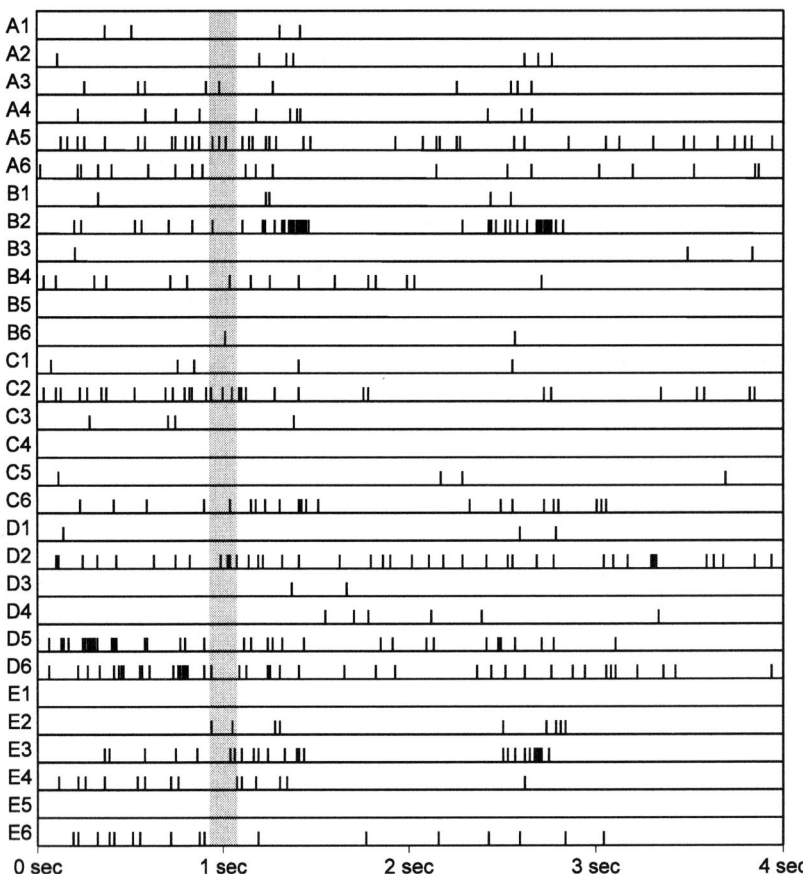

Fig. 6. Simultaneous recordings (over 4 seconds) of the firing times of 30 neurons from monkey striate cortex by Krüger and Aiple [Krüger and Aiple, 1988]. Each firing is denoted by a vertical bar, with a separate row for each neuron. For comparison we have shaded an interval of 150 msec. This time span is known to suffice for the completion of some complex multilayer cortical computations.

One completely different neural code that has been suggested as being relevant is the socalled *correlation code* (see [Recce, 1998] and [Maass, 1998b] for references). This code appears to be of particular interest from the point of view of *computer science logic*, since it hypothesizes that information is not transmitted between neurons in the form of numbers, but in the form of *second order objects*: (graded) *relations* or *sets*: Neurons whose firing time is statistically correlated during a neural computation communicate to other neurons the fact that they currently belong to the same set. A neuron can detect whether a critical number of presynaptic neurons belong to the same *set* because it can detect coincident firing times (as in our preceding discussion of the function ED_n, but now possibly with a higher firing threshold that requires coincident firing times of more than two presynaptic neurons).

3 On the Role of Space in Neural Computation

In the preceding section we had discussed examples for the different role that *time* plays in biological neural computation. In this section we will briefly discuss the role of another resource for neural computation: space, or more precisely the geometrical layout of neural circuits. The number of neurons to which a biological neuron has synaptic connections is by several orders of magnitude smaller than the number of neurons that participate in a typical computation. Hence "edges" or "wires" are a sparse resource in biological neural computation. This is not reflected in currently existing investigations of computational complexity issues for threshold circuits in theoretical computer science. There one typically wants to minimize the number of layers and the number of gates, with no charge for wires between adjacent layers. In addition in biological neural systems a large number of synapses connect neurons that are located quite close to each other. Obviously such architecture tends to keep *total wire length* small. This resource has also not yet been investigated in the context of threshold circuits in theoretical computer science. Its investigation appears to be of interest also from the point of view of related electronic hardware (for example cellular neural networks, see [Roska, 1997]). We refer to [Maass, 1998a] for some results in this direction.

[Valiant, 1994] addresses another important problem regarding the role of the spatial layout of neural circuits: Which local algorithms enable the computational units to carry out reliable information processing in a circuit whose layout is given by a random graph – hence not by a precise top-down design?

4 Outlook

Concurrently with the investigation of neural computation in living organisms one has started to design electronic hardware that captures particular aspects of the special role that time and space play in biological neural computation (see [Mead, 1989] and [Murray, 1998]). Examples are artificial retinas [Mead, 1989] schemes for low power analog communication between chips via pulses (address-event-representation, see [Douglas and Whatley, 1998, Mortara and Venier, 1998]) and programmable analog filters that employ pulses in a mix of analog and digital circuit techniques (see [Hamilton and Papathanasiou, 1998]) and cellular neural networks [Roska, 1997]. This approach is sometimes referred to as *Neuromorphic Engineering* (see [Smith, 1998] for the Proceedings of the first European Workshop on this topic). Obviously this area is still at a very early stage, and one might hope that theoretical computer science will play a role in its future development.

In principle theoretical computer science might be useful in modelling essential aspects of biological neural systems in a simplified mathematical framework, thereby providing a platform for extracting "portable" computational mechanisms and principles that can potentially be transported to novel *artificial* computing machinery. Unfortunately up to now, theoretically computer science has contributed very little in this direction (notable expections are for example [von Neumann, 1958] and [Valiant, 1994]). Perhaps one obstacle has been the difficulty for a non-expert to get an overview of the current state of the art in neurophysiology and neuromorphic engineering. This situation is now improving since a number of books with quite accessible surveys of most relevant topics have recently appeared (or will appear shortly): [Churchland and Sejnowski, 1992, Arbib, 1995] [Rieke et al., 1997, Ballard, 1997, Maass and Bishop, 1998, Koch, 1998].

Details to some of the specific models and results discussed in this article are available from http://www.cis.tu-graz.ac.at/igi/maass/ .

References

[Arbib, 1995] Arbib, M. A., editor (1995). *The Handbook of Brain Theory and Neural Networks*. MIT Press, Cambridge.

[Ballard, 1997] Ballard, D. H. (1997). *An Introduction to Natural Computation*. MIT-Press.

[Churchland and Sejnowski, 1992] Churchland, P. and Sejnowski, T. (1992). *The Computational Brain*. MIT Press, Cambridge.

[DasGupta and Schnitger, 1996] DasGupta, B. and Schnitger, G. (1996). Analog versus discrete neural networks. *Neural Computation*, 8(4):805–818.

[Dobrunz and Stevens, 1997] Dobrunz, L. and Stevens, C. (1997). Heterogenous release probabilities in hippocampal neurons. *Neuron*, 18:995–1008.

[Douglas and Whatley, 1998] Douglas, R. J. and Whatley, A. M. (1998). A pulse-coded communications infrastructure for neuromorphic systems. In Maass, W. and Bishop, C., editors, *Pulsed Neural Networks*. MIT-Press, Cambridge.

[Gerstner, 1998] Gerstner, W. (1998). Spiking neurons. In Maass, W. and Bishop, C., editors, *Pulsed Neural Networks*. MIT-Press, Cambridge.

[Gerstner and van Hemmen, 1994] Gerstner, W. and van Hemmen, L. (1994). How to describe neuronal activity: spikes, rates or assemblies? In *Advances in Neural Information Processing Systems*, volume 6, pages 463–470. Morgan Kaufmann.

[Hamilton and Papathanasiou, 1998] Hamilton, A. and Papathanasiou, K. (1998). Preprocessing for pulsed VLSI systems. In Maass, W. and Bishop, C., editors, *Pulsed Neural Networks*. MIT-Press, Cambridge.

[Hopcroft and Ullman, 1979] Hopcroft, J. E. and Ullman, J. D. (1979). *Introduction to automata theory, languages and computation*. Addison-Wesley, Reading Mas.

[Kleene, 1956] Kleene, S. C. (1956). Representation of events in nerve nets and finite automata. In *Automata Studies*, pages 3–42. Princeton University Press, Princeton N.J.

[Koch, 1998] Koch, C. (1998). *Biophysics of Computation: Information Processing in Single Neurons*. Oxford University Press, Oxford.

[Krüger and Aiple, 1988] Krüger, J. and Aiple, F. (1988). Multielectrode investigation of monkey stritate cortex: Spike train correlations in the infragranular layers. *Neurophysiology*, 60:798–828.

[Maass, 1997] Maass, W. (1997). Networks of spiking neurons: The third generation of neural network models. *Neural Networks*, 10:1659–1671.

[Maass, 1998a] Maass, W. (1998a). A model for universal analog computation in neural circuits with local connectivity. in preparation.

[Maass, 1998b] Maass, W. (1998b). A simple model for neural computation with firing rates and firing correlations. submitted for publication.

[Maass and Bishop, 1998] Maass, W. and Bishop, C., editors (1998). *Pulsed Neural Networks*. MIT-Press, Cambridge.

[Maass et al., 1991] Maass, W., Schnitger, G., and Sontag, E. (1991). On the computational power of sigmoid versus boolean threshold circuits. In *Proc. of the 32nd Annual IEEE Symposium on Foundations of Computer Science 1991*, pages 767–776.

[Maass and Zador, 1998a] Maass, W. and Zador, A. (1998a). Computing and learning with dynamic synapses. In Maass, W. and Bishop, C., editors, *Pulsed Neural Networks*. MIT-Press, Cambridge.

[Maass and Zador, 1998b] Maass, W. and Zador, A. M. (1998b). Dynamic stochastic synapses as computational units. In *Advances in Neural Processing Systems*, volume 10. MIT Press, Cambridge (to appear).

[McCulloch and Pitts, 1943] McCulloch, W. S. and Pitts, W. (1943). A logical calculus of the ideas immanent in nervous activity. *Bull. Math. Biophysics*, 5:115–133.

[Mead, 1989] Mead, C. (1989). *Analog VLSI and Neural Systems*. Addison-Wesley (Reading).

[Mortara and Venier, 1998] Mortara, A. and Venier, P. (1998). Analog VLSI pulsed networks for perceptive processing. In Maass, W. and Bishop, C., editors, *Pulsed Neural Networks*. MIT-Press, Cambridge.

[Murray, 1998] Murray, A. F. (1998). Pulse-based computation in VLSI neural networks. In Maass, W. and Bishop, C., editors, *Pulsed Neural Networks*. MIT-Press, Cambridge.

[Recce, 1998] Recce, M. (1998). Encoding information in neuronal activity. In Maass, W. and Bishop, C., editors, *Pulsed Neural Networks*. MIT-Press, Cambridge.

[Rieke et al., 1997] Rieke, F., Warland, D., Bialek, W., and de Ruyter van Steveninck, R. (1997). *SPIKES: Exploring the Neural Code*. MIT-Press, Cambridge.

[Roska, 1997] Roska, T. (1997). Implementation of cnn computing technology. In W. Gerstner, A. Germond, M. H. and Nicoud, J.-D., editors, *Proc. of ICANN 1997*, pages 1151–1155. Springer Verlag, Berlin.

[Smith, 1998] Smith, L. (1998). *Neuromorphic Systems: Engineering Silicon from Neurobiology*. World Scientific.

[Sontag, 1997] Sontag, E. D. (1997). Shattering all sets of 'k' points in "general position" requires (k-1)/2 parameters. *Neural Computation*, 9(2):337–348.

[Valiant, 1994] Valiant, L. G. (1994). *Circuits of the Mind*. Oxford University Press, Oxford.

[von Neumann, 1958] von Neumann, J. (1958). *The Computer and the Brain*. Yale University Press, New Haven.

From Algorithms to Working Programs: On the Use of Program Checking in LEDA

Kurt Mehlhorn[1] and Stefan Näher[2]

[1] Max-Planck-Insitut für Informatik, Im Stadtwald, D-66123 Saarbrücken, Germany
(mehlhorn@mpi-sb.mpg.de)
[2] Martin-Luther-Universität Halle-Wittenberg, FB Mathematik und Informatik,
Kurt-Mothes-Str. 1, D-06099 Halle (Saale), Germany
(naeher@infsn.informatik.uni-halle.de)

Abstract. We report on the use of program checking in the LEDA library of efficient data types and algorithms.

1 Introduction

LEDA [MN95, MNU97, MN98] is a collection of implementations of data structures and combinatorial algorithms. In the almost ten years of the project we translated hundreds of algorithms into programs. For the purpose of this paper an *algorithm* is the description of a problem solving method intended for a human reader and a *program* is a description intended for machine execution. Clearly, algorithms and programs are quite different animals; algorithms are formulated in natural language and are published in papers and books, and programs are written in computer languages and are executed on machines. We expected the process of implementation to be tedious and time-consuming, indeed, it was, but not intellectually challenging. We now believe that the implementation process is very difficult and challenging. We encountered the following difficulties.

- We and our co-workers are not perfect programmers. We make mistakes. We use program checking [BK89, SM90, WB97] to cope with the possibility of error.
- Geometric algorithms are usually designed for a hypothetical machine, the so-called *Real RAM*, which is equipped with arithmetic over the real numbers. The efficient realization of the Real RAM is non-trivial.
- The primary goal for algorithm design is asymptotic running time, the secondary design goal is elegance (remember that algorithms are intended for human readers). Actual running time is usually not a design goal. The actual running time of algorithms with the same asymptotic behavior may differ widely.

In this paper we concentrate on the first item. For the other two items we refer the reader to [MN98] and the references therein.

2 Program Checking

We start with an example and then generalize.

2.1 Planarity Testing

A graph is *planar* if it can be drawn in the plane without edge crossings. A planarity tester

```
bool Is_Planar(const graph& G)
```

takes a graph G and returns true if G is planar and returns false if G is non-planar[1]. The planarity test played a crucial role in the development of LEDA.

There are several linear time algorithms for planarity testing [HT74, LEC67, BL76]. An implementation of the Hopcroft and Tarjan algorithm was added to LEDA in 91. The implementation had been tested on a small number of graphs. In 93 we were sent a graph together with a planar drawing of it. However, our program declared the graph non-planar. It took us some days to discover the bug. More importantly, we realized that a complex question of the form "is this graph planar" deserves more than a yes-no answer. We adopted the thesis that

> *a program should justify (prove) its answers in a way that is easily checked by the user of the program.*

What does this mean for the planarity test?

If a graph is declared planar, a proof should be given in the form of a combinatorial embedding or a planar drawing. If a graph is declared non-planar, a proof should be given in the form of a Kuratowski subgraph[2]. Linear time algorithms for computing planar embeddings are described in [CNAO85, NC88, MM95] and linear algorithms for the computation of Kuratowski subgraphs are given in [Wil84, Kar90, HMN96]. The function

```
bool Is_Planar(graph& G, list<edge>& K)
```

returns true if G is planar and returns false otherwise. If G is planar, it also reorders the adjacency lists of G such that G becomes a plane map[3]. If G is non-planar, a set of edges forming a Kuratowski subgraph is returned in K. The function runs in linear time $O(n+m)$, where n and m are the number of nodes and edges of G, respectively. Its implementation is discussed in the chapter on

---

[1] We use the syntax of C++ in our examples. All functions mentioned in the paper are available in LEDA.
[2] Kuratowski [Kur30] has shown that every non-planar graph contains a subdivision of either K_5, the complete graph on five nodes, or $K_{3,3}$, the complete bipartite graph with three nodes on either side.
[3] A map is an undirected graph in which a cyclic order is imposed on the edges incident to any node. A map is plane if there is a planar embedding preserving the cyclic orders. We use plane map and combinatorial embedding as synonyms.

embedded graphs of [MN98]. If LEDA is installed on your computer system, you may want to exercise the planarity test demo before proceeding.

```
int GENUS(const graph& G)
{ if ( !Is_Map(G) ) error_handler(1,"Genus only applies to maps");
  int n = G.number_of_nodes();
  if ( n == 0 ) return 0;
  int nz = 0;
  node v;
  forall_nodes(v,G)
      if ( outdeg(v) == 0 )  nz++;
  int m = G.number_of_edges();
  node_array<int> cnum(G);
  int c = COMPONENTS(G,cnum);

  edge_array<bool> considered(G,false);
  int fc = 0;
  edge e;
  forall_edges(e,G)
  { if ( !considered[e] )
    { // trace the face to the left of e
      edge e1 = e;
      do { considered[e1] = true;
           e1 = G.face_cycle_succ(e1);
         }
      while (e1 != e);
      fc++;
    }
  }
  return (m/2 - n - nz - fc + 2*c)/2;
}
```

Fig. 1. A map is plane if and only if its genus is zero. The genus of a map is computed according to Euler's formula: one half of the number of undirected edges minus the number of nodes minus the number of isolated nodes minus the number of face cycles plus twice the number of connected components. The face cycle successor of an edge e is the next edge out of the target of e in the cyclic order of the edges around the target of e. In the LEDA representation of maps every undirected edge is represented by a pair of directed edges.

The crucial observation is now that the justifications, which Is_Planar gives for its answers, are easily checked. It is well-known that a connected map is plane if it satisfies the so-called Euler-relation $f - m + n - 2 = 0$, where f is the number of face cycles. It is also well-known that a graph is non-planar if it contains a Kuratowski subgraph. The functions

```
int GENUS(const graph& G)
bool CHECK_KURATOWSKI(const graph& G, const list<edge>& K)
```

compute the genus of a map G and check whether K is a Kuratowski subgraph of G, respectively. The implementation of the former function is shown in Figure 1, the implementation of the latter function is equally simple.

		BL_PLANAR			HT_PLANAR	
	G	T	T + E or K	C	T	T + E
P	0.25	0.52	0.53	0.08	0.84	1.32
	0.52	1.05	1.09	0.16	1.74	2.69
	1.07	2.13	2.22	0.33	3.59	5.77
P + $K_{3,3}$	0.309999	0.379999	1.77	0.0600014	0.83	–
	0.529999	0.640001	2.68	0.129997	1.7	–
	1.12	1.27	5.44	0.23	3.45	–
P + K_5	0.32	0.369999	1.82	0.0600014	0.869999	–
	0.549999	0.530003	2.57	0.119999	1.68	–
	1.1	1.3	7.09	0.229996	3.56001	–
MP	0.269997	0.720001	0.720001	0.110001	1.21	1.91
	0.459999	1.46	1.45	0.220001	2.50999	4.01
	0.93	2.94	2.92	0.440002	5.06	8.3
MP + e	0.269997	0.550003	2.05	0.0699997	0.330002	–
	0.449997	1.07001	3.81	0.139999	0.659996	–
	0.93	2.47	8.55	0.289993	1.35	–

Table 1. The running times of functions related to planarity: The first column shows the type of the input graph, the second column shows the time for the call $BL\_PLANAR(G)$, the third column shows the time for the call $BL\_PLANAR(G, K)$, the fourth column shows the time required to check the result of the computation in the third column, i.e, the time for the call $Genus(G) == 0$, if G is planar, and the call $CHECK\_KURATOWSKI(G, K)$ if G is non-planar, the fifth column shows the time for the call $HT\_PLANAR(G)$, and the last column shows the time for the call $HT\_PLANAR(G, K)$. The last call is only made when G is planar, since there there is no efficient Kuratowski finder implemented for the Hopcroft-Tarjan planarity test.

The meaning of the first column is as follows: P stands for a random planar map with n nodes and m uedges, P + $K_{3,3}$ stands for a random planar map with n nodes and m uedges plus a $K_{3,3}$ on six randomly chosen nodes, P + K_5 stands for a random planar mao with n nodes and m uedges plus a K_5 on five randomly chosen nodes, MP stands for a maximal planar map with n nodes, and MP + e stands for a maximal planar graph plus one additional edge between two random nodes that are not connected in G. In all cases the edges of the graph were permuted before the tests were started.

For each type of graph we used $n = 2^i \cdot 1000$, $m = 2^i \cdot 2000$ for $i = 0, 1$, and 2.

Table 1 shows the running times of several functions related to planarity. In this table BL_PLANAR stands for the planarity test of Lempel, Even, and Cederbaum with PQ-tree data structure of Booth and Luecker, the embedding algorithm of Chiba et al., and the Kuratowski finder of Hundack et al., and HT_PLANAR stands for the planarity test of Hopcroft and Tarjan and the embedding algorithm of Mehlhorn and Mutzel.

2.2 General Remarks

What have we achieved?

Verification for every Problem Instance: When a graph is declared planar, the resulting plane map is checked by testing whether its genus is zero, and if a graph is declared non-planar, the subgraph K is checked by CHECK_KURATOWSKI. In this way, the correctness of Is_Planar is established for each problem instance.

Trust with Minimal Investment: A user of Is_Planar does *not* have to understand the intricacies of the planarity test. It suffices to understand the functions GENUS and CHECK_KURATOWSKI. The implementation of either function is less than a page long and the underlying mathematics is simple compared to the mathematics underlying the planarity test. Observe that one only needs to understand that maps of genus zero are plane (about a two page proof) and that the existence of a Kuratowski subgraph implies non-planarity (again about a two page proof). There is, however, no need to understand why every non-planar graph contains a Kuratowski subgraph. The implementation proves this fact for every problem instance. In this way checkers allow to develop trust in an implementation with only minimal intellectual investment. It is even conceivable that checkers can be formally verified by means of automatic program verification. [BSM97] is a first example.

Program Libraries: Program libraries contain implemented algorithms. The implementor of a library may want to hide his code (after all, the source code of the programs constitutes his intellectual capital), but he may also want to make a convincing case that his code is correct. Program checkers resolve the conflict. We use the following guidelines for the specification and implementation of functions.

(1) Define the problem to be solved and what constitutes a justification for an answer.
(2) Prove that the suggested justification indeed proves correctness for any particular instance.
(3) Define the interface of the function.
(4) Define the interface of the checker and give its implementation.
(5) Give the implementation of the function. There is no need to make the implementation public.

A *matching* in a graph G is a subset M of the edges of G such that no two share an endpoint.

An odd-set cover OSC of G is a labeling of the nodes of G with non-negative integers such that every edge of G (which is not a self-loop) is either incident to a node labeled 1 or connects two nodes labeled with the same i, $i \geq 2$.

Let n_i be the number of nodes labeled i and consider any matching N. For i, $i \geq 2$, let N_i be the edges in N that connect two nodes labeled i. Let N_1 be the remaining edges in N. Then $|N_i| \leq \lfloor n_i/2 \rfloor$ and $|N_1| \leq n_1$ and hence

$$|N| \leq n_1 + \sum_{i \geq 2} \lfloor n_i/2 \rfloor$$

for any matching N and any odd-set cover OSC.

It can be shown that for a maximum cardinality matching M there is always an odd-set cover OSC with

$$|M| = n_1 + \sum_{i \geq 2} \lfloor n_i/2 \rfloor,$$

thus proving the optimality of M. In such a cover all n_i with $i \geq 2$ are odd, hence the name.

list<edge> MAX_CARD_MATCHING(*graph* G, *node_array*<int>& OSC,
    int heur = 0)

> computes a maximum cardinality matching M in G and returns it as a list of edges. The algorithm ([Edm65, Gab76]) has running time $O(nm \cdot \alpha(n, m))$. With *heur* = 1 the algorithm uses a greedy heuristic to find an initial matching. This seems to have little effect on the running time of the algorithm.
>
> An odd-set cover that proves the maximality of M is returned in OSC.

bool   CHECK_MAX_CARD_MATCHING(*graph* G, *list*<edge> M,
    node_array<int> OSC)

> checks whether M is a maximum cardinality matching in G and OSC is a proof of optimality. Aborts if this is not the case.

Fig. 2. The manual page for maximum cardinality matching. The first paragraph defines the problem, the second paragraph defines the notion of proof and the third and fourth paragraph establish that an odd-set cover constitutes a proof of optimality. Observe, that is is *not* necessary to understand why odd-set covers proving optimality exist.

Figures 2 and 3 illustrate items (1) to (4) for the case of maximum cardinality matchings in general graphs.

Debugging: Program checking amounts to a complete check of the post-condition of a program. It allows to assume that potentially incorrect programs are correct.

```
static bool False(string s)
{ cerr << "CHECK_MAX_CARD_MATCHING: " << s << "\n"; return false; }

bool CHECK_MAX_CARD_MATCHING(const graph& G, const list<edge>& M,
                             const node_array<int>& OSC)
{ int n = Max(2,G.number_of_nodes());
  int K = 1;
  array<int> count(n);
  for (int i = 0; i < n; i++) count[i] = 0;
  node v; edge e;

  forall_nodes(v,G)
  { if ( OSC[v] < 0 || OSC[v] >= n )
      return False("negative label or label larger than n - 1");
    count[OSC[v]]++;
    if (OSC[v] > K) K = OSC[v];
  }

  int S = count[1];
  for (int i = 2; i <= K; i++) S += count[i]/2;
  if ( S != M.length() )
    return False("OSC does not prove optimality");

  node_array<int> deg(G,0);
  forall(e,M) { deg[G.source(e)]++; deg[G.target(e)]++; }
  forall_nodes(v,G)
    if (deg[v] > 1) return False("M is not a matching");

  forall_edges(e,G)
  { node v = G.source(e); node w = G.target(e);
    if ( v == w || OSC[v] == 1 || OSC[w] == 1 ||
         ( OSC[v] == OSC[w] && OSC[v] >= 2) ) continue;
    return False("OSC is not a cover");
  }
  return true;
}
```

Fig. 3. The checker for maximum cardinality matchings.

If a program operates correctly on a particular instance, fine, and if it operates incorrectly, it is caught by the checker. Thus, if all subroutines of a function f are checked, no checker of a subroutine fires, and an error occurs during the execution of f, the error must be in f. This feature of program checking is extremely useful during the debugging phase of program development.

Testing: Program checking supports testing. Traditionally, testing is restricted to problem instances for which the solution is known by other means. Program checking allows to test on *any* instance. For example, we use the following program (among others) to check the matching algorithm.

```
for (int n = 0; n < 100; n++)
  for (int m = 0; m < 100; m++)
    { random_graph(G,n,m); // random graph with n nodes and m edges
      list<edge> M = MAX_CARD_MATCHING(G,OSC);
      CHECK_MAX_CARD_MATCHING(G,M,OSC);
    }
```

Hidden Assumptions: A checker can only be written if the problem at hand is rigorously defined. We noticed that some of our specifications contained hidden assumptions which were revealed during the design of the checker. For example, an early version of our biconnected components algorithm assumed that the graph contains no isolated nodes.

3 Conclusion

At the time of this writing LEDA contains checkers for most network algorithms (mostly based on linear programming duality), for planarity testing, for priority queues, and for the basic geometric algorithms (convex hulls, Delaunay diagrams, and Voronoi diagrams). Program checking has greatly increased our confidence in the correctness of our implementations. For further reading on program checking we refer the reader to [SM90, BS94, SM91, BSM97, BS95, BSM95, SWM95, BK89, BLR90, BW96, WB97], [MN98, AL94, OLPT97, MNS+96].

References

[AL94]   N.M. Amato and M.C. Loui. Checking linked data structures. In *Proceedings of the 24th Annual International Symposium on Fault-Tolerant Computing*, pages 164–173, 1994.

[BK89]   M. Blum and S. Kannan. Programs That Check Their Work. In *Proc. of the 21th Annual ACM Symp. on Theory of Computing*, 1989.

[BL76]   K.S. Booth and G.S. Luecker. Testing for the Consecutive Ones Property, Interval Graphs, and Graph Planarity Using *PQ*-tree Algorithms. *Journal of Comp. and Sys. Sciences*, 13:335–379, 1976.

[BLR90] M. Blum, M. Luby, and R. Rubinfeld. Self testing/correcting with applications to numerical problems. In *Proc. 22nd Annual ACM Symp. on Theory of Computing*, pages 73–83, 1990.

[BS94]   J. D. Bright and G. F. Sullivan. Checking mergeable priority queues. In *Proceedings of the 24th Annual International Symposium on Fault-Tolerant Computing*, pages 144–153, Los Alamitos, CA, USA, June 1994. IEEE Computer Society Press.

[BS95]   J. D. Bright and G. F. Sullivan. On-line error monitoring for several data structures. In *FTCS-25: 25th International Symposium on Fault Tolerant Computing Digest of Papers*, pages 392–401, Pasadena, California, 1995.

[BSM95]   J. D. Bright, G. F. Sullivan, and G. M. Masson. Checking the integrity of trees. In *FTCS-25: 25th International Symposium on Fault Tolerant Computing Digest of Papers*, pages 402–413, Pasadena, California, 1995.

[BSM97]   J.D. Bright, G. F. Sullivan, and G. M. Masson. A formally verified sorting certifier. *IEEE Transactions on Computers*, 46(12):1304–1312, 1997.

[BW96]   M. Blum and H. Wasserman. Reflections on the pentium division bug. *IEEE Trans. Comput.*, 45(4):385–393, April 1996.

[CNAO85] Norishige Chiba, Takao Nishizeki, Shigenobu Abe, and Takao Ozawa. A linear algorithm for embedding planar graphs using PQ-trees. *Journal of Computer and System Sciences*, 30(1):54–76, February 1985.

[Edm65]   J. Edmonds. Maximum matching and a polyhedron with 0,1 - vertices. *Journal of Research of the National Bureau of Standards*, 69B:125–130, 1965.

[Gab76]   H.N. Gabow. An efficient implementation of Edmond's algorithm for maximum matching on graphs. *JACM*, 23:221–234, 1976.

[HMN96]   C. Hundack, K. Mehlhorn, and S. Näher. A Simple Linear Time Algorithm for Identifying Kuratowski Subgraphs of Non-Planar Graphs. Manuscript, 1996.

[HT74]   J.E. Hopcroft and R.E. Tarjan. Efficient planarity testing. *Journal of the ACM*, 21:549–568, 1974.

[Kar90]   A. Karabeg. Classification and detetection of obstructions to planarity. *Linear and Multilinear Algebra*, 26:15–38, 1990.

[Kur30]   C. Kuratwoski. Sur le problème the courbes guaches en topologie. *Fundamenta Mathematicae*, 15:271–283, 1930.

[LEC67]   A. Lempel, S. Even, and I. Cederbaum. An Algorithm for Planarity Testing of Graphs. In P. Rosenstiehl, editor, *Theory of Graphs, International Symposium, Rome*, pages 215–232, 1967.

[MM95]   K. Mehlhorn and P. Mutzel. On the Embedding Phase of the Hopcroft and Tarjan Planarity Testing Algorithm. *Algorithmica*, 16(2):233–242, 1995.

[MN95]   K. Mehlhorn and S. Näher. LEDA, a platform for combinatorial and geometric computing. *Communications of the ACM*, 38:96–102, 1995.

[MN98]   K. Mehlhorn and S. Näher. *The LEDA Platform for Combinatorial and Geometric Computing*. Cambridge University Press, 1998. Draft versions of some chapters are available at http://www.mpi-sb.mpg.de/~mehlhorn.

[MNS+96] K. Mehlhorn, S. Näher, T. Schilz, S. Schirra, M. Seel, R. Seidel, and Ch. Uhrig. Checking Geometric Programs or Verification of Geometric Structures. In *Proc. of the 12th Annual Symposium on Computational Geometry*, pages 159–165, 1996.

[MNU97]   Kurt Mehlhorn, S. Näher, and Ch. Uhrig. The LEDA User Manual (Version R 3.5). Technical report, Max-Planck-Institut für Informatik, 1997. http://www.mpi-sb.mpg.de/LEDA/leda.html.

[NC88]   T. Nishizeki and N. Chiba. *Planar Graphs: Theory and Algorithms*. Annals of Discrete Mathematics (32). North-Holland Mathematics Studies, 1988.

[OLPT97] O. Devillers, G. Liotta, F.P. Preparata, and R. Tamassia. Checking the convexity of polytopes and the planarity of subdivisions. Technical report, Center for Geometric Computing, Department of Computer Science, Brown University, 1997.

[SM90] G. F. Sullivan and G. M. Masson. Using certification trails to achieve software fault tolerance. In Brian Randell, editor, *Proceedings of the 20th International Symposium on Fault-Tolerant Computing (FTCS '90)*, pages 423–433, Newcastle upon Tyne, UK, June 1990. IEEE Computer Society Press.

[SM91] G. F. Sullivan and G. M. Masson. Certification trails for data structures. In *Proceedings of the 21st International Symposium on Fault-Tolerant Computing*, pages 240–247, 1991.

[SWM95] G.F. Sullivan, D.S. Wilson, and G.M. Masson. Certification of computational results. *IEEE Transactions on Computers*, 44(7):833–847, 1995.

[WB97] Hal Wasserman and Manuel Blum. Software reliability via run-time result-checking. *Journal of the ACM*, 44(6):826–849, November 1997.

[Wil84] S.G. Williamson. Depth-First Search and Kuratowski Subgraphs. *JACM*, 31(4):681–693, 1984.

Computationally-Sound Checkers

Silvio Micali

Laboratory for Computer Science, MIT, Cambridge, MA 02139

Abstract. We show that CS proofs have important implications for validating one-sided heuristics for $\mathcal{NP}$. Namely, generalizing a prior notion of Blum's, we put forward the notion of a *CS checker* and show that special-type of CS proofs imply CS checkers for $\mathcal{NP}$-complete languages.

1 Introduction

Let us state the general problem of heuristic validation we want to solve, explain why prior notions of checkers may be inadequate for solving it, and discuss the novel properties we want from a checker.

1.1 The Problem of Validating One-Sided Heuristics for $\mathcal{NP}$

A GENERAL PROBLEM. $\mathcal{NP}$-complete languages contains very important and useful problems that we would love to solve. Unfortunately, it is extensively believed that $\mathcal{P} \neq \mathcal{NP}$ and $\mathcal{NP} \neq \mathcal{C}o\text{-}\mathcal{NP}$, and thus that our ability of successfully handling $\mathcal{NP}$-complete problems is severely limited. Indeed, if $\mathcal{P} \neq \mathcal{NP}$, then no efficient (i.e., polynomial-time) algorithm may decide membership in an $\mathcal{NP}$-complete language without making any errors. Moreover, if $\mathcal{NP} \neq \mathcal{C}o\text{-}\mathcal{NP}$, then no efficient algorithm may, in general, prove non-membership in an $\mathcal{NP}$-complete language by means of "short and easy to verify" strings.

In light of the above belief, the "best natural alternative" to deciding efficiently $\mathcal{NP}$-complete languages and conveying efficiently to others the results of our determinations, consists of tackling $\mathcal{NP}$-complete languages by means of efficient *heuristics* that are *one-sided*. Here by "heuristic" we mean a program (emphasizing that no claim is made about its correctness) and by "one-sided" we mean that such a program, on input a string x, outputs either (1) a proper $\mathcal{NP}$-witness, thereby *proving* that x is in the language, or (2) the symbol NO, thereby *claiming* (without proof) that x is not in the language.

But for an efficient one-sided heuristic to be really useful for tackling $\mathcal{NP}$-complete problems we should *know when it is right*. Of course, when such an heuristic outputs an $\mathcal{NP}$-witness, we can be confident of its correctness on the given input. However, when it outputs NO, skepticism is mandatory: even if the heuristic came with an a priori guarantee of returning the correct answer on most inputs, we might not know whether the input at hand is among those. Thus, in light of the importance of $\mathcal{NP}$-complete languages and in light of the many efficient one-sided heuristics suggested for these languages, a fundamental problem naturally arises:

Given an efficient one-sided heuristic H for an $\mathcal{NP}$-complete language, is there a meaningful and efficient way of using H so as to validate some of its NO-outputs?

INTERPRETING THE PROBLEM. The solvability of the above general problem critically depends on its specific interpretation.

One such interpretation was proposed by Manuel Blum when, a few years ago, he introduced the notion of a checker [3][1], and asked whether $\mathcal{NP}$-complete languages are checkable. His specific formulation of the general problem (which provided the motivation for our present work) is still open.

By contrast, in this paper we propose a new interpretation of the general problem and, under a complexity conjecture, provide its first (and positive) solution.

Thus, our result does not address Blum's open question, but a generalization of it. We shall immediately argue, however, that, without generalizing it, his original question may not possess a positive answer, nor enjoy some new but —to us— desirable properties.

1.2 Blum Checkers and Their Limitations

THE NOTION OF A BLUM CHECKER. Intuitively, a Blum checker for a given function f is an algorithm that either (a) determines with arbitrarily high probability that a given program, run on a given input, correctly returns the value of f at that input, or (b) determines that the program does not compute f correctly (possibly, at some other input). Let us quickly recall Blum's definition.

> **Informal Definition:** Let f be a function and C a probabilistic oracle-calling algorithm running in expected polynomial-time. Then, we say that C is a *Blum checker* for f if, on input an element x in f's domain and oracle access to any program P (allegedly computing f), the following two properties hold:
> 1. If $P(y) = f(y)$ for all y (i.e., if P correctly computes f for every input), then $C^P(x)$ outputs YES with probability 1; and
> 2. If $P(x) \neq f(x)$ (i.e., if P does not compute f correctly on the given input x), then $C^P(x)$ outputs YES with probability $\leq 1/2$.

The probabilities above are taken solely over the coin tosses of C whenever P is deterministic, or over the coin tesses of both algorithms otherwise.

The above notion of a Blum checker slightly differs from the original one.[2] In particular, according to our reformulation any correct program for computing f

---
[1] We shall call his notion a *Blum checker* to highlight its difference with ours.
[2] Disregarding minor issues, Blum's original formulation imposes an additional condition: roughly, that C run asymptotically faster than the fastest known algorithm for computing f —or asymptotically faster than P when checking P. This additional

immediately yields a checker for f, though not necessarily a useful one (because such a checker may be too slow, or because its correctness may be too hard to establish).[3]

Despite their stringent requirements, Blum checkers have been constructed for a variety of specific functions (see, in particular, the works of Blum, Luby, and Rubinfeld [4] and Lipton [18]).

Note that the notion of a checker is immediately extended to languages: an algorithm C is a Blum checker for a language L if it is a Blum checker for L's characteristic function. Indeed, the interactive proof-systems of [23] and [25] yield Blum checkers for, respectively, any $\#\mathcal{P}$- or $PSPACE$-complete language.[4]

BLUM CHECKERS VS. EFFICIENT HERISTICS FOR $\mathcal{NP}$-COMPLETE PROBLEMS. We believe that the question of whether Blum checkers for $\mathcal{NP}$-complete languages exist should be interpreted more broadly than originally intended. We in fact argue that, even if they existed, Blum checkers for $\mathcal{NP}$-complete languages might less relevant than desirable.

(Informal) Definition: We say that a Blum checker C for a function f is *irrelevant* if, for all efficient heuristic H for f, and for all x in f's domain, $C^H(x) = NO$ without ever calling H on input x.

Note that, if $\mathcal{P} \neq \mathcal{NP}$, then no efficient heuristic for an $\mathcal{NP}$-complete language is correct on all inputs. Thus, it is quite legitimate for a Blum checker for a $\mathcal{NP}$-complete language to output NO whenever its oracle is an efficient heuristic, without ever calling it on the specific input at hand: a NO-output simply indicates that the efficient heuristic is incorrect on some inputs (possibly different from the one at hand). However, constructing an irrelevant Blum checker for SAT's characteristic function under the assumption that $\mathcal{P} \neq \mathcal{NP}$ is not trivial. The difficulty lies in the fact that a checker does not know whether

---

   constraint aims at rebuffing a natural objection: who checks the checker? The condition is in fact an attempt to guarantee, in practical terms, that C is sufficiently different from (and thus "independent" of) P, so that the probability that both C and P make an error in a given execution is smaller than the probability that just P makes an error.

[3] Thus, running a checker C (as defined by us) with a program P may be useful only if C is much faster than P, or if C's correctness is much easier to prove —or believe— than that of P.

[4] In fact, the definition of a Blum checker for a language L is analogous to a restricted kind of interactive proof for L: one whose Prover is a probabilistic polynomial-time algorithm with access to an oracle for membership in L. Indeed, whenever a language L possesses such a kind of interactive proof-system, a checker C for L is constructed as follows. On inputs P (a program allegedly deciding membership in L) and x, the checker C simply runs both Prover and Verifier on input x, giving the Prover oracle access to program P. C outputs YES if the Verifier accepts, and rejects otherwise.

it is accessing a polynomial-time program (in which case, if $\mathcal{P} \neq \mathcal{NP}$, it could always output NO), or an exponential-time program that is correct on all inputs (in which case it should always output YES). We can, however, construct such an irrelevant Blum checker under the assumption that one-way functions exist. This assumption appears to be stronger than $\mathcal{P} \neq \mathcal{NP}$, but is widely believed and provides the basis of all modern cryptography.

(Informal) Definition. We say that a function f mapping binary strings to binary strings is *one-way* if it is length-preserving, polynomial-time computable, but not polynomial-time invertible in the following sense: for any polynomial-time algorithm A, if one generates at random a sufficiently long input z and computes $y = f(z)$, then the probability that $A(y)$ is a counter-image of f is negligible.

(Informal) Theorem 1: If one-way functions and Blum checkers for $\mathcal{NP}$-complete languages exist, then there exist irrelevant Blum checkers for $\mathcal{NP}$-complete languages.

(Informal) Proof: Let SAT be the $\mathcal{NP}$-complete language of all satisfiable formulae in conjunctive normal form, let P be a program allegedly deciding SAT, let C be a Blum checker for SAT, let f be a one-way function, and let $\mathcal{C}$ be the following oracle-calling algorithm.

On input an n-variable formula F in conjunctive normal form, and oracle access to P, $\mathcal{C}$ works in two stages. In the first stage, $\mathcal{C}$ randomly selects a (sufficiently long) string z and computes (in polynomial-time) $y = f(z)$. After that, $\mathcal{C}$ utilizes the completeness of SAT to construct, and feed to P, n formulae in conjunctive normal form, $F_1, \ldots, F_n$, whose satisfiability "encodes a counter-image of y under f".

(For instance, F_1 is constructed so as to be satisfiable if and only if there exists a counter-image of y whose first bit is 0. The checker feeds such an F_1 to P. If P outputs "F_1 is satisfiable," then $\mathcal{C}$ constructs F_2 to be a formula that is satisfiable if and only if there exists a counter-image of y whose 2-bit prefix is 00. If, instead, P responds "F_1 is not satisfiable," then $\mathcal{C}$ constructs F_2 to be a formula that is satisfiable if and only if there exists a counter-image of y whose 2-bit prefix is 10. And so on, until all formulae $F_1, \ldots, F_n$ are constructed and all outputs $P(F_1), \ldots, P(F_n)$ are obtained.)

Because string y is, by construction, guaranteed to be in the range of f, at the end of this process one either finds (a) a counter-image of y under f, or (b) a proof that P is wrong (because if no f-inverse of y has been found, then P must have provided a wrong answer for at least one of the formulae F_i). If event (b) occurs, $\mathcal{C}$ halts outputting NO. Else, in a second phase, $\mathcal{C}$ runs Blum checker C on input the original formula F and oracle access to P. When C halts so does $\mathcal{C}$, outputting the same YES/NO value that C does.

Let us now argue that $\mathcal{C}$ is a Blum checker for SAT. First, it is quite clear that $\mathcal{C}$ runs in probabilistic polynomial time. Then, there are two cases to consider.

1. *P correctly computes SAT's characteristic function.* In this case, a counter-image of y is found, and thus $\mathcal{C}^P$ does not halt in the first phase. Moreover,

in the second phase, $\mathcal{C}$ runs Blum checker C with the same correct program P. Therefore, by Property 1 of a Blum checker, C^P will output YES no matter what the original input formula F might be, and, by construction, so will $\mathcal{C}^P$. This shows that $\mathcal{C}$ enjoys Property 1 of a Blum checker for SAT.

2. $P(F)$ *provides the wrong answer about the satisfiability of F.* In this case, either $\mathcal{C}^P$ halts in Phase 1 outputting NO, or it executes Phase 2 by running $C^P(F)$, that is, the original Blum checker for SAT, C, on the same input F and the same oracle P. Therefore, by Property 2 of a Blum checker, the probability that $C^P(F)$ will halt outputting YES is no greater than 1/2. By construction, the same holds for $\mathcal{C}^P(F)$. This shows that $\mathcal{C}$ enjoys Property 2 of a Blum checker for SAT.

Finally, let us argue that, for any input F and any efficient P (no matter how well it may approximate SAT's characteristic function), almost always $\mathcal{C}^P(F) = NO$, without even calling P on F. In fact, because $\mathcal{C}$ runs in polynonial time, whenever P is polynomial-time, so is algorithm $\mathcal{C}^P$. Therefore, $\mathcal{C}^P$ has essentially no chance of inverting a one-way function evaluated on a random input. Therefore, $\mathcal{C}$ will output NO in Phase 1, where it does not call P on F). ∎

In sum, differently from many other contexts, the notion of a Blum checker may not be too useful for handling efficient heuristics for $\mathcal{NP}$-complete languages: either because no such checkers exist[5] or because they may exist but not be too useful.

BLUM CHECKERS ARE NOT COMPLEXITY-PRESERVING. The lesson we derive from the above sketched proof of Theorem 1 is that Blum's notion of a checker lacks a new property that we name *complexity preservation*. Intuitively, a Blum checker for Satisfiability, when given a "not-so-difficult" formula F, may ignore it altogether and instead call the to-be-tested efficient heuristic on *very special* and *possibly much harder* inputs, thus forcing the heuristic to make a mistake and justifying its own outputting NO (i.e., "the heuristic is wrong").

The possibility of calling a given heuristic H on inputs that are harder than the given one chills the chances of meaningfully validating H's answer whenever it happens to be correct.[6] Such possibility may not matter much if "the difference in computational complexity between any two inputs of similar length" is somewhat bounded. But it may matter a lot whenever if such a difference is enormous —which may be the case of $\mathcal{NP}$-complete languages, as they encode membership in both easy and hard languages. We thus wish to develop a notion of a "complexity-preserving" checker.

---

[5] Notice that this possibility does not contradict the fact that $\mathcal{NP}$ is contained in both $\#\mathcal{P}$ and $PSPACE$ and that $\#\mathcal{P}$- and $PSPACE$-complete languages are Blum checkable!

[6] Note that such possibility not only is present in the *definition* of a Blum checker, but also in all known *examples* of a Blum checker. Typically, in fact, a Blum checker works by calling its given heuristic on random inputs, and these may be more difficult than the specific, original one.

1.3 New Checkers for New Goals

THE OLD GOAL. Blum checkers are very useful to catch the occasional mistake of programs believed to be correct on all inputs. That is, they are ideally suited to check *fast programs for easy functions* (or slow program for hard functions). In fact, if f is an efficiently computable function, then we know a priori that there are efficient and correct programs for f. Therefore, if a reputable software company produces a program P for f, it might be reasonable to expect that P is correct. In this framework, by running a Blum checker for f, with oracle P, on a given input x we have nothing to lose[7] and something to gain. Indeed, if the checker answers YES, we have "verified our expectations" about the correctness of P at least on input x (a small knowledge gain), and if the checker answers NO, we have proved our expectations about P to be wrong (a big knowledge gain).

THE NEW GOAL. We instead want to develop checkers for a related, but different goal: validating efficient heuristics that are known to be incorrect on some inputs. That is, we wish to develop checkers suitable for handling *fast programs for hard functions*. Now, if f is a function hard to compute, then we know a priori that no efficient program correctly computes it. Therefore obtaining from a checker a proof that such an efficient program does not compute f correctly would be quite redundant. We instead want checkers that, at least occasionally, if an efficient heuristic for f happens to be correct on some input x, are capable of convincing us that this is the case.

INTERPRETING THE NEW GOAL. Several possible valid interpretations of this general constraint are possible. In this paper we focus on a single one: namely, we want checkers that are *complexity-preserving*. Let f be a function that is hard to compute (at least in the worst case). Then, intuitively, a complexity-preserving checker for f will, on input x, call a candidate program for f only on inputs for which evaluating f is essentially as difficult as for x.

Our point is that, while a given heuristic for satisfiability, H, may make mistakes on some formulae, it may return remarkably accurate answers on some class of formulae (e.g., those decidable in $O(2^{cn})$ time, for some constant $c < 1$, by a given deciding algorithm D). Intuitively, therefore, checkers should be defined (and built!) so that, if the input formula belongs to that class and $\mathcal{H}$ happens to be correct on the input formula, they call H only on additional formulae in that class.

2 Background on CS Proofs

Informally, a CS proof of a statement S consists of a short string, σ, which (1) is as easy to find as possible, (2) is very easy to verify, and (3) offers a strong

---

[7] Except for some amount of running time, but Blum checkers are often so fast (e.g., running in time sub-linear in that of the algorithm they check) that non even this is much of a concern.

computational guarantee about the verity of S. By "as easy to find as possible" we mean that a CS proof of a *true* statement (i.e., for the purposes of this paper, *derivable* in a given axiomatic theory) can be computed in a time essentially comparable to that needed to Turing-accept the statement. By "very easy to verify" we mean that the time necessary to inspect a CS Proof of a statement S is poly-logarithmic in the time necessary to Turing-accept S. Finally, by saying that the guarantee offered by a CS proof is "computational" we mean that false statements either do not have any CS proofs, or such "proofs" are practically impossible to find.

Let us now see how define these proof-systems more formally and then discuss their properties.

2.1 Common and Different Aspects of Various CS Proof-Systems

CS proof-systems consist of a pair of two algorithms, a Prover and a Verifier. However, we distinguish various types of such systems: all sharing a basic paradigm, but differing in (1) their "mechanics" and (2) the complexity assumptions underlying their implementations.

A COMMON PARADIGM: CONTROLLED INCONSISTENCY Though CS proofs differ from zero-knowledge arguments in important respects[8], they too allow the existence of false proofs, but ensure that these are computationally hard to find. That is,

False CS proofs may exist, but they will "never" be found.

Equivalently, *CS proof-systems are deliberately inconsistent, but practically indistinguishable from consistent systems.* Indeed, each CS proof specifies a security parameter, controlling the amount of computing resources necessary to "cheat" in the proof, so that these resources can be made arbitrarily high. Accordingly, CS proofs are meaningful only if we believe that the Provers who produced them, though more powerful than their corresponding Verifiers, are themselves computationally bounded.[9] ¿From a practical point of view, this is hardly a limitation. As long we restrict our attention to physically implementable processes, no prover in our Universe can perform $2^{1,000}$ steps of computation, at least during the existence of the human race. Thus, "practically speaking" all Provers are computationally bounded.

Besides being practically reasonable, CS proofs also are theoretically appealing: *mutatis mudandis*, they provide an answer to many of the oldest questions

---

[8] As we shall see, the latter may not enjoy *relative efficiency of proving*, nor *relative and ubiquitous* efficiency of verifying, nor *universality*.

[9] The transition from an interactive proof-system to a CS proof-system is analogous to the transition from perfect zero-knowledge proof-system to a computational zero-knowledge proof-system[22], which has proved to be a more flexible and powerful notion [12].

in complexity theory, and some new and fundamental ones as well. (After all, "the right notion is the one that allows us to prove the right theorem in the right way!")

THREE DIFFERENT RESOURCES. CS proof-systems differ in their use of the following three resources:

1. *Interaction.* CS proof-systems differ in the amount of interaction they require: informally, the number of rounds in which Prover and Verifier exchange messages until the Verifier accepts or reject. We call a CS proof-system non-interactive if it is 0-round. That is, if, on input the statement of a theorem T, the Prover outputs a CS proof of of T: a special string that the Verifier can check for correctness on its own.
2. *Shared randomness.* In a CS proof-system, the Prover, P, and the Verifier, V, may be probabilistic. We may thus distinguish various types of CS proof-systems according to the type of randomness source P and V share with each other: a random oracle (i.e., a long string of easily-accessible random bits), a (polynomially-long) random string, or no randomness at all.[10]
3. *Complexity assumptions.* CS proof-systems also differ in the complexity assumption necessary for concretly implement them. Such assumptions too could be considered a "resource" (to be used sparingly!).

These resources are quite interrelated, and typically one can decrease one at the cost of increasing another.

THREE TYPES OF CS PROOF-SYSTEMS. In this paper we consider only three types of CS proof-systems, briefly described below (in order of "decreasing complexity assumption"):

- *CS proof-systems sharing a random oracle.* In these proof-systems, when having a given statement as an input, both P and V have access to the same random oracle. A bit more precisely (given that a random oracle can be viewed as an infinite string of random bits), P and V have oracle access to a randomly selected function mapping $poly(k)$-bit strings to $poly(k)$-bit strings, where k is a security parameter.
  (We show that random oracles alone suffice for implementing these CS proof-systems. Such proof-system could be considered non-interactive.[11])
- *Interactive CS proof-systems.* In these systems Prover and Verifier exchange messages in arbitrarily many rounds.
  (Such systems can be implemented based on collision-free hash functions.)

---

[10] We may also consider the hybrid shared source of a hidden random string, that is, the image under a polynomial-time algorithm of a random string.
[11] In some sense, therefore, "maximizing" the second resources allows one to minimize the other two.

- *One-round CS proof-systems.* These are a special case of interactive CS proof-systems, where each of Prover and Verifier send a single message, the Verifier going first.
  (Such systems are implementable under (1) a concrete number-theoretic conjecture, or (2) under a generic conjecture. They are particularly important to us because they provide the first CS proof-system implying CS checkers.)
- *CS proof-systems with a random string.* These are a special case of one-round CS proof-systems, where the only message of the Verifier to the Prover consists of a random string (having length polynomial in a security parameter k).
  (Such systems are implementable based on a generic complexity conjecture —i.e., we provide a plausibility argument towards their existence. They can be considered non-interative —if a common random string is available.)

All mentioned types of CS proofs systems are designed to prove membership in a special (and yet quite "universal") language.

2.2 The CS Language.

PRELIMINARIES.

- *Encodings.* Throughout this paper, we assume usage of a standard binary encoding, and often identify an object with its encoding. (In particular, if A is an algorithm, we may —meaningfully if informally— give A as an input to another algorithm.)
  The length of an (encoded) object x is denoted by $|x|$.
  If q is a quadruple of binary strings, $q = (a, b, c, d)$, then our quadruple encoding is such that, for some positive constant c, $1 + |a| + |b| + |c| + |d| < |q| < c(1 + |a| + |b| + |c| + |d|)$.
- *(Steps.)* If M is a Turing machine and x an input, we denote by $\#M(x)$ the number of steps that M takes on input x.

Definition 1: We define the *CS language*, denoted by $\mathcal{L}$, to be the set of all quadruples $q = (M, x, y, t)$, such that M is (the description of) a Turing machine, x and y are a binary strings, and t a binary integer such that

1. $|x|, |y| \leq t$;
2. $M(x) = y$; and
3. $\#M(x) = t$.

Notice that, as long as M reads each bit of its inputs and writes each bit of its outputs, the above Property 1 is not a real restriction. Notice too that, due to our encoding, if $q = (M, x, y, t) \in \mathcal{L}$ then $t < 2^{|q|}$.

2.3 CS Proof-Systems with a Random Oracle

THE NOTION OF A CS PROOF-SYSTEM WITH A RANDOM ORACLE. By *n-call algorithm* we denote an oracle-calling algorithm that, in any possible execution, makes exactly n calls to its oracle.

Definition 2: Let $P^{(\cdot)}$ and $V^{(\cdot)}$ be two oracle-calling Turing machines, the second of which running in polynomial-time. We say that (P, V) is a *CS proof-system with a random oracle* if there exist a sequence of 6 positive constants, $c_1, \ldots, c_6$ (refered to as the *fundamental constants* of the system), such that the following two properties are satisfied:

1. *Feasible Completeness.* $\forall q = (M, x, y, t) \in \mathcal{L}$, $\forall k$, and $\forall f \in \Sigma^{k^{c_1}} \to \Sigma^{k^{c_1}}$,
   (1.1) $P^f(q, k)$ halts within $(|q|kt)^{c_2}$ computational steps, outputting a binary string $\mathcal{C}$ whose length is $\leq (|q|k)^{c_3}$, and
   (1.2) $V^f(q, k, \mathcal{C}) = YES$.
2. *Computational Soundness.* $\forall \tilde{q} \notin \mathcal{L}$, $\forall k$ such that $2^k > |q|^{c_4}$, and $\forall$ (cheating) deterministic $2^{c_5 k}$-call algorithm $\tilde{P}$, for a random oracle $\rho \in \Sigma^{k^{c_1}} \to \Sigma^{k^{c_1}}$,

$$Prob_\rho[V^\rho(\tilde{q}, k, \tilde{P}^\rho(\tilde{q}, k)) = YES] \leq 2^{-c_6 k}.$$

If $q = (M, x, y, t)$ and $V^f(q, k, \mathcal{C}) = YES$, we may call string $\mathcal{C}$ a *random-oracle CS proof (of security k) of $M(x) = y$ (in less than t steps)*. For variation of discourse, we may sometimes refer to such a $\mathcal{C}$ as a CS *witness* or a CS *certificate*.

THE CONSTRUCTABILITY OF CS PROOF-SYSTEMS WITH A RANDOM ORACLE.

Theorem 2 [24]: There exist CS proof-systems with a random oracle (without any other assumption).

2.4 One-Round CS Proof-Systems

Recall that a *circuit of size* $\leq s$ is a finite function computable by at most s Boolean gates, where each gate is either a NOT-gate (with one binary input and one binary output) or an AND-gate (with two binary inputs and one binary output). A circuit A may be taken to be deterministic, because it might have wired-in any finite lucky sequence of coin tosses. (In which case the probability that B is convinced in a random execution with A on input x solely depends on B's coin tosses.

Definition 4. Let (P, V) be a pair of algorithms, the second of which running in probabilistic polynomial time, which, on input x executes the following three phases: (1) V computes "a message" M on input x; (2) P computes a string y on inputs x and M; and (3) V outputs YES or NO by computing on input y and the internal state it reached at the end of phase 1. Let $\mathcal{L}$ be the CS language. We say that (P, V) is a *one-round CS proof-system* if there are four positive constants a, b, c, and d such that the following two properties are satisfied:

1′ *Feasible Completeness.* $\forall q = (M, x, y, t) \in \mathcal{L}$, and $\forall$ unary integers k, in every execution of (P, V) on inputs q and k,
(1.1′) P halts within $(nkt)^a$ computational steps, and
(1.2′) V outputs YES.

2′ *Computational Soundness.* $\forall \widetilde{q} \notin \mathcal{L}$, $\forall k > |q|^b$, and $\forall$ (cheating) circuit $\widetilde{P}$ of size $\leq 2^{ck}$, in a random execution of $\widetilde{P}$ with V on inputs $\widetilde{q}$ and k,
$$Prob[V \text{ outputs } YES] < 2^{-dk}.$$

We exhibit one-round CS proof-systems under either "We exhibit one-round CS proof-systems under either

1. *A concrete complexity assumption:* informally, the difficulty of deciding, given a prime p and an integer n (whose factorization is unknown), whether p divides $\phi(n)$
   (This assumption has been used by Cachin, Micali, and Stadler to construct computational private information retrieval systems with poly-logarithmic amount of communication [21]);
   or
2. *A generic, ad hoc conjecture:* informally, the replicability in the CS proof-system of Theorem 1 of the random oracle with a pseudo-random function

Theorem 3 [19]: Under Assumption 1 (properly formalized) there exist one-round CS proof-systems.

A more formal discussion of Conjecture 2 and why it suffices for implementing one-round CS proof-systems is presented in the next subsection. Indeed, CS proof-systems sharing a random string are a special case of one-round CS proof-systems.

2.5 CS Proof-Systems Sharing a Random String

THE NOTION OF OF A CS PROOF-SYSTEM SHARING A RANDOM STRING

In a CS proof-system sharing a random string, Prover and Verifier are ordinary (as opposed to oracle-calling) algorithms, sharing a short random string r. That is, whenever the security parameter is k, they share a string r that both believe to have been randomly selected among those having length k^c, where c is a positive constant. If string r is universally known, it can be shared by all Provers and Verifiers. (CS proof-systems sharing a random string are a special case of one-round CS proof-systems because r could be the message sent by the Verifier to the Prover.)

Definition 7: Let (P, V) be a pair of Turing machines, the second of which runs in polynomial-time. We say that (P, V) is a *CS proof-system sharing a random string* if there exists a sequence of 6 positive constants, $c_2, \ldots, c_6$ (refered to as the *fundamental constants* of the system[12]), such that the following two properties are satisfied:

---

[12] The "numbering" of these constants has been chosen to facilitate comparison with CS proof-systems with a random oracle.

1″ *Feasible Completeness.* $\forall q = (M, x, y, t) \in \mathcal{L}$, and $\forall$ binary string r,
(1.1″) On inputs q and r, P halts within $(|q| \cdot |r| \cdot t)^{c_2}$ computational steps outputting a binary string $\mathcal{C}$, whose length is $\leq (|q| \cdot |r|)^{c_3}$, such that
(1.2″) $V(q, r, \mathcal{C}) = YES$.

2″ *Computational Soundness.* $\forall \widetilde{q} \notin \mathcal{L}$, $\forall\ k > |q|^{c_4}$, and $\forall$ (cheating) circuits $\widetilde{P}$ whose size is $\leq 2^{c_5 k}$, for a random k^{c_1}-bit string r

$$Prob_r[\widetilde{P}(\widetilde{q}, r) = \widetilde{\mathcal{C}} \wedge V(\widetilde{q}, r, \widetilde{\mathcal{C}}) = YES] \leq 2^{c_6 k}.$$

We refer to the above strings r and $\mathcal{C}$ as, respectively, a *reference string* and a *CS certificate (of $q \in \mathcal{L}$, relative to r and (P, V))*.

THE CONSTRUCTABILITY OF CS PROOF-SYSTEMS SHARING A RANDOM STRING Though the "safest" conjecture sufficient for implying the existence of a new mathematical object simply is that "the object exists", it is often useful to prove that possibly stronger but longer-studied assumptions, such as the computational difficulty of *integer factorization* of that of finding *discrete logarithms*, are also sufficient.[13] Unfortunately, we have not been able to prove that some well-known complexity assumption suffices to imply the existence of CS proof systems with a random string, but their existence is nonetheless plausible. In particular, it is guaranteed by an ad hoc assumption: the replaceability, *in our context*, of random oracles with deterministic algorithms, possibly using short random seeds. (Such replacements are advocated, in more general contexts, by Bellare and Rogaway [2].)

To illustrate the details of such a replacement, denote by $(\mathcal{P}, \mathcal{V})$ the CS proof-system with a random oracle constructed in [24], and then consider substituting the random oracle of $(\mathcal{P}, \mathcal{V})$ with a specific pseudo-random function: the pseudo-random-oracle construction of Goldreich, Goldwasser and Micali [10].

Informally, their construction consists of a polynomial-time program that, on input a random and short (i.e., k-bit, where k is a security parameter) and *secret* seed and a query (binary string of a prescribed length), outputs an answer (binary string of another prescribed length). Assuming that unpredictable pseudo-random-bit generators exist[14], then no computationally-bounded algorithm (even if it chooses the queries) may distinguish the query-answer behavior of their program from that of a random oracle.

Syntactically, it is therefore possible to have the random seed of their construction be the reference string of a CS proof system sharing a random string, and have their polynomial-time program added to the specification of $(\mathcal{P}, \mathcal{V})$ so that, whenever $\mathcal{P}$ or $\mathcal{V}$ wish to query oracle f about a string x, they just run the

---

[13] For instance, the assumed computational difficulty of *integer factorization* and that of *discrete logarithms* have been proved sufficient for constructing, respectively, digital signature schemes unforgeable under an adaptive chosen message attack [11] and unpredictable pseudo-random number generators [5]

[14] And thus, thanks to result of Hastad, Impagliazzo, Levin, and Luby [14], if one-way functions exist

speficied polynomial-time construction on inputs the reference string and string x.

Notice that the so modified $(\mathcal{P}, \mathcal{V})$ is quite conceivably a CS proof-system with a random string, but this is not a consequence of the discussed indistinguishableness of the pseudo-random oracle of [10] from a truly-random oracle. In fact, such indistinguishability holds as long as the random seed of the pseudo-random oracle is kept secret, while identifying this seed with $(\mathcal{P}, \mathcal{V})$'s reference string makes it very public. And when its seed is made public, some statistical properties of the corresponding pseudo-random oracle are destroyed, while others are preserved.[15]

Let us emphasize that "random-oracle replacement does not always work": Canetti, Goldreich and Halevi [6] have proven that it is possible to construct algorithms so that they behave very differently when given access to a random oracle than when given access to *any* pseudo-random oracle with a public seed. In light of their result, it should be possible to construct special CS proof-systems sharing a random oracle so that they cannot be transformed into CS proof-systems sharing a random string by replacing the oracle with a pseudo-random function.[16]

On the other hand, it is known that, under traditional complexity assumptions, there are examples in which random oracles can be successfully substituted by pseudo-random ones.[17]

3 Defining CS Checkers

We start by defining an ideal version of CS checkers, and then show how to approximate it in a sufficiently close manner.

3.1 The Wishful Version of a CS Checker

The spirit of a CS checker is best conveyed wishfully assuming (for a second) that $\mathcal{NP}$ equalled $\mathit{Co\text{-}\mathcal{NP}}$. In that case, our CS checkers would take the following simple and appealing form.

WISHFUL CHECKERS. Define a *wishful checker* to be a polynomial-time algorithm C that, on input a Boolean formula F, outputs a Boolean Formula $\mathcal{F}$ satisfying the following two properties:

---

[15] Of course, for instance, if pseudo-random oracle PRO is indistinguishable from a random oracle when its seed is secret, the sequence $PRO(1), PRO(2), \cdots$ continues to contain roughly the same numbers of 0s and 1s after its seed is made public.

[16] Notice, however, that this does not imply that the same holds for every CS proof-system with a random oracle; in particular, for the mentioned $(\mathcal{P}, \mathcal{V})$.

[17] E.g., a random oracle provides a collision-free hash function, but such a function can be obtained with a "public-seed" construction under traditional complexity assumptions.

1. *(Membership Reversion)*: "$\mathcal{F} \in SAT$ iff $F \notin SAT$"; and
2. *(Complexity Preservation)*: "$\mathcal{F}$ is as hard to decide as F".

HOW TO USE A WISHFUL CHECKER. We interpret the above algorithm C as a kind of checker, because it immediately yields the following algorithm C' (more closely matching our intuition of a checker):

C' : Given an efficient one-sided heuristic for SAT, H, and an input formula, F, compute $\mathcal{F} = C(F)$. Then, call H so as to obtain the two values $H(F)$ and $H(\mathcal{F})$. If either value is different than NO, then a satisfying assignment has been computed either proving that $F \in SAT$ or that $F \notin SAT$. Else, $H(F) = H(\mathcal{F}) = NO$ prove that H is incorrect.

Note that, by the very definition of a wishful checker, the above proof that H is incorrect has been obtained without querying H on formulae not harder than the original input F.

3.2 The Informal Notion of a CS Checker

Let us now informally explain how, without assuming $\mathcal{NP} = Co\text{-}\mathcal{NP}$, CS checkers may approximate wishful ones to a sufficiently close extent. Renouncing to achieving greater generality, we discuss CS checkers solely in the context of $\mathcal{NP}$-complete languages, more particularly, of SAT.

CS CHECKERS. Informally speaking, a *CS checker* is a polynomial-time algorithm C that, on input a formula F, outputs a Boolean formula $\mathcal{F}$, called the *co-input*, satisfying the following properties:

1. *(Membership Semi-Reversion)*:
   1.1 At least one of F and $\mathcal{F}$ is satisfiable;
   1.2 If F is satisfiable, then no efficient algorithm has a non-negligible chance of finding a satisfying assignment for $\mathcal{F}$; and
2. *(Complexity Semi-Preservation)*: If $F \notin SAT$, then $\mathcal{F}$ is as hard to decide as F.

HOW TO USE A CS CHECKER. We interpret the above C as a checker because it immediately yields the following algorithm C' (that better matches what we may intuitively expect from a checker).

C' : Given an efficient one-sided heuristic for SAT, H, and an input formula, F, compute the co-input $\mathcal{F} = C(F)$. Then, call H so as to obtain $H(F)$. If $H(F) \neq NO$, HALT (a proof that $F \in SAT$ has been found). Else, call H so as to obtain $H(\mathcal{F})$ and HALT (if $H(\mathcal{F}) = NO$, then a proof that H is incorrect has been found; else, F can be interpreted to be unsatisfiable).

CS CHECKERS ARE GOOD ENOUGH. The computation of the above C' results in either (1) showing a satisfying assignment of F, or (2) showing a satisfying assignment of $\mathcal{F}$, or (3) showing that $H(F) = H(\mathcal{F}) = NO$.

A type-1 result clearly proves that $F \in SAT$.

A type-2 result is interpretable as saying that F is unsatisfiable. This is so because, if F belonged to SAT, then either a satisfying assignment of co-input $\mathcal{F}$ does not exist, or (by the very definition of a cs checker) the probability that it can be obtained in polynomial time is negligible. (Notice, in fact, that C' is efficient because both C and H are.)

A type-3 result proves that H is wrong. In fact, if $H(F) = NO$ is correct, then (by the very definition of a cs checker) $\mathcal{F} \notin SAT$, and thus $H(\mathcal{F}) = NO$ is incorrect. Moreover, *if H is correct on F*, our proof of H's incorrectness has been obtained in a complexity-preserving manner. We distinguish two cases:

1. If H is correct on F and $H(F) \neq NO$, then $H(F)$ is a (easy to verify) satisfying assignment of F, and thus C' does not call H on any co-input. Therefore, C vacuously does not call H on any $\mathcal{F}$ harder than F.
2. If H iscorrect on F and $H(F) = NO$, then $F \notin SAT$, in which case (again by the definition of a cs checker) $\mathcal{F}$ is guaranteed to have the same complexity of F.

If instead H is *not* correct about our original input F, then $H(F) = H(\mathcal{F}) = NO$ still is a proof of H's incorrectness, but not necessarily one obtained in a complexity-preserving manner. Notice, however, that lacking complexity preservation in this case is of no concern: if H happens wrong about our own original input, we are happy to prove that H wrong in any manner.[18]

THE COMPLEXITY PRESERVATION OF A CS CHECKER. To complete our informal discussion of CS checkers we must explain in what sense, whenever $F \notin SAT$, the complexity of F is close to that of $\mathcal{F}$. That is, we must explain (1) how we measure the complexity of the original input, and (2) how the co-input preserves this complexity.

1. *Complexity Meters.* The complexity of the original input F is defined to be the number of steps made by a chosen deciding algorithm for SAT, D, on input F. That is, when a CS checker for SAT is given an input formula F, it also given as an additional input the description of this chosen D. We refer to D as *the complexity meter*. In fact, by specifying D, we (implicitly) pin down the complexity of the original formula F. By insisting that D be a decider for SAT (i.e., that D be correct) we insist that the complexity of the original input be a "genuine" one.[19]

---

[18] Recall that in checking we care about our own original input x more than about H. Thus, if $H(x)$ is correct we aim at "proving" this fact, and we do not want to throw H away by calling it on much harder inputs. But if $H(x)$ is wrong, we do not mind dismissing H in any way. Least of all, we want to be convinced that $H(x)$ is right!

[19] In particular, if D were allowed to make errors, all formulae F could have constant complexity.

By properly choosing the complexity meter, one may be able to force the complexity of the original input to be small (and thus force the checker to query its given heuristic on a co-input of similarly small complexity). Choosing D to be the algorithm that tries all possible satisfying assignments for F is certainly legitimate, but not too meaningful. (Because any formula would have "maximum complexity" relative to such a complexity meter, the checker would essentially be free to call its given heuristic on any possible co-input.) Quite differently, if the original input F is known to belong to a class of formulae for which a given SAT algorithm performs very well (e.g., runs in sub-exponential time), by specifying that algorithm as our complexity meter, we force the checker to call its given heuristic only on a co-input of similarly low complexity.

Let us stress that we do not require that the checker, or someone choosing a complexity meter D, know how many steps D takes on the original input F. Nor do we require that one distinguish (somehow) for which inputs, if any, algorithm D (slow in the worst case) may be reasonably fast. Rather, we require that, if F happens to belong to those inputs on which D is fast, then it is this lower (and possibly unknown) complexity that should be preserved by a CS checker:

> *By specifying D one specifies **implicitly** the complexity of F, whatever it happens to be.*

For technical reasons, however, we require that D's running time be upper-bounded by 2^{2n}, a bound that essentially poses no real restrictions (in the sense that, within these bounds, any algorithm for satisfiability could alway be "timed-out" and then converted to an exhaustive search for a satisfying assignment).

2. *Complexity Co-meters.* The complexity of a co-input $\mathcal{F}$ is defined to be the number of steps taken on input $\mathcal{F}$ by a decider for SAT, $\mathcal{D}$, specified before hand. We refer to such a $\mathcal{D}$ as *the complexity co-meter*.

   Thus, a complexity co-meter is independent of the chosen complexity meter: the first is fixed once and for all (in fact, it could be made part of the very definition of a CS checker), while the second is chosen afresh each time a CS checker is run. Under these circumstances, at first glance, it may appear surprising that a CS checker may succeed in keeping the complexity of the co-input close to that of the original input. But the fixed co-meter $\mathcal{D}$ includes the code of the uniform algorithm, so that, in a sense, the complexity of a co-input is measured relative to a "decider for SAT that is easily constructed on input D".

   Notice that one could conceive stating complexity preservation by simply saying that the number of steps taken by a chosen D on the original input is polynomially close to the number of steps taken by the same D on the co-input. This is in fact a simpler way of having the co-meter easily depend on the meter. However, we needed to endow CS checkers with a bit more room to maneuver than that. In any case, we believe it preferable to have the meter that is a fixed component of the CS checker to be a *universal meter*.

3.3 The Notion of a CS Checker

PRELIMINARIES.

- We let CNF denote the language of all formulae in conjunctive normal form, and SAT the set of all satisfiable formulae in CNF. If $F \in SAT$, then we denote by $SAT(F)$ the set of all satisfiable assignments of F. For any positive integer n, CNF_n and SAT_n will denote, respectively, all formulae in CNF and SAT whose binary length is n.
- By a SAT *decider* we mean an algorithm that (correctly) decides the language SAT. (Thus a SAT decider D needs not to be one-sided, and may output just YES or NO.) We say that a SAT decider D is *reasonable* if, $\forall F \in CNF$, $\#D(F) \leq 2^{|F|}$.
- If A is a probabilistic algorithm, and E an event (involving executions of A on specified inputs), by $Prob_A[E]$ we denote the probability of E, taken over all possible coin tosses of A

Definition 8: Let Φ be a probabilistic polynomial-time algorithm, $\mathcal{D}$ a SAT decider, and $\mathcal{Q}(\cdot,\cdot,\cdot,\cdot)$ a positive polynomial. We say that $(\Phi, \mathcal{D}, \mathcal{Q})$ is a *CS checker* if, on input any CNF formula F, any reasonable SAT decider D, and any sufficiently long unary integer k, Φ outputs a formula $\mathcal{F}$ such that the following three properties hold:

1. $F \vee \mathcal{F} \in SAT$;
2. $F \in SAT \Rightarrow \forall$ poly(k)-size circuits A, $Prob_\Phi[A(\mathcal{F}) \in SAT(\mathcal{F})] < 2^k$; and
3. $F \notin SAT \Rightarrow \#\mathcal{D}(\mathcal{F}) < \mathcal{Q}(|F|, |D|, k, \#D(F))$.

If $\mathcal{C} = (\Phi, \mathcal{D}, \mathcal{Q})$ is a CS checker, we refer to Φ and $\mathcal{D}$ as, respectively, the *reducer* and the *complexity co-meter*.

REMARK: Note that even Properties 1 and 2 alone (i.e., leaving aside Property 3) constitute an surprising statement about SAT: informally, they state that from any formula F can be efficiently transformed to a formaul $\mathcal{F}$ such that (1) at least one of the two is satisfiable, but (2) no efficient algorithm can find a satisfying assignment for both. This is mighty close to saying that $\mathcal{NP} \neq Co\text{-}\mathcal{NP}$ without "crossing that line"! But the statement is even more interesting (and powerful) if (as expressed by property 3), $\mathcal{F}$ further has a complexity close to that of F.

4 Implementing CS Checkers for SAT

Let us recall some known properties of Cook's and Levin's $\mathcal{NP}$-completeness constructions.

KEY PROPERTIES OF COOK'S AND LEVIN'S CONSTRUCTIONS.
Given a polynomial-time predicate $A(\cdot, \cdot)$ and a positive constant b, these constructions consist of a polynomial-time algorithm that, on input a binary string

x, outputs a CNF formula ϕ that is satisfiable if and only if there is a binary string σ such that $|\sigma| \leq |x|^b$ and $A(x,\sigma) = YES$. We refer to such a string σ as a *witness (for x)*. The construction further enjoys the following well-known properties (which are actually required by Levin's definition of $\mathcal{NP}$-completeness):

(i) x is polynomial-time retrievable from ϕ;
(ii) a proper witness for x is polynomial-time computable from any satisfying assignment for ϕ (if one exists); and,
(iii) a satisfying assignment for ϕ is polynomial-time computable from any proper witness for x (if one exists).

Theorem 4: If one-round CS proof-systems exist, then CS checkers for SAT exist.

The proof of Theorem 4 actually needs a more detailed set up than we believe appropriate here. Accordingly, we shall prove it in the final paper, and prove instead below a weaker version of it, having very much in the same spirit but less details.

Theorem 4': If CS proof-systems sharing a random string exist, then CS checkers for SAT exist.

Proof: Let (P, V) be a CS proof-system sharing a random string with fundamental constants $c_2, \ldots, c_6$, and consider the following algorithm.

Algorithm Φ

Inputs: F, a CNF formula, D, a reasonable SAT solver, and k, a unary string
Subroutines: P and V

Code: Randomly select a k-bit (reference) string r for (P, V), and use Cook's [7] or Levin's [17] construction to compute a CNF formula $\mathcal{F}$ that is satisfiable if and only if there exist two binary strings t and σ such that, setting $q = (F, D, NO, t)$, the following three properties hold: (1) $t \leq 2|F|$, (2) $\sigma \leq (|q| \cdot k)^{c_3}$, and (3) $V(q, r, \sigma) = YES$.
{*Comment:* If it exists, σ is a CS certificate of $(D, F, NO, t) \in \mathcal{L}$, relative to (P, V) and reference string r. The existence of such a σ, however, does not guarantee that $D(F) = NO$.[20]}

Let us now show that there exist a SAT decider $\mathcal{D}$ and a positive polynomial $\mathcal{Q}$ such that $\mathcal{C} = (\Phi, \mathcal{D}, \mathcal{Q})$ is a CS checker.

---

[20] In fact, we "expect" that σ exists (and thus that F' is satisfiable) with "overwhelming probability", even when F is satisfiable.

To begin with, notice that, because of the polynomiality of V and of Cook's constuction, Φ is polynomial-time.[21]

Further, because properties 1 and 2 of a CS checker only depend on its reducer, let us show that they hold for our Φ prior to defining $\mathcal{D}$ and $\mathcal{Q}$.

Property 1 holds trivially if $F \in SAT$. Assume therefore that $F \notin SAT$. Then, because of the correctness and running time of the complexity meter D, we have $D(F) = NO$ within $t \leq 2^{2n}$ steps. Thus, by the (feasible) completeness of (P, V), for any possible reference string r there is a CS certificate σ of $q = (D, F, NO, t) \in \mathcal{L}$. Thus, $\mathcal{F} \in SAT$, proving that Property 1 holds in all cases.

Property 2 is established by contractiction. Assume that there exists an input formula $F \in SAT$ and a poly(k)-size circuit A that, with non-negligible probability, computes a satysfying assignment of a so constructed co-inputs $\mathcal{F}$. Then, by property (ii) of Cook's construction, from such a satisfying assignment (if it exists and is found) one computes in polynomial time both t and a CS certificate σ of $q = (D, F, NO, t) \in \mathcal{L}$. But if $F \in SAT$, then for no t is $q = (D, F, NO, t) \in \mathcal{L}$. Therefore, this contradicts the computational soundness of (P, V).

Let us finally show that there exist a SAT decider $\mathcal{D}$ and a positive polynomial $\mathcal{Q}$ such that, for all formulae $F \notin CNF$, for all complexity meters D, and for all security parameters k, if D, on input F, takes t ($\leq 2^{2|F|}$) steps to decide that no satisfying assignment for F exists, then, given any co-input $\mathcal{F}$ of F, $\mathcal{D}$ finds a satisfying assignment for $\mathcal{F}$ in at most $Q(|F|, |D|, k, t)$ steps.

Algorithm $\mathcal{D}$ works in four phases as follows:

$\mathcal{D}1$. Computes F, D, and r from $\mathcal{F}$.

(Due to property (i) of Cook's construction, $\mathcal{D}$ can execute this phase in time polynomial in $|\mathcal{F}|$. Thus, because $\mathcal{F}$ has been computed by $\mathcal{C}$ in time polynomial in $|F|$, $|D|$ and k, this phase is executable in time polynomial in $|F|$, $|D|$ and k.)

$\mathcal{D}2$. Runs D on input F to find the exact number of steps, t, taken by D to output NO on input F.

(Because D can be simulated with a slow-down polynomial in $|D|$, this phase takes time polynomial in $|D|$ and t.)

$\mathcal{D}3$. Run prover P on input $q = (D, F, NO, t)$ and reference string r to produce a CS certificate, σ, of $q \in \mathcal{L}$.

(Due to the feasible completeness of (P, V), this phase is executable in time polynomial in $|q|$, k and t; and thus in time polynomial in $|F|$, $|D|$, k, and t.)

$\mathcal{D}4$. Use σ to compute a satisfying assignment for $\mathcal{F}$.

(Due to property (iii) of Cook-Levin construction, also this phase can be implemented in time polynomial in $|F|$, $|D|$, and k.)

---

[21] Indeed, define $A(\cdot, \cdot)$ as follows: $A((F, D, r), (t, \sigma)) \stackrel{def}{=} V((D, F, NO, t), r, \sigma)$. Notice now that A is polynomial-time: in fact, V is the verifier of a CS proof system with a random string. Notice also that $|\sigma|$ is polynomially bounded in $|F|$, $|D|$ and $|r|$: in fact $q = (D, F, NO, t)$, $|t| \leq 2|F|$, and $|\sigma| \leq (|q|k)^{c_3}$.

Because each phase is implementable in time polynomial in $|F|$, $|D|$, k, and t, there exists a polynomial Q such that $\mathcal{D}(\mathcal{F})$ outputs a satisfying assignment of $\mathcal{F}$ in $Q(|F|,|D|,k,t)$ steps.

Finally notice that the above 4-phase procedure can be converted to a SAT decider by interleaving two different computations. In the first, an exhaustive search is conducted for deciding whether $\mathcal{F}$ is satisfiable. In the second, $\mathcal{F}$ is interpreted as a co-input of F, and the above 4-phase procedure is run. The so-modified $\mathcal{D}$ halts when either computation halts, and outputs what the halting computation does. ∎

4.1 Remarks

AN ALTERNATIVE FORMULATION. As we said, any CS checker $\mathcal{C}$ immediately yields an oracle-calling algorithm that, on input a formula F (a complexity meter D, and a security parameter k), and access to a one-sided efficient heuristic H, computes a co-input $\mathcal{F}$ and obtains $H(F)$ and $H(\mathcal{F})$.

With this is mind, we can rephrase Theorem 4' as follows (and obtain — implicitly— a definition of a CS checker that is more closely tailored to our implemetation).

Corollary 2': If CS proof-systems sharing a random string exist, then there exist (1) a polynomial-time oracle-calling algorithm $\mathcal{C}^{(\cdot)}(\cdot, \cdot, \cdot)$ that, whenever its first input is a CNF formula, F, queries its oracle at most twice: once about F, and possibly a second time about a second CNF formula $\mathcal{F}$, (2) a SAT decider $\mathcal{D}$, and a polynomial $Q(\cdot, \cdot, \cdot, \cdot)$ such that,

$\forall$ one-sided heuristics H for SAT, $\forall F \in CNF$, $\forall$ reasonable SAT deciders D solving F in $\leq 2^{2|F|}$ steps, $\forall$ sufficiently long unary integers k, the following two properties hold:

1. *Individual-Complexity Preservation.* If H is correct on F and $\mathcal{C}^H(F,D,r)$ queries H about a second CNF formula $\mathcal{F}$, then

$$\#\mathcal{D}(\mathcal{F}) \leq Q(|F|,|D|,|r|,\#D(F)).$$

2. *Computational Meaningfulness.* $\mathcal{C}^H(F,D,r)$ produces one of the following three outputs:

    (a) a satisfying assignment for F
    (i.e., a proof that F is satisfiable)
    (b) a CS proof, relative to (P,V) and reference string r, of $D(F) = NO$
    (i.e., evidence that F is not satisfiable)
    (c) a formula $\mathcal{F}$ such that, by construction, either F or $\mathcal{F}$ is satisfiable, and yet $H(F) = H(\mathcal{F}) = NO$
    (i.e., a proof that H is not correct).

Unlike Blum checkers, the above oracle-calling algorithm $\mathcal{C}$ does not provide answers that are correct with arbitrarily high probability (computed over its possible coin tosses). The type-(a) and type-(c) outputs of $\mathcal{C}$ are error-less, at least in the sense that any error here can be efficiently detected. But a type-(b) output of $\mathcal{C}$, interpreted as a (computationally) meaningful explanation that F is non-satisfiable, may be wrong in a non-easily detectable manner: if F is satisfiable, $\mathcal{C}$ could output a "false" CS proof of $D(F) = NO$ with positive probability. However, this probability is reasonably high only if an enormous amount of computation is performed, And in our application, all computation is performed by $\mathcal{C}$, which is polynomial-time, and by oracle H that is also polynomial-time. Therefore, the probability of a false type-(b) output is be absolutely negligible.

ANOTHER ADVANTAGE OF ONE-SIDED HEURISTICS. Our CS checkers only deal with one-sided heuristics for SAT. As already discussed, given the one-sided nature of $\mathcal{NP}$, this is a natural choice. On the other hand, could we have dealt with heuristics just outputting YES (i.e., "satisfiable, but with no proof") or NO?

So far, because of the self-reducibility property of $\mathcal{NP}$-complete problems, choosing between either types of heuristics has often been a matter of individual taste. Indeed, it is well known that a decision oracle for SAT can, in polynomial-time, be converted to a search oracle for SAT. As we explain below, however, this "equivalence" between decision and search relative to $\mathcal{NP}$-complete languages may cease to hold when one demands, as we do, that our reductions preserve the complexity of individual inputs, rather than just that of complexity classes.

When dealing with a one-sided efficient heuristics H for SAT, assuming that H is correct on F, we need only to take care of complexity preservation when $H(F) \neq NO$. In fact, if $H(F) \in SAT(F)$, then there is no need to call H on any co-input $\mathcal{F}$, and thus there is no complexity to be preserved. Presumably, however, if H outputs just YES and NO, we should care about preserving F's complexity also when $H(F)$ is correct and $H(F) = YES$. Now, to convince ourselves that $H(F) = YES$ is correct, we could run the self-reducibility algorithm, calling H on a sequence of formulae $F_1, \ldots, F_n$ (obtained by "fixing" a new variable each time), so as to find a satisfying assignment of F, or prove that H is wrong (on either F or some F_i). The problem is, however, that this self-reducibility process may not be complexity-preserving: It may be the case that our F is relatively easy, while some of the F_i's are very hard. Indeed, it is conceivable that it is the "degree of freedom" of the variables of the original formula F that make it easy to decide (without finding any $\mathcal{NP}$-proof of it) that F is satisfiable. However, after sufficiently many variables of F have been fixed, the difficulty of deciding satisfiability may grow dramatically high (though later on, when sufficiently many variables have been fixed, it will dramatically drop).

EXTRA COMPLEXITY PRESERVATION. Notice that, in the implementation of the proof of Theorem 4', the CS checker preserve the complexity of the original input F in a much closer manner than demanded by our definition. Indeed, a co-input

$\mathcal{F}$ in some sense, F' consists of the very encoding of the computation of the complexity meter D on input F.[22]

ONE ADDITIONAL APPLICATION. We believe that complexity preservation, in different formulations, will be useful to other contexts as well. In particular, it will enhance the meaningfulness of many reductions in a complexity setting. For instance, using complexity preservation, [13] present a more refined notion of a proof of knowledge [22] [20] [9] [1].

ONE OPEN PROBLEM. Is it possible to (define and) construct CS checkers that, when given an heuristic H and an input x, also receive a concise algorithmic representation of a ("non-trivial") set S, and call H only on x and elements of S? Such checkers could still be allowed to output a proof that the given heuristic H is wrong.[23] But, if H happens to be correct on S, and the given input happens to belong to S, they should output a "validation" for $H(x)$ (rather than a proof that H is wrong).

References

1. M. Bellare and O. Goldreich. *On Defining Proofs of Knowledge.* Proc. CRYPTO 92, Lecture Notes in Computer Science, Vol. 740, Springer Verlag, 1993, pp. 390-420.
2. M. Bellare and P. Rogaway. *Random Oracles are Practical: a Paradigm for Designing Efficient Protocols.* 1st Conference on Computer and Communications Security, ACM, pp. 62–73, 1993.
3. M. Blum and S. Kannan. *Designing Programs that check their work.* Proc. 21st Symposium on Theory of Computing, 1989, pp. 86-97.
4. M. Blum, M. Luby, and R. Rubinfeld. *Self-Testing and Self-Correcting Programs, With Applications to Numerical Problems.* Proc. 22nd ACM Symp. on Theory of Computing, 1990, pp. 73-83.
5. M. Blum and S. Micali. *How to Generate Cryptographically-Strong Sequences of Pseudo-Random Bits.* SIAM J. on Comp. vol 13, 1984
6. R. Canetti, O. Goldreich, and S. Halevi. In Preparation. 1998
7. S. Cook. *The Complexity of Theorem Proving Procedures.* Proc. 3rd Annual ACM Symposium on Theory of Computing, 1971, pp. 151-158.
8. U. Feige, A. Fiat, and A. Shamir. *Zero-knowledge Proofs of Identity.* Proc. of 19th Annual Symposium on Theory of Computing, 1987, pp. 1987.
9. A. Fiat and A. Shamir. *How to Prove Yourselves: Practical Solutions of Identification and Signature Problems.* Proc. Crypto 86, Springer- Verlag, 263, 1987, pp.186-194.
10. O. Goldreich, S. Goldwasser, and S. Micali. *How To Construct Random Functions.* J. of ACM 1986
11. S. Goldwasser, S. Micali, and R. Rivest, *A Digital Signature Scheme Secure Against Adaptive Chosen-Message Attacks,* SIAM J. Comput., Vol 17, No. 2, April 1988, pp. 281-308.

---

[22] We wonder whether this may yield a preferable formulation of complexity preservation.

[23] after all, an heuristic wrong everywhere should not be validated.

(A preliminary version of this article appeared with the title "A paradoxical solution to the signature problem" in Proc. of 25th Annual IEEE Symposium on the Foundations of Computer Science, FL, November 1984, pp. 464-479.)
12. O. Goldreich, S. Micali, and A. Wigderson.
13. S. Halevi and S. Micali. *A Stronger Notion of Proofs of Knowledge.* Unpublished Manuscript, 1997.
14. J. Hastad, R. Impagliazzo, L.A. Levin and M. Luby. *Construction of Pseudorandom Generators from any One-Way Function.* To appear is SIAM J. On Comp.
(This combines the works of Impagliazzo, Luby, and Levin, 21st Annual Symposium On Theory of Computing, 1989, and that of Hastad, 22nd Annual Symposium On Theory of Computing, 1990.)
15. R. Karp. *Reducibility among combinatorial problems.* Complexity of Compuyer Computations, R. Miller and J. Thatcher eds., Plenum, New York, 1972, pp. 85-103.
16. R. Impagliazzo, J. Hastad, L. Levin, and M. Luby. *Pseudo-Random Generation under uniform Assumptions.* STOC 1990.
17. L. Levin. *Universal Sequential Search Problems.* Problems Inform. Transmission, Vol. 9, No. 3, 1973, pp. 265-266.
18. R. Lipton. *New Directions in Testing.* Distributed Computing and Cryptography. (J. Feigembaum and M. Merritt Ed.) Vol. 2 of Dimacs Series in Discrete Mathematics and Theory of Computer Science. (Preliminary version: manuscript 1989.)
19. S. Micali, *A Concrete Construction Of Computationally Sound Checkers.* MIT-LC TM 579, 1998.
20. M. Tompa and H. Woll. *Random Self-Reducibility and Zero-knowledge Interactive Proofs of Possession of Information.* Proc. 28th Conference on Foundations of Computer Science, 1987, pp. 472-482.
21. C. Cachin, S. Micali and M. Stadler. *Computational Private Information Retrieval Systems With Poly-Logarithmic Amount of Communication.* Manuscript in preparation. 1998
22. S. Goldwasser and S. Micali and C. Rackoff. *The Knowledge Complexity of Interactive Proof Systems.* SIAM J. Comput., 18, 1989, pp. 186-208. An earlier version of this result informally introducing the notion of a proof of knowledge appeared in Proc. 17th Annual Symposium on Theory of Computing, 1985, pp. 291-304. (Earlier yet versions include "Knowledge Complexity," submitted to the 25th Annual Symposium on the Foundations of Computer Science, 1984.)
23. C. Lund and L. Fortnow and H. Karloff and N. Nisan. *Algebraic Methods for Interactive Proof Systems.* Proc. 22nd STOC, 1990.
24. S. Micali. *CS Proofs.* Proc. 35th Annual Symposium on Foundations of Computer Science, 1994, pp.
(An earlier version of this paper appeared as Technical Memo MIT/LCS/TM-510. Earlier yet versions were submitted to the 25th Annual Symposium on Theory of Computing, 1993, and the 34th Annual Symposium on Foundations of Computer Science, 1993.)
25. A. Shamir. *IP = PSPACE.* Proc. 31st IEEE Foundation of Computer Science Conference, 1990, pp. 11-15.

Reasoning About the Past

Mogens Nielsen

BRICS*
Department of Computer Science
University of Aarhus, Denmark

Abstract. In this extended abstract, we briefly recall the abstract (categorical) notion of bisimulation from open morphisms, as introduced by Joyal, Nielsen and Winskel. The approach is applicable across a wide range of models of computation, and any such bisimulation comes automatically with characteristic logics and games, which in their general formulations treat the future and the past (of computations) on an equal footing. This raises a number of questions concerning properties of such logics and games for concrete well known models from concurrency theory, in particular questions on the power of reasoning about the past.

1 Introduction

Concurrency theory is based on a number of different formal models of computation, e.g. labelled transition systems [8], synchronization trees [12], Petri nets [18], event structures [24], and trace structures [11], just to name a few. Similarly, concurrency theory deals with an abundance of notions for behavioural equivalence, e.g. bisimulation [12], trace equivalence [4], and pomset equivalence [19].

During the past decade, attempts have been made in order to understand the relationships between the confusingly many different concepts within concurrency theory. Here, we shall recall briefly the categorical approach of Winskel and Nielsen, with special emphasis on the abstract notion of bisimulation in terms of open morphisms [6], applicable to a wide range of models. Furthermore, we shall illustrate how to obtain automatically characteristic logics and games for such models, following [27].

The general idea is to equip a model of concurrency with morphisms, to be thought of as "simulations" between objects, so that it forms a category [10]. This idea has proven to be be useful in many different ways, as illustrated in [25]. For instance, many operations of process calculi can be understood as universal constructions (like product and coproduct), providing a way of understanding (in terms of morphisms) how the behaviour of process relates to its components. Also, categorical notions may be used in relating different models, typically in the form of (co)reflections, stating how one model may be "embedded" into

---

* Basic Research in Computer Science,
  Centre of the Danish National Research Foundation.

another, i.e. "viewed as a more abstract version", and allowing properties and constructions to be transferred between models via adjoints.

The "simulation" morphisms studied in [25] are generally too weak to yield useful abstract equivalences between processes. However, in [6], it is shown that the equivalences generated by special *open* morphisms (roughly speaking those morphisms which reflect as well as preserve behaviour), for general reasons, i.e. reasons applicable to all categories of models, yield notions of equivalence with a number of useful properties.

The definition of open morphisms is parameterised on a notion of computation paths for a given model. In this extended abstract, we present just two examples of model categories, tree structures (an example of an interleaving model), and event structures (an example of an independence model). For interleaving models like tree structures, we follow tradition and take as computation paths (or runs) sequences of consecutive transitions, which we can think of as picked out by a morphism from a string of action labels, with extension as standard extension of strings of action labels. For an independence model like event structures, we take computation paths as configurations, or more generally as a morphism from a pomset [19] to the event structure. Following Pratt, a computation path in the form of a pomset can be extended in "width" (adding concurrent events) as well as "height" (adding causally dependent events).

For the examples treated here, our notions of bisimulations from open morphisms specialise to familiar concepts; in particular, on tree structures with strings of actions as paths we obtain Park and Milner's strong bisimulation. For general reasons, i.e. from the existence of co-reflections between models, these carry over to other models, e.g. from tree structures to labelled transition systems, and from event structures to labelled Petri nets [16].

It should be noted that also other familiar behavioural equivalences are captured by the open morphism approach, e.g. Hoare's trace equivalence and Milner's weak bisimulation, both of which may be obtained by slightly changing the notion of path extension from the one presented here [14]. Also, the open morphism approach has been applied successfully to different categories of models, e.g. probabilistic systems [14], and timed systems [5].

Rather than having our bisimulations defined in terms of two parameters, a model and a path category, it was suggested in [6] to study presheaves as models derived directly from path categories. In recent work by Winskel and others, these presheaf models have been used successfully in dealing with higher-order models in concurrency [26,2]. Intuitively, a presheaf represents the effect of gluing together a set of computation paths to form a nondeterministic computation, and hence can be looked upon as labelled transition systems, in which the labels are morphisms of path extension. Following [27] this yields logical and game-theoretic characterisations of open morphisms and their bisimulations on presheaves. As we shall see, these operational characterisations in their general formulation treat the future and past of paths on an equal footing. Many models and their notion of bisimulation can be understood in a uniform way via their representation as presheaves, and via this our characterisations can be specialised to concrete models like tree structures and event structures.

The characteristic logics take the form of Hennessy-Milner like modal logics, with modalities indexed by path morphisms (path extensions, future modalities) and their inverses (path projections, past modalities). The idea of modal operators referring to the past is certainly not new, and it has been studied extensively also in the context of concurrency during recent years [9,23,17]. Here we focus on a few generalisations of the characteristic logics with past modalities arising form our general categorical setup, and we illustrate the power of past modalities in the setting of independence models with a few undecidability results.

2 Models, Morphisms, and Computation Paths

Here we quickly describe the models and notions of computation paths we'll use as illustrating examples. For the sake of presentation, we start out with the most general of our models, the *(labelled) event structures* of [13], and view the remaining notions as special instances of these. An event structure is to be thought of as describing the complete pattern of event occurrences of a nondeterministic and concurrent computation. It consists of a set of individual event occurrences, equipped with structure relating any two events as being either causally dependent, in conflict, or concurrent (sometimes also referred to as "independent").

Definition 1. *Define an* event structure *to be a structure* $(E, \leq, \#, L, l)$ *consisting of a set E, of* events *which are partially ordered by $\leq$, the* causal dependency *relation, a binary, symmetric, irreflexive relation $\# \subseteq E \times E$, the* conflict *relation, which satisfy for all $e, e', e'' \in E$*

$$\{e' \mid e' \leq e\} \text{ is finite}, \qquad e \# e' \leq e'' \Rightarrow e \# e'',$$

a set of labels, *L, and a labelling function $l : E \to L$. Say two events $e, e' \in E$ are* concurrent, *and write e co e', if $\neg(e \leq e'$ or $e' \leq e$ or $e \# e')$.*

The finiteness assumption restricts attention to discrete processes where an event occurrence depends only on finitely many previous occurrences. The axiom on the conflict relation expresses that if two events causally depend on events in conflict then they too are in conflict. The labelling function as usual represents information about the "type" of individual events.

Guided by our interpretation we can formulate a notion of computation state of an event structure $ES = (E, \leq, \#, L, l)$. Taking a computation state of a process to be represented by the set of events which have occurred in the computation, we define the states (usually called the *configurations* and denoted $\mathcal{C}(ES)$) to be those subsets $x \subseteq E$ satisfying

conflict-free: $\forall e, e' \in x. \neg(e \# e')$, and
downwards-closed: $\forall e, e'. e' \leq e \subset x \to e' \in x$.

Morphisms on event structures are defined in terms of a function telling how the events of one system may be simulated in another [24].

Definition 2. Let $ES = (E, \leq, \#, L, l)$ and $ES' = (E', \leq', \#', L, l')$ be event structures over a common labelling set L. A morphism from ES to ES' consists of a function η from E to E' on events which respects labelling, i.e. for all $e \in E$, $l(e) = l'(\eta(e))$, and satisfies

$$x \in \mathcal{C}(ES) \Rightarrow \eta x \in \mathcal{C}(ES') \,\&$$
$$\forall e_0, e_1 \in x.\ \eta(e_0) = \eta(e_1) \Rightarrow e_0 = e_1.$$

Let $\mathbf{E}_L$ denote the category of event structures over labelling set L with morphisms as described above and composition as composition of functions.

Notice, that configurations of an event structure can be thought of as *pomsets* in the sense of Pratt [19]. So, the subcategory of finite pomsets naturally play the role of finite computation paths of event structures. Pomsets are special event structures with empty conflict-relations. More precisely, define $\mathbf{Pom}_L$ to be the full subcategory of event structures with labels in L, and objects finite event structures with empty conflict-relation (i.e. pomsets).

Morphisms between pomsets, got by restricting those of event structures, are injective functions which send downwards-closed sets to downwards-closed sets. Thus a morphism from pomset P to pomset Q may not just extend P by extra events but also relax the causal dependency relation; two events causally related in P may have images no longer causally related in Q.

Similarly, we may pick out a subcategory of event structures as a model for sequential computations, by insisting on the co-relation being empty. In this way any two events are either causally ordered or in conflict, i.e. may be viewed as a *(labelled) tree structure* (also referred to as a synchronisation tree [12]. Define $\mathbf{T}_L$ to be the full subcategory of event structures with labels in L, and objects event structures with empty co-relation (i.e. tree structures).

The notion of a computation (or a run) of a tree structure is traditionally its (finite) paths, and hence we may choose naturally our notion of computation paths the subcategory of *strings* over L. Define $\mathbf{Str}_L$ to be the full subcategory of tree structures with labels in L, and objects finite event structures with empty co- and conflict-relation (i.e. strings).

Morphisms between such paths, inherited from event structures, correspond to extensions of the associated strings. So we can identify the category of such paths with the (partial-order) category of strings L^*, where a morphism from string s to string t corresponds to s being an initial prefix of t.

3 Open Morphisms and Bisimulation

Assume a category of models $\mathbf{M}$ (as examples, think of either $\mathbf{T}_L$ or $\mathbf{E}_L$ from the previous section), and a subcategory $\mathbf{P} \hookrightarrow \mathbf{M}$ of paths (as corresponding examples, think of either $\mathbf{Str}_L$ or $\mathbf{Pom}_L$).

Define a *path* in an object X of $\mathbf{M}$ to be a morphism $p: P \to X$, in $\mathbf{M}$, where P is an object in $\mathbf{P}$. A morphism $f: X \to Y$ in $\mathbf{M}$ takes such a path p in X to the path $f \circ p: P \to Y$ in Y. The morphism f expresses the sense in which Y

simulates X; any computation path in X is matched by the computation path $f \circ p$ in Y.

Our notion of *open* morphisms demand a stronger condition of a morphism $f : X \to Y$ expressed in the following *path-lifting condition*. For reasons which will become clear in the following, we define open morphisms with respect to a subclass of morphisms $\mathbf{P}_0$ of $\mathbf{P}$.

Whenever, for $m : P \to Q$ a morphism in $\mathbf{P}_0$, a "square"

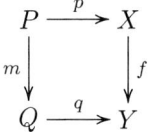

in $\mathbf{M}$ commutes, i.e. $q \circ m = f \circ p$, meaning the path $f \circ p$ in Y can be extended via m to a path q in Y, then there is a morphism p' such that in the diagram

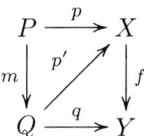

the two "triangles" commute, i.e. $p' \circ m = p$ and $f \circ p' = q$, meaning the path p can be extended via m to a path p' in X which matches q. When the morphism f satisfies this condition we shall say it is $\mathbf{P}_0$-*open*.

Say two objects X_1, X_2 of $\mathbf{M}$ are $\mathbf{P}_0$-*bisimilar* iff there is a *span* of $\mathbf{P}_0$-open morphisms f_1, f_2:

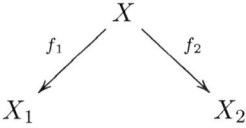

Proposition 1. *Two tree structures over the same labelling set L, are $\mathbf{Str}_L$-bisimilar iff they are strongly bisimilar in the sense of Milner.*

Two event structures over the same labelling set L, are $\mathbf{Pom}_L$-bisimilar iff they are strongly history preserving bisimilar in the sense of Bednarczyk.

For the exact definitions and proofs we refer here to [6,16] (and the characterisations provided in the following). From general categorical results, $\mathbf{P}$-bisimilarity is an equivalence relation for all models with pullbacks (including $\mathbf{T}_L$ and $\mathbf{E}_L$), which is preserved by co-reflections, e.g. unfoldings from labelled transition systems into $\mathbf{T}_L$ [6].

In checking whether a morphism is $\mathbf{P}$-open or for $\mathbf{P}$-bisimulation, for a path category $\mathbf{P}$, it suffices to consider a restricted class of morphisms, sufficient to generate the category $\mathbf{P}$.

Definition 3. *Let* **P** *be a category. Let* **P**$_0$ *consist of a subclass of morphisms of* **P**. *Say* **P**$_0$ *generates* **P** *iff the only subcategory of* **P** *which includes* **P**$_0$ *and all isomorphisms of* **P** *is* **P** *itself.*

Example 1. Morphisms of **Str**$_L$ and **Pom**$_L$ are generated as follows. *Strings:* the category **Str**$_L$ is generated by the set of morphisms representing the extension of a string by a single label.
Pomsets: The category **Pom**$_L$ is generated by the class of "atomic" morphisms of two kinds:

prefix: morphisms $m : P \to Q$ in **Pom**$_L$ expressing that pomset P is a prefix of pomset Q, where Q contains one more event than P; so m expresses that pomset Q consists of a copy of P with one additional event adjoined on top;

augmentation: morphisms $m : P \to Q$ in **Pom**$_L$ expressing that pomset P is an augmentation of pomset Q, with the causal dependency relation in P containing one more pair than that of Q.

Proposition 2. *Suppose* **P** *is generated by a subclass of morphisms* **P**$_0$.

1. *A morphism of* **M** *is* **P**-*open iff it is* **P**$_0$-*open.*
2. *Objects of* **M** *are* **P**-*bisimilar iff they are* **P**$_0$-*bisimilar.*

This proposition [27] allows us to limit the quantification over path extensions in the definition of open morphisms, but the definition of **P**-bisimilarity is still rather abstract. As we shall see in the following, we may obtain more operational and concrete characterisations via models of presheaves over **P** viewed as transition systems.

4 Presheaf Models and Transition Systems

Given a path category **P**, the category $\widehat{\mathbf{P}}$ of presheaves over **P** [10] consists of functors $\mathbf{P}^{op} \to \mathbf{Set}$ (where **Set** is the category of sets with functions) as objects, and natural transformations between functors as morphisms. Intuitively a presheaf $F : \mathbf{P}^{op} \to \mathbf{Set}$ can be thought of as specifying for a typical path object P the set $F(P)$ of paths from P. It acts on a morphism $m : P \to Q$ in **P** to give a function $F(m) : F(Q) \to F(P)$ saying how Q-paths restrict to P-paths.

In the following we assume that our path categories have an initial object I. This assumption is satisfied for our two examples of path categories, with the empty string (pomset) as initial object in **Str**$_L$ (**Pom**$_L$). A *rooted presheaf* is a presheaf F in which $F(I)$ is a singleton.

Let us see how a model, like a tree structure or an event structure, gives rise to a rooted presheaf. Consider a category of models **M** and a choice of path category forming a subcategory $\mathbf{P} \hookrightarrow \mathbf{M}$. There is a *canonical functor* from the category of models **M** to the category of presheaves $\widehat{\mathbf{P}}$. It takes an object X

of **M** to the presheaf $\mathbf{M}(-,X)$—more intuitively, it takes the model X to the presheaf which for each path object P yields the set of paths $\mathbf{M}(P,X)$ from P into X. The canonical functor takes a morphism $f : X \to Y$ in **M** to the natural transformation
$$\mathbf{M}(-,f) : \mathbf{M}(-,X) \to \mathbf{M}(-,Y)$$
whose component at an object P of **P** is the function $\mathbf{M}(P,X) \to \mathbf{M}(P,Y)$ taking p to $f \circ p$—intuitively, a path $p : P \to X$ in X is taken to a path $f \circ p : P \to Y$ in Y.

In a presheaf model $\widehat{\mathbf{P}}$ we can consider the image of **P** under the Yoneda embedding as its path category, and then apply the general definition of section 3, to obtain the class of **P**-open morphisms of the presheaf category. They form a category of *open maps* of the topos $\widehat{\mathbf{P}}$, in the sense of Joyal and Moerdijk. For general reasons (basically because the embeddings are full and dense), we get the following from [27].

Proposition 3. *(i) Two tree structures, over labelling set L, are $\mathbf{Str}_L$-bisimilar (i.e. strong bisimilar) iff their corresponding presheaves, under the canonical embedding, are related by a span of open morphisms in the full subcategory of rooted presheaves of $\widehat{\mathbf{Str}_L}$.*
(ii) Two event structures, over labelling set L, are $\mathbf{Pom}_L$-bisimilar (i.e. strong history-preserving bisimilar) iff their corresponding presheaves, under the canonical embedding, are related by a span of open morphisms in the full subcategory of rooted presheaves of $\widehat{\mathbf{Pom}_L}$.

Importantly, there is a simple and intuitive way of thinking of rooted presheaves as labelled transition systems. Formally, a *transition system* is a structure $(S, i, L, tran)$ where

- S is a set of *states*,
- $i \in S$ the *initial state*,
- L is a set of *labels*, and
- $tran \subseteq S \times L \times S$ is the *transition relation*. As usual, we write $s \xrightarrow{a} s'$ to indicate that $(s, a, s') \in tran$.

Assume that $\mathbf{P}_0$ is a subclass of morphisms of **P**. It will be helpful to think of a rooted presheaf over **P** as a transition system with labels taken from morphisms of $\mathbf{P}_0$:

Definition 4. *Let X be a rooted presheaf over **P**. Define its $\mathbf{P}_0$-transition system, denoted by $\mathcal{T}_{\mathbf{P}_0}(X)$ to consist of:*

- *states: (P,p) where P is an object of **P** and $p \in X(P)$,*
- *initial state: the unique member of $X(I)$,*
- *labels: $\mathbf{P}_0$,*
- *transitions: $(P,p) \xrightarrow{m} (Q,q)$ whenever $m : P \to Q$ in $\mathbf{P}_0$ and $(Xm)(q) = p$.*

The construction of $T_{\mathbf{P}_0}(X)$ allows us to view a rooted presheaf as a much more familiar model within computer science, a transition system. The states should be thought of as an abstract set of possible runs of the paths of $\mathbf{P}$, and the transitions simply represent how such runs extend each other. Notice that rooted presheaves over strings, $\widehat{\mathbf{Str}_L}$, correspond exactly to tree structures, $\mathbf{T}_L$, whereas $\widehat{\mathbf{Pom}_L}$ contains many structures not representable in $\mathbf{E}_L$, for details see [6].

5 Game and Logic Characterisations

Viewing presheaves as transition systems, we may lift existing notions of characteristic games and logics of the kind discussed in [12] (for logic) and [21] (for games and logic).

Let $T_0 = (S_0, i_0, L_0, tran_0)$ and $T_1 = (S_1, i_1, L_1, tran_1)$ be two transition systems. The game $G(T_0, T_1)$ played by two players (I and II) is defined as follows. The configurations of the game consists of pairs of states $(s_0 \in S_0, s_1 \in S_1)$ with (i_0, i_1) as the starting configuration. A *play* consists of a sequence of alternating moves by the two players (Player I making the first move), where a move consists of a choice of a transition from one of the systems, according to the following game rules:

At configuration (s_0, s_1)

- either Player I chooses a transition $s_0 \xrightarrow{a} s'_0$, after which Player II chooses a transition $s_1 \xrightarrow{a} s'_1$, and the game continues at configuration (s'_0, s'_1),
- or Player I chooses a transition $s_1 \xrightarrow{a} s'_1$, after which Player II chooses a transition $s_0 \xrightarrow{a} s'_0$, and the game continues at configuration (s'_0, s'_1)
- or Player I chooses a transition $s'_0 \xrightarrow{a} s_0$, after which Player II chooses a transition $s'_1 \xrightarrow{a} s_1$, and the game continues at configuration (s'_0, s'_1)
- or Player I chooses a transition $s'_1 \xrightarrow{a} s_1$, after which Player II chooses a transition $s'_0 \xrightarrow{a} s_0$, and the game continues at configuration (s'_0, s'_1).

Player I wins a play if Player II gets stuck, i.e. at some point cannot match a move by Player I according to the rules of the game. All other plays are won by Player II, i.e. all infinite plays and plays where Player at some point cannot make a move. A (history-free) *strategy* for a player is a set of rules which for each configuration tells the player how to proceed, i.e. for Player II a rule will associate to each configurations and a choice of back or forth transition in one of the systems by Player I, a set of matching transitions in the other system. A strategy is *winning* for a player, if he or she wins every play played according to the strategy.

Intuitively, the two players have different goals in game $G(T_0, T_1)$: Player I wants to show that the two transition systems are distinguishable, Player II that they are not. Generalising the results from [15] we get:

Proposition 4. *Let $\mathbf{P}_0$ generate $\mathbf{P}$. Two rooted presheaves in $\widehat{\mathbf{P}}$ are bisimilar iff Player II has a winning strategy in the game defined by their two $\mathbf{P}_0$-transition systems.*

Example 2. Games for synchronization trees and event structures are obtained from their canonical embeddings in presheaf categories. However, using properties of these concrete models, we may obtain even simpler game characterisations.
Tree structures: First of all, following the arguments of [3], in this case we can characterise bisimulation by restricting our games to only forwards moves, i.e. transitions labelled by extension morphisms. Secondly, we can apply Proposition 4 and further restrict the games to allow only moves involving extension with a single symbol, and finally such a morphism in the path category is determined by its domain and the label of the extended single symbol. Hence, we have obtained the original Stirling games characteristic for tree structures.
Event structures: Again, applying Proposition 4 strong history-preserving bisimulation between event structures is characterised by games with moves restricted to transitions labelled by "atomic" morphisms, i.e. prefix and augmentation morphisms. Furthermore, it can be shown that our games can be even further restricted to only forwards and backwards transitions labelled by "atomic" prefix morphisms [6,15].

Similarly, as shown in [6], we may obtain characteristic logics in our general setting of presheaf models.

Define $\mathbf{P}_0$-*assertions* by:

$$A ::= \overline{\langle m \rangle} A \mid \langle m \rangle A \mid \neg A \mid \bigwedge_{j \in J} A_j$$

where m is a morphism in $\mathbf{P}_0$, and J is an indexing set, possibly empty and *not* restricted to being finite. The modality $\overline{\langle m \rangle}$ is a "backwards" modality, while $\langle m \rangle$ is a "forwards" modality. We define the semantics with respect to a transition system with labelling set $\mathbf{P}_0$:

- $s \models \langle m \rangle A$ iff $\exists s'.\ s \xrightarrow{m} s'$ and $s' \models A$
- $s \models \overline{\langle m \rangle} A$ iff $\exists s'.\ s' \xrightarrow{m} s$ and $s' \models A$
- the boolean operations receive their expected meanings.

Notice that the logic takes the form of a kind of Hennessy-Milner logic, with modalities indexed by morphisms, and with a dual notion of backwards modalities. The following general characterisation result was shown in [27].

Proposition 5. *Let $\mathbf{P}_0$ generate $\mathbf{P}$. Two rooted presheaves in $\widehat{\mathbf{P}}$ are bisimilar iff their $\mathbf{P}_0$-transition systems satisfy the same assertions.*

Example 3. We determine a satisfaction relation for tree structures and event structures via their canonical embeddings in presheaf categories $\widehat{\mathbf{Str}_L}$, $\widehat{\mathbf{Pom}_L}$. Again, we may use properties of our concrete models in obtaining simplified characteristic logics.
Tree structures: Traditional Hennessy-Milner logic arises by reducing the seemingly richer logic based on all extension morphisms in $\mathbf{Str}_L$. Firstly, as remarked

11. Mazurkiewicz, A., Basic notions of trace theory. In de Bakker, de Roever and Rozenberg (eds.), Linear Time, Branching Time and Partial Orders in Logics and Models for Concurrency, Springer Lecture Notes in Computer Science 354, pp. 285–363, 1988.
12. Milner,A.J.R.G., *Communication and concurrency.* Prentice Hall, 1989.
13. Nielsen, M., Plotkin, G. and Winskel, G., Petri nets, Event structures and Domains, part 1. Theoretical Computer Science 13, pp. 85–108, 1981.
14. Nielsen, M., Cheng, A., Observing Behaviour Categorically. Proc. of FST&TCS'15, Springer Lecture Notes in Computer Science 1026, pp. 263–278,1995.
15. Nielsen, M., and Clausen, C., Bisimulations, Games, and Logic. Proc. of CONCUR'94, Springer Lecture Notes in Computer Science 836, pp. 385–400, 1994.
16. Nielsen, M., and Winskel, G., Petri nets and bisimulations. Theoretical Computer Science, 153, pp. 211–244, 1996.
17. Penczek, W., Kuiper, R., Traces and Logic. In *The Book of Traces*, eds. Diekert, V., Rozenberg, G., World Scientific, pp. 307–390, 1994.
18. Petri, C.A. Kommunikation mit Automaten. PhD thesis, Institut für Instrumentelle Mathematik, Bonn, Germany, 1962.
19. Pratt, V.R., Modelling concurrency with partial orders. International Journal of Parallel Programming, 15(1), pp. 33–71, 1986.
20. Shields, M.W., Concurrent Machines, Comput. J., 28, pp. 449–465, 1985.
21. Stirling, C., Games and Modal Mu-Calculus. Lecture Notes in Computer Science 1055, pp. 298 ff., 1996.
22. Thiagarajan, P.S., Elementary Net Systems. Springer Lecture Notes in Computer Science 254, pp. 26–59, 1986.
23. Vardi, M.Y., The Taming of the Converse: Reasoning about Two-way Computations. Springer Lecture Notes in Computer Science 193, pp. 413–424, 1985.
24. Winskel, G., Event structures. Springer Lecture Notes in Computer Science 255, pp. 325–392, 1987.
25. Winskel, G., and Nielsen, M., Models for concurrency. In the Handbook of Logic in Computer Science, vol. IV, ed. Abramsky, Gabbay and Maibaum, Oxford University Press, pp. 1–148, 1995.
26. Winskel, G., A Presheaf Semantics of Value-Passing Processes. In the proc. of CONCUR'96, Springer Lecture Notes in Computer Science 1119, pp. 98–114, 1996.
27. Winskel, G., and Nielsen, M., Presheaves as Transition Systems, DIMACS Series in Discrete Mathematics and Theoretical Computer Science, 29, pp. 129–140, 1997.

Satisfiability – Algorithms and Logic
(Extended Abstract)

Pavel Pudlák*

Mathematical Institute
Academy of Sciences
Prague, Czech Republic

Abstract. We present some recent results on algorithms for satisfiability of k-*CNF* formulas: fastest probabilistic algorithms. We mention some results in proof complexity that can be used to derive lower bounds on classes of algorithms for satisfiability.

1 Introduction

The satisfiability problem is to determine for a given boolean formula whether there exists a satisfying assignment. In general we can consider formulas in arbitrary basis, but most attention is given to CNF's, ie., conjunctions of disjunctions of variables and negated variables. The disjunctions are called *clauses*, the variables and negated variables are called *literals*. The set SAT of satisfiable CNF's was the first set for which NP-completeness was proved [2] and it is still the most important NP-complete set. Since it is an NP-complete set we do not expect that there can be found very efficient algorithms for it. Most researchers believe that there are only exponential algorithms for SAT, still we would like to know how much exponential they must be in general. This problem seems to have less structure than many other combinatorial problems. Therefore more restricted classes are considered, namely k-SAT, which is SAT restricted to conjunctions of disjunctions of size at most k, also called k-*CNF*'s. It seems that the best algorithms can only achieve running time which is a fixed root of the number of all assignments.

In this lecture we will describe some recent developments in design of algorithms for SAT. In particular we shall mention new probabilistic algorithms which use much less time than known deterministic ones. Then we shall consider the popular area of proof complexity. Results in this area can be used to prove lower bounds for some classes of algorithms. Finally we shall mention use of quantum computations for SAT. Also we present a new algorithm for 3-SAT which has worse running time than previously know ones, but it introduces a new idea in this field.

---

* The author was partially supported by grant A1019602 of the Academy of Sciences of the Czech Republic and by a joint grant of NSF (USA) and MŠMT (Czech Rep.) INT-9600919/ME-103.

2 Deterministic Algorithms

2.1

Most of the proposed algorithms for SAT can be viewed as instantiations of the Davis–Putnam procedure.[1] Davis–Putnam procedure is an algorithm which searches the tree of all assignments to propositional variables and at each step it checks if there is a clause which is false for the partial assignment so far constructed and it checks if all clauses are satisfied. If one clause is false, the search of the branch stops and the algorithm searches another branch. This is repeated until a satisfying assignment is found or all branches are excluded. When the clauses are short (as in a k-CNF, for k constant) the algorithm abandons the partial assignment quite often and in this way it saves a lot of time. The various applications of Davis-Putnam procedure differ in the strategy by which we choose next variable in the search process. We can also view the process as modifying the formula at each step when we assign new variable: we omit all clauses that are true and delete all literals that are false in the remaining clauses. We backtrack when an empty clause appears. Eg. $(x_1 \vee x_2) \wedge (\neg x_1 \vee x_3 \vee x_4)$, after assigning $x_1 := 1$, becomes $x_3 \vee x_4$.

2.2

The best known such algorithm was described by Monien and Speckenmeyer [11] (see **Algorithm A** in the table below). The strategy of this algorithm is to choose the shortest clause (in some fixed ordering of clauses) and the first variable in this clause (in some fixed ordering of variables). Suppose the first shortest clause is $C = x_1 \vee \ldots \vee x_k$. The following partial assignments make this clause true $1, 01, 001, \ldots 0 \ldots 01$ (with $k-1$ zeros), while $0 \ldots 00$ (k zeros) makes it false. Note that these partial assignments will be actually used, since after assigning several 0's the clause will not be satisfied and it will still be the shortest one. Let $F_k(n)$ be the number of steps needed in the worst case. We get the following recurrence $F_k(n) \leq F_k(n-1) + F_k(n-2) + \ldots + F_k(n-k)$. This is too rough and can be easily improved by considering *autark* variables. These are variables which occur in all clauses with the same sign (ie. all without negation, resp. all with negation). For such variables we know that one of the values can be always used whatever values we use for other variables (positive, if the occurrences are positive and vice versa). More generally, an autark partial assignment is an assignment which makes any clause true if the clause contains some assigned variable. When considering assignments to variables to the clause C, first we check if one of the assignments $1, 01, 001, \ldots 0 \ldots 01$ is autark (this can be done efficiently). If there is one, we take it and we do not have to consider others. If not, then any assignment hits another clause and does not make it true. Such a clause is then shortened. Thus, except possibly for the first clause,

---

[1] Apparently this procedure should rather be attributed to Davis, Logemann and Loveland [4]. In this paper we shall use the traditional name having in mind that it may not reflect the history quite accurately.

in each step we either have a clause of length at most $k-1$, or use an autark assignment, or we have a clause of length k and in the previous step we used an autark assignment. The autark assignments save more time than having clauses of shorter length, so we can upper bound the number of steps by assuming that in each step we consider a clause of length $k-1$. Hence the recurrence is better: $F_k(n) \leq F_k(n-1) + F_k(n-2) + \ldots + F_k(n-k+1)$. In case of $k=3$, this gives Fibonacci numbers.

2.3

Improvements have been considered mainly for the case of 3-*SAT*, see [16,17] (**Algorithm B**). The use of Davis-Putnam procedures in all these algorithms gives the impression that this is the only possible way of designing a nontrivial algorithm. Therefore we propose below a new algorithm for 3-*SAT* based on a different idea.

Let the number of variables be n, w.l.o.g. suppose n is even and that all clauses have length exactly 3. The algorithm divides the variables into two equal parts (arbitrarily). For each of the parts takes the clauses which contain variables only from this part and lists all satisfying assignments. This is done by considering all $2^{n/2}$ assignments one by one in each part. (We cannot save much by using some sophisticated subroutine here, as the number of satisfying assignments can be so big.) For each of the two parts we do the following. Take a satisfying assignment a for this part and look at the clauses which contain exactly two variables in this part, thus having one variable in the other part. If the clause is not satisfied, then the remaining variable is forced to have a particular value. It may happen that in this way a variable from the other part is forced to be 0 by one clause and 1 by another. Then the assignment a cannot be extended to all variables and we discard it. Otherwise we extend it by all the forced values in the other part. Thus we get partial assignments which assigns all variables on one part and some on the other. Clearly, there is a satisfying assignment for the whole formula iff there is a pair of consistent partial assignment one from one part the other from the other part.

Thus the task of finding a satisfying assignment reduces to the following question. We have two lists of partial assignments, each of size at most $2^{n/2}$. We have to find a consistent pair consisting of a partial assignment from one list and a partial assignment from the other list. We suspect that algorithms for such problems have been considered, but we were not able to find any related result. We will show that one can solve this problem faster than using the trivial procedure running in time 2^n. We shall use RAM as the computation model. The bound that one gets from the algorithm that we will sketch below is worse than the bounds on previously considered algorithms for 3-*SAT*, but very likely it can be significantly improved.

Suppose we have two sets of strings of $\{0, 1, *\}$ of length n denoted by A and B. In A the strings have no $*$'s in the first half and in B the strings have no $*$'s in the second half. We want to find an $a \in A$ and a $b \in B$ which are consistent, ie., on each position where both have a non-$*$ they coincide. Let α be a suitable constant such that $0 < \alpha < 1/2$, the optimal value can be computed later. For a $b \in B$ which has less than $(1/2 - \alpha)n$ *'s it is possible to find a consistent string in A, if there is any, in about $2^{(1/2-\alpha)n}$ steps. To this end we use a data structure to represent A which is a binary tree of depth $\leq n/2$ with branchings corresponding to the first $n/2$ coordinates and the leaves labelled by the remaining parts of the strings. In the same way we can search for a consistent string for every $a \in A$ which has less than $(1/2 - \alpha)n$ *'s. Altogether this will need at most $2.2^{n/2}.2^{(1/2-\alpha)n} = 2^{n-\alpha n+1}$ steps. If we do not find a consistent pair we discard such strings. Now each $a \in A$ has at most αn 0's and 1's in the second half and each $b \in B$ has at most αn 0's and 1's in the first half. For each $a \in A$ we replace the string a by strings constructed as follows. Take a subset of coordinates in the first half of size $\leq \alpha n$ and replace all entries in the other coordinates in the first half by *'s. Do the same with strings in B with the coordinates in the second half. Let the new sets be A^* resp. B^*. Then, clearly, there is a consistent pair (a,b), $a \in A$, $b \in B$, iff $A^* \cap B^* \neq \emptyset$. Finding the intersection can be done efficiently, eg., by sorting the union of the two lists A and B, hence the main term in the estimate will be the size of the two lists which is

$$2^{n/2} \sum_{i \leq \alpha n} \binom{n/2}{i} \approx 2^{n/2 + H(\alpha)n/2},$$

where H is the binary entropy function. Thus the optimal choice for α is given by the equation $1 - 2\alpha = H(\alpha)$ and the running time of the algorithm is $O(2^{0.83n})$.

There are several things that one may try for getting an improvement: find a faster algorithm for finding consistent pairs, use random partition into two sets of variables, but the most interesting would be to find a way to iterate this *divide and conquer* strategy, as that may lead to a substantial shortening of the running time.

3 Probabilistic Algorithms

3.1

Recently we proposed a simple algorithm which is essentially the Davis-Putnam procedure with tested variables chosen in random order and assigned random independent values [12] (we shall call it **Algorithm C**). For 3-*SAT* the best known deterministic algorithm Schiermeyer [17] is faster, but for $k > 3$ our algorithm beats Monien–Speckenmeyer [11], which is apparently the fastest algorithm for this case that had been published.

To explain the analysis of this algorithm, the best is to start with the special case where there is an isolated satisfying assignment to a k-*CNF* formula. This means that in Hamming distance 1 there is no satisfying assignment. We sketch an estimate on the probability that this assignment is found.

For a satisfying assignment a to a CNF formula Φ we say that C is a *critical clause* of Φ on a, if exactly one literal of C is true on a. That literal and its variable will also be called *critical*. It is easily seen that for an isolated satisfying assignment each variable is critical in some clause. More generally, if a is an arbitrary satisfying assignment and there is no satisfying assignment which differs from a only in the value of x_i, then x_i is a critical variable of some clause for a.

The critical clauses help us to save some bits when searching for a satisfying assignment. If we assign successively values to variables and it happens that we have so far assigned values according to some satisfying assignment a and it also happens that we have assigned all variables of a critical clause for a except for its critical variable, then the value of the critical variable is determined and we do not have to make choice. Fix an a and a critical clause C of length k. If we take a random order then with probability $1/k$ the critical variable will be the last among the variables of C. If a is isolated, then every variable is critical for some clause so we shall save in the average n/k bits. It is not difficult to turn this intuitive argument to a rigorous proof of an upper bound $2^{(1-1/k)n+o(n)}$ for the expected running time of the algorithm finding an isolated satisfying assignment. For the decision problem we get a bound of this form with superexponentially small probability of error.

In general we do not know that there is an isolated satisfying assignment. If there is no isolated satisfying assignment, then each satisfying assignment has a neighbour and the less it is isolated the more neighbours it has. When there are many satisfying assignments, then we are very likely to hit one by the random choice. It turns out that the trade-off between isolation and number of satisfying assignments works well and we get exactly the same bound in general. This can be seen as follows. Consider some ordering of variables and an assignment a. Let x_i be a variable and let a' be the part of a before x_i. We shall call a variable x_i *forced*, if a' forces the value for x_i, because x_i is the last variable of some clause which is not yet satisfied by a', ie., x_i appears as the last variable in some critical clause, where x_i is the critical variable. We shall call a variable *branching*, if both $a'0$ and $a'1$ can be extended to a satisfying assignment. Otherwise the variable is called *free*. For a random order the average ratio of free variables to forced variable is at most $(k-1):1$. To hit a satisfying assignment we need to guess the free variables only. Thus the worst case is really when there is only one satisfying assignment.

A modification of algorithm C gives apparently the best bound for polynomial size CNF's with unbounded lengths of clauses, namely $2^{n-\epsilon\sqrt{n}}$ for an $\epsilon > 0$. Here is a sketch of the algorithm and its analysis. First choose randomly half of the variables. Then try systematically all assignments to these variables. A standard simple argument shows that the probability that a clause of size $\geq \sqrt{n}$ is not satisfied by a random assignment to the chosen variables is at most $2^{-\Omega(\sqrt{n})}$, i.e., there are at most $2^{n/2-\Omega(\sqrt{n})}$ such partial assignments for which there are still some long clauses remaining. For such partial assignments we systematically search assignments for the remaining variables. For the others, we are left with a $\sqrt{n}$-CNF and we will find a satisfying assignment in the expected time $2^{(1-1/\sqrt{n})\frac{n}{2}}$ by Algorithm C.

3.2

Algorithm C is so simple that it surely must be possible to improve it. One possible direction has been considered and a substantial improvement has been achieved. The idea is to consider assignments which are more isolated – for some $d > 1$ there no other satisfying assignments in Hamming distance $\leq d$. In order to get a better performance it does not suffice to use the simple algorithm above. We shall use it, but first we have to pre-process the formula. This consists in deriving new clauses using resolution. Resolution is in some sense a dual procedure to Davis-Putnam, we shall say more about it later. Let us just recall that the basic rule of resolution produces from two clauses having a complementary literal a new clause which is the union of the two clauses less the two complementary literals. In the preprocessing phase we derive all clauses which can be derived without exceeding the clause size k^d. In this way we obtain new, slightly longer clauses which enhance the probability that a variable is forced in the random process.

In order explain the idea, let us consider only a single application of the resolution rule. Let $C_1 = x_1 \vee \neg x_2 \vee \neg x_3$ be a critical clause for the assignment $11111\dots 1$, which we assume to be isolated. In this clause x_1 is critical. Then x_2 must also be critical for some clause, say for $C_2 = x_2 \vee \neg x_i \vee \neg x_j$. We can resolve the two clauses and get $x_1 \vee \neg x_3 \vee \neg x_i \vee \neg x_j$. This is again a critical clause in case $1 \neq i, j$. Otherwise the clause is a tautology as it contains $x_1 \vee \neg x_1$. However there is no way to ensure the last condition, hence we have to do something else. If there is no satisfying assignment in distance 2, there must be a clause C_3 which is not satisfied by $00111\dots 1$. C_3 must contain some positive literals, hence it contains x_1 and/or x_2. If it contains x_2, we resolve C_3 with C_1 and we surely get a critical clause with the critical variable x_1. If not, then C_3 is already a critical clause with the critical variable x_1 and it is different from C_1. Thus we always have a new critical clause, therefore the value of x_1 is more likely to be forced.

Let us denote the *search problem* to find a satisfying assignment for k-*CNF*'s having exactly one satisfying assignment by *unique-k-SAT*. In particular for *unique*-3-*SAT* we get an algorithm (**Algorithm D**) with expected running time $O(2^{0.387n})$, while algorithm C gives only $2^{2n/3+o(n)}$.

The analysis of the above algorithm is not simple and it is even much harder to analyze the algorithm for general k-*CNF*'s, where no isolation is guaranteed. Until now it was possible to get asymptotically the same bounds for k-*SAT* as for *unique-k-SAT* only for $k > 4$. For $k = 3, 4$ the constants are worse than in the case of unique satisfying assignments. The best bound obtained so far for 3-*SAT* is $O(2^{0.446n})$, nevertheless, already this beats the best previous algorithm of Schiermeyer [17]! This proof was obtained by a computer search, but it is possible to beat the best previous record without a computer search if one is satisfied with only $O(2^{0.533n})$. It seems likely that the case of the unique satisfying assignments is the worst one also for $k = 3, 4$.

3.3

As usual, one would like to know if randomness is really needed. Put otherwise, can we derandomize these algorithms? In case of the algorithms for finding isolated satisfying assignments, it is fairly easy. A closer look at the proof reveals that we do not need to take the ordering completely randomly. For critical clauses of length $\leq k$ we need only that each ordering of a subset of k variables occurs equally likely (since we use the expectation of the number of forced variables). If k is a constant, there are polynomial size probability spaces which have this property and which can be constructed in polynomial time.

For general k-SAT it is a much more difficult task. Consider only the special case when the satisfying assignments have a lot of neighbours, so we can use only the fact that there are many of them. To derandomize this particular case we need to construct a small *hitting set* for such sets of assignments. The construction of such sets is not known. Instead of derandomizing algorithm C directly, we can use the following idea [12], (we shall call it **Algorithm E**). We shall look for *minimal* satisfying assignments, ie., those which have the minimal number of 1's. Then we use the following argument: either there is a minimal solution with few 1's and then we can find it by searching the small set of all such assignments, or a minimal assignment has a lot of 1's and then it has large isolation as it is isolated at least in the coordinates which have value 1. Unfortunately this gives worse constants in the exponent than the randomized algorithm C, see the table.

4 Circuit Complexity

The probabilistic algorithms were discovered when working on the complexity of depth 3 AND–OR–NOT circuits. Valiant observed long time ago that proving large lower bounds on such circuits would have interesting consequences. In particular a lower bound $2^{n/o(\log \log n)}$ on depth 3 circuits computing a boolean function of n variables implies that the function cannot be computed by a linear size log-depth circuit. Since 1986 [7] the best lower bounds are only of the form $2^{c\sqrt{n}}$ for a constant c. The progress has been achieved only in improving the constant c. A circuit of depth 3 with top gate OR is an OR of CNF's, thus better understanding of CNF's helps to improve lower bounds on such circuits. By considering isolated (Hamming distance 1) satisfying assignments to k-CNF's the complexity of the parity function has been determined up to a multiplicative constant: the minimal size of depth 3 circuits computing parity of n bits is of the order $n^{1/4}2^{\sqrt{n}}$, see [12] (previously even the best constant in the exponent was not known). In [13] a lower bound $\Omega(2^{c\sqrt{n}})$ with $c > 1$ was proved for BCH codes of small nonconstant minimal distance. It is rather paradoxical that improving *lower* bounds on circuit complexity is connected with improving *upper* bounds on algorithms for *SAT*.

5 Logic of Algorithms

5.1

A big achievement of proof complexity is the result that every algorithm for 3-SAT based on Davis-Putnam procedure has worst case complexity at least $2^{c \cdot n}$ for a positive constant c. This follows from a result of Urquhart [19] which uses ideas of Tseitin [18] and Haken [6]. It is worthwhile to explain this result in more details.

The propositional resolution calculus is the system based on the resolution rule described above. Successive applications of the rule produce new clauses from a given set of clauses. The system is complete in the sense that for any clause C which logically follows from a given set it is possible to derive a subclause of C. The system is sound, which means that it is possible to derive only clauses which logically follow from the given set. We consider also the empty clause which has no literals and which represents falsehood. Thus a set of clauses is unsatisfiable iff the empty clause can be derived. Hence we can prove that a CNF is unsatisfiable: take the set of clauses of the formula and derive the empty clause. Proving that a formula is unsatisfiable is the same as proving that the negation is satisfiable by all assignments, ie., the negation is a *tautology*. The size of a resolution proof is the number of clauses, including the initial ones, that are used in order to derive the empty clause.

Resolution can be used to prove that 2-SAT is in P. The point is that resolution of two clauses of length at most 2 is a clause of length at most 2 too. The number os such clauses is bounded by a polynomial, so we can systematically generate them all.

A connection with resolution and Davis-Putnam algorithms is given by the following proposition.

Proposition 1. *Suppose that a Davis-Putnam algorithm stops on an unsatisfiable CNF after N steps. Then there exists a resolution proof of unsatisfiability of size $\leq N$.*

Proof. Let T be the search tree on an unsatisfiable formula Φ. This means that for each leaf of T there is a clause of Φ which is false under the partial assignment given by the branch leading to the leaf. We shall pick one such clause for every leaf and extend this labelling to every vertex of the tree as follows. Let v be the parent of u and w, with a clause C the label of u and a clause D the label of w. Let x_i be the variable according to which the tree branches at v. If x_i does not occur in C (resp. D) we label v by C (resp. D). If both contain x_i, they contain it with different signs. Then we resolve C with D using x_i and use the result as the label of v. This labelling has the property that the clause C belonging to a vertex v is false under the partial assignment given by the path from the root to v. As the root determines the empty partial assignment, it can only be labelled by the empty clause. Thus the labelling is a resolution proof of unsatisfiability of Φ.

Now it suffices to use a lower bound on resolution proofs. Let us observe that Davis-Putnam algorithms produce resolution proofs in a tree form, while the lower bound on resolution proofs is for general proofs.

Theorem 1 ([19]). *There exists a sequence of tautologies Φ_n which are k-DNF's, contain d.n variables, for some constants k,d, and for some positive constant c, every resolution proof of Φ_n has size at least 2^{cn}.*

It follows that every Davis-Putnam algorithm must use at least 2^{cn} steps before it rejects the CNF obtained by negating Φ_n. The tautologies can be constructed explicitly. They express the easy fact that in a graph the number of vertices of odd degree must be even. This tautology cannot be expressed by a k-DNF with a constant k, therefore the statement is restricted to subgraphs of special graphs of constant degree where we also fix which vertices have odd degree (it suffices to have exactly one).

5.2

Another general framework for solving NP-problems is *integer linear programing*. The basic idea here is to represent the problem by linear inequalities with rational coefficients so that the solutions of the problem are encoded as *integer* solutions of the inequalities. The fact that solving linear inequalities in the domain of rational numbers (the *linear programing problem*) can be done efficiently helps in some cases, but not always. It is necessary to use some rules which are valid only for integer solutions. The most popular one among such systems is the *cutting plane* system. In this system one can derive new inequalities by taking positive linear combinations and round down the constant term in the inequality, if all coefficients at variables are integers. For this system unsolvable sets of inequalities have been constructed which do not have subexponential proofs of unsolvability [14]. It follows that any algorithm for k-SAT based on these rules has worst case running time at least 2^{n^ϵ}, for some absolute $\epsilon > 0$, which can be determined from the lower bound on cutting plane proofs. Of course, an algorithm for k-SAT with running time 2^{n^ϵ} with any constant $\epsilon < 1$ would be a sensational result. Thus there is a lot of room for improving, most likely the lower bounds.

For other systems for integer linear programing it is still an open problem to prove nontrivial lower bounds.

5.3

Another popular algorithm, especially in algebra, is Buchberger's algorithm for constructing a Gröbner basis of an ideal of polynomials. In logic we restrict ourselves to the domain $\{0,1\}$. In terms of polynomial equations this means that we assume equations $x_i^2 = x_i$ for every variable x_i. The natural logical framework for Buchberger's algorithm in the domain of $\{0,1\}$ is called the *polynomial calculus*. In this calculus we derive polynomials from a given set of polynomials by adding polynomials that have been derived and by multiplying a polynomial that have been derived by an arbitrary polynomial. A contradiction is reached, when we derive a constant nonzero polynomial. Exponential lower bounds on proofs in polynomial calculus have been recently obtained [15,9], hence we can again conclude that any direct use of the Gröbner basis algorithm cannot produce an algorithm for k-SAT running in subexponential time.

5.4

Fix some sufficiently general class of formulas, eg. DNF's, and let $TAUT$ be the set of tautologies in this class. A propositional proof system is, roughly speaking, a nondeterministic algorithm for $TAUT$. More precisely it is a polynomial time computable function $f : \Sigma^* \to TAUT$ which is onto and polynomial time computable, and Σ is a finite alphabet [3]. We say that w is a proof of $f(w)$ in the system determined by f. In concrete systems not all strings are proofs, but we can always modify the system by saying that "nonsensical" strings are proofs of some default tautology.

We have considered three concrete proof systems – resolution, cutting planes and polynomial calculus which cover certain types of algorithms. For a general algorithms we observe:

1. Every algorithm for SAT determines a proof system. Namely, for a $\Phi \in TAUT$, the proof of Φ is the computation of the algorithm on input $\neg\Phi$ (the computation that shows that $\neg\Phi$ is not satisfiable).
2. Every algorithm is based on an idea, a theory, assumptions etc., which are used to prove the soundness of the algorithm. This can be used to determine the logical framework, *the logic of the algorithm*, which can be turned into a proof system.

So far we have only one type of pairs consisting of an algorithm and a proof system: a Davis-Putnam algorithm and the resolution proof system. If we could find more such pairs, we could show that other types of algorithms have to run in exponential time, since exponential lower bounds have been proved for several other proof systems. In order to find a proof system for an algorithm, it is always possible to apply the idea in observation 2. One can express the mathematical assumption in a first order theory and then construct a propositional proof system from it using well-known means (see [10]). Unfortunately that would result in a very strong proof system for which we are not able to prove any lower bounds. A more promising way of finding a natural proof system for a given algorithm is to analyze the computations and try to turn them into proofs of some familiar proof systems. The following problems illustrate what I have in mind.

Problem 1. Do randomized algorithm provide some proofs? In particular, the algorithm C is just randomized Davis-Putnam procedure, does it provide resolution proofs on unsatisfiable formulas?

Problem 2. By derandomizing and slightly modifying the algorithm C, we obtained the deterministic algorithm E. Does this algorithm provide resolution proofs on inputs that it rejects? Maybe it gives at least bounded depth Frege proofs?

6 Quantum Speed-Up

It is well-known that SAT is a self reducible problem, therefore the search problem is equally difficult as the decision problem. It is not surprising then that all so far proposed algorithms are based on some search procedure. Recently Grover [5] proved an interesting general result on quantum computations. He showed that finding a unique element in a database of N elements can be done in expected time $O(\sqrt{N})$ using quantum computations. More precisely the time is $O(t\sqrt{N})$, where t is an upper bound on the time needed to check for an element if it belongs to the database (think of a database as a subroutine). It follows that some problems for which we use search, can be solved faster. For instance, finding a satisfying assignment to a formula with n variables can be done on quantum computers in expected time $2^{n/2+o(n)}$, where we can consider arbitrary polynomial size formulas, or even circuits. Let us recall that the best probabilistic algorithm that we know for the satisfiability of CNF's runs in expected time $2^{n-\epsilon\sqrt{n}}$ for some $\epsilon > 0$.

It seems unlikely that an NP-complete problem would have a polynomial time quantum algorithm. Still quantum algorithms may be substantially faster than the deterministic ones. An interesting question is when we can combine a nontrivial classical algorithm with Grover quantum search algorithm to get a faster algorithm. Let us analyze only the probabilistic algorithm C for the *unique*-3-*SAT* and leave others to the reader. For this it is very easy to use quantum search to get a square root speed-up (**Algorithm Q**). Take a random ordering of variables. In the favourable case, which occurs sufficiently often, 1/3 of the variables will be forced if we assign successively the values of the unique satisfying assignment. So we need only to find $\frac{2}{3}n$ bits. To find them we search the $2^{\frac{2}{3}n}$ strings using Grover's quantum search. Thus we get a quantum algorithm running in expected time $2^{n/3+o(n)}$.

7 Conclusions

In this short survey we have considered only the most popular discipline in the area of algorithms for SAT, namely the time complexity measured in terms of the number of variables of the formula and we have considered the worst inputs. A lot of research has been done on random instances of SAT, including bounds on the lengths of resolution proofs [1]. There are also results on algorithms for SAT where time is measured in terms of the number clauses or in terms of the length of the formula [8]. There has also been a lot of experimental work done in this area. While the experimental work may be useful for practical applications, say, when the inputs can be considered random, it does not reveal very much on the behaviour of the algorithms on worst inputs.

Consider for instance the popular idea of assigning the value to a variable which makes more clauses containing the variable true. More precisely, in some process of assigning values to variables, suppose that $x_i = 0$ will make k clauses containing x_i true and $x_i = 1$ will make l clauses containing x_i true. Then choose $x_i = 0$ iff $k > l$. The following formula shows that the such an algorithm would perform very badly on some instances. Suppose the number of variables is divisible by 3. Divide the variables in blocks of size 3 and on each block force the only satisfying assignment to be 111. Thus the only satisfying assignment of the formula will consist of all 1's. This formula has $7n/3$ clauses. Now add all clauses of size 3 which contain exactly one positive literal. Thus most variables will appear negatively, hence improving the satisfying assignment locally with the above rule will lead away from the satisfying assignment.

The most promising and rewarding area for future research seems to be the lower bounds for various classes of algorithms. As shown in Section 5 several tools for such lower bounds have been developed and now we should try to get tighter bounds for classes of algorithms for which we do have lower bounds and determine more classes to which the techniques can be applied.

8 Table

The table below gives the constant c in the upper bounds $2^{cn+o(n)}$ on the expected running time of some algorithms considered above for k-SAT. The number in parentheses is for $unique$-3-SAT.

k	A	B	C	D	E	Q
3	.695	.582	.667	.446 (.387)	.896	.334
4	.879		.75		.917	.375
5	.947		.8	.651	.931	.4

Acknowledgment

I would like to thank to Oleg Verbitski for reading the manuscript and pointing out several misprints and errors.

References

1. P. BEAME, R. KARP, T. PITASSI AND M. SAKS, On the complexity of unsatisfiability proofs for random k-CNF formulas, Proc. 30-th STOC, 1998, to appear.
2. S.A. COOK, The complexity of theorem proving procedures, *Proc. 3-rd STOC*, 1971, 151-158.
3. S.A. COOK AND A.R. RECKHOW, The relative efficiency of propositional proof systems, *J. of Symbolic Logic* 44(1), 1979, 36-50.

4. M. DAVIS, G. LOGEMANN AND D. LOVELAND, A machine program for theorem proving, *Communications of the ACM 5*, 1962, 394-397.
5. L.K. GROVER, A fast quantum mechanical algorithm for database search, *Proc. 28-th STOC 1996*, 212-218.
6. A. HAKEN, The intractability of resolution, *Theor. Computer Science*, 39, 1985, 297-308.
7. J. HÅSTAD, Almost optimal lower bounds for small depth circuits, *Proc. 18-th STOC*, 1986, 6-20.
8. E.A. HIRSCH, Two new upper bounds for *SAT*, *Proc. 9-th SODA*, 1998, to appear.
9. R. IMPAGLIAZZO, P. PUDLÁK AND J. SGALL, Lower Bounds for the Polynomial Calculus and the Groebner Basis Algorithm, to appear in *Computational Complexity*.
10. J. KRAJÍČEK, Bounded arithmetic, propositional logic, and complexity theory, Cambridge Univ. Press 1995.
11. B. MONIEN AND E. SPECKENMEYER, Solving satisfiability in less than 2^n steps, *Discrete Applied Math. 10*, 1985, 287-295.
12. R. PATURI, P. PUDLÁK AND F. ZANE, Satisfiability coding lemma, *Proc. 38-th FOCS*, 1997, 566-574.
13. R. PATURI, P. PUDLÁK, M.E. SAKS AND F. ZANE, An improved exponential-time algorithm for *k-SAT*, preprint, 1998.
14. P. PUDLÁK, Lower bounds for resolution and cutting planes proofs and monotone computations, *J. of Symb. Logic* 62(3), 1997, 981-998.
15. A. A. RAZBOROV, Lower bounds for the polynomial calculus, to appear in *Computational Complexity*.
16. I. SCHIERMEYER, Solving 3-Satisfiability in less than 1.579^n steps, *CSL'92*, LNCS 702, 1993, 379-394.
17. I. SCHIERMEYER, Pure literal look ahead: An $O(1.497^n)$ 3-satisfiability algorithm, preprint, 1996.
18. G.C. TSEITIN, On the complexity of derivations in propositional calculus, *Studies in mathematics and mathematical logic, Part II,* ed. A.O. Slisenko, 1968, 115-125.
19. A. URQUHART, Hard examples for resolution, J. of ACM 34, 1987, 209-219.

The Joys of Bisimulation

Colin Stirling

Department of Computer Science,
University of Edinburgh,
Edinburgh EH9 3JZ, UK,
cps@dcs.ed.ac.uk

1 Introduction

Bisimulation is a rich concept which appears in various areas of theoretical computer science. Its origins lie in concurrency theory, for instance see Milner [20], and in modal logic, see for example van Benthem [3].

In this paper we review results about bisimulation, from both the point of view of automata and from a logical point of view. We also consider how bisimulation has a role in finite model theory, and we offer a new undefinability result.

2 Basics

Labelled transition systems are commonly encountered in operational semantics of programs and systems. They are just labelled graphs. A transition system is a pair $\mathcal{T} = (\mathcal{S}, \{\stackrel{a}{\longrightarrow}: a \in \mathcal{A}\})$ where $\mathcal{S}$ is a non-empty set (of states), $\mathcal{A}$ is a non-empty set (of labels) and for each $a \in \mathcal{L}$, $\stackrel{a}{\longrightarrow}$ is a binary relation on $\mathcal{S}$. We write $s \stackrel{a}{\longrightarrow} s'$ instead of $(s, s') \in \stackrel{a}{\longrightarrow}$. Sometimes there is extra structure in a transition system, a set of atomic colours $\mathcal{Q}$, such that each colour $q \subseteq \mathcal{S}$ (the subset of states with colour q).

Bisimulations were introduced by Park [23] as a small refinement of the behavioural equivalence defined by Hennessy and Milner in [14] between basic CCS processes (whose behaviour is a transition system).

Definition 1 A binary relation $\mathcal{R}$ between states of a transition system is a *bisimulation* just in case whenever $(s, t) \in \mathcal{R}$ and $a \in \mathcal{A}$,

1. if $s \stackrel{a}{\longrightarrow} s'$ then $t \stackrel{a}{\longrightarrow} t'$ for some t' such that $(s', t') \in \mathcal{R}$ and
2. if $t \stackrel{a}{\longrightarrow} t'$ then $s \stackrel{a}{\longrightarrow} s'$ for some s' such that $(s', t') \in \mathcal{R}$.

In the case of an enriched transition system with colours there is an extra clause in the definition of a bisimulation that it preserves colours: if $(s,t) \in \mathcal{R}$ then

0. for all colours q, $s \in q$ iff $t \in q$

Simple examples of bisimulations are the identity relation and the empty relation. Two states of a transition system s and t are *bisimulation equivalent* (or *bisimilar*), written $s \sim t$, if there is a bisimulation relation $\mathcal{R}$ with $(s,t) \in \mathcal{R}$.

One can also present bisimulation equivalence as a game $\mathcal{G}(s_0,t_0)$, see for example [30,28], which is played by two participants, players I and II. A *play* of $\mathcal{G}(s_0,t_0)$ is a finite or infinite length sequence of the form $(s_0,t_0)\ldots(s_i,t_i)\ldots$. Player I attempts to show that the initial states are different whereas player II wishes to establish that they are equivalent. Suppose an initial part of a play is $(s_0,t_0)\ldots(s_j,t_j)$. The next pair (s_{j+1},t_{j+1}) is determined by one of the following two moves:

- Player I chooses a transition $s_j \xrightarrow{a} s_{j+1}$ and then player II chooses a transition with the same label $t_j \xrightarrow{a} t_{j+1}$,
- Player I chooses a transition $t_j \xrightarrow{a} t_{j+1}$ and then player II chooses a transition with the same label $s_j \xrightarrow{a} s_{j+1}$.

The play continues with further moves. Player I always chooses first, and then player II, with full knowledge of player I's selection, must choose a corresponding transition of the other state.

A play of a game continues until one of the players wins. In a position (s,t) if one of these states has an a transition and the other doesnt then s and t are clearly distinguishable (and in the case of an enriched transition systems if one of these states has a colour which the other doesnt have then again they are distinguishable). Consequently any position (s_n,t_n) where s_n and t_n are distinguishable counts as a win for player I, and are called I-wins. A play is won by player I if the play reaches a I-win position. Any play that fails to reach such a position counts as a win for player II. Consequently player II wins if the play is infinite, or if the play reaches the position (s_n,t_n) and neither state has an available transition.

Different plays of a game can have different winners. Nevertheless for each game one of the players is able to win any play irrespective of what moves her opponent makes. To make this precise, the notion of strategy is essential. A strategy for a player is a family of rules which tell the player how to move. However it turns out that we only need to consider *history-free* strategies whose rules do not depend on what happened previously in the play. For player I a rule is therefore of the form "at position (s,t) choose transition x" where x is $s \xrightarrow{a} s'$ or $t \xrightarrow{a} t'$ for some a. A rule for player II is "at position (s,t) when player I has chosen x choose y" where x is either $s \xrightarrow{a} s'$ or $t \xrightarrow{a} t'$ and y is a corresponding transition of the other state. A player uses the strategy π in a play if all her moves obey the rules in π. The strategy π is a *winning strategy* if the player wins every play in which she uses π.

Proposition 1 *For any game $\mathcal{G}(s,t)$ either player I or player II has a history-free winning strategy.*

Proposition 2 *Player II has a winning strategy for $\mathcal{G}(s,t)$ iff $s \sim t$.*

Transition systems are models for basic process calculi, such as CCS and CSP. Models for richer calculi capturing value passing, mobility, causality, time, probability and locations have been developed. The basic notion of bisimulation has been generalised, often in a variety of different ways, to cover these extra features. Bisimulation also has a nice categorical representation via co-algebras due to Aczel, see for example [25], which allows a very general definition. It is an interesting question whether all the different brands of bisimulation are instances of this categorical account. In this paper we shall continue to examine only the very concrete notion of bisimulation on transition systems.

3   Bisimulation Closure and Invariance

It is common to identify a root of a transition system (as some special start state). Above we defined a bisimulation on states of the same transition graph. Equally we could have defined it between states of different transition systems. When transition systems are rooted we can then say that two systems are bisimilar if their roots are.

A family Δ of rooted transition graphs is said to be *closed under bisimulation equivalence* when the following holds:

$$\text{if } \mathcal{T} \in \Delta \text{ and } \mathcal{T} \sim \mathcal{T}' \text{ then } \mathcal{T}' \in \Delta$$

Given a rooted transition system there is a "smallest" transition system which is bisimilar to it: this is its *canonical* transition graph which is the result of first removing any states which are not reachable from the root, and then identifying bisimilar states (using quotienting).

An alternative perspective on bisimulation closure is from the viewpoint of properties of transition systems. Properties whose transition systems are bisimulation closed are said to be *bisimulation invariant*. Over rooted transition graphs, property Φ is bisimulation invariant provided that:

$$\text{if } \mathcal{T} \models \Phi \text{ and } \mathcal{T} \sim \mathcal{T}' \text{ then } \mathcal{T}' \models \Phi$$

(By $\mathcal{T} \models \Phi$ we mean that Φ is true of the transition graph $\mathcal{T}$.) On the whole, "counting" properties are not bisimulation invariant, for example "has 32 states" or "has an even number of states". In contrast temporal properties are bisimulation invariant, for instance "will eventually do an a-transition" or "is never able to do a b-transition". Other properties such as "has an Hamiltonian circuit" or "is 3-colourable" are also not bisimulation invariant. Later we shall be interested in parameterised properties, that is properties of arbitrary arity. We say that an n-ary property $\Phi(x_1, \ldots, x_n)$ on transition systems is bisimulation invariant provided that:

$$\text{if } \mathcal{T} \models \Phi[s_1, \ldots, s_n] \text{ and } t_1, \ldots, t_n \text{ are states of } \mathcal{T}' \text{ and}$$
$$t_i \sim s_i \text{ for all } i : 1 \leq i \leq n \text{ then } \mathcal{T}' \models \Phi[t_1, \ldots, t_n]$$

(By $\mathcal{T} \models \Phi[s_1, \ldots, s_n]$ we mean that Φ is true of the states $s_1, \ldots, s_n$ of $\mathcal{T}$). An example of a property which is not bisimulation invariant is "$x_1 \ldots x_n$ is a cycle", and an example of a bisimulation invariant property is "x_1 is language equivalent to x_2".

The notions of bismulation closure and invariance have appeared independently in a variety of contexts, see for instance [2,3,4,7,22].

4 Caucal's Hierarchy

Bisimulation equivalence is a very fine equivalence between states. An interesting line of enquiry is to re-consider classical results in automata theory, replacing language equivalence with bismulation equivalence. These results concern definability, closure properties and decidability/undecidability.

Grammars can be viewed as generators of transition systems. Let Γ be a finite family of nonterminals and assume that $\mathcal{A}$ is a finite set (of terminals). A basic transition has the form $\alpha \xrightarrow{a} \beta$ where $\alpha, \beta \in \Gamma^*$ and $a \in \mathcal{A}$. A state is then any member of Γ^*, and the transition relations on states are defined as the least relations closed under basic transitions and the following prefix rule:

$$\text{PRE} \quad \text{if } \alpha \xrightarrow{a} \beta \text{ then } \alpha\delta \xrightarrow{a} \beta\delta$$

Given a state α we can define its rooted transition system whose states are just the ones reachable from α.

In the table below is a Caucal hierarchy of transition graph descriptions according to how the family of basic transitions is specified. In each case we assume a finite family of rules. Type 3 captures finite-state graphs, Type 2 captures context-free grammars in Greibach normal form, and Type $1\frac{1}{2}$, in fact, captures pushdown automata. For Type 0 and below this means that in each case there are finitely many basic transitions. In the other cases R_1 and R_2 are regular expressions over the alphabet Γ. The idea is that each rule $R_1 \xrightarrow{a} \beta$ stands for the possibly infinite family of basic transitions $\{\alpha \xrightarrow{a} \beta : \alpha \in R_1\}$ and $R_1 \xrightarrow{a} R_2$ stands for the family $\{\alpha \xrightarrow{a} \beta : \alpha \in R_1 \text{ and } \beta \in R_2\}$. For instance a Type -1 rule of the form $X^*Y \xrightarrow{a} Y$ includes for each $n \geq 0$ the basic transition $X^n Y \xrightarrow{a} Y$.

	Basic Transitions
Type -2	$R_1 \xrightarrow{a} R_2$
Type -1	$R_1 \xrightarrow{a} \beta$
Type 0	$\alpha \xrightarrow{a} \beta$
Type $1\frac{1}{2}$	$\alpha \xrightarrow{a} \beta$ where $|\alpha| = 2$ and $|\beta| > 0$
Type 2	$X \xrightarrow{a} \beta$
Type 3	$X \xrightarrow{a} Y$ or $X \xrightarrow{a} \epsilon$

This hierarchy is implicit in Caucal's work on understanding context-free graphs, and understanding when the monadic second-order theory of graphs is decidable [5,4,6]. With respect to language equivalence, the hierarchy collapses to just two levels, the regular and the context free. The families between, and including, Type 2 and Type -2 are equivalent.

The standard transformation from pushdown automata to context free grammars (Type $1\frac{1}{2}$ to Type 2) does not preserve bisimulation equivalence. In fact, with respect to bisimilarity pushdown automata is a richer family than context free grammars. For instance, normed[1] Type 2 transition systems are closed under canonical transition systems. Caucal and Monfort [7] show that this is not true for Type $1\frac{1}{2}$ transition systems: see [4] for further results about canonical transition graphs. Caucal showed in [5] that Type 0 transition systems coincide (up to isomorphism) with Type $1\frac{1}{2}$. There is a strict hierarchy between Type 0 and Type -2. Therefore, with respect to bisimulation equivalence there are five levels in the hierarchy.

Baeten, Bergstra and Klop proved that bisimulation equivalence is decidable on normed Type 2 transition systems [1]. The decidability result was generalized in [9] to encompass all Type 2 graphs. Groote and Hüttel proved that other standard equivalences (traces, failures, simulation, 2/3-bisimulation etc..,) on Type 2 graphs are all undecidable [13]. The most recent result is by Sénizergues [27], who shows that bisimulation equivalence is decidable on Type -1 transition systems (which generalises his proof of decidability of language equivalence for DPDA [26]). This leaves as an open question whether it is also decidable for Type -2 systems.

One can build an alternative hierarchy when a sequence $\alpha \in \Gamma^*$ is viewed as a *multiset*. In which case the rule PRE above is to be understood as if $\alpha \xrightarrow{a} \beta$ then $\alpha \cup \delta \xrightarrow{a} \beta \cup \delta$ where $\cup$ is multiset union. Christensen, Hirshfeld and Moller showed that bisimulation equivalence is decidable on Type 2 graphs [8]. Hüttel proved that other equivalences are undecidable [16]. Type 0 graphs are Petri nets. Jančar showed undecidability of bisimilarity on Petri nets [17]. Under this commutative interpretation, Type 0 and Type $1\frac{1}{2}$ transition systems are not equivalent. Hirshfeld (utilizing Jančar's technique) showed undecidability of bisimulation for Type $1\frac{1}{2}$ systems, for more details see the survey [21].

5 Logics

Bisimulations were independently introduced in the context of modal logic by van Benthem [2]. A variety of logics can be defined over transition graphs.

Let M be the following family of modal formulas where a ranges over $\mathcal{A}$:

$$\Phi ::= \text{tt} \mid \neg \Phi \mid \Phi_1 \vee \Phi_2 \mid \langle a \rangle \Phi$$

The inductive stipulation below defines when a state s has a modal property Φ, written $s \models_{\mathcal{T}} \Phi$, however we drop the index $\mathcal{T}$.

$$\begin{aligned}
s &\models \text{tt} \\
s &\models \neg \Phi & \text{iff} \quad s \not\models \Phi \\
s &\models \Phi \vee \Psi & \text{iff} \quad s \models \Phi \text{ or } s \models \Psi \\
s &\models \langle a \rangle \Phi & \text{iff} \quad \exists t.\, s \xrightarrow{a} t \text{ and } t \models \Phi
\end{aligned}$$

---

[1] A state t is terminal if it has no transitions. A state s is *normed* if for all s' such that $s \xrightarrow{w} s'$ for some $w \in A^*$, then there is a terminal t such that $s' \xrightarrow{u} t$ for some $u \in A^*$.

This modal logic is known as Hennessy-Milner logic [14]. In the context of an enriched transition system one adds propositions q for each colour $q \in Q$ to the logic, with semantic clause: $s \models q$ iff $s \in q$.

Bisimilar states have the same modal properties. Let $s \equiv_M t$ just in case s and t have the same modal properties.

Proposition 1 *If $s \sim t$ then $s \equiv_M t$.*

The converse of Proposition 1 holds for a restricted set of transition systems. A state s is immediately image-finite if for each $a \in \mathcal{A}$ the set $\{t : s \xrightarrow{a} t\}$ is finite. And s is *image-finite* if every member of $\{t : \exists w \in \mathcal{A}^*. s \xrightarrow{w} t\}$ is immediately image-finite.

Proposition 2 *If s and t are image-finite and $s \equiv_M t$ then $s \sim t$.*

These two results are known as the modal *characterisation* of bisimulation equivalence, due to Hennessy and Milner [14]. (There is also an unrestricted characterisation result for infinitary modal logic. And there are less restrictive notions than image-finiteness for when characterisation holds, see [12,15].)

The modal logic M is not very expressive. For instance it cannot define safety or liveness properties on transition systems which have been found to be very useful when analysing the behaviour of concurrent systems. Modal mu-calculus, μM, introduced by Kozen [19], has the rquired extra expressive power. The new constructs over and above those of M are:

$$\Phi ::= Z \mid \ldots \mid \mu Z.\Phi$$

where Z ranges over a family of propositional variables, and in the case of $\mu Z.\Phi$ there is a restriction that all free occurrences of Z in Φ are within the scope of an even number of negations (to guarantee monotonicity).

The semantics of M is extended to encompass the least fixed point operator μZ. Because of free variables valuations, $\mathcal{V}$, are used which assign to each variable Z a subset of states in $\mathcal{S}$. Let $\mathcal{V}[S/Z]$ be the valuation $\mathcal{V}'$ which agrees with $\mathcal{V}$ everywhere except Z when $\mathcal{V}'(Z) = S$. The inductive definition of satisfaction stipulates when a process E has the property Φ relative to $\mathcal{V}$, written $E \models_\mathcal{V} \Phi$, and the semantic clauses for the modal fragment are as before (except for the presence of $\mathcal{V}$).

$s \models_\mathcal{V} Z$ iff $s \in \mathcal{V}(Z)$
$s \models_\mathcal{V} \mu Z.\Phi$ iff $\forall S \subseteq \mathcal{S}.$ if $s \notin S$ then $\exists t \in \mathcal{S}. t \notin S$ and $t \models_{\mathcal{V}[S/Z]} \Phi$

The stipulation for the fixed point follows directly from the Tarski-Knaster theorem, as a least fixed point is the intersection of all prefixed points. (Again we would add atomic formulas q if we are interested in extended transition systems.)

The bisimulation characterisation result above, Propositions 1 and 2, remain true for closed formulas of μM.

Second-order propositional modal logic, 2M, is defined as an extension of M as follows:

$$\Phi ::= Z \mid \ldots \mid \Box \Phi \mid \forall Z.\Phi$$

The modality $\Box$ is the reflexive and transitive closure of $\bigcup \{[a] : a \in \mathcal{A}\}$, and is included so that 2M includes μM. As with modal mu-calculus we define when $s \models_\mathcal{V} \Phi$. The new clauses are:

$$s \models_\mathcal{V} \Box \Phi \quad \text{iff} \quad \forall t. \forall w \in \mathcal{A}^*. \text{ if } s \xrightarrow{w} t \text{ then } t \models_\mathcal{V} \Phi$$
$$s \models_\mathcal{V} \forall Z.\Phi \text{ iff } \forall S \subseteq \mathcal{S}.\ s \models_{\mathcal{V}[S/Z]} \Phi$$

The operator $\forall Z$ is a set quantifier, ranging over subsets of $\mathcal{S}$. There is a straightforward translation of μM into 2M. Let Tr be this translation. The important case is the fixed point: $\text{Tr}(\mu Z.\Phi) = \forall Z.(\Box(\text{Tr}(\Phi) \to Z) \to Z)$.

Formulas of M and closed formulas of μM are bisimulation invariant (from Proposition 1 and its generalisation to μM). This is not true in the case of 2M, for it is too rich for characterising bisimulation: for instance, a variety of "counting" properties are definable, such as "has at least two different successors under an a transition" . This means that two bisimilar states need not have the same 2M properties.

Besides modal logics we can also consider other logics over transition systems. First-order logic, FOL, over transition systems contains binary relations E_a for each $a \in \mathcal{A}$ (and monadic predicates $q(x)$ for each colour q if extended transition systems are under consideration). Formulas have the form:

$$\Phi ::= x E_a y \mid x = y \mid \neg \Phi \mid \Phi_1 \wedge \Phi_2 \mid \forall x.\Phi$$

A formula $\Phi(x_1, \ldots, x_n)$ with at most free variables $x_1, \ldots, x_n$ will be true or false of transition system $\mathcal{T}$ and states $s_1, \ldots, s_n$ in the usual way.

Richer logics include first-order logic with fixed points, μFOL, where there is the extra formulas:

$$\Phi ::= Z(x_1, \ldots, x_k) \mid \ldots \mid \mu Z(x_1, \ldots, x_k).\Phi(y_1, \ldots, y_k)$$

In the case of $\mu Z(\ldots).\Phi(\ldots)$, there is the same restriction as in μM that all free occurrences of Z in Φ lie within the scope of an even number of negations.

An alternative extension of first-order logic is monadic second-order logic, 2OL, with the extra formulas:

$$\Phi ::= Z(x_1) \mid \ldots \mid \forall Z.\Phi$$

Van Benthem's use of bisimulation was to identify which formulas of FOL are equivalent to modal formulas (to M formulas), see the survey [3]. A formula $\Phi(x)$ is equivalent to a modal formula Φ' provided that for any $\mathcal{T}$ and for any state s, $\mathcal{T} \models \Phi[s]$ iff $s \models_\mathcal{T} \Phi'$.

Proposition 3 *A FOL formula $\Phi(x)$ over transition systems is bisimulation invariant iff Φ is equivalent to an M formula.*

This result was generalised by Janin and Walukiewicz [18] to 2OL and μM, as follows:

Proposition 4 *A 2OL formula $\Phi(x)$ over transition systems is bisimulation invariant iff it is equivalent to a closed μM formula.*

One corollary of this result is that the bisimulation invariant (closed) formulas of 2M coincides with closed formulas of μM.

An interesting question is if there is also a characterisation of the bisimulation invariant formulas of μFOL. (See [22] for preliminary results but over finite models.)

6 Finite Model Theory

Finite model theory is concerned with relationships between complexity classes and logics over finite structures. It is interesting to consider bisimulation invariance in the context of finite model theory.

Rosen showed that Proposition 3 of the previous section remains true with the restriction to finite transition systems [24]. It is an open question whether Proposition 4 also remains true under this restriction.

Part of the interest in relationships between μM and 2M or 2OL with respect to finite transition systems is that within 2M and 2OL one can define NP-complete problems: examples include 3-colourability on finite connected undirected graphs. Consider such a graph. If there is an edge between two states s and t let $s \xrightarrow{a} t$ and $t \xrightarrow{a} s$. So in this case $\mathcal{A} = \{a\}$, and 3-colourability is given by:

$$\exists X. \exists Y. \exists Z. (\Phi \wedge \Box((X \to [a]\neg X) \wedge (Y \to [a]\neg Y) \wedge (Z \to [a]\neg Z)))$$

where Φ, which says that every vertex has a unique colour, is

$$\Box((X \wedge \neg Y \wedge \neg Z) \vee (Y \wedge \neg Z \wedge \neg X) \vee (Z \wedge \neg X \wedge \neg Y))$$

In contrast, μM formulas over finite transition systems can only express PTIME properties.

An interesting open question is whether there is a logic which captures exactly the PTIME properties of transition systems. Otto has shown that there is a logic for the PTIME properties that are bisimulation invariant [22]. The right setting is μFOL over canonical transition systems (where $=$ is $\sim$, and a linear ordering on states is thereby definable).

We now consider emaciated finite transition systems whose set $\mathcal{A}$ is a singleton. That is now $\mathcal{T} = (\mathcal{S}, \longrightarrow)$ where $\mathcal{S}$ is finite. We can define language equivalence on emaciated transition systems. Let $s \xrightarrow{n} t$, when $n \geq 0$, if there is a sequence of transitions of length n from s to t (and by convention $s \xrightarrow{0} s$). A state is terminal if it has no transitions. The language of state s is the set $L(s) = \{i \geq 0 : s \xrightarrow{i} t \text{ and } t \text{ is terminal}\}$. Consequently, s and s' are language equivalent if $L(s) = L(s')$. The property "x is language equivalent to y" as was noted earlier is bisimulation invariant. Notice that this is an example of a dyadic invariant property.

Proposition 1 *Language equivalence on (canonical) finite transition graphs is co-NP complete.*

Hence language equivalence over finite transition systems is definable in μFOL iff PTIME = NP. Dawar offers a different route to this observation [10].

A classical result (due to Immermann, Gurevich and Shelah) in a slightly normalised form is:

Proposition 2 *A μFOL formula $\Psi(y_1, \ldots, y_n)$ over finite transition systems is equivalent to a formula of the form $\exists u. (\mu Z(x_1, \ldots, x_m). \Phi(y_1, \ldots, y_n, \widetilde{u}))$ where Φ is first-order and contains at most $x_1, \ldots, x_m$ free.*

The argument places in the application (...) from $n+1$ to m are all filled by the same element u. This allows for the arity of the defining fixed point m to be larger than the arity of the μFOL formula n.

Consequently, if one can prove that "y is language equivalent to z", $\Psi(y, z)$, is not definable by a μFOL formula in normal form, $\exists u. (\mu Z(x_1, \ldots, x_m). \Phi(y, z, \widetilde{u}))$, then this would show that PTIME is different from NP. As a first step, we have proved the following using tableaux:

Theorem 1 *Language equivalence $\Psi(y, z)$ is not definable in μFOL by a normal formula of the form $\exists u. (\mu Z(x_1, x_2, x_3). \Phi(y, z, u))$.*

Acknowledgement: I would like to thank Julian Bradfield and Anuj Dawar for help in understanding finite model theory.

References

1. Baeten, J., Bergstra, J., and Klop, J. (1993). Decidability of bisimulation equivalence for processes generating context-free languages. *Journal of Association of Computing Machinery*, **40**, 653-682.
2. van Benthem, J. (1984). Correspondence theory. In *Handbook of Philosophical Logic*, Vol. II, ed. Gabbay, D. and Guenthner, F., 167-248, Reidel.
3. van Benthem, J. (1996). Exploring Logical Dynamics. *CSLI Publications*.
4. Burkart, O., Caucal, D., and Steffen, B. (1996). Bisimulation collapse and the process taxonomy. *Lecture Notes in Computer Science*, **1119**, 247-262.
5. Caucal, D. (1992). On the regular structure of prefix rewriting. *Theoretical Computer Science*, **106**, 61-86.
6. Caucal, D. (1996). On infinite transition graphs having a decidable monadic theory. *Lecture Notes in Computer Science*, **1099**, 194-205.
7. Caucal, D., and Monfort, R. (1990). On the transition graphs of automata and grammars. *Lecture Notes in Computer Science*, **484**, 311-337.
8. Christensen, S., Hirshfeld, Y., and Moller, F. (1993). Bisimulation is decidable for basic parallel processes. *Lecture Notes in Computer Science*, **715**, 143-157.
9. Christensen, S., Hüttel, H., and Stirling, C. (1995). Bisimulation equivalence is decidable for all context-free processes. *Information and Computation*, **121**, 143-148.

10. Dawar, A. (1997). A restricted second-order logic for finite structures, To appear in *Information and Computation*.
11. Emerson, E., and Jutla, C. (1988). The complexity of tree automata and logics of programs. *Extended version from FOCS '88*.
12. Goldblatt, R. (1995). Saturation and the Hennessy-Milner property. In *Modal Logic and Process Algebra*, ed. Ponse, A., De Rijke, M. and Venema, Y. *CSLI Publications*, 107-130.
13. Groote, J., and Hüttel, H. (1994). Undecidable equivalences for basic process algebra. *Information and Computation*, **115**, 354-371.
14. Hennessy, M. and Milner, R. (1985). Algebraic laws for nondeterminism and concurrency. *Journal of Association of Computer Machinery*, **32**, 137-162.
15. Hollenberg, M. (1995). Hennessy-Milner classes and process calculi. In *Modal Logic and Process Algebra*, ed. Ponse, A., De Rijke, M. and Venema, Y. *CSLI Publications*, 187-216.
16. Hüttel, H. (1994). Undecidable equivalences for basic parallel processes. *Lecture Notes in Computer Science*, **789**.
17. Jančar, P. (1994). Decidability questions for bisimilarity of Petri nets and some related problems. *Lecture Notes in Computer Science*, **775**, 581-594.
18. Janin, D. and Walukiewicz, I (1996). On the expressive completeness of the propositional mu-calculus with respect to the monadic second order logic. *Lecture Notes in Computer Science*, **1119**, 263-277.
19. Kozen, D. (1983). Results on the propositional mu-calculus. *Theoretical Computer Science*, **27**, 333-354.
20. Milner, R. (1989). *Communication and Concurrency*. Prentice Hall.
21. Moller, F. (1996). Infinite results. *Lecture Notes in Computer Science*, **1119**, 195-216.
22. Otto, M. (1997). Bisimulation-invariant ptime and higher-dimensional μ-calculus. *Preliminary report RWTH Aachen*.
23. Park, D. (1981). Concurrency and automata on infinite sequences. *Lecture Notes in Computer Science*, **154**, 561-572.
24. Rosen, E. (1995). Modal logic over finite structures. *Tech Report*, University of Amsterdam.
25. Rutten, J. (1995). A calculus of transition systems (towards universal coalgebra). In *Modal Logic and Process Algebra*, ed. Ponse, A., De Rijke, M. and Venema, Y. *CSLI Publications*, 187-216.
26. Sénizergues, G. (1997). The equivalence problem for deterministic pushdown automata is decidable. *Lecture Notes in Computer Science*, **1256**, 671-681.
27. Sénizergues, G. (1998). $\Gamma(A) \sim \Gamma(B)$? Draft paper.
28. Stirling, C. (1996). Modal and temporal logics for processes. *Lecture Notes in Computer Science*, **1043**, 149-237.
29. Stirling, C. (1996). Games and modal mu-calculus. *Lecture Notes in Computer Science*, **1055**, 298-312.
30. Thomas, W. (1993). On the Ehrenfeucht-Fraïssé game in theoretical computer science. *Lecture Notes in Computer Science*, **668**.

Towards Algorithmic Explanation of Mind Evolution and Functioning
(Extended Abstract)

Jiří Wiedermann *

Institute of Computer Science
Academy of Sciences of the Czech Republic
Pod vodárenskou věží 2 , 182 07 Prague 8, Czech Republic
e–mail `wieder@uivt.cas.cz`

'Any scientific theory of the mind has to treat it as an automaton.'
(P. Johnson–Laird [6], 1983, p. 477)

Abstract. The cogitoid is a computational model of cognition introduced recently by the author. In cogitoids, knowledge is represented by a lattice of concepts and associations among them. From computational point of view any cogitoid is an interactive transducer whose transitions from one configuration into the next one depend on the history of past transitions. Cogitoid's computational mechanism makes it possible for cogitoids to perform basic cognitive tasks such as abstraction formation, associative retrieval, causality learning, retrieval by causality, similarity–based behaviour, Pavlovian and operant conditioning, and reinforced learning. In addition, when a cogitoid is exposed to similar interaction as human brain during its existence, emergence of humanoid mind is to be expected. The respective development will subsequently feature emergence of various attentional mechanisms, essential living habits, development of abstract concepts, language understanding and acquisition, and, eventually, emergence of consciousness.

1 Introduction

The interest of computer science in answering questions related to minds and brains dates back to Turing who already by the end of forties came to the conclusion that operation of the brain can be modeled by digital computers [3]. Since then a number of models of the brain have been considered (cf. [9], [10]). Among them, the most popular models are those based on variations of the theme on artificial neurons. Within this framework a number of valuable specific problems related to cognition has been solved (for a recent overview cf. [1]). However, it seems that none of the respective approaches has lead to some

---

* This research was supported by GA ČR Grant No. 201/98/0717 and by an EU grant INCO–COOP 96–0195 'ALTEC–KIT' jointly with the accompanying grant of the MŠMT ČR No. OK–304

non–trivial computational, or algorithmical, explanation of mind functioning. In this respect an exception seems to be the pioneering work by L. Goldschlager who in 1984 in his work *'A Computational Theory of Higher Brain Function'* [5] initiated one possible line of attack towards understanding the operation of human mind. His novel approach, at least within computer science, was to forget about neuronal level that deals with primitive signals only, and instead to focus one's attention to a higher conceptual level where more complex entities are dealt with. In Goldschlager's 'memory surface' model formation of abstract concepts, association of ideas, train of thoughts, creativity, self and consciousness are explainable, at least to some extent. However, his computational model of the brain has not been formalized to a level that would allow a more rigorous reasoning when necessary. Also, memory surface model seems to neglect certain important mechanisms like those enabling a negative reinforcement of associations that seems to be a condition *sine qua non* in modeling of certain types of behaviour.

A further step towards a model of the brain that abstracts totally from the aspects *how* real brains might do what they do, and focuses onto the aspect *what* they do, has been recently done by the present author. This approach seems to be in the best spirit of computer science that keeps looking for machine independent models of any information–processing task. The author introduced a formal abstract model of the brain, the so–called *cogitoid* [8],[11]. The basic entities any cogitoid deals with are, similarly as with Goldschlager, concepts and associations among them. In contrast to memory surface model the cogitoid is a precisely defined algebraical structure — a lattice of concepts. In the course of computation new associations keep developing and strengthening among concepts.

In [8] it has been shown that cogitoids are able to realize basic behavioristic tasks. Besides behaviour elicited by the presentation of specific stimulus–response patterns (classical conditioning), the cogitoids are also able to acquire sequences of concepts, and even be the subjects of Pavlovian conditioning. Since the model allows both for positively and negatively reinforcing associations, operant conditioning, and delayed operand conditioning is within the reach of cogitoids also. The respective statements are formulated and proved as theorems.

In the subsequent paper [13] cogitoid's potential w.r.t. modeling of higher brain activities has been investigated. It appears that in a sufficiently large cogitoid that is equipped with similar sensors and effectors such as human brain is, and that is exposed to similar interaction as humans during their lives, emergence of humanoid mind can be expected.

The present paper reports the work in progress as far as cogitoids are concerned. It surveys the main results from author's works in this field. Due to the page limit the paper concentrates only on the most important or interesting issues. For more details, see the original papers by the author.

The structure of the paper at hand is as follows. In Section 2 an informal definition of a cogitoid is introduced. In Section 3 a brief account of basic results from [8] and [11] needed for the further explanation is given. In Section 4 the spontaneously emerging organizational structure of cogitoid's memory will be described. In Section 5 the evolution of mind, in several phases, is sketched.

The full version of the present paper is available as a technical report [13]. The book by Dennet 'Consciousness Explained' [4] presents a good companion reading. It offers an interesting orthogonal view of many topics treated in the paper at hand. This view is based on the most recent opinions and achievements in psychology, neurology, and philosophy.

2 The Cogitoid

Any cogitoid can be seen as a central part of a finite interactive computational device that interacts with its environment with the help of its sensors and effectors. The respective information flowing from sensors into a cogitoid and from a cogitoid to its effectors is represented by *concepts*. Each concept represents some 'event' as perceived by a cogitoid. It is assumed that there is only a finite (but huge) number of concepts. Over a set of concepts binary operations $\vee$ of concept join and $\wedge$ of concept meet are defined in such a way that the resulting algebraical structure forms a (finite) lattice.

A *lattice of concepts* is a lattice (cf. [2]) whose elements are concepts. For any two elements a and b of such a lattice, with $a \leq b$, we say that a is an *abstraction* of b, while b is a *concretization* of a. Then, a supremum of any two of its elements is the smallest concretization of these elements, while their infimum is the largest abstraction of these elements. We shall say that two concepts are *non–meeting* iff their largest abstraction is equal to the least element ℓ of the respective lattice.

With the help of the above mentioned two operations of concept meet and join, new concepts can be formed from existing ones. Especially, for any a and b $a \vee b$ is a concretization of either a or b, while $a \wedge b$ is their abstraction.

There is a special subset of concepts that is called *affects*, or *operant concepts*. Positive affects correspond to positive feelings, or emotions, of animals, while negative affects correspond to negative feelings, or emotions.

In a cogitoid, concepts may be explicitly related via associations. Associations emerge among concepts that occur in series or among similar concepts. Formally, an ordered pair of form (a, b) of concepts is called an *association*, denoted also as $a \rightarrow b$. We say that a is associated with b. There are two types of associations: *excitatory* and *inhibitory*. Among any pair of concepts both types of associations may occur.

Two concepts a and b resemble each other in the concept c iff $a \wedge b = c$ and $c \neq \ell$[1]. Since this is a symmetric relation this knowledge is represented as a pair of associations $a \rightarrow b$ and $b \rightarrow a$. We then write $a \approx b$.

At any time t any concept may be either *present* or *absent* in a cogitoid. If present, then a concept may be either in an *active* or in a *passive* state.

---

[1] Depending on the size of c we could introduce resemblance relations of a various degree of similarity; for simplicity reasons we abstain from such an idea. This is why c will not be mentioned in the sequel in the respective similarity relation.

Also, at each time t there are two quantities assigned to each concept: its *strength* and its *quality*. The strength of a present concept is always a non-negative integer while absent concepts have the strength zero. The quality of concepts can be positive, negative, or undefined. Positive affects have always positive quality, while negative affects have always negative quality. The quality of other concepts may be arbitrary and depends on the history of concept formation or on the context in which a concept is invoked (activated).

Similarly, the strength is also assigned to each excitatory or inhibitory association.

Currently passive concepts may be activated either directly from the environment or by internal stimuli via associations from other active concepts.

In the latter case, in order to activate, concepts should be sufficiently excited. The concepts get excited via associations. The strength of excitation depends on the strength and type of all associations leading from active concepts to the concept at hand. This concept is excited to the level that is proportional to the sum of strengths of all excitatory associations from currently active concepts decreased by the sum of strengths of all inhibitory associations from currently active concepts.

The cogitoid $\mathbb{C}$ is seen as an interactive transducer that reads an infinite sequence of input concepts. Each input concept i represents an event that is 'observed' by a cogitoid by its sensors.

The computation of $\mathbb{C}$ proceeds in rounds. At the end of each round a set of concepts is active. This set presents an output of the cogitoid — its behaviour, its actions, its reaction to the previous input. Let $\mathcal{A}_t$ be the set of concepts active at the end of the t-th computational round in a cogitoid $\mathbb{C}$.

Each round consists of six phases:

Phase 1: *Producing the output and reading the input:* The concepts in $\mathcal{A}_t$ are sent to the output. All concepts in the set $\mathcal{I}$ corresponding to all abstractions of i are activated. This models the formation of concepts by their simultaneous appearance.

Phase 2: *Activating new concepts by internal stimuli:* First, a single new concept o from among all currently passive concepts gets activated. This is done with the help of a *selection mechanism* which inspects the excitation of all currently passive concepts from concepts in $\mathcal{I} \vee \mathcal{A}_t$ and subsequently activates the most excited concept o.

Simultaneously with activating o, the set $\mathcal{O}$ of all abstractions of o gets activated also.

Phase 3: *Assigning quality to concepts.* The quality of affects is constant all the time and it will determine, via inheritance, the quality of all their currently active abstractions and concretizations. Should some concepts obtain in this way both positive and negative quality, their resulting quality remains undefined. The concepts whose quality cannot be determined by the preceding rule, get positive quality.

Phase 4: *Updating the Knowledge:* The strength of all currently activated concepts is increased by a small amount.

Similarly, the strength of associations between each concept in the set $\mathcal{A}_t$ and each in $\mathcal{O}$ is increased. This models the emergence of associations by cause and effect.

Finally, the associations by resemblance are updated by increasing the strength of associations between each active concept in $\mathcal{I} \cup \mathcal{A}_t$ and each resembling present passive concept, and vice versa.

In the above mentioned process, if the association to be strengthened is between the concepts a and b, then if the quality of a was positive or negative or undefined, respectively, then the excitatory or inhibitory association, or both associations, respectively, between a and b are strengthened.

Note that increasing the strength of associations in some cases means that new associations are established (since until that time associations can be seen as those with strength zero).

Phase 5: *Gradual forgetting:* If positive, then the strength of all concepts that are not currently active and the strength of all associations among them is decreased by a small amount.

Phase 6: *Deactivation:* The concepts in the set $\mathcal{A}_t$ are deactivated and the set $\mathcal{O}$ becomes the set $\mathcal{A}_{t+1}$ of all active concepts.

Note that the sequence $\{\mathcal{A}_t\}_{t \geq 0}$ models the 'train of thought' in our cogitoid.

The notion of the above described cogitoid can be formalized with the help of sets, mappings and constants that determine the amount of concepts and associations strengthening.

3 Basic Results

In [9] it is shown that for any cogitoids it is possible to perform basic cognitive tasks such as abstraction formation, associative retrieval, causality learning, retrieval by causality, and similarity–based behaviour. E.g., the latter behaviour can be acquired as follows. First, by presenting the cogitoid repeatedly two non–meeting concepts, a and b, one after the other, an association $a \to b$ will be established (this is called *classical conditioning*). Then, whenever $a' \approx a$ appears at cogitoid's input, in the next two steps the chain of activations $a' \to a \to b$ will be invoked.

As seen from the previous 'definition' of a cogitoid, all the previous basic cognitive tasks belong among cogitoid's built–in computational mechanisms.

The next domain of behaviour that can be acquired by cogitoids is that of *Pavlovian conditioning*. This is a phenomenon in which an animal can be conditioned (learned) to activate a concept as a response to an apparently unrelated stimulating concept (cf. [7], p. 217).

For instance, one may first 'train', by classical conditioning, a cogitoid to establish a strong association $s \to r$. Then, we may repeatedly confront such a cogitoid with a further, so far unseen concept a, with $a \wedge s = \ell$ that is presented to it jointly with s, as $s \vee a$. After a while we shall observe that a alone will elicit the response r. Nevertheless, after a few of such 'cheating' from our side, the cogitoid will abstain from eliciting r when seeing merely a (in psychology this is called *extinction*). More complicated instances of Pavlovian conditioning can be also observed in *arbitrary* cogitoids. The only condition is that cogitoids must be large enough to accommodate all the necessary concepts.

The respective proofs are not completely trivial. To a critical extent they depend on the setting of constants that govern the strengthening of concepts and associations.

In order to explain Pavlovian conditioning no use of negative operant concepts and related inhibitory associations are necessary.

Cogitoids are also able to realize so–called *operant behaviour*. This is a behaviour acquired, shaped, and maintained by stimuli occurring *after* the responses rather than before. Thus, the invocation of a certain response concept r is confirmed as a 'good one' (by invoking the positive operant concept p) or 'bad one' (the negative operant concept n) only after r has been invoked. It is the reward (p), or punishment (n) that act to enhance the likelihood of r being re–invoked under similar circumstances as before.

The real problem here is hidden in the last statement which says that r should be re–invoked (or not re–invoked) only under similar circumstances as before. Thus, inhibition, or excitation of r must not depend on s alone: in some contexts, r should be inhibited, while in others, excited. Such a context is called an *operant context*; it is represented by a concept that appears invariantly as the part of the input of a cogitoid during the circumstances at hand. Thanks to cogitoid learning abilities, this operant context gets tied to the respective operant concept (affect) which, later on, causes that all associations emerging from this pair will inherit the quality of the operant concept at hand. Therefore, in the future, these associations will inhibit or excite r as necessary.

It appears that by a similar mechanism that ties a certain operant concept to some temporarily prevailing operant context one can also explain a more complicated case of the so–called *delayed reinforcing* when the reinforcing stimulus — a punishment or a reward — does not necessarily appear immediately after the step that will be reinforced.

All of the latter statements concerning the learning abilities of cogitoids can be formalized and rigorously proven (see the original papers [9] and [11]). In the latter paper it is also shown that, after a suitable training, any cogitoid equipped with Turing machine tapes is able to simulate any Turing machine. The purpose of the training is to teach the cogitoid the transition function of the simulated Turing machine.

4 The Evolution of Cogitoid's Memory

The previous results show that any cogitoid has a potential to learn many cognitive tasks in parallel, intermixed in time one with the others in various ways. The key to efficient learning is rehearsal (classical conditioning) and operant conditioning. In order to master a task the cogitoid has to be repeatedly exposed to circumstances and interaction leading to the acquisition of the respective skills. A circumstance is characterized by the respective static operant context in which various objects can be used in numerous ways. Thanks to the computational properties of any cogitoid, circumstances get stored in the form of strengthening of the respective concepts, superimposing similar contexts one to the others. In this way the basic cogitoid's memory structures — the so–called *clusters*, evolve.

A cluster is a set of such concepts $b \in \mathcal{B}$ that share a common abstraction $a = \wedge_{b \in \mathcal{B}} b$. Thus, any $b \in \mathcal{B}$ resembles the remaining b's in a. The concept a is called the *center* of the cluster $C = \vee_{b \in \mathcal{B}} b$ while the sets b's are called the members of the cluster C. Members of a cluster are sometimes called 'episodic memories'. By the virtue of cogitoid's computational rules, the center of a cluster gets activated and strengthened each time when some of its members is activated.

Analogously, when the center a of the cluster $\mathcal{B}$ is activated at time t, all $b \in \mathcal{B}$ get excited. To activate a specific b, additional excitation from some other concepts is usually needed. Namely, assume that some concept $b \in \mathcal{B}$ is in the same time also a member of an other cluster $\mathcal{D}$, with its center e. Then the simultaneous activation of a and e can excite b to such a degree that the selection mechanism will activate b. Thus, a simultaneous activation of two or more centers of different clusters may activate the concept that is a member of all clusters at hand. This simple *discriminating mechanism* presents the basic mechanism that keeps automatically evolving in cogitoids for 'reminding' it what to do under not completely specified circumstances.

According to previously described general principles, in any cogitoid that interacts with its environment clusters and chains of associations keep developing automatically. From a structural point of view all these clusters and chains look alike. Nevertheless, they differ substantially as far as their semantic contents is concerned. This is because different circumstance lead to the development of structures with different semantics. Namely, in any cogitoid fundamental clusters evolve around three fundamental semantic categories. These categories correspond to specific operant contexts in which the interaction takes place, to objects that are involved in the interaction at hand, and to the way these objects are dealt with.

Contextual clusters evolve by a superimposition of episodic memories that are all pertinent to frequently occurring similar operant contexts, such as 'in the forest', 'on the street', 'christmas', 'winter', etc. Their centers are created by abstract concepts that correspond to objects that usually participate in these contexts. In the previous examples, this could be concepts corresponding to 'trees', 'paths', 'animals', or 'cars','houses', 'myself', etc. As explained in the previous part, when a particular context is activated in a cogitoid, the respective centers of contextual clusters get excited. Thus, this mechanism presents a kind of an *attentional mechanism* — the cogitoid is 'reminisced' of (i.e., excites concepts corresponding to) objects that used to play some important role at specific occasions.

Object clusters evolve around specific objects. The respective object presents the center of the respective cluster, while the members of the cluster provide the specific contexts, in which the object has frequently found its use in the past.

A specific object cluster will evolve e.g. around the concept 'key'. It can be used for unlocking or locking a door, a safe, a car, etc. When some object is activated, all the respective contexts in which the object at hand occurred frequently in the past will be excited. It is like offering all the possible occasions in which the object has been manipulated in the past. Thus, this mechanism presents some kind of *role assignment mechanism* for objects. To select some concrete role, additional excitation from other concepts is needed.

The previous two types of clusters are complemented by *functional clusters*. These are formed around frequently performed activities that are represented by previously mentioned specific contexts that are members of object clusters. A common abstraction of each of these activities presents the center of the respective cluster. Thus, there may be functional clusters for unlocking a door, a safe, etc. The respective cluster members then contain the starting operand contexts of a chain of 'algorithmic description' of the respective activities, inclusively the description of some elementary action that moves the activity towards the next step in its realization. In a sense, the respective mechanism plays a role of the so–called *frames* that have been known within AI for a while.

Note that while the first two types of clusters — contextual and object clusters — present a kind of static descriptions that are free of any action, functional clusters involve already some elementary actions. To push forward the actions of a cogitoid, a specific type of its memory organization evolves along with the previously mentioned clusters. This executive part of cogitoid's memory is given by algorithmic descriptions.

Algorithmic descriptions or *habits* are sequencies of clusters that are chained by associations among their centers. Each member in such sequences presents a further atomic stage in the process of realizing the algorithm at hand. By realizing one step in such a chain, the cogitoid finds itself in a new context. This new context may either activate the next step in the algorithmic chain at hand, or can trigger an other activity.

Initialization of the respective chaines starts at the level of corresponding concrete concepts. Namely, from the computational rules described in Phase 4 it follows that whenever in a cogitoid two concepts a and b are activated in two subsequent steps, an association $a \to b$ will emerge or strengthened. However, since both a and b are activated, all their abstractions get activated as well, by virtue of cogitoid's computational law. Thus, associations among all abstractions of a and all abstractions of b will also emerge, or will also be strengthened. This concerns especially the associations among centers of corresponding clusters to which a and b belong. If associations among different pairs of members of different clusters are strengthened, the association among the respective centers is strengthened at each such occasion. It follows that the respective centers are associated stronger than the individual pairs of members.

Thus, habits are present very strongly since they are continuously reinforced by their repeated execution under similar circumstances. Included is also some aspect of self stimulation since cogitoids behave as if actively seeking for opportunities to make use of habits that are appropriate to the given occasion. This is due to their discrimination mechanism that always selects some habit. At these opportunities habits are continuously shaped and therefore are becoming increasingly general.

We can conclude that the behaviour of a cogitoid is driven both by the chains of acquired associations as well as by the current context in which a cogitoid finds itself. The current context activates similar, more abstract concepts that 'trigger' the respective behaviour as dictated by the chain of the respective associations. Upon similar circumstances a cogitoid with a sufficiently evolved clusters and chaines of associations will behave similarly as in the past. Even upon some novel circumstance chains of abstraction at higher levels will be found that 'match' the current circumstance and will drive the cogitoid's behaviour. Thus, in practice a cogitoid can never find itself in a position when it does not 'know' what to do.

Note that in most cases, cogitoid's behaviour will unfold effortlessly, without the necessity of making use of some inference of rules.

5 Cogitoid's Mind Evolution

In order to trace mind evolution in cogitoids they have to be exposed to a proper training. It is quite difficult to describe the respective process 'in general', for arbitrary cogitoids in arbitrary environment. The difficulty lies in the fact that the environment must be cooperative, and, in some sense, patient enough to rise up the necessary abilities in the cogitoid. The corresponding 'educational' process should continue step by step, incrementally, from simple matters to more complicated ones. Bellow we shall describe such a process for a 'humanoid cogitoid' since this seems to be the only case where we can rely upon some experience and intuition.

Let us perform the following thought experiment: imagine a cogitoid being exchanged with one's brain, residing within the corresponding body. In such a case, we will assume that the cogitoid receives the same signals as the brain does. The opposite process also works: by sending the appropriate signals the cogitoid can service the same peripherals as a brain does.

Then we imagine that the resulting cogitoid 'lives' in a standard human environment during a standard human life span.

Under such circumstances we shall concentrate onto the evolution of cogitoid's memory structures mentioned in the previous sections.

In doing this experiment, from its very beginning there is one clear advantage of human beings over our cogitoids: there seems to be a certain amount of knowledge that is somehow present in human, or in general, in animal mind without being acquired by learning. This concerns various inherited, built–in, as it appears, instincts and reflexes, such as sucking or breathing. The corresponding activities are triggered in the appropriate situation without being ever 'trained' by the respective animal. To make the proposed thought experiment possible, we shall assume that cogitoids also have these innate abilities acquired by a suitable preprocessing that occurred prior to starting this experiment.

The mind evolution could be described as a process consisting of several phases. In order to proceed to the next one, the previous one should be passed (but a slight overlap in phases is possible).

The Dawn of Mind. The shaping of certain parts of minds in our humanlike cogitoid seems to already start in the prenatal stage. This is the first opportunity for an evolving mind to be exposed, and to get used and adjusted to stimuli coming from its evolving peripherals. Although the surrounding environment does not seem to be very stimulating, for a dawning mind, this is rather an advantage for it has to learn the essential, life functions preserving habits. Any unrelated intervention would be harmful to this process.

In a prenatal stage, a part of stimuli bears a continuos character — they do not change over time. Such stimuli are related to various 'system settings', such as blood pressure, body temperature, etc. This seems to be the right time for adjusting the respective control mechanisms to the correct values.

The mechanism that does the respective adjustment is very simple. By the uninterrupted stimulation of the respective concepts these concepts start to be present very strongly — in fact their strengths will never be exceeded by other concepts. Thanks to the mechanism of creating new concepts by the virtue of simultaneous occurrence the life function supporting concepts get bound to every other concept. Moreover, by the virtue of successive occurrence association emerge between these life supporting concepts, and other concepts, in both directions. Consequently, life supporting concepts present a pillar around which the rest of mind is built. The activation of the respective concepts means 'the system is running OK'. Any deviation of standard values will cause a kind of 'uncomfortable feeling' when the respective surveilling concepts will not be activated in their entirety. This can result into blocking of activities of some other concepts since a part of their excitation will be missing. Then the cogitoid can fall into unpredictable state.

Another part of prenatal stimuli bears a periodic character. They are indirectly mediated by reactions of mother organism to periodic changes between days and nights, and in general by the corresponding periodic activities, such as sleeping, awaking, etc. Various kinds of feelings are probably also projected into dawning mind of baby cogitoids — like fear, pain, sadness, pleasure, hunger, etc. At these occasions the mind also learns the right internal reactions, simply by copying the reactions of mother's organism.

The mechanism responsible for the respective learning is the same as before — the strengthening of the respective concepts, and the emergence of successor associations.

As a result, in a prenatal stage the foundations of essential living habits — 'run time support', so to speak, in computer science terms — are established in cogitoids.

Shaping the Mind. This is the period of life after the birth, including babyhood. The main task during this period is to learn the cogitoid to be good at interpreting its perception of various external and internal stimuli by responding to them with appropriate actions.

This has influence on shaping all three kinds of the cogitoid's memory.

First, based in its own perception the cogitoid constructs during this time the basic set of concepts corresponding to objects and space of the observable world. This is reflected in cogitoid's memory by strengthening of the respective concepts and along with it by emergence of the respective abstractions, by the virtue of the respective cogitoid's mechanisms. Establishing of first episodic memories begins. Consequently, contextual, object and functional clusters start to develop.

Next, causality is remembered via emergence of the corresponding successor associations. An increasingly coordinated linkage between own perception and own action is acquired as a result of behavioristic or operant learning. This is reflected in the ongoing shaping and improvement of the corresponding frames and roles via the formation of the respective clusters.

In the latter process, based on repeated occurrence of own experience with perception or own actions in many similar contexts the cogitoid's abstraction mechanism gives also rise to specific concepts that correspond to the concept of *self*. So far this concept is largely unrecognized by the cogitoid, nevertheless it is there and is heavily utilized. Namely, it is present in numerous roles centered around the object 'self'.

In addition, new habits are acquired along with the establishment of new attentional mechanisms.

Attentional mechanisms emerge simultaneously with establishment of habits by repeated exposition of a cogitoid to periodic events and by the automatic abstraction or generalization of them as explained in Section 4. To each operant context a specific, tailored to circumstances at hand, attentional mechanism will emerge. In the case of animals, some of these attentional mechanisms might be innate, but as seen from the above written the cogitoids are also able to learn to establish new attentional mechanisms.

Attentional mechanisms support concentration of a cogitoid to features that are important in the given operant context. By learning from experience, by rehearsal and reward, these features are grouped into one abstract concept whose activation helps in identification of the features in more complex concepts. Thus, any attentional mechanism may be viewed as a tool that amplifies the excitation of the respective features in other concepts. In this way it implements some kind of a filter through which the currently unimportant details are filtered out.

Any creature at this level of mental development possesses the basic abilities to survive in the respective environment. Besides instincts its basic behaviour is governed by habits acquired during its life. Making use of these, it is able to react to immediate environmental stimuli, or to stimuli provided by some internal sensors (such as hunger, cold, pain, etc.). It can hardly react to some internal mental stimuli (i.e., to stimuli other than those from sensors). Its attention span is limited to currently ongoing events. It has no long term intentions.

Language Acquisition, Understanding and Generation. When a cogitoid possesses powerful sensors and effectors that enable it to interact with its environment in an increasingly complex manner, and when its memory capacity is sufficient, and when subjected to the right training, a further development of mental abilities is to be expected.

Namely, the increased complexity of interaction leads to the development of an increased number of new concepts. If this is accompanied by a better mastering of, and extended sensitivity to, abstract internal stimuli then an advanced mind evolution results.

The respective algorithmic explanation is as follows. An animal has no other than indirect means to activate certain abstract concepts. For instance, it cannot activate an abstract concept 'hunger' without being really hungry or unless seeing some food. This is because it is more or less input driven, as explained at the close of the previous paragraph, and there are no stimuli, except those mentioned, that would activate exactly, and directly the abstract concept for hunger.

If there is such direct stimuli, then the cogitoid's mind would be able to treat them as any other direct stimuli. Consequently, habits, along with the corresponding attentional mechanisms, dealing only with abstract stimuli could develop in much the same way as they did in the case of concrete external stimuli.

These additional inputs that can directly activate so far unaccessible abstract concepts, are provided by the language. In the most general case a language need not be a spoken language, but for simplicity we shall concentrate to this particular case. Moreover, we shall consider only the case when there is already a language that a cogitoid has to learn, rather than the case when a language has to be invented.

In the former case, it appears that the language to be learned must be compatible with cogitoid's ability to generate the corresponding sounds. The generation of such sounds may be the subject of a specific training preceeding that of binding the sounds to some contexts.

Namely, when a cogitoid hears a spoken language along with perceiving respective visual stimuli, by the simultaneous occurrence composed concepts consisting of words (or sounds), and of the representation of their visual counterparts, start to emerge. By hearing the respective word the corresponding concepts will be activated by the virtue of resemblance. The same can be achieved by pronouncing the respective word by the cogitoid itself. In the course of such a self stimulation a specific attentional mechanism will emerge, as a part of a habit that may be called 'internal speaking'. The effect of this mechanism will be that a concept can be activated without actually hearing its name. This internal activation can in turn lead to the pronunciation of the respective word, in the right operant context. This seems to be the starting point of comprehending the algorithms underlying both language acquisition, understanding, and language generation.

In cogitoids, the hearing or utterance of each word is bound to a proper *semantic* operant context that is shaped in the process of language acquisition. In fact, it is the semantic operation context that provides the essential 'understanding' to cogitoids of what it is spoken about. In such cases semantic operant context may consists of complex abstract concepts that reflect the real linguistic context. What to hear and what to say in which semantic context must be acquired by rehearsal.

Fortunately, not everything what a cogitoid can ever hear, understand, or said must be literally learned. Due to its abstraction potential, along with semantic operant contexts corresponding to the current circumstances also more abstract, *syntactic* operant concepts start to emerge in cogitoid's memory.

Syntactic operant concepts are based on the syntactic similarity of sentences. Namely, during the acquisition of a language by a cogitoid the respective abstracting mechanisms will learn that certain categories of words play the role of nouns, while the other ones that of verbs, adjectives, etc. Each word gets associated with the corresponding syntactic class. Moreover, by the mechanism of learning sequencies cogitoids 'discover' that in sentences the respective words usually follow the same pattern. This will give rise to syntactic operant contexts that keep track on using the words in the proper order.

Both semantic and syntactic operant contexts take care of understanding and generating the language. Their proper coupling and ordering is maintained by the respective *speaking habits*, along with corresponding *semantic* and *syntactic attentional mechanisms*. The speaking habits trigger the respective speech understanding or production frames. A kind of an *acoustic attentional mechanism* also seems to play an important role in this process.

Eventually, a picture of some complex internal grammar that supports both understanding and generating of a language seems to emerge. Its emergence and utilization by cogitoids also explains an often discussed problem of the poverty of the stimuli [4]. This is a phenomenon that refers to the fact that, during the linguistic formative years, the child is not exposed to enough language to account for its linguistic abilities. Making use of this grammar one is able to generate and understand words and sentences never heard before.

Emergence of Consciousness. Language acquisition and generation seems to belong among the most difficult mental tasks. Once mastered, it allows for increased communication and thus, information exchange with other partners. This in turn calls for an immense development of the 'self' concept, and other abstract concepts related to it. The self becomes an important subject in various concept clusters. Especially, the self will become a center in an object cluster describing various activities in which the self plays a central role. Among these activities, there will be an abstract concept that corresponds to 'registering', or 'observing', in the widest sense (i.e., not necessarily visual observation). In the functional cluster centered around 'observing' there will be objects that can be observed. Next to more or less concrete objects from the outer world (such as 'house', or 'dog') there will also be abstract objects, like the 'self'. A prologue to consciousness is such a state of mind in which the 'self' excites 'observing' as a possible activity, and 'observing' excites the 'self' as a subject of observation. This mutual excitation can achieve such a degree that all the respective concepts will become active simultaneously. Of course, in our model this corresponds to the activation of a single larger encompassing concept that corresponds to consciousness. By a similar mechanism other related higher level mental notions can be also explained. For instance, introspection involves the self observing (thinking about) (it)self while thinking...

Once started, the feedback between the self and other concepts involved in consciousness will continuously strengthen the respective associations among the respective concepts. A habit of being conscious will emerge. Since that time, no cogitoid activity can take place without the participation of consciousness. Of course, in real brains there are states in which consciousness may be 'switched off'. In addition, there are concepts (mostly related to basic living or system functions) that cannot be included into consciousness. All this is caused by mechanisms that are not a part of our model.

Similarly as for any sufficiently often encountered operational contexts, attentional mechanism will automatically include consciousness into such contexts. Activation of the respective concepts corresponds to *conscious concentration.*

Consciousness jointly with the concentration in turn enables conscious 'focusing of mind' to various subjects, among them to abstract concepts. In this way consciousness acts as a kind of 'excitation amplifier' of the respective concepts. In this way thinking in abstract terms is enabled.

Development of Abstract Thinking. Abstract thinking is different from that, mostly about the observed world: it is a thinking about things that are non–existent, that have been invented in the process of thinking. A typical example of abstract thinking is mathematical thinking. In order it to arise a lot more mechanisms must develop in a cogitoid than in the case of everyday thinking.

In addition to the the respective abstract entities or concepts that have to be defined (i.e., understand) and named in order to be able to think about them, one has to develop specific aesthetic criteria. These are defined in terms of positive or negative operant concepts whose activation motivates further abstract thinking by bringing pleasurable or uncomfortable satisfaction from it.

New rules of handling these new concepts must be invented. By their frequent 'mental' application new habits must be acquired. A specific attentional mechanism corresponding to concentration to selected issues emerges. As a result, a corresponding 'computational' theory, with habits to think within its framework, will develop in cogitoid's memory.

In fact, the whole process of building such a theory is not unlike the process of langauge invention, language understanding and language mastering: in order to think about abstract things, one has to know their meaning, to know how to deal with them, and last but not least, one has to be able to speak about them.

In this way a cogitoid can develop many different abstract internal words. These worlds are governed by their own rules that may or may not correspond to the observed world. Examples of such worlds span from fairy tales, fantasy, religion up to mathematical theories.

6 Conclusion

The first results and intellectual experience with cogitoids point to the fact that cogitoids, or similar devices could provide an interesting framework for studying of cognition. This is because they are based on general principles that are consistent with theory of animal or human psychology. Building on two basic pillars, viz. classical and operant conditioning, cogitoids represent specific universal learning machines. The underlying algorithm enables them a continuous learning in the course of their potentially endless interaction with the environment. Within computer science this seems to represent a novel approach to brain and mind modeling. The results from [11] indicate that as long as we are able to formalize the cognitive task at hand we can prove theorems describing the respective behaviour of cogitoids. Results concerning higher brain function — such as mind development — bear so far a speculative character since we are not yet able to specify satisfactorily corresponding cognitive tasks. Nevertheless, even at this level of modeling the respective tools and results offer much more concrete paradigm for studying, discussing, and explaining such problems than it was possible until now. There is a lot of open ends both in the respective computational models and computational theory of the mind. It seems that the time has matured for computer science to introduce the respective issues as item No. 1 on its research agenda (cf. [12]).

References

1. Arbib, M. A. (Editor): The Handbook of Brain Theory and Neural Networks. The MIT Press, Cambridge — Massachusetts, London, England, 1995, 1118 p.
2. Birkhoff, G.: Lattice Theory. American Mathematical Society, New York, 1948
3. Davis, M.: Mathematical Logic and the Origin of Modern Computers. In: The Universal Turing Machine: A Half–Century Survey, R. Herken (ed.), Springer–Verlag Wien, New York, 1994, pp. 149–174
4. Dennet, D.C.: Consciousness Explained. Penguin Books, 1991, 511 p.
5. Goldschlager, L.G.: A Computational Theory of Higher Brain Function. Technical Report 233, April 1984, Basser Department of Computer Science, The University of Sydney, Australia, ISBN 0 909798 91 5
6. Johnson–Laird, P.: Mental Models: Towards a Cognitive Science of Language, Inference, and Consciousness. Cambridge University Press, Cambridge, 1983
7. Valiant, L.G.: Circuits of the Mind. Oxford University Press, New York, Oxford, 1994, 237 p., ISBN 0–19–508936–X
8. Wiedermann, J.: The Cogitoid: A Computational Model of Mind. Technical Report No. V–685, September 1996, Institute of Computer Science, Prague, September 1996, 17 p.
9. Wiedermann, J.: Towards Computational Models of the Brain: Getting Started. Neural Networks World, Vol 7., No.1, 1997, p.89–120
10. Wiedermann, J.: Towards Machines That Can Think (Invited Talk). In: Proceeding of the 24–th Seminar on Current Trends in Theory and Practice of Informatics SOFSEM'97, LNCS Vol. 1338, Springer Verlag, Berlin, 1997, pp.12–141
11. Wiedermann, J.: The Cogitoid: A Computational Model of Cognitive Behaviour (Revised Version). Institute of Computer Science, Prague, Technical Report V–743, 1998
12. Wiedermann, J.: Artificial Cognition: A Gauntlet Thrown to Computer Science. In: Proc. Cognitive Sciences, Slovak Technical University, May 1998; also as Technical Report V–742, Institute of Computer Science, Prague, 1998
13. Wiedermann, J.: Towards Algorithmic Explanation of Mind Evolution and Functioning. Full version of the present paper. Technical Report ICS AS ČR, 1998, to appear

Combinatorial Hardness Proofs for Polynomial Evaluation [*]

(Extended Abstract)

Mikel Aldaz[1], Joos Heintz[2,3], Guillermo Matera[3,4], José L. Montaña[1], and Luis M. Pardo[2]

[1] Universidad Pública de Navarra, Departamento de Matemática e Informática,
31006 Pamplona, Spain
`mikaldaz, pepe@upna.es`

[2] Universidad de Cantabria, Fac. de Ciencias, Depto. de Matemáticas, Est. y Comp.,
39071 Santander, Spain
`heintz, pardo@matesco.unican.es`

[3] Universidad de Buenos Aires, FCEyN, Departamento de Matemáticas,
(1428) Buenos Aires, Argentina
`joos, gmatera@dm.uba.ar`

[4] Universidad Nacional de Gral. Sarmiento, Instituto de Desarrollo Humano,
(1663) San Miguel, Argentina.

Abstract. We exhibit a new method for showing lower bounds for the time complexity of polynomial evaluation procedures. Time, denoted by L, is measured in terms of nonscalar arithmetic operations. The time complexity function considered in this paper is L^2. In contrast with known methods for proving lower complexity bounds, our method is purely combinatorial and does not require powerful tools from algebraic or diophantine geometry.

By means of our method we are able to verify the computational hardness of new natural families of univariate polynomials for which this was impossible up to now. By computational hardness we mean that the complexity function L^2 grows linearly in the degree of the polynomials of the family we are considering.

Our method can also be applied to classical questions of transcendence proofs in number theory and geometry. A list of (old and new) formal power series is given whose transcendency can be shown easily by our method.

1 Background and Results

The study of complexity issues for straight-line programs evaluating univariate polynomials is a standard subject in Theoretical Computer Science. One of the most fundamental tasks in this domain is the exhibition of *explicit* families of univariate polynomials which are "hard to compute" in the given context.

---

[*] Work partially supported by spanish DGCYT grant PB 96–0671–C02–02.

Following Motzkin ([1955]), Belaga ([1958]) and Paterson-Stockmeyer ([1973]) "almost all" univariate polynomials of degree d need for their evaluation at least $\Omega(d)$ additions/subtractions, $\Omega(d)$ scalar multiplications/divisions, and $\Omega(\sqrt{d})$ nonscalar multiplications/divisions. A family $(F_d)_{d \in \mathbb{N}}$ of univariate polynomials F_d satisfying the condition $\deg F_d = d$ is called *hard to compute* in a given complexity model if there exists a constant $c > 0$ such that any straight-line program evaluating the polynomial F_d requires the execution of at least $\Omega(d^c)$ arithmetic operations in the given model.

In the present contribution we shall restrict ourselves to the nonscalar complexity model. This model is well suited for lower bound considerations and does not represent any limitations for the generality of our statements.

Families of specific polynomials which are hard to compute where first considered by Strassen ([1974]). The method used in Strassen ([1974]) was later refined by Schnorr ([1978]) and Stoss ([1989]). Heintz & Sieveking ([1980]) introduced a considerably more adaptive method which allowed the exhibition of quite larger classes of specific polynomials which are hard to compute. However in its beginning the application of this new method was restricted to polynomials with *algebraic* coefficients. In Heintz & Morgenstern ([1993]) the method of Heintz-Sieveking was adapted to polynomials given by their roots and this adaption was considerably simplified in Baur ([1997]).

Finally the methods of Strassen ([1974]) and Heintz & Sieveking ([1980]) were unified to a common approach in Aldaz et al. ([1996]). This new approach was based on effective elimination and intersection theory with their implications for diophantine geometry (see e.g. Fitchas et al. ([1990]), Krick & Pardo ([1996]) and Puddu & Sabia ([1997])). This method allowed for the first time applications to polynomials having only integer roots.

The results of the present contribution are based on a new, considerably simplified version of the unified approach mentioned before. Geometric considerations are replaced by simple counting arguments which make our new method more flexible and adaptive. Our new method is inspired in Shoup & Smolensky ([1991]) and Baur ([1997]) and relies on a counting technique developed in Strassen ([1974]) (see also Schnorr ([1978]) and Stoss ([1989])). Except for this result (see Theorem 1) our method (Lemma 1) is elementary and requires only basic knowledge of algebra.

2 A General Lower Bound for the Nonscalar Complexity of Rational Functions

Let K be an algebraic closed field of characteristic zero. By $K[X]$ we denote the ring of univariate polynomials in the indeterminate X over K and by $K(X)$ its fraction field. Let α be a point of K. By $K[[X - \alpha]]$ we denote the ring of formal power series in $X - \alpha$ with coefficients in K and by $\mathcal{O}_\alpha$ the localization of $K[X]$ by the maximal ideal generated by the linear polynomial $X - \alpha$. This means that $\mathcal{O}_\alpha$ is the subring of $K(X)$ given by the rational functions $F := f/g$, with $f, g \in K[X]$ and $g(\alpha) \neq 0$.

Since K has characteristic zero there exists for every $\alpha \in K$ a natural embedding i_α from $\mathcal{O}_\alpha$ into $K[[X-\alpha]]$ given as follows: for any $F \in \mathcal{O}_\alpha$ let $i_\alpha(F)$ be the Taylor expansion of F in the point α, namely

$$i_\alpha(F) := \sum_{j \in \mathbb{N}} \frac{F^{(j)}(\alpha)}{j!}(X-\alpha)^j \ .$$

Here we denote by $F^{(j)}$, $j \in \mathbb{N}$, the j-th derivative of the rational function F.

Let F be an element of $K(X)$, i.e. a rational function over the field K. Let us recall the following standard notion of algebraic complexity theory (see Borodin & Munro ([1975]), von zur Gathen ([1988]), Heintz ([1989]), Stoss ([1989]), Strassen ([1990]), Pardo ([1995]) and Bürgisser et al. ([1997]), Chap. 4).

Let A be one of the following K-algebras: $K[X]$, $K(X)$ or $\mathcal{O}_\alpha$, where $\alpha \in K$.

Definition 1. *Let L be a natural number. A straight-line program of nonscalar length L in A is a sequence β of elements of A, namely $\beta = (Q_{-1}, Q_0, \ldots, Q_L)$, satisfying the following conditions:*

- $Q_{-1} := 1$.
- $Q_0 := X$.
- *For any ρ, $1 \le \rho \le L$, there exist $d_\rho \in \{0,1\}$ and $a_{\rho,j}, b_{\rho,j} \in K$, with $-1 \le j < \rho$, such that*

$$Q_\rho := \Bigl(\sum_{-1 \le j < \rho} a_{\rho,j} Q_j\Bigr) \cdot \Bigl(d_\rho \Bigl(\sum_{-1 \le j < \rho} b_{\rho,j} Q_j\Bigr) + (1-d_\rho)\Bigl(\sum_{-1 \le j < \rho} b_{\rho,j} Q_j\Bigr)^{-1}\Bigr)$$

holds.

Let F be an arbitrary element of the K-algebra A. We say that the straight-line program $\beta = (Q_{-1}, Q_0, \ldots, Q_L)$ computes F if there are field elements $c_l \in K$, with $-1 \le l \le L$, such that the following identity holds:

$$F = \sum_{-1 \le l \le L} c_l Q_l \ .$$

The nonscalar complexity $L_A(F)$ of an element F of A is defined as

$$L_A(F) := \min\{\text{nonscalar length of } \beta : \beta \text{ computes } F\} \ .$$

Now let F be a rational function belonging to the K-algebra $\mathcal{O}_\alpha$. Suppose that F is given by a straight-line program β in $\mathcal{O}_\alpha$. We are going to analyze how F depends on the parameters of the straight-line program β. To this end we use an idea going back to Strassen ([1974]) (see also Schnorr ([1978]) and Stoss ([1989])). The following analysis of the rational function F represents the main technical tool we use in this paper.

Let us first recall that the *height* of a given polynomial with integer coefficients is the maximum of the absolute values of its coefficients and the *weight* is the sum of the absolute values of these coefficients.

Theorem 1 (Representation theorem for rational functions). *Let L be a natural number and $N := (L+1)(L+2)$. Then there exists a family $(P_{L,j})_{j\in\mathbb{N}}$ of polynomials $P_{L,j} \in \mathbb{Z}[Z_1, \ldots, Z_N]$ with*

$$\deg P_{L,j} \leq j(2L-1) + 2 \qquad (1)$$

and

$$\text{weight } P_{L,j} \leq 2^{6((j+1)^L - 1)} \qquad (2)$$

such that for any $\alpha \in K$ and any $F \in \mathcal{O}_\alpha$ with $L_{\mathcal{O}_\alpha}(F) \leq L$ there exists a point $z_\alpha \in K^N$ satisfying the identity

$$i_\alpha(F) = \sum_{j \in \mathbb{N}} P_{L,j}(z_\alpha)(X - \alpha)^j .$$

For given natural numbers d and L let

$$\Phi_{d,L} : K^N \longrightarrow K^{d+1}$$

be the morphism of affine spaces defined by $\Phi_{d,L}(z) := (P_{L,d}(z), \ldots, P_{L,0}(z))$ for arbitrary $z \in K^N$. Let $W_{d,L} := \overline{im\,\Phi_{d,L}} \subseteq K^{d+1}$ be the Zariski closure over $\mathbb{Q}$ of the image $im\,\Phi_{d,L}$ of the morphism $\Phi_{d,L}$.

In the sequel we shall identify any polynomial $\sum_{0 \leq j \leq d} f_j X^j \in K[X]$ of degree d with its coefficient vector $(f_d, \ldots, f_0)$ which we consider as a point of the affine space K^{d+1}.

In order to state our technical lemma (namely Lemma 4 below) we need the following notion and notation.

Definition 2. *Let $U := (n_j)_{0 \leq j \leq d} \in \mathbb{N}^{d+1}$ be a given sequence of natural numbers. For fixed d we define a map*

$$\mu : K^{d+1} \longrightarrow \mathbb{N}$$

which is given in the following way: for any $(d+1)$-tuple $F := (f_j)_{0 \leq j \leq d}$ belonging to K^{d+1} let

$$\mu(F; U) := \#\left\{ \sum_{S \subseteq \{0,\ldots,d\}} \theta_S \prod_{j \in S} f_j^{v_j} : 1 \leq v_j \leq n_j, 0 \leq j \leq d, \theta_S \in \{0, 1\} \right\} .$$

For $U := (1)_{0 \leq j \leq d} \in \mathbb{N}^{d+1}$ we write simply

$$\mu(F) := \mu(F; U) = \mu(F; (1)_{0 \leq j \leq d}) .$$

Lemma 1 (Main Lemma). *Let d and L be given natural numbers. Let $U := (n_j)_{0 \leq j \leq d}$ be an arbitrary sequence of (positive) natural numbers and let $M_d := \sum_{0 \leq j \leq d} n_j$. Then for any polynomial F belonging to the algebraic variety $W_{d,L} \subseteq K^{d+1}$ we have*

$$\mu(F; U) \leq 2^{((d+1)^3 M_d)^{7L^2}} .$$

Sketch of proof. By continuity arguments it suffices to prove the statement of the lemma for an arbitrary $(d+1)$-tuple $F := (f_d, \ldots, f_0)$ which belongs to $\operatorname{im} \Phi_{d,L} = \operatorname{im}(P_{L,d}, \ldots, P_{L,0})$. Let $N := (L+1)(L+2)$. For fixed d, $L \in \mathbb{N}$ and $U := (u_j)_{0 \leq j \leq d}$ let us define the following set of polynomials of $\mathbb{Z}[Z_1, \ldots, Z_N]$:

$$\Gamma := \left\{ \sum_{S \subseteq \{0,\ldots,d\}} \theta_S \prod_{j \in S} P_{L,j}^{v_j} : 1 \leq v_j \leq n_j, 0 \leq j \leq d, \theta_S \in \{0,1\} \right\}.$$

For any $z \in K^N$ let us write $\Gamma(z) := \{P(z) : P \in \Gamma\} \subseteq K$. Clearly we have $\#\Gamma(z) \leq \#\Gamma$. From Definition 2 we conclude that for any $F := (f_d, \ldots, f_0)$ belonging to $\operatorname{im} \Phi_{d,L}$ there exists a point $z_F \in K^N$ such that the following holds:

$$\mu(F; U) = \#\Gamma(z_F) \leq \#\Gamma.$$

Therefore it suffices to show that $\#\Gamma \leq 2^{((d+1)^3 M_d)^{7L^2}}$ holds.

Let $D := \max\{\deg P : P \in \Gamma\}$ and $H := \max\{\operatorname{height} P : P \in \Gamma\}$. The number of monomials of any polynomial belonging to Γ is bounded from above by the combinatorial

$$\binom{D+N}{N} \leq (D+1)^N. \quad (3)$$

From the degree bound (1) we deduce the estimation:

$$D + 1 \leq 2(Ld+1)M_d. \quad (4)$$

From the weight bound (2) we infer the following bound for the height of any polynomial in Γ:

$$H \leq 2^{(6(d+1)^L - 5)M_d}. \quad (5)$$

Now, putting together (3), (4) and (5) we obtain the following estimation:

$$\#\Gamma \leq (2H+1)^{(D+1)^N} \leq 2^{(2(L+1)(d+1)M_d)^{7L^2}}.$$

From Horner's rule we deduce that we may suppose without loss of generality that $L \leq d$ holds. This implies finally

$$\#\Gamma \leq 2^{((d+1)^3 M_d)^{7L^2}}. \quad \square$$

From Lemma 1 we obtain easily a sufficient condition saying when a polynomial with integer coefficients is hard to compute. Let $p \in \mathbb{N}$ be a prime number. For any integer $q \in \mathbb{Z}$ let us denote by $\nu_p(q)$ the multiplicity of p in the prime factor decomposition of q.

Theorem 2. *There exists a universal constant $c > 0$ with the following property: let d and L be given natural numbers. Let $F := \sum_{0 \leq j \leq d} f_j X^j \in \mathbb{Z}[X]$ be a polynomial of degree at most d with integer coefficients such that F belongs to the algebraic variety $W_{d,L}$. Then for any prime number $p \in \mathbb{N}$ we have*

$$L^2 \geq c \cdot \frac{\log_2\left(\#\left\{\nu_p(\prod_{j \in S} f_j) : S \subseteq \{0,\ldots,d\}\right\}\right)}{\log_2 d}.$$

Sketch of proof. Let $p \in \mathbb{N}$ be a prime number and let $b := \#\left\{\nu_p\left(\prod_{j \in S} f_j\right) : S \subseteq \{0, \ldots, d\}\right\}$. Then it is not difficult to see that the following inequality holds

$$\#\left\{\sum_{S \subseteq \{0,\ldots,d\}} \theta_S \prod_{j \in S} f_j : \theta_S \in \{0,1\}, S \subseteq \{0,\ldots,d\}\right\} \geq 2^b . \tag{6}$$

From (6) we write $2^b \leq \mu(F)$ and from Lemma 1 we know that $\mu(F) \leq 2^{(d+1)^{28L^2}}$. Taking logarithms in both inequalities finishes the proof. □

3 Polynomials Which Are Hard to Compute

Theorem 2 yields hardness proofs for the following new families of polynomials:

1. Let φ be the Euler function and let $\pi(j)$ be the number of primes not exceeding j. Then the polynomials

$$\sum_{0 \leq j \leq d} 2^{2^{\varphi(j)}} X^j \quad \text{and} \quad \prod_{0 \leq j < d} (X - 2^{2^{\varphi(j)}}) ,$$

$$\sum_{0 \leq j \leq d} 2^{\pi(j)!} X^j \quad \text{and} \quad \prod_{0 \leq j < d} (X - 2^{\pi(j)!})$$

satisfy the complexity bound $L^2 = \Omega\left(\frac{d}{(\log_2 d)^2}\right)$.
2. Polynomials with integer coefficients of the form

$$\sum_{0 \leq j \leq d} 2^{2^{\lfloor \sqrt[n]{j} \rfloor}} X^j \quad \text{and} \quad \prod_{0 \leq j < d} (X - 2^{2^{\lfloor \sqrt[n]{j} \rfloor}})$$

satisfy the complexity bound $L^2 = \Omega\left(\frac{\sqrt[n]{d}}{\log_2 d}\right)$ for any natural number $n \geq 1$.
3. Let f_j be the j-th Fibonacci number. Then the polynomials

$$\sum_{0 \leq j \leq d} 2^{f_j} X^j \quad \text{and} \quad \prod_{0 \leq j < d} (X - 2^{f_j}) ,$$

$$\sum_{0 \leq j \leq d} 2^{j!} X^j \quad \text{and} \quad \prod_{0 \leq j < d} (X - 2^{j!})$$

satisfy the complexity bound $L^2 = \Omega\left(\frac{d}{\log_2 d}\right)$.

In all these examples L denotes the number of nonscalar multiplications/divisions necessary to evaluate the polynomials under consideration.

We remark that our method can also be applied to obtain new hardness proofs for the following known families of polynomials:

1. Polynomials with integer coefficients of the form $\sum_{0 \leq j \leq d} 2^{2^j} X^j$ (see Strassen ([1974])).

2. Polynomials with algebraic coefficients of the form $\sum_{1\leq j\leq d} e^{\frac{2\pi i}{j}} X^j$ (see Heintz & Sieveking ([1980])) or of the form $\sum_{1\leq j\leq d} \sqrt{p_j} X^j$, where p_j is the j-th prime number (see von zur Gathen & Strassen ([1980])).
3. Polynomials with algebraic coefficients given by their roots $\prod_{1\leq j\leq d}(X - \sqrt{p_j})$ (see Heintz & Morgenstern ([1993])).

4 Applications to Transcendental Function Theory

Our complexity method can also be applied to classical questions of transcendency in number theory and geometry. Using Lemma 1 and Newton's method to compute algebraic functions we obtain the following sufficient condition for the transcendency of formal power series.

Theorem 3. *Let K be an algebraically closed field of characteristic zero and let $\sigma := \sum_{j\in\mathbb{N}} f_j X^j \in K[[X]]$ be a given formal power series. For any $k \geq 0$ let us denote by σ_k the polynomial $\sigma_k := \sum_{0\leq j<2^k} f_j X^j$. Suppose that there exist constants $c > 0$, $\epsilon > 0$ and an infinite subset $M \subseteq \mathbb{N}$ such that for every $k \in M$ the following condition holds*

$$\sigma_k \notin W_{2^k-1, ck^{1+\epsilon}} \ .$$

Then the power series σ is transcendental over $K(X)$. In particular, if there exists a function $h : \mathbb{N} \to \mathbb{N}$ satisfying the inequality $h(k) \geq k^2$ for any $k \in \mathbb{N}$ and such that for infinitely many $k \in \mathbb{N}$ the polynomial σ_k does not belong to $W_{2^k-1, h(k)}$ then the power series σ is transcendental over $K(X)$.

Sketch of proof. Let us suppose that the power series σ is algebraic over $K(X)$. Then by Bochnak et al. ([1987]), Chap. 8, the power series σ can be interpreted as a holomorphic function. Using Newton's method (see Kung & Traub ([1978])) it is possible to show that there exist a Zariski open subset $U \subset K$ and a constant $c' > 0$ such that for every $\alpha \in U$ and for every $k \in \mathbb{N}$ the Taylor polynomial of (formal) degree $2^k - 1$ in the point α of the holomorphic function σ, namely

$$\sigma_k^\alpha := \sum_{0\leq j<2^k} \frac{\sigma^{(j)}(\alpha)}{j!}(X-\alpha)^j \ ,$$

is computable by a straight-line program of nonscalar length not exceeding $c'k$. This implies that for every $\alpha \in U$ and every $k \in \mathbb{N}$ we have

$$\sigma_k^\alpha \in W_{2^k-1, c'k} \ .$$

On the other hand, the Zariski open condition

$$\sigma_k \notin W_{2^k-1, ck^{1+\epsilon}}$$

implies that for any $k \in M$ there exists a neighbourhood V_k of zero in K such that the following holds: for $\alpha \in V_k$ the polynomial σ_k^α does not belong to $W_{2^k-1, ck^{1+\epsilon}}$.

Taking $k \in M$ large enough such that the inequality $c'k < ck^{1+\epsilon}$ is satisfied we reach a contradiction. Hence σ cannot be algebraic. ⊓

From Theorem 3 and Lemma 1 we deduce the following criterion of transcendency.

Corollary 1. *Let K be an algebraically closed field of characteristic zero and let $\sigma \in K[[X]]$ be a given power series. Suppose that there exist constants $c > 0$ and $\epsilon > 0$ such that for infinitely many $k \in \mathbb{N}$ the following inequality holds*

$$\mu(\sigma_k) > 2^{2^{2^{ck^{3+\epsilon}}}}.$$

Then the power series σ is transcendental over $K(X)$.

Theorem 3 and Corollary 1 imply the following (old and new) transcendency results.

Corollary 2. *The following power series which belong either to $\mathbb{Q}[[X]]$ or to $\mathbb{C}[[X]]$ are transcendental over the function field $\mathbb{C}(X)$:*

1. $\sum_{j \in \mathbb{N}} \frac{1}{2^{2^{\varphi(j)}}} X^j$.

2. $\sum_{j \in \mathbb{N}} \frac{1}{2^{\pi(j)!}} X^j$.

3. $\sum_{j \in \mathbb{N}} \frac{1}{2^{2^{\lfloor \sqrt[n]{j} \rfloor}}} X^j$, for $n \geq 1$.

4. $\sum_{j \in \mathbb{N}} \frac{1}{2^{2^{\lfloor (\log_2(j+1))^n \rfloor}}} X^j$, for $n \geq 4$.

5. $\sum_{j \in \mathbb{N}} \sqrt{p_{j+1}} X^j$, for p_j being the j-th prime number.

6. $\sum_{j \in \mathbb{N}} e^{\frac{2\pi i}{j+1}} X^j$.

References

[1996] Aldaz M., Heintz J., Matera G., Montaña J.L., Pardo L.M.: Time-space tradeoffs in algebraic complexity theory. J. of Complexity (submitted to), 1996.
[1997] Baur W.: Simplified lower bounds for polynomials with algebraic coefficients. J. of Complexity **13**(1) (1997) 38–41.
[1958] Belaga E.G.: Some problems involved in the computations of polynomials. Dokl. Akad. Nauk. SSSR **123** (1958) 775–777.
[1987] Bochnak J., Coste M., Roy M.-F.: Géométrie Algébrique Réelle. Ergebnisse der Mathematik und ihrer Grenzgebiete, 3. Folge, Vol. 12. Springer, 1987.
[1975] Borodin A., Munro I.: The Computational Complexity of Algebraic and Numeric Problems. American Elsevier, 1975.
[1997] Bürgisser P., Clausen M., Shokrollahi A.: Algebraic Complexity Theory. A Series of comprehensive studies in mathematics **315**. Springer, 1997.
[1990] Fitchas N., Galligo A., Morgenstern J.: Precise sequential and parallel complexity bounds for the quantifier elimination over algebraically closed fields. J. Pure Appl. Algebra **67** (1990) 1–14.

[1988] von zur Gathen J.: Algebraic complexity theory. Ann. Review of Comp. Sci. **3** (1988) 317–347.
[1980] von zur Gathen J., Strassen V.: Some polynomials that are hard to compute. Theoret. Comp. Sc. **11**(3) (1980) 331–335.
[1989] Heintz J.: On the computational complexity of polynomials and bilinear mappings. A survey. L. Huget and A. Poli, editors, Applied Algebra, Algebraic Algorithms and Error-Correcting Codes. Lectures notes in Computer Science **356**, 269–300. Springer, 1989.
[1993] Heintz J., Morgenstern J.: On the intrinsic complexity of elimination theory. J. of Complexity **9**(4) (1993) 471–498.
[1982] Heintz J., Schnorr C.P.: Testing polynomials which are easy to compute. Logic and Algorithmic: An International Symposium held in honor of Ernst Specker. l'Enseignement de Mathématiques **30**, 237–254. Genève, 1982.
[1980] Heintz J., Sieveking M.: Lower bounds for polynomials with algebraic coefficients. Theoret. Comp. Sc. **11**(3) (1980) 321–330.
[1996] Krick T., Pardo L.M.: A computational method for diophantine approximation. L. González-Vega and T. Recio, editors, Algorithms in Algebraic Geometry and Applications. Proceedings of MEGA'94, Progress in Mathematics **143**, 193–254. Birkhäuser, 1996.
[1978] Kung H.T., Traub J.F.: All algebraic functions can be computed fast. J. of the ACM **25**(2) (1978) 245–260.
[1955] Motzkin T.S.: Evaluation of polynomials and evaluation of rational functions. Bull. Amer. Math. Soc. **61** (1955) 163.
[1995] Pardo L.M: How lower and upper complexity bounds meet in elimination theory. G. Cohen, M. Giusti and T. Mora, editors, Applied Algebra, Algebraic Algorithms and Error-Correcting Codes. Proceedings AAECC-11, Lecture Notes in Computer Science **948**, 33–69. Springer, 1995.
[1973] Paterson M.S., Stockmeyer L.J.: On the number of nonscalar multiplications necessary to evaluate polynomials. SIAM J. of Computing **2**(1) (1973) 60–66.
[1997] Puddu S., Sabia J.: An effective algorithm for quantifier elimination over algebraically closed fields using straight-line programs. J. of Pure Appl. Algebra (to appear).
[1978] Schnorr C.P.: Improved lower bounds on the number of multiplications/divisions which are necessary to evaluate polynomials. Theoret. Comp. Sc. **7**(3) (1978) 251–261.
[1991] Shoup V., Smolensky R.: Lower bounds for polynomial evaluation and interpolation problems. Proceedings of the 32nd Annual Symposium FOCS, 378–383. IEEE Computer Society Press, 1991.
[1989] Stoss H.J.: On the representation of rational functions of bounded complexity. Theoret. Comp. Sc. **64**(1) (1989) 1–13.
[1974] Strassen V.: Polynomials with rational coefficients which are hard to compute. SIAM J. of Computing **3** (1974) 128–149.
[1990] Strassen V.: Algebraic Complexity Theory. Handbook of Theoretical Computer Science, Vol. A, Chap. 11, 635–672. Elsevier Science Publishers, 1990.

Minimum Propositional Proof Length is NP-Hard to Linearly Approximate

(Extended Abstract)

Michael Alekhnovich[1]*, Sam Buss[2]**,
Shlomo Moran[3]***, and Toniann Pitassi[4]†

[1] Moscow State University, Russia, michael@mail.dnttm.ru
[2] University of California, San Diego, sbuss@ucsd.edu
[3] Technion, Israel Institute of Technology, moran@cs.technion.ac.il
[4] University of Arizona, Tucson, toni@cs.arizona.edu

Abstract. We prove that the problem of determining the minimum propositional proof length is NP-hard to approximate within any constant factor. These results hold for all Frege systems, for all extended Frege systems, for resolution and Horn resolution, and for the sequent calculus and the cut-free sequent calculus. Also, if NP is not in $QP = DTIME(n^{\log^{O(1)} n})$, then it is impossible to approximate minimum propositional proof length within a factor of $2^{\log^{(1-\varepsilon)} n}$ for any $\varepsilon > 0$. All these hardness of approximation results apply to proof length measured either by number of symbols or by number of inferences, for tree-like or dag-like proofs. We introduce the Monotone Minimum (Circuit) Satisfying Assignment problem and prove the same hardness results for Monotone Minimum (Circuit) Satisfying Assignment.

1 Introduction

This paper proves lower bounds on the hardness of finding short propositional proofs of a given tautology and on the hardness of finding short resolution refutations. When considering Frege proof systems, which are textbook-style proof systems for propositional logic, the problem can be stated precisely as the following optimization problem:

Minimum Length Frege Proof:
Instance: A propositional formula φ which is a tautology.
Solution: A Frege proof P of φ.
Objective function: The number of symbols in the proof P.

---

* Supported in part by INTAS grant N96-753
** Supported in part by NSF grant DMS-9503247 and grant INT-9600919/ME-103 from NSF and MŠMT (Czech Republic)
*** Research supported by the Bernard Elkin Chair for Computer Science and by US-Israel grant 95-00238
† Supported in part by NSF grant CCR-9457782, US-Israel BSF grant 95-00238, and grant INT-9600919/ME-103 from NSF and MŠMT (Czech Republic)

For a fixed Frege system $\mathcal{F}$, let $\min_{\mathcal{F}}(\varphi)$ denote the minimum number of symbols in an $\mathcal{F}$-proof of φ. An algorithm M is said to approximate the Minimum Length Frege Proof problem within factor α, if for all tautologies φ, $M(\varphi)$ produces a Frege proof of φ of length $\leq \alpha \cdot \min_{\mathcal{F}}(\varphi)$. (Here, α may be a constant or may be a function of the length of φ.)

We are interested only in *polynomial time* algorithms for solving this problem. However, there is a potential pitfall here since the shortest proof of a propositional formula could be substantially longer than the formula itself,[1] and in this situation, an algorithm with runtime bounded by a polynomial of the length of the input could not possibly produce a proof of the formula. In addition, it seems reasonable that a "feasible" algorithm which is searching for a proof of a given length ℓ should be allowed runtime polynomial in ℓ, even if the formula to be proved is substantially shorter than ℓ. Therefore we shall only discuss algorithms that are polynomial time in the length of the shortest proof (or refutation) of the input.

Note that an alternative approach would be to consider a similar problem, **Minimum Length Equivalent Frege Proof**, an instance of which is a Frege proof of some tautology φ, and the corresponding solutions are (preferably shorter) proofs of φ. While our results are all stated in terms of finding a short proof to a given tautology, they hold also for that latter version where the instance is a proof rather than a formula.

A yet different approach could be studying algorithms which output the *size* (i.e., number of symbols) of a short proof of the input formula, rather than the proof itself. In this case it is possible for an algorithm to have run time bounded by a polynomial of the length of the input formula, even if the size of the shortest proof is exponential in the size of the formula. In the final section of this paper, we show that strong non-approximability results can be obtained for algorithms with run time bounded by a polynomial of the length of the formula for a variety of proof systems.

A related minimization problem concerns finding the shortest Frege proof when proof length is measured in terms of the number of steps, or lines, in the proof:

Minimum Step-Length Frege Proof:
Instance: A propositional formula φ which is a tautology.
Solution: A Frege proof P of φ.
Objective function: The number of steps in the proof P.

Resolution is a propositional proof system which is popular as a foundation for automated theorem provers. Since one is interested in finding resolution refutations quickly it is interesting to consider the following problem:

Minimum Length Resolution Refutation
Instance: An unsatisfiable set Γ of clauses.

---

[1] Is is known that $NP \neq coNP$ implies that some tautologies require superpolynomially long Frege proofs.

Solution: A resolution refutation R of Γ.
Objective function: The number of inferences (steps) in R.

The main results of this paper state that a variety of minimum propositional proof length problems, including the Minimum Length Frege Proof, the Minimum Step-Length Frege Proof and the Minimum Length Resolution Refutation problems, cannot be approximated to within a constant factor by any polynomial time algorithm unless $P = NP$. Furthermore, for these proof systems and for every constant ϵ, the Minimum Length Proof problems cannot be approximated to within a factor of $\epsilon \ln n$ unless $NP \subseteq DTIME(n^{O(\log \log n)})$ or to within a factor of $2^{\log^{(1-\epsilon)} n}$ unless $NP \subseteq QP$, where QP, quasi-polynomial time, is defined to equal $DTIME(2^{(\log n)^{O(1)}})$. Our results apply to all Frege systems, to all extended Frege systems, to resolution, to Horn clause resolution, to the sequent calculus, and to the cut-free sequent calculus; in addition, they apply whether proofs are measured in terms of symbols or in terms of steps (inferences), and they apply to either dag-like or tree-like versions of all these systems.

We let $\mathcal{F} \vdash^k \varphi$ mean that φ has an $\mathcal{F}$-proof of $\leq k$ symbols. One of the first prior results about the hardness of finding optimal length of Frege proofs was the second author's result [7] that, for a particular choice of Frege system $\mathcal{F}_1$ with the language $\wedge, \vee, \neg$ and $\rightarrow$, there is no polynomial time algorithm which, on input a tautology φ and a $k > 0$, can decide whether $\mathcal{F}_1 \vdash^k \varphi$, unless P equals NP. This result however applies only to a particular Frege system, and not to general Frege systems. It also did not imply the hardness of approximating Minimum Length Frege Proofs to within a constant factor.

A second related result, which follows from the results of Krajíček and Pudlák [13], is that if the RSA cryptographic protocol is secure, then there is no polynomial time algorithm for approximating the Minimum Step-Length Frege Proof problem to within a polynomial.

Another closely related prior result is the striking connection between the (non)automatizability of Frege systems and the (non)feasibility of factoring integers that was recently discovered by Bonet-Pitassi-Raz [6]. A proof system T is said be *automatizable* provided there is an algorithm M and a polynomial p such that whenever $T \vdash^n \varphi$ holds, $M(\varphi)$ produces some T-proof of φ in time $p(n)$ (see [8]). Obviously the automatizability of Frege systems is closely related to the solution of the Minimum Length Frege Proof problem. Our theorems give a linear or quasi-linear lower bound on the automatizability of the Minimum Proof Length problem based on the assumption that $P \neq NP$ or that $NP \nsubseteq QP$. It has recently been shown by Bonet-Pitassi-Raz [6] that Frege systems are not automatizable unless Integer Factorization is in P. Their result provides a stronger non-approximability conclusion, but requires assuming a much stronger complexity assumption.

For resolution, the first prior hardness result was Iwama-Miyano's proof in [11] that it is NP-hard to determine whether a set of clauses has a read-once refutation (which is necessarily of linear length). Subsequently, Iwama [10] proved that it is in NP-hard to find shortest resolution refutations; unlike us, he did not obtain an approximation ratio bounded away from 1.

2 Monotone Minimum Satisfying Assignment

The section introduces the Monotone Minimum Satisfying Assignment problem and shows it is harder to approximate than the Minimum Set Cover problem and the Minimum Label Cover. (The latter is needed for proving hardness of approximation within a superlinear factor). The reader can find a general introduction to and survey of the hardness of approximation and of probabilistically checkable proofs in [4] and [2]. Recall that an A-reduction, as defined by [12], is a polynomial-time Karp-reduction which preserves the non-approximating ratio to within a constant factor.

Consider the following NP-optimization problems:

Monotone Minimum Satisfying Assignment:
Instance: A monotone formula $\varphi(x_1, \ldots, x_n)$ over the basis $\{\vee, \wedge\}$.
Solution: An assignment $\langle v_1, \ldots, v_n \rangle$ such that $\varphi(v_1, \ldots, v_n) = \top$.
Objective function: The number of v_i's which equal $\top$.

We henceforth let $\rho(\varphi)$ denote the value of the optimal solution for the Monotone Minimum Satisfying Assignment problem for φ i.e., the minimum number of variables v_i which must be set *True* to force φ to have value *True*.

We will also consider the Monotone Minimum Circuit Satisfying Assignment problem which is to find the minimum number of variables which must be set *True* to force a given monotone circuit over the basis $\{\wedge, \vee\}$ evaluate *True*. It does not matter whether we consider circuits with bounded fanin or unbounded fanin since they can simulate each other. It is apparent that Monotone Minimum Circuit Satisfying Assignment is at least as hard as Monotone Minimum Satisfying Assignment.

Recall the Minimum Hitting Set problem, which is:

Minimum Hitting Set:
Instance: A finite collection $\mathcal{S}$ of nonempty subsets of a finite set U.
Solution: A subset V of U that intersects every member of $\mathcal{S}$.
Objective function: The cardinality of V.

It is easy to see that Monotone Minimum Satisfying Assignment is at least as hard as Minimum Hitting Set: namely Minimum Hitting Set can be reduced (via an A-reduction) to the special case of Monotone Minimum Satisfying Assignment where the propositional formula is in conjunctive normal form. Namely, given $\mathcal{S}$ and U, identify members of U with propositional variables and form a CNF formula which has, for each set in $\mathcal{S}$, a conjunct containing exactly the members of that set.

Lund and Yannakakis [14] noted that Minimum Hitting Set is equivalent to Minimum Set Cover (under A-reductions). Furthermore, it is known that the problem of approximating Minimum Set Cover to within any constant factor is not in polynomial time unless $P = NP$ [5]. If one makes a stronger complexity assumption, then one can obtain a better non-approximability result for Minimum Set Cover; namely, Feige [9] has proved that Minimum Set Cover cannot be approximated to within a factor of $(1-\epsilon)\ln n$ unless $NP \subseteq DTIME(n^{O(\log \log n)})$.

In fact, we can get stronger results than the above reduction of Minimum Set Cover to Monotone Minimum Satisfying Assignment. There are two ways to see this: firstly, we can use a construction due to S. Arora [private communication] to reduce Monotone Minimum Satisfying Assignment to the Minimum Label Cover problem, or alternatively we can use a "self-improvement" property of the Monotone Minimum Satisfying Assignment problem to directly prove better non-approximation results. Both approaches prove that Monotone Minimum Satisfying Assignment cannot be approximated to within a factor of $2^{(\log n)^{1-\epsilon}}$ unless $NP \subseteq QP$. The advantage of the first approach is that it gives a sharper result, namely, a reduction of Minimum Label Cover to Monotone Minimum Satisfying Assignment for Π_4-formula. The second approach is more direct in that it avoids the use of Label Cover. (We include details of the second approach in the full version of this paper, but not in this abstract.)

Minimum Label Cover: (see [2])
Instance: The input consists of: (i) a regular bipartite graph $G = (U, V, E)$, (ii) an integer N in unary, and (iii) for each edge $e \in E$, a partial function $\Pi_e : \{1, \ldots, N\} \to \{1, \ldots, N\}$ such that 1 is in the range of Π_e.

The integers in $\{1, \ldots, N\}$ are called *labels*. A *labeling* associates a nonempty set of labels with every vertex in U and V. A labeling *covers* an edge $e = (u, v)$ (where $u \in U$, $v \in V$) iff for every label ℓ assigned to v, there is some label t assigned to u such that $\Pi_e(t) = \ell$.

Solution: A labeling which covers all edges.
Objective function: The number of all labels assigned to vertices in U and V.

A Π_4-*formula* is a propositional formula which is written as an AND of OR's of AND's of OR's.

Theorem 1 (S. Arora) *There is an A-reduction from Minimum Label Cover to Monotone Minimum Satisfying Assignment such that the instances of Label Cover are mapped to Π_4 formulas.*

For space reasons, we omit the proof of this theorem.

It was proved in [1] that Minimum Label Cover is not approximable within a $2^{\log^{(1-\epsilon)} n}$ factor unless $NP \subseteq QP$. An immediate corollary of Theorem 1 is that Monotone Minimum Satisfying Assignment enjoys the same hardness of approximation, even when restricted to Π_4-formulas. Summarizing, we have

Theorem 2
(a) *If $P \neq NP$, then there is no polynomial time algorithm which can approximate Monotone Minimum Satisfying Assignment (and hence Monotone Minimum Circuit Satisfying Assignment) to within a constant factor.*
(b) *If $NP \not\subseteq DTIME(n^{O(\log \log n)})$, then Monotone Minimum Satisfying Assignment (and Monotone Minimum Circuit Satisfying Assignment) cannot be approximated to within a factor of $(1 - \epsilon) \ln n$ where n equals the number of distinct variables.*

(c) If $NP \not\subseteq QP$, then there is no polynomial time algorithm which can approximate Monotone Minimum Satisfying Assignment (or Monotone Minimum Circuit Satisfying Assignment) to within a factor of $2^{\log^{(1-\varepsilon)} n}$.

The main theorems of this paper are stated in the next section. Their proofs depend on the reduction of the Monotone Minimum Circuit Satisfying Assignment problem to problems on minimum T-proof length, for a variety of propositional proof systems T.

Open question. Is it possible to improve the non-approximation factor for Monotone Minimum Satisfying Assignment or Monotone Minimum Circuit Satisfying Assignment, or prove their hardness using just $P \neq NP$ as a complexity theory hypothesis?

In fact the known NP-hardness of Monotone Minimum Satisfying Assignment concerns Π_2 (CNF) formulae and Quasi NP-hardness uses Π_4 formulae. But in general the formula or circuit can have unbounded depth and thus it a priori has richer expressive abilities. Hence there could be some chance to prove its hardness by some other way without improving the corresponding factor of Label Cover, perhaps by using some extension of the self-improvement property.

3 Main Hardness Results

Our first main results state that it is hard to approximate the length of the shortest T-proof of a given tautology in a wide variety of propositional proof systems T.

Hardness Theorem 3 *Let T be one of the following propositional proof systems: (1) a Frege system, (2) an extended Frege system, (3) resolution, (4) Horn clause resolution, (5) the sequent calculus, or (6) the cut-free sequent calculus. Let T-proofs have length measured by either (a) number of symbols, or (b) number of steps (lines). Finally, for each system, we may either require proofs to be tree-like or allow them to be dag-like. (So overall, there are 24 possible choices for the system T.)*

(a) *If $P \neq NP$, then there is no polynomial time algorithm which can approximate Minimum Length T-Proof to within a constant factor.*

(b) *If $NP \not\subseteq DTIME(n^{O(\log \log n)})$, then there is a $c > 0$ such that there is no polynomial time algorithm which can approximate Minimum Length T Proof to within a factor of $c \cdot \log n$.*

(c) *If $NP \not\subseteq QP$, then there is no polynomial time algorithm which can approximate Minimum Length T Proof to within a factor of $2^{\log^{(1-\varepsilon)} n}$ for any ε.*

The proof of the Hardness Theorem 3 involves giving a reduction of the Monotone Minimum (Circuit) Satisfying Assignment problem to the Minimum Length T-Proof problem. Thus any hardness results for the Monotone Minimum

Satisfying Assignment or Monotone Minimum Circuit Satisfying Assignment problem immediately also apply to the Minimum Length Frege proof problem.

For space reasons, the proofs are omitted from this abstact, but they are already available in the full version of the paper.

4 Hardness Results for Long Proofs

In the previous sections we proved that it is NP-hard to approximate the minimal propositional proof length by any constant factor, and that if NP is not in QP then minimum proof-length cannot be approximated (in polynomial time) within a $2^{\log^{(1-\varepsilon)} n}$ factor. The tautologies used in the proofs of these results had "short" proofs (or refutations); that is, proofs whose length is polynomial in the size of the formula. However, if $NP \neq coNP$, then for any proof system $\mathcal{S}$, there are tautologies whose shortest $\mathcal{S}$-proof is of super-polynomial length. It is therefore interesting whether better non-approximability results can be achieved when the proof lengths are not bounded, and when the run time of the algorithm is required to be polynomial time in the length of the input formula only.

The following simple intuition implies that in this case, no polynomial time algorithm can guarantee a polynomial time approximation for the shortest refutation of a given unsatisfiable formula, unless $NP \not\subseteq P/poly$: [2]

Given an input formula ψ of length n, reduce it to a formula $\varphi = \psi \wedge \eta$, such that the size of η is polynomial in that of ψ, η is unsatisfiable, but its shortest refutation is longer than the shortest refutation of any unsatisfiable formula of length n by a super-polynomial factor. Then ψ is satisfiable iff on input φ, a supposed polynomially bounded approximation algorithm returns a number smaller than than the size of the shortest refutation of φ. This implies a polynomial time circuit for recognizing SAT. To make the above argument formal, we need few more definitions.

Definition 1. *For a proof system $\mathcal{S}$ and an unsatisfiable formula φ, $\min_{\mathcal{S}}(\varphi)$ is the minimum length of a refutation of φ in $\mathcal{S}$. For an integer n, $MAX_{\mathcal{S}}(n) = \max\{\min_{\mathcal{S}}(\varphi)\}$, where φ ranges over all unsatisfiable formulas of length $\leq n$.*

We say that a non-decreasing function f has *super-polynomial growth* if for every polynomial r, $f(n) > r(n)$ for almost all positive integers n. f has a *smooth super-polynomial growth* if in addition there is a constant D such that for each large enough n there is $1 < d < D$ such that $f(n^d) > f^d(n)$. [If we write $f(n) = n^{e(n)}$, then the first condition states that $e(n)$ is not bounded from above, and the second condition states that for each n there is m, $n < m < n^D$, such that $e(m) > e(n)$.]

Assume, for simplicity, that $\mathcal{S}$ contains the connective $\wedge$. Formulas ψ and η are said to be *disjoint* if their underlying sets of variables are disjoint.

---

[2] We present the results in terms of finding short refutations of unsatisfiable formulas, but equivalent definitions and results are easily obtained for finding short proofs of tautologies.

Theorem 4 *Assume that $NP \not\subseteq P/poly$, and let $\mathcal{S}$ be a proof system which satisfies:*

1. *For every pair of disjoint formulas ψ and η, where η is unsatisfiable, the following holds:*
   (a) *If ψ is unsatisfiable, then $\min_\mathcal{S}(\psi \wedge \eta) < \min_\mathcal{S}(\psi) + r(|\psi| + |\eta|)$ for some (fixed) polynomial r.*
   (b) *If ψ is satisfiable, than $\min_\mathcal{S}(\psi \wedge \eta) \geq \min_\mathcal{S}(\eta)$;*
2. *$MAX_\mathcal{S}(n)$ has a smooth super-polynomial growth.*

Then for any polynomial q, there is no polynomial time q-approximation algorithm for the minimum length proof in $\mathcal{S}$.

Observe that property 1 above holds trivially for all proof systems mentioned in this paper. Property 2 is known to hold for resolution, since in this case $MAX_\mathcal{S}(n) < 3^n$ for all n, and by [3], for each n there is an e, $1 < e < 3$, s.t. $MAX_\mathcal{S}(n^e) > 2^{\frac{n^e}{40}}$, thus property 2 holds for $D = 3$. We conjecture that this property holds for any known proof system in which the proof lengths are not polynomially bounded.

Proof. We show that the existence of a polynomial time q-approximation algorithm, AL, for $\mathcal{S}$, implies polynomial size circuits for solving SAT.

Let j be such that $q(n) < n^j$ for almost all n, and let D be the constant guaranteed by the smooth super-polynomial growth of $MAX_\mathcal{S}$. Since $MAX_\mathcal{S}$ has super-polynomial growth, for all large enough n it holds that $r(n + n^{2jD}) < MAX_\mathcal{S}(n)$. Fix an integer n_0 for which this inequality holds. Since the super-polynomial growth of $MAX_\mathcal{S}$ is smooth, there is a number d, $2j \leq d \leq 2jD$, such that $[MAX_\mathcal{S}(n_0)]^d < MAX_\mathcal{S}(m)$, where $m = n_0{}^d$. Let η_m be a formula of size $\leq m$ such that $\min_\mathcal{S}(\eta_m) = MAX_\mathcal{S}(m)$. An input formula ψ of size n_0 is reduced to $\varphi = \psi \wedge \eta_m$, where the variables of η_m are disjoint from these of ψ (note that φ is unsatisfiable and its size is polynomial in that of ψ). We claim that ψ is unsatisfiable if and only if AL on input φ will output a number $k < MAX_\mathcal{S}(m)$. To see this, observe that if ψ is unsatisfiable, then by property (1a) above, $\min_\mathcal{S}(\varphi) \leq \min_\mathcal{S}(\psi) + r(|\psi| + |\eta_m|) < 2MAX_\mathcal{S}(n_0)$. Hence, by the assumption on AL, AL must produce an output $k < (2MAX_\mathcal{S}(n_0))^j < MAX_\mathcal{S}(m) = \min_\mathcal{S}(\eta_m)$. On the other hand, if ψ is satisfiable, then, by property (1b), $\min_\mathcal{S}(\varphi) \geq \min_\mathcal{S}(\eta_m) = MAX_\mathcal{S}(m)$.

5 Acknowledgments

We are grateful to A.A. Razborov for extremely helpful discussions. We also would like to thank S. Arora for pointing out that Minimum Label Cover can be reduced to Monotone Minimum Satisfying Assignment.

References

1. S. ARORA, L. BABAI, J. STERN, AND Z. SWEEDYK, *The hardness of approximate optima in lattices, codes, and systems of linear equations*, Journal of Computer and System Sciences, 54 (1997), pp. 317–331. Earlier version in *Proc. 34th Symp. Found. of Comp. Sci.*, 1993, pp.724-733.
2. S. ARORA AND C. LUND, *Hardness of approximations*, in Approximation Algorithms for NP-hard Problems, D. S. Hochbaum, ed., PWS Publishing Co., Boston, 1996, p. ???
3. P. BEAME AND T. PITASSI, *Simplified and improved resolution lower bounds*, in Proceedings, 37th Annual Symposium on Foundations of Computer Science, Los Alamitos, California, 1996, IEEE Computer Society, pp. 274–282.
4. M. BELLARE, *Proof checking and approximation: Towards tight results*, SIGACT News, 27 (1996), pp. 2–13. Revised version at http://www-cse.ucsd.edu/users/mihir.
5. M. BELLARE, S. GOLDWASSER, C. LUND, AND A. RUSSELL, *Efficient probabalistically checkable proofs and applications to approximation*, in Proceedings of the Twenty-Fifth Annual ACM Symposium on Theory of Computing, Association for Computing Machinery, 1993, pp. 294–304.
6. M. L. BONET, T. PITASSI, AND R. RAZ, *No feasible interpolation for TC^0-Frege proofs*, in Proceedings of the 38th Annual Symposium on Foundations of Computer Science, Piscataway, New Jersey, 1997, IEEE Computer Society, pp. 264–263.
7. S. R. BUSS, *On Gödel's theorems on lengths of proofs II: Lower bounds for recognizing k symbol provability*, in Feasible Mathematics II, P. Clote and J. Remmel, eds., Birkhäauser-Boston, 1995, pp. 57–90.
8. M. CLEGG, J. EDMONDS, AND R. IMPAGLIAZZO, *Using the Groebner basis algorithm to find proofs of unsatisfiability*, in Proceedings of the Twenty-eighth Annual ACM Symposium on the Theory of Computing, Association for Computing Machinery, 1996, pp. 174–183.
9. U. FEIGE, *A threshold of $\ln n$ for approximating set cover*, in Proceedings of the Twenty-Eighth Annual ACM Symposium on Theory of Computing, Association for Computing Machinery, 1996, pp. 314–318.
10. K. IWAMA, *Complexity of finding short resolution proofs*, in Mathematical Foundations of Computer Science 1997, I. Prívara and P. Ruzicka, eds., Lecture Notes in Computer Science #1295, Springer-Verlag, 1997, pp. 309–318.
11. K. IWAMA AND E. MIYANO, *Intractibility of read-once resolution*, in Proceedings of the Tenth Annual Conference on Structure in Complexity Theory, Los Alamitos, California, 1995, IEEE Computer Society, pp. 29–36.
12. S. KHANNA, M. SUDAN, AND L. TREVISAN, *Constraint satisfaction: The approximability of minimization problems*, in Twelfth Annual Conference on Computational Complexity, IEEE Computer Society, 1997, pp. 282–296.
13. J. KRAJÍČEK AND P. PUDLÁK, *Some consequences of cryptographic conjectures for S_2^1 and EF*, in Logic and Computational Complexity, D. Leivant, ed., Berlin, 1995, Springer-Verlag, pp. 210–220.
14. C. LUND AND M. YANNAKAKIS, *On the hardness of approximating minimization problems*, Journal of the Association for Computing Machinery, 41 (1994), pp. 960–981.

Reconstructing Polyatomic Structures from Discrete X-Rays: NP-Completeness Proof for Three Atoms

(Extended Abstract)

Marek Chrobak[1]* and Christoph Dürr[2]

[1] Department of Computer Science, University of California, Riverside, CA 92521-0304. marek@cs.ucr.edu
[2] International Computer Science Institute, Berkeley, CA 94704-1198. cduerr@icsi.berkeley.edu, www.icsi.berkeley.edu/~cduerr/Xray

Abstract. We address a discrete tomography problem that arises in the study of the atomic structure of crystal lattices. A polyatomic structure T can be defined as an integer lattice in dimension $D \geq 2$, whose points may be occupied by c distinct types of atoms. To "analyze" T, we conduct ℓ measurements that we call *discrete X-rays*. A discrete X-ray in direction ξ determines the number of atoms of each type on each line parallel to ξ. Given ℓ such non-parallel X-rays, we wish to reconstruct T.

The complexity of the problem for $c = 1$ (one atom type) has been completely determined by Gardner, Gritzmann and Prangerberg [5], who proved that the problem is NP-complete for any dimension $D \geq 2$ and $\ell \geq 3$ non-parallel X-rays, and that it can be solved in polynomial time otherwise [8].

The NP-completeness result above clearly extends to any $c \geq 2$, and therefore when studying the polyatomic case we can assume that $\ell = 2$. As shown in another article by the same authors, [4], this problem is also NP-complete for $c \geq 6$ atoms, even for dimension $D = 2$ and axis-parallel X-rays. The authors of [4] conjecture that the problem remains NP-complete for $c = 3, 4, 5$, although, as they point out, the proof idea in [4] does not seem to extend to $c \leq 5$.

We resolve the conjecture from [4] by proving that the problem is indeed NP-complete for $c \geq 3$ in 2D, even for axis-parallel X-rays. Our construction relies heavily on some structure results for the realizations of 0-1 matrices with given row and column sums.

1 Introduction

The fundamental principle of the *transmission electron microscope* (TEM) is very similar to the more familiar optical microscope: it "shines" a focused beam of electrons towards a specimen, and the transmitted beam is projected onto a

---

* Research supported by NSF grant CCR-9503498 and conducted when the author was visiting International Computer Science Institute.

phosphor screen generating an image. The intensity represents the density and thickness of the specimen: denser or thicker areas of the specimen transmit fewer electrons and produce darker areas in the image. The development of the TEM in 1930's was necessitated by the limitations of the optical microscopes, whose magnification and resolution were insufficient to study the internal structure of organic cells or to find defects in bulk materials. Recently, new advancements in *high-resolution TEM* (HRTEM), led to development of instruments and techniques for studying biological molecules and for investigating the atomic structure of crystals. In particular, a technique called QUANTITEM [7,9] allows us to determine the number of atoms in atom columns of a crystal in certain directions. Given these numbers we wish to reconstruct the structure of the crystal. Problems of this nature are studied in *discrete tomography*, the area of mathematics and computer science that deals with inverse problems of reconstructing discrete density functions from a finite set of projections. The size of crystals that occur in the materials science applications is about 10^6 atoms, and thus efficient reconstruction algorithms would be of great interest.

The problem we address in this paper can be formulated as follows: Define a *polyatomic structure* T as an integer lattice in dimension $D \geq 2$, whose points may be occupied by c types of atoms. Each cell can be occupied by one atom or it could be empty. To "analyze" T, we conduct ℓ measurements that we refer to as *discrete X-rays*. (QANTITEM uses electron beams but, following [5], we use a more familiar term "X-ray" instead.) A discrete X-ray in direction ξ determines the number of atoms of each type on each line parallel to ξ. Given such ℓ non-parallel X-rays, we wish to reconstruct T.

The complexity of the problem for $c = 1$ (one atom type) has been completely determined by Gardner, Gritzmann and Prangerberg [5], who proved that the problem is NP-hard for any dimension $D \geq 2$ and $\ell \geq 3$ non-parallel X-rays, and that it can be solved in polynomial time otherwise [8].

The NP-hardness result above clearly extends to any $c \geq 2$, and therefore when studying the polyatomic case we can assume that $\ell = 2$. As shown in another article by the same authors, [4], this problem is also NP-hard for $c \geq 6$ atoms, even for dimension $D = 2$ and for the axis-parallel X-rays. The authors of [4] conjecture that the problem remains NP-hard for $c = 3, 4, 5$, and they point out that for these values of c "a substantially new technique will be needed, at least for the case $c = 3$".

We resolve the conjecture from [4] by proving that the problem is indeed NP-complete for $c = 3$ (and thus for any larger c as well) in 2D, even for the orthogonal case, that is, with the axis-parallel X-rays.

In the orthogonal case, the problem can be thought of as a reconstruction problem for four-valued matrices (three atom types and "holes") from given row and column sums for each atom. We will use capital letters A, B, C to denote the three atom types, and we will sometimes refer to these types as colors: *Azure*, *Beige*, and *Cyan*. For any atom type $a \in \{A, B, C\}$, denote by r_i^a (resp. s_j^a) the *row-sum* (resp. *column-sum*) of atom a, that is, the number of atoms of type a in row i (resp. in column j). Without loss of generality, we can concentrate on square $L \times L$ matrices. The vectors $\mathbf{r}^a = (r_1^a, \ldots, r_L^a)$ and $\mathbf{s}^a = (s_1^a, \ldots, s_L^a)$ are referred to, respectively, as the *column-sum vector* and the *row-sum vector*.

A *realization* of an instance $\mathcal{I} = (\mathbf{r}^A, \mathbf{s}^A, \mathbf{r}^B, \mathbf{s}^B, \mathbf{r}^C, \mathbf{s}^C)$ is a $L \times L$ matrix T with elements in $\{A, B, C, \Box\}$ such that $|\{j : T[i,j] = a\}| = r_i^a$ and $|\{j : T[j,i] = a\}| = s_i^a$, for $i = 1, \ldots, L$ and $a \in \{A, B, C\}$. We say $\mathcal{I}$ is *consistent* if it has a realization.

More specifically, we prove that the following *3-Color Consistency Problem (3CCP)* is NP-complete: "Given $\mathcal{I} = (\mathbf{r}^A, \mathbf{s}^A, \mathbf{r}^B, \mathbf{s}^B, \mathbf{r}^C, \mathbf{s}^C)$, is $\mathcal{I}$ consistent?".

If we restrict ourselves further to just one atom, the problem becomes equivalent to the reconstruction of 0-1 matrices from the row and column sums – a problem predating the discrete tomography research. The first efficient reconstruction algorithm was proposed in 1963 by Ryser [8], and a similar algorithm was later rediscovered in 1971 by Chang [2]. In addition to reconstruction, Ryser and others studied various structural properties of 0-1 matrices with given row and column sums. Our construction relies heavily on some results in this area. Interested readers are referred to an excellent survey by Brualdi [1] for more information on this topic.

2 The General Idea of the Proof

The proof is by a reduction from the Vertex Cover problem: "Given an undirected graph $G(V,E)$, and an integer K, is there a vertex cover in G of size K?".

Throughout this paper, we fix an undirected graph $G(V,E)$ with n vertices $V = \{1, \ldots, n\}$ and m edges $E = \{e_1, \ldots, e_m\}$, where $n, m > 0$. The proof is by constructing, in polynomial time, an instance $\mathcal{I}$ of 3CCP such that $\mathcal{I}$ is consistent iff G has a vertex cover of size K.

Suppose first that using some number d of atoms $C', D', \ldots$, we can force a unique realization, which has a form shown in Figure 1, that we refer to as a *frame*. In the frame, the empty entries form diagonally-oriented intervals of length n called *mirrors*. All other entries are filled with atoms $C', D', \ldots$. We have two rows of mirrors: m mirrors in the upper-left row, and $m+1$ mirrors in the lower-right row.

Use atom B' to create m copies of a candidate vertex cover U in the following way: The first row and column B'-sum is K and all other B'-sums are 1. (See Figure 1.) Then the pattern of B's in each lower-right mirror is the same, and is also the same as the pattern of holes in the upper-left mirrors. We associate U with this pattern: a vertex u is in U iff the uth cell in any upper-left mirror is a hole. We think of U as a

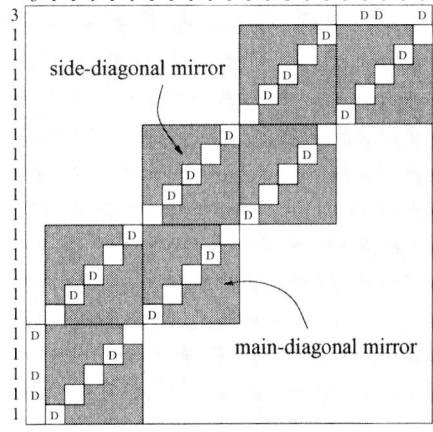

Fig. 1: *The frame and mirrors for $m = 3$.*

"beam" projected onto the last n cells in the first column, repeatedly reflected in a double-row of mirrors, and exiting through the last n cells in the first row. Finally, we use atom A' to verify that U is indeed a vertex cover. We convert the jth upper-left mirror into an *edge verifier* for edge $e_j = (u, v)$ (it may be necessary to add more rows and columns). Using appropriate sums for atom A', the realization of atoms $B', C', \ldots$ can be extended to a realization of all atoms iff either the uth cell or the vth cell in upper-left mirrors is a hole. Thus, either $u \in U$ or $v \in U$. (A similar approach was in [4].)

The main idea behind our proof is this: Define a partial order "$\preceq$" on all K-element vertex sets. The important property of "$\preceq$" is that its depth is polynomial, namely at most n^2. Further, "$\preceq$" has a unique minimum element U_{min}, and a unique maximum element U_{max}. Instead of using "perfect" mirrors, we use "skew" mirrors. These mirrors have the property that the reflected set is never smaller, with respect to "$\preceq$", than the set projected onto a skew mirror. These skew mirrors are also "flexible" – we know that they can reflect the same or a bigger set, but we cannot control what exactly the reflected set will be.

Now, instead of using m mirrors, we use n^2 segments, each having m skew mirrors in the upper row. In each segment, the jth skew mirror in the upper row is converted into an edge verifier for edge e_j. We "shine" U_{min} onto the first mirror in the bottom-left corner, and we make sure that the final set resulting from all reflections in the top-right corner is U_{max}. Since "$\preceq$" has depth n^2, there has to be a segment in which all mirrors reflect the same set U. Then the edge verifiers in this segment will verify that U is indeed a vertex cover.

3 0-1 Matrices with Given Row and Column Sums

By $\mathbf{x}, \mathbf{y}, \mathbf{z}$ we denote nonnegative integer vectors of length p, for example $\mathbf{x} = (x_1, \ldots, x_p)$. We now concentrate on the reconstruction problem for 0-1 matrices with given row and column sums: Given $\mathbf{x}$ and $\mathbf{y}$, is there a 0-1 matrix T that has x_i 1's in row i and y_j 1's in column j, for all $1 \leq i, j \leq p$? Again, in this case, we call T a *realization* and say that $\mathbf{x}, \mathbf{y}$ are *consistent*.

The structure function. Given a $p \times p$ matrix T, and integers $0 \leq k, l \leq p$ we partition T into four submatrices (which may have zero width or height): $T^{\blacksquare}_{k,l}, T^{\blacksquare}_{k,l}, T^{\blacksquare}_{k,l}$ and $T^{\blacksquare}_{k,l}$ defined by the intersections of the first k rows (resp. last $p - k$ rows) and the first l columns (resp. last $p - l$ columns). By $|T|_1$ and $|T|_0$ we denote the numbers of 1's and 0's in matrix T.

For a given instance $\mathbf{x}, \mathbf{y}$, let $\tau_{k,l} = (p-k)(p-l) + \sum_{j=1}^{l} y_j - \sum_{i=k+1}^{p} x_i$. We call τ the *structure function*. Then for any realization T we have

$$\tau_{kl} = \left(\left|T^{\blacksquare}_{kl}\right|_0 + \left|T^{\blacksquare}_{kl}\right|_1\right) + \left(\left|T^{\blacksquare}_{kl}\right|_1 + \left|T^{\blacksquare}_{kl}\right|_1\right) - \left(\left|T^{\blacksquare}_{kl}\right|_1 + \left|T^{\blacksquare}_{kl}\right|_1\right) = \left|T^{\blacksquare}_{kl}\right|_0 + \left|T^{\blacksquare}_{kl}\right|_1 \quad (1)$$

Consistent sums. A vector $\mathbf{z} = (z_1, \ldots, z_p)$ is *monotone* if $z_1 \leq \ldots \leq z_p$.

Lemma 1. *[1] Monotone vectors $\mathbf{x}, \mathbf{y}$ are consistent iff $\tau_{kl} \geq 0$ for all $k, l = 1, \ldots, p$.*

The implication ($\Rightarrow$) in Lemma 1 follows directly from Equation (1). The implication ($\Leftarrow$) can be proven constructively by giving an algorithm that produces a realization T for any pair $\mathbf{x}, \mathbf{y}$ for which the structure function is non-negative.

Decomposed realizations. We say that T is (k,l)-decomposed if $T_{kl}^{\blacksquare}$ consists only of 0's and $T_{k,l}^{\square\blacksquare}$ consists only of 1's. The following theorem follows immediately from Equation (1), and it will play a major role in this paper.

Lemma 2. *[1] For any k, l, $\tau_{k,l} = 0$ iff every realization (if any) is (k,l)-decomposed.*

Remark 1. Lemma 2 implies that if just one realization of $\mathbf{x}, \mathbf{y}$ is (k,l)-decomposed, then *all* realizations are (k,l)-decomposed as well.

4 The Skew-Mirror Lemma

0-1 Vectors and minorization. We use Greek letters $\boldsymbol{\alpha}, \boldsymbol{\beta}, \ldots$ for 0-1 vectors of length p, say $\boldsymbol{\alpha} = (\alpha_1, \ldots, \alpha_p)$. The *complement* $\bar{\boldsymbol{\alpha}}$ of $\boldsymbol{\alpha}$ is $\bar{\alpha}_i = 1 - \alpha_i$ for all $i = 1, \ldots, p$, and the *reverse* $\overleftarrow{\boldsymbol{\alpha}}$ is $\overleftarrow{\alpha}_i = \alpha_{p-i+1}$ for $i = 1, \ldots, p$.

We say that $\boldsymbol{\alpha}$ *minorizes* $\boldsymbol{\beta}$, denoted $\boldsymbol{\alpha} \preceq \boldsymbol{\beta}$, if $\sum_{i=1}^{k} \alpha_i \leq \sum_{i=1}^{k} \beta_i$ for all $k = 0, \ldots, p$. By straightforward verification, "$\preceq$" is a partial order.

The *total sum* of a 0-1 vector $\boldsymbol{\alpha}$ is $\sum_{i=1}^{p} \alpha_i$. If $\boldsymbol{\alpha}, \boldsymbol{\beta}$ are two 0-1 vectors with equal total sums, then the definitions above imply directly the following equivalences: $\qquad \boldsymbol{\alpha} \preceq \boldsymbol{\beta} \iff \bar{\boldsymbol{\alpha}} \succeq \bar{\boldsymbol{\beta}} \iff \overleftarrow{\boldsymbol{\alpha}} \succeq \overleftarrow{\boldsymbol{\beta}}.$

An important property of the minorization relation is that it is "shallow", that is its depth is only polynomial (unlike, for example, the lexicographic order). The next lemma gives a more accurate estimate on the depth of "$\preceq$".

Lemma 3. *Suppose that we have a strictly increasing sequence of 0-1 vectors $\boldsymbol{\alpha}^1 \prec \boldsymbol{\alpha}^2 \prec \ldots \prec \boldsymbol{\alpha}^q$ with total sums t. Then $q \leq t(p-t) + 1$.*

The 0-1 skew mirror. If T is a $p \times p$ matrix that realizes monotone vectors $\mathbf{x}, \mathbf{y}$, we say that T is a *perfect mirror* if $T[i,j] = 0$ for $i + j \leq p$ and $T[i,j] = 1$ for $i + j \geq p + 2$. In a perfect-mirror matrix the cells on the main diagonal $i + j = p + 1$ can be either 0 or 1, but all cells above it are 0, and all cells below it are 1. From Lemma 2 we immediately obtain the following corollary.

Corollary 1. *Let T be a realization of vectors $\mathbf{x}, \mathbf{y}$. Then T is a perfect mirror iff $\tau_{k,p-k} = 0$ for $k = 0, \ldots, p$.*

Lemma 4. *Let $\boldsymbol{\alpha}, \boldsymbol{\beta}$ be two 0-1 vectors of length p with equal total sums, and let $\mathbf{x}, \mathbf{y}$ be row and column sums defined by $x_i = i - \alpha_i$ and $y_i = i - \beta_i$, for $i = 1, \ldots, p$. Then*
(a) Vectors $\mathbf{x}, \mathbf{y}$ are consistent iff $\boldsymbol{\alpha} \preceq \overleftarrow{\boldsymbol{\beta}}$.
(b) Suppose that $\mathbf{x}, \mathbf{y}$ are consistent, and let T be any realization of $\mathbf{x}, \mathbf{y}$. Then T is a perfect mirror iff $\boldsymbol{\alpha} = \overleftarrow{\boldsymbol{\beta}}$.

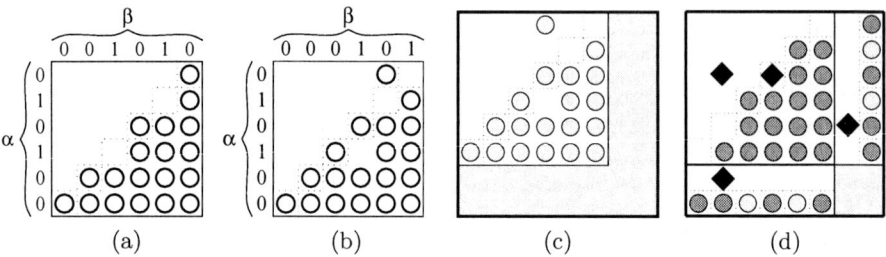

Fig. 2. *Examples of (a) a perfect mirror, (b) a skew mirror. Realizations of (c) BSM(010100, 000011) and (d) EV(010100, 001010, (2, 4)), (1 = ○, A = ●, B = ○ and C = ◆).*

5 Some Useful Gadgets

Beige skew mirror. Given two 0-1 vectors α, β of length n, we define the *beige skew mirror* as a $(n+2) \times (n+2)$ instance of 3CCP, $BSM(\alpha, \beta) = (\mathbf{x}^B, \mathbf{y}^B)$, with the following beige sums: $x_i^B = i - \alpha_i + 2$, $y_i^B = i - \beta_i + 2$, for $i = 1, \ldots, n$, and $x_i^B = y_i^B = n + 2$, for $i = n+1, n+2$. The azure and cyan sums are zero.

Lemma 5. *Let α, β be two 0-1 vectors with equal total sums. Then $BSM(\alpha, \beta)$ is consistent iff $\alpha \preceq \bar{\beta}$.*

Azure skew mirror. Given two 0-1 vectors γ, δ of length n, we define the *azure mirror* as a $(n+2) \times (n+2)$ instance of 3CCP, $ASM(\gamma, \delta) = (\mathbf{x}^A, \mathbf{y}^A, \mathbf{x}^B, \mathbf{y}^B)$, with the azure sums: $x_i^A = y_i^A = i$ for $i = 1, \ldots, n$, $x_{n+1}^A = y_{n+1}^A = 0$, $x_{n+2}^A = y_{n+2}^A = K$, and with the beige sums $x_i^B = \gamma_i$, $y_i^B = \delta_i$, for $i = 1, \ldots, n$, $x_{n+1}^B = y_{n+1}^B = 2$ and $x_{n+2}^B = y_{n+2}^B = n - K + 2$. The cyan sums are zero.

Lemma 6. *Let γ, δ be two 0-1 vectors of length n with total sums equal $n - K$. Then $ASM(\gamma, \delta)$ is consistent iff $\bar{\delta} \preceq \gamma$.*

Edge verifier. For 0-1 vectors γ, δ of length n and total sums equal $n - K$, and for an edge $e = (u, v)$ (with $u < v$) define the *edge verifier* for e, as a $(n+2) \times (n+2)$ instance of 3CCP, $EV(\gamma, \delta, e) = (\mathbf{x}^A, \mathbf{y}^A, \mathbf{x}^B, \mathbf{y}^B, \mathbf{x}^C, \mathbf{y}^C)$, where the azure and beige sums are the same as in $ASM(\gamma, \delta)$, and the cyan sums are: $x_u^C = 2$, $y_{n-u+1}^C = 1$, $x_v^C = 1$, $y_{n-v+1}^C = 2$, $x_{n+1}^C = 1$, $y_{n+1}^C = 1$.

Lemma 7. *Let γ be a 0-1 vector of length n with total sum $n - K$, and $e = (u, v)$ (with $u < v$) be an edge of G. Then $EV(\gamma, \bar{\gamma}, e)$ is consistent iff $\gamma_u = 0$ or $\gamma_v = 0$.*

Lemma 7 has the following interpretation: if we associate with γ the vertex set $U = \{u : \gamma_u = 0\}$, then $EV(\gamma, \bar{\gamma}, e)$ is consistent iff at least one endpoint of edge e belongs to U.

6 The Proof of NP-Completeness

The Reduction. Recall that G, K is the given instance of Vertex Cover, where $G = (V, E)$, $|V| = n$ and $|E| = m$. Define $J = K(n-K)+2$ and $L = (mJ+1)(n+2)$. We map G, K, into a $L \times L$ instance of 3CCP $\mathcal{I} = (\mathbf{r}^A, \mathbf{s}^A, \mathbf{r}^B, \mathbf{s}^B, \mathbf{r}^C, \mathbf{s}^C)$, where the row and column sums are defined as follows. (We only give the values of the non-zero sums.) We partition $L \times L$-matrices into $(n+2) \times (n+2)$-submatrices (*blocks*). A row or column is defined by a *block index* $a = 0, \ldots, mJ$ and an *offset* $i = 1, \ldots, n+2$. For $a \neq mJ$ the azure and beige sums are:

$$r^A_{a(n+2)+i} = s^A_{a(n+2)+i} = (mJ - a - 1)(n+2) + \begin{cases} i & i = 1, \ldots, n \\ 0 & i = n+1 \\ n - K & i = n+2 \end{cases}$$

$$r^B_{a(n+2)+i} = s^B_{a(n+2)+i} = a(n+2) + \begin{cases} i+2 & i = 1, \ldots, n \\ n+4 & i = n+1 \\ n+4+K & i = n+2 \end{cases}$$

and for $a = mJ$ the azure sums are zero and the beige sums are

$$r^B_{mJ(n+2)+i} = s^B_{mJ(n+2)+i} = \begin{cases} i+2 & i = 1, \ldots, K \\ i+1 & i = K+1, \ldots, n \\ n+2 & i = n+1, n+2 \end{cases}$$

Finally, we define the cyan sums. For $j = 0, \ldots, J-1$ and $k = 0, \ldots, m-1$ let $a = jm + k$ and $b = mJ - 1 - a$. If $e_k = (u, v)$ (with $u < v$), then

$$r^C_{a(n+2)+u} = 2 \qquad r^C_{a(n+2)+v} = 1 \qquad r^C_{a(n+2)+n+1} = 1$$
$$s^C_{b(n+2)+n-u+1} = 1 \qquad s^C_{b(n+2)+n-v+1} = 2 \qquad s^C_{b(n+2)+n+1} = 1$$

Realizations of Azure and Beige Atoms. Let $\mathcal{A}$ and $\mathcal{B}$ be $(n+2) \times (n+2)$ matrices filled with azure and beige atoms, respectively. We use notation $\mathcal{A}(\boldsymbol{\gamma}, \boldsymbol{\delta})$ for realizations of $ASM(\boldsymbol{\gamma}, \boldsymbol{\delta})$ and $\mathcal{B}(\boldsymbol{\alpha}, \boldsymbol{\beta})$ for realizations of $BSM(\boldsymbol{\alpha}, \boldsymbol{\beta})$. We define $\boldsymbol{\alpha}^0$ and $\boldsymbol{\beta}^{mJ+1}$ by $\boldsymbol{\alpha}^0 = \boldsymbol{\beta}^{mJ+1} = 0^K 1^{n-K}$.

For 0-1 vectors $\boldsymbol{\alpha}^1, \boldsymbol{\beta}^1, \ldots, \boldsymbol{\alpha}^{mJ}, \boldsymbol{\beta}^{mJ}$, each of total sum $n - K$, consider $L \times L$ azure and beige matrices of the following form:

$$\begin{bmatrix}
\mathcal{A} & \mathcal{A} & \mathcal{A} & \cdots & \mathcal{A}(\boldsymbol{\alpha}^{mJ}, \boldsymbol{\beta}^{mJ}) & \mathcal{B}(\boldsymbol{\alpha}^{mJ}, \boldsymbol{\beta}^{mJ+1}) \\
\mathcal{A} & \mathcal{A} & \mathcal{A} & & \mathcal{B}(\boldsymbol{\alpha}^{mJ-1}, \boldsymbol{\beta}^{mJ}) & \mathcal{B} \\
\mathcal{A} & \mathcal{A} & \mathcal{A} & & \mathcal{B} & \mathcal{B} \\
\vdots & & & & & \vdots \\
\mathcal{A} & \mathcal{A} & \mathcal{A}(\boldsymbol{\alpha}^3, \boldsymbol{\beta}^3) & & \mathcal{B} & \mathcal{B} \\
\mathcal{A} & \mathcal{A}(\boldsymbol{\alpha}^2, \boldsymbol{\beta}^2) & \mathcal{B}(\boldsymbol{\alpha}^2, \boldsymbol{\beta}^3) & & \mathcal{B} & \mathcal{B} \\
\mathcal{A}(\boldsymbol{\alpha}^1, \boldsymbol{\beta}^1) & \mathcal{B}(\boldsymbol{\alpha}^1, \boldsymbol{\beta}^2) & \mathcal{B} & & \mathcal{B} & \mathcal{B} \\
\mathcal{B}(\boldsymbol{\alpha}^0, \boldsymbol{\beta}^1) & \mathcal{B} & \mathcal{B} & \cdots & \mathcal{B} & \mathcal{B}
\end{bmatrix} \qquad (2)$$

Lemma 8. *Let $\mathcal{I}^{AB}$ be the restriction of $\mathcal{I}$ to the azure and beige sums. Then T is a realization of $\mathcal{I}^{AB}$ iff T has the form (2), where*

$$\boldsymbol{\alpha}^0 \preceq \overleftarrow{\boldsymbol{\beta}}^1 \preceq \boldsymbol{\alpha}^1 \preceq \overleftarrow{\boldsymbol{\beta}}^2 \preceq \ldots \preceq \boldsymbol{\alpha}^{mJ} \preceq \overleftarrow{\boldsymbol{\beta}}^{mJ+1}. \tag{3}$$

Theorem 1. *The problem 3CCP is NP-complete in the strong sense.*

Proof. (Sketch) We prove that G has a vertex cover of size K iff $\mathcal{I}$ is consistent.
($\Rightarrow$) Suppose U is a vertex cover of size K in G. Define $\boldsymbol{\gamma}$ by: $\gamma_u = 0$ iff $u \in U$. Take $\boldsymbol{\alpha}^i = \boldsymbol{\gamma}$, and $\boldsymbol{\beta}^i = \overleftarrow{\gamma}$ for $i = 1, \ldots, mJ$. By Lemma 8, this defines a realization of $\mathcal{I}^{AB}$. Furthermore, by Lemma 7, this realization can be extended to a realization of $\mathcal{I}$.
($\Leftarrow$) Let T be any realization of $\mathcal{I}$. By Lemma 8, T restricted to azure and beige atoms has the form (2). Using Lemma 3, there is j for which $\overleftarrow{\beta}^{mj+1} = \alpha^{mj+1} = \overleftarrow{\beta}^{mj+2} = \alpha^{mj+2} = \cdots = \overleftarrow{\beta}^{mj+m} = \alpha^{mj+m} = \overleftarrow{\beta}^{m(j+1)+1}$. Define $U = \{u : \alpha_u^{mj+1} = 0\}$. By Lemma 7, U is a vertex cover.

To conclude we note that the unary encoding of $\mathcal{I}$ has size $O(n^{10})$, and it can be computed easily in polynomial time. ∎

7 Final Comments

We proved that c-CCP is NP-complete for $c \geq 3$. Since 1-CCP can be solved efficiently in polynomial time (see [1]), the only unresolved case is for $c = 2$.

Relation to multicommodity flows. Consider the following problem: given a bipartite graph $H = (U, V, E)$ where E is the set of arcs directed from U to V, with each arc having capacity 1, we want to ship two commodities, from the vertices in U to the vertices in V, according to the given supplies in U and demands in V. More specifically, for each vertex $u_i \in U$ we are given a supply x_i^a of commodity a, and for each vertex $v_j \in V$ we are given a demand y_j^a of commodity a, where $a \in \{1, 2\}$. We wish to compute an integral 2-commodity flow from U to V of maximum total value. Let us call it *2-Commodity Integral 2-Layer Flow*, or *2-CI2LF*. It is known (see [3]) that the 2-commodity integral flow problem is NP-hard for directed networks. We can improve it to the 2-layer case. By modifying the argument outlined in Section 2, it is not difficult to show that 2-CI2LF is NP-hard as well: simply note that all but two atom types have unique realizations, and associate the entries not occupied by these atoms with the edges of the resulting graph H. (Another proof can be obtained by modifying the proof in [4] in a similar fashion.)

The argument above does not imply that 2-CCP is NP-complete, since the graphs corresponding to the 2-CCP problem are *complete* bipartite graphs. This leads to the following open problem: Can 2-CI2LF be solved in polynomial time for complete bipartite graphs?

Consequences to data security problems. Similarly as in [4], our result has consequences for problems in statistics and data security.

The reconstruction problem for contingency tables is similar to the 1-CCP problem, except that now we allow a realization to contain any non-negative integers. Our result implies that this problem is NP-hard even when we want to reconstruct a table whose entries are in the set $\{0, 1, \mu, \mu^2\}$, for some μ. (Modify the proof by representing each table entry in a μ-ary notation, where $\mu \geq L$, and associate color sums with the coefficients of 1, μ and μ^2.)

A related problem, arising in the 3D statistical data security problem, is to reconstruct a 3D table from its projections, which are called the row, column and file sums. Irving and Jerrum [6] proved that this problem is NP-hard even when all file sums are either 0 or 1. The work in [4] implies that the problem is NP-hard for $L \times L \times 7$ tables and all file sums equal 1. Our result implies that this problem remains NP-hard for tables of size $L \times L \times 4$ and file sums equal 1.

References

1. R.A. Brualdi. Matrices of zeros and ones with fixed row and column sum vectors. *Linear Algebra and Applications*, 33:159–231, 1980.
2. S.-K. Chang. The reconstruction of binary patters from their projections. *Communications of ACM*, 14:21–24, 1971.
3. S. Even, A. Itai, and A. Shamir. On the complexity of timetable and multicommodity flow problems. *SIAM Journal on Computing*, 5(4):691–703, 1976.
4. R.J. Gardner, P. Gritzmann, and D. Prangenberg. On the computational complexity of determining polyatomic structures by X-rays. To appear in *Theoretical Computer Science.*, 1997.
5. R.J. Gardner, P. Gritzmann, and D. Prangenberg. On the computational complexity of reconstructing lattice sets from their X-rays. Technical Report 970.05012, Techn. Univ. München, Fak. f. Math., 1997.
6. R.W. Irving and M.R. Jerrum. Three-dimensional statistical data security problems. *SIAM Journal on Computing*, 23:170–184, 1994.
7. C. Kisielowski, P. Schwander, F.H. Baumann, M. Seibt, Y. Kim, and A. Ourmazd. An approach to quantitate high-resolution transmission electron microscopy of crystalline materials. *Ultramicroscopy*, 58:131–155, 1995.
8. H.J. Ryser. *Combinatorial Mathematics*. Mathematical Association of America and Quinn & Boden, Rahway, New Jersey, 1963.
9. P. Schwander, C. Kisielowski, M. Seigt, F.H. Baumann, Y. Kim, and A. Ourmazd. Mapping projected potential, interfacial roughness, and composition in general crystalline solids by quantitative transmission electron microscopy. *Physical Review Letters*, 71:4150–4153, 1993.

Locally Explicit Construction of Rödl's Asymptotically Good Packings

Nikolai N. Kuzjurin

Institute for System Programming, Russian Academy of Sciences
B. Kommunisticheskaya 25, Moscow, 109004
nnkuz@ispras.ru

Abstract. We present a family of asymptotically good packings of l-subsets of an n-set by k-subsets and an algorithm that given a natural i finds the ith k-subset of this family. The bit complexity of this algorithm is almost linear in encoding length of i that is close to best possible complexity. A parallel NC-algorithm for this problem is presented as well.

1 Introduction

Research of the last decade has demonstrated the significance of probabilistic methods and algorithms (see [4,14]). By this reason, explicit constructions and derandomization techniques for objects whose existence had been proved by probabilistic arguments have been the focus of much research [2,14,15,19,22].

What we mean by explicit constructions? Say we wish to construct a family F of subsets of a finite set. By a *globally explicit* construction we mean an algorithm that lists all members of F in time polynomial in $|F|$. *Locally explicit* constructions would just ask for ith member of F to be evaluated in time polynomial in encoding length of i which is $O(\log |F|)$. Clearly, local is stronger than global, in analogy to the distinction between log-space and polynomial time.

One of the most significant achievements of the probabilistic method has been Rödl's probabilistic proof of the long-standing Erdős-Hanani conjecture, i.e. the proof of the existence of *asymptotically good* packings and coverings [4,17,24]. Our purpose here is to accomplish this result deterministically by means of the construction that is *locally explicit*.

We need to introduce some terminology and notation. Let $l < k < n$ be natural numbers and $[n] = \{0, 1, \ldots, n-1\}$ (the n-set). By a k-subset of $[n]$ we mean any subset g of $[n]$ such that $|g| = k$.

An (n, k, l)-packing is defined as a family P of k-subsets of $[n]$ such that every l-subset of $[n]$ is contained in at most one $g \in P$. The density of a packing P is defined as

$$\frac{|P|\binom{k}{l}}{\binom{n}{l}}.$$

It is easy to see that the density of any packing is at most 1. The sequences of packings of density tending to 1 as n tends to infinity are called *asymptotically good*.

P.Erdős and H.Hanani [6] conjectured that for all fixed l and k with $l < k$ the sequences of asymptotically good packings exist. In 1985 V.Rödl [17] using probabilistic methods proved this conjecture in full. Futher strenthenings and extensions were given in [3,7,8,11,13,16,18,23]. Recently D. Grable [9] found globally explicit construction of asymptotically good packings.

In this paper we present locally explicit construction of asymptotically good packings which has almost linear (in the input size) bit complexity. Our main contribution may be formulated as follows. We write $f = \tilde{O}(g)$ if there exists a constant $c > 0$ such that $f = O(g(\log g)^c)$.

Theorem 1. *For any natural numbers l and k with $l < k$ there is a sequence $P_n(k, l)$ of asymptotically good (n, k, l)-packings and enumeration of k-subsets in each $P_n(k, l)$ such that there is an algorithm of $\tilde{O}(\log n)$ bit complexity which given natural n, k, l and i finds the ith k-subset of $P_n(k, l)$.*

Moreover, we present a parallel NC-algorithm for this problem (Section 6). The proof of Theorem 1 will be given in the next 4 sections. Our result is not independent on Rödl's result. We use reccurrencies and algebraic construction to decrease exponentially the dimension and then apply Rödl's result along with exhaustive search for very small set.

2 Explicit Construction

We begin this section with a simple observation concerning the complexity of finding asymptotically good packings. If we compare all collections of k-subsets of an n-element set and choose one that is an (n, k, l)-packing of maximal size we will have found a maximal packing that is asymptotically good in view of [17]. The number of such collections is

$$O(2^{\binom{n}{k}})$$

This value characterizes the complexity of exhaustive search of finding a maximal packing as described above. In addition to form a basis of comparison for our final results, we shall use these facts in a modified situation in the sequel.

We begin with an outline of the construction. The deterministic construction consists of two parts. First of all, we choose some parameter t which will tend to infinity very slowly (as $n \to \infty$). This function will be defined more precisely later but certainly we shall always assume that $t < (1/2)\sqrt{n}$. The first part of construction is an explicit algebraic construction of asymptotically good (n, t, l)-packings. The second part comprises finding a maximal (t, k, l)-packing which is asymptotically good (as $t \to \infty$) because of Rödl's result [17]. The desired (n, k, l)-packing is the composition of the (n, t, l)-packing and the (t, k, l)-packing. This means that at the place of each t-subset of the (n, t, l)-packing we substitute a (t,k,l)-packing which is equivalent to the maximal packing $R(t,k,l)$. Clearly, the size of such a composed packing is the product of the sizes of the corresponding packings.

We describe now the first part of our construction in detail. Let p be the maximum prime such that

$$p < (\log \log n)^{1/5}. \tag{1}$$

We shall assume $p > t$. Let a natural r be such that

$$(\log \log n)^{1/10} < \frac{n}{p^r t} < (\log \log n)^{1/3}. \tag{2}$$

Let p_1 ($p_1 > t$) be the maximum prime such that

$$p_1 < \frac{n}{tp^r}. \tag{3}$$

It will be essential in the sequel that in view of [10] and (2-3) $p_1 \to \infty$ and $p_1 \sim \frac{n}{tp^r}$ as $n \to \infty$. It is easy to see that choosing appropriate r we can satisfy (2) because p is sufficiently small in view of (1).

Consider the system

$$\sum_{i=1}^{t} i^j x_i = 0 \bmod p, \quad j = 1, \ldots, t - l. \tag{4}$$

For arbitrarily fixed variables $x_{i_1}, \ldots, x_{i_l}$ in (4) we obtain a system with Vandermonde's determinant. For this reason the number of solutions is exactly p^l. Denote by $V(p,t,l)$ the set of solutions of the system (4). Note that the system (4) was used in [13] to prove some extensions of the Erdős-Hanani conjecture. In coding theory it is well-known as RS-codes over large alphabets.

Let $N = tp_1 p^r$, $[N] = \{0, 1, \ldots, N-1\}$. Consider a partition of $[N]$ in t parts

$$[N] = \bigcup_{i=1}^{t} S_i, \quad |S_i| = \frac{N}{t} = p_1 p^r,$$

where $S_i = \{(i-1)p_1 p^r, \ldots, ip_1 p^r - 1\}$. Define the function f which enumerates the Cartesian product $[p]^r \times [p_1]$ by natural numbers from $[p_1 p^r]$ in such a way that for any $z = (x_1 \ldots, x_r, y)$ from $[p]^r \times [p_1]$

$$f(z) = \sum_{i=1}^{r} x_i p^{r-i} + y p^r. \tag{5}$$

Note that f is a bijection. Define the mapping F, by componentwise application of f

$$F : V(p,t,l)^r \times V(p_1,t,l) \to [N/t]^t,$$

where $[N/t]^t$ denotes t^{th} Cartesian power of the set $[N/t] = \{0, 1, \ldots, N/t - 1\}$. Note that F defines t-tuples with elements belonging to $[p_1 p^r]$. For each such t-tuple we may obtain in an obvious way a t-subset with elements in S_i, $i = 1, \ldots, t$ (it is sufficient to modify x_i as follows $x_i = x_i + (i-1)p_1 p^r$). Denote this family of t-subsets by $S_n(t, l)$.

For the second part of our construction consider all (t, k, l)-packings and select one that is maximal denoting it by $R(t, k, l)$. Substitute (t, k, l)-packings equivalent to $R(t, k, l)$ in place of each t-tuple of $S_n(t, l)$ and denote such composed packing by $Q_t(n, k, l)$. Let $P_n(k, l) = Q_t(n, k, l)$ with $t = (\log \log \log n)^{1/3k}$.

3 Why the Construction is Asymptotically Good?

It is easy to verify that all t-subsets of $S_n(t, l)$ form (n, t, l)-packing. Indeed, fixing values of arbitrarily chosen l variables (for example, $x_1 = i_1, x_2 = i_2, \ldots, x_l = i_l$) for every i_j, $j = 1, \ldots, l$ the $r+1$ numbers $f^{-1}(i_j)$ are uniquely defined. Let $f^{-1}(i_j) = (z_1^{(j)}, \ldots, z_r^{(j)}, y^{(j)})$. For every l-tuple $(z_j^{(1)}, \ldots, z_j^{(l)})$, $j = 1, \ldots, r$ and $(y^{1)}, \ldots, y^{(l)})$ the solution of the system (4) with the values of l variables equal to numbers of this l-tuple is uniquely defined as well. It can be found by solving r systems of linear equations over $\mathbf{F}_p$ and one system of linear equations over $\mathbf{F}_{p_1}$. These $r+1$ solutions define unique t-subset by the mapping F.

Why the composition of (n, t, l)-packing P and (t, k, l)-packing Q is a (n, k, l)-packing? It is, of course, a family of k-subsets of $[n]$. Moreover, any two k-subsets from the same t-subset of P have no common l-subset because Q is (t, k, l)-packing. Any two k-subsets from different t-subsets have no common l-subset because P is (n, t, l)-packing. Thus, the resulting composed family is (n, k, l)-packing. Note, that such packings are the product of packings in the sence of [25].

In view of [10] $p_1 \sim \frac{n}{tp^r}$ as $n \to \infty$. The size of the packing $S_n(t, l)$ is

$$|S_n(t, l)| = p^{rl} p_1^l = p^{rl}((1 - o(1))\frac{n}{tp^r})^l = (1 - o(1))(\frac{n}{t})^l.$$

The size of the packing $P_n(k, l)$ is the product of the sizes of the corresponding packings

$$|P_n(k, l)| = |S_n(t, l)||R(t, k, l)|,$$

where $t = (\log \log \log n)^{1/3k}$. With our choice of t the following relation holds: $t^l \sim (t)_l$ as $n \to \infty$ and

$$|P_n(k, l)| = (1 - o(1))(\frac{n}{t})^l (1 - o(1))\frac{(t)_l}{(k)_l} = (1 - o(1))\frac{n^l}{(k)_l},$$

i.e. $P_n(k, l)$ is asymptotically good.

4 Algorithm

To show that our construction is locally explicit we must demonstrate how to number efficiently all k-subsets in $P_n(k,l)$. Let $P = p_1 p^r$ and $\mathbf{x} = (x_1, \ldots, x_t) \in S_n(t,l)$, $x_1 < \ldots < x_t$. The lexicographic number $L(\mathbf{x})$ of $\mathbf{x}$ is defined as follows. Let $y_i = x_i \pmod{P}$ and

$$L(\mathbf{x}) = \sum_{i=1}^{l} y_i P^{l-i}. \tag{6}$$

We number t-subsets of $S_n(t,l)$ in the lexicographic order, and k-subsets in $R(t,k,l)$ arbitrarily, thus, forming a list. At first, we number k-subsets in the first t-subset followed by those in the second one and so on. There are P^l t-subsets in $S_n(t,l)$ and there are $L = |R(t,k,l)|$ k-subsets in the list $R(t,k,l)$.

Recall that we know natural n, k, l and i as the input and wish to find efficiently the ith k-subset in $P_n(k,l)$. We describe now an algorithm and estimate its complexity. By the complexity we mean the bit complexity (see [1] for details). Algorithms for fast multiplication of two n bit integers and division of $2n$ bit integer by n bit integer of bit complexity $O(n \log n \log \log n)$ will be used [1].

Algorithm LocalSubset:
Input: n, k, l, i;
Output: the ith k-subset of $P_n(k,l)$.

1) Find the primes p and p_1 and natural r satisfying (1)-(3).
2) Find maximal (t,k,l)-packing $R(t,k,l)$.
3) Given i, $1 \leq i \leq P^l L$ represent it in the form

$$i = TL + d, \qquad 0 \leq d < L.$$

Then T is the number of t-subset and d is the number of k-subset in the list $R(t,k,l)$ corresponding to the Tth t-subset.

4) Transform T from the binary representation to the $P = p_1 p^r$ base representation of length l (denote it by $(y_1, \ldots, y_l)$).
5) Given $y_1, y_2, \ldots, y_l$ find l $(r+1)$-tuples $f^{-1}(y_j)$, $j = 1, \ldots, l$. Let $f^{-1}(y_j) = (z_1^{(j)}, \ldots, z_r^{(j)}, y^{(j)})$.
6) For each l-tuple of numbers $(z_j^{(1)}, \ldots, z_j^{(l)})$, $j = 1, \ldots, r$ and $(y^1), \ldots, y^{(l)})$ find the solution of system (4) with the first l variables equal to the values of this l-tuple. Thus, we obtain $r+1$ t-tuples.
7) Given this $r+1$ t-tuples of coefficients in representation (5), find the desired unique t-subset by the map F.
8) Given the list $R(t,k,l)$ and the number d select the dth k-subset in the t-subset that was found at the previous step.

5 Complexity

Now we estimate the complexity of each step.

1) It is known [10] that for any fixed $\alpha > 23/42$ and sufficiently large m the maximum prime p not greater than m is $p \geq m - m^\alpha$. Using this fact we estimate the complexity of finding the maximum prime $p < m$ as $m^\alpha T(p)$, where $T(p)$ is the complexity of primality test for p. It is sufficient to use the estimate $T(p) = O(\sqrt{p})$. Taking into account that in our construction we used only "small" primes p and p_1 (say, less than $m = (\log \log n)^{1/3}$) the bit complexity of finding the primes p and p_1 is $o(\log \log n)$.

2) To find maximal packings $R(t, k, l)$ and number its k-subsets in the list we use exhaustive search. By the observation which began this section we have that the complexity of this step is at most

$$O(2^{\binom{t}{k}}) = o(\log \log n),$$

where we have taken into account that $t = (\log \log \log n)^{1/3k}$.

3) It may be done using one division.

4) It may be done in $O(l)$ arithmetic operations with $O(\log n)$ bit integers. Indeed, we must divide T by P

$$T = T_1 P + y_l, \quad 0 \leq y_l < P,$$

then divide T_1 by P

$$T_1 = T_2 P + y_{l-1}, \quad 0 \leq y_{l-1} < P,$$

and so on until we define all y_j, $j = 1, 2, \ldots, l$. Thus, we do l divisions (e.g. constant) of integers of $O(\log n)$ bit each. Using fast division algorithm we may do it with bit complexity $\tilde{O}(\log n)$.

5) Given $y_1, y_2, \ldots, y_l$ we may find $r+1$ l-tuples $f^{-1}(y_j)$, $j = 1, \ldots, l$ as follows. Evaluate the powers of p: $p^{\frac{r}{2}}, p^{\frac{r}{4}}, \ldots, p^{\frac{r}{2^i}}, \ldots$ Because $r = o(\log n)$ there are $O(\log \log n)$ such powers and all such powers may be evaluated with bit complexity $\tilde{O}(\log n)$ using the fast multiplication algorithm.

To do this use the representation (see (5))

$$f(z) = P_1(z) + p^{\frac{r}{2}} P_2(z), \tag{7}$$

where $\deg P_i \leq r/2$, $i = 1, 2$.

Dividing $f(z)$ by $p^{r/2}$ we find P_1 and P_2. Continuing this recursive process we find all x_i and y in representation (5). Let $T(r)$ be the bit complexity of evaluating $f(z)$ which is the number less than $p_1 p^r$. The bit complexity is $\tilde{O}(\log n)$ because we have the recurrence:

$$T(r) \leq 2T(r/2) + \tilde{O}(\log p^r). \tag{8}$$

This implies the estimate
$$T(r) = O(\log r)\tilde{O}(\log p^r).$$
Taking into account that $r = o(\log n)$ we obtain the desired result.

6) We may find the solution of system (4) with the first l variables equal to the values of this l-tuple in $O(rt^3)$ operations in the field $\mathbf{F}_p$. To do this it is sufficient to solve r systems of linear equations over the finite field $\mathbf{F}_p$ and one system over $\mathbf{F}_{p_1}$. Note that with our choice of parameters r and t the relation $O(rt^3) = o(\log n)$ holds.

7) Given $r + 1$ t-tuples of coefficients in representation (5), we may find the desired unique t-tuple by the map F with $\tilde{O}(\log n)$ bit complexity. This is a consequence of the recurrences (7) and (8) which we use now for evaluations in opposite direction: from a t-tuple of numbers z to the number $f(z)$ by (5).

8) It may be done in time that is linear in the size of the list (i.e. in $o(\log \log n)$ operations). Finally, we must select this k-subset in the Tth t-subset in $S_n(t,l)$ that was found earlier.

What is the total complexity of the algorithm **LocalSubset**? All steps (as it was shown above) have bit complexity $\tilde{O}(\log n)$. Hence, we proved Theorem 1 claimed in the Introduction.

6 Parallel Complexity

The detailed analysis of the above algorithm shows that all operations may be efficiently parallelized and we obtain parallel algorithm which terminates in time $O((\log \log n)^{\text{const}})$ using $O((\log n)^{\text{const}})$ parallel processors.

Note that fast multiplication and division algorithms were parallelized (see [12]) and both parallel algorithms terminates in time $O((\log n)^{\text{const}})$ when operate with $O(n)$ bit numbers. In our case this parallel time is $O((\log \log n)^{\text{const}})$ because we operate with $O(\log n)$ bit numbers only.

Observe, briefly the main steps of the algorithm **LocalSubset**. Note that the sequential complexity of steps 1), 2) and 8) is $o(\log \log n)$ and it is nothing to parallelize within these steps. Steps 3–4 may be easily done in parallel time $O((\log \log n)^{\text{const}})$. At step 5 we evaluate $O(\log \log n)$ powers of p using fast parallel multiplication algorithm and recursively find all x_i and y in the representation (5). The parallel complexity is $O((\log \log n)^{\text{const}})$ because there are $\log r = o(\log \log n)$ recursive levels. Similar arguments show that this is true for step 7. At step 6 we solve for r l-tuples of numbers (independently) system (4) over the field $\mathbf{F}_p$ and for one l-tuple over the field $\mathbf{F}_{p_1}$. This may be done in parallel time $O((\log \log n)^{\text{const}})$ [12].

Thus, the total parallel time is $O((\log \log n)^{\text{const}})$ on $O((\log n)^{\text{const}})$ parallel processors. Noting that the class **NC** consists of problems solvable in deterministic time polynomial in the logarithm of the size of the input on polynomially-many parallel RAM processors (for details, see [12]) we obtain the following

Theorem 2. *There is an NC-algorithm which given arbitrary natural n, k, l and i finds the ith k-subset of $P_n(k,l)$.*

7 Discussion

There are three essential ideas to obtain our locally explicit construction:
1) the notion of composition of packings;
2) the algebraic construction of (n,t,l)-packings with slowly increasing t (as $n \to \infty$);
3) the use of refinement and direct products of packings to avoid large primes. The third part may be done in other manner using recent result of [20,21]. Instead of system (4) we may use its analogue over the field $\mathbf{F}_q$, where q is a prime power close to n. The efficient construction of such fields with $q \sim n$ was presented in [20,21].

This work was partially supported by the grant 98-01-00509 of the Russian Foundation for Fundamental Research. Part of this work was done while the author was visiting Bielefeld University.

References

1. A.V. Aho, J.E. Hopcroft and J.D. Ullman, The design and analysis of computer algorithms, Addison-Wesley, 1976.
2. N. Alon, J. Bruck, J. Naor, M. Naor and R. Roth, Construction of asymptotically good, low-rate error-correcting codes through pseudorandom graphs, IEEE Transactions on Information Theory, **38** (1992) 509-516.
3. N. Alon, J.H. Kim and J.H. Spencer, Nearly perfect matchings in regular simple hypergraphs, Preprint, 1996.
4. N. Alon and J.H. Spencer, The probabilistic method. John Wiley and Sons, New York, 1992.
5. P. Erdős and J. Spencer, Probabilistic methods in combinatorics, Akademic Press. New York, 1974.
6. P. Erdős and H. Hanani, On a limit theorem in combinatorial analysis. Publ. Math. Debrecen. **10** (1963) 10 - 13.
7. P. Frankl and V. Rödl, Near perfect coverings in graphs and hypergraphs, Europ. J. Combinatorics, **6** (1985), 317-326.
8. D.M. Gordon, O. Patashnik, G, Kuperberg and J.H. Spencer, Asymptotically optimal covering designs, J. Comb. Theory **A 75** (1996) 270 - 280.
9. D.A. Grable, Nearly-perfect hypergraph packing is in **NC**, Information Process. Letters, **60** (1997) 295-299.
10. H. Iwaniec and J. Pintz, Primes in short intervals, Monatsch. Math. **98** (1984) 115-143.
11. J. Kahn, A linear programming perspective on the Frankl-Rödl-Pippenger theorem, Random Structures and Algorithms, 8 (1996) 149-157.
12. R.M. Karp and V. Ramachandran, Parallel algorithms for shared-memory machines, In Handbook of Theoretical Computer Science (ed. J. van Leeuwen), Elsevier, 1990, 869-942.
13. N.N. Kuzjurin, On the difference between asymptotically good packings and coverings. - European J. Comb. **16** (1995) 35 - 40.
14. R. Motwani and P. Raghavan, Randomized Algorithms, Cambridge University Press, 1995.

15. M. Naor, L.J. Shulman and A. Srinivasan, Splitters and near-optimal derandomization, Proc. 36th Ann. IEEE FOCS, 1995, 182-191.
16. N. Pippenger and J. Spencer, Asymptotic behavior of the chromatic index for hypergraphs. J. Comb. Theory. Ser. **A51** (1989) 24 - 42.
17. V. Rödl, On a packing and covering problem. Europ. J. Combinatorics. **5** (1985) 69 - 78.
18. V. Rödl and L. Thoma, Asymptotic packing and the random greedy algorithm, Random Structures Algorithms **8** (1996) 161 - 177.
19. M. Saks, A. Srinivasan and S. Zhou, Explicit dispersers with polylog degree, Proc. Annu. 27th ACM STOC-95, 1995, 479-488.
20. I. Sparlinski, Approximate constructions in finite fields, In Finite Fields and Applications, London Math. Soc., Lect. Notes Ser., v. 233, Cambridge Univ. Press, Cambridge, 1996, 313-332.
21. I. Sparlinski, Finding irreducible and primitive polynomials, Appl. Algebra in Engin., Commun. and Computing, **4** (1993) 263-268.
22. J.H. Spencer, Ten Lectures on the Probabilistic Method, SIAM, Philadelphia, 1987.
23. J. Spencer, Asymptotic packing via a branching process, Random Structures Algorithms **7** (1995) 167 - 172.
24. J. Spencer, Asymptotically good coverings. Pacific J. Math. **118** (1985) 575 - 586.
25. V.A. Zinoviev, Cascade equal-weight codes and maximal packings, Problems of Control and Information Theory **12** (1983) 3 - 10.

Proof Theory of Fuzzy Logics: Urquhart's C and Related Logics *

Matthias Baaz[1], Agata Ciabattoni[2], Christian Fermüller[1], and Helmut Veith[1]

[1] Technische Universität Wien, Karlsplatz 13, A-1040 Austria
{baaz, chrisf}@@logic.at, veith@@dbai.tuwien.ac.at
[2] Dipartimento di Scienze Dell'Informazione Via Comelico, 39 Milano, Italy
ciabatto@@dotto.usr.dsi.unimi.it

Abstract. We investigate the proof theory of Urquhart's **C** and other logics underlying the most prominent fuzzy logics, such as Gödel, Product, and Łukasiewicz logic. All these logics share the property that their truth values are linearly ordered. We define hypersequent calculi for such logics, and show the following results: (1) Contraction-free counterparts of intuitionistic logic and Gödel logic (including **C**) admit cut-elimination. (2) Validity in these logics is decidable. (3) Hajek's basic fuzzy logic **BL** properly extends the contraction-free Gödel logic; the axiom for commutativity of the minimum is independent from the other axioms of **BL**. (4) All abovementioned logics are distinct from each other.

1 Introduction

Fuzzy logics are usually defined by arithmetic truth functions over the unit interval; this framework facilitates successful application of fuzzy formalisms to areas like control, scheduling, and AI, see e.g. [16]. Foundational investigations, too, have focused on algebraic and model theoretic aspects of fuzzy logics such as continuous t-norms (see [7] for the most thorough and recent treatment). While deep results have been obtained along this line of research, the proof theory of fuzzy logic is considerably less developed – with the notable exception of Avron's elegant cut-free formalisation [2] of infinite valued Gödel logic [6,4]. Methodologically, the key concept used in Avron's work are *hypersequents*, a generalization of Gentzen style sequents.

The aim of this paper is to gain a better proof theoretic understanding of the other eminent formalisations of fuzzy logic (e.g. Łukasiewicz logic [9,10] and product logic [8]) by studying logics that are contained in all these fuzzy logics; two outstanding examples of such *basic logics* are Urquhart's logic **C** as introduced in §3 of his handbook article on many-valued logic [15], and Hájek's Basic Fuzzy Logic **BL** [7]. Being of interest in their own right (since they express properties common to many fuzzy logics), they may in particular shed light

---
* Extended abstract, omitting most proofs; a full paper is in preparation. Work supported by the Austrian Science Foundation FWF (Grant P12652-MAT) and the COST Action # 15: *Many-valued Logics for Computer Science Applications*.

on long time open questions about the existence of cut-free (analytic) calculi for various fuzzy logics. (See Section 3 for some remarks on the importance of cut-elimination.)

Semantically, **C** has been characterized by model structures on ordered Abelian monoids and is therefore contained in Lukasiewicz logic which Urquhart characterized by ordered Abelian groups [15]. It has been noted earlier [11] that proof-theoretically, **C** corresponds to a Gödel logic without contraction. Here, we investigate *two different* Gödel logics without contraction that differ only in the axioms of residuation; the weaker of these logics coincides with **C**.

To obtain a cut-free calculus for such logics it is appropriate to proceed in a modular manner. In particular, linearity of truth values – a crucial property of all fuzzy logics – can be enforced on a given sequent calculus by transferring it to a hypersequent calculus in analogy to Avron's work on Gödel logic. It thus suffices to identify appropriate analytic calculi for the logics without linearity; in our case, they will be contraction-free fragments of intuitionistic logic.

Our main results can be summarized as follows:

- The contraction-free counterparts of intuitionistic logic and Gödel logic (including **C**) admit cut-elimination.
- Validity in these logics is decidable (and in fact in PSPACE).
- Hájek's **BL** properly extends the contraction-free Gödel logics. In particular, the axiom $[A \wedge (A \supset B)] \supset [B \wedge (B \supset A)]$ of **BL** (i.e., commutativity of the arithmetic minimum) is independent from the other axioms; this solves a question posed by Hájek.
- All abovementioned logics are distinct from each other.

Sections 2 present the Hilbert respectively Gentzen style and hypersequent calculi. Sections 3 and 4 contain the main results related to cut-elimination and **C**. Finally, Section 5 is devoted to **BL**.

2 Hilbert, Gentzen, and Hypersequent Calculi

Hilbert-Style Systems. The logics we are interested in will be defined via subsets of the following set of axioms.

Ax1 : $A \supset (B \supset A)$
Ax2 : $(A \supset B) \supset [(C \supset A) \supset (C \supset B)]$
Ax3 : $[A \supset (C \supset B)] \supset [(C \supset (A \supset B)]$
Ax4 : $(A \wedge B) \supset A$
Ax5 : $(A \wedge B) \supset B$
Ax6 : $A \supset [C \supset (A \wedge C)]$
Ax7 : $A \supset (A \vee B)$
Ax8 : $B \supset (A \vee B)$
Ax9 : $[(A \supset C) \wedge (B \supset C)] \supset [(A \vee B) \supset C]$

Res1 : $[(A \wedge B) \supset C] \supset [A \supset (B \supset C)]$
Res2 : $[(A \supset (B \supset C)] \supset [(A \wedge B) \supset C]$
Lin : $(A \supset B) \vee (B \supset A)$
Com : $[A \wedge (A \supset B)] \supset [B \wedge (B \supset A)]$
Abs : $\bot \supset A$
Contr : $[A \supset (A \supset B)] \supset (A \supset B)$

RULES: Modus Ponens

We shall investigate the following systems:

$$\mathbf{I}^- : \{\mathrm{Ax1}, \ldots, \mathrm{Ax9}, \mathrm{Abs}\} \qquad \mathbf{C} : \mathbf{I}^- \cup \{\mathrm{Lin}\}$$
$$\mathbf{I}^{-*} : \mathbf{I}^- \cup \{\mathrm{Res1}, \mathrm{Res2}\} \qquad \mathbf{C}^* : \mathbf{I}^{-*} \cup \{\mathrm{Lin}\}$$

For each of the above logics, one may also consider the fragments obtained by omitting the absurdity axiom Abs. As the presence of $\bot$ (and Abs) does not make any essential difference for the correspondences obtained, we will not explicitly state the results for those logics. The treatment of negation as a connective can be seen similarly, because $\neg A$ can be defined as $A \supset \bot$. It is then possible to derive the axioms presented in [11] for all systems considered.

We remark that Res1 is in fact derivable in $\mathbf{I}^-$.

Observe that both, $\mathbf{I}^-$ and $\mathbf{I}^{-*}$, are systems of intuitionistic logic without contraction. Urquhart's original formulation [15] is given by $\mathbf{C} - \{\mathrm{Abs}\}$. Since adding axiom Abs leaves the proof-theoretic properties of the logics unchanged, all results obtained about $\mathbf{C}$ also hold for $\mathbf{C} - \{\mathrm{Abs}\}$.

The starred systems are extensions of the above logics obtained by adding the residuation axioms Res1 and Res2; yet these axioms become redundant in presence of contraction. In Section 5, we show that Hájek's Basic Logic $\mathbf{BL}$ [7] corresponds to the logic obtained by adding axiom Com to $\mathbf{C}^*$. $\mathbf{C}, \mathbf{C}^*$, and $\mathbf{BL}$ turn into Gödel logic if we add axiom Contr (contraction).

To put our results into a broader context consider the following figure. It shows proper inclusions between various logics (see Corollary 7).

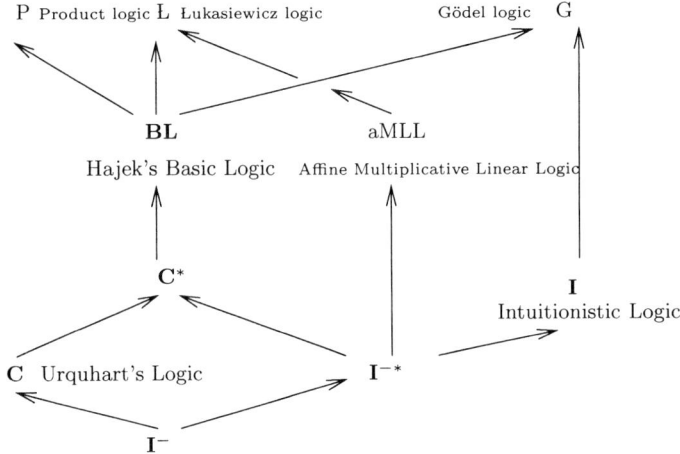

Sequent Calculi. For the purposes of this paper, it is convenient to treat sequents as *multisets* of formulas. Therefore, we do not need the exchange rule, and obtain the following calculus $\mathbf{LJ}$ for intuitionistic logic:

Axioms

$$A \Rightarrow A \qquad \bot \Rightarrow A \qquad \frac{\Gamma_1 \Rightarrow A \quad A, \Gamma_2 \Rightarrow B}{\Gamma_1, \Gamma_1 \Rightarrow B} \ (CUT)$$

Structural Rules

$$\frac{\Gamma \Rightarrow C}{\Gamma, A \Rightarrow C} \ (W) \qquad \frac{\Gamma, A, A \Rightarrow C}{\Gamma, A \Rightarrow C} \ (C)$$

Logical Rules

$$\frac{\Gamma, A \Rightarrow B}{\Gamma \Rightarrow A \supset B} \ (\supset\text{-right}) \qquad \frac{\Gamma_1 \Rightarrow A \quad B, \Gamma_2 \Rightarrow C}{\Gamma_1, \Gamma_2, A \supset B \Rightarrow C} \ (\supset\text{-left})$$

$$\frac{\Gamma_1 \Rightarrow A \quad \Gamma_2 \Rightarrow B}{\Gamma_1, \Gamma_2 \Rightarrow A \wedge B} \ (\wedge\text{-right}) \qquad \frac{\Gamma, A, B \Rightarrow C}{\Gamma, A \wedge B \Rightarrow C} \ (\wedge\text{-left})$$

$$\frac{\Gamma \Rightarrow A_i}{\Gamma \Rightarrow A_1 \vee A_2} \ (\vee_i\text{-right}) \qquad \frac{\Gamma, A \Rightarrow C \quad \Gamma, B \Rightarrow C}{\Gamma, A \vee B \Rightarrow C} \ (\vee\text{-left})$$

Remark: Note that in the above calculus we can replace the ($\vee$−left) rule by:

$$\frac{\Gamma_1, A \Rightarrow C \quad \Gamma_2, B \Rightarrow C}{\Gamma_1, \Gamma_2, A \vee B \Rightarrow C} \ (\vee'\text{-left})$$

This rule corresponds to the variant $(A \supset C) \supset [(B \supset C) \supset [(A \vee B) \supset C]]$ of Ax9 (in the sense of Definition 3, below). It is easy to see that, using cut, the ($\vee$−left) and ($\vee'$−left) rules are interderivable, while in a cut-free and contraction-free context this is not the case.

When investigating contraction-free fragments of **LJ**, we can consider alternative formulations for ($\wedge$−right) and ($\wedge$−left):

$$\frac{\Gamma \Rightarrow A \quad \Gamma \Rightarrow B}{\Gamma \Rightarrow A \wedge B} \ (\wedge'\text{-right}) \qquad \frac{\Gamma, A_i \Rightarrow C}{\Gamma, A_1 \wedge A_2 \Rightarrow C} \ (\wedge'_i\text{-left}) \quad \text{for } i = 1, 2$$

As is well known, using (C) and (W), these rules are interderivable with the corresponding ones in **LJ**, while in absence of contraction, they are distinct.

Thus one obtains four different calculi by substituting the rules for conjunction in contraction-free **LJ** by any combination of the above rules, i.e. ($\wedge$−left, $\wedge'$−right) or ($\wedge'_i$−left, $\wedge'$−right) or ($\wedge'_i$−left, $\wedge$−right) or ($\wedge$−left, $\wedge$−right). Let us discuss these cases. The first combination of these rules yields an undesirable result: it allows to derive contraction, and cuts are not eliminable see e.g. [14]. The **LJ** calculus without contraction containing the second combination of the above rules for conjunction corresponds to the Hilbert system given by axioms $\{Ax1 \ldots Ax9, \text{Res1}, \text{Abs}\}$ and axiom $(A \supset B) \supset ((A \supset C) \supset (A \supset (B \wedge C)))$, see [13,1]. We shall investigate the remaining two systems:

LJ$^-$
The rules for conjunction are ($\wedge'_i$−left) and ($\wedge$−right) and the remaining ones are those of **LJ** *except contraction* (C).
LJ^{-*}
Axioms and rules are those of **LJ** *except contraction* (C)

Hypersequent Calculi. Hypersequent calculi are a simple and natural generalization of ordinary Gentzen calculi [2,3].

Definition 1. *A* hypersequent *is a structure of the form:*

$$\Gamma_1 \Rightarrow \Delta_1 \mid \Gamma_2 \Rightarrow \Delta_2 \mid \cdots \mid \Gamma_n \Rightarrow \Delta_n$$

where every $\Gamma_i \Rightarrow \Delta_i$ is an ordinary sequent which is called component *of the hypersequent.*

The intended interpretation of the symbol "|" is usually disjunctive. As in sequent calculus, it is convenient to consider hypersequents as *multisets* of sequents. In the following, we use intuitionistic versions of hypersequents, i.e., the right hand side of components (i.e. sequents) contains at most one formula.

Like in ordinary sequent calculi, in a hypersequent calculus there are axioms and rules which are divided into two groups: *logical rules* and *structural rules*. The logical rules are essentially the same as those in sequent calculi, the only difference being the presence of dummy contexts H and H', called *side hypersequent* which are used as variables for (possibly empty) hypersequents.

The structural rules are divided into *internal* and *external rules*. The internal rules deal with formulas within components. If they are present, they are the usual weakening and contraction. The external rules manipulate whole components within a hypersequent. These are external weakening (EW) and external contraction (EC):

$$\frac{H}{H \mid \Gamma \Rightarrow A} \text{ (EW)} \qquad \frac{H \mid \Gamma \Rightarrow D \mid \Gamma \Rightarrow D}{H \mid \Gamma \Rightarrow D} \text{ (EC)}$$

In hypersequent calculi it is possible to define further structural rules which simultaneously act on several components of one or more hypersequent. It is this type of rule which increases the expressive power of hypersequent calculi with respect to ordinary sequent calculi. Indeed let us consider, for instance, the following rule (see [2])

$$\frac{H \mid \Pi_1, \Gamma_1 \Rightarrow A \quad H' \mid \Pi_2, \Gamma_2 \Rightarrow B}{H \mid H' \mid \Pi_1, \Pi_2 \Rightarrow A \mid \Gamma_1, \Gamma_2 \Rightarrow B} \text{ (Communication)}$$

Its intuitive meaning is that if we take a hypersequent as representing a multiprocess, then the above rule depicts an exchange of information between such multiprocesses. As shown in [2], the communication rule allows to prove axiom Lin.

For each of the sequent calculi discussed in the previous section, we define a corresponding hypersequent calculus:

HC, HC*, GLC (cf. [3])

Axioms, internal structural rules and logical rules are like **LJ⁻**, **LJ⁻*** and **LJ** respectively. (I.e., no internal contraction for **HC** and **HC***).

Further structural rules are (EW), (EC) and Communication

Our choice of the communication rule is not arbitrary. Two alternatives have been proposed in [3] for infinite valued Gödel logic. Neither

$$\frac{H \mid \Gamma_1, \Gamma_2 \Rightarrow A \quad H' \mid \Gamma_1, \Gamma_2 \Rightarrow B}{H \mid H' \mid \Gamma_1 \Rightarrow A \mid \Gamma_2 \Rightarrow B} \text{ (Com')}$$

is admissible in a contraction-free context nor the combination of

$$\frac{H \mid \Pi, \Gamma \Rightarrow A}{H \mid \Pi \Rightarrow A \mid \Gamma \Rightarrow A} \text{ (S}_I\text{)} \quad \text{and} \quad \frac{H \mid \Gamma_1 \Rightarrow A \quad H' \mid \Gamma_2 \Rightarrow B}{H \mid H' \mid \Gamma_2 \Rightarrow A \mid \Gamma_1 \Rightarrow B} \text{ (Com'')}:$$

Proposition 1. *Internal contraction is definable from either rule* (S_I) *or rule* (Com').

3 Cut Elimination

The question whether a logic enjoys cut-elimination (i.e., whether there is a cut-free Gentzen style system for this logic) is of eminent interest especially from the Computer Science point of view. Cut-free sequent and hypersequent calculi are "analytic" in the sense that a proof only contains formulas that occur as subformulas in the end sequent. This has important consequences. In our case, it implies that derivability is decidable. In fact it is not hard to see that – for the systems considered here – the problem is in PSPACE. More generally, we can say that cut-elimination is an essential prerequisite for efficient proof search.

Theorem 1. $\mathbf{LJ}^-, \mathbf{LJ}^{-*}$ *and* $\mathbf{LJ}$ *admit cut-elimination.*

Proof. (Sketch) For **LJ**, this is well-known. An inspection of the classical proof shows that the absence of contraction does not affect cut-elimination.

Theorem 2. *Whenever a sequent calculus admits cut-elimination then its hypersequent version with* Communication *as additional rule admits cut-elimination.*

Corollary 1. $\mathbf{HC}, \mathbf{HC}^*$ *and* **GLC** *admit cut-elimination.*

Corollary 2. *1. If* $\mathbf{LJ}^- \vdash A \wedge B, \Pi \Rightarrow C$ *from atomic axiom sequents, then* $\mathbf{LJ}^- \vdash A, \Pi \Rightarrow C$ *or* $\mathbf{LJ}^- \vdash B, \Pi \Rightarrow C$.
2. If $\mathbf{HC} \vdash A \wedge B, \Pi \Rightarrow C$ *from atomic axiom sequents, then* $\mathbf{HC} \vdash A, \Pi \Rightarrow C \mid B, \Pi \Rightarrow C$.

Proof. In the cut-free proof, omit the inference of ($\wedge'_i$-left) according to $A \wedge B$ in the end sequent.

Corollary 3. *1. In* $\mathbf{LJ}^{-*}$ *and* $\mathbf{HC}^*$, *non-atomic axioms are derivable from atomic axioms.*
2. In $\mathbf{LJ}^-$ *and* **HC**, *non-atomic axioms are* **not** *derivable from atomic axioms.*

Proof. $A \wedge B \Rightarrow A \wedge B$ is not derivable from atomic axioms in $\mathbf{LJ}^-$ and **HC**.

Corollary 4. *1.* $\mathbf{LJ}^{-*}$ *is a proper extension of* $\mathbf{LJ}^-$.
2. $\mathbf{HC}^*$ *is a proper extension of* **HC**.

Corollary 5. *Derivability in* **LJ**$^-$, **LJ**$^{-*}$, **LJ**, **HC**, **HC*** *and* **GLC** *is decidable.*

Proof. Proof search based decision algorithms for **LJ** (and **GLC**) are well known. For the sequent calculi without contraction it is even simpler to bound the number of sequents occurring in the cut-free proofs in terms of the size of the end sequent. In the case of hypersequents a bound for the number of hypersequents can be determined from the maximum number of components of the hypersequents in a proof where (EC) is applied whenever it is possible.

4 Correspondences

In this section we relate the various logics as given by their Hilbert-style axiomatizations to the sequent and hypersequent calculi introduced before.

Definition 2. *Let* $A_1, \ldots, A_n \Rightarrow B$ *be a sequent. Then the* generic interpretation $\mathfrak{I}$ *of* $A_1, \ldots, A_n \Rightarrow B$ *is defined as follows:* $\mathfrak{I}(\Rightarrow B) := B$,
$\mathfrak{I}(A_1, \ldots, A_n \Rightarrow B) := (A_1 \supset \ldots \supset (A_n \supset B) \ldots)$ *and*
$\mathfrak{I}(A_1, \ldots, A_n \Rightarrow) := (A_1 \supset \ldots \supset (A_n \supset \bot) \ldots)$

Definition 3. *Sequent rules*

$$\frac{S}{S^*} \quad \text{and} \quad \frac{T \quad T'}{T^*}$$

are called sound *for a (Hilbert-style) calculus* **H** *if* $\mathbf{H} \vdash \mathfrak{I}(S) \supset \mathfrak{I}(S^*)$ *and* $\mathbf{H} \vdash \mathfrak{I}(T) \supset (\mathfrak{I}(T') \supset \mathfrak{I}(T^*))$, *respectively. If all rules of a sequent calculus* **L** *are sound for* **H** *and* $\mathbf{H} \vdash \mathfrak{I}(S)$ *for all axioms* S *of* **L**, *then* **L** *is* sound *for* **H**.
A sequent calculus **L** *is* complete *for* **H** *if* $\mathbf{L} \vdash \Rightarrow A$ *whenever* $\mathbf{H} \vdash A$.

To prove the relative completeness and soundness theorems below we first observe that premises in chains of implications can be permuted arbitrarily already in **I**$^-$ (the weakest logic considered).

Proposition 2. *For any* $n \geq 2$ *and any permutation* π *of* $\{1, \ldots n\}$:

$$\mathbf{I}^- \vdash C[(A_1 \supset \ldots \supset (A_n \supset B) \ldots)] \text{ iff } \mathbf{I}^- \vdash C[(A_{\pi(1)} \supset \ldots \supset (A_{\pi(n)} \supset B) \ldots)]$$

Proof. It suffices to show that the formula

$$[(A_1 \supset \ldots \supset (A_n \supset B) \ldots)] \supset [(A_{\pi(1)} \supset \ldots \supset (A_{\pi(n)} \supset B) \ldots)]$$

is provable in **I**$^-$. For $n = 2$ this is Ax3; for $n > 2$ use Ax2. The generalization to arbitrary contexts C follows by induction on the size of C.

Lemma 1. *The rules of* **LJ**$^-$ *are sound for* **I**$^-$ *iff already the rules without side formulas are sound for* **I**$^-$.

Proof. Follows by induction on the number of side formulas, Prop. 2 and Ax2.

Theorem 3. $\mathbf{LJ}^-$ *is sound and complete for* $\mathbf{I}^-$.

Proof. (*Soundness*) The axioms of $\mathbf{LJ}^-$ translate into $A \supset A$ and $\bot \supset A$, respectively. The corresponding derivation in $\mathbf{I}^-$ is straightforward. Proving the soundness of the rules reduces to the derivation of single formulas by Lemma 1. For example, ($\supset$-left) translates into $A \supset ((B \supset C) \supset ((A \supset B) \supset C)$, which is derivable from Ax2 using Proposition 2. The other cases are similar.
(*Completeness*) Observe that Modus Ponens – the only rule of $\mathbf{I}^-$ – corresponds to the derivabilty of $A, A \supset B \Rightarrow B$ and the cut rule. It thus suffices to show that $\mathbf{LJ}^- \vdash$ Axi for all axioms Axi of $\mathbf{I}^-$. This is straightforward.

The following proposition states that for $\mathbf{I}^{-*}$ (and stronger systems) we may translate sequents by using conjunction instead of implication in the interpretation of the left hand side of a sequent. (This is not true for $\mathbf{I}^-$ by Corollary 4 and Theorem 3.)

Proposition 3. $\mathbf{I}^{-*} \vdash (A_1 \supset \ldots (A_n \supset B) \ldots)$ *iff* $\mathbf{I}^{-*} \vdash (A_1 \wedge \ldots \wedge A_n) \supset B$

Proof. Repeatedly apply Res1 and Res2.

Theorem 4. $\mathbf{LJ}^{-*}$ *is sound and complete for* $\mathbf{I}^{-*}$.

Proof. (*Soundness*) Lemma 1 also holds for $\mathbf{LJ}^{-*}$. Therefore it remains to check that the generic interpretation of the rule ($\wedge$-left), i.e. $(A \supset (B \supset C)) \supset (A \wedge B \supset C)$, is derivable in $\mathbf{I}^{-*}$. This follows directly from Proposition 3.
(*Completeness*) Analogous to the proof of Theorem 3.

To derive the linearity axiom Lin we have to "lift" sequents to hypersequents and use the communication rule.

Lemma 2. $\mathbf{HC} \vdash \Rightarrow$ Lin

We extend the generic interpretation of sequents to hypersequents:

Definition 4. *Let* $S_1 \mid \cdots \mid S_n$ *be a hypersequent. Then its generic interpretation is defined by* $\mathfrak{I}(S_1 \mid \cdots \mid S_n) := \mathfrak{I}(S_1) \vee \cdots \vee \mathfrak{I}(S_n)$

The definitions of relative soundness and completeness are extended to hypersequents in the obvious way. In analogy to Lemma 1 we may neglect the side sequents of all rules:

Lemma 3. *The* $\mathbf{HC}$-*rules are sound for* $\mathbf{C}$ *iff already the rules without side sequents are sound for* $\mathbf{C}$.

Proof. Observe that $(A \supset B) \supset (A \vee C \supset B \vee C)$ is derivable in $\mathbf{C}$. In fact it is already derivable in $\mathbf{I}^-$. The rest follows by induction on the number of side sequents.

Theorem 5. $\mathbf{HC}$ *is sound and complete for* $\mathbf{C}$.

Proof. (*Completeness*) Follows directly from Theorem 3 and Lemma 2.

The soundness and completeness of $\mathbf{LJ}^{-*}$ relative to $\mathbf{I}^{-*}$ can easily be "lifted" to the level of hypersequents by Lemma 3. By the above results, we thus obtain:

Theorem 6. $\mathbf{HC}^*$ *is sound and complete for* $\mathbf{C}^*$.

Corollary 6. *The logics* $\mathbf{I}^-$, $\mathbf{I}^{-*}$, $\mathbf{C}$, *and* $\mathbf{C}^*$ *are decidable.*

Proof. Follows from the respective completeness and soundness theorems and Corollary 5.

5 Applications to Hájek's Basic Logic

Basic Fuzzy Logic, **BL** for short, was introduced by Hájek in [7] as the logical counterpart of continuous T-norms. **BL** is given by the following axioms:

H1. $(A \supset B) \supset [(B \supset C) \supset (A \supset C)]$ **H5a.** $[A \supset (B \supset C)] \supset [(A \wedge B) \supset C]$
H2. $(A \wedge B) \supset A$ **H5b.** $[(A \wedge B) \supset C] \supset [A \supset (B \supset C)]$
H3. $(A \wedge B) \supset (B \wedge A)$ **H6.** $[(A \supset B) \supset C] \supset [[(B \supset A) \supset C] \supset C]$
H4. $[A \wedge (A \supset B)] \supset [B \wedge (B \supset A)]$ **H7.** $\bot \supset A$

Since the language of **BL** does not contain disjunction, ones defines disjunction as $A \vee B = [A \supset (A \supset B)] \sqcap [B \supset (B \supset A)]$ where $A \sqcap B = A \wedge (A \supset B)$.

Axioms H2, H4, H7, H5a, and H5b coincide with our axioms Ax4, Com, Abs, Res1, and Res2, respectively. Axiom H4 expresses the commutativity of the minimum in continuous T-norms and axiom H6 is a variant of "proof by cases".
Remark: By adding simple axioms to **BL** we can obtain three important logics: Lukasiewicz, Gödel and product logic. Indeed, if we add to **BL** axiom $\neg\neg A \supset A$ (involutivity of negation) we obtain **Lukasiewicz logic**, while if we add axiom Contr we obtain **Gödel logic**; finally, adding to **BL** axioms $\neg\neg A \supset ((B \wedge A \supset C \wedge A) \supset (B \supset C))$ and $A \wedge \neg A \supset \bot$ we obtain **product logic**.

Lemma 4. *All axioms of* $\mathbf{C}^*$ *are derivable in* **BL**.

To prove the next lemma, we need some additional notation:

Definition 5. *Let* α, β *be sets of hypersequents. Then* $\alpha \circ \beta$ *denotes the set of hypersequents* $H \mid G$ *such that* $H \in \alpha$ *and* $G \in \beta$.

Lemma 5. **BL** *strictly extends* $\mathbf{C}^*$.

Proof. We show that $\mathbf{HC}^* \not\vdash (A \wedge (A \supset B)) \supset (B \wedge (B \supset A))$ for atomic A, B. Assume the converse. This implies $\mathbf{HC}^* \vdash A, A \supset B \Rightarrow B \wedge (B \supset A)$ without cuts. We can rearrange the proof such that ($\wedge$-right) is the lowest inference. Consequently all hypersequents in

$$\left\{ \begin{array}{c} \Rightarrow B, \\ A, A \supset B \Rightarrow B \supset A \end{array} \right\} \circ \left\{ \begin{array}{c} A \Rightarrow B, \\ A \supset B \Rightarrow B \supset A \end{array} \right\} \circ \left\{ \begin{array}{c} A \supset B \Rightarrow B, \\ A \Rightarrow B \supset A \end{array} \right\} \circ \left\{ \begin{array}{c} A, A \supset B \Rightarrow B, \\ \Rightarrow B \supset A \end{array} \right\}$$

are provable in **HC***. Let us pick out H be $\Rightarrow B \mid A \supset B \Rightarrow B \supset A \mid A \supset B \Rightarrow B \mid \Rightarrow B \supset A$. H translates to $\mathfrak{I}(H)$ that is $B \vee (A \supset B) \supset (B \supset A) \vee (A \supset B) \supset B \vee (B \supset A)$. We have $v(\mathfrak{I}(H)) = v(B)$ in 3-valued Gödel logic for all valuations v with $1 > v(B) > v(A)$. Since for all sequents S which are derivable in **HC***, $\mathfrak{I}(S)$ must be valid in all Gödel logics, this concludes the proof.

Theorem 7. **BL** *is a proper extension of* **HC***.

Together with previously known results, this establishes the announced result:

Corollary 7. *All inclusions of logics of the figure in Section 2 are proper.*

References

1. Adillon, R. J., Verdú, V.: On the $\{\rightarrow, \wedge, \vee, \odot, 0\}$-contraction-less Intuitionistic Propositional Calculus. Draft. 1997.
2. Avron, A.: Hypersequents, Logical Consequence and Intermediate Logics for Concurrency. *Annals of Mathematics and Artificial Intelligence* Vol.4, 1991, 225-248.
3. Avron, A.: The Method of Hypersequents in Proof Theory of Propositional Non-Classical Logics. In *Logic: from foundations to applications. European logic colloquium, Keele, UK, July 20–29, 1993.* Oxford, Clarendon Press, 1996, 1-32.
4. Dummett, M.: A Propositional Logic with Denumerable Matrix. *Journal of Symbolic Logic* Vol. 24, 1959, 96-107.
5. Dyckhoff, R.: Contraction-Free Sequent Calculi for Intuitionistic Logic. *The Journal of Symbolic Logic*, Vol. 57/3, 1992, 795-807.
6. Gödel, K.: Zum Intuitionistischen Aussagenkalkül. *Ergebnisse eines mathematischen Kolloquiums* 4, 1933, 34-38.
7. Hájek, P.: Metamathematics of Fuzzy Logic. Kluwer, to appear.
8. Hájek, P., Godo L., Esteva, F.: A complete many-valued logic with product-conjunction. *Archive for Math. Logic*, Vol. 35 (1996), 191-208.
9. Łukasiewicz, J.: Zagadnienia prawdy (The problems of truth). In *Księga pamiątkowa XI zjazdu lekarzy i przyrodników polskich* 1922, 84-85,87.
10. Łukasiewicz, J.: Philosophische Bemerkungen zu mehrwertigen Systemen der Aussagenlogik. *Comptes Rendus de la Societe des Science et de Lettres de Varsovie*, cl.iii 23 (1930), 51-77.
11. Mendéz, J.M., Salto, F.: Urquhart's C with Intuitionistic Negation: Dummett's LC without the Contraction Axiom. *Notre Dame Journal of Formal Logic*, Vol. 36/3, 1995, 407-413.
12. Kiriyama, E., Ono, H.: The Contraction Rule in Decision Problems for Logics without Structural Rules. *Studia Logica*, Vol. 50/2, 1991, 299-319.
13. Ono, H., Komori, Y.: Logics without the Contraction Rule. *The Journal of Symbolic Logic*, Vol. 50/1, 1985, 169-201.
14. Troelstra, A. S., Schwichtenberg, H.: Basic Proof Theory. Cambridge University Press. 1996.
15. Urquhart, A.: Many-Valued Logic, in *Handbook of Philosophical Logic*, Vol III, ed. by D.Gabbay and F.Guenthner, Reidel, Dordrecht, 1984.
16. Zadeh, L.A.: Fuzzy Sets, Fuzzy Logic, and Fuzzy Systems. Selected Papers by Lotfi A. Zadeh ed. by G.J. Klir and B. Yuan. World Scientific Publishing, 1996.

Nonstochastic Languages as Projections of 2-Tape Quasideterministic Languages

Richard Bonner[1], Rūsiņš Freivalds[2], Jānis Lapiņš[3], and Antra Lukjanska[2]

[1] Department of Mathematics and Physics, Mälardalens University
[2] Institute of Mathematics and Computer Science, University of Latvia, Raiņa bulv. 29, Riga, Latvia[†]
[3] Department of Mathematics, University of Latvia, Zeļļu iela 8, Riga, Latvia

Abstract. A language $L^{(n)}$ of n-tuples of words which is recognized by a n-tape rational finite-probabilistic automaton with probability $1-\varepsilon$, for arbitrary $\varepsilon > 0$, is called quasideterministic. It is proved in [Fr 81], that each rational stochastic language is a projection of a quasideterministic language $L^{(n)}$ of n-tuples of words. Had projections of quasideterministic languages on one tape always been rational stochastic languages, we would have a good characterization of the class of the rational stochastic languages. However we prove the opposite in this paper. A two-tape quasideterministic language exists, the projection of which on the first tape is a nonstochastic language.

1 Introduction

Let N denote the set of all natural numbers. Let $n \in N$. By σ^n we denote a string consisting of n symbols σ. If Σ is a set, then Σ^n stands for the set of all the n-element strings over the alphabet Σ. A finite probabilistic automaton (FPA) is a system $\omega = (\Sigma, S, \Pi_0, M_\Sigma, F)$, where $\Sigma = \{\sigma_1, \sigma_2, \ldots, \sigma_e\}$ is a finite input alphabet, $S = \{s_1, s_2, \ldots, s_m\}$ is a finite set of states, $\Pi_0 = (p_1, p_2, \ldots, p_m)$ is a stochastic vector (the initial distribution of the probabilities of the states; $p_1 + p_2 + \ldots + p_m = 1$), M_Σ is a system of stochastic $m \times m$-matrices $M_{\sigma_1}, M_{\sigma_2}, \ldots, M_{\sigma_e}$ (the matrices of the probabilities for the transition from one state to another under the influence of the corresponding input symbol), and $F \subset S$ is a set of accepting states. Let $\eta_F = (\eta_1, \eta_2, \ldots, \eta_m)^T$ be a column matrix defined by $\eta_j = 1$ if $s_j \in F$ and $\eta_j = 0$ otherwise.

We say that a language L over the alphabet Σ is *acceptable* with cut-point γ ($0 \leq \gamma < 1$) by an automaton ω if the words $x_1 x_2 \ldots x_n$ in L are exactly the strings for which $\Pi_0 M_{x_1} M_{x_2} \ldots M_{x_n} \eta_F > \gamma$. In other words, we represent a language L in FPA ω with cut-point γ. For an arbitrary word x of L, if ω starts to work on x in a random state s_j distributed according to Π_0 then it stops in an accepting state with probability strictly larger than γ. A language L is called *stochastic* if it can be represented in some FPA with some cut-point

---

[†] Research supported by Grant No.96.0282 from the Latvian Council of Science

γ $(0 \le \gamma < 1)$. A FPA is called *rational* if all the components of its initial probability distribution and all elements its transition probability matrices are rational numbers. A language L is called *rational stochastic* if it can be represented in a rational FPA with a rational cut-point.

A FPA is called a finite *deterministic* automaton (FDA) if all the components of its initial distribution and all elements of its transition probability matrices are numbers from the set $\{0, 1\}$. A language L represented in a FDA is called *regular*.

We will consider in detail the case when a language L is represented in a FPA with cut-point $\gamma \ge \frac{1}{2}$ and so that $\Pi_0 M_{x_1} M_{x_2} \ldots M_{x_n} \eta_F \le 1 - \gamma$ for all strings $x_1 x_2 \ldots x_n$ *not* in L. In this case we say that the FPA *recognizes* the language L with probability γ.

Rabin and Scott [RS 59] introduced the concept of a multi-tape FDA, that is, a FDA that processes not words but tuples of words. Such an automaton has n tapes, over each of which a separate head can move in one direction, at most one unit at a time. Input words are written on tapes and every head observes a letter on the tape directly under it. It is presumed that the automaton can recognize the end of a word. This is provided by including in the alphabet Σ a special symbol # which is put on every tape immediately after an input word. So, the automaton is used to recognize sets of words in the alphabet $\Sigma \setminus \{\#\}$. It is assumed that when some head reaches a symbol #, further movement of this head becomes impossible. The work of the automaton ends when all the heads have observed the symbol #. We shall consider that the automaton has *accepted* the given n-tuple of words, if at this moment the automaton transits to an accepting state; otherwise, we consider the automaton to have *rejected* the input.

To formalise the definitions, let $n \in N$, denote by $W(n)$ the set of all the subsets of the set $\{1, 2, \ldots, n\}$, and let U_m denote the set of all m-dimensional stochastic vectors.

A *deterministic n-tape finite automaton* (n-FDA) is then a system

$$\omega = (\Sigma, S, s_1, \delta, \lambda, F)$$

where $\Sigma = \{\sigma_1, \sigma_2, \ldots, \sigma_e\}$ is a finite input alphabet containing the symbol #, $S = \{s_1, s_2, \ldots, s_m\}$ is a finite set of states with a singled out subset $F \subset S$ of accepting states and an initial state $s_1 \in S$, $\delta : S \times \Sigma^n \to S$ is a transition function from one state to another, and $\lambda : S \times \Sigma^n \to W(n)$ is a head movement function satisfying $i \notin \lambda(s_j, x_1, \ldots, x_n)$ if $x_i = \#$, $i = 1, \ldots, n$, $j = 1, \ldots, m$.

A *probabilistic n-tape finite automaton* (n-FPA) is a system

$$\omega = (\Sigma, S, \Pi_0, \delta, \lambda, F)$$

with Σ, S, F defined as above, and where $\Pi_0 = (p_1, p_2, \ldots, p_m) \in U_m$ is an initial probability distribution of states, $\delta : S \times \Sigma^n \to U_m$ is a state transition probability function, and $\lambda : S \times \Sigma^n \to U_{2^n}$ is a head movement function prescribing probabilities of subsets of tapes (points in $W(n)$) whereby subsets containing a tape in state # receive probability zero.

Let $y = (y_1, y_2, \ldots, y_n)$ be a n-tuple of strings over $\Sigma \setminus \{\#\}$. Denote by $P_\omega(y)$ the probability that n-FPA ω operates on y with initial distribution of the states Π_0 and stops operation in a state from the set F. We shall say that n-FPA ω *recognizes* a language $L = L^{(n)}$ of n-tuples of words over $\Sigma \setminus \{\#\}$ with probability γ ($\frac{1}{2} \leq \gamma < 1$), if for any n-tuple y of strings over $\Sigma \setminus \{\#\}$, we have $P_\omega(y) > \gamma$ if $y \in L$ while $P_\omega(y) \leq 1 - \gamma$ if $y \notin L$.

It was proved in [Fr 78] that there exists a language of pairs of words which cannot be recognized by any 2-tape FDA and cannot even be accepted by any 2-tape FNA; this language can however be recognized by a 2-tape FPA with probability $1 - \varepsilon$, for arbitrary $\varepsilon > 0$. It was proved in [Fr 91] that the class of languages that can be recognized by 2-tape FPA with probability $1 - \varepsilon$ for arbitrary $\varepsilon > 0$ is rather complex: in this class the emptiness problem is not decidable. In the present paper we characterize the complexity of this class in terms of projection languages. Here, the *projection* onto the first tape of a language $L = L^{(n)}$ of n-tuples of words $y = (y_1, y_2, \ldots, y_n)$ is defined as the language consisting of the words y_1 as y ranges over L.

We call a language L of n-tuples of word *quasideterministic* if for arbitrary $\varepsilon > 0$ there exists an n-FPA which recognizes L with probability $1 - \varepsilon$.

2 Results

It is known [RS 59] that the projection onto one of the tapes of a language of n-tuples of words that can be recognized by n-FDA, is a regular language. Indeed, more is true: the projection onto one of the tapes of an arbitrary language which is accepted by multi-tape finite nondeterministic automata is a regular language. For probabilistic automata, however, the situation is different. Freivalds [Fr 91] constructs a quasideterministic 3-language, the projection of which on the first tape is a nonstochastic language. The purpose of the present paper is to extend this result to 2-languages.

We begin by recording a simple fact, for reference; R denotes the set of real numbers.

Lemma 1. *Let $\xi_1, \xi_2, \ldots, \xi_t, \ldots$ be a sequence of random natural numbers. Then, for $t \geq 1$,*

$$\max_{c, c_1, \ldots, c_t \in R, c_t \neq 0} P\{\sum_{1 \leq i \leq t} c_i \xi_i = c\} \leq$$

$$\leq \max_{j_1, j_2, \ldots, j_t \in N} P\{\xi_t = j_t \mid \xi_1 = j_1, \ldots, \xi_{t-1} = j_{t-1}\}.$$

Proof: Denote by m the right side of the inequality, and let $c, c_1, \ldots, c_t \in R$, $c_t \neq 0$. Then

$$P\{\sum_{1 \leq i \leq t} c_i \xi_i = c\} = \sum_{j_1} \sum_{j_2} \cdots \sum_{j_{t-1}} P\{\xi_1 = j_1\} \times$$

$$\times P\{\xi_2 = j_2 \mid \xi_1 = j_1\} \times \ldots \times P\{\xi_{t-1} = j_{t-1} \mid \xi_1 = j_1, \ldots, \xi_{t-2} = j_{t-2}\} \times$$

$$\times P\left\{\xi_t = \frac{c - \sum_{i=1}^{t-1} c_i \xi_i}{c_t} \mid \xi_1 = j_1, \ldots, \xi_{t-1} = j_{t-1}\right\} \leq$$

$$\leq \sum_{j_1} P\{\xi_1 = j_1\} \times \sum_{j_2} P\{\xi_2 = j_2 \mid \xi_1 = j_1\} \times \ldots$$

$$\ldots \times \sum_{j_{t-1}} P\{\xi_{t-1} = j_{t-1} \mid \xi_1 = j_1, \ldots, \xi_{t-2} = j_{t-2}\} \times m \leq m.$$

Corollary 1. *Let M be a natural number and let $\xi_1, \xi_2, \ldots, \xi_t, \ldots$ be a sequence of independent random numbers uniformly distributed over the set $\{1, 2, \ldots, M\}$. Then $P\left\{\sum_{i=1}^{t} c_i \xi_i = c\right\} \leq \frac{1}{M}$ for arbitrary real numbers $c, c_1, \ldots, c_t$ ($c_t \neq 0$).*

We consider the following 2-language in the alphabet $\{1, 2, 3\}$:

$$B^{(2)} = \left\{(1^{s^2}, 1^1 2 1^3 2 1^5 2 \ldots 2 1^{2s-3} 3 1^{2s-1} 3) \mid s \in N\right\}.$$

Theorem 1. *For arbitrary $\varepsilon > 0$, there exists 2-FPA ω recognizing $B^{(2)}$ with probability $1 - \varepsilon$.*

Proof: Put $C^{(2)} = \left\{(1^j, 1^{i_1} 2 1^{i_2} \ldots 2 1^{i_s-1} 3 1^{i_s} 3) \mid j, s, i_1, \ldots, i_s \in N\right\}$. Let $\alpha = (\alpha_1, \alpha_2)$ denote a generic pair of words over the alphabet $\{1, 2, 3\}$, written on the tapes of an automaton ω. Observe that a 2-FPA ω recognizes $B^{(2)}$ with probability $1 - \varepsilon$ if it obeys the following rules.

1. If $\alpha \notin C^{(2)}$ then ω rejects α.
2. If $\alpha \in C^{(2)}$ then,
   (a) if $i_k + 2 = i_{k+1}$ for $1 \leq k \leq s-1$ but $j \neq \sum_{k=1}^{s} i_k$ or $i_1 \neq 1$, then ω rejects α,
   (b) if $i_1 = 1$ but $i_k + 2 \neq i_{k+1}$ for some k ($1 \leq k \leq s-1$), then ω rejects α with probability exceeding $1 - \varepsilon$,
   (c) if $i_1 = 1$ and $i_k + 2 = i_{k+1}$ for $1 \leq k \leq s-1$ and $j = \sum_{k=1}^{s} i_k = s^2$, then ω accepts α.

We now construct ω which obeys these rules. We present ω as a 2-FDA supplied with a generator of random equiprobable numbers. Fix a natural number $M > \frac{1}{\varepsilon}$ and let $\tau_1, \tau_2, \ldots, \tau_{s-1}$ be a sequence of independent uniformly distributed random numbers in the set $\{1, \ldots, M\}$. Denote by β_j the remainder from the division of i_j by M. For input $\alpha = (\alpha_1, \alpha_2) \in C^{(2)}$ we instruct ω to proceed as follows.

1. If the first letter of the word α_2 is 3, then if $\alpha = (1, 313)$ accept α, and if $\alpha \neq (1, 313)$ reject α.
2. If the first letter of the word α_2 is 2, then:
   (a) If $i_1 = 1$ then retain the number $\beta_1 = 1$ and move the first head by one unit. If $i_1 \neq 1$ then reject α.

(b) If the second head is located on the j:th number 2 ($2 \leq j < s-1$) and the retained number is β_{j-1}, then after moving of each M of following i_j units on the second tape, move the first head $M - \tau_{j-1} - \tau_j$ units. By the moving the last β_j units before $(j+1)$:th symbol 2 or before the first symbol 3 compare β_j with β_{j-1}. If $\beta_j = \beta_{j-1} + 2$ move the first head $\beta_{j-1} + 1$ units, while if $\beta_j = \beta_{j-1} + 2 - M$ move it $\beta_j + \tau_{j-1} - 1$ units. Retain the number β_j (instead of β_{j-1}). If $\beta_j \neq \beta_{j-1} + 2 (mod M)$ reject α.

(c) If the second head is located on the first symbol 3 and the retained number is β_{s-1}, then after moving each M following i_s units on the second tape, move the first head $M - \tau_{s-1}$ units. By going through the last β_s units before the second triple, compare the numbers β_s and β_{s-1}. If $\beta_s = \beta_{s-1} + 2$ move the first head $\beta_{s-1} + 1$ units, while if $\beta_s = \beta_{s-1} + 2 - M$ move it $\beta_s + \tau_{s-1} - 1$ units. If $\beta_s \neq \beta_{s-1} + 2 (mod M)$ reject α.

(d) If by moving the second head, the first head has moved a distance equal to the length of α_2, then accept α; otherwise reject α.

For input $\alpha \notin C^{(2)}$ we instruct ω to proceed as above until a difference from $C^{(2)}$ has been detected, and to reject α then.

Let $\alpha \notin B^{(2)}$. By moving the second head in $\alpha_2 = 12 1^3 21^5 2 \ldots 21^{2s-3} 31^{2s-1} 3$, the first head is moved all the $\sum_{k=1}^{s-1}(2k+1) = s^2$ units and ω outputs the correct answer $\alpha \notin B^{(2)}$ with probability one. If $\alpha = (1^j, 12 1^3 21^5 2 \ldots 21^{2s-3} 31^{2s-1} 3)$, then $j \neq s^2$ and, by the computations of the second head, the first head can be moved s^2 units. Since $s^2 \neq j$, the automaton ω outputs the correct answer $\alpha \notin B^{(2)}$ with probability one.

Let $K = \{k_1, k_2, \ldots, k_d\} = \{k \mid i_k + 2 \neq i_{k+1}\}$. We have two cases:

1. $\exists l \in K \quad i_l + 2 \neq i_{l+1} (mod M)$. Then $\beta_l + 2 \neq \beta_{l+1}$ and by 2 (b),(c) the automaton outputs a correct answer $\alpha \notin B^{(2)}$ with probability one.

2. $\forall l \in K \quad i_l + 2 = i_{l+1} (mod M)$. In this case, as a result of the movement of the second head, the first head moves a random number of τ units, $\tau = c_0 + \sum_{r=1}^{d} c_r \tau_r$, where c_r ($r = 0, 1, \ldots, d$) are integers such that $\sum_{r=1}^{d} |c_r| \neq 0$. The probability of a wrong answer will therefore be equal to the probability of the equality $\tau = j$. It now follows by Lemma 1 that if $P\{\tau = j\} < \varepsilon$ then ω outputs a correct answer with probability not less than $1 - \varepsilon$.

Theorem 2. *Let $P(x) = \sum_{j=0}^{l} c_j x^j$ be a polynomial of degree $l \geq 2$ with non-negative coefficients, mapping the set of natural numbers into itself. Then the language $L = \{1^{P(s)} \mid s \in N\}$ is nonstochastic.*

Before proving Theorem 2 we recall a known fact of Diophantine approximation.

Lemma 2. *Let $1, \psi_1, \psi_2, \ldots, \psi_t$ be real numbers forming a linearly independent system over the field of the rational numbers. Let $P(x)$ be a polynomial of positive degree with rational coefficients. Then the set of the fractional parts of the vectors $P(k)\Psi = (P(k)\psi_1, P(k)\psi_2, \ldots, P(k)\psi_t)$, $k \in N$, is everywhere dense in the unit t-dimensional hypercube.*

Proof: See, for example, [Ca 57] Ch. IV, Theorems III, IV, and VI. The sequence $P(k)\Psi$ of t-dimensional vectors is indeed uniformly distributed modulo 1 if (and only if) so is the 1-dimensional sequence $\tau_1 P(k)\psi_1 + \tau_2 P(k)\psi_2 + \ldots + \tau_t P(k)\psi_t = P(k)\sum_i \tau_i \psi_i$ for every non-zero vector $(\tau_1, \tau_2, \ldots, \tau_t)$ of integers. Since the numbers $1, \psi_1, \psi_2, \ldots, \psi_t$ are linearly independent over the rationals, the number $\psi = \sum_i \tau_i \psi_i$ is irrational, and hence the polynomial ψP has irrational coefficients. However, it is known that the values $Q(k)$, $k \in N$, of a polynomial Q having at least one irrational coefficient are uniformly distributed modulo 1.

Proof of Theorem 2. Assume, on the contrary, that the language L is stochastic. This means that there exist a FPA $\pounds = (\{1\}, S, \Pi_0, M_1, \eta_F)$ and a number γ such that

$$x = 1^k \in L \iff \Pi_0 M_1^k \eta_F > \gamma. \tag{1}$$

Let s be the cardinality of S and let $\tilde{\lambda}_j = |\tilde{\lambda}_j| e^{2\pi i \varphi_j}$, $j = 1, 2, \ldots, s$, the eigenvalues of the matrix M_1. Let $1, \psi_1, \psi_2, \ldots, \psi_t$ ($t \leq s$) be a system of linearly independent real numbers over the field of rational numbers such that $\varphi_j = r_{0j} + r_{1j}\psi_1 + \ldots + r_{tj}\psi_t$ with r_{uj} rational, and denote by R the common denominator of the numbers r_{uj}, $u = 0, 1, \ldots, t$, $j = 1, 2, \ldots, s$. Let $c < R$ be a nonnegative integer congruent to c_0 modulo R. Let the bold numeral **1** denote a generic column matrix consisting of 1's only, and put $M = M_1^R$ and $\eta = M_1^c (\eta_F - \gamma \cdot \mathbf{1})$. Since $\Pi_0 M_1^k \mathbf{1} = 1$, it follows from (1) that

$$1^{Rk+c} \in L \iff \Pi_0 M^k \eta > 0. \tag{2}$$

Put $|\lambda_j| = |\tilde{\lambda}_j|^R$, $\Theta_j = (\varphi_j - r_{0j})$ and $\lambda_j = |\lambda_j| e^{2\pi i \Theta_j}$, and let Z be a Jordan normal form of the matrix M, $M = T^{-1}ZT$. The numbers λ_j are clearly the eigenvalues of the matrix M. We have

$$F(k) = \Pi_0 M^k \eta = \Pi_0 T^{-1} Z^k T \eta = \sum_{m=1}^{s} |\lambda_m|^k \sum_{j=0}^{s-1} k^j a_{mj} e^{2\pi i(k\Theta_m + \alpha_{mj})}$$

where a_{mj}, α_{mj} are real, and, since $F(k)$ is also real, we may forthwith replace the complex exponential by the cosine. To simplify further, partition the set $J = \{\lambda_j \mid 1 \leq j \leq s\}$ of eigenvalues by their modulus: $J = \cup_{1 \leq \nu \leq N} J_\nu$, with J_ν consisting of all the eigenvalues of equal modulus ρ_ν and $\rho_1 < \ldots < \rho_N$. Write $F(k)$ in the form

$$F(k) = \sum_\nu \rho_\nu^k \sum_j k^j B_\nu(k, j)$$

where we have put

$$B_\nu(k, j) = \sum_{m \in J_\nu} a_{mj} \cos 2\pi(k\Theta_m + \alpha_{mj}).$$

Notice that the functions $k \mapsto B_\nu(k, j)$ cannot all vanish identically because F itself does not; indeed, since $P(Rk) \equiv c_0 \equiv c(\bmod R)$, we have by (2),

$$F(Q(l)) > 0 \ \ if \ \ Q(l) = \frac{P(Rl) - c}{R} \ \ and \ l \in N. \tag{3}$$

One may thus pick the largest index ν_0 among all ν for which $\sum_j k^j B_\nu(k,j)$ is not identically zero as a function of k, and then pick the largest index j_0 among all j for which $k \mapsto B_{\nu_0}(k,j)$ is not identically zero. Put for short $B(k) = B_{\nu_0}(k, j_0)$ and pick a value $k = k_0$ for which $B(k_0) \neq 0$.

Notice that if $k_v \to \infty$ is a sequence of integers for which $B(k_\nu) \to B(k_0)$ then
$$F(k) = \rho_{\nu_0}^k k^{j_0}(B(k) + o(1)) \ as \ k = k_\nu \to \infty, \qquad (4)$$
which for large ν forces the signs of $F(k_\nu)$ and $B(k_0)$ to coincide. By Lemma 2 we may first pick such a sequence k_ν of the form $Q(n_v)$ and conclude by (3) that $B(k_0) > 0$. Picking now k_ν a second time, this time of the form $Q(m_v) + 1$, we then see that $F(Q(m_v) + 1) > 0$ for large ν. This implies by (2) that for all such ν one has $R(Q(m_v) + 1) + c = P(l_\nu)$ for some $l_\nu \in N$, or, simplified, $P(Rm_v) + R = P(l_\nu)$. However, since P is of degree at least two and has positive coefficients, it is clear that this equation has no solution $l_\nu \in N$ if m_v is large enough.

The assumption that L is stochastic has lead us to a contradiction.

Theorem 3. *For each positive number ε there exists a language of pairs of words which is recognized by a finite probabilistic automaton with probability $1 - \varepsilon$, but the projection of which to one of the tapes is a nonstochastic language.*

Proof: Consider the language $B^{(2)}$. By Theorem 1, there exists a 2-FPA ω which recognizes this language with probability $1-\varepsilon$. On the other hand, the projection of $B^{(2)}$ to the first tape is the language $\{1^{s^2} \mid s \in N\}$, which is nonstochastic by Theorem 2.

References

[Fr 78] Rūsiņš Freivalds. *Recognition of languages with high probability by various types of automata.* "Dokladi AN SSSR", 1978, v. 239, No. 1, p. 60-62 (in Russian)
[Fr 81] Rūsiņš Freivalds. *Projections of languages recognizable by probabilistic and alternating finite multitape automata.* "Information Processing Letters", 1981, v. 13, No. 4/5, p. 195-198.
[Fr 91] Rūsiņš Freivalds. *Complexity of probabilistic versus deterministic automata.* "Lecture Notes in Computer Science", Springer, 1991, v. 502, p. 565-613
[RS 59] M.O. Rabin and D. Scott. *Finite automata and their decision problems.* "J. Res. Develop.", 1959, v. 3, No. 2, p. 114-125.
[Mi 66] B.T. Mirkin. *Towards the theory of multitape automata.* "Kibernetika", 1966, No 5, p. 12-18. (in Russian)
[Ca 57] J.W.S. Cassels. *An Introduction to Diophantine Approximation.* Cambridge Tracts in Mathematics and Mathematical Physics, vol. 45, 1957.

Flow Logic for Imperative Objects

Flemming Nielson and Hanne Riis Nielson

Department of Computer Science, Aarhus University, Denmark

Abstract. We develop a control flow analysis for the Imperative Object Calculus. We prove the correctness with respect to two Structural Operational Semantics that differ in *minor* technical ways, and we show that the proofs deviate in *major* ways as regards their use of proof techniques like coinduction and Kripke-logical relations.

1 Introduction

The advent of mobile computation renews the interest in static program analysis aimed at guaranteeing that software does not exhibit malicious or unintended behaviour. We consider here the problem in a pure form by studying a *control flow analysis* aimed at determining which software components might reach what places. When studying mobile computation one needs to be able to quickly adapt existing technologies for control and data flow analysis to the the variety of theoretical calculi designed for studying the problem. One such calculus is the *Imperative Object Calculus* [1] and this will be the one we study here. The control flow analysis will be expressed as an abstract *flow logic* in verbose form; this presentation focuses on the logical content of the analysis as opposed to the algorithmic techniques (that can then be added afterwards [2]), and is therefore particularly suited for quickly adapting existing technologies to novel calculi.

We briefly review the imperative object calculus and then specify the control flow analysis (Section 2). This specification must be interpreted *coinductively* because of the ability of the imperative object calculus to code recursion (in the manner of the fixed point combinator of the λ-calculus) and because of the abstract specificational style employed. The theoretical existence of best solutions, as opposed to a practical algorithm, is established by means of a *Moore Family* (or model intersection) property.

Semantic correctness of the specification is established by means of a *subject reduction* result with respect to a small-step structural operational semantics using environments (Section 3). The semantics is a small-step version of the one in [1]. Since the semantics introduces new intermediate syntax we need to specify the analysis also for these constructs. The proof of the subject reduction result employs *coinduction* as well as *Kripke-logical relations*.

Next we study the extent to which the structure of the correctness proof depends on the fine technical details of the operational semantics (Section 4). We do so by devising another semantics that only deviates from the former on some fine techical points; in fact we would claim that the semantics of [1] could

$$e ::= t^\ell \qquad \text{an expression is a labelled term}$$

$$
\begin{aligned}
t ::= \ & x & \text{variable} \\
| \ & [m_i = \varsigma(x_i).e_i{}^{i=1..n}] & \text{object (all } m_i \text{ distinct)} \\
| \ & e.m & \text{method invocation} \\
| \ & e_1.m := \varsigma(x_2).e_2 & \text{method udpdate} \\
| \ & \text{clone } e & \text{object cloning} \\
| \ & \text{let } x = e_1 \text{ in } e_2 & \text{local definition}
\end{aligned}
$$

Table 1. Imperative Object Calculus: expressions and terms.

equally well have been defined in a form resembling the modified semantics. We then observe that the formulation of the subject reduction result gets more complex in that a notion of "the signature of a state" seems to be needed; on the other hand, the subject reduction result can now be established *without* using coinduction and Kripke-logical relations.

We conclude by identifying the general principles that are illustrated by this study (Section 5). These insights are likely to be crucial for the ability to quickly and correctly devise static analyses for novel calculi for computation.

2 Control Flow Analysis for Imperative Objects

The Imperative Object Calculus is defined in Chapter 10 of [1]. It is an untyped but statically scoped calculus; a central term is the object, $[m_i = \varsigma(x_i).e_i{}^{i=1..n}]$, that is an ordered collection of n components, $m_i = \varsigma(x_i).e_i$, defining a method name, m_i, in terms of a method, $\varsigma(x_i).e_i$. The binders, $\varsigma(x_i)$, suspend the evaluation, and when an object is invoked, as in $[m_i = \varsigma(x_i).e_i{}^{i=1..n}].m_j$, the corresponding expression body, e_j, is evaluated in an environment where the formal parameter, x_j, is bound to the object itself, thereby permitting self-application and recursion. Method update, $e.m := \varsigma(x').e'$, redefines an *already existing* method name, m, to be the new method, $\varsigma(x').e'$, and returns the new object; this update takes place using a store (hence the name "imperative"). Cloning, clone e, produces a new object with the same method identifiers but using fresh locations. Finally, there is a construct, let $x = e_1$ in e_2, for local definitions.

The abstract syntax is summarised in Table 1. The main deviation from [1] is that we shall want to place labels (ranged over by ℓ) on all subexpressions in order to interface with the control flow analysis. To this end we formally distinguish between expressions, e, that are labelled terms, and terms, t, that are unlabelled expressions.

The control flow analysis of an expression aims at determining the sets of objects that can reach various points in the program. In the analysis, the presence of an object $[m_i = \varsigma(x_i).e_i{}^{i=1..n}]$ will be represented by the *abstract object* $\langle m_i{}^{i=1..n}\rangle$; this representation is stable under evaluation because in the Imperative Object Calculus, an object update is not allowed to introduce additional method names.

$(\hat{C}, \hat{\rho}, \hat{\sigma}) \models_\Sigma x^\ell$ iff $\hat{\rho}(x) \subseteq \hat{C}(\ell)$

$(\hat{C}, \hat{\rho}, \hat{\sigma}) \models_\Sigma [m_i = \varsigma(x_i).e_i\ ^{i=1..n}]^\ell$ iff
$\langle \{\varsigma(x_i).e_i\}^{i=1..n} \rangle \sqsubseteq \hat{\sigma}(\langle m_i{}^{i=1..n}\rangle) \wedge \langle m_i{}^{i=1..n}\rangle \in \hat{C}(\ell)$

$(\hat{C}, \hat{\rho}, \hat{\sigma}) \models_\Sigma (it_1^{\ell_1}.m)^\ell$ iff $(\hat{C}, \hat{\rho}, \hat{\sigma}) \models_\Sigma it_1^{\ell_1} \wedge \forall j, \langle m_i{}^{i=1..n}\rangle, \varsigma(x_0).t_0^{\ell_0}:$
$(((\langle m_i{}^{i=1..n}\rangle) \in \hat{C}(\ell_1) \wedge m_j = m \wedge \varsigma(x_0).t_0^{\ell_0} \in \hat{\sigma}(\langle m_i{}^{i=1..n}\rangle)|_j)$
$\Rightarrow (\langle m_i{}^{i=1..n}\rangle \in \hat{\rho}(x_0) \wedge \hat{C}(\ell_0) \subseteq \hat{C}(\ell) \wedge (\hat{C}, \hat{\rho}, \hat{\sigma}) \models_\Sigma t_0^{\ell_0}))$

$(\hat{C}, \hat{\rho}, \hat{\sigma}) \models_\Sigma (it_1^{\ell_1}.m := \varsigma(x_2).e_2)^\ell$ iff $(\hat{C}, \hat{\rho}, \hat{\sigma}) \models_\Sigma it_1^{\ell_1} \wedge \forall j, \langle m_i{}^{i=1..n}\rangle :$
$(((\langle m_i{}^{i=1..n}\rangle) \in \hat{C}(\ell_1) \wedge m_j = m) \Rightarrow (\langle m_i{}^{i=1..n}\rangle \in \hat{C}(\ell) \wedge \varsigma(x_2).e_2 \in \hat{\sigma}(\langle m_i{}^{i=1..n}\rangle)|_j))$

$(\hat{C}, \hat{\rho}, \hat{\sigma}) \models_\Sigma (\text{clone } it_1^{\ell_1})^\ell$ iff $(\hat{C}, \hat{\rho}, \hat{\sigma}) \models_\Sigma it_1^{\ell_1} \wedge \hat{C}(\ell_1) \subseteq \hat{C}(\ell)$

$(\hat{C}, \hat{\rho}, \hat{\sigma}) \models_\Sigma (\text{let } x = it_1^{\ell_1} \text{ in } it_2^{\ell_2})^\ell$ iff
$(\hat{C}, \hat{\rho}, \hat{\sigma}) \models_\Sigma it_1^{\ell_1} \wedge (\hat{C}, \hat{\rho}, \hat{\sigma}) \models_\Sigma it_2^{\ell_2} \wedge \hat{C}(\ell_1) \subseteq \hat{\rho}(x) \wedge \hat{C}(\ell_2) \subseteq \hat{C}(\ell)$

Table 2. Control Flow Analysis: the base part.

To be more precise, a proposed control flow analysis of an expression, e, is captured by the following three entities:

- The abstract cache, $\hat{C}$: here $\hat{C}(\ell)$ is (a superset of) the set of abstract objects that can result from the subexpression labelled ℓ.
- The abstract environment, $\hat{\rho}$: here $\hat{\rho}(x)$ is (a superset of) the set of abstract objects that the variable x might be instantiated to.
- The abstract store, $\hat{\sigma}$: here $\hat{\sigma}(\langle m_i{}^{i=1..n}\rangle)$ is an n-tuple of sets of methods; the jth component, $\hat{\sigma}(\langle m_i{}^{i=1..n}\rangle)|_j$, is (a superset of) the set of methods that might implement the method m_j within all objects of the form $[m_i = \varsigma(x_i).e_i{}^{i=1..n}]$ that are part of the program or that arise during evaluation.

Clearly the relation $\subseteq$ can be used to relate sets, and this can be extended in a pointwise manner to define a relation $\sqsubseteq$ that relates abstract states, and again in a componentwise manner to a relation $\sqsubseteq$ for relating triples of the form $(\hat{C}, \hat{\rho}, \hat{\sigma})$; it is immediate that this turns the set of triples (of form $(\hat{C}, \hat{\rho}, \hat{\sigma})$) into a complete lattice and we write $\sqcap$ for the greatest lower bound operation.

To verify that a proposed analysis, $(\hat{C}, \hat{\rho}, \hat{\sigma})$, is indeed an acceptable analysis of the program, e, we specify a judgement

$(\hat{C}, \hat{\rho}, \hat{\sigma}) \models_\Sigma e$

as shown in Table 2. The clause for variables is typical of the way the abstract value of a variable is included in the abstract value of a label. The clause for objects similarly includes the abstract object in the abstract value of the label; it also records the actual methods in the abstract store. The clause for method invocation performs a "recursive call" for verifying the analysis of the object; it then inspects each abstract object that might result and each possible method selected, and then takes care of the self-reference, includes the result of the method in the result of the call, and finally performs a "recursive call" for verifying the analysis of the method invoked. The clauses for method update, cloning and local definitions are less critical for understanding the basic features of the analysis.

Since the clauses of Table 2 are not compositional (due to the analysis of the expression $t_0^{\ell_0}$ in the clause for method invocation) we define $\models_\Sigma$ coinductively; we shall see below that this is more appropriate than an inductive definition.

Theorem 1. *For all expressions e, the set $S_e = \{(\hat{C}, \hat{\rho}, \hat{\sigma}) \mid (\hat{C}, \hat{\rho}, \hat{\sigma}) \models_\Sigma e\}$ is a Moore Family, i.e. $\forall Y \subseteq S_e : \sqcap Y \in S_e$.*

Proof. The proof of the theorem employs coinduction [4].

It is worth noting that the corresponding theorem (not just the proof) fails for the relation $\models'_\Sigma$ that is inductively defined by Table 2. It is a consequence of the Moore Family property (also called a model intersection property) that all expressions not only admit an acceptable analysis (take $Y = \emptyset$) but also a best acceptable analysis (take $Y = S_e$).

3 The First Approach

We now define our first semantics for the Imperative Object Calculus. It is a small-step semantics [5] corresponding to the big-step semantics of [1]. The overall (and slightly imprecise) idea is that a semantic judgement is of the form

$$\rho \vdash \langle e, \sigma \rangle \to \langle e', \sigma' \rangle$$

where the typical form of the finite environment, ρ, and the finite state, σ, is given by

$$\rho(x) = [m_i = \iota_i{}^{i=1..n}]$$
$$\sigma(\iota) = \varsigma(x).e$$

where ι ranges over a set of locations (just as x ranges over the variables).

However, this description does not take account of the *static scope* rules in the Imperative Object Calculus. In the manner of [5] this motivates introducing two new auxiliary expressions: the expression "close $\varsigma(x).e$ in ρ" that allows to encapsulate the environment at the point of definition, and "bind ρ in e" that allows to use a local environment for the evaluation of an expression e. To clarify the distinction between our original expressions, e, and the augmented ones, we shall term the latter *intermediate expressions*, *ie*. The precise details of the syntax of intermediate expressions follows from the definition of the semantics in Table 3. For lack of space, and since the definition is so close to the one in [1], we shall dispense with an explanation of the semantics.

To express semantic correctness of the analysis in terms of a subject reduction result, we need to extend the analysis of Table 2 to incorporate the intermediate expression "bind ρ in e" and the object denotation "$[m_i = \iota_i{}^{i=1..n}]$". This calls for adding the two clauses in Table 4; then, for the purposes of this section, $\models_\Sigma$ is defined coinductively by the Tables 2 and 4. One of the new clauses makes

$$\rho \vdash \langle x^\ell, \sigma \rangle \;\to\; \langle (\rho(x))^\ell, \sigma \rangle \text{ if } x \in \text{dom}(\rho)$$

$$\rho \vdash \langle [m_i = \varsigma(x_i).e_i{}^{i=1..n}]^\ell, \sigma \rangle \;\to\; \langle [m_i = \iota_i{}^{i=1..n}]^\ell, \sigma' \rangle$$
$$\text{if } \langle \iota_i{}^{i=1..n} \rangle \text{ are fresh } \wedge \sigma' = \sigma[\iota_i \mapsto (\text{close } \varsigma(x_i).e_i \text{ in } \rho)^{i=1..n}]$$

$$\frac{\rho \vdash \langle ie, \sigma \rangle \;\to\; \langle ie', \sigma' \rangle}{\rho \vdash \langle (ie.m)^\ell, \sigma \rangle \;\to\; \langle (ie'.m)^\ell, \sigma' \rangle}$$

$$\rho \vdash \langle ([m_i = \iota_i{}^{i=1..n}]^{\ell_1}.m)^\ell, \sigma \rangle \;\to\; \langle (\text{bind } \rho' \text{ in } e_j)^\ell, \sigma \rangle$$
$$\text{if } m_j = m \wedge \sigma(\iota_j) = (\text{close } \varsigma(x_j).e_j \text{ in } \rho_j) \wedge \rho' = \rho_j[x_j \mapsto [m_i = \iota_i{}^{i=1..n}]]$$

$$\frac{\rho \vdash \langle ie_1, \sigma \rangle \;\to\; \langle ie'_1, \sigma' \rangle}{\rho \vdash \langle (ie_1.m := \varsigma(x).e)^\ell, \sigma \rangle \;\to\; \langle (ie'_1.m := \varsigma(x).e)^\ell, \sigma' \rangle}$$

$$\rho \vdash \langle ([m_i = \iota_i{}^{i=1..n}]^{\ell_1}.m := \varsigma(x).e)^\ell, \sigma \rangle \;\to\; \langle [m_i = \iota_i{}^{i=1..n}]^\ell, \sigma' \rangle$$
$$\text{if } m_j = m \wedge \sigma' = \sigma[\iota_j \mapsto (\text{close } \varsigma(x).e \text{ in } \rho)]$$

$$\frac{\rho \vdash \langle ie, \sigma \rangle \;\to\; \langle ie', \sigma' \rangle}{\rho \vdash \langle (\text{clone } ie)^\ell, \sigma \rangle \;\to\; \langle (\text{clone } ie')^\ell, \sigma' \rangle}$$

$$\rho \vdash \langle (\text{clone } [m_i = \iota_i{}^{i=1..n}]^{\ell_1})^\ell, \sigma \rangle \;\to\; \langle [m_i = \iota'_i{}^{i=1..n}]^\ell, \sigma' \rangle$$
$$\text{if } \langle \iota'_i{}^{i=1..n} \rangle \text{ are fresh } \wedge \sigma' = \sigma[\iota'_i \mapsto \sigma(\iota_i)^{i=1..n}]$$

$$\frac{\rho \vdash \langle ie_1, \sigma \rangle \;\to\; \langle ie'_1, \sigma' \rangle}{\rho \vdash \langle (\text{let } x = ie_1 \text{ in } e_2)^\ell, \sigma \rangle \;\to\; \langle (\text{let } x = ie'_1 \text{ in } e_2)^\ell, \sigma' \rangle}$$

$$\rho \vdash \langle (\text{let } x = [m_i = \iota_i{}^{i=1..n}]^{\ell_1} \text{ in } e)^\ell, \sigma \rangle \;\to\; \langle (\text{bind } \rho' \text{ in } e)^\ell, \sigma \rangle$$
$$\text{if } \rho' = \rho[x \mapsto [m_i = \iota_i{}^{i=1..n}]]$$

$$\frac{\rho_1 \vdash \langle ie_2, \sigma \rangle \;\to\; \langle ie'_2, \sigma' \rangle}{\rho \vdash \langle (\text{bind } \rho_1 \text{ in } ie_2)^\ell, \sigma \rangle \;\to\; \langle (\text{bind } \rho_1 \text{ in } ie'_2)^\ell, \sigma' \rangle}$$

$$\rho \vdash \langle (\text{bind } \rho_1 \text{ in } [m_i = \iota_i{}^{i=1..n}]^{\ell_2})^\ell, \sigma \rangle \;\to\; \langle [m_i = \iota_i{}^{i=1..n}]^\ell, \sigma \rangle$$

Table 3. Operational Semantics: the first approach.

use of an auxiliary relation, $\mathcal{R}$, that relates the concrete enviroment, ρ, to the abstract environment, $\hat{\rho}$.

The other auxiliary relation, $\mathcal{S}$, is used to express the subject reduction result; since it is defined in terms of itself we shall employ a coinductive definition of the clause. To state the result we need the range of an environment, range(ρ), and the set of evaluated objects (i.e. $[m_i = \iota_i{}^{i=1..n}]$) in an intermediate expression, obj(ie); both definitions are straightforward and are therefore omitted.

Theorem 2. *If $\rho\, \mathcal{R}\, \hat{\rho}, (\hat{C}, \hat{\rho}, \hat{\sigma}) \models_\Sigma ie, \sigma\; \mathcal{S}_O\, (\hat{\rho}, \hat{\sigma})$ (for $O = \text{range}(\rho) \cup \text{obj}(ie)$) and if $\rho \vdash \langle ie, \sigma \rangle \to \langle ie', \sigma' \rangle$ then $\rho\, \mathcal{R}\, \hat{\rho}, (\hat{C}, \hat{\rho}, \hat{\sigma}) \models_\Sigma ie', \sigma'\; \mathcal{S}_{O'}\, (\hat{\rho}, \hat{\sigma})$ (for $O' = \text{range}(\rho) \cup \text{obj}(ie')$).*

Proof. Overall the proof is by induction on the shape of $\rho \vdash \langle ie, \sigma \rangle \to \langle ie', \sigma' \rangle$. However, to deal with the case of "bind" we use a stronger induction hypothesis using the notion of *Kripke-logical relations* [3]:

$(\hat{C}, \hat{\rho}, \hat{\sigma}) \models_\Sigma ([m_i = \iota_i{}^{i=1..n}])^\ell$ iff $\langle m_i{}^{i=1..n}\rangle \in \hat{C}(\ell)$

$(\hat{C}, \hat{\rho}, \hat{\sigma}) \models_\Sigma$ (bind ρ_1 in $it_2^{\ell_2})^\ell$ iff
$\rho_1 \mathcal{R} \hat{\rho} \wedge (\hat{C}, \hat{\rho}, \hat{\sigma}) \models_\Sigma it_2^{\ell_2} \wedge \hat{C}(\ell_2) \subseteq \hat{C}(\ell)$

$\rho \mathcal{R} \hat{\rho}$ iff $\forall x \in \text{dom}(\rho) \subseteq \text{dom}(\hat{\rho}) : \forall [m_i = \iota_i{}^{i=1..n}]:$
$((\rho(x) = [m_i = \iota_i{}^{i=1..n}]) \Rightarrow (\langle m_i{}^{i=1..n}\rangle \in \hat{\rho}(x)))$

$\sigma \mathcal{S}_O (\hat{\rho}, \hat{\sigma})$ iff $\forall [m_i = \iota_i{}^{i=1..n}] \in O : \forall j, \varsigma(x).e, \rho : ((\sigma(\iota_j) = (\text{close } \varsigma(x).e \text{ in } \rho))$
$\Rightarrow (\varsigma(x).e \in \hat{\sigma}(\langle m_i{}^{i=1..n}\rangle)|_j \wedge \rho \mathcal{R} \hat{\rho} \wedge \sigma \mathcal{S}_{\text{range}(\rho)} (\hat{\rho}, \hat{\sigma})))$

Table 4. Control Flow Analysis: the extensions for the first approach.

For all $\tilde{O}$:
  if $\rho \mathcal{R} \hat{\rho}, (\hat{C}, \hat{\rho}, \hat{\sigma}) \models_\Sigma ie, \sigma \mathcal{S}_O (\hat{\rho}, \hat{\sigma})$ (for $O = \text{range}(\rho) \cup \text{obj}(ie) \cup \tilde{O}$)
  and if $\rho \vdash \langle ie, \sigma \rangle \rightarrow \langle ie', \sigma' \rangle$
  then $\rho \mathcal{R} \hat{\rho}, (\hat{C}, \hat{\rho}, \hat{\sigma}) \models_\Sigma ie', \sigma' \mathcal{S}_{O'} (\hat{\rho}, \hat{\sigma})$ (for $O' = \text{range}(\rho) \cup \text{obj}(ie') \cup \tilde{O}$).

Also, to deal with the case of method update we proceed by *coinduction* on the definition of the auxiliary relation $\mathcal{S}$.

4 The Second Approach

We now define our second semantics for the Imperative Object Calculus. Here the overall (and slightly imprecise) idea is that a semantic judgement is still of the form

$$\rho \vdash \langle e, \sigma \rangle \rightarrow \langle e', \sigma' \rangle$$

but now the typical form of the finite environment, ρ, and the finite state, σ, is given by

$$\rho(x) = \iota$$
$$\sigma(\iota) = [m_i = \varsigma(x_i).e_i{}^{i=1..n}]$$

so that variables are mapped to locations that are mapped to vectors of methods (whereas before, variables were mapped to vectors of locations that were then mapped to methods). In our view this is a rather minor technical difference that could easily have been adopted when defining the semantics of the Imperative Object Calculus [1].

As before, this description does not take account of the static scope rules in the Imperative Object Calculus, and we therefore once more introduce two new auxiliary expressions. The resulting semantics is shown in Table 5.

To express semantic correctness of the analysis in terms of a subject reduction result, we once more need to extend the analysis of Table 2 to incorporate the intermediate expression "bind ρ in e" and the object denotation that now is simply "ι". This is done in Table 6; then, for the purposes of this section, $\models_\Sigma$ is defined coinductively by the Tables 2 and 6.

$\rho \vdash \langle x^\ell, \sigma \rangle \rightarrow \langle (\rho(x))^\ell, \sigma \rangle$ if $x \in \text{dom}(\rho)$

$\rho \vdash \langle [m_i = \varsigma(x_i).e_i{}^{i=1..n}]^\ell, \sigma \rangle \rightarrow \langle \iota^\ell, \sigma' \rangle$
  if ι is fresh $\wedge\, \sigma' = \sigma[\iota \mapsto [m_i = (\text{close } \varsigma(x_i).e_i \text{ in } \rho)^{i=1..n}]]$

$$\frac{\rho \vdash \langle ie, \sigma \rangle \rightarrow \langle ie', \sigma' \rangle}{\rho \vdash \langle (ie.m)^\ell, \sigma \rangle \rightarrow \langle (ie'.m)^\ell, \sigma' \rangle}$$

$\rho \vdash \langle (\iota^{\ell_1}.m)^\ell, \sigma \rangle \rightarrow \langle (\text{bind } \rho' \text{ in } e_j)^\ell, \sigma \rangle$
  if $\iota \in \text{dom}(\sigma) \wedge m_j = m \wedge \sigma(\iota) = [m_i = (\text{close } \varsigma(x_i).e_i \text{ in } \rho_i)^{i=1..n}] \wedge \rho' = \rho_j[x_j \mapsto \iota]$

$$\frac{\rho \vdash \langle ie_1, \sigma \rangle \rightarrow \langle ie'_1, \sigma' \rangle}{\rho \vdash \langle (ie_1.m := \varsigma(x).e)^\ell, \sigma \rangle \rightarrow \langle (ie'_1.m := \varsigma(x).e)^\ell, \sigma' \rangle}$$

$\rho \vdash \langle (\iota^{\ell_1}.m := \varsigma(x).e)^\ell, \sigma \rangle \rightarrow \langle \iota^\ell, \sigma[\iota \mapsto o] \rangle$
  if $\iota \in \text{dom}(\sigma) \wedge m_j = m \wedge \sigma(\iota) = [m_i = (\text{close } \varsigma(x_i).e_i \text{ in } \rho_i)^{i=1..n}] \wedge$
  $o = [m_i = \text{close } \varsigma(x_i).e_i \text{ in } \rho_i{}^{i=1..}, m_j = \text{close } \varsigma(x).e \text{ in } \rho, m_i = \text{close } \varsigma(x_i).e_i \text{ in } \rho_i{}^{i=..n}]]$

$$\frac{\rho \vdash \langle ie, \sigma \rangle \rightarrow \langle ie', \sigma' \rangle}{\rho \vdash \langle (\text{clone } ie)^\ell, \sigma \rangle \rightarrow \langle (\text{clone } ie')^\ell, \sigma' \rangle}$$

$\rho \vdash \langle (\text{clone } \iota^{\ell_1})^\ell, \sigma \rangle \rightarrow \langle (\iota')^\ell, \sigma' \rangle$
  if ι' is fresh $\wedge\, \sigma' = \sigma[\iota' \mapsto \sigma(\iota)]$

$$\frac{\rho \vdash \langle ie_1, \sigma \rangle \rightarrow \langle ie'_1, \sigma' \rangle}{\rho \vdash \langle (\text{let } x = ie_1 \text{ in } e_2)^\ell, \sigma \rangle \rightarrow \langle (\text{let } x = ie'_1 \text{ in } e_2)^\ell, \sigma' \rangle}$$

$\rho \vdash \langle (\text{let } x = \iota^{\ell_1} \text{ in } e_2)^\ell, \sigma \rangle \rightarrow \langle (\text{bind } \rho' \text{ in } e_2)^\ell, \sigma \rangle$
  if $\rho' = \rho[x \mapsto \iota]$

$$\frac{\rho_1 \vdash \langle ie_2, \sigma \rangle \rightarrow \langle ie'_2, \sigma' \rangle}{\rho \vdash \langle (\text{bind } \rho_1 \text{ in } ie_2)^\ell, \sigma \rangle \rightarrow \langle (\text{bind } \rho_1 \text{ in } ie'_2)^\ell, \sigma' \rangle}$$

$\rho \vdash \langle (\text{bind } \rho_1 \text{ in } \iota^{\ell_2})^\ell, \sigma \rangle \rightarrow \langle \iota^\ell, \sigma \rangle$

Table 5. Operational Semantics: the second approach.

One of the new clauses makes use of an auxiliary relation, $\mathcal{R}_\Sigma$, that relates the concrete enviroment, ρ, to the abstract environment, $\hat{\rho}$. However, now Σ is no longer a piece of mysterious notation, but is actually a *signature* of a state: this is a finite mapping from locations to tuples of the form $\langle m_i{}^{i=1..n} \rangle$. For obtaining the signature, $\text{sig}(\sigma)$, of a state, σ, we simply define $\text{sig}(\sigma)(\iota) = \langle m_i{}^{i=1..n} \rangle$ whenever $\sigma(\iota) = [m_i = \cdots{}^{i=1..n}]$.

The other auxiliary relation, $\mathcal{S}$, is used to express the subject reduction result; it no longer needs to be defined coinductively.

Theorem 3. If $\rho\, \mathcal{R}_{\text{sig}(\sigma)}\, \hat{\rho}, (\hat{C}, \hat{\rho}, \hat{\sigma}) \models_{\text{sig}(\sigma)} ie, \sigma\, \mathcal{S}\, (\hat{\rho}, \hat{\sigma}),\, \rho \vdash \langle ie, \sigma \rangle \rightarrow \langle ie', \sigma' \rangle$ then $\rho\, \mathcal{R}_{\text{sig}(\sigma')}\, \hat{\rho}, (\hat{C}, \hat{\rho}, \hat{\sigma}) \models_{\text{sig}(\sigma')} ie', \sigma'\, \mathcal{S}\, (\hat{\rho}, \hat{\sigma})$.

Proof. Overall the proof is by induction on the shape of $\rho \vdash \langle ie, \sigma \rangle \rightarrow \langle ie', \sigma' \rangle$ and it does not employ coinduction nor Kripke-logical relations. It merely makes use of three facts of which the first one is:

$(\hat{C}, \hat{\rho}, \hat{\sigma}) \models_\Sigma \iota^\ell$ iff $\Sigma(\iota) \in \hat{C}(\ell)$

$(\hat{C}, \hat{\rho}, \hat{\sigma}) \models_\Sigma$ (bind ρ_1 in $it_2^{\ell_2})^\ell$ iff
$\quad \rho_1 \mathcal{R}_\Sigma \hat{\rho} \wedge (\hat{C}, \hat{\rho}, \hat{\sigma}) \models_\Sigma it_2^{\ell_2} \wedge \hat{C}(\ell_2) \subseteq \hat{C}(\ell)$

$\rho \mathcal{R}_\Sigma \hat{\rho}$ iff $\forall x \in \text{dom}(\rho) \subseteq \text{dom}(\hat{\rho}) : \Sigma(\rho(x)) \in \hat{\rho}(x)$

$\sigma \mathcal{S} (\hat{\rho}, \hat{\sigma})$ iff $\forall \iota \in \text{dom}(\sigma) : \forall [m_i = (\text{close } \varsigma(x_i).e_i \text{ in } \rho_i)^{i=1..n}]$:
$\quad ((\sigma(\iota) = [m_i = (\text{close } \varsigma(x_i).e_i \text{ in } \rho_i)^{i=1..n}])$
$\quad \Rightarrow (\langle\{\varsigma(x_i).e_i\}^{i=1..n}\rangle \sqsubseteq \hat{\sigma}(\langle m_i^{i=1..n}\rangle)) \wedge \forall i : \rho_i \mathcal{R}_{\text{sig}(\sigma)} \hat{\rho}))$

Table 6. Control Flow Analysis: the extensions for the second approach.

If $\rho \vdash \langle ie, \sigma \rangle \to \langle ie', \sigma' \rangle$ then $\text{sig}(\sigma) \sqsubseteq \text{sig}(\sigma')$.

Here the partial ordering on signatures (which are just partial functions with a finite domain) is defined in the standard way, and the result just says that objects are never extended with new method names. The other two facts are:

If $\rho \mathcal{R}_{\Sigma_1} \hat{\rho}$ and $\Sigma_1 \sqsubseteq \Sigma_2$ then $\rho \mathcal{R}_{\Sigma_2} \hat{\rho}$.
If $(\hat{C}, \hat{\rho}, \hat{\sigma}) \models_{\Sigma_1} ie$ and $\Sigma_1 \sqsubseteq \Sigma_2$ then $(\hat{C}, \hat{\rho}, \hat{\sigma}) \models_{\Sigma_2} ie$.

These facts express a notion of "signature monotonicity".

5 Conclusion

In this paper we have proved two correctness theorems that exhibited some interesting differences despite studying the same analysis. We shall therefore conclude by identifying some of the general insights that we have gained from this work. A first observation is that:

> The abstract object used to identify an object in the analysis must be stable under evaluation.

This should hardly be surprising given that the semantic correctness results are formulated as subject reduction results. But to save work it is mandatory that this consideration be applied already when defining the analysis (and the semantics) rather than postponing it to the actual proof of semantic correctness (where it might fail). As an example, the object $[m_i = \varsigma(x_i).e_i^{\ i=1..n}]$ is identified by the tuple $\langle m_i^{i=1..n} \rangle$ of method identifiers; this works because the imperative object calculus does not allow to extend an object with new method identifiers. Hence a different way of identifying an object will be needed for an object calculus that allows extending an object with new method identifiers.

Let us briefly review the approach taken to specify the program analysis. One way of defining the specification would be in a *compositional* (or syntax-directed) manner (e.g. [2]), and this works well for *closed* systems; the approach taken in this paper was to use an *abstract* manner of specification in the sense that all method bodies are analysed when invoked rather than when defined, and this works well for *open systems*. Our second observation then is that:

The specification must be defined coinductively in case one takes the abstract approach, whereas it may be defined inductively in case one takes a compositional approach.

The need for coinduction shows up when establishing the Moore Family property but is *not* of concern when establishing the subject reduction result; also note that the coinductive and inductive definitions actually coincide in the case of compositional definitions.

Both of the operational semantics necessitated *extending the syntax* of the language with new intermediate constructs; for a subject reduction result to make sense, one then has to extend the analysis to these intermediate constructs as well. In both cases, this involved introducing an auxiliary relation, $\mathcal{R}$, for relating concrete environments to abstract environments, and an auxiliary relation, $\mathcal{S}$, for expressing the subject reduction result. Our third observation then is that:

> Coinduction is only needed for proving the subject reduction result, if one of the auxiliary relations is defined coinductively.

In particular, even though the specification of the analysis is defined coinductively (because we are taking an abstract approach) this in itself does not necessitate the use of coinduction for proving the subject reduction result.

In some cases the auxiliary relations, $\mathcal{R}$ and $\mathcal{S}$, are indexed by additional information; this can involve information about the state and information about the environment. Our fourth and final observation then is:

> Kripke-logical relations are needed for the induction hypothesis in the proof of the subject reduction result, if the index to one of the auxiliary relations can increase during the proof.

This clearly explains the rule of Kripke-logical relations in the first approach considered; there was no similar use of Kripke-logical relations in the second approach, although the result on "signature monotonicity" expresses a result about "Kripke-logical relations for free".

Acknowledgement This research was supported in part by the DART-AROS project funded by the Danish Research Councils.

References

1. M. Abadi and L. Cardelli. *A Theory of Objects*. Springer, 1996.
2. K. L. S. Gasser, F. Nielson, and H. R. Nielson. Systematic realisation of control flow analyses for CML. In *Proc. ICFP '97*, pages 38–51. ACM Press, 1997.
3. F. Nielson and H. R. Nielson. Layered predicates. In *Proc. REX'92 workshop on Semantics — foundations and applications*, volume 666 of *Lecture Notes in Computer Science*, pages 425–456. Springer, 1993.
4. F. Nielson and H. R. Nielson. Infinitary Control Flow Analysis: a Collecting Semantics for Closure Analysis. In *Proc. POPL '97*. ACM Press, 1997.
5. G. D. Plotkin. A structural approach to operational semantics. Technical Report FN-19, DAIMI, Aarhus University, Denmark, 1981.

Expressive Completeness of Temporal Logic of Action

Alexander Rabinovich

Department of Computer Science Raymond and Beverly Sackler Faculty of Exact
Sciences Tel Aviv University*, Tel Aviv 69978, Israel,
e.mail: rabino@math.tau.ac.il

Abstract. The paper compares the expressive power of monadic second order logic of order, a fundamental formalism in mathematical logic and theory of computation, with that of a fragment of Temporal Logic of Actions introduced by Lamport for specifying the behavior of concurrent systems.

1 Introduction

The Temporal Logic of Actions (TLA) was introduced by Lamport [3] as a logic for specifying concurrent systems and reasoning about them. One of the main differences of TLA from other discrete time temporal logics is its inability to specify that one state should immediately be followed by the other state, though it can be specified that one state is followed by the other state at some later time.

Lamport [2] argued in favor of this decision *'The number of steps in a Pascal implementation is not a meaningful concept when one gives an abstract, high level specification'*. For example, programs like $Pr_1 :: x := True; y := False$ and $Pr_2 :: x := True; Skip; y := False$ are not distinguishable by the TLA specifications, however, they are distinguishable in linear time temporal logic, one of the most popular temporal logics [4].

As a consequence of the decision not to distinguish between *'doing nothing and taking a step that produces no changes'* [3], the language of TLA contains the next time operator in a very restricted form. For the same reasons the TLA existential quantifier $\exists^{TLA}$ has a semantics different from the standard existential quantifier.

In this paper we consider the fragment of Lamport's Temporal Logic of Action where variables can only receive boolean values (BTLA). We compare the expressive power of BTLA with that of monadic second order logic of order.

One of the consequences of TLA design decision is that only *stuttering* closed languages are definable in TLA. We will show that

(1) if a stuttering closed ω-language is definable in monadic second order logic of order then it is definable in BTLA.

---

* Supported by a research grant of Tel Aviv University.

Together with Theorem 6 from [6] this shows that an ω-language is definable in BTLA if and only if it is stuttering closed and definable in monadic second order logic.

In [6] we proved that there is no compositional translation from BTLA into monadic second order logic. The proof of (1) provides a translation from monadic logic into BTLA. However, this translation is also not compositional.

A continuous time interpretation for TLA was suggested in [6] and it was shown there that this interpretation is more appropriate than the standard discrete time interpretation. A compositional translation from BTLA into monadic logic under the continuous time interpretation was given in [6]. Here we will show that

(2) there exists a compositional translation from monadic second order logic into BTLA under the continuous time interpretation.

Hence, under the continuous time interpretation, BTLA and second order monadic logic can be translated one into the other in a compositional way.

The paper is organized as follows. In section 2 we fix terminology and notations. Section 3 recalls the syntax and the semantics of monadic second order logic of order. Section 4 recalls the connection between automata on ω-strings and monadic second order logic (see [9,8] for a survey). We also provide here an automata theoretical characterization of the languages definable in the logic under a continuous time interpretation. The syntax and the semantics of BTLA is provided in section 5. Section 6 characterizes the expressive power of BTLA.

2 Terminology and Notations

Notations: $\mathbf{N}$ is the set of natural numbers; $\mathbf{R}$ is the set of real numbers, $\mathbf{R}^{\geq 0}$ is the set of non negative reals; **BOOL** is the set of booleans and Σ is a finite non-empty set.

A function from $\mathbf{N}$ to Σ is called an ω-string over Σ. A function h from the non-negative reals into a finite set Σ is called a *finitely variable signal* over Σ if there exists an unbounded increasing sequence $\tau_0 = 0 < \tau_1 < \tau_2 \ldots < \tau_n < \ldots$ such that h is constant on every interval (τ_i, τ_{i+1}). Below we will use 'signal' for 'finitely variable signal'. We say that a signal x is *right continuous* at t iff there is $t_1 > t$ such that $x(t) = x(t')$ for all t' which satisfies $t < t' < t_1$. We say that a signal is *right continuous* if it is right continuous at every t.

A set of ω-strings over Σ is called an ω-language over Σ. Similarly, a set of finitely variable (respectively, right continuous) signals over Σ is called a finitely variable (respectively, right continuous) Σ-signal language.

Let σ be an ω-string. We denote by $\sigma_{[n,\infty)}$ the ω-string $\langle s_n, s_{n+1}, \ldots \rangle$ and by $head(\sigma)$ its first letter s_0. For an ω-string σ and a letter s we denote by $s\sigma$ the ω-string $\langle s, s_0, s_1, \ldots \rangle$.

The collapse of an ω-string $\sigma = \langle s_0, s_1 \ldots s_n, \ldots \rangle$ is the ω-string $\sharp\sigma$ which is defined recursively as follows:

$$\sharp\sigma = \begin{cases} \sigma & \text{if } \forall i.\ s_i = s_0 \\ s_0 \sharp \sigma_{[i,\infty)} & \text{if } s_i \neq s_0 \text{ and } s_j = s_0 \text{ for all } j < i \end{cases}$$

Hence, operator $\sharp$ assigns to each ω-string σ the ω-string obtained by replacing every finite maximal subsequence $\langle s_i, s_{i+1} \ldots \rangle$ of identical letters in σ by a letter s_i. The ω-strings $\sigma = \langle s_0, s_1 \ldots s_n, \ldots \rangle$ and $\sigma' = \langle s'_0, s'_1 \ldots s'_n, \ldots \rangle$ are stuttering equivalent (notations $\sigma \simeq \sigma'$) if $\sharp\sigma = \sharp\sigma'$. Let L be an ω-language. We use the notation $Stutt(L)$ for the stuttering closure of L which is defined as $\{\sigma : \text{there exists } \sigma' \in L \text{ such that } \sigma \simeq \sigma'\}$. We say that an ω-language L is stuttering closed if $L = Stutt(L)$.

3 Monadic Second Order Theory of Order

3.1 Syntax

The language $L_2^{\leq}$ of monadic second order theory of order has individual variables, monadic second order variables, a binary predicate $<$, the usual propositional connectives and first and second order quantifiers $\exists^1$ and $\exists^2$. We use t, v for individual variables and x, y for second order variables.

The atomic formulas of $L_2^{\leq}$ are formulas of the form: $t < v$ and $x(t)$. The formulas are constructed from atomic formulas by logical connectives and first and second order quantifiers.

We write $\psi(x, y, t, v)$ to indicate that the free variables of a formula ψ are among x, y, t, v.

3.2 Semantics

A structure $K = \langle A, B, <_K \rangle$ for $L_2^{\leq}$ consists of a set A partially ordered by $<_K$ and a set B of monadic functions from A into $BOOL$. The satisfiability relation $K, \tau_1, \ldots \tau_m, \mathbf{x_1} \ldots \mathbf{x_n} \models \psi(t_1, \ldots t_m, x_1, \ldots x_m)$ is defined in a standard way.

We will be interested mainly in the following structures:

1. Structure $\omega = \langle \mathbf{N}, 2^{\mathbf{N}}, <_N \rangle$, where $2^{\mathbf{N}}$ is the set of all monadic functions from $\mathbf{N}$ into $BOOL$.
2. The signal structure Sig is defined as $Sig = \langle \mathbf{R}^{\geq 0}, SIG, <_R \rangle$, where SIG is the set of finitely variable boolean signals.
3. The right continuous signal structure is defined as $\langle \mathbf{R}^{\geq 0}, RSIG, <_R \rangle$, where $RSIG$ is the set of right continuous boolean signals.

3.3 Definability

Let $\phi(x)$ be an $L_2^{\leq}$ formula and $K = \langle A, B, <_K \rangle$ be a structure. We say that a set $C \subseteq B$ is definable by $\phi(x)$ if $\mathbf{x} \in C$ if and only if $K, \mathbf{x} \models \phi(x)$.
Example (Interpretations of Formulas).

1. The formula $\forall t_1 \forall t_2. \, t_1 < t_2 \land (\neg \exists t_3. \, t_1 < t_3 < t_2) \rightarrow (x(t_1) \leftrightarrow \neg x(t_2))$ defines the ω-language $\{(01)^\omega, (10)^\omega\}$ in the structure ω and defines the set of all signals in the signal and right continuous signal structures.

2. The formula $\exists y. \exists t'. y(t') \wedge \forall t. x(t) \rightarrow y(t) \wedge \forall t_1 \forall t_2. t_1 < t_2 \wedge y(t_1) \wedge y(t_2) \rightarrow \exists t_3. t_1 < t_3 < t_2 \wedge \neg y(t_3)$ defines in the structure ω the set of strings in which between any two occurrences of 1 there is an occurrence of 0. In the signal structure the above formula defines the set of signals that receive value 1 only at isolated points. The formula defines the empty language under right continuous signal interpretation.

In the above examples, all formulas have one free second order variable and they define languages over alphabet $\{0, 1\}$. A formula $\psi(x_1, \ldots x_n)$ with n free second order variables defines a language over alphabet $\{0, 1\}^n$.

4 Labeled Transition Systems

4.1 Syntax

A Labeled Transition System T is a triple $\langle Q, \Sigma, \rightarrow \rangle$ that consists of a set Q of states, a finite alphabet Σ of actions and a transition relation $\rightarrow$ which is a subset of $Q \times \Sigma \times Q$; we write $q \xrightarrow{a} q'$ if $\langle q, a, q' \rangle \in \rightarrow$; if Q is finite we say that the LTS is finite.

Sometimes the alphabet Σ of T will be the Cartesian product $\Sigma_1 \times \Sigma_2$ of other alphabets; in such a case we will write $q \xrightarrow{a,b} q'$ for the transition from q to q' labeled by the pair (a, b).

An **automaton** $\mathcal{A}$ over Σ is a triple $\langle T, INIT(\mathcal{A}), FAIR(\mathcal{A}) \rangle$, where $T = \langle Q, \Sigma, \rightarrow \rangle$ is an LTS over alphabet Σ; $INIT(\mathcal{A}) \subseteq Q$ - the initial states of $\mathcal{A}$. and $FAIR(\mathcal{A})$ - a collection of fairness conditions (subsets of Q).

4.2 Semantics

A run of an automaton $\mathcal{A}$ is an ω-sequence $q_0 a_0 q_1 a_1 \ldots$ such that $q_i \xrightarrow{a_i} q_{i+1}$ for all i. Such a run meets the initial conditions if $q_0 \in INIT(\mathcal{A})$. A run meets the fairness conditions if the set of states that occur in the run infinitely many times is a member of $FAIR(\mathcal{A})$.

An ω-string $a_0, a_1 \ldots$ over Σ is accepted by $\mathcal{A}$ if there is a run $q_0 a_0 q_1 a_1 \ldots$ that meets the initial and fairness conditions of $\mathcal{A}$. The ω-language *definable* (or accepted) by $\mathcal{A}$ is the set of all ω-strings acceptable by $\mathcal{A}$.

A right continuous signal x over Σ is accepted by $\mathcal{A}$ if there are an ω-string $a_0 a_1 \ldots a_n \ldots$ over alphabet Σ acceptable by $\mathcal{A}$ and an unbounded increasing sequence $0 = \tau_0 < \tau_1 < \ldots < \tau_i < \ldots$ of reals such that $x(\tau) = a_i$ for $\tau \in [\tau_i, \tau_{i+1})$.

A signal x over Σ is accepted by $\mathcal{A}$ if $\mathcal{A}$ is an automaton over the alphabet $\Sigma \times \Sigma$ and there are an ω-string $\langle a_0, b_0 \rangle \langle a_1, b_1 \rangle \ldots \langle a_n, b_n \rangle \ldots$ acceptable by $\mathcal{A}$ and an unbounded increasing sequence $0 = \tau_0 < \tau_1 < \ldots < \tau_i < \ldots$ of reals such that $x(\tau_i) = a_i$ and $x(\tau) = b_i$ for $\tau \in (\tau_i, \tau_{i+1})$.

It is clear that if the ω-languages acceptable by $\mathcal{A}$ and $\mathcal{B}$ are stuttering equivalent then $\mathcal{A}$ and $\mathcal{B}$ define the same right continuous signal language.

We say that an ω-language (respectively, finite variability signal language or right continuous language) is definable in monadic logic if the language is definable in monadic logic in the structure ω (respectively, the structure Sig or the structure of right continuous signals).

Theorem 1. *(Büchi [1]) An ω-language is definable by a finite state automaton iff it is definable by a monadic formula.*

Theorem 2. *[7] A finitely variable signal language is definable by a finite state automaton if and only if it is definable by a monadic formula.*

Theorem 3. *A right continuous signal language is definable by a finite state automaton if and only if it is definable by a monadic formula.*

Proof. The only if direction is obtained by a direct formalization of the behavior of a finite state automaton.

The if direction is obtained by the method of interpretation [5] as follows. Let $\phi(x_1, \ldots x_n)$ be a monadic formula. First, we construct a monadic formula $\phi^*(x_1, \ldots x_n)$ such that the language definable by ϕ under finitely variable interpretation coincides with the language definable by ϕ under right continuous signal interpretation.

Let $rsignal(x)$ be the formula $\forall t \exists t'.\ t' > t \wedge \forall t''.t < t'' < t' \rightarrow x(t) = x(t'')$. It is clear that a signal satisfies $rsignal(x)$ iff it is right continuous.

Let ϕ' be obtained from ϕ by relativizing all the second order quantifiers to the right continuous signals, i.e., by replacing "$\forall x.\ \ldots$" (respectively "$\exists x.\ \ldots$") by "$\forall x.\ rsignal(x) \rightarrow \ldots$" (respectively, "$\exists x.\ \wedge \ldots$"). It is easy to see that a right continuous signal satisfies ϕ under right continuous interpretation iff it satisfies ϕ' under finitely variable interpretation. Hence, the required formula $\phi^*(x_1, \ldots x_n)$ can be defined as $rsignal(x_1) \wedge rsignal(x_2) \wedge \ldots \wedge rsignal(x_n) \wedge \phi'$.

By Theorem 2, there exists an automaton over $\Sigma \times \Sigma$ such that the finitely variable signal language definable by $\mathcal{A}$ is the same as the language definable by ϕ^*. Let $\mathcal{B}$ be the automaton over Σ defined as follows: (1) remove from $\mathcal{A}$ all transitions of the form $q_1 \xrightarrow{a,b} q_2$ for $a \neq b$. (2) replace transitions of the form $q_1 \xrightarrow{a,a} q_2$ by $q_1 \xrightarrow{a} q_2$. It is not difficult to verify that the right continuous signal language definable by ϕ is the same as the right continuous signal language definable by $\mathcal{B}$. □

5 Temporal Logic of Actions

We consider the fragment of Lamport's [3] Temporal Logic of Action where variables can only receive boolean values (BTLA).

5.1 Syntax

The symbol set of BTLA consists of:

1. A set Var of variables;
2. A set Var' of primed version for variables; $Var' = \{x' : x \in Var\}$.
3. Logical connectives $\wedge$ and $\neg$.
4. TLA existential quantifier $\exists^{TLA}$.
5. Modal operator $\Box$.
6. The special operator **Enabled**.

The syntax of BTLA formulas is summarized in Fig. 1.

$\langle\text{formula}\rangle \triangleq \langle\text{elementary formula}\rangle \mid \neg\,\langle\text{formula}\rangle \mid \langle\text{formula}\rangle \wedge \langle\text{formula}\rangle$
$\qquad\qquad\quad \mid \Box\langle\text{formula}\rangle \mid \exists^{TLA}x.\langle\text{formula}\rangle$
$\langle\text{elementary formula}\rangle \triangleq \langle\text{simple state formula}\rangle \mid \langle\text{enabled formula}\rangle \mid \langle\text{action formula}\rangle$
$\langle\text{enabled formula}\rangle \triangleq \textbf{Enabled}(\langle\text{action}\rangle)$
$\langle\text{action formula}\rangle \triangleq \Box[\langle action\rangle]_{\langle simple\ state\ formula\rangle}$
$\langle\text{action}\rangle \triangleq$ boolean combination of variables and primed variables.
$\langle\text{simple state formula}\rangle \triangleq$ boolean combination of variables.

Fig. 1. Syntax of BTLA

Remark: (Primed variables) Priming a variable in TLA 'corresponds' to applying the next operator in temporal logic (see definition 1 (2) below). One can see that this next operator is used in BTLA in very restricted form.

Remark: (Free and bound occurrences of variables) A variable x occurs free in x and in x'. The only binding operator of BTLA is the existential quantifier. $\exists^{TLA}x.\psi$ binds all free occurrences of x in ψ.

5.2 Semantics of BTLA

A state is a function from a set Var of variables into the boolean set **BOOL**. A state sequence σ is an ω-sequence $\langle s_0, s_1, \ldots, s_n \ldots\rangle$ of states. State sequences $\sigma = \langle s_0, s_1 \ldots s_n, \ldots\rangle$ and $\sigma' = \langle s'_0, s'_1 \ldots s'_n, \ldots\rangle$ are equivalent up to a variable x (notation $\sigma =_x \sigma'$) if for every n, the states s_n and s'_n coincide on all variables distinct from x.

There is a natural correspondence between the states for variables $v_1, \ldots, v_k$ and the letters of the alphabet $\{0, 1\}^k$. This correspondence is lifted to the correspondence between state sequences (respectively, sets of state sequences) and ω-strings (respectively, ω-languages) over the alphabet $\{0, 1\}^k$. We will say that state sequences σ and σ' are stuttering equivalent up to x (notations $\sigma \simeq_x \sigma'$) if there exist σ_1, σ'_1 such that $\sigma =_x \sigma_1$, $\sigma' =_x \sigma'_1$ and σ_1 is stuttering equivalent to σ'_1.

We are going to recall the definition of the satisfaction relation between state sequences and a superset of BTLA formulas, which was called raw TLA by Lamport [3]. In the following definition x denotes a BTLA variables and A denotes an action.

Definition 1. *The satisfaction relation $\models$ is defined as follows:*

1. $\sigma \models x$ *if* $head(\sigma)(x)$ *is equal to* TRUE.
2. $\sigma \models x'$ *if* $\sigma_{[1,\infty)} \models x$.
3. $\sigma \models \psi_1 \wedge \psi_2$ *if* $\sigma \models \psi_1$ *and* $\sigma \models \psi_2$
4. $\sigma \models \neg\psi$ *if not* $\sigma \models \psi$.
5. $\sigma \models \mathbf{Enabled}(A)$ *if there exists* σ' *such that* $head(\sigma)\sigma' \models A$.
6. $\sigma \models \Box\psi$ *if* $\sigma_{[n,\infty)} \models \psi$ *for every* n.
7. $\sigma \models \exists^{TLA} x.\psi$ *if there is* σ' *such that* $\sigma \simeq_x \sigma'$ *and* $\sigma' \models \psi$.

For an action A and a simple state formula p, the BTLA action formula $\Box[A]_p$ is considered as an abbreviation of the raw TLA formula $\Box(A \vee (p \leftrightarrow p'))$, where p' the formula obtained from p by replacing every variable x by its primed version x'. We will also use $\Box[A]_{p,q}$ as a shortand for a BTLA formula which is equivalent to the raw BTLA formula $\Box(A \vee ((p \leftrightarrow p') \wedge (q \leftrightarrow q')))$. As ususal $\Diamond\phi$ is an abbreviation for $\neg\Box\neg\phi$.

Note that the set of sequences which satisfies a BTLA formula is closed under stuttering, i.e., $\sigma \models \psi$ and $\sigma \simeq \sigma'$ imply $\sigma' \models \psi$.

6 Expressive Completeness of BTLA

6.1 Discrete Time

Theorem 4. *An ω-language is definable in BTLA if and only if it is stuttering closed and definable in $L_2^{\leq}$.*

Proof. The only if direction is Theorem 5 of [6]. In order to show the if direction it is enough to show that for every ω-language accepted by a finite state automaton its stuttering closure is definable in BTLA.

For an automaton $\mathcal{A} = \langle T, INIT(\mathcal{A}), FAIR(\mathcal{A}) \rangle$, where $T = \langle Q, \Sigma, \rightarrow \rangle$ is an LTS over alphabet Σ we first define the formulas $Init_\mathcal{A}, Next_\mathcal{A}, Fair_\mathcal{A}$ as follows

$$Init_\mathcal{A} \triangleq \bigvee_{q \in Init(\mathcal{A})} state = q$$

$$Next_\mathcal{A} \triangleq \Box[\bigvee_{q_1 \xrightarrow{a} q_2} (state = q_1 \wedge state' = q_2 \wedge x = a)]_{q,x}$$

$$Fair_\mathcal{A} \triangleq \bigvee_{Q \in FAIR(\mathcal{A})} Fair_Q, \text{ where}$$

for a set $Q = \{q_1, \ldots q_n\}$ the formula $Fair_Q$ is

$$Fair_Q \triangleq (\Diamond\Box \bigvee_{q \in Q} state = q) \wedge (\bigwedge_{q \in Q} \Box\Diamond state = q)$$

Let $\phi_{\mathcal{A}}$ be the formula $\exists^{TLA} state.\ (Init_{\mathcal{A}} \wedge Next_{\mathcal{A}} \wedge Fair_{\mathcal{A}})$.

In these definitions *state* and x range over ω-strings over the alphabets Q and Σ respectively. It is well-known that the strings over an alphabet of size $n < 2^k$ can be coded by k-tuples of strings over the binary alphabet. So, the above formulas are the shorthands of the BTLA formulas.

It is easy to check that an ω-string $a_0 a_1 \ldots a_n \ldots$ satisfies $\phi_{\mathcal{A}}$ if either it is stuttering equivalent to an ω-string acceptable by $\mathcal{A}$ or there are ω-strings $b_0 b_1 \ldots b_n \ldots$ and $q_0 q_1 \ldots q_n \ldots$ such that (1) $a_0 a_1 \ldots a_n \ldots$ and $b_0 b_1 \ldots b_n \ldots$ are stuttering equivalent and (2) there is n such that $\{q_{n+1}\} \in FAIR(\mathcal{A})$ and $\forall i > n.\ b_i = b_n \wedge q_i = q_{n+1}$ and $\forall i \leq n.\ q_i \xrightarrow{b_i} q_{i+1}$.

Therefore, if all fairness conditions of $\mathcal{A}$ contain at least two states then the language definable by $\phi_{\mathcal{A}}$ is the stuttering closure of the language accepted by $\mathcal{A}$.

It is not difficult to show that every automaton is equivalent to an automaton with all fairness conditions of cardinality at least two. Hence, every stuttering closed ω-language definable in $L_2^<$ is definable in BTLA. □

6.2 Continuous Time

We say that an ω-string $s = a_0 \ldots a_n \ldots$ represents a right continuous signal x if there is an unbounded ω-sequence $0 = \tau_0 < \tau_1 \ldots \tau_n \ldots$ such that $x(\tau) = a_i$ for $\tau \in [\tau_i, \tau_{i+1})$. It is clear that the set of ω-strings that represents a right continuous signal is stuttering closed.

A signal (a right continuous signal) language L is *speed-independent* if for every bijective increasing function $\rho : \mathbf{R}^{\geq 0} \to \mathbf{R}^{\geq 0}$ the following condition holds: $x \in L$ iff $x \circ \rho \in L$. It is clear that the set of right continuous signals representable by an ω-string is speed independent. Representability induces one-one correspondence between stuttering closed ω-languages and speed-independent right continuous signal languages. Through this correspondence we associate with every BTLA formula ϕ the set of right continuous signals that are representable by the ω-strings which satisfy ϕ. This set of right continuous signals is said to be definable by a BTLA formula ϕ under right continuous interpretation. It was shown in [6] that right continuous signal interpretation for BTLA is more appropriate than the standard discrete time interpretation.

Proposition 5. *For every monadic formula ϕ there exists a BTLA formula ϕ^* such that the language definable by ϕ under right continuous interpretation is the same as the right continuous signal language definable by ϕ^*.*

Proof. Let ϕ be a monadic formula and let L be a language definable by ϕ under right continuous interpretation. By Theorem 3, there exists an automaton $\mathcal{A}$ such that L is the right continuous signal language definable by $\mathcal{A}$. Let L' be the ω-language definable by $\mathcal{A}$. Let L'' be the stuttering closure L'. Note that the ω-languages L' and L'' represent the right continuous signal language L. From the proof of Theorem 4 it follows that L'' is definable by a BTLA formula ϕ^*. Therefore, ϕ and ϕ^* define the same language under right continuous signal interpretation. This completes the proof of the proposition. □

Proposition 5 implies the *if direction* of the following theorem; the *only-if direction* is Corollary 18 of [6].

Theorem 6. *A right continuous signal language is definable in BTLA if and only if it is definable in monadic second order logic of order.*

Observe that the proof of Theorem 6 is constructive. In particular, one can extract from the proof translation algorithms $Alg : L_2^< \to BTLA$ and $Alg' : BTLA \to L_2^<$ such that the formulas ϕ and $Alg(\phi)$ (respectively, ϕ and $Alg'(\phi)$) define the same right continuous signal language.

Let us comment first on the translation algorithm from monadic logic into BTLA. In the proof of Proposition 5 we first translated monadic formulas into automata and then translated automata into a BTLA formulas. The size of the automaton obtained from ϕ might be much larger than the size of ϕ. In fact, for every k and every n there is a formula of the size $m > n$ such that the corresponding automaton has at least $exp_k(m)$ states, where $exp_k(n)$ is the k time iterated exponential function (e.g. $exp_2(n) = 2^{2^n}$). Obviously, the above translation algorithm from monadic logic into BTLA is not compositional.

On the other hand, the translation from BTLA into monadic logic that is extracted from the proof of the *only-if direction* of Theorem 6 is compositional (see Section 8 in [6]). Below an alternative proof of the *if direction* of Theorem 6 is provided in which we define a compositional translation Tr from monadic logic into BTLA. In the full version of the paper we show that the translation has the following property: the right continuous signal languages definable by a formula $\phi(x_1, \ldots x_n)$ without free first order variables and by BTLA formula $Tr(\phi)$ are the same.

Hence, under the right continuous signal interpretation, BTLA and second order monadic logic can be translated one into the other in a compositional way.

Let $Sing(t)$ be a BTLA formula that defines the ω-language $0^*11^*0^\omega$.
Let $ORDER(t_1, t_2)$ be the formula

$$\Box(t_1 \leftrightarrow t_2)$$
$$\vee \Diamond(t_1 \wedge \neg t_2 \wedge \Diamond(\neg t_1 \wedge \neg t_2 \wedge \Diamond t_2))$$
$$\vee \Diamond(t_2 \wedge \neg t_1 \wedge \Diamond(\neg t_1 \wedge \neg t_2 \wedge \Diamond t_1))$$

Let $Contains(x, t)$ be the formula

$$(\Diamond(x \wedge t)) \to ((\Box(t \to x)) \wedge \sqcup[(t \to x') \vee (t \leftrightarrow t' \wedge x \leftrightarrow x')]_x)$$

The translation is defined in Fig. 2. We use $free_1(\phi)$ (respectively, $free_2(\phi)$ for the set of first order (respectively, second order) variables which are free in ϕ.

$$Tr(t_1 < t_2) \triangleq \Box(t_1 \to \Diamond t_2)$$
$$Tr(x(t)) \triangleq \Box(t \to x)$$
$$Tr(\phi_1 \wedge \phi_2) \triangleq Tr(\phi_1) \wedge Tr(\phi_2)$$
$$Tr(\neg \phi) \triangleq \neg Tr(\phi)$$
$$Tr(\exists^1 t.\ \phi) \triangleq \exists^{BTLA} t.\ Sing(t) \wedge Tr(\phi) \wedge$$
$$\bigwedge_{t_i \in free_1(\phi)} Ordered(t, t_i) \wedge \bigwedge_{x \in free_2(\phi)} (Contains(x, t) \vee Contains(\neg x, t))$$
$$Tr(\exists^2 x.\ \phi) \triangleq \exists^{BTLA} x.\ Tr(\phi) \wedge \bigwedge_{t \in free_1(\phi)} (Contains(x, t) \vee Contains(\neg x, t))$$

Fig. 2. Translation from $L_2^{\leq}$ into BTLA

References

1. J. R. Büchi. On a decision method in restricted second order arithmetic In *Proc. International Congress on Logic, Methodology and Philosophy of Science*, E. Nagel at al. eds, Stanford University Press, pp 1-11, 1960.
2. L. Lamport. What good is temporal logic. In R. E. A. Manson editor *Information Processing 83, Proceedings of IFIP 9th World Congress*, Paris pp. 657-668. IFIP, North Holland.
3. L. Lamport. The Temporal Logic of Actions. ACM Transactions on Programming Languages and Systems, 16(3), pp. 872-923, 1994.
4. Z. Manna and A. Pnueli. The Temporal Logic of Reactive and Concurrent Systems, Springer Verlag, 1992.
5. M. O. Rabin. Decidable theories. In J. Barwise editor *Handbook of Mathematical Logic*, North-Holland, 1977.
6. A. Rabinovich. On Translations of Temporal Logic of Actions into Monadic Second Order Logic. Theoretical Computer Science 193 (1998), 197-214.
7. A. Rabinovich and B. A. Trakhtenbrot. From Finite Automata to Hybrid Systems. Fundamentals of Computation Theory 1997, LNCS vol. 1279, pages 411-422, 1997.
8. W. Thomas. Automata on Infinite Objects. In J. van Leeuwen editor *Handbook of Theoretical Computer Science*, The MIT Press, 1990.
9. B. A. Trakhtenbrot and Y. M. Barzdin. Finite Automata, North Holland Amsterdam, 1973.

Reducing AC-Termination to Termination

Maria C. F. Ferreira[1], Delia Kesner[2], and Laurence Puel[2]

[1] Dep. de Informática, Fac. de Ciências e Tecnologia, Univ. Nova de Lisboa,
Quinta da Torre, 2825 Monte da Caparica, Portugal, cf@di.fct.unl.pt.
[2] CNRS and Laboratoire de Recherche en Informatique, Bât 490, Université de
Paris-Sud, 91405 Orsay Cedex, France, {kesner,puel}@lri.fr.

Abstract. We present a new technique for proving AC-termination. We show that if certain conditions are met, AC-termination can be reduced to termination, i. e., termination of a TRS S modulo an AC-theory can be inferred from termination of another TRS R with no AC-theory involved. This is a new perspective and opens new possibilities to deal with AC-termination.

1 Introduction

Termination of term rewriting systems (TRS's) is crucial for the use of rewriting in proofs and computations, and many theories have been developed in this field. However, many interesting and useful systems have operators which are *associative* and *commutative* (AC), and most techniques developed for proving termination of TRS's do not carry over to AC-rewriting so that they need to be adapted to this case. Along these lines, a lot of work has been done on the development of suitable AC-compatible orderings, as for example [4, 11, 12, 14, 15, 19, 18], exploring the possibilities of adapting the *recursive path ordering (rpo)* [5, 10] to the AC-case. It is well-known that the rpo technique cannot handle all terminating TRS's and the same remark applies to AC-extensions of it wrt AC-termination; we see an example.

Example 1. Take $R : f(f(a(x,y),x),y) \to f(a(f(a(x,y),x),f(a(x,y),y)),y)$, where a is an AC-symbol.

Let $>_{AC}$ be any AC-compatible order having the subterm property (1) and being closed under contexts (2). Then $a(f(a(x,y),x),f(a(x,y),y)) >_{AC} f(a(x,y),x)$ by (1); $f(a(f(a(x,y),x),f(a(x,y),y)),y) >_{AC} f(f(a(x,y),x),y)$, by (2), which is not useful for our purpose since we rather need to have the inequality the other way around. Note that rpo-like AC-extensions, as for example those in [11, 18], cannot deal with this example since they do enjoy properties (1) and (2); furthermore most rpo-like AC-extensions are designed for ground terms and their application to open terms together with the property of closedness under substitutions is not always easy to obtain; this example suggests that other techniques are necessary to deal with AC-function symbols.

A useful approach to termination consists in using sound transformations such that the transformed systems are somehow easier to deal with, wrt termination proofs, than the original ones. Examples of this approach can be found in [1, 2, 3, 20, 21, 22, 9]. Most transformation techniques were originally meant for standard term rewriting, which raises the pertinent question of validity of the transformations in an equational setting. In [7], a transformation of equational TRS's into equational TRS's is proposed in such a way that termination of the resulting equational system implies termination of the original equational system. In particular, this transformation changes the rewriting system but leaves the equational part intact. However, proving equational termination and in particular, AC-termination is not an easy task (AC-termination is even undecidable for terminating systems [16]), and so the technique proposed in [7] while simplifying the problem still has the disadvantage that we are left, after the transformation, with an equational (AC) system.

In this paper, we propose a new technique that allows us to reduce AC-termination to termination, i. e., termination of a TRS S modulo an AC-theory can be inferred from termination of another TRS R with *no theory* involved. We use the dummy-elimination technique defined in [7], but we completely eliminate the equational part of the system along with the elimination of the AC-symbols; as a consequence, the proof of soundness of our transformation is not a particular case of that in [7] (we show by examples why that cannot be) so that we have to define a new interpretation of terms to achieve the desired results. It is also pertinent to remark that our soundness proof is just based on the *existence* of an AC-compatible order having some additional properties but not on any particular definition of such orders.

Coming back to Ex. 1 we transform the AC-system R into the following system R', by eliminating the symbol a and introducing a fresh constant $\diamond$:

$$R' : f(f(\diamond,x),y) \to f(\diamond,y) \qquad\qquad f(f(\diamond,x),y) \to x$$
$$f(f(\diamond,x),y) \to f(\diamond,x) \qquad\qquad f(f(\diamond,x),y) \to y$$

Since the system R' is terminating (that can be proved for example by the rpo technique), then AC-termination of R will follow.

For the sake of simplicity and clarity, we present our technique for systems having just one AC-symbol. However, the technique is also applicable whenever more AC-symbols are present (see sec. 3).

The rest of the paper is organized as follows. Section 2 is devoted to explaining the transformation which allows to reduce AC-termination to termination and to proving its soundness. In sec. 3 we discuss the details concerning the application of our technique, we consider when many AC-symbols are present, the possibility of eliminating only some AC-symbols while keeping others; we also discuss the weak and strong points of this technique, and present some conclusions. Due to space restrictions proofs are omitted, but we refer the reader to [8] for a full and detailed version of this paper.

2 The Transformation and Its Soundness

We assume the reader is familiar with the basic notion pertaining to partial orders, quasi-orders, rewriting and rewriting modulo AC-theories and, due to lack of space, will only present some needed notions; for more information the reader is referred to [6, 8, 13, 17].

A TRS is *terminating* if it admits no infinite rewrite sequence. If EQ is an equational system and R a TRS, we say that R is E-terminating (or that R/EQ is terminating) if the relation $\to_{R/EQ}$ is terminating, i. e., if there are no infinite sequences of the form: $s_0 =_{EQ} s'_0 \to_R s_1 =_{EQ} s'_1 \to_R s_2 =_{EQ} s'_2 \to_R s_3 \ldots$, where $=_{EQ}$ represents the equational theory generated by the set of equations EQ, and $\to_R$ represents the rewrite relation generated by R.

An equational rewrite system R/EQ (TRS S) is *compatible* with a quasi-order $\succeq = \succ \cup \sim$ (on $\mathcal{T}(\mathcal{F}, \mathcal{X})$) (resp. partial order $>$) if $=_{EQ} \subseteq \sim$ and $\to_R \subseteq \succ$ (resp. $\to_R \subseteq >$). It is well-known that a TRS is terminating if and only if it is compatible with a reduction (thus well-founded) order, and that an equational rewrite system is terminating if and only if it is compatible with a reduction quasi-order. Furthermore if a TRS R is terminating then $\to_R^+$ is a reduction order on $\mathcal{T}(\mathcal{F}, \mathcal{X})$.

In AC-rewriting it is common to use the flattened version of terms. A term is *flattened* wrt an AC-function symbol f if it does not contain any nested occurrences of this symbol. Note that for a flattened term to make sense we need to admit that the AC-symbol can have any arity ≥ 2. Given a term t, we denote its flattened version wrt an AC-function symbol f, by $\bar{t}^f$ or simply $\bar{t}$.

For our purposes we need the *existence* of an AC-compatible order having some additional properties, namely *subterm compatibility*, *closedness under contexts*, *well-foundedness* when taken with respect to a well-founded precedence, and *AC-compatibility*. We will present our results making use of Kapur and Sivakumar's order [11], which enjoys these properties. In the following, let $>_{ac}$ and $\sim_{ac}$ denote respectively the order and compatible congruence relation defined in [11], and let $\geq_{ac} = >_{ac} \cup \sim_{ac}$.

As we mentioned before, we present our results for the case where we have only one AC-symbol. The technique is however applicable if we have a (possibly infinite) collection of AC-symbols we wish to eliminate (see sec. 3).

Let a be an AC-function symbol *not* occurring in signature $\mathcal{F}$; a is the function symbol to be eliminated. Since we will work with flattened terms, we consider that the symbol a has variable arity ≥ 2. Flattening will be done always with respect to this function symbol. Let $\diamond$ be a constant also not occurring in $\mathcal{F}$. We denote by $\mathcal{F}_a$ and $\mathcal{F}_\diamond$ resp. the sets $\mathcal{F} \cup \{a\}$ and $\mathcal{F} \cup \{\diamond\}$. In $\mathcal{T}(\mathcal{F}_a, \mathcal{X})$, we consider the relation $\to_{R/AC}$, where AC is the set of the associative and commutative equations for a. We define a transformation on terms that induces a transformation $\mathcal{H}$ on the AC-systems, and then show that termination of $\to_{R/AC}$ can be inferred from termination of $\to_{\mathcal{H}(R/AC)}$. The relevant point of this transformation is that the system $\mathcal{H}(R/AC)$ does *not* contain any equation, thus we are in fact reducing AC-termination to termination.

The main idea behind the term algebra transformation is to recursively break a term t into pieces, $\texttt{cap}(t), \texttt{dec}(t)$, that do not contain the function symbol to be eliminated; one of these blocks, namely the one above all occurrences of the function symbol "a", is denoted by cap(t) and treated especially. This transformation was introduced in [9], from which we take the definition:

Definition 1. *Functions* cap: $T(\mathcal{F}_a, \mathcal{X}) \to T(\mathcal{F}_\diamond, \mathcal{X})$ *and* dec: $T(\mathcal{F}_a, \mathcal{X}) \to \mathcal{P}(T(\mathcal{F}_\diamond, \mathcal{X}))$ *are defined inductively as follows:*

- cap(x) = x and dec(x) = $\emptyset$, *for any* $x \in \mathcal{X}$,
- cap($f(\ldots t_i \ldots)$) = $f(\ldots \text{cap}(t_i) \ldots)$, *and* dec($f(\ldots t_i \ldots)$) = $\bigcup_i \text{dec}(t_i)$,
- cap($a(\ldots t_i \ldots)$) = $\diamond$, *and* dec($a(\ldots t_i \ldots)$) = $\bigcup_i (\{\text{cap}(t_i)\} \cup \text{dec}(t_i))$.

For example, the term $t = f(a(g(a(x, y), z), a(x, s(x))), x, h(a(x, h(y))))$ has cap(t) = $f(\diamond, x, h(\diamond))$ and dec(t) = $\{g(\diamond, z), x, y, \diamond, s(x), h(y)\}$.

We can extend both the function cap and the notion of flattening to substitutions as follows.

Definition 2. *Let* $\sigma : \mathcal{X} \to T(\mathcal{F}_a, \mathcal{X})$ *be an arbitrary substitution. The substitutions* cap(σ) : $\mathcal{X} \to T(\mathcal{F}_\diamond, \mathcal{X})$ *and* $\overline{\sigma} : \mathcal{X} \to T(\mathcal{F}_a, \mathcal{X})$ *are defined respectively by* cap(σ)(x) = cap($\sigma(x)$), *and* $\overline{\sigma}(x) = \overline{\sigma(x)}$, *for all* $x \in \mathcal{X}$.

We now define the transformation on TRS's. As can be expected we will transform the lhs and rhs's of the rules in R, creating new rules and simultaneously getting rid of the AC-equations.

Definition 3. *Given an AC-rewriting system R/AC over $T(\mathcal{F}_a, \mathcal{X})$ such that the function symbol a is AC, $\mathcal{H}(R/AC)$ is a TRS over $T(\mathcal{F}_\diamond, \mathcal{X})$ given by*

$$\mathcal{H}(R/AC) = \{\text{cap}(l) \to u \mid l \to r \in R \text{ and } u \in \{\text{cap}(r)\} \cup \text{dec}(r)\}$$

Note that in some cases $\mathcal{H}(R/AC)$ may not be a TRS in the usual sense, since cap(l) may eliminate variables needed in the rhs's of the transformed rules. From the definition of $\mathcal{H}$, we see that in general the TRS $\mathcal{H}(R/AC)$ has more rules but is syntactically and furthermore semantically simpler than the original one. Since in the transformed version no equations are involved, proving termination becomes an easier task provided the transformation is sound. This is exactly the original characteristic of our technique which consists in reducing AC-termination to termination.

In contrast with some other techniques to show termination, our transformation is not complete, i. e., there exist AC-terminating systems R such that $\mathcal{H}(R/AC)$ are not terminating. To see that consider the TRS $R: f(x, x) \to f(a(x, x), x)$, where a is an AC-symbol. R is AC-terminating, while $\mathcal{H}(R/AC) = \{f(x, x) \to f(\diamond, x), f(x, x) \to x\}$ is clearly non-terminating: the term $f(\diamond, \diamond)$ reduces to itself.

However, complete techniques are usually difficult or impossible to be implemented, and our aim is to provide a *new* tool to deal with and *simplify* the problem of AC-termination.

We now show that the transformation $\mathcal{H}$ is sound, i. e., termination of $\mathcal{H}(R/AC)$ implies AC-termination of R/AC. The proof proceeds along the following general lines. To each term in $T(\mathcal{F}_a, \mathcal{X})$ we associate a term over a different signature. For that we consider the set of terms $T(\Sigma)$ where $\Sigma = T(\mathcal{F}_\diamond, \mathcal{X})$,

i. e., *each term* in $T(\mathcal{F}_\diamond, \mathcal{X})$ is seen *as a function symbol* in $T(\Sigma)$; note that variables in $\mathcal{X}$ are now interpreted as constants in this new algebra of terms. Furthermore, while $\diamond$ in $T(\mathcal{F}_\diamond, \mathcal{X})$ is defined to be a constant and $\mathcal{F}_\diamond$ does not contain any AC-symbol, $\diamond$ is an AC-symbol in Σ, the only one. If $\mathcal{H}(R/AC)$ is terminating, there is a well-founded order $>$ on $T(\mathcal{F}_\diamond, \mathcal{X})$ compatible with $\mathcal{H}(R/AC)$; such an order provides a well-founded precedence on Σ upon which we will consider $\geq_{ac}$. As a consequence $>_{ac}$ will be well-founded on $T(\Sigma)$, so we can conclude AC-termination of R/AC if we show that the interpretation of the terms from $T(\mathcal{F}_a, \mathcal{X})$ in $T(\Sigma)$ is compatible with both the AC-theory and the rewriting relation defined by R. In other words we only need to ensure that if $s =_{AC} u \to_R v =_{AC} t$, with $s, u, v, t \in T(\mathcal{F}_a, \mathcal{X})$, then $S \sim_{ac} U >_{ac} V \sim_{ac} T$, where $S, U, V, T \in T(\Sigma)$ are the interpretations of, respectively, s, u, v, t.

Because we want to mark the distinction between the more traditional terms and terms in $T(\Sigma)$, in which function symbols are themselves terms, we will use a slightly different notation for the terms in $T(\Sigma)$.

Notation 4 *A term in $T(\Sigma)$ will be denoted as $s(\!(s_1, \ldots, s_k)\!)$, where $s \in \Sigma$ and $s_i \in T(\Sigma)$, for all $1 \leq i \leq k$, $k \geq 0$. $\Sigma_{AC} = \{\diamond\}$ and all function symbols in $T(\mathcal{F}_\diamond, \mathcal{X})$ are of fixed arity except $\diamond$ and symbols t for which its root has varyadic arity in $\mathcal{F}$. As usual, we will represent constants $s(\!(\,)\!)$ simply by s.*

A good start point to define an interpretation of terms is to use that in [7].

Definition 5. *A term $t \in T(\mathcal{F}_a, \mathcal{X})$ is mapped to a term $\mathtt{tree}(t) \in T(\Sigma)$, by the function $\mathtt{tree} : T(\mathcal{F}_a, \mathcal{X}) \to T(\Sigma)$, defined inductively as:*

- $\mathtt{tree}(x) = x(\!(\,)\!)$, *for any* $x \in \mathcal{X}$.
- $\mathtt{tree}(f(s_1, \ldots, s_k)) = \mathtt{cap}(f(s_1, \ldots, s_k))(\!(t^1_1, \ldots, t^1_{n_1}, \ldots t^k_1, \ldots, t^k_{n_k})\!)$, *where* $\mathtt{tree}(s_i) = \mathtt{cap}(s_i)(\!(t^i_1, \ldots, t^i_{n_i})\!)$ *for all* $1 \leq j \leq k$.
- $\mathtt{tree}(a(s_1, \ldots, s_k)) = \mathtt{cap}(a(s_1, \ldots, s_k))(\!(\mathtt{tree}(s_1), \ldots, \mathtt{tree}(s_k))\!)$.

However, the interpretation $\mathtt{tree}(\_)$ does not allow to show soundness of our transformation $\mathcal{H}$ as it looses vital information about the structure of terms having AC-symbols. The problems encountered are similar to the ones posed when one tries to extend rpo to AC-flattened terms. We illustrate those problems in an example showing that $\mathtt{tree}(\_)$ does not work with non-flattened terms nor with flattened terms.

Example 2. Consider the terms $s = f(a(x, a(y, z)))$ and $t = f(a(a(x, y), z))$, where a is AC. We have $s =_{AC} t$ but $\mathtt{tree}(s) = f(\diamond)(\!(x, \diamond(\!(y, z)\!))\!)$ is not AC-equal to $\mathtt{tree}(t) = f(\diamond)(\!(\diamond(\!(x, y)\!), z)\!)$. Now consider $R : a(0, 1) \to f(0, 1)$, where a is AC. Since $s = h(a(a(0, 1), 2)) \to_{R/AC} h(a(f(0, 1), 2)) = t$, one would like to show that $\mathtt{tree}(\bar{s})$ is in some sense greater than $\mathtt{tree}(\bar{t})$.

Now, $\mathcal{H}(R/AC)$ is the rule $\diamond \to f(0, 1)$, $\bar{s} = h(a(0, 1, 2))$ and $\bar{t} = t$ so that $\mathtt{tree}(\bar{s}) = h(\diamond)(\!(0, 1, 2)\!)$ and $\mathtt{tree}(\bar{t}) = h(\diamond)(\!(f(0, 1), 2)\!)$. Using $>_{ac}$ (or RPO on flattened terms) one obtains $\mathtt{tree}(\bar{t}) >_{ac} \mathtt{tree}(\bar{s})$ which is exactly the contrary of what one wants.

So we need a different interpretation and propose the following.

Definition 6. *A term $t \in \mathcal{T}(\mathcal{F}_a, \mathcal{X})$ is mapped to a term $\mathcal{I}(t) \in \mathcal{T}(\Sigma)$, by the function $\mathcal{I}: \mathcal{T}(\mathcal{F}_a, \mathcal{X}) \to \mathcal{T}(\Sigma)$, defined inductively as:*

- $\mathcal{I}(x) = x(\!(\,)\!)$, *for any $x \in \mathcal{X}$,*
- $\mathcal{I}(f(s_1, \ldots, s_k)) = \mathtt{cap}(f(s_1, \ldots, s_k))(\!(\mathcal{I}(s_1), \ldots, \mathcal{I}(s_k))\!)$, *for any $f \in \mathcal{F}_a$.*

The term $f(g(a(a(0, x), y)))$ is interpreted by $f(g(\diamond))(\!(g(\diamond)(\!(\diamond(\!(\diamond(\!(0, x)\!), y)\!))\!))\!)$.

Remark 7 *From now on we assume that $\mathcal{H}(R/AC)$ is well-defined and terminating. This means, in particular, that for any rule $l \to r$ in R we must have $var(r) \subseteq var(\mathtt{cap}(l))$; this fact will be of use later.*

Since $\mathcal{H}(R/AC)$ is terminating, $\to^+_{\mathcal{H}(R/AC)}$ is a well-founded partial order on $\mathcal{T}(\mathcal{F}_\diamond, \mathcal{X})$, which is closed under contexts and substitutions. In general this order will not possess the subterm property, which we will need at a later stage, but fortunately that is not a problem since, as was noted by Kamin and Lévy [10], we can easily extend such an order to another one enjoying that property (at the expense of losing closedness under contexts). Before proceeding further, we recall this order and its properties.

Definition 8. *We define a relation $\gg$ on $\mathcal{T}(\mathcal{F}_\diamond, \mathcal{X})$ as follows: $s \gg t$ iff $(s \neq t$ and $\exists C: s \to^*_{\mathcal{H}(R/AC)} C[t])$.*

Lemma 1. *In the conditions of def. 8, if $\mathcal{H}(R/AC)$ is terminating then $\gg$ is a well-founded partial order on $\mathcal{T}(\mathcal{F}_\diamond, \mathcal{X})$ extending $\to_{\mathcal{H}(R/AC)}$ (i. e., $\to^+_{\mathcal{H}(R/AC)} \subseteq \gg$), closed under substitutions and satisfying the subterm property.*

From now on we take $\gg$ (def. 8) as a precedence in $\mathcal{T}(\Sigma)$ and consider the quasi-order $\geq_{ac}$ associated to it. As was previously noted it is not necessary to make $\gg$ total since we do not require $>_{ac}$ to be total.

We want to prove that if $s \to_{R/AC} t$ then $\geq_{ac}$ decreases the interpretations of these terms. However $\geq_{ac}$ compares flattened terms and so some flattening operation has to be performed either on s and t or on their interpretations. It turns out that if we flatten a term in $\mathcal{T}(\mathcal{F}_a, \mathcal{X})$ wrt a before interpreting it, the interpretation of it in $\mathcal{T}(\Sigma)$ will be in a flattened form wrt to $\diamond$, so we flatten terms before interpreting them.

So now we prove that if $s \to_{R/AC} t$ then $\mathcal{I}(\bar{s}) >_{ac} \mathcal{I}(\bar{t})$, and we proceed in two steps showing that $s =_{AC} t \Rightarrow \mathcal{I}(\bar{s}) \sim_{ac} \mathcal{I}(\bar{t})$ and that $s \to_R t \Rightarrow \mathcal{I}(\bar{s}) >_{ac} \mathcal{I}(\bar{t})$.

Definition 9. *For any terms $s, t \in \mathcal{T}(\mathcal{F}_a, \mathcal{X})$, $s \equiv_{AC} t$ if and only if either $s = t$, or $\bar{s} = f(s_1, \ldots, s_k)$, $\bar{t} = f(t_1, \ldots, t_k)$, and $s_i \equiv_{AC} t_{\pi(i)}$, for all $1 \leq i \leq k$, and some permutation π, such that π is the identity whenever $f \notin AC$.*

The relation $\equiv_{AC}$ is used in order to translate the equality $=_{AC}$, which is defined on the set of ordinary terms, into the set of flattened terms. It is also worth to notice that on $\mathcal{T}(\mathcal{F}_a, \mathcal{X})$, $\equiv_{AC} \subset \sim_{ac}$. The following results are straightforward.

Lemma 2. *The relation $\equiv_{AC}$ is an equivalence relation on $\mathcal{T}(\mathcal{F}_a, \mathcal{X})$; furthermore, for all $s, t \in \mathcal{T}(\mathcal{F}_a, \mathcal{X})$, $s =_{AC} t \Rightarrow \overline{s} \equiv_{AC} \overline{t}$, and $\overline{s} \equiv_{AC} \overline{t} \Rightarrow \mathcal{I}(\overline{s}) \sim_{ac} \mathcal{I}(\overline{t})$.*

Corollary 1. *Let $s, t \in \mathcal{T}(\mathcal{F}_a, \mathcal{X})$. Then $s =_{AC} t$ implies $\mathcal{I}(\overline{s}) \sim_{ac} \mathcal{I}(\overline{t})$.*

We turn now to the case of inequality.

Lemma 3. *Let $s \in \mathcal{T}(\mathcal{F}_a, \mathcal{X}) \setminus \mathcal{X}$, $t \in \mathcal{T}(\mathcal{F}_a, \mathcal{X})$ be terms such that $\mathrm{var}(t) \subseteq \mathrm{var}(\mathtt{cap}(s))$ and $\mathtt{cap}(s) \gg v$ for all $v \in \mathtt{dec}(t) \cup \{\mathtt{cap}(t)\}$; let $\sigma : \mathcal{X} \to \mathcal{T}(\mathcal{F}_a, \mathcal{X})$ be any substitution. Then $\mathcal{I}(\overline{s\sigma}) >_{ac} \mathcal{I}(\overline{t\sigma})$.*

From the definition of $\mathcal{H}(R/AC)$ (def. 3), and the assumption that $\mathcal{H}(R/AC)$ is terminating, it is easy to check that we can replace s and t in the above lemma by, respectively, l and r, for any rule $l \to r \in R$, so we can state that the interpretation is compatible with the rules of R/AC.

Corollary 2. *Suppose that $\mathcal{H}(R/AC)$ is terminating, and let $l \to r$ be any rule in R and let $\sigma : \mathcal{X} \to \mathcal{T}(\mathcal{F}_a, \mathcal{X})$ be any substitution. Then $\mathcal{I}(\overline{l\sigma}) >_{ac} \mathcal{I}(\overline{r\sigma})$.*

We still have to check that if a reduction occurs within a non-trivial context, the same result holds, i. e., $l \to r \in R$ implies $\mathcal{I}(\overline{C[l\sigma]}) >_{ac} \mathcal{I}(\overline{C[r\sigma]})$.

Theorem 1. *Under the assumption that $\mathcal{H}(R/AC)$ is terminating, let $s, t \in \mathcal{T}(\mathcal{F}_a, \mathcal{X})$ such that $s \to_R t$. Then $\mathcal{I}(\overline{s}) >_{ac} \mathcal{I}(\overline{t})$.*

We can now prove our main result.

Theorem 2. *If $\mathcal{H}(R/AC)$ is terminating then R/AC is AC-terminating.*

Proof. Suppose that R/AC does not terminate. Then we have an infinite sequence of the form $s_0 =_{AC} s'_0 \to_R s_1 =_{AC} s'_1 \to_R s_2 =_{AC} s'_2 \ldots$ Using corollary 1 and theorem 1 this translates to the following sequence on $\mathcal{T}(\Sigma)$:

$$\mathcal{I}(\overline{s_0}) \sim_{ac} \mathcal{I}(\overline{s'_0}) >_{ac} \mathcal{I}(\overline{s_1}) \sim_{ac} \mathcal{I}(\overline{s'_1}) >_{ac} \mathcal{I}(\overline{s_2}) \ldots$$

where $>_{ac}$ is taken over the well-founded precedence $\gg$ in $\mathcal{T}(\mathcal{F}_\diamond, \mathcal{X})$. Since $>_{ac}$ and $\sim_{ac}$ are compatible and $>_{ac}$ is well-founded (the precedence is well-founded), this is a contradiction.

3 Discussion and Conclusions

For the sake of simplicity we presented our technique for AC-systems with only one AC-symbol. The technique is however valid in the presence of more (possibly infinite) AC-symbols, but its application can be done in different ways, namely map all AC-symbols to the same constant $\diamond$, or map groups of AC-symbols to different constants (in the extreme case each AC-symbol a is associated to a different constant $\diamond_a$). These different forms of applying the transformation are not equivalent, being the last one the finer one.

It may also be interesting to eliminate only some AC-symbols while keeping others (for example when the application of the previous technique is not possible); in this case the resulting system will still be an AC-system. Soundness of this transformation can be shown along the same general lines; the technical details become however much more unpleasant.

We would also like to point out that this technique is not always appropriate whenever the symbol we want to eliminate occurs in the lhs of rewrite rules since then all variables occurring in the rhs must also occur in the lhs **above** (or parallel to) the to-be-eliminated symbol. This means that our technique can hardly cope with systems having defined AC-symbols, since in those cases the restriction required on the variables will usually not be fulfilled. A possible way to deal with more cases where the symbol is defined has been recently pointed out in [1], and it seems that a similar solution could be applied for AC-symbols (this is currently under investigation). However, the interesting property of our technique is that it can be used to eliminate just some symbols of the system, treating the rest of them with more classical techniques to prove termination. Thus, this technique is not just an alternative to other techniques to prove AC-termination, but also a complementary tool that allows to reduce AC-systems to systems without AC-symbols.

The idea of extending dummy elimination to equational rewriting had already been explored in [7], but with the restriction that the equational part of the system remained unchanged. This is in line with most works dealing with AC-termination that either define new techniques or try to extend existing ones to the AC-setting, but without ever questioning the setting itself. The motivation of our work consists precisely in changing this setting, and we do so by eliminating the equational part of the system. This presents a totally different way of looking at AC-termination and is, as far as we know, the first technique which allows to show termination of AC-systems while ignoring the AC-equations, i. e. by showing termination of some system which has no associated equations.

In the future we would like to pursue this line of research by studying the possible application of transformations defined in the literature and/or define new ones. Also, we would like to investigate what kind of equational theories are amenable to a similar treatment as the one presented here for AC.

References

[1] T. Aoto and Y. Toyama. Termination transformation by tree lifting ordering. In *Proc. of the 9th Int. Conf. on Rewriting Techniques and Applications - RTA 98*, volume 1379 of *LNCS*. Springer, 1998.
[2] T. Arts. *Automatically proving termination and innermost normalisation of term rewriting systems*. PhD thesis, Universiteit Utrecht, May 1997.
[3] F. Bellegarde and P. Lescanne. Termination by completion. *Applicable Algebra in Engineering, Communication and Computing*, 1(2):79–96, 1990.
[4] C. Delor and L. Puel. Extension of the associative path ordering to a chain of associative commutative symbols. In *Proc. of the 5th Int. Conf. on Rewrite Techniques and Applications (RTA)*, number 690 in LNCS, pages 389–404. Springer, 1993.

[5] N. Dershowitz. Orderings for term rewriting systems. *Theoretical Computer Science*, 17(3):279–301, 1982.
[6] N. Dershowitz and J.-P. Jouannaud. Rewrite systems. In J. van Leeuwen, editor, *Handbook of Theoretical Computer Science*, volume B. Elsevier, 1990.
[7] M. C. F. Ferreira. Dummy Elimination in Equational Rewriting. In *Proc. of the 7th Int. Conf. on Rewriting Techniques and Applications*, volume 1103 of *LNCS*, pages 78–92. Springer, 1996.
[8] M. C. F. Ferreira and D. Kesner and L. Puel. Reducing AC-Termination to Termination. Technical Report 1175, Université Paris-Sud. 1998.
[9] M. C. F. Ferreira and H. Zantema. Dummy elimination: making termination easier. In *Fundamentals of Computation Theory, 10th Int. Conference FCT'95*, volume 965 of *LNCS*, pages 243–252. Springer, 1995.
[10] S. Kamin and J. J. Lévy. Two generalizations of the recursive path ordering. University of Illinois, 1980.
[11] D. Kapur and G. Sivakumar. A Total, Ground Path Ordering for Proving Termination of AC-Rewrire Systems. In *Proc. of the 8th Int. Conf. on Rewriting Techniques and Applications - RTA'97*, volume 1232 of *LNCS*. Springer, 1997.
[12] D. Kapur, G. Sivakumar, and H. Zhang. A new method for proving termination of ac-rewrite systems. In *Proc. of the 10th Conf. on Foundations of Software Technology and Theoretical Computer Science*, volume 472 of *LNCS*, pages 133 – 148. Springer, 1990.
[13] J. W. Klop. Term rewriting systems. In S. Abramsky, D. M. Gabbay, and T. S. E. Maibaum, editors, *Handbook of Logic in Computer Science*, volume II, pages 1–116. Oxford University Press, 1992.
[14] C. Marché. Normalized rewriting and normalized completion. In *Proc. of the 9th IEEE Symposium on Logic in Computer Science*, pages 394–403, 1994.
[15] P. Narendran and M. Rusinowitch. Any ground associative-commutative theory has a finite canonical system. In *Proc. of the 4th Int. Conf. on Rewriting Techniques and Applications*, volume 488 of *LNCS*, pages 423–434. Springer, 1991.
[16] H. Osaki and A. Middeldorp. Type introduction for equational rewriting. In *Proc. of the 4th Int. Symposium Logical Foundations of Computer Science - LFCS 97*, volume 1234 of *LNCS*, pages 283–293. Springer, 1997.
[17] D. A. Plaisted. Equational reasoning and term rewriting systems. In D. Gabbay, C. J. Hogger, and J. A. Robinson, editors, *Handbook of Logic in Artificial Intelligence and Logic Programming*, volume 1 - Logical Foundations, pages 273–364. Oxford Science Publications, Clarendon Press - Oxford, 1993.
[18] A. Rubio. A total AC-compatible ordering with RPO scheme, 1997. Draft.
[19] A. Rubio and R. Nieuwenhuis. A precedence-based total AC-compatible ordering. *Theoretical Computer Science*, 142:209–227, 1995.
[20] H. Xi. Towards automated termination proofs through "freezing". In *Proc. of the 9 th Int. Conf. in Rewriting Techniques and Applications - RTA 98*, volume 1379 of *LNCS*. Springer, 1998.
[21] H. Zantema. Termination of term rewriting: interpretation and type elimination. *Journal of Symbolic Computation*, 17:23–50, 1994.
[22] H. Zantema. Termination of term rewriting by semantic labelling. *Fundamenta Informaticae*, 24:89–105, 1995.

On One-Pass Term Rewriting *

Zoltán Fülöp[1], Eija Jurvanen[2], Magnus Steinby[3], and Sándor Vágvölgyi[4]

[1] József Attila University, Department of Computer Science, H-6701 Szeged, P. O. Box 652, Hungary, `fulop@inf.u-szeged.hu`
[2] Turku Centre for Computer Science, DataCity, Lemminkäisenkatu 14 A, FIN-20520 Turku, Finland, `jurvanen@utu.fi`
[3] Turku Centre for Computer Science, and Department of Mathematics, University of Turku, FIN-20014 Turku, Finland, `steinby@utu.fi`
[4] József Attila University, Department of Applied Informatics, H-6701 Szeged, P. O. Box 652, Hungary, `vagvolgy@inf.u-szeged.hu`

Reducing a term with a term rewriting system (TRS) is a highly nondeterministic process and usually no bound for the lengths of the possible reduction sequences can be given in advance. Here we consider two very restrictive strategies of term rewriting, *one-pass root-started rewriting* and *one-pass leaf-started rewriting*. If the former strategy is followed, rewriting starts at the root of the given term t and proceeds continuously towards the leaves without ever rewriting any part of the current term which has been produced in a previous rewrite step. When no more rewriting is possible, a *one-pass root-started normal form* of the term t has been reached. The leaf-started version is similar, but the rewriting is initiated at the leaves and proceeds towards the root. The requirement that rewriting should always concern positions immediately adjacent to parts of the term rewritten in previous steps distinguishes our rewriting strategies from the IO and OI rewriting schemes considered in [5] or [2]. It also implies that the top-down and bottom-up cases are different even for a linear TRS.

Let $\mathcal{R} = (\Sigma, R)$ be a TRS over a ranked alphabet Σ. For any Σ-tree language T, we denote the sets of one-pass root-started sentential forms, one-pass root-started normal forms, one-pass leaf-started sentential forms and one-pass leaf-started normal forms of trees in T by $1r\mathrm{S}_{\mathcal{R}}(T)$, $1r\mathrm{N}_{\mathcal{R}}(T)$, $1\ell\mathrm{S}_{\mathcal{R}}(T)$ and $1\ell\mathrm{N}_{\mathcal{R}}(T)$, respectively. We show that the following inclusion problems, where $\mathcal{R} = (\Sigma, R)$ is a left-linear TRS and T_1 and T_2 are two regular Σ-tree languages, are decidable.

The one-pass root-started sentential form inclusion problem: $1r\mathrm{S}_{\mathcal{R}}(T_1) \subseteq T_2$?
The one-pass root-started normal form inclusion problem: $1r\mathrm{N}_{\mathcal{R}}(T_1) \subseteq T_2$?
The one-pass leaf-started sentential form inclusion problem: $1\ell\mathrm{S}_{\mathcal{R}}(T_1) \subseteq T_2$?
The one-pass leaf-started normal form inclusion problem: $1\ell\mathrm{N}_{\mathcal{R}}(T_1) \subseteq T_2$?

In [9] the inclusion problem for ordinary sentential forms is called the second-order reachability problem and the problem is shown to be decidable for a TRS $\mathcal{R}$ which preserves recognizability, i.e. if the set of sentential forms of the trees of

---

* This research was supported by the exchange program of the University of Turku and the József Attila University, and by the grants MKM 665/96 and FKFP 0095/97.

any recognizable tree language T is also recognizable. In our problems the sets of normal forms or sentential forms are not necessarily regular.

Many questions concerning term rewriting systems have been studied using tree automata; cf. [2], [4], [8], [9], [10], [11], [12], for example. We also prove the decidability of the four inclusion problems by reducing them to the emptiness problem of certain finite tree recognizers.

We thank the referees for useful comments.

1 Preliminaries

Here we introduce the basic notions used in the paper, but for more about term rewriting and tree automata, we refer the reader to [1], [3], [6] and [7].

In what follows Σ is a *ranked alphabet*. For each $m \geq 0$, the set of m-ary symbols in Σ is denoted by Σ_m, and Σ is *unary* if $\Sigma = \Sigma_1$. If Y is an alphabet disjoint with Σ, the set $T_\Sigma(Y)$ of Σ-*terms with variables* in Y is the smallest set U including Y such that $f(t_1, \ldots, t_m) \in U$ whenever $m \geq 0$, $f \in \Sigma_m$ and $t_1, \ldots, t_m \in U$. If $c \in \Sigma_0$, we write just c for $c()$. The set $T_\Sigma(\emptyset)$ of *ground* Σ-*terms* is denoted by T_Σ. Terms are also called *trees*. Ground Σ-terms and subsets of T_Σ are called Σ-*trees* and Σ-*tree languages*, respectively. The *height* hg(t) of a tree $t \in T_\Sigma(Y)$ is defined so that hg$(t) = 0$ for $t \in Y \cup \Sigma_0$, and hg$(t) = \max\{\text{hg}(t_1), \ldots, \text{hg}(t_m)\} + 1$ for $t = f(t_1, \ldots, t_m)$. The set var(t) ($\subseteq Y$) of variables appearing in t is also defined as usual (cf. [7]).

Let $X = \{x_1, x_2, \ldots\}$ be a set of variables. For each $n \geq 0$, we put $X_n = \{x_1, \ldots, x_n\}$ and abbreviate $T_\Sigma(X_n)$ to $T_{\Sigma,n}$. A tree $t \in T_{\Sigma,n}$ is *linear* if no variable appears twice in t. The subset $\widetilde{T}_{\Sigma,n}$ of $T_{\Sigma,n}$ is defined so that $t \in T_{\Sigma,n}$ belongs to $\widetilde{T}_{\Sigma,n}$ if and only if each of $x_1, \ldots, x_n$ occurs in t exactly once and their left-to-right order is $x_1, \ldots, x_n$. Also, let $\widetilde{T}_{\Sigma,X} = \bigcup_{n=0}^\infty \widetilde{T}_{\Sigma,n}$. If $f \in \Sigma_m$, $m \geq 1$ and $t_1, \ldots, t_m \in \widetilde{T}_{\Sigma,X}$, then $\|f(t_1, \ldots, t_m)\|$ is the tree in $\widetilde{T}_{\Sigma,X}$ obtained from $f(t_1, \ldots, t_m)$ by renaming the variables. If $t \in T_{\Sigma,n}$ and $\sigma \colon X \to T_\Sigma(X)$ is a substitution such that $\sigma(x_i) = t_i$ ($i = 1, \ldots, n$), we write $\sigma(t) = t[t_1, \ldots, t_n]$.

A *term rewriting system* (TRS) over Σ is a system $\mathcal{R} = (\Sigma, R)$, where R is a finite set of *rewrite rules* $p \to r$ such that $p, r \in T_\Sigma(X)$, var$(r) \subseteq$ var(p) and $p \notin X$. A rule $p \to r$ is *ground* if $p, r \in T_\Sigma$. The rewrite relation $\Rightarrow_\mathcal{R}$ on T_Σ induced by $\mathcal{R}$ is defined so that $t \Rightarrow_\mathcal{R} u$ if u is obtained from t by replacing an occurrence of a subtree of t of the form $p[t_1, \ldots, t_n]$ by $r[t_1, \ldots, t_n]$, where $p \to r \in R$, $p, r \in T_{\Sigma,n}$ and $t_1, \ldots, t_n \in T_\Sigma$. The reflexive, transitive closure of $\Rightarrow_\mathcal{R}$ is denoted by $\Rightarrow_\mathcal{R}^*$. Hence $s \Rightarrow_\mathcal{R}^* t$ iff there exists a *reduction sequence*

$$t_0 \Rightarrow_\mathcal{R} t_1 \Rightarrow_\mathcal{R} \ldots \Rightarrow_\mathcal{R} t_n$$

in $\mathcal{R}$ such that $n \geq 0$, $t_0 = s$ and $t_n = t$. Note that we apply a TRS to ground terms only. For any TRS $\mathcal{R} = (\Sigma, R)$, let lhs$(\mathcal{R}) = \{p \mid (\exists r)\, p \to r \in R\}$. The TRS $\mathcal{R}$ is *left-linear* if every p in lhs$(\mathcal{R})$ is linear, and it is then *in standard form* if lhs$(\mathcal{R}) \subseteq \widetilde{T}_{\Sigma,X}$. A tree $s \in T_\Sigma$ is *irreducible* with respect to $\mathcal{R}$ if $s \Rightarrow_\mathcal{R} u$ for no u, and it is a *normal form* of a Σ-tree t if it is irreducible and $t \Rightarrow_\mathcal{R}^* s$.

In a *top-down Σ-recognizer* $\mathcal{A} = (A, \Sigma, P, a_0)$ (1) A is a (finite) unary ranked alphabet of *states* such that $A \cap \Sigma = \emptyset$, (2) P is a finite set of *transition rules*, each of the form $a(f(x_1, \ldots, x_m)) \to f(a_1(x_1), \ldots, a_m(x_m))$, also written simply $a(f) \to f(a_1, \ldots, a_m)$, where $m \geq 0$, $f \in \Sigma_m$ and $a, a_1, \ldots, a_m \in A$, and (3) $a_0 \in A$ is the *initial state*. We treat $\mathcal{A}$ as the TRS $(\Sigma \cup A, P)$ and the rewrite relation $\Rightarrow_\mathcal{A} \subseteq T_{\Sigma \cup A} \times T_{\Sigma \cup A}$ is defined accordingly. For each $a \in A$, let $T(\mathcal{A}, a) = \{t \in T_\Sigma \mid a(t) \Rightarrow^*_\mathcal{A} t\}$. The tree language *recognized* by $\mathcal{A}$ is the set $T(\mathcal{A}) = T(\mathcal{A}, a_0)$. A tree language $T \subseteq T_\Sigma$ is *recognizable*, or *regular*, if $T(\mathcal{A}) = T$ for a top-down Σ-recognizer $\mathcal{A}$.

In a *generalized top-down Σ-recognizer* $\mathcal{A} = (A, \Sigma, P, a_0)$ the rewrite rules of P are of the form $a(t(x_1, \ldots, x_n)) \to t[a_1(x_1), \ldots, a_n(x_n)]$, where $n \geq 0$, $a, a_1, \ldots, a_n \in A$, and $t \in \widetilde{T}_{\Sigma,n}$. The relations $\Rightarrow_\mathcal{A}$, $\Rightarrow^*_\mathcal{A}$, and the set $T(\mathcal{A})$ are defined as in a top-down Σ-recognizer.

A *bottom-up Σ-recognizer* is a quadruple $\mathcal{A} = (A, \Sigma, P, A_f)$, where (1) A is a finite set of *states* of rank 0, $\Sigma \cap A = \emptyset$, (2) P is a finite set of *transition rules*, of the form $f(a_1, \ldots, a_m) \to a$ with $m \geq 0$, $f \in \Sigma_m$, $a_1, \ldots, a_m, a \in A$, and (3) A_f ($\subseteq A$) is the set of *final states*. We say that $\mathcal{A}$ is *total deterministic* if for all $f \in \Sigma_m$, $m \geq 0$, $a_1, \ldots, a_m \in A$, there is exactly one rule of the form $f(a_1, \ldots, a_m) \to a$. We treat $\mathcal{A}$ as the rewriting system $(\Sigma \cup A, P)$, and the tree language recognized by it can be defined as the set $T(\mathcal{A}) = \{t \in T_\Sigma \mid (\exists a \in A_f)\ t \Rightarrow^*_\mathcal{A} a\}$. For any bottom-up Σ-recognizer $\mathcal{A}$, one can effectively construct a total deterministic bottom-up Σ-recognizer $\mathcal{B}$ such that $T(\mathcal{A}) = T(\mathcal{B})$.

In a *generalized bottom-up Σ-recognizer* $\mathcal{A} = (A, \Sigma, P, A_f)$ P is a finite set of rewrite rules $t[a_1, \ldots, a_n] \to a$, where $n \geq 0$, $t \in \widetilde{T}_{\Sigma,n}$ and $a_1, \ldots, a_n, a \in A$. The tree language recognized by $\mathcal{A}$ is $T(\mathcal{A}) = \{t \in T_\Sigma \mid (\exists a \in A_f)\ t \Rightarrow^*_\mathcal{A} a\}$.

It is easy to see that both generalized top-down and bottom-up Σ-recognizers recognize exactly the regular Σ-tree languages. Moreover, the emptiness problem "$T(\mathcal{A}) = \emptyset$?" is obviously decidable for both types of automata.

2 One-Pass Term Rewriting

The first of our two modes of one-pass rewriting may be described as follows.

Let $\mathcal{R} = (\Sigma, R)$ be a TRS and t the Σ-tree to be rewritten. The portion of t first rewritten should include the root. Rewriting then proceeds towards the leaves so that each rewrite step applies to a root segment of a maximal unprocessed subtree but never involves any part of the tree produced by a previous rewrite step. For the formal definition we associate with $\mathcal{R}$ a TRS in which a new special symbol forces this mode of rewriting.

Definition 2.1. *The* one-pass root-started TRS *associated with a given TRS* $\mathcal{R} = (\Sigma, R)$ *is the TRS* $\mathcal{R}_\# = (\Sigma \cup \{\#\}, R_\#)$, *where* $\#$ *is a new unary symbol, the* separator mark, *and* $R_\#$ *is the set of all rewrite rules*

$$\#(p(x_1, \ldots, x_n)) \to r[\#(x_1), \ldots, \#(x_n)]$$

obtained from a rule $p \to r$ *in* R, *where* $p, r \in T_{\Sigma,n}$, *by adding* $\#$ *to the root of the left-hand side and above the variables in the right-hand side.*

Example 2.1. If $R = \{f(g(x_1), x_2) \to f(x_1, g(x_2)), g(x_1) \to g(c)\}$, where $f \in \Sigma_2$, $g \in \Sigma_1$, and $c \in \Sigma_0$, then

$$R_\# = \{\#(f(g(x_1), x_2)) \to f(\#(x_1), g(\#(x_2))), \#(g(x_1)) \to g(c)\} \ .$$

For any TRS $\mathcal{R}$, the associated one-pass root-started TRS $\mathcal{R}_\#$ is terminating. For recovering the one-pass root-started reduction sequences of $\mathcal{R}$ from the reduction sequences of $\mathcal{R}_\#$, we introduce the tree homomorphism $\delta\colon T_{\Sigma \cup \{\#\}} \to T_\Sigma$ which just erases the separator marks. If

$$\#(t) \Rightarrow_{\mathcal{R}_\#} t_1 \Rightarrow_{\mathcal{R}_\#} t_2 \Rightarrow_{\mathcal{R}_\#} \cdots \Rightarrow_{\mathcal{R}_\#} t_k$$

is a reduction sequence with $\mathcal{R}_\#$ starting from some $t \in T_\Sigma$, then

$$t \Rightarrow_\mathcal{R} \delta(t_1) \Rightarrow_\mathcal{R} \delta(t_2) \Rightarrow_\mathcal{R} \cdots \Rightarrow_\mathcal{R} \delta(t_k)$$

is a *one-pass root-started reduction sequence* with $\mathcal{R}$. The terms $t, \delta(t_1), \ldots, \delta(t_k)$ are called *one-pass root-started sentential forms* of t in $\mathcal{R}$. If t_k is irreducible in $\mathcal{R}_\#$, then $\delta(t_k)$ is a *one-pass root-started normal form* of t in $\mathcal{R}$. The sets of all one-pass root-started sentential forms and normal forms of a Σ-tree t are denoted by $1rS_\mathcal{R}(t)$ and $1rN_\mathcal{R}(t)$, respectively. This notation is extended to sets of Σ-trees in the natural way.

Note that for any TRS $\mathcal{R} = (\Sigma, R)$ and any $t \in T_\Sigma$, the sets $1rS_\mathcal{R}(t)$ and $1rN_\mathcal{R}(t)$ are finite and effectively computable but that $1rS_\mathcal{R}(T)$ and $1rN_\mathcal{R}(T)$ are not necessarily regular even for a regular Σ-tree language T.

The one-pass TRS used for defining the one-pass leaf-started rewriting mode of a given TRS is constructed in two stages.

Definition 2.2. *Let $\mathcal{R} = (\Sigma, R)$ be a TRS. First we extend R to the set R_e of all rules*

$$p[y_1, \ldots, y_n] \to r[y_1, \ldots, y_n]$$

such that $p \to r \in R$ with $p, r \in T_{\Sigma, n}$, and for each i, $1 \leq i \leq n$, either $y_i \in X$ or $y_i \in \Sigma_0$, and $p[y_1, \ldots, y_n] \in \widetilde{T}_{\Sigma, X}$. Now let $\Sigma' = \{f' \mid f \in \Sigma\}$ be a disjoint copy of Σ such that for any $f \in \Sigma$, f and f' have the same rank. The one-pass leaf-started TRS associated with $\mathcal{R}$ is the TRS $\mathcal{R}^\# = (\Sigma \cup \Sigma' \cup \{\#\}, R^\#)$, where $\#$ is a new unary symbol, the separator mark, and $R^\#$ consists of all rules

$$p[\#(x_1), \ldots, \#(x_n)] \to \#(r'(x_1, \ldots, x_n)) \ ,$$

where $p \to r \in R_e$, with $p, r \in T_{\Sigma, n}$, and r' is obtained from r by replacing every symbol $f \in \Sigma$ by the corresponding symbol f' in Σ'.

Example 2.2. Let $R = \{f(g(x_1), x_2) \to f(x_1, c), g(c) \to c\}$, where $\Sigma = \{f, g, c\}$, $f \in \Sigma_2$, $g \in \Sigma_1$, and $c \in \Sigma_0$. Then $\Sigma' = \{f', g', c'\}$ and the one-pass leaf-started TRS associated with $\mathcal{R} = (\Sigma, R)$ is the TRS $\mathcal{R}^\# = (\Sigma \cup \Sigma' \cup \{\#\}, R^\#)$ where $R^\#$ consists of the five rules

$$f(g(\#(x_1)), \#(x_2)) \to \#(f'(x_1, c')), \ f(g(c), \#(x_1)) \to \#(f'(c', c')),$$
$$f(g(\#(x_1)), c) \to \#(f'(x_1, c')), \ f(g(c), c) \to \#(f'(c', c')), \ g(c) \to \#(c') \ .$$

Clearly, $\Rightarrow_{\mathcal{R}_e} = \Rightarrow_{\mathcal{R}}$. The reduction sequences of $\mathcal{R}^\#$ represent reduction sequences of $\mathcal{R}$ which start at the leaves of a term and proceed towards the root of it so that symbols introduced by a previous rewrite step never form a part of the left-hand side of the rule applied next. Moreover, $\mathcal{R}^\#$ passes only once over the term because the left-hand sides and the right-hand sides of its rules share only the symbol $\#$. The corresponding one-pass reduction sequence of $\mathcal{R}$ is recovered by applying the tree homomorphism $\delta : T_{\Sigma \cup \Sigma' \cup \{\#\}} \to T_\Sigma$ which erases the $\#$-marks and the primes from the symbols $f' \in \Sigma'$. Then each reduction sequence

$$t \Rightarrow_{\mathcal{R}^\#} t_1 \Rightarrow_{\mathcal{R}^\#} t_2 \Rightarrow_{\mathcal{R}^\#} \ldots \Rightarrow_{\mathcal{R}^\#} t_k$$

with $\mathcal{R}^\#$ yields the *one-pass leaf-started reduction sequence*

$$t \Rightarrow_\mathcal{R} \delta(t_1) \Rightarrow_\mathcal{R} \delta(t_2) \Rightarrow_\mathcal{R} \ldots \Rightarrow_\mathcal{R} \delta(t_k)$$

with $\mathcal{R}$. The terms $t, \delta(t_1), \ldots, \delta(t_k)$ are called *one-pass leaf-started sentential forms* of t in $\mathcal{R}$. If t_k is irreducible in $\mathcal{R}^\#$, then $\delta(t_k)$ is a *one-pass leaf-started normal form* of t in $\mathcal{R}$. The sets of all one-pass leaf-started sentential forms and normal forms of a Σ-tree t are denoted by $1\ell\,\mathrm{S}_\mathcal{R}(t)$ and $1\ell\,\mathrm{N}_\mathcal{R}(t)$, respectively. This notation is extended to sets of Σ-trees in the natural way.

Note that without the new rules of the extended TRS $\mathcal{R}_e$ many natural one-pass leaf-started rewriting sequences of $\mathcal{R}$ could be missed.

3   The One-Pass Root-Started Inclusion Problems

First we consider the one-pass root-started normal form inclusion problem. It is assumed that the tree languages are given as tree recognizers.

Theorem 3.1. *For any left-linear TRS $\mathcal{R} = (\Sigma, R)$, the following* one-pass root-started normal form inclusion problem *is decidable.*
  Instance: *Recognizable Σ-tree languages T_1 and T_2.*
  Question: $1r\mathrm{N}_\mathcal{R}(T_1) \subseteq T_2$?

For proving Theorem 3.1, we need the following auxiliary notation. For a set A of unary symbols such that $A \cap \Sigma = \emptyset$ and any alphabet Y, let $T_\Sigma(A(Y))$ be the least subset T of $T_{\Sigma \cup A}(Y)$ for which (1) $a(y) \in T$ for all $a \in A$, $y \in Y$, and (2) $m \geq 0$, $f \in \Sigma_m$, $t_1, \ldots, t_m \in T$ implies $f(t_1, \ldots, t_m) \in T$.

Let $\mathcal{A} = (A, \Sigma, P, a_0)$ be a top-down Σ-recognizer. For any $a \in A$, $n \geq 0$ and any $t \in T_{\Sigma,n}$, the set $\mathcal{A}(a,t)$ ($\subseteq T_\Sigma(A(X_n))$) is defined so that (1) for $x_i \in X_n$, $\mathcal{A}(a, x_i) = \{a(x_i)\}$, (2) for $c \in \Sigma_0$, $\mathcal{A}(a,c) = \{c\}$ if $a(c) \to c \in P$, and $\mathcal{A}(a,c) = \emptyset$ otherwise, and (3) for $t = f(t_1, \ldots, t_m)$, $\mathcal{A}(a,t) =$

$$\{f(s_1, \ldots, s_m) \mid s_1 \in \mathcal{A}(a_1, t_1), \ldots, s_m \in \mathcal{A}(a_m, t_m),\ a(f) \to f(a_1, \ldots, a_m) \in P\}\ .$$

For any $s \in T_\Sigma(A(X))$ and any variable $x_i \in X$, we denote by $\mathrm{st}(s, x_i)$ the set of states $b \in A$ such that $b(x_i)$ appears as a subterm in s.

Clearly, $\mathcal{A}(a,t) \neq \emptyset$ iff there is a computation of $\mathcal{A}$ which starts in state a at the root of t, continues to the leaves of t, and if $\mathcal{A}$ reaches in a state b a leaf labelled by a nullary symbol c, then $b(c) \to c$ is in P. Each $s \in \mathcal{A}(a,t)$ represents the situation when such a successful computation has been completed so that all leaves labelled with a nullary symbol have also been processed. If $t \in \widetilde{T}_{\Sigma,n}$, then every $s \in \mathcal{A}(a_0,t)$ is of the form $s = t[a_1(x_1),\ldots,a_n(x_n)]$ and for any $t_1, \ldots, t_n \in T_\Sigma$, the tree s appears in a computation of $\mathcal{A}$ on $t[t_1,\ldots,t_n]$ of the form

$$a_0(t[t_1,\ldots,t_n]) \Rightarrow^*_\mathcal{A} t[a_1(t_1),\ldots,a_n(t_n)] = s[t_1,\ldots,t_n] \Rightarrow^*_\mathcal{A} \ldots$$

in which each subterm t_i is processed starting in the corresponding state a_i. However, if t is not linear, then a variable x_i may appear in a term $s \in \mathcal{A}(a_0,t)$ together with more than one state symbol, and then the corresponding subterm t_i should be accepted by a computation starting with each $a \in \text{st}(s,x_i)$.

Proof (of Theorem 3.1). Consider a left-linear TRS $\mathcal{R} = (\Sigma, R)$ and any recognizable Σ-tree languages T_1 and T_2. Let $\mathcal{A} = (A, \Sigma, P_1, a_0)$ and $\mathcal{B} = (B, \Sigma, P_2, b_0)$ be top-down Σ-recognizers for which $T(\mathcal{A}) = T_1$ and $T(\mathcal{B}) = T_2^c \; (= T_\Sigma \setminus T_2)$. We construct a generalized top-down Σ-recognizer $\mathcal{C}$ such that for any $t \in T_\Sigma$,

$$t \in T(\mathcal{C}) \quad \text{iff} \quad t \in T(\mathcal{A}) \text{ and } s \in T(\mathcal{B}) \text{ for some } s \in 1r\text{N}_\mathcal{R}(t) \;. \tag{1}$$

Then $1r\text{N}_\mathcal{R}(T_1) \subseteq T_2$ iff $T(\mathcal{C}) = \emptyset$, and the latter condition is decidable.

Let $\mathcal{C} = (C, \Sigma, P, (a_0, \{b_0\}))$ be the generalized top-down Σ-recognizer with the state set $C = (A \times \wp(B)) \cup (\bar{A} \times \wp(B))$, where $\wp(B)$ is the power set of B and $\bar{A} = \{\bar{a} \mid a \in A\}$ is a disjoint copy of A, and the set P of transition rules is defined as follows. The rules are of three different types.

Type 1. If $p \to r$ is a rule in R and $(a, H) \in A \times \wp(B)$, where $H = \{b_1, \ldots, b_k\}$, we include in P any rule

$$(a,H)(p(x_1,\ldots,x_n)) \to p[(a_1, H_1)(x_1), \ldots, (a_n, H_n)(x_n)] \;,$$

where $p[a_1(x_1),\ldots,a_n(x_n)] \in \mathcal{A}(a,p)$ and there are terms $s_1 \in \mathcal{B}(b_1,r), \ldots, s_k \in \mathcal{B}(b_k,r)$ such that $H_i = \text{st}(s_1, x_i) \cup \ldots \cup \text{st}(s_k, x_i)$ for all $i = 1, \ldots, n$. For $H = \emptyset$ ($k = 0$), this is interpreted to mean that $H_1 = \ldots = H_n = \emptyset$ should hold, and if $p \to r$ is a ground rule ($n = 0$), we include $(a, H)(p) \to p$ in P iff $a(p) \Rightarrow^*_\mathcal{A} p$ and $b_i(r) \Rightarrow^*_\mathcal{B} r$ for all $i = 1, \ldots, k$.

Type 2. Let NI be the set of all terms $q \in \widetilde{T}_{\Sigma,X}$ such that (1) $\text{hg}(q) \leq \max\{\text{hg}(p) \mid p \in \text{lhs}(\mathcal{R})\} + 1$, and (2) $\sigma(q) \neq \sigma'(p)$ for all $p \in \text{lhs}(\mathcal{R})$ and all substitutions σ and σ'. For each $p(x_1, \ldots, x_n) \in$ NI and any $(a,H) \in A \times \wp(B)$ with $H = \{b_1, \ldots, b_k\}$, we include in P any rule

$$(a,H)(p(x_1,\ldots,x_n)) \to p[(\bar{a}_1, H_1)(x_1), \ldots, (\bar{a}_n, H_n)(x_n)] \;,$$

where $p[a_1(x_1),\ldots,a_n(x_n)] \in \mathcal{A}(a,p)$, and there are terms $s_1 \in \mathcal{B}(b_1,p), \ldots, s_k \in \mathcal{B}(b_k,p)$ such that $H_i = \text{st}(s_1, x_i) \cup \ldots \cup \text{st}(s_k, x_i)$ for all $i = 1, \ldots, n$. The cases $H = \emptyset$ and $n = 0$ are treated similarly as above.

Type 3. For each $(\bar{a}, H) \in \bar{A} \times \wp(B)$, where $H = \{b_1, \ldots, b_k\}$, we add to P rules as follows.

(i) For $c \in \Sigma_0$, we include in P the rule $(\bar{a}, H)(c) \to c$ iff $a(c) \to c$ is in P_1 and P_2 contains $b_i(c) \to c$ for every $b_i \in H$.

(ii) For $f \in \Sigma_m$, $m > 0$, we add to P all rules

$$(\bar{a}, H)(f(x_1, \ldots, x_m)) \to f((\bar{a}_1, H_1)(x_1), \ldots, (\bar{a}_m, H_m)(x_m)) ,$$

where $a(f(x_1, \ldots, x_m)) \to f(a_1(x_1), \ldots, a_m(x_m))$ is in P_1, and there are rules $b_i(f(x_1, \ldots, x_m)) \to f(b_{i1}(x_1), \ldots, b_{im}(x_m))$ ($i = 1, \ldots, k$) in P_2 such that $H_j = \{b_{1j}, \ldots, b_{kj}\}$ for each $j = 1, \ldots, m$.

We can show that $\mathcal{C}$ has the property described in (1). If $t \in T(\mathcal{C})$, then $(a_0, \{b_0\})(t) \Rightarrow_\mathcal{C}^* t$ and this derivation can be split into two parts

$$(a_0, \{b_0\})(t) \Rightarrow_\mathcal{C}^* \tilde{t}[(a_1, H_1)(t_1), \ldots, (a_n, H_n)(t_n)] \Rightarrow_\mathcal{C}^* \tilde{t}[t_1, \ldots, t_n] = t , \quad (2)$$

where $n \geq 0$, $t \in \widetilde{T}_{\Sigma,n}$ and, for every $1 \leq i \leq n$, $t_i \in T_\Sigma$ and $(a_i, H_i) \in A \times \wp(B)$. In the first part of (2) only Type 1 rules are used, and hence $\tilde{t}[a_1(x_1), \ldots, a_n(x_n)] \in \mathcal{A}(a_0, \tilde{t})$. Moreover, for some $k \geq 0$, $\tilde{s} \in \widetilde{T}_{\Sigma,k}$, and $s_1, \ldots, s_k \in T_\Sigma$,

$$\#(t) = \#(\tilde{t}[t_1, \ldots, t_n]) \Rightarrow_{\mathcal{R}_\#} \cdots \Rightarrow_{\mathcal{R}_\#} \tilde{s}[\#(s_1), \ldots, \#(s_k)] = s ,$$

where every s_j is a copy of exactly one of the t_i. (Of course, s_j may be equal to more than one t_i.) For each $i = 1, \ldots, n$, let $K(i) = \{ j \mid s_j \text{ is a copy of } t_i \}$. Then for some $u \in \mathcal{B}(b_0, \tilde{s})$, $H_i = \bigcup\{ \text{st}(u, x_j) \mid j \in K(i) \}$ for all $i = 1, \ldots, n$.

In the second part of (2), it is first checked using Type 2 rules that $\tilde{s}[s_1, \ldots, s_k] \in 1r\mathrm{N}_\mathcal{R}(t)$, and the computations $(a_i, H_i)(t_i) \Rightarrow_\mathcal{C}^* t_i$ are finished using Type 3 rules. That means for every $i = 1, \ldots, n$, that (a) $t_i \in T(\mathcal{A}, a_i)$ and (b) $t_i \in T(\mathcal{B}, b)$ for all $b \in H_i$. Therefore

$$a_0(t) \Rightarrow_\mathcal{A}^* \tilde{t}[a_1(t_1), \ldots, a_n(t_n)] \Rightarrow_\mathcal{A}^* \tilde{t}[t_1, \ldots, t_n] = t$$

and there are $b_1, \ldots, b_k \in B$ such that

$$b_0(\tilde{s}[s_1, \ldots, s_k]) \Rightarrow_\mathcal{B}^* \tilde{s}[b_1(s_1), \ldots, b_k(s_k)] \Rightarrow_\mathcal{B}^* \tilde{s}[s_1, \ldots, s_k] .$$

The converse of (1) can be proved similarly. □

The corresponding result for sentential forms can be proved by modifying suitably the definition of the recognizer $\mathcal{C}$.

Theorem 3.2. *For any left-linear TRS* $\mathcal{R} = (\Sigma, R)$, *the following* one-pass root-started sentential form inclusion problem *is decidable.*
  Instance: *Recognizable Σ-tree languages T_1 and T_2.*
  Question: $1r\mathrm{S}_\mathcal{R}(T_1) \subseteq T_2$? □

4 The One-Pass Leaf-Started Inclusion Problems

Now we consider the one-pass leaf-started sentential form inclusion problem. Again the tree languages are assumed to be given in the form of tree recognizers.

Theorem 4.1. *For any left-linear TRS* $\mathcal{R} = (\Sigma, R)$, *the following* one-pass leaf-started sentential form inclusion problem *is decidable.*
  Instance: *Recognizable Σ-tree languages T_1 and T_2.*
  Question: $1\ell\, S_{\mathcal{R}}(T_1) \subseteq T_2$?

Proof. Let $\mathcal{A} = (A, \Sigma, P_1, A_f)$ and $\mathcal{B} = (B, \Sigma, P_2, B_f)$ be bottom-up Σ-recognizers that recognize T_1 and T_2, respectively. We may assume that $\mathcal{B}$ is total deterministic. We construct a generalized bottom-up Σ-recognizer $\mathcal{C} = (C, \Sigma, P, C_f)$ such that $T(\mathcal{C}) = \emptyset$ iff $1\ell\, S_{\mathcal{R}}(T_1) \subseteq T_2$ as follows.
  Let $C = (A \times B) \cup (\bar{A} \times \bar{B})$, where $\bar{A} = \{\bar{a} \mid a \in A\}$ and $\bar{B} = \{\bar{b} \mid b \in B\}$, and let $C_f = \{\bar{a} \mid a \in A_f\} \times \{\bar{b} \mid b \in (B \setminus B_f)\}$. The set P consists of the following rules which are of three different types.
  Type 1. For every $p \to r \in R_e$ with $p, r \in T_{\Sigma,n}$, $n \geq 0$, and for all $a_1, \ldots, a_n$, $a \in A$, $b_1, \ldots, b_n$, $b \in B$ such that $p[a_1, \ldots, a_n] \Rightarrow^*_{\mathcal{A}} a$ and $r[b_1, \ldots, b_n] \Rightarrow^*_{\mathcal{B}} b$, let P contain the rule $p[(a_1, b_1), \ldots, (a_n, b_n)] \to (a, b)$.
  Type 2. For all $a \in A$ and $b \in B$, let $(a, b) \to (\bar{a}, \bar{b})$ be in P.
  Type 3. For all $f \in \Sigma_m$, $m \geq 0$, $f(a_1, \ldots, a_m) \to a \in P_1$ and $f(b_1, \ldots, b_m) \to b \in P_2$, let P contain $f((\bar{a}_1, \bar{b}_1), \ldots, (\bar{a}_m, \bar{b}_m)) \to (\bar{a}, \bar{b})$.
  The way $\mathcal{C}$ processes a Σ-tree t can be described as follows. First $\mathcal{C}$, using rules of Type 1, follows some one-pass leaf-started rewriting sequences by $\mathcal{R}$ on subtrees of t computing in the first components of its states the evaluations by $\mathcal{A}$ of these subtrees and in the second components the evaluations by $\mathcal{B}$ of the translations of the subtrees produced by these one-pass leaf-started rewriting sequences. At any time $\mathcal{C}$ may switch by rules of Type 2 to a mode in which it by rules of Type 3 computes in the first components of its states the evaluation by $\mathcal{A}$ of t and in the second components the evaluation by $\mathcal{B}$ of the one-pass leaf-started sentential form of t produced by $\mathcal{R}$ when the rewriting sequences on the subtrees are combined. This means that for any $t \in T_\Sigma$, $a \in A$ and $b \in B$,

$$t \Rightarrow^*_{\mathcal{C}} (\bar{a}, \bar{b}) \quad \text{iff} \quad t \Rightarrow^*_{\mathcal{A}} a \text{ and } s \Rightarrow^*_{\mathcal{B}} b \text{ for some } s \in 1\ell\, S_{\mathcal{R}}(t) ,$$

which, by recalling the definition of C_f, implies immediately that $T(\mathcal{C}) = \emptyset$ iff $1\ell\, S_{\mathcal{R}}(T_1) \subseteq T_2$, as required. $\square$

  Finally, we turn to one-pass leaf-started normal forms.

Theorem 4.2. *For any left-linear TRS* $\mathcal{R} = (\Sigma, R)$, *the following* one-pass leaf-started normal form inclusion problem *is decidable.*
  Instance: *Recognizable Σ-tree languages T_1 and T_2.*
  Question: $1\ell\, N_{\mathcal{R}}(T_1) \subseteq T_2$?

Proof. Let $\mathcal{A} = (A, \Sigma, P_1, A_f)$ and $\mathcal{B} = (B, \Sigma, P_2, B_f)$ be total deterministic bottom-up Σ-recognizers such that $T(\mathcal{A}) = T_1$ and $T(\mathcal{B}) = T_2$. We construct a generalized bottom-up Σ-recognizer $\mathcal{C} = (C, \Sigma, P, C_f)$ such that $T(\mathcal{C}) = \emptyset$ iff $1\ell\, N_{\mathcal{R}}(T_1) \subseteq T_2$ as follows.
  Let $mx = \max\{\operatorname{hg}(p) \mid p \in \operatorname{lhs}(R_e)\}$ and $T_{mx} = \{t \in \widetilde{T}_{\Sigma,X} \mid \operatorname{hg}(t) \leq mx\}$. Now let $C = (A \times B) \cup (\bar{A} \times \bar{B} \times (T_{mx} \cup \{ok\}))$, where $\bar{A} = \{\bar{a} \mid a \in A\}$ and

$\bar{B} = \{\bar{b} \mid b \in B\}$, and $C_f = \{\bar{a} \mid a \in A_f\} \times \{\bar{b} \mid b \in (B \setminus B_f)\} \times (T_{mx} \cup \{ok\})$.
The set P consists of the following rules of five different types.

Type 1. For every rule $p \to r \in R_e$ with $p, r \in T_{\Sigma,n}$, $n \geq 0$, and any states $a_1, \ldots, a_n, a \in A, b_1, \ldots, b_n, b \in B$ such that $p[a_1, \ldots, a_n] \Rightarrow_{\mathcal{A}}^* a$ and $r[b_1, \ldots, b_n] \Rightarrow_{\mathcal{B}}^* b$, let P contain the rule $p[(a_1, b_1), \ldots, (a_n, b_n)] \to (a, b)$.

Type 2. For all $a \in A$ and $b \in B$, let $(a, b) \to (\bar{a}, \bar{b}, x_1)$ be in P.

Type 3. For all $f \in \Sigma_m$, $m \geq 0$, $u_1, \ldots, u_m$, $u \in T_{mx}$, $f(a_1, \ldots, a_m) \to a \in P_1$ and $f(b_1, \ldots, b_m) \to b \in P_2$ such that $u = \|f(u_1, \ldots, u_m)\|$ and $u \in (T_{mx} \setminus \text{lhs}(\mathcal{R}_e))$, let P contain $f((\bar{a}_1, \bar{b}_1, u_1), \ldots, (\bar{a}_m, \bar{b}_m, u_m)) \to (\bar{a}, \bar{b}, u)$. For $m = 0$, we get $f \to (\bar{a}, \bar{b}, f)$.

Type 4. For any $f \in \Sigma_m$, $m \geq 0$, $f(a_1, \ldots, a_m) \to a \in P_1$, $f(b_1, \ldots, b_m) \to b \in P_2$ and $u_1, \ldots, u_m \in T_{mx}$ such that $\|f(u_1, \ldots, u_m)\| \notin T_{mx}$, let P contain the rule $f((\bar{a}_1, \bar{b}_1, u_1), \ldots, (\bar{a}_m, \bar{b}_m, u_m)) \to (\bar{a}, \bar{b}, ok)$.

Type 5. For any $f \in \Sigma_m$ with $m \geq 1$, $a_1, \ldots, a_m, a \in A$, $b_1, \ldots, b_m, b \in B$, and sequence $y_1, \ldots, y_m \in T_{mx} \cup \{ok\}$ such that $ok \in \{y_1, \ldots, y_m\}$, $f(a_1, \ldots, a_m) \to a \in P_1$, $f(b_1, \ldots, b_m) \to b \in P_2$, let P contain the rule $f((\bar{a}_1, \bar{b}_1, y_1), \ldots, (\bar{a}_m, \bar{b}_m, y_m)) \to (\bar{a}, \bar{b}, ok)$.

It can now be shown that for any $t \in T_\Sigma$, $a \in A$, $b \in B$ and $y \in T_{mx} \cup \{ok\}$,

$$t \Rightarrow_\mathcal{C}^* (\bar{a}, \bar{b}, y) \quad \text{iff} \quad t \Rightarrow_\mathcal{A}^* a \text{ and } s \Rightarrow_\mathcal{B}^* b \text{ for some } s \in 1\ell\mathrm{N}_\mathcal{R}(t) ,$$

and hence $T(\mathcal{C}) = \emptyset$ iff $1\ell\mathrm{N}_\mathcal{R}(T_1) \subseteq T_2$. □

References

1. J. Avenhaus. *Reduktionssysteme.* Springer, 1995.
2. M. Dauchet and F. De Comite. A gap between linear and non-linear term-rewriting systems. In *RTA-87*, LNCS **256**. Springer, 1987, 95–104.
3. N. Dershowitz and J.-P. Jouannaud. *Rewrite Systems*, volume B of *Handbook of Theoretical Computer Science*, chapter 6, pages 243–320. Elsevier, 1990.
4. A. Deruyver and R. Gilleron. The reachability problem for ground TRS and some extensions. In *TAPSOFT'89*, LNCS **351**. Springer, 1989, 227–243.
5. J. Engelfriet and E. M. Schmidt. IO and OI. Part I. *J. Comput. Syst. Sci.*, 15(3):328–353, 1977. Part II. *J. Comput. Syst. Sci.*, 16(1):67–99, 1978.
6. F. Gécseg and M. Steinby. *Tree automata.* Akadémiai Kiadó, Budapest, 1984.
7. F. Gécseg and M. Steinby. *Tree Languages*, volume 3 of *Handbook of Formal Languages*, chapter 1, pages 1–68. Springer, 1997.
8. R. Gilleron. Decision problems for term rewriting systems and recognizable tree languages. In *STACS'91*, LNCS **480**. Springer, 1991, 148–159.
9. R. Gilleron and S. Tison. Regular tree languages and rewrite systems. *Fundam. Inf.*, 24(1,2):157–175, 1995.
10. D. Hofbauer and M. Huber. Linearizing term rewriting systems using test sets. *J. Symb. Comput.*, 17(1):91–129, 1994.
11. G. Kucherov and M. Tajine. Decidability of regularity and related properties of ground normal form languages. *Inf. Comput.*, 118(1):91–100, 1995.
12. S. Vágvölgyi and R. Gilleron. For a rewrite system it is decidable whether the set of irreducible, ground terms is recognizable. *Bull. EATCS*, 48:197–209, 1992.

On the Word, Subsumption, and Complement Problem for Recurrent Term Schematizations*
(Extended Abstract)

Miki Hermann[1] and Gernot Salzer[2]

[1] LORIA (CNRS), BP 239, 54506 Vandœuvre-lès-Nancy, France.
hermann@loria.fr
[2] Technische Universität Wien, Karlsplatz 13, 1040 Wien, Austria.
salzer@logic.at

Abstract. We investigate the word and the subsumption problem for recurrent term schematizations, which are a special type of constraints based on iteration. By means of unification, we reduce these problems to a fragment of Presburger arithmetic. Our approach is applicable to all recurrent term schematizations having a finitary unification algorithm. Furthermore, we study a particular form of the complement problem. Given a finite set of terms, we ask whether its complement can be finitely represented by schematizations, using only the equality predicate without negation. The answer is negative as there are ground terms too complex to be represented by schematizations with limited resources.

1 Introduction

Infinite sets of first-order terms with structural similarities appear frequently in several branches of automated deduction, like logic programming, model building, term rewriting, equational unification, or clausal theorem proving. They are usually produced by saturation-based procedures, like equational completion or hyper-resolution. A usual requirement for effective use of such sets is the possibility to handle them by finite means. There exist several approaches to cope with this phenomenon, like lazy evaluation, set constraints, or term schematizations. Lazy evaluation usually does not combine well with unification or other operations. Set constraints allow to describe regular sets of first-order terms, using the potential of regular tree grammars and tree automata, and having the good properties of regular tree languages. Schematizations exploit the recurring term structure in infinite sets, as produced by self-resolving clauses or by self-overlapping rewrite rules.

Several formalisms for recurrent term schematizations were introduced within the last years. They rely on the same principle, namely the iteration of first-order contexts, but differ in the expressive power. The main concern in this

---

* Full version is at http://www.loria.fr/~hermann/publications/redelim.ps.gz. This work was done while the second author was visiting LORIA and was funded by Univeristé Henri Poincaré, Nancy 1.

work is the decidability of unification and the construction of finite complete sets of unifiers. Formalisms satisfying these requirements are ρ-terms [CH95], I-terms [Com95], R-terms [Sal92], and primal grammars [HG97], all of them with a finitary unification algorithm. Set operations were studied in [AHL97].

Applications of recurrent schematizations are quite rare and mostly theoretical, like in model building [Pel97] or cycle unification [Sal94]. One reason is that there are still some open problems to be solved prior to a successful implementation. A *sine qua non* of automated deduction is redundancy elimination. The elementary tools in this respect are testing for equality and subsumption. In other words, we need to solve the word problem and the subsumption problem for recurrent term schematizations. Moreover, only positive set operations were studied in [AHL97] without considering the complement. Complement building is interesting from the algebraic and logic point of view, e.g., during construction of counter-examples or for quantifier elimination.

In the first part of the paper, we investigate the word and the subsumption problem for primal grammars. By means of unification, we reduce them to a problem in Presburger arithmetic. Our approach is applicable to all recurrent term schematizations having a finitary unification algorithm. In the second part, we study a particular form of the complement problem. Given a finite set of terms, we ask whether its complement can be represented finitely by schematizations, using only the equality predicate without negation. The answer is negative as there are ground first-order terms too complex to be represented by primal grammars with limited resources.

2 Term Schematizations

2.1 Syntax

The language of primal terms is based on four kinds of symbols: first-order variables $\mathcal{V}$, counter variables $\mathcal{C}$, function symbols $\mathcal{F}_p$ of arities $p \geq 0$, and defined symbols $\mathcal{D}_{q,p}$ of counter arities $q \geq 1$ and first-order arities $p \geq 0$. Nullary function symbols are called constants. The set of all function and defined symbols is denoted by $\mathcal{F}$ and $\mathcal{D}$, respectively.

Let $\mathbb{N}$ be the set of natural numbers. The set of *counter expressions* $\mathcal{L}$ is the set of linear expressions over $\mathcal{C}$ with coefficients in $\mathbb{N}$. Two counter expressions are considered equal if they are equivalent with respect to the usual equalities of addition and multiplication. Furthermore, we drop parentheses where possible and do not distinguish between natural numbers and their symbolic representation. The set of *primal terms* $\mathcal{P}$ is defined inductively as the smallest set satisfying the following conditions: $\mathcal{V} \subseteq \mathcal{P}$; $f(\boldsymbol{t}) \in \mathcal{P}$ if $f \in \mathcal{F}_p$ and $\boldsymbol{t} \in \mathcal{P}^p$; $\hat{f}(\boldsymbol{l};\boldsymbol{t}) \in \mathcal{P}$ if $\hat{f} \in \mathcal{D}_{q,p}$, $\boldsymbol{l} \in \mathcal{L}^q$, and $\boldsymbol{t} \in \mathcal{P}^p$. The sets of counter variables and first-order variables of a primal term t are denoted by $\mathcal{C}\mathit{Var}(t)$ and $\mathit{Var}(t)$, respectively.

2.2 Semantics

In the sequel, we assume that the reader is familiar with the basic notions of term rewriting. With each defined symbol $\hat{f} \in \mathcal{D}_{q,p}$, we associate two rewrite rules $\hat{f}(0, \boldsymbol{n}; \boldsymbol{x}) \to r_1^{\hat{f}}$ and $\hat{f}(m+1, \boldsymbol{n}; \boldsymbol{x}) \to r_2^{\hat{f}}[\hat{f}(m, \boldsymbol{n}+\boldsymbol{\delta}; \boldsymbol{x})]_A$, where $m, \boldsymbol{n}$ and $\boldsymbol{x}$ are counter variables and first-order variables, respectively; $r_1^{\hat{f}}$ and $r_2^{\hat{f}}$ are primal terms, whose variables are among those of the left hand sides of the rules; all defined symbols in $r_1^{\hat{f}}$ and $r_2^{\hat{f}}$ are smaller than $\hat{f}$ with respect to a given precedence relation on the defined symbols; A is a set of independent first-order positions of $r_2^{\hat{f}}$ without the root position; $\boldsymbol{\delta}$ is either the null vector or a k-dimensional unit vector, i.e., all components of $\boldsymbol{\delta}$ are zero except one which may be zero or one. The first-order positions are those not below a defined symbol. Two positions are independent if none is a prefix of the other.

Let $\mathcal{R}$ be the set of all rewrite rules associated with the defined symbols. The rewrite relation $\longrightarrow_\mathcal{R}$ generated by $\mathcal{R}$ is the smallest relation that contains $\mathcal{R}$, and is closed under congruence and substitution. By $t\downarrow_\mathcal{R}$ we denote the normal form of t with respect to $\mathcal{R}$. Note that $t\downarrow_\mathcal{R}$ is a first-order term if t contains no counter variables. The first-order terms represented by a primal term t are defined as $L(t) = \{t\xi\downarrow_\mathcal{R} \mid \xi: \mathcal{C} \longrightarrow \mathbb{N}\}$. Two primal terms s and t are *equivalent*, denoted by $s \doteq t$, if $s\xi\downarrow_\mathcal{R} = t\xi\downarrow_\mathcal{R}$ holds for all substitutions $\xi: \mathcal{C} \longrightarrow \mathbb{N}$.

2.3 Unification

A substitution is a mapping $\sigma: (\mathcal{V} \cup \mathcal{C}) \longrightarrow (\mathcal{P} \cup \mathcal{L})$, which is well-typed and whose domain is finite, i.e., $\sigma(x) \in \mathcal{P}$ for $x \in \mathcal{V}$, $\sigma(n) \in \mathcal{L}$ for $n \in \mathcal{C}$, and $\text{dom}(\sigma) = \{v \in (\mathcal{V} \cup \mathcal{C}) \mid \sigma(v) \neq v\}$ is finite. The application of σ to a term t is written as $t\sigma$; the composition of two substitutions σ, τ is written as $\sigma\tau$ with the understanding that $t\sigma\tau = (t\sigma)\tau$ for all terms t. We denote σ by the set $\{v \mapsto v\sigma \mid v \in \text{dom}(\sigma)\}$. Normalization is extended to substitutions in the natural way, i.e., $\sigma\downarrow_\mathcal{R} = \{v \mapsto v\sigma\downarrow_\mathcal{R} \mid v \in \text{dom}(\sigma)\}$.

A substitution σ is a unifier of two primal terms s and t iff for all $\xi: \mathcal{C} \longrightarrow \mathbb{N}$ the first-order substitution $\sigma\xi\downarrow_\mathcal{R}$ unifies the first-order terms $s\xi\downarrow_\mathcal{R}$ and $t\xi\downarrow_\mathcal{R}$. A set of unifiers Σ is complete iff for every counter substitution ξ there exists $\sigma \in \Sigma$, such that $\sigma\xi\downarrow_\mathcal{R}$ is a most general unifier of $s\xi\downarrow_\mathcal{R}$ and $t\xi\downarrow_\mathcal{R}$. Note that σ is a unifier of s and t iff $s\sigma \doteq t\sigma$, i.e., our notion of unifiability corresponds to the standard one in unification theory. This is not true for completeness: a unifier need not be an instance of any substitution in a given complete set of unifiers. Unification of primal terms is decidable and finitary, i.e., for any pair of primal terms there exists a finite set of unifiers which is complete. Moreover, complete sets of unifiers can be effectively computed [HG97].

2.4 First-Order Formulas

In this paper, we use first-order formulas to define the word problem in a concise way and to compare different notions of subsumption. Quantified counter variables are interpreted over the domain of natural numbers, quantified first-order variables over the Herbrand universe with respect to the underlying set of function symbols. Free variables are treated as constants.

Additionally, we use vectors and notations from linear algebra as a compact representation of similar objects. For example, $\boldsymbol{x} \doteq \boldsymbol{s}(\boldsymbol{k})$ stands for a set of equations of the form $x \doteq s(\boldsymbol{k})$, where x is a variable from $\boldsymbol{x}$ and $s \in \boldsymbol{s}$ is a term containing variables $k_1, k_2, \ldots$ from $\boldsymbol{k}$. Furthermore, $\{\boldsymbol{n} \mapsto \boldsymbol{Ck} + \boldsymbol{c}\}$ represents the substitution replacing each variable in $\boldsymbol{n}$ by the corresponding row in the vector of linear expressions, which is obtained by multiplying the matrix $\boldsymbol{C}$ of natural numbers by the vector $\boldsymbol{k}$ of counter variables and adding the vector $\boldsymbol{c}$.

Let s and t be primal terms containing the variables $\boldsymbol{x} = \mathcal{V}ar(s)$, $\boldsymbol{y} = \mathcal{V}ar(t)$, $\boldsymbol{m} = \mathcal{CV}ar(s)$ and $\boldsymbol{n} = \mathcal{CV}ar(t)$. A complete set of unifiers for s and t can be considered as a solved form of the equation $s \doteq t$ in the following way. A unifier $\sigma = \{\boldsymbol{x} \mapsto \boldsymbol{s}'(\boldsymbol{k}), \boldsymbol{y} \mapsto \boldsymbol{t}'(\boldsymbol{k}), \boldsymbol{m} \mapsto \boldsymbol{Ck} + \boldsymbol{c}, \boldsymbol{n} \mapsto \boldsymbol{Dk} + \boldsymbol{d}\}$, where $\boldsymbol{k}$ are auxiliary counter variables introduced during unification, corresponds to the formula $\phi_\sigma(\boldsymbol{x}, \boldsymbol{y}, \boldsymbol{m}, \boldsymbol{n}) = \exists \boldsymbol{k}\big(\boldsymbol{x} \doteq \boldsymbol{s}'(\boldsymbol{k}) \wedge \boldsymbol{y} \doteq \boldsymbol{t}'(\boldsymbol{k}) \wedge \boldsymbol{m} = \boldsymbol{Ck} + \boldsymbol{c} \wedge \boldsymbol{n} = \boldsymbol{Dk} + \boldsymbol{d}\big)$. Note that unification does not introduce auxiliary first-order variables. However, s' and t' may contain variables from $\boldsymbol{x}$ and $\boldsymbol{y}$; in this case these variables do not occur in the domain of the substitution. The formula associated with a complete set of unifiers Σ is the disjunction of the formulas corresponding to the single unifiers: $\phi_\Sigma(\boldsymbol{x}, \boldsymbol{y}, \boldsymbol{m}, \boldsymbol{n}) = \bigvee_{\sigma \in \Sigma} \phi_\sigma(\boldsymbol{x}, \boldsymbol{y}, \boldsymbol{m}, \boldsymbol{n})$. Therefore the formulas $s \doteq t$ and $\phi_\Sigma(\boldsymbol{x}, \boldsymbol{y}, \boldsymbol{m}, \boldsymbol{n})$ are equivalent.

2.5 Miscellaneous Notations

If t is a primal term and $A \subseteq \mathcal{P}os(t)$ is a set of independent first-order positions, then $t[\circ]_A$ is called a *context*. If s is a context and t is a context or primal term, then the concatenation of s and t, denoted by $s \cdot t$, is the context or primal term $s\{\circ \mapsto t\}$. Concatenation is associative, hence we drop parentheses where possible. The empty context $\circ$ serves as unit element with respect to concatenation. Exponentiation is defined by $s^0 = \circ$ and $s^{i+1} = s \cdot s^i$.

The depth of a primal term t, denoted by $depth(t)$, is recursively defined as $depth(t) = 0$ for $t \in (\mathcal{V} \cup \mathcal{F}_0)$, and $depth(f(\boldsymbol{t})) = depth(\hat{f}(\boldsymbol{l}; \boldsymbol{t})) = 1 + depth(\boldsymbol{t})$ for $f \in \mathcal{F}_p$ ($p > 0$) and $\hat{f} \in \mathcal{D}$. The depth of a set or vector of terms $\boldsymbol{t}$ is defined as $depth(\boldsymbol{t}) = \max\{depth(t) \mid t \in \boldsymbol{t}\}$. The depth of the set of rewrite rules $\mathcal{R}$ associated with $\mathcal{D}$ is the depth of the set of all right hand sides: $depth(\mathcal{R}) = depth(\{r_1^{\hat{f}}, r_2^{\hat{f}}[\hat{f}(\boldsymbol{m}, \boldsymbol{n} + \boldsymbol{\delta}; \boldsymbol{x})]_A \mid \hat{f} \in \mathcal{D}\})$.

3 Redundancy Elimination

Recurrent term schematizations are of potential use in all areas concerned with first-order terms, mostly in automated deduction, like term rewriting with equational completion and proofs by consistency, or clausal theorem proving. An ubiquitous problem appearing there is the duplication of objects. Redundancy elimination plays therefore a vital role. In the simplest case, we need to maintain the set property, where no element (term, clause, literal) must occur twice. Another case of redundancy is the presence of two elements, where one is an instance of the other. In the first case we have to solve the *word problem*, i.e., to determine whether two terms s and t represent the same object in the underlying theory. The latter case is usually referred to as the *subsumption problem*.

3.1 Word Problem

Definition 1. *The **word problem** for two primal terms s and t is the question whether the formula $\forall \boldsymbol{n}\ (s \doteq t)$ is valid in the equational theory generated by $\mathcal{R}$, where $\boldsymbol{n} = \mathcal{CV}ar(s) \cup \mathcal{CV}ar(t)$.*

One possibility to solve the word problem is to reduce s and t to unique normal forms, followed by a check whether the latter are syntactically equal. This approach is described for R-strings in [Sal91]. In this paper, we choose a different approach: we transform the word problem to a unification problem and a subsequent problem in Presburger arithmetic. The first method is efficient but works only if we can define a unique normal form. In general, there is no obvious way of defining the normal form of a primal term. Our approach does not depend on a specific syntactic representation for schematizations, but requires only the existence of a finitary and terminating unification algorithm. Therefore, our method is applicable to all known recurrent schematizations, i.e., to ρ-terms, I-terms, R-terms, and primal grammars.

We proceed in three steps.

1. *Elimination of first-order variables:* replace all first-order variables by new constants. Observe that the formula $\forall \boldsymbol{n}(s \doteq t)$ is valid if and only if the corresponding formula $\forall \boldsymbol{n}(s^* \doteq t^*)$ is valid, where the terms s^*, t^* are obtained from the terms s, t by replacing each first-order variable x by a new constant c_x.
2. *Unification:* solve the equation $s^* \doteq t^*$. We solve the equation $s^* \doteq t^*$ by means of unification. Note that a finitary and terminating unification algorithm exists for all four known recurrent schematizations. This means that the output of the unification algorithm is a finite disjunction of formulas $\exists \boldsymbol{k}(\boldsymbol{n} = \mathbf{N}_i \boldsymbol{k} + \boldsymbol{d}_i)$, where $\mathbf{N}_i$ and $\boldsymbol{d}_i$ is a matrix and a vector of non-negative integers, respectively, and $\boldsymbol{k}$ are new counter variables introduced during unification. The resulting formula $\phi(\boldsymbol{n}) = \exists \boldsymbol{k} \bigvee_i (\boldsymbol{n} = \mathbf{N}_i \boldsymbol{k} + \boldsymbol{d}_i)$ contains only counter variables, since there are no first-order variables in s^* and t^*.
3. *Validity check:* check whether the formula $\forall \boldsymbol{n}\ \phi(\boldsymbol{n})$ is valid. The formula $\phi(\boldsymbol{n})$ represents a complete set of unifiers, one per disjunct, of the problem $s^* \doteq t^*$. To show that the universally quantified formula $\forall \boldsymbol{n}(s^* \doteq t^*)$ is valid, we need to prove that the unifiers from $\phi(\boldsymbol{n})$ cover the whole Cartesian product $\mathbb{N}^{|\boldsymbol{n}|}$. By correctness of the applied unification algorithm, the formulas $\forall \boldsymbol{n}(s^* \doteq t^*)$ and $\forall \boldsymbol{n}\ \phi(\boldsymbol{n})$ are equivalent. The latter expression is a Π_2-formula of Presburger arithmetic and can be solved by usual methods [Coo72].

3.2 Subsumption Problem

In the first-order case, a term s subsumes a term t if there exists a substitution σ, such that $s\sigma = t$. In the free algebra, this is equivalent to $\exists \boldsymbol{x}(s \doteq t)$, where $\boldsymbol{x} = \mathcal{V}ar(s)$. An alternative definition is that the formula $\forall \boldsymbol{y} \exists \boldsymbol{x}(s \doteq t)$ is valid, where $\boldsymbol{x} = \mathcal{V}ar(s)$ and $\boldsymbol{y} = \mathcal{V}ar(t)$. These two definitions are equivalent, except for singular signatures, since in the empty theory (without axioms) validity in the equational theory is equivalent to validity in the inductive theory.

For schematizations, there are several possibilities to define subsumption. Let s and t be two primal terms from a schematization G, where $\boldsymbol{m} = \mathcal{CV}ar(s)$, $\boldsymbol{n} = \mathcal{CV}ar(t)$, $\boldsymbol{x} = \mathcal{V}ar(s)$, and $\boldsymbol{y} = \mathcal{V}ar(t)$. Recall that we check the validity of formulas in the equational theory of $\mathcal{R}$, i.e., the free algebra generated by $\mathcal{R}$. The possibilities to define that s subsumes t are: (1) the formula $\exists \boldsymbol{m} \exists \boldsymbol{x}(s \doteq t)$ is valid; (2) the formula $\forall \boldsymbol{n} \forall \boldsymbol{y} \exists \boldsymbol{m} \exists \boldsymbol{x}(s \doteq t)$ is valid; (3) the formula $\forall \boldsymbol{n} \exists \boldsymbol{m}(s \doteq t)$ is valid; (4) the formula $\forall \boldsymbol{n} \exists \boldsymbol{m} \exists \boldsymbol{x}(s \doteq t)$ is valid. The first two approaches are straightforward extensions of the first-order concept. The second approach does not meet a natural requirement for subsumption, namely independence of the underlying signature. Subsumption should be a local test on two terms independent of other elements. There exist two terms s, t, such that s subsumes t (according to the second definition) over a signature $\mathcal{F}$, but not over an extended signature $\mathcal{F}' \supset \mathcal{F}$ [AHL97, Example 14]. The same terms also show that the first two subsumption concepts are not equivalent, since there is no substitution σ, such that $s\sigma \doteq t$, as required by the first concept. The problems with the second concept originate from quantification over first-order variables. One possibility to avoid them is to quantify only the counter variables, as in the third approach. This concept is not satisfactory either, since it does not capture usual first-order subsumption. When we extend the third concept with usual equational first-order subsumption, we get the fourth concept.

Hence, we have two suitable concepts for subsumption: the first and the last one. Intuitively, the first concept expresses that there is a uniform mapping σ, relating the term s and t in the equational theory of the schematization. In particular, for the counter variable vectors $\boldsymbol{m}$ and $\boldsymbol{n}$, this means that $\boldsymbol{m}$ is a linear expression of $\boldsymbol{n}$. In contrast, the fourth concept requires this uniformity only on the first-order level; the vectors $\boldsymbol{m}$ and $\boldsymbol{n}$ need not be related by a linear function. Clearly, the first concept implies the fourth concept. The converse is not true.

The last subsumption concept encompasses the first one. Moreover, the last concept corresponds to the natural view that schematizations are just a finite representation of infinite sets of first-order terms: s subsumes t if every term represented by t is subsumed by a term represented by s. Therefore we adopt the last concept of subsumption.

Definition 2. *Let s and t be primal terms, where $\boldsymbol{m} = \mathcal{CV}ar(s)$, $\boldsymbol{n} = \mathcal{CV}ar(t)$, and $\boldsymbol{x} = \mathcal{V}ar(s)$. The term s **subsumes** t if the formula $\forall \boldsymbol{n} \exists \boldsymbol{m} \exists \boldsymbol{x}(s \doteq t)$ is valid. A set S subsumes a set T if for each term $t' \in T$ there exists a term $s' \in S$, such that s' subsumes t'.*

A primal term s subsumes a primal term t iff the set $L(s)$ subsumes the set $L(t)$.

Similar to the word problem, we want to reduce subsumption to unification. We proceed in four steps: we replace certain first-order variables by new constants, apply the unification algorithm, simplify the resulting formula, and check its validity in Presburger arithmetic.

1. *Elimination of first-order variables in t:* replace all first-order variables in t by new constants, producing the term t^*. The formula $\forall \boldsymbol{n} \exists \boldsymbol{m} \exists \boldsymbol{x}(s \doteq t)$ is valid iff $\forall \boldsymbol{n} \exists \boldsymbol{m} \exists \boldsymbol{x}(s \doteq t^*)$ holds by the way how we interpret free variables.
2. *Unification:* solve the equation $s = t^*$ by means of a unification algorithm. Its output can be written as the finite formula $\phi(\boldsymbol{m}, \boldsymbol{n}, \boldsymbol{x}) = \exists \boldsymbol{k} \bigvee_i (\boldsymbol{x} = \boldsymbol{u}_i(\boldsymbol{k}) \wedge \boldsymbol{m} = \mathbf{M}_i \boldsymbol{k} + \boldsymbol{c}_i \wedge \boldsymbol{n} = \mathbf{N}_i \boldsymbol{k} + \boldsymbol{d}_i)$, where $\boldsymbol{k}$ are the new counter variables introduced during unification, $\mathbf{M}_i$, $\mathbf{N}_i$ are matrices of non-negative integers, and $\boldsymbol{c}_i$, $\boldsymbol{d}_i$ are vectors of non-negative integers.
3. *Simplification:* remove the equations $\boldsymbol{x} = \boldsymbol{u}_i(\boldsymbol{k})$ and $\boldsymbol{m} = \mathbf{M}_i \boldsymbol{k} + \boldsymbol{c}_i$ from the formula $\phi(\boldsymbol{m}, \boldsymbol{n}, \boldsymbol{x})$, producing $\phi'(\boldsymbol{n})$. Note that $\exists \boldsymbol{m} \exists \boldsymbol{x}\, \phi(\boldsymbol{m}, \boldsymbol{n}, \boldsymbol{x})$ is equivalent to $\phi'(\boldsymbol{n})$, since the variables $\boldsymbol{m}$ and $\boldsymbol{x}$ are existentially quantified and appear only once and separated on the left-hand side of equations.
4. *Validity check:* check if $\forall \boldsymbol{n}\, \phi'(\boldsymbol{n})$ is valid. The result $\forall \boldsymbol{n} \exists \boldsymbol{k} \bigvee_i (\boldsymbol{n} = \mathbf{N}_i \boldsymbol{k} + \boldsymbol{d}_i)$ belongs to the Π_2-fragment of Presburger arithmetic.

3.3 Complexity Issues

Both the word problem and the subsumption problem reduce in the last step to a Π_2-formula in Presburger arithmetic. While the complexity of full Presburger arithmetic is at least doubly exponential and Cooper presents in [Coo72] an algorithm of triple exponential complexity, the Π_2-fragment is only coNP-complete, as it was proved by Grädel [Grä88] and Schöning [Sch97]. Our formulas are quite simple and do not cover the whole Π_2-fragment: they are of the form $\forall \boldsymbol{n} \exists \boldsymbol{k} \bigvee_i (\boldsymbol{n} = \mathbf{N}_i \boldsymbol{k} + \boldsymbol{d}_i)$, i.e., the formula is in disjunctive normal form and the variables $\boldsymbol{n}$ appear only once separated on the left-hand side. Therefore we can ask whether our special problems are still coNP-complete. The lower bound reductions used by Grädel and Schöning require more complex formulas. However, following an idea in [Sch97], due to Grädel, we can prove the coNP-hardness of our problems by a reduction from SIMULTANEOUS INCONGRUENCES [GJ79]. This NP-complete problem is defined as follows: given a set $\{(a_1, b_1), \ldots, (a_p, b_p)\}$ of ordered pairs of positive integers, with $a_i \leq b_i$, the problem asks whether there is an integer n such that $n \not\equiv a_i \pmod{b_i}$ holds for all i. We use the dual problem to show coNP-hardness. Encoding $n \equiv a_i \pmod{b_i}$ as $\exists k (n = b_i k + a_i)$, we obtain the disjunction $\exists k \bigvee_{i=1}^{p}(n = b_i k + a_i)$. The final formula is $\forall n \exists k \bigvee_i (n = b_i k + a_i)$, which is of the same type as the formulas obtained from word and subsumption problems. Note that in both cases only the problem solved in the last step is coNP-complete. The overall complexity of our algorithms is determined by the complexity of unification. In particular, the cardinality of a minimal complete set of unifiers can be at least exponential [Sal91]; and we have to compute all solutions to obtain the formula. Hence, the formula in the last step can be exponentially longer than the input of the original problem.

4 Complement Problem

If t is a first-order term, its Herbrand universe is $\mathcal{H}(t) = \{t\sigma \mid \sigma\colon \mathcal{X} \longrightarrow \mathcal{T}(\mathcal{F})\}$, the set of the ground instances of t with respect to the underlying signature $\mathcal{F}$. Similarly, if T is a set of first-order terms, its Herbrand universe $\mathcal{H}(T)$ is the union of the Herbrand universes $\mathcal{H}(t)$ for each $t \in T$. For a primal term t, its Herbrand universe is the set $\mathcal{H}(L(t))$, i.e., the Herbrand universe of the schematized set. Finally, the Herbrand universe of a set of primal terms T is obtained as the union of the Herbrand universes $\mathcal{H}(t)$ for each $t \in T$.

Given a set of first-order or primal terms T, its *complement* is the set $T^c = \mathcal{T}(\mathcal{F}) \setminus \mathcal{H}(T)$. A class $\mathbb{C}$ is a collection of sets of terms satisfying a common property. For a given class $\mathbb{C}$, the *complement problem* is the question whether for each finite set of terms $T \in \mathbb{C}$ there exists a finite set of terms $T' \in \mathbb{C}$, such that $\mathcal{H}(T') = T^c$ holds. The set T' is called a finite complement representation.

For first-order terms, Lassez and Marriott proved that finite sets of linear terms always have a finite complement representation [LM87]. On the other hand, they showed that this is not true for arbitrary finite sets of first-order terms. Since schematizations were introduced to increase the expressive power of first-order terms, we might expect to be able to represent the complements of non-linear terms by a finite set of primal terms. However, as we show in the sequel, already the very simple non-linear term $f(x,x)$ has no finite complement representation by primal terms.

The potential of primal terms resides in the possibility to generate arbitrarily deep terms by iterating contexts. The expressive power of iteration is limited by the fact that the number of contexts must be finite. The maximal number of consecutive iterations during a reduction of a primal term is measured by the iteration depth. Each iteration terminates with the application of the base rule $\hat{f}(0,\ldots) \to r_1^{\hat{f}}$ for some defined symbol $\hat{f}$. Therefore we can determine the iteration depth by counting the occasions when a variable gets decremented to 0. The iteration depth of a primal term is then the maximum over all reductions. Inspection of the rewrite system $\mathcal{R}$ reveals that there is a correspondence between the application of base rules and the number of counter positions present in the primal term: each iteration consumes a counter position.

Definition 3. *The **iteration depth** of a primal term is the function τ defined recursively as follows:*

- $\tau(x) = \tau(a) = 0$ *for a first-order variable x and a constant a,*
- $\tau(f(t_1,\ldots,t_n)) = \max\{\tau(t_i) \mid i = 1,\ldots,n\}$ *for an n-ary function symbol f,*
- $\tau(\hat{f}(c; t_1,\ldots,t_n)) = |c| + \max\{\tau(t_i) \mid i = 1,\ldots,n\}$ *for a defined symbol $\hat{f}$.*

The iteration depth naturally extends to a set of primal terms T, defined by $\tau(T) = \max\{\tau(t) \mid t \in T\}$.

This definition emphasizes the static aspect by looking at the primal term only. The operational aspect, namely counting the occasions when a variable is decremented to 0, is expressed by the equalities $\tau(\hat{f}(0,\ldots)\theta) = 1 + \tau(r_1^{\hat{f}}\theta)$ and $\tau(\hat{f}(n+1,\ldots)\theta) = \tau(r_2^{\hat{f}}\theta)$ for each defined symbol $\hat{f}$ and substitution θ.

Iteration of contexts consumes resources of the primal term. On one hand, a single iteration can produce an arbitrarily deep term. On the other hand, there are ground first-order terms that require a certain iteration depth. We use two different contexts, $f(\circ, a)$ and $f(a, \circ)$, to force a consumption of resources. Consider the ground term $s = f(\circ, a)^m \cdot a$. If the value of m is sufficiently large, then a primal term t representing s must contain a defined symbol through which we iterate the context $f(\circ, a)$, and the iteration depth of t must be at least 1. If we simply concatenate two blocks of the same context, like in $f(\circ, a)^m \cdot f(\circ, a)^m \cdot a$, we do not necessarily need to increase the iteration depth of the primal term. However, if we insert the context $f(a, \circ)$ between the two blocks, producing the term $s = f(\circ, a)^m \cdot f(a, \circ) \cdot f(\circ, a)^m \cdot a$, we force a primal term t representing s to have an iteration depth of at least 2. Repeating the step, this idea leads to an upper bound on the number of context blocks $f(\circ, a)^m \cdot f(a, \circ)$ that can be represented by a given primal term t.

Lemma 1. *Let t be a primal term without first-order variables and let $s = w \cdot (f(\circ, a)^m \cdot f(a, \circ))^n \cdot a$ be a ground first-order term, where w is a proper subcontext of $f(\circ, a)^m \cdot f(a, \circ)$. If $s \in L(t)$ and $m > \tau(t) \times depth(\mathcal{R}) + depth(t)$ then $n \leq \tau(t)$.*

The lemma indicates that if we choose the value of n in the term s larger than the iteration depth $\tau(t)$ of the primal term t, then we cannot represent s by t using iteration only. Therefore, the term t must contain variables.

Corollary 1. *If $s = (f(\circ, a)^m \cdot f(a, \circ))^n \cdot a$ is an instance of a primal term t with $\tau(t) < n$ and $m > \tau(t) \times depth(\mathcal{R}) + depth(t)$, then t must end with a variable.*

We show by contradiction that the complement of the first-order term $f(x,x)$ has no finite representation. The underlying idea is to choose a ground term $s = f(s_1, s_2)$ from the complement, such that both s_1 and s_2 are too complex to be produced by iteration alone, and s_2 is twice as deep as s_1. Therefore a term representing s must be of the form $f(u, v)$, where both u and v end with variables y and z, respectively. If $y \neq z$ then the terms $f(u, v)$ and $f(x, x)$ are unifiable, contradicting the assumption that $f(u, v)$ represents (part of) the complement of $f(x, x)$. If $y = z$, then there is no substitution σ, such that $u\sigma\downarrow_\mathcal{R} = s_1$ and $v\sigma\downarrow_\mathcal{R} = s_2$ hold.

Theorem 1. *The complement of a finite set of first-order terms cannot be represented in general by a finite set of primal terms.*

5 Conclusion

We presented general algorithms for solving the word and the subsumption problem for primal terms that also work for ρ-terms, I-terms, and R-terms. The algorithms require a finitary unification algorithm for the schematization formalisms, implies multiply recursive derivation length (Hofbauer [6]), and termination under the lexicographic path ordering implies multiply recursive derivation length (Weiermann [11]). What is known in the general case of total termination? More generally, what is the expressivity of Kruskal's theorem when applied to finite rewrite systems? Weiermann has produced a theoretical upper bound for the complexity of terminating rewrite systems by Kruskal's theorem, using the Hardy hierarchy: the length of a derivation is dominated by the Hardy function $(s^\omega)^{\bar{\phi}_{\Omega^\omega}(0)}$, where $\bar{\phi}_{\Omega^\omega}(0)$ is an ordinal notation from Bachmann's system for the small Veblen ordinal. To give a proof theoretic intuition about this measure, primitive recursion corresponds to the provably total functions of the $\Sigma_0^1 - Ind$ fragment of Peano arithmetic and multiple recursion corresponds to the $\Sigma_0^2 - Ind$ fragment. However $(s^\omega)^{\bar{\phi}_{\Omega^\omega}(0)}$ is not even provably total in ATR_0. So there is a huge gap between the upper bound formulated by Weiermann and the observed complexity of common rewrite systems. Weiermann concluded his article by emphasising that "it is an *open problem* to prove or disprove that there are always multiply recursive bounds on the derivation lengths of a finite rewrite system $\mathcal{R}$ over a finite signature, for which the rewrite relation $\to_\mathcal{R}$ is contained in a simplification ordering (...)."

In addition to the practical interest of knowing the expressivity of total termination orderings, there is a theoretical issue. The study of known total termination orderings tells us that it is possible to classify the derivation lengths with the order type of the ordering. More precisely, the derivation length is connected to the order type through the so called slow-growing hierarchy. Can this result extend to all totally terminating rewrite systems, or even to all systems reducing by Kruskal's theorem, as suggested by Cichon in [2]? For the homeomorphic embedding of Kruskal's theorem, the maximal order type was studied by Schmidt [9]: it corresponds to the multiply recursive functions in the slow-growing hierarchy.

The purpose of this article is to present a "negative" result. We produce an example of a totally terminating finite rewrite system, which goes above multiple recursion. So this furnishes a new lower bound for the complexity of totally terminating rewrite systems and for rewrite systems that reduces by Kruskal's theorem. This contradicts Cichon's conjecture too. Our construction relies on the famous combinatorial game of the *Hydra battle* [7], which can be seen as a geometrical representation of the Hardy hierarchy. The paper is organised as follows: in the first section, we recall standard notions of term rewriting theory and termination. The second section is devoted to the presentation of the Hydra battle and the third section to the construction the rewrite system $\mathcal{H}$ which encodes the Hydra battle. The proof of total termination for $\mathcal{H}$ is based on a new characterisation of total termination.

1 Rewriting Background

This article assumes some familiarity with term rewriting theory. We recall here some useful basic notions. A comprehensive survey is to be found in Dershowitz-Jouannaud [3].

Let $\mathcal{F}$ be a finite signature whose function symbols have fixed arity. Given a set of variables $\mathcal{V}$, $\mathcal{T}(\mathcal{F}, \mathcal{V})$ denotes the term algebra built up from $\mathcal{V}$ and $\mathcal{F}$, and $\mathcal{T}(\mathcal{F})$ the set of closed terms of $\mathcal{T}(\mathcal{F}, \mathcal{V})$. For a rewrite system $\mathcal{R}$, we write $\xrightarrow{+}_\mathcal{R}$ for the associated rewrite relation. $\mathcal{R}$ *terminates* if $\xrightarrow{+}_\mathcal{R}$ is Noetherian. The complexity of a terminating rewrite system is measured by the *derivation length* function $Dl_\mathcal{R}$, which is the longest derivation allowed by the rewrite system.

Definition 1 (Derivation length). *Let $\mathcal{T}(\mathcal{F}, \mathcal{V})$ be a term algebra and $\mathcal{R}$ a terminating rewrite system over $\mathcal{T}(\mathcal{F}, \mathcal{V})$. Define the* derivation length *functions $dl_\mathcal{R}$ and $Dl_\mathcal{R}$:*

$$dl_\mathcal{R} : \mathcal{T}(\mathcal{F}) \to \mathbb{N}$$
$$t \mapsto \max\{dl_\mathcal{R}(u),\ t \to_\mathcal{R} u\} + 1$$

$$Dl_\mathcal{R} : \mathbb{N} \to \mathbb{N}$$
$$m \mapsto \max\{n \in \mathbb{N},\ \exists t \in \mathcal{T}(\mathcal{F}), dl_\mathcal{R}(t) = n \ \wedge\ |t| \leq m\}$$

where $|t|$ is the height of t.

Given a well-ordered set $(\mathcal{A}, \prec)$, an *interpretation* for a rewrite system $\mathcal{R}$ on $\mathcal{A}$ is a morphism $[\ ] : \mathcal{T}(\mathcal{F}) \to \mathcal{A}$ such that

$$\forall u \forall v \in \mathcal{T}(\mathcal{F})\ \ u \xrightarrow{+}_\mathcal{R} v \Rightarrow [u] \succ [v].$$

Since $(\mathcal{A}, \prec)$ is well-founded, the interpretation ensures termination.

Definition 2. *Let $\mathcal{T}(\mathcal{F}, \mathcal{V})$ be a term algebra and $(\mathcal{A}, \prec)$ be a well-ordered set. For any morphism $[\ ]$ of $\mathcal{T}(\mathcal{F}) \to \mathcal{A}$, we say that*

$[\ ]$ *is* strictly monotone *if for all $u, v, t_1, \ldots, t_n \in \mathcal{T}(\mathcal{F})$, for all $f \in \mathcal{F}$*

$$[u] \prec [v] \Rightarrow [f(t_1, \ldots, u, \ldots, t_n)] \prec [f(t_1, \ldots, v, \ldots, t_n)].$$

$[\ ]$ *is* monotone *if for all $u, v, t_1, \ldots, t_n \in \mathcal{T}(\mathcal{F})$, for all $f \in \mathcal{F}$*

$$[u] \preceq [v] \Rightarrow [f(t_1, \ldots, u, \ldots, t_n)] \preceq [f(t_1, \ldots, v, \ldots, t_n)],$$

where $\preceq$ is the reflexive closure of $\prec$.

$[\ ]$ *has the* subterm property *if for all $u_1 \ldots u_n \in \mathcal{T}(\mathcal{F})$, for all $f \in \mathcal{F}$*

$$\forall i\ 1 \leq i \leq n\ \ [u_i] \prec [f(u_1, \ldots, u_n)].$$

Most of the time, interpretations are defined in a compositional way: each symbol of the signature is assigned a function on $\mathcal{A}$ of the same arity. In this case, the interpretation is monotone if each function is increasing, strictly monotone if each function is strictly increasing and it has the subterm property if the result of each function is strictly greater than each of its arguments. We now come to the definition of total termination, due to Ferreira and Zantema [4].

Definition 3 (Total termination). *A rewrite system is totally terminating if there exists a well-ordered algebra $(\mathcal{A}, \prec)$ and a strictly monotone interpretation for $\mathcal{R}$ on $(\mathcal{A}, \prec)$.*

In other words, if there exists a well-ordered algebra $(\mathcal{A}, \prec)$ and a strictly monotone morphism $[\ ] : \mathcal{T}(\mathcal{F}) \to \mathcal{A}$ such that

$$\forall l \to r \in \mathcal{R} \ \forall \sigma : \mathcal{V} \to \mathcal{T}(\mathcal{F}) \ [l\sigma] \succ [r\sigma]$$

then $\mathcal{R}$ is totally terminating. It is a well-known result that any totally terminating rewrite system on a finite signature with fixed arity is compatible with the homeomorphic embedding relation of Kruskal's theorem (see [4] for instance). We now give another characterisation of total termination, which requires only monotonicity, instead of strict monotonicity.

Proposition 1. *Let $\mathcal{T}(\mathcal{F}, \mathcal{V})$ be a term algebra and let $\mathcal{R}$ be a rewrite system on $\mathcal{T}(\mathcal{F}, \mathcal{V})$. If there exists a well-ordered algebra $(\mathcal{A}, \prec)$ and a morphism $[\ ] : \mathcal{T}(\mathcal{F}) \to \mathcal{A}$ such that*
  (i) *for all $l \to r$ in $\mathcal{R}$, for all substitutions $\sigma : \mathcal{V} \to \mathcal{T}(\mathcal{F})$, $[l\sigma] \succ [r\sigma]$,*
  (ii) $[\ ]$ *has the subterm property,*
  (iii) $[\ ]$ *is monotone,*

then $\mathcal{R}$ is totally terminating.

Proof. We construct a strictly monotone interpretation $\mathcal{I}$ for $\mathcal{R}$ on the well-ordered algebra $(\mathsf{mul}(\mathcal{A}), \mathsf{mul}(\prec))$ (we write $\mathsf{mul}(\mathcal{A})$ to mean the set of finite multisets on $\mathcal{A}$ and $\mathsf{mul}(\prec)$ the multiset extension of $\prec$ on $\mathsf{mul}(\mathcal{A})$). Let $\cup$ denote the union of multisets. For each term u in $\mathcal{T}(\mathcal{F})$, define $\mathcal{I}(u)$ as the multiset of $\mathsf{mul}(\mathcal{A})$ containing the interpretations of u and its subterms:

$$\mathcal{I}(c) = \{[c]\} \ \text{ whenever } c \text{ is a constant symbol,}$$
$$\mathcal{I}(f(t_1, \ldots, t_n)) = \{[f(t_1, \ldots, t_n)]\} \cup \mathcal{I}(t_1) \ldots \cup \mathcal{I}(t_n).$$

  – $\mathcal{I}$ is compatible with $\mathcal{R}$: let $l \to r$ in $\mathcal{R}$ and $\sigma : \mathcal{V} \to \mathcal{T}(\mathcal{F})$, a substitution. By (i), $[l\sigma] \succ [r\sigma]$, which with (ii) implies $\{[l\sigma]\}\mathsf{mul}(\succ)\mathcal{I}(r\sigma)$. Hence $\mathcal{I}(l\sigma)\mathsf{mul}(\succ)\mathcal{I}(r\sigma)$.
  – $\mathcal{I}$ is strictly monotone: let $u, v \in \mathcal{T}(\mathcal{F})$ such that $\mathcal{I}(u)\mathsf{mul}(\prec)\mathcal{I}(v)$ and let $f \in \mathcal{F}$ of arity $n + 1$. For all $t_1, \ldots, t_n \in \mathcal{T}(\mathcal{F})$, we have

$$\mathcal{I}(f(t_1, \ldots, u, \ldots, t_n)) = \{[f(t_1, \ldots, u, \ldots, t_n)]\} \cup \mathcal{I}(u) \cup \mathcal{I}(t_1) \cup \ldots \cup \mathcal{I}(t_n)$$
$$\mathcal{I}(f(t_1, \ldots, v, \ldots, t_n)) = \{[f(t_1, \ldots, v, \ldots, t_n)]\} \cup \mathcal{I}(v) \cup \mathcal{I}(t_1) \cup \ldots \cup \mathcal{I}(t_n).$$

By hypothesis, we have $\mathcal{I}(u)\mathsf{mul}(\prec)\mathcal{I}(v)$, which implies

$$\mathcal{I}(u) \cup \mathcal{I}(t_1) \cup \ldots \cup \mathcal{I}(t_n) \ \mathsf{mul}(\prec) \ \mathcal{I}(v) \cup \mathcal{I}(t_1) \cup \ldots \cup \mathcal{I}(t_n).$$

So it remains to show that $[f(t_1, \ldots, u, \ldots, t_n)] \preceq [f(t_1, \ldots, u, \ldots, t_n)]$. Suppose $[u] \succ [v]$. By hypothesis (ii) on $[\ ]$, this would imply $\{[u]\}\mathsf{mul}(\succ)\mathcal{I}(v)$, which contradicts the hypothesis $\mathcal{I}(u)\mathsf{mul}(\prec)\mathcal{I}(v)$. So $[u] \preceq [v]$, which with (iii) ensures $[f(t_1, \ldots, u, \ldots, t_n)] \preceq [f(t_1, \ldots, v, \ldots, t_n)]$. This concludes the proof. □

2 The Hydra Battle

The rewrite system we present is based on the *Battle of Hercules and the Hydra* of [7]. Let us recall the general principle. A Hydra is a finite tree, each leaf corresponding to a head. At each step of the game, Hercules chops off one of the heads of the Hydra and the monster grows in turn: if the cut leaf has a grandparent in the tree, then the branch issued from this grandparent is multiplied. The number of copy equals the rank of the step in the game. This implies that the multiplication rate of the Hydra is increasing during the game. Hercules wins when the Hydra is reduced to the empty tree.

A battle may easily be interpreted by a decreasing sequence of ordinals. Associate to each node n in the tree the ordinal $\langle n \rangle = \omega^{\langle n_1 \rangle \oplus \cdots \oplus \langle n_i \rangle}$, where $n_1, \ldots, n_i$ are the children of n and $\oplus$ denotes the ordinal natural sum. The whole tree is interpreted by $\langle r_1 \rangle \oplus \cdots \oplus \langle r_n \rangle$, where $r_1, \ldots, r_n$ are the children of the root. Here is an example of battle, with the associated ordinal labelling. So every strategy is a winning strategy for Hercules.

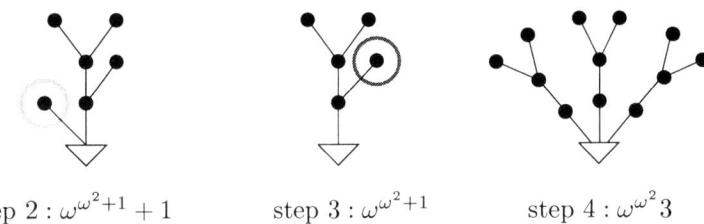

step 2: $\omega^{\omega^2+1} + 1$ step 3: ω^{ω^2+1} step 4: $\omega^{\omega^2} 3$

We now concentrate on a particular strategy, which we call "standard". We describe it from the ordinal point of view. Let $\mathcal{CNF}(\varepsilon_0)$ denote the set of notations in Cantor Normal Form for ordinals below ε_0. Given a limit ordinal λ, a *fundamental sequence* $(\lambda_n)_{n \in \mathbb{N}}$ for λ is simply a strictly increasing sequence whose supremum is λ. A canonical assignment of fundamental sequences for $\mathcal{CNF}(\varepsilon_0)$ is defined recursively as follows:

$$\omega_n = n$$
$$(\alpha + \lambda)_n = \alpha + \lambda_n$$
$$(\omega^{\beta+1})_n = \omega^\beta n$$
$$(\omega^\lambda)_n = \omega^{\lambda_n}.$$

Definition 4 (The standard Hydra battle). *For all n in $\mathbb{N}$, define the function $h_n : \mathcal{CNF}(\varepsilon_0) \to \mathcal{CNF}(\varepsilon_0)$ by*

$$h_n(0) = 0$$
$$h_n(\alpha + 1) = \alpha$$
$$h_n(\lambda) = \lambda_n, \text{ if } \lambda \text{ is a limit ordinal.}$$

Given an initial ordinal α_0, the battle is a sequence $(\alpha_n, n)_{n \in \mathbb{N}}$ of $\mathcal{CNF}(\varepsilon_0) \times \mathbb{N}$ such that for all n in $\mathbb{N}$ $\alpha_{n+1} = h_n(\alpha_n)$. In a pair (α_n, n), the ordinal α_n is the Hydra. The second element n is the rank of the step in the game.

For any initial configuration, the standard battle is finite. This fact is however not provable in Peano arithmetic. Indeed, given an initial configuration α, the length of this battle is greater than $s^\alpha(0)$, the αth element of the Hardy hierarchy applied to 0. This can be established using standard tools of number theoretic functions. We do not go into technical details here and we invite the interested reader to consult some classical texts, such as Cichon [1] and Wainer [10]. The only result we need for our construction is the following proposition.

Proposition 2. *The function of* $\mathbb{N} \to \mathbb{N}$ *which associates to each integer* n *the length of the Hydra Battle starting from* $(\omega^{\omega^\omega}, n)$ *with standard strategy is not multiply recursive.*

Proof. The Hardy function $s^{\omega^{\omega^\omega}}$ is not multiply recursive (Robbin [8]). □

It follows that any rewrite system $\mathcal{R}$ encoding the Hydra Battle for trees of height 4, that is ordinals below ω^{ω^ω}, admits a derivation length function which is not multiply recursive. For each n in $\mathbb{N}$, $(\omega^{\omega^\omega}, n)$ reduces in $(\omega^{\omega^n}, n+1)$ in one step. For each $\alpha < \omega^{\omega^\omega}$ and for each $m \in \mathbb{N}$, $\mathcal{R}$ encodes the battle with initial configuration (α, m). In particular, it encodes the battle with initial configuration $(\omega^{\omega^n}, n+1)$.

3 Encoding the Hydra Battle as a Rewrite System

3.1 Construction of the Rewrite System $\mathcal{H}$

We now model the process of the Hydra battle by the rewrite system $\mathcal{H}$. A first system for the Hydra battle appears in [3], but its termination cannot be established by Kruskal's theorem. The version we present here is totally terminating. The underlying idea for the transcription is very different. The intuition is as follows. The ordinals of ω^{ω^ω} in Cantor Normal Form are interpreted by terms built up from the constant 0 and the binary function symbol H. For this, define $\mathcal{O}$ by

$$\mathcal{O}: \quad \omega^{\omega^\omega} \to \mathcal{T}(0, \mathsf{H})$$
$$0 \mapsto 0$$
$$\omega^\alpha \mapsto \mathsf{H}(\mathcal{O}(\alpha), 0)$$
$$\beta + \omega^\alpha \mapsto \mathsf{H}(\mathcal{O}(\alpha), \mathcal{O}(\beta))$$

To deal with the rank of the step in a battle, we introduce two unary function symbols, $[]$ and $\bullet$. Each step (α, n) of the battle will be encoded by the term $\bullet\, []^n\, \mathcal{O}(\alpha)$. For each ordinal α in ω^{ω^ω}, the system $\mathcal{H}$ should then allow us to derive

$$\bullet\, []^n\, \mathcal{O}(\alpha) \xrightarrow{+}_{\mathcal{H}} \bullet\, []^{n+1}\, \mathcal{O}(h_n(\alpha)).$$

Let's have a closer look on the definition of the ordinal function h_n. Given an ordinal α in ω^{ω^ω}, we distinguish three cases for the computation of $h_n(\alpha)$:

Case 1 : if α is a successor ordinal of the form $s(\beta)$, then $h_n(\alpha) = \beta$,
Case 2 : if α is a limit ordinal of the form $\gamma + \omega^{s(\beta)}$, then $h_n(\alpha) = \gamma + \omega^\beta n$.
Case 3 : if α is a limit ordinal of the form $\gamma + \omega^{\beta + \omega^{s(a)}}$, $h_n(\alpha) = \gamma + \omega^{\beta + \omega^a n}$.

So if we write $t = \mathcal{O}(\beta)$, $u = \mathcal{O}(\gamma)$ and $v = \mathcal{O}(a)$, $\mathcal{H}$ should allow us to derive

Case 1 : $\bullet \, []^n \, \mathsf{H}(0,t) \xrightarrow{+}_{\mathcal{H}} \bullet \, []^{n+1} \, t$,

Case 2 : $\bullet \, []^n \, \mathsf{H}(\mathsf{H}(0,t),u) \xrightarrow{+}_{\mathcal{H}} \bullet \, []^{n+1} \, \mathsf{H}(t, \mathsf{H}(t, \ldots \mathsf{H}(t,u) \ldots))$ (n occurrences of t),

Case 3 : $\bullet \, []^n \, \mathsf{H}(\mathsf{H}(\mathsf{H}(0,t),u),v) \xrightarrow{+}_{\mathcal{H}} []^{n+1} \mathsf{H}(\mathsf{H}(t, \ldots \mathsf{H}(t,u) \ldots), v)$ (n occurrences of t).

The first case can be handled directly by a single rewrite rule. For the two last cases, we need to introduce three intermediate function symbols: $\circ$, c^1 (for case 2) and c^2 (for case 3). Consider finally the signature $\mathcal{F} = \{\circ, \bullet, [], 0, \mathsf{H}, c^1, c^2\}$, where 0 is a constant symbol, $\circ$, $\bullet$, $[]$ are unary function symbols, H, c_1 are of arity 2 and c^2 is of arity 3. $\mathcal{H}$ is defined on $\mathcal{T}(\mathcal{F}, \mathcal{V})$ by

$$\mathcal{H} \begin{cases}
\circ x \to \bullet \, [] \, x & (1) \\
\bullet \, [] \, x \to [] \, \bullet \bullet x & (2) \\
\mathsf{H}(0, x) \to \circ x & (3) \\
\bullet \mathsf{H}(\mathsf{H}(0, y), z) \to c^1(y, z) & (4) \\
\bullet \mathsf{H}(\mathsf{H}(\mathsf{H}(0, x), y), z) \to c^2(x, y, z) & (5) \\
\bullet c^1(x, y) \to c^1(x, \mathsf{H}(x, y)) & (6) \\
\bullet c^2(x, y, z) \to c^2(x, \mathsf{H}(x, y), z) & (7) \\
c^1(y, z) \to \circ z & (8) \\
c^2(x, y, z) \to \circ \mathsf{H}(y, z) & (9) \\
[] \circ x \to \circ \, [] \, x & (10) \\
\bullet x \to x & (11)
\end{cases}$$

3.2 Complexity of $\mathcal{H}$

We verify that $\mathcal{H}$ simulates the Hydra battle.

Lemma 1. *Let $\alpha \in \omega^{\omega^\omega}$. For all $n \geq 1$, $\bullet \, []^n \, \mathcal{O}(\alpha) \xrightarrow{+}_{\mathcal{H}} \bullet \, []^{n+1} \, \mathcal{O}(h_n(\alpha))$.*

Proof. We consider the three cases mentioned above.

Case 1 :

$$\begin{aligned}
\bullet \, []^n \, \mathsf{H}(0, t) &\xrightarrow{+}_{\mathcal{H}} []^n \mathsf{H}(0, t) & (11) \\
&\xrightarrow{+}_{\mathcal{H}} []^n \circ t & (3) \\
&\xrightarrow{+}_{\mathcal{H}} \circ []^n \, t & (10)^n \\
&\xrightarrow{+}_{\mathcal{H}} \bullet \, []^{n+1} \, t & (1)
\end{aligned}$$

Case 2 :

$$\begin{aligned}
\bullet \, []^n \, \mathsf{H}(\mathsf{H}(0,t), u) &\xrightarrow{+}_{\mathcal{H}} []^n \, \bullet^{2^n} \, \mathsf{H}(\mathsf{H}(0,t), u) & (2)^n \\
&\xrightarrow{+}_{\mathcal{H}} []^n \, \bullet^{n+1} \, \mathsf{H}(\mathsf{H}(0,t), u) & (11)^* \\
&\xrightarrow{+}_{\mathcal{H}} []^n \, \bullet^n \, c^1(t, u) & (4) \\
&\xrightarrow{+}_{\mathcal{H}} []^n c^1(t, \mathsf{H}(t, \mathsf{H}(t, \ldots \mathsf{H}(t, u) \ldots))) & (6)^n \\
&\xrightarrow{+}_{\mathcal{H}} []^n \circ \mathsf{H}(t, \mathsf{H}(t, \ldots \mathsf{H}(t, u) \ldots)) & (8) \\
&\xrightarrow{+}_{\mathcal{H}} \circ []^n \, \mathsf{H}(t, \mathsf{H}(t, \ldots \mathsf{H}(t, u) \ldots)) & (10)^n \\
&\xrightarrow{+}_{\mathcal{H}} \bullet \, []^{n+1} \, \mathsf{H}(t, \mathsf{H}(t, \ldots \mathsf{H}(t, u) \ldots)) & (1)
\end{aligned}$$

Case 3 :

- $[]^n\, \mathsf{H}(\mathsf{H}(\mathsf{H}(0,t),u),v) \xrightarrow{+}_{\mathcal{H}} []^n\, \bullet^{2^n}\, \mathsf{H}(\mathsf{H}(\mathsf{H}(0,t),u),v)$ $(2)^n$
  $\xrightarrow{+}_{\mathcal{H}} []^n\, \bullet^{n+1}\, \mathsf{H}(\mathsf{H}(\mathsf{H}(0,t),u),v)$ $(11)^*$
  $\xrightarrow{+}_{\mathcal{H}} []^n\, \bullet^n\, c^2(t,u,v)$ (5)
  $\xrightarrow{+}_{\mathcal{H}} []^n c^2(t,\mathsf{H}(t,\ldots \mathsf{H}(t,u)\ldots),v)$ $(7)^n$
  $\xrightarrow{+}_{\mathcal{H}} []^n \circ \mathsf{H}(\mathsf{H}(t,\ldots \mathsf{H}(t,u)\ldots),v)$ (9)
  $\xrightarrow{+}_{\mathcal{H}} \circ []^n\, \mathsf{H}(\mathsf{H}(t,\ldots \mathsf{H}(t,u)\ldots),v)$ $(10)^n$
  $\xrightarrow{+}_{\mathcal{H}} \bullet []^{n+1}\, \mathsf{H}(\mathsf{H}(t,\ldots \mathsf{H}(t,u)\ldots),v)$ (1)

□

Corollary 1. *$Dl_{\mathcal{H}}$ is not multiply recursive.*

Proof. Consequence of proposition 2. □

3.3 $\mathcal{H}$ Is Totally Terminating

The proof of total termination is based on proposition 1: we associate to each function symbol appearing in $\mathcal{H}$ a monotone function which enjoys the subterm property. Our starting point is the intentional meaning of the symbols 0 and H: each term t built up from 0 and H may simply be interpreted by the ordinal $\mathcal{O}^{-1}(t)$. For c_1 and c_2, we shall use the function f, defined by

$$f : \mathcal{CNF}(\varepsilon_0) \times \mathcal{CNF}(\varepsilon_0) \to \mathcal{CNF}(\varepsilon_0)$$
$$(x,y) \mapsto y + \omega^{x+1}$$

Note that the definition of f uses the ordinal sum $+$, which is not strictly monotonic. For instance, $f(2, \omega^2 3 + \omega + 7) = \omega^3 + \omega + 7$.

Lemma 2. *For all α, β in $\mathcal{CNF}(\varepsilon_0)$*

(i) $f(\alpha, \beta \oplus \omega^\alpha) = f(\alpha, \beta)$,
(ii) $f(\alpha, \beta) \leq \beta \oplus \omega^\alpha \omega$,
(iii) $f(\alpha, \beta) > \alpha$ and $f(\alpha, \beta) > \beta$,
(iv) f is an increasing function.

Proof. Straightforward. □

For the symbols $\bullet$, $\circ$ and $[]$, consider the sub-system

$$\begin{cases} \circ x \to \bullet []\, x \\ \bullet []\, x \to []\, \bullet \bullet x \\ []\, \circ x \to \circ []\, x \end{cases}$$

This admits an interpretation on $\omega \times \omega$: interpret $\circ$ by $(m,n) \mapsto (2m+3, n)$, $[]$ by $(m,n) \mapsto (2m+2, n)$ and $\bullet$ by $(m,n) \mapsto (m, n+m+1)$. Combining the interpretations for 0, H, c_1 and c_2 on $\mathcal{CNF}(\varepsilon_0)$ and the interpretations for $[]$, $\circ$ and $\bullet$ on $\omega \times \omega$, we finally define $[\ ]$ on $\mathcal{CNF}(\varepsilon_0) \times \omega \times \omega$ as follows:

$$[0] = (0,0,0)$$
$$[\mathsf{H}] = (\alpha,m,n),(\beta,m',n') \mapsto (\omega^\alpha \oplus \beta, 0, 0)$$
$$[c^1] = (\alpha,m,n),(\beta,m',n') \mapsto (f(\alpha,\beta),0,0)$$
$$[c^2] = (\alpha,m,n),(\beta,m',n'),(\gamma,m'',n'') \mapsto (\gamma \oplus \omega^{f(\alpha,\beta)},0,0)$$
$$[\bullet] = (\alpha,m,n) \mapsto (\alpha, m, n+m+1)$$
$$[\circ] = (\alpha,m,n) \mapsto (\alpha, 2m+3, n)$$
$$[[]] = (\alpha,m,n) \mapsto (\alpha, 2m+2, n)$$

$\mathcal{CNF}(\varepsilon_0) \times \omega \times \omega$ is ordered by the lexicographic combination of $(\mathcal{CNF}(\varepsilon_0), \in)$ and $(\omega \times \omega, \in)$. We write $<$ for this ordering.

Lemma 3.
  (i) $[\ ]$ *has the subterm property,*
  (ii) $[\ ]$ *is monotone,*
  (iii) *for all* $l \to r \in \mathcal{H}$, *for all substitutions* $\sigma : \mathcal{V} \to \mathcal{T}(\mathcal{F})$, $[l\sigma] > [r\sigma]$.

Proof. (i) and (ii) are direct, using lemma 2 for f. For (iii), we examine each rule:

(1) $\mathsf{H}(0,t) \to \circ t : (\alpha+1, 0, 0) > (\alpha, 2m+3, n)$
(2) $c^1(u,v) \to \circ v : (f(\beta,\gamma), 0, 0) > (\gamma, 2m''+3, n'')$
(3) $c^2(t,u,v) \to \circ \mathsf{H}(u,v) : (\gamma \oplus \omega^{f(\alpha,\beta)}, 0, 0) > (\gamma \oplus \omega^\beta, 3, 0)$
(4) $\circ t \to \bullet [] t : (\alpha, 2m+3, n) > (\alpha, 2m+2, n+2m+3)$
(5) $\bullet [] t \to [] \bullet \bullet t : (\alpha, 2m+2, n+2m+3) > (\alpha, 2m+2, n+2m+2)$
(6) $[] \circ t \to \circ [] t : (\alpha, 4m+8, n) > (\alpha, 4m+7, n)$
(7) $\bullet \mathsf{H}(\mathsf{H}(0,u),v) \to c^1(u,v) : (\gamma \oplus \omega^{\beta+1}, 0, 1) > (f(\beta,\gamma), 0, 0)$
(8) $\bullet \mathsf{H}(\mathsf{H}(\mathsf{H}(0,t),u),v) \to c^2(t,u,v) : (\gamma \oplus \omega^{\beta+\omega^{\alpha+1}}, 0, 1) > (\gamma \oplus \omega^{f(\alpha,\beta)}, 0, 0)$
(9) $\bullet c^1(t,u) \to c^1(t, \mathsf{H}(t,u)) : (f(\alpha,\beta), 0, 1) > (f(\alpha,\beta), 0, 0)$
(10) $\bullet c^2(t,u,v) \to c^2(t, \mathsf{H}(t,u), v): (\gamma \oplus \omega^{f(\alpha,\beta)}, 0, 1) > (\gamma \oplus \omega^{f(\alpha,\beta)}, 0, 0)$
(11) $\bullet t \to t : (\alpha, m, n+n+1) > (\alpha, m, n)$.

(t, u, v are terms of $\mathcal{T}(\mathcal{F})$ whose interpretations are (α,m,n), (β,m',n') and (γ,m'',n'') respectively). $\square$

Proposition 3. $\mathcal{H}$ *is totally terminating.*

Proof. Consequence of lemma 3 and proposition 1. $\square$

3.4 Extension of $\mathcal{H}$

The rewrite system $\mathcal{H}$ models a restrained version of Hydra battle with ordinals below ω^{ω^ω}. It may easily be extended to deal with higher ordinals, below ε_0. To reach $\omega^{\omega^{\omega^\omega}}$, one adds a 4-ary function symbol c^3 and so on. In this way one exhausts the provably total functions of Peano arithmetic.

Perspectives

We have exhibited a totally terminating rewrite system which departs from multiple recursion. What still remains open is what complexity can be achieved via total termination or termination by Kruskal's theorem. Moreover, our example rekindles the debate on the relationship between order type and length of derivation for a rewrite system. Our construction is interesting from a proof-theoretical point of view. We have shown that it is possible to encode the Hardy hierarchy by a finite rewrite system. So it can be directly connected with the work of Weiermann in [12], which uses the Hardy hierarchy too. Unfortunately, our construction is restrained to ordinals below ε_0. Is it possible to describe higher ordinals and reach $\phi_{\Omega^\omega}(0)$, the maximal order type of homeomorphic embedding of Kruskal's theorem ? This then would imply that the bound formulated by Weiermann is, surprisingly, a least upper bound.

References

1. E.A. Cichon, *A short proof of two recently discovered independence results using recursion theoretic methods.* Proceedings of the American Mathematical Society, vol 97 (1983), p.704-706.
2. E.A. Cichon, *Termination proofs and complexity characterisations.* Proof theory, P. Aczel, H. Simmons and S. Wainer Eds, Cambridge university press (1992), p.173-193.
3. N. Dershowitz and J.P. Jouannaud, *Rewrite systems.* Handbook of Theoretical Computer Science, J. Van Leeuwen Ed., north-Holland 1990, p.243-320.
4. M.C.F. Ferreira and H. Zantema, *Total termination of term rewriting.* Proceedings of RTA-93, Lecture Notes in Computer Science 690, p. 213-227.
5. D. Hofbauer, *Termination proofs with multiset path orderings imply primitive recursive derivation lengths.* Theoretical Computer Science 105-1 (1992), p.129-140.
6. D. Hofbauer, *Termination proofs and derivation lengths in term rewriting systems* Dissertation, Technische Universität Berlin, 1991 (also available as Technical Report: TU Berlin, Forschungsberichte des Fachbereichs Informatik 92-46, 1992).
7. L. Kirby and J. Paris, *Accessible independence results for Peano arithmetic.* Bull. London Math. Soc. 14 (1982), p.285-225.
8. J.W. Robbin, *Subrecursive Hierarchies.* Ph.D. Princeton
9. D. Schmidt, *Well-partial orderings and their maximal order types.* Habilitationsschrift, Fakultät für Mathematik der Ruprecht-Karl-Universität, Heidelberg (1977).
10. S.S. Wainer, *Ordinal recursion, and a refinement of the extented Grzegorczyk hierarchy.* Journal of Symbolic Logic 37-2 (1972), p.281-292.
11. A. Weiermann, *Termination proofs by lexicographic path orderings yield multiply recursive derivation lengths.* Theoretical Computer Science 139 (1995), p.355-362.
12. A. Weiermann, *Complexity bounds for some finite forms of Kruskal's theorem.* Journal of Symbolic Computation 18 (1994), p.463-488.

Computing ϵ-Free NFA from Regular Expressions in $O(n \log^2(n))$ Time[*]

Christian Hagenah and Anca Muscholl

Institut für Informatik, Universität Stuttgart,
Breitwiesenstr. 20-22, 70565 Stuttgart, Germany

Abstract. The standard procedure to transform a regular expression to an ϵ-free NFA yields a quadratic blow-up of the number of transitions. For a long time this was viewed as an unavoidable fact. Recently Hromkovič et.al. [5] exhibited a construction yielding ϵ-free NFA with $O(n \log^2(n))$ transitions. A rough estimation of the time needed for their construction shows a cubic time bound. The known lower bound is $\Omega(n \log(n))$. In this paper we present a sequential algorithm for the construction described in [5] which works in time $O(n \log(n) + \text{size of the output})$. On a CREW PRAM the construction is possible in time $O(\log(n))$ using $O(n + (\text{size of the output})/\log(n))$ processors.

1 Introduction

Among various descriptions of regular languages regular expressions are especially interesting because of their succinctness. On the other hand, the high degree of expressiveness leads to algorithmically hard problems, for example testing equivalence is PSPACE-complete. Given a regular expression we are often interested in computing an equivalent nondeterministic finite automaton *without ϵ-transitions (NFA)*. This conversion is of interest due to some operations which can be easily performed on NFA, as for example intersection.

In this paper we present efficient sequential and parallel algorithms for converting regular expressions into small NFA. For a regular expression E we take the number of letters as the size of E, whereas the size of an NFA is measured as the number of transitions. It is known that the translation from NFA to regular expressions can yield an exponential blow-up, [3]. The other direction however can be achieved in polynomial time. One classical method for constructing NFA from regular expressions is based on position automata. This construction yields NFA of quadratic size, see e.g. [1,2]. A substantial improvement on this construction was achieved in [5], where a refinement of position automata was shown to yield NFA with $O(n \log^2(n))$ transitions. This is optimal up to a possible $\log(n)$ factor, as shown in [5] by proving a $O(n \log(n))$ lower bound. However, the precise complexity of the conversion proposed in [5] was not investigated. A trivial estimation of the construction of [5] leads to a cubic algorithm.

---

[*] Research was partly supported by the French-German project PROCOPE.

Performing the conversion form regular expressions to NFA efficiently is important from a practical viewpoint. The best one can hope for is to perform the construction in time proportional to the output size. In the present paper we propose efficient sequential and parallel algorithms for converting regular expressions to NFA. Our approach is based on the construction proposed in [5], but using a slightly different presentation. This allows us to obtain an algorithm which works in time $O(n \log(n))$ + size of the output). Therefore, our algorithm has worst-case complexity of $O(n \log^2(n))$. In the parallel setting we are able to perform the construction on a CREW PRAM in $O(\log(n))$ time by using $O(n)$ processors for computing the description of the states of the NFA, resp. $O(n \log(n))$ processors in the worst-case for the output NFA. Previously known was an $O(\log(n))$ time algorithm using $O(n/\log(n))$ processors, which computes an NFA *with* ϵ-transitions, see [4]. The paper is organized as follows. The sequential algorithm is presented in Sect. 4. Basic notions on position automata are recalled in Sect. 2, whereas Sect. 3 deals with the common follow sets construction of [5].

2 Preliminaries

Let A denote a finite alphabet. We consider non-empty regular expressions over A, i.e. (bracketed) expressions E built from ϵ and the letters in A, using concatenation $\cdot$, union $+$ and Kleene star $*$. The regular language defined by a regular expression E is denoted $\mathcal{L}(E)$. Finite automata are denoted as usual as $\mathcal{A} = (Q, A, q_0, \delta, F)$, with Q as set of states, $\delta \subseteq Q \times A \times Q$ as transition relation, q_0 as initial state and F as set of final states. The language recognized by $\mathcal{A}$ is denoted $L(\mathcal{A})$.

For algorithmic purposes a regular expression E over A is given by some syntax tree t_E. The syntax tree t_E has leaves labelled by ϵ or $a \in A$, and the inner nodes are either binary and labelled by $+$ or $\cdot$, or they are unary and labelled by $*$. The inner nodes of a syntax tree will be named $F, G, \ldots$ and we will denote them as subexpressions of E. For two subexpressions F, G of E we write $F \leq G$ ($F < G$, resp.) if F is an ancestor (a proper ancestor, resp.) of G. For a subexpression F let firststar(F) denote the largest subexpression G with $G \leq F$ such that G^* is the parent node of G.

A subtree t of t_E is a connected subgraph (i.e. a tree) of t_E. A subtree t is called *full subtree* if it contains all descendants of its root. This means that a full subtree of t_E corresponds to a subexpression of E.

We may suppose without loss of generality that the leaves of t_E are labelled with pairwise distinct letters. This allows to identify the leaves of t_E labelled by A uniquely by their labelling. For example, for $E = (a^* + b)^* a b^*$ we replace A by $\{a_1, a_2, b_1, b_2\}$ and E by $(a_1^* + b_1)^* a_2 b_2^*$.

3 Position Automata

In this section we recall some basic notions related to the construction of position automata from regular expressions. We follow the definitions from [2,5].

3.1 Positions and Sets of Positions

Given a regular expression E, the set $\mathrm{pos}(E)$ comprises all positions of E which are labelled by letters from A. According to our convention, $\mathrm{pos}(E) \subseteq A$. Positions of E will be named $x, y, \ldots$.

Lemma 1. *Let E be a regular expression, $n = |\mathrm{pos}(E)|$. Then we can compute in linear time an equivalent expression E', $\mathcal{L}(E) = \mathcal{L}(E')$, such that E' has length $O(n)$.*

The size $|E|$ of the expression E is defined as $|\mathrm{pos}(E)|$. Moreover, $\mathrm{pos}(t)$ and $|t|$ are defined analogously for a subtree t of t_E. Throughout the paper we denote by n the size $|E|$ of E. The lemma above says that we may assume that the size of the syntax tree t_E satisfies $|t_E| \in O(n) = O(|\mathrm{pos}(t_E)|)$.

For a regular expression E we consider two distinguished subsets of positions, $\mathrm{first}(E)$ and $\mathrm{last}(E)$. The set $\mathrm{first}(E) \subseteq \mathrm{pos}(E)$ contains all positions which can occur as first letter in some word in $\mathcal{L}(E)$. Similarly, $\mathrm{last}(E)$ contains all positions which can occur as last letter in some word in $\mathcal{L}(E)$. Formally:

$$\mathrm{first}(E) = \{x \in \mathrm{pos}(E) \mid xA^* \cap \mathcal{L}(E) \neq \emptyset\},$$
$$\mathrm{last}(E) = \{x \in \mathrm{pos}(E) \mid A^*x \cap \mathcal{L}(E) \neq \emptyset\}.$$

The sets $\mathrm{first}(E), \mathrm{last}(E)$ can be computed inductively by noting that e.g. $\mathrm{first}(F+G) = \mathrm{first}(F) \cup \mathrm{first}(G)$, $\mathrm{first}(F^*) = \mathrm{first}(F)$ and $\mathrm{first}(F \cdot G) = \mathrm{first}(F)$ if $\epsilon \notin \mathcal{L}(F)$, resp. $\mathrm{first}(F \cdot G) = \mathrm{first}(F) \cup \mathrm{first}(G)$ if $\epsilon \in \mathcal{L}(F)$. For a given position $x \in \mathrm{pos}(E)$ let $\mathrm{follow}(x) \subseteq \mathrm{pos}(E)$ contain all positions y which are immediate successors of x in some word of $\mathcal{L}(E)$:

$$\mathrm{follow}(x) = \{y \in \mathrm{pos}(E) \mid A^*xyA^* \cap \mathcal{L}(E) \neq \emptyset\}.$$

As above, $\mathrm{follow}(x)$ can be defined recursively by means of $\mathrm{follow}(x, F) = \mathrm{follow}(x) \cap \mathrm{pos}(F)$. We omit the definition here, since anyway we will not compute the sets $\mathrm{follow}(x)$ globally.

3.2 Automata

First, last and follow sets are the basic components of an NFA $\mathcal{A}_E$ recognizing $\mathcal{L}(E)$, called *position automaton* in [5]. Let $\mathcal{A}_E = (Q, A, \delta, q_0, F)$ be defined by

$$Q = \mathrm{pos}(E) \,\dot{\cup}\, \{q_0\}$$
$$\delta = \{(q_0, x, x) \mid x \in \mathrm{first}(E)\} \cup \{(x, y, y) \mid y \in \mathrm{follow}(x)\}$$
$$F = \begin{cases} \mathrm{last}(E) & \text{if } \epsilon \notin \mathcal{L}(E) \\ \mathrm{last}(E) \cup \{q_0\} & \text{otherwise} \end{cases}$$

Recall for the above definition that $\mathrm{pos}(E) \subseteq A$. The following equivalence is easy to check:

Proposition 2. *For every regular expression E we have $\mathcal{L}(\mathcal{A}_E) = \mathcal{L}(E)$.*

The construction above yields ε-free automata with $n+1$ states and $O(n^2)$ transitions. In [5] a refined construction was presented, based on the idea of a *system of common follow sets (CFS system)*, which is defined as follows:

Definition 3 ([5]). *Let E be a regular expression. A CFS system S for E is given as $S = (dec(x))_{x \in pos(E)}$, where $dec(x) \subseteq \mathcal{P}(pos(E))$ is a decomposition of $follow(x)$:*

$$follow(x) = \bigcup_{C \in dec(x)} C.$$

Let $\mathcal{C}_S = \{first(E)\} \cup \bigcup_{x \in pos(E)} dec(x)$. The CFS automaton $\mathcal{A}_S$ associated with S is defined as $\mathcal{A}_S = (Q, A, q_0, \delta, F)$ where

$Q = \mathcal{C}_S \times \{0, 1\}$

$q_0 = \begin{cases} (first(E), 1) & \text{if } \epsilon \in \mathcal{L}(E) \\ (first(E), 0) & \text{otherwise} \end{cases}$

$\delta = \{(C, f), x, (C', f')) \mid x \in C, \ C' \in dec(x) \text{ and } f' = 1 \Leftrightarrow x \in last(E)\}$

$F = \mathcal{C}_S \times \{1\}$

Lemma 4. *Let E be a regular expression and let S be a CFS system for E. Then the CFS automaton $\mathcal{A}_S$ recognizes $\mathcal{L}(E)$.*

It is shown in [5] how to obtain a CFS system S for a given regular expression E such that $|\mathcal{C}_S| \in O(n)$, $\sum_{C \in \mathcal{C}_S} |C| \in O(n \log n)$ and $|dec(x)| \in O(\log n)$ for all $x \in pos(E)$. This yields a CFS automaton with $O(n)$ states and $O(n \log^2(n))$ transitions.

4 Computing a Common Follow Sets System

4.1 Properties of Follow Sets

The running time of our algorithm relies heavily on some structural properties of follow sets which are discussed in the following.

Lemma 5. *Let E be a regular expression and let F, G be subexpressions with $E \leq F \leq G$. Then we have:*

1. *$first(F) \cap first(G) \neq \emptyset$ implies $first(G) \subseteq first(F)$.*
2. *$F \leq H \leq G$ and $\emptyset \neq first(G) \subseteq first(F)$ implies $first(G) \subseteq first(H) \subseteq first(F)$.*
3. *$x \in pos(G) \setminus first(G)$ implies $x \notin first(F)$.*

The proof of the lemma is a straightforward application of the inductive definition. An analogous lemma can be also stated for last sets.

The next lemma deals with the relation between follow sets and a decomposition of the syntax tree, which will be used recursively in the definition of the CFS system. For simplifying the notation we will denote for $x \in pos(E)$, $E \leq F$, the set $follow(x) \cap pos(F)$ by $follow_F(x)$. Analogously, $follow_t(x)$ denotes the set $follow(x) \cap pos(t)$ for a subtree t.

Lemma 6. *Given a regular expression E, a syntax tree t_E and subexpressions F, G with $E < F < G$. Let t, t' be subtrees of t_E such that $pos(t) \subseteq pos(F) \setminus pos(G)$ and $pos(t') \subseteq pos(G)$. Then we have for all $x, x', y \in pos(E)$:*

1. $follow_F(x) = \emptyset$ for all $x \in pos(t') \setminus last(G)$;
2. $follow_F(x) = follow_F(x')$ for all $x, x' \in pos(t') \cap last(G)$;
3. $follow_{t'}(y) = first(G) \cap pos(t')$ for all $y \in pos(t)$ with $follow_{t'}(y) \neq \emptyset$.

4.2 Recursive Definition of CFS Systems

The CFS system defined in [5] is based on a divide-and-conquer construction. Consider a subtree t of t_E and let F denote the root of t. Let $x \in pos(t)$. If $|t| = 1$ then we define

$$C_0 = follow_t(x) = follow(x) \cap \{x\}, \quad dec(x,t) = \{C_0\}.$$

Suppose now that $|t| > 1$. Then let t_1 be a subtree of t such that $1/3|t| \leq |t_1| \leq 2/3|t|$ and let $t_2 = t \setminus t_1$. Let F_1 denote the root of t_1. Clearly, for every position $x \in pos(t)$ we have $follow_t(x) = follow_{t_1}(x) \dot\cup follow_{t_2}(x)$. We distinguish two cases, depending on $x \in pos(t_1)$ or $x \in pos(t_2)$.

i) Let $x \in pos(t_1)$. If $x \notin last(F_1)$ then by Lem. 6 we have $follow_{t_2}(x) = \emptyset$. Otherwise, for $x \in last(F_1)$ then again by Lem. 6 we have $follow_{t_2}(x) = follow_{t_2}(x')$ for all $x' \in last(F_1) \cap pos(t_1)$.
Let $C_1 = follow_{t_2}(x')$ for some $x' \in pos(t_1) \cap last(F_1)$ and define $dec(x,t)$ as

$$dec(x,t) = \begin{cases} dec(x,t_1) & \text{if } x \notin last(F_1) \\ dec(x,t_1) \cup \{C_1\} & \text{otherwise} \end{cases}$$

ii) Let $x \in pos(t_2)$. If $follow_{t_1}(x) \neq \emptyset$ then we have $follow_{t_1}(x) = first(F_1) \cap pos(t_1)$ by Lem. 6.
Let $C_2 = first(F_1) \cap pos(t_1)$ and define $dec(x,t)$ as

$$dec(x,t) = \begin{cases} dec(x,t_2) & \text{if } follow_{t_1}(x) = \emptyset \\ dec(x,t_2) \cup \{C_2\} & \text{otherwise} \end{cases}$$

It can be easily verified that $dec(x,t)$ is a decomposition[1] of $follow_t(x)$, i.e. $follow_t(x) = \bigcup_{C \in dec(x,t)} C$. Hence, we obtain a CFS system $\mathcal{C}(t)$ restricted to t, where

$$\mathcal{C}(t) = \bigcup \{dec(x,t) \mid x \in pos(t)\} = \{C \mid C \in dec(x,t) \text{ for some } x \in pos(t)\}.$$

Note that $|\mathcal{C}(t)| \leq |\mathcal{C}(t_1)| + |\mathcal{C}(t_2)| + 2$. This yields $|\mathcal{C}(t)| \leq 3|t| - 2$. Similarly, the following estimations can be easily verified (see also Lem. 4 of [5]):

$\sum_{C \in \mathcal{C}(t)} |C| \leq 2|t| \log(|t|) + 1$ and
$|dec(x,t)| \leq 2 \log(|t|) + 1$, for all $x \in pos(t)$.

---

[1] In [5] the corresponding set $\bigcup_{C \in dec(x,t)} C$ is just a subset of $follow_t(x)$. Having equality here simplifies the recursive definition and the correctness proof of the decomposition.

5 A Sequential $O(n \log(n))$ Algorithm for Computing a Common Follow Sets System

We consider now the computation of the sets defined in the previous section. For $C_0 = \text{follow}(x) \cap \{x\}$ we can determine whether $x \in \text{follow}(x)$ by checking whether $x \in \text{last}(S) \cap \text{first}(S)$ for $S = \text{firststar}(x)$. For the recursion step we have to determine C_1, C_2 with

$$C_1 = \text{follow}_{t_2}(x) \quad \text{and} \quad C_2 = \text{first}(F_1) \cap \text{pos}(t_1).$$

We want to compute both C_1, C_2 and the positions $x \in \text{pos}(t)$ to which C_1 or C_2 is added in linear time, i.e. in time $O(|t|)$. As shown below, the computation of C_1 reduces to computing a union of first sets restricted to $\text{pos}(t_2)$. This yields two problems: First we need an efficient way to compute intersections of first sets with a given set of positions. Second, the union of restricted first sets has to be disjoint. The solution to both problems will rely on a suitable data structure for first sets. Before discussing the data structure let us consider the set C_1 in more detail[2].

Definition 7. *Let $E \leq F$ be regular expressions. We define $\text{fnext}(F) \subseteq \text{pos}(E)$ as*

$$\text{fnext}(F) = \begin{cases} \text{first}(G) & \text{if } F \cdot G \text{ is the parent node of } F \\ \text{first}(F) & \text{if } F^* \text{ is the parent node of } F \\ \emptyset & \text{otherwise} \end{cases}$$

Analogously, $\text{lprev}(F)$ is defined by replacing first by last and by requiring that $G \cdot F$ is the parent node of F.

Using the fnext operator we are able to express follow sets as unions of first sets. Compared with Lem. 3 in [5] we need for expressing $\text{follow}_{t_2}(x)$ at most *one* first set which is not contained in F. Of course, this is necessary in order to be able to determine C_1 in time $O(|t|)$:

Proposition 8. *Let E be a regular expression with $E \leq F < F_1$ and let t_E be a syntax tree of E. Let t_2 be a subtree with root F and $\text{pos}(t_2) \cap \text{pos}(F_1) = \emptyset$ and consider a position $x \in \text{last}(F_1)$. Then we have*

$$\text{follow}_{t_2}(x) = \bigcup_{G \in \mathcal{G}} (\text{fnext}(G) \cap \text{pos}(t_2))$$

where the union is taken over

$$\mathcal{G} = \{G \mid \text{last}(G) \supseteq \text{last}(F_1) \text{ and } (F < G \leq F_1 \text{ or } G = \text{firststar}(F))\}.$$

---

[2] In the definition below $\text{fnext}(F)$ corresponds to $\text{first}(\text{fnext}(F))$ in [5].

Proof: Note that for every $G \leq F_1$ with $\text{last}(F_1) \subseteq \text{last}(G)$ we have $\text{fnext}(G) \cap \text{pos}(t_2) \subseteq \text{follow}_{t_2}(x)$. Conversely, consider a position $y \in \text{follow}_{t_2}(x)$ with $y \notin \text{fnext}(G)$, for all $F < G \leq F_1$ with $\text{last}(F_1) \subseteq \text{last}(G)$. Hence, there exists some node G, $E \leq G \leq F$, with $y \in \text{fnext}(G)$ and $\text{last}(F_1) \subseteq \text{last}(G)$. Clearly, the parent node of G is G^* (otherwise, $\text{fnext}(G) \cap \text{pos}(t_2) = \emptyset$), thus $y \in \text{first}(G) \cap \text{pos}(t_2)$. If $G = \text{firststar}(F)$ then we are done. Otherwise $G < H = \text{firststar}(F)$. In this case it is not difficult to verify using Lem. 5 that for all $G < H$ with $\text{first}(G) \cap \text{pos}(t_2) \neq \emptyset$ we also have $\text{first}(G) \cap \text{pos}(t_2) = \text{first}(H) \cap \text{pos}(t_2)$. Therefore, $y \in \text{first}(H) \cap \text{pos}(t_2)$. □

Our algorithm is based on a suitable order on positions of E, which allows manipulating first sets efficiently. We use an array called firstdata such that for each subexpression F of E the set $\text{first}(F)$ is a subinterval of firstdata. The crucial point is the order of positions within firstdata. Consider a fixed syntax tree t_E of E. We first define a forest $\mathcal{F}$ by deleting all edges from nodes labelled $F \cdot G$ to the child labelled G, whenever $\epsilon \notin \mathcal{L}(F)$. Let $\mathcal{F} = \{T_1, \ldots, T_k\}$ be the forest thus obtained, then we denote the trees T_i as *first-trees*. Note that each $\text{first}(F)$ is the union of all $\text{first}(F')$ with $F < F'$ where F' belongs to the same first-tree as F.

We define a total order on $\mathcal{F}$ as follows. For $1 \leq i \neq j \leq k$ let $T_i \prec T_j$ whenever the roots F_i, F_j of T_i, resp. T_j satisfy

- either $F_j < F_i$, i.e. F_j is an ancestor of F_i,
- or F_i and F_j are incomparable w.r.t. $<$ and F_i lies to the right of F_j.

The order $\prec$ corresponds thus to a reversed preorder traversal of t_E, i.e. right child—left child—parent node.

Suppose that after renaming $\mathcal{F} = \{T_1, \ldots, T_k\}$ with $T_i \prec T_j$ for all $i < j$. The array firstdata is given as $\text{fdata}(T_1) \cdots \text{fdata}(T_k)$, with $\text{fdata}(T_i)$ being the list of positions corresponding to the yield of T_i. Moreover, by a preorder traversal of each T_i we can determine for each subexpression F of T_i the subinterval of $\text{fdata}(T_i)$ corresponding to $\text{first}(F)$. The set $\text{first}(F)$ is described by its starting position $\text{fstart}(F)$ within $\text{fdata}(T_i)$ and its length $\text{flength}(F) = |\text{first}(F)|$.

Remark 9. (i) Let F, G be subexpressions of E. Then we have $\emptyset \neq \text{first}(F) \subseteq \text{first}(G)$ if and only if $\text{fstart}(G) \leq \text{fstart}(F)$ and $\text{fstart}(G) + \text{flength}(G) \geq \text{fstart}(F) + \text{flength}(F)$, i.e. if the subinterval corresponding to $\text{first}(G)$ includes the subinterval corresponding to $\text{first}(F)$. Moreover, firstdata allows to determine the intersection $\text{first}(F) \cap \text{pos}(t)$ in $O(|t|)$ time, where F is a subexpression and t is subtree of t_E (described as set of positions in increasing order).

(ii) A similar data structure lastdata can be defined for the last sets.

We are now ready to describe an algorithm UnionFirst for the following problem. Given subexpressions F, F_1 of E with $F < F_1$ and subtrees t, t_2 of F, resp. a subtree t_1 of F_1, where $t = t_1 \dot\cup t_2$, $\text{pos}(F_1) \cap \text{pos}(t_2) = \emptyset$, and a position $x \in \text{last}(t_1)$. We want to compute the set $C = \text{follow}_{t_2}(x)$. Recall from Prop. 8 that

$$C = \bigcup_{G \in \mathcal{G}} (\text{fnext}(G) \cap \text{pos}(t_2)),$$

with $G \in \mathcal{G}$ if and only if $\text{last}(F_1) \subseteq \text{last}(G)$, and either $F < G \leq F_1$ or $G = \text{firststar}(F)$.

function UnionFirst (node F_1, tree t_2) : nodelist;
    var rootlist, tocheck: nodelist; G: node;
    begin
    rootlist := nil;
    tocheck := nil;
    $G := F_1$;
    while $(G \neq \text{root}(t_2)$ and $\text{last}(F_1) \subseteq \text{last}(G))$ **do**
        begin
        $A :=$ parent expression of G;
        if $A = G^*$ **then**
            rootlist := rootlist $\setminus \{H \mid H \in$ tocheck and $\text{first}(H) \subseteq \text{first}(G)\}$;
            rootlist := rootlist $\circ\, G$;
            tocheck := $\{G\}$;
        else if $A = G \cdot H$ **then**
            if $\epsilon \in \mathcal{L}(G)$ **then** rootlist := rootlist $\circ H$;
            else rootlist := $H\circ$ rootlist **endif**;
            tocheck := tocheck $\cup \{H\}$;
        endif;
        $G := A$;
    endwhile;
    return(rootlist);
    end

The proof of the next proposition is omitted for lack of space.

Proposition 10. *Let F, F_1 be subexpressions of E, $E \leq F < F_1$, and let t_E be a syntax tree. Let t_1 be a subtree with root F_1 and let t_2 be a subtree with root F and $\text{pos}(F_1) \cap \text{pos}(t_2) = \emptyset$. Let $x \in \text{last}(t_1)$ be a position and let $\mathcal{G}$ be defined as above. Then UnionFirst(F_1, t_2) yields a list* rootlist $= (H_1, \ldots, H_l)$ *of (names of) subexpressions of E satisfying the following:*

1. $\bigcup_{i=1}^{l} \text{first}(H_i) = \bigcup_{G \in \mathcal{G}, G \neq \text{firststar}(F)} \text{fnext}(G)$.
   Moreover, $\text{first}(H_i) \cap \text{first}(H_j) = \emptyset$ for all $i \neq j$.
2. *Let $T(H_i)$ denote the first-tree in the forest $\mathcal{F}$ containing H_i. Then $T(H_i) \preceq T(H_j)$ for all $1 \leq i < j \leq k$. Moreover, if $T(H_i) = T(H_j)$ then H_i precedes H_j w.r.t. preorder (in t_E).*
3. *UnionFirst(F_1, t_2) runs in $O(|t_2|)$ steps.*

Remark 11. Given the assertion of Prop. 10 it is not hard to verify that the set $\bigcup_{G \in \mathcal{G}} (\text{fnext}(G) \cap \text{pos}(t_2))$ can be computed in $O(|t_2|)$ steps using rootlist and firstdata. More precisely, we can precompute in time $O(|t_2|)$ a list fdata(t_2) corresponding to pos(t_2) sorted as firstdata. Next, we scan rootlist and fdata(t_2) in parallel, building the intersection. Hereby we use Rem. 9 in order to determine in constant time whether a position belongs to a set first(G). Note that rootlist has at most $|t_2|$ elements, since the while loop in UnionFirst is executed at most $|t_2|$ times. Finally, if $S = \text{firststar}(F)$ is defined we can check whether $\text{last}(F_1) \subseteq \text{last}(S)$ in $O(1)$ time and compute $\text{first}(S) \cap \text{pos}(t_2)$ in time $O(|t_2|)$.

Theorem 12. *Given a regular expression E and a syntax tree t_E for E of size $O(|E|) = O(n)$. We can compute a CFS system S for E in time $O(n \log(n))$. Therefore, we can compute an NFA $\mathcal{A}_S$ for E of size $|\mathcal{A}_S|$ in time $O(n \log(n) + |\mathcal{A}_S|)$. The worst-case complexity of the algorithm is thus $O(n \log^2(n))$.*

Proof: Recall the recursive definition of $\text{dec}(x,t)$ given in Sect. 4.2. For a position $x \in \text{pos}(t_1)$ we test whether $x \in \text{last}(F_1)$ in constant time (using lastdata), whereas $C_1 = \text{follow}_{t_2}(x)$ can be computed in time $O(|t_2|)$. The case where $x \in \text{pos}(t_2)$ is dual. Here, the set $C_2 = \text{first}(F_1) \cap \text{pos}(t_1)$ can be determined in constant time, whereas determining which $x \in \text{pos}(t_2)$ satisfy $\text{follow}_{t_1}(x) \neq \emptyset$ requires $O(|t_1|)$ steps. To see this, note that for a position $y \in C_2$ we have $\{x \in \text{pos}(t_2) \mid \text{follow}_{t_1}(x) \neq \emptyset\} = \text{precede}(y) \cap \text{pos}(t_2)$, where $\text{precede}(y)$ is defined as the dual of $\text{follow}(y)$, i.e., $\text{precede}(y) = \{x \mid A^*xyA^* \cap \mathcal{L}(E) \neq \emptyset\}$. Moreover, by duality we have $\text{precede}(y) \cap \text{pos}(t_2) = \bigcup_{G \in \mathcal{G}} (\text{lprev}(G) \cap \text{pos}(t_2))$, with $G \in \mathcal{G}$ if $\text{first}(F_1) \subseteq \text{first}(G)$, and either $F < G \leq F_1$ or $G = \text{firststar}(F)$.

Therefore, we can compute in time $O(|t|)$ the sets $\text{dec}(x,t)$ from $\text{dec}(x,t_1)$ and $\text{dec}(x,t_2)$ for all positions x of t. Hence, our algorithm runs in time $O(n \log(n))$. Finally, outputting the transitions of the NFA $\mathcal{A}_S$ is possible in time $O(|\mathcal{A}_S|)$.
□

In the parallel setting we have again an output-size optimal algorithm on a CREW PRAM, as stated below. For lack of space we omit the proofs.

Theorem 13. *Given a regular expression E and a syntax tree t_E for E of size $O(|E|) = O(n)$. We can compute a CFS system S for E on a CREW PRAM in time $O(\log(n))$ using $O(n)$ processors. Therefore, we can compute an NFA $\mathcal{A}_S$ for E of size $|\mathcal{A}_S|$ in time $O(\log(n))$ using $O(n + |\mathcal{A}_S|/\log(n))$ processors (i.e., $O(n \log(n))$ processors in the worst case).*

Acknowledgment: We thank Volker Diekert for many comments and contributions and to the anonymous referees for suggestions which helped improving the presentation.

References

1. G. Berry and R. Sethi. From regular expressions to deterministic automata. *Theoretical Computer Science*, 48:117–126, 1986.
2. A. Brüggemann-Klein. Regular expressions into finite automata. *Theoretical Computer Science*, 120:197–213, 1993.
3. A. Ehrenfeucht and P. Zeiger. Complexity measures for regular expressions. *Journal of Computer and System Sciences*, 12:134–146, 1976.
4. A. Gibbons and W. Rytter. *Efficient Parallel Algorithms*. Cambridge University Press, 1989.
5. J. Hromkovič, S. Seibert, and T. Wilke. Translating regular expressions into small ϵ-free nondeterministic finite automata. In *Proc. of the 14th Ann. Symp. on Theor. Aspects of Comp. Sci. (STACS'97)*, no. 1200 in LNCS, p. 55–66, 1997. Springer.

Iterated Length-Preserving Rational Transductions*
(Extended Abstract)

Michel Latteux, David Simplot, and Alain Terlutte

C.N.R.S. U.R.A. 369, L.I.F.L. Université de Lille I, Bât. M3, Cité Scientifique,
59655 Villeneuve d'Ascq Cedex, France

Abstract. The purpose of this paper is the study of the smallest family of transductions containing length-preserving rational transductions and closed under union, composition and iteration. We give several characterizations of this class using restricted classes of length-preserving transductions, by showing the connections with "context-sensitive transductions" and transductions associated with recognizable picture languages. We also study the class obtained by only using length-preserving rational functions and we show the relations with "deterministic context-sensitive transductions".

1 Introduction

The family of rational languages turns out to be one of the most important classes within the Chomsky hierarchy. Finite automata that are the main object for studying rational languages are now used in most domains of computer science. Rational transductions introduced by C. C. Elgot and J. E. Mezei [4] are a natural extension of rational languages and were very useful to represent several kinds of computations. The theory of rational transductions was mainly developed by M. P. Schtzenberger, S. Eilenberg and M. Nivat (see [3,11,12]). This theory is now well established and its basic results can be found in [2,3]. More recently, some representation theorems were achieved in terms of compositions of morphisms and inverse morphisms [13].

At the contrary, there is only a few papers dealing with iteration of rational transductions (see [6,14]). Since several mechanisms of computation are actually iterations of rational transductions, it seems that this study deserves to be undertaken. For instance, finitely generated congruences, derivations in a grammar, partial commutation — as well as semi-commutation —, L-systems are examples of such mechanisms.

The set of transductions equipped with the operation of composition has a semigroup structure closed under iteration. The subset of rational transductions is closed under composition but not under iteration. We are mainly interested by the rational closure of the set of rational transductions, that is the smallest set

---

* This work was partially supported by the group MOSYDIS of the PRC/GDR AMI

of transductions closed under union, composition and iteration and containing the rational transductions. Indeed several interesting transductions need to compose rational transductions and iterated rational transductions. For instance, the mirror operation is shown in the preliminaries to be such a transduction. It is neither a rational transduction nor an iterated rational transduction but it can be realized by composition of these two kinds of transductions.

In this paper, we shall restrict ourself to iterated length-preserving transductions, more precisely, we shall study the rational closure of the class of length-preserving rational transductions, that is the smallest family of transductions containing length-preserving rational transductions and closed under union, composition and iteration. There are two main reasons for this choice. First, one easily verifies that iterations of arbitrary rational transductions can be obtained by composition of arbitrary rational transductions with iterated length-preserving rational transductions. At reverse, arbitrary rational transductions can be achieved by composition of length-preserving rational transductions and iterations of faithful rational transductions. In this way the projection from A^* onto B^* with $B \subseteq A$ is equal to the composition of the iteration of the rational function which erases only the first occurrence of a letter of $A \setminus B$ with the length-preserving rational function which corresponds to the intersection with B^*.

For lack of space we only give rough sketchs of proofs, but they can be found in the full version of this abstract which is available as Technical Report [9].

2 Preliminaries

We assume the reader to be familiar with basic formal language theory (see [2,3] for more precisions). The goal of this section is to fix notations and terminology.

2.1 Words, Languages, and Transductions

For a finite alphabet Σ, we denote by Σ^* the free monoid generated by Σ. The neutral element of this monoid is the empty word, which is denoted by ε. The size of the alphabet Σ is denoted by $||\Sigma||$ and is equal to its number of letters. The length of a word u is denoted by $|u|$.

The classes of regular, deterministic context-sensitive and context-sensitive languages over Σ are denoted respectively by $\text{Rec}(\Sigma^*)$, $\text{CS}_\text{d}(\Sigma^*)$ and $\text{CS}(\Sigma^*)$.

Now we give some basic definitions about transductions. A transduction is a subset of $X^* \times Y^*$ where X and Y are two finite alphabets. For a word u, the set of images of u by a transduction τ is denoted by $u\tau$ and is defined by: $u\tau = \{v \mid (u,v) \in \tau\}$.

The set of transductions has a semigroup structure according to the composition operation. Let τ and σ be two transductions. The composition of τ and σ is the transduction defined by:

$$\tau\sigma = \{(u,w) \mid \exists v \text{ such that } (u,v) \in \tau \wedge (v,w) \in \sigma\}.$$

A transduction τ from X^* into Y^* is rational if and only if it is a rational part of $X^* \times Y^*$ (according to the usual concatenation product in this monoid). It is the class of transductions which can be realized by a finite transducer — that is a finite automaton where edges are labelled by an input and an output word.

We say that a transduction τ is functional if for each word u, $u\tau$ contains at most one word. When we deal with a function τ we will write $u\tau = v$ instead of $u\tau = \{v\}$.

A transduction τ is length-preserving (l.p.) if and only if for each couple $(u,v) \in \tau$ we have $|u| = |v|$. In the remainder of the paper, we consider only l.p. transductions. The class of all l.p. rational transductions is denoted by $\mathcal{T}$ and the class of l.p. rational functions is denoted by $\mathcal{F}$.

In our proofs, we shall use several particular kinds of letter-to-letter morphisms — the class of letter-to-letter morphisms is a particular class of l.p. functions denoted by $\mathcal{H}$. For an arbitrary alphabet A, the identity over A^* is denoted by I_A — notice that I_A is equivalent to the intersection with A^* which is denoted by $(\cap A^*)$. When we consider an alphabet Σ which is the cartesian product of n alphabets, $\Sigma = X_1 \times X_2 \times \ldots X_n$ (with $n \geq 1$), the morphism Π_i, with $1 \leq i \leq n$ is the projection onto the ith component.

2.2 A Introductory Example

Let us start with a simple example to explain the use of iterated length-preserving rational transductions. Let X be an arbitrary alphabet. We consider the function f which associates the mirror image with each word of X^*: $\forall w \in X^*$, $wf = \widetilde{w}$. Although this function is not rational, we show that f can be obtained by composition of l.p. rational functions and iterated l.p. rational functions.

Let Σ be the alphabet $X \cup \dot{X}$ containing non-marked and marked letters of X. We define a rational function τ from Σ^* into Σ^*: for each word $w = auv$ with $a \in X$, $u \in X^*$ and $v \in \dot{X}^*$, we have $w\tau = u\dot{a}v$; the image of the empty word is the empty word, $\varepsilon\tau = \varepsilon$, and the transduction is undefined in the other cases.

For instance, the successive applications of τ on a word over X of length five are:

$$w = a_1 a_2 a_3 a_4 a_5, \quad w\tau = a_2 a_3 a_4 a_5 \dot{a}_1, \quad w\tau^2 = a_3 a_4 a_5 \dot{a}_2 \dot{a}_1,$$
$$\ldots \quad w\tau^5 = \dot{a}_5 \dot{a}_4 \dot{a}_3 \dot{a}_2 \dot{a}_1, \quad w\tau^6 = \emptyset \ .$$

It is clear that for any word $w \in X^*$, the set $w\tau^+(\cap \dot{X}^*)$ — where τ^+ denotes the iteration of τ — contains a single word which is the marked mirror of w. If we denote by φ the morphism from $\dot{X}^*$ into X^* which gives the unmarked image, we have $f = (\cap X^*)\tau^+\varphi$.

3 Iterations of Rational Transductions

In this section we study the rational closure of $\mathcal{T}$, denoted by $\text{Rat}(\mathcal{T})$, which is the smallest class of transductions which contains $\mathcal{T}$ and closed under union, composition and iteration.

The iteration is the natural extension of the Kleene operator to the semigroup of length-preserving transductions. Let τ be a transduction. The iteration of τ, denoted by τ^+, is defined by $\tau^+ = \cup_{i \geq 1} \tau^i$. Let $\mathcal{C}$ be a class of transductions. The class $\mathcal{C}_+$ denotes the class of iterated transductions of $\mathcal{C}$: $\mathcal{C}_+ = \{\tau^+ \mid \tau \in \mathcal{C}\}$.

For the sake of simplicity of our proofs, we introduce a new class of l.p. transductions called one-step transductions which is denoted by $\mathcal{O}$. A one-step transduction is defined by a couple (X, P) where X is a finite alphabet and P a finite set of l.p. rules of $X^* \times X^*$. The transduction τ realizes one step of the rewriting system defined by P: $\tau = \{(u\alpha u', u\beta u') \in X^* \times X^* \mid (\alpha, \beta) \in P\}$.

The last characterization we give in the next theorem concerns recognizable picture languages. A picture over an alphabet Σ is a matrix of elements of Σ. In 1992, D. Giammarresi and A. Restivo give a definition of recognizable picture languages in terms of projection of local picture languages which are the natural extension of local string languages to the two-dimensional case (see [5] for complete definitions).

The first row of a picture p is denoted by $\text{fr}_\top(p)$ and the last row is denoted by $\text{fr}_\bot(p)$. The transduction associated to a given recognizable picture language L, denoted by τ_L is the set of couples of words which appear on the first and last rows of a picture of L: $\tau_L = \{(\text{fr}_\top(p), \text{fr}_\bot(p)) \mid p \in L\}$. The idea is to consider pictures as computations over the words which occur on the first lines. The class of transductions associated to recognizable picture languages is denoted by $\mathcal{T}_{\text{Rec}(LF)}$.

Here is the main result of this section which gives a characterization of the class $\text{Rat}(\mathcal{T})$.

Theorem 1 (Representation theorem). *Let τ be a transduction. The following properties are equivalent:*

i. *the transduction τ can be defined by using union, composition and iteration of l.p. rational transductions ($\tau \in \text{Rat}(\mathcal{T})$),*
ii. *there exist three l.p. rational transductions σ_1, σ_2 and σ_3 such that $\tau = \sigma_1 \sigma_2^+ \sigma_3$ ($\tau \in \mathcal{T T_+ T}$),*
iii. *there exist two l.p. rational transductions σ_1 and σ_3 and a one-step transduction σ_2 such that $\tau = \sigma_1 \sigma_2^+ \sigma_3$ ($\tau \in \mathcal{T O_+ T}$),*
iv. *there exist two letter-to-letter morphisms φ and ψ and a context-sensitive language A such that $\tau = \varphi^{-1}(\cap A)\psi$ ($\tau \in \mathcal{T}_{\text{CS}}$),*
v. *there exists a recognizable picture language L such that $\tau \setminus \{(\varepsilon, \varepsilon)\} = \tau_L$ ($\tau \in \mathcal{T}_{\text{Rec}(LF)}$).*

Sketch of proof. We can suppose that all the transductions we use do not contain the couple $(\varepsilon, \varepsilon)$ and that all context-sensitive languages are ε-free. We show the different implications.

($i. \Rightarrow ii.$) Clearly, $\mathcal{TT_+T}$ contains $\mathcal{T}$. Then, it suffices to show that $\mathcal{TT_+T}$ is closed under union, composition and iteration. Let $\tau = \sigma_1 \sigma_2^+ \sigma_3$ and $\tau' = \sigma_1' \sigma_2'^+ \sigma_3'$ be two transductions where the σ_i and the σ_i' are l.p. rational transductions. By choosing carefully the different alphabets used by these transductions, we can assume that the following property holds

$$\sigma_1 \sigma_2' = \sigma_1' \sigma_2 = \sigma_2 \sigma_3' = \sigma_2' \sigma_3 = \sigma_1 \sigma_3 = \sigma_2 \sigma_2' = \sigma_2' \sigma_2 = \sigma_1' \sigma_3' = \sigma_1' \sigma_3 = \emptyset \ .$$

The expected closure properties hold since we have:

$$\tau + \tau' = (\sigma_1 + \sigma_1')(\sigma_2 + \sigma_2')^+(\sigma_3 + \sigma_3') \ ,$$
$$\tau \tau' = \sigma_1 (\sigma_2 + \sigma_3 \sigma_1' + \sigma_2')^+ \sigma_3' \ ,$$
$$\tau^+ = \sigma_1 (\sigma_2 + \sigma_3 \sigma_1)^+ \sigma_3 \ .$$

($ii. \Rightarrow i.$) This implication is obvious.

($ii. \Rightarrow iii.$) It is clear that it suffices to prove that $\mathcal{T}_+$ is included in $\mathcal{TO_+T}$. Let $\tau \in \mathcal{T}$ be a l.p. rational transduction realized by a finite transducer T whose transitions are letter-to-letter. The idea is to simulate several computations of T over the input word.

The first rational transduction marks the first and the last letter of the word. The one-step transduction contains rules which (1) put an initial state on the first letter in a non-deterministic way, (2) realize a transition over two successive letters, and (3) delete the state on the last letter if this state leads to a final state by reading this letter. The last rational transduction verifies that the word contains no state.

($iii. \Rightarrow ii.$) Since $\mathcal{O}$ is included in $\mathcal{T}$, this implication is trivial.

($iii. \Rightarrow iv.$) One-step transductions have been introduced because it is easy to prove the following assertion.

Assertion. *The class* CS *is closed under* $\mathcal{O}_+$.

In order to show this assertion, we consider a ε-free context-sensitive language $L \in \mathrm{CS}(X^*)$ and a one-step transduction τ associated with the couple (Y^*, P). The language $L' = L(\cap Y^*)$ is also a ε-free context-sensitive language and then, is generated by a grammar $G = (Y, V, Q, S)$ in Kuroda normal form [7]. It is clear that the language $L\tau^+ = L'\tau^+$ is generated by the length-increasing grammar $G' = (Y, P \cup Q, S)$ and then is also context-sensitive. This ends the proof of the assertion.

Let τ be a transduction of $\mathcal{TO_+T}$. The implication $iii. \Rightarrow iv.$ is easily deduced by showing that the language A defined by

$$A = \{u \in (X \times Y)^* \mid u\Pi_2 \in u\Pi_1 \tau\}$$

is context-sensitive. Indeed, A is the image of the regular language $\{(a, a) \mid a \in X\}^*$ by the transduction τ' which applies τ on the second component. Hence, we have $\tau = \Pi_1^{-1}(\cap A)\Pi_2$.

($iv. \Rightarrow iii.$) Since inverse morphisms and morphisms are rational transductions, we just have to show that the intersection with a context-sensitive language A is a transduction of $\mathcal{TO_+T}$. We use the assertion that follows.

Assertion. *Let $A \in \mathrm{CS}(\Sigma^*)$ be a context-sensitive language. There exist a regular language R and a one-step transduction σ such that $A = R\sigma^+(\cap \Sigma^*)$.*

The idea is to simulate the derivations of a length-increasing grammar $G = (\Sigma, V, P, S)$ which generates A by using only l.p. rules. The rules realized by the one-step transduction σ are $u\$^{|v|-|u|} \to v$ – where \$ is a new symbol – for every $u \to v \in P$, and the rules allowing to move the symbol \$: $a\$ \to \$a$ for $a \in \Sigma \cup V$. It is easy to see that $A = (S\$^*)\sigma^+(\cap \Sigma^*)$.
So, the intersection with A is realized by the transduction $\tau = \sigma_1 \sigma_2^+ \sigma_3$ where the σ_i are defined by (X denotes the alphabet used in σ):

$$\sigma_1 = \{(u,v) \in \Sigma^* \times (\Sigma \times X)^*) \mid v\Pi_1 = u \wedge v\Pi_2 \in R\} \ ,$$
$$\sigma_2 = \{(u,v) \in ((X \times X)^*)^2 \mid u\Pi_1 = v\Pi_1 \wedge v\Pi_2 \in u\Pi_2\sigma\} \ ,$$
$$\sigma_3 = \{(u,v) \in (\Sigma \times \Sigma)^* \times \Sigma^* \mid u\Pi_1 = u\Pi_2 = v\} \ .$$

It is easy to see that the transductions σ_1 and σ_3 are rational and that σ_2 is a one-step transduction.

($ii. \Rightarrow v.$) We construct a recognizable picture language whose transduction realizes the iterated l.p. transduction σ_2 which is supposed to be included in $X^* \times Y^*$.
The idea is to show that the language K which contains all the pictures p such that if a word u is on the ith row ($i > 1$) then the above row ($i - 1$) contains a word v such that $u \in v\sigma_2$ is a recognizable language. We use a variant of the Nivat's theorem [11] which states that $\sigma_2 = \Pi_1^{-1}(\cap R)\Pi_2$ where R is a regular language of $(X \times Y)^*$. It brings out that all this controls can be made locally and it is easy to see that K is recognizable.
We have $\tau_K = \sigma_2^+$. The recognizable picture language L is obtained from K by adding a line at the top which corresponds to the application of σ_1 and a line at the bottom which realizes σ_3.

($v. \Rightarrow i.$) The recognizable picture language L is the image by a projection π (letter-to-letter morphism) of a local picture language K. The principle is to show that for a local picture language, the set of authorized rows below a given row is obtained by applying a l.p. rational transduction. In the same way, the set of words authorized to appear on the top lines (respectively bottom lines) is a regular set T (resp. B). Then we have $\tau_K = (\cap T)(\sigma^+ \cup I)(\cap B)$. Hence, we have $\tau_L = \pi^{-1}\tau_K \pi$ which belongs to $\mathrm{Rat}(\mathcal{T})$.
□

Because of Point $iv.$, by analogy to rational transductions, the transductions of $\mathrm{Rat}(\mathcal{T})$ are called "context-sensitive transductions". Moreover, by using the fact that the class of context-sensitive languages is closed under intersection and complement, we easily deduce the following closure properties.

Proposition 1. *The class of transductions* $\text{Rat}(\mathcal{T})$ *is closed under intersection and difference.*

The representation theorem allows us to deduce easily a result concerning recognizable picture language theory [8] which is similar to the theorem which states that the frontier of recognizable tree languages is the class of context-free languages [10].

Corollary 1 (Latteux Simplot 1997, [8]). *The family of frontiers of recognizable picture languages is exactly the family of ε-free context-sensitive languages.*

In this corollary, "frontiers" means the "bottom lines" of the pictures.

4 Iterations of Rational Functions

In this section we show that we can obtain all the transductions of $\text{Rat}(\mathcal{T})$ by using only functions. More precisely, we show that $\text{Rat}(\mathcal{T})$ and $\text{Rat}(\mathcal{F})$ coincide. We introduce a new class, denoted by $\text{Fin}(\mathcal{F}_+)$, which is the smallest class of transductions containing $\mathcal{F}_+$ and closed under union and composition. This class is interesting since it contains $\mathcal{T}$ and corresponds to "deterministic context-sensitive transduction" as shown below.

First, we remark that $\mathcal{F}$ is included in $\text{Fin}(\mathcal{F}_+)$. Indeed, when the image of a function uses the same alphabet as the domain, we cannot forbid the iteration of the function. But, using disjoint alphabets, we can use composition of two iterations in order to simulate a function. Thus the family of length-preserving rational functions is included in $\mathcal{F}_+\mathcal{F}_+$.

Like in Theorem 1 (*i.* $\Leftrightarrow$ *ii.*), we show that one iteration suffices in $\text{Fin}(\mathcal{F}_+)$.

Proposition 2. *The family* $\text{Fin}(\mathcal{F}_+)$ *is equal to the family* $\mathcal{F}\mathcal{F}_+\mathcal{F}$. *That means, a transduction* τ *belongs to* $\text{Fin}(\mathcal{F}_+)$ *if and only if* $\tau = f_1 f_2^+ f_3$ *for some f_1, f_2 and f_3 in $\mathcal{F}$.*

Sketch of proof. Since $\mathcal{F}\mathcal{F}_+\mathcal{F}$ is obviously included in $\text{Fin}(\mathcal{F}_+)$, it suffices to show that $\mathcal{F}\mathcal{F}_+\mathcal{F}$ is closed under union and composition.

Let τ_1 and τ_2 in $\mathcal{F}$ $\mathcal{F}_+$ $\mathcal{F}$. We have $\tau_1 = f_1 g_1^+ h_1$ and $\tau_2 = f_2 g_2^+ h_2$ for some $f_i, g_i, h_i \in \mathcal{F}$. We shall suppose that $(\varepsilon, \varepsilon)$ does not belong to all this functions.

An application of $\tau_1 + \tau_2$ looks like the following:

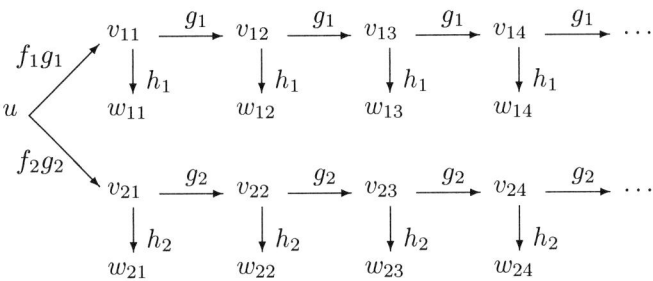

The idea is to apply alternatively g_1 and g_2 like shown in the next figure:

In order to avoid to break the computation the functions f_i and g_i are transformed in complete functions (each word has an image) by adding a new symbol which is never selected by the functions h_i. Hence $\tau_1 + \tau_2$ belongs to $\mathcal{F}\mathcal{F}_+\mathcal{F}$.

For the closure under composition, we remark that for a rational function f and a word $w \in \Sigma^n$ with $||\Sigma|| = k$, we have $uf^+ = \cup_{1 \leq i \leq k^n} uf^i$. Hence, by inserting a counter (with a word of length n over an alphabet of size k, one can count from 0 to $k^n - 1$), it is possible by iteration of a rational function g to cover the words of $ug_1h_1f_2g_2^+$, followed by the words of $ug_1^2h_1f_2g_2^+$ and so on. The applications of this function g look like the following:

$$u \to ug_1, ug_1h_1f_2g_2, 0 \to ug_1, ug_1h_1f_2g_2^2, 1 \to ug_1, ug_1h_1f_2g_2^3, 2$$
$$\to \ldots \to ug_1, ug_1h_1f_2g_2^{k^n}, k^n - 1 \to ug_1^2, ug_1^2h_1f_2g_2, 0$$
$$\to ug_1^2, ug_1^2h_1f_2g_2^2, 1 \to \ldots \to$$

As previously, h_1, f_2 and g_2 have to be complete. The selection consists in taking the second component. Hence we just have to apply f_1, to iterate g, to select the second component and to apply h_2 and we obtain $\tau_1\tau_2$. □

We have seen that $\mathcal{F} \subset \mathrm{Fin}(\mathcal{F}_+)$, the next result states that $\mathcal{T}$ is also included in this family.

Proposition 3. *The class of l.p. rational transductions is included in* $\mathrm{Fin}(\mathcal{F}_+)$.

Sketch of proof. Let τ be a l.p. rational transduction. We know that the transduction τ is equal to $\varphi^{-1}(\cap R)\psi$ where φ and ψ are two letter-to-letter morphisms and R is a regular language. Since letter-to-letter morphisms and the intersections with regular languages are l.p. rational functions, it suffices to show that inverse morphisms are in $\mathrm{Fin}(\mathcal{F}_+)$. Let φ be a letter-to-letter morphism from X^* into Y^*. For a word $u \in Y^*$ of length n, we enumerate in lexical order all the words over Y of length n and a word is selected if its image by φ is u. All these operations are rational and the result holds. □

Since $\mathrm{Fin}(\mathcal{F}_+)$ is included in $\mathrm{Rat}(\mathcal{F})$, the following equality clearly holds.

Proposition 4. *The classes* $\mathrm{Rat}(\mathcal{T})$ *and* $\mathrm{Rat}(\mathcal{F})$ *coincide.*

As we have shown that $\mathrm{Rat}(\mathcal{T})$ corresponds to the class of context-sensitive transductions, we show the connection between $\mathrm{Fin}(\mathcal{F}_+)$ and deterministic context-sensitive transductions.

Theorem 2. *Let τ be a transduction. It belongs to $\mathrm{Fin}(\mathcal{F}_+)$ if and only if there exist two letter-to-letter morphisms φ and ψ and a deterministic context-sensitive language A such that $\tau = \varphi^{-1}(\cap A)\psi$.*

Sketch of proof. It suffices to show that, for a l.p. rational function $f \subseteq X^* \times X^*$, the language $A = \{u \in (X \times X)^* \mid u\Pi_2 \in u\Pi_1 f^+\}$ is a deterministic context-sensitive language. We use the "decomposition theorem" [1] which states that a rational function is the composition of a right sequential transduction σ_r with a left sequential function σ_l. It is easy to build a deterministic linear bounded automaton (LBA) which simulates the successive applications of σ_r and σ_l on the first component of words over $X \times X$ and turns to a final state if the two components are identical.

At reverse, since inverse letter-to-letter morphisms are l.p. rational transductions – which belong to $\mathrm{Fin}(\mathcal{F}_+)$ by Proposition 3 –, we just have to show that the intersection with a deterministic context-sensitive language of Σ^* is in $\mathrm{Fin}(\mathcal{F}_+)$. A step of computation of a deterministic LBA is a l.p. rational function if we omit transitions which read the last letter and lead to a non-final state. Hence, it suffices to duplicate the word into $(X \times X)^*$ and to return the first component if a computation on the second component of the LBA leads to a final state. □

We deduce the following closure properties.

Proposition 5. *The class of transduction $\mathrm{Fin}(\mathcal{F}_+)$ is closed under inverse, intersection and difference.*

5 Conclusion

We have two classes $\mathrm{Rat}(\mathcal{T})$ and $\mathrm{Fin}(\mathcal{F}_+)$. The first class coincides also with $\mathrm{Rat}(\mathcal{F})$ and with the class of context-sensitive transductions. The second class corresponds to the class of deterministic context-sensitive transductions. Then the class $\mathrm{Fin}(\mathcal{F}_+)$ is included in $\mathrm{Rat}(\mathcal{T})$. To prove the reverse inclusion is equivalent to prove $\mathrm{CS_d} = \mathrm{CS}$, and is also equivalent to $SPACE(n) = NSPACE(n)$, which is a well-known open problem.

We can also consider the classes obtained by iteration of l.p. sub-sequential functions, l.p. sequential functions or deterministic l.p. functions (sub-sequential functions without output function associated with final states). Do we obtain new classes of transductions? Are these classes helpful to distinguish CS_d and CS?

References

1. Arnold, A., and Latteux, M.: A new proof of two theorems about rational transductions. *Theoretical Computer Science 8*, 2 (1979), 261–263.
2. Berstel, J.: *Transductions and Context-Free Languages*. Teubner Studienbücher, Stuttgart, 1979.
3. Eilenberg, S.: *Automata, Languages and Machines*, vol. A. Academic Press, New York, 1974.
4. Elgot, C. C., and Mezei, J. E.: On relations defined by generalized finite automata. *IBM Journal of Research and Development 9* (1965), 47–68.
5. Giammarresi, D., and Restivo, A.: Two-dimensional languages. In *Handbook of Formal Languages*, A. Salomaa and G. Rozenberg, Eds., vol. 3. Springer-Verlag, Berlin, 1997, pp. 215–267.
6. Greibach, S. A.: Full AFL's and nested iterated substitution. *Information and Control 16*, 1 (1970), 7–35.
7. Kuroda, S.-Y.: Classes of languages and linear bounded automata. *Information and Control 7*, 2 (1964), 207–223.
8. Latteux, M., and Simplot, D.: Context-sensitive string languages and recognizable picture languages. *Information and Computation 138*, 2 (1997), 160–169.
9. Latteux, M., Simplot, D., and Terlutte, A.: Iterated length-preserving transductions. Tech. Rep. it-312, L.I.F.L., Univ. Lille 1, France, March 1998.
10. Mezei, J, and Wright, J. B.: Algebraic Automata and Context-Free Sets. *Information and Control 11*, 2-3 (1967), 3–29.
11. Nivat, M.: Transductions des langages de Chomsky. *Ann. de l'Inst. Fourier 18* (1968), 339–456.
12. Schützenberger, M. P.: Sur les relations rationnelles entre monoides libres. *Theoretical Computer Science 3*, 2 (1976), 243–259.
13. Turakainen, P.: Transducers and compositions of morphisms and inverse morphisms. In *Studies in honour of Arto Kustaa Salomaa on the occasion of his fiftieth birthday* (1984), vol. 186 of *Ann. Univ. Turku. Ser. A I*, pp. 118–128.
14. Wood, D.: Iterated a-NGSM maps and Γ systems. *Information and Control 32*, 1 (1976), 1–26.

The Head Hierarchy for Oblivious Finite Automata with Polynomial Advice Collapses

Holger Petersen

Institut für Informatik, Universität Stuttgart
Breitwiesenstr. 20–22, D-70565 Stuttgart
petersen@informatik.uni-stuttgart.de

Abstract. We show that the hierarchy of classes of languages accepted by finite multi-head automata with oblivious head movements that receive polynomial advice strings collapses to the fifth level. A characterization of nondeterministic logarithmic space with polynomial advice is simplified. In the presence of polynomial advice, the question whether deterministic and nondeterministic logarithmic space are equivalent can be reduced to the question whether simple nondeterministic automata can be simulated deterministically. Polynomial time can be characterized by a one-head device.

For automata without advice we prove that multi-head counter automata, stack automata, and non-erasing stack automata do not lose power by the restriction to oblivious head movements.

1 Introduction

The investigation of finite multi-head automata goes back at least to the work of Kosmidiadi and Marchenkov [7]. Already in this early reference a connection between multi-head automata and space complexity of Turing machine computations was established. The characterizations of the complexity classes

$$\mathbf{L} = \bigcup_{k \geq 1} \mathbf{DFA}(k), \qquad \mathbf{NL} = \bigcup_{k \geq 1} \mathbf{NFA}(k)$$

are well-known, see e.g. [3] (here $\mathbf{DFA}(k)$ ($\mathbf{NFA}(k)$) denotes the class of languages accepted by deterministic (nondeterministic) finite automata with k two-way input heads). Other complexity classes can be characterized with the help of two-way multi-head automata as well; as an example we mention that $\mathbf{P}$ is the class of languages accepted by deterministic or nondeterministic multi-head pushdown automata [2] or alternating finite multi-head automata [6]. See [11] for more equivalences between complexity classes and classes based on multi-head automata.

Holzer [4] defined oblivious (data-independent) finite multi-head automata, in order to obtain multi-head devices characterizing logspace uniform $\mathbf{NC}^1$, the class of languages accepted by logspace-uniform families of circuits with logarithmic depth and bounded fan-in. He also investigated the class of languages accepted by these devices if a polynomial advice in the sense of Karp and Lipton [5] is supplied to the automata. He left open whether the infinite union over all possible numbers of input heads collapses to a fixed level, as it does for non-oblivious automata. The main result of this paper gives a positive answer to this question.

Despite their apparent weakness, oblivious finite multi-head automata with a small number of heads are able to perform tasks like string matching, i.e., decide whether a pattern occurs in a text. The straightforward quadratic algorithm for a device with two heads can be adapted to the oblivious variant. Instead of aborting a comparison after the first mismatch, both heads move right until the one reading the text reaches the end-marker. Only then are the heads reset for the next comparison. A match is recorded in the finite control and the process is continued until all positions have been checked.

The present paper is organized as follows. We introduce informally the concepts and notation used. Then we show the collapse of the hierarchy of language classes accepted by oblivious finite multi-head automata with polynomial advice. This is a result analogous to Theorem 14 of [4] for non-oblivious automata. However, as pointed out in [4], a new technique has to be developed since the simulation of non-oblivious automata relies on the input being well-formed, i.e., composed of the input and a corresponding advice string. Checking this property in a straight-forward manner spoils the obliviousness that should hold for every input.

For finite multi-head automata Holzer could reduce the number of input heads to two in the presence of a polynomial advice. We show that this characterization of **NL**/*poly* by two-head finite automata can be improved to one-head bounded counter automata.

We give strengthened versions of the result from [4] linking simple automata to the relation between **L**/*poly* and **NL**/*poly*.

For two-way pushdown automata a single head simulates any finite number of heads in the analogous setting, thus leading to a characterization of **P**/*poly*.

In the concluding remarks of [4] other devices than multi-head finite automata are mentioned in connection with oblivious and non-oblivious computations. Here we investigate counter automata as well as some variants of stack automata and show that obliviousness is no restriction for these devices.

In a final section we discuss some unresolved questions.

2 Notation

We consider one-way and two-way devices equipped with a finite control and a finite number of input heads that can be moved independently. The read-only input tape is bordered by special symbols, the end-markers. These devices start their computation in a fixed initial state with their input heads next to the left end-marker and accept by final state. Each step depends on the internal state of the automaton and the symbols read by the heads. A counter automaton has in addition access to a counter that can be incremented, decremented, and tested for being zero. A stack is a sequential storage similar to a pushdown store with the additional option to move the stack pointer in a read-only mode into the stack contents. It is non-erasing if no symbol is ever removed once it has been written. For more detailed definitions see [11,4].

An automaton is called *oblivious* (data-independent) if the movements of its *input heads* depend on the length of the input only. We also require that all computations on words of a given length terminate after the same number of steps, either by accepting or by reaching a configuration that admits no further computation. Note that oblivious counter and stack automata may still modify their storage contents in an arbitrary manner.

The classes of languages accepted by k-head deterministic resp. nondeterministic two-way finite automata and deterministic resp. nondeterministic counter automata will be denoted by **DFA**(k), **NFA**(k), **DCA**(k), and **NCA**(k). If the counters are bounded by the input-length we call the respective classes **DBCA**(k) and **NBCA**(k). The class of languages accepted by k-head two-way deterministic pushdown automata is **DPDA**(k).

A prefix **Di** indicates that the heads move in an oblivious fashion.

Let C be a class of languages. Then $C/poly$ is the class of languages $L' = \{w \mid \alpha_{|w|}w \in L\}$ where $L \in C$ and $\alpha = (\alpha_i)$ is a sequence of polynomially length-bounded advice strings. Up to easy transformations of the input this definition is equivalent to the ones in [5,1].

3 Automata with Advice

The first result of this section shows that the "syntactic" hierarchy of language classes defined in terms of oblivious automata with an increasing number of heads receiving a polynomial advice collapses to a fixed level. The idea is to construct a simulator with a fixed number of heads for a given multi-head automaton that uses a modified advice in order to compensate for the lack of heads. Note however that it is not sufficient to design a simulator that works in an oblivious way on input strings containing the intended advice. Rather the simulator has to work in this way on every input string.

Our first goal is an arithmetical encoding of the positions of many heads with the help of a fixed number that does not depend on the contents, but only on the length of a given input string.

Lemma 1. *Let m be a number stored as the distance of the first head of a multi-head automaton to the right end-marker. Then the automaton can compute m^k as the distance of another head for every $k > 0$ such that m^k does not exceed the input length. At most three additional heads are required and m is still encoded by the first head.*

Proof. In order to preserve m the automaton uses a second head that moves in the opposite direction of the first one. Two further heads alternate in storing $1, m, m^2, \ldots$. Suppose m^i is currently stored. This number is repeatedly decremented and for every decrement operation m is added to the position of the head storing the next result with the help of the first two heads. Starting from 0 the resulting value is m^{i+1}. □

Theorem 1. *The infinite union $\bigcup_{k\geq 1} \mathbf{DiDFA}(k)/poly$ collapses to its fifth level,*

$$\bigcup_{k\geq 1} \mathbf{DiDFA}(k)/poly = \mathbf{DiDFA}(5)/poly.$$

Proof. Let $L \in \mathbf{DiDFA}(k)/poly$ be accepted by an oblivious k-head finite automaton A via a sequence of advice strings $\alpha = (\alpha_n)$, and let x be a fixed symbol from the alphabet of A. Define a new advice $\alpha' = (\alpha'_n)$ by $\alpha'_n = x^{(n+|\alpha_n|+2)^k - n - |\alpha_n|} \alpha_n$, i.e., the new advice incorporates the old one and adds padding symbols that will result in an input string that has length m^k, where m is the number of symbols accessible to A (including end-markers). Clearly α' is polynomially bounded.

We give an oblivious algorithm for a finite automaton B equipped with five heads receiving advice α' that simulates A with advice α. First B stores as the distance of its first head to the right end-marker $m = 1, 2, \ldots$ and in turn attempts to compute m^k according to the preceding lemma. If eventually m^k is equal to the input length for some m this first phase terminates and m is kept fixed. If during the computation B determines that the input-length is not a k-th power it rejects its input.

Now B starts a step-by-step simulation of A on a suffix of length m of B's input. The head positions of A can be expressed as k distances to the right end-marker that may vary between 0 and $n + |\alpha_n| + 1$. These distances will be encoded as a single k-digit, m-ary number p for $m = n + |\alpha_n| + 2$. Notice that the input for B has length m^k (assuming that the "matching" advice is supplied). This number p is stored on head 3 (heads 1 and 2 store m). Initially $p = \sum_{i=0}^{k-1}(n+|\alpha_n|)m^i$.

In order to simulate a single step of A the automaton B first determines the symbols scanned by the heads of A. It divides p by m, storing the result as the distance of head 4 and keeping the remainder r as the position of head 1. Now an input symbol can be read by head 1 and remembered in B's finite control. Then B copies r onto head 5 and adds rm^{k-1} to the number stored by head 4. It achieves this by moving head 3 over the entire input string (that has length m^k) and incrementing the number stored on head 4 for every m-th symbol (using heads 1 and 2). This sequence of operations is repeated r times using head 5 as a counter. Notice that the left end-marker will not be available at the proper position because of the padding. Therefore B in its simulation of A substitutes the left end-marker for every symbol read at distance $m - 1$, which will be determined with the help of the value m stored by heads 1 and 2.

By repeatedly computing the remainders and k times rotating the number stored on head 4, all symbols scanned by the heads of A can be determined. If the "correct" advice is presented to B it has assembled all the information necessary to simulate a step of A. By repeating the rotation process described above it can update the encoding of the head positions and then record the new internal state of A. The input is accepted by B if and only if A accepts.

It might happen that the input length is a k-th power although it has not been composed properly of the advice and a corresponding input. This will do no harm because B always simulates a computation of A on an input of length $m - 2$. Since A is oblivious the head movements are the same for every string of this length. If B's input length is not a k-th power it always rejects after the first (oblivious) stage. Therefore B is oblivious on every input. We have (A accepts $\alpha_{|w|}w$) if and only if (B accepts $\alpha'_{|w|}w = x^{(|w|+|\alpha_{|w|}|+2)^k - |w| - |\alpha_{|w|}|}\alpha_{|w|}w$). □

With the equality $\mathbf{L}/poly = \mathbf{DFA}(2)/poly$ from [4] we obtain:

Corollary 1.

$$\mathbf{L}/poly = \bigcup_{k \geq 1} \mathbf{DiDFA}(k)/poly \quad \text{iff} \quad \mathbf{DFA}(2)/poly = \mathbf{DiDFA}(5)/poly.$$

Turning now to non-oblivious automata, we improve the characterization

$$\mathbf{NL}/poly = \mathbf{NFA}(2)/poly$$

(the nondeterministic analogue of Theorem 14 in [4]) from two-head finite automata to one-head counter automata, where the counter is bounded by the input length. A bounded counter can easily be simulated with the help of a two-way input head.

Theorem 2. $\mathbf{NL}/poly = \mathbf{NBCA}(1)/poly$.

Proof. By the equality $\mathbf{NL}/poly = \mathbf{NFA}(2)/poly$ it suffices to give a simulation of a nondeterministic two-head automaton A with polynomial advice by a nondeterministic bounded counter automaton B. Let A's advice be $\alpha = (\alpha_n)$ and introduce a new symbol $\#$. We design an advice $\beta = (\beta_n)$ by letting $m = n + |\alpha_n| + 2$ and set

$$\beta_n = $$
$$0^{(m-1)m+(m-1)}10^{(m-1)m^2+(m-1)m+(m-1)}1$$
$$0^{(m-1)m^2+(m-1)m+(m-1)}10^{(m-1)m^2+(m-1)m}1\ldots$$
$$\ldots 0^{im+j}10^{im^2+jm+i}10^{im^2+jm+j}10^{im^2+jm}1\ldots$$
$$\ldots 0^{0m+0}10^{0m^2+0m+0}10^{0m^2+0m+0}10^{0m^2+0m}1\#\alpha_n,$$

where i and j run through $(m-1)\ldots 0$.

If there are two or more symbols $\#$ in its input string B rejects. Therefore words in the language defined will not contain $\#$.

Now we describe the step-by-step simulation of A by B. If the distances of A's heads from the right end-marker are i and j respectively, B's counter will store the number $im + j$. The blocks of 0's in B's advice are arranged in groups of four. Counting modulo four starting at the left end-marker, B locates a group such that the length of the first block matches the counter contents.

From this group B copies the length of the second block onto the counter, which now stores $im^2 + jm + i$, and moves its input head to the right end-marker. Decrementing the counter for every step of the input head, B moves its head nondeterministically on position i and remembers the input symbol read by A's head. The head is never moved over $\#$ in these operations, thus bounding the number subtracted from the counter by $m - 1$. Then B nondeterministically returns to the group and checks that the counter contents agree with the fourth block. This comparison verifies that the correct position has been read by the input head and that the initial group has been reached. The length of the third block in the group is copied onto the counter and the process is repeated for the second head of A. After restoring the original counter contents the encoded head positions are updated according to the state transition by adding or subtracting one or m respectively. The last operation can be carried out with the help of the input head. If A enters an accepting state, B accepts as well. None of the phases of the simulation stores a number exceeding the input length on the counter. □

From the characterization in [4] follows, that the classes of languages accepted by deterministic and nondeterministic Turing machines with logarithmic space bound and polynomial advice coincide if and only if a corresponding inclusion relation holds for two-way finite automata with two heads, $\mathbf{NL}/poly = \mathbf{L}/poly$ if and only if $\mathbf{NFA}(2)/poly \subseteq \mathbf{DFA}(2)/poly$. With the previous simulation we obtain a strengthened version of Corollary 15 in [4].

Corollary 2.

$\mathbf{NL}/poly = \mathbf{L}/poly$ *if and only if* $\mathbf{NBCA}(1)/poly = \mathbf{DFA}(2)/poly$.

In the case of automata without advice it is known that the equivalence between determinism and nondeterminism can be reduced to a question about one-way automata, $\mathbf{NL} = \mathbf{L}$ if and only if $\mathbf{1NFA}(2) \subseteq \mathbf{DFA}(k)$ for some k [10]. We observe that an analogous relation holds for automata with advice, thus giving another variant of Corollary 15 from [4].

Proposition 1.

$\mathbf{NL}/poly = \mathbf{L}/poly$ *if and only if* $\mathbf{1NFA}(2)/poly \subseteq \mathbf{DFA}(2)/poly$.

With the same technique as in [4], two-head alternating automata [6] characterize languages accepted with a polynomial time bound in the presence of a polynomial advice. The next result gives a characterization of this class $\mathbf{P}/poly$ using a pushdown automaton with a single head.

Theorem 3. $\mathbf{DPDA}(1)/poly = \mathbf{P}/poly$.

Proof (sketch). The linear time simulation of deterministic one-head pushdown automata on the RAM shows that any language in $\mathbf{DPDA}(1)/poly$ is in $\mathbf{P}/poly$ as well.

Conversely, $\mathbf{P}/poly$ can be characterized as the class of languages having small circuits, i.e., families of polynomial size circuits. The circuit value problem can be decided by a deterministic two-way pushdown automaton, see [8]. The advice for input length n is simply the encoding of the circuit for length n, where the separation of the block of input symbols from the code of the circuit does not pose a problem for the pushdown automaton. □

The preceding proof requires a substantial transformation of a given advice when going from a polynomial time Turing machine to a two-way pushdown automaton. The following brief discussion shows that by exploiting the power of pushdown automata a simple padding actually suffices. Monien [9] has shown that any set accepted by a successor RAM (the set of instructions includes increment by one, a comparison with zero, and memory transfer operations with indirect addressing) in time bound $t(n)$ can be accepted by a counter pushdown automaton with a counter bounded by $t(n)\log^2 t(n)$, provided that the bound $t(n)$ is constructible by the pushdown automaton. A counter pushdown automaton is a deterministic two-way device with a pushdown storage and a single input head that may read blank symbols beyond the end of its input string. If the counter is bounded by $s(n)$ it may access at most $s(n)$ cells on its input tape. Clearly it does not matter whether the blank symbols are read before or after the input string. By providing a sufficient number of blank symbols before the original advice, a new advice that admits a simulation of any polynomially time bounded successor RAM (and therefore of any polynomially time bounded Turing machine) can be constructed.

4 Oblivious Automata Without Advice

Holzer [4] suggests to study oblivious counter machines or stack automata. We show here that obliviousness or data-independence is no restriction for these multi-head devices if the storage may be used freely.

Lemma 2. *Every k-head finite automaton can be simulated by a $2k$-head oblivious counter automaton,* **NFA**$(k) \subseteq$ **DiNCA**$(2k)$ *and* **DFA**$(k) \subseteq$ **DiDCA**$(2k)$.

Proof. Note that for a k-head automaton the length of a shortest accepting computation can be bounded by a polynomial $p(n)$ in $O(n^k)$. Add a second set of heads and modify the finite control in such a way that the automaton stops after exactly $p(n)$ steps for every input word. The resulting finite automaton will be simulated by an oblivious counter automaton that keeps an encoding of all head positions as an $(n+2)$-ary number on its counter, similar to the proof of Theorem 1. It uses its $2k$ input heads to cycle through all possible combinations of head positions and decrement the counter in every step. When the counter reaches zero, the automaton records the symbols read by its heads. Then it continues to cycle through the head positions until it has exhausted all possibilities, now incrementing the counter. It repeats this process, interchanging increment and decrement, thus recovering the initial counter contents. In order to simulate a step of the finite automaton the counter automaton determines the next state from the previous one and the symbols read. Here the mode depends on the simulated automaton being deterministic or nondeterministic. In order to adjust the encoded head positions the automaton repeatedly cycles through head configurations generating intervals of length $(n+2)^i$ for all $0 \leq i < 2k$. The encoded positions are updated by either incrementing or decrementing the counter along with each step in the interval. The head movements are the same for every encoding and thus the counter automaton is oblivious. □

Theorem 4. *Every deterministic or nondeterministic multi-head one-counter automaton can be simulated by an oblivious device of the same kind.*

Proof. First we notice that every k-head counter automaton can be simulated by a $2k$-head finite automaton in the deterministic case and by a $3k$-head finite automaton in the nondeterministic case [11, Theorem 13.8]. Now apply the preceding lemma to the resulting finite automata to prove the claim. □

Theorem 5. *Every non-erasing multi-head stack automaton is equivalent to an oblivious device of the same kind.*

Proof (sketch). The class characterized by non-erasing multi-head stack automata in their nondeterministic as well as in their deterministic variant is **PSPACE** [11, Theorem 13.29]. It thus suffices to describe the oblivious simulation of a $p(n)$ space-bounded deterministic Turing machine M by a non-erasing multi-head stack automaton A, where p is a polynomial. Without loss of generality we assume that M's computations on input strings of length n have the same length. We equip A with a number of heads that suffices to count up to $p(n)$. Initially A writes the first configuration of M on the stack, adds $p(n) - n$ blanks and a marker symbol. Then it repeatedly generates the successor configuration terminated by the marker-symbol from the old configuration until M either accepts or rejects.

We note that the proof of [11, Theorem 13.29] uses the input heads to store information during the copying process and thus does not give an oblivious simulation. □

Theorem 6. *Every multi-head stack automaton is equivalent to an oblivious device of the same kind.*

Proof (sketch). Multi-head stack automata characterize the time complexity class **DTIME**(2^{Pol}) [11, Theorem 13.35]. We describe an oblivious stack automaton A simulating a deterministic Turing machine M that accepts in $O(2^{p(n)})$ steps for some polynomial p. Without loss of generality we require M to be a single tape Turing machine such that computations on inputs of size n have identical length. Analogous to the proof of Theorem 13.20.3 of [11] the head movements of M can be normalized with quadratic overhead to facilitate the calculation of M's head position. Stack automaton A makes use of its heads to copy, compare, and modify binary strings of polynomial length at the top of its stack. In this way A computes tuples encoding M's state, tape contents, and head position in a similar way as the pushdown-automaton in the proof mentioned above, using binary instead of unary encoding of numbers. By the normalization A works in an oblivious way. □

5 Open Problems

Unlike the situation for finite multi-head automata with advice, oblivious or non-oblivious, we do not have a hierarchy or collapse result for nonuniform finite multi-head automata as defined in [4]. In the case of our collapse result for oblivious automata the number of heads is not known to be optimal and, using refined simulation techniques, it seems possible to reduce this number. The lower bound is two since, as pointed out in [4], one-head automata are not sufficient for this type of simulation. In the case of nondeterministic automata the characterization of **NL**/*poly* has been simplified to counter automata, but the technique does not carry over to deterministic machines. Does some other simulation give an analogous result, or is this an inherent weakness of deterministic devices?

Acknowledgement. I wish to thank Markus Holzer for several helpful discussions.

References

1. J. L. Balcázar, J. Díaz, and J. Gabarró. *Structural Complexity I*, volume 11 of *EATCS Monographs on Theoretical Computer Science*. Springer, Berlin-Heidelberg-New York, 1988.
2. S. A. Cook. Characterizations of pushdown machines in terms of time-bounded computers. *Journal of the Association for Computing Machinery*, 18:4–18, 1971.
3. J. Hartmanis. On non-determinancy in simple computing devices. *Acta Informatica*, 1:336–344, 1972.
4. M. Holzer. Multi-head finite automata: Data-independent versus data-dependent computations. In I. Prívara and P. Ružička, editors, *Proceedings of the 22nd Symposium on Mathematical Foundations of Computer Science (MFCS), Bratislava, 1997*, number 1295 in Lecture Notes in Computer Science, pages 299–308. Springer, 1997.
5. R. M. Karp and R. J. Lipton. Turing machines that take advice. *L'Enseignement Mathématique*, 28:191–209, 1982.
6. K. N. King. Alternating multihead finite automata. *Theoretical Computer Science*, 61:149–174, 1988.
7. V. A. Kosmidiadi and S. S. Marchenkov. On multihead automata. *Systems Theory Research*, 21:124–156, 1971. Translation of Probl. Kib. 21:127–158, 1969, in Russian.
8. R. E. Ladner. The Circuit Value Problem is log space complete for P. *SIGACT News (ACM Special Interest Group on Automata and Computability Theory)*, 7:18–20, 1975.
9. B. Monien. Characterizations of time-bounded computations by limited primitive recursion. In *Proceedings of the 2nd International Colloquium on Automata, Languages and Programming (ICALP), Saarbrücken, 1974*, number 14 in Lecture Notes in Computer Science, pages 280–293. Springer, 1974.
10. I. H. Sudborough. On tape-bounded complexity classes and multihead finite automata. *Journal of Computer and System Sciences*, 10:62–76, 1975.
11. K. Wagner and G. Wechsung. *Computational Complexity*. Mathematics and its Applications. D. Reidel Publishing Company, Dordrecht, 1986.

The Equivalence Problem for Deterministic Pushdown Transducers into Abelian Groups

Géraud Sénizergues

LaBRI
Université de Bordeaux I
351, Cours de la Libération 33405 Talence, France
ges@labri.u-bordeaux.fr; fax: 05-56-84-66-69
http://www.labri.u-bordeaux.fr/~ges

Abstract. The equivalence problem for deterministic pushdown transducers with inputs in a free monoid X^* and outputs in an abelian group H is shown to be *decidable*. The result is obtained by constructing a *complete formal system* for equivalent pairs of deterministic rational series on the variable alphabet associated with the dpdt $\mathcal{M}$ with coefficients in the monoid H^0 (the monoid obtained by adjoining a zero to the group H).

1 Introduction

We show here that, given two deterministic pushdown *transducers* (dpdt's for short) A, B from a free monoid X^* into an abelian group H, one can decide whether $S(A) = S(B)$ or not (i.e. whether A, B compute the same function $f : X^* \to H$). This result generalizes the decidability of the equivalence problem for deterministic pushdown *automata* ([13]) and can be also considered as a step towards the solution of the equivalence problem for dpdt's from a free monoid X^* into another free monoid Y^*. This last problem has been addressed in [9,10,16] and remains open (see section 6 for other related problems).
Our solution leans on the methods developed in [13] and our exposition will often refer the reader to this article. Complete proofs can be found in [14, section 11, p.108-143], an example is treated in [14, section 12, p.153-158].

2 Preliminaries

2.1 Semi-Rings

The reader is refered to [1] for formal power series. We just review here some basic vocabulary and properties.

Semi-Ring $\mathsf{K}\langle\langle\ W\ \rangle\rangle$ Let us consider a semi-ring $(\mathsf{K}, +, \cdot, 0_K, 1_K)$ and an alphabet W. By $(\mathsf{K}\langle\langle\ W\ \rangle\rangle, +, \cdot, \emptyset, \epsilon)$ we denote the semi-ring of *series* over the set of non-commutative undeterminates W, with coefficients in K. The sum and product are defined as usual.

A map $\psi : K\langle\langle\ W\ \rangle\rangle \to K\langle\langle\ W'\rangle\rangle$ which is a semi-ring homomorphism, a σ-additive map and which fixes every element of K, will be called a *substitution*. The *support* of S is the language: $\mathrm{supp}(S) = \{w \in W^* \mid S_w \neq 0_K\}$.

Semi-Ring B$\langle\langle\ H\ \rangle\rangle$ Let $(\mathsf{B}, +, \cdot, 0, 1)$ where $\mathsf{B} = \{0,1\}$ denote the semi-ring of "booleans" and let $(H, \cdot)$ be a group. By $\mathsf{B}\langle\langle\ H\ \rangle\rangle$ we denote the semi-ring of formal power series with undeterminates in the group H and coefficients in B. It is isomorphic with the semi-ring of subsets of H.

2.2 Automata

Finite H-Automata Let $(H, \cdot)$ be some group. We call a finite H-automaton over the alphabet W any 5-tuple

$$\mathcal{M} = <W, Q, \delta, h_0, q_0, Q'>$$

such that Q is the finite set of states, δ, the set of transitions, is a finite subset of $Q \times H \times W \times Q$, $h_0 \in H$, $q_0 \in Q$ and $Q' \subseteq Q$. As H is embedded in the semi-ring $K = \mathsf{B}\langle\langle\ H\ \rangle\rangle$ such an automaton can be seen as a finite automaton with multiplicities in K and the series recognized by $\mathcal{M}$, $\mathrm{S}(\mathcal{M})$, is defined as usual. It can be defined, for example, by $\mathrm{S}(\mathcal{M}) = h_0 \cdot A \cdot B^* \cdot C$, where $A \in K_{1,Q}\langle\langle\ W\ \rangle\rangle, B \in K_{Q,Q}\langle\langle\ W\ \rangle\rangle$, and $C \in K_{Q,1}\langle\langle\ W\ \rangle\rangle$ are given by: $A_{q_0} = \epsilon$, $A_q = \emptyset$ (for $q \in Q - \{q_0\}$), $B_{q,q'} = \sum_{(q,h,v,q')\in\delta} h \cdot v$, $C_{q,1} = \emptyset$ (if $q \notin Q'$), $C_{q,1} = \epsilon$ (if $q \in Q'$). $\mathcal{M}$ is said W-*deterministic* iff,

$$\forall q \in Q, \forall v \in W, \mathrm{Card}(\{(r, h, v, r') \in \delta | q = r\}) \leq 1. \tag{1}$$

Finite m-H-Automata Let $n, m \in \mathbb{N} - \{0\}$ be positive integers. By $\mathsf{B}\langle\langle\ H\ \rangle\rangle_{n,m}\langle\langle\ W\ \rangle\rangle$ we denote the set of matrices of dimension (n, m) with entries in the semi-ring $\mathsf{B}\langle\langle\ H\ \rangle\rangle\langle\langle\ W\ \rangle\rangle$. We call a finite m-H-automaton over the alphabet W any 5-tuple $\mathcal{M} = <W, Q, \delta, h_0, q_0, (Q_j)_{1\leq j\leq m}>$ such that $<W, Q, \delta, h_0, q_0, Q>$ is a finite H-automaton and for every $j \in [1, m]$, $Q_j \subseteq Q$. For every $j \in [1, m]$ we denote by $\mathcal{M}_j$ the finite H-automaton $\mathcal{M}_j = <W, Q, \delta, h_0, q_0, Q_j>$. The vector recognized by $\mathcal{M}$, $\mathrm{S}(\mathcal{M})$, is defined by:

$$\mathrm{S}(\mathcal{M}) = (\mathrm{S}(\mathcal{M}_1), \ldots, \mathrm{S}(\mathcal{M}_j), \ldots, \mathrm{S}(\mathcal{M}_m)).$$

$\mathcal{M}$ is said W-*deterministic* iff it fulfills the above condition (1).

Pushdown H-Automata We call a *pushdown H-automaton* on the alphabet X any 6-tuple

$$\mathcal{M} = <X, Z, Q, \delta, q_0, z_0>$$

where Z is the finite stack-alphabet, Q is the finite set of states, $q_0 \in Q$ is the initial state, z_0 is the initial stack-symbol and $\delta : QZ \times (X \cup \{\epsilon\}) \to \mathcal{P}_f(H \times QZ^*)$, is the transition mapping. Let $q, q' \in Q, \omega, \omega' \in Z^*, z \in Z, h \in H, u \in X^*$ and

$a \in X \cup \{\epsilon\}$; we note $(qz\omega, h, au) \longmapsto_{\mathcal{M}} (q'\omega'\omega, h \cdot h', u)$ if $(h', q'\omega') \in \delta(qz, a)$.
$\overset{*}{\longmapsto}_{\mathcal{M}}$ is the reflexive and transitive closure of $\longmapsto_{\mathcal{M}}$. For every $q\omega, q'\omega' \in QZ^*$
and $h \in H, u \in X^*$, we note $q\omega \overset{(h,u)}{\longrightarrow}_{\mathcal{M}} q'\omega'$ iff $(q\omega, 1_H, u) \overset{*}{\longmapsto}_{\mathcal{M}} (q'\omega', h, \epsilon)$. $\mathcal{M}$
is said *deterministic* iff for every $z \in Z, q \in Q$:

$$\text{either Card}(\delta(qz, \epsilon)) = 1 \text{ and for every } x \in X, \text{Card}(\delta(qz, x)) = 0, \quad (2)$$

$$\text{or Card}(\delta(qz, \epsilon)) = 0 \text{ and for every } x \in X, \text{Card}(\delta(qz, x)) \leq 1. \quad (3)$$

The mode qz is said ϵ-bound (resp. ϵ-free) when condition (2) (resp. (3)) is true.
A H-dpda $\mathcal{M}$ is said *normalized* iff, for every $q \in Q, z \in Z, x \in X$:

$$q'\omega' \in \delta_2(qz, x) \Rightarrow |\omega'| \leq 2, \text{ and } q'\omega' \in \delta_2(qz, \epsilon) \Rightarrow |\omega'| = 0, \quad (4)$$

where $\delta_2 : QZ \times (X \cup \{\epsilon\}) \to \mathcal{P}_f(QZ^*)$, is the second component of the map δ. Given some finite set $F \subseteq QZ^*$ of configurations, the *series recognized by* $\mathcal{M}$ *with final configurations* F is defined by

$$S(\mathcal{M}, F) = \sum_{c \in F} \sum_{q_0 z_0 \overset{hu}{\longrightarrow}_{\mathcal{M}} c} h \cdot u.$$

One can see the coefficient $S_u \in \mathbb{B}\langle\!\langle\ H\ \rangle\!\rangle$ of a word $u \in X^*$ in the series $S(\mathcal{M}, F)$ either as the "multiplicity" with which the word u is recognized , or as the "output" of the automaton $\mathcal{M}$ on the "input" u.
We suppose that Z contains a special symbol e subject to the property:

$$\forall q \in Q, \delta(qe, \epsilon) = \{(1_H, q)\} \text{ and im}(\delta_2) \subseteq \mathcal{P}_f(Q(Z - \{e\})^*). \quad (5)$$

2.3 Free Monoids Acting on Semi-Rings

Actions of Monoids The general notions of right-action and σ-right-action of a monoid over a semi-ring is the same as in [13, §2.3.2].

The Action of $H \times W^*$ on $\mathbb{B}\langle\!\langle\ H\ \rangle\!\rangle\langle\!\langle\ W\ \rangle\!\rangle$ A σ-right-action of the monoid $H \times W^*$ over $\mathbb{B}\langle\!\langle\ H\ \rangle\!\rangle\langle\!\langle\ W\ \rangle\!\rangle$ is defined by: $\forall S \in \mathbb{B}\langle\!\langle\ H\ \rangle\!\rangle\langle\!\langle\ W\ \rangle\!\rangle, \forall h \in H, \forall w \in W^*, T = S \bullet (h, w)$ is the series:

$$\forall v \in W^*, T_v = h^{-1} \cdot S_{w \cdot v}.$$

In words, $S \bullet (h, w)$ is the left-quotient of S by the monomial $h \cdot w$. (From now on, we identify the pair $(h, w) \in H \times W^*$ with the monomial $h \cdot w \in \mathbb{B}\langle\!\langle\ H\ \rangle\!\rangle\langle\!\langle\ W\ \rangle\!\rangle$).

The Action of $H \times X^*$ on $\mathbb{B}\langle\!\langle\ H\ \rangle\!\rangle\langle\!\langle\ V\ \rangle\!\rangle$ Let $\mathcal{M}$ be some H-dpda (for sake of simplicity , we suppose here that $\mathcal{M}$ is normalized). The *variable* alphabet $V_{\mathcal{M}}$ associated with $\mathcal{M}$ is defined as: $V_{\mathcal{M}} = \{[p, z, q] | p, q \in Q, z \in Z\}$ (from now on, we abbreviate $V_{\mathcal{M}}$ by just V). Let us consider the set $P_{\mathcal{M}}$ of all the pairs of one of the following forms:

$$([p,z,q], h \cdot x \cdot [p',z_1,p''][p'',z_2,q]) \text{ or } ([p,z,q], h \cdot x \cdot [p',z',q]) \text{ or } ([p,z,q], h \cdot a) \quad (6)$$

where $p, q, p', p'' \in Q, x \in X, a \in X \cup \{\epsilon\}, (h, p'z_1z_2) \in \delta(pz, x), (h, p'z') \in \delta(pz, x), (h, q) \in \delta(pz, a)$.

We define a σ-right-action $\otimes$ of the monoid $H \times (X \cup \{\epsilon\})^*$ over the semi-ring $(\mathsf{B}\langle\!\langle\ H\ \rangle\!\rangle)\langle\!\langle\ V\ \rangle\!\rangle$ by: for every $p, q \in Q, z \in Z, x \in X, h \in H, k \in \mathsf{B}\langle\!\langle\ H\ \rangle\!\rangle$:

$$[p,z,q] \otimes x = (\sum_{([p,z,q],m) \in P_{\mathcal{M}}} m) \bullet (1_H, x), \quad [p,z,q] \otimes e = h \text{ iff } ([p,z,q], h) \in P_{\mathcal{M}},$$
(7)

$$[p,z,q] \otimes e = \emptyset \text{ iff } (\{[p,z,q]\} \times H \cdot V^*) \cap P_{\mathcal{M}} = \emptyset, \quad (8)$$

$$k \otimes x = \emptyset, \quad k \otimes e = \emptyset. \quad (9)$$

The action is extended to all monomials by: for every $k \in \mathsf{B}\langle\!\langle\ H\ \rangle\!\rangle, \beta \in V^*, y \in X \cup \{\epsilon\}, S \in \mathsf{B}\langle\!\langle\ H\ \rangle\!\rangle\langle\!\langle\ V\ \rangle\!\rangle, h \in H$,

$$(k \cdot [p,z,q] \cdot \beta) \otimes y = k \cdot ([p,z,q] \otimes y) \cdot \beta, \quad S \otimes h = h^{-1} \cdot S. \quad (10)$$

Action $\odot$ We define a map $\rho_\epsilon : \mathsf{B}\langle\!\langle\ H\ \rangle\!\rangle\langle\!\langle\ V\ \rangle\!\rangle \to \mathsf{B}\langle\!\langle\ H\ \rangle\!\rangle\langle\!\langle\ V\ \rangle\!\rangle$ as the unique σ-additive map such that,

$$\rho_\epsilon(\emptyset) = \emptyset, \quad \rho_\epsilon(\epsilon) = \epsilon,$$

and for every $p \in Q, z \in Z, q \in Q, \beta \in V^*, k \in \mathsf{B}\langle\!\langle\ H\ \rangle\!\rangle, S \in \mathsf{B}\langle\!\langle\ H\ \rangle\!\rangle\langle\!\langle\ V\ \rangle\!\rangle$,

$$\rho_\epsilon([p,z,q] \cdot \beta) = \rho_\epsilon(([p,z,q] \otimes e) \cdot \beta) \text{ if } pz \text{ is } \epsilon - \text{bound},$$

$$\rho_\epsilon([p,z,q] \cdot \beta) = [p,z,q] \cdot \beta \text{ if } pz \text{ is } \epsilon - \text{free},$$

$$\rho_\epsilon(k \cdot S) = k \cdot \rho_\epsilon(S).$$

The right-action $\odot$ of the monoid $H \times X^*$ over the semi-ring $\mathsf{B}\langle\!\langle\ H\ \rangle\!\rangle\langle\!\langle\ V\ \rangle\!\rangle$ is then the unique monoid-action fulfilling: for every $S \in \mathsf{B}\langle\!\langle\ H\ \rangle\!\rangle\langle\!\langle\ V\ \rangle\!\rangle, h \in H, x \in X$,

$$S \odot (h, x) = \rho_\epsilon(\rho_\epsilon(S) \otimes (h, x)).$$

Case where H is abelian Let us consider the case where H is abelian. Let $\varphi : \mathsf{B}\langle\!\langle\ H\ \rangle\!\rangle \cup V \to \mathsf{B}\langle\!\langle\ H\ \rangle\!\rangle\langle\!\langle\ X\ \rangle\!\rangle$ defined by:

$$\forall k \in \mathsf{B}\langle\!\langle\ H\ \rangle\!\rangle, \varphi(k) = k; \quad \forall v \in V, \varphi(v) = \sum_{v \odot (h \cdot u) = \epsilon} h \cdot u.$$

One can check that, as H is supposed abelian, there exists a unique σ-additive semi-ring homomorphism $\tilde{\varphi} : \mathsf{B}\langle\!\langle\ H\ \rangle\!\rangle\langle\!\langle\ V\ \rangle\!\rangle \to \mathsf{B}\langle\!\langle\ H\ \rangle\!\rangle\langle\!\langle\ X\ \rangle\!\rangle$ which extends φ. Let us denote by the same letter the original φ and its extension $\tilde{\varphi}$.

Lemma 21 *For every $S \in \mathsf{B}\langle\!\langle\ H\ \rangle\!\rangle\langle\!\langle\ V\ \rangle\!\rangle, h \in H, u \in X^*$,*

1. $\varphi(S) = \varphi(\rho_\epsilon(S))$,
2. $\varphi(S \odot (h, u)) = \varphi(S) \bullet (h, u)$ *(i.e. φ is a morphism of right-actions).*

We denote by $\equiv$ the kernel of φ i.e.: for every $S, T \in \mathsf{B}\langle\!\langle\ H\ \rangle\!\rangle\langle\!\langle\ V\ \rangle\!\rangle, S \equiv T \Leftrightarrow \varphi(S) = \varphi(T)$.

2.4 Length-Functions

Let us suppose now that H admits a presentation over a finite alphabet $\hat{Y}$: $\varphi_H :\hat{Y}^* \to H$ is a surjective monoid-homomorphism. We suppose the presentation φ_H is "symmetric" in the sense that $\hat{Y} = Y \cup \bar{Y}$, $Y \cap \bar{Y} = \emptyset$ and for every $y \in Y$ (resp. $y \in \bar{Y}$), there exists a unique $\bar{y} \in \bar{Y}$ (resp. $\bar{y} \in Y$) such that $\varphi_H(y \cdot \bar{y}) = 1_H$. For every $h \in H$, the *length* of h, relative to the presentation φ_H, is defined by:

$$\ell(h) = \min\{|u| \mid u \in \hat{Y}^*, \varphi_H(u) = h\}.$$

One can notice that the map $(h, h') \mapsto \ell(h^{-1} \cdot h')$ is a distance over H. In the case of the free group $\mathsf{F}(W)$ with basis W, by $\ell(*)$ we denote the length-function associated with the standard presentation of $\mathsf{F}(W)$ over the set of generators $W \cup W^{-1}$.

3 Deterministic Series, Vectors, and Matrices

3.1 Determinism

Let us fix a group $(H, \cdot)$ and a structured alphabet $(W, \smile)$. (We recall it just means that $\smile$ is an equivalence relation over the set W).

W-Deterministic Rational Matrices For every $n, m \geq 1$ we define an equivalence relation $\sim$ over $\mathsf{B}\langle\!\langle\, H\, \rangle\!\rangle_{1,m}\langle\!\langle\, W\, \rangle\!\rangle$ by: $S \sim T \Leftrightarrow \exists h \in H, S = h \cdot T$. The right-action $\bullet$ is extended componentwise to $\mathsf{B}\langle\!\langle\, H\, \rangle\!\rangle_{n,m}\langle\!\langle\, W\, \rangle\!\rangle$ by: for every $S \in \mathsf{B}\langle\!\langle\, H\, \rangle\!\rangle_{n,m}\langle\!\langle\, W\, \rangle\!\rangle, h \in H, u \in W^*, (S \bullet (h, u))_{i,j} = S_{i,j} \bullet (h, u)$. For every $S \in \mathsf{B}\langle\!\langle\, H\, \rangle\!\rangle_{n,m}\langle\!\langle\, W\, \rangle\!\rangle$ we define the set of residuals of S, $\mathsf{Q}(S)$ and the set of row-residuals of S, $\mathsf{Q}_r(S)$, by:

$$\mathsf{Q}(S) = \{S \bullet (h, u) \mid h \in H, u \in W^*\}, \quad \mathsf{Q}_r(S) = \cup_{1 \leq i \leq n} \mathsf{Q}(S_{i,*}).$$

Let us denote by $(\mathsf{H}^0, \cdot, 1_H)$ the submonoid of $(\mathsf{B}\langle\!\langle\, H\, \rangle\!\rangle, \cdot, 1_H)$ consisting of the empty series and all the singletons $\{h\}$ for $h \in H$. H^0 can be seen as the monoid obtained by "adjoining a zero" to the group H. We sometimes use the symbol 0 for the element $\emptyset \in \mathsf{H}^0$ and we identify every $h \in H$ with the corresponding $\{h\} \in \mathsf{H}^0$. By $\mathsf{H}^0\langle\!\langle\, W\, \rangle\!\rangle$ we denote the subset of series in $\mathsf{B}\langle\!\langle\, H\, \rangle\!\rangle\langle\!\langle\, W\, \rangle\!\rangle$ whose coefficients are all in H^0.

Proposition 31 *Let $m \geq 1$, $S \in \mathsf{B}\langle\!\langle\, H\, \rangle\!\rangle_{1,m}\langle\!\langle\, W\, \rangle\!\rangle$. The following properties are equivalent:*
(1) S is recognized by some W-deterministic finite m-H-automaton
(2) $\forall j \in [1, m], \forall u \in W^*, ((S_j)_u \in \mathsf{H}^0)$ and $\mathsf{Q}(S)/\sim$ is finite

This proposition has been established in [3, prop.4, p.93] in the case where $m = 1$ and H is a free-group. The extension to $m \geq 1$ and to *any* group H is straightforward.

Definition 32 *Let $S \in \mathsf{B}\langle\!\langle\, H\, \rangle\!\rangle_{1,m}\langle\!\langle\, W\, \rangle\!\rangle$. S is said W-deterministic rational iff it fulfills one of points (1)(2) of proposition 31.*

Length and Norm Let us consider a W-deterministic, finite, m-H-automaton $\mathcal{M} =< W, Q, \delta, h_0, q_0, (Q_j)_{1 \leq j \leq m} >$. We define the length of $\mathcal{M}$, $\bar{k}(\mathcal{M})$, the initial length of $\mathcal{M}$, $k_0(\mathcal{M})$ and the norm of $\mathcal{M}$, $\|\mathcal{M}\|$ as:

$$\bar{k}(\mathcal{M}) = \sup\{\ell(h) \mid \exists q \in Q, v \in W, r \in Q, (q, h, v, r) \in \delta\}; \quad k_0(\mathcal{M}) = \ell(h_0) \text{ and}$$

$$\|\mathcal{M}\| = \operatorname{Card}(Q).$$

Let us consider now a vector $S \in \mathsf{H}^0_{1,m}\langle\!\langle\ W\ \rangle\!\rangle$. We define the length of S, $\bar{\ell}(S)$, the initial length of S, $\ell_0(S)$, and the norm of S, $\|S\|$ by:

$$\bar{\ell}(S) = \inf\{\mu \in \mathsf{R}_+ \mid \forall i, j \in [1, m], \forall u, v \in W^*, S_{i,u} \neq 0 \Rightarrow$$

$$\ell((S_{i,u})^{-1} \cdot S_{j,v}) \leq \mu \cdot \ell(u^{-1} \cdot v)\},$$

$$\ell_0(S) = \ell_0(S_{u_0}), \|S\| = \operatorname{Card}(\mathsf{Q}(S)/\sim),$$

where $S_{j,u}$ denotes the coefficient of S_j on the word u, u_0 is the minimum word of $\cup_{j=1}^m \operatorname{supp}(S_j)$ and we define $\ell_0(\emptyset^m) = 0$.

Lemma 33 *For every W-deterministic rational vector $S \in \mathsf{B}\langle\!\langle\ H\ \rangle\!\rangle_{1,m}\langle\!\langle\ W\ \rangle\!\rangle$, there exists some m-W-dfa $\mathcal{M}$ such that $\mathsf{S}(\mathcal{M}) = S$ and $\bar{k}(\mathcal{M}) \leq 2 \cdot \bar{\ell}(S) \cdot \|S\|$, $k_0(\mathcal{M}) \leq \ell_0(S)$, $\|\mathcal{M}\| \leq \|S\|$.*

Let us consider now a matrix $S \in \mathsf{H}^0_{n,m}\langle\!\langle\ W\ \rangle\!\rangle$. We define the length of S, $\bar{\ell}(S)$, the initial length of S, $\ell_0(S)$, and the norm of S, $\|S\|$ by:

$$\bar{\ell}(S) = \max\{\bar{\ell}(S_{i,*}), 1 \leq i \leq n\}, \quad \ell_0(S) = \max\{\ell_0(S_{i,*}), 1 \leq i \leq n\}, \text{ and}$$

$$\|S\| = \operatorname{Card}(\mathsf{Q}_r(S)/\sim).$$

In general, $\bar{\ell}(S), \|S\|$ belong to $\mathbb{N} \cup \{\infty\}$ and $\ell_0(S)$ belongs to $\mathbb{N}$.

$\smile$-Determinism Let $n, m \geq 1, S \in \mathsf{B}_{n,m}\langle\!\langle\ W\ \rangle\!\rangle$. S is said $\smile$-*deterministic* iff it is *deterministic* in the sense of definition 3.5 of [13] (this definition was a straightforward extension of [7, definition 3.2 p.188]). A finite m-H-automaton $\mathcal{M} =< W, Q, \delta, h_0, q_0, (Q_j)_{1 \leq j \leq m} >$ will be said $\smile$-*deterministic* if and only if, for every $q \in Q, A, A' \in W, h, h' \in H, r, r' \in Q, j, j' \in [1, m]$:

$$((q, h, A, r) \in \delta \text{ and } (q, h', A', r') \in \delta) \Rightarrow A \smile A' \text{ and } j \neq j' \Rightarrow Q_j \cap Q_{j'} = \emptyset. \tag{11}$$

Deterministic Rational Matrices

Proposition 34 *Let $m \geq 1, S \in \mathsf{B}\langle\!\langle\ H\ \rangle\!\rangle_{1,m}\langle\!\langle\ W\ \rangle\!\rangle$. The following properties are equivalent:*
(1) S is W-deterministic rational and $\operatorname{supp}(S)$ is $\smile$-deterministic.
(2) $\forall j \in [1, m], \forall u \in W^*, S_{i,u} \in H^0$, $\mathsf{Q}(S)/\sim$ is finite and $\operatorname{supp}(S)$ is $\smile$-deterministic.
(3) S is recognized by some finite m-H-automaton which is both W-deterministic and $\smile$-deterministic.

Definition 35 Let $m \geq 1, S \in \mathsf{B}\langle\!\langle\ H\ \rangle\!\rangle_{1,m}\langle\!\langle\ W\ \rangle\!\rangle$. The vector S is said fully deterministic rational (deterministic rational, for short) iff it fulfills one of points (1)(2)(3) of proposition 34.

Definition 36 Let $n, m \geq 1, S \in \mathsf{B}\langle\!\langle\ H\ \rangle\!\rangle_{n,m}\langle\!\langle\ W\ \rangle\!\rangle$. The matrix S is said fully deterministic rational (deterministic rational, for short) iff every row-vector $S_{i,*}$, for $1 \leq i \leq n$, is fully deterministic rational.

We denote by $\mathsf{DRH}^0_{n,m}\langle\!\langle\ W\ \rangle\!\rangle$ the set of Deterministic Rational matrices of dimension (n, m), with coefficients in $\mathsf{B}\langle\!\langle\ H\ \rangle\!\rangle\langle\!\langle\ W\ \rangle\!\rangle$.

Ordering We define a partial ordering on $\mathsf{B}\langle\!\langle\ H\ \rangle\!\rangle\langle\!\langle\ W\ \rangle\!\rangle$ by: for every $S, T \in \mathsf{B}\langle\!\langle\ H\ \rangle\!\rangle\langle\!\langle\ W\ \rangle\!\rangle$, $S \sqsubseteq T \Leftrightarrow (\forall u \in W^*, S_u = 0 \text{ or } S_u = T_u)$. Given $S, T \in \mathsf{B}\langle\!\langle\ H\ \rangle\!\rangle\langle\!\langle\ W\ \rangle\!\rangle$ such that $S \sqsubseteq T$ we define $T - S \in \mathsf{B}\langle\!\langle\ H\ \rangle\!\rangle\langle\!\langle\ W\ \rangle\!\rangle$ by: $\forall u \in W^*, (T - S)_u = T_u(\text{ if } S_u = 0); (T - S)_u = 0(\text{ if } S_u = T_u)$.

3.2 Algebraic Properties

Let us fix now some *abelian* group $(H, \cdot)$, some normalized H-dpda $\mathcal{M}$ and consider the structured alphabet $(V, \smile)$ associated with $\mathcal{M}$. As B is embedded into $\mathsf{B}\langle\!\langle\ H\ \rangle\!\rangle$ the notations introduced in [13, §1.3.1] are still valid here. Some *new* statements concerning the functions $\bar{\ell}, \ell_0$ are introduced in the two next lemmas.

Lemma 37 Let $n, m, s \geq 1, S \in \mathsf{DH}^0_{n,m}\langle\!\langle\ V\ \rangle\!\rangle, T \in \mathsf{H}^0_{m,s}\langle\!\langle\ V\ \rangle\!\rangle$. Then
(1) $\bar{\ell}(S \cdot T) \leq \max\{\bar{\ell}(S), \bar{\ell}(T)\} + 2 \cdot \ell_0(T) + 2 \cdot \bar{\ell}(T) \cdot \|T\|$.
(2) $\ell_0(S \cdot T) \leq \ell_0(S) + \ell_0(T)$. (3) $\|S \cdot T\| \leq \|S\| + \|T\|$.

Lemma 38 Let $m \geq 1, S \in \mathsf{DH}^0_{1,m}\langle\!\langle\ V\ \rangle\!\rangle, u \in X^*$. Then
(1) $\bar{\ell}(S \odot u) \leq \bar{\ell}(S)$.
(2) $\ell_0(S \odot u) \leq \ell_0(S) + \|S\| \cdot \bar{k}(\mathcal{M}) \cdot |u| + |Q| \cdot \bar{k}(\mathcal{M}) \cdot |u|^2$.
(3) $\|S \odot u\| \leq \|S\| + |Q| \cdot |u|$.

Deterministic Spaces The notions of linear combination, d-space and generating set are defined as in [13, §3.2] except that B is replaced by H^0 everywhere.

Lemma 39 Let $S_1, \ldots, S_j, \ldots, S_m \in \mathsf{DRH}^0\langle\!\langle\ V\ \rangle\!\rangle$. The following are equivalent

1. $\exists \boldsymbol{\alpha}, \boldsymbol{\beta} \in \mathsf{DRH}^0_{1,m}\langle\!\langle\ V\ \rangle\!\rangle, \boldsymbol{\alpha} \not\equiv \boldsymbol{\beta}$, such that $\sum_{1 \leq j \leq m} \alpha_j \cdot S_j \equiv \sum_{1 \leq j \leq m} \beta_j \cdot S_j$,

2. $\exists j_0 \in [1, m], \exists \boldsymbol{\gamma} \in \mathsf{DRH}^0_{1,m}\langle\!\langle\ V\ \rangle\!\rangle, (\gamma_{j_0} \not\equiv \epsilon)$, such that $S_{j_0} \equiv \sum_{1 \leq j \leq m} \gamma_j \cdot S_j$,

3. $\exists j_0 \in [1, m], \exists \boldsymbol{\gamma}' \in \mathsf{DRH}^0_{1,m}\langle\!\langle\ V\ \rangle\!\rangle, \gamma'_{j_0} \equiv \emptyset$, such that $S_{j_0} \equiv \sum_{1 \leq j \leq m} \gamma'_j \cdot S_j$.

(This lemma generalizes [13, lemma 3.7] which generalized the idea of [12, lemma 11 p.589]).

4 Deduction System

4.1 General Systems

We use here a notion of *deduction system* which was inspired by [4]. The reader is referred to [13, section 4] for a precise definition of this notion and of the related notion of *strategy*.

4.2 The System $\mathcal{H}_0$

We define here a particular deduction system $\mathcal{H}_0$ "Taylored for the equivalence problem for H-dpda's". Given a fixed H-dpda $\mathcal{M}$ over the terminal alphabet X, we consider the variable alphabet V associated to $\mathcal{M}$ (see section 2.3) and the set $\mathsf{DRH}^0 \langle\!\langle\, V\, \rangle\!\rangle$ (the set of Deterministic Rational series over V^*, with coefficients in H^0). The set of assertions is defined by :

$$\mathcal{A} = \mathbb{N} \times \mathsf{DRH}^0 \langle\!\langle\, V\, \rangle\!\rangle \times \mathsf{DRH}^0 \langle\!\langle\, V\, \rangle\!\rangle$$

i.e. an assertion is here a *weighted equation* over $\mathsf{DRH}^0 \langle\!\langle\, V\, \rangle\!\rangle$. The "cost-function" $J : \mathcal{A} \to \mathbb{N} \cup \{\infty\}$ is defined by : $J(n, S, S') = n + 2 \cdot \mathrm{Div}(S, S')$, where $\mathrm{Div}(S, S')$, the *divergence* between S and S', is defined by: $\mathrm{Div}(S, S') = \inf\{|u|, (\varphi(S))_u \neq (\varphi(S'))_u\}$. (Notice that, $J(n, S, S') = \infty \iff S \equiv S'$).

We define a binary relation $\mid\!\!\vdash\;\; \subset \mathcal{P}_f(\mathcal{A}) \times \mathcal{A}$, the *elementary deduction relation*, as the set of all the pairs having one of the following forms:

$(H0)$ $\{(p, S, T)\}$ $\mid\!\!\vdash (p+1, S, T)$
$(H1)$ $\{(p, S, T)\}$ $\mid\!\!\vdash (p, T, S)$
$(H2)$ $\{(p, S, S'), (p, S', S'')\}$ $\mid\!\!\vdash (p, S, S'')$
$(H3)$ $\emptyset$ $\mid\!\!\vdash (0, S, S)$
$(H'3)$ $\emptyset$ $\mid\!\!\vdash (0, S, T)$ for $T \in \{\emptyset, \epsilon\}, S \equiv T$
$(H4)$ $\{(p+1, S \odot x, T \odot x) \mid x \in X\}$ $\mid\!\!\vdash (p, S, T)$
  where $(\forall h \in H, S \not\equiv h \wedge T \not\equiv h)$
$(H5)$ $\{(p, S, S')\}$ $\mid\!\!\vdash (p+2, S \odot x, S' \odot x)$
  for $x \in X$
$(H6)$ $\{(p, S_1 \cdot T + S_2, T)\}$ $\mid\!\!\vdash (p, S_1^* \cdot S_2, T)$
  where $(\forall h \in H, S_1 \not\equiv h)$
$(H7)$ $\{(p, S_1, T_1), (p, S_2, T_2)\}$ $\mid\!\!\vdash (p, S_1 + S_2, T_1 + T_2)$
$(H8)$ $\{(p, S, S')\}$ $\mid\!\!\vdash (p, S \cdot T, S' \cdot T)$
$(H9)$ $\{(p, T, T')\}$ $\mid\!\!\vdash (p, S \cdot T, S \cdot T')$
$(H10)$ $\emptyset$ $\mid\!\!\vdash (0, S, \rho_\epsilon(S))$
$(H11)$ $\emptyset$ $\mid\!\!\vdash (0, S, \rho_e(S))$,

where $p \in \mathbb{N}, S, S', T, T' \in \mathsf{DRH}^0 \langle\!\langle\, V\, \rangle\!\rangle, (S_1, S_2), (T_1, T_2) \in \mathsf{DRH}^0_{1,2} \langle\!\langle\, V\, \rangle\!\rangle$. The map ρ_ϵ involved in rule $(H10)$ was defined in §2.3 and we define the new map ρ_e involved in rule (H11) as the unique substitution $\mathsf{B}\langle\!\langle\ H\ \rangle\!\rangle\langle\!\langle\ V\ \rangle\!\rangle \to \mathsf{B}\langle\!\langle\ H\ \rangle\!\rangle\langle\!\langle\ V\ \rangle\!\rangle$ such that, for every $p, q \in Q, z \in Z$,

$$\rho_e([p,e,q]) = \emptyset(\text{ if } p \neq q), \rho_e([p,e,q]) = \epsilon(\text{ if } p = q), \rho_e([p,z,q]) = [p,z,q](\text{ if } z \neq e),$$

where e is the "dummy" symbol introduced in (5). ρ_e maps every $S \in \mathsf{DRH}^0\langle\!\langle\ V\ \rangle\!\rangle$ into an image $\rho_e(S) \in \mathsf{DRH}^0\langle\!\langle\ V\ \rangle\!\rangle$. A series $S \in \mathsf{DRH}^0\langle\!\langle\ V\ \rangle\!\rangle$ is called marked (resp. unmarked) iff its support has at least one occurence (resp. no occurence) of a letter of the form $[p, e, q]$. Let us define $\vdash\!\!-$ by : for every $P \in \mathcal{P}_f(\mathcal{A}), A \in \mathcal{A}$,

$$P \vdash\!\!- A \iff P \ \|\!\!\stackrel{<*>}{\vdash\!\!-}\ \circ\ \|\!\!\stackrel{[1]}{\vdash\!\!-}_{0,3,4,10,11} \circ\ \|\!\!\stackrel{<*>}{\vdash\!\!-}\ \{A\}.$$

where $\|\!\!\vdash\!\!-_{0,3,4,10,11}$ is the relation defined by $H0, H3, H'3, H4, H10, H11$ only. We let $\mathcal{H}_0 = <\mathcal{A}, J, \vdash\!\!->$.

Lemma 41 : $\mathcal{H}_0$ *is a deduction system.*

5 Strategies

5.1 Triangulations

Let us consider a sequence $\mathcal{S}$ of n "weighted" linear equations :

$$(\mathcal{E}_i) : p_i, \sum_{j=1}^{d} \alpha_{i,j} S_j\ ,\ \sum_{j=1}^{d} \beta_{i,j} S_j$$

where $p_i \in \mathbb{N}$, and $A = (\alpha_{i,j}), B = (\beta_{i,j})$ are deterministic rational matrices of dimension (n, d), with indices $m \leq i \leq m+n-1, 1 \leq j \leq d$. As in [13, section 5], we associate to such a system $\mathcal{S}$, another system of equations $\mathrm{INV}(\mathcal{S})$ and two integers $\mathrm{D}(\mathcal{S}), \mathrm{W}(\mathcal{S})$, which depend on the matrices A, B only (and, essentially, not on the series $S_1, S_2, \ldots, S_d$).

5.2 Constants

Let us fix a normalized H-dpda $\mathcal{M}$ and an initial equation

$$A_0 = (\Pi_0, S_0^-, S_0^+) \in \mathbb{N} \times \mathsf{DRH}^0\langle\!\langle\ V\ \rangle\!\rangle \times \mathsf{DRH}^0\langle\!\langle\ V\ \rangle\!\rangle.$$

Some *constants*, i.e. integers depending on $(\mathcal{M}, A_0)$ only, are defined. The integers $k_0, k_1, D_1, k_2, K_1, K_2, \bar{K}_3, K_3^0, K_3, \bar{K}_4, K_4^0, K_4, d_0, D_2, N_0$ and the sequences $(\delta_i, \ell_i, L_i, \bar{s}_i, s_i^0, s_i, S_i, \Sigma_i)_{1 \leq i \leq d_0}$ are defined by formulas similar to those of [13, section 6]. We then introduce four new constants: $\bar{K}_2, \bar{L}_2, K_7, K_8$ ([14, section 6]).

5.3 Strategies

By some slight adapatations of the strategies devised for the system $\mathcal{D}_0$ (see [13, section 7]), we obtain strategies for the particular system $\mathcal{H}_0$.

T_{cut}: $T_{cut}(A_1 \cdots A_n) = B_1 \cdots B_m$ iff $\exists i \in [1, n-1], \exists S_i, S'_i, S_n, S'_n \in \mathsf{DRH}^0\langle\!\langle\, V\, \rangle\!\rangle$, $h \in H$, such that

$$O_i \sqsubseteq S_i, O'_i \sqsubseteq S'_i, O_n \sqsubseteq S_n, O'_n \sqsubseteq S'_n, \quad O_i \equiv O'_i \equiv O_n \equiv O'_n \equiv \emptyset,$$

$$A_i = (p_i, S_i, S'_i), A_n = (p_n, S_n, S'_n), p_i < p_n,$$

$$S_i - O_i = h \cdot (S_n - O_n), S'_i - O'_i = h \cdot (S'_n - O'_n), \text{ and } m = 0$$

T_H: $T_H(A_1 \cdots A_n) = B_1 \cdots B_m$ iff $A_n = (p, S, T), p \geq 0, \exists h \in H, S \equiv T \equiv h$

and $m = 0$.

The strategies $T_\emptyset, T_A, T_B^+, T_B^-, T_C$ and the compound strategies $\mathcal{S}_{AB}, \mathcal{S}_{ABC}$ are then defined similarly as in [13, end of section 7].

Lemma 51 : $T_{cut}, T_\emptyset, T_H, T_A, T_B^+, T_B^-, T_C$ are $\mathcal{H}_0$-strategies. Moreover, $\mathcal{S}_{AB}$, $\mathcal{S}_{ABC}$ are closed $\mathcal{H}_0$-strategies.

6 Completeness of $\mathcal{H}_0$

Let us fix a tree $\tau = \mathcal{T}(\mathcal{S}_{AB}, (\pi_0, U_0^-, U_0^+))$ (i.e. τ is the proof tree associated with the assertion (π_0, U_0^-, U_0^+) by the strategy $\mathcal{S}_{AB}$). We suppose that, for every $\alpha \in \{-, +\}$

$$\bar{\ell}(U_0^\alpha) \leq \bar{L}_2, \;\; \mathrm{rd}(U_0^\alpha) \leq D_2 \;\; \text{and} \;\; U_0^-, U_0^+ \text{ are both unmarked}. \qquad (12)$$

A careful analysis of such a tree τ shows that, for every infinite branch with sequence of labels $A_1, \cdots, A_i, \cdots$, there exists some $n \geq 1$ such that $A_1 \cdots A_n \in \mathrm{dom}(T_C)$ (see [14, section 11.12], the ideas follow those of [13, section 8], combined with the new lemmas 37 and 38). We can then deduce, along the same lines as in [13, lemma 9.1], the lemma below.

Lemma 61 : Let A_0 be some true assertion which is supposed unmarked. Then the tree $\mathcal{T}(\mathcal{S}_{ABC}, A_0)$ is finite.

Theorem 62 The system $\mathcal{H}_0$ is complete.

Theorem 63 The equivalence problem for deterministic pushdown H-automata is decidable.

Other related results can be found in [8],[11],[6]. The decidability results [15, theorem p.203],[5, corollary 2 p.549] raise the problem of whether theorem 63 still holds for any commutative *monoid* H ?

References

1. J. Berstel and C. Reutenauer. *Rational Series and their Languages*. Springer, 1988.
2. A.P. Biryukov. Some algorithmic problems for finitely defined commutative semi-groups. *Siberian Math. Journal 8*, pages 384–391, 1967.
3. C. Choffrut. A generalization of Ginsburg and Rose's characterisation of gsm mappings. In *Proceedings ICALP 79*, pages 88–103. LNCS, Springer-Verlag, 1979.
4. B. Courcelle. An axiomatic approach to the Korenjac-Hopcroft algorithms. *Math. Systems theory*, pages 191–231, 1983.
5. R.H. Gilman. Presentations of groups and monoids. *Journal of Algebra 57*, pages 544–554, 1979.
6. T. Harju and J. Karhumäki. The equivalence problem of multitape finite automata. *TCS 78*, pages 347–355, 1991.
7. M.A. Harrison, I.M. Havel, and A. Yehudai. On equivalence of grammars through transformation trees. *TCS 9*, pages 173–205, 1979.
8. O. Ibarra. The unsolvability of the equivalence problem for ϵ-free ngsm's with unary input (output) alphabet and applications. In *Proceedings FOCS 78*, pages 74–81. IEEE, 1978.
9. O.H. Ibarra and L. Rosier. On the decidability of equivalence problem for deterministic pushdown transducers. *Information Processing Letters 13*, pages 89–93, 1981.
10. K. Kulik II and J. Karhumäki. Synchronizable deterministic pushdown automata and the decidability of their equivalence. *Acta Informatica 23*, pages 597–605, 1986.
11. W. Kuich and D. Raz. On the multiplicity equivalence problem for context-free grammars. In *Important Results and trends in Theoretical Computer Science (Colloquium in Honor of Aarto Salomaa)*, pages 232–250. Springer-Verlag, LNCS 812, 1994.
12. Y.V. Meitus. The equivalence problem for real-time strict deterministic pushdown automata. *Cybernetics and Systems analysis*, pages 581–594, 1990. Original article (in russian) in Kibernetika 5, p.14-25, 1989.
13. G. Sénizergues. L(A) = L(B)? In *Proceedings INFINITY 97*, pages 1–26. Electronic Notes in Theoretical Computer Science 9, URL: http://www.elsevier.nl/locate/entcs/volume9.html, 1997.
14. G. Sénizergues. L(A) = L(B)? Technical report, corrected and extended version of nr 1161-97, LaBRI, Université Bordeaux I, can be accessed at URL:http://www.labri.u-bordeaux.fr/~ges, 1998.
15. M. Taiclin. Algorithmic problems for commutative semigroups. *Soviet Math. Dokl.*, pages 201–204, 1968.
16. E. Tomita and K. Seino. A direct branching algorithm for checking the equivalence of two deterministic pushdown transducers, one of which is real-time strict. *Theoretical Computer science*, pages 39–53, 1989.

The Semi-Full Closure of Pure Type Systems

Gilles Barthe

Institutionen för Datavetenskap, Chalmers Tekniska Högskola, Göteborg, Sweden
Departamento de Informatica, Universidade do Minho, Braga, Portugal
gilles@di.uminho.pt

Abstract. We show that every functional Pure Type System may be extended to a semi-full Pure Type System. Moreover, the extension is conservative and preserves weak normalization. Based on these results, we give a new, conceptually simple type-checking algorithm for functional Pure Type Systems.

1 Introduction

Pure Type Systems (PTSs) [1] capture in a unified setting many typed λ-calculi that form the basis of typed functional languages and type-theory based proof-development systems. One central issue in the theory of PTSs is the problem of type-checking, which consists in deciding whether a judgment $\Gamma \vdash M : A$ is derivable according to the rules of a given PTS λS. Although type-checking is undecidable in general, most systems of interest have decidable type-checking. For such systems, the question remains whether it is possible to find reasonable, sound and complete, algorithms for type-checking. The existence of such algorithms is not obvious and indeed the completeness of the most natural type-checking algorithm, due to R. Pollack [8], remains an open problem. In a nutshell, the problem is caused by the second premise of the abstraction rule, which makes it difficult to prove completeness by induction on the structure of derivations.

Nevertheless several authors have proposed type-checking algorithms that are sound and complete for some specific classes of PTSs. In the early 90s, R. Pollack [7, 8] introduced the class of semi-full PTSs—informally a PTS is semi-full if it has "enough rules"—and gave a sound and complete type-checking algorithm for PTSs in that class. Unfortunately, many PTSs of interest are not semi-full. Later L.S. van Benthem Jutting, J. McKinna and R. Pollack [3, 8] gave an alternative algorithm that is sound and complete for functional PTSs, a large class of PTSs that comprises most of the systems that appear in the literature. In order to check for the second premise of the abstraction rule, their algorithm invokes a complex derivability relation with Π-application and Π-conversion, as given by the application rule

$$\frac{\Gamma \vdash M : \Pi x\colon A.\ B \quad \Gamma \vdash N : A}{\Gamma \vdash M\ N : (\Pi x\colon A.\ B)\ N}$$

and the reduction rule $(\Pi x\colon A.\ B)\ N \to_\pi B\{x := N\}$. Their algorithm is not fully satisfactory in the sense that it requires to consider an extended framework.

More recently, P. Severi [9] has suggested another algorithm that appeals to Pure Type Systems without the Π-condition (PTSWs). Those are a variant of PTSs in which the abstraction rule is

$$\frac{\Gamma, x : A \vdash M : B}{\Gamma \vdash \lambda x{:}A.\ M : \Pi x{:}A.\ B}$$

Again PTSWs are used to check for the second premise of the abstraction rule. While Severi's algorithm eliminates the need for considering new reduction relations, it still introduces a new framework. As a result, Severi needs to prove numerous properties for PTSWs before proving the soundness and completeness of the algorithm. Finally there are other algorithms that are concerned with the smaller class of (weakly) injective PTSs [2, 6]. These algorithms are simpler but do not cover all existing systems. For example some of the languages of the Automath family [4] and predicative F^ω [5] are not weakly injective.

The purpose of this paper is to present a new sound and complete typechecking algorithm for functional PTSs. The novelty of our algorithm is to remain within the framework of PTSs. It is an improvement over [3, 8, 9]: our algorithm is conceptually clearer and suppresses the need for introducing new frameworks such as the ones of [3, 8, 9]. In order to define our algorithm and prove it correct, we show that every functional PTS may be extended conservatively to a semi-full PTS, its *semi-full closure*. This result makes it possible to check, using Pollack's algorithm for semi-full PTSs, the second clause in the abstraction rule in the semi-full closure of the PTS under consideration.

Contents The paper is organized as follows. Section 2 briefly reviews the definition of PTSs. Section 3 introduces the semi-full closure of a PTS. Section 4 shows that the semi-full closure of a PTS is a conservative extension of the original PTS provided the latter is functional. In Section 5, we use that result to prove the soundness and completeness of a type-checking algorithm for functional PTSs.

2 Pure Type Systems

In this section, we present the syntax of PTSs and refer to standard texts, see e.g. [1], for examples and motivations.

Definition 1 (Specification). *A specification is a triple* $\mathbf{S} = (\mathcal{S}, \mathcal{A}, \mathcal{R})$ *where* $\mathcal{S}$ *is a set of sorts,* $\mathcal{A} \subseteq \mathcal{S} \times \mathcal{S}$ *is a set of axioms and* $\mathcal{R} \subseteq \mathcal{S} \times \mathcal{S} \times \mathcal{S}$ *is a set of rules. A specification* $\mathbf{S} = (\mathcal{S}, \mathcal{A}, \mathcal{R})$ *is functional if for every* $s_1, s_2, s_2', s_3, s_3' \in \mathcal{S}$,

$$\begin{array}{l}(s_1, s_2) \in \mathcal{A} \ \ \wedge\ (s_1, s_2') \in \mathcal{A} \ \ \Rightarrow s_2 \equiv s_2' \\ (s_1, s_2, s_3) \in \mathcal{R} \wedge (s_1, s_2, s_3') \in \mathcal{R} \Rightarrow s_3 \equiv s_3'\end{array}$$

Every specification $\mathbf{S}$ yields a PTS $\lambda \mathbf{S}$ as specified below. Throughout this section, $\mathbf{S} = (\mathcal{S}, \mathcal{A}, \mathcal{R})$ is a fixed specification.

Definition 2 (Pure Type System).

1. The set $\mathcal{T}$ of pseudo-terms is given by the abstract syntax

$$\mathcal{T} = V \mid \mathcal{S} \mid \mathcal{T}\mathcal{T} \mid \lambda V : \mathcal{T}.\mathcal{T} \mid \Pi V : \mathcal{T}.\mathcal{T}$$

   where V is a fixed countably infinite set of variables.
2. β-reduction $\to_\beta$ is defined as the compatible closure of the contraction

$$(\lambda x{:}A.\ M)\ N \to_\beta M\{x := N\}$$

   where $\bullet\{\bullet := \bullet\}$ is the standard substitution operator. The reflexive-transitive and reflexive-symmetric-transitive closures of $\to_\beta$ are denoted by $\twoheadrightarrow_\beta$ and $=_\beta$ respectively.
3. A pseudo-context is a finite ordered list $x_1 : A_1, \ldots, x_n : A_n$ where $x_1, \ldots, x_n \in V$ and $A_1, \ldots, A_n \in \mathcal{T}$. The empty context is denoted by $\langle\rangle$ and the set of pseudo-contexts is denoted by $\mathcal{G}$. If $\Gamma \in \mathcal{G}$, we let $\text{dom}(\Gamma) = \{x \mid \exists t \in \mathcal{T}.\ x : t \in \Gamma\}$.
4. A judgment is a triple $\Gamma \vdash M : A$ where $\Gamma \in \mathcal{G}$ and $M, A \in \mathcal{T}$. The rules of Pure Type Systems are given in Figure 1. If $\Gamma \vdash M : A$ is derivable according to those rules, then Γ and M are legal.
5. $\lambda \mathbf{S} = (\mathcal{E}, \mathcal{G}, \to_\beta, \vdash)$ is the Pure Type System (PTS) induced by $\mathbf{S}$.

Some of the results of this paper are concerned with normalization.

(axiom) $\langle\rangle \vdash s_1 : s_2$ if $(s_1, s_2) \in \mathcal{A}$

(start) $\dfrac{\Gamma \vdash A : s}{\Gamma, x : A \vdash x : A}$ if $x \in V \setminus \text{dom}(\Gamma)$

(weakening) $\dfrac{\Gamma \vdash A : B \quad \Gamma \vdash C : s}{\Gamma, x : C \vdash A : B}$ if $x \in V \setminus \text{dom}(\Gamma)$ and $A \in V \cup \mathcal{S}$

(product) $\dfrac{\Gamma \vdash A : s_1 \quad \Gamma, x : A \vdash B : s_2}{\Gamma \vdash (\Pi x{:}A.\ B) : s_3}$ if $(s_1, s_2, s_3) \in \mathcal{R}$

(application) $\dfrac{\Gamma \vdash F : (\Pi x{:}A.\ B) \quad \Gamma \vdash a : A}{\Gamma \vdash F\ a : B\{x := a\}}$

(abstraction) $\dfrac{\Gamma, x : A \vdash b : B \quad \Gamma \vdash (\Pi x{:}A.\ B) : s}{\Gamma \vdash \lambda x{:}A.\ b : \Pi x{:}A.\ B}$

(conversion) $\dfrac{\Gamma \vdash A : B \quad \Gamma \vdash B' : s}{\Gamma \vdash A : B'}$ if $B =_\beta B'$

Fig. 1. RULES FOR PURE TYPE SYSTEMS

Definition 3. *We write* $\lambda\mathbf{S} \models \mathsf{WN}(\beta)$ *and* $\lambda\mathbf{S} \models \mathsf{SN}(\beta)$ *respectively if every legal term in* $\lambda\mathbf{S}$ *is β-weakly normalizing and β-strongly normalizing respectively.*

We conclude this section with a list of properties of PTSs.

Lemma 1 (Closure properties).

1. Substitution. *If* $\Gamma, x : A, \Delta \vdash B : C$ *and* $\Gamma \vdash a : A$, *then*[1] $\Gamma, \Delta\{x := a\} \vdash A\{x := a\} : B\{x := a\}$.
2. Correctness of Types. *If* $\Gamma \vdash A : B$ *then either* $B \in \mathcal{S}$ *or there exists* $s \in \mathcal{S}$ *such that* $\Gamma \vdash B : s$.
3. Correctness of Contexts. *If* $\Gamma, x : C, \Delta \vdash A : B$ *then there exists* $s \in \mathcal{S}$ *such that* $\Gamma \vdash C : s$.
4. Subject Reduction. *If* $\Gamma \vdash M : A$ *and* $M \twoheadrightarrow_\beta N$ *then* $\Gamma \vdash N : A$.
5. Predicate Reduction. *If* $\Gamma \vdash M : A$ *and* $A \twoheadrightarrow_\beta A'$ *then* $\Gamma \vdash M : A'$.

Lemma 2 (Uniqueness of Types). *Assume* $\mathbf{S}$ *is functional.*

$$\Gamma \vdash M : A \quad \wedge \quad \Gamma \vdash M : A' \quad \Rightarrow \quad A =_\beta A'$$

3 The Semi-Full Closure of a Specification

Semi-fullness is a technical condition ensuring that a PTS "has enough rules". This is to be contrasted with negative notions such as functionality or injectivity which ensure that a PTS "does not have too many rules". Because of the nature of semi-fullness, every PTS may be extended to a semi-full one—while a non-functional or non-injective PTS may not be extended to a functional or an injective one. In fact, there are several ways to extend a PTS into a semi-full one. The next definition suggests two possibilities: the stratified closure, which is layered so as to facilitate reasoning, and the compact closure, which is more suited for type-checking purposes.

Definition 4 (Semi-full, semi-full closure). *Let* $\mathbf{S} = (\mathcal{S}, \mathcal{A}, \mathcal{R})$ *be a specification. Define*

$$\mathcal{O} = \{s \in \mathcal{S} \mid \exists s', s'' \in \mathcal{S}.\ (s, s', s'') \in \mathcal{R}\}$$
$$\mathcal{P} = \{(s_1, s_2) \in \mathcal{O} \times \mathcal{S} \mid \forall s \in \mathcal{S}.\ (s_1, s_2, s) \notin \mathcal{R}\}$$

1. $\mathbf{S}$ *is semi-full if* $\mathcal{P} = \emptyset$.
2. *The* compact semi-full closure *of* $\mathbf{S}$ *is the specification* $\mathbf{S}^\bullet = (\mathcal{S}^\bullet, \mathcal{A}, \mathcal{R}^\bullet)$ *where* $\mathcal{S}^\bullet = \mathcal{S} \cup \{\bullet\}$ *and*

$$\mathcal{R}^\bullet = \mathcal{R} \cup \{(s_1, s_2, \bullet) \mid (s_1, s_2) \in \mathcal{P}\} \cup \{(s, \bullet, \bullet) \mid s \in \mathcal{O}\}$$

---

[1] Substitution is extended from pseudo-terms to pseudo-contexts in the usual way.

3. The stratified semi-full closure of **S** is the specification $\mathbf{S}^\omega = (\mathcal{S}^\omega, \mathcal{A}, \mathcal{R}^\omega)$ where $\mathcal{S}^\omega = \mathcal{S} \cup \{\bullet_i \mid i \in \mathbb{N}\}$ and

$$\mathcal{R}^\omega = \mathcal{R} \cup \{(s_1, s_2, \bullet_0) \mid (s_1, s_2) \in \mathcal{P}\} \cup \{(s, \bullet_i, \bullet_{i+1}) \mid s \in \mathcal{O}, i \in \mathbb{N}\}$$

4. Let $j \in \mathbb{N}$. The j-closure of **S** is the specification $\mathbf{S}^j = (\mathcal{S}^j, \mathcal{A}, \mathcal{R}^j)$ where $\mathcal{S}^j = \mathcal{S} \cup \{\bullet_i \mid i \leq j\}$ and

$$\mathcal{R}^j = \mathcal{R} \cup \{(s_1, s_2, \bullet_0) \mid (s_1, s_2) \in \mathcal{P}\} \cup \{(s, \bullet_i, \bullet_{i+1}) \mid s \in \mathcal{O} \wedge i < j\}$$

5. By convention, we set $\mathbf{S}^{-1} = \mathbf{S}$.

The next result provides an alternative characterization of semi-fullness; in fact it corresponds to Pollack's original definition of semi-fullness.

Lemma 3. Let $\mathbf{S} = (\mathcal{S}, \mathcal{A}, \mathcal{R})$ be a specification.

1. **S** is semi-full if for every $s_2' \in \mathcal{S}$ and $(s_1, s_2, s_3) \in \mathcal{R}$, there exists $s_3' \in \mathcal{S}$ such that $(s_1, s_2', s_3') \in \mathcal{R}$.
2. $\mathbf{S}^\bullet$ and $\mathbf{S}^\omega$ are semi-full.
3. If **S** is functional so are $\mathbf{S}^\bullet$, $\mathbf{S}^\omega$ and $\mathbf{S}^j$ for every $j \in \mathbb{N}$.

We conclude this section by relating $\mathbf{S}^\bullet$, $\mathbf{S}^\omega$ and the $\mathbf{S}^j$s.

Lemma 4.

1. If $\Gamma \vdash^x M : A$ for $x \in \{\bullet, \omega, i\}$ then $M \in \mathcal{T}$ and $A \in \mathcal{T} \cup \{\bullet\} \cup \{\bullet_i \mid i \in \mathbb{N}\}$.
2. Let $A \in \mathcal{T}$. Then $\Gamma \vdash^\bullet M : A \Leftrightarrow \Gamma \vdash^\omega M : A$.
3. Let $M \in \mathcal{T}$. Then $\Gamma \vdash^\bullet M : \bullet \Leftrightarrow \exists j \in \mathbb{N}.\ \Gamma \vdash^\omega M : \bullet_j$.
4. $\Gamma \vdash^\omega M : A \Leftrightarrow \exists j \in \mathbb{N}.\ \Gamma \vdash^j M : A$.

4 Conservativity

The purpose of this section is to prove that $\lambda \mathbf{S}^\bullet$ is conservative—in a sense to be made precise below— over $\lambda \mathbf{S}$. We proceed in three steps: first, we prove that $\lambda \mathbf{S}^i$ is conservative over $\lambda \mathbf{S}^{i-1}$ for every $i \in \mathbb{N}$. Second, we conclude that $\lambda \mathbf{S}^\omega$ is conservative over $\lambda \mathbf{S}$. Third, we derive that $\lambda \mathbf{S}^\bullet$ is conservative over $\lambda \mathbf{S}$. The development of this section is similar to [9] but the use of PTSs and stratification simplify the proofs. Throughout this section, we let $\mathbf{S} = (\mathcal{S}, \mathcal{A}, \mathcal{R})$ be a functional specification.

Definition 5 (Contracting functions).

1. M is i-constrained in Γ, written $!^i_\Gamma(M)$, if one of the conditions below holds:
   (a) $M \in \mathcal{S} \cup \{\bullet_0, \ldots, \bullet_{i-1}\}$—in case $i = 0$ this clause is simply $M \in \mathcal{S}$;
   (b) there exists $A \in \mathcal{T}$ such that $\Gamma \vdash^i M : A$ and $!^i_\Gamma(A)$.

2. The contracting functions $\phi_\Gamma^i : \mathcal{T} \to \mathcal{T}$ are defined simultaneously for all $\Gamma \in \mathcal{G}$:

$$\phi_\Gamma^i(s) = s \qquad\qquad s \in \mathcal{S}$$
$$\phi_\Gamma^i(x) = x \qquad\qquad x \in \mathcal{V}$$
$$\phi_\Gamma^i(\Pi x{:}A.\ B) = \Pi x{:}\phi_\Gamma^i(A).\ \phi_{\Gamma,x:A}^i(B)$$
$$\phi_\Gamma^i(\lambda x{:}A.\ M) = \lambda x{:}\phi_\Gamma^i(A).\ \phi_{\Gamma,x:A}^i(M)$$
$$\phi_\Gamma^i(M\ N) = \begin{cases} (\phi_{\Gamma,x:A}^i(P))\{x := \phi_\Gamma^i(N)\} & \text{if } \neg !_\Gamma^i(M) \text{ and } M \equiv \lambda x{:}A.\ P \\ \phi_\Gamma^i(M)\ \phi_\Gamma^i(N) & \text{otherwise} \end{cases}$$

3. Γ is i-constrained, written $!^i(\Gamma)$, is defined as follows:

$$!^i(\langle\rangle) = \text{true} \qquad !^i(\Gamma, x : A) = !^i(\Gamma) \wedge !_\Gamma^i(A)$$

4. $!_\Gamma^i(M)$ is defined as $!^i(\Gamma) \wedge !_\Gamma^i(M)$.
5. The contracting function $\phi^i : \mathcal{G} \to \mathcal{G}$ is defined as follows:

$$\phi^i(\langle\rangle) = \langle\rangle \qquad \phi^i(\Gamma, x : A) = \phi^i(\Gamma), x : \phi_\Gamma^i(A)$$

There are alternative definitions of $!_\Gamma^i(M)$ that are not recursive but we prefer our compact definition. The next result is the key to our development.

Proposition 1 (Conservativity of $\lambda \mathbf{S}^i$ over $\lambda \mathbf{S}^{i-1}$). If $\Gamma \vdash^i M : D$ and $!_\Gamma^i(M)$ then there exists $B \in \mathcal{T}$ such that $D =_\beta B$ and $\phi^i(\Gamma) \vdash^{i-1} \phi_\Gamma^i(M) : B$.

Proof. By induction on the structure of derivations.

Next we derive the conservativity of $\mathbf{S}^i$ over $\mathbf{S}$.

Definition 6. Let $\Gamma \in \mathcal{G}$ and $M \in \mathcal{T}$ For every $i \in \mathbb{N}$, define

$$\Phi_\Gamma^0(M) = \phi_\Gamma^0(M) \qquad \Phi_\Gamma^{i+1}(M) = \Phi_{\phi^{i+1}(\Gamma)}^i$$
$$??_\Gamma^0(M) = !_\Gamma^0(M) \qquad ??_\Gamma^{i+1}(M) = !_\Gamma^{i+1}(M) \wedge ??_{\phi^{i+1}(\Gamma)}^i(\phi_\Gamma^{i+1}(M))$$

We have:

Lemma 5. If $\Gamma \vdash^i M : A$ and $??_\Gamma^i(M)$ then there exists $B \in \mathcal{T}$ such that $B =_\beta A$ and $\Phi^i(\Gamma) \vdash \Phi_\Gamma^i(M) : B$.

Proof. By a straightforward induction on i.

To conclude we need to 'take the limit' of the conservativity results.

Definition 7.

1. For every $\Gamma \in \mathcal{G}$ and $M \in \mathcal{T}$, let $\mu(\Gamma, N)$ denote, when it exists, the smallest number such that $\exists A \in \mathcal{T}.\ \Gamma \vdash^i M : A$.
2. Define $??_\Gamma^\omega(M) = \mu(\Gamma, M) \downarrow \wedge ??_\Gamma^{\mu(\Gamma,M)}(M)$ and $\Phi_\Gamma^\omega(M) = \Phi_\Gamma^{\mu(\Gamma,M)}(M)$.

3. For every $\Gamma \in \mathcal{G}$, let $\mu(\Gamma)$ denote, when it exists, the smallest number such that $\exists M, A \in \mathcal{T}. \ \Gamma \vdash^i M : A$.
4. Define $??^\omega(\Gamma) = \mu(\Gamma) \downarrow \ \wedge \ ??^{\mu(\Gamma)}(\Gamma)$ and $\Phi^\omega(\Gamma) = \Phi^{\mu(\Gamma)}(\Gamma)$.

Lemma 6. *If $\Gamma \vdash^\omega M : A$ and $??^\omega_\Gamma(M)$ then there exists $B \in \mathcal{T}$ such that $B =_\beta A$ and $\Phi^\omega(\Gamma) \vdash \Phi^\omega_\Gamma(M) : B$.*

Proof. Follows from Lemmas 4 and 5.

$\lambda \mathbf{S}^\omega$ and $\lambda \mathbf{S}^\bullet$ type the same terms, hence we can transfer the conservativity result to $\lambda \mathbf{S}^\bullet$.

Corollary 1 (Conservativity of $\lambda \mathbf{S}^\bullet$ over $\lambda \mathbf{S}$). *If $\Gamma \vdash^\bullet M : A$ and $??^\omega_\Gamma(M)$ then there exists $B \in \mathcal{T}$ such that $B =_\beta A$ and $\Phi^\omega(\Gamma) \vdash \Phi^\omega_\Gamma(M) : B$.*

Proof. Follows from Corollary 6 and Lemma 4.

For the purpose of type-checking, the crucial property is the one given in the next corollary.

Corollary 2. *If $\Gamma \vdash^\bullet \Pi x{:}A.\ B : s$ and $\Gamma, x : A \vdash B : s'$ and $s \in \mathcal{S}$ then $\Gamma \vdash \Pi x{:}A.\ B : s$.*

Proof. Necessarily $??^\omega_\Gamma(\Pi x{:}A.\ B)$ hence there exists $C \in \mathcal{T}$ such that $C =_\beta s$ and $\Phi^\omega(\Gamma) \vdash \Phi^\omega_\Gamma(\Pi x{:}A.\ B) : C$. Now $C \twoheadrightarrow_\beta s$ hence by Predicate Reduction $\phi^\omega(\Gamma) \vdash \Phi^\omega_\Gamma(\Pi x{:}A.\ B) : s$. To conclude, observe that $\Phi^\omega(\Gamma) = \Gamma$ and $\Phi^\omega_\Gamma(\Pi x{:}A.\ B) = \Pi x{:}A.\ B$ since $\Gamma, x : A \vdash B : s'$.

In order to type-check, one sometimes needs to check the convertibility of types. Hence it is important for the PTS to be β-normalizing. Fortunately, $.^\bullet$ preserves normalization.

Proposition 2. $\lambda \mathbf{S} \models \mathsf{WN}(\beta) \Rightarrow \lambda \mathbf{S}^\bullet \models \mathsf{WN}(\beta)$.

Proof. It is enough to show that $\lambda \mathbf{S}^{i-1} \models \mathsf{WN}(\beta) \Rightarrow \lambda \mathbf{S}^i \models \mathsf{WN}(\beta)$. Assume $\Gamma \vdash^i M : A$. To show that $M \in \mathsf{WN}(\beta)$. We proceed by induction on the length (and not on the structure) of terms and reason by cases (whether or not $!^i(\Gamma)$).

5 Application to Type-Checking

In this section, we exploit the results of the previous section and the decidability of type-checking for semi-full normalizing PTSs [3, 8] to establish the decidability of type-checking for functional normalizing PTSs.

An important step towards decidability of type-checking is to provide a syntax-directed presentation of the rules of PTSs. In a nutshell, a set of rules is syntax-directed if the premises of a rule are determined—up to inessential details—by its conclusion. The next definition provides such a set of rules. It uses an auxiliary relation $\vdash^\bullet_\mathsf{sfsd}$ which instantiates the derivability relation $\vdash_\mathsf{sfsd}$ of [3, 8] to $\lambda \mathbf{S}^\bullet$.

(axiom) $\quad \langle\rangle \vdash_{\mathsf{sfsd}}^{\bullet} s_1 : s_2 \quad\quad$ if $(s_1, s_2) \in \mathcal{A}$

(start) $\quad \dfrac{\Gamma \vdash_{\mathsf{sfsd}}^{\bullet} A :\twoheadrightarrow_{wh} s}{\Gamma, x : A \vdash_{\mathsf{sfsd}}^{\bullet} x : A} \quad$ if $x \in V \setminus \mathrm{dom}(\Gamma)$

(weakening) $\quad \dfrac{\Gamma \vdash_{\mathsf{sfsd}}^{\bullet} A : B \quad \Gamma \vdash_{\mathsf{sfsd}}^{\bullet} C :\twoheadrightarrow_{wh} s}{\Gamma, x : C \vdash_{\mathsf{sfsd}}^{\bullet} A : B} \quad$ if $x \in V \setminus \mathrm{dom}(\Gamma)$ and $A \in V \cup \mathcal{S}$

(product) $\quad \dfrac{\Gamma \vdash_{\mathsf{sfsd}}^{\bullet} A :\twoheadrightarrow_{wh} s_1 \quad \Gamma, x : A \vdash_{\mathsf{sfsd}}^{\bullet} B :\twoheadrightarrow_{wh} s_2}{\Gamma \vdash_{\mathsf{sfsd}}^{\bullet} (\Pi x{:}A.\, B) : s_3} \quad$ if $(s_1, s_2, s_3) \in \mathcal{R}^{\bullet}$

(application) $\quad \dfrac{\Gamma \vdash_{\mathsf{sfsd}}^{\bullet} F :\twoheadrightarrow_{wh} (\Pi x{:}A'.\, B) \quad \Gamma \vdash_{\mathsf{sfsd}}^{\bullet} a : A}{\Gamma \vdash_{\mathsf{sfsd}}^{\bullet} F\, a : B\{x := a\}} \quad$ if $A =_\beta A'$

(abstraction) $\quad \dfrac{\Gamma, x : A \vdash_{\mathsf{sfsd}}^{\bullet} b : B \quad B \in \mathcal{S} \Rightarrow B \in \mathcal{S}^\tau}{\Gamma \vdash_{\mathsf{sfsd}}^{\bullet} \lambda x{:}A.\, b : \Pi x{:}A.\, B}$

Fig. 2. SYNTAX-DIRECTED RULES FOR SEMI-FULL CLOSURES

Definition 8 (Syntax-directed Rules).

1. Weak-head reduction $\rightarrow_{wh}$ is defined as the closure[2] of the contraction

$$(\lambda x : A.\, P)\, Q\, R_1 \ldots R_n \quad \rightarrow_{wh} \quad P\{x := Q\}\, R_1 \ldots R_n$$

2. A sort s is a typed sort, written $s \in \mathcal{S}^\tau$, if $\exists s' \in \mathcal{S}.\, (s, s') \in \mathcal{A}$.
3. The derivability relation $\Gamma \vdash_{\mathsf{sfsd}}^{\bullet} M : A$ is given by the rules of Figure 2 where we write $\Gamma \vdash_{\mathsf{sfsd}}^{\bullet} M :\twoheadrightarrow_{wh} A$ if $\exists A' \in \mathcal{T}.\, \Gamma \vdash_{\mathsf{sfsd}}^{\bullet} M : A' \wedge A' \twoheadrightarrow_{wh} A$.
4. The derivability relation $\Gamma \vdash_{\mathsf{nat}} M : A$ is given by the rules of Figure 3 where we write $\Gamma \vdash_{\mathsf{nat}} M :\twoheadrightarrow_{wh} A$ if $\exists A' \in \mathcal{T}.\, \Gamma \vdash_{\mathsf{nat}} M : A' \wedge A' \twoheadrightarrow_{wh} A$.

The soundness and completeness of $\vdash_{\mathsf{sfsd}}^{\bullet}$ over $\vdash^{\bullet}$ is already known.

Proposition 3 ([3]). *For every specification $\mathcal{S}$,*

1. *Soundness:* $\Gamma \vdash_{\mathsf{sfsd}}^{\bullet} M : A \quad \Rightarrow \quad \Gamma \vdash^{\bullet} M : A.$
2. *Completeness:* $\Gamma \vdash^{\bullet} M : A \quad \Rightarrow \quad \exists A' \in \mathcal{T}.\, \Gamma \vdash_{\mathsf{sfsd}}^{\bullet} M : A' \wedge A =_\beta A'$

Using the above proposition, we conclude that $\vdash_{\mathsf{nat}}$ is sound and complete with respect to $\vdash$.

---

[2] We insist on the closure not being compatible so weak-head reduction differs from β-reduction by applying only at the top-level.

(axiom) $\langle \rangle \vdash_{\mathsf{nat}} s_1 : s_2$ if $(s_1, s_2) \in \mathcal{A}$

(start) $\dfrac{\Gamma \vdash_{\mathsf{nat}} A \mathrel{:\twoheadrightarrow_{wh}} s}{\Gamma, x : A \vdash_{\mathsf{nat}} x : A}$ if $x \in V \setminus \mathrm{dom}(\Gamma)$

(weakening) $\dfrac{\Gamma \vdash_{\mathsf{nat}} A : B \quad \Gamma \vdash_{\mathsf{nat}} C \mathrel{:\twoheadrightarrow_{wh}} s}{\Gamma, x : C \vdash_{\mathsf{nat}} A : B}$ if $x \in V \setminus \mathrm{dom}(\Gamma)$ and $A \in V \cup \mathcal{S}$

(product) $\dfrac{\Gamma \vdash_{\mathsf{nat}} A \mathrel{:\twoheadrightarrow_{wh}} s_1 \quad \Gamma, x : A \vdash_{\mathsf{nat}} B \mathrel{:\twoheadrightarrow_{wh}} s_2}{\Gamma \vdash_{\mathsf{nat}} (\Pi x{:}A.\, B) : s_3}$ if $(s_1, s_2, s_3) \in \mathcal{R}$

(application) $\dfrac{\Gamma \vdash_{\mathsf{nat}} F \mathrel{:\twoheadrightarrow_{wh}} (\Pi x{:}A'.\, B) \quad \Gamma \vdash_{\mathsf{nat}} a : A}{\Gamma \vdash_{\mathsf{nat}} F\, a : B\{x := a\}}$ if $A =_\beta A'$

(abstraction) $\dfrac{\Gamma, x : A \vdash_{\mathsf{nat}} b : B \quad \Gamma \vdash^{\bullet}_{\mathsf{sfsd}} \Pi x{:}A.\, B : s}{\Gamma \vdash_{\mathsf{nat}} \lambda x{:}A.\, b : \Pi x{:}A.\, B}$ if $s \in \mathcal{S}$

Fig. 3. SYNTAX-DIRECTED RULES FOR FUNCTIONAL PURE TYPE SYSTEMS

Theorem 1. *If* **S** *is functional, then*

1. *Soundness:* $\Gamma \vdash_{\mathsf{nat}} M : A \;\Rightarrow\; \Gamma \vdash M : A.$
2. *Completeness:* $\Gamma \vdash M : A \;\Rightarrow\; \exists A' \in \mathcal{T}.\; \Gamma \vdash_{\mathsf{nat}} M : A' \wedge A =_\beta A'$

Proof. Soundness is proved by induction on the derivations, using Corollary 2 in the (abstraction) rule. Completeness is also proved by induction on the structure of derivations, using soundness. The proofs are routine.

Corollary 3 (Decidability of type-checking). *If* $\mathbf{S} = (\mathcal{S}, \mathcal{A}, \mathcal{R})$ *is a functional specification, then type-checking is decidable provided* $\mathcal{S}$, $\mathcal{A}$ *and* $\mathcal{R}$ *are recursive and* $\lambda\mathbf{S} \models \mathsf{WN}(\beta)$.

Proof. We need to prove that all side-conditions are decidable. Weak normalization is needed in order to decide β-convertibility. Details are omitted.

Acknowledgments The author is supported by a European TMR Fellowship.

References

[1] H. Barendregt. Lambda calculi with types. In S. Abramsky, D. Gabbay, and T. Maibaum, editors, *Handbook of Logic in Computer Science*, pages 117–309. Oxford Science Publications, 1992. Volume 2.

[2] G. Barthe. Type checking injective pure type systems. Manuscript, 1997.

[3] L.S. van Benthem Jutting, J. McKinna, and R. Pollack. Checking algorithms for pure type systems. In H. Barendregt and T. Nipkow, editors, *Proceedings of TYPES'93*, volume 806 of *Lecture Notes in Computer Science*, pages 19–61. Springer-Verlag, 1994.

[4] R. Nederpelt, H. Geuvers, and R. de Vrijer, editors. *Selected papers on Automath*, volume 133 of *Studies in Logic and the Foundations of Mathematics*. North-Holland, Amsterdam, 1994.

[5] S. Peyton Jones and E. Meijer. Henk: a typed intermediate language. Proceedings of the ACM Workshop on Types in Compilation, 1997.

[6] E. Poll. A typechecker for bijective pure type systems. Technical Report CSN93/22, Technical University of Eindhoven, June 1993.

[7] R. Pollack. Typechecking in pure type systems. In B. Nordström, editor, *Informal proceedings of LF'92*, pages 271–288, 1992. Available from http://www.dcs.ed.ac.uk/lfcsinfo/research/types-bra/proc/index.html.

[8] R. Pollack. *The Theory of LEGO: A Proof Checker for the Extended Calculus of Constructions*. PhD thesis, University of Edinburgh, 1994.

[9] P. Severi. *Normalisation in lambda calculus and its relation to type inference*. PhD thesis, Technical University of Eindhoven, 1996.

Predicative Polymorphic Subtyping

Marcin Benke

Institute of Informatics
Warsaw University
ben@mimuw.edu.pl

Abstract. We consider a version of the Mitchell's polymorphic type containment (aka *subsumption*). Here, it is not suited for the system $\mathcal{F}$ but for the *predicative* version of it introduced by Leivant. The aim of the approach is to propose a new notion of polymorphic functional subsumption which may be decidable unlike Mitchell's one. We define a system for predicative subsumption in the style of Mitchell and then prove its equivalence with a Gentzen-style definition. Next, we study the notion of bicoercibility. This is followed by a reduction of the typability in the Leivant's system with subsumption to subsumption itself.

1 Introduction

The notion of polymorphism formalised as system $\mathcal{F}$ ([Gir72, Rey74]) is widely used in functional programming community. This notion was refined in many different directions in order to obtain a strongly expressible system with user friendly facilities. The ML type system introduced in [Mil78] was a successful milestone in this direction with its decidable type inference [DM82]. On the other hand it came out that system $\mathcal{F}$ itself is less convenient as it has undecidable type inference [Wel94].

People studied different variations of type systems that are closely connected with these two main systems. There was some effort spent in order to enrich the ML system with less restricted quantification. Existential quantification in algebraic datatypes was introduced by Perry in [Per90] and by Läufer and Odersky in [LO94]. This extension is in fact Milner style version of Mitchell and Plotkin abstract types [MP88]. Then the system with existential quantification was extended with universal quantification by [Rém94] which gave a possibility to declare objects with polymorphic methods. Läufer and Odersky introduced in [OL96] a type system with type schemes that may have quantification in any place, but type variables may be instantiated by types with no quantifiers only. All the aforementioned systems have decidable type inference.

Another direction was to change system $\mathcal{F}$ so that one can get decidable type inference. One way consisted in restriction on the form of types. Types of rank at most 2 (rank is the maximal number of steps made to the left on a path leading in a type to a quantifier) lead to decidable type inference while the rank 3 leads to undecidability [KT92]. The idea of changing the system itself was the crucial point in the work of Mitchell [Mit88], where additional relation of *containment* (known also as *subsumption*) was introduced, as well as in the one by Leivant [Lei91], where stratification on type variables was imposed (variable may be substituted for by a type with quantifiers on a lower level).

The Mitchell's system has already its own story. Unfortunately, type inference is undecidable for this system as shown in [Wel96]. Moreover, the containment relation is undecidable, too ([TU96, Wel95]). This was surprising in the light of the fact that the bicoercibility relation (σ is bicoercible with τ iff $\sigma \leq \tau$ and $\tau \leq \sigma$) is decidable [Tiu95].

By contrast the Leivant's system has not been thoroughly studied yet.

Contributions of the paper This paper presents a version of the Mitchell's subsumption relation suited for Leivant's stratified system $\mathcal{F}$. This subsumption relation does not cover the rule for distribution of quantifier, however. This choice was made because proofs are getting very involved in presence of such rule (see [LMS95]). Moreover, it seems that the presence of this rule does not add any substantial complexity as the same mechanism gives undecidability for the Mitchell's system with this rule and without [Chr98]. At last Läufer and Odersky use exactly this subsumption in [OL96].

The notion of subsumption is given in the Mitchell style. We characterise this notion in the spirit of [LMS95] proving its equivalence with a Gentzen-style definition. Next, we study the notion of bicoercibility obtaining very straightforward characterisation of it. A reduction of the typability problem to the problem of deciding subsumption is studied then. The technique is taken from [Jim95]. We end with an algorithm for deriving subsumption in the restricted case for types that have all possible zero-level quantification (variables may be substituted for only by quantifier free types) and may have a vector of level one quantifiers in the root position (variables may be substituted for by terms with zero-level quantification). This result extends the result by Läufer and Odersky [OL96].

Organisation of the paper Section 2 contains preliminary definitions. Section 3 presents the notion of subtyping in the spirit of Mitchell. In Section 4 we introduce the sequent system in the sense of [LMS95] and prove its equivalence with the system defined in Section 3. Section 5 introduces a technical tool of syntax-driven system. The idea of this system is taken from [Wel95]. Bicoercions are studied in Section 6.

Due to the page limit, some proofs have been omitted. They may be found in the full version, tentatively published as [Ben98].

2 Definitions

2.1 Stratification

Let t be a natural number. Systems of tier[1] t have the following components:
Type expressions τ and their levels $L(\tau)$ are defined inductively:

---

[1] while we use the word *level* to characterize types, the word *tier* is used to denote systems; a sytem of tier t may involve types of levels $0, \ldots, t$.

- For each level $p = 0, 1, ..., t$ there is a denumerable supply of type variables of level p: $\alpha^p, \alpha_1^p, \ldots$ (we omit the level superscript when it is irrelevant or clear from context). A type variable of level p is also a type expression of level p.
- If σ and τ are type expressions then $\sigma \to \tau$ is a type expression of level $\max(L(\sigma), L(\tau))$
- If τ is a type expression of level p then $\forall \alpha^q.\tau$ is a type expression of level $\max(p, q+1)$

If T is a set of types, then by $L(T)$ we shall mean the multiset of levels

$$\{L(\tau) \mid \tau \in T\}$$

The relation $\leq_{Mul}$ will denote the usual multiset ordering.

Let τ be a type with $L(\tau) = p+1$ by *rank* of τ we shall mean the maximum length of a path leading in τ to a quantifier binding a variable of level p.

Note that by erasing all level indices from a predicative polymorphic type we get an ordinary "system F" type. We shall denote the result of such erasure performed on type τ by $F(\tau)$.

2.2 Representing Types: Trees and Canonical Form

Let $\mathcal{P} = \{1,2\}^*$ and $\mathcal{P}^\omega = \{1,2\}^\omega$ be sets of respectively finite and infinite words over $\{1,2\}$ (which may be viewed as paths in binary trees). We shall use capital greek letters (e.g. Π, Δ) to denote such paths.

We may view a type as a tree (with domain being a prefix-closed subset of $\mathcal{P}$) where each arrow corresponds to an internal node, type variables are leaf nodes, and each node may be labelled with a set of quantifiers. We shall denote such a set labelling a node at address Π in type τ by $Q(\tau, \Pi)$. The domain of such a tree shall be called the *tree skeleton* of a type.

Definition 1. *We say that τ is in the canonical form, if*

1. *It has no redundant quantifiers.*
2. *$BV(\tau) = \{\alpha_1, \ldots, \alpha_n\}$ (for some $n \in N$).*
3. *names of bound variables are different from the names of free variables and no variable is bound more than once.*
4. *If for some paths Δ, Π where Δ is lexicographically earlier than Π, $\tau_\Delta = \alpha_i$, $\tau_\Pi = \alpha_j$ (and there is no lexicographically earlier path leading to α_j) then $i < j$.*
5. *At every node the quantifiers are ordered according to indices of the variables they bound.*

In [Wel96], Wells introduced another convenient representation for types, which he called *leaf groups*: a type σ is a set of leaf groups $\{G_1, \ldots, G_n\}$. Each leaf group denotes the set of occurrences of a particular variable — free or bound. A group referring to a free variable is of the form $\langle \alpha, P \rangle$ where α is the name of the variable and $P \subseteq \mathcal{P}$ is the set of paths leading to leaves labelled with α; a group referring to a variable bound at the node with address Π is of the form $\langle \Pi, P \rangle$.

The connection between leaf-group representation and canonical forms is established by the following

Proposition 1. *Two types have the same canonical form if and only if they have the same leaf-group representation.*

The tree representatiuon allows us to define some useful (partial) functions, used later in the paper:

- sign(Π), defined for $\Pi \in \mathcal{P}$ as $+1$ for positive paths (even number of 1's) and -1 for negative ones.
- leaf(σ, Π) — longest prefix of Π leading to a leaf in σ.
- quant(σ, Π) — path from root to the quantifier binding the variable at leaf(σ, Π), if any.
- leafvar(σ, Π) — name of the variable at leaf(σ, Π) if it is free, undefined otherwise.
- quantsign(σ, Π) = sign(quant(σ, Π)) if quant(σ, Π) is defined, 0 if the variable is free, undefined if leaf(σ, Π) is undefined.

3 Subtyping

3.1 A Hilbert-Style System

Axioms:

(refl) $\sigma \leq \sigma$

(inst) $\forall \alpha^q . \sigma \leq \forall \beta^p . \sigma[\rho/\alpha^q]$ $L(\rho) \leq q$ $\beta \notin \mathrm{FV}(\forall \alpha^q . \sigma)$

Rules:

$$(\rightarrow) \; \frac{\rho' \leq \rho \quad \tau \leq \tau'}{\rho \rightarrow \tau \leq \rho' \rightarrow \tau'} \qquad (\forall) \; \frac{\sigma \leq \sigma'}{\forall \alpha . \sigma \leq \forall \alpha . \sigma'}$$

$$(\mathrm{trans}) \; \frac{\sigma \leq \sigma_1 \quad \sigma_1 \leq \sigma'}{\sigma \leq \sigma'}$$

We shall sometimes use the notation $\sigma \leq_t \tau$ to denote the fact that this inequality is derivable in the system of tier t. We shall also use the notation $\sigma \sqsubset \tau$ to indicate the fact that there is a derivation of $\vdash \sigma \leq \tau$ such that its last rule is other than (trans). It is easy to see that

$$\vdash \sigma \leq \tau \quad \text{iff} \quad \sigma \sqsubset^* \tau$$

3.2 Relationship to the Mitchell's System

Our system differs from the original system of Mitchell [Mit88] (which operates on the types without level annotations) in two points: firstly, the (inst) axiom is restricted to predicative instances. Besides, the system of Mitchell contains additionally an axiom of *distributivity*:

$$\forall \alpha. \sigma_1 \to \sigma_2 \leq \forall \alpha. \sigma_1 \to \forall \alpha. \sigma_2$$

Let $\vdash^M$ denote derivability in the Mitchell's system. We have the following

Lemma 1. *If $\vdash \sigma \leq \tau$ then $\vdash^M F(\sigma) \leq F(\tau)$*

4 A Sequent System for Predicative Subtyping

$$\sigma \vdash \sigma$$

$$\frac{\rho' \vdash \rho \quad \tau \vdash \tau'}{\rho \to \tau \vdash \rho' \to \tau'} \ (\to)$$

$$\frac{\sigma[\rho/\alpha^q] \vdash \tau \quad L(\rho) \leq q}{\forall \alpha^q.\sigma \vdash \tau} \ (\forall L)$$

$$\frac{\sigma \vdash \tau}{\sigma \vdash \forall \alpha^q.\tau} \ (\forall R) \qquad \alpha \notin \mathrm{FV}(\sigma)$$

This system differs from [LMS95] in two points: first in the rule $(\forall L)$ instance is predicative, second, the rule $(\forall R)$ is restricted to the case called $(\forall_0\text{-}R)$ in [LMS95]. The latter change reflects the fact that we have no axiom of distributivity in the system defined in the previous section.

4.1 Cut Elimination

The relation $\vdash$ is transitive, i.e. the following rule is admissible:

$$\frac{\sigma \vdash \rho \quad \rho \vdash \tau}{\sigma \vdash \tau} \ (cut)$$

Lemma 2. *1. $(\forall R)$ permutes with $(\forall L)$, that is for every derivation ending with the consecutive applications of rules $(\forall R)$ and $(\forall L)$ there exists a derivation of the same sequent ending with $(\forall L)$ followed by $(\forall R)$.*
2. $(\forall R)$ on the right permutes with (cut).
3. $(\forall L)$ on the left permutes with (cut).

Theorem 1. *For any sequent derivable in the system with the rule (cut) there exists an equivalent cut-free derivation*

Cut elimination allows us to state the equivalence of Hilbert- and sequent-style systems:

Theorem 2. *For any types σ and τ we have*

$$\sigma \leq \tau \Leftrightarrow \sigma \vdash \tau$$

5 A Syntax-Driven System

Lemma 3. $\forall \boldsymbol{\alpha}.\sigma_1 \to \sigma_2 \leq \forall \boldsymbol{\gamma}.\tau_1 \to \tau_2$ *(with $\boldsymbol{\gamma} \notin \mathrm{FV}(\forall\boldsymbol{\alpha}.\sigma_1 \to \sigma_2)$) if and only if there exist $\boldsymbol{\rho}$ such that $L(\rho_i) \leq L(\alpha_i)$ as well as $\tau_1 \leq \sigma_1[\boldsymbol{\rho}/\boldsymbol{\alpha}]$ and $\sigma_2[\boldsymbol{\rho}/\boldsymbol{\alpha}] \leq \tau_2$*

Proof. The right-to-left implication is easy to check:

$$\cfrac{\forall\boldsymbol{\alpha}.\sigma_1 \to \sigma_2 \leq \forall\boldsymbol{\gamma}.(\sigma_1 \to \sigma_2)[\boldsymbol{\rho}/\boldsymbol{\alpha}] \quad \cfrac{\cfrac{\tau_1 \leq \sigma_1[\boldsymbol{\rho}/\boldsymbol{\alpha}] \quad \sigma_2[\boldsymbol{\rho}/\boldsymbol{\alpha}] \leq \tau_2}{(\sigma_1 \to \sigma_2)[\boldsymbol{\rho}/\boldsymbol{\alpha}] \leq \tau_1 \to \tau_2}}{\forall\boldsymbol{\gamma}.(\sigma_1 \to \sigma_2)[\boldsymbol{\rho}/\boldsymbol{\alpha}] \leq \forall\boldsymbol{\gamma}.\tau_1 \to \tau_2}(\forall)}{\forall\boldsymbol{\alpha}.\sigma_1 \to \sigma_2 \leq \forall\boldsymbol{\gamma}.\tau_1 \to \tau_2}(trans)$$

On the other hand, by Theorem 2 to prove the converse implication it suffices to prove that if

$$\forall\boldsymbol{\alpha}.\sigma_1 \to \sigma_2 \vdash \forall\boldsymbol{\gamma}.\tau_1 \to \tau_2 \qquad (1)$$

then there exist $\boldsymbol{\rho}$ such that $\tau_1 \vdash \sigma_1[\boldsymbol{\rho}/\boldsymbol{\alpha}]$ and $\sigma_2[\boldsymbol{\rho}/\boldsymbol{\alpha}] \vdash \tau_2$

Let as assume 1. It is easy to see that its derivation must have ended with a sequence of intermixed rules $(\forall L)$ and $(\forall R)$ respectively for α-s and γ-s. However, by Lemma 2 all applications of $(\forall R)$ can be moved down. Thus we have

$$\forall\boldsymbol{\alpha}.\sigma_1 \to \sigma_2 \vdash \tau_1 \to \tau_2 \qquad (2)$$

Now, unless $\boldsymbol{\alpha}$ is empty, the only rule yielding such sequent is $(\forall L)$. By applying this argument repeatedly until α is empty, we may conclude that there are $\boldsymbol{\rho}$ (of levels compatible with $\boldsymbol{\alpha}$) such that

$$(\sigma_1 \to \sigma_2)[\boldsymbol{\rho}/\boldsymbol{\alpha}] \vdash \tau_1 \to \tau_2$$

from which the thesis follows, since unless this is an axiom instance, the only rule yielding such a sequent is $(\to)$.

5.1 Polarity

Lemma 4 (Wells). *If $\sigma \leq \tau$ then for all $\Pi \in \mathcal{P}^\omega$ the following properties hold:*
1. *If $\mathrm{quantsign}(\sigma, \Pi) = 0 = \mathrm{quantsign}(\tau, \Pi)$ then $\mathrm{leaf}(\sigma, \Pi) = \mathrm{leaf}(\tau, \Pi)$ and $\mathrm{leafvar}(\sigma, \Pi) = \mathrm{leafvar}(\tau, \Pi)$.*
2. *$\mathrm{quantsign}(\sigma, \Pi) \geq \mathrm{quantsign}(\tau, \Pi)$.*
3. *If $\mathrm{leaf}(\sigma, \Pi) < \mathrm{leaf}(\tau, \Pi)$ then $\mathrm{quantsign}(\sigma, \Pi) = +1$.*
4. *If $\mathrm{leaf}(\sigma, \Pi) > \mathrm{leaf}(\tau, \Pi)$ then $\mathrm{quantsign}(\tau, \Pi) = -1$.*

Lemma 5. *If $\sigma \leq \tau \leq \sigma'$ then for every path $\Pi \in \mathrm{dom}(\sigma) \cap \mathrm{dom}(\sigma')$ we have that $\Pi \in \mathrm{dom}(\tau)$.*

Proof. Assume the contrary, i.e. that there is a path Π such that

$$\mathrm{leaf}(\sigma, \Pi) > \mathrm{leaf}(\tau, \Pi) < \mathrm{leaf}(\sigma', \Pi)$$

By Lemma 4, since $\sigma \leq \tau$ and $\mathrm{leaf}(\sigma, \Pi) > \mathrm{leaf}(\tau, \Pi)$, we have that $\mathrm{quantsign}(\tau, \Pi) = -1$. On the other hand, since $\tau \leq \sigma'$ and $\mathrm{leaf}(\tau, \Pi) < \mathrm{leaf}(\sigma', \Pi)$, we obtain that $\mathrm{quantsign}(\tau, \Pi) = +1$. Hence, by contradiction, the thesis is proved.

6 Bicoercions

Bicoercion is an important relation introduced by Tiuryn [Tiu95].

Two types σ and τ are said to be *bicoercible* if $\vdash \sigma \leq \tau$ and $\vdash \tau \leq \sigma$ holds.

Proposition 2. *If σ and τ are bicoercible then $F(\sigma)$ and $F(\tau)$ are bicoercible in $\vdash^M$.*

Proof. By Lemma 1 $\vdash \sigma \leq \tau$ implies $\vdash^M F(\sigma) \leq F(\tau)$ and $\vdash \tau \leq \sigma$ implies $\vdash^M F(\tau) \leq F(\sigma)$

In our system, however, bicoercibility is a stronger notion: two types are bicoercible if and only if they have the same canonical form, i.e. are identical up to α conversion, quantifier reordering and redundant quantifiers.

6.1 Proof System

The system given below is for deriving expressions of the form $\sigma \equiv \tau$, where σ and τ are polymorphic types.

Axioms:
(A1) $\quad \sigma \equiv \sigma$
(A2) $\quad \forall \alpha \forall \beta. \sigma \equiv \forall \beta \forall \alpha. \sigma$
(A3) $\quad \forall \alpha. \sigma \equiv \sigma \quad$ if $\alpha \notin FV(\sigma)$

Rules:

$$\frac{\sigma \equiv \sigma' \quad \tau \equiv \tau'}{\sigma \to \sigma' \equiv \tau \to \tau'} \; (\to) \qquad \frac{\sigma \equiv \sigma'}{\forall \alpha. \sigma \equiv \forall \alpha. \sigma'} \; (\forall)$$

$$(\text{trans}) \; \frac{\sigma \equiv \rho \quad \rho \equiv \tau}{\sigma \equiv \tau} \qquad (\text{symm}) \; \frac{\sigma \equiv \tau}{\tau \equiv \sigma}$$

We write $\vdash \sigma \equiv \tau$ to indicate that there is a derivation of $\sigma \equiv \tau$ in the above proof system.

Lemma 6 ([Tiu95]). *The types $\forall \alpha \beta. \sigma$ and $\forall \beta \alpha. \sigma$ are bicoercible.*

Proof. The proof from [Tiu95] carries over to the predicative case without any changes.

Lemma 7. *If $\vdash \sigma \equiv \tau$ then σ and τ are bicoercible.*

Proof. Soundness of **A2** follows from the previous lemma. Soundness of other axioms and rules is easily verified, since they correspond to respective axioms and rules for subtyping (apart from *symm*, but this rule is obvious anyway).

Corollary 1. *Every type is bicoercible with its canonical form.*

Lemma 8. *Bicoercible types have identical tree skeletons.*

Lemma 9. *Let $\sigma = \forall \alpha.\sigma', \tau = \forall \beta.\tau'$ (where σ', τ' have no quantifiers on top level) be types with no redundant quantifiers and such that for every path $\Pi \in \mathrm{dom}(\tau)$ such that $\mathrm{quant}(\tau, \Pi) = \varepsilon$, $\Pi \in \mathrm{dom}(\sigma)$. If $\sigma \leq \tau$ then*

$$L(\alpha) \geq_{Mul} L(\beta)$$

Theorem 3. *If σ and τ are bicoercible types in canonical form then they are identical.*

Proof. If σ and τ are bicoercible then there exist types $\sigma_1, \ldots \sigma_n, \tau_1, \ldots \tau_m$ such that

$$\sigma \sqsubset \sigma_1 \sqsubset \cdots \sigma_n \sqsubset \tau \sqsubset \tau_1 \sqsubset \cdots \sqsubset \tau_m \sqsubset \sigma$$

Such sequence of types will be called a *circle*. We may also assume that all the types are in canonical form. Since all the types in the circle are bicoercible, they all have the same tree skeleton. Now, by inspection of the circle it is easy to prove that if $\forall \alpha.\eta, \forall \beta.\theta$ are members of the circle (and in canonical form, with η and θ having no quantifiers at the top level) then

1. $L(\alpha) \leq_{Mul} L(\beta)$,
2. hence $L(\alpha) = L(\beta)$ (since this is a circle),
3. hence $\alpha = \beta$ by the definition of the canonical form
4. hence η and θ are bicoercible

Obviously, if $\sigma_1 \to \sigma_2$ and $\tau_1 \to \tau_2$ are bicoercible then so are σ_1 and τ_1 as well as σ_2 and τ_2. This completes the proof.

Theorem 4. *σ and τ are bicoercible iff $\vdash \sigma \equiv \tau$ holds.*

7 Typability

In this section we define a predicative variant of system F with subsumption. Following an idea of Trevor Jim [Jim95, Jim96] typability in this system can be reduced to the satisfiability problem and hence also to the subsumption problem (albeit at the price of increasing level by one).

The set of types is the set of stratified types of level not greater than t defined in section 1.1.

A *type environment* is a function from term variables to types. If α is the set of all type variables of level less then t which are free in σ but not in E, then $Gen(E, \sigma) = \forall \alpha.\sigma$

The typing judgements $E \vdash M : \sigma$ of predicative system Γ with subsumption are defined by the following rules:

$$E(x:\sigma) \vdash x:\sigma$$

$$\frac{E(x:\tau) \vdash M:\rho}{E \vdash \lambda x.M : \tau \to \rho}$$

$$\frac{E \vdash M : \tau \to \rho \quad E \vdash N : \tau}{E \vdash MN : \rho}$$

$$\frac{E \vdash M : \tau}{E \vdash M : \sigma} \quad Gen(E, \sigma) \leq \tau$$

Note that the last rule has as special cases the rules for instantiation and generalisation:

$$\frac{E \vdash M : \tau}{E \vdash M : \forall \alpha.\tau} \quad \alpha \notin FV(E)$$

$$\frac{E \vdash M : \forall \alpha^q.\tau}{E \vdash M : \tau[\rho/\alpha^q]} \quad L(\rho) \leq \alpha$$

The following theorem, formulated by Jim for system $\mathcal{F}$, carries over to the predficative case:

Theorem 5 (Jim95). *For a given closed term M there exist types σ, τ such that M is typable if and only if $\sigma \leq \tau$.*

8 Future Work

We have proven some preliminary results on predicative system $\mathcal{F}$ with subsumption. This work was done in order to make necessary background for studying the type inference problem in this interesting type theory. The basic aim would be to prove decidability in general case — this would give very strong extension of the ML polymorphism. Some partial decidability results going beyond the Läufer and Odersky approach are interesting, too.

References

[Ben98] Marcin Benke. Predicative polymorphic subtyping. Technical Report TR98-02(251), Institute of Informatics, Warsaw University, March 1998. Available from http://zls.mimuw.edu.pl/~ben/Papers/.

[Chr98] Jacek Chrząszcz. Polimorphic subtyping without distributivity. MFCS'98 (in this volume), 1998.

[DM82] Luis Damas and Robin Milner. Principal type-schemes for functional programs. In *Conf. Rec. ACM Symp. Principles of Programming Languages*, pages 207–211, 1982.

[Gir72] J.-Y. Girard. *Interprétation fonctionelle et élimination des coupures dans l'arithmetique d'ordre supérieur.* PhD thesis, Université Paris VII, 1972.
[Jim95] Trevor Jim. System F plus subsumption reduces to Mitchell's subtyping relation. Manuscript, 1995.
[Jim96] Trevor Jim. What are principal typings and what are they good for? In *Conf. Rec. ACM Symp. Principles of Programming Languages*, pages 42–53, 1996.
[KT92] A. J. Kfoury and J. Tiuryn. Type reconstruction in finite-rank fragments of the second-order λ-calculus. *Information and Computation*, 2(98):228–257, June 1992 1992.
[Lei91] Daniel Leivant. Finitely stratified polymorphism. *Information and Computation*, 93:93–113, 1991.
[LMS95] G. Longo, K. Milsted, and S. Soloviev. A logic of subtyping. In *Proc. IEEE Symp. on Logic in Computer Science*, pages 292–299, 1995.
[LO94] Konstantin Läufer and Martin Odersky. Polymorphic type inference and abstract data types. *ACM Transactions on Programming Languages and Systems*, 16(5):1411–1430, September 1994.
[Mil78] Robin Milner. A theory of type polymorphism in programming. *Journal of Computer and System Sciences*, 17(14):348–375, December 1978.
[Mit88] John C. Mitchell. Polymorphic type inference and containment. *Information and Computation*, 76(2/3):211–249, 1988. Reprinted in *Logical Foundations of Functional Programming,* ed. G. Huet, Addison-Wesley (1990) 153–194.
[MP88] John Mitchell and Gordon Plotkin. Abstract types have existential types. *ACM Transactions on Programming Languages and Systems*, 10(3):470–502, 1988.
[OL96] Martin Odersky and Konstantin Läufer. Putting type annotations to work. In *Conf. Rec. ACM Symp. Principles of Programming Languages*, pages 54–67, 1996.
[Per90] N. Perry. *The Implementation of Practical Functional Programming Languages.* PhD thesis, Imperial College of Science, Technology, and Medicine, University of London, 1990.
[Rém94] Didier Rémy. Programming objects with ml-art, and extension to ml with abstract and record types. In *Proceedings of Theoretical Aspects of Programming Languages*, number 789 in LNCS, pages 321–346. Springer, 1994.
[Rey74] J. C. Reynolds. Mathematical foundations of software development. volume 19 of *Lecture Notes in Computer Science*, chapter Towards a theory of type structure., pages 408–425. Springer, 1974.
[Tiu95] Jerzy Tiuryn. Equational axiomatization of bicoercibility for polymorphic types. In Ed. P.S. Thiagarajan, editor, *Proc. 15th Conference Foundations of Software Technology and Theoretical Computer Science*, volume 1026 of *Lecture Notes in Computer Science*, pages 166–179. Springer Verlag, 1995.
[TU96] Jerzy Tiuryn and Paweł Urzyczyn. The subtyping problem for second-order types is undecidable. In *Proceedings of 11th LICS*, 1996.
[Wel94] J. Wells. Typability and type checking in the second-order λ-calculus are equivalent and undecidable. In *Proc. 9th Ann. IEEE Symp. Logic in Comput. Sci.*, pages 176–185, 1994.
[Wel95] J. Wells. The undecidability of Mitchell's subtyping relation. Technical report, Computer Sci. Dept., Boston University, December 1995.
[Wel96] J. B. Wells. Typability is undecidable for F+eta. Technical Report 96-022, Boston University, March 9, 1996.

A Computational Interpretation of the $\lambda\mu$-Calculus

G.M. Bierman

University of Cambridge

Abstract. This paper proposes a simple computational interpretation of Parigot's $\lambda\mu$-calculus. The $\lambda\mu$-calculus is an extension of the typed λ-calculus which corresponds via the Curry-Howard correspondence to classical logic. Whereas other work has given computational interpretations by translating the $\lambda\mu$-calculus into other calculi, I wish to propose here a direct computational interpretation. This interpretation is best given as a single-step semantics which, in particular, leads to a relatively simple, but powerful, operational theory.

1 Introduction

It is well-known that the typed λ-calculus can be viewed as a term assignment for natural deduction proofs in intuitionistic logic (**IL**). Consequently the set of types of all closed λ-terms enumerates all intuitionistic tautologies. This is known as the Curry-Howard correspondence, or the formulae-as-types principle. Thus one can talk of a computational interpretation of **IL**. A natural question is whether there is such a computational interpretation of classical logic (**CL**). A first step is to devise a well behaved natural deduction formulation for **CL** and give a term assignment. A number of proposals have been made but recently Parigot [9] introduced a extension of the typed λ-calculus, which he called the $\lambda\mu$-calculus. The set of types of all closed $\lambda\mu$-terms enumerates all classical tautologies and the calculus is particularly well behaved, satisfying both strong normalisation and confluence.

However two questions remain. First, what does the extension to the $\lambda\mu$-calculus mean computationally? Secondly, if the $\lambda\mu$-calculus is extended in much the same way as the λ-calculus is extended to yield PCF, what is its operational theory? Of course the answer to the second question is heavily dependent upon the answer to the first. In this paper I suggest that the $\lambda\mu$-calculus has a natural computational reading: it is a λ-calculus which is able to manipulate the runtime environment via indexed catch and throw operators. This can easily be expressed using evaluation contexts which are common in work on control operators.

Morris-style contextual equivalence is commonly accepted as the natural notion of equivalence for functional languages. There has been significant effort in devising alternative characterisations of contextual equivalence which are more amenable for constructing proofs. For PCF the common solution is to use some form of (applicative) bisimilarity [3]. However these techniques do not often extend to languages with control. In §6 I give a simple notion of program equivalence, based on transitions in an abstract machine, which coincides with contextual equivalence.

2 Parigot's $\lambda\mu$-Calculus

In his seminal paper Parigot introduced an extension of the typed λ-calculus, which he called the $\lambda\mu$-calculus. The extension is such that terms no longer have a single type but a *sequence* of types, one of which is said to be the active type and the rest which are said to be passive. I shall not go into great detail here—the reader is referred to any one of a number of good introductions [1, 7, 8].

Types are given by the grammar $\phi ::= \bot \mid \phi \to \phi$, and raw $\lambda\mu$-terms are given by

$$
\begin{aligned}
M ::=~ & x & & \text{Variable} \\
\mid~ & \lambda x{:}\phi.M & & \text{Abstraction} \\
\mid~ & MM & & \text{Application} \\
\mid~ & [a{:}\phi]M & & \text{Passification} \\
\mid~ & \mu a{:}\phi.M & & \text{Activation;}
\end{aligned}
$$

where x is taken from a countable set of λ-variables, ϕ is a well-formed type (formula) and a is taken from a countable set of μ-variables.

Typing judgements are of the form, $\Gamma \triangleright M{:}\phi, \Sigma$, where Γ is a set of pairs of λ-variables and types written $x{:}\psi$, M is a term from the above grammar and Σ denotes a set of pairs of μ-variables and types written $a{:}\varphi$ (thus ϕ is the active type). The typing rules are as follows.

$$\frac{}{\Gamma, x{:}\phi \triangleright x{:}\phi, \Sigma}\text{ Identity}$$

$$\frac{\Gamma, x{:}\phi \triangleright M{:}\psi, \Sigma}{\Gamma \triangleright \lambda x{:}\phi.M{:}\phi \to \psi, \Sigma}\to_\mathcal{I} \qquad \frac{\Gamma \triangleright M{:}\phi \to \psi, \Sigma \quad \Gamma \triangleright N{:}\phi, \Sigma}{\Gamma \triangleright MN{:}\psi, \Sigma}\to_\mathcal{E}$$

$$\frac{\Gamma \triangleright M{:}\phi, \Sigma}{\Gamma \triangleright [a{:}\phi]M{:}\bot, a{:}\phi, \Sigma}\text{ Passify} \qquad \frac{\Gamma \triangleright M{:}\bot, a{:}\phi, \Sigma}{\Gamma \triangleright \mu a{:}\phi.M{:}\phi, \Sigma}\text{ Activate}$$

The new rules are the Passify and Activate. The former takes a term whose active type is ϕ (where ϕ is not $\bot$) and passifies it, i.e. ϕ becomes a passive type (and is hence labelled with a). The resulting term has an active type of $\bot$.[1] The Activate rule works similarly but in the reverse direction.

There are a number of reduction rules associated with the $\lambda\mu$-calculus. In full they are as follows.

$$
\begin{aligned}
(\lambda x{:}\phi.M)N &\leadsto_\beta M[x := N] \\
\mu a{:}\phi.[a{:}\phi]M &\leadsto_\beta M & & \text{where } a \notin \mu\text{FV}(M) \\
(\mu a{:}\phi \to \psi.M)N &\leadsto_c \mu a{:}\psi.M[[a{:}\phi \to \psi]P \Leftarrow [a{:}\psi]PN] \\
\lambda x.Mx &\leadsto_\eta M & & \text{where } x \notin \text{FV}(M) \\
[a{:}\phi]\mu b{:}\phi.M &\leadsto_\eta M[a/b]
\end{aligned}
$$

In the second β-rule, $\mu\text{FV}(M)$ denotes the set of free μ-variables in the term M (I shall omit its rather obvious definition). In the commuting conversion ($\leadsto_c$) I have used the

---

[1] This ensures that every term has an active type. It is possible to give a formulation where terms need not have an active type.

notation $M[N \Leftarrow P]$ to denote the term M where *all* occurrences of the subterm N have been replaced by the term P. In the last η-rule, $M[a/b]$ denotes the term M where all free occurrences of the μ-variable b are replaced with a. All forms of substitution are assumed to be non-capturing.

3 A Computational Interpretation

As it stands it is unclear what this move to **CL** has given us—clearly we have terms at new types and new terms at old types, but what does this mean computationally? In order to find an answer I shall consider the operational behaviour of the calculus, namely the execution of closed terms (programs) to canonical values.

Before presenting the operational behaviour I need first to introduce some standard terminology from work on control operators, e.g. [2]. To formalise the notion of an evaluation order, Felleisen [*op. cit.*], defined an *evaluation context*. This is essentially a term with a single 'hole' in it, written $E[\bullet]$ (this will be defined formally in the next section). The result of placing a term, M, in that hole is written $E[M]$. Evaluation contexts are devised so that every closed term, M, is either a canonical value or can be written *uniquely* as $E[R]$, where R is a redex. The context $E[\bullet]$ can be thought of as representing the rest of the computation that remains to be done after R has been reduced. In this sense it can be seen as the *continuation* of R or, more simply, the *current continuation*.

Evaluation is then written as $(E[R], \mathcal{E}) \Rightarrow (M', \mathcal{E}')$, where $\mathcal{E}$ is a function from μ-variables to evaluation contexts—the need for this will become clear. The important evaluation rules are

$$(E[\mu a.M], \mathcal{E}) \Rightarrow (M, \mathcal{E} \uplus (a \mapsto E[\bullet]))$$
$$(E[[a]M], \mathcal{E} \uplus (a \mapsto E'[\bullet])) \Rightarrow (E'[M], \mathcal{E} \uplus (a \mapsto E'[\bullet]));$$

where $\mathcal{E} \uplus (a \mapsto E[\bullet])$ denotes the extension of the function $\mathcal{E}$ with the mapping $a \mapsto E[\bullet]$. Thus in the first reduction rule the current continuation is captured ('catch'), added to $\mathcal{E}$ and indexed with a. In the second reduction rule the appropriate indexed continuation is taken from $\mathcal{E}$, replacing the current continuation (i.e. the term M is 'thrown back' to an earlier continuation). In summary, the Activate and Passify rules are interpreted as indexed catch and throw operators, respectively.

4 μPCF

Rather than develop an operational theory for the $\lambda\mu$-calculus, I shall first enrich it with natural numbers, a conditional, pairs and recursion. This is essentially what Ong and Stewart call μPCF [8]. The next step is to choose an evaluation strategy. Most work on control operators has considered a *call-by-value* strategy and to aid comparison I shall adopt the same. It is important to note that what is developed in this section can easily be adjusted to reflect a call-by-name strategy; some details are sketched in §7. This is in contrast with Ong and Stewart's framework, which requires significant changes to move from call-by-name to call-by-value (some details are in their paper [8]). For completeness the typing rules for the new constructors are given below.

$$\frac{\Gamma \triangleright M\colon \mathsf{int}, \Sigma}{\Gamma \triangleright \underline{n}\colon \mathsf{int}, \Sigma} \quad \frac{\Gamma \triangleright M\colon \mathsf{int}, \Sigma}{\Gamma \triangleright \mathsf{suc}(M)\colon \mathsf{int}, \Sigma} \quad \frac{\Gamma, f\colon \phi \to \phi, x\colon \phi \triangleright M\colon \phi, \Sigma \quad \Gamma, f\colon \phi \to \phi \triangleright N\colon \psi, \Sigma}{\Gamma \triangleright \mathsf{letrec}\ f = \lambda x.M\ \mathsf{in}\ N\colon \psi, \Sigma}$$

$$\frac{\Gamma \triangleright M\colon \mathsf{int}, \Sigma \quad \Gamma \triangleright N\colon \phi, \Sigma \quad \Gamma \triangleright P\colon \phi, \Sigma}{\Gamma \triangleright \mathsf{ifz}\ M\ \mathsf{then}\ N\ \mathsf{else}\ P\colon \phi, \Sigma}$$

$$\frac{\Gamma \triangleright M\colon \phi, \Sigma \quad \Gamma \triangleright N\colon \psi, \Sigma}{\Gamma \triangleright \langle M, N\rangle\colon \phi \times \psi, \Sigma} \quad \frac{\Gamma \triangleright M\colon \phi \times \psi, \Sigma}{\Gamma \triangleright \mathsf{fst}(M)\colon \phi, \Sigma} \quad \frac{\Gamma \triangleright M\colon \phi \times \psi, \Sigma}{\Gamma \triangleright \mathsf{snd}(M)\colon \psi, \Sigma}$$

The syntactic classes of values, evaluation contexts and redexes are defined as follows.

$$\begin{aligned}
\text{Values} \quad & v ::= \underline{n} \mid \lambda x.M \mid \langle v, v\rangle \\
\text{Evaluation Contexts}\ E ::= & \bullet \mid vE \mid EM \\
& \mid \langle E, M\rangle \mid \langle v, E\rangle \mid \mathsf{fst}(E) \mid \mathsf{snd}(E) \\
& \mid \mathsf{suc}(E) \mid \mathsf{ifz}\ E\ \mathsf{then}\ M\ \mathsf{else}\ M \\
\text{Redexes} \quad & R ::= vv \mid \mathsf{fst}(v) \mid \mathsf{snd}(v) \\
& \mid \mathsf{suc}(v) \mid \mathsf{ifz}\ v\ \mathsf{then}\ M\ \mathsf{else}\ M \\
& \mid \mathsf{letrec}\ f = \lambda x.M\ \mathsf{in}\ N \mid [a]M \mid \mu a.M
\end{aligned}$$

The fundamental property of evaluation contexts is the following.

Lemma 1. *Every closed term, M, is either a value, v, or is uniquely of the form $E[R]$, where $E[\bullet]$ is an evaluation context and R is a redex.*

We can now write out the (single-step) reduction rules in full, which are as follows.

$$\begin{aligned}
(E[(\lambda x.M)v], \mathcal{E}) &\Rightarrow (E[M[x := v]], \mathcal{E}) \\
(E[\mathsf{fst}(\langle v, w\rangle)], \mathcal{E}) &\Rightarrow (E[v], \mathcal{E}) \\
(E[\mathsf{snd}(\langle v, w\rangle)], \mathcal{E}) &\Rightarrow (E[w], \mathcal{E}) \\
(E[\mathsf{suc}(\underline{n})], \mathcal{E}) &\Rightarrow (E[\underline{n+1}], \mathcal{E}) \\
(E[\mathsf{ifz}\ \underline{0}\ \mathsf{then}\ M\ \mathsf{else}\ N], \mathcal{E}) &\Rightarrow (E[M], \mathcal{E}) \\
(E[\mathsf{ifz}\ (\underline{n+1})\ \mathsf{then}\ M\ \mathsf{else}\ N], \mathcal{E}) &\Rightarrow (E[N], \mathcal{E}) \\
(E[\mathsf{letrec}\ f = \lambda x.M\ \mathsf{in}\ N], \mathcal{E}) &\Rightarrow (E[N[f := \lambda x.\mathsf{letrec}\ f = \lambda x.M\ \mathsf{in}\ M]], \mathcal{E}) \\
(E[\mu a.M], \mathcal{E}) &\Rightarrow (M, \mathcal{E} \uplus (a \mapsto E[\bullet])) \\
(E[[a]M], \mathcal{E} \uplus (a \mapsto E'[\bullet])) &\Rightarrow (E'[M], \mathcal{E} \uplus (a \mapsto E'[\bullet]))
\end{aligned}$$

5 Examples

To demonstrate the expressive power of this computational interpretation I shall show the dynamics of particular ML-like exception handling and 'callcc' primitives are preserved by their encodings into μPCF (the encodings are due to Ong and Stewart [8]).

5.1 Exception Handling

ML can be extended with exceptions in a number of ways. One such method was given by Gunter *et al.* [5] and simplified by Ong and Stewart [8]. Typed exceptions arc identified with names, thus typing judgements (for ML) are now of the form $\Gamma; \Delta \triangleright M\colon \phi$

where Γ is the usual typing environment and Δ is the typing environment for the exception names. Two new operators are added to ML whose typing rules are as follows.

$$\frac{\Gamma; \Delta \triangleright M: A}{\Gamma; \Delta, a: A \triangleright \mathit{raise}(a, M): B} \qquad \frac{\Gamma; \Delta, a: A \triangleright M: A \to B \qquad \Gamma; \Delta, a: A \triangleright N: B}{\Gamma; \Delta \triangleright \mathit{handle}(a, M, N): B}$$

The intended interpretation is that the first rule evaluates M to a value v and then raises an exception named a associated with v. The second rule evaluates M to a value (say v) and then evaluates N. If N evaluates to a value w then this is the overall result, but if it raises an exception named a with a value u, then this is applied to v. Given as reduction rules the intended interpretation is as follows.

$$\begin{aligned} \mathit{handle}(a, v, w) &\leadsto w & (a \notin \mathrm{FN}(w)) \\ \mathit{handle}(a, v, E[\mathit{raise}(a, u)]) &\leadsto vu & (a \notin \mathrm{FN}(v, u)) \end{aligned}$$

These operators can be translated into μPCF as follows (where b is a fresh μ-variable).

$$\begin{aligned} [\![\mathit{raise}(a, M)]\!] &\stackrel{\mathrm{def}}{=} (\lambda x.\mu b.[a]x)[\![M]\!] \\ [\![\mathit{handle}(a, M, N)]\!] &\stackrel{\mathrm{def}}{=} \mu b.[b][\![M]\!](\mu a.[b][\![N]\!]) \end{aligned}$$

It is relatively easy to show that this translation preserves the operational behaviour, e.g.

$$\begin{aligned} &([\![\mathit{handle}(a, M, E[\mathit{raise}(a, N)])]\!], \mathcal{E}) \\ &\stackrel{\mathrm{def}}{=} (\mu b.[b][\![M]\!](\mu a.[b]E[(\lambda x.\mu c.[a]x)[\![N]\!]]), \mathcal{E}) \\ &\Rightarrow^2 ([\![M]\!](\mu a.[b]E[(\lambda x.\mu c.[a]x)[\![N]\!]]), \mathcal{E} \uplus \{b \mapsto \bullet\}) \\ &\Rightarrow^* (v(\mu a.[b]E[(\lambda x.\mu c.[a]x)[\![N]\!]]), \mathcal{E} \uplus \{b \mapsto \bullet\}) \\ &\Rightarrow ([b]E[(\lambda x.\mu c.[a]x)[\![N]\!]], \mathcal{E} \uplus \{a \mapsto (v\bullet), b \mapsto \bullet\}) \\ &\Rightarrow (E[(\lambda x.\mu c.[a]x)[\![N]\!]], \mathcal{E} \uplus \{a \mapsto (v\bullet), b \mapsto \bullet\}) \\ &\Rightarrow^+ (E[\mu c.[a]u)], \mathcal{E} \uplus \{a \mapsto (v\bullet), b \mapsto \bullet\}) \\ &\Rightarrow ([a]u, \mathcal{E} \uplus \{a \mapsto (v\bullet), b \mapsto \bullet, c \mapsto E[\bullet]\}) \\ &\Rightarrow (vu, \mathcal{E} \uplus \{a \mapsto (v\bullet), b \mapsto \bullet, c \mapsto E[\bullet]\}) \end{aligned}$$

5.2 Call-with-Current-Continuation (callcc)

ML can be extended with operators to manipulate first-class continuations in a number of ways. I shall consider a proposal again due to Gunter *et al.* [5] and simplified by Ong and Stewart [8]. Here (typed) continuations are associated with names, and so typing judgements are of the form $\Gamma; \Delta \triangleright M: A$, where Δ is the typing environment for continuation names. Three new operators are added to ML, whose typing rules are as follows.

$$\frac{\Gamma; \Delta \triangleright M: (A \to B) \to A}{\Gamma; \Delta \triangleright \mathit{callcc}(M): A} \qquad \frac{\Gamma; \Delta \triangleright M: A}{\Gamma; \Delta, a: A \triangleright \mathit{abort}(a, M): B} \qquad \frac{\Gamma; \Delta, a: A \triangleright M: A}{\Gamma; \Delta \triangleright \mathit{set}(a, M): A}$$

The *callcc* operator applies the term M to an abstraction of the current continuation. The *set* serves as a delimiter for continuations, and the *abort* discards the current continuation (delimited by a). Their intended operational behaviour is as follows.

$$\begin{aligned} \mathit{set}(a, E[\mathit{abort}(a, M)]) &\leadsto M & (a \notin \mathrm{FN}(M)) \\ \mathit{set}(a, v) &\leadsto v & (a \notin \mathrm{FN}(v)) \\ E[\mathit{callcc}(M)] &\leadsto \mathit{set}(a, E[M(\lambda x.\mathit{abort}(a, E[x]))]) \end{aligned}$$

Ong and Stewart provided a translation of these operators into μPCF, which is as follows.

$$[\![callcc(M)]\!] \stackrel{\text{def}}{=} \mu a.[a]([\![M]\!](\lambda x.\mu b.[a]x))$$
$$[\![abort(a, M)]\!] \stackrel{\text{def}}{=} \mu b.[a][\![M]\!] \quad \text{where } b \notin \mu\text{FV}([\![M]\!])$$
$$[\![set(a, M)]\!] \stackrel{\text{def}}{=} \mu a.[a][\![M]\!]$$

Again it is simple to check that this translation preserves the operational behaviour, e.g.

$$([\![set(a, E[abort(a, M)])]\!], \mathcal{E})$$
$$\stackrel{\text{def}}{=} (\mu a.[a]E[\mu b.[a][\![M]\!]], \mathcal{E})$$
$$\Rightarrow^2 (E[\mu b.[a][\![M]\!]], \mathcal{E} \uplus \{a \mapsto \bullet\})$$
$$\Rightarrow ([a][\![M]\!], \mathcal{E} \uplus \{a \mapsto \bullet, b \mapsto E[\bullet]\})$$
$$\Rightarrow ([\![M]\!], \mathcal{E} \uplus \{a \mapsto \bullet, b \mapsto E[\bullet]\})$$

5.3 Pairing

It is easy to verify that $\phi \times \psi \equiv \neg(\phi \to \neg\psi)$ in **CL**. This logical equivalence can be used to simulate pairing in μPCF. The constructor and deconstructors are encoded as follows.[2]

$$pair \stackrel{\text{def}}{=} \lambda m{:}\phi.\lambda n{:}\psi.\lambda f{:}(\phi \to (\psi \to \bot)).f\, m\, n$$
$$fst \stackrel{\text{def}}{=} \lambda p.\mu a.p(\lambda x.\mu b.[a]x)$$
$$snd \stackrel{\text{def}}{=} \lambda p.\mu a.p(\lambda y.\lambda x.[a]x)$$

It is left to the reader to verify that these encodings satisfy the expected operational behaviour.

6 Operational Theory

An implementation based on the reduction rules given in §4 would work as follows. Take a term M: if it is a value then we are done; if not it can be given uniquely as $E[R]$. One takes the relevant reduction step (determined by R)—the resulting term is either a value, in which case we are done, or it has to be re-written again as an evaluation context and a redex. This process is repeated until a value is reached. The continual intermediate step of rewriting a term into an evaluation context and a redex would be inefficient in practice and is quite cumbersome theoretically. Consequently I shall give a new set of reduction rules where the context and the redex are actually separated. Reduction rules are now of the form $(S, M, \mathcal{E}) \longrightarrow (S', M', \mathcal{E}')$, where S is a stack of *evaluation frames*, which are defined as follows.

$$F ::= \bullet M \mid v \bullet \mid \langle \bullet, M \rangle \mid \langle v, \bullet \rangle$$
$$\mid \mathsf{fst}(\bullet) \mid \mathsf{snd}(\bullet) \mid \mathsf{suc}(\bullet) \mid \mathsf{ifz}\, \bullet \, \mathsf{then}\, M \, \mathsf{else}\, M$$

(Clearly $\mathcal{E}$ is now a function from μ-variables to stacks.) The reduction rules essentially describe the transitions of a simple abstract machine.[3] In full they are as follows.

---
[2] A similar encoding using control operators was given by Griffin [4].
[3] Harper and Stone [6] give similar transition rules in their analysis of SML and Pitts [10] has used similar rules in work on functional languages with dynamic allocation of store.

$$
\begin{aligned}
(F[\bullet] :: S, v, \mathcal{E}) &\longrightarrow (S, F[v], \mathcal{E}) \\
(S, MN, \mathcal{E}) &\longrightarrow ((\bullet N) :: S, M, \mathcal{E}) & M \text{ not a value} \\
(S, vN, \mathcal{E}) &\longrightarrow ((v\bullet) :: S, N, \mathcal{E}) & N \text{ not a value} \\
(S, (\lambda x.M)v, \mathcal{E}) &\longrightarrow (S, M[x := v], \mathcal{E}) \\
(S, \langle M, N \rangle, \mathcal{E}) &\longrightarrow ((\langle \bullet, N \rangle) :: S, M, \mathcal{E}) & M \text{ not a value} \\
(S, \langle v, N \rangle, \mathcal{E}) &\longrightarrow ((\langle v, \bullet \rangle) :: S, N, \mathcal{E}) & N \text{ not a value} \\
(S, \mathsf{fst}(M), \mathcal{E}) &\longrightarrow (\mathsf{fst}(\bullet) :: S, M, \mathcal{E}) & M \text{ not a value} \\
(S, \mathsf{fst}(\langle v, w \rangle), \mathcal{E}) &\longrightarrow (S, v, \mathcal{E}) \\
(S, \mathsf{snd}(M), \mathcal{E}) &\longrightarrow (\mathsf{snd}(\bullet) :: S, M, \mathcal{E}) & M \text{ not a value} \\
(S, \mathsf{snd}(\langle v, w \rangle), \mathcal{E}) &\longrightarrow (S, w, \mathcal{E}) \\
(S, \mathsf{suc}(M), \mathcal{E}) &\longrightarrow (\mathsf{suc}(\bullet) :: S, M, \mathcal{E}) & M \text{ not a value} \\
(S, \mathsf{suc}(\underline{n}), \mathcal{E}) &\longrightarrow (S, \underline{n+1}, \mathcal{E}) \\
(S, \mathsf{ifz}\, M \text{ then } N \text{ else } P, \mathcal{E}) &\longrightarrow ((\mathsf{ifz}\, \bullet \text{ then } N \text{ else } P) :: S, M, \mathcal{E}) M \text{ not a value} \\
(S, \mathsf{ifz}\, \underline{0} \text{ then } M \text{ else } N, \mathcal{E}) &\longrightarrow (S, M, \mathcal{E}) \\
(S, \mathsf{ifz}\, \underline{(n+1)} \text{ then } M \text{ else } N, \mathcal{E}) &\longrightarrow (S, N, \mathcal{E}) \\
(S, \mathsf{letrec}\, f = \lambda x.M \text{ in } N, \mathcal{E}) &\longrightarrow (S, N[f := \lambda x.\mathsf{letrec}\, f = \lambda x.M \text{ in } M], \mathcal{E}) \\
(S, \mu a.M, \mathcal{E}) &\longrightarrow ([], M, \mathcal{E} \uplus (a \mapsto S)) \\
(S, [a]M, \mathcal{E} \uplus (a \mapsto T)) &\longrightarrow (T, M, \mathcal{E} \uplus (a \mapsto T))
\end{aligned}
$$

An example may make these reduction rules clearer. Consider an instance of the 'callcc' reduction rule given in §5.2.

$$set(a, (\lambda x.N)(abort(a, M))) \rightsquigarrow M$$

The left hand term is translated to the μPCF-term $\mu a.[a](\lambda x.\llbracket N \rrbracket)(\mu b.[a]\llbracket M \rrbracket)$, which reduces as follows.

$$
\begin{aligned}
&(S, & \mu a.[a](\lambda x.\llbracket N \rrbracket)(\mu b.[a]\llbracket M \rrbracket), & \mathcal{E}) \\
\longrightarrow &([], & [a](\lambda x.\llbracket N \rrbracket)(\mu b.[a]\llbracket M \rrbracket), & \mathcal{E} \uplus \{a \mapsto S\}) \\
\longrightarrow &(S, & (\lambda x.\llbracket N \rrbracket)(\mu b.[a]\llbracket M \rrbracket), & \mathcal{E} \uplus \{a \mapsto S\}) \\
\longrightarrow &(((\lambda x.\llbracket N \rrbracket)\bullet) :: S, & \mu b.[a]\llbracket M \rrbracket, & \mathcal{E} \uplus \{a \mapsto S\}) \\
\longrightarrow &([], & [a]\llbracket M \rrbracket, & \mathcal{E} \uplus \{a \mapsto S, b \mapsto ((\lambda x.\llbracket N \rrbracket)\bullet) :: S\}) \\
\longrightarrow &(S, & \llbracket M \rrbracket, & \mathcal{E} \uplus \{a \mapsto S, b \mapsto ((\lambda x.\llbracket N \rrbracket)\bullet) :: S\})
\end{aligned}
$$

It is easy to define a function $\ulcorner E \urcorner$ which converts a given evaluation context, E to a stack of frames, and a function $S@M$ which takes a stack of frames, S, and a term, M, and converts the stack back to an evaluation context before inserting M. For example

$$\ulcorner(((\lambda x.M)\bullet)P)Q\urcorner \stackrel{\mathrm{def}}{=} ((\lambda x.M)\bullet) :: ((\bullet P) :: ((\bullet Q) :: []))$$
$$((\lambda x.M)\bullet) :: ((\bullet P) :: ((\bullet Q) :: []))@N \stackrel{\mathrm{def}}{=} (((\lambda x.M)N)P)Q$$

The two sets of reduction rules can be related in the following sense.

Proposition 1 $(S@M, \mathcal{E}) \Rightarrow (N, \mathcal{E}')$ iff $\exists S', M'.N = S'@M', (S, M, \ulcorner \mathcal{E} \urcorner) \longrightarrow^* (S', M', \ulcorner \mathcal{E}' \urcorner)$

An important fact (first discovered by Pitts [10] in a different setting) is that the set

$$\searrow \stackrel{\mathrm{def}}{=} \{(S, M, \mathcal{E}) \mid \exists v, \mathcal{E}'.(S, M, \mathcal{E}) \longrightarrow^* ([], v, \mathcal{E}')\}$$

has a direct, inductive definition which is as follows.

$([], v, \mathcal{E}) \searrow$

$$\frac{(S, F[v], \mathcal{E}) \searrow}{(F[\bullet] :: S, v, \mathcal{E}) \searrow}$$

$$\frac{((\bullet N) :: S, M, \mathcal{E}) \searrow}{(S, MN, \mathcal{E}) \searrow} M \text{ not a value} \quad \frac{((v\bullet) :: S, N, \mathcal{E}) \searrow}{(S, vN, \mathcal{E}) \searrow} M \text{ not a value}$$

$$\frac{(S, M[x := v], \mathcal{E}) \searrow}{(S, (\lambda x.M)v, \mathcal{E}) \searrow} \quad \frac{(S, N[f := \lambda x.\mathsf{letrec}\ f = \lambda x.M \text{ in } M], \mathcal{E}) \searrow}{(S, \mathsf{letrec}\ f = \lambda x.M \text{ in } N, \mathcal{E}) \searrow}$$

$$\frac{(\langle \bullet, N \rangle :: S, M, \mathcal{E}) \searrow}{(S, \langle M, N \rangle, \mathcal{E}) \searrow} M \text{ not a value} \quad \frac{(\langle v, \bullet \rangle :: S, N, \mathcal{E}) \searrow}{(S, \langle v, N \rangle, \mathcal{E}) \searrow} N \text{ not a value}$$

$$\frac{(\mathsf{fst}(\bullet) :: S, M, \mathcal{E}) \searrow}{(S, \mathsf{fst}(M), \mathcal{E}) \searrow} M \text{ not a value} \quad \frac{(S, v, \mathcal{E}) \searrow}{(S, \mathsf{fst}(\langle v, w \rangle), \mathcal{E}) \searrow}$$

$$\frac{(\mathsf{snd}(\bullet) :: S, M, \mathcal{E}) \searrow}{(S, \mathsf{snd}(M), \mathcal{E}) \searrow} M \text{ not a value} \quad \frac{(S, w, \mathcal{E}) \searrow}{(S, \mathsf{snd}(\langle v, w \rangle), \mathcal{E}) \searrow}$$

$$\frac{(T, M, \mathcal{E} \uplus (a \mapsto T)) \searrow}{(S, [a]M, \mathcal{E} \uplus (a \mapsto T)) \searrow} \quad \frac{([], M, \mathcal{E} \uplus (a \mapsto S)) \searrow}{(S, \mu a.M, \mathcal{E}) \searrow}$$

Given two terms M and N such that $\emptyset \triangleright M : \phi, \Sigma$ and $\emptyset \triangleright N : \phi, \Sigma$, they are said to be *ciu-similar*, written $M \leq_{\phi, \Sigma} N$, just when $\forall S, \mathcal{E}$. if $(S, M, \mathcal{E}) \searrow$ then $(S, N, \mathcal{E}) \searrow$. They are said to be *ciu-equivalent*, written $M \simeq_{\phi, \Sigma} N$ just when $M \leq_{\phi, \Sigma} N$ and $N \leq_{\phi, \Sigma} M$. Both these relations are extended to open terms in the obvious way.

This notion of equivalence is quite refined, consider the following terms (where Ω is a looping term, which can be defined using the recursion operator).

$$T_1 \stackrel{\text{def}}{=} \mu a.[a](\lambda y.\mu c.[a](\lambda x.\mathsf{ifz}\ y\ \mathsf{then}\ \Omega\ \mathsf{else}\ \underline{0}))$$
$$T_2 \stackrel{\text{def}}{=} \lambda z.\mu b.[b]((\lambda y.\mu c.[b]((\lambda x.\mathsf{ifz}\ y\ \mathsf{then}\ \Omega\ \mathsf{else}\ \underline{0})z))z)$$

It is easy to verify that $T_1 \underline{n} \simeq_{\mathsf{int}} T_2 \underline{n}$ for all natural numbers n. However they are *not* ciu-equivalent as

$$([(\lambda s.s(s\underline{1}))\bullet], T_1, \emptyset) \searrow,$$

but it is *not* the case that

$$([(\lambda s.s(s\underline{1}))\bullet], T_2, \emptyset) \searrow .$$

This is an important example as T_1 and T_2 *are* equivalent given the definition of applicative bisimilarity by Ong and Stewart [8]. (Their notion of bisimilarity is hence not a congruence.)

We can make the following definitions.

$$(M, \mathcal{E}) \Downarrow (v, \mathcal{E}') \stackrel{\text{def}}{=} (M, \mathcal{E}) \Rightarrow^* (v, \mathcal{E}') \text{ and} (v, \mathcal{E}') \not\Rightarrow$$
$$(M, \mathcal{E}) \Downarrow \stackrel{\text{def}}{=} \exists v, \mathcal{E}'.(M, \mathcal{E}) \Downarrow (v, \mathcal{E}')$$

Let C be a context, which is a μPCF-term with (possibly many) hole(s) in it (not to be confused with an evaluation context). We say that two terms M and N are *contextually equivalent*, written $M \approx N$, when $\forall \mathcal{C}, \mathcal{E}.(\mathcal{C}[M], \mathcal{E}) \Downarrow$ iff $(\mathcal{C}[N], \mathcal{E}) \Downarrow$. In other words, two terms are contextually equivalent if no larger program can tell them apart.

The two terms given above (T_1 and T_2) are not contextually equivalent, as the context $(\lambda s.s(s\underline{1}))\bullet$ distinguishes them. Clearly this notion of contextual equivalence is highly desirable but awkward to work with given the quantification over all contexts. However the notion of ciu-equivalence is more usable and an interesting question is in what sense they are related. In fact we find that they coincide!

Theorem 1. $\forall M, N. M \approx N$ iff $M \simeq N$.

Proof. The proof is adapted from the standard one for purely functional languages (see, for example, the chapter by Pitts [11]). It uses a variant of Howe's method.

This means that to prove two terms contextually equivalent we need only to show that they are ciu-equivalent, which is significantly easier. For example, it is simple to show the following ciu-equivalences.

$$(\lambda x.M)v \simeq M[x := v]$$
$$\mu a.[a]M \simeq M \qquad a \notin \mu\text{FV}(M)$$
$$(\mu a.M)N \simeq \mu b.M[[a]P \Leftarrow [b]PN]$$

For example, the second equivalence holds by the assumption that $a \notin \mu\text{FV}(M)$ and by observing

$$\frac{\dfrac{(S, M, \mathcal{E} \uplus (a \mapsto S))\searrow}{([], [a]M, \mathcal{E} \uplus (a \mapsto S))\searrow}}{(S, \mu a.[a]M, \mathcal{E})\searrow}$$

7 Call-by-Name

This paper has so far considered only call-by-value computation. However it is very simple to provide a computational interpretation for a call-by-name evaluation strategy. The main difference is in the (new) definition of values, evaluation contexts and redexes, which are as follows.

Values $\qquad v ::= \underline{n} \mid \lambda x.M \mid \langle M, M \rangle$

Evaluation Contexts $E ::= \bullet \mid EM \mid \text{fst}(E) \mid \text{snd}(E) \mid \text{suc}(E) \mid \text{ifz } E \text{ then } M \text{ else } M$

Redexes $\qquad R ::= vM \mid \text{fst}(v) \mid \text{snd}(v) \mid \text{suc}(v) \mid \text{ifz } v \text{ then } M \text{ else } M$
$\qquad\qquad\qquad \mid \text{rec } x.M \mid [a]M \mid \mu a.M$

The evaluation rules are as before except for the following.

$$(E[(\lambda x.M)N], \mathcal{E}) \Rightarrow (E[M[x := N]], \mathcal{E})$$
$$(E[\text{fst}(\langle M, N \rangle)], \mathcal{E}) \Rightarrow (E[M], \mathcal{E})$$
$$(E[\text{snd}(\langle M, N \rangle)], \mathcal{E}) \Rightarrow (E[N], \mathcal{E})$$
$$(E[\text{rec } x.M], \mathcal{E}) \Rightarrow (E[M[x := (\text{rec } x.M)]], \mathcal{E})$$

The development of the corresponding operational theory follows closely that outlined in §6. The differs sharply from the treatment given by Ong and Stewart [8] who have to introduce completely new reduction rules to move from a call-by-name to a call-by-value setting.

8 Conclusion

In this paper I have given a simple computation interpretation of the $\lambda\mu$-calculus: it is a λ-calculus which is extended with indexed operators to manipulate the runtime environment. This is maybe not too surprising as Griffin [4] has shown the close relationship between classical logic and languages with control. This interpretation can be expressed as a single-step reduction semantics using environment contexts. In turn I gave an equivalent semantics expressed as steps of a simple abstract machine, which eliminated the need for the evaluation contexts. Using this simple abstract machine it is possible to define a notion of program equivalence based on a termination relation which coincides with a natural definition of contextual equivalence.

Clearly the work by Ong and Stewart [8] is most closely related to that reported here. Their thesis is that μPCF is a foundational language for call-by-value functional computation with control and this paper can be seen as further evidence to that claim. However I would claim that the operational treatment given here is more intuitive, more flexible (in that different calling mechanisms can be handled easily) and leads to a more refined notion of program equivalence.

References

[1] G.M. BIERMAN. A classical linear λ-calculus. Technical Report 401, Cambridge Computer Laboratory 1996.
[2] M. FELLEISEN. The theory and practice of first-class prompts. POPL 1988.
[3] A.D. GORDON. Bisimilarity as a theory of functional programming: Mini-course. Technical Report NS–95–2, BRICS, Department of Computer Science, University of Århus, July 1995.
[4] T.G. GRIFFIN. A formulae-as-types notion of control. POPL 1990.
[5] C.A. GUNTER, D. RÉMY, AND J.G. RIECKE. A generalisation of exceptions and control in ML-like languages. FPCA 1995.
[6] R. HARPER AND C. STONE. An interpretation of Standard ML in type theory. Technical Report CMU–CS–97–147, School of Computer Science, Carnegie Mellon University, June 1997.
[7] M. HOFMANN AND T. STREICHER. Continuation models are universal for $\lambda\mu$-calculus. LICS 1997.
[8] C.-H.L. ONG AND C.A. STEWART. A Curry-Howard foundation for functional computation with control. POPL 1997.
[9] M. PARIGOT. $\lambda\mu$-calculus: an algorithmic interpretation of classical natural deduction. LPAR 1992. LNCS 624.
[10] A.M. PITTS. Operational semantics for program equivalence. Slides from talk given at MFPS, 1997.
[11] A.M. PITTS. Operationally-based theories of program equivalence. In *Semantics and Logics of Computation*, CUP, 1997.

Polymorphic Subtyping Without Distributivity

Jacek Chrząszcz*

Institute of Informatics,
Warsaw University,
ul. Banacha 2, 02-097 Warsaw, Poland.
email: chrzaszcz@mimuw.edu.pl

Abstract. The subtyping relation in the polymorphic second-order λ-calculus was introduced by John C. Mitchell in 1988. It is known that this relation is undecidable, but all known proofs of this fact strongly depend on the distributivity axiom. Nevertheless it has been conjectured that this axiom does not influence the undecidability. The paper shows undecidability of subtyping when we remove distributivity from its definition. Furthermore, the full equational axiomatisation of the corresponding equivalence relation is given. Both results follow from an analysis of rewriting-style subtyping derivations.

1 Introduction

Polymorphism is one of the most important issues in modern programming. When we go beyond the limits of shallow polymorphism imposed by common functional languages like ML or Haskell, we encounter (among others) the following question: can values of a given type σ replace values of type τ without a type clash, or, differently speaking, is σ a *subtype* of τ?

In the higher-order polymorphic functional language, system **F**, the subtyping is characterized by $\forall \alpha_1 \ldots \alpha_n \, \sigma \;\sqsubseteq_\mathbf{F}\; \forall \beta_1 \ldots \beta_m \, \sigma \{\alpha_1 \mapsto \rho_1, \ldots, \alpha_n \mapsto \rho_n\}$ where $\beta_1, \ldots, \beta_m$ are not free in $\forall \alpha_1 \ldots \alpha_n \, \sigma$, and $\rho_1, \ldots, \rho_n$ are arbitrary types. But this relation is not strong enough. One can see for example that when $\sigma \sqsubseteq_\mathbf{F} \sigma'$ then terms of type $\tau \to \sigma$ can very well replace those of type $\tau \to \sigma'$. Therefore John C. Mitchell in [7] extended the notion of subtyping to a containment relation $\sqsubseteq_\mathrm{M}$. It is defined by four axioms and three rules. The axioms are: reflexivity, quantifier instantiation, generalization (adding an empty quantifier), and quantifier distributivity over an arrow. The rules ensure transitivity and a particular closure by context. The particularity lies within the rule concerning the arrow symbol: $\sigma_1 \to \tau_1 \sqsubseteq_\mathrm{M} \sigma_2 \to \tau_2$ if $\sigma_2 \sqsubseteq_\mathrm{M} \sigma_1$ and $\tau_1 \sqsubseteq_\mathrm{M} \tau_2$. Adding a subsumption rule for $\sqsubseteq_\mathrm{M}$

$$\frac{E \vdash M : \sigma \quad \sigma \sqsubseteq_\mathrm{M} \tau}{E \vdash M : \tau}$$

properly extends System **F**. Unfortunately the resulting system $\mathbf{F}_\eta$ has all the bad undecidability properties of system **F**. Typability and type-checking are

---
* The author is partly supported by Polish KBN Grant 8 T11C 034 10.

undecidable in both systems (see [10] for **F** and [12] for typability and [9] for type-checking in $\mathbf{F}_\eta$) and so is Mitchell subtyping relation [9] (see also [11] for alternative proof).

The solution for designers of functional languages with full polymorphism is to seek for non-trivial decidable subrelations of $\sqsubseteq_M$. And to do this it is essential to understand what exactly makes this relation undecidable.

The present paper analyses the subtyping relation when we remove distributivity from its definition. This axiom seems to be the least intuitive while the subtyping relation without it remains quite powerful and useful.

Both known undecidability proofs strongly depend on the distributivity axiom. Tiuryn and Urzyczyn [9] encode computations of a certain device called *stack register machine* with undecidable halting problem. The encoding is based on subtyping derivations in Longo-Milsted-Soloviev system [6], which deeply incorporates the distributivity axiom.

Another approach is taken by Wells [11], who reduces the well known semi-unification problem to subtyping. Having two pairs of types σ_1, τ_1 and σ_2, τ_2, the author constructs a universal type $\Pi(\delta_1, \zeta_1, \delta_2, \zeta_2)$ such that if γ are all variables in $\sigma_1, \tau_1, \sigma_2, \tau_2$ then $\forall \beta \, (\Pi(\beta, \beta, \beta, \beta) \to \bot) \sqsubseteq_M (\forall \gamma \, \Pi(\sigma_1, \tau_1, \sigma_2, \tau_2)) \to \bot$ if and only if σ_1, τ_1 and σ_2, τ_2 are semi-unifiable. The distributivity axiom plays a crucial role in this proof. Intuitively after performing common substitution, it is the only way to separate quantifiers and therefore apply two separate substitutions yielding $\sigma_1 \theta \theta_1 = \tau_1 \theta$ and $\sigma_2 \theta \theta_2 = \tau_2 \theta$.

Our work proves that removing distributivity does not suffice to obtain a decidable subtyping relation.

The undecidability result is based on a reduction similar to that in the work of Tiuryn and Urzyczyn. In fact their encoding depends only on the notion of weight of an inequality and on two properties of the subtyping relation $\sqsubseteq_M$ (Lemmas 12 and 14). The present paper, instead of basing the notion of weight on Longo-Milsted-Soloviev derivations, defines a type rewriting system which is a straightforward translation of original Mitchell axioms. The notion of weight is defined using this system, and its further analysis enables us to show that the subtyping without distributivity has all the properties of $\sqsubseteq_M$ (see Lemmas 10 and 11) on which depends the encoding from [9].

The organization of the paper is as follows. After the preliminaries, the rewriting and its correspondence with the subtyping axiomatization is given in Section 3. Then it is shown that rewriting derivations can be normalized, i.e. the steps can be rearranged so that all positive steps precede the negative ones. Using normal derivations we characterize in Section 6 the equivalence relation induced by the new subtyping preorder. Its equational axiomatization is given, similar to the one given by Tiuryn [8] for full Mitchell subtyping relation.

Section 7 introduces yet another class of derivations, ordered ones, which are normal derivations where rewriting takes place from the root towards the leaves in the positive phase, and the other way in the negative phase. It is somewhat similar to Wells's syntax-driven system [11], but using rewriting helps keeping all proof information in one piece.

Ordered derivations constitute a very useful tool to prove subtyping properties necessary to encode computations of a stack register machine (Section 8). Finally, in Section 9 we conclude the paper with some suggestions on how else a decidable subtyping relation could be found.

2 Notation

We work with polymorphic types (like in system **F**), defined over the denumerable set of variables V by the grammar $\mathcal{T} ::= V \mid \forall V\, \mathcal{T} \mid \mathcal{T} \to \mathcal{T}$.

Because of quantifiers, variables occurring in a given type may be either bound or free, that is not bound by any quantifier. For a given type σ, the set of all free variables occurring in it is denoted by $FV(\sigma)$. Types which differ only in the names of bound variables are considered to be equal. Furthermore we assume that in each type the names of bound variables are distinct and different from the names of free ones.

Variables appearing in types are denoted by the initial Greek letters α, β, γ, sequences of quantified variables by $\boldsymbol{\delta}$, $\boldsymbol{\zeta}$, $\boldsymbol{\vartheta}$, and types by other Greek letters like σ, τ, ρ... Often we abbreviate $\forall \alpha \forall \beta\, \tau$ as $\forall \alpha \beta\, \tau$. We admit that quantifiers bind stronger than $\to$, and we use the symbol $\bot$ to denote the type $\forall \alpha\, \alpha$.

The number of arrows appearing in a given type is called its *weight*.

Substitutions are denoted $\{\alpha_1 \mapsto \xi_1, \ldots, \alpha_n \mapsto \xi_n\}$ and we use postfix notation for their application to types. Substitutions bind stronger than quantifiers, so in the type $\forall \beta\, \sigma\{\alpha \mapsto \xi\}$ the free occurrences of β in the type ξ will be bound by the quantifier $\forall \beta$.

Polymorphic types may also be regarded as labelled trees. Leaves are labelled by variables, internal nodes by $\to$ symbols and each node is labelled by a finite sequence of quantifiers. The set of positions in such a tree can be identified with a finite set of finite strings over the alphabet $\{0, 1\}$. We define the sign of a given position p to be positive if p contains even number of zeros, and negative otherwise. The empty string is denoted by ϵ, and given two strings p, r their concatenation is written as pr.

Positions will be compared by two different partial orders. The prefix order is denoted $\leq$, and the notation $p \not\geq r$ means that p and r are not comparable. The other order is the lexicographic order $\preceq$ induced by $0 \prec 1$.

If σ is a type and p a position in the tree of σ, then we denote by $\sigma|_p$ a subtree at the position p. If σ and ϱ are types and p is a position in the tree of σ then $\sigma[\varrho]_p$ stands for σ with the subtree at position p replaced by ϱ. Unlike a substitution, this operation may make some free variables in ϱ bound by quantifiers in σ.

3 Subtyping Relation

The relation of subtyping $\sqsubseteq$ is defined by 3 axioms and 3 rules:

(refl) $\sigma \sqsubseteq \sigma$

(inst) $\forall \alpha\, \sigma \sqsubseteq \sigma\{\alpha \mapsto \rho\}$

(quant) $\sigma \sqsubseteq \forall \alpha\, \sigma \quad$ if $\alpha \notin FV(\sigma)$

Rules:

$$(\to\text{-context}) \quad \frac{\sigma_2 \sqsubseteq \sigma_1 \quad \tau_1 \sqsubseteq \tau_2}{\sigma_1 \to \tau_1 \sqsubseteq \sigma_2 \to \tau_2}$$

$$(\forall\text{-context}) \quad \frac{\sigma \sqsubseteq \tau}{\forall\alpha\,\sigma \sqsubseteq \forall\alpha\,\tau}$$

$$(\text{trans}) \quad \frac{\sigma \sqsubseteq \rho \quad \rho \sqsubseteq \tau}{\sigma \sqsubseteq \tau}$$

Our definition differs from the original one introduced by Mitchell in [7] in the lack of the distributivity axiom:

$$(\text{distr}) \quad \forall\alpha(\sigma \to \tau) \sqsubseteq_M \sigma \to \forall\alpha\,\tau \quad \text{if } \alpha \notin FV(\sigma)$$

4 Type Rewriting

For technical reasons instead of derivations in the above system we use type rewriting according to the following rules:

$$C^+[\forall\alpha\,\sigma]_p \stackrel{p}{\Longrightarrow}{}_i^+ C^+[\sigma\{\alpha \mapsto \rho\}]_p \qquad C^-[\sigma\{\alpha \mapsto \rho\}]_r \stackrel{r}{\Longrightarrow}{}_i^- C^-[\forall\alpha\,\sigma]_r$$

$$C^+[\sigma]_p \stackrel{p}{\Longrightarrow}{}_q^+ C^+[\forall\alpha\,\sigma]_p \qquad C^-[\forall\alpha\,\sigma]_r \stackrel{r}{\Longrightarrow}{}_q^- C^-[\sigma]_r$$
$$\text{if } \alpha \notin FV(\sigma) \qquad\qquad\qquad \text{if } \alpha \notin FV(\sigma)$$

where $C^+[\,]_p$ and $C^-[\,]_r$ denote contexts with a hole at a positive position p and negative position r respectively. When we do not want to precise what step is considered, we omit the appropriate decoration and write for example $\Longrightarrow_i$ or $\Longrightarrow^+$. Any finite number of steps may be denoted by a double arrow $\Longrightarrow$. Given a derivation step $\sigma \Longrightarrow \tau$, the absolute value of the difference of weights of σ and τ is called the *weight* of this step. Intuitively, it is the number of arrows introduced (or removed) by this step.

Lemma 1. *For any types σ and τ we have $\sigma \sqsubseteq \tau$ if and only if $\sigma \Longrightarrow \tau$.*

5 Normal Derivations

The goal of the present section is to show that every subtyping inequality can be proved by a *normal* derivation, that is a sequence of positive steps (which increase the weight of a type) followed by a sequence of negative steps.

Definition 1 (normal derivation).
A derivation $\sigma \Longrightarrow \tau$ is called normal, *if it is of the form $\sigma \Longrightarrow^+ \rho \Longrightarrow^- \tau$ for certain type ρ.*

The *weight* of a normal derivation is the sum of weights of steps in it. It is easy to see that it is equal $2 \times weight(\rho) - weight(\sigma) - weight(\tau)$.

Before showing how to normalize derivations, define the notion of safe steps.

Definition 2 (safe step).
A rewriting step $\sigma \Longrightarrow \tau$ is safe, if one of the following conditions is satisfied:

1. *it is a $\Longrightarrow_q$ step;*
2. *it is a $\Longrightarrow_i$ step, in which the inserted type is $\bot$;*
3. *it is a $\stackrel{p}{\Longrightarrow}_i$ step, in which the inserted type is a variable β bound at a position $p' \leq p$, of the same sign as the sign of p.*

The second condition above covers particularly the instantiation of empty quantifiers, i.e. the steps $C^+[\forall \alpha\, \sigma] \Longrightarrow_i^+ C^+[\sigma]$ and respectively $C^-[\sigma] \Longrightarrow_i^- C^-[\forall \alpha\, \sigma]$ for $\alpha \notin FV(\sigma)$.

The last condition of Definition 2 expresses the following situations:

$$C^+[\forall \beta\, C^+[\forall \alpha\, \sigma]] \Longrightarrow_i^+ C^+[\forall \beta\, C^+[\sigma\{\alpha \mapsto \beta\}]]$$

$$C^-[\forall \beta\, C^+[\sigma\{\alpha \mapsto \beta\}]] \Longrightarrow_i^- C^-[\forall \beta\, C^+[\forall \alpha\, \sigma]]$$

All safe steps have the weight 0. In addition, a safe step can be inserted in a normal derivation without changing its weight. The formal statement of this property, together with an example of a step of weight 0 which is not safe will be given by the end of the section.

The following lemma shows how to permute negative and positive steps,[1] which is necessary to normalize a derivation.

Lemma 2. *Let σ and τ be types such that $\sigma \Longrightarrow^- \rho \Longrightarrow^+ \tau$. There exists a type ρ' such that $\sigma \Longrightarrow^+ \rho' \Longrightarrow^- \tau$. In addition, if the step $\sigma \Longrightarrow^- \rho$ is safe then after the permutation the step $\rho' \Longrightarrow^- \tau$ will remain safe and similarly if $\rho \Longrightarrow^+ \tau$ is a safe step then so will be $\sigma \Longrightarrow^+ \rho'$.*

Proof. Let $\sigma \stackrel{p}{\Longrightarrow}^- \rho \stackrel{r}{\Longrightarrow}^+ \tau$. Depending on p and r we have three possibilities to consider: $p \not\geq r$, $p < r$, and $p > r$. The first one is easy, the other two, dual to each other, are also easy when any of the rewriting steps is a q-step. Otherwise both steps are instantiations and we use the fact that $\upsilon\{\beta \mapsto \xi_\beta\{\alpha \mapsto \xi_\alpha\}\} = \upsilon\{\beta \mapsto \xi_\beta\}\{\alpha \mapsto \xi_\alpha\}$ for any types υ, ξ_α, and ξ_β. □

Lemma 3. *Given two types σ and τ such that $\sigma \sqsubseteq \tau$, there exists a normal derivation $\sigma \Longrightarrow\!\!\!\!\Rightarrow^+ \Longrightarrow\!\!\!\!\Rightarrow^- \tau$.*

Now we define the *weight* of an inequality and explain the meaning of safe steps.

Definition 3 (weight of inequality).
The weight of an inequality $\sigma \sqsubseteq \tau$ is the minimal weight of a normal derivation $\sigma \Longrightarrow\!\!\!\!\Rightarrow^+ \rho \Longrightarrow\!\!\!\!\Rightarrow^- \tau$.

---

[1] A permutation the other way is not always possible. One cannot for example change the order of the steps $\forall \alpha\, \alpha \stackrel{\epsilon}{\Longrightarrow}_i^+ (\beta \to \beta) \to \beta \stackrel{0}{\Longrightarrow}_i^- (\forall \alpha\, \alpha) \to \beta$

Proposition 4. Let $\sigma_1 \sqsubseteq \tau$ and let $\sigma_0 \Longrightarrow \sigma_1$ be a safe step. Then the weight of the inequality $\sigma_0 \sqsubseteq \tau$ is no greater than the weight of the inequality $\sigma_1 \sqsubseteq \tau$.

Let us give an example of a rewriting step of weight 0, which is not safe.

$$\forall \alpha \left((\bot \to \alpha) \to \alpha \right) = \forall \alpha \left((\bot \to \beta)\{\beta \mapsto \alpha\} \to \alpha \right) \overset{0}{\Longrightarrow}{}_{i}^{-} \forall \alpha \left(\forall \beta (\bot \to \beta) \to \alpha \right)$$

It is not safe, because the variable α is bound at a positive position, unlike the variable β. If we put this step in front of a one-step derivation of weight $weight(\xi)$:

$$\forall \alpha \left(\forall \beta (\bot \to \beta) \to \alpha \right) \overset{\epsilon}{\Longrightarrow}{}_{i}^{+} \forall \beta (\bot \to \beta) \to \xi$$

and permute the steps according to Lemma 2, we get a normal derivation of weight $3 \times weight(\xi)$:

$$\forall \alpha \left((\bot \to \alpha) \to \alpha \right) \overset{\epsilon}{\Longrightarrow}{}_{i}^{+} ((\bot \to \xi) \to \xi) \overset{0}{\Longrightarrow}{}_{i}^{-} \forall \beta (\bot \to \beta) \to \xi$$

6 Bicoercible Types

Types σ and τ are called *bicoercible* if $\sigma \sqsubseteq \tau$ and $\tau \sqsubseteq \sigma$. We show in the present section, that the bicoercibility relation may be identified with the least congruence containing the axioms

(B1) $\forall \alpha \forall \beta \, \sigma \sim \forall \beta \forall \alpha \, \sigma$

(B2) $\forall \alpha \, \sigma \sim \sigma\{\alpha \mapsto \bot\}$ (if all occurrences of α in σ are positive[2])

The equivalence relation induced by the original Mitchell system was characterized by Tiuryn in [8]. It is defined by axioms **(B1)**, **(B2)** and the axiom

(B3) $\forall \alpha \, (\sigma_1 \to \sigma_2) \sim_M \sigma_1 \to \forall \alpha \, \sigma_2$ (if $\alpha \notin FV(\sigma_1)$)

which is not true in our system.[3] The rest of the present section is devoted to the proof that the relation $\sim$, defined as the least congruence containing axioms **(B1)** and **(B2)**, is the equivalence relation induced by the quasi-order $\sqsubseteq$.

Lemma 5. *If $\sigma \Longrightarrow \tau$ and $\tau \sqsubseteq \sigma$ then the statement $\sigma \sim \tau$ can be derived from the axioms* **(B1)** *and* **(B2)**.

Proof. Since both situations are similar, assume $\sigma \Longrightarrow \tau$ is a positive step. By analyzing what changes introduced by the first step can be undone by further steps of a normal derivation $\sigma \Longrightarrow^+ \tau \Longrightarrow^+ \rho \Longrightarrow^- \sigma$, we conclude that inserted types can only be variables, that the sign of a variable binding must not change, and finally that when a variable bound at a positive position has negative occurrences (or vice versa) the changes to its binding can only be trivial. If follows that all kinds of acceptable changes can be simulated by **(B1)** and **(B2)**. □

---
[2] It covers also the situation $\alpha \notin FV(\sigma)$
[3] It is not true that $\forall \alpha \, (\beta \to (\alpha \to \beta)) \sim \beta \to \forall \alpha \, (\alpha \to \beta)$

Lemma 6. *If the statement $\sigma \sim \tau$ can be proved by a single use of the axiom* **(B1)** *or* **(B2)** *then $\sigma \Longrightarrow\!\!\!\!\!\twoheadrightarrow \tau$ and $\tau \Longrightarrow\!\!\!\!\!\twoheadrightarrow \sigma$. Moreover these derivations use only safe steps.*

Proof. There are four possibilities, depending on what axiom was used and what was the position sign. Since most of the reasoning is simple we analyze only the use of **(B2)** at a positive position: $\sigma = \sigma[\forall \boldsymbol{\delta}\alpha\,\varrho]_p \sim \sigma[\forall \boldsymbol{\delta}\,\varrho\{\alpha \mapsto \bot\}]_p = \tau$.

The derivation from left to right can be done by a single i-step. Constructing a derivation from right to left is more complicated. Let $r_1, \ldots, r_n$ be all occurrences of α in ϱ. For all i we have $\varrho\{\alpha \mapsto \bot\}|_{r_i} = \forall \zeta_i \bot = \forall \zeta_i \beta\, \beta$, so

$$\sigma[\forall \boldsymbol{\delta}\,\varrho\{\alpha \mapsto \bot\}]_p \xrightarrow[q]{p}{}^+ \sigma[\forall \boldsymbol{\delta}\gamma\,\varrho\{\alpha \mapsto \bot\}]_p$$
$$\xrightarrow[i]{pr_1}{}^+ \sigma[\forall \boldsymbol{\delta}\gamma\,\varrho\{\alpha \mapsto \bot\}[\forall \zeta_1\,\gamma]_{r_1}]_p \xrightarrow[i]{pr_2}{}^+ \cdots$$
$$\cdots \xrightarrow[i]{pr_n}{}^+ \sigma[\forall \boldsymbol{\delta}\gamma\,\varrho\{\alpha \mapsto \bot\}[\forall \zeta_1\,\gamma]_{r_1}\cdots[\forall \zeta_n\,\gamma]_{r_n}]_p$$
$$= \sigma[\forall \boldsymbol{\delta}\gamma\,\varrho\{\alpha \mapsto \gamma\}]_p = \sigma[\forall \boldsymbol{\delta}\alpha\,\varrho]_p$$

where γ is fresh. All steps $\xrightarrow[i]{pr_i}{}^+$ are correct and safe because every r_i is positive, and the inserted variable γ is bound at a positive position p. □

Theorem 1 (characterization of bicoercibility).
Inequalities $\sigma \sqsubseteq \tau$ and $\tau \sqsubseteq \sigma$ are both true if and only if $\sigma \sim \tau$.

For the rest of the proof the most important is the following proposition which follows from Lemma 6 and Proposition 4.

Proposition 7. *If $\sigma_0 \sim \sigma_1$ and $\sigma_1 \sqsubseteq \tau$ then the inequality $\sigma_0 \sqsubseteq \tau$ also holds and the weights of both inequalities are the same.*

7 Decomposing Inequalities

The present section shows the properties of $\sqsubseteq$ that will be directly used in the proof of undecidability. To this end we introduce ordered derivations. Intuitively, positive rewriting steps will be ordered by their positions compared with the lexicographic ordering (induced by $0 \prec 1$). At one position, all q-steps will precede i-steps ordered from outermost to innermost. Negative steps will be ordered dually. The rewriting will then take place first at positive positions from the root towards the leaves, and then at negative positions, from the leaves upwards.

Definition 4 (ordered derivations).
A sequence of positive steps $\sigma_0 \xrightarrow{p_1}{}^+ \cdots \xrightarrow{p_n}{}^+ \sigma_n$ is ordered if:

1. *for all i, j if $i < j$ then $p_i \preceq p_j$;*
2. *for any given p, $\xrightarrow[q]{p}{}^+$ steps precede $\xrightarrow[i]{p}{}^+$ steps;*

3. there are no pairs of steps

$$\sigma_j[\forall \boldsymbol{\delta}\,\alpha\,\boldsymbol{\zeta}\beta\,\varrho]_p \stackrel{p}{\Longrightarrow}{}_i^+ \sigma_j[\forall \boldsymbol{\delta}\,\alpha\,\boldsymbol{\zeta}\,\varrho\{\beta \mapsto \xi_\beta\}]_p$$
$$\stackrel{p}{\Longrightarrow}{}_i^+ \sigma_j[\forall \boldsymbol{\delta}\,\boldsymbol{\zeta}\,\varrho\{\beta \mapsto \xi_\beta\}\{\alpha \mapsto \xi_\alpha\}]_p$$

A sequence of negative steps $\tau_0 \Longrightarrow^- \cdots \Longrightarrow^- \tau_n$ is ordered, if the dual sequence of positive steps $\tau_n \to \gamma \Longrightarrow^+ \cdots \Longrightarrow^+ \tau_0 \to \gamma$ (for γ fresh) is ordered.

A normal derivation $\sigma \Longrightarrow\!\!\!\Longrightarrow^+ \rho \Longrightarrow\!\!\!\Longrightarrow^- \tau$ is called ordered if both sequences $\sigma \Longrightarrow\!\!\!\Longrightarrow^+ \rho$ and $\rho \Longrightarrow\!\!\!\Longrightarrow^- \tau$ are ordered.

Lemma 8. *If $\sigma \sqsubseteq \tau$, then there exists an ordered derivation $\sigma \Longrightarrow\!\!\!\Longrightarrow^+ \Longrightarrow\!\!\!\Longrightarrow^- \tau$, of weight equal to the weight of the inequality.*

Proof. Consider a normal derivation $\sigma \Longrightarrow\!\!\!\Longrightarrow^+ \rho \Longrightarrow\!\!\!\Longrightarrow^- \tau$ of minimal weight. We observe that every two consecutive steps which are not ordered can be permuted. Since each phase is sorted separately, the middle type ρ does not change and neither does the weight of the derivation. □

Lemma 9. *Let $\sigma \sqsubseteq \tau$ and let $\sigma|_p = \varrho_\sigma$ and $\tau|_p = \varrho_\tau$ be subterms such that no variable that is free in ϱ_σ (resp. in ϱ_τ) is bound in σ (resp. τ). If p is positive, then $\varrho_\sigma \sqsubseteq \varrho_\tau$ and if p is negative then $\varrho_\tau \sqsubseteq \varrho_\sigma$.*

Proof. Let $\sigma \Longrightarrow\!\!\!\Longrightarrow^+ \rho \Longrightarrow\!\!\!\Longrightarrow^- \tau$ be an ordered derivation. Only those steps which have p as a prefix can influence a subterm at position p. Since the derivation is ordered, those steps form two groups, 'in the middle' of both phases. It turns out that those groups of steps can easily be transformed into the desired derivation $\varrho_\sigma \Longrightarrow\!\!\!\Longrightarrow \varrho_\tau$ (or $\varrho_\tau \Longrightarrow\!\!\!\Longrightarrow \varrho_\sigma$ if p is negative). □

Lemma 10. *The inequality $\sigma' \to \sigma'' \sqsubseteq \tau' \to \tau''$ holds if and only if both inequalities $\sigma' \sqsupseteq \tau'$ and $\sigma'' \sqsubseteq \tau''$ hold. Moreover, if all inequalities are true then the weight of the first one is equal to the sum of weights of the others.*

Lemma 11. *Let σ and τ be types. Assume that there exist two paths: positive p and negative r such that $\sigma|_p = \sigma|_r = \alpha$ and $\tau|_p = \tau|_r = \alpha$ for a free variable α. Suppose furthermore that τ is not of the form $\forall \gamma\, \tau'$. Then*

1. *$\forall \alpha\, \sigma \sqsubseteq \tau\{\alpha \mapsto \xi\}$ if and only if $\sigma\{\alpha \mapsto \xi\} \sqsubseteq \tau\{\alpha \mapsto \xi\}$;*
2. *if both inequalities hold then the weight of the latter is no smaller than the weight of the former;*
3. *moreover if $\mathrm{weight}(\xi) > 0$ then the weight of the first incquality is strictly greater than the weight of the second.*

Proof. If $\sigma\{\alpha \mapsto \xi\} \sqsubseteq \tau\{\alpha \mapsto \xi\}$, then since $\forall \alpha\, \sigma \overset{\epsilon}{\Longrightarrow}_i^+ \sigma\{\alpha \mapsto \xi\}$, we have

$$\forall \alpha\, \sigma \sqsubseteq \tau\{\alpha \mapsto \xi\} \tag{1}$$

In order to show the opposite implication, suppose inequality (1) holds. Consider its ordered derivation of minimal weight. It begins with some meaningless $\overset{\epsilon}{\Longrightarrow}^+$ steps. All they can do is to add some empty quantifiers after the initial $\forall \alpha$. Now the latter must be instantiated:

$$\forall \alpha\, \sigma' \overset{\epsilon}{\Longrightarrow}_i^+ \sigma'\{\alpha \mapsto \chi\} \tag{2}$$

where σ' stands for $\forall \beta_1 \ldots \beta_n\, \sigma$, and $\beta_1, \ldots, \beta_n$ are the empty quantifiers. By Lemma 9, $\xi \sim \chi$, so also $\sigma\{\alpha \mapsto \xi\} \sim \sigma'\{\alpha \mapsto \chi\}$. By Proposition 7 this yields the desired inequality. Its weight is no greater than the weight of (1) minus the weight of (2), which is strictly positive, when $weight(\xi)$ is so. Since $weight(\xi) = weight(\chi)$ we have our claim. □

8 Encoding a Machine

A stack register machine $\mathcal{M}$ is a deterministic computing device with two main registers V_1 and V_2 and a finite number of auxiliary registers $v_1, \ldots, v_n$. The registers can hold nonempty words over the finite alphabet Σ, containing instruction labels with one special end label e. A machine step consists of taking the top label ℓ from the first main register and executing the instruction assigned to ℓ. If the read label is e the machine halts.

The encoding idea is the following. For each label $\ell \in \Sigma$ we construct a type σ_ℓ embodying the associated instruction. Then we give the label composition rules to encode words contained by the machine's registers. So a given configuration yields $n+2$ types, $\sigma_{V_1}, \sigma_{V_2}, \sigma_{v_1}, \ldots, \sigma_{v_n}$, containing one special free variable γ.

Proposition 12. *Given a stack register machine $\mathcal{M}$, it halts for an instantaneous description $(V_1, V_2, v_1, \ldots, v_n)$ if and only if the encoding inequality holds*

$$\sigma_{V_1} \sqsubseteq (\sigma_{v_1} \to \sigma_{v_1}) \to \cdots \to (\sigma_{v_n} \to \sigma_{v_n}) \to \sigma_{V_2} \to \gamma$$

Proof. The proof is based on two properties of $\sqsubseteq$ which are exactly Lemmas 10 and 11. They correspond to Lemmas 12 and 14 in [9]. See the latter paper for all the details. □

Theorem 2 (Main result).
The relation $\sqsubseteq$ of polymorphic subtyping without distributivity is undecidable.

Proof. This follows from our Proposition 12 and Lemma 16 from the paper of Tiuryn and Urzyczyn [9] that reduces the halting problem of deterministic two-counter automata to the halting problem of stack register machines. □

9 Conclusions

The same proof would also be valid if we removed both (**distr**) and (**quant**) from the definition of subtyping. The only difference would emerge in Section 6, but Proposition 7 remains true which enables us to continue with the undecidability proof.

Another direction of searching for decidable subtyping relation is the predicative version of System **F**, presented by Daniel Leivant [5]. Some work has already been done in this field (see [1]) and we believe that further investigation should lead to interesting results.

References

1. M. Benke, "Predicative Polymorphic Subtyping", these proceedings.
2. J. Chrząszcz, "Polymorphic Subtyping Without Distributivity" Technical Report TR98-03(252), Institute of Informatics, Warsaw University, May 1998. URL: http:\\zls.mimuw.edu.pl\~chrzaszc\papers
3. J.-Y. Girard, Y. Lafont, P. Taylor. *Proofs and Types*. Cambridge Tracts in Theoretical Computer Science. Cambridge University Press, 1989.
4. T. Jim, "System F plus Subsumption Reduces to Mitchell's Subtyping Relation", Manuscript (1995).
5. D. Leivant, "Finitely Stratified Polymorphism", *Information and Computation*, **93** (1), 1991, 93-113.
6. G. Longo, K. Milsted, S. Soloviev, "A Logic of Subtyping", *Proc. 10th IEEE Symp. Logic in Computer Science*, 1995, pp. 292-299.
7. J.C. Mitchell, "Polymorphic Type Inference and Containment", *Information and Computation*, **76** (2-3), 1988, pp. 211-249.
8. J. Tiuryn, "Equational Axiomatization of Bicoercibility for Polymorphic Types", *Proc. Conf. Foundations of Software Technology and Theoretical Computer Science'95*, Bangalore, India; 18-20 Dec 1995. Lecture Notes in Computer Science 1026, pp. 166-179.
9. J. Tiuryn, P. Urzyczyn, "The Subtyping Problem for Second-Order Types is Undecidable" *Proc. 11th IEEE Symp.Logic in Computer Science*, 1996, pp. 74-85.
10. J.B. Wells, "Typability and Type Checking in the Second-Order λ-calculus are Equivalent and Undecidable", *Proc. 9th IEEE Symp. Logic in Computer Science*, 1994, pp. 176-185.
11. J.B. Wells, "The Undecidability of Mitchell's Subtyping Relationship", Technical Report, Computer Science Department, Boston University, Number 95-019, December 10 1995.
12. J.B. Wells, "Typability is Undecidable for F+Eta" Technical Report, Computer Science Department, Boston University, Number 96-022, March 9 1996.

A (Non-elementary) Modular Decision Procedure for LTrL

Paul Gastin[1], Raphaël Meyer[2], and Antoine Petit[2]

[1] LIAFA, Université Paris 7, 2, place Jussieu, F-75251 Paris Cedex 05
{Paul.Gastin}@liafa.jussieu.fr
[2] LSV, URA 2236 CNRS, ENS de Cachan, 61, av. du Prés. Wilson, F-94235 Cachan Cedex
{rmeyer,petit}@lsv.ens-cachan.fr

Abstract. Thiagarajan and Walukiewicz [18] have defined a temporal logic LTrL on Mazurkiewicz traces, patterned on the famous propositional temporal logic of linear time LTL defined by Pnueli. They have shown that this logic is equal in expressive power to the first order theory of finite and infinite traces.

The hopes to get an "easy" decision procedure for LTrL, as it is the case for LTL, vanished very recently due to a result of Walukiewicz [19] who showed that the decision procedure for LTrL is non-elementary.

However, tools like Mona [8] or Mosel [7] show that it is possible to handle non-elementary logics on significant examples.

Therefore, it appears worthwhile to have a direct decision procedure for LTrL; in this paper we propose such a decision procedure, in a modular way. Since the logic LTrL is not pure future, our algorithm constructs by induction a finite family of Büchi automata for each LTrL-formula. As expected by the results of [19], the main difficulty comes from the "Until" operator.

Topics: logic in computer science, automata and formal languages, theory of parallel and distributed computation, model-checking

1 Introduction

A run of a distributed system can be viewed, in many settings, as a partial order between the events of the system. Two events are ordered if and only if their executions depend causally one of the other. The partial orders that arise in this fashion, are frequently Mazurkiewicz traces [9, 4]. A major interest of this model lies on the fact that a trace can be seen either as a labelled ordered graph expressing directly the partial order or as an equivalence class of sequences, each of them representing a linearization of the partial order.

In a natural way, and in order to exploit directly the partial order underlying to a trace, a good amount of research have focused on developing temporal logics that can be directly interpreted over traces seen as labelled partially ordered graphs (rather than as sets of sequences). Recently, Thiagarajan and

Walukiewicz [18] have defined a new temporal logic, denoted by LTrL, patterned on the propositional temporal logic of linear time LTL defined by Pnueli [12] with exactly the expressive power of $FO(<)$, the first order theory of finite and infinite traces (see also [10]). This work was the outcome of a long sequence of papers on logics on traces [17, 5, 1, 13].

One of the most important property of LTL is to have a PSPACE-complete decision procedure for LTL is [15] (whereas the decision procedure for the equivalent logic $FO(<)$ is non-elementary [16]). This logic LTL has thus been used successfully in model-checking, see e.g. [3]. So very naturally, people were interested in the complexity of the decision procedure for LTrL which was expected to be of "low" complexity. These hopes vanish after the very recent work of Walukiewicz who shows that the decision procedure for LTrL is non-elementary [19].

This result could be seen as an irremediable drawback for the use of LTrL for "practical" model-checking. Nevertheless, recent tools such as Mona [8] or Mosel [7] have been proposed to handle non-elementary logics. Moreover, significant "real" problems have been solved using these tools. Therefore, it appears worthwhile to have a decision procedure for LTrL. Up to now, the only existing one lies on the transformation of a LTrL-formula into an equivalent $FO(<)$-formula used by Thiagarajan and Walukiewicz to prove their main theorem [18].

To achieve this goal, we propose as the main result of this paper, a modular direct decision procedure for LTrL. Precisely we construct for any LTrL-formula α, a Büchi automaton recognizing the set of (the linearizations of) the models of α. From the well-known decidability of the emptiness of the language recognized by a Büchi automaton, we get our decision procedure. Our construction is performed in a modular way from the structure of α.

In fact, since the logic LTrL is not pure future but contains some "present" operators, we need to construct by induction not a simple automaton but a sequence of automata indexed by some alphabetic information. The constructions for the boolean operators and the local next operators are classical and do not present any new difficulty (note nevertheless that the negation requires of course a complementation of a Büchi automaton). The crucial step is the construction for the "Until" operator. From [19], we know that the operator is responsible of the non-elementariness and therefore the construction can not be simple. We use the notion of alphabetic automaton (classical from finite automata theory) in order to simplify the presentation as much as possible.

Our paper is organized as follows. In Section 2 we set the basic definitions for the trace theory and the temporal logic LTrL and we introduce the problem of Model-Checking in this logic. Section 3 is devoted to the presentation of the main tools we will be using: Büchi automata and alphabetic automata. These tools allow us to describe our modular constructions in Section 4 in a more concise manner. Finally, we give some conclusions about those constructions in Section 5, and devise some possible paths for future research.

2 Model-Checking on LTrL

2.1 Traces

A *dependence alphabet* is a pair (Σ, I) where Σ is a finite set and $I \subseteq \Sigma \times \Sigma$ is a symmetric and irreflexive relation called the *independence relation*. Elements of Σ are called *actions*; two actions $a, b \in \Sigma$ are said *independent* if $(a, b) \in I$. The complementary relation $D = \Sigma \times \Sigma \setminus I$ is called the *dependence relation*. If $a \in \Sigma$, then $I(a)$ and $D(a)$ denote respectively the sets of letters independent and dependent with a.

A (Mazurkiewicz) *trace* on (Σ, I) is a Σ-labelled partially ordered set (poset) that respects the dependence relation [9]. More formally, let $(E, \leq, \lambda)$ be a Σ-labelled poset, that is: E is a finite or infinite set, $\leq$ is a partial order on E and λ is a labelling function from E into Σ. For every subset Y of E, we define $\downarrow Y = \{x \in E \mid \exists y \in Y, x \leq y\}$. If Y is a singleton $\{y\}$ we shall write $\downarrow y$ instead of $\downarrow \{y\}$. We also define the *covering relation* $\lessdot$ on $E \times E$: $x \lessdot y$ if $x < y$ and $\forall z \in E, x \leq z \leq y$ implies $z = x$ or $z = y$.

Then a *trace* over (Σ, I) is a Σ-labelled poset $u = (E, \leq, \lambda)$ satisfying:

$(T1)$ $\forall e \in E,\ \downarrow e$ is a finite set
$(T2)$ $\forall e, e' \in E,\ e \lessdot e' \Rightarrow (\lambda(e), \lambda(e')) \in D$
$(T3)$ $\forall e, e' \in E,\ (\lambda(e), \lambda(e')) \in D \Rightarrow e \leq e'$ or $e' \leq e$

Elements of E are called *events*. A *configuration* of a trace $u = (E, \leq, \lambda)$ is a finite subset $c \subseteq E$ satisfying $\downarrow c = c$. It can be viewed as a finite trace which is a prefix of u. The set of letters of *maximal* events in c (for the partial order induced by u) is denoted by $max(c)$. We denote by C_u the set of configurations of a trace u. Remark that $\emptyset \in C_u$ and $max(\emptyset) = \emptyset$. The *transition relation* $\rightarrow_u \subseteq C_u \times \Sigma \times C_u$ is given by: $c \xrightarrow{a}_u c'$ iff there exists $e \in E$ such that $\lambda(e) = a$, $e \notin c$ and $c' = c \cup \{e\}$.

2.2 LTrL

Thiagarajan and Walukiewicz[18] have proposed a temporal logic on (finite and infinite) traces, called LTrL, that is expressively complete, i.e. equivalent to the first-order logic $FO(<)$ on (finite and infinite) traces.

The set of formulas belonging to LTrL is defined inductively by: [1]

$$\text{LTrL}(\Sigma, I) ::= \underline{tt} \mid \neg \alpha \mid \alpha \vee \beta \mid @\alpha \mid \alpha\,\mathcal{U}\,\beta \mid \underline{a}$$

The semantics of these formulas are defined inductively: let $u = (E, \leq, \lambda)$ be a trace and c a configuration of u, then

- $u, c \models \underline{tt}$.
  Furthermore, the boolean connectives $\neg$ and $\vee$ have the usual interpretations.

---
[1] Our notations are slightly different from the ones of the original article.

- $u, c \models @\underline{a} \alpha$ iff $\exists c' \in C_u$, $c \xrightarrow{a}_u c'$ and $u, c' \models \alpha$.
- $u, c \models \alpha \, \mathcal{U} \, \beta$ iff $\exists c' \in C_u$, $c \subseteq c'$, such that $u, c' \models \beta$ and $\forall c'' \in C_u$, $c \subseteq c'' \subset c'$ implies $u, c'' \models \alpha$.
- $u, c \models \underline{a}$ iff $\exists c' \in C_u$ such that $c' \xrightarrow{a}_u c$, i.e. $a \in max(c)$.

Note that the formulas of the form $\underline{a}$ describe the present configuration at which they are evaluated. For this reason, they will be referred to as *present formulas*. Thus, if c and c' are two configurations such that $max(c) = max(c')$, then for any trace u and any formula α of LTrL it holds:

$$cu, c \models \alpha \Leftrightarrow c'u, c' \models \alpha$$

Denote by $\mathbf{R}(\Sigma, I)$ the set of finite and infinite traces on the dependence alphabet (Σ, I), then every formula α of LTrL defines a trace language $L(\alpha)$ in the following way:

$$L(\alpha) = \{u \in \mathbf{R}(\Sigma, I) \mid u, \emptyset \models \alpha\}$$

2.3 The Model-Checking Problem

The *global* nature of LTrL's temporal operators makes it easy to specify global liveness and safety properties. However, this global nature has a counterpart regarding the complexity of verification tasks such as satisfiability or model-checking. Actually, Walukiewicz has shown[19] that the satisfiability problem is non-elementary in LTrL, and that this non-elementarity is due to the possible nesting of "until" operators. However, this should not prevent us from looking for direct algorithms for verification tasks. Indeed, such algorithms would allow to check at least formulas with a low "until depth".

Our approach is to start with a formula α of LTrL and build an automaton recognizing the language defined by α inductively on the structure of that formula. With these constructions, the satisfiability and model-checking problems are reduced to checking for emptiness of automata.

Our main result can be stated as follows:

Theorem 1. *Let α be a formula of the temporal logic LTrL for a given dependence alphabet (Σ, I). A Büchi automaton $\mathcal{A}_\alpha$ on Σ^∞ that accepts exactly the linearizations of traces satisfying α, can be constructed inductively on the structure of α.*

3 Tools

3.1 Words and Traces

Although we want to recognize *trace* languages, our approach is to construct automata that will recognize I-closed sets of words. More formally, let Σ^∞ be the set of finite and infinite words on the alphabet Σ, and recall that $\mathbf{R}(\Sigma, I)$ is the set of finite and infinite traces on (Σ, I). Now let φ denote the canonical morphism from Σ^∞ into $\mathbf{R}(\Sigma, I)$. A word language $L \subseteq \Sigma^\infty$ is said to be I-closed if it is closed by the morphism φ, i.e. if $L = \varphi^{-1}(\varphi(L))$.

Note that $\mathbf{R}(\Sigma, I)$ can also be defined as $\Sigma^\infty / \sim_I$, where $\sim_I$ is the congruence induced by I on Σ^∞ (see e.g. [4] for further details).

3.2 Büchi Automata

In the sequel, we shall use some *Büchi automata*, that is non-deterministic automata on words with a particular *repeating* acceptance condition for infinite words [2, 11].

Definition A Büchi automaton on Σ is a t-uple $\mathcal{A} = (Q, \rightarrow, S, F, R)$ where:

- Q is the finite set of *states*
- $\rightarrow \subseteq Q \times \Sigma \times Q$ the *transition relation*. If $(q, a, q') \in \rightarrow$, we shall write $q \xrightarrow{a} q'$. The transitive closure of $\rightarrow$ will be denoted by $\rightarrow^*$.
- $S \subseteq Q$ is the set of *initial* states of $\mathcal{A}$.
- F is the set of *final* states.
- $R \subseteq Q$ is the set of *repeated* states.

Executions A finite *execution* of $\mathcal{A}$ on a finite word $u = a_1 a_2 \ldots a_n$, with $\forall i, a_i \in \Sigma$ is a sequence $\rho = q_0, q_1, q_2, \ldots, q_n$ of states such that $q_0 \in S$ and for all $0 \leq i \leq n-1$, we have $q_i \xrightarrow{a_{i+1}} q_{i+1}$. If $u = a_1 a_2 \ldots a_k \ldots$ is an infinite word, an *execution* of $\mathcal{A}$ on u is an infinite sequence $\rho = q_0, q_1, q_2, \ldots, q_k, \ldots$ of states such that $q_0 \in S$ and for all $i \leq 0$, it holds $q_i \xrightarrow{a_{i+1}} q_{i+1}$. Denote by $\text{rep}(\rho)$ the set of states $q \in Q$ such that $\{i \geq 0 \mid q = q_i\}$ is infinite. Note that since $\mathcal{A}$ is non-deterministic, it can have different executions on the same word.

Acceptance Conditions If u is a finite word, an execution ρ of $\mathcal{A}$ on u is *accepting* if the last state of ρ is in F. If u is a infinite word, an execution ρ of $\mathcal{A}$ on u is *accepting* if $\text{rep}(\rho) \cap R \neq \emptyset$.

Finally, the language $L(\mathcal{A})$ of words *accepted* or *recognized* by $\mathcal{A}$ is the set of (finite and infinite) words w such that there exists an accepting execution of $\mathcal{A}$ on w.

3.3 Alphabetic Automata

For our constructions we will need the notion of *alphabetic* automata: an automaton is alphabetic if all the words that can be used to reach a particular state have the same alphabet.

Formally, a Büchi automaton $\mathcal{A} = (Q, \rightarrow, S, F, R)$ is *alphabetic* if it satisfies the following property:

$$\forall q \in Q, \exists B \subseteq \Sigma, \forall q_0 \in S, \forall u \in \Sigma^*, q_0 \xrightarrow{u}{}^* q \Rightarrow \text{alph}(u) = B$$

If this is indeed the case, then we set $alph(q) = B$.

Our constructions use intensively the following result which can be considered as "folklore" in automata theory:

Proposition 1. *Every Büchi automaton can be transformed into an alphabetic Büchi automaton recognizing the same language.*

Proof. Let $\mathcal{A} = (Q, \rightarrow, S, F, R)$ be a Büchi automaton on the alphabet Σ. Let $\mathcal{B} = (Q_1, \rightarrow_1, S_1, F_1, R_1)$ such that:

- $Q_1 = Q \times 2^\Sigma$,
- $(q, E) \xrightarrow{a}_1 (q', E')$ iff $q \xrightarrow{a} q'$ and $E' = E \cup \{a\}$,
- $S_1 = S \times \{\emptyset\}$,
- $F_1 = \bigcup_{G \subseteq \Sigma} F \times G$,
- and $R_1 = \bigcup_{G \subseteq \Sigma} R \times G$.

Then a word u is accepted by $\mathcal{A}$ iff it is accepted by $\mathcal{B}$. Moreover, $\mathcal{B}$ is alphabetic, since if a state (q, E) is reachable from an initial state $(q_0, \emptyset)$ in $\mathcal{B}$ through a word u, then $E = alph(u)$.

4 Modular Constructions

4.1 Methodology

If α is an LTrL formula, then $L(\alpha)$ is a *trace* language. Our goal is to build a Büchi automaton on words that recognizes the word language $\varphi^{-1}(L(\alpha))$. We achieve this construction in a modular way, inductively on the structure of the formula α.

One of the problems we encounter while attempting to make such constructions comes from the fact that the logic LTrL is not *pure future-oriented*, since it features present formulas. For instance, if $\alpha = \underline{a}$, then $L(\alpha) = \emptyset$, but $L(@ \underline{a}) = a\mathbf{R}(\Sigma, I)$. This example shows that $L(\alpha)$ is not a sufficient information for our modular approach. In order to overcome this problem, we shall define, for every formula $\alpha \in$ LTrL a finite family of automata $(\mathcal{A}_\alpha^B)_B$ for every set B *compatible* with I, that is such that $B \times B \subseteq I$. For a *compatible* set B, the automata $\mathcal{A}_\alpha^B$ will recognize the word language $\varphi^{-1}(L_B(\alpha))$ where $L_B(\alpha)$ is the trace language $\{u \mid \forall c, max(c) = B \Rightarrow cu, c \models \alpha\}$. Intuitively, our idea is that $L_B(\alpha)$ will be the set of traces u that satisfy α starting at a configuration whose maximal letters are described by B. This trick will allow us to deal with $\underline{a}$ formulas more smoothly.

With these notations, the word language $\varphi^{-1}(L(\alpha))$ will be recognized by the automaton $\mathcal{A}_\alpha^\emptyset$.

Subsection 4.2 deals with the easy cases for the formula α, while the difficult case of formulas of the form $\beta \mathcal{U} \gamma$ is presented in Subsection 4.3.

4.2 Boolean Connectors and Next-Step Operator

The Constant Formula Assume $\alpha = \underline{tt}$. Then $L(\alpha) = \mathbf{R}(\Sigma, I)$, so $\varphi^{-1}(L(\alpha)) = \Sigma^\infty$. Moreover, for every compatible subset $B \subseteq \Sigma$, the langage $L_B(\alpha)$ is $\mathbf{R}(\Sigma, I)$, so $\varphi^{-1}(L_B(\alpha)) = \Sigma^\infty$. We set $\mathcal{A}^B_{\underline{tt}} = (Q, \to, S, F, R)$ where:

- $Q = \{0\}$,
- $0 \xrightarrow{a} 0$ for all $a \in \Sigma$,
- $S = \{0\}$,
- and $F = R = \{0\}$.

In that way, it is obvious that every (finite or infinite) word is accepted by $\mathcal{A}^B_{\underline{tt}}$.

The Present Formulas Assume that $\alpha = \underline{a}$ for some $a \in \Sigma$. Now, depending on the set B we choose, we will construct two different automata. Indeed, if $a \in B$, then $L_B(\underline{a}) = \mathbf{R}(\Sigma, I)$, while if $a \notin B$, then $L_B(\underline{a}) = \emptyset$.

- For $B \ni a$, we define $\mathcal{A}^B_\alpha = \mathcal{A}^B_{\underline{tt}}$;
- For $B \not\ni a$, we define $\mathcal{A}^B_\alpha = \mathcal{A}^B_{\neg \underline{tt}}$, this automaton being the same as $\mathcal{A}^B_{\underline{tt}}$ except that $F = R = \emptyset$.

In both cases, the automaton we construct recognizes exactly $\varphi^{-1}(L_B(\alpha))$.

Negation Assume that $\alpha = \neg \beta$, and that the family of automata $(\mathcal{A}^B_\beta)_B$ is given. For a fixed B, the automaton $\mathcal{A}^B_\alpha$ recognizing exactly the language $\Sigma^\infty \setminus \varphi^{-1}(L_B(\beta))$, which is precisely $\varphi^{-1}(L_B(\alpha))$, can be obtained by the classical method of Safra[14]. Note that this step involves an exponential blow-up in the size of the automata.

Disjunction Assume that we know the automata $\mathcal{A}^B_\beta$ and $\mathcal{A}^B_\gamma$ recognizing respectively the languages $\varphi^{-1}(L_B(\beta))$ and $\varphi^{-1}(L_B(\gamma))$.

We want to construct the automata $(\mathcal{A}^B_\alpha)_B$ for the formula $\alpha = \beta \vee \gamma$. We can assume that $\mathcal{A}^B_\beta$ and $\mathcal{A}^B_\gamma$ have disjoint sets of states, and define $\mathcal{A}^B_\alpha$ as the disjoint union of these two automata. It is clear that $\mathcal{A}^B_\alpha$ recognizes $\varphi^{-1}(L_B(\beta)) \cup \varphi^{-1}(L_B(\gamma))$, i.e. $\varphi^{-1}(L_B(\beta) \cup L_B(\gamma))$, which is precisely $\varphi^{-1}(L_B(\alpha))$.

The Indexed Next-Step Operator Assume that $\alpha = @\beta$ for some $a \in \Sigma$ and some $\beta \in \text{LTrL}$. Assume that we have a fixed compatible set B. Then it is easy to see that $L_B(@\beta) = aL_{B'}(\beta)$ with $B' = (B \cap I(a)) \cup \{a\}$.

Therefore we define the automaton $\mathcal{A}^B_\alpha = (Q_1, \to_1, S_1, F_1, R_1)$ from the automaton $\mathcal{A}^{B'}_\beta = (Q, \to, S, F, R)$ (with $B' = (B \cap I(a)) \cup \{a\}$ as above) as follows:

- $Q_1 = Q \times \{0, 1, 2\}$,
- $S_1 = S \times \{0\}$,

- $F_1 = F \times \{1\}$,
- $R_1 = R \times \{1\}$,

and $\to_1$ in the following way:

- $(p,0) \xrightarrow{b}_1 (p',0)$ iff $b \in I(a)$ and $p \xrightarrow{b} p'$;
- $(p,0) \xrightarrow{b}_1 (p',2)$ iff $b \in D(a) \setminus \{a\}$ and $p \xrightarrow{b} p'$;
- $(p,0) \xrightarrow{b}_1 (p',1)$ iff $b = a$ and $p \xrightarrow{b} p'$;
- $(p,1) \xrightarrow{b}_1 (p',1)$ iff $p \xrightarrow{b} p'$;
- $(p,2) \xrightarrow{b}_1 (p',2)$ iff $p \xrightarrow{b} p'$.

It should be clear that $\mathcal{A}_\alpha^B$ recognizes exactly $\varphi^{-1}(L_B(\alpha))$ from the preceding remark.

4.3 A Construction for the Until Operator

We first explain the idea of the simple construction in the word case. If we want to check some property of the form $\beta \, \mathcal{U} \, \gamma$ on some word w, each time we read a letter a we have to start a new verification for β, until some time in the future when we will start to check the property γ. So every new verification started for β will be added to some set P, but the size of this set is bounded since if two different verifications reach the same state $p \in P$, then we know that we only have to carry on with one verification.

The situation with traces is much more involved. We also have to check β for all suffixes up to the suffix which satisfies γ. The problem is that when a trace t is represented by a word w, the suffixes of t are represented by certain subwords of w and not only suffixes of w. Therefore, each time we read a letter a we might either skip it or apply it to the current verification of β and γ depending on whether or not it is part of the considered trace suffix. Moreover, when we start checking γ, we might still need to start new verifications of β.

Finally, since we deal with infinite computation we need to take care of the acceptance conditions. Since the states that code the various verifications of β are put together in a set structure, we cannot know directly what are the repeated states of each verification. To overcome this problem, we will take an *ordering* transition that will transform this set into an ordered one, thus allowing to trace each verification individually. Note that this problem is not specific to traces.

Notations Assume that $\alpha = \beta \, \mathcal{U} \, \gamma$, and that the families of automata $(\mathcal{A}_\beta^B)_B$ and $(\mathcal{A}_\gamma^B)_B$ are given. We set $\mathcal{A}_\beta^B = (Q_1^B, \to_1^B, S_1^B, F_1^B, R_1^B)$ and $\mathcal{A}_\gamma^B = (Q_2^B, \to_2^B, S_2^B, F_2^B, R_2^B)$. We can assume, without loss of generality, that the sets $(Q_i^B)_B$ are pairwise disjoint for a fixed $i = 1, 2$. Now let $\mathcal{A}_1$ (resp. $\mathcal{A}_2$) be the disjoint union of the automata $\mathcal{A}_\beta^B$ (resp. $\mathcal{A}_\gamma^B$) for every compatible set $B \subseteq \Sigma$.

From Proposition 1, we can assume without loss of generality that $\mathcal{A}_1$ and $\mathcal{A}_2$ are alphabetic. We set, for $i = 1, 2$, $\mathcal{A}_i = (Q_i, \to_i, S_i, F_i, R_i)$.

Let's fix some total order on the set Q_1. This order induces a function $f : 2^{Q_1} \to Q_1^{\leq n}$ with $n = |Q_1|$, which maps every subset P of Q_1 to its corresponding ordered set $\underline{P}$.

Set of States and Transitions Recall that $\alpha = \beta \,\mathcal{U}\, \gamma$. For a fixed compatible set B, we define $\mathcal{A}_\alpha^B = (Q, \rightarrow, S, F, R)$ as follows:

- $Q = (2^{Q_1} \times Q_2 \times 2^\Sigma) \cup (Q_1^{\leq n} \times Q_2 \times 2^\Sigma)$,
- $S = \{\emptyset\} \times S_2 \times \{B\}$,
- $F = 2^{F_1} \times F_2 \times 2^\Sigma$,
- $R = R_1^{\leq n} \times R_2 \times 2^\Sigma$,

and let the transition relation $\rightarrow$ be defined in the following way:

- $(P, q, E) \xrightarrow{a} (P', q, E')$ iff $alph(q) \subseteq I(a)$ and $P' = P_a \cup (P \cap I(a)) \cup \{q_E\}$ such that $\exists q \in S_1^E, q \xrightarrow{a}_1 q_E$; and P_a satisfies $\forall q \in P, \exists q' \in P_a, q \xrightarrow{a}_1 q'$ and $\forall q' \in P_a, \exists q \in P, q \xrightarrow{a}_1 q'$; and $P \cap I(a) = \{q \in P \mid alph(q) \subseteq I(a)\}$; and finally $E' = (E \cap I(a)) \cup \{a\}$. These transitions are called β-transitions.
- $(P, q, E) \xrightarrow{a} (P', q', E')$ iff $P' = P_a$, $q \xrightarrow{a}_2 q'$, and $E' = (E \cap I(a)) \cup \{a\}$. These transitions are called γ-transitions.
- $(P, q, E) \xrightarrow{a} (\underline{P}, q', E')$ iff $\underline{P} = f(P_a)$, $q \xrightarrow{a}_2 q'$ and $E' = (E \cap I(a)) \cup \{a\}$. These transitions are called *ordering* transitions.
- $(\underline{P}, q, E) \xrightarrow{a} (\underline{P}', q', E')$ iff $\exists n_0 \leq n \mid \underline{P} \in Q_1^{n_0}$ and $\underline{P}' \in Q_1^{n_0}$ and for all $1 \leq i \leq n_0$, $\underline{P}_i \xrightarrow{a}_1 \underline{P}'_i$; $q \xrightarrow{a}_2 q'$; and $E' = (E \cap I(a)) \cup \{a\}$. These transitions are called *finishing* transitions.

Proposition 2. *The automaton $\mathcal{A}_\alpha^B$ recognizes exactly the word language $\varphi^{-1}(L_B(\beta \,\mathcal{U}\, \gamma))$.*

Due to lack of space, we do not provide here a proof of this proposition. The interested reader shall find the complete proof in a technical report[6].

5 Conclusion

We have shown that it is possible to have a direct construction of a Büchi automaton for every formula of LTrL and to make these constructions in a modular way, which allows us to reuse the automata for a formula α in any construction for a formula β where α would be a subformula of β. The two constructions involving an exponential blow-up are for the negation and the until operator. We know that the blow-up for the until operator is unavoidable. As far as the negation is concerned, we could use some transformations on the formulas in order to push every negation inside as far as possible, and describe an easy construction for a conjunction operator. What remains to be done is to find efficient algorithms for fragments of LTrL that are elementary, as Walukiewicz has open the way in [19], and find good characterizations of the expressive power of these fragments.

References

[1] R. Alur, D. Peled, and W. Penczek. Model-checking of causality properties. In *Proceedings of LICS'95*, pages 90–100, 1995.
[2] J.R. Büchi. Weak second-order arithmetic and finite automata. *Z. Math Logik Grundlag. Math.*, 6:66–92, 1960.
[3] C. Courcoubetis, M. Y. Vardi, P. Wolper, and M. Yannakakis. Memory efficient algorithms for the verification of temporal properties. *formal Methods in System Design*, 1:275–288, 1992.
[4] V. Diekert and G. Rozenberg, editors. *The Book of Traces*. World Scientific, Singapore, 1995.
[5] W. Ebinger. *Charakterisierung von Sprachklassen unendlicher Spuren durch Logiken*. Dissertation, Institut für Informatik, Universität Stuttgart, 1994.
[6] P. Gastin, R. Meyer, and A. Petit. A (non-elementary) modular decision procedure for LTrL. Technical report, LSV, ENS de Cachan, June 1998.
[7] P. Kelb, T. Margaria, M. Mendler, and C. Gsottberger. Mosel: a flexible toolset for monadic second-order logic. In *Proceedings of CAV'97, LNCS 1254*, 1997.
[8] N. Klarlund. Mona & Fido: The logic-automaton connection in practice. In *Proceedings of CSL'97, LNCS*, 1998.
[9] A. Mazurkiewicz. Concurrent program schemes and their interpretations. DAIMI Rep. PB 78, Aarhus University, Aarhus, 1977.
[10] R. Meyer and A. Petit. Expressive completeness of LTrL on finite traces: an algebraic proof. In *Proceedings of STACS'98*, number 1373 in LNCS, pages 533–543, 1998.
[11] D. Perrin and J. E. Pin. Infinite words. Technical report, LITP, Avril 1997.
[12] A. Pnueli. The temporal logics of programs. In *Proceedings of the 18th IEEE FOCS, 1977*, pages 46–57, 1977.
[13] R. Ramanujam. Locally linear time temporal logic. In *Proceedings of LICS'96*, pages 118–128, 1996.
[14] S. Safra. On the complexity of ω-automata. In *Proceedings of the 29th annual IEEE Symp. on Foundations of Computer Science*, pages 319–327, 1988.
[15] A. Sistla and E. Clarke. The complexity of propositional linear time logic. *J. ACM*, 32:733–749, 1985.
[16] L. Stockmeyer. The complexity of decision problems in automata theory and logic. PhD thesis, TR 133, M.I.T., Cambridge, 1974.
[17] P. S. Thiagarajan. A trace based extension of linear time temporal logic. In *Proceedings of the 9th Annual IEEE Symposium on Logic in Computer Science (LICS'94)*, pages 438–447, 1994.
[18] P. S. Thiagarajan and I. Walukiewicz. An expressively complete linear time temporal logic for Mazurkiewicz traces. In *Proceedings of the 12th Annual IEEE Symposium on Logic in Computer Science (LICS'97)*, 1997.
[19] I. Walukiewicz. Difficult configurations - on the complexity of LTrL. In *Proceedings of ICALP'98*, 1998.

Complete Abstract Interpretations Made Constructive

Roberto Giacobazzi[1], Francesco Ranzato[2], and Francesca Scozzari[1]

[1] Dipartimento di Informatica, Università di Pisa, Italy
{giaco,scozzari}@di.unipi.it
[2] Dipartimento di Matematica Pura ed Applicata, Università di Padova, Italy
franz@math.unipd.it

Abstract. Completeness is a desirable, although uncommon, property of abstract interpretations, formalizing the intuition that, relatively to the underlying abstract domains, the abstract semantics is as precise as possible. We consider here the most general form of completeness, where concrete semantic functions can have different domains and ranges, a case particularly relevant in functional programming. In this setting, our main contributions are as follows. (i) Under the weak and reasonable hypothesis of dealing with continuous semantic functions, a constructive characterization of complete abstract interpretations is given. (ii) It turns out that completeness is an abstract domain property. By exploiting (i), we therefore provide explicit constructive characterizations for the least complete extension and the greatest complete restriction of abstract domains. This considerably extends previous work by the first two authors, who recently proved results of mere existence for more restricted forms of least complete extension and greatest complete restriction. (iii) Our results permit to generalize, from a natural perspective of completeness, the notion of quotient of abstract interpretations, a tool introduced by Cortesi et al. for comparing the expressive power of abstract interpretations. Fairly severe hypotheses are required for Cortesi et al.'s quotients to exist. We prove instead that continuity of the semantic functions guarantees the existence of our generalized quotients.

1 Introduction and Motivation

Within the classical and widely adopted Cousot and Cousot framework for approximating generic semantic definitions [7,8], it is well known that *completeness* for an abstract interpretation is a much richer property than plain mandatory soundness. In fact, roughly speaking, a complete abstract interpretation turns out to be as precise as possible, relatively to its underlying abstract domains where approximate computations are encoded. This simple intuition explains why, although being a rather uncommon property in practice, notably in static program analysis, completeness is a highly desirable feature for an abstract interpretation, especially in abstract model checking (indeed, some authors arguably term it "optimality"). Examples of complete abstract interpretations can be found, e.g., when comparing algebraic polynomial systems [10] and program semantics [9].

In recent years, there has been a number of papers dealing with various theoretical issues related to completeness in abstract interpretation (cf. [6,12], [16,17,18,20]). Among them, Giacobazzi and Ranzato's paper [12] points out that completeness for an abstract interpretation only depends on the underlying abstract domain, and therefore is an abstract domain property. In view of this basic observation, the following problem is then considered: Given an abstract interpretation with underlying abstract domain A, do there exist the least extension and the greatest restriction of A making the whole abstract interpretation complete? Giacobazzi and Ranzato [12] give an affirmative answer, by showing that greatest complete restrictions (called *complete kernels*) always exist, and, for continuous concrete semantic operations, *least complete extensions* exist as well. According to [11], these two operators on abstract domains are, resp., instances of generic abstract domain simplifications and refinements. Following the standard notation, let us denote resp. by $\alpha_{X,Y}$ and $\gamma_{Y,X}$ the abstraction and concretization maps for a concrete domain X and an abstract domain Y. In [12], given a semantic operation $f : C^n \to C$, an abstract interpretation $I = \langle A, f^\sharp \rangle$, with $f^\sharp : A^n \to A$, is complete w.r.t. $\langle C, f \rangle$ when $\alpha_{C,A} \circ f = f^\sharp \circ \alpha_{C^n, A^n}$. Thus, functions of generic type $C \to D$, occurring frequently in denotational semantics for functional programming, cannot be handled. Moreover, Giacobazzi and Ranzato's results, in general, only prove the existence of least complete extensions and complete kernels, and give a constructive iterative methodology for obtaining least complete extensions only when the semantic operations are additive. However, additivity is a fairly restrictive hypothesis to be widely applicable in practice. By contrast, the present work deals with the most general formulation of completeness for abstract interpretations – no hypothesis on the type of semantic functions is assumed – and fully solves the limitations of Giacobazzi and Ranzato's approach, in particular on the side of complete domain construction.

Let us explain more in detail the general approach pursued in this paper. Firstly, given any concrete domain C, we denote by $\mathcal{L}_C$ the so-called *lattice of abstract interpretations* of C [7,8]. Let $f : C \to D$ be any concrete semantic function occurring in some complex semantic specification, and assume that an abstract semantics is given by $f^\sharp : A \to B$, where $A \in \mathcal{L}_C$ and $B \in \mathcal{L}_D$. The concept of soundness is standard and well-known: $\langle A, B, f^\sharp \rangle$ is a sound abstract interpretation – or $f^\sharp$ is a correct approximation of f relatively to A and B – when $\alpha_{D,B} \circ f \sqsubseteq f^\sharp \circ \alpha_{C,A}$ ($\sqsubseteq$ denotes pointwise ordering). On the other hand, $\langle A, B, f^\sharp \rangle$ is complete when equality holds, i.e. $\alpha_{D,B} \circ f = f^\sharp \circ \alpha_{C,A}$. Since $\alpha_{D,B} \circ f \sqsubseteq f^\sharp \circ \alpha_{C,A} \Leftrightarrow \alpha_{D,B} \circ f \circ \gamma_{A,C} \sqsubseteq f^\sharp$, the canonical best correct approximation $f^{b_{A,B}} : A \to B$ of f relatively to the abstract domains A and B is defined by $f^{b_{A,B}} \stackrel{\text{def}}{=} \alpha_{D,B} \circ f \circ \gamma_{A,C}$. In this scenario, the following observation still holds: Given A and B, there exists $f^\sharp$ such that $\langle A, B, f^\sharp \rangle$ is complete iff $\langle A, B, f^{b_{A,B}} \rangle$ is complete. This means that, even in this general context, *completeness is an abstract domain property*, and gives rise to the question whether abstract domains can be minimally refined and/or simplified so that completeness is achieved. Let us give a simple example concerning Mycroft's strictness analysis for functional programs [3,15]. Consider the following function F of type Nat × Nat → Bool:

$$F(\langle x,y\rangle) \stackrel{\text{def}}{=} \text{if } (x = 3 \text{ and } y = 3) \text{ then } \textit{true} \text{ else } \bot$$

Following Burn et al. [3], from F one gets in the most natural way its denotational "collecting" semantics $f : \mathbf{P}(\mathbb{N}_\bot \times \mathbb{N}_\bot) \to \mathbf{P}(\textit{Bool}_\bot)$, where $\mathbf{P}$ is the Hoare powerdomain operator and $\bot$ denotes undefinedness (i.e., both nontermination and error). Let $S = \{0 < 1\}$ be the basic strictness domain, abstracting both $\mathbf{P}(\mathbb{N}_\bot)$ and $\mathbf{P}(\textit{Bool}_\bot)$, and such that $S \times S$ abstracts $\mathbf{P}(\mathbb{N}_\bot \times \mathbb{N}_\bot)$. Concretization and abstraction maps are the usual ones, e.g. $\gamma(\langle 0,0\rangle) = \{\langle\bot,\bot\rangle\}$ and $\gamma(\langle 0,1\rangle) = \{\langle\bot,x\rangle \mid x \in \mathbb{N}_\bot\}$. Then, the best correct approximation $f^b : S \times S \to S$ of f is as follows: $f^b = \{\langle 0,0\rangle \mapsto 0, \langle 0,1\rangle \mapsto 0, \langle 1,0\rangle \mapsto 0, \langle 1,1\rangle \mapsto 1\}$. Clearly, f^b is not complete: For instance, $\alpha(f(\{\langle\bot,\bot\rangle,\langle 4,5\rangle\})) = \alpha(\{\bot\}) = 0$, whilst $f^b(\alpha(\{\langle\bot,\bot\rangle,\langle 4,5\rangle\})) = f^b(\langle 1,1\rangle) = 1$. These phenomena of incompleteness in strictness analysis are analyzed in depth in [17,18], which, however, do not investigate the issue of achieving completeness by minimally modifying the abstract domains. Moreover, because the range and domain of f are different, the method of [12] is not applicable here. Instead, the methodology proposed here allows to constructively derive the least extension $\mathcal{E}(S \times S)$ of $S \times S$ which induces a complete abstract interpretation. It should be clear that by adding a point to $S \times S$ which is able to represent the information that the first and second components are surely not simultaneously equal to $3 \in \mathbb{N}_\bot$, one gets a domain inducing a complete abstract interpretation. Indeed, our methodology allows to constructively derive that $\mathcal{E}(S \times S) = (S \times S) \cup \{\langle\neq,\neq\rangle\}$, where $\gamma(\langle\neq,\neq\rangle) = (\mathbb{N}_\bot \times \mathbb{N}_\bot) \setminus \{\langle 3,3\rangle\}$. In this way, one gets a best correct approximation $f^{b*} : \mathcal{E}(S \times S) \to S$ such that $f^{b*}(\langle\neq,\neq\rangle) = 0$, and therefore completeness has been achieved.

Let us illustrate the main contributions of the paper. In Section 3, the concept of completeness is formalized by resorting to the Cousot and Cousot *closure operator* approach to abstract interpretation [5,8]. This allows us to be independent from specific representations of abstract domain's objects. It is shown that completeness is an abstract domain property, which gives rise to a mathematically compact equation between closures, studied in later sections. Moreover, we observe that if an abstract interpretation $f^\sharp : A \to B$ is complete, and therefore $f^\sharp = f^{b_{A,B}}$, then for all the abstract domains A' more concrete than A and B' more abstract than B, it turns out that $f^{b_{A',B'}} : A' \to B'$ is still complete. This implies that it is not meaningful to search for the complete kernel of A and the least complete extension of B, because, e.g., if the complete kernel of A would exist then A itself would already be complete. Instead, one should try to solve the converse problems. Under the working hypothesis of dealing with continuous semantic functions, a key constructive characterization of the domains inducing complete abstract interpretations is given in Section 4. More precisely, given a continuous semantic function $f : C \to D$, we show that $f^{b_{A,B}} : A \to B$ is complete iff A is more concrete than a certain domain $R_f(B)$ depending on B iff B is more abstract than a certain domain $L_f(A)$ depending on A. Thus, the mappings $L_f : \mathcal{L}_C \to \mathcal{L}_D$ and $R_f : \mathcal{L}_D \to \mathcal{L}_C$ form an adjunction. By exploiting these results, we are able to characterize: (1) the least complete extension of A relative to B as the least domain which contains both A and $R_f(B)$, and (2) the complete kernel of B relative to A as the greatest domain contained in both B and $L_f(A)$. As a further consequence, we subsume the more restrictive notions of least complete extension and greatest complete restriction studied in [12] and the corresponding results of existence as well as the constructive characterization given for additive semantic functions. In Section 5, we investigate the relationship between completeness and the concept of *quotient* of an abstract interpretation, recently introduced by Cortesi et al. [4] for comparing the precision of abstract interpretations in computing a given property. Informally, the quotient of a complex abstract domain A w.r.t. a property P of A (i.e., a further abstraction of A) represents which part of A contributes in computing the property P. We show that, in general, Cortesi et al.'s quotients do not always exist: In particular, the basic assumption of continuity of the semantic functions does not ensure their existence. However, we observe that quotients, when they exist, turn out to be certain least complete extensions, which naturally formalize the intuition behind the notion of quotient. Thus, a simple and natural generalization of the notion of quotient is proposed, which retains the advantage of being always well-defined, under the hypothesis of continuity of the semantic functions.

2 Preliminaries

Basic Notation. If S is any set, P a poset, and $f, g : S \to P$ then we write $f \sqsubseteq g$ if for all $x \in S$, $f(x) \leq_P g(x)$. If $S \subseteq P$ then $max(S) \stackrel{\text{def}}{=} \{s \in S \mid \forall t \in S.\, s \leq t \Rightarrow s = t\}$. Given two posets C and D, $C \xrightarrow{m} D$, $C \xrightarrow{c} D$, and $C \xrightarrow{a} D$ denote, resp., the set of all monotone, continuous (i.e. preserving lub's of chains), and (completely) additive (i.e. preserving all lub's, empty set included) functions from C to D. ω denotes the first infinite ordinal. For a complete lattice C, given $f : C \to C$, for any $i \in \mathbb{N}$, the i-th power $f^i : C \to C$ of f is inductively defined, for any $x \in C$, as x if $i = 0$, and as $f(f^{i-1}(x))$ if i is a successor.

The Lattice of Abstract Interpretations. In standard Cousot and Cousot's abstract interpretation theory, abstract domains can be equivalently specified either by Galois connections, i.e. adjunctions, or by closure operators (see [5,8]). In the first case, the concrete domain C and the abstract domain A are related by an adjunction (α, C, A, γ). It is generally assumed that (α, C, A, γ) is a Galois insertion (GI), i.e. α is onto or, equivalently, γ is 1-1. In the second case instead, an abstract domain is specified as an (*upper*) *closure operator* (shortly uco or closure) on the concrete domain C, i.e., a monotone, idempotent and extensive operator on C. These two approaches are equivalent, modulo isomorphic representations of domain's objects. In the following, $\langle uco(C), \sqsubseteq \rangle$ denotes the poset of all uco's on C. Let us recall that each $\rho \in uco(C)$ is uniquely determined by the set of its fixpoints, which is its image, i.e. $\rho(C) = \{x \in C \mid \rho(x) = x\}$, and that $\rho \sqsubseteq \eta$ iff $\eta(C) \subseteq \rho(C)$. Also, when $\langle C, \leq, \vee, \wedge, \top, \bot \rangle$ is a complete lattice, $\langle uco(C), \sqsubseteq, \sqcup, \sqcap, \lambda x.\top, \lambda x.x \rangle$ is a complete lattice, and $X \subseteq C$ is the set of fixpoints of a uco iff X is meet-closed, i.e. $X = \mathcal{M}(X) \stackrel{\text{def}}{=} \{\wedge Y \mid Y \subseteq X\}$ (where $\wedge \emptyset = \top \in X$). Moreover, given $\rho \subset uco(C)$, $\langle \rho(C), \leq \rangle$ is a complete meet subsemilattice of C. Hence, for a concrete domain C which is a complete lattice, we will identify $uco(C)$ with the lattice $\mathcal{L}_C$ of abstract interpretations of C, i.e. the complete lattice of all possible abstract domains of C. Often, we will find convenient to identify closures with their sets of fixpoints. This does not give rise to ambiguity, since one can distinguish their use as functions or sets according to the context. The ordering on $uco(C)$ corresponds precisely to the standard order used in abstract interpretation to compare abstract domains with regard to their precision: A_1 is more precise than A_2 (i.e., A_1 is more concrete than A_2 or A_2 is more abstract than A_1) iff $A_1 \sqsubseteq A_2$ in $uco(C)$. Lub and glb of $uco(C)$ have therefore the following reading as operators on abstract domains. Let $\{A_i\}_{i \in I} \subseteq uco(C)$: (i) $\sqcup_{i \in I} A_i$ is the most concrete among the domains which are abstractions of all the A_i's, i.e. it is their least (w.r.t. $\sqsubseteq$) common abstraction; (ii) $\sqcap_{i \in I} A_i$ is the most abstract among the domains (abstracting C) which are more concrete than every A_i; this domain is known as reduced product of all the A_i's.

3 Completeness by Closures

Let $f : C \xrightarrow{m} D$ be any monotone semantic function, where C and D are complete lattices playing the rôle of concrete semantic domains. Let an abstract interpretation $\langle A, B, f^\sharp \rangle$ of $\langle C, D, f \rangle$ be specified by the GIs $(\alpha_{C,A}, C, A, \gamma_{A,C})$ and $(\alpha_{D,B}, D, B, \gamma_{B,D})$, and by an abstract function $f^\sharp : A \xrightarrow{m} B$. It is known [8] that $f^\sharp$ is a correct approximation of f, i.e. $\alpha_{D,B} \circ f \sqsubseteq f^\sharp \circ \alpha_{C,A}$, if and only if $\alpha_{D,B} \circ f \circ \gamma_{A,C} \sqsubseteq f^\sharp$. Thus, $f^{b_{A,B}} \stackrel{\text{def}}{=} \alpha_{D,B} \circ f \circ \gamma_{A,C} : A \to B$ is called the canonical best correct approximation of f relatively to the abstract domains A and B. $\langle A, B, f^\sharp \rangle$ is called *complete* when $\alpha_{D,B} \circ f = f^\sharp \circ \alpha_{C,A}$. In this case, $f^\sharp = f^\sharp \circ \alpha_{C,A} \circ \gamma_{A,C} = \alpha_{D,B} \circ f \circ \gamma_{A,C} = f^{b_{A,B}}$, i.e. $f^\sharp$ indeed is the best correct approximation $f^{b_{A,B}}$. This means that, given two abstract domains A and B, there exists $f^\sharp$ such that $\langle A, B, f^\sharp \rangle$ is complete iff $\langle A, B, f^{b_{A,B}} \rangle$ is complete. Since $f^{b_{A,B}}$ only depends on A and B, we get that *completeness is an abstract domain property*. Thus, given A and B, we refer to completeness of A and B in order to refer to completeness of the whole abstract interpretation $\langle A, B, f^{b_{A,B}} \rangle$. By using closure operators, if $\rho = \gamma_{A,C} \circ \alpha_{C,A} \in uco(C)$ and $\eta = \gamma_{B,D} \circ \alpha_{D,B} \in uco(D)$ are the uco's associated, resp., with A and B, one can extend an analogous result in [12] by showing that A and B are complete iff $\eta \circ f = \eta \circ f \circ \rho$. This justifies the following general definition of completeness.

Definition 1. Let C and D be complete lattices, $f : C \xrightarrow{m} D$, $\rho \in uco(C)$, and $\eta \in uco(D)$. Then, the pair $\langle \rho, \eta \rangle$ is *complete* for f if $\eta \circ f = \eta \circ f \circ \rho$. Also, if $F \subseteq C \xrightarrow{m} D$ then $\langle \rho, \eta \rangle$ is complete for F whenever $\forall f \in F$. $\eta \circ f = \eta \circ f \circ \rho$. □

First, let us notice that, equivalently, one can define $\langle \rho, \eta \rangle$ complete for f when $f \circ \rho \sqsubseteq \eta \circ f$. Further, it is worth remarking that our definition encompasses the case where $f : C \to C$ and one is interested in two different abstractions of input and output, i.e. $\rho, \eta \in uco(C)$ with $\rho \neq \eta$. Whenever $f : C \to C$ and $\rho = \eta$, the above definition of completeness boils down to the equation $\rho \circ f = \rho \circ f \circ \rho$ considered in [7,12]. Also, it would not be too difficult (although notationally heavy) to develop the whole theory by considering semantic functions of type $C^n \to D^m$.

For any given set of functions $F \subseteq C \xrightarrow{m} D$, we will use the following helpful notation: $\Gamma(C, D, F) \stackrel{\text{def}}{=} \{\langle\rho, \eta\rangle \in uco(C) \times uco(D) \mid \forall f \in F. \ \eta \circ f = \eta \circ f \circ \rho\}$. Whenever $F = \{f\}$, we simply write $\Gamma(C, D, f)$. The following result lists some interesting properties of completeness, where points (i)–(iii) generalize an analogous result given in [12].

Proposition 1.
(i) $\langle\lambda x.x, \eta\rangle, \langle\rho, \lambda x.\top_D\rangle \in \Gamma(C, D, F)$.
(ii) $\forall d \in D. \ \Gamma(C, D, \lambda x.d) = uco(C) \times uco(D)$.
(iii) If $\langle\rho, \eta\rangle \in \Gamma(C, D, f)$ and $\langle\eta, \mu\rangle \in \Gamma(D, E, g)$ then $\langle\rho, \mu\rangle \in \Gamma(C, E, g \circ f)$.
(iv) If $\langle\rho, \eta\rangle \in \Gamma(C, D, F)$, $\delta \sqsubseteq \rho$ and $\beta \sqsupseteq \eta$, then $\langle\delta, \beta\rangle \in \Gamma(C, D, F)$.

Given $F \subseteq C \xrightarrow{m} D$ and $\eta \in uco(D)$, let us now introduce the following operators transforming abstract domains of C (as usual, we follow the standard conventions $\sqcap \emptyset = \top_{uco(C)}$ and $\sqcup \emptyset = \bot_{uco(C)}$).

- $\mathcal{K}_F^\eta(\rho) \stackrel{\text{def}}{=} \sqcap \{\varphi \in uco(C) \mid \rho \sqsubseteq \varphi, \ \langle\varphi, \eta\rangle \in \Gamma(C, D, F)\}$;
- $\mathcal{E}_F^\eta(\rho) \stackrel{\text{def}}{=} \sqcup \{\varphi \in uco(C) \mid \varphi \sqsubseteq \rho, \ \langle\varphi, \eta\rangle \in \Gamma(C, D, F)\}$.

Also, given $\rho \in uco(C)$, analogous operators $\mathcal{K}_F^\rho$ and $\mathcal{E}_F^\rho$ of type $uco(D) \to uco(D)$ are introduced. Thus, e.g., $\mathcal{E}_F^\eta(\rho)$ is the least common abstraction of all the domains φ more concrete than ρ and such that $\langle\varphi, \eta\rangle$ is complete for F. As a consequence of Proposition 1 (iv), one can draw the following two important remarks: (i) If $\langle\mathcal{K}_F^\eta(\rho), \eta\rangle \in \Gamma(C, D, F)$ then $\mathcal{K}_F^\eta(\rho) = \rho$; (ii) If $\langle\rho, \mathcal{E}_F^\rho(\eta)\rangle \in \Gamma(C, D, F)$ then $\mathcal{E}_F^\rho(\eta) = \eta$. This means that it does not make sense to search for the greatest restriction ρ^g of $\rho \in uco(C)$ such that $\langle\rho^g, \eta\rangle$ is complete, and, dually, the least extension η^l of $\eta \in uco(D)$ such that $\langle\rho, \eta^l\rangle$ is complete, because either they coincide with their arguments or they do not exist. That is why we introduce just the following notions.

Definition 2. If $\langle\rho, \mathcal{K}_F^\rho(\eta)\rangle \in \Gamma(C, D, F)$ then $\mathcal{K}_F^\rho(\eta)$ is called the *complete kernel* of η relative to ρ. Dually, if $\langle\mathcal{E}_F^\eta(\rho), \eta\rangle \in \Gamma(C, D, F)$ then $\mathcal{E}_F^\eta(\rho)$ is called the *least complete extension* of ρ relative to η. □

As far as complete kernels are concerned, it is an easy task to show that they always exist, although no explicit characterization can be given.

Proposition 2. *Let $F \subseteq C \xrightarrow{m} D$, $\rho \in uco(C)$ and $\eta \in uco(D)$. There exists the complete kernel of η relative to ρ.*

Let us now consider the case where $C = D$ and $\rho = \eta$. It is important to remark that if $\langle\mathcal{E}_F^\rho(\rho), \rho\rangle \in \Gamma(C, C, F)$, then $\mathcal{E}_F^\rho(\rho)$ is the most abstract among the domains $\varphi \sqsubseteq \rho$ such that $\rho \circ f = \rho \circ f \circ \varphi$. Thus, we stress that ρ thought of as output abstraction is considered fixed. We will see in Section 5 how this concept can be usefully exploited. Moreover, let us recall that in Giacobazzi and Ranzato's approach [12], the least complete extension of ρ, when it exists, is instead defined as the most abstract among the domains $\varphi \sqsubseteq \rho$ such that $\varphi \circ f = \varphi \circ f \circ \varphi$. Hence, by defining $E_F(\rho) \stackrel{\text{def}}{=} \sqcup \{\varphi \in uco(C) \mid \varphi \sqsubseteq \rho,\ \langle \varphi, \varphi \rangle \in \Gamma(C,C,F)\}$, this latter least complete extension exists whenever $\langle E_F(\rho), E_F(\rho) \rangle \in \Gamma(C,C,F)$. Therefore, in this case, ρ considered as output abstraction is not fixed. Thus, we remark that this latter concept of least complete extension is different from that introduced in Definition 2. In the next section, we will study both these interesting notions. In order to distinguish them, when $\langle E_F(\rho), E_F(\rho) \rangle \in \Gamma(C,C,F)$, we will call $E_F(\rho)$ the *absolute* least complete extension of ρ. Moreover, analogous dual considerations hold for complete kernels: We will call them absolute complete kernels.

4 Constructive Characterization of Completeness

The following key result characterizes complete abstract interpretations in a "constructive" way: In fact, it shows that a completeness equation $\eta \circ f = \eta \circ f \circ \rho$ holds iff ρ contains a certain set of points depending on η, and, in a dual fashion, iff η is contained in a certain set of points which depends on ρ. The proof makes use of a variant of the axiom of choice, known as Hausdorff's Maximal Principle [2, pag. 192]. We will exploit largely the following compact notation: For any $f : C \to D$ and $y \in D$, $H_y^f \stackrel{\text{def}}{=} \{x \in C \mid f(x) \leq y\}$.

Theorem 1. *Let $F \subseteq C \stackrel{c}{\hookrightarrow} D$, $\rho \in uco(C)$ and $\eta \in uco(D)$. Then,*
$\langle \rho, \eta \rangle \in \Gamma(C,D,F) \Leftrightarrow \eta \subseteq \{y \in D \mid \cup_{f \in F} max(H_y^f) \subseteq \rho\} \Leftrightarrow \cup_{f \in F, y \in \eta} max(H_y^f) \subseteq \rho$.
Moreover, $\{y \in D \mid \cup_{f \in F} max(H_y^f) \subseteq \rho\} \in uco(D)$.

It is then useful to observe that, for any arbitrary set of points S and any uco ρ, the following equivalence holds: $S \subseteq \rho \Leftrightarrow \rho \sqsubseteq \mathcal{M}(S)$. Thus, as the above theorem suggests, given any set of continuous functions $F \subseteq C \stackrel{c}{\hookrightarrow} D$, we define two mappings $L_F : uco(C) \to uco(D)$ and $R_F : uco(D) \to uco(C)$ as follows:

$$L_F(\rho) \stackrel{\text{def}}{=} \{y \in D \mid \cup_{f \in F} max(H_y^f) \subseteq \rho\}; \qquad R_F(\eta) \stackrel{\text{def}}{=} \mathcal{M}(\cup_{f \in F, y \in \eta} max(H_y^f)).$$

In this way, Theorem 1 can be restated as follows:

$$\langle \rho, \eta \rangle \in \Gamma(C,D,F) \Leftrightarrow L_F(\rho) \sqsubseteq \eta \Leftrightarrow \rho \sqsubseteq R_F(\eta).$$

In particular, $(L_F, uco(C), uco(D), R_F)$ is an adjunction. Consequently, for any $\rho \in uco(C)$ and $\eta \in uco(D)$, one gets the following characterizations for the operators $\mathcal{K}_F^\rho$ and $\mathcal{E}_F^\eta$:

- $\mathcal{K}_F^\rho(\beta) = \sqcap \{\mu \in uco(D) \mid \beta, L_F(\rho) \sqsubseteq \mu\} = \beta \sqcup L_F(\rho);$
- $\mathcal{E}_F^\eta(\delta) = \sqcup \{\varphi \in uco(C) \mid \varphi \sqsubseteq \delta, R_F(\eta)\} = \delta \sqcap R_F(\eta).$

Hence, since $L_F(\rho) \sqsubseteq \beta \sqcup L_F(\rho)$ and $\delta \sqcap R_F(\eta) \sqsubseteq R_F(\eta)$, by Theorem 1, we obtain that $\langle \rho, \beta \sqcup L_F(\rho) \rangle, \langle \delta \sqcap R_F(\eta), \eta \rangle \in \Gamma(C,D,F)$, and therefore, according to Definition 2, we can draw the following consequences:

$$F(\langle x, y \rangle) \stackrel{\text{def}}{=} \text{if } (x = 3 \text{ and } y = 3) \text{ then true else } \bot$$

Following Burn et al. [3], from F one gets in the most natural way its denotational "collecting" semantics $f : \mathbf{P}(\mathbb{N}_\bot \times \mathbb{N}_\bot) \to \mathbf{P}(\mathit{Bool}_\bot)$, where $\mathbf{P}$ is the Hoare powerdomain operator and $\bot$ denotes undefinedness (i.e., both nontermination and error). Let $S = \{0 < 1\}$ be the basic strictness domain, abstracting both $\mathbf{P}(\mathbb{N}_\bot)$ and $\mathbf{P}(\mathit{Bool}_\bot)$, and such that $S \times S$ abstracts $\mathbf{P}(\mathbb{N}_\bot \times \mathbb{N}_\bot)$. Concretization and abstraction maps are the usual ones, e.g. $\gamma(\langle 0, 0 \rangle) = \{\langle \bot, \bot \rangle\}$ and $\gamma(\langle 0, 1 \rangle) = \{\langle \bot, x \rangle \mid x \in \mathbb{N}_\bot\}$. Then, the best correct approximation $f^b : S \times S \to S$ of f is as follows: $f^b = \{\langle 0, 0 \rangle \mapsto 0, \langle 0, 1 \rangle \mapsto 0, \langle 1, 0 \rangle \mapsto 0, \langle 1, 1 \rangle \mapsto 1\}$. Clearly, f^b is not complete: For instance, $\alpha(f(\{\langle \bot, \bot \rangle, \langle 4, 5 \rangle\})) = \alpha(\{\bot\}) = 0$, whilst $f^b(\alpha(\{\langle \bot, \bot \rangle, \langle 4, 5 \rangle\})) = f^b(\langle 1, 1 \rangle) = 1$. These phenomena of incompleteness in strictness analysis are analyzed in depth in [17,18], which, however, do not investigate the issue of achieving completeness by minimally modifying the abstract domains. Moreover, because the range and domain of f are different, the method of [12] is not applicable here. Instead, the methodology proposed here allows to constructively derive the least extension $\mathcal{E}(S \times S)$ of $S \times S$ which induces a complete abstract interpretation. It should be clear that by adding a point to $S \times S$ which is able to represent the information that the first and second components are surely not simultaneously equal to $3 \in \mathbb{N}_\bot$, one gets a domain inducing a complete abstract interpretation. Indeed, our methodology allows to constructively derive that $\mathcal{E}(S \times S) = (S \times S) \cup \{\langle \neq, \neq \rangle\}$, where $\gamma(\langle \neq, \neq \rangle) = (\mathbb{N}_\bot \times \mathbb{N}_\bot) \setminus \{\langle 3, 3 \rangle\}$. In this way, one gets a best correct approximation $f^{b*} : \mathcal{E}(S \times S) \to S$ such that $f^{b*}(\langle \neq, \neq \rangle) = 0$, and therefore completeness has been achieved.

Let us illustrate the main contributions of the paper. In Section 3, the concept of completeness is formalized by resorting to the Cousot and Cousot *closure operator* approach to abstract interpretation [5,8]. This allows us to be independent from specific representations of abstract domain's objects. It is shown that completeness is an abstract domain property, which gives rise to a mathematically compact equation between closures, studied in later sections. Moreover, we observe that if an abstract interpretation $f^\sharp : A \to B$ is complete, and therefore $f^\sharp = f^{b_{A,B}}$, then for all the abstract domains A' more concrete than A and B' more abstract than B, it turns out that $f^{b_{A',B'}} : A' \to B'$ is still complete. This implies that it is not meaningful to search for the complete kernel of A and the least complete extension of B, because, e.g., if the complete kernel of A would exist then A itself would already be complete. Instead, one should try to solve the converse problems. Under the working hypothesis of dealing with continuous semantic functions, a key constructive characterization of the domains inducing complete abstract interpretations is given in Section 4. More precisely, given a continuous semantic function $f : C \to D$, we show that $f^{b_{A,B}} : A \to B$ is complete iff A is more concrete than a certain domain $R_f(B)$ depending on B iff B is more abstract than a certain domain $L_f(A)$ depending on A. Thus, the mappings $L_f : \mathcal{L}_C \to \mathcal{L}_D$ and $R_f : \mathcal{L}_D \to \mathcal{L}_C$ form an adjunction. By exploiting these results, we are able to characterize: (1) the least complete extension of A relative to B as the least domain which contains both A and $R_f(B)$, and (2) the

In recent years, there has been a number of papers dealing with various theoretical issues related to completeness in abstract interpretation (cf. [6,12], [16,17,18,20]). Among them, Giacobazzi and Ranzato's paper [12] points out that completeness for an abstract interpretation only depends on the underlying abstract domain, and therefore is an abstract domain property. In view of this basic observation, the following problem is then considered: Given an abstract interpretation with underlying abstract domain A, do there exist the least extension and the greatest restriction of A making the whole abstract interpretation complete? Giacobazzi and Ranzato [12] give an affirmative answer, by showing that greatest complete restrictions (called *complete kernels*) always exist, and, for continuous concrete semantic operations, *least complete extensions* exist as well. According to [11], these two operators on abstract domains are, resp., instances of generic abstract domain simplifications and refinements. Following the standard notation, let us denote resp. by $\alpha_{X,Y}$ and $\gamma_{Y,X}$ the abstraction and concretization maps for a concrete domain X and an abstract domain Y. In [12], given a semantic operation $f : C^n \to C$, an abstract interpretation $I = \langle A, f^\sharp \rangle$, with $f^\sharp : A^n \to A$, is complete w.r.t. $\langle C, f \rangle$ when $\alpha_{C,A} \circ f = f^\sharp \circ \alpha_{C^n, A^n}$. Thus, functions of generic type $C \to D$, occurring frequently in denotational semantics for functional programming, cannot be handled. Moreover, Giacobazzi and Ranzato's results, in general, only prove the existence of least complete extensions and complete kernels, and give a constructive iterative methodology for obtaining least complete extensions only when the semantic operations are additive. However, additivity is a fairly restrictive hypothesis to be widely applicable in practice. By contrast, the present work deals with the most general formulation of completeness for abstract interpretations – no hypothesis on the type of semantic functions is assumed – and fully solves the limitations of Giacobazzi and Ranzato's approach, in particular on the side of complete domain construction.

Let us explain more in detail the general approach pursued in this paper. Firstly, given any concrete domain C, we denote by $\mathcal{L}_C$ the so-called *lattice of abstract interpretations* of C [7,8]. Let $f : C \to D$ be any concrete semantic function occurring in some complex semantic specification, and assume that an abstract semantics is given by $f^\sharp : A \to B$, where $A \in \mathcal{L}_C$ and $B \in \mathcal{L}_D$. The concept of soundness is standard and well-known: $\langle A, B, f^\sharp \rangle$ is a sound abstract interpretation – or $f^\sharp$ is a correct approximation of f relatively to A and B – when $\alpha_{D,B} \circ f \sqsubseteq f^\sharp \circ \alpha_{C,A}$ ($\sqsubseteq$ denotes pointwise ordering). On the other hand, $\langle A, B, f^\sharp \rangle$ is complete when equality holds, i.e. $\alpha_{D,B} \circ f = f^\sharp \circ \alpha_{C,A}$. Since $\alpha_{D,B} \circ f \sqsubseteq f^\sharp \circ \alpha_{C,A} \Leftrightarrow \alpha_{D,B} \circ f \circ \gamma_{A,C} \sqsubseteq f^\sharp$, the canonical best correct approximation $f^{b_{A,B}} : A \to B$ of f relatively to the abstract domains A and B is defined by $f^{b_{A,B}} \stackrel{\text{def}}{=} \alpha_{D,B} \circ f \circ \gamma_{A,C}$. In this scenario, the following observation still holds: Given A and B, there exists $f^\sharp$ such that $\langle A, B, f^\sharp \rangle$ is complete iff $\langle A, B, f^{b_{A,B}} \rangle$ is complete. This means that, even in this general context, *completeness is an abstract domain property*, and gives rise to the question whether abstract domains can be minimally refined and/or simplified so that completeness is achieved. Let us give a simple example concerning Mycroft's strictness analysis for functional programs [3,15]. Consider the following function F of type Nat × Nat → Bool:

- The complete kernel of η rel. to ρ is the least common abstraction of η and $L_F(\rho)$;
- The least complete extension of ρ rel. to η is the reduced product of ρ and $R_F(\eta)$.

For any $\rho \in uco(C)$ and $\eta \in uco(D)$, it is helpful to define two dual mappings $\mathcal{F}_F^\rho : uco(D) \to uco(C)$ and $\mathcal{G}_F^\eta : uco(C) \to uco(D)$ as follows:

$$\mathcal{F}_F^\rho(\mu) \stackrel{\text{def}}{=} \rho \sqcap R_F(\mu); \qquad \mathcal{G}_F^\eta(\varphi) \stackrel{\text{def}}{=} \eta \sqcup L_F(\varphi).$$

Summing up, we have shown the following result, which explicitly states what one must add to ρ in order to get its least complete extension relative to η and, dually, what one must subtract from η in order to get its complete kernel relative to ρ.

Theorem 2. Let $F \subseteq C \xrightarrow{c} D$, $\rho \in uco(C)$ and $\eta \in uco(D)$.

- $\mathcal{F}_F^\rho(\eta) = \mathcal{M}(\rho \cup (\cup_{f \in F, y \in \eta} max(H_y^f)))$
  is the least complete extension of ρ rel. to η;
- $\mathcal{G}_F^\eta(\rho) = \eta \cap \{y \in D \mid \cup_{f \in F} max(H_y^f) \subseteq \rho\}$
  is the complete kernel of η rel. to ρ.

Example 1. Consider the example sketched in Section 1. Let $\rho \in uco(\mathbf{P}(\mathbb{N}_\perp \times \mathbb{N}_\perp))$ be the uco associated to the input abstract domain $S \times S$, and $\eta \in uco(\mathbf{P}(Bool_\perp))$ be the uco associated to the output abstract domain S. Therefore, $\rho = \{\{\langle \perp, \perp \rangle\}, \{\perp\} \times \mathbb{N}_\perp, \mathbb{N}_\perp \times \{\perp\}, \mathbb{N}_\perp \times \mathbb{N}_\perp\}$ and $\eta = \{\{\perp\}, Bool_\perp\}$. The semantic function f is obviously continuous and hence, by Theorem 2, the least complete extension of ρ for f relative to η does exist, and it is given by the reduced product $\mathcal{F}_f^\rho(\eta) = \rho \sqcap R_f(\eta)$. Thus, for $y \in \eta$, let us compute $max(H_y^f)$. We have that:

- $max(H_{Bool_\perp}^f) = \{\mathbb{N}_\perp \times \mathbb{N}_\perp\}$;
- $max(H_{\{\perp\}}^f) = max(\{Z \in \mathbf{P}(\mathbb{N}_\perp \times \mathbb{N}_\perp) \mid f(Z) \subseteq \{\perp\}\})$
  $= max(\{Z \in \mathbf{P}(\mathbb{N}_\perp \times \mathbb{N}_\perp) \mid \langle 3, 3 \rangle \notin Z\}) = \{(\mathbb{N}_\perp \times \mathbb{N}_\perp) \setminus \{\langle 3, 3 \rangle\}\}$.

Hence, $\mathcal{F}_F^\rho(\eta) = \mathcal{M}(\rho \cup \{(\mathbb{N}_\perp \times \mathbb{N}_\perp) \setminus \{\langle 3, 3 \rangle\}\}) = \rho \cup \{(\mathbb{N}_\perp \times \mathbb{N}_\perp) \setminus \{\langle 3, 3 \rangle\}\}$. Thus, as announced in Section 1, and as one naturally expects, this shows that the least complete extension of $S \times S$ can be obtained by adding a point $\langle \neq, \neq \rangle$ with concrete meaning $(\mathbb{N}_\perp \times \mathbb{N}_\perp) \setminus \{\langle 3, 3 \rangle\}$, i.e. denoting that first and second components are surely not simultaneously equal to the value 3. It should be clear that this refined input abstract domain induces now a complete abstract interpretation. □

Let us now turn to absolute complete kernels and absolute least complete extensions, as formally introduced at the end of Section 3. What follows generalizes the results in [12, Section 6], where the hypothesis consisted of dealing with additive semantic functions. Assume that $C = D$, i.e. $F \subseteq C \xrightarrow{c} C$, and let $\rho \in uco(C)$. By Theorem 1, for any $\varphi \in uco(C)$, we have that:

- $\varphi \sqsubseteq \mathcal{F}_F^\rho(\varphi) \Leftrightarrow \varphi \sqsubseteq \rho$ and $\langle \varphi, \varphi \rangle \in \Gamma(C, C, F)$;
- $\mathcal{G}_F^\rho(\varphi) \sqsubseteq \varphi \Leftrightarrow \rho \sqsubseteq \varphi$ and $\langle \varphi, \varphi \rangle \in \Gamma(C, C, F)$.

Therefore, for the operator E_F introduced at the end of Section 3, we obtain that $E_F(\rho) = \sqcup \{\varphi \in uco(C) \mid \varphi \sqsubseteq \mathcal{F}_F^\rho(\varphi)\}$. Then, since $\mathcal{F}_F^\rho : uco(C) \to uco(C)$ is clearly monotone for any ρ, and hence admits the greatest fixpoint, we get $E_F(\rho) = gfp(\mathcal{F}_F^\rho)$. Moreover, by Theorem 2, $\langle gfp(\mathcal{F}_F^\rho), gfp(\mathcal{F}_F^\rho) \rangle \in \Gamma(C, C, F)$. This means that the absolute least complete extension of ρ exists, and it is $gfp(\mathcal{F}_F^\rho)$. Dual considerations hold for complete kernels. Thus, we get the following constructive characterization for absolute completeness.

Theorem 3. *Let* $F \subseteq C \xrightarrow{c} C$ *and* $\rho \in uco(C)$. *Then*, $gfp(\mathcal{F}_F^\rho)$ *and* $lfp(\mathcal{G}_F^\rho)$ *are, resp., the absolute least complete extension and absolute complete kernel of* ρ.

5 Generalized Quotients of Abstract Interpretations

The concept of *quotient* of an abstract interpretation has been recently introduced by Cortesi et al. [4] in order to formalize the least amount of information of a complex abstract domain A that is useful for computing some property that A is able to represent. Cortesi et al. [4] show how to exploit this notion for comparing the precision of two abstract interpretations in computing a given common property. Notably, they compare the well-known Jacobs and Langen *Sharing* [13] and Marriott and Søndergaard *Pos* [14] Prolog abstract interpretations, by demonstrating that *Pos* is strictly more precise than *Sharing* for computing variable groundness information. Further, Bagnara et al. [1] show, also experimentally, that in order to compute pair-sharing information, the use of the quotient of *Sharing* w.r.t. the pair-sharing Søndergaard domain [19] leads to remarkable gains of efficiency, when compared with the full domain *Sharing*.

Let us recall from [4] the definition of quotient. Let A be any complete lattice, $f : A \xrightarrow{m} A$ be a monotone semantic function on A, and $\rho \in uco(A)$ be an abstraction of A. Here, $\langle A, f \rangle$ models any abstract interpretation of some reference semantic definition, while ρ plays the rôle of the property (i.e. the abstraction of A) one is interested in. The equivalence relation $r_\rho \subseteq A \times A$ is defined as follows:[1]

$$\langle a_1, a_2 \rangle \in r_\rho \text{ iff } \forall i \in \omega. \ \rho(f^i(a_1)) = \rho(f^i(a_2)).$$

Roughly speaking, $\langle a_1, a_2 \rangle \in r_\rho$ when A views a_1 and a_2 as "equivalent" w.r.t. the computation of the property ρ. Thus, according to this intuition, the quotient $\mathcal{Q}_\rho(A)$ of A w.r.t. ρ is defined (cf. [4, Definition 3.5]) as the subset of A of the lub's of all equivalence classes of r_ρ: That is, if $[a]$ denotes a generic equivalence class for r_ρ, then $\mathcal{Q}_\rho(A) \stackrel{\text{def}}{=} \{\vee[a] \mid a \in A\}$, and the ordering is that inherited from A. Cortesi et al. [4, Theorem 3.6] show that if the equivalence r_ρ is additive, i.e. $\forall i \in I. \ \langle a_i, b_i \rangle \in r_\rho \Rightarrow \langle \vee_{i \in I} a_i, \vee_{i \in I} b_i \rangle \in r_\rho$, then $\mathcal{Q}_\rho(A)$ is well-defined, namely

---

[1] This definition considers the case of the first limit ordinal ω for practical purposes – a generalization to any (possibly transfinite) ordinal would be straightforward.

it is in turn an abstraction of A, i.e. the set of fixpoints of a uco on A, and ρ is an abstraction of $\mathcal{Q}_\rho(A)$.

Cortesi et al.'s results can be sharpened as follows. Firstly, it is useful to recall (see [8, Section 6.3]) that, in general, given an equivalence relation R on a complete lattice L, R is additive iff $\lambda x. \vee_L [x]_R \in uco(L)$. Thus, the hypothesis that the equivalence relation r_ρ is additive is indeed equivalent to the fact that $\lambda a. \vee [a] \in uco(A)$, i.e. that the quotient $\mathcal{Q}_\rho(A)$ is well-defined. In this case, that $\mathcal{Q}_\rho(A) \sqsubseteq \rho$ (i.e. [4, Theorem 3.6 (ii)]) is an immediate consequence: In fact, if $a = \rho(a)$ and $b \in [a]$ then $a = \rho(a) = \rho(f^0(a)) = \rho(f^0(b)) = \rho(b)$, and hence $b \leq a$, i.e. $\vee[a] = a$.

En passant, we observe that the additivity of f is an obvious sufficient condition guaranteeing that the quotient exists. Actually, the quotients presented in [1,4] exist just because the involved semantic functions are additive.

Lemma 1. *If $f : A \xrightarrow{a} A$ and $\rho \in uco(A)$ then $\mathcal{Q}_\rho(A) \in uco(A)$.*

It turns out that the quotient abstract domain satisfies the following remarkable property of "minimality": When a quotient $\mathcal{Q}_\rho(A)$ exists, if $\phi_\rho = \lambda a. \vee [a] \in uco(A)$ is the uco associated to $\mathcal{Q}_\rho(A)$, then ϕ_ρ is the most abstract solution in $uco(A)$ of the system of equations $\{\rho \circ f^i = \rho \circ f^i \circ \psi \mid i \in \omega\}$.

Lemma 2. *Let $\rho \in uco(A)$ such that $\mathcal{Q}_\rho(A) \in uco(A)$. Then, $\forall i \in \omega. \rho \circ f^i = \rho \circ f^i \circ \mathcal{Q}_\rho(A)$, and for any $\psi \in uco(A)$, $\forall i \in \omega. \rho \circ f^i = \rho \circ f^i \circ \psi$ implies $\psi \sqsubseteq \mathcal{Q}_\rho(A)$.*

Since the system of equations $\{\rho \circ f^i = \rho \circ f^i \circ \psi \mid i \in \omega\}$ is clearly equivalent to the system $\{\psi \sqsubseteq \rho\} \cup \{\rho \circ f^i = \rho \circ f^i \circ \psi \mid i > 0\}$, the above lemma says that, when a quotient $\mathcal{Q}_\rho(A)$ exists (i.e. $\mathcal{Q}_\rho(A) \in uco(A)$), it is characterized as follows:

$$\mathcal{Q}_\rho(A) = \sqcup\{\psi \in uco(A) \mid \psi \sqsubseteq \rho, \forall i > 0. \rho \circ f^i = \rho \circ f^i \circ \psi\}.$$

Of course, in the terminology of this paper, this means that when it exists, $\mathcal{Q}_\rho(A)$ is the least complete extension of ρ for the set of functions $\{f^i\}_{i>0}$ relative to ρ itself. However, it may well happen that, for some A, f and ρ, such least complete extension exists, whilst the quotient $\mathcal{Q}_\rho(A)$ does not exist, as the following simple example shows.

Example 2. Let A be the lattice depicted in the figure. Also, let $f : A \to A$ be defined as $f = \{a \mapsto a, b \mapsto a, c \mapsto e, d \mapsto e, e \mapsto e\}$, and let $\rho \in uco(A)$ such that $\rho(A) = \{a, b, e\}$. Trivially, f is monotone (and therefore continuous) but not additive. Moreover, f is idempotent, and therefore, for any $i \geq 1$, $f^i = f$. It turns out that, for any $i \geq 1$, $\rho \circ f^i = f$. As a consequence, r_ρ is not an additive equivalence relation. In fact, for any $i \geq 0$, $\rho(f^i(c)) = \rho(f^i(d))$: If $i = 0$ then, $\rho(c) = \rho(d) = b$; if $i \geq 1$ then, $\rho(f^i(c)) = f(c) = e = f(d) = \rho(f^i(d))$. But, $\rho(f(c \vee d)) = \rho(f(b)) = f(b) = a$. Hence, this means

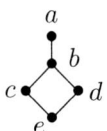

that the quotient $Q_\rho(A)$ does not exist. Instead, as each f^i is monotone, by Theorem 2, the least complete extension of ρ for $\{f^i\}_{i>0}$ relative to ρ does exist. Moreover, this is given by the following reduced product: $\rho \sqcap (\cup_{i>0, y \in \rho} max(H_y^{f^i}))$. It is then a routine task to check that this is the domain A itself, i.e. the identity uco $\lambda x.x$. □

Then, Lemma 2 and Example 2 hint to generalize the notion of quotient as the least complete extension of ρ for $\{f^i\}_{i>0}$ relative to ρ, whenever this exists.

Definition 3. Given a complete lattice A, $f : A \xrightarrow{m} A$, and $\rho \in uco(A)$, the *generalized quotient* of A w.r.t. ρ is well-defined when there exists the least complete extension $\Phi_\rho(A)$ of ρ for $\{f^i\}_{i>0}$ relative to ρ; in such a case, the generalized quotient is defined to be $\Phi_\rho(A)$. □

It is here worth noting that the above definition naturally extends the intuitive meaning of the concept of quotient: In fact, the abstract domain $\Phi_\rho(A)$ is the most abstract domain which is more concrete than the property ρ and which is as good as A for propagating the information through the semantic function f. In other words, $\Phi_\rho(A)$ encodes exactly the least amount of information of A that is useful for computing the property ρ. Thus, this exactly formalizes the clear intuition behind the concept of quotient. As an immediate consequence of Theorem 2, we are then able to give the following theorem ensuring that, when the semantic function f is continuous, generalized quotients always exist.

Theorem 4. *If $f : A \xrightarrow{c} A$ then, for any $\rho \in uco(A)$, the generalized quotient $\Phi_\rho(A)$ exists.*

Acknowledgments. We wish to thank Enea Zaffanella for his helpful remarks on quotients of abstract interpretations and an anonymous referee for his/her useful comments. The work of Francesco Ranzato has been supported by an individual grant no. 202.12199 from Comitato 12 "Scienza e Tecnologie dell'Informazione" of Italian CNR.

References

1. R. Bagnara, P.M. Hill, and E. Zaffanella. Set-sharing is redundant for pair-sharing. In *Proc. 4th Int. Static Analysis Symp.*, LNCS 1302:53–67, 1997.
2. G. Birkhoff. *Lattice Theory.* AMS Colloq. Publications vol. XXV, 3rd ed., 1967.
3. G.L. Burn, C. Hankin, and S. Abramsky. Strictness analysis for higher-order functions. *Sci. Comput. Program.*, 7:249–278, 1986.
4. A. Cortesi, G. Filé, and W. Winsborough. The quotient of an abstract interpretation. *Theor. Comput. Sci.*, 202(1-2):163–192, 1998.
5. P. Cousot. *Méthodes itératives de construction et d'approximation de points fixes d'opérateurs monotones sur un treillis, analyse sémantique des programmes.* PhD thesis, Université Scientifique et Médicale de Grenoble, 1978.
6. P. Cousot. Completeness in abstract interpretation (Invited Lecture). In *Proc. 1995 Joint Italian-Spanish Conference on Declarative Programming*, pp. 37–38, 1995.

7. P. Cousot and R. Cousot. Abstract interpretation: a unified lattice model for static analysis of programs by construction or approximation of fixpoints. In *Proc. 4th ACM POPL*, pp. 238–252, 1977.
8. P. Cousot and R. Cousot. Systematic design of program analysis frameworks. In *Proc. 6th ACM POPL*, pp. 269–282, 1979.
9. P. Cousot and R. Cousot. Inductive definitions, semantics and abstract interpretation. In *Proc. 19th ACM POPL*, pp. 83–94, 1992.
10. P. Cousot and R. Cousot. Abstract interpretation of algebraic polynomial systems. In *Proc. 6th AMAST Conf.*, LNCS 1349:138–154, 1997.
11. R. Giacobazzi and F. Ranzato. Refining and compressing abstract domains. In *Proc. 24th ICALP*, LNCS 1256:771–781, 1997.
12. R. Giacobazzi and F. Ranzato. Completeness in abstract interpretation: a domain perspective. In *Proc. 6th AMAST Conf.*, LNCS 1349:231–245, 1997.
13. D. Jacobs and A. Langen. Static analysis of logic programs for independent AND-parallelism. *J. Logic Program.*, 13(2-3):154–165, 1992.
14. K. Marriott and H. Søndergaard. Precise and efficient groundness analysis for logic programs. *ACM Lett. Program. Lang. Syst.*, 2(1-4):181–196, 1993.
15. A. Mycroft. *Abstract interpretation and optimising transformations for applicative programs*. PhD thesis, CST-15-81, Univ. of Edinburgh, 1981.
16. A. Mycroft. Completeness and predicate-based abstract interpretation. In *Proc. ACM PEPM Conf.*, pp. 179–185, 1993.
17. U.S. Reddy and S.N. Kamin. On the power of abstract interpretation. *Computer Languages*, 19(2):79–89, 1993.
18. R.C. Sekar, P. Mishra, and I.V. Ramakrishnan. On the power and limitation of strictness analysis. *J. ACM*, 44(3):505–525, 1997.
19. H. Søndergaard. An application of abstract interpretation of logic programs: occur check reduction. In *Proc. ESOP '86*, LNCS 213:327–338, 1986.
20. B. Steffen. Optimal data flow analysis via observational equivalence. In *Proc. 14th MFCS Symp.*, LNCS 379:492–502, 1989.

Timed Bisimulation and Open Maps

Thomas Hune and Mogens Nielsen

BRICS*, Department of Computer Science, University of Aarhus, Denmark,
{baris,mn}@brics.dk

Abstract. Open maps have been used for defining bisimulations for a range of models, but none of these have modelled real-time. We define a category of timed transition systems, and use the general framework of open maps to obtain a notion of bisimulation. We show this to be equivalent to the standard notion of timed bisimulation. Thus the abstract results from the theory of open maps apply, e.g. the existence of canonical models and characteristic logics. Here, we provide an alternative proof of decidability of bisimulation for finite timed transition systems in terms of open maps, and illustrate the use of open maps in presenting bisimulations.

1 Introduction

During the past decade, a number of formalisms for real-time systems have been introduced and studied, e.g. the timed automata [AD90] and timed process algebras [Wan90]. A great deal of the theory of untimed systems has been lifted successfully to the setting of formalisms modelling real-time behaviour of systems. As examples, many results from automata theory apply also to timed automata, [AD90, AD94, ACM97], and a number of timed versions of classical specification logics have been studied, [AH91, LLW95].

In this paper we study the notion of bisimulation [Mil89] for timed transition systems. The notion of bisimulation for timed models has already been introduced and studied by many researchers, e.g. in [Wan90, AKLN95, NSY93]. Timed bisimulation was shown decidable for finite timed transition systems by Čerāns in [Čer92], and since then more efficient algorithms have been discovered [LLW95, WL97] and implemented in tools for automatic verification[KN94]. These results like most other results concerning verification of real-time systems build on the region construction [AD90, AD94] which makes it possible to express the uncountable behaviour of a real-time system in a finite way.

One of the main advantages of Milners notion of bisimulation for untimed transition systems, is the fact that for two transition systems, the property of being bisimilar may be expressed in terms of presenting an explicit bisimulation between the two systems, i.e. a relation on the states of the two systems. Unfortunately, this property does not generalise to the setting of timed transition

---

* Basic Research in Computer Sciencs, Centre of the Danish National Research Foundation

systems, where bisimulations are defined in terms of the uncountable unfolded version of given timed transition systems, and where the decision procedures from e.g. [Čer92] produce relations over nontrivial regional constructions

The contribution of this paper is first and foremost to show the applicability of the general categorical framework of bisimulations in terms of open maps from [JNW96]. This framework has already been applied successfully to (re)define a number of observational equivalences [CN96]. Here we define a category of timed transition systems, where the morphisms are to be thought of as simulations, and an accompanying path (sub)category of timed words, which, following [JNW96], provides us with a notions of open maps and a bisimulation with a number of useful properties, like a canonical (presheaf) model, and a characteristic (modal) logic.

We show this notion of bisimulation to coincide with the standard timed bisimulation from [Čer92], and hence we may apply the general results from [JNW96] to this standard notion. Furthermore, we show within the framework of open maps that bisimilarity is decidable for finite timed transition systems. More importantly, for two bisimilar systems, our decision procedure will produce a span of open maps, i.e. a representation of bisimilarity within the framework of timed transition systems, matching the internal representation of bisimulations for untimed transition systems.

In Section 2 a category of timed transition systems and a path subcategory are defined, and they are shown to have the required properties for applying the approach of [JNW96]. Next, in Section 3 the resulting notion of bisimulation is studied, and shown to coincide with the standard notion of timed bisimulation. Finally, in Section 4 we provide a new proof of the decidability of timed bisimulation and illustrate the use of open maps to express bisimulations. Section 5 contains conclusions and future work.

2 A Category of Timed Transition Systems

As a model for real time systems we use timed transition systems. These will be the objects of our category. A timed transition system is basically a timed automata without a set of accepting states and acceptance condition. R.Alur and D.L.Dill [AD94] call this a timed transition table.

Definition 1 (Timed Transition Systems) *A timed transition system is a quintuple* (S, Σ, s_0, X, T) *where*

- *S is a set of states and s_0 is the initial state.*
- *Σ is a finite alphabet of actions.*
- *X is a set of clock variables.*
- *T is the set of transitions such that $T \subseteq S \times \Sigma \times \Delta \times 2^X \times S$ where Δ is a clock constraint generated by the grammar $\Delta ::= c \sharp x \mid x + c \sharp y \mid \Delta \wedge \Delta$ in which $\sharp \in \{\leq, <, \geq, >\}$, c is an integer constant and x, y are clock variables. A transition $(s, \sigma, \delta, \lambda, s')$ is written $s \xrightarrow[\delta,\lambda]{\sigma} s'$.*

Before looking at an example of a timed transition system and discussing how we interpret the behaviour of such a system, we will define our notion of paths for timed transition systems.

Definition 2 (Timed Words) *A timed word over an alphabet Σ is a finite sequence of pairs $\alpha = (\sigma_1, \tau_1)(\sigma_2, \tau_2)(\sigma_3, \tau_3) \cdots (\sigma_n, \tau_n)$, where for all $0 \leq i \leq n$ $\sigma_i \in \Sigma, \tau_i \in \mathbf{R}_+$ and furthermore $\tau_i < \tau_{i+1}$.*

A pair (σ, τ) represents an occurrence of action σ at time τ relative to the starting time (0) of the execution.

Example 1 *The timed transition system in Figure 1 has two clocks x and y, and three actions a,b,c. The state s_0 is the initial state.*

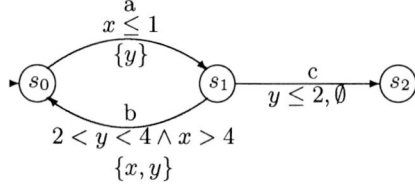

Fig. 1. The timed transition system from Example 1.

To keep track of values of clocks during an execution, we introduce the notion of a clock evaluation.

Definition 3 (Clock Evaluation) *A clock evaluation is a function $\nu : X \to \mathbf{R}_+$ which assigns times to the clock variables of a system. We define $(\nu+c)(x) := \nu(x)+c$ for all clock variables x. If λ is a set of clock variables then $\nu[\lambda \mapsto 0](x) := 0$ if $x \in \lambda$, and $\nu(x)$ otherwise.*

For a constraint δ to be satisfied in a clock evaluation ν we require that the expression $\delta[\nu(x)/x]^1$ evaluates to true. A constraint δ defines a subset of $\mathbf{R}^n$ where n is the number of clocks in X. We will speak of this subset as the meaning of δ and write it $[\![\delta]\!]_X$. A clock evaluation defines a point in $\mathbf{R}^n$ which we shall denote by $[\![\nu]\!]_X$, so the constraint δ is satisfied for the clock evaluation ν if and only if $[\![\nu]\!]_X \in [\![\delta]\!]_X$.

Definition 4 *Let T be a timed transition system. A configuration is a pair $\langle s, \nu \rangle$, where s is a state and ν is a clock evaluation. T can make the run $\langle s_0, \nu_0 \rangle \xrightarrow{\sigma_1}_{\tau_1} \langle s_1, \nu_1 \rangle \xrightarrow{\sigma_2}_{\tau_2} \cdots \xrightarrow{\sigma_n}_{\tau_n} \langle s_n, \nu_n \rangle$ iff for all $i > 0$ there is a transition $s_{i-1} \xrightarrow{\sigma_i}_{\delta_i, \lambda_i} s_i$ such that $[\![\nu_{i-1} + (\tau_i - \tau_{i-1})]\!]_X \in [\![\delta_i]\!]_X$ and $\nu_i = (\nu_{i-1} + (\tau_i - \tau_{i-1}))[\lambda_i \mapsto 0]$. The state s_0 is the initial state of T and ν_0 is the constant 0 function. We define τ_0 to be 0. The timed word $(\sigma_1, \tau_1)(\sigma_2, \tau_2)(\sigma_3, \tau_3) \cdots (\sigma_n, \tau_n)$ is generated by this run.*

---

[1] $\delta[y/x]$ is syntactic substitution of y for x in δ.

The morphisms of our category will be simulation morphisms following the approach of [JNW96]. This leads to the following definition of a morphism.

Definition 5 *A morphism (m, η) between timed transition systems T and T' consists of two components; a map $m : S \to S'$ between the states and a map $\eta : X' \to X$ between the clocks. These maps must satisfy that $m(s_0) = s'_0$ and whenever there is a transition in T of the form $s \xrightarrow{\sigma}_{\delta, \lambda} s'$ then there is a transition $m(s) \xrightarrow{\sigma}_{\delta', \lambda'} m(s')$ in T' satisfying the following two constraints:*

1. $\lambda' = \eta^{-1}(\lambda)$ where $\eta^{-1}(\lambda) = \{x' \in X' \mid \eta(x') \in \lambda\}$
2. $[\![\delta]\!]_X \subseteq [\![\delta'[\eta(x)/x]]\!]_X$

Example 2 *There is a morphism from the timed transition system in Figure 2 to the one in Figure 1 mapping states t_0 and t_2 to s_0, t_1 and t_3 to s_1, and t_4 to s_2. The clock variable x is mapped to z and y to u. It should be easy to check that the two constraints in Definition 5 are satisfied.*

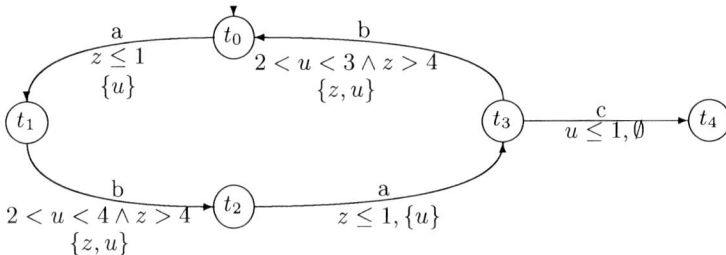

Fig. 2. A timed transition system with a morphism to the system in Figure 1.

Definition 6 *For a function $\eta : X' \to X$ and a clock evaluation $\nu : X \to \mathbf{R}_+$ we define $\eta^{-1}(\nu) : X' \to \mathbf{R}_+$, as $\eta^{-1}(\nu)(x) := \nu(\eta(x))$*

Theorem 1 *Given two timed transition systems T and T' and a morphism (m, η) from T to T', we have that if T can make the run $\langle s_0, \nu_0 \rangle \xrightarrow{\sigma_1}_{\tau_1} \langle s_1, \nu_1 \rangle \xrightarrow{\sigma_2}_{\tau_2} \cdots \xrightarrow{\sigma_n}_{\tau_n} \langle s_n, \nu_n \rangle$ which generates the timed word $(\sigma_1, \tau_1)(\sigma_2, \tau_2)(\sigma_3, \tau_3) \cdots (\sigma_n, \tau_n)$, then T' can make the run $\langle m(s_0), \eta^{-1}(\nu_0) \rangle \xrightarrow{\sigma_1}_{\tau_1} \langle m(s_1), \eta^{-1}(\nu_1) \rangle \xrightarrow{\sigma_2}_{\tau_2} \cdots \xrightarrow{\sigma_n}_{\tau_n} \langle m(s_n), \eta^{-1}(\nu_n) \rangle$ generating the same timed word.*

Definition 7 *The category CTTS_Σ has timed transition systems with alphabet Σ as objects, and the morphisms from Definition 5 as arrows. For morphisms $T \xrightarrow{(m, \eta)} T'$ and $T' \xrightarrow{(m', \eta')} T''$ composition is defined as $(m', \eta') \circ (m, \eta) := (m' \circ m, \eta \circ \eta')$. The identity morphism is the morphism where both m and η are the identity function.*

CTTS$_\Sigma$ has a number of useful properties. For our purpose here we only need the following.

Theorem 2 *CTTS$_\Sigma$ has pullbacks and products.*

The pullback construction is a combination of the pullback and pushout constructions in sets with functions, for m and η respectively.

2.1 A Path Category

We need to represent our observations (timed words) as a subcategory of CTTS$_\Sigma$ to use the framework of open maps.

Definition 8 *Given a timed word* $\alpha = (\sigma_1, \tau_1)(\sigma_2, \tau_2)(\sigma_3, \tau_3) \cdots (\sigma_n, \tau_n)$ *we define a timed transition system* $\mathcal{T}_\alpha$: $s_0 \xrightarrow[\delta_1, \lambda_1]{\sigma_1} s_1 \xrightarrow[\delta_2, \lambda_2]{\sigma_2} \cdots \xrightarrow[\delta_n, \lambda_n]{\sigma_n} s_n$. *There are $n + 1$ states in the timed transition system and a clock variable for each of the 2^n subsets of states* $\{s_1, s_2, \ldots, s_n\}$. *We define λ_i and δ_i as*

$$\lambda_i = \{x_j \mid s_i \in x_j\} \text{ and } \delta_i = \bigwedge_{x_j \in X}(x_j = \tau_i - \tau_{I(s_i, x_j)})$$

where $I(s_i, x_j) = \max(k : k < i \wedge s_k \in x_j)$. *If there is no such k then $I(s_i, x_j) = 0$ and we define $\tau_0 := 0$. The index returned by $I(s_i, x_j)$ is the index of the last state at which x_j was rest. This defines a subclass* TTS$_{\text{TW}}$ *of timed transition systems. We write $\mathcal{T}_\alpha$ for the transition system in* TTS$_{\text{TW}}$ *representing* α.

The only purpose of this seemingly ad hoc construction is that it allows us to identify runs of α in $\mathcal{T}$ with morphisms from $\mathcal{T}_\alpha$ to $\mathcal{T}$, as expressed formally in the following two theorems.

Theorem 3 *The full subcategory of* CTTS$_\Sigma$ *with objects from* TTS$_{\text{TW}}$, *denoted* CTTS$_{\text{TW}}$, *is isomorphic to the category of timed words (as objects) with word extensions (as morphisms).*

Theorem 4 *Given a timed word α and a timed transition system $\mathcal{T}$, there is a one to one correspondence between runs of α and morphisms* $\mathcal{T}_\alpha \xrightarrow{(m, \eta)} \mathcal{T}$.

3 Timed Bisimulation

Given our category of timed transition systems and the path category we can use the general framework from [JNW96] to define our notions of open maps and bisimulation.

Definition 9 (Open Map [JNW96]) *A morphism* $\mathcal{T} \xrightarrow{(m, \eta)} \mathcal{T}'$ *is open iff for all* $\mathcal{T}_\alpha \xrightarrow{(p, \eta_p)} \mathcal{T}$ *with* $\mathcal{T}_\alpha \in$ CTTS$_{\text{TW}}$ *and all morphisms* $\mathcal{T}_\alpha \xrightarrow{(f, \eta_f)} \mathcal{T}_{\alpha'}$

in CTTS$_{TW}$ such that the square $\mathcal{T}_\alpha \xrightarrow{(p,\eta_p)} \mathcal{T}$ $\xrightarrow{(f,\eta_f)\downarrow \quad \downarrow (m,\eta)}$ $\mathcal{T}_{\alpha'} \xrightarrow{(q,\eta_q)} \mathcal{T}'$ commutes there exists a morphism $(p',\eta_{p'}) : \mathcal{T}_{\alpha'} \to \mathcal{T}$ such that the in the diagram
$$\mathcal{T}_\alpha \xrightarrow{(p,\eta_p)} \mathcal{T}$$
$$(f,\eta_f)\downarrow \ \ (p',\eta_{p'})\nearrow \ \ \downarrow(m,\eta)$$
$$\mathcal{T}_{\alpha'} \xrightarrow{(q,\eta_q)} \mathcal{T}'$$
the two triangles commute.

Definition 10 *Two timed transition systems $\mathcal{T}_1$ and $\mathcal{T}_2$ are $\mathcal{TW}$-bisimilar iff there exists a span $\mathcal{T}_1 \xleftarrow{(m,\eta)} \mathcal{T} \xrightarrow{(m',\eta')} \mathcal{T}_2$ with vertex $\mathcal{T}$ of open morphisms.*

It follows from [JNW96] and Theorem 2 that $\mathcal{TW}$-bisimulation is the equivalence generated by open maps.

The 'standard' notion of timed bisimulation is defined in terms of configurations.

Definition 11 (Timed Bisimulation) *Two timed transition systems are bisimilar iff there exists a relation R over configurations $(\langle s, \nu_s\rangle, \langle t, \nu_t\rangle)$ of the two systems satisfying $(\langle s^{in}, \nu_s^0\rangle, \langle t^{in}, \nu_t^0\rangle) \in R$ and for all $(\langle s, \nu_s\rangle, \langle t, \nu_t\rangle) \in R$*

- *whenever $\langle s,\nu_s\rangle \xrightarrow[\tau]{\sigma} \langle s', \nu'_s\rangle$ then $\langle t,\nu_t\rangle \xrightarrow[\tau]{\sigma} \langle t',\nu'_t\rangle$ with $(\langle s',\nu'_s\rangle, \langle t',\nu'_t\rangle) \in R$ for some $\langle t',\nu'_t\rangle$.*
- *whenever $\langle t,\nu_t\rangle \xrightarrow[\tau]{\sigma} \langle t',\nu'_t\rangle$ then $\langle s,\nu_s\rangle \xrightarrow[\tau]{\sigma} \langle s',\nu'_s\rangle$ with $(\langle s',\nu'_s\rangle, \langle t',\nu'_t\rangle) \in R$ for some $\langle s',\nu'_s\rangle$.*

Theorem 5 *Two timed transition systems $\mathcal{T}$ and $\mathcal{T}'$ are $\mathcal{TW}$-bisimilar iff they are bisimilar according to Definition 11.*

Example 3 *Since identity morphisms are open we have that two timed transition systems are bisimilar if we can find an open map from one to the other. In Figure 3 the (only) morphism from $\mathcal{T}$ to $\mathcal{T}'$ is open. From Theorem 7 it should be easy for the reader to check that this indeed is an open morphism.*

In this setup, one can present the fact that two timed transition systems are timed bisimilar by providing a concrete span of open maps. In the next section we will show how one can effectively construct a finite span of open maps for any given pair of bisimilar finite timed transition systems.

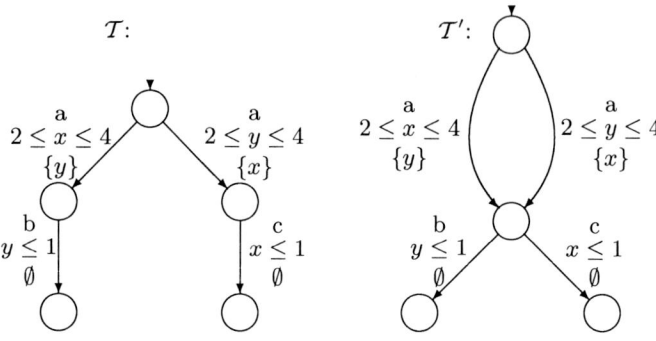

Fig. 3. Two bisimilar timed transition systems.

4 Decidability

Showing the decidability of timed bisimulation amounts to deciding whether there exists a span of open maps between two finite timed transition systems. Our approach is first to show that openness of a morphism between two finite timed transition systems is decidable, and next to show an upper bound on the size of the vertex of a span for two bisimilar finite timed transition systems.

Definition 12 *A configuration $\langle s, \nu \rangle$ is reachable iff there is a run that has $\langle s, \nu \rangle$ as a configuration.*

We will now characterise the open maps in terms of runs and configurations.

Theorem 6 *A morphism $\mathcal{T}_1 \xrightarrow{(m,\eta)} \mathcal{T}_2$ is open iff for all reachable configurations $\langle s_1, \nu \rangle$ in $\mathcal{T}_1$, and for all $\nu' = \nu + \tau$ whenever there is a transition $m(s_1) \xrightarrow[\delta_2, \lambda_2]{\sigma} s'_2$ such that $[\![\eta^{-1}(\nu')]\!]_{X_2} \in [\![\delta_2]\!]_{X_2}$, then there exists a transition $s_1 \xrightarrow[\delta_1, \lambda_1]{\sigma} s'_1$ such that $m(s'_1) = s'_2$ and $[\![\nu']\!]_{X_1} \in [\![\delta_1]\!]_{X_1}$.*

To get a decidable characterisation of openness we introduce the notion of regions, [AD94].

Definition 13 (Region[AD94]) *Given a finite set of clocks X and a constant c_X a region is an equivalence class of valuations such that $\nu \cong \nu'$ iff*

- *For each $x \in X$: $\lfloor \nu(x) \rfloor = \lfloor \nu'(x) \rfloor$[2] or both $\nu(x) > c_X$ and $\nu'(x) > c_X$.*
- *For every pair of clock variables $x, y \in X$ where both $\nu(x) \leq c_X$ and $\nu(y) \leq c_X$ we have that $fract(\nu(x)) \leq fract(\nu(y))$ iff $fract(\nu'(x)) \leq fract(\nu'(y))$.*
- *For every clock variable $x \in X$ where $\nu(x) \leq c_X$ we have $fract(\nu(x)) = 0$ iff $fract(\nu'(x)) = 0$.*

---

[2] We use $\lfloor x \rfloor$ for the largest integer smaller than or equal to x and $fract(x) := x - \lfloor x \rfloor$.

The region to which ν belongs is denoted by $[\nu]$. For a finite timed transition system $\mathcal{T}$ let $\mathcal{R}_\mathcal{T}$ be the set of regions associated with the set of clock variables of $\mathcal{T}$ and the largest constant referred to in the constraints of $\mathcal{T}$ transitions. A pair $\langle s, reg \rangle$ where $reg \in \mathcal{R}_\mathcal{T}$, is called an extended state.

Our operations on clock evaluations can be extended to regions which will be used below. We can now give a characterisation of open maps in terms of extended states.

Theorem 7 *For finite $\mathcal{T}_1$ and $\mathcal{T}_2$ a morphism $(m, \eta) : \mathcal{T}_1 \to \mathcal{T}_2$ is open iff for all reachable extended states $\langle s_1, reg \rangle$ in $\mathcal{T}_1$, and for all $reg' \in Reach(reg)^3$, whenever there is a transition $m(s_1) \xrightarrow[\delta_2, \lambda_2]{\sigma} s_2'$ such that $[\![\eta^{-1}(reg')]\!]_{X_2} \subseteq [\![\delta_2]\!]_{X_2}$, then there exists a transition $s_1 \xrightarrow[\delta_1, \lambda_1]{\sigma} s_1'$ such that $m(s_1') = s_2'$ and $[\![reg']\!]_{X_1} \subseteq [\![\delta_1]\!]_{X_1}$.*

Notice that Theorem 7 implies immediately the decidability of openness of a morphism between two finite timed transition systems.

Next, looking for the existence of a finite vertex of a span of open maps between two finite timed transition systems, a first attempt could be to look for a subsystem of their product. Unfortunately this is not enough in all cases, however we still have the following theorem.

Theorem 8 *Given two finite timed transition systems $\mathcal{T}_1$ and $\mathcal{T}_2$ if there exists a vertex $\mathcal{T}$ such that $\mathcal{T}_1 \xleftarrow{(m,\eta)} \mathcal{T} \xrightarrow{(m',\eta')} \mathcal{T}_2$ are open maps then there is a finite vertex $\mathcal{T} \times_\mathcal{R}$ and open morphisms $\mathcal{T}_1 \xleftarrow{(p, \eta_p)} \mathcal{T} \times_\mathcal{R} \xrightarrow{(q, \eta_q)} \mathcal{T}_2$.*

We prove this by constructing the finite vertex $\mathcal{T} \times_\mathcal{R}$. The clocks of $\mathcal{T} \times_\mathcal{R}$ are the disjoint union of the clocks of the two systems. We will write a clock evaluation over this set of clocks as $(\nu \uplus \nu')$, acting as ν for the clocks in $\mathcal{T}$ and as ν' otherwise. The states are triples of the form $\langle s_1, s_2, reg \rangle$, where reg is a region over the new set of clocks. For a reachable configuration $\langle s, \nu \rangle$ in $\mathcal{T}$, there is a state $\langle m(s), m'(s), reg \rangle$ in $\mathcal{T} \times_\mathcal{R}$, where $[\![\eta^{-1}(\nu) \uplus \eta'^{-1}(\nu)]\!]_{X_\mathcal{R}} \in [\![reg]\!]_{X_\mathcal{R}}$. There is a transition $\langle m(s), m'(s), reg \rangle \xrightarrow[\delta_\mathcal{R}, \lambda_\mathcal{R}]{\sigma} \langle m(s'), m'(s'), reg' \rangle$, if there is a run in $\mathcal{T}$ which uses the transition $s \xrightarrow[\delta, \sigma]{\sigma} s'$ with clock evaluations ν and ν' before and after the transition respcetively, such that $[\![\eta^{-1}(\nu) \uplus \eta'^{-1}(\nu)]\!]_{X_\mathcal{R}} \in [\![reg]\!]_{X_\mathcal{R}}$ and $[\![\eta^{-1}(\nu') \uplus \eta'^{-1}(\nu')]\!]_{X_\mathcal{R}} \in [\![reg']\!]_{X_\mathcal{R}}$. Here $\lambda_\mathcal{R} = \eta^{-1}(\lambda) \cup \eta'^{-1}(\lambda)$, and $\delta_\mathcal{R}$ is the logical expression for the region to which $\eta^{-1}(\nu'') \uplus \eta'^{-1}(\nu'')$ belongs, where ν'' is the clock evaluation enabling then transition in $\mathcal{T}$ during the run.

Importantly, this construction defines a system of bounded size in the number of states of $\mathcal{T}_1$ and $\mathcal{T}_2$ and the number of regions over the disjoint union of the clocks of the two systems. The morphisms (p, η_p) and (q, η_q) are projections which can easily be shown to be morphisms and using Theorem 7 can be shown to be open.

From the proof of Theorem 8, we have the following corollary.

Corollary 1 *For finite timed transition systems timed bisimulation is decidable.*

---

[3] The function *Reach* returns a set of regions reachable from its argument.

5 Conclusion

We have shown that the general framework of open maps may also be applied to the setting of timed systems, providing a way of expressing a bisimulation purely within the framework of timed transition systems. Furthermore, a decision procedure for bisimulation was presented within this framework.

We see our main contribution as extending the open maps approach to the setting of timed systems. This opens up a number of possibilities of applying general results from the categorical setting to concrete timed bisimulations, like the one studied here. One particularly interesting example is the characteristic path logic obtained from [JNW96]. Properties of this logic and its relation to other timed logics will be subject to future work.

We also propose the span of open maps idea as a useful way of expressing timed bisimulations. On the other hand, we do not claim that our alternative decision procedure as presented here is more efficient than existing ones, e.g. [LLW95, WL97].

We have used the same method for timed transition systems extended by invariants on the states [HN98], and the method proved to be robust under this kind of extension.

References

[ACM97] E. Asarin, P. Caspi, and O. Maler. A Kleene theorem for timed automata. *Proc. of LICS'97*, 1997.

[AD90] R. Alur and D.L. Dill. Automata for modelling real-time systems. *Proc. of ICALP'90*, LNCS 433:pages 322–335, 1990.

[AD94] R. Alur and D.L. Dill. A theory of timed automata. *Theoretical Computer Science*, 126, 1994.

[AH91] R. Alur and T.A. Henzinger. Logics and models for real time: A survey. *Real-Time:Theory in Practice*, LNCS 600:pages 74–106, 1991.

[AKLN95] J.H. Andersen, K.J. Kristoffersen, K.G. Larsen, and J. Niedermann. Automatic synthesis of real time systems. *Proc. of ICALP'95*, LNCS 944:pages 535–546, 1995.

[Čer92] K. Čerāns. Decidability of bisimulation equivalence for parallel timer processes. *Proc. of CAV'92*, LNCS 663, 1992.

[CN96] A. Cheng and M. Nielsen. Open maps (at) work. *Proc. of FST&TCS '95*, LNCS 1026, 1996.

[HN98] T. Hune and M. Nielsen. Timed bisimulation and open maps. Technical Report RS-98-4, BRICS, 1998.

[JNW96] A. Joyal, M. Nielsen, and G. Winskel. Bisimulation from open maps. *Information and Computation*, 127,2:pages 164–185, 1996.

[KN94] K.J. Kristoffersen and J. Niedermann. User's manual for Epsilon. Available via anonymous ftp at cs.auc.dk, December 1994.

[LLW95] F. Laroussinie, K. G. Larsen, and C. Weise. From timed automata to logic – and back. *Proc. of MFCS'95*, LNCS 969:pages 529–539, 1995.

[Mil89] R. Milner. *Communication and Concurrency*. Prentice Hall International Series in Computer Science, 1989.

[NSY93] X. Nicollin, J. Sifakis, and S. Yovine. From ATP to timed graphs and hybrid systems. *Acta Informatica*, 30:pages 181–202, 1993.

[Wan90] Y. Wang. Real-time behaviour of asynchronous agents. *Proc. of CONCUR'90*, LNCS 458, 1990.

[WL97] C. Weise and D. Lenzkes. Efficient scaling-invariant checking of timed bisimulation. *Proc. of STACS'97*, LNCS 1200:pages 177–188, 1997.

Deadlocking States in Context-Free Process Algebra

Jiří Srba*

Faculty of Informatics MU,
Botanická 68a, 60200 Brno, Czech Republic
srba@fi.muni.cz

Abstract. Recently the class of BPA (or context-free) processes has been intensively studied and bisimilarity and regularity appeared to be decidable (see [CHS95, BCS95, BCS96]). We extend these processes with a deadlocking state into BPA$_\delta$ systems. Bosscher has proved that bisimilarity and regularity remain decidable [Bos97]. We generalise his approach introducing strict and nonstrict version of bisimilarity. We show that the BPA$_\delta$ class is more expressive w.r.t. (both strict and nonstrict) bisimilarity but it remains language equivalent to BPA. Finally we give a characterization of those BPA$_\delta$ processes which can be equivalently (up to bisimilarity) described within the 'pure' BPA syntax.

1 Introduction

This paper deals with BPA processes (Basic Process Algebra) extended with deadlocks. BPA represents the class of processes introduced by Bergstra and Klop (see [BK85]), which corresponds to the transition systems associated with Greibach normal form (GNF) context-free grammars in which only left-most derivations are permitted. For detailed description of the relation between language and process theory we refer to [HM96]. We define the class BPA$_\delta$ of BPA processes extended with deadlocks and introduce two alternative definitions – strict and nonstrict – of bisimilarity within this class.

The definition of BPA$_\delta$ systems is based on a special variable δ (we call it a deadlock). In the usual presentation every variable used in a BPA system is supposed to be defined but for the deadlock variable we allow no definition. This causes that if the system reaches a state where the first variable is δ, the system sticks at this state and no more actions can be performed.

Bosscher has proved in [Bos97] that decidability of bisimilarity and regularity in BPA systems extends to the BPA$_\delta$ systems. The trick used for this extention is based on the idea that δ can be simulated by an unnormed variable.

The main topic this article deals with is the issue of the language equivalence and of describing BPA$_\delta$ in bisimilar BPA syntax. We show in Section 3 that extending the BPA systems with deadlocks does not yield any language extension.

---

* The author is supported by the Grant Agency of the Czech Republic, grant No. 201/97/0456

On the other hand the class of BPA$_\delta$ systems is larger with regard to bisimilarity. An interesting question explored in this paper (Section 4) is concerned with deciding whether there exists an alternative description of a BPA$_\delta$ system in bisimilar BPA syntax. We show that it is decidable for the strict bisimilarity and we find a nice semantic characterization of the situation in the nonstrict case. Moreover we show that the corresponding BPA syntax can be effectively constructed.

Several proofs in this paper are just sketched and their full version can be obtained in [Srb98].

2   Basic Definitions

When dealing with processes we need some structure to describe their operational semantics. As the most suitable structure transition systems are widely used. We introduce the labelled transition system in the extended version with the set of final states as can be found e.g. in [Mol96].

Definition 1. (labelled transition system) *A labelled transition system is a tuple $(S, Act, \longrightarrow, \alpha_0, F)$ where S is a set of states; Act is a set of actions (or labels); $\longrightarrow \subseteq S \times Act \times S$ is a transition relation, written $\alpha \xrightarrow{a} \beta$, for $(\alpha, a, \beta) \in \longrightarrow$; $\alpha_0 \in S$ is the root (or start state) of the transition system; $F \subseteq S$ is the set of final states which are terminal: for each $\alpha \in F$ there is no $a \in Act$ and $\beta \in S$ such that $\alpha \xrightarrow{a} \beta$.*

As usual we extend the transition relation to the elements of Act^*. We also write $\alpha \longrightarrow^* \beta$ instead of $\alpha \xrightarrow{w} \beta$ if $w \in Act^*$ is irrelevant.

Definition 2. (language generation) *Let $(S, Act, \longrightarrow, \alpha_0, F)$ be a labelled transition system and suppose that $\alpha \in S$. The language generated by the state α is $L(\alpha) \stackrel{def}{=} \{w \in Act^* \mid \exists \alpha' \in F : \alpha \xrightarrow{w} \alpha'\}$. We say that two states α and β are language equivalent, written $\alpha =_L \beta$, iff $L(\alpha) = L(\beta)$. Two labelled transition systems are language equivalent iff their roots are language equivalent.*

Definition 3. (bisimilarity) *Let $(S, Act, \longrightarrow, \alpha_0, F)$ be a labelled transition system. A binary relation $R \subseteq S \times S$ is a bisimulation iff whenever $(\alpha, \beta) \in R$ then for each $a \in Act$:*

- *if $\alpha \xrightarrow{a} \alpha'$ then $\exists \beta' \in S : \beta \xrightarrow{a} \beta' \wedge (\alpha', \beta') \in R$*
- *if $\beta \xrightarrow{a} \beta'$ then $\exists \alpha' \in S : \alpha \xrightarrow{a} \alpha' \wedge (\alpha', \beta') \in R$*
- *$\alpha \in F \Leftrightarrow \beta \in F$*

States $\alpha, \beta \in S$ are bisimilar ($\alpha \sim \beta$), iff $(\alpha, \beta) \in R$ for some bisimulation R.

2.1 BPA and BPA$_\delta$ Systems

Assume that $\mathcal{V}ar$ and $\mathcal{A}ct$ are finite sets of *variables* and *actions* such that $\mathcal{V}ar \cap \mathcal{A}ct = \emptyset$. We define the class $\mathcal{E}_{\text{BPA}}$ of *BPA expressions* as the union of ϵ *(empty process)* and a set $\mathcal{E}_{\text{BPA}}^+$, which is defined by the following abstract syntax:

$$E ::= a \mid X \mid E_1.E_2 \mid E_1 + E_2$$

Here a ranges over $\mathcal{A}ct$ and X ranges over $\mathcal{V}ar$. We state $\mathcal{E}_{\text{BPA}} \stackrel{\text{def}}{=} \{\epsilon\} \cup \mathcal{E}_{\text{BPA}}^+$.

We call the BPA expressions as processes and later on we assume fixed sets $\mathcal{V}ar$ and $\mathcal{A}ct$ if no confusion is caused. As usual, we restrict our attention to *guarded* expressions: a BPA expression is guarded iff every variable occurrence is within the scope of an atomic action.

Definition 4. (BPA system) *A* BPA system *is a quadruple* $(\mathcal{V}ar, \mathcal{A}ct, \Delta, X_1)$ *where $\mathcal{V}ar$ and $\mathcal{A}ct$ are finite sets of distinct variables* $(\mathcal{V}ar = \{X_1, \ldots, X_n\})$ *resp. actions; $X_1 \in \mathcal{V}ar$ is the* leading variable; *Δ is a finite set of recursive equations* $\Delta = \{X_i \stackrel{\text{def}}{=} E_i \mid i = 1, \ldots, n\}$ *where each $E_i \in \mathcal{E}_{\text{BPA}}^+$ is a guarded BPA expression with variables drawn from the set $\mathcal{V}ar$ and actions from $\mathcal{A}ct$.*

Speaking about variables and actions used in the system $(\mathcal{V}ar, \mathcal{A}ct, \Delta, X_1)$ we use the notation $\mathcal{V}ar(\Delta)$ and $\mathcal{A}ct(\Delta)$ and for shorter referring to the BPA system we often identify the system $(\mathcal{V}ar, \mathcal{A}ct, \Delta, X_1)$ with Δ.

Assume that we have a BPA system $(\mathcal{V}ar, \mathcal{A}ct, \Delta, X_1)$. This system determines a labelled transition system $(S, \mathcal{A}ct, \longrightarrow, X_1, \{\epsilon\})$ whose states are BPA expressions built over $\mathcal{V}ar$ and $\mathcal{A}ct$, $\mathcal{A}ct$ is the set of labels, the transition relation is the least relation satisfying the following SOS rules, X_1 is the root and ϵ is the only final state.

$$\frac{}{a \stackrel{a}{\longrightarrow} \epsilon} \qquad \frac{E \stackrel{a}{\longrightarrow} E'}{E.F \stackrel{a}{\longrightarrow} E'.F} \text{ if } E' \neq \epsilon \qquad \frac{E \stackrel{a}{\longrightarrow} \epsilon}{E.F \stackrel{a}{\longrightarrow} F}$$

$$\frac{E \stackrel{a}{\longrightarrow} E'}{E+F \stackrel{a}{\longrightarrow} E'} \qquad \frac{F \stackrel{a}{\longrightarrow} F'}{E+F \stackrel{a}{\longrightarrow} F'} \qquad \frac{E \stackrel{a}{\longrightarrow} E'}{X \stackrel{a}{\longrightarrow} E'} \text{ if } X \stackrel{\text{def}}{=} E \in \Delta$$

We now define the class BPA$_\delta$ of BPA systems with deadlock. The definition is very similar to the definition of BPA systems except for a new distinct variable δ. There is no operational rule for δ in the BPA$_\delta$ systems.

Definition 5. (BPA$_\delta$ system) *A* BPA$_\delta$ system *is a quadruple* $(\mathcal{V}ar, \mathcal{A}ct, \Delta, X_1)$ *where $\mathcal{V}ar = \{X_1, \ldots, X_n, \delta\}$ (δ is a special variable called* deadlock*), $\mathcal{A}ct$ is a finite set of actions and Δ is a finite set of recursive equations* $\Delta = \{X_i \stackrel{\text{def}}{=} E_i \mid i = 1, \ldots, n\}$ *where each $E_i \in \mathcal{E}_{\text{BPA}}^+$ is a guarded BPA expression with variables drawn from the set $\mathcal{V}ar$ and actions from $\mathcal{A}ct$.*

It is obvious that any BPA system is trivially a BPA_δ system. BPA_δ labelled (strict or nonstrict) transition system is defined as in the case of BPA systems. If $F = \{\epsilon\}$ is the only final state we call the labelled transition system *strict* and if the final states are $F = \{\epsilon, \delta\} \cup \{\delta.E | E \in \mathcal{E}_{BPA}^+\}$ we call it *nonstrict*.

This means that the relation of bisimulation differs for both these approaches. Similarly, we call the bisimulation *strict* resp. *nonstrict* (and write $\overset{s}{\sim}$ resp. $\overset{n}{\sim}$) according to the type of the labelled transition system we take into account. These two notions of bisimilarity imply that $\delta \overset{n}{\sim} \epsilon$ but $\delta \overset{s}{\not\sim} \epsilon$. An easy consequence of decidability of bisimilarity in BPA_δ [Bos97] is that both $\overset{s}{\sim}$ and $\overset{n}{\sim}$ are decidable. Following lemma results from the definition of $\overset{s}{\sim}$ and $\overset{n}{\sim}$.

Lemma 1. $\overset{s}{\sim} \subseteq \overset{n}{\sim}$

Let $X \in \mathcal{V}ar$. We define the *norm* of X as $\|X\| \overset{\text{def}}{=} \min\{ \operatorname{length}(w) \mid \exists E : X \overset{w}{\longrightarrow} E \not\rightarrow\}$, if such w exists; or $\|X\| \overset{\text{def}}{=} \infty$ otherwise. We call the variable X *normed* iff $\|X\| < \infty$. A process Δ is normed iff its leading variable is normed.

Definition 6. *A BPA (resp. BPA_δ) system Δ is said to be in* Greibach Normal Form (GNF) *iff all its defining equations are of the form* $X \overset{\text{def}}{=} \sum_{j=1}^{m} a_j \alpha_j$ *where* $m > 0$, $a_j \in \mathcal{A}ct(\Delta)$ *and* $\alpha_j \in \mathcal{V}ar(\Delta)^*$. *If* $\operatorname{length}(\alpha_j) < k$ *for each j then Δ is said to be in k–GNF.*

Following theorem justifies the usage of 3–GNF.

Theorem 1. *Let Δ be a BPA_δ system. We can effectively find a BPA_δ system Δ' in 3–GNF such that $\Delta' \overset{s}{\sim} \Delta$ resp. $\Delta' \overset{n}{\sim} \Delta$.*

Proof. The proof is based on the proof of 3–GNF for BPA systems (see e.g. [Hüt91]), which had to be modified to capture the behaviour of deadlocks. In fact we had to use some additional transformations exploiting (from left to right) the rules $\delta + E \sim E$ and $\delta.E \sim \delta$. □

3 Expressibility of BPA_δ Systems

In this section we justify the importance of introducing a deadlocking state into the BPA systems. We show that deadlocks enlarge the descriptive power of BPA systems w.r.t. both strict and nonstrict bisimilarity. On the other hand introducing deadlocks does not allow to generate more languages.

Theorem 2. *There exists a BPA_δ system such that no BPA system is strictly bisimilar to it.*

Proof. No BPA system can be strictly bisimilar to the system $\{X \overset{\text{def}}{=} a\delta\}$ since δ is reachable in this system and there is no match for δ in any BPA system. □

Theorem 3. *There exists a BPA_δ system such that no BPA system is non-strictly bisimilar to it.*

Proof. We define a BPA_δ system Δ and show that there is no BPA system Δ' such that $\Delta \stackrel{n}{\sim} \Delta'$. Consider $\Delta = \{X \stackrel{\text{def}}{=} aXX + b + c\delta\}$ and suppose that there is a BPA system Δ' in 3–GNF, $\Delta' = \{Y_i \stackrel{\text{def}}{=} E_i \mid i = 1, \ldots, n\}$, such that $\Delta \stackrel{n}{\sim} \Delta'$. Then there are infinitely many states reachable from the leading variable X of the system Δ. They are of the form X^n for $n \geq 1$ and for each such state there must be reachable a state E from Δ' such that $X^n \stackrel{n}{\sim} E$. The state X^n still has norm 1 whereas norm 1 for BPA processes implies that it must be a single variable. Thus Δ is nonstrictly bisimilar to a system with finitely many reachable states, which is contradiction – Δ is a system where infinitely many nonstrictly nonbisimilar states are reachable. □

In what follows we show that the classes of BPA and BPA_δ systems are equivalent w.r.t. language generation. We will consider just the nonstrict case ($F = \{\epsilon, \delta\} \cup \{\delta.E \mid E \in \mathcal{E}_{BPA}^+\}$) since it is obvious that the strict case cannot bring any language extension.

Definition 7. *We define classes of languages generated by BPA resp. BPA_δ systems as following:* $\mathcal{L}(BPA) \stackrel{\text{def}}{=} \{L(\Delta) \mid \Delta \text{ is a BPA system}\}$ *and* $\mathcal{L}(BPA_\delta) \stackrel{\text{def}}{=} \{L(\Delta_\delta) \mid \Delta_\delta \text{ is a } BPA_\delta \text{ system}\}$.

Theorem 4. *It holds that* $\mathcal{L}(BPA) = \mathcal{L}(BPA_\delta)$.

Proof. We show that for a BPA_δ system Δ_δ there exists a BPA system Δ such that $L(\Delta_\delta) = L(\Delta)$. The other direction is obvious.

Our proof will be constructive. For each variable $X \in \Delta_\delta$ we define a couple of new variables X^ϵ, X^δ. The first one will simulate the language behaviour of X when reaching the state ϵ, the second one will simulate ending in the suffix of the form $\delta\alpha$. We use the notation $a\alpha \in Y$ meaning that $a\alpha$ is a summand in the defining equation of the variable Y. W.l.o.g. let Δ_δ be a BPA_δ system in 3–GNF. The variables of the system Δ will be $\mathcal{V}ar(\Delta) \stackrel{\text{def}}{=} \cup_{X \in \mathcal{V}ar(\Delta_\delta) - \{\delta\}} \{X^\epsilon, X^\delta\} \cup \{X_1^{\epsilon\delta}\}$ where X^ϵ, X^δ are distinct fresh variables and $X_1^{\epsilon\delta}$ is the leading variable, supposing that X_1 was the leading variable of Δ_δ. Next we realize that the summands of the defining equation for $X \in \mathcal{V}ar(\Delta_\delta) - \{\delta\}$ are exactly of one of the following form (because of 3–GNF):

(a) aAB **(b)** bC **(c)** c **(d)** $dD\delta$ **(e)** $e\delta$ (1)

where $a, b, c, d, e \in \mathcal{A}ct(\Delta_\delta)$ and $A, B, C, D \in \mathcal{V}ar(\Delta_\delta)$ such that $A, B, C, D \neq \delta$. Notice that we can suppose that there is no summand of the form $a\delta A$ because it can be replaced with $a\delta$.

We now define the variables from Δ. For each $X \in \mathcal{V}ar(\Delta_\delta) - \{\delta\}$ and for the summands of the variables X^ϵ and X^δ will hold:

if $aAB \in X$ then $aA^\epsilon B^\epsilon \in X^\epsilon$ and $aA^\epsilon B^\delta + aA^\delta \in X^\delta$
if $bC \in X$ then $bC^\epsilon \in X^\epsilon$ and $bC^\delta \in X^\delta$
if $c \in X$ then $c \in X^\epsilon$
if $dD\delta \in X$ then $dD^\epsilon + dD^\delta \in X^\delta$
if $e\delta \in X$ then $e \in X^\delta$
if $X_1^\epsilon \stackrel{\text{def}}{=} E$ and $X_1^\delta \stackrel{\text{def}}{=} F$ then $X_1^{\epsilon\delta} \stackrel{\text{def}}{=} E + F$

If it is the case that there is a variable $Y \in Var(\Delta)$ such that Y does not have any summand we define $Y \stackrel{\text{def}}{=} aY$. (This variable cannot generate any nonempty language because it is unnormed). Finally we state $X_1^{\epsilon\delta}$ to be the leading variable of the system Δ.

Example 1. Let us have a BPA$_\delta$ system $\Delta_\delta = \{X \stackrel{\text{def}}{=} aXX + b + c\delta + bY,\ Y \stackrel{\text{def}}{=} b\}$. The corresponding language equivalent BPA system Δ looks as following:
$\Delta = \{X^\epsilon \stackrel{\text{def}}{=} aX^\epsilon X^\epsilon + b + bY^\epsilon,\ X^\delta \stackrel{\text{def}}{=} aX^\epsilon X^\delta + aX^\delta + c + bY^\delta,\ Y^\epsilon \stackrel{\text{def}}{=} b,\ Y^\delta \stackrel{\text{def}}{=} a.Y^\delta,\ X^{\epsilon\delta} \stackrel{\text{def}}{=} aX^\epsilon X^\epsilon + b + bY^\epsilon + aX^\epsilon X^\delta + aX^\delta + c + bY^\delta\}$.

It is not difficult to see that the newly defined system Δ is in 3–GNF and we show that $L(\Delta_\delta) = L(\Delta)$. For this we need one lemma using following notation.

Definition 8. *Let Δ' be a BPA (resp. BPA$_\delta$) system in 3–GNF, $n \geq 1$ and $Y \in Var(\Delta')$. We define $L_n^\epsilon(Y)$ and $L_n^\delta(Y)$ as following:*

$$L_n^\epsilon(Y) \stackrel{\text{def}}{=} \{w \in Act(\Delta')^* \mid Y \stackrel{w}{\longrightarrow} \epsilon \wedge length(w) \leq n\}$$
$$L_n^\delta(Y) \stackrel{\text{def}}{=} \{w \in Act(\Delta')^* \mid \exists \alpha \in Var(\Delta')^* : Y \stackrel{w}{\longrightarrow} \delta\alpha \wedge length(w) \leq n\}.$$

Lemma 2. *For all $n \geq 1$ and $X \in Var(\Delta_\delta) - \{\delta\}$ holds that $L_n^\epsilon(X) = L_n^\epsilon(X^\epsilon)$ and $L_n^\delta(X) = L_n^\epsilon(X^\delta)$.*

Proof. The proof is led by induction on n, following the subcases from (1). □

To finish the proof of our theorem let us define for $n \geq 1$ the set $L_n(Y) \stackrel{\text{def}}{=} \{w \in L(Y) \mid length(w) \leq n\}$. Notice that because of the Lemma 2 we get $L_n(X_1) = L_n^\epsilon(X_1) \cup L_n^\delta(X_1) = L_n^\epsilon(X_1^\epsilon) \cup L_n^\epsilon(X_1^\delta) = L_n(X_1^{\epsilon\delta})$ for all $n \geq 1$.

Now it is clear that $L(X_1) = L(X_1^{\epsilon\delta})$ since if $w \in L(X_1)$ then $\exists n : w \in L_n(X_1)$ and so $w \in L_n(X_1^{\epsilon\delta})$ which implies that $w \in L(X_1^{\epsilon\delta})$. The other direction is similar. We have shown that $L(\Delta_\delta) = L(\Delta)$ and our proof is complete. □

4 Describing BPA$_\delta$ in BPA Syntax

We have shown that w.r.t. bisimilarity the class of BPA$_\delta$ systems is strictly larger than that of BPA. This challenges the question whether a given BPA$_\delta$ system can be equivalently described in BPA syntax.

Theorem 5. Let $(\mathcal{V}ar, \mathcal{A}ct, \Delta, X_1)$ be a BPA_δ system. It is decidable whether there exists a BPA system Δ' such that $\Delta \overset{s}{\sim} \Delta'$. Moreover if the answer is positive, the system Δ' can be effectively constructed.

Proof. The proof is standard and is based on the fact that $\delta \overset{s}{\not\sim} \epsilon$. Suppose w.l.o.g. that the system Δ is in 3–GNF. The notation $\alpha \in E$ means again that α is a summand in the expression E.

We will construct the sets $M_0, M_1, \ldots$ of variables from which the deadlock is reachable as following: $M_0 \overset{\text{def}}{=} \{\delta\}$ and for $i \geq 0$ the sets M_{i+1} are defined as $M_{i+1} \overset{\text{def}}{=} M_i \cup \{X \in \mathcal{V}ar \mid \exists a \in \mathcal{A}ct, \exists Y \in \mathcal{V}ar, \exists Z \in M_i : (X \overset{\text{def}}{=} E) \in \Delta,\ a.Z \in E \lor a.Z.Y \in E \lor (a.Y.Z \in E \text{ and } \|Y\| < \infty)\}$.

We remind that the norm of a variable can be effectively computed. Let us denote the fixed point of this construction as M. We can see that for each $X \in \mathcal{V}ar: X \longrightarrow^* \delta.\alpha$ for some $\alpha \in \mathcal{V}ar^*$ iff $X \in M$. If $X_1 \in M$ then Δ cannot be expressed by a BPA syntax since the deadlocking state is reachable from X_1. If $X_1 \notin M$ we can naturally transform Δ into a BPA system. □

The situation for the nonstrict case will be nicely characterised by the Corollary 1. In what follows, the set of variables from which a deadlocking state is reachable will be of great importance. Hence we define the set $\mathcal{V}ar_\delta$ of such variables: $\mathcal{V}ar_\delta \overset{\text{def}}{=} \{X \in \mathcal{V}ar \mid X \longrightarrow^* \delta \text{ or } \exists E \in \mathcal{E}^+_{\text{BPA}} : X \longrightarrow^* \delta.E\} - \{\delta\}$ and we state $\mathcal{V}ar_\epsilon \overset{\text{def}}{=} \mathcal{V}ar - \{\delta\} - \mathcal{V}ar_\delta$. The sets $\mathcal{V}ar_\delta$ and $\mathcal{V}ar_\epsilon$ can be effectively constructed as we have demonstrated in the proof of the Theorem 5. In what follows let the variables U, V, X, Y, Z range over $\mathcal{V}ar_\delta$ and A, B, C over $\mathcal{V}ar_\epsilon$.

Theorem 6. Let $(\mathcal{V}ar, \mathcal{A}ct, \Delta, X_1)$ be a BPA_δ system in 3–GNF. Suppose that there are only finitely many pairwise nonstrictly nonbisimilar $Y\alpha \in \mathcal{V}ar_\delta.\mathcal{V}ar^*$ such that $X_1 \longrightarrow^* Y\alpha$. Then there exists a BPA system $(\mathcal{V}ar', \mathcal{A}ct', \Delta', X_1')$ such that $\Delta \overset{n}{\sim} \Delta'$.

Proof. Let us suppose that $X_1 \in \mathcal{V}ar_\epsilon$. Then the system Δ can be trivially transformed into bisimilar BPA system Δ'. Thus assume that $X_1 \in \mathcal{V}ar_\delta$.

We may suppose w.l.o.g. that each summand of every defining equation in Δ does not contain an unnormed variable (resp. δ) followed by another variable. We define functions f_α for each $\alpha \in \mathcal{V}ar^*$. These functions take an expression from $\mathcal{E}^+_{\text{BPA}}$ in 3–GNF and transform it into another expression. Our goal is following. We want to achieve $f_\alpha(E) \overset{n}{\sim} E\alpha$ and there should be no deadlock in $f_\alpha(E)$. For each $\alpha \in \mathcal{V}ar^*$ let us also define a function r_α which returns the set of the new variables added by the function f_α. Let us assume that $X, Y, U \in \mathcal{V}ar_\delta$, $A, B, C \in \mathcal{V}ar_\epsilon$ with $\|C\| = \infty$, $\beta \in \mathcal{V}ar_\epsilon^*$ such that $\|\beta\| < \infty$ and $\gamma \in \mathcal{V}ar^*$.

$$f_\alpha(\sum_{i=1}^{n} a_i\alpha_i) = \sum_{i=1}^{n} f_\alpha(a_i\alpha_i) \qquad r_\alpha(\sum_{i=1}^{n} a_i\alpha_i) = \bigcup_{i=1}^{n} r_\alpha(a_i\alpha_i)$$

$$\begin{aligned}
f_\alpha(aXY) &= aX^{Y\alpha} & r_\alpha(aXY) &= \{X^{Y\alpha}\} \\
f_\alpha(aX\delta) &= aX^\epsilon & r_\alpha(aX\delta) &= \{X^\epsilon\} \\
f_\alpha(a\delta) &= a & r_\alpha(a\delta) &= \emptyset \\
f_\alpha(aX) &= aX^\alpha & r_\alpha(aX) &= \{X^\alpha\} \\
f_\alpha(aAB) &= aAB\beta U^\gamma & r_\alpha(aAB) &= \{U^\gamma\} & \text{if } \alpha = \beta U\gamma \\
 &= aAB\beta C & &= \emptyset & \text{if } \alpha = \beta C\gamma \\
 &= aAB\alpha & &= \emptyset & \text{otherwise} \\
f_\alpha(a) &= a\beta U^\gamma & r_\alpha(a) &= \{U^\gamma\} & \text{if } \alpha = \beta U\gamma \\
 &= a\beta C & &= \emptyset & \text{if } \alpha = \beta C\gamma \\
 &= a\alpha & &= \emptyset & \text{otherwise} \\
f_\alpha(aA\delta) &= aA & r_\alpha(aA\delta) &= \emptyset \\
f_\alpha(aA) &= aA\beta U^\gamma & r_\alpha(aA) &= \{U^\gamma\} & \text{if } \alpha = \beta U\gamma \\
 &= aA\beta C & &= \emptyset & \text{if } \alpha = \beta C\gamma \\
 &= aA\alpha & &= \emptyset & \text{otherwise} \\
f_\alpha(aXA) &= aX^{A\alpha} & r_\alpha(aXA) &= \{X^{A\alpha}\} \\
f_\alpha(aAX) &= aAX^\alpha & r_\alpha(aAX) &= \{X^\alpha\}
\end{aligned}$$

Let us now construct the nonstrictly bisimilar BPA system Δ' where $\mathsf{Var}' \stackrel{\text{def}}{=} \mathsf{Var}_\epsilon \cup \mathsf{Added}$; $\mathsf{Act}' \stackrel{\text{def}}{=} \mathsf{Act}$; $\Delta' \stackrel{\text{def}}{=} \Delta_\epsilon \cup \Gamma$; $X_1' \stackrel{\text{def}}{=} X_1^\epsilon$. The sets Added and Γ are outputs of the following algorithm and $\Delta_\epsilon \subseteq \Delta$ contains exactly the defining equations for variables from Var_ϵ.

Algorithm 1

1 Solve$:=\{X_1^\epsilon\}$
2 Added$:=\{X_1^\epsilon\}$
3 $\Gamma := \emptyset$
4 **while** Solve $\neq \emptyset$ **do**
5   Let us fix $X^\alpha \in$ Solve with $(X \stackrel{\text{def}}{=} E) \in \Delta$
6      $\Gamma := \Gamma \cup \{X^\alpha \stackrel{\text{def}}{=} f_\alpha(E)\}$
7   Add$:=\{Y^\beta \in r_\alpha(E) \mid \forall Z^\omega \in$ Added $: Y\beta \stackrel{n}{\not\sim} Z\omega\}$
8   **while** $\exists Y^\beta, Z^\omega \in$ Add $: Y^\beta \neq Z^\omega \wedge Y\beta \stackrel{n}{\sim} Z\omega$ **do**
9   Add$:=$ Add $- \{Y^\beta\}$
10   **endwhile**
11   Solve$:=$ (Solve $-\{X^\alpha\}) \cup$ Add
12   Added$:=$ Added $\cup$ Add
13   **for** $\forall Y^\beta \in r_\alpha(E) -$ Add **do**
14   replace all occurences of Y^β in Γ with Z^ω
15   where $Z^\omega \in$ Added $: Y\beta \stackrel{n}{\sim} Z\omega$
16   **endfor**
17 **endwhile**

In the following lemmas we demonstrate that the algorithm is correct and yields a BPA system Δ' such that $\Delta \stackrel{n}{\sim} \Delta'$.

Lemma 3. *For the loop 4–17 of Alg.1 holds the following invariant:* $\forall Y^\beta, Z^\omega \in$ Added $: Y^\beta \neq Z^\omega \Rightarrow Y\beta \not\sim^n Z\omega$.

Proof. An easy observation. □

Lemma 4. *Whenever during the execution of Alg.1 we have* $Y^\alpha \in$ Added *then* $Y \in \mathcal{V}ar_\delta$.

Proof. All variables in Added had to be produced by the function r_α (see line 7 and 12). It is easily seen that $\{Y \mid Y^\beta \in r_\alpha(E)\} \subseteq \mathcal{V}ar_\delta$ for any $\alpha \in \mathcal{V}ar^*$ and $E \in \mathcal{E}^+_{\text{BPA}}$ such that E is in 3–GNF. □

Lemma 5. *Whenever during the execution of Alg.1 we have* $Y^\beta \in$ Added *then* $X_1 \longrightarrow^* Y\beta$.

Proof. By induction on the number of repetitions of the loop 4–17.
Basic step: The only variable in the set Added before the execution of the loop 4–17 started is X_1^ϵ. However $X_1\epsilon = X_1$ and so $X_1 \longrightarrow^* X_1\epsilon$.
Induction step: Suppose that at line 12 we have added a new variable Y^β into Added. So at line 7 we had to have $Y^\beta \in r_\alpha(E)$ for some $X^\alpha \in$ Solve and $(X \stackrel{\text{def}}{=} E) \in \Delta$. The induction hypothesis says that $X_1 \longrightarrow^* X\alpha$ (X^α had to be added in some previous repetition of the main loop). It must hold that $\alpha\gamma Y^\beta \in f_\alpha(E)$ where $\gamma \in \mathcal{V}ar^*_\epsilon$ and $\|\gamma\| < \infty$. From the construction of f_α we can also see that $X\alpha \longrightarrow^* Y\beta$. Thus we get $X_1 \longrightarrow^* X\alpha \longrightarrow^* Y\beta$. □

Lemma 6. *Alg.1 cannot loop forever (under the assumption of the Theorem 6).*

Proof. Suppose that the algorithm loops forever which means that the set Solve is never empty. But in every loop we remove exactly one element from the set Solve (line 11). This implies that the set Added will grow arbitrarily because the set Add is infinitely often unempty (otherwise the algorithm would stop). The contradiction is immediate from the Lemmas 3, 4 and 5. □

The following lemma is crucial for the proof of our theorem.

Lemma 7. *After the execution of Alg.1 we have* $V^\alpha \stackrel{n}{\sim} V\alpha$ *for all* $V^\alpha \in$ Added.

Proof. We use the *stratified bisimulation relations* [Mil89] $\sim_k$. By induction on k we show that $V^\alpha \sim_k V\alpha$ for all $k \geq 0$. This implies that $V^\alpha \stackrel{n}{\sim} V\alpha$. This straightforward but also long and technical proof can be found in [Srb98]. □

Lemma 8. *The system Δ' is a BPA system and moreover* $X_1 \stackrel{n}{\sim} X_1^\epsilon$.

Proof. Immediately from the Lemma 7. □

We have constructed a BPA system Δ' such that $\Delta \stackrel{n}{\sim} \Delta'$. □

Theorem 7. Let $(\mathcal{V}ar, \mathcal{A}ct, \Delta, X_1)$ be a BPA_δ system. Suppose that there are infinitely many pairwise nonstrictly nonbisimilar $Y\alpha \in \mathcal{V}ar_\delta.\mathcal{V}ar^*$ such that $X_1 \longrightarrow^* Y\alpha$. Then there is no BPA system Δ' such that $\Delta \stackrel{n}{\sim} \Delta'$.

Proof. The proof is based on the fact that $\alpha \stackrel{n}{\sim} \beta$ implies $\|\alpha\|=\|\beta\|$. Let us assume w.l.o.g. that there exists Δ' in 3–GNF such that $\Delta \stackrel{n}{\sim} \Delta'$. We show that this is not possible. Since there are infinitely many reachable states $Y_1\alpha_1, Y_2\alpha_2,\ldots$ of Δ which are pairwise nonstrictly nonbisimilar there must be corresponding states $\beta_1, \beta_2, \ldots$ of the system Δ' such that $Y_i\alpha_i \stackrel{n}{\sim} \beta_i$ for $i = 1, 2, \ldots$. Let us now define a constant N_{max} as $N_{max} \stackrel{def}{=} \max\{\|Y_i\|_\delta \mid i = 1, 2, \ldots\}$ where $\|Y\|_\delta \stackrel{def}{=} \min\{length(w) \mid Y \stackrel{w}{\longrightarrow} \delta$ or $\exists E \in \mathcal{E}_{\text{BPA}}^+ : Y \stackrel{w}{\longrightarrow} \delta.E\}$. Notice that the definition of N_{max} is correct since for all i $\|Y_i\|_\delta < \infty$ (because $Y_i \in \mathcal{V}ar_\delta$) and there are only finitely many different Y_i's.

Clearly $\|Y_i\alpha_i\| \leq N_{max}$ for all i. This implies that the norm of β_i is also less or equal N_{max} for all i. However, Δ' is a BPA system and all variables in Δ' are guarded. This means that there are only finitely many different states of Δ' such that their norm is less or equal N_{max}. Hence there must be two states β_k and β_l with $k \neq l$ such that $\beta_k = \beta_l$. This implies that $\beta_k \stackrel{n}{\sim} \beta_l$. Then also $Y_k\alpha_k \stackrel{n}{\sim} Y_l\alpha_l$, which is contradiction. □

Suppose that we have a BPA_δ system and that there are infinitely many nonbisimilar states from which, after some 'short' sequence of actions, a deadlocking state is reachable. Then the corresponding (nonstrictly bisimilar) BPA system does not exists. This condition appears to be both necessary and sufficient as is illustrated by the following corollary.

Corollary 1. Let $(\mathcal{V}ar, \mathcal{A}ct, \Delta, X_1)$ be a BPA_δ system. There are only finitely many pairwise nonstrictly nonbisimilar $Y\alpha \in \mathcal{V}ar_\delta.\mathcal{V}ar^*$ such that $X_1 \longrightarrow^* Y\alpha$ if and only if there exists a BPA system $(\mathcal{V}ar', \mathcal{A}ct', \Delta', X_1')$ such that $\Delta \stackrel{n}{\sim} \Delta'$.

Proof. An immediate consequence of the Theorems 6 and 7. □

5 Conclusion Remarks

In this paper we have focused on the class of BPA processes extended with deadlocks. We have shown that for language equivalence the extention is no acquisition. On the other hand the BPA_δ class is larger with regard to the relation of bisimulation. We introduce two notions of bisimilarity to capture the different understanding of deadlock behaviour. If we do not distinguish between ϵ and δ, we speak about nonstrict bisimilarity and if we do, we call the appropriate bisimulation as strict. We have solved the question whether, given a BPA_δ system Δ, there is an equivalent description (with regard to bisimilarity) of Δ in terms of BPA syntax. The solution for the strict bisimilarity is straightforward. However, the answer to the problem dealing with the nonstrict bisimilarity exploited a nice semantic characterization of the subclass of BPA_δ processes bisimilarly describable in BPA syntax: a BPA_δ system can be transformed into a BPA system (preserving nonstrict bisimilarity) if and only if finitely many nonbisimilar states starting with some in δ-ending variable are reachable. There is still an open problem whether this semantic characterization is syntactically checkable.

Acknowledgements: First of all, I would like to thank Ivana Černá for her help and encouragement throughout the work. I am very grateful for her advise and valuable discussions. My warm thanks go also to Mojmír Křetínský and Antonín Kučera for their constant support and comments.

References

[BCS95] O. Burkart, D. Caucal, and B. Steffen. An elementary decision procedure for arbitrary context-free processes. In *Proceedings of MFCS'95*, volume 969 of *LNCS*, pages 423–433, 1995.

[BCS96] O. Burkart, D. Caucal, and B. Steffen. Bisimulation collapse and the process taxonomy. In *Proceedings of CONCUR'96* [Con96], pages 247–262.

[BK85] J.A. Bergstra and J.W. Klop. Algebra of communicating processes with abstraction. *Theoretical Computer Science*, 37:77–121, 1985.

[Bos97] D. Bosscher. *Grammars Modulo Bisimulation.* PhD thesis, CWI, University of Amsterdam, 1997.

[CHS95] S. Christensen, H. Hüttel, and C. Stirling. Bisimulation equivalence is decidable for all context-free processes. *Information and Computation*, 121:143–148, 1995.

[Con96] *Proceedings of CONCUR'96*, volume 1119 of *LNCS*. Springer-Verlag, 1996.

[HM96] Y. Hirshfeld and F. Moller. Decidability results in automata and process theory. In *Logics for Concurrency: Automata vs Structure*, volume 1043 of *LNCS*, pages 102–148. Faron Moller and Graham Birtwistle, 1996.

[Hüt91] H. Hüttel. *Decidability, Behavioural Equivalences and Infinite Transition Graphs.* PhD thesis, The University of Edinburgh, 1991.

[Mil89] R. Milner. *Communication and Concurrency.* Prentice-Hall, 1989.

[Mol96] F. Moller. Infinite results. In *Proceedings of CONCUR'96* [Con96], pages 195–216.

[Srb98] Jiří Srba. Comparing the classes BPA and BPA with deadlocks. Technical Report FIMU-RS-98-05, Faculty of Informatics, Masaryk University, 1998.

A Superpolynomial Lower Bound for a Circuit Computing the Clique Function with At Most $(1/6) \log \log n$ Negation Gates

Kazuyuki Amano and Akira Maruoka

Graduate School of Information Sciences, Tohoku University,
Sendai, 980-8579 JAPAN
{ama|maruoka}@ecei.tohoku.ac.jp

Abstract. We investigate about a lower bound on the size of a Boolean circuit that computes the clique function with a limited number of negation gates. To derive strong lower bounds on the size of such a circuit we develop a new approach by combining the three approaches: the restriction applied for constant depth circuits[Has], the approximation method applied for monotone circuits[Raz2] and boundary covering developed in the present paper. Based on the approach the following statement is established: If a circuit C with at most $\lfloor (1/6) \log \log m \rfloor$ negation gates detects cliques of size $(\log m)^{3(\log m)^{1/2}}$ in a graph with m vertices, then C contains at least $2^{(1/5)(\log m)^{(\log m)^{1/2}}}$ gates. In addition, we present a general relationship between negation-limited circuit size and monotone circuit size of an arbitrary monotone function.

1 Introduction

Recently there has been substantial progress in obtaining strong lower bounds for restricted Boolean circuits that compute some function, in particular for those circuit models such as constant depth circuits or monotone circuits. Exponential lower bounds are derived for the size of constant depth circuits computing the parity function[Has] and for the size of monotone circuits computing the clique function[Raz],[AB],[AM]. It is natural to ask if we could make use of the approaches developed to obtain these bounds so as to derive strong lower bounds for more generalized model. As such a generalized model, we consider circuits with a limited number of negation gates. In fact, it remains open so far to derive non-trivial lower bounds on the size of a circuit computing some monotone function with, say, a constant number of negation gates[SW]. Fischer[Fis] showed that for any function f, the size of the smallest circuit computing f with an arbitrary number of NOT gates and the one with at most $\lceil \log(n+1) \rceil$ NOT gates are polynomially related(See also [BNT]). So if one can prove superpolynomial lower bounds on the size of circuits with at most $\lceil \log(n+1) \rceil$ NOT gates computing some problem in NP, then we have that P$\neq$NP. So we try to obtain superpolynomial lower bounds on the size of circuits, with $O(\log \log n)$ NOT gates rather than $O(\log n)$ NOT gates, computing some problem in NP. More precisely we prove the following: If a circuit C with at most $\lfloor (1/6) \log \log m \rfloor$ NOT gates detects cliques of size $(\log m)^{3(\log m)^{1/2}}$ in a graph with m vertices, then C contains at least $2^{(1/5)(\log m)(\log m)^{1/2}}$ gates (Theorem 8). The problem of detecting a clique in a graph with m vertices will be written as a Boolean function of $n = \binom{m}{2}$ variables. In addition, we present a general relationship between negation-limited circuit size and monotone circuit size of an arbitrary monotone function (Theorem 2).

To achieve the main results, we develop a new approach by combining the three approaches: the restriction applied for constant depth circuits[Has], the approximation method applied for monotone circuits[Raz] and boundary covering developed in the present paper. A Boolean function f can be viewed as dividing the Boolean cube into two regions: The one is written as $\{v \in \{0,1\}^n \mid f(v) = 1\}$ and the other as $\{v \in \{0,1\}^n \mid f(v) = 0\}$. So we can think of the boundary between the two regions, which is defined as the collection of pairs of vectors (w, w') such that $f(w) \neq f(w')$ and the Hamming distance between w and w' is 1. The idea of the proof of the main theorem is as follows. Firstly, we verify in Section 3 a theorem that says the problem of proving lower bounds on the negation-limited circuit size of a monotone function f can be reduced to the one of proving lower bounds on the maximum of monotone circuit sizes of monotone functions such that the union of boundaries of the latter monotone functions covers the boundary of the former monotone function f. Secondly, we analyze carefully in Section 4 the proof due to Amano and Maruoka[AM] for an exponential lower bound on the monotone circuit size of the clique function, and verify a statement that we still need superpolynomial number of gates in a monotone circuit that computes even a certain small fraction of the boundary of the clique function (Theorem 3). Finally, we verify in Section 5 a statement (Theorem 8) that, no matter what collection of monotone functions we take to cover the boundary of the clique function, the largest fraction of the boundary covered by some monotone function in the collection is more than what is needed to apply the result (Theorem 3) in the second part. This is the most difficult part of the proof.

Throughout this paper, the function $\log x$ denotes logarithm base 2 of x.

2 Preliminaries

For w in $\{0,1\}^n$, let w_i denote the value of the ith bit of w. Let w and w' be in $\{0,1\}^n$. We denote $w \leq w'$ if $w_i \leq w'_i$ for all $1 \leq i \leq n$, and $w < w'$ if $w \leq w'$ and $w \neq w'$. Let $\text{Ham}(w, w')$ denote the Hamming distance between w and w', written as, $\text{Ham}(w, w') = |\{i \in \{1, \ldots, n\} \mid w_i \neq w'_i\}|$, where $|S|$ denotes the size of set S.

A *Boolean circuit* is a directed acyclic graph with gate nodes (or, simply gates) and input nodes. Operation AND or OR is associated with each gate whose in-degree is 2, whereas NOT is associated with each gate whose in-degree is 1. A Boolean variable or a constant, namely, 0 or 1, is associated with each input node whose in-degree is 0. In particular, a circuit with no NOT gates is called *monotone*. A Boolean function of n variables is called *monotone* if $f(w) \leq f(w')$

holds for any $w, w' \in \{0,1\}^n$ such that $w \leq w'$. Let $\mathcal{M}^n$ denote the set of all monotone functions of n variables. The *size* of a circuit C, denoted size(C), is the number of gates in the circuit C. The *circuit complexity* (respectively, *monotone circuit complexity*) of a function f, denoted size(f) (respectively, size$_{mon}(f)$), is the size of the smallest circuit (respectively, monotone circuit) computing f. For a function f and a positive integer t, the *circuit complexity with t limited negation* (negation limited complexity, for short) of a function f, denoted size$_t(f)$, is the size of the smallest circuit C that computes f and includes at most t NOT gates. If the function f cannot be computed with only t NOT gates, then size$_t(f)$ is undefined. Let $C_g(w)$ denote the output of the gate g in circuit C that has as input w. We say a gate g in C separates a pair of vectors (w, w') (or simply, a gate g separates (w, w') when no confusion arises) if $C_g(w) = 0$, $C_g(w') = 1$ and Ham$(w, w') = 1$. In particular, when g is taken to be the output gate in circuit C, we simply say that the circuit C separates such a pair (w, w'). Similarly, when a circuit separates a pair we say the function computed by the circuit separates the pair. Although we can generalize the notion of separation by dropping the condition that the Hamming distance between vectors being 1, we don't need such generalization for the purpose of our argument.

3 Relating Negation Limited and Monotone Circuit Complexity

In this section, we establish a relationship between negation-limited complexity and monotone circuit complexity for a monotone Boolean function.

Definition 1 *Let f be a Boolean function of n variables. A sensitive graph of f, denoted $G(f)$, is defined as follows: $G(f) = (V, E)$ is a directed graph with $V = \{0,1\}^n$ and $E = \{(w, w') \mid \text{Ham}(w, w') = 1, f(w) = 0 \text{ and } f(w') = 1\}$.*

So a sensitive graph of f is a graph whose edge set consists of pairs that are separated by the function f. Let G_1 and G_2 be two graphs on the same set V of vertices, that is, $G_1 = (V, E_1)$ and $G_2 = (V, E_2)$. Then the *union* $G_1 \cup G_2$ is defined to be the graph $(V, E_1 \cup E_2)$. Furthermore we say G_1 contains G_2 if $E_1 \supseteq E_2$ holds, which is denoted by $G_1 \supseteq G_2$.

Theorem 2. *Let f be a monotone function of n variables. For any positive integer t,*

$$\text{size}_t(f) \geq \min_{F' = \{f_1, \ldots, f_\alpha\} \subseteq \mathcal{M}^n} \left\{ \max_{f' \in F'} \{\text{size}_{mon}(f')\} \,\Big|\, \bigcup_{f' \in F'} G(f') \supseteq G(f) \right\},$$

where $\alpha = 2^{t+1} - 1$. □

This theorem shows that the problem of deriving lower bounds on the negation-limited circuit complexity of a monotone function can be reduced to the one of deriving the maximum lower bound for monotone circuit complexity among monotone functions whose sensitive graphs cover that of the original function.

Note that the set of all variables $F' = \{x_1, \ldots, x_n\}$ satisfies the condition $\cup_{f' \in F'} G(f') \supseteq G(f)$ for any monotone function f. Hence, if $\alpha \geq n$, that is, $t \geq \log(n+1) - 1$, the right-hand side of the inequality in Theorem 2 does not give any non-trivial lower bound.

Proof (of Theorem 2). Let f be a monotone function of n variables. Let C denote the circuit of the smallest size that computes f using no more than t NOT gates. That is, size(C) =size$_t(f)$. Just for simplicity of notation, we assume the number of NOT gates in C is given by t. Furthermore, without loss of generality we assume that the output gate of C is not a NOT gate. Let $g_1, \ldots, g_t$ be a list of NOT gates of C arranged in topological order. For $0 \leq i \leq t$ and $u = (u_1, \ldots, u_i) \in \{0,1\}^i$, let C_u denote the subcircuit of C obtained by restricting the output of the NOT gates g_j to constant u_j for $1 \leq j \leq i$ and making the input to g_{i+1} in C the output of entire circuit C_u, where g_{t+1} is supposed to denote the output gate of the entire circuit C. In particular, for the empty sequence λ, C_λ denote the circuit obtained by making the input to g_1 in C the output of entire circuit C_λ. Then it is easy to see that, for any $(w, w') \in \{0,1\}^n \times \{0,1\}^n$ separated by circuit C, there exists $0 \leq i \leq t$ and $u \in \{0,1\}^i$ such that the circuit C_u separates the (w, w') or (w', w). This is because as such an i we can simply take i such that g_{i+1} is the first gate in the sequence $(g_1, \ldots, g_{t+1})$ such that $C_{g_{i+1}}(w) \neq C_{g_{i+1}}(w')$, and put $u_j = g_j(w) (= g_j(w'))$ for $1 \leq j \leq i$. The number of the circuits represented as C_u for $u \in \{0,1\}^*$ such that $|u| \leq t$ is given by $\sum_{j=1}^{t} 2^j + 1 = 2^{t+1} - 1 = \alpha$. Hence, denoting by f_j's functions computed by circuit C_u's, we have $\cup_{f' \in F'} G(f') \supseteq G(f)$ for $F' = \{f_1, \ldots, f_\alpha\}$. Thus, since size$_t(f)(=$ size$(C)) \geq$ size$_{mon}(f')$ for any $f' \in F'$, the proof is completed. □

4 Hardness of Approximating Clique Function

The clique function, denoted CLIQUE(m, s), of $m(m-1)/2$ variables $\{x_{i,j} \mid 1 \leq i < j \leq m\}$ is defined to take the value 1 if and only if the undirected graph on m vertices represented in the obvious way by the input contains a clique of size s.

A graph on m vertices is called *good* if, for some positive integer s_2, it corresponds to a clique on some set of s_2 vertices, and having no other edges. Let $I(m, s_2)$ denote the set of such good graphs. A graph on m vertices is called *bad* if, for some positive integer s_1, there exists a partition of the vertices into $m \bmod (s_1 - 1)$ sets of size $\lceil m/(s_1 - 1) \rceil$ and $s_1 - 1 - (m \bmod (s_1 - 1))$ sets of size $\lfloor m/(s_1 - 1) \rfloor$ such that any two vertices chosen from different sets have an edge between them, and no other edges exist. Let $O(m, s_1)$ denote the set of such bad graphs.

For $1 \leq s_1 \leq s_2 \leq m$, let $F(m, s_1, s_2)$ denote the set of all monotone functions f of $\binom{m}{2}$ variables representing a graph G on m vertices such that function f takes the value 0 if G contains no clique of size s_1, the value 1 if G contains a clique of size s_2, and an arbitrary value otherwise. For any function f in $F(m, s_1, s_2)$, the value of f is 1 for any good graph in $I(m, s_2)$, and is 0 for any bad graph in $O(m, s_1)$. The following theorem will be needed in the next section to prove the main theorem.

Theorem 3. *Let s_1 and s_2 be positive integers such that $64 \leq s_1 \leq s_2$ and $s_1^{1/3} s_2 \leq m/200$. Suppose that C is a monotone circuit and that the fraction of good graphs in $I(m, s_2)$ such that C outputs 1 is at least $h = h(s_2)$. Then at least one of the followings holds: (i) The number of gates in C is at least $(h/2)2^{s_1^{1/3}/4}$, (ii) The fraction of bad graphs in $O(m, s_1)$ such that C outputs 0 is at most $2/s_1^{1/3}$.* □

The proof of Theorem 3 is done by using the same argument as in the proof of Theorem 1 in [AM]. In [AM], Amano and Maruoka presented a simplified proof of an exponential lower bound on the monotone complexity of the clique function based on the Razborov's approximation method. The key of the proof is to define the approximate operations $\overline{\vee}$ (which approximates an OR gate) and $\overline{\wedge}$ (which approximates an AND gate) in terms of DNF and CNF formulas such that the size of terms and clauses in the formulas is limited appropriately. For the purpose of the argument of the current paper, adopt the same definition for the approximate operations as in [AM] except for the values of the parameters l and r in their definitions, and follow their arguments to obtain Theorem 3. See [AM] for the detail.

Choose $l = \lfloor s_1^{1/3}/4 \rfloor$ and $r = \lfloor 30 s_1^{1/3} \rfloor$. Put $w = m \bmod (s_1 - 1)$. By the analogous argument to the proof of Theorem 1 in [AM], we can get

Fact 4 $|I(m, s_2)| = (m!)/(s_2!(m - s_2)!)$ and
$$|O(m, s_1)| = \frac{m!}{(\lceil m/(s_1-1) \rceil!)^w (\lfloor m/(s_1-1) \rfloor!)^{s_1-1-w} w!(s_1-1-w)!}.$$

Lemma 5 *Let C be a monotone circuit. An approximator circuit $\overline{C}$ (i.e., $\overline{C}$ is a circuit obtained by replacing every OR and AND gates in C by $\overline{\vee}$ and $\overline{\wedge}$ gates, respectively) outputs identically 0 or the fraction of bad graphs in $O(m, s_1)$ such that $\overline{C}$ outputs 1 is at least $1 - s_1^{1/3}$.*

Proof. Replace the formula in the proof of Lemma 1 in [AM] with $\prod_{k=1}^{l-1}(1 - (k\lceil m/(s_1-1)\rceil)/m) > 1 - s_1^{1/3}$. The rest of the proof is analogous to that of Lemma 1 in [AM] □

Lemma 6 *The number of bad graphs in $O(m, s_1)$ for which the OR and $\overline{\vee}$ gates produce different outputs (the OR gate produces 0, whereas the $\overline{\vee}$ gate produces 1) is at most*
$$\frac{(m/s_1^{1/6})^{r+1}(m-r-1)!}{(\lceil m/(s_1-1)\rceil!)^w (\lfloor m/(s_1-1)\rfloor!)^{s_1-1-w} w!(s_1-1-w)!}.$$

Proof. Replace the formula in the middle of the proof of Lemma 2 in [AM] with $(l(l-1)/2)(m/(s_1-1)) + \sqrt{(l(l-1)/2)(2m^2/(s_1-1))} < m/s_1^{1/6}$. The rest of the proof is analogous to that of Lemma 2 in [AM]. □

Lemma 7 *The number of good graphs in $I(m, s_2)$ for which the AND and $\overline{\wedge}$ gates produce different outputs (the AND gate produces 1, whereas the $\overline{\wedge}$ gate produces 0) is at most $((2rs_2)^{l+1}(m - l - 1)!)/(s_2!(m - s_2)!)$.*

Proof. Replace the formula in the middle of the proof of Lemma 3 in [AM] with $r(s_2 - 1) + \sqrt{(r(r-1)/2)s_2(s_2-1)} < 2rs_2$. The rest of the proof is analogous to that of Lemma 3 in [AM].

Proof (of Theorem 3). Let C be a monotone circuit such that $\Pr_{v \in I(m,s_2)}[C(v) = 1] \geq h(s_2)$ and $\Pr_{u \in O(m,s_1)}[C(u) = 0] > 2/s_1^{1/3}$ hold. To prove the theorem, we show that the circuit C must satisfy the condition (i) in Theorem 3. From Lemma 5, the approximator circuit $\overline{C}$ satisfies $\Pr_{v \in I(m,s_2)}[C(v) \neq \overline{C}(v)] \geq h(s_2)$ or $\Pr_{u \in O(m,s_1)}[C(u) \neq \overline{C}(u)] > 1/s_1^{1/3}$. Thus in view of Fact 4, Lemmas 5, 6 and 7, the size of C is at least

$$\frac{1}{2}\min\left(\frac{h(s_2)m!}{(2rs_2)^{l+1}(m-l-1)!}, \frac{m!}{s_1^{1/3}(m/s_1^{1/6})^{r+1}(m-r-1)!}\right). \quad (1)$$

The coefficient $1/2$ here is caused by taking into account the extra restriction of the $\overline{\vee}$ and $\overline{\wedge}$ alternation mentioned in [AM]. An elementary calculation completes the proof. □

5 Proof of the Main Theorem

The goal of this section is to prove Theorem 8, which says that $\lfloor (1/6) \log \log m \rfloor$ NOT gates are not enough to compute the clique function feasibly. We don't intend here to optimize the constant $1/6$ in the number of NOT gates.

Theorem 8. *For any sufficiently large integer m,*

$$size_{\lfloor (1/6)\log\log m \rfloor}(CLIQUE(m, (\log m)^{3(\log m)^{1/2}})) > 2^{(1/5)(\log m)^{(\log m)^{1/2}}}.$$

Before proceeding to the proof, we describe an idea of the proof.

Let f be $CLIQUE(m, s)$. Suppose to the contrary that a small circuit C with t NOT gates computes f. By Theorem 2, there are $2^{t+1} - 1 (= \alpha)$ monotone functions $f_1, \ldots, f_\alpha$ such that each of them has small monotone complexity and that $\cup_{i \in \{1,\ldots,\alpha\}} G(f_i) \supseteq G(f)$. Let $l_0 < l_1 < \cdots < l_\alpha$ be some monotone increasing sequence with $l_0 = s$ and $l_\alpha = m$. For $1 \leq i \leq \alpha$, a graph v is called *good in the i-th layer* if v consists of a clique of size l_{i-1}, and has no other edges. For $1 \leq i \leq \alpha$, a graph u is called *bad in the i-th layer* if there exists a partition of some l_i vertices into $s - 1$ sets with equal size such that any two vertices chosen from different sets have an edge between them, and no other edges exist. In other words, a bad graph in the i-th layer is a $(s - 1)$-partite complete graph on some subset of vertices with size l_i. Note that for any good graph v in the first layer, since deleting an edge from v breaks the clique in v, an edge ending at v is in $G(f)$. For any bad graph u in any layer, since adding an appropriate edge to u produces a clique with size s, an edge starting at u is in $G(f)$.

A Superpolynomial Lower Bound for a Circuit 405

Since $\cup_{i\in\{1,...,\alpha\}} G(f_i) \supseteq G(f)$, there exists a function f' in $\{f_1,\ldots,f_\alpha\}$ such that $G(f')$ contains at least $1/\alpha$ fraction of edges of $G(f)$ ending at good graphs in the first layer, and hence f' outputs 1 on at least $1/\alpha$ fraction of good graphs in the first layer. W.l.o.g., let the function f' be denoted by f_1. Since Theorem 3 says that a small monotone circuit can not separate not small number of good graphs in the first layer from not small number of bad graphs in the same layer, there are not small number of bad graphs u in the first layer on which the function f_1 takes the value 1. But since by adding an edge appropriately to such a u we get a graph u^+ which contains a clique of size s. Hence there are many edges, denoted (u, u^+), that are not included in the edges in $G(f_1)$. Since $\cup_{i\in\{1,...,\alpha\}} G(f_i) \supseteq G(f)$, there exists a function f'' in $\{f_2,\ldots,f_\alpha\}$ such that $G(f'')$ contains at least $1/\alpha$ fraction of such edges (u, u^+). W.l.o.g., let the function f'' be denoted by f_2. On the other hands, since f_2 is monotone, f_2 takes the value 1 on the good graph v in the second layer such that $u^+ \leq v$ and that $f_2(u^+) = 1$. Applying Theorem 3 again, we can conclude that f_2 outputs 1 for not too small fraction of bad graphs in the second layer. It can be shown that f_1 also outputs 1 for such bad graphs.

By continuing above argument, we can conclude that every functions $f_1,\ldots,f_\alpha$ outputs 1 on some bad graph u in the last layer, contradicting the fact that $\cup_{i\in\{1,...,\alpha\}} G(f_i) \supseteq G(f)$. This is an outline of the proof.

Proof (of Theorem 8). Let m be a sufficiently large integer. Put $t = \lfloor (1/6)\log\log m \rfloor$, $s = (\log m)^{3(\log m)^{1/2}}$, $M = 2^{(1/5)(\log m)^{(\log m)^{1/2}}}$ and $\alpha = 2^{t+1} - 1$. We suppose to the contrary that a circuit C with at most t NOT gates computes CLIQUE(m, s) and that size$(C) \leq M$. From Theorem 2, there exist monotone functions $f_1,\ldots,f_\alpha \in \mathcal{M}^n$ such that size$_{mon}(f_i) \leq M$ for any $1 \leq i \leq \alpha$, and $\cup_{i\in\{1,...,\alpha\}} G(f_i) \supseteq G(\text{CLIQUE}(m, s))$.

Let $l_0 = s$, $l_\alpha = m$ and for every $j = 1,\ldots,\alpha-1$, let $l_j = m^{1/10 + (1/3)(j-1)/(\log m)^{1/6}}$. Since $l_{\alpha-1} \leq m^{1/10+(1/3)(2^{(1/6)\log\log m + 1})/(\log m)^{1/6}} = m^{1/10 + 2/3} < m^{9/10}$, we have $l_0 < l_1 < \cdots < l_\alpha$. Let V be a set of m vertices of the graph associated with CLIQUE. For $j \in \{0,\ldots,\alpha\}$, let $\mathcal{L}_j$ denote $\{L \subseteq V \mid |L| = l_j\}$ and let $\mathcal{L}_j(L)$ denote $\{L' \subseteq L \mid L' \in \mathcal{L}_j\}$. For $i \in \{1,\ldots,\alpha\}$ and $L_i \in \mathcal{L}_i$, a graph v is called *good on the set L_i in the i-th layer* if it corresponds to a clique of size l_{i-1} on some $L_{i-1} \in \mathcal{L}_{i-1}(L_i)$ (i.e., $|L_{i-1}| = l_{i-1}$ and $L_{i-1} \subseteq L_i$), having no other edges. For $i \in \{1,\ldots,\alpha\}$ and $L_i \in \mathcal{L}_i$, a graph u is called *bad on the set L_i in the i-th layer* if there exists a partition of L_i into $V_1,\ldots,V_{s-1}$ such that (i)$|V_i| \in \{\lfloor |L_i|/(s-1) \rfloor, \lceil |L_i|/(s-1) \rceil\}$ for $i = 1,\ldots,s-1$, and (ii) a graph u has an edge (w, w') iff $w \in V_i$ and $w' \in V_j$ such that $i \neq j$.

Let I_{L_i} (respectively, O_{L_i}) denote the set of all good (respectively, bad) graphs on the set L_i in the i-th layer. Note that a good graph in the first layer (respectively, a bad graph in the last layer) is a minterm (respectively, a maxterm) of CLIQUE(m, s). We also note that there are one to one corresponding between I_{L_i} and $I(l_i, l_{i-1})$, and that between O_{L_i} and $O(l_i, s)$, where $I(l_i, l_{i-1})$ and $O(l_i, s)$ are defined in Section 4, and hence separating I_{L_i} and O_{L_i} is equivalent to computing a function in $F(l_i, s, l_{i-1})$. Since $s^{1/3}l_{i-1} \leq l_i/200$ holds, the following claim is straightforward from Theorem 3.

Claim 9 *Let $i \in \{1, \ldots, \alpha\}$ and $L_i \in \mathcal{L}_i$. Suppose that C is a monotone circuit and the fraction of good graphs in I_{L_i} such that C outputs 1 is at least h. Then at least one of the followings holds: (i) The number of gates in C is at least $(h/2)2^{s^{1/3}/4}$, (ii) The fraction of bad graphs in O_{L_i} such that C outputs 0 is at most $2/s^{1/3}$.* □

Proof (of Theorem 8 (continued)). For $L \subseteq V$, let v_L denote a graph corresponding to a clique on the set L, having no other edges. Recall that $\mathcal{L}_0 = \{L \subseteq V \mid |L| = s\}$. Thus for any $L_0 \in \mathcal{L}_0$, there exists $u < v_{L_0}$ such that the edge (u, v_{L_0}) is in $G(\text{CLIQUE}(m, s))$. Hence there exists i_1 in $\{1, \ldots, \alpha\}$ such that $\Pr_{L_0 \in \mathcal{L}_0}[\exists u < v_{L_0} \ (u, v_{L_0}) \in G(f_{i_1})] \geq 1/\alpha > 1/2^{t+1}$ holds, and this implies $\Pr_{L_0 \in \mathcal{L}_0}[f_{i_1}(v_{L_0}) = 1] \geq 1/2^{t+1}$. Then it is easy to see that

$$\Pr_{L_1 \in \mathcal{L}_1}\left[\Pr_{v \in I_{L_1}}[f_{i_1}(v) = 1] \geq \frac{1}{2^{t+2}}\right] \geq \frac{1}{2^{t+2}}. \quad (2)$$

Now we call a $L_1 \in \mathcal{L}_1$ *dense* if $\Pr_{v \in I_{L_1}}[f_{i_1}(v) = 1] \geq 1/2^{t+2}$ holds. Put $h = 1/2^{t+2}$. An easy calculation shows $h \geq 1/m$. Thus by applying Claim 9 to every dense L_1, we have $\text{size}_{mon}(f_{i_1}) \geq (1/2m)2^{s^{1/3}/4} = 2^{(1/4)(\log m)(\log m)^{1/2} - \log m - 1} > M$ or $\Pr_{u \in O_{L_1}}[f_{i_1}(u) = 1] \geq 1 - 2/s^{1/3} \geq 1/2$ for any dense L_1. Since the former contradicts the assumption $\text{size}_{mon}(f_{i_1}) \leq M$, we have $\Pr_{u \in O_{L_1}}[f_{i_1}(u) = 1] \geq 1/2$ for any dense L_1. By (2), we have

$$\Pr_{L_1 \in \mathcal{L}_1}\left[\Pr_{u \in O_{L_1}}[f_{i_1}(u) = 1] \geq \frac{1}{2}\right] \geq \frac{1}{2^{t+2}}. \quad (3)$$

The proof is done by induction on a level of the layers. We use (3) as the induction basis and the induction step is as follows. For a proof of this claim, see the appendix.

Claim 10 *Suppose $c_1 > 1$ and $c_2 > 1$. Put $c_3 = \alpha$. Let $f_1, \ldots, f_{c_3}$ be the monotone functions such that $\cup_{i \in \{1,\ldots,c_3\}} G(f_i) \supseteq G(\text{CLIQUE}(m, s))$ and $\text{size}_{mon}(f_i) \leq M$ for any $1 \leq i \leq c_3$. Suppose that for distinct indices $i_1, \ldots, i_k \in \{1, \ldots, c_3\}$ $\Pr_{L_k \in \mathcal{L}_k}\left[\Pr_{u \in O_{L_k}}[f_{i_1}(u) = \cdots = f_{i_k}(u) = 1] \geq 1/c_1\right] \geq 1/c_2$. If $c_1 c_2 c_3 \leq s^{1/3}/8$, then there exists $i_{k+1} \in \{1, \ldots, c_3\} \setminus \{i_1, \ldots, i_k\}$ such that*

$$\Pr_{L_{k+1} \in \mathcal{L}_{k+1}}\left[\Pr_{u \in O_{L_{k+1}}}[f_{i_1}(u) = \cdots = f_{i_k}(u) = f_{i_{k+1}}(u) = 1] \geq \frac{1}{4c_1 c_2 c_3}\right] \geq \frac{1}{2c_2 c_3}.$$

Proof (of Theorem 8 (continued)). First we claim that for any $k \in \{1, \ldots, \alpha\}$, there are k distinct indices $i_1, \ldots, i_k \in \{1, \ldots, \alpha\}$ such that

$$\Pr_{L_k \in \mathcal{L}_k}\left[\Pr_{u \in O_{L_k}}[f_{i_1}(u) = \cdots = f_{i_k}(u) = 1] \geq \frac{1}{2^{k^2(t+2)}}\right] \geq \frac{1}{2^{k(t+2)}} \quad (4)$$

holds. The claim is proved by induction on k. The basis, $k = 1$, is trivial from (3). Now we suppose the claim holds for any $k \leq l$ and let $k = l + 1$. By the induction hypothesis, we have

$$\Pr_{L_l \in \mathcal{L}_l}\left[\Pr_{u \in O_{L_l}}[f_{i_1}(u) = \cdots = f_{i_l}(u) = 1] \geq \frac{1}{2^{l^2(t+2)}}\right] \geq \frac{1}{2^{l(t+2)}}.$$

Putting $c_1 = 2^{l^2(t+2)}$, $c_2 = 2^{l(t+2)}$ and $c_3 = \alpha$, we have $4c_1c_2c_3 \leq 2^{2+l^2(t+2)+l(t+2)+(t+1)} \leq 2^{(l+1)^2(t+2)}$, $2c_2c_3 \leq 2^{1+l(t+2)+t+1} = 2^{(l+1)(t+2)}$ and $c_1c_2c_3$
$\leq 2^{(l+1)^2(t+2)}/4 \leq 2^{(2^{t+1})^2(t+2)}/4 \leq 2^{2^{3t}}/8 \leq 2^{2^{(1/2)\log\log m}}/8 = 2^{\sqrt{\log m}}/8$
$< (\log m)^{\sqrt{\log m}}/8 = s^{1/3}/8$. Thus by Claim 10,
$$\Pr_{L_{l+1}\in\mathcal{L}_{l+1}}\left[\Pr_{u\in O_{L_{l+1}}}\left[f_{i_1}(u) = \cdots = f_{i_{l+1}}(u) = 1\right] \geq \frac{1}{4c_1c_2c_3}\right] \geq \frac{1}{2c_2c_3}$$
holds. Therefore
$$\Pr_{L_{l+1}\in\mathcal{L}_{l+1}}\left[\Pr_{u\in O_{L_{l+1}}}\left[f_{i_1}(u) = \cdots = f_{i_{l+1}}(u) = 1\right] \geq \frac{1}{2^{(l+1)^2(t+2)}}\right] \geq \frac{1}{2c_2c_3} \geq \frac{1}{2^{(l+1)(t+2)}}.$$

This completes the induction step and hence the proof of the claim.

Recalling $\mathcal{L}_\alpha = \{V\}$ and setting k in (4) to α, we have $\Pr_{u\in O_V}[\forall i \in \{1,\ldots,\alpha\}$ $f_i(u) = 1] > 0$. Thus there exists $u \in O_V$ and $u^+ \in \text{CLIQUE}(m,s)^{-1}(1)$ such that $(u, u^+) \in G(\text{CLIQUE}(m,s))$ and $(u, u^+) \notin G(f_i)$ for any $i \in \{1,\ldots,\alpha\}$. This implies $\cup_{i\in\{1,\ldots,\alpha\}} G(f_i) \not\supseteq G(\text{CLIQUE}(m,s))$, a contradiction. This completes the proof. □

References

[AB] N. Alon and R. B. Boppana, "The Monotone Circuit Complexity of Boolean Functions", *Combinatorica*, Vol. 7, No. 1, pp. 1–22, 1987.
[AM] K. Amano and A. Maruoka, "Potential of the Approximation Method", *Proc. 37th FOCS*, pp. 431–440, 1996.
[BNT] R. Beals, T. Nishino and K. Tanaka, "More on the Complexity of Negation-Limited Circuits", *Proc. 27th STOC*, pp. 585–595, 1995.
[Fis] M.J. Fischer, "The Complexity of Negation-Limited Networks–a Brief Survey", *Lecture Notes in Computer Science 33*, Springer-Verlag, Berlin, pp. 71-82, 1974.
[Has] J. Håstad, "Almost Optimal Lower Bounds for Small Depth Circuits", *Proc. 18th STOC*, pp. 6–20, 1986.
[Raz] A.A. Razborov, "Lower Bounds on the Monotone Complexity of Some Boolean Functions", *Soviet Math. Dokl.*, Vol. 281, pp. 798–801, 1985.
[Raz2] A.A. Razborov, "On the Method of Approximations", *Proc. 21st STOC*, pp. 167–176, 1989.
[SW] M. Santha and C. Wilson, "Limiting Negations in Constant Depth Circuits", *SIAM J. Comput.*, Vol. 22, No. 2, pp. 294–302, 1993.

Appendix

Proof (of Claim 10). Let $\mathcal{L}_k^{bad}$ denote the collection of sets $L_k \in \mathcal{L}_k$ with $\Pr_{u\in O_{L_k}}[f_{i_1}(u) = \cdots = f_{i_k}(u) = 1] \geq 1/c_1$. By the assumption of Claim 10, $\Pr_{L_k\in\mathcal{L}_k}[L_k \in \mathcal{L}_k^{bad}] \geq 1/c_2$(A1). Let $u \in O_{L_k}$ be chosen arbitrarily so that $f_{i_1}(u) = \cdots = f_{i_k}(u) = 1$. By the definition of a sensitive graph, any of $G(f_{i_1}),\ldots,G(f_{i_k})$ does not contain an edge from u. Let u^+ be a graph obtained from u by adding an arbitrary edge whose both endpoints are in L_k. Clearly, $(u, u^+) \in G(\text{CLIQUE}(m, s))$. Since $u^+ \leq v_{L_k}$, we have

$$\forall L_k \in \mathcal{L}_k^{bad} \exists u \in O_{L_k} \exists u^+ \leq v_{L_k} \quad (u, u^+) \in \bigcup_{j \in \{1,\ldots,c_3\} \setminus \{i_1,\ldots,i_k\}} G(f_j).$$

Therefore there exists $l \in \{1, \ldots, c_3\} \setminus \{i_1, \ldots, i_k\}$ with $\Pr_{L_k \in \mathcal{L}_k^{bad}}[\exists u \in O_{L_k} \exists u^+ \leq v_{L_k} \ (u, u^+) \in G(f_l)] \geq 1/c_3$. If $(u, u^+) \in G(f_l)$ for $u \in O_{L_k}$, then $f_l(u^+) = 1$, which together with $u^+ \leq v_{L_k}$ implies $f_l(v_{L_k}) = 1$ by the monotonicity of f_l. Thus we can conclude that $\Pr_{L_k \in \mathcal{L}_k}[f_l(v_{L_k}) = 1 \mid L_k \in \mathcal{L}_k^{bad}] \geq 1/c_3$. From this and (A1), there exists $l \in \{1, \ldots, c_3\}/\{i_1, \ldots, i_k\}$ such that $\Pr_{L_k \in \mathcal{L}_k}[L_k \in \mathcal{L}_k^{bad}$ and $f_l(v_{L_k}) = 1] \geq 1/c_2c_3(A2)$. Now we choose an index l arbitrarily satisfying the above inequality and let $i_{k+1} = l$. Letting $\mathcal{L}_k^{target}$ denote a collection of sets $L_k \in \mathcal{L}_k$ such that $L_k \in \mathcal{L}_k^{bad}$ and that $f_{i_{k+1}}(v_{L_k}) = 1$, we have $\Pr_{L_k \in \mathcal{L}_k}[L_k \in \mathcal{L}_k^{target}] \geq 1/c_2c_3$. By a similar argument to the derivation of (2), we have

$$\Pr_{L_{k+1} \in \mathcal{L}_{k+1}} \left[\Pr_{L_k \in \mathcal{L}_k(L_{k+1})} [L_k \in \mathcal{L}_k^{target}] \geq \frac{1}{2c_2c_3} \right] \geq \frac{1}{2c_2c_3}. \tag{5}$$

Now we call a $L_{k+1} \in \mathcal{L}_{k+1}$ *dense* if $\Pr_{L_k \in \mathcal{L}_k(L_{k+1})}[L_k \in \mathcal{L}_k^{target}] \geq 1/2c_2c_3(A3)$ holds, and let $\mathcal{L}_{k+1}^{dense}$ denote a collection of all dense sets in $\mathcal{L}_{k+1}$. Note that $\Pr_{v \in I_{L_{k+1}}}[f_{i_{k+1}}(v) = 1] \geq 1/2c_2c_3$, for any dense $L_{k+1} \in \mathcal{L}_{k+1}^{dense}$. Put $h = 1/2c_2c_3 > 1/m$. Thus by applying Claim 9 to every dense L_{k+1}, we have size $mon(f_{i_{k+1}}) > (1/2m)2^{s^{1/3}/4} > M$ or $\Pr_{u \in O_{L_{k+1}}}[f_{i_{k+1}}(u) = 0] \leq 2/s^{1/3} \leq 1/4c_1c_2c_3$, for any dense L_{k+1}. (We use the assumption $c_1c_2c_3 \leq s^{1/3}/8$ in Claim 10 here.) Since the former contradicts the assumption $size_{mon}(f_{i_{k+1}}) \leq M$, we have $\Pr_{u \in O_{L_{k+1}}}[f_{i_{k+1}}(u) = 0] \leq 1/4c_1c_2c_3(A4)$, for any $L_{k+1} \in \mathcal{L}_{k+1}^{dense}$. By (A3), we have

$$\Pr_{L_k \in \mathcal{L}_k(L_{k+1})} \left[\Pr_{u \in O_{L_k}} [f_{i_1}(u) = \cdots = f_{i_k}(u) = 1] \geq \frac{1}{c_1} \right] \geq \frac{1}{2c_2c_3},$$

which, together with the fact that all O_{L_k}'s are disjoint, implies $\Pr_{u \in O'_k}[f_{i_1}(u) = \cdots = f_{i_k}(u) = 1] \geq 1/2c_1c_2c_3$, where $O'_k = \cup_{L_k \in \mathcal{L}_k(L_{k+1})} O_{L_k}$. It is not difficult to see that $\Pr_{u \in O_{L_{k+1}}}[f_{i_1}(u) = \cdots = f_{i_k}(u) = 1] \geq 1/2c_1c_2c_3(A5)$ holds. By (A4) and (A5), for any $L_{k+1} \in \mathcal{L}_{k+1}^{dense}$, we have $\Pr_{u \in O_{L_{k+1}}}[f_{i_1}(u) = \cdots = f_{i_{k+1}}(u) = 1] \geq 1/4c_1c_2c_3$. Now Claim 10 is straightforward from this and (5). □

On Counting AC^0 Circuits with Negative Constants

Andris Ambainis[1], David Mix Barrington[2], and Hương LêThanh[3]

[1] Computer Science Division, University of California at Berkeley
ambainis@cs.berkeley.edu
[2] Computer Science Department, University of Massachusetts
barring@cs.umass.edu
[3] Laboratoire de Recherche en Informatique, Université de Paris-Sud
huong@lri.fr

Abstract. Continuing the study of the relationship between TC^0, AC^0 and arithmetic circuits, started by Agrawal et al. [1], we answer a few questions left open in this paper. Our main result is that the classes DiffAC^0 and GapAC^0 coincide, under poly-time, log-space, and log-time uniformity. From that we can derive that under logspace uniformity, the following equalities hold:

$$C_=AC^0 = PAC^0 = TC^0.$$

1 Introduction

The study of counting complexity classes was started by the pioneering work of Valiant [16] on the class $\#P$. It consists of functions which associate to a string x the number of accepting computations of an NP-machine on x. A well-known complete problem for this class is the computing of the permanent of an integer matrix. The class $\#L$ was defined later analogously with respect to NL-computation [3,18,14]. Each of these classes can be defined equivalently either by counting the number of accepting subtrees of the corresponding class of uniform circuits, or by computing functions via the arithmetized versions of these circuit classes [17,18,14].

These counting classes contain functions which take only natural numbers as values. Counting classes computing functions which might also take negative values were introduced via the so-called Gap-classes. The class Gap$P$ was defined by Fenner, Fortnow and Kurtz [8], and the class Gap$L$ was introduced by analogy in [18]. For both classes there are two equivalent definitions. They can either be defined as the set of functions computable as the difference of two functions from the corresponding counting class, or as functions which are computed by the corresponding arithmetic circuits augmented by the constant -1.

Recently, counting classes related to circuit model based language classes were also defined. The class $\#NC^1$ was introduced by Caussinus et al. in [7], and the class $\#AC^0$ by Agrawal et al. in [1]. The corresponding Gap-classes were also defined in these papers. The two definitions for GapNC^1 are again easily seen to be equivalent, the principal reason for this being the fact that the PARITY language is in NC^1. The same argument fails to work for the two definitions of GapAC^0 since PARITY can not be computed in AC^0. In fact, one of the problems left open in [1] was the exact relationship between these two classes (GapAC^0 and DiffAC^0 in the notation of the paper).

The main result of our paper is that GapAC^0 and DiffAC^0 actually coincide. We will prove this in the log-time uniform setting, thus showing that it also holds in the log-space uniform and P-uniform settings. As a consequence of this result, we can simplify the relationships among the various boolean complexity classes defined in terms of these arithmetic classes, resolving several open problems of [1]. For example, under log-space uniformity, the classes TC^0, $C_=AC^0$ and PAC^0 are all equal. (This result was proven in [1] only under P-uniformity.) Under log-time uniformity, we have the new series of containments $TC^0 \subseteq C_=AC^0 \subseteq PAC^0$.

2 Preliminaries

Following [1], we will consider three notions of uniformity for circuit families. A family $\{C_n\}_{n\geq 1}$ of circuits is said to be *P uniform* (*log-space uniform*) if there exists a Turing machine $\mathcal{M}$ and a polynomial $T(n)$ (a function $S(n) = O(\log n)$) such that $\mathcal{M}$, given n in unary, produces a description of the circuit C_n within time $T(n)$ (using space $S(n)$).

The definition of *log-time uniformity* [5] is a bit more complicated. A family $\{C_n\}$ of circuits is said to be *log-time uniform* if there is a Turing machine that can answer queries in its *direct connection language* in time $O(\log n)$. The direct connection language consists of all tuples $\langle i, j, t, y \rangle$ where i is the number of a gate of C_n, j is the number of one of its children (or the number of the referenced input x_j, if gate i is an input gate), t gives the type of gate i, and y is any string of length n. The log-time Turing machine has a read-only random-access input tape, so that it can determine the length of its input by binary search. As shown in [5] and [4], log-time uniform circuits are equivalent in power to circuits given by *first-order formulas* with variables for input positions and atomic predicates for order, equality, and binary arithmetic on these variables.

By De Morgan's law, it is sufficient to consider circuits in which negations occur only on the input level, and all the other gates are OR- or AND-gates. For such circuits the notion of *subtree* was introduced in [17]. Let C be a Boolean circuit and let $T(C)$ be the circuit obtained from C by duplicating all gates whose fan-out is greater than one, until the underlying graph of $T(C)$ is a tree. Let x be an input of C. A subtree H of C on input x is a subtree of $T(C)$ defined as follows: the output gate of the circuit $T(C)$ belongs to H; for each non-input gate g already belonging to H, if g is an AND-gate then all its input gates belong to H; if g is an OR-gate then exactly one of its input gates belongs to H. A subtree on input x is an *accepting* subtree if all its leaves evaluate to 1.

We now define how to *arithmetize* a Boolean circuit. The input variables $x_1, x_2, \ldots, x_n$ take as values the natural numbers 0 or 1, and the negated input variables $\bar{x}_i$ take the values $1 - x_i$. Each OR-gate is replaced by a $+$-gate and each AND-gate by a $\times$-gate. It was shown in [17] that the number of accepting subtrees of the circuit C on input $(x_1, \bar{x}_1, \ldots, x_n, \bar{x}_n)$ is equal to the output of its arithmetized circuit on the same input.

Note that the output of such an arithmetic circuit is always non-negative. If the constant -1 is allowed in the circuit, functions with negative values can also be computed.

Let $\#\mathcal{C}$ be a class of functions from $\{0,1\}^*$ to $\mathbb{N}$. By definition, $\#\mathcal{C} - \#\mathcal{C}$ is the class of functions expressible as the difference of two functions from $\#\mathcal{C}$.

Definition 1 *[1] Let U be any of three uniformity definitions: P, log-space, or log-time. For any $k > 0$, U-uniform $\#AC_k^0$ ($GapAC_k^0$) is the class of functions computed by depth k, polynomial size, U-uniform circuits with $+,\times$-gates having unbounded fan-in, where inputs of the circuits are from $\{0, 1, x_i, 1 - x_i\}$ (from $\{0, 1, -1, x_i, 1 - x_i\}$) and $x_i \in \{0,1\}$ for all $i = 1, \ldots, n$. Let*

$$\#AC^0 = \bigcup_{k>0} \#AC_k^0,$$
$$DiffAC^0 = \#AC^0 - \#AC^0,$$
$$GapAC^0 = \bigcup_{k>0} GapAC_k^0.$$

It is easy to see that under all three uniformity conditions, Diff$AC^0 \subseteq$ GapAC^0. A very natural question, left open by Agrawal et al. [1], is whether DiffAC^0=GapAC^0.

Let PARITY denote the usual 0-1 parity function which computes the sum of its inputs modulo 2, and let F-PARITY be its *Fourier representation*, that is F-PARITY $(x_1, \ldots, x_n) = \prod_{i=1}^{n}(1 - 2x_i)$ (this function takes value 1 or -1). It is clear that F-PARITY is in GapAC^0. Another open question was whether this function belongs to DiffAC^0.

In the next sections we will give a positive answer to both questions. By a $\#AC^0$ circuit we mean an arithmetized AC^0 circuit in the above sense. Throughout this paper we will need the following fact.

Fact 1 *For each integer N of m bits there exists a $\#AC^0$ circuit with $O(m^2)$ gates, which on input 1^m computes N. This circuit is log-time uniform if the binary representation of N is given as input.*

Proof. Let $N = N_{m-1}N_{m-2}\ldots N_1N_0$ be the binary representation of N. The formula

$$N = \sum_{i=0}^{m-1} N_i \cdot 2^i = \sum_{i=0}^{m-1} N_i \cdot \underbrace{(1+1) \cdot (1+1) \cdots (1+1)}_{i \text{ times}}$$

will give a $\#AC^0$ circuit of depth 3 and size $O(m^2)$ computing N, using the circuit C_r of depth 2 and size $(3r + 1)$ introduced in [1], whose number of accepting subtrees on input 1^{2r} is 2^r. Note that the family of circuits $\{C_r\}_{r \geq 1}$ is both P-uniform and logspace-uniform. A log-time Turing machine can use its random access input tape to reference the single bit of N needed to answer any particular query. □

3   DiffAC^0 = GapAC^0

Theorem 1 *DiffAC^0 = GapAC^0 for log-time uniform circuits (and hence for P-uniform and log-space uniform circuits as well).*

Proof. It is enough to show that Gap$AC^0 \subseteq$ DiffAC^0 under each uniformity condition. We will first describe our general construction, and then show that it can be carried out preserving log-time uniformity (and hence the other two conditions as well). Given an arithmetic circuit C for the inputs of length n, which uses the constant -1, we will construct two other arithmetic circuits A and B, each with only positive constants, such that $C(x) = A(x) - B(x)$ for all input x of length n. We will show, by induction on the depth of C, that for each gate g, we can build two $\#AC^0$ circuits A^g and B^g such that $g(x) = A^g(x) - B^g(x)$.

The construction is trivial for gates of depth 0. Consider now a gate g of depth $d \geq 1$ having as input gates $g_1, g_2, \ldots, g_m$. Suppose that for each $i = 1, \ldots, m$, we have already constructed two $\#AC^0$ circuits A_i^g and B_i^g satisfying $g_i(x) = A_i^g(x) - B_i^g(x)$. If g is a $+$-gate, the construction of A^g and B^g is straightforward. The interesting case is when g is a $\times$-gate. For ease of notation we set $a_i = A_i^g(x)$ and $b_i = B_i^g(x)$.

Without any negative constants, we can compute the product $\prod_{i=1}^m (a_i + b_i)$ which is of no immediate help in getting $\prod_{i=1}^m (a_i - b_i)$. The key idea is to notice that we can also compute some other products of positive linear combinations of the a_i's and b_i's as well, such as $\prod_{i=1}^m (a_i + 2b_i)$, and use linear algebra to solve for the combination we want.

Specifically, we will find a sequence of integers $c_1(m), c_2(m), \ldots, c_{m+1}(m)$, each of which depends only on m and has $O(m)$ bits, such that

$$\prod_{i=1}^m (a_i - b_i) \tag{1}$$

$$= \sum_{k=1}^{m+1} c_k(m) \cdot \prod_{i=1}^m (a_i + k \cdot b_i) \tag{2}$$

$$= \sum_{k:c_k(m)>0} c_k(m) \cdot \prod_{i=1}^m (a_i + k \cdot b_i) - \sum_{k:c_k(m)<0} (-c_k(m)) \cdot \prod_{i=1}^m (a_i + k \cdot b_i). \tag{3}$$

We must show that this sequence of integers exists, and that each $c_k(m)$ is computable by a log-time uniform $\#AC^0$ circuit. Then the log-time uniform circuits A^g and B^g can easily be constructed to calculate the two sums in expression (3), and the difference will be the value of g as desired.

References

1. M. Agrawal, E. Allender, S. Datta, *On TC^0, AC^0 and Arithmetic Circuits*. In Proceedings of the 12th Annual IEEE Conference on Computational Complexity, pp:134–148, 1997.
2. E. Allender, R. Beals, M. Ogihara, *The complexity of matrix rank and feasible systems of linear equations*. In Proceedings of the 28th ACM Symposium on Theory of Computing (STOC), pp:161–167, 1996.
3. C. Álvarez, B. Jenner, *A very hard logspace counting class*. Theoretical Computer Science, 107:3–30, 1993.
4. D. A. M. Barrington, N. Immerman, *Time, Hardware, and Uniformity*. In L. A. Hemaspaandra and A. L. Selman, eds., *Complexity Theory Retrospective II*, Springer Verlag, pp:1–22, 1997.
5. D. A. M. Barrington, N. Immerman, and H. Straubing, *On Uniformity Within NC^1*. Journal of Computer and System Science, 41:274–306, 1990.
6. P. Beame, S. Cook, H. J. Hoover, *Log depth circuits for division and related problems*. SIAM Journal on Computing, 15:994–1003, 1986.
7. H. Caussinus, P. McKenzie, D. Thérien, H. Vollmer, *Nondeterministic NC^1*. In Proceedings of the 11th Annual IEEE Conference on Computational Complexity, pp:12–21, 1996.
8. S. A. Fenner, L. J. Fortnow, S. A. Kurtz, *Gap-definable counting classes*. Journal of Computer and System Science, 48(1):116–148, 1995.
9. J. Köbler, U. Schöning, J. Torán, *On counting and approximation*. Acta Informatica, 26:363–379, 1989.
10. B. Litow, *On iterated integer product*. Information Processing Letters, 42(5):269–272, 1992.
11. M. Mahajan, V. Vinay, *Determinant: Combinatorics, Algorithms and Complexity*. In Proceedings of SODA'97. ftp://ftp.eccc.uni-trier.de/pub/eccc/reports/1997/TR97-036/index.html
12. A. A. Razborov, *Lower bound on size of bounded depth networks over a complete basis with logical addition*. Mathematicheskie Zametli, 41:598–607, 1987. English translation in Mathematical Notes of the Academy of Sciences of the USSR, 41:333–338, 1987.
13. R. Smolensky, *Algebraic methods in the theory of lower bounds for Boolean circuit complexity*. In Proceedings of the 19th ACM Symposium on the Theory of Computing (STOC), pp:77–82, 1987.
14. S. Toda, *Classes of arithmetic circuits capturing the complexity of computing the determinant*. IEICE Transactions, Informations and Systems, E75-D:116–124, 1992.
15. S. Toda, *Counting problems computationally equivalent to the determinant*. Manuscript.
16. L. Valiant, *The complexity of computing the permanent*. Theoretical Computer Science, 8:189–201, 1979.
17. H. Venkateswaran, *Circuit definitions of non-deterministic complexity classes*. SIAM Journal on Computing, 21:655–670, 1992.
18. V. Vinay, *Counting auxiliary pushdown automata and semi-unbounded arithmetic circuits*. In Proceedings of the 6th IEEE Structure in Complexity Theory Conference, pp:270–284, 1991.

A Second Step Towards Circuit Complexity-Theoretic Analogs of Rice's Theorem*

Lane A. Hemaspaandra[1] and Jörg Rothe[2]

[1] Dept. of Computer Science, University of Rochester, Rochester, NY 14627, USA
[2] Inst. für Informatik, Friedrich-Schiller-Universität Jena, 07740 Jena, Germany

Abstract. Rice's Theorem states that every nontrivial language property of the recursively enumerable sets is undecidable. Borchert and Stephan [BS97] initiated the search for circuit complexity-theoretic analogs of Rice's Theorem. In particular, they proved that every nontrivial counting property of circuits is UP-hard, and that a number of closely related problems are SPP-hard.

The present paper studies whether their UP-hardness result itself can be improved to SPP-hardness. We show that their UP-hardness result cannot be strengthened to SPP-hardness unless unlikely complexity class containments hold. Nonetheless, we prove that every P-constructibly bi-infinite counting property of circuits is SPP-hard. We also raise their general lower bound from unambiguous nondeterminism to constant-ambiguity nondeterminism.

1 Introduction

Rice's Theorem [Ric53,Ric56] states that every nontrivial language property of the recursively enumerable sets is undecidable.

Theorem 1 (Rice's Theorem, Version I). *Let $\mathcal{A}$ be a nonempty proper subset of the class of recursively enumerable sets. Then the following problem is undecidable: Given a Turing machine M, is $L(M) \in \mathcal{A}$?*

In fact, the theorem can be stated in the following more provocative form ([Ric53], see [BS97]).

Theorem 2 (Rice's Theorem, Version II). *Let $\mathcal{A}$ be a nonempty proper subset of the class of recursively enumerable sets. Then either the halting problem or its complement many-one reduces to the problem: Given a Turing machine M, is $L(M) \in \mathcal{A}$?*

This theorem conveys quite a bit of information about the nature of programs and their semantics. Programs are completely nontransparent. One can (in general) decide

---

* Supported in part by grants NSF-CCR-9322513 and NSF-INT-9513368/DAAD-315-PRO-foab, and a NATO Postdoctoral Science Fellowship from the DAAD ("Gemeinsames Hochschulsonderprogramm III von Bund und Ländern"). Work done in part while the first author was visiting FSU Jena, and while the second author was visiting the Univ. of Rochester. Email: lane@cs.rochester.edu, rothe@informatik.uni-jena.de. Current address of second author: Dept. of Comp. Sci., Univ. of Rochester, Rochester, NY 14627, USA.

nothing—emptiness, nonemptiness, infiniteness, etc.—about the languages of given programs other than the trivial fact that each accepts some language and that language is a recursively enumerable language.[1] Recently, Kari [Kar94] has proven, for cellular automata, an analog of Rice's Theorem: All nontrivial properties of limit sets of cellular automata are undecidable.

A bold and exciting paper of Borchert and Stephan [BS97] proposes and initiates the search for *complexity-theoretic* analogs of Rice's Theorem. Borchert and Stephan note that Rice's Theorem deals with properties of programs, and they suggest as a promising complexity-theoretic analog properties of boolean circuits. In particular, they focus on counting properties of circuits. Let $\mathbb{N}$ denote $\{0, 1, 2, \ldots\}$. Boolean functions are functions that for some n map $\{0,1\}^n$ to $\{0,1\}$. Circuits built over boolean gates (and encoded in some standard way—in fact, for simplicity of expression, we will often treat a circuit and its encoding as interchangeable) are ways of representing boolean functions. As Borchert and Stephan point out, the parallel is a close one. Programs are concrete objects that correspond in a many-to-one way with the semantic objects, languages. Circuits (encoded into Σ^*) are concrete objects that correspond in a many-to-one way with the semantic objects, boolean functions. Given an arity n circuit c, $\#(c)$ denotes under how many of the 2^n possible input patterns c evaluates to 1.

Definition 1. *1.* [BS97] *Each $A \subseteq \mathbb{N}$ is a* counting property of circuits. *If $A \neq \emptyset$, we say it is a* nonempty property, *and if $A \neq \mathbb{N}$, we say it is a* proper property.

2. [BS97] *Let A be a counting property of circuits. The* counting problem for A, Counting(A), *is the set of all circuits c such that $\#(c) \in A$.*

3. (see [GJ79]) *For each complexity class $\mathcal{C}$ and each set $B \subseteq \Sigma^*$, we say B is $\mathcal{C}$-hard if $(\forall L \in \mathcal{C})\,[L \leq^{\mathrm{P}}_{\mathrm{T}} B]$, where as is standard $\leq^{\mathrm{P}}_{\mathrm{T}}$ denotes polynomial-time Turing reducibility.*

4. (following usage of [BS97]) *Let A be a counting property and let $\mathcal{C}$ be a complexity class. By convention, we say that counting property A is $\mathcal{C}$-hard if the counting problem for A,* Counting(A), *is $\mathcal{C}$-hard. (Note in particular that by this we do not mean $\mathcal{C} \subseteq \mathrm{P}^A$—we are speaking just of the complexity of A's counting problem.)*

For succinctness and naturalness, and as it introduces no ambiguity here, we throughout this paper use "counting" to refer to what Borchert and Stephan originally referred to as "absolute counting." For completeness, we mention that their sets Counting(A) are not entirely new: For each A, Counting(A) is easily seen (in light of the fact that circuits can be parsimoniously simulated by Turing machines, which themselves, as per the references cited in the proof of Theorem 4 (see the full version of this paper [HR97]), can be parsimoniously transformed into boolean formulas) to be

---

[1] One must stress that Rice's Theorem refers to the languages accepted by the programs (Turing machines) rather than to machine-based actions of the programs (Turing machines)—such as whether they run for at least seven steps on input 1776 (which is decidable) or whether for some input they infinite loop (which is not decidable, but Rice's Theorem does not speak directly to this issue, that is, Rice's Theorem does not address the computability of the set $\{M \mid \text{there is some input } x \text{ on which } M(x) \text{ infinite loops}\}$).

many-one equivalent to the set, known in the literature as SAT_A or A-SAT, $\{f \mid$ the number of satisfying assignments to boolean formula f is an integer contained in the set $A\}$ [GW87,CGH$^+$89]. Thus, Counting(A) inherits the various properties that the earlier papers on SAT_A established for SAT_A, such as completeness for certain counting classes. We will at times draw on this earlier work to gain insight into the properties of Counting(A).

The results of Borchert and Stephan that led to the research reported on in the present paper are the following. Note that Theorem 3 is a partial analog of Theorem 2, and Corollary 1 is a partial analog of Theorem 1.

Theorem 3. [BS97] *Let A be a nonempty proper subset of $\mathbb{N}$. Then one of the following three classes is $\leq_{\mathrm{m}}^{\mathrm{P}}$-reducible to* Counting($A$): *NP, coNP, or* UP $\oplus$ coUP.

Corollary 1. [BS97] *Every nonempty proper counting property of circuits is UP-hard.*

Borchert and Stephan's paper proves a number of other results—regarding an artificial existentially quantified circuit type yielding NP-hardness, definitions and results about counting properties over rational numbers and over $\mathbb{Z}$, and so on—and we highly commend their paper to the reader. They also give a very interesting motivation. They show that, in light of the work of Valiant and Vazirani, any nontrivial counting property of circuits is hard for either NP or coNP, with respect to *randomized reductions*. Their paper and this one seek to find to what extent or in what form this behavior carries over to deterministic reductions.

The present paper makes the following contributions. First, we extend the above-stated results of Borchert and Stephan, Theorem 3 and Corollary 1. Regarding the latter, from the same hypothesis as their Corollary 1 we derive a stronger lower bound—$\mathrm{UP}_{\mathcal{O}(1)}$-hardness. That is, we raise their lower bound from *unambiguous* nondeterminism to *low-ambiguity* nondeterminism. Second, we show that our improved lower bound cannot be further strengthened to SPP-hardness unless an unlikely complexity class containment—SPP $\subseteq \mathrm{P}^{\mathrm{NP}}$—occurs. Third, we nonetheless under a very natural hypothesis raise the lower bound on the hardness of counting properties to SPP-hardness. The natural hypothesis strengthens the condition on the counting property to require not merely that it is nonempty and proper, but also that it is infinite and coinfinite in a way that can be certified by polynomial-time machines.

2 The Complexity of Counting Properties of Circuits

All the notations and definitions in this paragraph are standard in the literature. Fix the alphabet $\Sigma = \{0, 1\}$. FP denotes the class of polynomial-time computable functions from Σ^* to Σ^*. Given any two sets $A, B \subseteq \Sigma^*$, we say A polynomial-time many-one reduces to B ($A \leq_{\mathrm{m}}^{\mathrm{P}} B$) if $(\exists f \in \mathrm{FP})\,(\forall x \in \Sigma^*)\,[x \in A \iff f(x) \in B]$. For each set A, $\|A\|$ denotes the number of elements in A. The length of each string $x \in \Sigma^*$ is denoted by $|x|$. We use DPTM (respectively, NPTM) as a shorthand for deterministic polynomial-time Turing machine (nondeterministic polynomial-time Turing machine). Turing machines and their languages (with or without oracles) are denoted as

A Second Step Towards Circuit Complexity-Theoretic Analogs of Rice's Theorem

is standard, as are complexity classes (with or without oracles), e.g., M, M^A, $L(M)$, $L(M^A)$, P, and P^A. We allow both languages and functions to be used as oracles. In the latter case, the model is the standard one, namely, when a query q is asked to a function oracle f the answer is $f(q)$. For each $k \in \mathbb{N}$, a "$[k]$" denotes a restriction of at most k oracle questions (in a sequential—i.e., "adaptive" or "Turing"—fashion). For example, $\mathrm{P}^{\mathrm{FP}[2]}$ denotes $\{L \mid (\exists \mathrm{DPTM}\ M)(\exists f \in \mathrm{FP})\,[L = L(M^f) \wedge (\forall x \in \Sigma^*)\,[M^f(x)$ makes at most two oracle queries$]]\}$, which happens to be merely an ungainly way of describing the complexity class P. A "$[\mathcal{O}(1)]$" denotes that, for some constant k, a "$[k]$" restriction holds.

We will define, in a uniform way via counting functions, some standard ambiguity-limited classes and counting classes. To do this, we will take the standard "#" operator ([Tod91] for the concept and [Vol94] for the notation, see the discussion in [HV95]) and will make it flexible enough to describe a variety of types of counting functions that are well-motivated by existing language classes. In particular, we will add a general restriction on the maximum value it can take on. (For the specific case of a polynomial restriction such an operator, $\#_{\mathrm{few}}$, was already introduced by Hemaspaandra and Vollmer [HV95], see below.)

Definition 2. *For each function $g : \mathbb{N} \to \mathbb{N}$ and each class $\mathcal{C}$, define $\#_g \cdot \mathcal{C} = \{f : \Sigma^* \to \mathbb{N} \mid (\exists L \in \mathcal{C})\,(\exists\ \mathrm{polynomial}\ s)\,(\forall x \in \Sigma^*)\,[f(x) \leq g(|x|) \wedge \|\{y \mid |y| = s(|x|) \wedge \langle x, y \rangle \in L\}\| = f(x)]\}$.*

Note that for the very special case of $\mathcal{C} = \mathrm{P}$, which is the case of importance in the present paper, this definition simply yields classes that speak about the number of accepting paths of Turing machines that obey some constraint on their number of accepting paths. In particular, the following clearly holds for each g: $\#_g \cdot \mathrm{P} = \{f : \Sigma^* \to \mathbb{N} \mid (\exists\ \mathrm{NPTM}\ N)\,(\forall x \in \Sigma^*)\,[N(x)\text{ has exactly } f(x) \text{ accepting paths and } f(x) \leq g(|x|)]\}$.

In using Definition 2, we will allow a bit of informality regarding describing the functions g. For example, we will write $\#_1$ when formally we should write $\#_{\lambda x.1}$, and so on in similar cases. Also, we will now define some versions of the $\#_g$ operator that focus on collections of bounds of interest to us.

Definition 3. *1. For each class $\mathcal{C}$, $\#_{\mathrm{const}} \cdot \mathcal{C} = \{f : \Sigma^* \to \mathbb{N} \mid (\exists k \in \mathbb{N})\,[f \in \#_k \cdot \mathcal{C}]\}$. 2. [HV95] For each class $\mathcal{C}$, $\#_{\mathrm{few}} \cdot \mathcal{C} = \{f : \Sigma^* \to \mathbb{N} \mid (\exists\ \mathrm{polynomial}\ s)\,[f \in \#_s \cdot \mathcal{C}]\}$.*

Definition 4. *1. [Val79] $\#\mathrm{P} = \{f : \Sigma^* \to \mathbb{N} \mid (\exists\ \mathrm{NPTM}\ N)\,(\forall x \in \Sigma^*)\,[N(x)\text{ has exactly } f(x) \text{ accepting paths}]\}$. 2. [Val76] $\mathrm{UP} = \{L \mid (\exists f \in \#_1 \cdot \mathrm{P})\,(\forall x \in \Sigma^*)\,[x \in L \iff f(x) > 0]\}$. 3. ([Bei89], see also [Wat88]) For each $k \in \mathbb{N} - \{0\}$, $\mathrm{UP}_{\leq k} = \{L \mid (\exists f \in \#_k \cdot \mathrm{P})\,(\forall x \in \Sigma^*)\,[x \in L \iff f(x) > 0]\}$. 4. ([HZ93], see also [Bei89]) $\mathrm{UP}_{\mathcal{O}(1)} = \{L \mid (\exists f \in \#_{\mathrm{const}} \cdot \mathrm{P})\,(\forall x \in \Sigma^*)\,[x \in L \iff f(x) > 0]\}$. (Equivalently, $\mathrm{UP}_{\mathcal{O}(1)} = \bigcup_{k \geq 1} \mathrm{UP}_{\leq k}$.) 5. [All86,AR88] $\mathrm{FewP} = \{L \mid (\exists f \in \#_{\mathrm{few}} \cdot \mathrm{P})\,(\forall x \in \Sigma^*)\,[x \in L \iff f(x) > 0]\}$. 6. [CH90] $\mathrm{Few} = \mathrm{P}^{(\#_{\mathrm{few}} \cdot \mathrm{P})[1]}$. 7. $\mathrm{Const} = \mathrm{P}^{(\#_{\mathrm{const}} \cdot \mathrm{P})[\mathcal{O}(1)]}$.[2] 8. [FFK94,OH93] $\mathrm{SPP} = \{L \mid (\exists f \in \#\mathrm{P})\,(\exists g \in \mathrm{FP})\,(\forall x \in \Sigma^*)\,[(x \notin L \iff f(x) = 2^{|g(x)|}) \wedge (x \in L \iff f(x) = 2^{|g(x)|} + 1)]\}$.*

---

[2] As noted in the proof of Theorem 4 (see the full version of this paper [HR97]), $\mathrm{P}^{(\#_{\mathrm{const}} \cdot \mathrm{P})[\mathcal{O}(1)]} = \mathrm{P}^{(\#_{\mathrm{const}} \cdot \mathrm{P})[1]}$. Thus, the definition of Const is more analogous to the definition of Few than one might realize at first.

It is well-known that $UP = UP_{\leq 1} \subseteq UP_{\leq 2} \subseteq \cdots \subseteq UP_{\mathcal{O}(1)} \subseteq FewP \subseteq Few \subseteq SPP$ (the final containment is due to Köbler et al. [KSTT92], see also [FFK94] for a more general result), and clearly $UP_{\mathcal{O}(1)} \subseteq Const \subseteq Few$. SPP plays a central role in much of complexity theory (see [For97]), and in particular is closely linked to the closure properties of #P [OH93]. Regarding relationships with the polynomial hierarchy, $P \subseteq UP \subseteq FewP \subseteq NP$, and $Few \subseteq P^{FewP}$ (so $Few \subseteq P^{NP}$). It is widely suspected that $SPP \not\subseteq PH$ (where PH denotes the polynomial hierarchy), though this is an open research question. UP, $UP_{\mathcal{O}(1)}$, and FewP are tightly connected to the issue of whether one-way functions exist [GS88,AR88,HZ93], and Watanabe [Wat88] has shown that $P = UP$ if and only if $P = UP_{\mathcal{O}(1)}$.

Intuitively, UP captures the notion of unambiguous nondeterminism, FewP allows polynomially ambiguous nondeterminism and, most relevant for the purposes of the present paper, $UP_{\mathcal{O}(1)}$ allows constant-ambiguity nondeterminism. Corollary 2 raises the UP lower bound of Borchert and Stephan (Corollary 1) to a $UP_{\mathcal{O}(1)}$ lower bound. This is obtained via the even stronger bound provided by Theorem 4, which itself extends Theorem 3.

Theorem 4. *Let A be a nonempty proper subset of $\mathbb{N}$. Then one of the following three classes is $\leq_m^p$-reducible to* Counting(A): *NP, coNP, or Const.*

Corollary 2. *Every nonempty proper counting property of circuits is $UP_{\mathcal{O}(1)}$-hard (indeed, is even $UP_{\mathcal{O}(1)}$-$\leq_{1\text{-}tt}^p$-hard).*

Our proof, which can be found in the full version of this paper [HR97], applies a constant-setting technique that Cai and Hemaspaandra (then Hemachandra) [CH90] used to prove that $FewP \subseteq \oplus P$, and that Köbler et al. [KSTT92] extended to show that $Few \subseteq SPP$.

Corollary 2 raised the lower bound of Corollary 1 from UP to $UP_{\mathcal{O}(1)}$. It is natural to wonder whether the lower bound can be raised to SPP. This is especially true in light of the fact that Borchert and Stephan obtained SPP-hardness results for their notions of "counting problems over $\mathbb{Z}$" and "counting problems over the rationals"; their UP-hardness result for standard counting problems (i.e., over $\mathbb{N}$) is the short leg of their paper. However, we note that extending the hardness lower bound to SPP under the same hypothesis seems unlikely. Let BH denote the boolean hierarchy [CGH+88]. It is well-known that $NP \subseteq BH \subseteq P^{NP} \subseteq PH$.

Proposition 1. *If $A \subseteq \mathbb{N}$ is finite or cofinite, then* Counting(A) $\in BH$.

This result needs no proof, as it follows easily from [CGH+89, Lemma 3.1 and Theorem 3.1.1(a)] (those results exclude the case $0 \in A$ but their proofs clearly apply also to that case) or from [GW87, Theorem 15], in light of the relationship between Counting(A) and SAT$_A$ mentioned earlier in the present paper. Similarly, from earlier work one can conclude that, though for all finite and cofinite A it holds that Counting(A) is in the boolean hierarchy, these problems are not good candidates for complete sets for that hierarchy's higher levels—or even its second level. In particular, from the approach of the theorem and proof of [CGH+89, Theorem 3.1.2] (see also [GW87, Theorem 15]) it is not too hard to see that $(\exists B) [(\forall \text{ finite } A) [\text{Counting}(A)$ is not $\leq_m^{p,B}$-hard for $\text{NP}^B] \wedge (\forall \text{ cofinite } A) [\text{Counting}(A)$ is not $\leq_m^{p,B}$-hard for $\text{coNP}^B]]$.

In light of the fact that SPP-hardness means SPP-$\leq_T^p$-hardness, the bound of Proposition 1 yields the following result (one can equally well state the stronger claim that no finite or cofinite counting property of circuits is SPP- $\leq_m^p$ -hard unless SPP $\subseteq$ BH).

Corollary 3. *No finite or cofinite counting property of circuits is* SPP-*hard unless* SPP $\subseteq$ P^{NP}.

Though we have not in this paper discussed models of relativized circuits and relativized formulas to allow this work to relativize cleanly (and we do not view this as an important issue), we mention in passing that there is a relativization in which SPP is not contained in P^{NP} (indeed, relative to which SPP strictly contains the polynomial hierarchy) [For97].

Corollary 3 makes it clear that if we seek to prove the SPP-hardness of counting properties, we must focus only on counting properties that are simultaneously infinite and coinfinite. Even this does not seem sufficient. The problem is that there are infinite, coinfinite sets having "gaps" so huge as to make the sets have seemingly no interesting usefulness at many lengths (consider, e.g., the set $\{i \mid (\exists j) [i = \texttt{AckermannFunction}(j,j)]\}$). Of course, in a recursion-theoretic context this would be no problem, as a Turing machine in the recursion-theoretic world is free from time constraints and can simply run until it finds the desired structure (which we will see is a boundary event). However, in the world of complexity theory we operate within (polynomial) time constraints. Thus, we consider it natural to add a hypothesis, in our search for an SPP-hardness result, requiring that infiniteness and coinfiniteness of a counting property be constructible in a polynomial-time manner.

Recall that a set of nonnegative integers is infinite exactly if it has no largest element. We will say that a set is P-constructibly infinite if there is a polynomial-time function that yields elements at least as long as each given (unary) input length.

Definition 5. *Let* $B \subseteq \Sigma^*$. *We say that* B *is* P-constructibly infinite *if* $(\exists f \in \text{FP}) (\forall n \in \mathbb{N}) [f(0^n) \in B \wedge |f(0^n)| \geq n]$.

Let us adopt the standard bijection between Σ^* and $\mathbb{N}$—the natural number i corresponds to the lexicographically $(i+1)$st string in Σ^*: $0 \leftrightarrow \epsilon, 1 \leftrightarrow 0, 2 \leftrightarrow 1, 3 \leftrightarrow 00$, etc. If $A \subseteq \mathbb{N}$, we say that A is P-constructibly infinite if A, viewed as a subset of Σ^* via this bijection, is P-constructibly infinite according to Definition 5. If $A \subseteq \Sigma^*$ and $\overline{A}$ (or $A \subseteq \mathbb{N}$ and $\mathbb{N}-A$) are P-constructibly infinite, we will say that A is P-*constructibly bi-infinite*.

Note that some languages that are infinite (respectively, bi-infinite) are not P-constructibly infinite (respectively, bi-infinite), e.g., languages with huge gaps between successive elements.

Borchert and Stephan [BS97] also study "counting problems over the rationals," and in this study they use a root-finding-search approach to establishing lower bounds. In the following proof, we apply this type of approach (by which we mean the successive interval contraction of the same flavor used when trying to capture the root of a function on $[a, b]$ when one knows initially that, say, $f(a) > 0$ and $f(b) < 0$) to counting problems (over $\mathbb{N}$). In particular, we use the P-constructibly bi-infinite hypothesis to "trap" a boundary event of $\overline{A}$.

Theorem 5. *Each P-constructibly bi-infinite counting property of circuits is* SPP-*hard.*

Proof: Let $A \subseteq \mathbb{N}$ be any P-constructibly bi-infinite counting property of circuits. Let L be any set in SPP. Since $L \in$ SPP, there are functions $f \in \#$P and $g \in$ FP such that, for each $x \in \Sigma^*$: $(x \in L \iff f(x) = 2^{|g(x)|} + 1) \land (x \notin L \iff f(x) = 2^{|g(x)|})$. Let h and $\overline{h}$ be FP functions certifying that A and $\overline{A}$ are P-constructibly infinite. We will describe a DPTM N that $\leq_T^P$-reduces L to Counting(A). For clarity, let $\widehat{w}$ henceforth denote the natural number that in the above bijection between $\mathbb{N}$ and Σ^* corresponds to the string w. For convenience, we will sometimes view A as a subset of $\mathbb{N}$ and sometimes as a subset of Σ^* (and in the latter case we implicitly mean the transformation of A to strings under the above-mentioned bijection).

Since clearly $A \leq_m^P$ Counting(A),[3] we for convenience will sometimes informally speak as if the set A (viewed via the bijection as subset of Σ^*) is an oracle of the reduction. Formally, when we do so, this should be viewed as a shorthand for the complete $\leq_T^P$-reduction that consists of the $\leq_T^P$-reduction between L and A followed by the $\leq_m^P$-reduction between A and Counting(A).

We now describe N. On input x, $|x| = n$, N proceeds in three steps. (As a shorthand, we will consider x fixed and will write N rather than $N^{\text{Counting}(A)}(x)$.)

(1) N runs $\overline{h}$ and h on suitable inputs to find certain sufficiently large strings in $\overline{A}$ and A. In particular, let $\overline{h}(0^{|g(x)|+1}) = y$. So we have $y \notin A$ and $|y| \geq |g(x)| + 1$, and thus $\widehat{y} \geq 2^{|g(x)|+1} - 1 \geq 2^{|g(x)|}$. Recall that $|x| = n$. Since both $\overline{h}$ and g are in FP, there exists a polynomial p such that $|y| \leq p(n)$, and thus certainly $\widehat{y} < 2^{p(n)+1}$. So let $h(0^{p(n)+2}) = z$, which implies $z \in A$ and $|z| \geq p(n) + 2$. Thus, $\widehat{z} \geq 2^{p(n)+2} - 1 > 2^{p(n)+1} > \widehat{y}$. Since $h \in$ FP, there clearly exists a polynomial q such that $\widehat{z} < 2^{q(n)}$. To summarize, N has found in time polynomial in $|x|$ two strings $y \notin A$ and $z \in A$ such that $2^{|g(x)|} \leq \widehat{y} < \widehat{z} < 2^{q(n)}$.

(2) N performs a search on the interval $[\widehat{y}, \widehat{z}] \subseteq \mathbb{N}$ to find some $\widehat{u} \in \mathbb{N}$ that is a boundary event of $\overline{A}$. That is, $\widehat{u}$ will satisfy: (a) $\widehat{y} \leq \widehat{u} \leq \widehat{z}$, (b) $\widehat{u} \notin A$, and (c) $\widehat{u} + 1 \in A$. Since $\widehat{z} < 2^{q(n)}$, the search will terminate in time polynomial in $|x|$. We state the standard search algorithm (searching to find a boundary event of $\overline{A}$):

Input $\widehat{y}$ and $\widehat{z}$ satisfying $\widehat{y} < \widehat{z}$, $\widehat{y} \notin A$, and $\widehat{z} \in A$.
Output $\widehat{u}$, a boundary event of $\overline{A}$ satisfying $\widehat{y} \leq \widehat{u} \leq \widehat{z}$.

$\widehat{u} := \widehat{y}$;

while $\widehat{z} > \widehat{u} + 1$ **do**

  $\widehat{a} := \lfloor \frac{\widehat{u}+\widehat{z}}{2} \rfloor$; **if** $\widehat{a} \notin A$ **then** $\widehat{u} := \widehat{a}$ **else** $\widehat{z} := \widehat{a}$

end while

---

[3] Either one can encode a string n (corresponding to the number $\widehat{n}$ in binary) directly into a circuit c_n such that $\#(c_n) = \widehat{n}$ (which is easy to do), or one can note the following indirect transformation: Let N' be an NPTM that on input n produces exactly $\widehat{n}$ accepting paths. Using a parsimonious Cook/Karp/Levin reduction (as described earlier), we easily obtain a family of circuits $\{\widetilde{c}_n\}_{n \in \Sigma^*}$ such that, for each $n \in \Sigma^*$, $\#(\widetilde{c}_n) = \widehat{n}$.

(3) Now consider the #P function $e(\langle m, x \rangle) = m + f(x)$ and the underlying NPTM E witnessing that $e \in$ #P. Let d_E be the parsimonious Cook/Karp/Levin reduction that on each input $\langle m, x \rangle$ outputs a circuit (representation) $\widetilde{c}_{\langle m,x \rangle}$ such that $\#(\widetilde{c}_{\langle m,x \rangle}) = e(\langle m, x \rangle)$. Recall that N has already computed $\widehat{u}$ (which itself depends on x and the oracle). N, using d_E to build its query, now queries its oracle, Counting(A), as to whether $\widetilde{c}_{\langle \widehat{u}-2^{|g(x)|}, x \rangle} \in$ Counting(A), and N accepts its input x if and only if the answer is "yes." This completes the description of N.

As argued above, N runs in polynomial time. We have to show that it correctly $\leq_T^P$-reduces L to Counting(A). Assume $x \notin L$. Then $f(x) = 2^{|g(x)|}$, and thus $e(\langle \widehat{u} - 2^{|g(x)|}, x \rangle) = \widehat{u} \notin A$. This implies that the answer to the query "$\widetilde{c}_{\langle \widehat{u}-2^{|g(x)|}, x \rangle} \in$ Counting(A)?" is "no," and so N rejects x. Analogously, if $x \in L$, then $f(x) = 2^{|g(x)|} + 1$, and thus $e(\langle \widehat{u} - 2^{|g(x)|}, x \rangle) = \widehat{u} + 1 \in A$, and so N accepts x. ∎

Finally, though we have stressed ways in which hypotheses that we feel are natural yield hardness results, we mention that for a large variety of complexity classes (amongst them R, coR, BPP, PP, and FewP) one can state somewhat artificial hypotheses for A that ensure that Counting(A) is many-one hard for the given class. For example, if A is any set such that either $\{i \mid i$ is a boundary event of $A\}$ is P-constructibly infinite or $\{i \mid i$ is a boundary event of $\overline{A}\}$ is P-constructibly infinite, then Counting(A) is SPP-$\leq_m^P$-hard.

Acknowledgments: We are grateful to L. Fortnow, K. Regan, and H. Vollmer for helpful literature pointers and history, and to B. Borchert, E. Hemaspaandra, and G. Wechsung for helpful discussions and suggestions. We thank J. Hartmanis for commending to us the importance of finding complexity-theoretic analogs of index sets, and we commend to the reader, as J. Hartmanis did to us, the open issue of whether a crisp complexity-theoretic analog can be found to the early work of Hartmanis and Lewis [HL71].

References

[All86] E. Allender. The complexity of sparse sets in P. In *Proceedings of the 1st Structure in Complexity Theory Conference*, pages 1–11. Springer-Verlag *Lecture Notes in Computer Science #223*, June 1986.

[AR88] E. Allender and R. Rubinstein. P-printable sets. *SIAM Journal on Computing*, 17(6):1193–1202, 1988.

[Bei89] R. Beigel. On the relativized power of additional accepting paths. In *Proceedings of the 4th Structure in Complexity Theory Conference*, pages 216–224. IEEE Computer Society Press, June 1989.

[BS97] B. Borchert and F. Stephan. Looking for an analogue of Rice's Theorem in circuit complexity theory. In *Proceedings on the 1997 Kurt Gödel Colloquium*, pages 114–127. Springer-Verlag *Lecture Notes in Computer Science #1289*, 1997.

[CGH+88] J. Cai, T. Gundermann, J. Hartmanis, L. Hemachandra, V. Sewelson, K. Wagner, and G. Wechsung. The boolean hierarchy I: Structural properties. *SIAM Journal on Computing*, 17(6):1232–1252, 1988.

[CGH+89] J. Cai, T. Gundermann, J. Hartmanis, L. Hemachandra, V. Sewelson, K. Wagner, and G. Wechsung. The boolean hierarchy II: Applications. *SIAM Journal on Computing*, 18(1):95–111, 1989.

[CH90]   J. Cai and L. Hemachandra. On the power of parity polynomial time. *Mathematical Systems Theory*, 23(2):95–106, 1990.
[FFK94] S. Fenner, L. Fortnow, and S. Kurtz. Gap-definable counting classes. *Journal of Computer and System Sciences*, 48(1):116–148, 1994.
[For97] L. Fortnow. Counting complexity. In L. Hemaspaandra and A. Selman, editors, *Complexity Theory Retrospective II*, pages 81–107. Springer-Verlag, 1997.
[GJ79]   M. Garey and D. Johnson. *Computers and Intractability: A Guide to the Theory of NP-Completeness*. W. H. Freeman and Company, 1979.
[GS88]   J. Grollmann and A. Selman. Complexity measures for public-key cryptosystems. *SIAM Journal on Computing*, 17(2):309–335, 1988.
[GW87]   T. Gundermann and G. Wechsung. Counting classes with finite acceptance types. *Computers and Artificial Intelligence*, 6(5):395–409, 1987.
[HL71]   J. Hartmanis and F. Lewis. The use of lists in the study of undecidable problems in automata theory. *Journal of Computer and System Sciences*, 5(1):54–66, 1971.
[HR97]   L. Hemaspaandra and J. Rothe. Complexity-theoretic analogs of Rice's Theorem. Technical Report TR-662, Department of Computer Science, University of Rochester, Rochester, NY, July 1997.
[HV95]   L. Hemaspaandra and H. Vollmer. The Satanic notations: Counting classes beyond #P and other definitional adventures. *SIGACT News*, 26(1):2–13, 1995.
[HZ93]   L. Hemaspaandra and M. Zimand. Strong forms of balanced immunity. Technical Report TR-480, Department of Computer Science, University of Rochester, Rochester, NY, December 1993. Revised May, 1994.
[Kar94] J. Kari. Rice's Theorem for the limit sets of cellular automata. *Theoretical Computer Science*, 127(2):229–254, 1994.
[KSTT92] J. Köbler, U. Schöning, S. Toda, and J. Torán. Turing machines with few accepting computations and low sets for PP. *Journal of Computer and System Sciences*, 44(2):272–286, 1992.
[OH93]   M. Ogiwara and L. Hemachandra. A complexity theory for closure properties. *Journal of Computer and System Sciences*, 46(3):295–325, 1993.
[Ric53] H. Rice. Classes of recursively enumerable sets and their decision problems. *Transactions of the AMS*, 74:358–366, 1953.
[Ric56] H. Rice. On completely recursively enumerable classes and their key arrays. *Journal of Symbolic Logic*, 21:304–341, 1956.
[Tod91] S. Toda. *Computational Complexity of Counting Complexity Classes*. PhD thesis, Department of Computer Science, Tokyo Institute of Technology, Tokyo, Japan, 1991.
[Val76] L. Valiant. The relative complexity of checking and evaluating. *Information Processing Letters*, 5(1):20–23, 1976.
[Val79] L. Valiant. The complexity of enumeration and reliability problems. *SIAM Journal on Computing*, 8(3):410–421, 1979.
[Vol94] H. Vollmer. *Komplexitätsklassen von Funktionen*. PhD thesis, Institut für Informatik, Universität Würzburg, Würzburg, Germany, 1994.
[Wat88] O. Watanabe. On hardness of one-way functions. *Information Processing Letters*, 27:151–157, 1988.

Model Checking Real-Time Properties of Symmetric Systems *

E. Allen Emerson and Richard J. Trefler

Computer Sciences Department and
Computer Engineering Research Center
University of Texas, Austin, TX, 78712, USA
http://www.cs.utexas.edu/users/emerson/

Abstract. We develop efficient algorithms for model checking quantitative properties of symmetric reactive systems in the general framework of a Real-Time Mu-calculus. Previous work has been limited to qualitative correctness properties. Our work not only permits handling of quantitative correctness, but it provides a strictly more expressive framework for qualitative correctness since the Mu-calculus strictly subsumes, e.g, CTL*. Unlike the previous "group-theoretic" approaches of [CE96] and [ES96] and the technical "automata-theoretic" approach of [ES97], our new approach may be viewed as "model-theoretic".

1 Introduction

Model checking [CE81] (c.f. [QS82], [LP85]) is an algorithmic method for determining whether a given finite state system M satisfies a temporal logic specification f. Lichtenstein and Pnueli [LP85] argued that in practice the complexity of model checking will be dominated by $|M|$, the size of M. Unfortunately, $|M|$ can be of size exponential in the program text. For example, a system with n processes running in parallel, each having just 3 local states, can have 3^n global states.

Symmetry reduction is a technique designed to substantially ameliorate this state explosion problem by exploiting the fact that many such systems are symmetric in their design and operation (cf. [JR91], [ID96], [ES96], [CE96], [ES97], [GS97]). Symmetry is a form of redundancy that can be factored out. Many synchronization and coordination protocols are the parallel composition of n processes which are identical up to renaming. The state graph M of such a system may reflect considerable symmetry. For example, states (C_1, T_2) and (T_1, C_2) may be present in a solution to the mutex problem. By clustering together such symmetry equivalent states, we can form the symmetry reduced quotient structure $\hat{M}$. $\hat{M}$, whose states are named by representatives of the clusters, may be exponentially smaller than M. Then the temporal formula f may be model

---

* The authors' work was supported in part by NSF grants CCR-941-5496 and CCR-980-4736 and SRC contract 97-DP-388. The authors can be reached at {emerson,trefler}@cs.utexas.edu

checked over $\hat{M}$ to determine if f holds of M. In practice, $\hat{M}$ is typically constructed incrementally from the program text, avoiding the self defeating task of first building M.

Work on symmetry reduction in model checking originally reduced M to an 'unannotated' symmetry reduced quotient structure $\overline{M}$ [ES96], [CE96]. However, that work, due to certain technical provisos regarding the internal symmetry of the specifications, was unable to handle fairness despite otherwise catering for CTL*. To remedy this, [ES97] introduced the annotated quotient structure $\hat{M}$ where the transitions between representative states are labeled with permutations indicating how the meaning of all coordinates shift from representative to representative. [ES97] also introduced a threaded quotient structure M^* indicating how the meaning of individual coordinates shift. By combining automata with these quotient graphs in an automata-theoretic [VW86] treatment, [ES97] developed a technical approach that allowed fairness properties to be checked efficiently.

In this paper we investigate model checking quantitative, discrete real-time properties over the quotients $\hat{M}$ and M^* in the framework of the Real-Time μ-calculus (RTLμ) (c.f. [Ko83], [Em92], [Se96]) which strictly subsumes the logics considered in previous work. We define a new notion of "twisted truth" or permuted satisfaction of a formula over annotated structures, $\hat{M}, \hat{s} \models f$, and prove that this permuted truth corresponds to the usual one over unannotated structures $M, \hat{s} \models f$, that is $\hat{M}, \hat{s} \models f$ iff $M, \hat{s} \models f$. This new notion leads to an efficient model checking algorithm for a formulation of an Indexed Real-Time Mu-calculus, IRTLμ. In particular, we give an $O(|\hat{M}||f|n)$ algorithm, which actually operates on M^*, for evaluating IRTLμ formulae of alternation depth 1 over $\hat{M}$. This algorithm can be generalized to work on arbitrary formulae of the μ-calculus. Our treatment of these problems, providing an alternative means of handling fairness properties, is done without appeal to automata. Instead, our techniques show how expressive model checking over the annotated quotient structure can be accomplished in a model-theoretic framework.

Interestingly, quantitative temporal properties of the structure M are preserved in $\hat{M}$ even though $\hat{M}$ may be exponentially smaller than M. For example, if the number of states of $\hat{M} < k <$ the number of states of M, then checking for the existence of a path no longer than k steps to a state where symmetric assertion P is true takes time proportional to k in M but proportional to the size of $\hat{M}$ in the symmetry reduced structure. This is not so for arbitrary boolean assertions f and is complicated in the annotated $\hat{M}$ by the shifting meaning of coordinates. A subtlety that arises is the fact that cycles in the annotated quotient may not correspond to cycles in the original structure. The extent to which this subtlety must be clarified in order to solve the model checking problem is a key issue in this paper.

Finally, we present results which relate to the difficulty of model checking temporal formulae of symmetric systems. We show that model checking certain temporal modalities over annotated structures is NP-hard. Furthermore, the model checking problem for certain quantitatively bounded fairness problems is NP-hard even over unreduced structures, in contrast with the polynomial algorithms for checking unbounded fairness. Against the background of these somewhat negative results we identify some classes of formulae and structures for which symmetry can reduce NP-hard problems to problems which can be solved in polynomial time.

The rest of the paper is organized as follows. In section 2 the logics discussed in the paper are introduced. Section 3 discusses the definition of symmetry reduced structures. The precise correspondence between structures, their symmetry reduced annotated quotient structures and temporal formulae is given in section 4. An IRTLμ model checking algorithm is given in section 5. Section 6 discusses the complexity of model checking symmetric structures and some applications of reasoning about symmetry in structures. Finally, section 7 contains a short conclusion.

2 RTLμ

Let LP denote a finite set of local propositions. $\mathcal{I}$ denotes an index set $[1..n]$ for some $n \in \mathbf{N}$ which denotes the set of natural numbers.

RTLμ is formed thusly: the set of atomic propositions is $LP \times \mathcal{I}$; we will write $(P, i) \in LP \times \mathcal{I}$ as P_i. We assume a set of variables over sets of states Var $= V \cup (V' \times \mathcal{I})$, where V and V' are unindexed and disjoint. RTLμ is the set of formulae defined by $P_i \in LP \times \mathcal{I}$ and $Y, Y_i \in$ Var are atomic formulae; if f and g are formulae then so are $\neg f$, $f \wedge g$ and $\langle R \rangle f$. Finally, suppose $f(Y)$ is a formula syntactically monotone in Y, $Y \in$ Var, that is, every occurrence of Y falls within an even number of $\neg$ occurrences, then $\mu Y.f(Y)$ and $\mu k Y.f(Y)$, $k \in \mathbf{N}$, are formulae. $\nu Y.f(Y)$ and $\nu k Y.f(Y)$ are abbreviations for $\neg \mu Y.\neg f(\neg Y)$ and $\neg \mu k Y.\neg f(\neg Y)$ respectively and $[R]f$ is an abbreviation for $\neg \langle R \rangle \neg f$.

Let iRTLμ be the sub-logic whose atomic formulae are propositions in $LP \times \{i\}$ and variables in V and V' $\times \{i\}$. Then the indexed mu-calculus, IRTLμ, is the logic whose atomic formulae are variables in V and formulae of the form $\vee_i f_i$ or $\wedge_i f_i$ where the f_i are isomorphic formulae of iRTLμ. Formulae of IRTLμ are also formed from the connectives $\wedge, \neg, \langle R \rangle, \mu Y$ and $\mu k Y$. Furthermore, if $f(Y)$ and g are formulae of IRTLμ, where Y is an unindexed variable not appearing within any μY in f, then the formula $f(g)$, which is obtained by replacing each of occurrence of Y in $f(Y)$ by g is a formula of IRTLμ.

Formulae of RTLμ are given semantics relative to finite structures $M = (S, R)$ where $S \subseteq LP^{\mathcal{I}}$ and $R \subseteq S \times S$. For $s \in S$ we say that $P_i \in s$ iff the ith element of s is $P \in LP$. Given formula f and structure M, the meaning of f, written f^M, is a mapping from valuations, $\sigma \in [\text{Var} \rightarrow 2^S]$, into 2^S and is defined below. We say that state s in M satisfies f if $s \in f^M$ (alternatively written $M, s \models f$).

- For atomic formulae, $P_i^M(\sigma) = \{s \in S \mid P_i \in s\}$ and $Y^M(\sigma) = \sigma(Y)$.
- For boolean combinations, $(f \wedge g)^M(\sigma) = f^M(\sigma) \cap g^M(\sigma)$ and $(\neg f)^M(\sigma) = S \setminus f^M(\sigma)$.
- $(\langle R \rangle f)^M(\sigma) = \{s \in S \mid \exists t \in S, t \in f^M(\sigma) \text{ and } (s, t) \in R\}$.

- $(\mu Y.f(Y))^M(\sigma) = \cap\{S' \subseteq S \mid (f(Y))^M(\sigma[S'/Y]) = S'\}$.
  $(\mu 0 Y.f(Y))^M(\sigma) = \emptyset$.
  $(\mu(k+1)Y.f(Y))^M(\sigma) = (f(Y))^M(\sigma[(\mu k Y.f(Y))^M(\sigma)/Y])$.

For valuation σ, $\sigma[A/Y]$ is the valuation everywhere equal to σ except that $\sigma[A/Y](Y) = A$. We say that variable Y is free (not bound) in formula f if it does not occur syntactically within a μY. operator. A sentence is a formula with no free variables and we assume that all bound variables are bound uniquely in sentences. Altdepth(f), is the alternation depth of the formula f, that is, $1 +$ the depth of nesting of alternating μ's and ν's when f is put in positive normal form.

We will at times make use of Computation Tree Logic (CTL) [CE81] and CTL* [EH86] formulae to explicate results. Both logics use the universal A and existential E path quantifiers together with the standard X, F, G and U temporal operators over paths. CTL is restricted to formulae where each path quantifier is matched with a single path operator while CTL* allows arbitrary nesting and boolean combinations of temporal operators over paths to be combined with a single path quantifier. It is known that CTL formulae can be seen as simple macros for μ-calculus formulae of alternation depth 1, while CTL* can be translated into the μ-calculus but with an exponential blowup in formula length. We note that the CTL* fairness formula $\mathsf{E}(\wedge_i \mathsf{GF} P_i)$, which says that there is a path along which for each i, P_i is satisfied infinitely often, can be expressed in RTLμ as $\nu Z. \wedge_i \mathsf{EXEF}(P_i \wedge Z)$ which is expressed in IRTLμ as $\nu Z. \wedge_i (\mu Y_i.\langle R\rangle((P_i \wedge Z) \vee Y_i))$.

3 Symmetry of Structures

Sym $\mathcal{I} = \{\pi \mid \pi$ is a permutation on $\mathcal{I}\ \}$. *Sym* $\mathcal{I}$ together with the function composition operator, $\circ$, is a group, where the inverse of π is denoted by π^{-1}. Given state s and permutation π, we say that π acts on s, written $\pi(s)$, in the following way $\pi(s) = \{P_{\pi(j)} \in LP \times \mathcal{I} \mid P_j \in s\}$. For example, let $s = (C_1, T_2)$ and π be the permutation which flips 1 and 2. Then $\pi(s) = (T_1, C_2)$. Similarly for structures, $\pi(M)$, π acting on M, is the structure (S', R') where $S' = \{\pi(s) \in LP^{\mathcal{I}} \mid s \in S\}$ and $\pi(s) \to \pi(t) \in R'$ iff $s \to t \in R$. Then π is an automorphism of M if $\pi(M) = M$ and $Aut(M)$ is the set $\pi \in$ *Sym* $\mathcal{I}$ such that π is an automorphism of M. $Aut(M)$ is a subgroup of *Sym* $\mathcal{I}$ which will be denoted by $Aut(M) \preceq$ *Sym* $\mathcal{I}$.

Let f be a formula of RTLμ, $\pi(f)$ is the formula g which is identical to f except that each occurrence of $P_i \in LP \times \mathcal{I}$ ($Y_i \in V' \times \mathcal{I}$) is replaced by $P_{\pi(i)}$ ($Y_{\pi(i)}$). Then $Aut(f) = \{\pi \in$ *Sym* $\mathcal{I} \mid \pi(f) \equiv f\}$ [ES96]. $Auto(f)$ [1] is the set of $\pi \in$ *Sym* $\mathcal{I}$ such that for each maximal propositional sub-formula, g, of f, $\pi(g) \equiv g$. Both $Aut(f)$ and $Auto(f)$ are subgroups of *Sym* $\mathcal{I}$ and $Auto(f) \preceq Aut(f)$.

---

[1] A more general definition of $Auto(f)$ can be found in [ES96].

Given structure $M = (S, R)$, let G be a subgroup of $Aut(M)$. Two states, $s, s' \in S$ are equivalent with respect to G, written $s \equiv_G s'$, if there exists $\pi \in G$ such that $\pi(s) = s'$ [ES96], [CE96]. Because G is a (sub)group it is clear that $\equiv_G$ is an equivalence relation. Then for each equivalence class in $\equiv_G$ we choose an arbitrary member state to represent that class and refer to that state as $\overline{s}$ and its equivalence class as $[\overline{s}]$. That is, given $\overline{s} \in S$, $\overline{s}$ is the representative of $[\overline{s}] = \{t \in S \mid t \equiv_G \overline{s}\}$. $\overline{M} = M/\equiv_G = (\overline{S}, \overline{R})$, is the unannotated quotient structure where $\overline{S}$ is the set of $\overline{s}$ such that $\overline{s}$ is the representative of an equivalence class of $\equiv_G$. $\overline{R} \subseteq \overline{S} \times \overline{S}$ is the set of transitions $(\overline{s}, \overline{t})$ such that there exists $s' \equiv_G \overline{s}$, $t' \equiv_G \overline{t}$ and $s' \to t' \in R$.

$\hat{M} = M/\equiv_G = (\hat{S}, \hat{R})$, is the annotated quotient structure where $\hat{S}$ is the set of $\hat{s}$ such that, as above, $\hat{s}$ is the representative of an equivalence class of $\equiv_G$. $\hat{R} \subseteq \hat{S} \times G \times \hat{S}$ is defined by the restrictions : (i) if $\hat{s} \to t \in R$ then there is a unique π such that $\hat{s} \xrightarrow{\pi} \hat{t} \in \hat{R}$ and $\pi(\hat{t}) = t$; (ii) $\hat{s} \xrightarrow{\pi} \hat{t} \in \hat{R}$ only if $\hat{s} \to \pi(\hat{t}) \in R$.

Model checking for RTLμ could be carried out on either $\hat{M}$ of M^*. We choose to make use of M^* which can be seen as a data structure which, for a modest increase in the number of states in $\hat{M}$, can help organize the model checking algorithm. However, the fact that the number of states in M^* is larger, by a factor of $|\mathcal{I}|$, over $\hat{M}$ can be misleading. Although it may be unusual two states $\hat{s}$ and $\hat{t}$ may have an exponential number of labeled arcs between them in $\hat{M}$ but in M^* those same two states will have at most a quadratic number of arcs between them. Technically $M^* = (\hat{S} \times (\{0\} \cup \mathcal{I}), R^*, RED)$ where R^* and RED are transition relations. For $i, j \in \mathcal{I}$, $\langle \hat{s}, i \rangle \to \langle \hat{t}, j \rangle \in R^*$ iff there exists π such that $\hat{s} \xrightarrow{\pi} \hat{t} \in \hat{R}$ and $\pi^{-1}(i) = j$. For $i \in \mathcal{I}$, $\langle \hat{s}, 0 \rangle \to \langle \hat{s}, i \rangle \in RED$ and $\langle \hat{s}, i \rangle \to \langle \hat{s}, 0 \rangle \in RED$. Finally, $\langle \hat{s}, 0 \rangle \to \langle \hat{t}, 0 \rangle \in R^*$ iff there is a π such that $\hat{s} \xrightarrow{\pi} \hat{t} \in \hat{R}$.

4 Temporal Formulae on Annotated Structures

We define the meaning of an RTLμ formula, f, on an annotated structure $\hat{M}$. We say that state $\hat{s}$ in $\hat{M}$ satisfies f if $\hat{s} \in f^{\hat{M}}$ (alternatively written $\hat{M}, \hat{s} \models f$). For the purposes of this definition $\hat{M}$ could, in general, be any annotated structure and need not correspond to the symmetry reduced quotient of any particular unannotated structure. $f^{\hat{M}} : [\text{Var} \to 2^{\hat{S}}] \to 2^{\hat{S}}$ and valuation $\hat{\sigma} \in [\text{Var} \to 2^{\hat{S}}]$.

- For P_i, $P_i^{\hat{M}}(\hat{\sigma})$ is the set of states $\hat{s} \in \hat{S}$ such that $P_i \in \hat{s}$. For $Y \in \text{Var}$, $Y^{\hat{M}}(\hat{\sigma})$ is the set of states $\hat{s} \in \hat{S}$ such that $\hat{s} \in \hat{\sigma}(Y)$.
- $(f \wedge g)^{\hat{M}}(\hat{\sigma}) = f^{\hat{M}}(\hat{\sigma}) \cap g^{\hat{M}}(\hat{\sigma})$.
  $(\neg f)^{\hat{M}}(\hat{\sigma}) = \hat{S} \setminus f^{\hat{M}}(\hat{\sigma})$.
- $(\langle \hat{R} \rangle f)^{\hat{M}}(\hat{\sigma}) = \{\hat{s} \in \hat{S} \mid \text{there exists } \hat{s} \xrightarrow{\pi} \hat{t} \in \hat{R} \text{ and } \hat{t} \in (\pi^{-1}(f))^{\hat{M}}(\hat{\sigma})\}$.
- $(\mu Y. f(Y))^{\hat{M}}(\hat{\sigma}) = \cap \{\hat{S}' \subseteq \hat{S} \mid (f(Y))^{\hat{M}}(\hat{\sigma}[\hat{S}'/Y]) = \hat{S}'\}$.
  $(\mu 0 Y. f(Y))^{\hat{M}}(\hat{\sigma}) = \emptyset$.
  $(\mu(k+1)Y. f(Y))^{\hat{M}}(\hat{\sigma}) = f(Y)^{\hat{M}}(\hat{\sigma}[(\mu k Y. f(Y))^{\hat{M}}(\hat{\sigma})/Y])$.

The following theorem relates the meaning of RTLμ sentence f over a structure M and the meaning of f over the annotated structure $\hat{M}$. For $\sigma \in [\text{Var} \to 2^S]$ and $\hat{\psi} \in [\text{Var} \to 2^{\hat{S}}]$ we say that σ and $\hat{\psi}$ correspond iff $\sigma(Y) = \{t \mid t \equiv_G \hat{s}$, for some $\hat{s} \in \hat{\psi}(Y)\}$.

Theorem 1. *Let $\hat{M} = M/\equiv_G$, for $G \preceq Aut(M)$, and let σ and $\hat{\psi}$ be corresponding valuations. Then for any* RTLμ *sentence f, $\hat{s} \in f^M(\sigma)$ iff $\hat{s} \in f^{\hat{M}}(\hat{\psi})$.*

5 Real-Time Mu-Calculus Model Checking

We show how to reduce the problem of model checking IRTLμ formula f over $\hat{M}$ to the problem of checking the transformed formulae $T(f)$, over the threaded structure M^*. This reduction implies an algorithm for model checking f over M, by checking $T(f)$ over M^*.

We proceed as follows, firstly, we define a translation from formulae of IRTLμ over $LP \times \mathcal{I}$, $V \cup (V' \times \mathcal{I})$ and $\langle R \rangle$ to formulae of RTLμ over LP, V, V', $\langle R \rangle$ and $\langle RED \rangle$. The intuition behind this transformation is that the states of M^*, which are of the form $\langle \hat{s}, i \rangle$, only record the satisfaction of propositions P_i which are true at $\hat{s}$. Therefore the subscript i can be dropped in these 'local' states. We then use the state $\langle \hat{s}, 0 \rangle$ as a 'global' state to collect information about all the $\langle \hat{s}, i \rangle$'s. It is then possible to trade the universal quantification over i in formulae of the form $\wedge_i f_i$ for a modal operator $[RED]$ at state $\langle \hat{s}, 0 \rangle$ and check that all the $\langle \hat{s}, i \rangle$'s satisfy f. We then model check the transformed formulae over the structure M^* [EL86]. Since M^* is an unannotated structure, μ-calculus model checking algorithms may be applied directly to the problem of checking whether the transformed formula is satisfied by M^*. For the purposes of this model checking we define the meaning of $P \in LP$ over M^* as follows $P^{M^*}(\psi) = \{\langle \hat{s}, i \rangle \mid i \neq 0 \text{ and } P_i \in \hat{s}\}$. The meaning of a compound formula or variable is defined by its standard meaning as given in Section 2.

Technically, we distinguish between global and local IRTLμ formulae. Global formulae are those where all indexed propositions and variables appear within the scope of an $\wedge_i$ or $\vee_i$ quantifier. Local formulae have at least one indexed proposition or variable which does not appear within the scope of any $\wedge_i$ or $\vee_i$. Then for formula, f, of IRTLμ we define the transform of f, $T(f)$ as follows.

- $T(P_i) = P$. $T(Y_i) = Y$. $T(Z) = Z$.
- For f and g both local or both global: $T(f \wedge g) = T(f) \wedge T(g)$.
  $T(f \vee g) = T(f) \vee T(g)$.
- For f global and g_i local: $T(f \wedge g_i) = (\langle RED \rangle T(f)) \wedge T(g_i)$.
  $T(f \vee g_i) = (\langle RED \rangle T(f)) \vee T(g_i)$.
- $T(\neg f) = \neg T(f)$.
- $T(\wedge_i f_i) = [RED]T(f_1)$. Because the f_i's are isomorphic, $T(f_1) = T(f_2)$, we need only check for $T(f_1)$.
- $T(\vee_i f_i) = \langle RED \rangle T(f_1)$.
- $T(\mu Z.f(Z)) = \mu Z.T(f(Z))$.
  $T(\mu k Z.f(Z)) = \mu k Z.T(f(Z))$.

- $\mathcal{T}(\mu Y_i.f(Y_i)) = \mu Y.\mathcal{T}(f(Y))$.
  $\mathcal{T}(\mu k Y_i.f(Y_i)) = \mu k Y.\mathcal{T}(f(Y))$.

The idea behind $\mathcal{T}(\wedge_i f_i) = [RED]\mathcal{T}(f_1)$ is that $\hat{s}$ satisfies $\wedge_i f_i$ in $\hat{M}$ iff for all $i \in \mathcal{I}$, $\langle \hat{s}, i \rangle$ satisfies f in M^* and we use $\langle \hat{s}, 0 \rangle$ to check whether in fact this is the case. Recall from the definition of M^* that the only transitions in RED are from $\langle \hat{s}, 0 \rangle$ to $\langle \hat{s}, i \rangle$ and vice versa.

Let $\hat{\sigma}$ be a valuation over $\hat{M}$ and ψ be a valuation over M^*. Then we say that $\hat{\sigma}$ and ψ correspond when for global variable Z, $\psi(Z) = \{\langle \hat{s}, 0 \rangle \mid \hat{s} \in \hat{\sigma}(Z)\}$ and for local variable $Y \in V'$, $\psi(Y) = \{\langle \hat{s}, i \rangle \mid \hat{s} \in \hat{\sigma}(Y_i)\}$.

Proposition 1. *Let f be a global formula of* IRTLμ *while $\hat{\sigma}$ and ψ are corresponding valuations. Then $\hat{s} \in f^{\hat{M}}(\hat{\sigma})$ iff $\langle \hat{s}, 0 \rangle \in (\mathcal{T}(f))^{M^*}(\psi)$.*

Let f_i be a local formula of IRTLμ *while $\hat{\sigma}$ and ψ are corresponding valuations. Then $\hat{s} \in f_i^{\hat{M}}(\hat{\sigma})$ iff $\langle \hat{s}, i \rangle \in (\mathcal{T}(f))^{M^*}(\psi)$.*

Theorem 2. *For global sentence f of* IRTLμ, $\mathcal{T}(f)$ *can be model checked over the structure M^* in time $\mathcal{O}((|M^*||f|)^{\mathrm{altdepth}(f)})$.*

Remark : This time bound can be improved to $\mathcal{O}((|M^*||f|)^{\lfloor(\mathrm{altdepth}(f)+1)/2\rfloor})$ [LB94] and in general we may take advantage of any μ-calculus algorithm with a better time bound.

Corollary 1. *The model checking problem 'does* IRTLμ *formula f hold in M' can be solved in time $\mathcal{O}((|M^*||f|)^{\mathrm{altdepth}(f)})$.*

6 Applications

Certain problems that are in general NP-hard become solvable in polynomial time in the special case of symmetric structures. We now discuss the complexity of model checking symmetric structures and symmetric formulae. Our first results show that model checking certain basic temporal logic formulae over annotated symmetry reduced structures is NP-hard. Secondly, we show that model checking the indexed bounded fairness formula $\mathsf{E} \wedge_i \mathsf{GF}^{\leq k} P_i$ over unannotated structures is NP-hard. This implies that it is unlikely that there is a 'short' IRTLμ formula expressing this property and hence is an indication that bounded fairness may be exponentially harder to check than more standard fairness notions.

Theorem 3. $\hat{M}, \hat{s} \models \mathsf{EF}p$ *is NP-hard.* $\hat{M}, \hat{s} \models \mathsf{EG}p$ *is NP-hard.*

Proof Idea: Let $q(P_1, \ldots, P_n)$ be a boolean formula over the propositions P_1 through P_n. Define p as $q(Q_1, Q_3, \ldots, Q_{2n-1})$ where the $2n$, Q_i's are fresh propositional symbols. $\hat{M}$ is the annotated structure which consists of the single state $\hat{s}$ which is labeled with the propositions $Q_1, Q_3, \ldots, Q_{2n-1}$. There are two transitions from $\hat{s}$ to $\hat{s}$, the first is labeled by the rotation permutation $(1\,2\ldots 2n)$ and the second by the transposition permutation $(1\,2)$. Arbitrary composition of these two permutations is enough to create any permutation in $Sym\,2n$. It can then be shown that $q(P_1, \ldots, P_n)$ is satisfiable iff $\hat{M}, \hat{s} \models \mathsf{EF}p$. □

Theorem 4. $M, s \models \mathsf{E} \bigwedge_i \mathsf{GF}^{\leq k} P_i$ is NP-hard.

The following theorems relate the CTL* formula $\mathsf{E}(\bigwedge_i \mathsf{F} P_i)$ to any equivalent translation into the μ-calculus. $\mathsf{E}(\bigwedge_i \mathsf{F} P_i)$ says that there is a path such that for all i, eventually P_i is true.

Theorem 5. *[SC85] (c.f. [CE81])* $M, s \models \mathsf{E}(\bigwedge_i \mathsf{F} P_i)$ is NP-complete.

However, when $Aut(M) = Aut(s) = Sym\ \mathcal{I}$, where $Aut(s) = \{\pi \in Sym\ \mathcal{I} \mid \pi(s) = s\}$, then the model checking problem for $\mathsf{E}(\bigwedge_i \mathsf{F} P_i)$ can be solved efficiently.

Proposition 2. $M, s \models [\mathsf{E}(\bigwedge_i \mathsf{F} P_i)]$ iff $M, s \models \bigvee_{\pi \in Sym\ \mathcal{I}} \mathsf{EF}(P_{\pi(1)} \wedge \mathsf{EF}(P_{\pi(2)} \wedge \ldots \wedge \mathsf{EF} P_{\pi(n)})))$

Theorem 6. $Aut(M) = Aut(s) = Sym\ \mathcal{I}$ implies that $M, s \models \mathsf{E}(\bigwedge_i \mathsf{F} P_i)$ iff $M, s \models \mathsf{EF}(P_1 \wedge \mathsf{EF}(P_2 \wedge \ldots \wedge \mathsf{EF} P_n))$.

Proof idea: Right to left follows from the existence of a path through each of the P_i's. Suppose s satisfies $\mathsf{E}(\bigwedge_i \mathsf{F} P_i)$ then s also satisfies $\mathsf{EF}(P_{\pi(1)} \wedge \mathsf{EF}(P_{\pi(2)} \wedge \ldots \wedge \mathsf{EF} P_{\pi(n)}))$ for some π. By state symmetry [ES96] s also satisfies $\mathsf{EF}(P_1 \wedge \mathsf{EF}(P_2 \wedge \ldots \wedge \mathsf{EF} P_n))$. □

Because $\mathsf{EF}(P_1 \wedge \mathsf{EF}(P_2 \wedge \ldots \wedge \mathsf{EF} P_n))$ is a CTL formula and can be translated into a μ-calculus formula of alternation depth 1 it can be model checked on a structure M in time linear in the size of the structure and the formula as opposed to the presumed exponential time algorithm for model checking $\mathsf{E} \bigwedge_i \mathsf{F} P_i$.

We can extend this reasoning as follows.

Theorem 7. Suppose $M, s \models \mathsf{AGEF} s$ and $Aut(s) = Aut(M) = Sym\ \mathcal{I}$. Then $M, s \models \mathsf{E}(\bigwedge_i \mathsf{F} P_i)$ iff $M, s \models \bigwedge_i \mathsf{EF} P_i$ iff $M, s \models \mathsf{EF} P_1$.

The point being that $\bigwedge_i \mathsf{EF} P_i$ can be translated into an IRTLμ formula of alternation depth 1 and hence can be model checked on the symmetry reduced structure $\hat{M}$ quickly where as it seems that $\mathsf{EF}(P_1 \wedge \mathsf{EF}(P_2 \wedge \ldots \wedge \mathsf{EF} P_n))$ cannot be. It is, in general, interesting to consider the classes of linear time formula h_i and structures M for which s satisfies $\mathsf{E} \bigwedge_i h_i$ is equivalent to s satisfies $\bigwedge_i \mathsf{E} h_i$ because the latter formula can be checked much more quickly on both the large and the symmetry reduced structures.

7 Conclusion

This paper has described a general framework for performing model checking for formulae of the μ-calculus on symmetric systems. We have given efficient model checking algorithms for indexed sub-logics of the Real-Time μ-calculus over annotated structures. These real-time logics are useful for describing the quantitative and qualitative properties of a large class of programs that operate in real-time environments, such as network communication protocols and embedded real-time control systems. Furthermore, our framework subsumes indexed formulations of RTCTL [EM92] and CTL* [ES97].

We have also shown that the threaded graph construction of [ES97], used in a different form in [ES96], is more general and thus more applicable than previously thought. M^* supports general μ-calculus model checking. But that leaves the question, 'where did the automata go?' The answer is that M^* may be viewed as an automaton of a particularly simple nature, one whose job it is to steadily keep track of shifting indices. We remark that for checking fairness our method requires essentially quadratic time in $|\hat{M}|$ for weak fairness versus linear time in $|\hat{M}|$ for [ES97]; but this is an artifact of using the more general μ-calculus of alternation depth 2 (c.f. [EL86] vs [EL87]).

The work presented here deals with quantitative, discrete real-time logics. These logics are exponentially more succinct but not strictly more expressive than their untimed counterparts. An interesting area for further research is reasoning about symmetry on explicitly timed structures which model dense or discrete time as discussed in [AC90], [Al91] and [He91]. We have also identified an interesting open problem in the realm of model checking symmetric structures, that is to fully characterize the relationship between formulae of the form $\wedge_i \mathsf{E} h_i$ and $\mathsf{E}(\wedge_i h_i)$ over symmetric structures.

References

[AC90] Alur, R., Courcoubetis, C., and Dill, D., Model Checking for Real-Time Systems. In *Proceedings of the Fifth Annual Symposium on Logic in Computer Science*, pp. 414-425, IEEE Computer Society Press, 1990.
[Al91] Alur, R., *Techniques for Automatic Verification of Real-Time Systems*. PhD thesis, Stanford University, 1991.
[CE81] Clarke, E. M., and Emerson, E. A., Design and Verification of Synchronization Skeletons using Branching Time Temporal Logic, Logics of Programs Workshop, IBM Yorktown Heights, New York, Springer LNCS no. 131., pp. 52-71, May 1981.
[CE96] Clarke, E. M., Filkorn, T., and Jha, S., Exploiting Symmetry in Temporal Logic Model Checking. In *Fifth International Conference on Computer Aided Verification*, Crete, Greece, June 1993. Journal version appears as: Clarke, E. M., Enders, R. Filkorn, T. and Jha, S., Exploiting Symmetry in Temporal Logic Model Checking. In *Formal Methods in System Design*, Kluwer, vol. 9, no. 1/2, August 1996.
[Em92] E. Allen Emerson Real–Time and the μ–Calculus. In *Proceedings of Real-Time: Theory in Practice*, LNCS, Vol. 600, pp. 176-194, Springer, June 1992.
[EH86] Emerson, E. A., and Halpern, J. Y., 'Sometimes' and 'Not Never' Revisited: On Branching versus Linear Time Temporal Logic, *JACM*, vol. 33, no. 1, pp. 151-178, Jan. 86.
[EL86] Emerson, E. A., and Lei, C.-L., Efficient Model Checking in Fragments of the Mu-Calculus, IEEE Symp. on Logic in Computer Science (LICS), Cambridge, Mass., 1986.

[EL87] Emerson, E. A., and Lei, C.-L.m Modalities for Model Checking: Branching Time Strikes Back, pp. 84-96, ACM POPL85; journal version appears in Sci. Comp. Prog. vol. 8, pp 275-306, 1987.

[EM92] Emerson, E. A., Mok, A. K., Sistla, A. P., and Srinivasan, J., Quantitative Temporal Reasoning. In *Journal of Real Time Systems*, vol. 4, pp. 331-352, 1992.

[ES96] Emerson, E. A. and Sistla, A. P., Symmetry and Model Checking. In *Fifth International Conference on Computer Aided Verification*, Crete, Greece, June 1993. Journal Version appeared in *Formal Methods in System Design*, Kluwer, vol. 9, no. 1/2, August 1996.

[ES97] Emerson, E. A. and Sistla, A. P., Utilizing Symmetry when Model Checking under Fairness Assumptions. In *Seventh International Conference on Computer Aided Verification* Springer-Verlag, 1995. Journal version, *TOPLAS* 19(4): 617-638 (1997).

[GS97] Gyuris, V. and Sistla, A. P., On-the-Fly Model checking under Fairness that Exploits Symmetry. In *Proceedings of the 9th International Conference on Computer Aided Verification, Haifa, Israel*, 1997.

[He91] Henzinger, T., The Temporal Specification and Verification of Real-Time Systems, Ph.D. Thesis, Stanford University, 1991, report number STAN-CS-91-1380.

[ID96] Ip, C-W. N., Dill, D. L., Better Verification through Symmetry. In *Proc. 11th International Symposium on Computer Hardware Description Languages(CHDL)*, April, 1993. Journal version appeared in *Formal Methods in System Design*, Kluwer, vol. 9, no. 1/2, August 1996.

[JR91] Jensen, K. and Rozenberg, G. (eds.), High-Level Petri Nets: Theory and Application, Springer- Verlag, 1991.

[Ko83] Kozen, D., Results on the Propositional Mu-Calculus, Theor. Comp. Sci., pp. 333-354, Dec. 83.

[LP85] Litchtenstein, O., and Pnueli, A., Checking That Finite State Concurrent Programs Satisfy Their Linear Specifications, POPL85, pp. 97-107, Jan. 85.

[LB94] Long, D., Browne, A., Clarke, E. Jha, S. and Marrero, W., An Improved Algorithm for the Evaluation of Fixpoint Expressions. In *Proc. of the 6th Inter. Conf. on Computer Aided Verification, Stanford, Springer LNCS no. 818*, June 1994.

[QS82] Queille, J. P., and Sifakis, J., Specification and verification of concurrent programs in CESAR, Proc. 5th Int. Symp. Prog., Springer LNCS no. 137, pp. 195-220, 1982.

[Se96] Seidl, H., A Modal μ-Calculus for Durational Transition Systems. In *Eleventh Annual IEEE Symposium on Logic In Computer Science*, IEEE Computer Society Press, 1996.

[SC85] Sistla, A. P., and Clarke, E. M., The Complexity of Propositional Linear Temporal Logic, J. ACM, Vol. 32, No. 3, pp.733-749, 1985.

[VW86] Vardi, M., and Wolper, P. , An Automata-theoretic Approach to Automatic Program Verification, Proc. IEEE LICS, pp. 332-344, 1986.

Locality of Order-Invariant First-Order Formulas

Martin Grohe[1] and Thomas Schwentick[2]

[1] Institut für mathematische Logik, Eckerstr. 1, 79104 Freiburg, Germany
[2] Institut für Informatik, Johannes-Gutenberg-Universität Mainz, 55099 Mainz, Germany

Abstract. A query is *local* if the decision of whether a tuple in a structure satisfies this query only depends on a small neighborhood of the tuple. We prove that all queries expressible by *order-invariant* first-order formulas are local.

1 Introduction

One of the fundamental properties of first-order formulas is their *locality*, which means that the decision of whether in a fixed structure a formula holds at some point (or at a tuple of points) only depends on a small neighborhood of this point (tuple). This result, proved by Gaifman [5], provides very convenient proofs that certain queries cannot be expressed by a first-order formula. For example, to decide whether there is a path between two vertices of a graph it clearly does not suffice to look at small neighborhoods of these vertices. Hence by locality, s-t-connectivity is not expressible in first-order logic. Recently, Libkin and others [3,8,9,10] systematically started to explore locality as tool for proving inexpressibility results. The ultimate goal of this line of research would have been to separate complexity classes, in particular to separate TC_0 from LOGSPACE. However, a recent result of Hella [7], showing that even uniform AC_0 contains non-local queries, has destroyed these hopes.

Nevertheless, locality remains an important tool for proving inexpressibility results for query languages. In database theory, one often faces a situation where the physical representation of the database, which we consider as a relational structure, induces an order on the structure, but this order is hidden to the user. The user may use the order in her queries, but the result of the query should not depend on the given order. It may seem that this does not help her, but actually there are first-order formulas that use the order to express order-invariant queries that cannot be expressed without the order. This is an unpublished result due to Gurevich [6]; for examples of such queries we refer the reader to [1,2] and Example 3 (due to [4]).

Formally, we say that a first-order formula $\varphi(\bar{x})$ whose vocabulary contains the order symbol $\leq$ is *order-invariant* on a class $\mathcal{C}$ of structures if for all structures $\mathfrak{A} \in \mathcal{C}$, tuples $\bar{a}$ of elements of $\mathfrak{A}$, and linear orders $\leq_1, \leq_2$ on $\mathfrak{A}$ we have: $\varphi(\bar{a})$ holds in $(\mathfrak{A}, \leq_1)$ if, and only if, $\varphi(\bar{a})$ holds in $(\mathfrak{A}, \leq_2)$. It is an easy consequence of the interpolation theorem that if a formula is order-invariant on the class of all structures, it is equivalent to a first-order formula that does not use the ordering. This is no longer true when restricted to the class of all finite structures, or to a class consisting of a single infinite structure. Unfortunately, these are the cases showing up naturally in applications to computer science. We prove that for *all* classes $\mathcal{C}$ of structures the first-order formulas that are order-invariant on $\mathcal{C}$ can only define queries that are local on all structures in $\mathcal{C}$. This gives us a good intuition about the expressive power of order-invariant first-order formulas and a simple method to prove inexpressibility results.

The paper is organized as follows: After the preliminaries, we prove the locality of order-invariant first-order formulas with one free variable in Section 3. In Section 4 we reduce the case of formulas with arbitrarily many variables to the one-variable case. Due to space limitations, we can only sketch most of the proofs. The full paper is available via `http://www.Informatik.Uni-Mainz.DE/~tick/` and `http://logimac.mathematik.uni-freiburg.de/preprints/grohe/pub.html`.

We would like to thank Juha Nurmonen for pointing us to the problem and Clemens Lautemann for fruitful discussions about its solution.

2 Preliminaries

A *vocabulary* is a set τ containing finitely many relation and constant symbols. A τ-*structure* $\mathfrak{A}$ consists of a set A, called the *universe* of $\mathfrak{A}$, an interpretation $R^{\mathfrak{A}} \subseteq A^r$ for each r-ary relation symbol $R \in \tau$, and an interpretation $c^{\mathfrak{A}} \in A$ of each constant symbol $c \in \tau$. For example, a graph can be considered as an $\{E\}$-structure $\mathfrak{A} = (A, E^{\mathfrak{A}})$, where E is a binary relation symbol. An *ordered structure* is a structure whose vocabulary contains the distinguished binary relation symbol $\leq$ which is interpreted as a linear order of the universe.

$[i, j]$ always denotes the set $\{i, i+1, \ldots, j\}$ of integers.

Occasionally, we need to consider strings as finite structures. For each $l \geq 1$, we let τ_l denote the vocabulary $\{\leq, P_1, \ldots, P_l, \underline{\min}, \underline{\max}\}$ with unary relation symbols P_j and constant symbols $\underline{\min}$ and $\underline{\max}$. We represent a string $s = s_1 \cdots s_n$ over an l-letter alphabet $\Sigma = \{\alpha_1, \ldots, \alpha_l\}$ by the ordered τ_l-structure with universe $[1, n]$, where P_j is interpreted as $\{i \mid s_i = \alpha_j\}$, for every j, and $\underline{\min} = 1$, $\underline{\max} = n$. In our notation we do not distinguish between the string s and its representation as a finite structure s.

If $\mathfrak{A}$ is a structure and $B \subseteq A$ a subset that contains all constants of $\mathfrak{A}$, then the (induced) substructure of $\mathfrak{A}$ with universe B is denoted by $\langle B \rangle^{\mathfrak{A}}$.

Let $\sigma \subset \tau$ be vocabularies. The σ-*reduct* of a τ-structure $\mathfrak{A}$, denoted by $\mathfrak{A}|_\sigma$, is the σ-structure with universe A in which all symbols of σ are interpreted as in $\mathfrak{A}$. On the other hand, each τ-structure $\mathfrak{A}$ such that $\mathfrak{A}|_\sigma = \mathfrak{B}$ is called a τ-*expansion* of $\mathfrak{B}$. For a σ-structure $\mathfrak{B}$, relations $R_1 \subseteq B^{k_1}, \ldots, R_l \subseteq B^{k_l}$, and $b_1, \ldots, b_m \in B$, by $(\mathfrak{B}, R_1, \ldots, R_l, b_1, \ldots, b_m)$ we denote the expansion of $\mathfrak{B}$ of a suitable vocabulary $\tau \supset \sigma$ that contains in addition to the symbols in σ a new k_i-ary relation symbol for each $i \leq l$ and m new constant symbols.

Let $k \geq 1$ and $\mathcal{C}$ a class of τ-structures. A k-*ary query on* $\mathcal{C}$ is a mapping ρ that assigns a k-ary relation on A to each structure $\mathfrak{A} \in \mathcal{C}$ such that for isomorphic τ-structures $\mathfrak{A}, \mathfrak{B} \in \mathcal{C}$ each isomorphism f between $\mathfrak{A}$ and $\mathfrak{B}$ is also an isomorphism between the expanded structures $(\mathfrak{A}, \rho(\mathfrak{A})), (\mathfrak{B}, \rho(\mathfrak{B}))$.

2.1 Types and Games

Equivalence in first-order logic can be characterized in terms of the following *Ehrenfeucht-Fraïssé game*:

Definition 1. *Let $r \geq 0$ and $\mathfrak{A}, \mathfrak{A}'$ structures of the same vocabulary. The r-round EF-game on $\mathfrak{A}, \mathfrak{A}'$ is played by two players called the spoiler and the duplicator. In each of the r rounds of the game the spoiler either chooses an element v_i of $\mathfrak{A}$ or an element v'_i of $\mathfrak{A}'$. The duplicator answers by choosing an element v'_i of $\mathfrak{A}'$ or an element v_i of $\mathfrak{A}$, respectively. The duplicator wins the game if the mapping that maps v_i to v'_i (for $i \leq r$) and each constant $c^{\mathfrak{A}}$ to the corresponding constant $c^{\mathfrak{A}'}$ is a partial isomorphism, that is, an isomorphism between the substructure of $\mathfrak{A}$ generated by its domain and the substructure of $\mathfrak{A}'$ generated by its image. It is clear how to define the notion of a winning strategy for the duplicator in the game.*

The *quantifier-depth* of a first-order formula is the maximal number of nested quantifiers in the formula. The *r-type* of a structure $\mathfrak{A}$ is the set of all first-order sentences of quantifier-depth at most r satisfied by $\mathfrak{A}$. It is a well-known fact that for each vocabulary τ there is only a finite number of distinct r-types of τ-structures (simply because there are only finitely many inequivalent first-order formulas of vocabulary τ and quantifier-depth at most r). We write $\mathfrak{A} \sim_r \mathfrak{A}'$ to denote that $\mathfrak{A}$ and $\mathfrak{A}'$ have the same r-type.

Theorem 1. *Let $r \geq 0$ and $\mathfrak{A}, \mathfrak{A}'$ structures of the same vocabulary. Then $\mathfrak{A} \sim_r \mathfrak{A}'$ if, and only if, the duplicator has a winning strategy for the r-round EF-game on $\mathfrak{A}, \mathfrak{A}'$.*

The following two simple examples, both needed later, may serve as an exercise for the reader in proving non-expressibility results using the EF-game.

Example 1. Let $r \geq 1$ and $m = 2^r + 1$. Using the r-round EF-game, it is not hard to see that the strings $1^m 0^m$ and $1^{m-1} 0^{m+1}$ have the same r-type. This implies, for example, that the class $\{1^n 0^n \mid n \geq 1\}$ cannot be defined by a first-order sentence.

Example 2. We may consider Boolean algebras as structures of vocabulary $\{\sqcup, \sqcap, \neg, 0, 1\}$. In particular, let $\mathfrak{P}(n)$ denote the power-set algebra over $[1, n]$. It is not hard to prove that for each $r \geq 1$ there exists an n such that $\mathfrak{P}(n) \sim_r \mathfrak{P}(n+1)$. Thus the class $\{\mathfrak{P}(n) \mid n \text{ even}\}$ cannot be defined by a first-order sentence.

In some applications, it is convenient to modify the EF-game as follows: Instead of choosing an element in a round of the game, the spoiler may also skip the round. In this case, v_i and v'_i remain undefined; we may also write $v_i = v'_i = \bot$. Of course undefined v_is are not considered in the decision whether the duplicator wins. It is obvious that the duplicator has a winning strategy for the r-round *modified EF-game* on $\mathfrak{A}, \mathfrak{A}'$ if, and only if, she has a winning strategy for the original r-round EF-game on $\mathfrak{A}, \mathfrak{A}'$.

2.2 Order Invariant First-Order Logic

Definition 2. *Let τ be a vocabulary that does not contain $\leq$ and $\mathcal{C}$ a class of τ-structures. A formula $\varphi(x_1, \ldots, x_k)$ of vocabulary $\tau \cup \{\leq\}$ is* order-invariant *on $\mathcal{C}$ if for all $\mathfrak{A} \in \mathcal{C}$, $a_1, \ldots, a_k \in A$, and linear orders $\leq_1, \leq_2$ of A we have*

$$(\mathfrak{A}, \leq_1) \models \varphi(a_1, \ldots, a_k) \iff (\mathfrak{A}, \leq_2) \models \varphi(a_1, \ldots, a_k).$$

If φ is order invariant on the class $\{\mathfrak{A}\}$ we also say that φ is order-invariant on $\mathfrak{A}$.

To simplify our notation, if a $\tau \cup \{\leq\}$-formula $\varphi(\bar{x})$ is order-invariant on a class $\mathcal{C}$ of τ-structures and $\mathfrak{A} \in \mathcal{C}, \bar{a} \in A$ we write $\mathfrak{A} \models_{\text{inv}} \varphi(\bar{a})$ to denote that for some, hence for all orderings $\leq$ on A we have $(\mathfrak{A}, \leq) \models \varphi(\bar{a})$. Furthermore, we say that $\varphi(\bar{x})$ defines the query $\mathfrak{A} \mapsto \{\bar{a} \mid \mathfrak{A} \models_{\text{inv}} \varphi(\bar{a})\}$ on $\mathcal{C}$.[1] Let us emphasize that, although *order-invariant first-order logic* sounds like a restriction of pure first-order logic, it is actually an extension: There are queries on the class of all finite structures that are definable by an order-invariant first-order formula, but not by a pure first-order formula [6]. The following example can be found in [4].

Example 3. There is an order-invariant first-order sentence φ of vocabulary $\{\sqcup, \sqcap, \neg, 0, 1, \leq\}$ that defines the query $\{\mathfrak{P}(n) \mid n \text{ even}\}$ on the class of all finite Boolean algebras. By Example 2, this query is not definable in first-order logic.

2.3 Local Formulas

Let $\mathfrak{A}$ be a τ-structure. The *Gaifman graph* of $\mathfrak{A}$ is the graph with universe A where $a, b \in A$ are adjacent if they occur in a tuple $\bar{c}$ of some relation of $\mathfrak{A}$. The distance $d^{\mathfrak{A}}(a, b)$ between two elements $a, b \in A$ is defined to be the length of a shortest path from a to b in the Gaifman graph of $\mathfrak{A}$; if no such path exists we let $d^{\mathfrak{A}}(a, b) = \infty$. The δ-*ball* around $a \in A$ is defined to be the set $B_{\delta}^{\mathfrak{A}}(a) = \{b \in A \mid d^{\mathfrak{A}}(a, b) \leq \delta\}$, and the δ-*sphere* is the set $S_{\delta}^{\mathfrak{A}}(a) = \{b \in A \mid d^{\mathfrak{A}}(a, b) = \delta\}$. If $\mathfrak{A}$ is clear from the context, we usually omit the superscript $\mathfrak{A}$. For sets $B, C \subseteq A$ we let $d(B, C) = \min\{d(b, c) \mid b \in B, c \in C\}$ and $B_{\delta}(B) = \bigcup_{b \in B} B_{\delta}(b)$, $S_{\delta}(B) = \bigcup_{b \in B} S_{\delta}(b)$. For tuples $\bar{a} = a_1 \ldots a_k, \bar{b} = b_1 \ldots b_l \in A$ we let $d(\bar{a}, \bar{b}) = d(\{a_1, \ldots, a_k\}, \{b_1, \ldots, b_l\})$, $B_{\delta}(\bar{a}) = B_{\delta}(\{a_1, \ldots, a_k\})$, and $S_{\delta}(\bar{a}) = S_{\delta}(\{a_1, \ldots, a_k\})$.

Definition 3. *(1) A k-ary query ρ on a class $\mathcal{C}$ is* local *if there exists a $\lambda \geq 0$ such that for all $\mathfrak{A} \in \mathcal{C}$ and $\bar{a}, \bar{b} \in A^k$ we have*

$$\langle B_{\lambda}(\bar{a})\rangle^{\mathfrak{A}} \cong \langle B_{\lambda}(\bar{b})\rangle^{\mathfrak{A}} \implies (\bar{a} \in \rho(\mathfrak{A}) \iff \bar{b} \in \rho(\mathfrak{A})).$$

The least such λ is called the locality rank *of ρ.*
(2) A formula $\varphi(\bar{x})$ that is order-invariant on a class $\mathcal{C}$ is local, *if the query it defines is local. The* locality rank *of $\varphi(\bar{x})$ is the locality rank of this query.*

It should be emphasized that, in the definition of local order-invariant formulas, neither the isomorphisms nor the distance function refer to the linear order.

Gaifman [5] has proved that first-order formulas can only define local queries.

3 Locality of Invariant Formulas with One Free Variable

In this section we are going to show the locality of order-invariant first-order formulas with one free variable. Before we formally state and prove this result, we need some preparation.

---

[1] This is ambiguous because $\varphi(\bar{x})$ also defines a query on the class of all $\tau \cup \{\leq\}$-structures. But if we speak of a query defined by an order-invariant formula, we always refer to the query defined in the text.

Lemma 1. For all $l, r \in \mathbb{N}$ there are $m, n \in \mathbb{N}$ such that for all l-strings s of size at least n there are unary relations P and P' on s such that (1) $|P| = m$, (2) $|P'| = m - 1$, and (3) $(s, P) \sim_r (s, P')$.

Proof. Let $l, r \in \mathbb{N}$ be fixed and t the number of r-types of vocabulary τ_l. We let $m = 2^r + 1$ and choose n large enough such that whenever the edges of a complete graph with n vertices are colored with t colors, there is an induced subgraph of size $2m + 1$ all of whose edges have the same color.

Let $s = s_1 \cdots s_{n'}$ be an l-string of length $n' \geq n$. For $i < j \leq n'$ we let $\langle i, j \rangle$ denote the l-substring $s_i \cdots s_j$.

For $i < j \leq n'$ we color the pair $\{i, j\}$ (that is, the edge $\{i, j\}$ of the complete graph on $[1, n]$) with the r-type of (the representation of) $\langle i, j \rangle$. By the choice of n we find $2m + 1$ vertices $p_1 < \ldots < p_{2m+1} \leq n'$ such that all structures $\langle p_i, p_j \rangle$, for $i < j \leq 2m + 1$, have the same r-type. We let $P = \{p_1, \ldots, p_m\}$ and $P' = \{p_1, \ldots, p_{m-1}\}$.

We claim that $(s, P) \sim_r (s, P')$. Intuitively, we prove this claim by carrying over a winning strategy for the duplicator on the strings $u = 1^m 0^m$ and $u' = 1^{m-1} 0^{m+1}$ to our structures. Recall from Example 1 that such a strategy exists. Formally, we proceed as follows: We define a mapping $f : [1, n'] \to [1, 2m] \cup \{\bot\}$ by

$$f(x) = \begin{cases} i & \text{if } p_i \leq i < p_{i+1} \\ \bot & \text{otherwise} \end{cases}$$

Consider the r-round EF-game on (s, P), (s, P'). As usual, let v_i and v_i' be the elements chosen in round i. It is not too difficult to prove, by induction on i, that the duplicator can play in such a way that for every $i \leq r$ one of the following conditions holds:

(1) $v_i < p_1$ and $v_i' = v_i$.
(2) $v_i > p_{2m+1}$ and $v_i = v_i'$.
(3) $p_1 \leq v_i \leq p_{2m+1}$ and the following two subconditions hold:
   (a) The duplicator has a winning strategy for the $(r - i)$-round modified EF-game on $(u, f(v_1), \ldots, f(v_i))$ and $(u', f(v_1'), \ldots, f(v_i'))$.
   (b) The duplicator has a winning strategy for the $(r - i)$-round modified EF-game on $(\langle p_{f(v_i)}, p_{f(v_i)+1} \rangle, g(v_1), \ldots, g(v_i))$ and $(\langle p_{f(v_i')}, p_{f(v_i')+1} \rangle, g'(v_1'), \ldots, g'(v_i'))$, where g is the identity on $\langle p_{f(v_i)}, p_{f(v_i)+1} \rangle$ and $\bot$ everywhere else and g' is the identity on $\langle p_{f(v_i')}, p_{f(v_i')+1} \rangle$ and $\bot$ everywhere else.

Clearly, this implies the claim and thus the statement of the lemma. □

Lemma 2. *If a first-order formula $\varphi(x)$ is order-invariant on a class $\mathcal{C}$ of structures then it is local on $\mathcal{C}$.*

Proof. Let $\varphi(x)$ be a first-order formula of quantifier-depth r that is order-invariant on a class $\mathcal{C}$ of τ-structures.

Let l_0 be the number of different r-types of vocabulary $\tau \cup \{Q_0, \ldots, Q_{2^r}\}$, where the Q_i are new unary relation symbols and let $l := l_0^2$. Let m and n be given by Lemma 1 above w.r.t. r and l. Let $\kappa := n(2^r + 1) + 2^r$ and let $\lambda := 5\kappa + 1$.

Let $\mathfrak{A} \in \mathcal{C}$. Let $a, b \in A$, where $B_\lambda^{\mathfrak{A}}(a) \cong B_\lambda^{\mathfrak{A}}(b)$ via an isomorphism π.

Our goal is to show that there are linear orders $\leq_1$ and $\leq_2$ on A such that $(\mathfrak{A}, \leq_1, a) \sim_r (\mathfrak{A}, \leq_2, b)$. From this we can conclude

$$\mathfrak{A} \models_{\text{inv}} \varphi(a) \iff (\mathfrak{A}, \leq_1) \models \varphi(a) \iff (\mathfrak{A}, \leq_2) \models \varphi(b) \iff \mathfrak{A} \models_{\text{inv}} \varphi(b)$$

In order to prove the existence of such linear orders, we first show that, w.l.o.g., we can assume the following.

(∗) There is a set $W \supseteq \{a, b\}$, and an automorphism ρ on $\langle B_\kappa(W) \rangle$ such that $\rho(a) = b$. To show this, we distinguish the following two cases.

- Case 1: $d(a, b) > 2\kappa$. In this case we simply set $W := \{a, b\}$ and define ρ by

$$\rho(x) := \begin{cases} \pi(x) & \text{if } x \in B_\kappa(a) \\ \pi^{-1}(x) & \text{if } x \in B_\kappa(b) \end{cases}$$

- Case 2: $d(a, b) \leq 2\kappa$. Assume first that $d(a, \pi^i(a)) > 4\kappa$, for some $i > 0$. Then we also have $d(b, \pi^i(a)) > 2\kappa$. Furthermore, by the choice of λ, $B_\kappa(a) \cong B_\kappa(\pi^i(a)) \cong B_\kappa(b)$. We can conclude from the proof given below that

$$\mathfrak{A} \models_{\text{inv}} \varphi(a) \iff \mathfrak{A} \models_{\text{inv}} \varphi(\pi^i(a)) \iff \mathfrak{A} \models_{\text{inv}} \varphi(b).$$

If, on the other hand, $d(a, \pi^i(a)) \leq 4\kappa$, for every i, we set $W = \{\pi^i(a) \mid i \in \mathbb{Z}\}$ and $\rho = \pi$.

Hence, we can assume (∗). In the following we only make use of $B_\kappa(a) \cong B_\kappa(b)$ (as opposed to $B_\lambda(a) \cong B_\lambda(b)$).

It is easy to see that every sphere $S_i(W)$ is a disjoint union of orbits of ρ, i.e. a disjoint union of sets of the form $O(v) = \{\rho^j(v) \mid j \in \mathbb{Z}\}$, for some v. We fix, for every i, some linear order of the orbits of the sphere $S_i(W)$. Next we fix a preorder $\prec$ on A with the following properties.

- $\prec$ is a linear order on $A - B_\kappa(W)$,
- $c \prec c'$, whenever $c \in B_\kappa(W)$ and $c' \in A - B_\kappa(W)$,
- $c \prec c'$, whenever $c, c' \in B_\kappa(W)$ and $d(c, W) < d(c', W)$,
- $c \prec c'$, whenever $c, c' \in B_\kappa(W)$, c and c' are in the same sphere $S_i(W)$ but the orbit of c comes before the orbit of c in the order of the orbits of $S_i(W)$, and
- c and c' are not related with respect to $\prec$, whenever $c, c' \in B_\kappa(W)$ and c and c' are in the same orbit.

Both linear orders $\leq_1$ and $\leq_2$ will be refinements of $\prec$. They will only differ inside some of the orbits.

We can assume that no sphere $S_i(W)$, with $i \leq \kappa$, is empty. Otherwise, $B_\kappa(W)$ would be a union of connected components of $\mathfrak{A}$, hence we could fix any linear order $\leq$ on the orbits of $B_\kappa(W)$ and define $\leq_1$ by combining $\leq$ with $\prec$ and $\leq_2$ by combining the image of $\leq$ under ρ with $\prec$.

For each orbit O, we fix a vertex $v(O)$ and define a linear order $\leq^0$ on O by $v(O) \leq^0 \rho(v(O)) \leq^0 \rho^2(v(O)) \leq^0 \cdots$, if O is finite and by $\cdots \leq^0 \rho^{-1}(v(O)) \leq^0 v(O) \leq^0$

$\rho(v(O)) \leq^0 \cdots$, if O is infinite. For every k, we denote by $\leq^k$ the image of $\leq^0$ under ρ^k. It is easy to see that $(S_i(W), \leq^j) \cong (S_i(W), \leq^{j'})$, for all i, j, j'.

To catch the intuitive idea of the proof, the reader should picture the spheres $S_i(W)$ (for $0 \leq i \leq \kappa$) as a sequence of concentric cycles, W itself being innermost. Outside these cycles is the rest of the structure $\mathfrak{A}$, fixed once and for all by the order $\prec$. The automorphism ρ is turning the cycles, say, clockwise. In particular, it turns the cycle W far enough to map a to b. Each cycle is ordered clockwise by $\leq^0$. The ordering $\leq^k$ is the result of turning the cycle k-steps. (Unfortunately, all this is not exactly true, because usually the orbits do not form whole spheres. They may form small cycles or "infinite cycles". But essentially it is the right picture.)

To define the orders $\leq_1$ and $\leq_2$ we proceed as follows. We determine two sequences $1 \leq j_1 \leq \cdots \leq j_m \leq \kappa - 1$ and $1 \leq k_1 \leq \cdots \leq k_{m-1} \leq \kappa - 1$. We define $\leq_1$ sphere-wise. On W we let $\leq_1 = \leq^0$ and $\leq_2 = \leq^1 = \rho(\leq^0)$. Then $\leq_1$ looks from a as $\leq_2$ looks from b, and this is how it should be. For all $j < j_1$ we leave $\leq_1 = \leq^0$ on $S_j(W)$ but once we reach j_1 we turn it one step. That is, we let $\leq_1 = \leq^1$ on $S_{j_1}(W)$. We stick with this, until we reach $S_{j_2}(W)$, and there we turn again and let $\leq_1 = \leq^2$. We go on like this, and after the last turn at $S_{j_m}(W)$ we have $\leq_1 = \leq^m$, and that is what we wanted. Similarly, we define $\leq_2$ by starting with $\leq^1$ and taking turns at all spheres k_i, for $1 \leq i \leq m-1$. Again we end up with $\leq_2 = \leq^m$ on all spheres $S_k(W)$ for $k \geq k_{m-1}$, hence, on the outermost cycle $S_\kappa(W)$ both orderings are the same.

But of course the turns can be detected, so how can we hide that we took one more turn in defining $\leq_1$? The idea is to consider the sequence of spheres as a long string, whose letters are the types of the spheres. The positions where a turn is taken can be considered as a unary predicate on this string. By Lemma 1, we can find unary predicates of sizes m and $m-1$, respectively, such that the expansions of our string by these predicates are indistinguishable. This is exactly what we need.

Essentially, this is what we do. But of course there are nasty details ...

Let $h = 2^r + 1$. For every i with $1 \leq i \leq n$ and $j < j' \leq 2^r$ let $T^i_{(j,j')}$ denote the substructure of $\mathfrak{A}$ that is induced by the spheres $S_{ih-j}(W), \ldots, S_{ih+j'}(W)$. Furthermore let, for every i, $1 \leq i \leq n$, the i-th *super-sphere* T^i be the structure $T^i_{(2^r, 2^r)}$.

Let the linear order $\leq^j$ on T^i be defined by combining the orders $\leq^j$ on the spheres of T^i with $\prec$. Finally let $\trianglelefteq^j$ be the linear order on T^i that is obtained by combining $\prec$ with $\leq^j$, for the spheres $S_q(W)$ with $q \leq ih$, and with $\leq^{j+1}$ for the spheres $S_q(W)$ with $q > ih$. For every i, j, j' it holds $(T^i, \leq^j) \cong (T^i, \leq^{j'})$ and $(T^i, \trianglelefteq^j) \cong (T^i, \trianglelefteq^{j'})$.

For every i, we define the unary relations $Q_0, \ldots, Q_{2^r}$ on T_i by $Q_j = S_{ic-j}(W) \cup S_{ic+j}(W)$, i.e., a vertex v is in Q_j, if its distance from the central sphere in T^i is j. Now we define an l-string $s = s_1 \cdots s_n$ as follows. Let $z_1, \ldots, z_l$ be an enumeration of all pairs of r-types of $\tau \cup \{Q_0, \ldots, Q_{2^r}\}$-structures. We set $s_i = \alpha_j$ whenever z_j is the pair (r-type of $(T^i, \overline{Q}, \leq^0)$, r-type of $(T^i, \overline{Q}, \trianglelefteq^0)$).

By Lemma 1 and our choice of the parameters l, r, m, n there exist unary relations P and P' such that $|P| = m$, $|P'| = m - 1$ and the duplicator has a winning strategy in the r-round game on (s, P) and (s, P'). Now we are ready to define the linear orders $\leq_1$ and $\leq_2$ on A. For every i, let $u(i) = |\{j < i \mid j \in P\}|$ and $u'(i) = |\{j < i \mid j \in P\}|$.

- $\leq_1$ is defined on T^i as $\leq^{u(i)}$, if $i \notin P$ and as $\trianglelefteq^{u(i)}$, if $i \in P$.
- $\leq_2$ is defined on T^i as $\leq^{u(i)}$, if $i \notin P'$ and as $\trianglelefteq^{u(i)}$, if $i \in P'$.

Observe that, although T^i and T^{i+1} are not disjoint, these definitions are consistent. It remains to show that the duplicator has a winning strategy in the r-round game on $(\mathfrak{A}, \leq_1, a)$ and $(\mathfrak{A}, \leq_2, b)$. The proof of this fact is given in the full version. The winning strategy of the duplicator is obtained by transferring the winning strategy on (s, P) and (s, P'), making use of the gap preserving technique that was invented in [11]. □

4 Locality of Invariant Formulas with Arbitrarily Many Free Variables

Lemma 3. *Let τ be a vocabulary and $r \geq 0, k \geq 1$. Then there exists a $\kappa = \kappa(\tau, r, k)$ such that the following holds: If $\varphi(x_1, \ldots, x_k)$ is a first-order formula of vocabulary τ and quantifier-depth at most r that is order-invariant on a τ-structure $\mathfrak{A}$, then for all $\bar{a} = a_1 \ldots a_k, \bar{b} = b_1 \ldots b_k \in A^k$ with $d(a_i, a_j), d(b_i, b_j) > 2\kappa$ (for $1 \leq i < j \leq k$) we have*

$$\langle B_\kappa(\bar{a})\rangle^{\mathfrak{A}} \cong \langle B_\kappa(\bar{b})\rangle^{\mathfrak{A}} \implies (\mathfrak{A} \models \varphi(\bar{a}) \iff \mathfrak{A} \models \varphi(\bar{b})).$$

Proof. We only sketch the proof. Details are given in the full version.

The proof is by induction on k. For $k = 1$ the lemma just restates the locality of order-invariant first-order formulas with one free variable, proved in Lemma 2.

For $k > 1$, we assume that we have k-tuples $\bar{a}, \bar{b}$ in $\mathfrak{A}$ such that all the a_i, a_j and b_i, b_j are far apart (as the hypothesis of the Lemma requires) and we have an isomorphism $\pi : \langle B_\kappa(\bar{a})\rangle^{\mathfrak{A}} \cong \langle B_\kappa(\bar{b})\rangle^{\mathfrak{A}}$ for a sufficiently large κ. We prove that $\bar{a}$ and $\bar{b}$ cannot be distinguished by order-invariant formulas of vocabulary τ and quantifier-depth at most r. We distinguish between three cases:

The first is that some b_i, say, b_k, is far away from $\bar{a}$. Then we can treat $a_1, \ldots, a_{k-1}$ as constants and apply Lemma 2 to show that a_k and b_k cannot be distinguished in the expanded structure $(\mathfrak{A}, a_1, \ldots, a_{k-1})$. (Here we use the hypothesis $d(a_i, a_j) > 2k$ for all $i \leq k - 1$). Then we treat b_k as a constant and apply the induction hypothesis to prove that the $(k-1)$-tuples $a_1 \ldots a_{k-1}$ and $b_1 \ldots b_{k-1}$ cannot be distinguished in the expanded structure $(\mathfrak{A}, b_k)$. (This requires our hypothesis that $d(b_i, b_k) > 2\kappa$ for all $i \leq k - 1$.)

The second case is similar, we assume that for some $h \geq 1$ the iterated partial isomorphism π^h maps some a_i far away from $\bar{a}$. Then we first show that $\bar{a}$ and $\pi^h(\bar{a})$ cannot be distinguished and then that $\pi^h(\bar{a})$ and $\bar{b}$ cannot be distinguished.

The third case is that for all $h \geq 1$ the entire tuple $\pi^h(\bar{a})$ is close to $\bar{a}$. Then some restriction of π is an automorphism of a substructure of $\mathfrak{A}$ that maps $\bar{a}$ to $\bar{b}$. We can modify this substructure in such a way that the tuples $\bar{a}$ and $\bar{b}$ can be encoded by single elements and then apply Lemma 2. □

Theorem 2. *Every first-order formula that is order-invariant on a class $\mathcal{C}$ of structures is local on $\mathcal{C}$.*

Proof. Again we only sketch the proof. For more details we refer the reader to the full version. The proof is by induction on the number k of free variables of a formula. We have already proved that formulas with one free variable are local.

So let $\varphi(x_1, \ldots, x_k)$ be invariant on $\mathcal{C}$, $\mathfrak{A} \in \mathcal{C}$, and $\bar{a}, \bar{b} \in A^k$ such that $\langle B_\lambda(\bar{a})\rangle^{\mathfrak{A}} \cong \langle B_\lambda(\bar{b})\rangle^{\mathfrak{A}}$ for a sufficiently large λ. Either all the a_i, a_j and b_i, b_j are far apart, then we can apply Lemma 3, or some of them are close together. In the latter case, we define a new structure where we encode pairs of elements of $\mathfrak{A}$ that are close together by new elements. This does not spoil the distances too much, and we can encode our k-tuples by smaller tuples that still have isomorphic neighborhoods. On these we apply the induction hypothesis. □

5 Further Research

The obvious question following our result is: What else can be added to first-order logic such that it remains local. Hella [7] proved that invariant first-order formulas that do not only use an order, but also addition and multiplication, are *not* local. On the other hand, we conjecture that just adding order and addition does not destroy locality.

However, the fact that invariant formulas with built-in addition and multiplication are not local is more relevant to complexity theory, since first-order logic with built-in addition and multiplication captures uniform AC_0. One way to apply locality techniques to complexity theoretic questions in spite of Hella's non-locality result is to weaken the notion of locality. For example, it is conceivable that all invariant AC_0 or even TC_0-queries are local in the sense that if two points of a structure of size n have isomorphic neighborhoods of radius $O(\log n)$, then they are indistinguishable.

This would still be sufficient to separate LOGSPACE from these classes.

References

1. S. Abiteboul, R. Hull, and V. Vianu. *Foundations of Databases*. Addison-Wesley, 1995.
2. O. Belegradek, A. Stolboushkin, and M. Taitslin. Extended order-generic queries, 1997. Submitted for publication.
3. G. Dong, L. Libkin, and L. Wong. Local properties of query languages. In *Proceedings of the 6th International Conference on Database Theory*, volume 1186 of *Lecture Notes in Computer Science*, pages 140–154. Springer-Verlag, 1997.
4. H.-D. Ebbinghaus and J. Flum. *Finite Model Theory*. Springer-Verlag, 1995.
5. H. Gaifman. On local and non-local properties. In *Proceedings of the Herbrand Symposium, Logic Colloquium '81*. North Holland, 1982.
6. Y. Gurevich. Private communication.
7. L. Hella. Private communication.
8. L. Hella, L. Libkin, and Y. Nurmonen. Notions of locality and their logical characterizations over finite models, 1998. unpublished.
9. L. Libkin. On forms of locality over finite models. In *Proceedings of the 12th IEEE Symposium on Logic in Computer Science*, pages 204–215, 1997.
10. L. Libkin. On counting and local properties. To appear in *Proceedings of the 13th IEEE Symposium on Logic in Computer Science*, 1998.
11. T. Schwentick. Graph connectivity and monadic NP. In *Proceedings of the 35th Annual IEEE Symposium on Foundations of Computer Science*, pages 614–622, 1994.

Probabilistic Concurrent Constraint Programming: Towards a Fully Abstract Model

Alessandra Di Pierro and Herbert Wiklicky

{adp,herbert}@cs.city.ac.uk
City University, Northampton Square, London EC1V OHB

Abstract. This paper presents a Banach space based approach towards a denotational semantics of a probabilistic constraint programming language. This language is based on the concurrent constraint programming paradigm, where randomness is introduced by means of a probabilistic choice construct. As a result, we obtain a declarative framework, in which randomised algorithms can be expressed and formalised. The denotational model we present is constructed by using functional-analytical techniques. As an example, the existence of fixed-points is guaranteed by the Brouwer-Schauder Fixed-Point Theorem. A concrete fixed-point construction is also presented which corresponds to a notion of observables capturing the exact results of both finite and infinite computations.

1 Introduction

Probabilistic Concurrent Constraint Programming (PCCP) was introduced in [4] in order to allow the formulation of randomised algorithms within the declarative framework of Concurrent Constraint Programming (CCP) [13]. The main feature of this language is a construct for probabilistic choice expressing a kind of nondeterminism which allows a program to make stochastic moves during its execution. An operational semantics describing such a behaviour was also given in [4]. The ultimate aim of this work is to provide PCCP with a denotational semantics which is fully abstract with respect to the notion of observables introduced in [4], and corresponding to the exact results of both finite and infinite computations.

One major problem that makes this task difficult is the presence in the language of nondeterminism (though in its *probabilistic*, thus more refined, version) in combination with synchronisation. In fact, any model reflecting these two aspects cannot be 'too abstract'; information about the branching structure and the synchronisation cannot be ignored, for which relatively complex structures are usually required like, for instance, the *reactive sequences* [3] or the *bounded traces operators* [13] adopted for CCP.

Another problem, somehow orthogonal to the first one, arises from the combination of (probabilistic) nondeterminism and infinite computations, in that it is difficult to find the appropriate structure of the domain where limits can be characterised by a fixed-point operator.

As a first step, we concentrate in this paper on the second problem and we abstract from the problem of synchronisation. We therefore define a denotational semantics capturing both probabilistic nondeterminism and infinite limit results for a sub-language of PCCP which has no suspension mechanism (all the guards are *true*). This language corresponds to Constraint Logic Programming where the or-nondeterminism is replaced by a probabilistic choice among the input clauses. Thus we call it Probabilistic Constraint Logic Programming (PCLP).

The domains we will consider for the semantics of PCLP are based on linear structures, that is on vector spaces and their structure preserving morphisms: linear mappings and operators. Vector spaces provide a common and most widely used model for various sciences, ranging from physics to economics, but they are much less popular in computer science. In the context of our investigations they come into considerations because they combine quantitative and qualitative concepts. This is useful in PCCP (PCLP), as a computation in this paradigm incorporates some quantitative information, besides the usual qualitative one, in the form of probabilities associated to the choice. Furthermore, vector spaces are very well studied mathematical structures, which makes it possible to utilise a great number of well established results.

We argue that the introduction of quantitative aspects in the semantics of CCP plays a fundamental role in modelling the program behaviour. Thanks to the ability to measure the "strength" of a constraint (quality) by means of the probability (quantity) assigned to it, our denotational model succeeds in capturing some observable behaviours that the more classical powerdomain or metric based approaches fail to capture. More specifically, while the Smyth powerdomain and the metric approaches to the semantics of constraint programming have been shown unable to model the *exact* (infinite) results of a computation [1], the probabilistic model we define in this paper perfectly matches this behaviour. Moreover, it can be used also for the standard (non-probabilistic) version of Constraint Logic Programming.

2 Probabilistic Constraint Logic Programming

In [4] we introduce the language PCCP, which is essentially CCP where the nondeterministic choice is replaced by a probabilistic one[1]. This allows us to see the execution of a program as a random walk on the transition graph. Probabilistic Constraint Logic Programming is the sub-language of PCCP obtained by replacing all the guards in the probabilistic choice construct by *true*. This eliminates the aspect of synchronisation from the language, thus allowing us to abstract (for the time being) from this problem.

The syntax of PCLP is given in Table 1. Successful termination is expressed by the agent **stop**; the agent **tell**, the hiding operator $\exists_x$ and the procedure call $p(x)$ are the usual ones (of CCP). The operator $\|$ expresses the parallel

---

[1] Another approach to incorporate probabilistic aspects into CCP languages was introduced later in [7]. It is based on the use of random variables and is substantially different in both the aim and the method from our approach.

$$P ::= D.A$$

$$D ::= \epsilon \mid D.D \mid p(x) : -A$$

$$A ::= \textbf{stop} \mid \textbf{tell}(c) \mid \square_{i=1}^{n} \ true \mid p_i \to A_i \mid A \parallel A \mid \exists_x A \mid p(x)$$

Table 1. The syntax for PCLP.

composition of two agents. Additionally we provide a "probabilistic" choice $\square$. Operationally this construct expresses the choice of one of the agents A_i according to the assigned probabilities p_i. The intended meaning of this is the usual interpretation of probability in probability theory: if the choice is repeated (under the same condition and sufficiently often) the relative frequency of executions of an agent A_i is exactly p_i. For the definition of the (cylindric) constraint system underlying the language we refer to [13].

3 Operational Semantics for PCLP

For the operational semantics of PCLP we essentially use the probabilistic transition system introduced in [4]. Randomness is expressed by labels representing the probability that a transition takes place.

A *configuration* represents the state of the system at a certain moment, namely the agent A which has still to be executed, and the current store d. We denote a configuration by $<A,d>$. The probabilistic transition system for PCLP consists of a pair $(Conf, \longrightarrow_p)$, where $Conf$ is a set of configurations and $\longrightarrow_p \subseteq Conf \times \mathbb{R} \times Conf$ is the transition relation defined in Table 2. We denote the transitive closure of transition relation $\longrightarrow_p$ by $\longrightarrow_{p'}^*$, where p' is the product of the probabilities associated to each single step.

3.1 The Observables

The notion of *observables* we consider captures the *exact* results of both finite and infinite computations together with their associated probabilities.

Given a program P, we define the result $\mathcal{R}_P$ of an agent A and an initial store d as the (multi-)set of all pairs $<c,p>$, where c is the least upper bound of the partial constraints accumulated during a computation starting from d; and p is the probability of reaching that result.

$$\mathcal{R}_P(A,d) = \{<c,p> \mid <A,d> \longrightarrow_p^* <B,c> \not\longrightarrow\} \cup \\ \{<\bigsqcup_i d_i, \prod_i p_i> \mid <A,d_0> \longrightarrow_{p_0} \ldots\}.$$

The first term describes the results of finite computations, where the least upper bound of the partial store corresponds to the final store. The second term

$$
\boxed{\begin{array}{l}
\mathbf{R1} \ <\mathbf{tell}(c),d> \to_1 <\mathbf{stop}, c \sqcup d> \\[4pt]
\mathbf{R2} \ <\square_{i=1}^n \ true \mid p_i \to A_i, d> \to_{p_j} <A_j, d> \qquad j \in [1,n] \\[6pt]
\mathbf{R3} \ \dfrac{<A,c> \to_p <A',c'>}{\begin{array}{l}<A \parallel B, c> \to_p <A' \parallel B, c'> \\ <B \parallel A, c> \to_p <B \parallel A', c'>\end{array}} \\[12pt]
\mathbf{R4} \ \dfrac{<A, d \sqcup \exists_x c> \to_p <B, d'>}{<\exists_x^d A, c> \to_p <\exists_x^{d'} B, c \sqcup \exists_x d'>} \\[10pt]
\mathbf{R5} \ <p(y), c> \to_1 <\exists_\alpha(\delta_{y\alpha} \wedge \exists_x(\delta_{\alpha x} \wedge A)), c> \qquad p(x) :-A \in P
\end{array}}
$$

Table 2. The transition system for PCLP.

covers the infinite results. The probability of obtaining a certain result depends on the probabilities p associated to the possible paths which lead to it. To capture the true behaviour of an agent we have to identify different computational paths leading to the same result as well as to collect the accumulated probabilities associated with different interleavings. In order to do this in a precise way we define the following operation. By $c_i j$ we denote the jth occurrence of the constraint c_i in the multi-set of all results.

$$\mathcal{K}(\{<c_{ij}, p_{ij}>\}_{i,j}) = \{<c_i, P_{c_i}> \mid \ P_{c_i} = \textstyle\sum_j p_{ij}\}_i.$$

Another operation normalises the probabilities. This is necessary as the probabilities in each interleaving add up to one such that the overall sum of probabilities is exactly the number of possible interleavings. This process of renormalisation effectively implies that all interleavings are equally likely.

$$\mathcal{N}(\{<c_i, p_i>\}_i) = \{<c_i, \tfrac{p_i}{P}> \mid \ P = \textstyle\sum_i p_i\}.$$

With these two operations we can define the observables associated to an agent A and an initial store d as:

$$\mathcal{O}_P(A,d) = \mathcal{N}(\mathcal{K}(\mathcal{R}_P(A,d))).$$

Note that this notion of observables differs from the classical notion of input/output behaviour in CCP. In the classical case a constraint c belongs to the input/output observables of a given agent A if *at least* one path leads from the initial store d to the final result c. In the probabilistic case we have to consider *all* possible paths leading to the same result c and combine the associated probabilities.

4 A Denotational Semantics for PCLP

To simplify our presentation we will assume in the following that the set of agents $\mathcal{A} = \{A_1, \ldots, A_{|\mathcal{A}|}\}$ is finite, and that the set of constraints $\mathcal{C} = \{c_i\}_{i=0}^{\infty}$ is countable. In the case of uncountable constraint systems we can generalise our approach, replacing sums by integrals, l^1 by L^1, etc. Then most of the results presented here can be transferred into an appropriate measure-theoretic setting.

We assign to each agent a (probability) distribution on the set of constraints.

Definition 1. *A distribution ρ on the set of constraints $\mathcal{C}$ is a map from $\mathcal{C}$ into the real interval $[0,1]$ satisfying the normalisation condition: $\sum_{c \in \mathcal{C}} \rho(c) = 1$. The set of distributions on $\mathcal{C}$ is denoted by $\mathcal{D}(\mathcal{C})$ or simply $\mathcal{D}$.*

We define an *interpretation* $I : \mathcal{A} \mapsto \mathcal{D}$ as a function from the set of agents $\mathcal{A}$ into the set of distributions $\mathcal{D}(\mathcal{C})$ on $\mathcal{C}$. The set of all possible interpretations is denoted by $\mathcal{I}$. For an agent $A \in \mathcal{A}$ we represent its interpretation by $I(A) = \{<c_i, p_i>\}_i$, where $c_i \in \mathcal{C}$ and $p_i = I(A)(c_i)$. We will omit those pairs where the probability vanishes.

The set of possible interpretations of an agent A, $\mathcal{I}(A)$, forms a subset of the (real) *Banach space* $l^1(\mathcal{C})$. The elements of the Banach space $l^1(\mathcal{C})$ are given by sequences of real numbers indexed by the elements of a (countable) constraint system such that the sum of their absolute values exists:

$$l^1(\mathcal{C}) = \{<x_i, c_i> \mid x_i \in \mathbb{R}, c_i \in \mathcal{C} \text{ and } \sum_{c_i \in \mathcal{C}} |x_i| < \infty\}.$$

On the space of sequences $l^1(\mathcal{C})$ we define a scalar product and vector addition *pointwise* and the norm as the usual l^1-norm (with $q, p \in \mathbb{R}$) by:

$$q \cdot \{<c_i, p_i>\}_i = \{<c_i, qp_i>\}_i,$$
$$\{<c_i, p_i>\}_i + \{<c_i, q_i>\}_i = \{<c_i, p_i + q_i>\}_i,$$
$$\|\{<c_i, p_i>\}_i\| = \sum_{c_i \in \mathcal{C}} |p_i|.$$

In order to model all constructs of our language we define two additional operations on this space: a tensor product $\otimes$ and a *pointwise* hiding operator:

$$\{<c_i, p_i>\}_i \otimes \{<c_j, q_j>\}_j = \{<c_i \sqcup d_j, p_i q_j>\}_{i,j},$$
$$\exists_x \{<c_i, p_i>\}_i = \{<\exists_x c_i, p_i>\}_i.$$

We can embed the set of all possible interpretations, $\mathcal{I}$, in a similarly defined Banach space $l^1(\mathcal{C})^{|\mathcal{A}|}$, i.e. the (finite) cartesian product of $|\mathcal{A}|$ copies of $l^1(\mathcal{C})$.

Proposition 1. *The space of interpretations $\mathcal{I}$ forms a convex, closed (non-empty) subset of the Banach space $l^1(\mathcal{C})^{|\mathcal{A}|}$.*

$\Phi(I)(\mathbf{stop})$	$= \{<true, 1>\}$
$\Phi(I)(\mathbf{tell}(c))$	$= \{<c, 1>\}$
$\Phi(I)(\Box_{i=1}^n\ true\ \|\ p_i \to A_i)$	$= \sum_{i=1}^n p_i \cdot \Phi(I)(A_i)$
$\Phi(I)(A_1\ \|\ A_2)$	$= \Phi(I)(A_1) \otimes \Phi(I)(A_2)$
$\Phi(I)(\exists_x A)$	$= \exists_x \Phi(I)(A)$
$\Phi(I)(p(x))$	$= I(\Delta_y^x A)$

Table 3. The compositional definition of $\Phi : \mathcal{I} \to \mathcal{I}$.

On the set of interpretations $\mathcal{I}$ we define inductively the fixed-point operator Φ as in Table 3 (where $\Delta_y^x A$ is a shorthand notation for $\exists_\alpha(\delta_{y\alpha} \wedge \exists_x(\delta_{\alpha x} \wedge A))$ as in **R5** in Table 2). Some useful properties of this operator are stated in the following proposition.

Proposition 2. *The operator Φ is well-defined on $\mathcal{I} \subset l^1(\mathcal{C})^{|\mathcal{A}|}$ and has the following properties: (i) Φ is continuous, and (ii) Φ is compact, i.e. the closure of $\Phi(\mathcal{I})$ is compact.*

Proof. (Idea) Ad (i): Φ is linear and bound and therefore continuous, ad (ii): Φ is the limit of finite-dimensional operators as PCLP is finitely branching. □

To guarantee the existence of a fixed-point of Φ we use a classical theorem from functional analysis [5, Theorem 18.10'].

Theorem 1. *(Brouwer-Schauder Theorem) Let $F : K \mapsto K$ be a continuous mapping from a non-empty closed, convex set K in a Banach space into itself, with the closure of $F(K)$ compact. Then there exists a fixed-point of F, i.e. a point $c \in K$ such that $F(c) = c$.*

By Theorem 1 and Propositions 1 and 2, we can guarantee:

Theorem 2. *The operator Φ has a fixed-point.*

4.1 Construction of a Fixed-Point

In order to concretely construct a fixed-point of Φ we will mimic the classical fixed-point construction: Starting with the initial interpretation I_0 assigning to each agent A the distribution $I_0(A) = \{<true, 1>\}$, we iteratively apply Φ in order to construct a (pointwise) limit of the sequence of interpretations $\{I_n\}_n =$

$I_0, \Phi(I_0), \Phi^2(I_0), \ldots, \Phi^n(I_0), \ldots$. This limit will be a fixed-point of Φ because of continuity. To show convergence we need some auxiliary constructions.

We introduce the notion of *volume* of a constraint with respect to a distribution. This is roughly the probability concentrated in the upward closure $\uparrow c$ of a constraint $c \in \mathcal{C}$ (with respect to $\sqsubseteq$ in the constraint system) and will be essential in the construction of the limit interpretation defining the meaning of our programs.

Definition 2. *Given a distribution $\rho \in \mathcal{D}$, we define the volume of a constraint c with respect to ρ as*

$$vol_\rho(c) = \sum_{d \in \uparrow c} \rho(d).$$

There is a one-to-one correspondence between the original distribution ρ and the distribution of volumes vol_ρ. Using a general inclusion-exclusion principle, e.g. [6, Eqn. 3.3], we can show the following lemma.

Lemma 1. *Given the volume $vol_\rho(c)$ of each constraint $c \in \mathcal{C}$ with respect to a distribution $\rho \in \mathcal{D}$ it is possible to reconstruct the distribution ρ uniquely, by*

$$\rho(c) = vol(c) - \sum_{d>c} vol(d) + \sum_{d>e>c} vol(d \sqcup e) - \sum_{f>d>e>c} vol(d \sqcup e \sqcup f) + \ldots$$

The sequence $\{I_n\}_n$ in general is not pointwise monotone (e.g. example 2 below), therefore it is not obvious how to prove its convergence directly. However, it is easy to see that the corresponding sequence of volume distributions does converge.

Lemma 2. *Let $A \in \mathcal{A}$, $c \in \mathcal{C}$ and $\{I_n\}_n$ the sequence defined above. Then the sequence $\{vol_{I_n(A)}(c)\}_n$ converges.*

Proof. (Sketch) For each constraint $c \in \mathcal{C}$ the following holds $\forall n \in \mathbb{N}$:

- $vol_{I_n(A)}(c) \leq 1$, i.e. the volume of each constraint is bound by one in each interpretation, because Φ is "normalised", i.e. maps $\mathcal{I}(A)$ into $\mathcal{I}(A)$.
- $vol_{I_n(A)}(c) \leq vol_{I_{n+1}(A)}(c)$, i.e. the sequence of volumes is monotone (increasing).

Therefore, the limit $\lim_{n\to\infty} vol_{I_n(A)}(c)$ of a monotone and bound sequence of real numbers $vol_{I_n(A)}(c)$ exists. □

By Lemma 1 and continuity of Φ we can reconstruct the pointwise limit of distributions from the pointwise limit of volumes of the constraints.

Theorem 3. *The sequence $\{<c, I_n(A)(c)>\}_n$ converges pointwise to a fixed-point of Φ.*

We are now in a position to define a semantics for PCLP.

Definition 3. For each agent $A \in \mathcal{A}$ we define its semantics $Q(A)$ as the point-wise limit of $\{I_n(A)\}_n$,

$$Q(A) = \lim_{n \to \infty} I_n(A) = \lim_{n \to \infty} \Phi^n(I_0(A)).$$

For this semantics we can establish the correspondence with the observables defined in Section 3 by structural induction.

Theorem 4. For all agents $A \in \mathcal{A}$ the fixed-point semantics $Q(A)$ coincides with the observables

$$\mathcal{O}_P(A, true) = Q(A).$$

We would like to point out that alternative fixed-point constructions can be defined which model different notions of observables.

4.2 Examples

Example 1. Consider the following PCLP program for computing the natural numbers:
$$nat(x) :- \; true|\tfrac{1}{2} \to \mathbf{tell}(x = 0)$$
$$\square \; true|\tfrac{1}{2} \to \exists_y(\mathbf{tell}(x = s(y)) \; \| \; nat(y))$$

The sequence of interpretations $I_n(nat(x))$ converges pointwise to

$$Q(nat(x)) = \{<x = 0, 1/2>, <x = s(0), 1/4>, \ldots, <x = s^n(0), 1/2^{n+1}>, \ldots\}.$$

This clearly coincides with the observables $\mathcal{O}_P(nat(x), true)$. Note that, contrary to the classical approach, these observables not only tell us that all numbers may be computed but also that the probability of computing larger numbers decreases.

Example 2. The following declarations have been used in [1] (in their CCP formulation) as an example of the inapplicability of metric and order-theoretic approaches to modelling the exact results observables in constraint programming.
$$p(x) :- \; q(x)$$
$$q(x) :- \; true|\tfrac{1}{2} \to p(x)$$
$$\square \; true|\tfrac{1}{2} \to (r(x) \; \| \; \mathbf{tell}(c))$$
$$r(x) :- \; \mathbf{tell}(\mathit{false}).$$

In [1] it was shown that the interpretations defined by using an analogous of the Φ operator do not converge with respect to any metric or order. In our quantitative semantics we get convergence as the limit $\lim_n I_n$ exists for all three agents,
$$Q(p(x)) = Q(q(x)) = Q(r(x)) = \{<\mathit{false}, 1>\}.$$

5 Future Work

We plan to extend the denotational semantics developed here for PCLP to the full language PCCP. To this purpose it will be necessary to add an appropriate encoding of the branching structure and the synchronisation for dealing with global choice. It seems that we can still use an underlying Banach space structure; however it will be necessary to replace vectors (distributions) by matrices (operators) in order to keep track of the computational traces. We expect this model to be the 'quantitative' counterpart of the various (equivalent) fully abstract models developed until now for CCP. The only two attempts to describe the exact results of infinite computations in CCP we are aware of are [11] and [2], whereas the two approaches [3, 13] we already mentioned in the introduction have been shown to be fully abstract only with respect to the results of finite computations.

Additional investigations will compare our construction to other approaches towards the semantics of probabilistic programming languages [12, 9, 8], probabilistic predicate transformers [10] and logics and stochastic processes.

References

[1] F. S. de Boer, A. Di Pierro, and C. Palamidessi. Nondeterminism and Infinite Computations in Constraint Programming. *Theoretical Computer Science*, 151(1), 1995. Selected Papers of the Workshop on Topology and Completion in Semantics, Chartres, France.

[2] F. S. de Boer and M. Gabbrielli. Infinite Computations in Concurrent Constraint Programming. *Electronic Notes in Theoretical Computer Science*, 6:16, 1997.

[3] F.S. de Boer and C. Palamidessi. A Fully Abstract Model for Concurrent Constraint Programming. In S. Abramsky and T.S.E. Maibaum, editors, *TAPSOFT/CAAP*, volume 493, pages 293–319. Springer Verlag, 1991.

[4] A. Di Pierro and H. Wiklicky. An operational semantics for Probabilistic Concurrent Constraint Programming. In P. Iyer, Y. Choo, and D. Schmidt, editors, *ICCL'98 – International Conference on Computer Languages*, pages 174–183. IEEE Computer Society and ACM SIGPLAN, IEEE Computer Society Press, May 1998.

[5] K. Goebel and W.A. Kirk. *Topics in Metric Fixed Point Theory*, volume 28 of *Cambridge studies in advanced mathematics*. Cambridge University Press, Cambridge, 1990.

[6] C. M. Grinstead and J. L. Snell. *Introduction to Probability*. American Mathematical Society, Providence, Rhode Island, second revised edition, 1997.

[7] V. Gupta, R. Jagadeesan, and V. A. Saraswat. Probabilistic concurrent constraint programming. In *Proceedings of CONCUR 97*. Springer Verlag, 1997.

[8] Claire Jones. *Probabilistic Non-Determinism*. PhD thesis, University of Edinburgh, Edingburgh, 1993.

[9] Dexter Kozen. Semantics for probabilistic programs. *Journal of Computer and System Sciences*, 22:328–350, 1981.

[10] C. Morgan, A. McIver, K. Seidel, and J.W. Sanders. Probabilistic predicate transformers. Technical Report PRG-TR-4-95, Programming Research Group, Oxford University Computing Laboratory, 1995.

[11] S. O. Nyström and B. Jonsson. Indeterminate Concurrent Constraint Programming: A Fixpoint Semantics for Non-Terminating Computations. In D. Miller, editor, *Proc. of the 1993 International Logic Programming Symposium*, Series on Logic Programming, pages 335–352. The MIT Press, 1993.

[12] N. Saheb-Djahromi. CPO's of measures for nondeterminism. *Theoretical Computer Science*, 12:19–37, 1980.

[13] V. A. Saraswat, M. Rinard, and P. Panangaden. Semantics foundations of concurrent constraint programming. In *Proceedings of POPL*, pages 333–353. ACM, 1991.

Lazy Functional Algorithms for Exact Real Functionals

Alex K. Simpson

LFCS, Department of Computer Science, University of Edinburgh,
JCMB, King's Buildings, Edinburgh, EH9 3JZ, Scotland
<Alex.Simpson@dcs.ed.ac.uk>

Abstract. We show how functional languages can be used to write programs for real-valued functionals in exact real arithmetic. We concentrate on two useful functionals: definite integration, and the functional returning the maximum value of a continuous function over a closed interval. The algorithms are a practical application of a method, due to Berger, for computing quantifiers over streams. Correctness proofs for the algorithms make essential use of domain theory.

1 Introduction

In exact real number computation, infinite representations of reals are employed to avoid the usual rounding errors that are inherent in floating point computation [4,5,6,17]. For certain real number computations that are highly sensitive to small variations in the input, such rounding errors become inordinately large and the use of floating-point algorithms can lead to completely erroneous results [1,14]. In such situations, exact real number computation provides guaranteed correctness, although at the (probably inevitable) price of a loss of efficiency. How to improve efficiency is a field of active research [9].

Lazy functional programming provides a natural implementational style for exact real algorithms. One reason is that lazy functional languages support lazy infinite data structures, such as streams, which can be coveniently used to represent real numbers. The efficient management of such infinite data structures (for example, using call-by-need to avoid repeated computations) can be entrusted to the language implementer, leaving the programmer free to concentrate on the essentials of the algorithms being developed. Also, functional programming naturally supports the recursive definition of functions, which is the most useful method of defining exact functions on real numbers. Such considerations were important motivating factors in [4,5,17,6,7,10].

One principal distinguishing feature of functional languages is their acceptance of functions as first-class values, and the associated possibility of passing functions as arguments to other function(al)s. In the context of exact real number computation, this raises the question of whether it is possible to write functional algorithms to implement useful functionals on real numbers. In [8], Edalat and Escardó show how to extend Real PCF [10] with primitive functionals for definite integration, and for the maximum value attained by a continuous function over a closed interval. However, their operational semantics is nondeterministic, and requires a parallel evaluation strategy which is not readily supported within the context of the standard *sequential* functional languages. The problem of whether such algorithms are possible sequentially was originally posed in Di Gianantonio's PhD thesis [6], where it was conjectured that they are not.

In this paper we show that Di Gianantonio's conjecture is false. We provide sequential functional algorithms for the specific and useful functionals of integration and maximum. The algorithms rely on a clever, but little known, idea of Berger, who showed how to compute quantifiers over predicates on streams sequentially [2]. Berger's algorithms deserve to be better known, especially in the light of their possible applications.

The work of Berger, Di Gianantonio, Escardó and Edalat, referred to above, was carried out in the context of the minimal functional language PCF [15] (and extensions of it). It would be fully possible to write this paper in the same setting, but we prefer instead to adopt a less spartan approach. The goal of this paper is to describe and verify particular functional algorithms. We therefore use an easily readable, although not formally defined, functional pseudocode for expressing algorithms (just as an informal imperative pseudocode is used to specify algorithms throughout computer science). We also make use of a simple type discipline to specify the domains and codomains of functions.

Not only does the type discipline improve the readability of the code, it also serves a more significant purpose. The statements of correctness of the algorithms and their verification make essential use of a denotational semantics defined in terms of the type structure. Indeed it is a further benefit of using a functional language that a denotational semantics is easily obtained using standard constructions on complete partial orders. Because we have not formally defined the language, we cannot formally define its semantics either. Nonetheless, the denotational semantics of functional programming languages is now well enough understood that it is possible to use such semantics in an informal way with full mathematical rigour. Our approach is to use denotational semantics as one more mathematical tool for verifying informally specified algorithms, alongside all the other tools available from the body of mathematics as a whole.

Perhaps what is most interesting about the use of denotational semantics in this paper is that it goes beyond the mere existence of fixed-points and their basic properties. Instead, the correctness proofs make use of topological properties (moduli of continuity) of the denotations of higher-order functions. Understanding the denotational semantics is helpful even to appreciate the correctness of the algorithms informally. In order to verify the algorithms rigorously, some use of domain theory appears to be essential.

2 Types and Their Denotations

In our functional pseudocode, we assume basic datatypes like `int`, the type of integers, `bool`, the type of booleans, as well as some convenient finite types:

```
type two = {0,1}
type three = {-1,0,1}
```

We assume that two is a subtype of three in the obvious way (so we shall not bother to include explicit coercions between them). The type constructors we use are A → B, function space, A × B, cartesian product, and A stream. Function application is assumed to be lazy. Mainly for denotational simplicity, we interpret × as a lazy product (thus a pair may converge in one component but not the other). The behaviour of streams is best explained via the denotational semantics.

For the denotational semantics, we use directed-complete partial orders with least element (henceforth cpos) for interpreting datatypes, and continuous functions between them for interpreting programs (see e.g. [12]). In Sec. 3 we refer to cpos as topological spaces, understanding them as carrying the Scott topology.

Given a set X, we write $X_\perp$ for the flat cpo with least element $\perp$ and with all other elements taken from the set X. Basic types are interpreted as flat cpos by: $[\![\texttt{int}]\!] = \mathbb{Z}_\perp$; $[\![\texttt{bool}]\!] = \mathbb{B}_\perp$ where $\mathbb{B} = \{\textit{true}, \textit{false}\}$; $[\![\texttt{two}]\!] = \mathbf{2}_\perp$ where $\mathbf{2} = \{0, 1\}$; and $[\![\texttt{three}]\!] = \mathbf{3}_\perp$ where $\mathbf{3} = \{-1, 0, 1\}$. The function space is interpreted as the cpo, $[\![\texttt{A} \to \texttt{B}]\!]$, of all continuous functions from $[\![\texttt{A}]\!]$ to $[\![\texttt{B}]\!]$ ordered pointwise. The interpretation of the product type, $[\![\texttt{A} \times \texttt{B}]\!]$, is the straightforward cartesian product of $[\![\texttt{A}]\!]$ and $[\![\texttt{B}]\!]$ (as partially ordered sets).

Streams will be denoted by possibly infinite sequences, so we develop some notation for these. For a set X we write: X^* for the set of finite sequences of its elements; X^ω for the set of infinite sequences; and X^∞ for the set of all sequences, i.e. $X^\infty = X^* \cup X^\omega$. For any sequence α, we write $|\alpha|$ for the (possibly infinite) length of α and $\alpha(i)$ (where $0 \le i < |\alpha|$) for the $(i+1)$-th element of α. We use textual juxtaposition, $\alpha\beta$, for the concatenation of a finite sequence α with an arbitrary sequence β. We write $\alpha\lceil_n$ for the largest finite prefix, β, of α such that $|\beta| \le n$. For $x \in X$ we write $\vec{x}$ for the infinite constant sequence. In the paper, we shall only ever use streams formed from base types. These have a straightforward interpretation. If $[\![\texttt{A}]\!] = X_\perp$ then $[\![\texttt{A stream}]\!] = X^\infty$ with $\alpha \le \beta$ if and only if α is a prefix of β. We write $hd : X^\infty \to X_\perp$ and $tl : X^\infty \to X^\infty$ for the evident head and tail functions. The "cons" operation on streams (written :: in our pseudocode) is the evident left-strict function from $X_\perp \times X^\infty$ to X^∞.

3 Real Numbers: Representation and Semantics

In order to write algorithms for functions and functionals on the reals, we first need to choose a representation for real numbers. It is well known that the standard base n notation for reals does not provide an adequate representation, as many simple functions (e.g. addition) are not computable exactly. However, many alternative choices of adequate representation are available. There are discussions of these issues in e.g. [6,9].

We shall use one of the simplest possible representations: a modification of the standard binary representation using negative digits. We consider an infinite sequence $\alpha \in \mathbf{3}^\omega$ (recall that $\mathbf{3} = \{-1, 0, 1\}$) as representing the real number

$$q(\alpha) = \sum_{i=0}^{\infty} \alpha(i) \times 2^{-(i+1)}$$

This defines a surjective function q from $\mathbf{3}^\omega$ to $\mathbb{I}$, where we write $\mathbb{I}$ for the closed interval $[-1, 1]$. The whole real line can be represented using a mantissa from $\mathbf{3}^\omega$ and an exponent from $\mathbb{Z}$, thus $(z, \alpha) \in \mathbb{Z} \times \mathbf{3}^\omega$ represents the real number $2^z \times q(\alpha)$. This representation will be used in the full version of the paper, but, for lack of space, is not considered further in this conference version.

We use the natural type definition to implement the representation.

```
type interval = three stream
```

There is, however, a mismatch between the datatype and the representation of reals. We have that $[\![\mathtt{interval}]\!] = \mathbf{3}^\infty$, whereas only elements in the subset $\mathbf{3}^\omega$ have been given interpretations as real numbers.

Just as not all values of type interval represent real numbers, neither do all functions of type interval $\to$ interval represent functions on real numbers. We use the denotational semantics to distinguish those that do. For greater generality we work with n-ary functions.

An arbitrary function $\phi : (\mathbf{3}^\infty)^n \to \mathbf{3}^\infty$ is said to be *total* if it restricts to a function $\bar\phi : (\mathbf{3}^\omega)^n \to \mathbf{3}^\omega$. Clearly $\bar\phi$, when it exists, is unique. Similarly, a function $\theta : (\mathbf{3}^\omega)^n \to \mathbf{3}^\omega$ is said to be *real* if there exists a function $\tilde\theta : \mathbb{I}^n \to \mathbb{I}$ such that, for all $\alpha_1, \ldots, \alpha_n \in \mathbf{3}^\omega$, it holds that $q(\theta(\alpha_1, \ldots, \alpha_n)) = \tilde\theta(q(\alpha_1), \ldots, q(\alpha_n))$. Again $\tilde\theta$ is uniquely determined (because q is surjective). Putting the two together, we say that $\phi : (\mathbf{3}^\infty)^n \to \mathbf{3}^\infty$ is *real-total* if it is total and $\bar\phi$ is real, in which case we write $\tilde\phi : \mathbb{I}^n \to \mathbb{I}$ for the unique induced function.

A functional program of type interval $\to$ interval will always be denoted by a *continuous* $\phi : \mathbf{3}^\infty \to \mathbf{3}^\infty$. By topological trivialities, if we endow $\mathbf{3}^\omega$ with the subspace topology of the Scott topology on $\mathbf{3}^\infty$, and we endow $\mathbb{I}$ with the quotient topology of $\mathbf{3}^\omega$ under q, then, for any continuous real-total ϕ, we have that $\bar\phi$ and $\tilde\phi$ are continuous. The proposition below makes this observation more interesting.

Proposition 1.

1. The induced topologies on $\mathbf{3}^\omega$ and $\mathbb{I}$ are the product and Euclidean topologies respectively.
2. For any continuous $f : \mathbb{I}^n \to \mathbb{I}$, there exists a real $\theta : (\mathbf{3}^\omega)^n \to \mathbf{3}^\omega$ such that $f = \tilde\theta$.
3. For any continuous $\theta : (\mathbf{3}^\omega)^n \to \mathbf{3}^\omega$ there exists a total $\phi : (\mathbf{3}^\infty)^n \to \mathbf{3}^\infty$ such that $\theta = \bar\phi$.

In the full version of the paper the definitions and results in this section will be related to work on totality in domain theory [2,3,16], and to topological injectivity (and projectivity) results [11].

4 Moduli of Continuity and Stream Quantifiers

Consider any continuous function $\phi : \mathbf{2}^\infty \to X_\perp$ where X is any set. We say that f is *total* if, for all $\alpha \in \mathbf{2}^\omega$, it holds that $\phi(\alpha) \in X$.

Proposition 2. *For any total $\phi : \mathbf{2}^\infty \to X_\perp$ there exists $n \in \mathbb{N}$ such that, for all $\alpha \in \mathbf{2}^\infty$, it holds that $\phi(\alpha) = \phi(\alpha\lceil_n)$.*

We call the least n satisfying the property stated in the proposition the *intensional modulus of continuity* of ϕ, and we write $imc(\phi)$ for it.

Corollary 1. *For any total $\phi : \mathbf{2}^\infty \to X_\perp$ there exists $n \in \mathbb{N}$ such that, for all $\alpha, \beta \in \mathbf{2}^\omega$, it holds that $\alpha\lceil_n = \beta\lceil_n$ implies $\phi(\alpha) = \phi(\beta)$.*

We call the smallest such n the *extensional modulus of continuity* of ϕ, and we write $emc(\phi)$ for it. Obviously $emc(\phi) \leq imc(\phi)$. In the full version of the paper there will be a discussion of the relative benefits of the two notions of modulus.

Our first application, due to Berger [2], is to provide a universal quantifier for total predicates on `two stream`. The algorithm is presented in Fig. 1 below.

```
witness-not: (two stream → bool)  → two stream
witness-not (P) =
  lazylet w = witness-not (λv. P(0 :: v))
    in if P(0 :: w) then 1 :: witness-not (λv. P(1 :: v))
                    else 0 :: w

forall : (two stream → bool) → bool
forall (P) = P (witness-not P)
```

Fig. 1. Algorithms for the stream quantifier

Proposition 3. *For any total $\phi : \mathbf{2}^\infty \to \mathbb{B}_\perp$:*

$$[\![\mathtt{forall}]\!](\phi) = \begin{cases} true & \text{if, for all } \alpha \in \mathbf{2}^\omega,\ \phi(\alpha) = true \\ false & \text{otherwise} \end{cases}$$

Proof. One proves, by induction on $imc(\phi)$, that, for all total $\phi : \mathbf{2}^\infty \to \mathbb{B}_\perp$: if there exists $\alpha \in \mathbf{2}^\omega$ such that $\phi(\alpha) = \mathit{false}$ then $[\![\mathtt{witness\text{-}not}]\!](\phi)$ is one such α; otherwise $[\![\mathtt{witness\text{-}not}]\!](\phi) = \widetilde{1}$. The proposition follows easily. □

5 Functional Algorithms for Maximum and Integration

The denotation of every program of type `interval → interval` will be a continous function ϕ from $\mathbf{3}^\infty$ to $\mathbf{3}^\infty$. If ϕ is real-total then there is a corresponding continuous $\tilde{\phi} : \mathbb{I} \to \mathbb{I}$. Our goal in this paper is to show how Berger's algorithms can be applied to the practical problem of computing the values of functionals acting on continuous functions on $\mathbb{I}$.

We shall concentrate on two basic and useful functionals: the functional that finds the maximum value attained by a continuous function over the closed interval $[0, 1]$, and the function that computes the definite integral of a continuous function over $[0, 1]$. That such maximum values and definite integrals exist for all continuous functions are very basic results in analysis. Observe that both operations return values in $\mathbb{I}$.

5.1 Maximum

The algorithm for the functional `max-fun` is presented in Fig. 2. A first lemma states the important properties of the main auxiliary function defined there.

```
sub-one: interval → interval
sub-one (1 :: r)    = −1 :: r
sub-one (0 :: r)    = −1 :: sub-one(r)
sub-one (−1 :: r) = $\overrightarrow{-1}$

max-real: interval × interval → interval
max-real (d₁ :: r₁, d₂ :: r₂) =
  let d = d₁ − d₂ in case d of   2 then d₁ :: r₁
                                 1 then d₁ :: max-real(r₁, sub-one(r₂))
                                 0 then d₁ :: max-real(r₁, r₂)
                                −1 then d₂ :: max-real(sub-one(r₁), r₂)
                                −2 then d₂ :: r₂

max-fun: (interval → interval) → interval
max-fun (f) =
  let d = head (f($\overrightarrow{1}$)) in if forall (λv. head(f(v)) = d)
                     then d :: (max-fun(λv. tail(f(v))))
                     else max-real (max-fun (λv. f(0 :: v)),
                                    max-fun (λv. f(1 :: v)) )
```

Fig. 2. Maximum-value algorithm

Lemma 1. $[\![$max-real$]\!]$ *is real-total with:* $\widetilde{[\![\text{max-real}]\!]}(x, y) = max(x, y)$ *Moreover, for all* $\alpha, \beta \in \mathbf{3}^{\infty}$, $|[\![$max-real$]\!](\alpha, \beta)| \geq min(|\alpha|, |\beta|)$.

Observe that the lemma includes the intensional information that `max-real` only examines n digits of the input streams in order to produce n digits of output. This is crucial in the proof of the proposition below, which states the correctness of `max-fun`.

Proposition 4. *For any real-total* ϕ, *it holds that* $[\![$max-fun$]\!](\phi) \in \mathbf{3}^{\omega}$ *and*

$$q([\![\text{max-fun}]\!](\phi)) = max\{\tilde{\phi}(x) \mid 0 \leq x \leq 1\}.$$

To prove Proposition 4, we prove, by induction on $n \in \mathbb{N}$, that, for all real-total $\phi : \mathbf{3}^{\infty} \to \mathbf{3}^{\infty}$, it holds that $[\![$max-fun$]\!](\phi)\lceil_n = d_1 \ldots d_n \in \mathbf{3}^n$ such that:

$$\left| max\{\tilde{\phi}(x) \mid 0 \leq x \leq 1\} - \sum_{i=1}^{n} d_i.2^{-i} \right| \leq 2^{-n} \qquad (1)$$

The base case, $n = 0$, is trivial. When $n > 0$, consider $h(\phi) : \mathbf{2}^{\infty} \to \mathbf{3}_{\perp}$ defined by $h(\phi)(\alpha) = hd(\phi(\alpha))$. Because ϕ is real-total, we have that $h(\phi)$ is total. The required inequality (1) is now proved by an inner induction on $emc(h(\phi))$.

Briefly, if $emc(h(\phi)) = 0$ then (1) is proved using the outer induction hypothesis on n and the general equality, valid for any continuous $f : \mathbb{I} \to \mathbb{I}$:

$$max\{f(x) + c \mid 0 \leq x \leq 1\} = max\{f(x) \mid 0 \leq x \leq 1\} + c. \tag{2}$$

When $emc(h(\phi)) > 0$ then (1) is proved using the induction hypothesis on the extensional modulus of continuity (the intensional information of Lemma 1 is needed) together with the general equality, valid for any continuous $f : \mathbb{I} \to \mathbb{I}$:

$$max\{f(x) \mid 0 \leq x \leq 1\} = max(\ max\{f(x/2) \mid 0 \leq x \leq 1\}, \tag{3}$$
$$max\{f((x+1)/2) \mid 0 \leq x \leq 1\}\ ).$$

5.2 Integration

Integration can be performed by much the same method. Observe that integration enjoys the following equalities, for any continuous $f : \mathbb{I} \to \mathbb{I}$:

$$\int_0^1 f(x) + c \, \mathrm{d}x = \int_0^1 f(x) \, \mathrm{d}x + c$$

$$\int_0^1 f(x) \, \mathrm{d}x = \int_0^1 f(x/2) \, \mathrm{d}x \oplus \int_0^1 f((x+1)/2) \, \mathrm{d}x,$$

where $\oplus : \mathbb{I} \times \mathbb{I} \to \mathbb{I}$ computes the average of two reals. The above equations are wholly analogous to (2) and (3) for maximum. Indeed we shall obtain an integration algorithm by replacing the binary `max-real` used in `max-fun` with a function computing the average of two reals. However, the translation is not completely straightforward. Recall that the intensional information of Lemma 1 was crucial to the proof of Proposition 4. This contrasts with the easy:

Proposition 5. *There is no real-total $\phi : \mathbf{3}^\infty \to \mathbf{3}^\infty$ such that, for all $x, y \in \mathbb{I}$, $\tilde{\phi}(x, y) = x \oplus y$ and, for all $\alpha, \beta \in \mathbf{2}^\infty$, $|\phi(\alpha, \beta)| \geq min(|\alpha|, |\beta|)$.*

The observed problem is a quirk of the particular representation of real numbers we are using. A neat way of solving it is to use a second representation. Recall that the set of dyadic rationals is $\mathbb{Q}_d = \{m/2^n \mid m, n \in \mathbb{Z}\}$. We write $\mathbb{D}$ for the set $\mathbb{Q}_d \cap [-1, 1]$, which we call the set of *dyadic digits*. We consider an infinite sequence $\gamma \in \mathbb{D}^\omega$ as representing the real number $q'(\gamma) = \sum_{i=0}^\infty \gamma(i) \times 2^{-(i+1)}$. This defines a surjective function $q' : \mathbb{D}^\omega \to [-1, 1]$ extending $q : \mathbf{3}^\omega \to [-1, 1]$.

In order to write algorithms working with dyadic digits we assume an implemented datatype `dyadic` of dyadic digits, complete with the associated operations for the basic arithmetic operations on dyadic rationals. Then we simply define a new datatype for the interval $[-1, 1]$ in terms of dyadic streams:

```
type q-interval = dyadic stream
```

Semantically we assume that $[\![\mathtt{dyadic}]\!] = \mathbb{D}_\bot$, so $[\![\mathtt{q\text{-}interval}]\!] = \mathbb{D}^\infty$. The notions of a function $\phi : \mathbb{D}^\infty \to \mathbb{D}^\infty$ being *total* and *real-total* are defined entirely analogously to the cases for $\mathbf{3}^\infty$.

The full algorithm for integration is presented in Fig. 3. For convenience we assume that `three` is a subtype of `dyadic` and (hence) `interval` is a subtype of `q-interval`.

```
coerce: q-interval → interval
coerce (qd₁ :: qd₂ :: qr) =
  let qc = (2×qd₁) + qd₂ in case  qc < -1       then  -1::coerce((qc + 2) :: qr)
                                  qc > 1        then   1::coerce((qc - 2) :: qr)
                                  otherwise then  0::coerce(qc :: qr)

q-avg: q-interval × q-interval → q-interval
q-avg (qd₁ :: qr₁, qd₂ :: qr₂) = (qd₁+qd₂)/2 :: q-avg (qr₁, qr₂)

q-int: (interval → interval) → q-interval
q-int (f) = let d = head (f(1⃗)) in if forall (λv. head(f(v)) = d)
                                    then d :: (q-int(λv. tail(f(v))))
                                    else q-avg ( q-int (λv. f(0 :: v)),
                                                 q-int (λv. f(1 :: v)) )

integrate: (interval → interval) → interval
integrate (f) = coerce (q-int(f))
```

Fig. 3. Integration algorithm

Lemma 2.

1. For any $\gamma \in \mathbb{D}^\omega$, it holds that $[\![\text{coerce}]\!](\gamma) \in \mathbf{3}^\omega$ and $q([\![\text{coerce}]\!](\gamma)) = q'(\gamma)$.
2. The function $[\![\text{q-avg}]\!] : \mathbb{D}^\infty \to \mathbb{D}^\infty$ is real-total with $\widetilde{[\![\text{q-avg}]\!]}(x, y) = x \oplus y$. Moreover, for all $\gamma, \gamma' \in \mathbb{D}^\infty$, $|[\![\text{q-avg}]\!](\gamma, \gamma')| \geq min(|\gamma|, |\gamma'|)$.

Proposition 6. For any real-total ϕ, it holds that $[\![\text{integrate}]\!](\phi) \in \mathbf{3}^\omega$ and

$$q([\![\text{integrate}]\!](\phi)) = \int_0^1 \tilde{\phi}(x) \, dx.$$

The proof structure closely follows that of Proposition 4.

6 Further Developments

In the full version of the paper an extension of the integration algorithm will be presented that integrates, over any closed interval, functions defined from the interval to the whole real line. This makes use of the mantissa-exponent representation of the real line mentioned briefly in Sec. 3.

The algorithms in this extended abstract were implemented by Reinhold Heckmann in Gofer in summer 1997. The extensions to functions from an arbitrary closed interval to the full real line have recently been implemented in Haskell by David Plume. The integration algorithm performs abysmally on any interesting functions. The maximum algorithm performs a little better. A partial quantitative analysis of this situation will appear in the full version of the paper. The intrinsic intractability of the operations of integration and finding maximum values is to be expected from the work of Ko [13].

Acknowledgements

I have benefited from discussions with Pietro Di Gianantonio, Gordon Plotkin and, especially, Martín Escardó. I thank Ieke Moerdijk, Jaap van Oosten and Harold Schellinx for their hospitality in Utrecht, where the paper was written with financial support from an NWO Pionier Project.

References

1. J.-C. Bajard, D. Michelucci, J.-M. Moreau, and J.-M. Muller. Introduction to special issue: "Real Numbers and Computers". *Journal of Universal Computer Science*, 1(7):436–438, 1995.
2. U. Berger. *Totale Objecte und Mengen in Bereichstheorie*. PhD Thesis, University of Munich, 1990.
3. U. Berger. Total objects and sets in domain theory. *Journal of Pure and Applied Logic*, 60:91–117, 1993.
4. H.J. Boehm and R. Cartwright. Exact real arithmetic: Formulating real numbers as functions. In D. Turner, editor, *Research Topics in Functional Programming*, pages 43–64. Adison-Wesley, 1990.
5. H.J. Boehm, R. Cartwright, M. Riggle, and M.J. O'Donnel. Exact real arithmetic: a case study in higher order programming. In *ACM Symposium on LISP and Functional Programming*, 1986.
6. P. Di Gianantonio. *A Functional Approach to Computability on Real Numbers*. PhD Thesis, University of Pisa, 1993.
7. P. Di Gianantonio. An abstract data type for real numbers. In *Proceedings of ICALP-97*, pages 121–131. Springer LNCS 1256, 1997.
8. A. Edalat and M.H. Escardó. Integration in Real PCF. *Information and Computation*, To appear, 1998.
9. A. Edalat and P.J. Potts. *Exact Real Computer Arithmetic*. Presented at workshop: New Paradigms for Computation on Classical Spaces, Birmingham, 1997.
10. M.H. Escardó. PCF extended with real numbers. *Theoretical Computer Science*, 162(1):79–115, 1996.
11. M.H. Escardó. Properly injective spaces and function spaces. *Topology and its Applications*, To appear, 1998.
12. C.A. Gunter. *Semantics of Programming*. MIT Press, 1992.
13. Ker-I Ko. *Complexity Theory of Real Functions*. Birkhauser, Boston, 1991.
14. V. Menissier-Morain. Arbitrary precision real arithmetic: Design and algorithms. *Journal of Symbolic Computation*, Submitted, 1996.
15. G.D. Plotkin. LCF considered as a programming language. *Theoretical Computer Science*, 5(1):223–255, 1977.
16. G.D. Plotkin. Full abstraction, totality and PCF. *Math. Struct. in Comp. Sci.*, To appear, 1998.
17. J. Vuillemin. Exact real arithmetic with continued fractions. *IEEE Transactions on Computers*, 39(8):1087–1105, 1990.

Randomness vs. Completeness: On the Diagonalization Strength of Resource-Bounded Random Sets *

Klaus Ambos-Spies[1], Steffen Lempp[2], and Gunther Mainhardt[1]

[1] Mathematisches Institut, Universität Heidelberg
ambos,mainhard@math.uni-heidelberg.de
[2] Department of Mathematics, University of Wisconsin, Madison
lempp@math.wisc.edu

Abstract. We show that the question of whether the p-tt-complete or p-T-complete sets for the deterministic time classes **E** and **EXP** have measure 0 in these classes in the sense of Lutz's resource-bounded measure cannot be decided by relativizable techniques. On the other hand, we obtain the following absolute results if we bound the norm, i.e., the number of oracle queries of the reductions: For $r = tt, T$,

$$\mu_p(\{C : C \ p\text{-}r(kn)\text{-complete for } \mathbf{E}\}) = 0 \text{ and}$$
$$\mu_{p_2}(\{C : C \ p\text{-}r(n^k)\text{-complete for } \mathbf{EXP}\}) = 0.$$

In the second part of the paper we investigate the diagonalization strength of random sets in an abstract way by relating randomness to a new genericity concept. This provides an alternative, quite elegant and powerful approach for obtaining results on resource-bounded measures like the ones in the first part of the paper.

1 Introduction

Lutz's resource-bounded measure provides a framework for the quantitative analysis of complexity classes (see Lutz [13]). The most interesting results of this theory have been obtained for the deterministic exponential time classes $\mathbf{E} = \mathbf{DTIME}(2^{lin})$ and $\mathbf{EXP} = \mathbf{DTIME}(2^{poly})$, which are captured by the p- and p_2-measure, respectively. Here, the question of determining the measure of the complete sets for these classes under the various types of polynomial-time reducibilities became a challenging problem which in part is still unsolved.

---

* Research supported in part by the Human Capital and Mobility Program of the European Community under grant CHRX-CT93-0415 (COLORET). The second author would like to acknowledge partial support by National Science Foundation grant DMS-9504474 and a grant of the British Engineering and Physical Sciences Research Council. The main results of Section 3 were obtained by the first and second author when they visited the University of Leeds in the spring of 1996. In Section 4 some recent work by the first and third author is reported.

Mayordomo [14] has shown that the class of p-m-complete sets for **E** (or **EXP**) has p-measure 0. Ambos-Spies, Neis and Terwijn [5] extended Mayordomo's theorem to bounded truth-table completeness, by showing certain relations between genericity and randomness. The somewhat weaker form of this result for p_2-measure was independently obtained by Buhrman and Mayordomo [8] by looking at resource-bounded Kolmogorov complexity.

For reducibilities with nonconstant norm, however, this question remained open. In fact, Allender and Strauss [1] have shown that, assuming **BPP** = **EXP**, the class of the p-T-hard sets has p-measure 1, whence the class of p-T-complete sets does not have p-measure 0, and their result can be easily extended to p-tt-completeness. Since Heller [12] has constructed an oracle A relative to which **BPP**^{A} = **EXP**^{A}, this shows that it is impossible to use relativizable techniques to extend the result of Ambos-Spies et al. from bounded truth-table to truth-table or even Turing reducibility.

The first goal of our paper is to show that the p-measure (and p_2-measure) of the classes of the p-tt-complete and p-T-complete sets for **E** and **EXP** is in fact oracle dependent. This is shown by complementing the result of Allender and Strauss as follows: Assuming **P** = **PSPACE** (or at least **PSPACE** $\subseteq$ **DTIME**(2^{kn}) for some k), we show that the p-measure of the class of p-T-complete sets for **EXP** is 0. (Recently, this result was independently proved by Buhrman et al. [10] by using a new nonmonotone martingale concept suggested by Regan.)

By analyzing the proof of our theorem, we can extend the absolute smallness results for the classes of complete sets as follows: For arbitrary but fixed k, the class $\{C : C \; p\text{-}T(kn)\text{-complete for } \mathbf{EXP}\}$ has p-measure 0 and $\{C : C \; p\text{-}T(n^k)\text{-complete for } \mathbf{EXP}\}$ has p_2-measure 0, where $p\text{-}T(l(n))$ refers to p-T-reductions of norm $l(n)$, i.e., to reductions having the number of oracle queries on an input of length n bounded by $l(n)$ for arbitrary but fixed oracle. (These results were obtained independently by Buhrman and Van Melkebeek [9].) Note that by the theorem of Allender and Strauss, the latter is the best possible result provable by relativizable techniques.

By expressing resource-bounded measure in terms of resource-bounded randomness, the above results can be viewed as consequences of the diagonalizations "built" into random sets. In the second part of the paper we address the question of which types of diagonalizations are subsumed by randomness. Since the most common types of diagonalizations in complexity theory have been formalized by corresponding genericity notions (see [2] for details), this question can be answered by isolating the genericity notions which are implied by randomness. First results in this direction were obtained by Ambos-Spies et al. in [5], where the compatibility of the genericity concept of [3] with randomness was shown. This genericity concept, however, is too weak for dealing with reducibilities of nonconstant norm. Here, we introduce a new genericity concept compatible with randomness, which captures the diagonalizations of the type required for establishing our smallness results in the first part of the paper. Though this genericity approach is somewhat technical, once the required relations to randomness are established it becomes a quite powerful tool for obtaining results on resource-bounded measure.

Our notation is mainly standard and follows [4]. We let $\Sigma = \{0,1\}$ be the binary alphabet, Σ^* be the set of finite binary strings, and Σ^∞ be the set of infinite binary strings. Sometimes, we identify strings with numbers, and subsets A of Σ^* with their characteristic sequences $A(0)A(1)A(2)\ldots$. The initial segment of A of length n is denoted by $A\restriction n = A(0)\ldots A(n-1)$. We assume the reader to be familiar with the polynomial-time reducibilities m (many-one), btt (bounded truth-table), tt (truth-table) and T (Turing). For $r = tt, T$, $r(l(n))$ will denote that the norm of the reduction is bounded by $l(n)$.

2 Resource-Bounded Measure and Randomness

In this section, we introduce the fragment of Lutz's resource-bounded measure theory required for the following. For more details, we refer to Lutz [13].

Lutz's theory is defined in terms of martingales. A characterization of classical measure by martingales was given by Ville in 1939, while Schnorr [15] was the first to look at computable martingales. He also defined resource-bounded randomness in these terms.

A *martingale* is a function $d : \Sigma^* \to Q_+$, (where Q_+ is the set of nonnegative rationals) which satisfies the so-called *martingale condition* $d(x0)+d(x1) = 2d(x)$ for all strings x. A martingale d *succeeds on a set* X if $\limsup_{n\to\infty} d(X\restriction n) = \infty$, and d *succeeds on a class* **C** if d succeeds on all sets $X \in \mathbf{C}$. By Ville, a class **C** has (classical) measure 0 iff some martingale succeeds on **C**.

A $t(n)$-*martingale* is a martingale $d \in \mathbf{DTIME}(t(n))$, and d is called a p-*martingale* [p_2-*martingale*] if d is an n^k-martingale [$2^{(\log n)^k}$-*martingale*] for some $k \geq 1$. A class **C** has p-*measure* 0, $\mu_p(\mathbf{C}) = 0$, if some p-martingale succeeds on **C**. The p_2-*measure* is defined correspondingly. Note that martingales operate on initial segments $X\restriction x$. Since $|X\restriction x|^k \approx 2^{k|x|}$ and $2^{(\log(|X\restriction x|))^k} \approx 2^{|x|^k}$, this implies that p-measure corresponds to $\mathbf{E} = \mathbf{DTIME}(2^{lin})$ while p_2-measure corresponds to $\mathbf{EXP} = \mathbf{DTIME}(2^{poly})$. Lutz has shown that $\mu_p(\mathbf{DTIME}(2^{kn})) = 0$ for all k but $\mu_p(\mathbf{E}) \neq 0$, whence a measure on **E** can be defined as follows: A class **C** has *measure* 0 *in* **E** if $\mu_p(\mathbf{C} \cap \mathbf{E}) = 0$, and **C** has *measure* 1 *in* **E** if the complement $\overline{\mathbf{C}}$ of **C** has measure 0 in **E**. Similarly, we obtain a measure on **EXP** based on the p_2-measure.

The resource-bounded measure can also be defined in terms of resource-bounded randomness (see, e.g., [4]): A set R is $t(n)$-*random* [p-*random*, p_2-*random*] if no $t(n)$-martingale [p-martingale, p_2-martingale] succeeds on R. For any k, there is an n^k-random set R in **E** but there are no such sets in $\mathbf{DTIME}(2^{kn})$. Hence there is no p-random set in **E** but such sets exist in **EXP**. Moreover, a class **C** has p-measure 0 iff **C** does not contain any n^k-random set for some $k \geq 1$. In a similar way, $2^{(\log n)^k}$-random sets characterize the p_2-measure and the measure on **EXP**.

3 Randomness vs. Completeness

In this section we show that the measure of the p-Turing and p-truth-table complete sets in **E** and **EXP** cannot be determined by relativizable techniques. Allender and Strauss [1] have shown that, assuming **BPP** = **EXP**, the p-T-complete sets have measure 1 in **E** and **EXP**, and their result easily extends to the p-tt-complete sets. We complement this result by showing that the p-T-complete sets, and hence the p-tt-complete sets, have measure 0 in **E** and **EXP** if we assume that **PSPACE** $\subseteq$ **P**. Oracles A and B relative to which **BPP**^{A} = **EXP**^{A} and **PSPACE**^{B} = **P**^{B} have been constructed by Heller [12] and Baker, Gill and Solovay [6], respectively. Our proof also yields the following absolute result: The class of the p-$T(kn)$-complete sets for **E** (or **EXP**) has p-measure 0 and the class of the p-$T(n^k)$-complete sets for **EXP** has p_2-measure 0 (for arbitrary but fixed k).

Theorem 1. *(Allender and Strauss [1]) Let A be n^2-random. Then A is p-tt-hard for **BPP**.*

Corollary 1. *Assume **BPP** = **EXP**. Every n^2-random set is p-tt-hard for **EXP**. Hence, in particular, for $r \in \{tt, T\}$, the class $\{C : C$ p-r-complete for **E**$\}$ has measure 1 in **E**, and the class $\{C : C$ p-r-complete for **EXP**$\}$ has measure 1 in **EXP**.*

Theorem 1 extends the result of Bennett and Gill in [7] that the p-T-hard sets for **BPP** have classical measure 1. The proof of Allender and Strauss uses results on pseudo-random number generators, and in [1], the result is only claimed for p-T-reducibility. In 1996, the third author obtained an alternative, elementary proof based on the original proof of Bennett and Gill, which also yields the result for p-tt-reducibility. This proof will appear in Mainhardt's Ph.D. thesis.

Corollary 1 shows that if **BPP** = **EXP**, i.e., if **BPP** is "large", then the p-tt- and p-T-complete sets for **E** and **EXP** are abundant in the sense of Lutz's measure. We now complement this observation by showing that if **PSPACE** (hence **BPP**) is "small", in particular if **PSPACE** = **P**, then the p-tt- and p-T-complete sets for **E** and **EXP** are scarce.

Theorem 2. *Assume that **PSPACE** $\subseteq$ **DTIME**(2^{kn}). There is no n^{k+2}-random set which is p-T-complete for **E** or **EXP**. Hence*

$$\mu_p(\{C : C \text{ p-T-complete for } \mathbf{E} \text{ } (\mathbf{EXP})\}) = 0,$$

*whence $\{C : C$ p-T-complete for **E**$\}$ and $\{C : C$ p-T-complete for **EXP**$\}$ have measure 0 in **E** and **EXP**, respectively.*

Proof (sketch). Let R be n^{k+2}-random and, for any set X, let

$$L(X) = \{x : \| \{xy : |x| = |y| \text{ \& } xy \in X\} \| \text{ even}\}.$$

Then, for $X \in \mathbf{E}$ (**EXP**), we also have $L(X) \in \mathbf{E}$ (**EXP**), whence it suffices to show that $L(R) \not\leq_T^P R$. So, given a p-T-reduction M, we will define an n^{k+2}-martingale d which succeeds on $\mathbf{C} = \{B : L(B) = M^B\}$.

Fix a polynomial time bound p for M where w.l.o.g. $p(n) > 2n$, and let $q(n)$ be a polynomial which bounds the norm of M, i.e., the number of oracle queries in the computation of $M^X(x)$ for any string x of length n and any oracle X. Note that $q(n) \leq p(n)$. Finally, fix an easily recognizable infinite sequence $x_0 < x_1 < x_2 < \ldots$ of strings such that $p(|x_m|) < 2^{|x_m|} < |x_{m+1}|$ for $m \geq 0$. On the interval $[x_m, x_{m+1})$ the martingale d will be defined in such a way that, for any set B,

(3.1) $\qquad L(B)(x_m) = M^B(x_m) \Rightarrow d(B\restriction x_{m+1}) \geq \frac{3}{2} d(B\restriction x_m)$

will hold. Obviously this will make d succeed on $\mathbf{C}$.

For the definition of d, fix $x = x_m$, $x' = x_{m+1}$ and some initial segment $\sigma = X \restriction x$. Assume that $d(\sigma)$ is given, and let $n = |x|$ and $n' = |x'|$. We will define $d(X \restriction y)$ for all proper extensions $X \restriction y$ of σ with $y < x'$.

Note that, for any set X, $L(X)(x)$ is determined by $X \cap I_x$, where $I_x = \{xy : |y| = |x|\}$. Similarly, $M^X(x)$ only depends on $X \restriction x'$. So, for $\tau = X \restriction x'$, τ determines $L(X)(x)$ and $M^X(x)$ whence we may denote these values by $L(\tau)(x)$ and $M^\tau(x)$, respectively.

We call a proper extension $\tau = X \restriction y$ of σ a *complete* extension if $y = x'$ and a *partial* extension if $y < x'$. A complete extension τ is called *positive* if $L(\tau)(x) = M^\tau(x)$ and *negative* otherwise. For a partial extension σ', let $pos(\sigma')$ be the number of positive extensions of σ'. Note that, for B as in the premise of (3.1), $B\restriction x'$ are positive, and that one half of the complete extensions of σ are positive. So, in order to guarantee (3.1) it suffices to define d on the interval $[x, x')$ in such a way that the capital $d(\sigma)$ is uniformly distributed among the positive extensions (while for the negative extensions $X \restriction x'$, $d(X \restriction x') = 0$). This is achieved by letting

$$\frac{d((X\restriction y)0)}{d((X\restriction y)1)} = \frac{pos((X\restriction y)0)}{pos((X\restriction y)1)}$$

for any partial extension $X \restriction y$.

It remains to show that the martingale d is n^{k+2}-time bounded. For this, it suffices to show that $pos(X\restriction y)$ can be computed in $2^{(k+1)m}$ steps for any partial extension $X \restriction y$ of σ, $m = |y|$. To do so, we distinguish three cases.

Let v be the unique element of I_x such that there are exactly $q(n)$ elements in I_x greater than v, and let w be the greatest element of I_x. Then, for $y < v$, one half of the total extensions are positive. For $y > w$, $L(X\restriction y)$ is already determined by $X\restriction y$. So here we can compute $pos(X\restriction y)$ by looking at the query tree of the computation $M(x)$ and counting the (appropriately weighted) paths giving output $L(X\restriction y)$ which are consistent with $X\restriction y$. Note that this query tree has depth at most $q(n)$ where q is the polynomial norm of the reduction M. So this procedure can be carried out in $poly < (2^{q(n)})$ steps. For $q(n) > kn$, however, this exceeds the time $|X\restriction y|^{kn}$ available to d. This problem is overcome by our assumption that **PSPACE** $\subseteq$ **DTIME**(2^{kn}), since the above search of a tree of polynomial depth requires only polynomial space.

The case of $v \le y \le w$ is similar. Here, in addition, we have to cycle through the (at most $2^{q(n)+1}$ many) extensions $X \restriction w$ of $X \restriction y$ in order to determine $L(X \restriction x)$. Then, for each $X \restriction w$, the positive extensions are counted as in the second case. □

By relativizing the proofs of Corollary 1 and Theorem 2, we obtain

Corollary 2. *For $r = tt, T$, the measure of $\{C : C$ p-r-complete for $\mathbf{E}$ ($\mathbf{EXP}$)$\}$ in $\mathbf{E}$ ($\mathbf{EXP}$) is oracle dependent.*

Corollary 2 has been obtained independently by Buhrman et al. [10] by investigating a nonmonotone variant of resource-bounded martingales introduced by Regan.

By analyzing how the complexity of the martingale d defined in the proof of Theorem 2 depends on the norm q of the reductions M we consider (without the assumption that **PSPACE** $\subseteq$ **DTIME**(2^{kn})), we obtain the following absolute results, which have been independently obtained by Buhrman and Van Melkebeek [9].

Theorem 3. *(a) For any k, there is a number k' such that no $n^{k'}$-random set is p-T(kn)-complete for $\mathbf{E}$ or $\mathbf{EXP}$. Hence, for fixed but arbitrary k,*

$$\mu_p(\{C : C \text{ p-}T(kn)\text{-complete for } \mathbf{E}\ (\mathbf{EXP})\}) = 0.$$

(b) For any k there is a number k' such that no $2^{(\log n)^{k'}}$-random set is p-T(n^k)-complete for $\mathbf{EXP}$. Hence, for fixed but arbitrary k,

$$\mu_{p_2}(\{C : C \text{ p-}T(n^k)\text{-complete for } \mathbf{EXP}\}) = 0.$$

Note that, by Corollary 1, the second part of the theorem cannot be improved by relativizable techniques. Also note that, by the first part of Theorem 3, no p-random set is p-T(lin)-complete for **EXP**. The proof of the second part of the theorem can be easily modified to yield the following resource-bounded random separation for the classes **P** and **PSPACE**, in fact for **P** and $\oplus$ **P**.

Theorem 4. *For any p_2-random set R, $\mathbf{P}^R \ne \oplus \mathbf{P}^R$, so $\mathbf{P}^R \ne \mathbf{PSPACE}^R$.*

Again by Corollary 1, the resource-bound in Theorem 4 cannot be improved by relativizable techniques.

4 Genericity Compatible with Randomness

Guided by the results of the preceding section, we now introduce a new resource-bounded genericity concept compatible with measure. This concept will yield simpler proofs of the results above and of related results.

The theorems in Section 3 have been obtained by exploiting the "built-in diagonalizations" in a random set. The construction of an incomplete set is a quite simple exercise in diagonalization. So, once we have isolated the diagonalization arguments subsumed by a randomness concept, properties of the random set which can be forced by this type of diagonalization can be established quite easily. Formalizations of different types of diagonalization techniques have been given in terms of genericity, where a generic set is a set having all properties which can be forced by diagonalizations of this type (see [2] for a survey of genericity concepts introduced in complexity theory). Unfortunately, however, most of the genericity concepts in the literature are too strong for being compatible with randomness.

The first successful attempt to isolate some diagonalizations built into random sets was made by Ambos-Spies, Neis and Terwijn [5] by showing that the genericity concept of Ambos-Spies, Fleischhack and Huwig [3] is compatible with randomness. In particular they showed that every n^{k+1}-random set is AFH-n^k-generic (in the sense of [3]), whence any property shared by all AFH-n^k-generic sets (for any fixed k) has p-measure 1 and measure 1 in **E**. Since no AFH-n^2-generic set is p-btt-complete for **E**, in [5] Ambos-Spies et al. concluded that the class of the p-btt-complete sets for **E** has p-measure 0, hence measure 0 in **E**.

The genericity concept of [3], however, is tailored for diagonalizations over bounded query reductions: As shown in [5], for any unbounded nondecreasing polynomial-time computable function f there are AFH-n^k-generic sets (for $k \geq 1$) which are p-tt($f(n)$)-complete for **E**. So the diagonalization strength of this genericity concept does not suffice to obtain results on reducibilities of unbounded norm as in Theorem 3.

Our new genericity concept, which will be sufficiently strong to cope with this situation and which still is subsumed by randomness, refines the concept of [3] by adding a device allowing look-aheads. This additional feature was inspired by Regan's new concept of a nonmonotone martingale introduced in [10]. The look-aheads give us extra strength similar to nonmonotonicity but – in the context of genericity – our approach is technically simpler. (The difference between the common genericity concepts and our new look-ahead genericity notion parallels the difference between self-reducibility and auto-reducibility. This will be made more explicit in the full version of this paper.)

Definition 1. *A prediction machine M is an oracle Turing machine where, whenever $M^X(x)$ is defined, then $M^X(x) = (y, i)$ for some string $y \geq x$ and some $i \in \Sigma$. Moreover, the computation of $M^X(x)$ is subject to the following two constraints:*

(4.1) *If $M^X(x) = (y, i)$ then $M^{X \cup \{y\}}(x) = M^{X - \{y\}}(x)$.*
(4.2) *If in the computation of $M^X(x)$ the oracle is queried for some string $z \geq x$ then $M^X(x)$ is defined.*

A prediction function f is the functional computed by a prediction machine. f predicts A at x if $f^A(x) = (y, A(y))$, and f predicts A iff f predicts A at some x. f is dense along A if $f^A(x)$ is defined for infinitely many x.

Note that (4.1) is a necessary fairness condition while (4.2) is an optional condition expressing that additional information on X for strings $\geq x$ can be only required if actually a prediction is made at x. In order to get corresponding resource-bounded genericity concepts, we will introduce time bounds (which, in order to make the bounds compatible with those for martingales, will be exponentially blown up) and bounds on the size of the look-ahead.

Definition 2. *A $t(n)$-prediction function f is a functional computed by a $t(2^{n+1})$-time bounded prediction machine M. If, moreover, $M^X(x)$ queries at most $l(|x|)$ strings $\geq x$ then the function f is an $l(n)$-l.a. $t(n)$-prediction function. A set G is l.a. $t(n)$-generic [$l(n)$-l.a. $t(n)$-generic] if every $t(n)$-prediction [$l(n)$-l.a. $t(n)$-prediction] function f which is dense along G predicts G.*

Note that AFH-genericity coincides with 0-l.a. genericity in the above sense. On the other hand, one can easily show that look-ahead genericity is weaker than general genericity in the sense of [2], whence it induces resource-bounded category concepts on **E** and **EXP**. In particular, there are l.a. n^k-generic sets in **E** but, for any length bound l, there is no $l(n)$-l.a. n^k-generic set in **DTIME**(2^{kn}). The following theorem shows the compatibility of the new concept with resource-bounded measure if we appropriately bound the norm of the look-ahead. We omit the proof, which resembles the proof of Theorem 3.

Theorem 5. *Every n^{2k+3}-random set is (kn)-l.a. n^k-generic. Furthermore, every $2^{(\log n)^{k+1}}$-random set is n^k-l.a. $2^{(\log n)^k}$-generic.*

Now, in order to obtain alternative proofs of the results in Section 3 based on our new genericity concept, it suffices to prove the corresponding results for generic sets. For instance, in order to obtain Theorem 3(a) from Theorem 5, it suffices to show that no $(2kn)$-l.a. n^2-generic set is complete for **E** under p-T-reductions of norm kn. This can be shown by expressing a straightforward diagonalization in terms of prediction functions, which guarantees that, for a $(2kn)$-l.a. n^2-generic set G, $L(G) \not\leq^P_{T(kn)} G$, where $L(G)$ is defined as in the proof of Theorem 2.

References

1. E. Allender, M. Strauss. Measure on small complexity classes with applications for BPP. In *Proceedings of the 35th Symposium on Foundations of Computer Science*, 867-818, IEEE Computer Society Press, 1994.
2. K. Ambos-Spies. Resource-bounded genericity. In *Computability, Enumerability, Unsolvability* (S. B. Cooper et al., Eds.), London Mathematical Society Lecture Notes Series 224, 1-59, Cambridge University Press, 1996.
3. K. Ambos-Spies, H. Fleischhack, H. Huwig. Diagonalizations over deterministic polynomial time. In *Proceedings of the First Workshop on Computer Science Logic*, CSL'87, Lecture Notes in Computer Science 329, 1-16, Springer Verlag, 1988.
4. K. Ambos-Spies, E. Mayordomo. Resource-bounded measure and randomness. In *Complexity, Logic and Recursion Theory*, Lecture Notes in Pure and Applied Mathematics 187, 1-47, Dekker, 1997.

5. K. Ambos-Spies, H.-C. Neis, S. A. Terwijn. Genericity and measure for exponential time. Theoretical Computer Science 168 (1996) 3-19.
6. T. Baker, J. Gill, R. Solovay. Relativizations of the $P =? NP$ question. SIAM Journal on Computing 5 (1975) 431-442.
7. C. Bennett, J. Gill. Relative to a random oracle $P^A \neq NP^A \neq co\text{-}NP^A$ with probability 1. SIAM Journal on Computing 10 (1981) 96-113.
8. H. Buhrman, E. Mayordomo. An excursion to the Kolmogorov random strings. In *Proceedings of the 10th IEEE Structure in Complexity Theory Conference*, 197-205, IEEE Computer Society Press, 1995.
9. H. Buhrman, D. v. Melkebeek. Hard Sets are Hard to Find. In *Proceedings of the 13th IEEE Conference on Comput. Complexity*, IEEE Computer Society Press, 1998.
10. H. Buhrman, D. v. Melkebeek, K. W. Regan, D. Sivakumar, M. Strauss. A generalization of resource-bounded measure with an application. In *Proceedings of the Symposium on Theoretical Aspects of Computer Science*, Lecture Notes in Computer Science, Springer Verlag, 1998.
11. S. A. Fenner. Notions of resource-bounded category and genericity. In *Proceedings of the 6th IEEE Structure in Complexity Theory Conference*, 196-212, IEEE Computer Society Press, 1991.
12. H. Heller. On relativized exponential and probabilistic complexity classes. Information and Control 71 (1986) 231-243.
13. J. H. Lutz. The quantitative structure of exponential time. In *Complexity Theory Retrospective II* (L.A. Hemaspaandra, A.L. Selman, eds.), Springer-Verlag, 1997.
14. E. Mayordomo. Almost every set in exponential time is P-bi-immune. Theoretical Computer Science 136 (1994) 487-506.
15. C. P. Schnorr. Zufälligkeit und Wahrscheinlichkeit. Lecture Notes in Mathematics 218, Springer-Verlag, 1971.

Positive Turing and Truth-Table Completeness for NEXP Are Incomparable

Levke Bentzien

Mathematisches Institut
Universität Heidelberg
Im Neuenheimer Feld 294
D-69120 Heidelberg
bentzien@math.uni-heidelberg.de

Abstract. Using ideas introduced by Buhrman et al. ([2], [3]) to separate various completeness notions for **NEXP** = **NTIME**(2^{poly}), positive Turing complete sets for **NEXP** are studied. In contrast to many-one completeness and bounded truth-table completeness with norm 1 which are known to coincide on **NEXP** ([3]), whence any such set for **NEXP** is positive Turing complete, we give sets A and B such that

(1) A is $\leq^P_{bT(2)}$-complete but not $\leq^P_{posT}$-complete for **NEXP**

(2) B is $\leq^P_{posT}$-complete but not $\leq^P_{tt}$-complete for **NEXP**.

These results come close to optimality since a further strengthening of (1), as was done by Buhrman in [1] for **EXP** = **DTIME**(2^{poly}), seems to require the assumption **NEXP** = co-**NEXP**.

1 Introduction

Polynomial time reductions and the corresponding completeness notions are central concepts of complexity theory. Since 1975, when Ladner, Lynch and Selman ([4]) studied the different types of polynomial time reductions on **E** = **DTIME**(2^{lin}), the relation between the corresponding completeness notions for deterministic and nondeterministic time classes have become of interest. Watanabe [5] proved separation results for the most important completeness notions for **EXP** exploiting special structural properties. These results were extended by Buhrman, Homer, Spaan and Torenvliet ([2], [3]) to **NEXP**. They constructed sets witnessing the separation of a great variety of completeness notions, including optimal results for bounded truth-table and bounded Turing completeness. Most of their constructions make use of some kind of surplus of queries to enable diagonalizations.

For positive Turing complete sets, this surplus arises only when constructing a positive Turing complete set which is not truth-table complete, whereas other techniques have to be applied for a separation in the other direction.

The goal of this paper is to investigate the mutual relation between positive Turing and truth-table completeness for **NEXP**. We show that

1. there is a set which is $\leq_{bT(2)}^{P}$-complete but not $\leq_{posT}^{P}$-complete for **NEXP**
2. there is a set which is $\leq_{posT}^{P}$-complete but not $\leq_{tt}^{P}$-complete for **NEXP**.

The second result strengthens the separation of Turing and truth-table completeness given in [2], while the first one settles an open question stated by Buhrman in [1] — namely, for which k there is a set that is $\leq_{btt(k)}^{P}$-complete but not $\leq_{posT}^{P}$-complete for **NEXP**.

In both cases, the proof consists of three steps. In the first step, we define a reduction M (of appropriate type) we want to use to code some fixed many-one-complete set K for **NEXP** into the desired set A. In the second step we construct sets A and W in stages such that $K = L(M, A)$ and $W \not\leq_{r}^{P} A$ where r is the reduction type we want to diagonalize against. Finally, we show that A and W are members of **NEXP**. Thus the central task will be to coordinate the coding of K into A with diagonalization against reductions of type r.

As mentioned above, in proving the second result we will exploit the fact that a positive Turing reduction may contain super-polynomial many queries in its entire computation tree, which cannot be covered by a polynomial time bounded truth-table reduction. Though the situation is different in proving the first result, there we will be able to code K into A since the necessary adding (or removing) of strings will not change the behavior of the positive Turing reductions considered there.

We close this section by introducing some notation.

Let $\Sigma = \{0, 1\}$. Strings are elements of Σ^*, and denoted by lowercase letters $x, y, z, \ldots$. Other lowercase letters usually denote natural numbers. For a string x, $x[i]$ denotes the $(i+1)$th bit in x, i.e. $x = x[0] \ldots x[n-1]$, where $n = |x|$ is the length of x. We will use the binary representation of numbers to code $\mathbb{N}$ into Σ^* and some convenient pairing function $\langle x, y \rangle$ on Σ^*. The concatenation of x and y will be denoted by xy and 0^n will denote the string of length n consisting only of zeroes.

Languages (also called sets) are subsets of Σ^*, and denoted by capital letters $A, B, C, O, X, \ldots$. For any set S the cardinality of S is denoted by $|S|$. For sets A and B, $A \oplus B = \{0x : x \in A\} \cup \{1x : x \in B\}$ is the join of A and B. We identify a set with its infinite characteristic sequence, i.e. $x \in A$ iff $A(x) = 1$ and $x \notin A$ iff $A(x) = 0$.

The reader is assumed to be familiar with the standard (oracle) Turing machine model. In this paper, reductions are characterized in this model, whence we distinguish between adaptive oracle Turing machines – where queries may depend on the answers to previous queries – and non-adaptive oracle Turing machines. Since we will simulate oracle Turing machines given several different oracles, we use the notation

$$sim(M, x, O)$$

for the computation of machine M on input x given oracle O besides the more commonly used notation $M^O(x)$. Moreover, at one point we will replace the oracle set O by some finite string α such that $sim(M, x, \alpha)$ will denote the computation of M on input x where the i-th query is answered by $\alpha[i-1]$, the i-th bit of α. In addition we use $sim(M, x, O) = 0$ for rejecting, and $sim(M, x, O) = 1$ for accepting computations. The language accepted by the oracle Turing machine M equipped with oracle A will be denoted by $L(M, A)$. An oracle Turing machine M is *positive* if $L(M, A) \subseteq L(M, B)$ whenever $A \subseteq B$. For sets A and B we say that

- A *Turing reduces to* B ($A \leq_T^P B$) if $A = L(M, B)$ for some polynomial time bounded oracle Turing machine M
- A *positive Turing reduces to* B ($A \leq_{posT}^P B$) if $A \leq_T^P B$ via some positive oracle Turing machine
- A *truth-table reduces to* B ($A \leq_{tt}^P B$) if $A \leq_T^P B$ via some non-adaptive oracle Turing machine
- A *bounded Turing reduces to* B *with norm* k ($A \leq_{bT(k)}^P B$) if $A \leq_T^P B$ via some oracle Turing machine making at most k queries
- A *bounded truth-table reduces to* B *with norm* k ($A \leq_{btt(k)}^P B$) if $A \leq_T^P B$ via some non-adaptive oracle Turing machine making at most k queries
- A *many-one reduces to* B ($A \leq_m^P B$) if there exists a polynomial time bounded function f such that $x \in A$ iff $f(x) \in B$.

For $r \in \{T, posT, tt, bT(k), btt(k), m\}$ we say that A is $\leq_r^P$-*hard* for **NEXP** if $B \leq_m^P A$ for all $B \in$ **NEXP** and A is $\leq_r^P$-*complete* for **NEXP** if A is $\leq_r^P$-hard for **NEXP** and $A \in$ **NEXP**. Finally, for use as standard $\leq_m^P$-complete set for **NEXP** we define a set $K \in$ **NTIME**(2^n) by

$$K = \{\langle e, x, l\rangle : N_e \text{ accepts } x \text{ within } l \text{ steps}\}$$

where N_e is the e-th nondeterministic Turing machine.

2 Being $\leq_{bT(2)}^P$-Complete Without Being $\leq_{posT}^P$-Complete

In [1] Buhrman gave a construction of a set which is $\leq_{btt(2)}^P$-complete but not $\leq_{posT}^P$-complete for **EXP**. That construction uses the parity function to obtain $\leq_{btt(2)}^P$-completeness, whence this approach cannot be carried out for **NEXP** unless we assume **NEXP** = co-**NEXP**. The construction given below avoids this difficulty by using a $\leq_{bT(2)}^P$-reduction, where the answer to the first query indicates which one out of two eventually relevant informations is in fact relevant. That the resulting set is $\leq_{bT(2)}^P$-hard but not $\leq_{posT}^P$-hard for **NEXP** will be quite clear from the construction, while some work has to be done on showing that A is in **NEXP** itself.

Theorem 1. *There is a $\leq^P_{bT(2)}$-complete set A for* **NEXP** *which is not $\leq^P_{posT}$-complete for* **NEXP**.

Proof. First define a sequence $\{b_n : n \in \mathbb{N}\}$ of natural numbers and a sequence of intervals I_n by

$$b_n = \begin{cases} 1 & \text{if } n \leq 1 \\ 2^{b_{n-1}^{n-1}} + 1 & \text{if } n > 1 \end{cases} \quad (1)$$

$$I_n = \begin{cases} \{y : |y| \leq 1\} & \text{if } n = 1 \\ \{y : b_{n-1}^{n-1} < |y| \leq b_n^n\} & \text{if } n > 1 \end{cases}. \quad (2)$$

Let $\{M_i : i \geq 1\}$ be an enumeration of polynomial time bounded Turing reductions where the running time of M_i is bounded by n^i. We will construct sets A and W in stages such that $W \subseteq \{0^{b_n} : n \geq 1\}$ and $M_n^A(0^{b_n}) = 1 - W(0^{b_n})$ if M_n is a positive reduction. On the other hand, we will guarantee that $K \leq^P_{bT(2)} A$ by

$$x \in K \iff 00^{\log(b_n)} \in A \ \& \ 1\langle x,0\rangle \in A$$
$$\text{or } 00^{\log(b_n)} \notin A \ \& \ 1\langle x,1\rangle \in A \quad (3)$$

for n such that $b_{n-1}^{n-1} < |\langle x,0\rangle| \leq b_n^n$. Note that $\log(b_n) = b_{n-1}^{n-1}$ whence the intended reduction from K to A will be computable in polynomial time.

The construction is as follows:

stage 0: $B_0 = C_0 = \emptyset$

stage n: First we have to check whether the reduction M_n behaves in a positive way on input 0^{b_n}. I.e., we have to compute

$$pos(n) = \begin{cases} 1 & \text{if } M_n^X(0^{b_n}) \leq M_n^Y(0^{b_n}) \text{ whenever } X \subseteq Y \\ 0 & \text{otherwise} \end{cases}$$

Let $A_{<n} = \left(\bigcup_{i<n} B_i\right) \oplus \left(\bigcup_{i<n} C_i\right)$ and

$$C'_n = \{\langle x,1\rangle : \langle x,1\rangle \in I_n\} \quad (4)$$
$$A'_n = A_{<n} \cup (\emptyset \oplus C'_n) \quad (5)$$

Compute $s_n = sim(M_n, 0^{b_n}, A'_n)$ and fix B_n and C_n by

$$B_n = \begin{cases} \{0^{\log(b_n)}\} & \text{if } s_n = 1 \lor pos(n) = 0 \\ \emptyset & \text{otherwise} \end{cases} \quad (6)$$

$$C_n = \begin{cases} C'_n \cup \{\langle x,0\rangle : x \in K \ \& \ \langle x,1\rangle \in I_n\} & \text{if } B_n \neq \emptyset \\ \{\langle x,1\rangle : x \in K \ \& \ \langle x,1\rangle \in I_n\} & \text{if } B_n = \emptyset \end{cases} \quad (7)$$

end of stage n.

Finally, let $B = \bigcup_{n \geq 1} B_n$, $C = \bigcup_{n \geq 1} C_n$, $W = \{0^{b_n} : n \geq 1 \ \& \ s_n = 0\}$ and

$$A = B \oplus C = \bigcup_{n \geq 1} A_{<n}.$$

Note that A'_n, the oracle used in the simulation during stage n, provides information about $A_{<n}$, but queries in the interval I_n are answered according to a mere syntactical property (see (4),(5)). Nevertheless, since M_n on input 0^{b_n} only queries strings of length $\leq b_n^n$, the adding (resp. removing) of strings during stage n in (7) and (6) doesn't affect the outcome of this simulation if M_n is positive, in which case

$$s_n = sim(M_n, 0^{b_n}, A'_n) = sim(M_n, 0^{b_n}, A_{<n+1}) = sim(M_n, 0^{b_n}, A). \quad (8)$$

Since $W(0^{b_n}) = 1 - s_n$, (8) implies that W is not polynomial time positive Turing reducible to A. Moreover, by (7), C_n provides information about K but only B_n tells whether the pair $\langle x, 0\rangle$ or $\langle x, 1\rangle$ carries information about x. To be more precise, for n such that $b_{n-1}^{n-1} < |\langle x, 1\rangle| \leq b_n^n$, $x \in K$ iff either $B_n \neq \emptyset$ and the string $\langle x, 0\rangle$ entered C_n or $B_n = \emptyset$ and the string $\langle x, 1\rangle$ still is a member of C_n. But this is equivalent to (3) whence A is $\leq^P_{bT(2)}$-hard for **NEXP**. It therefore remains to show that A and W are members of **NEXP**.

Claim. $A_{<n}(x)$ can be nondeterministically computed in time $n \cdot 2^{3n|x|^2}$.

Proof. (By induction on n.) For $n = 1$, $A_{<n} = \emptyset$. So assume that the claim holds for $A_{<n}$. Then, for $|x| \leq b_{n-1}^{n-1}$ or $|x| > b_n^n$, $A_{<n+1}(x) = A_{<n}(x)$ whence it suffices to consider $x \in B_n \oplus C_n$. Note that for $x \in I_n$

$$b_n^n < 2^{n \cdot b_{n-1}^{n-1}+1} \leq 2^{|x|^2}.$$

Case 1: $x = 00^{\log(b_n)}$. By definition, $A_{<n+1}(x) = B_n(0^{\log(b_n)}) = 1$ iff $s_n = 1 \vee pos(n) = 0$. By guessing two sequences of oracle answers β_0, β_1 of length $2^{|x|^2}$ witnessing that M_n does not behave positively on input 0^{b_n} we can nondeterministically check that $pos(n) = 0$ in time $2^{3|x|^2}$.

Though we cannot directly compute $s_n = sim(M_n, 0^{b_n}, A'_n)$, in a second algorithm we can guess the oracle answers in this computation and check that this guess is (more or less) correct. That is, nondeterministically guess a string α of length $2^{|x|^2}$ and check that $sim(M_n, 0^{b_n}, \alpha) = 1$. If this is the case and q_i denotes the i-th query to the oracle during this simulation, then accept iff

$$\alpha[i-1] = 1 \Rightarrow A'_n(q_i) = 1 \quad (9)$$

for all $i \leq l$ where l denotes the number of queries made in $sim(M_n, 0^{b_n}, \alpha)$. By (5) and the induction hypothesis, all this can be done nondeterministically in time

$$2 \cdot 2^{|x|^2} + 2^{|x|^2} \cdot n \cdot 2^{3n|x|^2} < (n+1) \cdot 2^{3(n+1)|x|^2}.$$

Combining these two algorithms disjunctively we can nondeterministically compute $A_{<n+1}(x)$, since $A_{<n+1}(x) = 1$ implies that at least one of the two

computations has an accepting path. On the other hand, if $A_{<n+1}(x) = 0$ then M_n behaves positively on input 0^{b_n} and $s_n = sim(M_n, 0^{b_n}, A'_n) = 0$, whence no α satisfying (9) can lead to an accepting path in the second computation. Therefore, none of the two algorithms can have an accepting path.

Case 2: $x = 1\langle y, j\rangle$ for some string y and $j \leq 1$. By definition of C_n in (7), in this case $x \in C_n$ iff at least two out of the following assertions hold:

$$j = 1, \quad y \in K, \quad B_n(0^{\log(b_n)}) = 1.$$

Obviously, using a nondeterministic computation for K and the algorithm given in case 1, this can be nondeterministically decided in time $(n+1) \cdot 2^{3(n+1)|x|^2}$, too.

For other x in the given interval, $A_{<n+1}(x) = 0$ by definition of $A_{<n+1}$. □

Since $n < \log(b_n)$, the above claim shows that $A \in \text{NTIME}(2^{|x|^3})$, whence A is $\leq^P_{bT(2)}$-complete for **NEXP**.

Claim. For $n \geq 4$, $W(0^{b_n})$ can be computed in time 2^{2b_n}.

Proof. Since $W(0^{b_n}) = 1 - s_n$, we have to compute $s_n = sim(M_n, 0^{b_n}, A'_n)$. The running time of M_n on input 0^{b_n} is bounded by b_n^n, whence we have to compute $A'_n(q)$ for at most b_n^n many queries q of length $\leq b_n^n$. For queries q such that $|q| \leq b_{n-1}$ this can be done in time $2^{2^{|q|^3}} \leq 2^{2^{b_{n-1}^3}} \leq 2^{b_n}$ by the previous claim. For queries q such that $|q| > b_{n-1}$, one of the following cases applies.

1. $q = 00^{\log(b_n)}$.
   Then $A'_n(q) = 0$
2. $q = 1y$ and $|y| > b_{n-1}^{n-1}$.
   Then $A'_n(q) = 1 \iff y = \langle z, 1\rangle$ for some z.
3. $q = 1y$ and $b_{n-2}^{n-2} < |y| \leq b_{n-1}^{n-1}$.
   Then $A'_n(q) = 1 \iff$
   (a) $A(00^{\log(b_{n-1})}) = 1$ and $[y = \langle z, 1\rangle \vee (y = \langle z, 0\rangle \ \& \ z \in K)]$, or
   (b) $A(00^{\log(b_{n-1})}) = 0$ and $y = \langle z, 1\rangle \ \& \ z \in K$) for some z.
4. For other q, $A'_n(q) = 0$.

Note that in the third case the computation of $A(00^{\log(b_{n-1})})$ can be done in time $2^{2^{(\log(b_{n-1})+1)^3}} \leq 2^{b_n}$ by the previous claim and the computation of $K(z)$ can be done deterministically in time $2^{2^{|y|}} \leq 2^{b_n}$. Therefore, $A'_n(q)$ can be computed in time 2^{b_n} for all queries q made by M_n on input 0^{b_n}, whence $W(0^{b_n}) = 1 - s_n$ can be computed in time $2^{b_n} \cdot b_n^n \leq 2^{2b_n}$ for $n \geq 4$. □

Since the claim implies that $W \in \text{DTIME}(2^{2n})$ and W is not $\leq^P_{posT}$-reducible to A, A is not $\leq^P_{posT}$-complete for **NEXP**.

This completes the proof of Theorem 1. □

3 Being $\leq_{posT}^{P}$-Complete Without Being $\leq_{tt}^{P}$-Complete

Considering the set A given in [2] to separate $\leq_{T}^{P}$- from $\leq_{tt}^{P}$-completeness for **EXP**, we easily obtain a separation of $\leq_{posT}^{P}$-completeness from $\leq_{tt}^{P}$-completeness for **EXP** by the set $A \oplus \overline{A}$. Though the construction of A may be carried out in **NEXP**, this second step fails for **NEXP** since $A \oplus \overline{A}$ will not be a member of **NEXP** unless we assume **NEXP** = co-**NEXP**. Therefore, to obtain the desired separation, we give a new set B in **NEXP** with the required properties by a more direct construction.

To a great extent, the proof of the next theorem relies on the construction of A in [2]. Once the question how to code K into B in a positive way is settled, the necessary steps to diagonalize against polynomial time truth-table reductions can be carried out just the same way as in the construction of A in [2].

Theorem 2. *There is a set B which is $\leq_{posT}^{P}$-complete but not $\leq_{tt}^{P}$-complete for* **NEXP**.

Proof. B will be constructed in such a way that K is recognized by a positive Turing machine M with oracle B which acts as follows.

On input x of length n, M performs $n^2 + 1$ many rounds of querying B. In the first n^2 rounds M chooses two queries depending on the answers to the previous queries and accepts (rejects) if both answers are 1 (0). Otherwise, M starts the next round. If the final round is reached, only one query is chosen and M accepts iff the answer is 1.

It es easy to see that M is indeed positive and that the computation tree of M on input x may contain $2(2^{|x|^2} - 1) + 2^{|x|^2}$ many different queries. This fact provides enough flexibility to diagonalize against polynomial time truth-table reductions. As before, B will be constructed in stages together with a witness set W. During the n-th stage we will fix B on a suitable interval J_n and diagonalize against the n-th truth-table reduction M_n, where $\{M_i : i \geq 1\}$ is some enumeration of polynomial time truth-table reductions and the running time of M_i is bounded by n^i. We will only sketch this construction here, since all the technical details can be taken from Theorem 6 in [2].

First we define for every string z the tree $T(z)$ of queries to be chosen by M. Consider a balanced tree of depth $|z|^2$ where the root is labeled by $\{\langle z,1\rangle, \langle z,2\rangle\}$ and if a node is labeled by $\{\langle z,i-1\rangle, \langle z,i\rangle\}$ then its left son is labeled by $\{\langle z, 2i-1\rangle, \langle z, 2i\rangle\}$ and its right son by $\{\langle z, 2i+1\rangle, \langle z, 2i+2\rangle\}$. (I. e. take all the pairs $\langle z,1\rangle, \ldots \langle z, 2(2^{|z|^2+1} - 1)\rangle$ and, starting from the root and proceeding for every level from left to right, attach to each node the next two pairs.) Finally, we remove from the leaves the pairs $\langle z,i\rangle$ for which i is odd. W.l.o.g. we assume that for given z, all the pairs $\langle z,i\rangle$ occurring in $T(z)$ are of the same length.

The resulting labeled tree will guide M in the following way: the first round performed by M on input z will consist of querying $\langle z,1\rangle, \langle z,2\rangle$, i.e. the strings that are labeled to the root of $T(z)$. As described above, M accepts (rejects) in round m if all queries made in this round are answered by 1 (0). Otherwise, if M has reached the node n_m of $T(z)$ in this round, then M chooses the left (right) son of n_m if only the first (resp. the second) answer is 1 and starts the next round by querying the strings that are labeled to that node.

An easy computation shows that for any set S with $|S| < 2 \cdot 2^{|z|^2} - 1$, there exists a node n in $T(z)$ such that S and the label of n are disjoint. Let $N_z(S)$ denote the first such node and $P_z(S)$ the path in $T(z)$ that leads to it.

Now we are ready to sketch the construction of B and W. Again we will use a suitable sequence of diagonalization points $\{v_n : n \geq 1\}$ such that $v_{n+1} > 2^{v_n^n}$, and of intervals $J_n = \{y : v_{n-1}^{n-1} < |y| \leq v_n^n\}$. During the n-th stage of the construction we do the following:

1. simulate M_n on input 0^{v_n} and let Q_n be the set of queries made by M_n on this input. (Note that M_n is a truth-table reduction whence Q_n is not oracle dependent.)
2. For every $y = \langle z, i \rangle \in J_n$ fix $B(y)$ according to the following rules:
   (a) If $N_z(Q_n)$ exists then

   $B(y) = 1 \iff$
   y is a member of the label of $N_z(Q_n)$ & $z \in K$
   or
   there is a node on $P_z(Q_n)$ such that y is a member of its label and
   (i is odd and $P_z(Q_n)$ proceeds to the left) $\vee$
   (i is even and $P_z(Q_n)$ proceeds to the right) (10)

   (b) If $N_z(Q_n)$ does not exist then

   $$B(y) = 1 \iff (z \in K \ \& \ i \in \{1, 2\})$$ (11)

3. Let $W(0^{v_n}) = 1 - s_n$ where $s_n = sim(M_n, 0^{v_n}, B)$

By this construction it is straightforward to see that $K = L(M, B)$ for the positive Turing machine M described above, that W is not polynomial time truth-table reducible to B, and that $B \in \mathbf{NEXP}$ if we make sure that $v_n^n < 2^{|y|^2}$ for all $y \in J_n$.

To see that $W \in \mathbf{EXP}$, note that any query $q = \langle z, i \rangle$ made by M_n on input 0^{v_n} is either of length $< v_{n-1}^{n-1}$, in which case we can compute $K(z)$ deterministically in time 2^{v_n} and therefore $B(q)$ in time exponential in v_n, or $q \in Q_n \cap J_n$, in which case $B(q)$ is computed according (10) or (11). If (10) applies, $B(q)$ is decided by comparing Q_n and $T(z)$. If (11) applies, the length of z is small compared to v_n^n. Therefore, if we choose the sequence $\{v_n : n \geq 1\}$ such that $K(z)$ can be deterministically decided for those z in time exponential in v_n, $W(0^{v_n}) = 1 - sim(M_n, 0^{v_n}, B)$ will be computable in time exponential in v_n, too. (For details on the choice of v_n and the distinction between "small" and "non-small" z see [2]).

This completes the proof of Theorem 2. □

4 Conclusions

In Section 2 we proved that positive and non-positive completeness notions differ for **NEXP** from a quite early stage, i.e. the difference can be shown for $\leq^P_{bT(2)}$-complete, i.e. $\leq^P_{btt(3)}$-complete sets for **NEXP**. A strengthening of this result to $\leq^P_{btt(2)}$-complete sets for **NEXP**, as can be done for **EXP**, seems unlikely unless we assume **NEXP** = co-**NEXP**. An oracle which gives evidence for the oracle dependence of such a stronger result would be an interesting further step.

In Section 3 we strengthened the separation of Turing and truth-table completeness for **NEXP** given in [2] to positive Turing complete sets. Taken together, these two results prove the incomparability of $\leq^P_{posT}$-completeness and $\leq^P_{tt}$-completeness for **NEXP**.

If we consider bounded reductions for which the number of queries is bounded by some constant, $\leq^P_{bT(k)}$-completeness and $\leq^P_{btt(l)}$-completeness for **NEXP** are incomparable for $k < l < 2^k - 1$ (see [3]). For a similar result in the case of *positive* bounded reductions, we may adapt the construction given in Section 3 to obtain (for $k \geq 2$) a set C which is *positive bounded Turing complete with norm k* but not *positive bounded truth-table complete with norm l* ($\leq^P_{pos-bT(k)}$- and $\leq^P_{pos-btt(l)}$-complete for short) for $l < 2^{\lfloor k/2 \rfloor} + 2^{\lfloor (k-1)/2 \rfloor} - 1$. Moreover, in [3] a set is given which is *disjunctive* $\leq^P_{btt(k+1)}$-complete but not $\leq^P_{bT(k)}$-complete for **NEXP** (for $k \geq 1$).

Therefore, if we let $g(k) = 2^{\lfloor k/2 \rfloor} + 2^{\lfloor (k-1)/2 \rfloor} - 1$, $\leq^P_{pos-bT(k)}$-completeness and $\leq^P_{pos-btt(l)}$-completeness for **NEXP** are incomparable for $k < l < g(k)$. If we replace the positive Turing reduction M used in Section 3 to code K into B by a more complex one, we can improve the bound $g(k)$ to $g'(k)$ where $g' \in \Theta(3^{n/2})$.

Acknowledgments: I would like to thank Frank Stephan and Klaus Ambos-Spies for discussing with me the results presented in this paper and providing many fruitful ideas to the proof in Section 2. I would also like to thank the anonymous referees for their helpful comments on an earlier version of this paper.

References

1. H. Buhrman. *Resource Bounded Reductions.* PhD thesis, Universiteit van Amsterdam, Amsterdam, 1993.
2. H. Buhrmann, S. Homer, and L. Torenvliet. Completeness for nondeterministic complexity classes. *Mathematical Systems Theory* **24**, 179–200, 1991.
3. H. Buhrmann, E. Spaan, and L. Torenvliet. Bounded reductions. In K. Ambos-Spies, S. Homer, and U. Schöning, editors, *Complexity Theory*, pages 83–99. Cambridge University Press, 1993.
4. R. Ladner, N. Lynch, and A. Selman. A comparison of polynomial-time reducibilities. *Theoretical Computer Science* **1**, 103–123, 1975.
5. O. Watanabe. A comparison of polynomial time completeness notions. *Theoretical Computer Science* **54**, 249–265, 1987.

Tally NP Sets and Easy Census Functions

Judy Goldsmith[1] *, Mitsunori Ogihara[2] **, and Jörg Rothe[3] ***

[1] Department of Computer Science, University of Kentucky, Lexington, KY 40506, USA
   goldsmit@cs.engr.uky.edu
[2] Department of Computer Science, University of Rochester, Rochester, NY 14627, USA
   ogihara@cs.rochester.edu
[3] Institut für Informatik, Friedrich-Schiller-Universität Jena, 07740 Jena, Germany
   rothe@informatik.uni-jena.de

Abstract. We study the question of whether every P set has an easy (i.e., polynomial-time computable) census function. We characterize this question in terms of unlikely collapses of language and function classes such as $\#P_1 \subseteq FP$, where $\#P_1$ is the class of functions that count the witnesses for tally NP sets. We prove that every $\#P_1^{PH}$ function can be computed in $FP^{\#P_1^{\#P_1}}$. Consequently, every P set has an easy census function if and only if every set in the polynomial hierarchy does. We show that the assumption $\#P_1 \subseteq FP$ implies $P = BPP$ and $PH \subseteq MOD_k P$ for each $k \geq 2$, which provides further evidence that not all sets in P have an easy census function. We also relate a set's property of having an easy census function to other well-studied properties of sets, such as rankability and scalability (the closure of the rankable sets under P-isomorphisms). Finally, we prove that it is no more likely that the census function of any set in P can be approximated (more precisely, can be n^α-enumerated in time n^β for fixed α and β) than that it can be precisely computed in polynomial time.

1 Introduction

Does every P set have an easy (i.e., polynomial-time computable) census function? Many important properties similar to this one were studied during the past decades to gain more insight into the nature of feasible computation. Among the questions that were previously studied are the question of whether or not every P set has an easy to compute ranking function [GS91,HR90], whether every P set is P-isomorphic to some rankable set [GH96], whether every sparse set in P is P-printable [HY84,AR88], whether there exists an infinite set in P having no infinite

---

* Supported in part by NSF grant CCR-9315354.
** Supported in part by the National Science Foundation under grants CCR-9701911 and INT-9726724.
*** Supported in part by grants NSF-INT-9513368/DAAD-315-PRO-fo-ab and NSF-CCR-9322513 and by a NATO Postdoctoral Science Fellowship from the Deutscher Akademischer Austauschdienst ("Gemeinsames Hochschulsonderprogramm III von Bund und Ländern"). Current address: Department of Computer Science, University of Rochester, Rochester, NY 14627, USA. Work done in part while visiting the University of Kentucky and the University of Rochester.

P-printable subset [AR88,HRW97], whether every P-printable set is P-isomorphic to some tally set in P [AR88], and whether every P set admits easy certificate schemes [HRW97], to name just a few. Some of those questions arise in the field of data compression and are related to Kolmogorov complexity, some are linked to the question of whether one-way functions exist. Extending this line of research, the present paper studies the complexity of computing the census functions of sets in P. Census functions have proven to be a particularly important and useful notion in complexity theory, and their use has had a profound impact upon almost every area of the field (see the extensive literature related to the isomorphism conjecture of Berman and Hartmanis [BH77] or, for instance, [KL80,HY84,HIS85,KS85,LS86,BBS86,AR88,GH96] for other topics).

Valiant, in his seminal papers [Val79a,Val79b], introduced $\#P$, the class of functions that count the solutions of NP problems, and its tally version $\#P_1$ for which the inputs are given in unary. Although $\#P_1$ has not become as prominent as $\#P$, it contains a number of quite interesting and important problems such as the problem Self-Avoiding Walk (see [Wel93]): Given an integer n in unary, compute the number of self-avoiding walks on the square lattice having length n and rooted at the origin. Self-Avoiding Walk is a well-known classical problem of statistical physics and polymer chemistry, and it is an intriguing open question whether Self-Avoiding Walk is $\#P_1$-complete (see [Wel93]).

We characterize the question of whether every P set has an easy census function in terms of collapses of language and function classes that are considered to be unlikely. In particular, every P set has an easy census function if and only if $\#P_1 \subseteq FP$. The main technical contribution in Section 3 is that $\#P_1^{PH}$ is contained in $FP^{\#P_1^{\#P_1}}$. An immediate consequence of this result are upward collapse results of the form: the collapse $\#_1 \cdot P \subseteq FP$ implies the collapse $\#_1 \cdot PH \subseteq FP$. Thus, every P set has an easy census function if and only if every set in PH does. Note that the corresponding upward collapse for the $\#$ operator applied to the levels of PH follows immediately from the upward collapse property of the polynomial hierarchy itself: $\# \cdot P \subseteq FP$ implies $NP = P$ and thus $PH = P$; so, $\# \cdot PH = \# \cdot P \subseteq FP$. However, for the $\#_1$ operator this is not so clear, since the assumption $\#_1 \cdot P \subseteq FP$ merely implies that all *tally* NP sets are in P (equivalently, $NE = E$), from which one cannot immediately conclude that $\#_1 \cdot NP$ or even $\#_1 \cdot PH$ is contained in FP. In fact, Hartmanis et al. [HIS85] show that in some relativized world, $NE = E$ and yet the (weak) exponential-time hierarchy does not collapse. In light of this result, it is quite possible that the assumption of all tally NP sets being in P does not force all tally sets from higher levels of the polynomial hierarchy into P. We also show that the assumption $\#P_1 \subseteq FP$ implies both $P = BPP$ and $PH \subseteq MOD_k P$ for each $k \geq 2$, which provides further evidence that not all sets in P have a census function computable in polynomial time. We also relate a set's property of having an easy census function to other well-studied properties of sets, such as rankability [GS91] and scalability [GH96]. In particular, though every rankable set has an easy census function, we show that (even when restricted to the sets in P) the converse is not true unless $P = PP$. This expands the result of Hemaspaandra and Rudich that every P set is rankable if and only if $P = PP$ [HR90] by showing that $P = PP$ is already implied by the apparently weaker hypothesis that every P set *with an easy census function* is rankable.

Cai and Hemaspaandra [CH89] introduced the notion of enumerative counting as a way of approximating the value of a #P function deterministically in polynomial time. Hemaspaandra and Rudich [HR90] show that every P set is k-enumeratively rankable for some fixed k in polynomial time if and only if #P = FP. They conclude that it is no more likely that one can enumeratively rank all sets in P than that one can exactly compute their ranking functions in polynomial time. In Section 4, we similarly characterize the question of whether the census function of all P sets is n^α-enumerable in time n^β for fixed constants α and β, or equivalently, whether every #P$_1$ function is n^α-enumerable in time n^β. We show that this implies #P$_1 \subseteq$ FP, and we thus conclude that it is no more likely that one can n^α-enumerate the census function of every P set in time n^β than that one can precisely compute its census function in polynomial time. Finally, Section 5 provides a number of relativization results.

2 Notation and Definitions

Fix the alphabet $\Sigma = \{0,1\}$. Σ^* denotes the set of all strings over Σ. For any string $x \in \Sigma^*$, we denote the length of x by $|x|$. For any set $L \subseteq \Sigma^*$, the number of strings in L is denoted $|L|$, and the complement of L in Σ^* is denoted $\overline{L}$. Let $L^{=n}$ (respectively, $L^{\leq n}$) denote the set of strings in L of length n (respectively, of length at most n). As a shorthand, we use Σ^n to denote $(\Sigma^*)^{=n}$. For any set L, the *census function of L*, $census_L : \Sigma^* \to \mathbb{N}$, is defined by $census_L(1^n) \stackrel{\text{df}}{=} |L^{=n}|$,[1] and χ_L denotes the *characteristic function of L*. A set S is said to be *sparse* if there is a polynomial p such that for each length n, $census_S(1^n) \leq p(n)$. A set T is said to be *tally* if $T \subseteq \{1\}^*$.

The definition of Turing machines and their languages, Turing transducers and the functions they compute, relativized (i.e., oracle) computations, (relativized) complexity classes, etc. is standard in the literature. We briefly recall the complexity classes most important in this paper. FP denotes the class of polynomial-time computable functions. FE is the class of functions that can be computed by deterministic transducers running in time 2^{cn} for some constant c. FP$_1$ is the class of functions computable in polynomial time by deterministic transducers with a unary input alphabet. An unambiguous Turing machine is a nondeterministic Turing machine that on each input has at most one accepting path. UP [Val76] (respectively, UE) is the class of all languages accepted by some unambiguous Turing machine running in polynomial time (respectively, in time 2^{cn} for some constant c). For any nondeterministic Turing machine M and any input $x \in \Sigma^*$, let $\text{acc}_M(x)$ denote the number of accepting paths of $M(x)$. A *spanP machine* [KST89] is an NP machine that has a special output device on which some output is printed for each accepting path. For any spanP machine M and any input $x \in \Sigma^*$, $\text{span}_M(x)$ is defined to be the number of different outputs of $M(x)$ if $M(x)$ has at least

---

[1] The census function of L at n is often defined as the number of elements in L of length up to n. This definition and our definition are compatible as long as our computability admits subtraction. We let $census_L$ map strings 1^n (as opposed to numbers n in binary notation) to $|L^{=n}|$ to emphasize that the input to the transducer computing $census_L$ is given in unary.

one accepting path, and 0 otherwise. A *tally NP machine* (respectively, a *tally spanP machine*) is an NP (respectively, a spanP) machine with a unary input alphabet.

Valiant introduced the function classes $\#\mathrm{P} \stackrel{\mathrm{df}}{=} \{\mathrm{acc}_M \mid M \text{ is an NP machine}\}$ [Val79a,Val79b] and $\#\mathrm{P}_1 \stackrel{\mathrm{df}}{=} \{\mathrm{acc}_M \mid M \text{ is a tally NP machine}\}$ [Val79b]. The class spanP $\stackrel{\mathrm{df}}{=} \{\mathrm{span}_M \mid M \text{ is a spanP machine}\}$ [KST89] can analogously be restricted to tally sets: $\mathrm{spanP}_1 \stackrel{\mathrm{df}}{=} \{\mathrm{span}_M \mid M \text{ is a tally spanP machine}\}$. We will use the common operator notation at times in order to generalize function classes such as $\#\mathrm{P}$ and $\#\mathrm{P}_1$. For any language class $\mathcal{C}$, define $\# \cdot \mathcal{C}$ (respectively, $\#_1 \cdot \mathcal{C}$) to be the class of all functions f for which there exist a set $A \in \mathcal{C}$ and a polynomial p such that $(\forall x \in \Sigma^*)[f(x) = |\{y \mid |y| = p(|x|) \text{ and } \langle x, y \rangle \in A\}|]$ (respectively, $(\forall n \in \mathbb{N})[f(1^n) = |\{y \mid |y| = p(n) \text{ and } \langle 1^n, y \rangle \in A\}|])$. Let $\mathrm{E} \stackrel{\mathrm{df}}{=} \bigcup_{c>0} \mathrm{DTIME}[2^{cn}]$ and $\mathrm{NE} \stackrel{\mathrm{df}}{=} \bigcup_{c>0} \mathrm{NTIME}[2^{cn}]$, and define $\#\mathrm{E} \stackrel{\mathrm{df}}{=} \{\mathrm{acc}_M \mid M \text{ is an NE machine}\}$. For the other classes we consider, we simply give a reference to the paper in which the class is defined: The polynomial hierarchy PH [MS72,Sto77], PP [Gil77], BPP [Gil77], $\mathrm{MOD}_k\mathrm{P}$ [CH90] for fixed $k \geq 2$ (if $k = 2$, we write $\oplus\mathrm{P}$ [PZ83,GP86] instead of $\mathrm{MOD}_2\mathrm{P}$), and SPP [OH93,FFK94]. We also consider nonuniform classes $\mathcal{C}/\mathrm{poly}$ [KL80] for any language class $\mathcal{C}$, and we analogously define nonuniform function classes $\mathcal{F}/\mathrm{poly}$ for any function class $\mathcal{F}$.

Definition 1. *A bijection $\phi : \Sigma^* \to \Sigma^*$ is a P-isomorphism if ϕ is computable and invertible in polynomial time. A P-isomorphism ϕ is* length-preserving *if for all $x \in \Sigma^*$, $|\phi(x)| = |x|$. A P-isomorphism ϕ mapping set $A \subseteq \Sigma^*$ to set $B \subseteq \Sigma^*$ is* order-preserving *if for any two strings x and y satisfying either $x, y \in A$ or $x, y \notin A$, if $x \leq y$, then $\phi(x) \leq \phi(y)$.*

Definition 2. [GS91] *The* ranking function *of a language $A \subseteq \Sigma^*$ is the function $r : \Sigma^* \to \mathbb{N}$ that maps each $x \in \Sigma^*$ to $|\{y \leq x \mid y \in A\}|$. A language A is* rankable *if its ranking function is computable in polynomial time.*

Generalizing rankability, Goldsmith and Homer [GH96] introduced the property of scalability. They showed that the scalable sets are precisely those that are P-isomorphic to some rankable set. The definition below is based on this characterization.

Definition 3. [GH96] *A language A is* scalable *if it is P-isomorphic to a rankable set. For any oracle X, the X-scalable sets are those that are P^X-isomorphic to some set rankable in FP^X.*

3 Does P Have Easy Census Functions?

We start by exploring the relationships between the properties of a set being rankable, being scalable, and having an easy census function. Let A be any set (not necessarily in P). Consider the following conditions:

(i) A is rankable.
(ii) A has an easy census function.
(iii) A is P-isomorphic to some rankable set (i.e., A is scalable).
(iv) A is P-isomorphic to some rankable set via some length-preserving isomorphism.
(v) A is P-isomorphic to some rankable set via some order-preserving isomorphism.

It is immediately clear that for any set A, (i) implies each of (ii), (iv), and (v), and each of (iv) and (v) implies (iii). The next proposition shows that the rankable sets are closed under order-preserving P-isomorphisms (thus, conditions (i) and (v) in fact are equivalent) and that the class of sets having an easy census function is closed under length-preserving P-isomorphisms. The latter fact gives that (iv) implies (ii), since every rankable set has an easy census function. Due to space constraints, all proofs of this paper except the proof of Theorem 5 are omitted; they can be found in the full version of this paper [GOR98].

Proposition 1. *(1) The class of all rankable sets is closed under order-preserving P-isomorphisms. (2) The class of sets having an* FP*-computable census function is closed under length-preserving* P*-isomorphisms.*

So we are left with only the four conditions (i) to (iv). Since there are nonrecursive sets with an FP-computable census function, but any set satisfying one of (i), (iii), or (iv) is in P, condition (ii) in general cannot imply any of the other three conditions. On the other hand, when we restrict our attention to the sets in P having easy census functions, we can show that (ii) implies (i) if and only if P = PP. Thus, even when restricted to P sets, it is unlikely that (ii) is equivalent to (i).

Theorem 1. *All* P *sets with an easy census function are rankable if and only if* P = PP.

Corollary 1. *All* P *sets are rankable if and only if all sets in* P *with an easy census function are rankable.*

One might ask whether or not all P sets outright have an easy census function (which, if true, would make Corollary 1 trivial). The following characterization of this question in terms of unlikely collapses of certain function and language classes suggests that this probably is not true. Thus, Corollary 1 is nontrivial with the same certainty with which we believe that for instance not all $\#P_1$ functions are in FP.

Theorem 2. *The following are equivalent.*

1. *Every* P *set has an* FP-*computable census function.*
2. $\#P_1 \subseteq FP$.
3. $\#E = FE$.
4. $P^{\#P_1} = P$.
5. *For every language L accepted by a logspace-uniform depth 2 AND-OR circuit family of bottom fan-in 2, census$_L$ is in* FP.

Theorem 2 can as well be stated for more general classes than $\#P_1 = \#_1 \cdot P$. In particular, this comment applies to $\#_1 \cdot \mathcal{C}$, where for instance $\mathcal{C} = NP$ or $\mathcal{C} = PH$. Noticing that spanP$_1 = \#_1 \cdot NP$ and focusing on the first two conditions of Theorem 2, this observation is exemplified as follows.

Theorem 3. *(1) Every* NP *set has an* FP-*computable census function if and only if* spanP$_1 \subseteq$ FP. *(2) Every set in* PH *has an* FP-*computable census function if and only if* $\#_1 \cdot$ PH $\subseteq$ FP.

We will show later that the conditions of Theorem 2 in fact are equivalent to the two conditions stated in either part of Theorem 3. Next, we give some more evidence that the collapse $\#P_1 \subseteq$ FP is unlikely to hold.

Theorem 4. *If* $\#P_1 \subseteq$ FP, *then* PH $\subseteq$ MOD$_k$P *for any fixed* $k \geq 2$, *and* P = BPP.

Now we show that the conditions of Theorem 2 in fact are equivalent to the two conditions stated in either part of Theorem 3. To this end, we establish the following theorem, which is interesting in its own right.

Theorem 5. $\#P_1^{\text{PH}} \subseteq \text{FP}^{\#P_1^{\#P_1}}$.

Remark 1. Note that Toda's result PH $\subseteq$ P$^{\#P}$ [Tod91] immediately gives that $\#P^{\text{PH}} \subseteq \#P^{\#P}$ and $\#P_1^{\text{PH}} \subseteq \#P_1^{\#P}$. Observe that the oracle is a $\#P$ function. In contrast to the inclusion $\#P_1^{\text{PH}} \subseteq \#P_1^{\#P}$, Theorem 5 establishes containment of $\#P_1^{\text{PH}}$ in a class in which only $\#P_1$ oracles occur. Though our proof also applies the techniques of [Tod91,TO92], the result we obtain seems to be incomparable with the above-mentioned immediate consequence of Toda's Theorem. Note also that it is unlikely that Theorem 5 can be improved to even $\#P^{\text{PH}}$ being contained in $\text{FP}^{\#P_1^{\#P_1}}$, since this would imply that $\#P^{\text{PH}} \subseteq$ FP/poly and thus, in particular, would collapse the polynomial hierarchy. Though the proof of Theorem 5 in fact establishes the statement of Theorem 5 (and its corollaries) even for the class $\oplus P^{\text{PH}} =$ BPP$^{\oplus P}$ in place of PH, we focus on the PH case.

Proof of Theorem 5. Let f be any function in $\#P_1^{\text{PH}}$. Note that $\#P_1^{\text{PH}} = \#_1 \cdot$ PH. Thus, there exist a set $L \in$ PH and a polynomial p such that for each length n, $f(1^n) = |\{y \in \{0,1\}^{p(n)} \mid 1^n \# y \in L\}|$, where for convenience we assume that $p(n)$ is a power of 2 for each n. By Toda and Ogihara's result that PH $\subseteq \oplus$P/poly [TO92], there exist a set $A \in \oplus$P, an advice function h computable in $\text{FP}_1^{\#P_1}$ (see [GOR98] for a proof of this claim), and a polynomial q such that for each length m and each x of length m, $|h(1^m)| = q(m)$, and $x \in L$ if and only if $\langle x, h(1^m)\rangle \in A$. Let M be a machine witnessing that $A \in \oplus$P, i.e., for every string z, $z \in A$ if and only if acc$_M(z)$ is odd.

Toda [Tod91] defined inductively the following sequence of polynomials: For $j \in \mathbb{N}$, let $s_0(j) \stackrel{\text{df}}{=} j$, and for each $j \in \mathbb{N}$ and $i > 0$, let $s_i(j) \stackrel{\text{df}}{=} 3(s_{i-1}(j))^4 + 4(s_{i-1}(j))^3$. One very useful property of this sequence of polynomials is that for all $i, j \in \mathbb{N}$, $s_i(j) = c \cdot 2^{2^i}$ for some $c \in \mathbb{N}$ if j is even, and $s_i(j) = d \cdot 2^{2^i} - 1$ for some $d \in \mathbb{N}$ if j is odd (see [Tod91] for the induction proof).

We describe a polynomial-time oracle transducer T that, on input 1^n, invokes its $\#P_1^{\#P_1}$ function oracle g and then prints in binary the number $f(1^n)$. Fix the input 1^n. First, T transfers the input to the oracle g. Formally, function g is defined by

$$g(1^n) \stackrel{\text{df}}{=} \sum_{y \in \{0,1\}^{p(n)}} \left(s_{\ell_n}(\text{acc}_M(\langle 1^n \# y, h(1^{n+1+p(n)})\rangle))\right)^2,$$

where $\ell_n \stackrel{\mathrm{df}}{=} \log p(n)$. Informally speaking, that g is in $\#\mathrm{P}_1^{\#\mathrm{P}_1}$ follows from the properties of the Toda polynomials, from the closure of $\#\mathrm{P}$ under addition and multiplication, and from the fact that advice function h is computable in $\mathrm{FP}_1^{\#\mathrm{P}_1}$. For a formal proof of $g \in \#\mathrm{P}_1^{\#\mathrm{P}_1}$, the reader is referred to the full version of this paper [GOR98].

We use the shorthands $a_n = h(1^{n+1+p(n)})$ for the advice string for length n strings, and $j_y = \mathrm{acc}_M(\langle 1^n \# y, a_n\rangle)$ for each fixed y, $|y| = p(n)$. By the above properties of the Toda polynomials, it follows that for each y of length $p(n)$, if j_y is even then $s_{\ell_n}(j_y) = c \cdot 2^{2^{\ell_n}}$ for some $c \in \mathbb{N}$, and if j_y is odd then $s_{\ell_n}(j_y) = d \cdot 2^{2^{\ell_n}} - 1$ for some $d \in \mathbb{N}$. Thus, recalling that $2^{\ell_n} = p(n)$, we have that $(s_{\ell_n}(j_y))^2 = (c^2 \cdot 2^{p(n)-1})2^{p(n)+1}$ if j_y is even, and $(s_{\ell_n}(j_y))^2 = (d^2 \cdot 2^{p(n)-1} - d)2^{p(n)+1} + 1$ if j_y is odd. Defining $\widehat{c}(n) \stackrel{\mathrm{df}}{=} c^2 \cdot 2^{p(n)-1}$ and $\widehat{d}(n) \stackrel{\mathrm{df}}{=} d^2 \cdot 2^{p(n)-1} - d$, we obtain

$$(s_{\ell_n}(j_y))^2 = \begin{cases} \widehat{c}(n) \cdot 2^{p(n)+1} & \text{if } j_y \text{ is even} \\ \widehat{d}(n) \cdot 2^{p(n)+1} + 1 & \text{if } j_y \text{ is odd}. \end{cases}$$

Thus, since $f(1^n) \leq 2^{p(n)}$ and since j_y is odd if and only if $1^n \# y \in L$, the rightmost $p(n) + 1$ bits of the binary representation of $g(1^n)$ represent the value of $f(1^n)$. Hence, after the value $g(1^n)$ has been returned by the oracle, T can output $f(1^n)$ by printing the $p(n) + 1$ rightmost bits of $g(1^n)$. This completes the proof. □

Since $\#\mathrm{P}_1 \subseteq \mathrm{FP}$ implies $\mathrm{FP}^{\#\mathrm{P}_1^{\#\mathrm{P}_1}} \subseteq \mathrm{FP}$, we have from Theorem 5:

Corollary 2. $\#\mathrm{P}_1 \subseteq \mathrm{FP}$ *if and only if* $\#\mathrm{P}_1^{\mathrm{PH}} \subseteq \mathrm{FP}$, *and in particular,* $\#\mathrm{P}_1 \subseteq \mathrm{FP}$ *if and only if* $\mathrm{spanP}_1 \subseteq \mathrm{FP}$.

Corollary 3. *Every* P *set has an easy census function if and only if every set in* PH *has an easy census function.*

Köbler et al. [KST89] proved that $\mathrm{spanP} = \#\mathrm{P}$ if and only if $\mathrm{NP} = \mathrm{UP}$. Using the analogous result for tally sets, we can show that spanP_1 and $\#\mathrm{P}_1$ are different classes unless $\mathrm{NE} = \mathrm{UE}$, or unless every sparse set in NP is low for SPP. A set S is *$\mathcal{C}$-low* for some class $\mathcal{C}$ if $\mathcal{C}^S = \mathcal{C}$. In particular, it is known that every sparse NP set is low for P^{NP} [KS85] and for PP [KSTT92], but it is not known whether all sparse NP sets are low for SPP. There are oracles known for which some sparse NP set is not SPP-low.

Theorem 6. *If* $\mathrm{spanP}_1 = \#\mathrm{P}_1$, *then* $\mathrm{NE} = \mathrm{UE}$ *and every sparse* NP *set is* SPP-*low.*

4 Enumerative Approximation of Census Functions

Definition 4. [CH89] *Let* $f : \Sigma^* \to \Sigma^*$ *and* $g : \mathbb{N} \to \mathbb{N}$ *be two functions. A Turing transducer* E *is a* $g(n)$-*enumerator of* f *if for all* $n \in \mathbb{N}$ *and* $x \in \Sigma^n$, *(1)* E *on input* x *prints a list* $\mathcal{L}_x$ *with at most* $g(n)$ *elements, and (2)* $f(x)$ *is a member of list* $\mathcal{L}_x$. *A function* f *is* $g(n)$-*enumerable in time* $t(n)$ *if there exists a* $g(n)$-*enumerator of* f *that runs in time* $t(n)$. *A set is* $g(n)$-*enumeratively rankable in time* $t(n)$ *if its ranking function is* $g(n)$-*enumerable in time* $t(n)$.

Recall from the introduction Hemaspaandra and Rudich's result that every P set is k-enumeratively rankable for some fixed k (and indeed, even $\mathcal{O}(n^{1/2-\epsilon})$-enumeratively rankable for some $\epsilon > 0$) in polynomial time if and only if $\#\text{P} = \text{FP}$ [HR90]. We similarly characterize the question of whether the census function of all P sets is n^α-enumerable in time n^β for fixed constants α and β. By the analog of Theorem 2 for fixed time n^β, this is equivalent to asking whether every $\#\text{P}_1$ function is n^α-enumerable in time n^β. We show that this implies $\#\text{P}_1 \subseteq \text{FP}$, and we thus conclude that it is no more likely that one can n^α-enumerate the census function of every P set in time n^β than that one can precisely compute its census function in polynomial time. It would be interesting to know if this result can be improved to hold for polynomial time instead of time t for some fixed polynomial $t(n) = n^\beta$.

Theorem 7. *Let $\alpha, \beta > 0$ be constants. If every $\#\text{P}_1$ function is n^α-enumerable in time n^β, then $\#\text{P}_1 \subseteq \text{FP}$.*

5 Oracle Results

Theorem 8. *There exists an oracle D such that $\#\text{P}_1^D \subseteq \text{FP}^D \neq \#\text{P}^D$.*

Corollary 4. *There exists an oracle D such that all sets in P^D have a census function computable in FP^D, yet some set in P^D is not rankable by any function in FP^D.*

Theorem 9. *There exists an oracle A such that there exists an A-scalable set B whose census function is not in FP^A.*

Theorem 10. *There exists an oracle D such that $D \in \text{P}^D$ is not D-scalable and its census function is not in FP^D.*

Theorem 11. *There exists an oracle A such that $A \in \text{P}^A$ is not A-scalable and its census function is in FP^A.*

Acknowledgments. We are deeply indebted to Lance Fortnow, Lane Hemaspaandra, and Gabriel Istrate for interesting discussions and for helpful comments and suggestions, and we thank Eric Allender and Lane Hemaspaandra for pointers to the literature.

References

[AR88] E. Allender and R. Rubinstein. P-printable sets. *SIAM Journal on Computing*, 17(6):1193–1202, 1988.

[BBS86] J. Balcázar, R. Book, and U. Schöning. The polynomial-time hierarchy and sparse oracles. *Journal of the ACM*, 33(3):603–617, 1986.

[BH77] L. Berman and J. Hartmanis. On isomorphisms and density of NP and other complete sets. *SIAM Journal on Computing*, 6(2):305–322, 1977.

[CH89] J. Cai and L. Hemachandra. Enumerative counting is hard. *Information and Computation*, 82(1):34–44, 1989.

[CH90] J. Cai and L. Hemachandra. On the power of parity polynomial time. *Mathematical Systems Theory*, 23(2):95–106, 1990.

[FFK94] S. Fenner, L. Fortnow, and S. Kurtz. Gap-definable counting classes. *Journal of Computer and System Sciences*, 48(1):116–148, 1994.
[GH96] J. Goldsmith and S. Homer. Scalability and the isomorphism problem. *Information Processing Letters*, 57(3):137–143, 1996.
[Gil77] J. Gill. Computational complexity of probabilistic Turing machines. *SIAM Journal on Computing*, 6(4):675–695, 1977.
[GOR98] J. Goldsmith, M. Ogihara, and J. Rothe. Tally NP sets and easy census functions. Technical Report TR 684, University of Rochester, Rochester, NY, March 1998.
[GP86] L. Goldschlager and I. Parberry. On the construction of parallel computers from various bases of boolean functions. *Theoretical Computer Science*, 43(1):43–58, 1986.
[GS91] A. Goldberg and M. Sipser. Compression and ranking. *SIAM Journal on Computing*, 20(3):524–536, 1991.
[HIS85] J. Hartmanis, N. Immerman, and V. Sewelson. Sparse sets in NP−P: EXPTIME versus NEXPTIME. *Information and Control*, 65(2/3):159–181, 1985.
[HR90] L. Hemachandra and S. Rudich. On the complexity of ranking. *Journal of Computer and System Sciences*, 41(2):251–271, 1990.
[HRW97] L. Hemaspaandra, J. Rothe, and G. Wechsung. Easy sets and hard certificate schemes. *Acta Informatica*, 34(11):859–879, 1997.
[HY84] J. Hartmanis and Y. Yesha. Computation times of NP sets of different densities. *Theoretical Computer Science*, 34(1/2):17–32, 1984.
[KL80] R. Karp and R. Lipton. Some connections between nonuniform and uniform complexity classes. In *Proceedings of the 12th ACM Symposium on Theory of Computing*, pages 302–309, April 1980. An extended version has also appeared as: Turing machines that take advice, *L'Enseignement Mathématique*, 2nd series 28, 1982, pages 191–209.
[KS85] K. Ko and U. Schöning. On circuit-size complexity and the low hierarchy in NP. *SIAM Journal on Computing*, 14(1):41–51, 1985.
[KST89] J. Köbler, U. Schöning, and J. Torán. On counting and approximation. *Acta Informatica*, 26(4):363–379, 1989.
[KSTT92] J. Köbler, U. Schöning, S. Toda, and J. Torán. Turing machines with few accepting computations and low sets for PP. *Journal of Computer and System Sciences*, 44(2):272–286, 1992.
[LS86] T. Long and A. Selman. Relativizing complexity classes with sparse oracles. *Journal of the ACM*, 33(3):618–627, 1986.
[MS72] A. Meyer and L. Stockmeyer. The equivalence problem for regular expressions with squaring requires exponential space. In *Proceedings of the 13th IEEE Symposium on Switching and Automata Theory*, pages 125–129, 1972.
[OH93] M. Ogiwara and L. Hemachandra. A complexity theory for feasible closure properties. *Journal of Computer and System Sciences*, 46(3):295–325, 1993.
[PZ83] C. Papadimitriou and S. Zachos. Two remarks on the power of counting. In *Proceedings of the 6th GI Conference on Theoretical Computer Science*, pages 269–276. Springer-Verlag *Lecture Notes in Computer Science #145*, 1983.
[Sto77] L. Stockmeyer. The polynomial-time hierarchy. *Theoretical Computer Science*, 3(1):1–22, 1977.
[TO92] S. Toda and M. Ogiwara. Counting classes are at least as hard as the polynomial-time hierarchy. *SIAM Journal on Computing*, 21(2):316–328, 1992.
[Tod91] S. Toda. PP is as hard as the polynomial-time hierarchy. *SIAM Journal on Computing*, 20(5):865–877, 1991.
[Val76] L. Valiant. The relative complexity of checking and evaluating. *Information Processing Letters*, 5(1):20–23, 1976.

[Val79a] L. Valiant. The complexity of computing the permanent. *Theoretical Computer Science*, 8(2):189–201, 1979.

[Val79b] L. Valiant. The complexity of enumeration and reliability problems. *SIAM Journal on Computing*, 8(3):410–421, 1979.

[Wel93] D. Welsh. *Complexity: Knots, Colourings and Counting*. Cambridge University Press, 1993.

Average-Case Intractability vs. Worst-Case Intractability

Johannes Köbler and Rainer Schuler

Theoretische Informatik, Universität Ulm, D-89069 Ulm, Germany

Abstract. We use the assumption that all sets in NP (or other levels of the polynomial-time hierarchy) have efficient average-case algorithms to derive collapse consequences for $\mathcal{MA}$, $\mathcal{AM}$, and various subclasses of $\mathcal{P}/\text{poly}$. As a further consequence we show for $\mathcal{C} \in \{\mathcal{P}(\mathcal{PP}), \mathcal{PSPACE}\}$ that $\mathcal{C}$ is not tractable in the average-case unless $\mathcal{C} = \mathcal{P}$.

1 Introduction

In general, the average-case complexity of an algorithm depends (by definition) on the distribution on the inputs. In fact, there exist certain (so called malign or universal) distributions relative to which the average-case complexity of any algorithm coincides with its worst-case complexity [26]. Fortunately, these distributions are not recursive. Even for the class of polynomial-time bounded algorithms, malign distributions are not computable in polynomial time [31].

In recent literature, it has been shown that several $\mathcal{NP}$-complete problems are solvable efficiently on average (i.e., in time polynomial on μ-average) with respect to certain natural distributions μ on the instances. However, this is not true in general unless $\mathcal{E} = \mathcal{NE}$ [10]. In fact, some natural $\mathcal{NP}$ problems A are complete for $\mathcal{NP}$ in the sense that with respect to a particular distribution, A is not efficiently solvable on average unless any $\mathcal{NP}$ problem is efficiently solvable on average with respect to any polynomial-time computable distribution [25]. It is therefore one of the main open problems in average-case complexity theory whether $\mathcal{NP}$ problems can be solved efficiently on average with respect to natural, i.e. polynomial-time computable, distributions.

Let $\mathcal{AP}_{\mathcal{FP}}$ denote the class of sets that are decidable in time polynomial on average with respect to every polynomial-time computable distribution. As noted above, $\mathcal{NP} \subseteq \mathcal{AP}_{\mathcal{FP}}$ implies that $\mathcal{E} = \mathcal{NE}$ [10]. This result provides an interesting connection between average-case complexity and worst-case complexity. Namely, if all $\mathcal{NP}$ problems can be decided in time polynomial on average, then all sets in $\mathcal{NE}$ can be decided in (worst-case) exponential time.

Similarly, as observed in [14], any random self-reducible set which can be decided in time polynomial on average (under the distribution induced by the random self-reduction) can be decided by a randomized algorithm in (worst-case) polynomial time. For example, Lipton [27] used an idea of Beaver and Feigenbaum [8] to show that multivariate polynomials of low degree are (functionally) random self-reducible. In particular, it follows from Lipton's result that if there is an algorithm computing the permanent efficiently for all but a sufficiently small (polynomial) fraction of all $n \times n$ matrices (over GF(p) where $p > n+1$ is prime), then it is possible to compute the permanent of any $n \times n$ matrix in expected polynomial time. Using this property it is not hard to show that $\mathcal{P}(\mathcal{PP}) \not\subseteq \mathcal{AP}_{\mathcal{FP}}$ unless $\mathcal{PP} = \mathcal{ZPP}$. From Corollary 1 below, $\mathcal{P}(\mathcal{PP}) \subseteq \mathcal{AP}_{\mathcal{FP}}$ even implies that $\mathcal{PP} = \mathcal{P}$ (in fact, it is easy to verify that $\mathcal{PP} = \mathcal{P}$ already follows from the assumption that the middle bit class $\mathcal{MP}$ [16] is contained in $\mathcal{AP}_{\mathcal{FP}}$). This means that for $\mathcal{C} = \mathcal{P}(\mathcal{PP})$, $\mathcal{C}$ is not tractable on the average unless $\mathcal{C}$ is tractable in the worst-case. As shown in Corollary 5, the same holds for $\mathcal{C} = \mathcal{PSPACE}$. Hence, the question arises whether a similar relationship holds for other classes $\mathcal{C}$ as, e.g., $\mathcal{C} = \mathcal{NP}$ or, more generally, for $\mathcal{C} = \Sigma_k^p$.

In contrast to worst-case complexity, where $\mathcal{NP} \subseteq \mathcal{P}$ implies that $\mathcal{PH} \subseteq \mathcal{P}$, it is not known whether $\mathcal{NP} \subseteq \mathcal{AP}_{\mathcal{FP}}$ implies that all sets in $\Delta_2^p = \mathcal{P}(\mathcal{NP})$ are contained in $\mathcal{AP}_{\mathcal{FP}}$ (see [19] for an exposition). Consider for example an $\mathcal{NP}$ search problem. It is not known whether an efficient average-case algorithm for the corresponding decision problem can be used to compute solutions efficiently on average. To see the difficulty consider the computation of a deterministic Turing machine M with oracle A, where the distribution on the inputs of M is computable in polynomial time. Since the oracle queries can be adaptive, it depends on the oracle set A which queries are actually made. Hence, the distribution induced on the oracle queries is not necessarily computable in polynomial time. On the other hand, it is known that $\mathcal{NP} \subseteq \mathcal{AP}_{\mathcal{FP}}$ implies that $\Theta_2^p = \mathcal{P}_{\text{tt}}(\mathcal{NP})$ is contained in $\mathcal{AP}_{\mathcal{FP}}$ (cf. Theorem 5). We refer the reader to [19,36] for further discussions of this and related questions. As shown in [35], the class $\mathcal{AP}_{\mathcal{FP}}$ is not closed under Turing reducibility, moreover, $\mathcal{AP}_{\mathcal{FP}}$ even contains Turing complete sets for $\mathcal{EXP}$ (note that $\mathcal{EXP}$ is not contained in $\mathcal{AP}_{\mathcal{FP}}$).

We use the assumption that all sets in NP (or higher levels of the polynomial-time hierarchy) are efficiently solvable on average to derive collapse consequences for $\mathcal{MA}$, $\mathcal{AM}$, and various subclasses of $\mathcal{P}/\text{poly}$. Our results are based on the following special properties of any set $A \in \mathcal{AP}_{\mathcal{FP}}$: Firstly, for any $\mathcal{P}$-printable domain D there is an algorithm that decides A efficiently on all inputs in the domain D. Secondly, since A is efficiently decidable on average with respect to the standard distribution μ_{st} (which is uniform on Σ^n), there is an algorithm for A that is polynomial in the worst case for all but a polynomial fraction of the strings of each length. Roughly speaking, we exploit these two properties in the following context: A serves as an oracle in a computation that generates oracle queries in such a way that it is sufficient to answer these queries either on some $\mathcal{P}$-printable domain or on any domain which contains a large fraction of the strings of each length. In particular, we get the following collapse consequences. (The notion of instance complexity and the class IC[log,poly] of sets of strings with low instance complexity were introduced in [32].)

- If $\mathcal{NP} \subseteq \mathcal{AP}_{\mathcal{FP}}$ then $\mathcal{MA} = \mathcal{NP}$ and $\mathcal{NP} \cap \mathcal{P}/\log = \mathcal{P}$.
- If $\Delta_2^p \subseteq \mathcal{AP}_{\mathcal{FP}}$ then $\Delta_2^p \cap \text{IC}[\log, \text{poly}] = \mathcal{P}$ and every self-reducible set in $\mathcal{P}/\text{poly}$ is in $\mathcal{ZPP}$.

– If $\Sigma_2^p \subseteq \mathcal{AP}_{\mathcal{FP}}$ then $\mathcal{AM} = \mathcal{NP}$ and all sets in $\Sigma_2^p \cap \Pi_2^p$ that conjunctively, disjunctively, or bounded truth-table reduce to some sparse set are in $\mathcal{P}$.
– If $\Delta_3^p \subseteq \mathcal{AP}_{\mathcal{FP}}$ then $\Sigma_2^p \cap \Pi_2^p \cap \mathcal{P}/\text{poly} = \mathcal{P}$.
– If $\Sigma_3^p \subseteq \mathcal{AP}_{\mathcal{FP}}$ then $\Sigma_3^p \cap \Pi_3^p \cap \mathcal{P}/\text{poly} = \mathcal{P}$.

Recently a series of plausible consequences, not known to follow from the assumption $\mathcal{P} \neq \mathcal{NP}$, have been derived from the assumption that $\mathcal{NP}$ is not small in $\mathcal{EXP}$, see, e.g., [30,28,29,1]. It is interesting to note that the assumption $\mathcal{NP} \subseteq \mathcal{AP}_{\mathcal{FP}}$ is contradictory to Lutz' hypothesis that $\mathcal{NP}$ is not small in $\mathcal{EXP}$, as follows directly from the fact that $\mathcal{AP}_{\mathcal{FP}}$ is small in $\mathcal{EXP}$ [37,13].

In this extended abstract proofs are omitted; see [22] for a full version.

2 Preliminaries

All languages are over the binary alphabet $\Sigma = \{0, 1\}$. The *length* of a string $x \in \Sigma^*$ is denoted by $|x|$. For a language A, let $A^{=n}$ denote the set of all strings in A of length n. Strings in 1^* are called *tally* and a set T is *tally* if $T \subseteq 1^*$. A set S is called *sparse* if the cardinality of $S^{=n}$ is bounded above by a polynomial in n. TALLY denotes the class of all tally sets, and SPARSE denotes the class of all sparse sets. The cardinality of a finite set A is denoted by $\|A\|$. The *join* of two sets A and B is $A \oplus B = \{0x \mid x \in A\} \cup \{1x \mid x \in B\}$. The join of language classes is defined analogously. To encode pairs (or tuples) of strings we use a standard polynomial-time computable pairing function denoted by $\langle \cdot, \cdot \rangle$ whose inverses are also computable in polynomial time. We assume that this function encodes tuples of tally strings again as a tally string. IN denotes the set of non-negative integers and by log we denote the function $\log n = \max\{1, \lceil \log_2 n \rceil\}$.

We assume that the reader is familiar with fundamental complexity theoretic concepts such as (oracle) Turing machines and the polynomial-time hierarchy (see, for example, [6,34]). Let $\mathcal{C}$ be a complexity class. A set A is $\mathcal{P}^{\mathcal{C}}$-printable if there exists a set $C \in \mathcal{C}$ and a polynomial-time bounded oracle Turing transducer T such that the output of T with oracle C and input 1^n is an enumeration of all strings in A of length n. An oracle Turing machine T is non-adaptive, if for all oracles C and all inputs x, the queries of T on input x are independent of C. T is honest if there exists a constant c such that $|x| \leq |y|^c$ for all x and for all oracle queries y of T on input x. A set A is $\mathcal{P}_{\text{tt,honest}}(\mathcal{C})$-printable if A is $\mathcal{P}(\mathcal{C})$-printable and the respective Turing transducer is honest and non-adaptive.

Next we review the notion of advice functions introduced by Karp and Lipton [20] to characterize non-uniform complexity classes. A function $h: 0^* \to \Sigma^*$ is called a *polynomial-length function* if for some polynomial p and for all $n \geq 0$, $|h(0^n)| = p(n)$. For a class $\mathcal{C}$ of sets, let $\mathcal{C}/\text{poly}$ be the class of sets L such that there is a set $I \in \mathcal{C}$ and a polynomial-length function h such that for all n, and for all x in Σ^n: $x \in L \Leftrightarrow \langle x, h(0^n) \rangle \in I$. The function h is called an *advice function* for L, whereas I is the corresponding *interpreter set*.

In the following we will also make use of multi-valued advice functions. A (total) multi-valued function h maps every string x to a non-empty subset of Σ^*, denoted by set-$h(x)$. We say that g is a *refinement* of h if for all x, set-$g(x) \subseteq$ set-$h(x)$. A *multi-valued* advice function h has the property that for some polynomial p and all n, set-$h(0^n) \subseteq \Sigma^{p(n)}$. Furthermore, for all $w \in$ set-$h(0^n)$ and for all x in Σ^n it holds that $x \in L \Leftrightarrow \langle x, w \rangle \in I$. Let $\mathcal{F}$ be a class of (possibly multi-valued) functions and let $L \in \mathcal{C}/\text{poly}$. Then L is said to *have an advice function in $\mathcal{F}$* (with respect to interpreter class $\mathcal{C}$) if some $h \in \mathcal{F}$ is an advice function for L with respect to some interpreter set $I \in \mathcal{C}$.

Let μ be a probability distribution on Σ^*. Associated with μ are a distribution function that we also denote by μ and a density function, denoted by μ'. Both μ and μ' are functions from Σ^* to the interval $[0,1]$, with the property that $\sum_x \mu'(x) = 1$ and $\mu(x) = \sum_{y \leq x} \mu'(y)$ where, as usual, $\leq$ denotes the lexicographic ordering on Σ^*. Let t be a function from $\mathbb{N}$ to $\mathbb{N}$. A distribution μ *t-dominates* a distribution ν, if $\mu'(x) \cdot t(|x|) \geq \nu'(x)$ for all x. If t is a constant, then we say that μ dominates ν by a constant factor, similarly, if t is bounded by a polynomial, then we say that μ polynomially dominates ν.

Let μ be a distribution. A function $f : \Sigma^* \to \mathbb{N}$ is polynomial on μ-average [25], if there exists a constant $\epsilon > 0$ such that $\sum_{x \neq \lambda} \frac{f^\epsilon(x)}{|x|} \mu'(x) < \infty$.

The class of functions polynomial on μ-average has many closure properties that are known for polynomials [25,17]. A further important property is robustness under the polynomial domination of distributions [25,17], i.e., any function that is polynomial on ν-average is also polynomial on μ-average provided that ν dominates μ.

A distribution is said to be $\mathcal{P}$-computable if its distribution function μ is $\mathcal{P}$-computable, i.e., there exists a polynomial-time deterministic Turing transducer M such that for all x and all k it holds that $|M(x, 1^k) - \mu(x)| \leq 2^{-k}$. Here the output of M is interpreted as a rational number, in some appropriate way. For example, if $M(x, 1^k) = \langle p, q \rangle$, then $M(x, 1^k)$ computes the number p/q.

As usual let $\mathcal{FP}$ denote the set of polynomial-time computable functions. An important subclass of the class of $\mathcal{P}$-computable distributions is the class of so-called $\mathcal{FP}$-computable distributions for which μ can be efficiently computed without error. For a complexity class $\mathcal{C}$, we say that a distribution is $\mathcal{FP}(\mathcal{C})$-computable (in symbols: $\mu \in \mathcal{FP}(\mathcal{C})$) if its distribution function μ is $\mathcal{FP}(\mathcal{C})$-computable, i.e., there exist functions $f \in \mathcal{FP}(\mathcal{C})$ and $g \in \mathcal{FP}$ such that for all x, $\mu(x) = f(x)/g(x)$.

As the following theorem shows, a problem is solvable in time polynomial on ν-average for every $\mathcal{FP}$-computable distribution ν if and only if it is solvable in time polynomial on μ-average for every $\mathcal{P}$-computable distribution μ.

Theorem 1. *[17] Every $\mathcal{P}$-computable distribution μ is dominated by a $\mathcal{FP}$-computable distribution ν by a constant factor. Furthermore, for all x, the binary representation of $\nu(x)$ is of length linear in the length of x.*

Following [25,17] we assume that all natural distributions are either $\mathcal{P}$-computable or dominated by a $\mathcal{P}$-computable distribution. In this sense, a set is efficiently decidable on average (under natural distributions) if it is decidable in time polynomial on μ-average with respect to every distribution $\mu \in \mathcal{FP}$.

Definition 1. *[38] Let $\mathcal{F}$ be a set of distributions. A set A is decidable in average polynomial time under distributions in $\mathcal{F}$ (in symbols, $A \in \mathcal{AP}_\mathcal{F}$) if for every distribution $\mu \in \mathcal{F}$ there exists a deterministic Turing machine M such that $A = L(M)$ and the running time of M is polynomial on μ-average.*

As noted by Ben-David et al. [10], all sets in $\mathcal{AP}_{\mathcal{FP}}$ are decidable in polynomial time on tally inputs. In [33], Schapire shows that a function f is polynomial on μ-average if and only if there exists a polynomial p such that for all m, $\mu\{x \mid f(x) > p(|x|, m)\} \le \frac{1}{m}$. From this characterization it follows immediately that any function f that is polynomial on μ-average is in fact polynomially bounded on $\Sigma^{\le n}$, except for a subset which has low probability under μ.

Proposition 1. *Let f be polynomial on μ-average. For every polynomial p there is a polynomial p' such that for all n, $\mu\{x \in \Sigma^{\le p(n)} \mid f(x) > p'(n)\} \le \frac{1}{p(n)}$.*

Proposition 2. *Let f be polynomial on μ-average.*

1. *Then for every polynomial p there exists a polynomial p' such that $f(x) \le p'(|x|)$ holds for all x with $\mu'(x) \ge 1/p(|x|)$.*
2. *If $\mu = \mu_{st}$ is the standard distribution (where $\mu'_{st}(x) = \frac{1}{|x|(|x|+1)} \cdot 2^{-|x|}$ for all $x \ne \lambda$), then for every polynomial p there exists a polynomial p' such that for all $n > 0$, $\|\{x \in \Sigma^n \mid f(x) > p'(n)\}\| \le \frac{2^n}{p(n)}$.*

3 Eliminating Tally and Printable Oracle Queries

The following consequence was the first that has been derived from the assumption that all $\mathcal{NP}$ problems are decidable in time polynomial on μ-average for any distribution $\mu \in \mathcal{FP}$.

Theorem 2. *[10] If $\mathcal{NP} \subseteq \mathcal{AP}_{\mathcal{FP}}$, then $\mathcal{E} = \mathcal{NE}$ (or, equivalently, $\mathcal{NP} \cap$ TALLY $\subseteq \mathcal{P}$).*

Put in other words, if $\mathcal{NP}$ problems have efficient average-case decision algorithms, then $\mathcal{P}(\mathcal{NP} \cap \text{TALLY})$, a subclass of $\mathcal{P}/\text{poly}$, collapses downto $\mathcal{P}$. We observe that similar collapse consequences downto $\mathcal{P}$ can be derived for other subclasses of $\mathcal{P}/\text{poly}$ (see Corollary 1). Some of these collapse consequences follow immediately from recent results investigating the complexity of sparse and tally descriptions for sets in $\mathcal{P}/\text{poly}$ [7,21,15,2]. For the others we can exploit an interesting connection between the worst-case complexity of a set L and the average-case complexity of oracles used in the computation of an advice function for L.

The following theorem shows that if an advice function h for some set L can be efficiently computed relative to some oracle which is efficiently decidable on average, then h is computable in polynomial time.

Theorem 3. *1. If $\mathcal{P}(D) \cap$ TALLY $\subseteq \mathcal{AP}_{\mathcal{FP}}$ then any advice function that is computable in $\mathcal{FP}(D)$ is computable in $\mathcal{FP}$.*

2. If $\{A \mid A \leq^p_m D\} \cap \text{TALLY} \subseteq \mathcal{AP}_{\mathcal{FP}}$ then any advice function that is computable in $\mathcal{FP}_{tt}(D)$ is computable in $\mathcal{FP}$.

Now, using results from [7,2,21,15], we can state similar collapse consequences as in Theorem 2 for several subclasses of $\mathcal{P}/\text{poly}$. We note that by using a different proof technique it has been shown in [5] that $\mathcal{BPP} = \mathcal{P}$ follows from the assumption that every tally set in Σ_4^p is contained in $\mathcal{P}$.

Corollary 1.
1. If $\mathcal{NP} \cap \text{TALLY} \subseteq \mathcal{AP}_{\mathcal{FP}}$ then $\mathcal{NP} \cap \mathcal{P}/\log = \mathcal{P}$.
2. If $\Delta_2^p \cap \text{TALLY} \subseteq \mathcal{AP}_{\mathcal{FP}}$ then $\Delta_2^p \cap \text{IC}[\log, \text{poly}] = \mathcal{P}$.
3. If $\Sigma_2^p \cap \text{TALLY} \subseteq \mathcal{AP}_{\mathcal{FP}}$ then all sets in $\Sigma_2^p \cap \Pi_2^p$ that conjunctively, disjunctively, or bounded truth-table reduce to some sparse set are in $\mathcal{P}$.
4. If $\Delta_3^p \cap \text{TALLY} \subseteq \mathcal{AP}_{\mathcal{FP}}$ then $\Sigma_2^p \cap \Pi_2^p \cap \mathcal{P}/\text{poly} = \mathcal{P}$ and hence $\mathcal{BPP} = \mathcal{P}$.
5. If $\Sigma_3^p \cap \text{TALLY} \subseteq \mathcal{AP}_{\mathcal{FP}}$ then $\Sigma_3^p \cap \Pi_3^p \cap \mathcal{P}/\text{poly} = \mathcal{P}$.

In our next theorem we consider the complexity of sets in $\mathcal{P}/\text{poly}$ that have advice functions which can be computed by nondeterministic transducers under some oracle. The proof makes use of the following proposition which shows as a special case that any set $A \in \mathcal{AP}_{\mathcal{FP}}$ is efficiently decidable on any $\mathcal{P}$-printable domain B.

Proposition 3. Let $A \in \mathcal{AP}_{\mathcal{FP}(C)}$ and let B be a $\mathcal{P}(C)$-printable set for some oracle C. Then there exists a set $D \in \mathcal{P}$ such that $B \subseteq D$ and $D \cap A \in \mathcal{P}$.

For an oracle B, a (multivalued) function h is in $\mathcal{NPMV}(B)$ if there exists a non-deterministic polynomial-time transducer T such that set-$h(x)$ consists of all output values of T^B on input x. h is in $\mathcal{NPMV}_{\text{honest}}(B)$ if, additionally, there exists a constant c such that $|y|^c > |x|$ for all oracle queries y of T^B on input x. If $B = \emptyset$ then we simply write $\mathcal{NPMV}$ instead of $\mathcal{NPMV}(\emptyset)$.

Theorem 4. Assume that $A \in \mathcal{AP}_{\mathcal{FP}(\mathcal{NP}(A))}$. Then any advice function $h \in \mathcal{NPMV}_{\text{honest}}(A)$ has a refinement in $\mathcal{NPMV}$, implying that any set $L \in (\mathcal{NP} \cap \text{co-}\mathcal{NP})/\text{poly}$ that has an advice function in $\mathcal{NPMV}_{\text{honest}}(A)$ belongs to $\mathcal{NP} \cap \text{co-}\mathcal{NP}$.

As an application we get the following consequence for the class IP[$\mathcal{P}/\text{poly}$] that contains all sets having an interactive proof with prover complexity restricted to $\mathcal{P}/\text{poly}$ [4,3].

Corollary 2. If $\mathcal{NP} \subseteq \mathcal{AP}_{\mathcal{FP}(\Sigma_2^p)}$ then IP[$\mathcal{P}/\text{poly}$] $\subseteq \mathcal{NP} \cap \text{co-}\mathcal{NP}$.

As mentioned in the introduction, if a decision problem L_2 is decidable in time polynomial on μ_2-average for any $\mathcal{FP}$-computable distribution μ_2, then this does not necessarily imply that also any set L_1 in $\mathcal{P}(L_2)$ is efficiently decidable on average with respect to any $\mathcal{FP}$-computable distribution μ_1. If however for any $\mathcal{FP}(\mathcal{NP})$-computable distribution μ_2, L_2 is decidable in time polynomial on μ_2-average, then we can show that indeed L_1 is efficiently solvable on average with respect to any $\mathcal{FP}$-computable distribution μ_1. For this it suffices to show that the distribution on the oracle queries induced by μ_1 and by the reduction of L_1 to L_2 is $\mathcal{FP}(\mathcal{NP})$-computable.

Theorem 5. *Let $\mathcal{C}$ and $\mathcal{D}$ be language classes where $\mathcal{C}$ is closed under polynomial-time many-one equivalence. Then $\mathcal{C} \subseteq \mathcal{AP}_{\mathcal{FP}(\mathcal{C} \oplus \mathcal{D})}$ implies $\mathcal{P}(\mathcal{C}) \subseteq \mathcal{AP}_{\mathcal{FP}(\mathcal{C} \oplus \mathcal{D})}$ and $\mathcal{C} \subseteq \mathcal{AP}_{\mathcal{FP}(\mathcal{D})}$ implies $\mathcal{P}_{tt}(\mathcal{C}) \subseteq \mathcal{AP}_{\mathcal{FP}(\mathcal{D})}$.*

Corollary 3. *1. If $\mathcal{NP} \subseteq \mathcal{AP}_{\mathcal{FP}(\mathcal{NP})}$ then $\Delta_2^p \cap \mathrm{IC}[\log, \mathrm{poly}] = \mathcal{P}$.*
2. If $\Sigma_2^p \subseteq \mathcal{AP}_{\mathcal{FP}(\Sigma_2^p)}$ then $\Sigma_2^p \cap \Pi_2^p \cap \mathcal{P}/\mathrm{poly} = \mathcal{P}$ and in particular, $\mathcal{BPP} \subseteq \mathcal{P}$.

4 Eliminating Random Oracle Queries

As mentioned in the introduction, any random self-reducible set which can be decided in time polynomial on average (under the distribution induced by the random self-reduction) can be decided by a randomized algorithm in expected polynomial time. As shown by Feigenbaum and Fortnow, many complexity classes like $\mathcal{PP}$, $\mathcal{Mod}_k\mathcal{P}$, and $\mathcal{PSPACE}$ have complete sets that are random self-reducible. By combining the results stated in [14] with Corollary 4 below, it is not hard to verify that for $\mathcal{K} \in \{\mathcal{P}(\mathcal{PP}), \mathcal{MP}, \mathcal{Mod}_k\mathcal{P}, \mathcal{Mod}\mathcal{P}, \mathcal{PSPACE}\}$, $\mathcal{K}$ is not contained in $\mathcal{AP}_{\mathcal{FP}}$ unless $\mathcal{K} \subseteq \mathcal{ZPP}$ where the middle bit class $\mathcal{MP}$, the classes $\mathcal{Mod}_k\mathcal{P}$, $k \geq 2$, and the generalized Mod class $\mathcal{Mod}\mathcal{P}$ have been introduced and studied in [16], [12,18,9], and [23], respectively.

In Theorem 7 below we show a similar collapse for the subclass of $\mathcal{P}/\mathrm{poly}$ consisting of all sets L for which a multivalued advice function can be computed by a randomized algorithm under an oracle that is easily decidable on average. Let $h \in \mathcal{NPMV}(B)$. Then we say that $h \in \mathcal{FZPP}(B)$ if h is computable by an $\mathcal{NPMV}(B)$ transducer that, when considered as a probabilistic Turing machine, on any input x produces with probability at least $1/2$ some output y.

Let M be a randomized Turing machine. If we fix a sequence $r \in \{0,1\}^*$ of the probabilistic choices of M, then the computation of M on input x is deterministic. We use $M_r(x)$ to denote the output of M on input x and computation path r. Assuming that M uses a functional oracle $f : \Sigma^* \to \Sigma^*$ and p is a polynomial bounding the running time of M, we define for any input x the distribution $\mu_{M,f,x}$ induced by M on input x with oracle f,

$$\mu'_{M,f,x}(y) = \frac{\|R_y(x)\|}{\|R(x)\|}$$

where $R_y(x) = \{(r,i) \mid r \in \{0,1\}^{p(n)}$ and y is the ith query of $M_r^f(x)\}$ and $R(x) = \bigcup_{y \in \Sigma^*} R_y(x)$.

Assume that $g \in \mathcal{FZPP}^f$ via some transducer M. Then we say that the computation of M^f is dominated by a distribution μ (in symbols: $g \in \mathcal{FZPP}^f_\mu$ via M) if μ dominates the ensemble $(\mu_{M,f,x})_{x \in \Sigma^*}$, i.e., there exists a polynomial s such that for all x and y, $\mu'(y) \geq \mu'_{M,f,x}(y)/s(|x|)$. By $\mathcal{ZPP}^f_\mu$ we denote the class of all languages whose characteristic function belongs to $\mathcal{FZPP}^f_\mu$.

For positive integers i and m, $\mathrm{bin}_m(i)$ denotes the binary representation of i, padded to length m. For any function f, we define the set

$$A_f - \{y\mathrm{bin}_{q(|y|)}(i) \mid 1 \leq i \leq |f(y)|+1 \text{ and the } i\text{th bit of } f(y)1 \text{ is one}\}$$

that contains for any string y all strings yz such that $z = \text{bin}_{q(|y|)}(i)$ for some $i \in \{1, \ldots, |f(y)| + 1\}$ and the ith bit of $f(y)1$ is one. Here, $q(n)$ is a fixed polynomial bounding the length of the strings $f(y)$, $y \in \Sigma^n$.

Lemma 1. *Let f be a function and q be a polynomial such that $|f(y)| < q(|y|)$. Then $\mathcal{FZPP}^f_\mu \subseteq \mathcal{FZPP}^{A_f}_\nu$ where ν is the distribution defined as*

$$\nu'(u) = \begin{cases} \mu'(y)/q(|y|), & \text{if } u = y\text{bin}_{q(|y|)}(i) \text{ for some } i \in \{1, \ldots, q(|y|)\}, \\ 0, & \text{otherwise.} \end{cases}$$

Theorem 6. *If $A_f \in \mathcal{AP}_{\mathcal{FP}}$ and if μ is a distribution in $\mathcal{FP}$, then $\mathcal{FZPP}^f_\mu$ has a refinement in $\mathcal{FZPP}$ and, in particular, $\mathcal{ZPP}^f_\mu \subseteq \mathcal{ZPP}$.*

Since the standard distribution μ_{st} is easily seen to be in $\mathcal{FP}$, we immediately get the following corollary.

Corollary 4. *If $A_f \in \mathcal{AP}_{\mathcal{FP}}$, then any function $h \in \mathcal{FZPP}^f_{\mu_{\text{st}}}$ has a refinement in $\mathcal{FZPP}$ and, in particular, $\mathcal{ZPP}^f_{\mu_{\text{st}}} \subseteq \mathcal{ZPP}$.*

Now we are ready to show that any advice function which can be computed by a randomized algorithm under an oracle that is easily decidable on average is computable in the same way without the help of an oracle.

Theorem 7. *Any advice function that is computable in $\mathcal{FZPP}(D)$ where $\mathcal{P}(D) \subseteq \mathcal{AP}_{\mathcal{FP}}$ has a refinement in $\mathcal{FZPP}$.*

By using results from [11,24,27,14,16,23] it is easy to derive the following corollary.

Corollary 5. 1. *[39] If $\Delta^p_2 \subseteq \mathcal{AP}_{\mathcal{FP}}$ then every self-reducible set in $\mathcal{P}/\text{poly}$ is in $\mathcal{ZPP}$.*
2. *If $\mathcal{NP} \subseteq \mathcal{AP}_{\mathcal{FP}(\mathcal{NP})}$ then every self-reducible set in $\mathcal{P}/\text{poly}$ is in $\mathcal{ZPP}$.*
3. *For $\mathcal{K} \in \{\mathcal{P}(\mathcal{PP}), \mathcal{MP}, \mathcal{M}od\mathcal{P}, \mathcal{PSPACE}\}$, $\mathcal{K}$ is not contained in $\mathcal{AP}_{\mathcal{FP}}$ unless $\mathcal{K} = \mathcal{P}$.*

Finally, by applying a technique used to show that $\mathcal{MA} \subseteq \mathcal{ZPP}(\mathcal{NP})$ [1] we extend a result in [19] showing that $\mathcal{NP} \subseteq \mathcal{AP}_{\mathcal{FP}}$ implies $\mathcal{BPP} = \mathcal{ZPP}$. More specifically, we derive under the same assumption $\mathcal{NP} \subseteq \mathcal{AP}_{\mathcal{FP}}$ that $\mathcal{MA}$ can be derandomized, i.e., $\mathcal{MA} = \mathcal{NP}$, whereas under the stronger assumption $\Sigma^p_2 \subseteq \mathcal{AP}_{\mathcal{FP}}$ also $\mathcal{AM}$ can be derandomized, i.e., $\mathcal{AM} = \mathcal{NP}$. Note that $\mathcal{AM} = \mathcal{NP}$ has some immediate strong implications as, for example, that Graph Isomorphism is in $\mathcal{NP} \cap \text{co-}\mathcal{NP}$.

Theorem 8. 1. *If $\mathcal{NP} \subseteq \mathcal{AP}_{\mathcal{FP}}$ then $\mathcal{MA} = \mathcal{NP}$.*
2. *If $\Sigma^p_2 \subseteq \mathcal{AP}_{\mathcal{FP}}$ then $\mathcal{AM} = \mathcal{NP}$.*

We notice that in order to derive $\mathcal{MA} = \mathcal{NP}$ ($\mathcal{AM} = \mathcal{NP}$) it suffices to assume that for any set L in co-$\mathcal{NP}$ (respectively, Π^p_2) and any $\mathcal{FP}$-computable distribution μ there is some nondeterministic Turing machine for L whose running time is polynomial on μ-average.

References

1. V. Arvind and J. Köbler. On resource-bounded measure and pseudorandomness. In *Proc. 17th Conference on Foundations of Software Technology and Theoretical Computer Science*, volume 1346 of *Lecture Notes in Computer Science*, pages 235–249. Springer-Verlag, 1997.
2. V. Arvind, J. Köbler, and M. Mundhenk. Upper bounds for the complexity of sparse and tally descriptions. *Mathematical Systems Theory*, 29(1):63–94, 1996.
3. V. Arvind, J. Köbler, and R. Schuler. On helping and interactive proof systems. *International Journal of Foundations of Computer Science*, 6(2):137–153, 1995.
4. L. Babai, L. Fortnow, and C. Lund. Non-deterministic exponential time has two-prover interactive protocols. *Computational Complexity*, 1:1–40, 1991.
5. L. Babai, L. Fortnow, N. Nisan, and A. Wigderson. BPP has subexponential time simulations unless EXPTIME has publishable proofs. *Computational Complexity*, 3:307–318, 1993.
6. J. L. Balcázar, J. Díaz, and J. Gabarró. *Structural Complexity I*. EATCS Monographs on Theoretical Computer Science. Springer-Verlag, second edition, 1995.
7. J.L Balcázar and U. Schöning. Logarithmic advice classes. *Theoretical Computer Science*, 99:279–290, 1992.
8. D. Beaver and J. Feigenbaum. Hiding instances in multioracle queries. In *Proc. 7th Symposium on Theoretical Aspects of Computer Science*, volume 415 of *Lecture Notes in Computer Science*, pages 37–48. Springer-Verlag, 1990.
9. R. Beigel and J. Gill. Counting classes: thresholds, parity, mods, and fewness. *Theoretical Computer Science*, 103:3–23, 1992.
10. S. Ben-David, B. Chor, O. Goldreich, and M. Luby. On the theory of average case complexity. *Journal of Computer and System Sciences*, 44:193–219, 1992.
11. N. Bshouty, R. Cleve, R. Gavaldà, S. Kannan, and C. Tamon. Oracles and queries that are sufficient for exact learning. *Journal of Computer and System Sciences*, 52:421–433, 1996.
12. J. Cai and L. A. Hemachandra. On the power of parity polynomial time. *Mathematical Systems Theory*, 23:95–106, 1990.
13. J.-Y. Cai and A. Selman. Fine separation of average time complexity classes. In *Proc. 13th Symposium on Theoretical Aspects of Computer Science*, volume 1046 of *Lecture Notes in Computer Science*, pages 331–343. Springer-Verlag, 1996.
14. J. Feigenbaum and L. Fortnow. Random-self-reducibility of complete sets. *SIAM Journal on Computing*, 22:994–1005, 1993.
15. R. Gavaldà. Bounding the complexity of advice functions. *Journal of Computer and System Sciences*, 50(3):468–475, 1995.
16. F. Green, J. Köbler, K. Regan, T. Schwentick, and J. Torán. The power of the middle bit of a #P function. *Journal of Computer and System Sciences*, 50(3):456–467, 1995.
17. Y. Gurevich. Average case completeness. *Journal of Computer and System Sciences*, 42(3):346–398, 1991.
18. U. Hertrampf. Relations among MOD-classes. *Theoretical Computer Science*, 74:325–328, 1990.
19. R. Impagliazzo. A personal view of average-case complexity. In *Proc. 10th Structure in Complexity Theory Conference*, pages 134–147. IEEE Computer Society Press, 1995.
20. R. M. Karp and R. J. Lipton. Some connections between nonuniform and uniform complexity classes. In *Proc. 12th ACM Symposium on Theory of Computing*, pages 302–309. ACM Press, 1980.

21. J. Köbler. Locating P/poly optimally in the extended low hierarchy. *Theoretical Computer Science*, 134(2):263–285, 1994.
22. J. Köbler and R. Schuler. Average-case intractability vs. worst-case intractability. See ftp://theorie.informatik.uni-ulm.de/pub/papers/ti/ap.ps.gz.
23. J. Köbler and S. Toda. On the power of generalized MOD-classes. *Mathematical Systems Theory*, 29(1):33–46, 1996.
24. J. Köbler and O. Watanabe. New collapse consequences of NP having small circuits. In *Proc. 22nd International Colloquium on Automata, Languages, and Programming*, volume 944 of *Lecture Notes in Computer Science*, pages 196–207. Springer-Verlag, 1995.
25. L. Levin. Average case complete problems. *SIAM Journal on Computing*, 15:285–286, 1986.
26. M. Li and P.M.B. Vitányi. Average case complexity under the universal distribution equals worst-case complexity. *Information Processing Letters*, 42:145–149, 1992.
27. R. J. Lipton. New directions in testing. In J. Feigenbaum and M. Merritt, editors, *Distributed Computing and Cryptography*, volume 2 of *DIMACS Series in Discrete Mathematics and Theoretical Computer Science*. American Mathematical Society, 1991.
28. J. H. Lutz. Observations on measure and lowness for Δ_2^P. *Theory of Computing Systems*, 30:429–442, 1997.
29. J. H. Lutz. The quantitative structure of exponential time. In L. A. Hemaspaandra and A. L. Selman, editors, *Complexity Theory Retrospective II*, pages 225–260. Springer-Verlag, 1997.
30. J. H. Lutz and E. Mayordomo. Cook versus Karp-Levin: separating reducibilities if NP is not small. *Theoretical Computer Science*, 164:141–163, 1996.
31. P.B. Milterson. The complexity of malign measures. *SIAM Journal on Computing*, 22(1):147–156, 1993.
32. P. Orponen, K. Ko, U. Schöning, and O. Watanabe. Instance complexity. *Journal of the ACM*, 41(1):96–121, 1994.
33. R. E. Schapire. The emerging theory of average-case complexity. Technical Report TM-431, Massachusetts Institut of Technology, 1990.
34. U. Schöning. *Complexity and Structure*, volume 211 of *Lecture Notes in Computer Science*. Springer-Verlag, 1986.
35. R. Schuler. Truth-table closure and Turing closure of average polynomial time have different measures in EXP. In *Proc. 11th Annual IEEE Conference on Computational Complexity*, pages 190–197. IEEE Computer Society Press, 1996.
36. R. Schuler and O. Watanabe. Towards average-case complexity analysis of NP optimization problems. In *Proc. 10th Structure in Complexity Theory Conference*, pages 148–159. IEEE Computer Society Press, 1995.
37. R. Schuler and T. Yamakami. Sets computable in polynomial time on average. In *Proc. 1st International Computing and Combinatorics Conference*, volume 959 of *Lecture Notes in Computer Science*, pages 400–409. Springer-Verlag, 1995.
38. R. Schuler and T. Yamakami. Structural average case complexity. *Journal of Computer and System Sciences*, 52:308–327, 1996.
39. O. Watanabe, 1996. Personal communication.

Shuffle on Trajectories: The Schützenberger Product and Related Operations *

Tero Harju[1], Alexandru Mateescu[2], and Arto Salomaa[1]

[1] Turku Centre for Computer Science and Department of Mathematics,
University of Turku, 20014 Turku, Finland
[2] Turku Centre for Computer Science and Faculty of Mathematics,
University of Bucharest, Romania.
{harju, mateescu, asalomaa}@utu.fi

Abstract. We investigate the problem of finding monoids that recognize languages of the form $L_1 \shuffle_T L_2$, where T is an arbitrary set of trajectories. Thereby, we describe two such methods: one based on the so-called trajectories monoids and the other based on monoids of matrices. Many well-known operations such as catenation, bi-catenation, shuffle, literal shuffle and insertion are just particular instances of the operation $\shuffle_T$. Hence, our results offer a uniform treatment for classical methods, notably the Schützenberger product. We also investigate some other related operations.

1 Preliminaries

In this paper we investigate the problem of finding monoids that recognize languages of the form $L_1 \shuffle_T L_2$, where T is an arbitrary set of trajectories.

The solution offers a uniform method to find monoids that recognize a number of operations with languages such as, for instance: catenation, bi-catenation, shuffle, literal shuffle, balanced literal shuffle, insertion, etc. Also, we compare our solution with other well-known constructions, notably with the Schützenberger product. Some other operations with languages are considered, too.

The operation of shuffle on trajectories of words and languages was introduced in [3]. The operations considered below are defined using the notion of the trajectory. A trajectory defines the general strategy to switch from one word to another word when carrying out the shuffle operation. The operation is extended to sets T of trajectories.

Let Σ be an alphabet. The set of all words over Σ is denoted by Σ^*. The *empty word* is denoted by λ. If $w \in \Sigma^*$, then $|w|$ denotes the length of w and $|w|_a$ denotes the number of occurrences of a in w, where $a \in \Sigma$. Note that $|\lambda| = 0$. If A is a set, then the set of all subsets of A is denoted by $\mathcal{P}(A)$.

The *shuffle* operation, denoted by $\shuffle$, is defined recursively by:

$$(ax \shuffle by) = a(x \shuffle by) \cup b(ax \shuffle y) \quad \text{and} \quad x \shuffle \lambda = \lambda \shuffle x = \{x\},$$

---

* This work has been partially supported by the Project 137358 of the Academy of Finland.

where $x, y \in \Sigma^*$ and $a, b \in \Sigma$. Other operations with words and languages that we consider in this paper are:

literal shuffle, denoted by $\sqcup\!\sqcup_{lit}$: if $x, y \in \Sigma^*$, $x = x_1 x_2 \ldots x_n$, $y = y_1 y_2 \ldots y_m$, where $x_i, y_j \in \Sigma$, $1 \leq i \leq n$, $1 \leq j \leq m$, then

$$x \sqcup\!\sqcup_{lit} y = \begin{cases} x_1 y_1 x_2 y_2 \ldots x_m y_m x_{m+1} x_{m+2} \ldots x_n, & \text{if } n > m, \\ x_1 y_1 x_2 y_2 \ldots x_n y_n y_{n+1} y_{n+2} \ldots y_m, & \text{if } n \leq m; \end{cases}$$

balanced literal shuffle, denoted by $\sqcup\!\sqcup_{blit}$, and defined as $\sqcup\!\sqcup_{lit}$, but only for words of the same length;
insertion, denoted by $\longleftarrow$: $\quad x \longleftarrow y = \{x'yx'' \mid x'x'' = x\}$;
bi-catenation, denoted by $\odot$: $\quad x \odot y = \{xy, yx\}$;
anti-catenation, denoted by $\bullet$: $\quad x \bullet y = yx$.

All the above operations are extended in a natural way to operations with languages, i.e., if $\bowtie$ is an operation with words, $\bowtie \in \{\cdot, \sqcup\!\sqcup, \sqcup\!\sqcup_{lit}, \sqcup\!\sqcup_{blit}, \longleftarrow, \odot, \bullet\}$ and L_1, L_2 are languages, then

$$L_1 \bowtie L_2 = \bigcup_{x \in L_1, y \in L_2} x \bowtie y.$$

2 Shuffle on Trajectories

In this section we introduce the notions of the trajectory and shuffle on trajectories. Let $V = \{r, u\}$ be the set of *versors* in the plane: r stands for the *right* direction, whereas u stands for the *up* direction.

Definition 1. *A trajectory is an element t, $t \in V^*$.*

Let Σ be an alphabet and let t be a trajectory, let d be a versor, $d \in V$, and $\alpha, \beta \in \Sigma^*$.

Definition 2. *The shuffle of α with β on the trajectory dt, denoted $\alpha \sqcup\!\sqcup_{dt} \beta$, is recursively defined as follows:*
if $\alpha = ax$ and $\beta = by$, where $a, b \in \Sigma$ and $x, y \in \Sigma^*$, then

$$ax \sqcup\!\sqcup_{dt} by = \begin{cases} a(x \sqcup\!\sqcup_t by), & \text{if } d = r, \\ b(ax \sqcup\!\sqcup_t y), & \text{if } d = u; \end{cases}$$

if $\alpha = ax$ and $\beta = \lambda$, where $a \in \Sigma$ and $x \in \Sigma^*$, then

$$ax \sqcup\!\sqcup_{dt} \lambda = \begin{cases} a(x \sqcup\!\sqcup_t \lambda), & \text{if } d = r, \\ \emptyset, & \text{if } d = u; \end{cases}$$

if $\alpha = \lambda$ and $\beta = by$, where $b \in \Sigma$ and $y \in \Sigma^*$, then

$$\lambda \sqcup\!\sqcup_{dt} by = \begin{cases} \emptyset, & \text{if } d = r, \\ b(\lambda \sqcup\!\sqcup_t y), & \text{if } d = u. \end{cases}$$

Finally,

$$\lambda \sqcup\!\sqcup_t \lambda = \begin{cases} \lambda, & \text{if } t = \lambda, \\ \emptyset, & \text{otherwise.} \end{cases}$$

Comment. Note that if $|\alpha| \neq |t|_r$ or $|\beta| \neq |t|_u$, then $\alpha \shuffle_t \beta = \emptyset$.

Example 1. Let α and β be the words $\alpha = a_1a_2a_3a_4a_5a_6a_7a_8$, $\beta = b_1b_2b_3b_4b_5$ and assume that $t = r^3u^2r^3ururu$. The shuffle of α with β on the trajectory t is

$$\alpha \shuffle_t \beta = \{a_1a_2a_3b_1b_2a_4a_5a_6b_3a_7b_4a_8b_5\}.$$

Remark 1. Here we show that a number of operations with words and languages are particular cases of the operation of shuffle on trajectories.

1. Let T be the set $T = \{r, u\}^*$. Observe that $\shuffle_T = \shuffle$, the shuffle operation.
2. Assume that $T = r^*u^*$. It follows that $\shuffle_T = \cdot$, the catenation operation.
3. Define $T = r^*u^*r^*$ and note that $\shuffle_T = \hookleftarrow$, the insertion operation.
4. Consider $T = (ru)^*$ and observe that $\shuffle_T = \shuffle_{blit}$, the balanced literal shuffle.
5. Assume that $T = (ru)^*(r^* \cup u^*)$. Note that in this case $\shuffle_T = \shuffle_{lit}$, the literal shuffle.
6. Let T be the set $T = r^*u^* \cup u^*r^*$. In this case $\shuffle_T = \odot$, i.e., it is the bi-catenation operation.
7. Consider $T = u^*r^*$ and observe that $\shuffle_T = \bullet$, the anti-catenation operation.

3 The Problem of Recognition and Trajectories Monoids

We now go into some general facts about languages recognized by monoids. Of a special interest is the case of languages of the form $L_1 \shuffle_T L_2$, where T is a set of trajectories.

A monoid M_1 is *embedded* in a monoid M_2 iff there exists an injective morphism from M_1 to M_2. A monoid M_1 *divides* a monoid M_2, denoted $M_1 < M_2$, iff M_1 is isomorphic with a quotient of a submonoid of M_2. Clearly, if M_1 is embedded in M_2, then M_1 divides M_2. The division relation is transitive.

The unit element of a monoid is denoted by 1. If M is a monoid, then the set $\mathcal{P}(M)$ is a monoid with the multiplication defined by $AB = \{xy \mid x \in A, y \in B\}$, where $A, B \subseteq M$.

Definition 3. *Let L be a language, $L \subseteq \Sigma^*$. A monoid M recognizes L iff there exists a morphism $\varphi : \Sigma^* \longrightarrow M$, a subset F of M, $F \subseteq M$, such that $L = \varphi^{-1}(F)$. If, additionally, $\varphi(w) = 1$ iff $w = \lambda$, then we say that M unit-separately recognizes L.*

For each language L, $L \subseteq \Sigma^*$, there exists a monoid M that recognizes L. An example of such a monoid is the syntactic monoid of L. The *syntactic congruence* defined by L is the congruence $\approx_L$ on Σ^* defined as: $x \approx_L y$ iff for all $\alpha, \beta \in \Sigma^*$: $\alpha x \beta \in L$ iff $\alpha y \beta \in L$, where $x, y \in \Sigma^*$. The *syntactic monoid* of L, denoted by M_L, is the quotient monoid $\Sigma^*/\approx_L$. One can easily verify that M_L recognizes L. A monoid M recognizes L iff M_L divides M. If a monoid M_1 recognizes L and if M_1 divides M_2, then M_2 recognizes L, too.

Assume that a monoid M recognizes L, but M does not unit-separately recognize L. In this case we can adjoin a new unit element, say $1'$, to M and the new monoid M' unit-separately recognizes L. The morphism φ is extended to a new morphism φ' such that $\varphi'(\lambda) = 1'$. Moreover, $F' = F \cup \{1'\}$, if $\lambda \in L$, and $F' = F$, otherwise. Hence, for each language L there exists a monoid M that unit-separately recognizes L. However, there are monoids that do not unit-separately recognize any language, for instance the monoid $\{1\}$.

The following theorem goes back to Kleene:

Theorem 1. *A language L is regular iff L is recognized by some finite monoid.*

Let L_1, L_2 be languages $L_1, L_2 \subseteq \Sigma^*$. Let M_1, M_2 be monoids, such that M_i recognizes L_i, $i = 1, 2$. Assume that $\bowtie$ is an operation with languages such that $L_1 \bowtie L_2 \subseteq \Sigma^*$.

The following problem has been widely investigated: find a function $\Psi_\bowtie$ such that the language $L_1 \bowtie L_2$ is recognized by $\Psi_\bowtie(M_1, M_2)$.

For more details on this problem, as well as for a large bibliography, the reader is referred to [1], [4], or more recently, [5].

In the sequel we solve this problem for the operation $\sqcup_T$, where T is an arbitrary set of trajectories. The solution offers a uniform method to find monoids that recognize a large number of operations with languages. Also, we compare our solution with other well-known constructions, mainly with the Schützenberger product.

Definition 4. *Let L_1, L_2 be languages, $L_1, L_2 \subseteq \Sigma^*$, and let M_1, M_2 be monoids such that M_i recognizes L_i, $i = 1, 2$. Assume that T is a set of trajectories, $T \subseteq \{r, u\}^*$ and let M_T be a monoid that recognizes T. The trajectories monoid associated to (M_1, M_2, M_T), denoted by $T(M_1, M_2, M_T)$, is by definition the monoid $\mathcal{P}(M_1 \times M_2 \times M_T)$, i.e., $T(M_1, M_2, M_T) = \mathcal{P}(M_1 \times M_2 \times M_T)$.*

Theorem 2. *Let L_1, L_2 be languages, $L_1, L_2 \subseteq \Sigma^*$, and let M_1, M_2 be monoids such that M_i recognizes L_i, $i = 1, 2$. Assume that T is a set of trajectories, $T \subseteq \{r, u\}^*$ and let M_T be a monoid that recognizes T. The language $L_1 \sqcup_T L_2$ is recognized by the trajectories monoid $T(M_1, M_2, M_T)$.*

Proof. Let $\varphi_i : \Sigma^* \longrightarrow M_i$ be morphisms such that $\varphi_i^{-1}(F_i) = L_i$, for some $F_i \subseteq M_i$, $i = 1, 2$. Assume that $\varphi_T : \Sigma^* \longrightarrow M_T$ is a morphism such that $\varphi_i^{-1}(F_T) = T$, for some $F_T \subseteq M_T$.

Define the morphism $\varphi : \Sigma^* \longrightarrow T(M_1, M_2, M_T)$, $\varphi(a) = \{(\varphi_1(a), 1, \varphi_T(r)), (1, \varphi_2(a), \varphi_T(u))\}$, where $a \in \Sigma$. It is easy to see that the morphism φ has the following remarkable property:

$$\varphi(x) = \{(\varphi_1(\alpha), \varphi_2(\beta), \varphi_T(t)) \mid x \in \alpha \sqcup_t \beta, \text{ where } \alpha, \beta \in \Sigma^*, t \in V^*\}.$$

Consider the set $F = \{K \subseteq M_1 \times M_2 \times M_T \mid K \cap (F_1 \times F_2 \times F_T) \neq \emptyset\}$. Using the above property of φ, one can easily show that $\varphi^{-1}(F) = L_1 \sqcup_T L_2$. Hence, the trajectories monoid $T(M_1, M_2, M_T)$ recognizes the language $L_1 \sqcup_T L_2$. □

Note that the monoid $T(M_1, M_2, M_T)$ unit-separately recognizes the language $L_1 \sqcup_T L_2$. Also, note that if M_1, M_2 and M_T are finite monoids, then also the trajectories monoid $T(M_1, M_2, M_T)$ is finite. Hence,

Corollary 1. *If L_1, L_2 and T are regular languages, then $L_1 ⊔\!\!\!⊔_T L_2$ is a regular language.*

Consider now the case $T = \{r, u\}^*$. Therefore the operation $⊔\!\!\!⊔_T$ is the shuffle, $⊔\!\!\!⊔$. Note that M_T is the trivial monoid, i.e., $M_T = \{1\}$. Hence, the monoid $M_1 \times M_2 \times M_T$ is isomorphic with the monoid $M_1 \times M_2$ and, consequently, in this case the trajectories monoid is $\mathcal{P}(M_1 \times M_2)$. Thus we obtain

Corollary 2. *If L_1, L_2 are languages, then the language $L_1 ⊔\!\!\!⊔ L_2$ is recognized by the monoid $\mathcal{P}(M_1 \times M_2)$. Moreover, if L_1 and L_2 are regular languages, then $L_1 ⊔\!\!\!⊔ L_2$ is a regular language.*

See [4], Proposition 1.3, for an entirely different proof of the above corollary.

Similar results can be obtained for the operations of bi-catenation, literal shuffle, balanced literal shuffle, and insertion. However, we do not enter this discussion in this paper.

4   Catenation and the Schützenberger Product

Of a special interest is the case of the catenation operation. We quote from Eilenberg, [1], vol. B, page 249: "The catenation product AB of two recognizable subsets A and B of Σ^*, turns out to be a rather complicated operation when looked at from the point of view of the syntactic invariants. It requires a new operation on semigroups due to Schützenberger."

The Schützenberger product of two monoids M_1 and M_2, denoted by $M_1 \diamond M_2$, is the submonoid of $[\mathcal{P}(M_1 \times M_2)]^{2\times 2}$ generated by all matrices of the following form:

$$\begin{pmatrix} (m_1, 1) & N \\ 0 & (1, m_2) \end{pmatrix}$$

where $m_i \in M_i$, $i = 1, 2$, and $N \subseteq M_1 \times M_2$.

The following theorem goes back to Schützenberger, [6], see also [1], vol. B, Theorem 2.1, or [4], Theorem 1.4.

Theorem 3. *If M_i are monoids such that M_i recognizes L_i, $i = 1, 2$, then the monoid $M_1 \diamond M_2$ recognizes $L_1 L_2$.*

By Remark 1, for $T = r^* u^*$, $⊔\!\!\!⊔_T$ is the catenation operation. We start by considering M_T as being the syntactic monoid of T, denoted by M_{cat}. Since T is a regular language, it follows that M_{cat} is a finite monoid. Moreover, by a classical method, one can obtain that $M_{cat} = \{1, \alpha, \beta, \gamma, 0\}$, where 1 is the unit element, 0 is the zero element, $\alpha^2 = \alpha$, $\beta^2 = \beta$, $\alpha\beta = \gamma\beta = \gamma$ and $\beta\alpha = \beta\gamma = \gamma\alpha = \gamma\gamma = 0$. The morphism $\varphi_T : \Sigma^* \longrightarrow M_T$, defined by $\varphi_T(r) = \alpha$ and $\varphi_T(u) = \beta$, has the property that $\varphi_T^{-1}(F_T) = T$, where $F_T = \{1, \alpha, \beta, \gamma\}$.

From Theorem 2 it follows in a straightforward manner:

Theorem 4. *If M_i are monoids such that M_i recognizes L_i, $i = 1, 2$, then the trajectories monoid $\mathcal{T}(M_1, M_2, M_{cat})$ recognizes $L_1 L_2$.*

In the remainder of this section we establish the interrelation between the trajectories monoid $\mathcal{T}(M_1, M_2, M_{cat})$ and the Schützenberger product $M_1 \diamond M_2$.
Notations. Let $\varphi : \Sigma^* \longrightarrow \mathcal{T}(M_1, M_2, M_{cat})$ be the morphism from the above theorem, with the property that $\varphi^{-1}(F) = L_1 L_2$, for some $F \subseteq \mathcal{T}(M_1, M_2, M_{cat})$. We denote by $M_1[cat]M_2$ the monoid $\varphi(\Sigma^*)$.

Note that $M_1[cat]M_2$ is a submonoid of $\mathcal{T}(M_1, M_2, M_{cat})$. Moreover, the monoid $M_1[cat]M_2$ recognizes the language $L_1 L_2$, too.

Let m be in $M_1[cat]M_2$, $m \neq 1$. From the definition of the monoid $M_1[cat]M_2$ there exists a word $x \in \Sigma^*$, such that $m = \varphi(x)$. One can easily verify that x is a nonempty word. Consider the mapping $\psi : M_1[cat]M_2 \longrightarrow M_1 \diamond M_2$ defined as

$$\psi(m) = \begin{pmatrix} (\varphi_1(x), 1) & \{(\varphi_1(x'), \varphi_2(x'')) \mid x = x'x''\} \\ 0 & (1, \varphi_2(x)) \end{pmatrix}$$

and $\psi(1) = 1$. One can prove that ψ is well defined.

The following theorem shows the relation between the trajectories monoids and the Schützenberger product.

Theorem 5. *The mapping ψ is a morphism and, moreover, if M_i unit-separately recognizes L_i, $i = 1, 2$, then ψ is injective.*

Corollary 3. *The monoid $M_1[cat]M_2$ divides the monoid $M_1 \diamond M_2$, if M_i unit-separately recognizes L_i, $i = 1, 2$.*

Note that, using the above corollary and Theorem 4 we obtain a new proof of the classical Theorem 3 in the case that M_i unit-separately recognizes L_i, $i = 1, 2$.

5 Monoids of Matrices

The Schützenberger product $M_1 \diamond M_2$ is a monoid of matrices. Consequently, the following natural problem arises: does there exist for every set of trajectories T and for all languages L_1 and L_2 a monoid of matrices that recognizes the language $L_1 \sqcup_T L_2$? The answer to this question is positive and the first part of this section is dedicated to this problem. The section ends with some Schützenberger-like products for some other operations: literal shuffle, bi-catenation, insertion, etc.

In this section we restrict our attention to regular sets of trajectories. This is not a major restriction mathematically. However, nonregular sets of trajectories lead to infinite matrices.

We start with some general facts concerning regular languages and finite automata. Let R be a regular language, $R \subseteq \Sigma^*$. There exists a finite automaton $A = (Q, \Sigma, \delta, Q_{in}, Q_{fin})$ such that the language accepted by A, denoted by $L(A)$ is R, i.e., $L(A) = R$. Note that we do not assume that A is a deterministic automaton. Without loss of generality, we may assume that the set Q of states, is of the form $Q = \{1, 2, \ldots, n\}$, for some $n \geq 1$. Let Δ_A be the transition matrix associated to A, i.e., Δ_A is an $n \times n$ matrix $\Delta_A = (d_{ij})_{1 \leq i,j \leq n}$ such that $d_{ij} = \{x \in \Sigma \mid \delta(i, x) = j\}$. Note that the entries in Δ_A are subsets of Σ. Let Δ_A^k be the kth power of the matrix Δ_A, where $k \geq 1$. By definition, Δ_A^0 is the unit matrix of size $n \times n$. Moreover, denote by Δ_A^* the matrix $\sum_{k \geq 0} \Delta_A^k$. Note that Δ_A^* does always exist.

Let $\eta_{in} = (i_j)_{1 \leq j \leq n}$ be the row matrix of size $1 \times n$, where $i_j = 1$, if $j \in Q_{in}$ and $i_j = 0$, otherwise. Similarly, let $\mu_{fin} = (f_j)_{1 \leq j \leq n}$ be the column matrix of size $n \times 1$, where $f_j = 1$, if $j \in Q_{fin}$ and $f_j = 0$, otherwise.

The following theorem is well known, see [1], [2].

Theorem 6. *Using the above notations:*
(i) if $\Delta_A^k = (\alpha_{ij})_{1 \leq i,j \leq n}$, then $\alpha_{ij} = \{w \in \Sigma^* \mid \delta(i, w) = j \text{ and } |w| = k\}$, where $k \geq 0$.
(ii) the language accepted by the automaton A is $L(A) = \eta_{in} \Delta_A^* \mu_{fin}$.

Let $A_T = (Q_T, \{r, u\}, \delta_T, Q_{T,in}, Q_{T,fin})$ be a finite automaton such that $L(A_T) = T$. Assume that $Q_T = \{1, 2, \ldots, n\}$ and $\Delta_T = (d_{ij})_{1 \leq i,j \leq n}$ is the transition matrix associated to A_T.

Let Σ be an alphabet, and let L_1, L_2 be languages, $L_i \subseteq \Sigma^*$, $i = 1, 2$. Assume that L_i is recognized by the monoid M_i, $i = 1, 2$. Let $\varphi_i : \Sigma^* \longrightarrow M_i$ be a morphism such that $L_i = \varphi_i^{-1}(F_i)$, for some $F_i \subseteq M_i$, $i = 1, 2$.

Let a be in Σ and let $\theta_a : \{r, u\}^* \longrightarrow M_1 \times M_2$ be the morphism defined by $\theta_a(r) = (\varphi_1(a), 1)$ and $\theta_a(u) = (1, \varphi_2(a))$.

For each $a \in \Sigma$, we define the matrix $\theta_a(\Delta_T)$ as the $n \times n$ matrix obtained from the transition matrix $\Delta_T = (d_{ij})_{1 \leq i,j \leq n}$ by replacing each d_{ij} with $\theta_a(d_{ij})$.

Let $\mathcal{M}_T(M_1, M_2)$ be the monoid generated by the following set of matrices: $\{\theta_a(\Delta_T) \mid a \in \Sigma\}$. Note that $\mathcal{M}_T(M_1, M_2)$ is a submonoid of the monoid $[\mathcal{P}(M_1 \times M_2)]^{n \times n}$.

The following theorem gives a positive answer to the question considered at the beginning of this section.

Theorem 7. *Let $L_1, L_2 \subseteq \Sigma^*$ be languages and assume that L_i is recognized by the monoid M_i, $i = 1, 2$, and let $T \subseteq \{r, u\}^*$ be a regular set of trajectories. The language $L_1 \sqcup\!\sqcup_T L_2$ is recognized by the monoid of matrices $\mathcal{M}_T(M_1, M_2)$.*

Having a language of the form $L = L_1 \sqcup\!\sqcup_T L_2$ we proved that L is recognized by the trajectories monoid $\mathcal{T}(M_1, M_2, M_T)$ and also by the monoid of matrices $\mathcal{M}_T(M_1, M_2)$. Hence, a natural question occurs: what is the interrelation between these two monoids?

Let $\varphi : \Sigma^* \longrightarrow \mathcal{T}(M_1, M_2, M_T)$ be the morphism such that $\varphi^{-1}(F) = L_1 \sqcup\!\sqcup_T L_2$. Denote by $M_1[T]M_2$ the monoid $\varphi(\Sigma^*)$. Note that $M_1[T]M_2$ is a submonoid of $\mathcal{T}(M_1, M_2)$ and, moreover, $M_1[T]M_2$ recognizes $L_1 \sqcup\!\sqcup_T L_2$.

Consider the mapping $\psi : M_1[T]M_2 \longrightarrow \mathcal{M}_T(M_1, M_2)$ defined as follows: $\psi(1) = 1$ and, for each $m \in M_1[T]M_2$, such that $m \neq 1$, there exists a nonempty word $x \in \Sigma^*$ such that $m = \varphi(x)$. Assume that $x = a_1 a_2 \ldots a_m$, where $a_i \in \Sigma$, $1 \leq i \leq m$. Then, by definition,

$$\psi(m) = \theta_{a_1}(\Delta_T)\theta_{a_2}(\Delta_T)\ldots\theta_{a_m}(\Delta_T).$$

One can show that ψ is well defined.

Theorem 8. *The mapping ψ is a morphism and, moreover, if M_i unit-separately recognizes L_i, $i = 1, 2$, then ψ is injective.*

Corollary 4. *The monoid $M_1[T]M_2$ divides the monoid $\mathcal{M}_T(M_1, M_2)$, if M_i unit-separately recognizes L_i, $i = 1, 2$.*

Now we briefly describe the products corresponding to the operations listed in Remark 1. All these products are similar to the Schützenberger product. Matrices in the product monoid $\mathcal{M}_T(M_1, M_2)$ are referred to as *product matrices*.

1. *The shuffle operation:* the transition matrix is $\Delta_T = (\{r, u\})$ and the product matrices are of the form (N), where $N \subseteq M_1 \times M_2$.
2. *The catenation operation:* the transition matrix is:

$$\Delta_T = \begin{pmatrix} r & u \\ 0 & u \end{pmatrix} \quad \text{Product matrices:} \quad \begin{pmatrix} (m_1, 1) & N \\ 0 & (1, m_2) \end{pmatrix}$$

where, $N \subseteq M_1 \times M_2$. Hence, we obtain again the Schützenberger product.
3. *The insertion operation:* the transition matrix is:

$$\Delta_T = \begin{pmatrix} r & u & 0 \\ 0 & u & r \\ 0 & 0 & r \end{pmatrix} \quad \text{Product matrices:} \quad \begin{pmatrix} (m_1, 1) & N_1 & N_2 \\ 0 & (1, m_2) & N_3 \\ 0 & 0 & (n_1, 1) \end{pmatrix}$$

where, $N_1, N_2, N_3 \subseteq M_1 \times M_2$ and $m_1, n_1 \in M_1$, $m_2 \in M_2$.
4. *The balanced literal shuffle operation:* the transition matrix is:

$$\Delta_T = \begin{pmatrix} 0 & r \\ u & 0 \end{pmatrix} \quad \text{Product matrices:} \quad \begin{pmatrix} m & 0 \\ 0 & n \end{pmatrix}, \begin{pmatrix} 0 & m' \\ n' & 0 \end{pmatrix},$$

where, $m, n, m', n' \in M_1 \times M_2$.
5. *The literal shuffle operation:* the transition matrix is:

$$\Delta_T = \begin{pmatrix} 0 & r & 0 & u \\ u & 0 & r & 0 \\ 0 & 0 & r & 0 \\ 0 & 0 & 0 & u \end{pmatrix} \quad \text{Product matrices:} \quad \begin{pmatrix} m & n & N_1 & N_2 \\ p & q & N_3 & N_4 \\ 0 & 0 & (m_1, 1) & 0 \\ 0 & 0 & 0 & (1, m_2) \end{pmatrix}$$

where, $N_1, N_2, N_3, N_4 \subseteq M_1 \times M_2$, $m_1 \in M_1$, $m_2 \in M_2$ and, moreover, $m = q = 0, n, p \in M_1 \times M_2$ or $n = p = 0, m, q \in M_1 \times M_2$.

6. *The bi-catenation operation:* the transition matrix is:

$$\Delta_T = \begin{pmatrix} r & u & 0 & 0 \\ 0 & u & 0 & 0 \\ 0 & 0 & u & r \\ 0 & 0 & 0 & r \end{pmatrix} \quad \text{Product matrices:} \quad \begin{pmatrix} (m_1,1) & N_1 & 0 & 0 \\ 0 & (1,m_2) & 0 & 0 \\ 0 & 0 & (1,m_2) & N_2 \\ 0 & 0 & 0 & (m_1,1) \end{pmatrix}$$

where, $N_1, N_2 \subseteq M_1 \times M_2$, $m_1 \in M_1$, and $m_2 \in M_2$.

7. *The anti-catenation operation:* the transition matrix is:

$$\Delta_T = \begin{pmatrix} u & r \\ 0 & r \end{pmatrix} \quad \text{Product matrices:} \quad \begin{pmatrix} (1,m_2) & N \\ 0 & (m_1,1) \end{pmatrix}, \text{ where } N \subseteq M_1 \times M_2.$$

6 Conclusion

We introduced a uniform method to find monoids that recognize languages of the form $L_1 \shuffle_T L_2$ and consequently, our results are applicable to many of the most important operations with languages. Many details concerning this method remain to be clarified. The above problems can be formulated for all associative operations $\shuffle_T$ and languages of the form $L_1 \shuffle_T L_2 \shuffle_T \ldots \shuffle_T L_n$, $n \geq 3$.

Acknowledgements. The authors are grateful to Volker Diekert for his comments that have improved the initial version of this paper.

References

[1] S. Eilenberg, *Automata, Languages and Machines*, Academic Press, New York, vol. A, 1974, vol. B, 1976.
[2] W. Kuich and A. Salomaa, *Semirings, Automata, Languages*, Springer-Verlag, Berlin, 1986.
[3] A. Mateescu, G. Rozenberg and A. Salomaa, "Shuffle on Trajectories: Syntactic Constraints", *Theoretical Computer Science*, 197, 1-2, (1998) 1-56.
[4] J.E. Pin, *Varieties of Formal Languages*, North Oxford Academic, 1986.
[5] J.E. Pin, "Syntactic Semigroups", in *Handbook of Formal Languages*, eds. G. Rozenberg and A. Salomaa, Vol. 1, Springer, 1997, 679-746.
[6] M.P. Schützenberger, "On finite monoids having only trivial subgroups", *Information and Control*, 8, (1965) 190-194.

Gaußian Elimination and a Characterization of Algebraic Power Series

Werner Kuich[*]

Abteilung für Theoretische Informatik
Institut für Algebra und Diskrete Mathematik
Technische Universität Wien
kuich@tuwien.ac.at

Abstract. We show first how systems of equations can be solved by Gaußian elimination. This yields a characterization of algebraic power series and of $\mathfrak{Alg}(A')$, $A' \subseteq A$, A a continuous semiring. In the case of context-free languages this characterization coincides with the characterization given by Gruska [7].

1 Introduction and Basic Results

In 1971, Gruska [7] characterized context-free languages by certain expressions that are similar to regular expressions.

Let Σ_∞ be an infinite alphabet. Let L be a formal language over Σ, $\Sigma \subset \Sigma_\infty$, and $x \in \Sigma_\infty$. Define the morphism $h_x^L : \Sigma_\infty^* \to \mathfrak{P}(\Sigma_\infty^*)$ by $h_x^L(x) = L$, $h_x^L(x') = \{x'\}$, $x' \neq x$, and extend it to a substitution $h_x^L : \mathfrak{P}(\Sigma_\infty^*) \to \mathfrak{P}(\Sigma_\infty^*)$.

Define $L^{x,1} = h_x^\emptyset(L)$, $L^{x,j+1} = h_x^{L^{x,j}}(L)$, $j \geq 1$, and $L^x = \bigcup_{j \geq 1} L^{x,j}$. We call a set of languages $\mathfrak{E} \subseteq \mathfrak{P}(\Sigma_\infty^*)$ *equationally closed* iff $\mathfrak{E}$ is a semiring closed under the following operation: if $L \in \mathfrak{E}$ and $x \in \Sigma_\infty$ then $L^x \in \mathfrak{E}$.

Gruska [7] proved that the set of context-free languages over Σ, $\Sigma \subset \Sigma_\infty$, coincides with the least equationally closed semiring containing the finite languages.

Consider the language equation $x = L$ and its approximation sequence $(\sigma^j)_{j \geq 0}$ (see Autebert, Berstel, Boasson [1] and Kuich [10]). Then we have $\sigma^0 = \emptyset$, $\sigma^1 = h_x^\emptyset(L)$, $\sigma^{j+1} = h_x^{\sigma^j}(L)$, $j \geq 1$. Hence, $\sigma^j = L^{x,j}$, $j \geq 1$, and L^x is the least solution of $x = L$. Denote the least solution (i. e., the least fixpoint) of $x = L$ by $\mu x.L$. Then a set of languages $\mathfrak{E} \subseteq \mathfrak{P}(\Sigma_\infty^*)$ is equationally closed iff it is a semiring closed under least solutions $\mu x.L$ of language equations $x = L$, where $L \in \mathfrak{E}$ and $x \in \Sigma_\infty$. It is this formulation of Gruska's result which we will generalize in our paper.

Earlier, in 1968, a similar theorem on recognizable trees was proved by Thatcher and Wright [15] and, via the yield of trees, was projected to the context-free languages. (See Gecseg, Steinby [5], Example 14.9.) Bozapalidis [3] extended the characterization of recognizable trees to rational formal power series on trees

---

[*] Supported by Stiftung Aktion Österreich-Ungarn

and projected it via the yield to algebraic power series (see Bozapalidis [3], Sections 5 and 6). We give a complete and direct proof of Bozapalidis' result on algebraic power series in Section 2 and generalize it in Section 3.

We put Gaußian elimination into the center of our consideration and prove by it characterizations of context-free languages, algebraic power series and algebraic elements of continuous semirings. Here Gaußian elimination means the step-by-step elimination of variables (or sets of variables) in the process of solving algebraic systems of equations (see Autebert, Berstel, Boasson [1], Theorem 2.4).

It is assumed that the reader is familiar with the basics of semiring theory. Notions and notations that are not defined are taken from Kuich [10]. In the sequel, A will always be a *continuous* semiring. This is a complete and naturally ordered semiring such that, for all index sets I and all families $(a_i \mid i \in I)$ the following condition is satisfied:

$$\sum_{i \in I} a_i = \sup\{\sum_{i \in E} a_i \mid E \subseteq I, \ E \text{ finite}\}.$$

Here "sup" denotes the least upper bound with respect to the natural order (see Goldstern [6], Sakarovitch [13], and Karner [8]).

A subsemiring $\bar{A}$ of A is called *rationally closed* iff for all $a \in \bar{A}$, we have $a^* := \sum_{i \geq 0} a^i \in \bar{A}$. By definition, $\mathfrak{Rat}(A')$ is the smallest rationally closed subsemiring of A containing $A' \subseteq A$. Furthermore, the collection of the components of the least solutions of all A'-algebraic systems, where A' is a fixed subset of A, is denoted by $\mathfrak{Alg}(A')$. Here, an A'-*algebraic system* is a system of formal equations $y_i = p_i$, $1 \leq i \leq n$, where $p_1, \ldots, p_n$ are semiring-polynomials in the polynomial semiring over A (with variables $y_1, \ldots, y_n$) with coefficients in A'. See also Lausch, Nöbauer [12], Chapter 1, §4.)

In case the basic semiring is the semiring of formal power series $A\langle\!\langle(\Sigma \cup Y)^*\rangle\!\rangle$, where A is a commutative continuous semiring and Σ and $Y = \{y_1, \ldots, y_n\}$ are disjoint alphabets, we consider algebraic systems. Here an *algebraic system* is a system of formal equations $y_i = p_i$, $1 \leq i \leq n$, where $p_1, \ldots, p_n$ are polynomials in $A\langle(\Sigma \cup Y)^*\rangle$. The collection of the components of the least solutions of all algebraic systems as defined above is denoted by $A^{\text{alg}}\langle\!\langle \Sigma^* \rangle\!\rangle = \mathfrak{Alg}(\{aw \mid a \in A, \ w \in \Sigma^*\})$.

We need a more general framework for our considerations. In the sequel, A will denote a *continuous* semiring, Σ_∞ an infinite alphabet, and $\Sigma \subset \Sigma_\infty$ a (finite) alphabet. Additionally, in the remainder of this section, A will be *commutative*.

Let $h : \Sigma_\infty^* \to A\langle\!\langle \Sigma_\infty^* \rangle\!\rangle$ be a monoid morphism. Extend it in the usual manner to a semiring morphism $h : A\langle\!\langle \Sigma_\infty^* \rangle\!\rangle \to A\langle\!\langle \Sigma_\infty^* \rangle\!\rangle$ by $h(r) = \sum_{\alpha \in \Sigma_\infty^*} (r, \alpha) h(\alpha)$, $r \in A\langle\!\langle \Sigma_\infty^* \rangle\!\rangle$. Our first result is that h is a complete morphism, i.e., $h(\sum_{i \in I} r_i) = \sum_{i \in I} h(r_i)$ for all $r_i \in A\langle\!\langle \Sigma_\infty^* \rangle\!\rangle$, $i \in I$, and all index sets I.

Theorem 1 *Let $h : \Sigma_\infty^* \to A\langle\!\langle \Sigma_\infty^* \rangle\!\rangle$ be a monoid morphism. Then the extended semiring morphism $h : A\langle\!\langle \Sigma_\infty^* \rangle\!\rangle \to A\langle\!\langle \Sigma_\infty^* \rangle\!\rangle$ is a complete morphism.*

Proof. We obtain

$$h(\sum_{i \in I} r_i) = h(\sum_{i \in I} \sum_{\alpha \in \Sigma_\infty^*} (r_i, \alpha)\alpha) =$$
$$h(\sum_{\alpha \in \Sigma_\infty^*} (\sum_{i \in I} (r_i, \alpha))\alpha) = \sum_{\alpha \in \Sigma_\infty^*} \sum_{i \in I} (r_i, \alpha)h(\alpha) =$$
$$\sum_{i \in I} \sum_{\alpha \in \Sigma_\infty^*} (r_i, \alpha)h(\alpha) = \sum_{i \in I} h(r_i).$$

□

Corollary 2 *Let $h : \Sigma_\infty^* \to A\langle\!\langle \Sigma_\infty^* \rangle\!\rangle$ be a monoid morphism. Then the extended semiring morphism $h : A\langle\!\langle \Sigma_\infty^* \rangle\!\rangle \to A\langle\!\langle \Sigma_\infty^* \rangle\!\rangle$ is an ω-continuous mapping.*

Proof. Since $A\langle\!\langle \Sigma_\infty^* \rangle\!\rangle$ is continuous, each ω-chain has the form $(\sum_{0 \leq i \leq n} r_i \mid n \in \mathbb{N})$, $r_i \in A\langle\!\langle \Sigma_\infty^* \rangle\!\rangle$, $i \in \mathbb{N}$, and its least upper bound is $\sum_{i \in \mathbb{N}} r_i$. The equality $h(\sum_{i \in \mathbb{N}} r_i) = \sum_{i \in \mathbb{N}} h(r_i)$ is now implied by Theorem 1. □

We use the following notation: If $r \in A\langle\!\langle \Sigma_\infty^* \rangle\!\rangle$, and $y_1, \ldots, y_n \in \Sigma_\infty$ are distinguished variables, we write $r(y_1, \ldots, y_n)$. If $h : \Sigma_\infty^* \to A\langle\!\langle \Sigma_\infty^* \rangle\!\rangle$ is a morphism such that $h(x) = x$, $x \in \Sigma_\infty - \{y_1, \ldots, y_n\}$, then we write for $h(r(y_1, \ldots, y_n))$ simply $r(h(y_1), \ldots, h(y_n))$. Hence, $r(y_1, \ldots, y_n)$ induces a mapping from $(A\langle\!\langle \Sigma_\infty^* \rangle\!\rangle)^n$ into $A\langle\!\langle \Sigma_\infty^* \rangle\!\rangle$ whose value at $\sigma_1, \ldots, \sigma_n$ is given by $r(\sigma_1, \ldots, \sigma_n)$. By Corollary 2, this mapping is ω-continuous.

A *system (of equations)* with variables $y_1, \ldots, y_n \in \Sigma_\infty$ is given by

$$y_i = r_i(y_1, \ldots, y_n), \quad 1 \leq i \leq n, \quad r_i \in A\langle\!\langle \Sigma_\infty^* \rangle\!\rangle.$$

Let $\hat{\Sigma}_\infty = \Sigma_\infty - \{y_1, \ldots, y_n\}$. A *solution* to this system is given by $(\sigma_1, \ldots, \sigma_n) \in (A\langle\!\langle \hat{\Sigma}_\infty^* \rangle\!\rangle)^n$ such that

$$\sigma_i = r_i(\sigma_1, \ldots, \sigma_n), \quad 1 \leq i \leq n.$$

A solution $(\sigma_1, \ldots, \sigma_n)$ is termed *least solution* iff $\sigma_i \sqsubseteq \tau_i$, $1 \leq i \leq n$, for all solutions $(\tau_1, \ldots, \tau_n)$. Hence, the least solution of the system $y_i = r_i(y_1, \ldots, y_n)$, $1 \leq i \leq n$, is nothing else than the least fixpoint of the mapping $(r_1, \ldots, r_n) : (A\langle\!\langle \hat{\Sigma}_\infty^* \rangle\!\rangle)^n \to (A\langle\!\langle \hat{\Sigma}_\infty^* \rangle\!\rangle)^n$ defined by $(r_1, \ldots, r_n)(\sigma_1, \ldots, \sigma_n) = (r_1(\sigma_1, \ldots, \sigma_n), \ldots, r_n(\sigma_1, \ldots, \sigma_n))$. Since this mapping is ω-continuous, we can apply the Fixpoint Theorem (see Wechler [16]) to achieve our next result. We use a vectorial notation in Theorem 3.

Theorem 3 *Let $y = r$, $r \in (A\langle\!\langle \Sigma_\infty^* \rangle\!\rangle)^n$, be a system of equations, where $y = (y_1, \ldots, y_n)$ and $r = (r_1, \ldots, r_n)$. Then the least solution of $y = r(y)$ exists in $(A\langle\!\langle \hat{\Sigma}_\infty^* \rangle\!\rangle)^n$, $\hat{\Sigma}_\infty = \Sigma_\infty - \{y_1, \ldots, y_n\}$, and equals*

$$\sup(r^i(0) \mid i \in \mathbb{N}),$$

where r^i is the i-th iterate of the mapping r.

Theorem 3 indicates how we can compute an approximation to the least solution of a system of equations $y = r$. The *approximation sequence* $(\sigma^j)_{j \geq 0}$, $\sigma^j \in (A\langle\!\langle \hat{\Sigma}_\infty^* \rangle\!\rangle)^n$, $j \geq 0$, associated to $y = r(y)$ is defined as follows:

$$\sigma^0 = 0, \quad \sigma^{j+1} = r(\sigma^j), \quad j \geq 0.$$

Clearly, $(\sigma^j \mid j \in \mathbb{N})$ is an ω-chain and its least upper bound $\sup(\sigma^j \mid j \in \mathbb{N})$ is the least solution of $y = r$.

The following method for the resolution of algebraic systems is called Gaußian elimination by Autebert, Berstel, Boasson [1], Theorem 2.4. In the case of commutative continuous semirings it was used by Kuich [9], Theorem 4.9. Bekic [2] has proved a more general result. For a full treatment of least fixpoint and least pre-fixpoint solutions see Thorem 6.1 of Esik [4].

Consider disjoint alphabets $\{y_1, \ldots, y_n\}$ and $\{z_1, \ldots, z_m\}$ of variables and let $\hat{\Sigma}_\infty = \Sigma_\infty - \{y_1, \ldots, y_n, z_1, \ldots, z_m\}$. Let $p_i(z_1, \ldots, z_m, y_1, \ldots, y_n)$, $1 \leq i \leq n$, and $q_j(z_1, \ldots, z_m, y_1, \ldots, y_n)$, $1 \leq j \leq m$, be power series in $A\langle\langle \Sigma_\infty^* \rangle\rangle$ and consider the system of equations

$$z_j = p_j(z_1, \ldots, z_m, y_1, \ldots, y_n), \quad 1 \leq j \leq m,$$
$$y_i = q_i(z_1, \ldots, z_m, y_1, \ldots, y_n), \quad 1 \leq i \leq n.$$

Let $(t_1(z_1, \ldots, z_m), \ldots, t_n(z_1, \ldots, z_m)) \in (A\langle\langle (\hat{\Sigma}_\infty \cup \{z_1, \ldots, z_m\})^* \rangle\rangle)^n$ and $(r_1, \ldots, r_m) \in (A\langle\langle \hat{\Sigma}_\infty^* \rangle\rangle)^n$ be the least solutions of the systems $y_i = q_i(z_1, \ldots, z_m, y_1, \ldots, y_n)$, $1 \leq i \leq n$, and $z_j = p_j(z_1, \ldots, z_m, t_1(z_1, \ldots, z_m), \ldots, t_n(z_1, \ldots, z_m))$, $1 \leq j \leq m$, respectively. Then $(r_1, \ldots, r_m, t_1(r_1, \ldots, r_m), \ldots, t_n(r_1, \ldots, r_m))$ is the least solution of the original system.

In the next theorem and its proof, we use a vectorial notation: $z = (z_1, \ldots, z_m)$, $y = (y_1, \ldots, y_n)$, $p = (p_1, \ldots, p_m)$, $q = (q_1, \ldots, q_n)$, etc.

Theorem 4 (Bekic [2]) *Consider the system of equations*

$$z = p(z, y), \quad y = q(z, y).$$

Let $t(z)$ and r be the least solutions of the systems $y = q(z, y)$ and $z = p(z, t(z))$, respectively. Then $(r, t(r))$ is the least solution of the system $z = p(z, y)$, $y = q(z, y)$.

Moreover, r is the least solution of the system $z = p(z, t(r))$.

2 Algebraic Expressions for Algebraic Power Series

In this section, A denotes a *commutative continuous* semiring. We introduce the following notation: Let $r(y_1, \ldots, y_i, \ldots, y_n) \in A\langle\langle \Sigma_\infty^* \rangle\rangle$, where $y_1, \ldots, y_i, \ldots, y_n$ are variables. We denote the least $\sigma \in A\langle\langle (\Sigma_\infty - \{y_i\})^* \rangle\rangle$ such that $r(y_1, \ldots, \sigma, \ldots, y_n) = \sigma$ by $\mu y_i.r(y_1, \ldots, y_i, \ldots, y_n)$, $1 \leq i \leq n$. This means that σ is the least fixpoint of the equation $y_i = r(y_1, \ldots, y_i, \ldots, y_n)$ and μy_i is a fixpoint operator. Observe that $\mu y_i.r(y_1, \ldots, y_i, \ldots, y_n) \in A\langle\langle (\Sigma_\infty - \{y_i\})^* \rangle\rangle$.

Lemma 5 *Let $r(y_1, \ldots, y_n, y) \in A\langle\langle \Sigma_\infty^* \rangle\rangle$ and $\sigma_i \in A\langle\langle (\Sigma_\infty - \{y\})^* \rangle\rangle$, $1 \leq i \leq n$. Let $s(y_1, \ldots, y_n) = \mu y.r(y_1, \ldots, y_n, y)$. Then*

$$s(\sigma_1, \ldots, \sigma_n) = \mu y.r(\sigma_1, \ldots, \sigma_n, y).$$

Proof. Let $(\tau^j(y_1,\ldots,y_n))_{j\geq 0}$ be the approximation sequence of $y = r(y_1,\ldots,y_n, y)$. Then $(\tau^j(\sigma_1,\ldots,\sigma_n))_{j\geq 0}$ is the approximation sequence of $y = r(\sigma_1,\ldots,\sigma_n, y)$. By Corollary 2, we infer that $\mu y.r(\sigma_1,\ldots,\sigma_n, y) = \sup(\tau^j(\sigma_1,\ldots,\sigma_n) \mid j \in \mathbb{N}) = s(\sigma_1,\ldots,\sigma_n)$. □

A subsemiring $\bar{A}$ of $A\langle\!\langle \Sigma_\infty^* \rangle\!\rangle$ is called *equationally closed* iff, for all $r \in \bar{A}$ and $y \in \Sigma_\infty$ the power series $\mu y.r$ is again in $\bar{A}$.

Let $A\{\Sigma_\infty^*\} = \{r \in A\langle \Sigma^* \rangle \mid \Sigma \subset \Sigma_\infty \text{ finite}\}$ and $A^{\text{alg}}\{\!\{\Sigma_\infty^*\}\!\} = \{r \in A^{\text{alg}}\langle\!\langle \Sigma^* \rangle\!\rangle \mid \Sigma \subset \Sigma_\infty \text{ finite}\}$. Denote by $\bar{A}\{\!\{\Sigma_\infty^*\}\!\}$ the least equationally closed semiring containing $A\{\Sigma_\infty^*\}$. We will show in this section that $\bar{A}\{\!\{\Sigma_\infty^*\}\!\} = A^{\text{alg}}\{\!\{\Sigma_\infty^*\}\!\}$.

Theorem 6 *Let $t(y_1,\ldots,y_n)$, $\sigma_j \in \bar{A}\{\!\{\Sigma_\infty^*\}\!\}$, $1 \leq j \leq n$. Then $t(\sigma_1,\ldots,\sigma_n) \in \bar{A}\{\!\{\Sigma_\infty^*\}\!\}$.*

Proof. The proof is by induction on the number of applications of the operations $+, \cdot$ and μ to generate $t(y_1,\ldots,y_n)$.

(i) Let $t(y_1,\ldots,y_n) \in A\{\Sigma_\infty^*\}$, i.e., $t(y_1,\ldots,y_n) \in A\langle \Sigma^* \rangle$ for some $\Sigma \subset \Sigma_\infty$. Since $t(\sigma_1,\ldots,\sigma_n)$ is generated from $\sigma_1,\ldots,\sigma_n$ by applications of sum, product and scalar product, we infer that $t(\sigma_1,\ldots,\sigma_n) \in \bar{A}\{\!\{\Sigma_\infty^*\}\!\}$.

(ii) We only prove the case of the operator μ. Let $\sigma_1,\ldots,\sigma_n \in \bar{A}\{\!\{\Sigma_\infty^*\}\!\} \cap A\langle\!\langle \Sigma^* \rangle\!\rangle$ for some Σ and choose a $y \in \Sigma_\infty$ that is not in $\Sigma \cup \{y_1,\ldots,y_n\}$. Without loss of generality assume that $t(y_1,\ldots,y_n) = \mu y.r(y_1,\ldots,y_n,y)$ (the variable y is "bound"), where $r(y_1,\ldots,y_n,y) \in \bar{A}\{\!\{\Sigma_\infty^*\}\!\}$. By induction hypothesis, we obtain $r(\sigma_1,\ldots,\sigma_n,y) \in \bar{A}\{\!\{\Sigma_\infty^*\}\!\}$. Hence, $t(\sigma_1,\ldots,\sigma_n) = \mu y.r(\sigma_1,\ldots,\sigma_n,y) \in \bar{A}\{\!\{\Sigma_\infty^*\}\!\}$ by Lemma 5. □

Theorem 7 $A^{\text{alg}}\{\!\{\Sigma_\infty^*\}\!\} \subseteq \bar{A}\{\!\{\Sigma_\infty^*\}\!\}$.

Proof. The proof is by induction on the number of variables of algebraic systems. We use the following induction hypothesis: If $\tau \in (A^{\text{alg}}\{\!\{\Sigma_\infty^*\}\!\})^n$, $n \geq 1$, is the least solution of an algebraic system $y_i = q_i(y_1,\ldots,y_n)$, $1 \leq i \leq n$, with n variables $y_1,\ldots,y_n$ where $q_i \in A\{\Sigma_\infty^*\}$, then $\tau_i \in \bar{A}\{\!\{\Sigma_\infty^*\}\!\}$.

(1) Let $n = 1$ and assume that r is the least solution of the algebraic system $z = p(z)$. Then $r = \mu z.p(z) \in \bar{A}\{\!\{\Sigma_\infty^*\}\!\}$.

(2) Let $z, y_1,\ldots,y_n$ be variables and $p, q_1,\ldots,q_n$ be polynomials in $A\{\Sigma_\infty^*\}$, and consider the algebraic system $z = p(z,y)$, $y = q(z,y)$, where $y = (y_1,\ldots,y_n)$ and $q = (q_1,\ldots,q_n)$. Let $t(z) \in (A^{\text{alg}}\{\!\{\Sigma_\infty^*\}\!\})^n$ be the least solution of $y = q(z,y)$. By our induction hypothesis we obtain $t(z) \in (\bar{A}\{\!\{\Sigma_\infty^*\}\!\})^n$. Since $p(z,y)$ is a polynomial, it is in $\bar{A}\{\!\{\Sigma_\infty^*\}\!\}$. Hence, by Theorem 6, $p(z,t(z))$ is in $\bar{A}\{\!\{\Sigma_\infty^*\}\!\}$. This implies $\mu z.p(z,t(z)) \in \bar{A}\{\!\{\Sigma_\infty^*\}\!\}$. Again, by Theorem 6, $t(\mu z.p(z,t(z)) \in (\bar{A}\{\!\{\Sigma_\infty^*\}\!\})^n$. By Theorem 4, $(\mu z.p(z,t(z)), t(\mu z.p(z,t(z))))$ is the least solution of the algebraic system $z = p(z,y)$, $y = q(z,y)$. Hence the components of the least solution of this algebraic system are in $\bar{A}\{\!\{\Sigma_\infty^*\}\!\}$. □

We now show the converse to Theorem 7.

Theorem 8 $\bar{A}\{\{\Sigma_\infty^*\}\} \subseteq A^{\mathrm{alg}}\{\{\Sigma_\infty^*\}\}$.

Proof. We show that $A^{\mathrm{alg}}\{\{\Sigma_\infty^*\}\}$ is an equationally closed semiring that contains $A\{\Sigma_\infty^*\}$. Easy constructions (see Theorem 3.11 of Kuich [10]) show that $A^{\mathrm{alg}}\{\{\Sigma_\infty^*\}\}$ is a semiring containing $A\{\Sigma_\infty^*\}$. Hence we have only to show that $\mu z.r$, $r \in A^{\mathrm{alg}}\{\{\Sigma_\infty^*\}\}$ and $z \in \Sigma_\infty$, is in $A^{\mathrm{alg}}\{\{\Sigma_\infty^*\}\}$.

Let $r \in A^{\mathrm{alg}}\{\{\Sigma_\infty^*\}\}$ be the first component of the least solution of the algebraic system $y_i = p_i(y_1, \ldots, y_n, z)$, $1 \leq i \leq n$. Then, by Theorem 4, $\mu z.r$ is the z-component of the least solution of the algebraic system $z = y_1$, $y_i = p_i(y_1, \ldots, y_n, z)$, $1 \leq i \leq n$. □

We have now achieved the main result of this section.

Theorem 9 $\bar{A}\{\{\Sigma_\infty^*\}\} = A^{\mathrm{alg}}\{\{\Sigma_\infty^*\}\}$ and $\bar{A}\{\{\Sigma_\infty^*\}\} \cap A\langle\!\langle \Sigma^* \rangle\!\rangle = A^{\mathrm{alg}}\langle\!\langle \Sigma^* \rangle\!\rangle$, $\Sigma \subset \Sigma_\infty$, Σ *finite.*

Analogous to the regular expressions (see Salomaa [14]) and similar to the context-free expressions of Gruska [7], we define algebraic expressions.

Assume that Σ_∞, A and $U = \{+, \cdot, \mu, [,]\}$ are mutually disjoint. A word E over $\Sigma_\infty \cup A \cup U$ is an *algebraic expression* over $A\{\Sigma_\infty^*\}$ iff

(i) E is a symbol of A, or
(ii) E is a symbol of Σ_∞, or else
(iii) E is of one of the forms $[E_1 + E_2]$, $[E_1 \cdot E_2]$, or $\mu y.E_1$, where E_1 and E_2 are algebraic expressions and y is a symbol of Σ_∞.

Each algebraic expression E over $A\{\Sigma_\infty^*\}$ denotes a formal power series $|E|$ in $A\{\{\Sigma_\infty^*\}\}$ according to the following conventions:

(i) The power series denoted by $a \in A$ is $a\varepsilon$ in $A\{\Sigma_\infty^*\}$.
(ii) The power series denoted by $x \in \Sigma_\infty$ is x in $A\{\Sigma_\infty^*\}$.
(iii) For algebraic expressions E_1, E_2 over $A\{\Sigma_\infty^*\}$ and $y \in \Sigma_\infty$, $|[E_1 + E_2]| = |E_1| + |E_2|$, $|[E_1 \cdot E_2]| = |E_1| \cdot |E_2|$, $|\mu y.E_1| = \mu y.|E_1|$.

Let ϕ be the mapping from the set of algebraic expressions over $A\{\Sigma_\infty^*\}$ into the finite sets of $\mathfrak{P}(\Sigma_\infty)$ defined by

(i) $\phi(a) = \emptyset$, $a \in A$.
(ii) $\phi(x) = \{x\}$, $x \in \Sigma_\infty$.
(iii) $\phi([E_1 + E_2]) = \phi([E_1 \cdot E_2]) = \phi(E_1) \cup \phi(E_2)$, $\phi(\mu y.E_1) = \phi(E_1) - \{y\}$, for algebraic expressions E_1, E_2 and $y \in \Sigma_\infty$.

Given an algebraic expression E, $\phi(E)$ contains the "free symbols" of E. This means that $|E|$ is a formal power series in $A\langle\!\langle \phi(E)^* \rangle\!\rangle$. Theorem 9 and the above definitions yield some corollaries.

Corollary 10 *A power series r is in $A^{\mathrm{alg}}\{\{\Sigma_\infty^*\}\}$ iff there exists an algebraic expression E over $A\{\Sigma_\infty^*\}$ such that $r = |E|$.*

Corollary 11 *A power series r is in $A^{\mathrm{alg}}\langle\!\langle \Sigma^* \rangle\!\rangle$ iff there exists an algebraic expression E over $A\{\Sigma_\infty^*\}$, where $\phi(E) \subseteq \Sigma$, such that $r = |E|$.*

Observe that $\mathbb{B}\{\{\Sigma_\infty^*\}\}$ is isomorphic to the semiring $\mathfrak{L}(\Sigma_\infty) = \{L \mid L \subseteq \Sigma^*, \Sigma \subset \Sigma_\infty\}$ of formal languages over Σ_∞. Hence, if $A = \mathbb{B}$ then each algebraic expression E over $\mathbb{B}\{\Sigma_\infty^*\}$ denotes by this isomorphism a formal language in $\mathfrak{L}(\Sigma_\infty)$ according to the following conventions:

(i) The language denoted by 0 or 1 is $\emptyset$ or $\{\varepsilon\}$, respectively.
(ii) The language denoted by $x \in \Sigma_\infty$ is $\{x\}$.
(iii) For algebraic expressions E_1, E_2 over $\mathbb{B}\{\Sigma_\infty^*\}$ and $y \in \Sigma_\infty$, $|[E_1 + E_2]| = |E_1| \cup |E_2|$, $|[E_1 \cdot E_2]| = |E_1| \cdot |E_2|$, $|\mu y.E_1| = |E_1|^y$.

Corollary 12 (Gruska [7]) *A formal language L in $\mathfrak{L}(\Sigma_\infty)$ is context-free iff there exists an algebraic expression E over $\mathbb{B}\{\Sigma_\infty^*\}$ such that $L = |E|$.*

Corollary 13 *A formal language L over Σ is context-free iff there exists an algebraic expression E over $\mathbb{B}\{\Sigma_\infty^*\}$, where $\phi(E) = \Sigma$, such that $L = |E|$.*

3 Algebraic Expressions for Algebraic Elements

In this section A will denote a continuous semiring (*not* necessarily commutative) and $Y = \{y, y_1, y_2, \ldots, y_n, \ldots\}$ will denote a countably infinite set of variables disjoint with A. Consider the continuous semiring $\mathbb{N}^\infty \langle\!\langle A^* \rangle\!\rangle$ and let $a \in A$. The polynomial $\bar{p} \in \mathbb{N}^\infty \langle A^* \rangle$ defined by $(\bar{p}, a) = 1$, $(\bar{p}, \bar{w}) = 0$ for all $\bar{w} \in A^* - \{a\}$, will be denoted by $\bar{a}$. Let $\varphi : A^* \to A$ be the monoid morphism defined by $\varphi(\bar{a}) = a$, $a \in A$. Then, by Goldstern [6], φ extends uniquely to a complete semiring morphism $\varphi : \mathbb{N}^\infty \langle\!\langle A^* \rangle\!\rangle \to A$ defined by $\varphi(\bar{r}) = \sum_{\bar{w} \in A^*} (\bar{r}, \bar{w}) \varphi(\bar{w})$ for $\bar{r} \in \mathbb{N}^\infty \langle\!\langle A^* \rangle\!\rangle$. Each $\bar{r}(y_1, \ldots, y_n) \in \mathbb{N}^\infty \langle\!\langle (A \cup \{y_1, \ldots, y_n\})^* \rangle\!\rangle$ induces a mapping $\bar{r} : (\mathbb{N}^\infty \langle\!\langle A^* \rangle\!\rangle)^n \to \mathbb{N}^\infty \langle\!\langle A^* \rangle\!\rangle$ as follows. Let $h : (A \cup \{y_1, \ldots, y_n\})^* \to \mathbb{N}^\infty \langle\!\langle A^* \rangle\!\rangle$ be a monoid morphism defined by $h(\bar{a}) = \bar{a}$, $a \in A$, $h(y_i) = \bar{s}_i \in \mathbb{N}^\infty \langle\!\langle A^* \rangle\!\rangle$, $1 \leq i \leq n$. Then h extends uniquely to a complete semiring morphism $h : \mathbb{N}^\infty \langle\!\langle (A \cup \{y_1, \ldots, y_n\})^* \rangle\!\rangle \to \mathbb{N}^\infty \langle\!\langle A^* \rangle\!\rangle$ defined by $h(\bar{r}) = \sum_{\bar{\alpha} \in (A \cup \{y_1, \ldots, y_n\})^*} (\bar{r}, \bar{\alpha}) h(\bar{\alpha})$. Now the value of $\bar{r}$ at $(\bar{s}_1, \ldots, \bar{s}_n)$ is defined to be $h(\bar{r})$, i. e., $\bar{r}(\bar{s}_1, \ldots, \bar{s}_n) = h(\bar{r})$. Consider now a system with one equation

$$y = \bar{r}(y_1, \ldots, y_n, y), \quad \bar{r} \in \mathbb{N}^\infty \{\{(A \cup \{y_1, \ldots, y_n, y\})^*\}\}.$$

Its least solution (i. e., its least fixpoint) in $\mathbb{N}^\infty \{\{(A \cup \{y_1, \ldots, y_n, y\})^*\}\}$ is denoted by $\mu y.\bar{r}(y_1, \ldots, y_n, y)$.

By Lausch, Nöbauer [12] (see also Kuich [10], Section 3) each semiring-polynomial in the polynomial semiring $A(\{y_1, \ldots, y_n\})$ over the semiring A has a representation as a finite sum of product terms, where a product term has the form $t(y_1, \ldots, y_n) = a_0 y_{i_1} a_1 \cdots a_{k-1} y_{i_k} a_k$, $a_j \in A$, $1 \leq i_1, \ldots, i_k \leq n$, i. e., a semiring-polynomial p has a representation

$$p(y_1, \ldots, y_n) = \sum_{1 \leq j \leq m} t_j(y_1, \ldots, y_n),$$

where t_j is a product term, $1 \leq j \leq m$.

A product term $t(y_1, \ldots, y_n)$ as above induces the mapping $t : A^n \to A$ defined by $t(s_1, \ldots, s_n) = a_0 s_{i_1} a_1 \cdots a_{k-1} s_{i_k} a_k$ for all $s_1, \ldots, s_n \in A$. A semiring-polynomial $p(y_1, \ldots, y_n)$ represented as above induces the mapping $p : A^n \to A$ defined by $p(s_1, \ldots, s_n) = \sum_{1 \le j \le m} t_j(s_1, \ldots, s_n)$ for all $s_1, \ldots, s_n \in A$.

Lemma 14 *Let $p \in A\langle\{y_1, \ldots, y_n\}\rangle$. Then there exists $\bar{p} \in \mathbb{N}\langle(A \cup \{y_1, \ldots, y_n\})^*\rangle$ such that $p(\varphi(\bar{s}_1), \ldots, \varphi(\bar{s}_n)) = \varphi(\bar{p}(\bar{s}_1, \ldots, \bar{s}_n))$ for all $\bar{s}_1, \ldots, \bar{s}_n \in \mathbb{N}^\infty \langle\!\langle A^* \rangle\!\rangle$.*

Proof. If $t(y_1, \ldots, y_n) = a_0 y_{i_1} a_1 \cdots a_{k-1} y_{i_k} a_k$, $a_j \in A$, is a product term, define $\bar{t}(y_1, \ldots, y_n) = \bar{a}_0 y_{i_1} \bar{a}_1 \cdots \bar{a}_{k-1} y_{i_k} \bar{a}_k$. If the semiring-polynomial $p(y_1, \ldots, y_n)$ has a representation $p(y_1, \ldots, y_n) = \sum_{1 \le j \le m} t_j(y_1, \ldots, y_n)$, where $t_j(y_1, \ldots, y_n)$, $1 \le j \le m$, is a product term, define $\bar{p}(y_1, \ldots, y_n) = \sum_{1 \le j \le m} \bar{t}_j(y_1, \ldots, y_n)$. But observe that the definition of $\bar{p}(y_1, \ldots, y_n)$ depends on the representation choosen for $p(y_1, \ldots, y_n)$, while the mapping induced by $p(y_1, \ldots, y_n)$ does not depend on the representation choosen for $p(y_1, \ldots, y_n)$. Let now $\bar{s}_1, \ldots, \bar{s}_n \in \mathbb{N}^\infty \langle\!\langle A^* \rangle\!\rangle$. Then $t(\varphi(\bar{s}_1), \ldots, \varphi(\bar{s}_n)) = a_0 \varphi(\bar{s}_{i_1}) a_1 \cdots a_{k-1} \varphi(\bar{s}_{i_k}) a_k$, $\bar{t}(\bar{s}_1, \ldots, \bar{s}_n) = \bar{a}_0 \bar{s}_{i_1} \bar{a}_1 \cdots \bar{a}_{k-1} \bar{s}_{i_k} \bar{a}_k$ and $\varphi(\bar{t}(\bar{s}_1, \ldots, \bar{s}_n)) = a_0 \varphi(\bar{s}_{i_1}) a_1 \cdots a_{k-1} \varphi(\bar{s}_{i_k}) a_k$. Hence, $t(\varphi(\bar{s}_1), \ldots, \varphi(\bar{s}_n)) = \varphi(\bar{t}(\bar{s}_1, \ldots, \bar{s}_n))$. Moreover,

$$p(\varphi(\bar{s}_1), \ldots, \varphi(\bar{s}_n)) = \sum_{1 \le j \le m} t_j(\varphi(\bar{s}_1), \ldots, \varphi(\bar{s}_n)),$$
$$\bar{p}(\bar{s}_1, \ldots, \bar{s}_n) = \sum_{1 \le j \le m} \bar{t}_j(\bar{s}_1, \ldots, \bar{s}_n) \text{ and}$$
$$\varphi(\bar{p}(\bar{s}_1, \ldots, \bar{s}_n)) = \sum_{1 \le j \le m} \varphi(\bar{t}_j(\bar{s}_1, \ldots, \bar{s}_n)).$$

Hence, $p(\varphi(\bar{s}_1), \ldots, \varphi(\bar{s}_n)) = \varphi(\bar{p}(\bar{s}_1, \ldots, \bar{s}_n))$. □

We are now ready for the characterization of $\mathfrak{Alg}(A')$, $A' \subseteq A$.

Theorem 15 *Let $A' \subseteq A$. Then $\mathfrak{Alg}(A') = \{\varphi(\bar{r}) \mid \bar{r} \in \mathbb{N}^{\infty\,\mathrm{alg}}\{\!\{A^*\}\!\}\}$.*

Proof. Consider an A'-algebraic system

$$y_i = p_i(y_1, \ldots, y_n), \quad p_i \in A'\langle\{y_1, \ldots, y_n\}\rangle, \quad 1 \le i \le n,$$

with least solution $\sigma \in \mathfrak{Alg}(A')^n$. Construct according to Lemma 14 the algebraic system

$$y_i = \bar{p}_i(y_1, \ldots, y_n), \quad \bar{p}_i \in \mathbb{N}^\infty \langle(A' \cup \{y_1, \ldots, y_n\})^*\rangle, \quad 1 \le i \le n,$$

where, for all $\bar{s}_1, \ldots, \bar{s}_n \in \mathbb{N}^\infty \langle\!\langle A^* \rangle\!\rangle$, $p(\varphi(\bar{s}_1), \ldots, \varphi(\bar{s}_n)) = \varphi(\bar{p}(\bar{s}_1, \ldots, \bar{s}_n))$. Let $\bar{\tau} \in (\mathbb{N}^{\infty\,\mathrm{alg}}\{\!\{A'^*\}\!\})^n$ be the least solution of the algebraic system $y_i = \bar{p}_i$, $1 \le i \le n$. Let $(\sigma^j)_{j \ge 0}$ and $(\bar{\tau}^j)_{j \ge 0}$ be the approximation sequences of the systems $y_i = p_i$ and $y_i = \bar{p}_i$, $1 \le i \le n$, respectively. We claim that $\sigma^j = \varphi(\bar{\tau}^j)$, $j \ge 0$, and show it by induction on j. We obtain $\sigma^0 = 0 = \varphi(\bar{0}) = \varphi(\tau^0)$ and, for $j \ge 0$, $\sigma^{j+1} = p(\sigma^j) = p(\varphi(\bar{\tau}^j)) = \varphi(\bar{p}(\bar{\tau}^j)) = \varphi(\bar{\tau}^{j+1})$. Here the second equality follows by the induction hypothesis and the third equality by Lemma 14. Since φ is an ω-continuous mapping, we infer that $\sigma = \varphi(\bar{\tau})$.

Hence, for each $a \in \mathfrak{Alg}(A')$ there exists an $\bar{r} \in \mathbb{N}^{\infty\,\mathrm{alg}} \langle\!\langle A'^* \rangle\!\rangle$ and vice versa such that $a = \varphi(\bar{r})$. □

We now restate our definitions of Section 2 to achieve a characterization of $\mathfrak{Alg}(A')$ by algebraic expressions.

Let $A' \subseteq A$ and let U be defined as in Section 2. A word E over $A' \cup Y \cup U$ is an *algebraic expression* over $\mathbb{N}^\infty\{(A' \cup Y)^*\}$ iff

(i) E is a symbol of $\mathbb{N}^\infty$, or
(ii) E is a symbol of $A' \cup Y$, or else
(iii) E is of one of the forms $[E_1 + E_2]$, $[E_1 \cdot E_2]$ or $\mu y.E_1$, where E_1 and E_2 are algebraic expressions and y is a symbol of Y.

Each algebraic expression E over $A\{\Sigma_\infty^*\}$ denotes a formal power series $|E|$ in $\mathbb{N}^\infty\{\{(A' \cup Y)^*\}\}$ according to the following conventions:

(i) The power series denoted by $n \in \mathbb{N}^\infty$ is $n\varepsilon$ in $\mathbb{N}^\infty\{(A' \cup Y)^*\}$.
(ii) The power series denoted by $a \in A'$ or $y \in Y$ is $\bar{a} \in \mathbb{N}^\infty\{(A' \cup Y)^*\}$ or $y \in \mathbb{N}^\infty\{(A' \cup Y)^*\}$, respectively.
(iii) For algebraic expressions E_1, E_2 over $\mathbb{N}^\infty\{(A' \cup Y)^*\}$ and $y \in Y$ we set $|[E_1 + E_2]| = |E_1| + |E_2|$, $|[E_1 \cdot E_2]| = |E_1| \cdot |E_2|$, $|\mu y.E_1| = \mu y.|E_1|$.

Let ϕ be the mapping from the set of algebraic expressions over $\mathbb{N}^\infty\{(A' \cup Y)^*\}$ into the finite sets of $\mathfrak{P}(Y)$ defined by:

(i) $\phi(n) = \emptyset$ for each $n \in \mathbb{N}^\infty$.
(ii) $\phi(a) = \emptyset$ for each $a \in A'$; $\phi(y) = \{y\}$ for each $y \in Y$.
(iii) $\phi([E_1 + E_2]) = \phi([E_1 \cdot E_2]) = \phi(E_1) \cup \phi(E_2)$, $\phi(\mu y.E_1) = \phi(E_1) - \{y\}$ for algebraic expressions E_1, E_2 and $y \in \Sigma_\infty$.

Theorem 15 and the above definitions yield the following corollary.

Corollary 16 *Let $A' \subseteq A$. An element $a \in A$ is in $\mathfrak{Alg}(A')$ iff there exists an algebraic expression E over $\mathbb{N}^\infty\{(A' \cup Y)^*\}$, where $\phi(E) = \emptyset$, such that $a = \varphi(|E|)$.*

References

1. Autebert, J.-M., Berstel, J., Boasson, L.: Context-free languages and pushdown automata. In: Handbook of Formal Languages (Eds.: G. Rozenberg and A. Salomaa), Springer, 1997, Vol. 1, Chapter 3, 111–174.
2. Bekic, H.: Definable operations in general algebras, and the theory of automata and flowcharts. Tech. Report, IBM Labor, Wien, 1967.
3. Bozapalidis, S.: Equational elements in additive algebras. Technical Report, Aristotle University of Thessaloniki, 1997.
4. Esik, Z.: Completeness of Park induction. Theor. Comput. Sci. 177(1997) 217–283.
5. Gecseg, F., Steinby, M.: Tree Languages. In: Handbook of Formal Languages (Eds.: G. Rozenberg and A. Salomaa), Springer, 1997, Vol. 3, Chapter 1, 1–68.
6. Goldstern, M.: Vervollständigung von Halbringen. Diplomarbeit, Technische Universität Wien, 1985.

7. Gruska, J.: A characterization of context-free languages. Journal of Computer and System Sciences 5(1971) 353–364.
8. Karner, G.: On limits in complete semirings. Semigroup Forum 45(1992) 148–165.
9. Kuich, W.: The Kleene and the Parikh theorem in complete semirings. ICALP87, Lect. Notes Comput. Sci. 267(1987) 212–225.
10. Kuich, W.: Semirings and formal power series: Their relevance to formal languages and automata theory. In: Handbook of Formal Languages (Eds.: G. Rozenberg and A. Salomaa), Springer, 1997, Vol. 1, Chapter 9, 609–677.
11. Kuich, W., Salomaa, A.: Semirings, Automata, Languages. EATCS Monographs on Theoretical Computer Science, Vol. 5. Springer, 1986.
12. Lausch, H., Nöbauer, W.: Algebra of Polynomials. North-Holland, 1973.
13. Sakarovitch, J.: Kleene's theorem revisited. Lect. Notes Comput. Sci. 281(1987) 39–50.
14. Salomaa, A.: Formal Languages. Academic Press, 1973.
15. Thatcher, J. W., Wright, J. B.: Generalized finite automata theory with an application to a decision problem of second-order logic. Math. Systems Theory 2(1968) 57–81.
16. Wechler, W.: Universal Algebra for Computer Scientists. EATCS Monographs on Computer Science, Vol. 25. Springer, 1992.

D0L-Systems and Surface Automorphisms

Luis-Miguel Lopez[1] and Philippe Narbel[2]

[1] IGM, Univ. Marne-La-Vallée. 2, Butte Verte, 93166 Noisy-le-Grand, France
   `lopez@univ-mlv.fr`
[2] LABRI, Univ. Bordeaux I. 351, Cours de la Libération 33405 Talence, France
   `narbel@labri.u-bordeaux.fr`

Abstract. We introduce a new relationship between formal language theory and surface theory. More specifically, we show how substitutions on words can represent automorphisms of surfaces. This correspondance is applied to construct and analyze non-periodic irreducible automorphisms. We use results about D0L-systems, mainly the decidability of the non-repetitiveness of a D0L-language.

1 Introduction

This paper shows how results in formal language theory, specifically iterated substitutions, can be used to solve a problem in surface theory. Its first main point is to introduce how substitutions on words can represent surfaces automorphisms, i.e. bijective bicontinuous maps of a surface onto itself. Iterating an automorphism conjugates then to iterating a substitution [10], and automorphisms can be studied in the scope of substitutions and D0L-systems theory [20]. This approach is applied here to the classification of the automorphisms of compact oriented surfaces, which are known for a long time to be either periodic, or reducible, or non-periodic irreducible [16]. Since, this classification into three families has been made more precise [7, 13, 8, 4, 22]. In particular non-periodic irreducible automorphisms have the striking property to setwise fix two closed sets of infinite pairwise disjoint simple curves. They are difficult to obtain, and much effort has been dedicated to finding systematic constructions [1, 18, 22, 4, 17, 11, 21]; in [22] Thurston indicated one and Penner generalized it in [18]. Nevertheless the proofs rely on intricate geometric manipulations of graphs, and do not lead to an effective description of the stable set of curves. By using D0L-systems results [5, 6, 19, 14, 15], we obtain a simpler and constructive proof.

Thurston's method uses two mutually transverse sets of simple closed curves $C \cup D$ on a surface Σ, to which is associated a basis of simple automorphisms, the so-called Dehn twists. Penner's generalization states that any positive composition involving each of these twists at least once is non-periodic and irreducible. This is the result we prove here. The first step we take is to make such a system of curves $C \cup D$ an oriented labelled graph Γ. This graph can be proved invariant under an associated set of twists, i.e. the twists can be interpreted as graph maps [2]. Next, curves on the surface Σ are associated by homotopy to admissible paths of Γ, and can be coded into words by concatenating the visited edges labels. Applying a twist is then the same as applying the corresponding graph map which is the same as applying the corresponding substitution to the coding words. Iteration of a composition of twists h corresponds to iteration of the corresponding substitution θ. Hence, the fixed point set obtained by infinite iterations of h can be studied through the fixed point set of θ, i.e. its boundary language [3, 15]. Now, deciding irreducibility of h is the same as deciding whether the boundary of θ does not contain periodic words. This can be solved by deciding non-repetitiveness for the D0L-language corresponding to θ [6] (see also [9]). Since the boundary of θ is constructive [15], we also obtain a constructive representation of the sets of curves fixed by the constructed non-periodic irreducible automorphisms.

The main technical steps of the proof are the following: First, we must obtain substitutions, i.e. endomorphisms over *free* monoids. Hence, relations among the coding words are dealt with by a recoding, such that the corresponding monoid becomes free. Second, the properties required to apply [6] are proved for the whole set of the considered substitutions, i.e. primitivity, elementarity [5], strongly closedness, and non-cyclicity [6]. All the missing proofs will appear in a full version of this article.

2 Automorphisms as Substitutions

An **admissible path** in an oriented graph Γ is an oriented edge-path, i.e. an oriented indexing map from an interval of $\mathbb{Z}$ towards the set of edges of Γ. According to if the interval is finite, half-infinite or all of $\mathbb{Z}$, the path is respectively said to be **finite, one-way infinite** or **two-way infinite**. A path is said to be **closed** if it can be defined by a two-way infinite periodic indexing map. Let Γ be embedded in a surface Σ. A **simple** curve on Σ, i.e. a non-crossing curve, which is **homotopic**, i.e. which can be continuously deformed, to an admissible path in Γ is said to be **carried** by Γ (for a finite curve, this deformation must take place with endpoints fixed). Such a carried curve is called a **leaf**. In Figure 22(ii) is pictured a deformation of a leaf (normal style) to an admissible path of a graph (dashed style). A **lamination** carried by Γ on Σ is a maximal set of pairwise non-homotopic, pairwise disjoint two-way infinite carried leaves. Assigning a distinct label to every edge of Γ gives the **edge alphabet** A of Γ. The **coding** of a path is the word obtained by concatenating its edges' labels according to its indexing map. Hence, to each leaf carried by Γ corresponds a word. If this word is unique, we say that Γ gives a **well-defined coding over the edge alphabet**. If Γ contains homotopic distinct edge-paths this is no more true. So more generally, a coding is said to be **well-defined** if 1) there is a set of finite admissible paths in Γ, called the **coding paths**, 2) a **path alphabet** B with one label for each coding path, and 3) a **coding rule** such that every leaf carried by Γ has a unique coding over B. In the sequel, to save notation, the same symbol denotes a coding path and its corresponding label.

A **homeomorphism** is a bijective bicontinuous map $X \to Y$ where X, Y are two topological spaces. When $X = Y$, it is called an **automorphism**. An automorphism h of a surface is called **periodic** when there is $n > 0$ such that h^n is homotopic to the identity. It is called **reducible** when it fixes a finite set of pairwise disjoint simple closed curves. Non-periodic irreducible automorphisms have been proved to be **pseudo-Anosov**, i.e. they setwise fix two laminations $\mathcal{L}_1$ and $\mathcal{L}_2$ (see [7, 13, 8, 4, 22]). Putting a transverse measure on each of them, such an automorphism is respectively expanding on $\mathcal{L}_1$ and shrinking on $\mathcal{L}_2$ by factors inverse of each other. A **Dehn twist** is a basic automorphism non homotopic to the identity: Consider a simple closed curve c on an oriented surface, such that c is homotopically non-zero, i.e. cannot be continuously deformed into a single point; next, cut off the surface along c, and apply one (or more) turn(s) to one of the separated parts; finally paste back along c. Figure 21 shows a cylinder-like piece of a surface where one can observe the local effect of a twist along c (bold style) on a transverse leaf (normal style) intersecting at p. Applying the twist drags the leaf along c from p (left part of the figure).

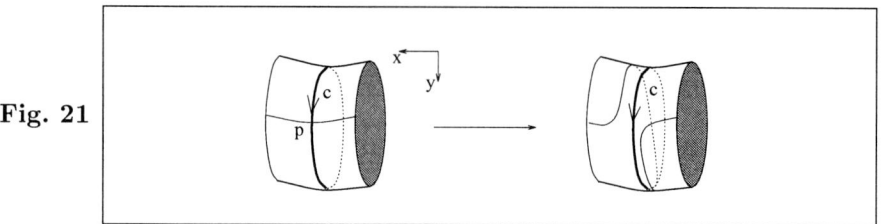

Fig. 21

More formally, this map applies to a parametrized annulus $\{(r, e^{i\alpha}) | r \in [r_0, r_1], 0 < r_0 < r_1, \alpha \in [0, 2\pi)\}$ and is homotopic to $f_n : (r, e^{i\alpha}) \to (r, e^{2i\pi n \frac{r - r_0}{r_1 - r_0}} e^{i\alpha})$ for some $n \in \mathbb{Z}$. According to n's sign, the twist is said **positive** or **negative**. The annulus can be described as a neighborhood of the above homotopically non-zero simple closed curve c. In this case we say that a Dehn twist has been defined **along** c. This is well-defined since f_n's restriction to the boundary is the identity, and different annuli neighbourhoods yield homotopic maps. Thus when performing a Dehn twist on a surface along a curve, we always suppose that its neighbourhood has been fixed.

An **oriented graph map** between two oriented graphs Γ to Γ' is a map which sends vertices of Γ to vertices of Γ', and oriented edges of Γ to oriented paths of Γ'. If Γ and Γ' are embedded oriented graphs in Σ, an automorphism h of Σ sends Γ to Γ' if h induces an oriented graph map up to homotopy rel. endpoints. If $\Gamma' = \Gamma$, we say that Γ is **invariant** under h. Note that if Γ is an invariant graph, then every leaf carried by Γ has its image by h still carried by Γ. As an instance, Figure 22(i) shows a cylinder-like piece of surface where a twist along c intersects at p a single leaf (normal style) carried by Γ (dashed style). Applying the twist results in Figure 22(ii) showing that the leaf is still carried by Γ.

Fig. 22

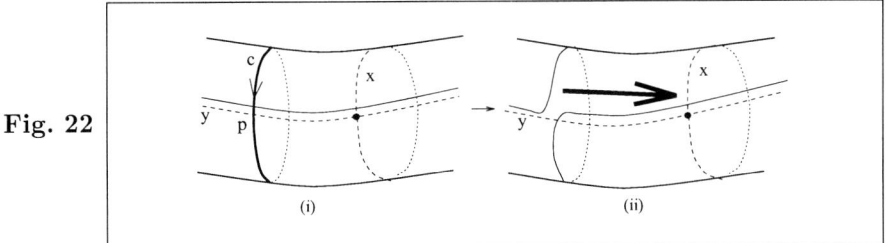

Given an alphabet A, a **substitution** θ over A is a transformation which sends any letter a of A to some word over A, and which is extended to any word $w = ...w_i w_{i+1}...$ with $w_i \in A$ by sending it to $...\theta(w_i)\theta(w_{i+1})....$ If finite words are considered, a substitution is an endomorphism on the free monoid A^* and if some word to apply the substitution is specified, one speaks of a **D0L-system**. In what follows we denote by cod the coding map from leaves to words.

Lemma 21 *Suppose there is a well-defined coding on Γ. Let Γ be invariant under h. Then h induces a unique substitution θ_h.*

Proof. Let g_h be the graph map induced by h; take the coding paths, calculate their images by g_h and rewrite the result into the path alphabet: this is a substitution associated to h. Well-definedness of cod implies uniqueness. ◇

For instance, assuming that the edge alphabet is on use in Figure 22 where the twist along c intersects the edge labelled y, then the corresponding substitution is the one which sends y to xy leaving all the other letters fixed. We say that h **preserves the coding** iff $cod \circ h = \theta_h \circ cod$, i.e. application of θ_h on a leaf coding gives the coding of the image leaf by h. Note that since the graph map induced by h preserves concatenation of oriented edges, if the current alphabet is the edge one, preservation of the coding trivially holds. If coding preservation holds only for a subset $\mathcal{L}$ of all the carried leaves we say that h **preserves the coding on $\mathcal{L}$**. This latter condition is automatically realized in the following case: let $\mathcal{L}$ be a set of leaves on which the coding is injective up to homotopy, and assume that up to homotopy too $h(\mathcal{L}) \subset \mathcal{L}$, then h preserves the coding on $\mathcal{L}$.

Lemma 22 *Suppose there is a well-defined coding on Γ. Let H be the semi-group of automorphisms leaving Γ invariant and preserving the coding. Then H is homomorphic to a semi-group of substitutions.*

Proof. The sets of maps in the statement are clearly semi-groups. So we have $cod \circ (h \circ h') = \theta_{h \circ h'} \circ cod$. On the other hand $cod \circ h \circ h' = \theta_h \circ cod \circ h' = \theta_h \circ \theta_{h'} \circ cod$ because h and h' preserve codings. Hence, $\theta_{h \circ h'} = \theta_h \circ \theta_{h'}$. ◇

If the coding is preserved only on a subset $\mathcal{L}$ of all the leaves, the above equalities hold on $\mathcal{L}$. So the lemma remains true if we define H as the semi-group of automorphisms leaving Γ invariant and $\mathcal{L}$ setwise fixed up to homotopy.

3 Non-periodic Irreducible Automorphisms Construction

Consider a closed oriented surface Σ with negative Euler characteristic (i.e. there is a triangulation of Σ which graph has a negative Euler characteristic). Let C

and D be two sets of pairwise non-parallel and disjoint oriented simple closed curves on Σ and assume that each component of the complement of $C \cup D$ in Σ is a disk (i.e. $C \cup D$ **fills** Σ) without being a bigon (i.e. C **hits** D **efficiently**), and that the orientation given by C and D at any point of $C \cap D$ agrees with that of Σ. In Figure 31 we show such a system of curves $C \cup D$ (dashed style) on a two-holed torus. The set $\Gamma = C \cup D$ as a set of points can be considered as an oriented graph: its vertex set being $C \cap D$ and its oriented edges being the segments of $C \cup D$ between two vertices. Now, consider another copy of $\Gamma = C \cup D$, denoted by **T**, in general position with respect to Γ slightly under Γ ("under" with respect to Γ's orientation). In Figure 32, we show **T** (bold style) relatively to the $C \cup D$ of Figure 31.

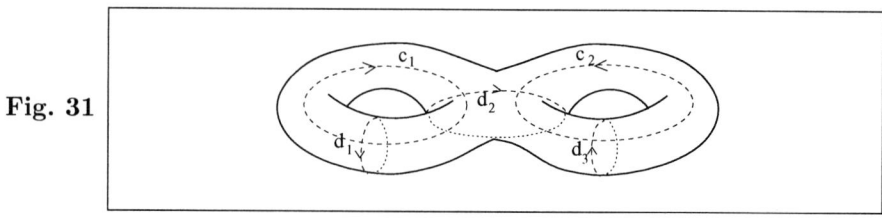

Fig. 31

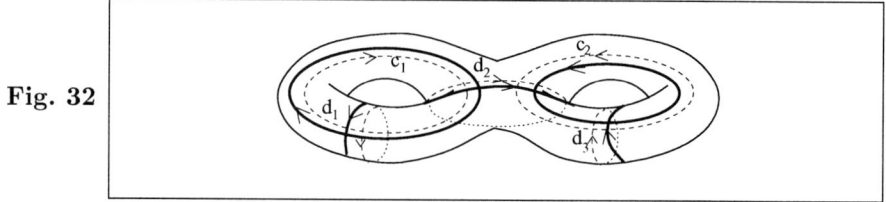

Fig. 32

We denote by H_+ the semi-group of automorphisms generated by compositions of Dehn twists along **T**, where each curve is involved at least once. The sequel is dedicated to prove the announced following result due to Thurston [22], and to Penner [18] in its strengthened version:

Theorem 31 *Consider a system of curves $C \cup D$ as described above. Then each automorphism in H_+ is non-periodic and irreducible (i.e. pseudo-Anosov).*

First, according to 21 and 22, the next result indicates that we are close to interpret the automorphisms of H_+ as substitutions:

Lemma 32 *Γ is invariant under every $h \in H_+$.*

However, we also need a well-defined coding. The coding over the edge alphabet induces a well-defined coding if carried leaves with different associated admissible paths are not pairwise homotopic. Nevertheless, the above construction of the graph Γ may lead to such cases. We show indeed in Figure 33 a case for a three-holed torus where a subset of a system $C \cup D$ (dashed style) generates a square (gray region) in its complement: clearly, carried leaves may be homotopic, though having different associated admissible paths.

Fig. 33

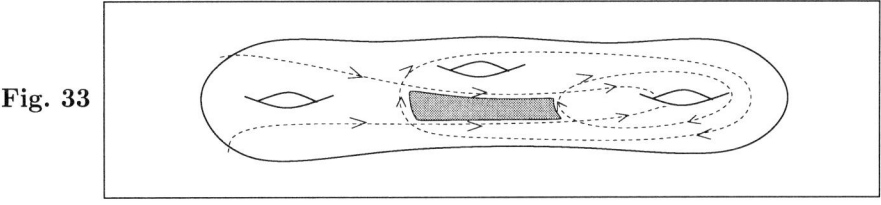

The next lemma shows that one must deal only with such squares:

Lemma 33 *If there is no square among the polygon components of $\Sigma \setminus \Gamma$ then Γ gives a well-defined coding over the edge alphabet.*

Hence, in view of Lemmas 21 to 33, if there is no square in $\Sigma \setminus \Gamma$, the semi-group of substitutions generated by the twists along **T** is the semi-group of substitutions associated to H_+, henceforth denoted by Θ_{H_+}. As a full example let us revisit the two-holed torus and its system of curves $C \cup D = \{c_1, c_2, d_1, d_2, d_3\}$ given in Figure 31. In Figure 34, the same surface is shown, but with the edges of Γ (dashed style) labelled by an alphabet of 8 letters $A = \{x_1, y_1, y_2, s_1, s_2, t_1, z_1, z_2\}$. Considering the set of twists curves **T** given in Figure 32, the associated substitutions over A are the following (θ_γ denotes the substitution associated to the twist along γ, and we indicate only the labels with non-trivial images):

$\theta_{c_1}(x_1) = y_1 y_2 x_1$, $\theta_{c_2}(z_2) = s_2 s_1 z_2$, $\theta_{d_2}(y_2) = z_1 z_2 y_2$, $\theta_{d_1}(y_1) = x_1 y_1$, $\theta_{d_3}(s_1) = t_1 s_1$.
$\theta_{c_1}(z_1) = y_2 y_1 z_1$, $\theta_{c_2}(t_1) = s_1 s_2 t_1$, $\theta_{d_2}(s_2) = z_2 z_1 s_2$.

Fig. 34

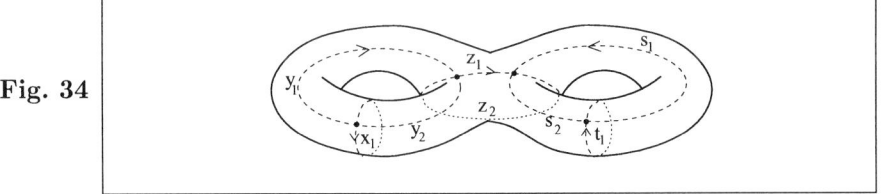

Let us deal now with the case where squares occur in $\Sigma \setminus \Gamma$. Two squares in $\Sigma \setminus \Gamma$ are said to be **in relation** iff they have a side in common. This is extended by transitivity to an equivalence relation, and a **square-region** is the subset of Σ given by the union of all the squares of a class. Note that no twist curve can lie in the interior of a square-region because there are no parallel curves in **T**. So each twist curve intersecting a square-region intersects its frontier. Accordingly, we locate a set of **entering points** and **exiting points** on the edges of the frontiers. We push these points towards the end vertices of the edges they lie on, and we obtain two sets of vertices En and Ex. For each square-region $\mathcal{R}$, let $\mathcal{P}_\mathcal{R}$ be a set of representatives of the homotopy classes of admissible paths from En to Ex. For instance, Figure 35 shows a square-region $\mathcal{R}$ made of 6 squares (gray region) (dashed style is used for Γ and bold style for the twists curves **T**). For this square-region, $En = \{a, d, h, i, j\}$ and $Ex = \{b, c, e, f, g\}$. Denoting by (x, y) an homotopy class in $\mathcal{P}_\mathcal{R}$ where $x \in En$, $y \in Ex$, then (i, c), (a, f) are such instances. We take the union of the $\mathcal{P}_\mathcal{R}$'s over all the square-regions, and add to it the subset of the edge alphabet formed by the edges outside or on the frontier of the square-regions provided they do not already belong to some $\mathcal{P}_\mathcal{R}$. We denote by A_H the alphabet obtained by assigning a distinct label to each of these paths and edges. Note however that a square-region does not need to be **simply connected**, i.e. not all its edge-paths with common endpoints are necessarily pairwise homotopic. In this case, carried leaves may stay forever in a square-region. We say that an admissible path in Γ is in **normal position** if it does not run the boundary of any square along the left edge (necessarily upwards), then along the top one (necessarily rightwards). It is obvious that any admissible path can be put in normal position in a unique way.

Fig. 35

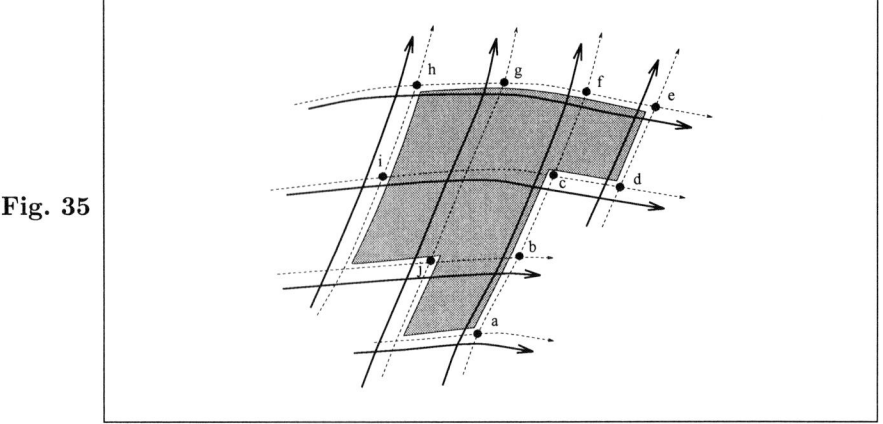

Now, a coding rule on Γ can be defined assuming that admissible paths are in normal position. Once a path is in this position we push it slightly under Γ (for example on **T**). If the path does not meet the square-regions then either it is contained in one of them and then is coded by the empty word, or it lies outside and then is coded over the edge alphabet. If the path meets square-regions it is followed from some vertex towards the positive direction, and points in En and Ex are successively marked as the path crosses frontier edges. The same thing is done backwards from the chosen vertex. We pull the path marked in this way back to Γ. Now: 1) a subpath from a vertex of En to one of Ex is coded by its label in A_H; 2) a subpath from one in Ex to one in En is coded in the edge-alphabet; 3) a subpath ending at a vertex in Ex or beginning at a vertex in En is coded by the empty word; 4) a subpath ending at En or beginning at Ex is coded in the edge alphabet. Uniqueness of the normal position and of the coding paths labelling imply well-definedness of the coding on Γ over A_H. Note that this coding may be not injective. There may even exist infinite leaves with the empty word as coding word (when for instance square-regions are not simply connected). However, the following lemma gives injectivity for a subset of leaves on which we shall later restrict the discussion. We denote by $\mathcal{I}$ the set of all the two-way infinite leaves carried by Γ:

Lemma 34 *The above coding is injective on the subset of $\mathcal{I}$ meeting the square-regions along finite segments.*

Two problems remain to be solved in order to be able to apply Lemmas 21 and 22: preservation of the coding and the possible infinite cardinality of A_H. A substitution θ over an alphabet A is said to be: - **strongly growing** if for every $s \in A$, $length(\theta^n(s))$ with $n \in \mathbb{N}$ becomes arbitrarily large; - **primitive** if there exists a finite power K such that for all $s \in A$, $alph(\theta^K(s)) = A$, where $alph(w)$ is the set of labels that occur in w; - **strongly closed** if for all $s \in A$, $alph(\theta(s)) = A$; - **cyclic** if there exists a cyclic permutation π of the alphabet A, such that for each $s \in A$, $\theta(s) = s_1...s_m$, then $s_{i+1} = \pi(s_i)$, with $1 \leq i \leq m-1$; and for each pair $s, t \in A$ such that $\pi(s) = t$, then $\pi(last(\theta(s))) = first(\theta(t))$, where $first(w)$ extracts the first label of w, and $last(w)$ its last one; - **simplifiable** if there exists an alphabet B, such that $|B| < |A|$, where $|.|$ denotes the cardinality, and substitutions $f : A^* \to B^*$ and $g : B^* \to A^*$ such that $\theta = gf$. This amounts to say that there is a set of words $u_1, \ldots, u_k$ with $k < |A|$ such that θ's image words can be rewritten in terms of the u_i's [5]; - **elementary** if it cannot be simplified; - **repetitive** if for each $k > 1$ there is a non-empty word u such that $u^k = uuu...u$ is a subword of a word in $L_\theta = \{\theta^n(s), s \in A, n \in \mathbb{N}\}$; - **strongly repetitive** if there is a non-empty word u such that for each $k > 1$, $u^k = uuu...u$ is a subword of a word in L_θ. In the sequel, θ_h denotes the substitution associated to the automorphism h.

Lemma 35 *Suppose there is a well-defined coding on Γ over the edge alphabet. Then every substitution $\theta_h \in \Theta_{H_+}$ is primitive.*

Lemma 36 *Suppose $\Sigma \setminus \Gamma$ contains square-regions and let $h \in H_+$. Then there is a finite path-alphabet $A_h \subset A_H$ and some $k > 0$ such that, for all $n \geq k$ and every admissible path γ, the coding word $cod(h^n(\gamma))$ only uses labels in A_h.*

Lemma 37 *Let h, Γ and A_h be as given in the above lemma. Then h preserves the coding over A_h on $h^{|\mathbf{T}|}(\mathcal{I})$.*

Proof. This follows from the above lemma and from Lemma 34. ◇

Hence similarly to the case with no square-regions, and according to Lemmas 21 and 22, the semi-group of substitutions over A_h generated by the twists along $\mathbf{T}$, also denoted Θ_{H_+}, is homomorphic to H_+.

Lemma 38 *Let θ be a primitive substitution on a finite alphabet A. Then for every $n > 0$, θ is primitive on all the subwords of length n of the words in L_θ.*

Lemma 39 *Let h, Γ and A_h be as given in the above lemma. Then θ_h is primitive over an alphabet $A'_h \subseteq A_h$.*

Lemma 310 *Every $\theta_h \in \Theta_{H_+}$ over A'_h can effectively be made elementary.*

Note that by Lemma 38, a simplified primitive substitution can be seen to be still primitive. Moreover Lemma 37 is still valid for the alphabet after simplification. In the sequel, every θ_h is therefore considered elementary.

Corollary 311 *Every substitution $\theta_h \in \Theta_{H_+}$ is strongly growing.*

Lemma 312 *Suppose there is a well-defined coding on Γ. Then every substitution in $\theta_h \in \Theta_{H_+}$ is non-cyclic.*

Lemma 313 *Suppose there is a well-defined coding on Γ. Then for every substitution $\theta_h \in \Theta_{H_+}$, there is $m \geq 1$ such that θ_h^m is not strongly repetitive.*

Proof. According to Lemmas 35 and 39, θ_h is primitive over A_h. Hence the sequence of $alph(\theta_h(s)), alph(\theta_h^2(s)), \ldots$ with $s \in A_h$, is constant and equal to A_h from some power $m \geq 1$. Hence, θ_h^m is strongly closed. Since $\theta_h^m \in \Theta_{H_+}$, it is also non-cyclic and elementary. According to the main theorem in ([6], Section 6), any such substitution is non repetitive and therefore non strongly repetitive. $\Diamond$

The **pointed language** $\widehat{L_\theta}$ associated to $L_\theta = \{\theta^n(s), s \in A, n \in \mathbb{N}\}$ is obtained by giving every possible origin to each of its words and indexing its letters accordingly. Finite words are assumed padded to both infinities with some dummy symbol δ, so that $\widehat{L_\theta}$ is a subset of $(A \cup \{\delta\})^{\mathbb{Z}}$, and inherit the topology of this set. The boundary of $\widehat{L_\theta}$, called in short **boundary of** θ, is a set of one-way and two-way infinite words if θ is strongly growing (see [15]). Considering only the two-way infinite words, we identify together all the words that are equal up to a translation of indexes in $\mathbb{Z}$, i.e. up to applying a shift. The resulting set of words is denoted by $\mathbf{L}_\theta$.

Now, recall that a lamination on Σ is a maximal set of pairwise non-homotopic, pairwise disjoint two-way infinite carried leaves:

Lemma 314 *There is a lamination Λ_h associated to $\mathbf{L}_{\theta_h}$.*

We are now in position to prove Theorem 31: According to Lemma 37, h preserves the coding and $\mathbf{L}_{\theta_h}$ is by construction setwise fixed by θ_h. Thus Λ_h is setwise fixed by h up to homotopy. Moreover, since for every $m \geq 1$, the same set of subwords appears in L_θ and L_{θ^m}, the boundary of θ is equal to the boundary of θ^m. So Λ_h is homotopic to Λ_{h^m}. Now Λ_h contains no compact leaf. Indeed, closed ones correspond to periodic words in $\mathbf{L}_{\theta_h}$ and according to Lemma 313 there are none: if there is a periodic word ...$uuuu$... in $\mathbf{L}_{\theta_h}$, for every $n > 1$, u^n must appear in some word of L_{θ_h}, i.e. θ_h should be strongly repetitive. Hence every leaf is dense in Λ_h. So Γ can be naturally measured thanks to the ergodic theorem [12]: for a generic two-way infinite leaf the limit of the average number of occurences of any letter in any subword converges as the length of the subword goes to infinity, and the measure at each edge is the limit of its corresponding label. The effect of h on Λ_h is expanding on the measure since θ_h

is strongly growing. Now consider another graph Γ' obtained like Γ, but with reversed orientations of the curves in $C \cup D$. Similarily, Γ' is invariant under h^{-1}. We denote by B its edge alphabet, by $B_{h^{-1}}$ its path alphabet, by $\mathbf{L}_{\theta^{-1}}$ and $\Lambda_{h^{-1}}$ the language and the fixed lamination corresponding to h^{-1}. However, no more substitution is associated to h acting on $\Lambda_{h^{-1}}$ but instead a transformation which inverts the substitution $\theta_{h^{-1}}$ (see [19, 14, 15]) by slicing the words belonging to $\mathbf{L}_{\theta^{-1}}$ into subwords in $\{\theta_{h^{-1}}(s),\ s \in B_{h^{-1}}\}$, and by replacing these occurences by the labels s. Existence and uniqueness of such a slicing for two-way infinite words generated by a non strongly repetitive and injective substitution on the letters is proved in [14]. Now since elementarity implies injectivity (see [5]), $\theta_{h^{-1}}$ has all the required properties: its inverse $\theta_{h^{-1}}^{-1}$ is well-defined on $\mathbf{L}_{\theta^{-1}}$. Hence, by construction, $\Lambda_{h^{-1}}$ is also fixed by h, and h is shrinking on the transverse measure associated to $\mathbf{L}_{\theta^{-1}}$: subwords lengths are shrunk by a coefficient inverse of the expanding one. That Λ_h and $\Lambda_{h^{-1}}$ are transverse is immediate. $\diamondsuit$

References

[1] P. Arnoux and JC. Yoccoz. Construction de difféomorphismes pseudo-Anosov. *C.R. Acad. Sci., Paris, Ser. I*, 292:75–78, 1981.
[2] M. Bestvina and M. Handel. Train-tracks for surface homeomorphisms. *Topology*, 34(1):109–140, 1995.
[3] L. Boasson and N. Nivat. Adherence of languages. *J. Comp. Syst. Sc.*, 20:285–309, 1980.
[4] A.J. Casson and S. Bleiler. *Automorphisms of Surfaces after Nielsen and Thurston*. Number 9 in Student Text. London Mathematical Society, 1988.
[5] A. Ehrenfeucht and G. Rozenberg. Simplifications of homomorphisms. *Information and Control*, 38:298–309, 1978.
[6] A. Ehrenfeucht and G. Rozenberg. Repetitions of subwords in D0L languages. *Information and Control*, 59:13–35, 1983.
[7] A. Fathi, F. Laudenbach, and V. Poenaru, editors. *Travaux de Thurston sur les surfaces*. Soc. Math. de France, 1979. Astérisque, Volume 66-67.
[8] M. Handel and W. P. Thurston. New proofs of some results of Nielsen. *Adv. in Math.*, 56:173–191, 1985.
[9] Y. Kobayashi and Otto F. Repetitiveness of D0L languages is decidable in polynomial time. In *Proceedings of MFCS'97, Slovakia*, pages 337–346. Springer Verlag, 1997. Lecture Notes in Comp. Sci., 1295.
[10] L-M. Lopez and Ph. Narbel. Generalized sturmian languages. In Z. Fueloep and F. Gecseg, editors, *ICALP'95*, pages 336–347. Springer, 1995. Lecture Notes in Computer Science 944.
[11] J. E. Los. Pseudo-Anosov maps and invariant train tracks in the disc: a finite algorithm. *Proc. of London Math. Soc.*, 66(2):400–430, 1993.
[12] R. Mañé. *Ergodic Theory and Differentiable Dynamics*. Springer-Verlag, 1983.
[13] R.T. Miller. Geodesic laminations from Nielsen's viewpoint. *Adv. in Math.*, 45:189–212, 1982.
[14] B. Mossé. Puissances de mots et reconnaissabilité des points fixes d'une substitution. *Theoretical Computer Science*, 99:327–334, 1992.
[15] Ph. Narbel. The boundary of iterated morphisms on free semi-groups. *Intern. J. of Algebra and Computation*, 6(2):229–260, 1996.

[16] J. Nielsen. Untersuchungen zur topologie des geschlossenen zweiseitigen flächen. *Acta Math.*, 50:189–358, 1927.
[17] A. Papadopoulos and R.C. Penner. Enumerating pseudo-Anosov foliations. *Pacific Journ. of Math.*, 142(1):159–173, 1990.
[18] R.C. Penner. A construction of pseudo-Anosov homeomorphisms. *Transc. of the Amer. Math. Soc.*, 310(1):179–197, 1988.
[19] M. Quéffelec. *Substitution Dynamical Systems. Spectral Analysis.* Number 1294 in Lecture Notes in Mathematics. Springer-Verlag, 1987.
[20] G. Rozenberg and A. Salomaa. *The mathematical theory of L systems.* Academic press, 1980.
[21] I. Takarajima. On a construction of pseudo-Anosov diffeomorphism by sequences of train tracks. *Pacific Journ. of Math.*, 166(1):123–191, 1994.
[22] William P. Thurston. On the geometry and dynamics of diffeomorphisms of surfaces. *Bull. Am. Math. Soc., New Ser.*, 19(2):417–431, 1988.

About Synchronization Languages

Isabelle Ryl, Yves Roos, and Mireille Clerbout

C.N.R.S. U.R.A. 369, L.I.F.L. Université de Lille I, Bât. M3, Cité Scientifique
59655 Villeneuve d'Ascq Cedex, FRANCE

Abstract. Synchronization languages are a model used to describe the behaviors of distributed applications whose synchronization constraints are expressed by synchronization expressions. Synchronization languages were conjectured by Guo, Salomaa and Yu to be characterized by a rewriting system. We have shown that this conjecture is not true. This negative result has led us to extend the rewriting system and Salomaa and Yu to extend the definition of synchronization languages. The aim of this paper is to establish the link between these two extensions, we show that the behaviors expressed by the two families of synchronization languages are only separated by morphisms.

1 Introduction

Synchronization languages, introduced in [6], are regular languages which correspond to synchronization expressions introduced by Govindarajan, Guo, Yu and Wang [5] within the framework of the *ParC* project. These expressions allow a programmer to express minimal synchronization constraints of a program in a distributed context. A synchronization language can be seen as the set of correct executions of a distributed application where each action is split in two atomic actions, its start and its termination. In this sense, synchronization languages take place in interleaving semantics (see [9] for a comparison between interleaving semantics and non-interleaving semantics) with split of actions [10]. Guo, Salomaa and Yu have defined in [6] a rewriting system named R in order to characterize synchronization languages. One part of this system is a semi-commutation [4], using this part we can put in sequence actions which occur in parallel. The second part of the system is a generalized partial commutation [4], using this part, we can rewrite a word corresponding with a parallel execution in a word with the same parallelism degree. The main interest of R is that synchronization languages are closed under R and Guo et al. have conjectured that the converse holds (see [6]).

We have shown in [1] that this conjecture is true in the particular case of languages defined over alphabets of two actions but not in the general case. In order to bypass this negative result, Salomaa and Yu have chosen to extend the definition of synchronization languages [8] and we have chosen to keep the first definition and to extend the system R [7]. This paper makes the link between the two definitions of synchronization languages and we show that the behaviors expressed by the two families of synchronization languages are only separated by morphisms.

2 Preliminaries

In the following, we shall denote by Π_Y the *projection* onto the sub-alphabet Y, i.e. the image of w by the morphism Π_Y defined by: for each letter x, if $x \in Y$ then $\Pi_Y(x) = x$, else $\Pi_Y(x) = \varepsilon$, where ε denotes the empty word. The *shuffle product* of two words u and v is $u \sqcup\!\sqcup v = \{u_1v_1u_2v_2...u_nv_n \mid u_i \in \Sigma^*, v_i \in \Sigma^*, u = u_1u_2...u_n, v = v_1v_2...v_n\}$.

Let us consider a rewriting system R. We shall write $u \xrightarrow{R} v$ if there is a rule $\alpha \longrightarrow \beta$ in R and two words w and w' such that $u = w\alpha w'$ and $v = w\beta w'$ and we shall write $\xrightarrow{*}_R$ if there are words $w_0, w_1, \ldots, w_n$, $(n \geq 0)$, such that $w_0 = u, w_n = v$, and for each $i < n, w_i \xrightarrow{R} w_{i+1}$. We denote $f_R(u) = \{v \in \Sigma^* \mid u \xrightarrow{*}_R v\}$ and $f_R(L) = \bigcup_{u \in L} f_R(u)$.

3 Synchronization Languages

Synchronization expressions are a high-level tool which allows a programmer to express the synchronization constraints his distributed application has to respect. The statements are tagged and, during the execution, a statement can be executed immediately if it satisfies the constraints described by the expression, if it does not, the execution is delayed.

A synchronization expression may be:

- a statement tag or ε for no action,
- if e_1 and e_2 are synchronization expressions:
  - $(e_1 \rightarrow e_2)$ which imposes that the execution of e_2 starts only after the end of the execution of e_1,
  - $(e_1 \parallel e_2)$ which allows the executions of e_1 and e_2 to overlap. Because of the definition of $\parallel$, the same statement tag cannot appear in both operands,
  - $(e_1 \mid e_2)$ which specifies that either e_1 or e_2 can be executed but not both,
  - $(e_1 \& e_2)$ which imposes that the execution satisfies both expressions e_1 and e_2,
  - (e_1^*) which allows the execution of e_1 to be repeated an arbitrary number of times.

Example 1. The expression $(a \parallel b) \rightarrow (c \parallel d)$ represents the following constraints: statements c and d can be executed only after the end of a and b (there is no synchronization constraint between a and b and between c and d).

With a synchronization expression, we associate a language describing all the executions which respect the constraints expressed by the expression. So, the language corresponds with the possible execution traces. Moreover, each action a is split in two parts, the beginning of the execution of a, a_s and its termination a_t in order to obtain words which show the real concurrency of actions. So, from an expression e over Σ, we construct $L(e) \subseteq (\Sigma_s \cup \Sigma_t)^*$ which is an st-language:

Definition 1. *Let Σ be a finite alphabet. The alphabets Σ_s and Σ_t are defined by the relation:*
$$(a \in \Sigma) \Leftrightarrow (a_s \in \Sigma_s) \Leftrightarrow (a_t \in \Sigma_t).$$
A word $u \in (\Sigma_s \cup \Sigma_t)^*$ is an st-word if and only if for each $x \in \Sigma$, $\Pi_{\{x_s,x_t\}}(u) \in (x_s x_t)^*$. We extend this definition in a canonical way to languages. We denote by ST_Σ the language which contains all st-words over the alphabet $\Sigma_s \cup \Sigma_t$.

Definition 2. *Let Σ be the alphabet of actions (or tags). The language $L(e) \subseteq (\Sigma_s \cup \Sigma_t)^*$ associated with an expression e over Σ is inductively defined by:*

- *$L(\varepsilon) = \varepsilon$,*
- *for each action a, $L(a) = a_s a_t$,*
- *if $e = e_1 \to e_2$ then $L(e) = L(e_1).L(e_2)$,*
- *if $e = e_1 \mid e_2$ then $L(e) = L(e_1) \cup L(e_2)$,*
- *if $e = e_1 \& e_2$ then $L(e) = L(e_1) \cap L(e_2)$,*
- *if $e = e_1 \parallel e_2$ then $L(e) = L(e_1) \sqcup L(e_2)$,*
- *if $e = e_1^*$ then $L(e) = (L(e_1))^*$.*

We denote by LS *the family of synchronization languages.*

Notice that we obtain st-languages because we only compute shuffle product of languages defined over disjoint alphabets. Moreover, by construction, synchronization languages are clearly regular languages.

Guo, Salomaa and Yu [6] have defined the rewriting system R in order to characterize synchronization languages.

Definition 3. *Let Σ be an alphabet of actions. The rewriting system R_Σ is the union of the semi-commutation*
$$\theta_\Sigma = \bigcup_{x \neq y} \{x_s y_s \leftrightarrow y_s x_s, x_t y_t \leftrightarrow y_t x_t, x_s y_t \to y_t x_s\}$$
and the set of rules
$$a_{1t} \ldots a_{mt} a_{1s} \ldots a_{ms} b_{1t} \ldots b_{nt} b_{1s} \ldots b_{ns}$$
$$\updownarrow$$
$$b_{1t} \ldots b_{nt} b_{1s} \ldots b_{ns} a_{1t} \ldots a_{mt} a_{1s} \ldots a_{ms}$$
for each sequence $a_1, \ldots, a_m, b_1, \ldots, b_n$ of pairwise distinct elements of Σ with $m, n \geq 1$.

Example 2. *Let $u = a_s b_s a_t a_s b_t b_s a_t b_t$. Using the second part of R, we have:*
$$a_s b_s \underbrace{a_t a_s}\, \underbrace{b_t b_s}\, a_t b_t \xrightarrow{R} a_s b_s \underbrace{b_t b_s}\, \underbrace{a_t a_s}\, a_t b_t.$$

Guo et al. have shown in [6] that each synchronization language is closed under R and they have conjectured that an arbitrary regular st-language closed under R is a synchronization language. We have shown in [1] that this conjecture is true in the particular case of languages defined over alphabets of two actions but not in the general case. So, we have extended the system R in order to find a characterization of synchronization languages. Salomaa and Yu have chosen another way, they have extended the definition of synchronization expressions.

4 Extensions

4.1 Extension of Salomaa and Yu [8]

Salomaa and Yu have proposed a new definition of synchronization expressions (that we call generalized synchronization expressions), they allow to use a parallel operator between two expressions defined over non-disjoint alphabets, for example, $a \parallel (a \to (b \mid c))$ is a generalized expression. These expressions lead to generalized synchronization languages:

Definition 4. *Generalized synchronization languages (LS$_G$) are built like synchronization languages except for the parallel operator: if $e = e_1 \parallel e_2$ then $L(e) = (L(e_1) \sqcup L(e_2)) \cap \text{ST}_\Sigma$. This operation is called st-shuffle.*

Clearly, we have LS $\subseteq$ LS$_G$. Salomaa and Yu have shown that each generalized synchronization language is closed under θ and they have conjectured that an arbitrary regular st-language closed under θ is a generalized synchronization language.

4.2 Extension of R

We have chosen to keep the first definition of LS to remain closed to the implementation in *ParC*. The problem was that R does not give a characterization of synchronization languages. The synchronization expressions semantics leads us to conjecture that some projection properties are missing to R. Therefore, we define a new system as follow:

Definition 5. *Let Σ be an alphabet of actions. The rewriting system R'_Σ is defined, for each u in $(\Sigma_s \cup \Sigma_t)^*$, by:*

$$(u \xrightarrow[R'_\Sigma]{} v) \Leftrightarrow (\forall a, b \in \Sigma, \Pi_{\{a_s, a_t, b_s, b_t\}}(u) \xrightarrow[R_\Sigma]{*} \Pi_{\{a_s, a_t, b_s, b_t\}}(v)).$$

The system R' is an extension of R, in particular, it is easy to see that for each word u, we have $f_R(u) \subseteq f_{R'}(u)$. The idea is to extend the second part of R. For example, for any integer $n > 1$, a factor $a_t(c_t c_s)^n a_s b_t b_s$ can be rewritten in $b_t b_s a_t (c_t c_s)^n a_s$ using R' but not using R. We have shown that each synchronization language is closed under R' [7] and we have shown that the system R' is best suited to the study of synchronization languages:

Lemma 1 (Clerbout, Roos and Ryl [3]). *For each rewriting system S such that each synchronization language defined over a compatible alphabet of actions is closed under S, we have: $\forall L \subseteq \text{ST}_\Sigma$, $(L = f_{R'_\Sigma}(L)) \Rightarrow (f_S(L) = L)$.*

This lemma shows that R' is a good choice. Nevertheless, there exists a language, $f_{R'}(b_s(a_s a_t)^* c_s b_t (a_s a_t)^* c_t)$, which is regular and which is shown not to be a synchronization language [1]. Therefore, we get the proposition:

Proposition 1 (Clerbout et al. [3]). *There does not exist any rewriting system S such that each synchronization language is closed under S and each regular st-language closed under S is a synchronization language.*

5 Relations Between θ and R'

The aim of the rest of this paper is to establish a link between synchronization languages and generalized synchronization languages. In order to do this, we will use morphisms because of the following ideas. First the difference between LS and LS_G comes only from the alphabets of the operands of the $\|$, so it could be useful to rename some letters. Secondly, the difference between θ and R' comes from the second part of the system R and we will see that we can remove this difference using a renaming. So, we define st-morphisms:

Definition 6. *When Σ and X are two alphabets of actions, a strictly alphabetical morphism from Σ^* into X^* is called action morphism. An action morphism φ from Σ^* into X^* is extended in a natural way to obtain φ from $(\Sigma_s \cup \Sigma_t)^*$ into $(X_s \cup X_t)^*$. With each action morphism φ from Σ^* into X^*, we associate a rational function $\hat{\varphi}$ called st-morphism:*

$$\hat{\varphi} = \{(u, \varphi(u)) \mid u \in \text{ST}_\Sigma \text{ and } \varphi(u) \in \text{ST}_X\}.$$

Note that $\hat{\varphi}$ is equal to $(\cap \text{ST}_X) \circ \varphi \circ (\cap \text{ST}_\Sigma)$. We denote by Φ_{st} the st-morphism family.

Let us show a property of st-morphisms which will be useful in the rest of the paper.

Lemma 2. *The composition of two st-morphisms is an st-morphism.*

Proof. Let $\hat{\varphi}, \hat{\psi}$ be in Φ_{st} respectively associated with the action morphisms φ from X^* into Υ^* and ψ from Υ^* into Σ^*. We have:

$$\hat{\varphi} = \{(u, \varphi(u)) \mid u \in \text{ST}_X \text{ and } \varphi(u) \in \text{ST}_\Upsilon\},$$
$$\hat{\psi} = \{(u, \psi(u)) \mid u \in \text{ST}_\Upsilon \text{ and } \psi(u) \in \text{ST}_\Sigma\}.$$

The composition of these two st-morphisms is:

$$\hat{\psi} \circ \hat{\varphi} = \{(u, \psi \circ \varphi(u)) \mid u \in \text{ST}_X, \varphi(u) \in \text{ST}_\Upsilon \text{ and } \psi \circ \varphi(u) \in \text{ST}_\Sigma\}.$$

Since $\psi \circ \varphi(u) \in \text{ST}_\Sigma$ implies that $\varphi(u) \in \text{ST}_\Upsilon$, $\hat{\psi} \circ \hat{\varphi}$ is an st-morphism.

Now, we will consider the families of languages closed under the rewriting systems we have defined and their closures under st-morphisms.

Definition 7. *We denote respectively by L_θ, R_θ and $\text{R}_{R'}$, the family of θ-closed st-languages, regular θ-closed st-languages and regular R'-closed st-languages.*

Lemma 3. *The family of θ-closed st-languages is closed under st-morphism:* $\Phi_{\text{st}}(\text{L}_\theta) = \text{L}_\theta$.

Proof. Clearly we have $L_\theta \subseteq \Phi_{st}(L_\theta)$. Let $L \subseteq (X_s \cup X_t)^*$ be a θ_X-closed st-language. Let φ be an action morphism from an alphabet containing X into Σ and $\hat{\varphi}$ be the associated st-morphism. Let us show that $\hat{\varphi}(L) = f_{\theta_\Sigma}(\hat{\varphi}(L))$. Let $u \in \hat{\varphi}(L)$. It suffices to show that: $(u \in \hat{\varphi}(L)$ and $u \xrightarrow[\theta_\Sigma]{} v) \Rightarrow (v \in \hat{\varphi}(L))$.

Let us set $u = u_1 xy u_2$ and $v = u_1 yx u_2$ with $(x, y) \in \theta_\Sigma$. Since $u \in \hat{\varphi}(L)$, there exists $w = w_1 ztw_2 \in L$ such that $\varphi(w_1) = u_1$, $\varphi(z) = x$, $\varphi(t) = y$ and $\varphi(w_2) = u_2$. We have $(z, t) \in \theta_X$ because $(\varphi(z), \varphi(t)) \in \theta_\Sigma$. Since L is closed, the word $w_1 tzw_2$ belongs to L so, we have $\varphi(w_1 tzw_2) = v \in \varphi(L)$ and since θ_X and θ_Σ preserve the st-property, v belongs to $\hat{\varphi}(L)$.

We deduce from the previous lemma the corollary:

Corollary 1. *The family of regular θ-closed st-languages is closed under st-morphism:* $\Phi_{st}(R_\theta) = R_\theta$.

Proof. Clearly we have $R_\theta \subseteq \Phi_{st}(R_\theta)$ and $\Phi_{st}(R_\theta) \subseteq L_\theta$. Since the closure under st-morphism preserves regularity of st-languages, we have immediately $\Phi_{st}(R_\theta)) \subseteq R_\theta$.

Now we will state the main result of this section:

Proposition 2. *The family of the images by st-morphisms of regular R'-closed st-languages and the family of regular θ-closed st-languages coincide:* $\Phi_{st}(R_{R'}) = R_\theta$.

The proof of this proposition is not obvious but we can see the main idea with an example (a complete proof can be found in [2]). Let us consider the word $u = a_s b_s a_t a_s b_t b_s a_t b_t$, clearly, $f_\theta(u) \neq f_{R'}(u)$ because $a_s b_s b_t b_s a_t a_s a_t b_t$ belongs to $f_{R'}(u)$ but not to $f_\theta(u)$. Using some renaming (we mark alternatively the occurrences of an action with 1 and 2), we convert u into $v = a_{1s} b_{1s} a_{1t} a_{2s} b_{1t} b_{2s} a_{2t} b_{2t}$ clearly, we have $f_\theta(v) = f_{R'}(v)$. The idea is that in each projection over subalphabets of two actions of a marked word, we can never find a factor which is the left part of a rule of the second part of R.

6 Relation Between LS and LS$_G$

The aim of this section is to compare LS and LS$_G$. We start with some interesting properties of the family of synchronization languages.

Lemma 4. *Let L be an st-language and $\hat{\varphi}$ be an st-morphism. We have:*

$$(L \in \text{LS}) \Leftrightarrow (\hat{\varphi}^{-1}(L) \in \text{LS}).$$

Proof. First, we show the implication from left to right by induction on the construction of L as union, product, star, intersection and shuffle of synchronization languages. The only non-trivial case is the case when $L = L_1 \sqcup L_2$. Since L_1 and L_2 are defined over disjoint alphabets, $\hat{\varphi}^{-1}(L_1)$ and $\hat{\varphi}^{-1}(L_2)$ are also defined over disjoint alphabets. Since these two languages are st-languages, their shuffle is an st-language. Thus, we have $\hat{\varphi}^{-1}(L) = \hat{\varphi}^{-1}(L_1 \shuffle L_2) = \hat{\varphi}^{-1}(L_1) \shuffle \hat{\varphi}^{-1}(L_2)$. By induction hypothesis, $\hat{\varphi}^{-1}(L_1)$ and $\hat{\varphi}^{-1}(L_2)$ are synchronization languages, therefore $\hat{\varphi}^{-1}(L)$ is a synchronization language.

Conversely, let us consider an st-language L defined over an alphabet of actions Υ and an action morphism φ defined from X^* into Υ^* such that $\hat{\varphi}^{-1}(L)$ is a synchronization language. Let us consider the partition of the alphabet X, $X = X_1 \cup \ldots \cup X_n$ with: $(\exists 1 \leq k \leq n \mid x, y \in X_k) \Leftrightarrow (\varphi(x) = \varphi(y))$. Let x_1 be in X_1, ..., x_n be in X_n. Let us set:

$$M = \hat{\varphi}^{-1}(L) \cap [(x_{1s}x_{1t})^* \shuffle \ldots \shuffle (x_{ns}x_{nt})^*].$$

From this equality, we deduce $M \in$ LS. Since the restriction of φ from the alphabet of actions of M into the alphabet of actions of L establishes a bijection between these two alphabets, and since $L = \varphi(M)$, L is a synchronization language.

Lemma 5. *The family of the images by st-morphism of synchronization languages is closed under intersection.*

Proof. Let L_1 and L_2 be two synchronization languages defined over the alphabets of actions X_1 and X_2, respectively. According to Lemma 2, the composition of two st-morphisms is an st-morphism so we can always distinguish these alphabets: without loss of generality, we can take $X_1 \cap X_2 = \emptyset$. Let φ_1 and φ_2 be two action morphisms respectively defined from X_1^* into Υ_1^* and from X_2^* into Υ_2^*. Let us consider the following alphabets:

$$X_1' = \{x \in X_1 \mid \varphi_1(x) \notin \varphi_2(X_2)\},$$
$$X_2' = \{x \in X_2 \mid \varphi_2(x) \notin \varphi_1(X_1)\},$$
$$\Sigma = \{(x,y) \in X_1 \times X_2 \mid \varphi_1(x) = \varphi_2(y)\},$$

(note that $X_1' \cap X_2' = \emptyset$) and the following action morphisms:

$$\begin{array}{ll}
\psi_1 : (X_1' \cup \Sigma)^* \longrightarrow X_1^* & \psi_2 : (X_2' \cup \Sigma)^* \longrightarrow X_2^* \\
\quad x \in X_1' \longmapsto x & \quad x \in X_2' \longmapsto x \\
\quad (x,y) \in \Sigma \longmapsto x & \quad (x,y) \in \Sigma \longmapsto y \\
\psi : \Sigma^* \longrightarrow (\Upsilon_1 \cap \Upsilon_2)^* \\
\quad (x,y) \longmapsto \varphi_1(x) = \varphi_2(y)
\end{array}$$

We denote by $\hat{\varphi}_1$, $\hat{\varphi}_2$, $\hat{\psi}$, $\hat{\psi}_1$ and $\hat{\psi}_2$ the st-morphisms induced by these action morphisms. We will show that we have: $\hat{\varphi}_1(L_1) \cap \hat{\varphi}_2(L_2) = \hat{\psi}(\hat{\psi}_1^{-1}(L_1) \cap \hat{\psi}_2^{-1}(L_2))$. We first remark the two following properties:

Assertion. Let $u_1 \in L_1$ and $u_2 \in L_2$. We have:

$$(\hat{\varphi}_1(u_1) = \hat{\varphi}_2(u_2)) \Leftrightarrow (\exists v \in ST_\Sigma \mid \hat{\psi}_1(v) = u_1 \text{ and } \hat{\psi}_2(v) = u_2).$$

For the implication from left to right, it suffices to note that if $u_1 = x_1 \ldots x_k$ and $u_2 = y_1 \ldots y_k$, the word $v = (x_1, y_1) \ldots (x_k, y_k)$ belongs to ST_Σ with $\hat{\psi}_1(v) =$

u_1 and $\hat{\psi}_2(v) = u_2$. The implication from right to left can be deduced from the definitions of the alphabets and of the action morphisms.

Assertion. For each $v \in \mathrm{ST}_\Sigma$, we have: $\hat{\psi}(v) = \hat{\varphi}_1(\hat{\psi}_1(v)) = \hat{\varphi}_2(\hat{\psi}_2(v))$.

From these two assertions, we deduce the following equivalences:

$$w \in \hat{\varphi}_1(L_1) \cap \hat{\varphi}_2(L_2)$$
$$\Leftrightarrow \exists u_1 \in L_1, u_2 \in L_2 \mid \hat{\varphi}_1(u_1) = \hat{\varphi}_2(u_2) = w$$
$$\Leftrightarrow \exists v \in \mathrm{ST}_\Sigma \mid \hat{\psi}_1(v) = u_1 \in L_1, \hat{\psi}_2(v) = u_2 \in L_2,$$
$$w = \hat{\varphi}_1(\hat{\psi}_1(v)) = \hat{\varphi}_2(\hat{\psi}_2(v)) = \hat{\psi}(v)$$
$$\Leftrightarrow \exists v \in \hat{\psi}_1^{-1}(L_1) \cap \hat{\psi}_2^{-1}(L_2) \mid w = \hat{\varphi}_1(\hat{\psi}_1(v)) = \hat{\varphi}_2(\hat{\psi}_2(v)) = \hat{\psi}(v)$$
$$\Leftrightarrow w \in \hat{\psi}(\hat{\psi}_1^{-1}(L_1) \cap \hat{\psi}_2^{-1}(L_2))$$

Thus, we have $\hat{\varphi}_1(L_1) \cap \hat{\varphi}_2(L_2) = \hat{\psi}(\hat{\psi}_1^{-1}(L_1) \cap \hat{\psi}_2^{-1}(L_2))$.

The synchronization languages family is closed under inverse st-morphism (Lemma 4), therefore $\hat{\psi}_1^{-1}(L_1)$ and $\hat{\psi}_2^{-1}(L_2)$ are synchronization languages so, we have $\hat{\varphi}_1(L_1) \cap \hat{\varphi}_2(L_2) \in \Phi_{\mathrm{st}}(\mathrm{LS})$.

Lemma 6. *The family of the images by st-morphism of synchronization languages is closed under st-shuffle.*

Proof. Let us recall that the st-shuffle of two languages is the set of st-words of their shuffle product. Let L_1 and L_2 be two synchronization languages respectively defined over the alphabets of actions X_1 and X_2. We can suppose that $X_1 \cap X_2 = \emptyset$. Let φ_1 and φ_2 be two action morphisms respectively defined from X_1^* into Υ_1^* and from X_2^* into Υ_2^*. We consider the action morphism:

$$\begin{aligned}\varphi_3 : (X_1 \cup X_2)^* &\longrightarrow (\Upsilon_1 \cup \Upsilon_2)^* \\ x \in X_1 &\longmapsto \varphi_1(x) \\ x \in X_2 &\longmapsto \varphi_2(x)\end{aligned}$$

Since L_1 and L_2 are defined over disjoint alphabets, $L_1 \shuffle L_2$ is an st-language and we have: $[\hat{\varphi}_1(L_1) \shuffle \hat{\varphi}_2(L_2)] \cap \mathrm{ST}_{\Upsilon_1 \cup \Upsilon_2} = \hat{\varphi}_3(L_1 \shuffle L_2)$.

The closure properties of family of the images by st-morphisms of synchronization languages we have shown lead us to the main result of this section which establishes the link between synchronization languages and generalized synchronization languages:

Proposition 3. *The family of the images by st-morphism of generalized synchronization languages and the family of the images by st-morphism of synchronization languages coincide:* $\Phi_{\mathrm{st}}(\mathrm{LS_G}) = \Phi_{\mathrm{st}}(\mathrm{LS})$.

Proof. The family $\Phi_{\mathrm{st}}(\mathrm{LS})$ contains all the finite languages which represent the execution of one action (like $a_s a_t$) and it is closed under union, product, star, intersection (Lemma 5) and st-shuffle (Lemma 6) so, it contains the generalized synchronization languages: $(\mathrm{LS_G} \subseteq \Phi_{\mathrm{st}}(\mathrm{LS})) \Rightarrow (\Phi_{\mathrm{st}}(\mathrm{LS_G}) \subseteq \Phi_{\mathrm{st}}(\mathrm{LS}))$. Moreover, the family of synchronization languages is included in the family of generalized synchronization languages, thus: $(\mathrm{LS} \subseteq \mathrm{LS_G}) \Rightarrow (\Phi_{\mathrm{st}}(\mathrm{LS}) \subseteq \Phi_{\mathrm{st}}(\mathrm{LS_G}))$.

7 Conclusion

We have established a link between synchronization languages and their extensions and between the rewriting systems used to try to characterize them. The following diagram shows the present situation: some inclusions miss to complete the study of synchronization languages families. We have already shown that a non-obvious family included in R_θ (the family of well-formed languages defined in [6]) is included in $\Phi_{st}(LS)$, so we would like to extend this result and we conjecture that $R_\theta \subseteq \Phi_{st}(LS)$ that is to say, we conjecture the equality $\Phi_{st}(R_{R'}) = R_\theta = \Phi_{st}(LS) = \Phi_{st}(LS_G)$. Since $R_{R'} \subseteq R_\theta$, we also conjecture the inclusion $R_{R'} \subseteq \Phi_{st}(LS)$. The inclusion $R_\theta \subseteq LS_G$ is conjectured by Salomaa and Yu [8], this conjecture leads to the following: $\Phi_{st}(R_{R'}) = R_\theta = \Phi_{st}(LS) = \Phi_{st}(LS_G) = LS_G$. Nevertheless, a characterization of generalized synchronization languages may be more difficult because of the very little structural relations between a generalized synchronization language and one of its generalized synchronization expressions.

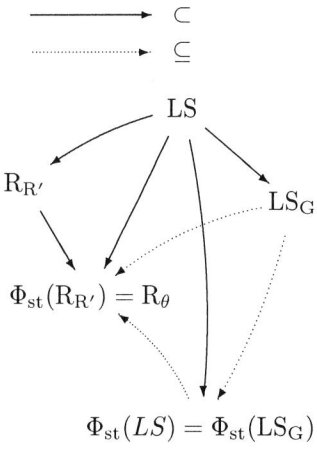

References

1. CLERBOUT, M., ROOS, Y., AND RYL, I. Synchronization languages. *Theoretical Computer Science*. to appear.
2. CLERBOUT, M., ROOS, Y., AND RYL, I. About synchronization languages (full version). Tech. Rep. IT-98-313, Université des Sciences et Technologies de Lille, 1998.
3. CLERBOUT, M., ROOS, Y., AND RYL, I. Langages de synchronisation et systèmes de réécriture. Tech. Rep. IT-98-311, Université des Sciences et Technologies de Lille, 1998.
4. DIEKERT, V., AND ROZENBERG, G., Eds. *The Book of Traces*. World Scientific, Singapore, 1995.
5. GOVINDARAJAN, R., GUO, L., YU, S., AND WANG, P. ParC project: Practical constructs for parallel programming languages. In *Proc. IEEE 15th Annual Internationnal Computer Software & Applications Conference* (1991), pp. 183–189.
6. GUO, L., SALOMAA, K., AND YU, S. On synchronization languages. *Fundamenta Informaticae 25* (1996), 423–436.
7. RYL, I., ROOS, Y., AND CLERBOUT, M. Partial characterization of synchronization languages. In *Proc. 22nd International Symposium on Mathematical Foundations of Computer Science (MFCS'97)* (Bratislava, Slovakia, 1997), I. Prívara and P. Ružička, Eds., vol. 1295 of *Lecture Notes in Computer Science*, Springer-Verlag, Berlin, pp. 209–218.
8. SALOMAA, K., AND YU, S. Rewriting rules for synchronization languages. In *Structures in Logic and Computer Science, a Selection of Essays in Honor of A.*

Ehrenfeucht, J. Mycielski, G. Rozenberg, and A. Salomaa, Eds., vol. 1261 of *Lecture Notes in Computer Science*. Springer-Verlag, Berlin, 1997, pp. 322–338.
9. SASSONE, V., NIELSEN, M., AND WINSKEL, G. Models for concurrency: Towards a classification. *Theoretical Computer Science 170*, 1-2 (1996), 297–348.
10. VAN GLABBECK, R. J., AND VAANDRAGER, F. The difference between splitting in n and $n+1$. *Information and Computation 136*, 2 (1997), 109–142.

őr# When Can an Equational Simple Graph Be Generated by Hyperedge Replacement?

Klaus Barthelmann

Johannes Gutenberg-Universität Mainz, Institut für Informatik,
D-55099 Mainz, Germany
barthel@informatik.uni-mainz.de

Abstract. Infinite hypergraphs with sources arise as the canonical solutions of certain systems of recursive equations written with operations on hypergraphs. There are basically two different sets of such operations known from the literature, HR and VR. VR is strictly more powerful than HR on simple hypergraphs. Necessary conditions are known ensuring that a VR-equational simple hypergraph is also HR-equational. We prove that two of them, namely having finite tree-width or not containing the infinite bipartite graph, are also sufficient. This shows that equational hypergraphs behave like context-free sets of finite hypergraphs.
Using an alternate characterization of VR-equational simple hypergraphs [3], this result provides a (necessary and) sufficient condition and an effective procedure to translate a language-theoretic definition of an infinite hypergraph [9] into an operational one based on substitution [11].

1 Introduction

Operations on hypergraphs have been defined in the literature in essentially two different ways. *Hyperedge replacement* (*HR*) rewrites a hypergraph by substituting a hypergraph for a hyperedge. The inserted hypergraph contains some distinguished vertices (often called "sources") which are glued with the former attachment vertices of the removed hyperedge. A rewriting rule, therefore, has to specify the label of the hyperedge to replace (which in turn determines the number of attachment vertices) and the hypergraph (including sources) to be inserted. This operational description can readily be turned into algebraic operations [7, 12]. *Vertex replacement* (*VR*) rewrites a hypergraph by removing a vertex together with its adjacent hyperedges, inserting a new hypergraph, and connecting it with the former neighbours of the vertex by new hyperedges. Only the labels of the former neighbours and the labels of the removed hyperedges are taken into account when creating hyperedges. A rewriting rule, therefore, has to specify the label of the vertex to replace, the hypergraph to be inserted, and the embedding transformation. This operational description is modeled by algebraic operations rather indirectly [18, 15]. These VR operations only make sense for simple hypergraphs, that is, hypergraphs without parallel hyperedges with the same label. On the other hand, it is well-known that the VR operations are more powerful than the HR operations on simple hypergraphs. The reader can find comparisons of both approaches in [20, 17]. Suffice it to say that the duality is actually deeper than one might expect from the preceding remarks.

Algebraic operations are mainly used in systems of recursive equations. One is often interested in "polynomial" equations which have sets of finite hypergraphs as their (least) solutions, those that can be generated by the operational description above. But a system of recursive equations can also be used to describe a single, usually infinite hypergraph as its least solution. This is the aspect we consider in this paper. Such hypergraphs are called "equational". They arise naturally in denotational semantics, because they generalize regular trees (see [10]). In many cases, they allow to verify properties of a computation. For example, equational hypergraphs defined with HR operations (*HR-equational* hypergraphs for short) can be used to analyse recursive applicative program schemes [11]. Caucal [9, 3] has given a pure language-theoretic description of equational hypergraphs defined with VR operations (*VR-equational* hypergraphs for short). Every equational hypergraph has a decidable monadic second-order (MSO) theory [11, 9, 3], because it can be defined inside the (infinite) complete binary tree by formulas of MSO logic. It is also known that, vice versa, every hypergraph that can be defined inside the (infinite) complete binary tree by formulas of MSO logic is equational [13, 3]. Although equational hypergraphs certainly do not constitute the largest class of hypergraphs with a decidable MSO theory, they have perhaps the most natural description.

This paper investigates the relationship between the classes of equational hypergraphs defined with HR and VR operations. It is clear that every HR-equational simple hypergraph is also VR-equational. Since an HR-equational hypergraph has finite tree-width [14], there are VR-equational hypergraphs which are not HR-equational (see also [9]). Bounded tree-width and equivalent properties are known to be necessary and sufficient for a VR-equational set of finite hypergraphs to be HR-equational [16, 2]. We establish a similar relationship between VR-equational and HR-equational hypergraphs, thereby answering a question raised in [3]. We directly transform the corresponding systems of equations and do not need to build a tree decomposition (which would also give the result according to [14]) or consider logical properties of the hypergraphs (like in [16]). The paper is organized as follows. We first recall the notion of hypergraphs and operations and show how to solve systems of regular equations for hypergraphs. Section 3 states the main theorem. Since space is short, all proofs are omitted. (The full version is available as a technical report [4].) We conclude this paper with some remarks.

2 Preliminaries

First we fix some notation. $\mathbb{N}$ is the set of nonnegative integers and $\mathbb{N}_+$ the set of positive ones. We set $[n] = \{1, \ldots, n\}$ for $n \in \mathbb{N}$; $[0] = \emptyset$. The set of finite sequences (words) of elements from a set A is called A^*. If $f: A \to B$ is a mapping, we write $f^*: A^* \to B^*$ for its unique extension to A^*. $[a]_R$ is the equivalence class of $a \in A$ modulo R and $A/R = \{[a]_R \mid a \in A\}$. Accordingly,

$[\ ]_R \colon A \to A/R$ is the canonic surjection. If a formula $\phi(x_1, \dots, x_n)$ contains at most n free variables $x_1, \dots, x_n$ in some specified order and $d_1, \dots, d_n$ are appropriate values for them in a structure S then $S, d_1, \dots, d_n \models \phi(x_1, \dots, x_n)$ means that S satisfies ϕ under the assignment $x_i := d_i$, $i \in [n]$.

2.1 Hypergraphs and Operations

Let us fix a (sufficiently large) finite set L of *labels* together with a *rank* mapping $\rho \colon L \to \mathbb{N}_+$.

Definition 1 (hypergraph, homomorphism). *A hypergraph G of sort $n \in \mathbb{N}$ consists of a set V of vertices, a set E of hyperedges, a vertex mapping vert $\colon E \to V^*$, a label mapping lab $\colon E \to L$, and a source mapping src $\colon [n] \to V$, such that $|\mathrm{vert}(e)| = \rho(\mathrm{lab}(e))$ for all $e \in E$. $\mathrm{src}(i)$ is called the ith source.*

In a simple hypergraph, the conditions $\mathrm{vert}(e) = \mathrm{vert}(e')$ and $\mathrm{lab}(e) = \mathrm{lab}(e')$ together imply $e = e'$ for all $e, e' \in E$. Therefore, $E \subseteq L \times V^*$ can always be assumed, and $\mathrm{vert}((l, \boldsymbol{v})) = \boldsymbol{v}$, $\mathrm{lab}((l, \boldsymbol{v})) = l$.

A homomorphism $h \colon G \to G'$ between hypergraphs of sort n consists of two mappings $h_V \colon V \to V'$ and $h_E \colon E \to E'$ such that $\mathrm{vert}' \circ h_E = h_V^* \circ \mathrm{vert}$, $\mathrm{lab}' \circ h_E = \mathrm{lab}$ and $\mathrm{src}' = h_V \circ \mathrm{src}$. For simple hypergraphs, h_E is determined by h_V.

As usual, an isomorphism is a homomorphism with an inverse.

The following three definitions introduce operations on isomorphism classes of hypergraphs. They are taken from [15] with small modifications. (The restricted operations in [18, 16, 17] are sufficient for ordinary graphs.)

Definition 2 (constants). *The constants 0_n, $n \in \mathbb{N}$, denote discrete hypergraphs of sort n: $V = [n]$, $E = \emptyset$ (which also determines vert and lab) and $\mathrm{src}(i) = i$ for $i \in [n]$. The constants 1_n, $n \in \mathbb{N}$, denote one-vertex hypergraphs of sort n: $V = [1]$, $E = L$, $\mathrm{vert}(l) = 1^{\rho(l)}$, $\mathrm{lab}(l) = l$ for $l \in L$ and $\mathrm{src}(i) = 1$ for $i \in [n]$. The constants $l \in L$ denote hypergraphs of sort $\rho(l)$ containing exactly one hyperedge: $V = [\rho(l)]$, $E = \{e\}$, $\mathrm{vert}(e) = 1 \dots \rho(l)$, $\mathrm{lab}(e) = l$ and $\mathrm{src}(i) = i$ for $i \in [\rho(l)]$.*

Definition 3 (parallel composition). *Parallel composition $\|_{n_1,n_2}$ is a binary operation on hypergraphs G_1 and G_2 of respective sorts n_1 and n_2. Its result has sort $\max\{n_1, n_2\}$. We can assume that V_1 is disjoint from V_2 and E_1 is disjoint from E_2. Let $\approx$ be the least equivalence relation on $V_1 \cup V_2$ such that $\mathrm{src}_1(i) \approx \mathrm{src}_2(i)$ for every $i \in [\min\{n_1, n_2\}]$. Then $G_1 \|_{n_1, n_2} G_2$ is determined as follows: $V = (V_1 \cup V_2)/{\approx}$, $E = E_1 \cup E_2$, $\mathrm{vert} = [\ ]_\approx^* \circ (\mathrm{vert}_1 \cup \mathrm{vert}_2)$, $\mathrm{lab} = \mathrm{lab}_1 \cup \mathrm{lab}_2$, and $\mathrm{src} = ([\ ]_\approx \circ \mathrm{src}_1) \cup ([\ ]_\approx \circ \mathrm{src}_2)$.*

Note that all constants denote simple hypergraphs. However, the parallel composition of two simple hypergraphs is not always simple. The definition has to be modified a little in this case: We assume $E_i \subseteq L \times V_i^*$ ($i = 1, 2$) as usual and set $E = \{(l, [\boldsymbol{v}]_\approx^*) \mid (l, \boldsymbol{v}) \in E_1 \cup E_2\}$. That is, parallel hyperedges with the same label, whose attachment vertices are all among the sources, are fused

Simple hypergraphs of sort n are essentially relational structures with n constants. Their domain is V. They include predicates edge_l of arity $\rho(l)$ for every $l \in L$, such that $\mathrm{edge}_l(v_1, \ldots, v_n)$ means $(l, v_1 \ldots v_n) \in E$. The constant s_i for every $i \in [n]$ represents the vertex $\mathrm{src}(i)$. Adding equality and a finite number of variables $x_1, \ldots, x_k$, atomic formulas are built as usual. The set of *quantifier-free* (*qf*) formulas is obtained from atomic formulas with the truth values true, false and the Boolean connectives $\wedge$, $\vee$ and $\neg$. The set of *positive qf* (*pqf*) formulas consists of those qf formulas that can be formed without negation. We denote by $\mathrm{QF}_n(x_1, \ldots, x_k)$ and $\mathrm{PQF}_n(x_1, \ldots, x_k)$ the respective sets of qf and pqf formulas up to (tauto)logical equivalence. (The axioms for equality are included, namely reflexivity, symmetry, transitivity and substitution of equals for equals in formulas. Note that logical equivalence and logical implication are decidable for (p)qf formulas, because they are even decidable for the fragment of first-order logic with equality, where all formulas are in prefix normal form and contain only universal quantifiers.) $\mathrm{QF}_n(x_1, \ldots, x_k)$ and $\mathrm{PQF}_n(x_1, \ldots, x_k)$ are finite for every n and k. It is more convenient to speak about their elements as formulas, and this should not cause any confusion.

Definition 4 (qfd operations). *A qf definition scheme Δ from m to n consists of a vertex formula $\delta \in \mathrm{QF}_m(x_1)$ of the form $\delta'(x_1) \vee \bigvee_{i \in [n]} x_1 = s_{\sigma_i}$, hyperedge formulas $\eta_l \in \mathrm{QF}_m(x_1, \ldots, x_{\rho(l)})$ for each $l \in L$, and source specifications $\sigma_i \in [m]$ for each $i \in [n]$. A pqf definition scheme is defined similarly with PQF_m instead of QF_m.*

Δ determines a unary operation Def_Δ on simple hypergraphs G_1 of sort m, which is called (positive) quantifier-free definable *(qfd or pqfd). Its result $\mathrm{Def}_\Delta(G_1)$ has sort n and is determined as follows: $V = \{v \in V_1 \mid G_1, v \models \delta(x_1)\}$, $E = \{(l, v_1 \ldots v_{\rho(l)}) \mid l \in L, v_1, \ldots, v_{\rho(l)} \in V, G_1, v_1, \ldots, v_{\rho(l)} \models \eta_l(x_1, \ldots, x_{\rho(l)})\}$, $\mathrm{src}(i) = \mathrm{src}_1(\sigma_i)$ for every $i \in [n]$.*

One pqfd operation deserves a special name because it works on hypergraphs in general. ren_f is an operation of type $m \to n$, where $f\colon [n] \to [m]$ is a mapping. Its definition scheme contains $\delta(x_1) \equiv \mathrm{true}$, $\eta_l(x_1, \ldots, x_{\rho(l)}) \equiv \mathrm{edge}_l(x_1, \ldots, x_{\rho(l)})$ for each $l \in L$, and $\sigma_i \equiv f(i)$ for each $i \in [n]$. ($\mathrm{ren}_f(G)$ is like G except that its source mapping is $\mathrm{src} \circ f$.)

The *HR operations* are 0_n, $n \in \mathbb{N}$, all l, $l \in L$, ren_f for all mappings $f\colon [n] \to [m]$, $m, n \in \mathbb{N}$, and $\|_{n_1,n_2}$, $n_1, n_2 \in \mathbb{N}$. This set of operations is obviously equivalent to the one introduced in [7, 12], where two other operations are used instead of parallel compositions, one for forming the disjoint union of two hypergraphs and one for fusing vertices. The *VR operations* are 0_n and 1_n, $n \in \mathbb{N}$, all pqfd operations Def_Δ and $\|_{n_1,n_2}$, $n_1, n_2 \in \mathbb{N}$. VR operations are defined on simple hypergraphs; the interpretation of parallel composition is slightly different from the HR case, as said above.

2.2 Equations in ω-Complete Categories

Solving equations amounts to forming fixpoints in some algebraic structure. However, no topological or order-theoretic definition of convergence for sequences of hypergraphs is known. There is no choice but using a more general framework based on homomorphisms [5, 6]. It will require a basic understanding of category theory (categories, functors, colimits and their construction for sets), which we assume from the reader. Let us review only a few less common facts (see also [3]).

Hypergraphs of sort $n \in \mathbb{N}$ and their homomorphisms form (the respective objects and arrows of) a category $\mathcal{G}_n$. The same holds for simple hypergraphs; the corresponding category is called $\mathcal{SG}_n$. $\mathcal{G}_n$ has all colimits. They can be obtained from the corresponding constructions in the category of sets, applying them to vertices and hyperedges separately. Similarly, $\mathcal{SG}_n$ has all colimits. Let us remark that 0_n is the initial object in $\mathcal{G}_n$ (and $\mathcal{SG}_n$). $G_1 \|_{n,n} G_2$ is the coproduct of G_1 and G_2 in $\mathcal{G}_n$ and, under the modified definition mentioned earlier, also in $\mathcal{SG}_n$.

Every hypergraph operation induces a functor. (Here the restriction to *positive* quantifier-free definable operations is essential.) Moreover, all these functors and arbitrary compositions preserve colimits. A special case of this fact deserves a particular name. A category is ω-*complete* iff every diagram of the form

$$A_0 \xrightarrow{h_0} A_1 \xrightarrow{h_1} \cdots \xrightarrow{h_{i-1}} A_i \xrightarrow{h_i} \cdots$$

has a colimit. A functor is ω-*continuous* iff it preserves all colimits of this form.

Fact 5 ([1, 5]). *If $F: \mathcal{G}_{m_1} \times \cdots \times \mathcal{G}_{m_n} \to \mathcal{G}_{m_1} \times \cdots \times \mathcal{G}_{m_n}$ is an ω-continuous functor then its fixpoints (the tuples of hypergraphs $\mathbf{H}$ such that $F(\mathbf{H})$ is componentwise isomorphic to $\mathbf{H}$) form a category with an initial object. This initial object $\mathbf{G}$ (the initial fixpoint) is the colimit of the diagram*

$$\mathbf{0} \xrightarrow{\mathbf{f}} F(\mathbf{0}) \xrightarrow{F(\mathbf{f})} \cdots \xrightarrow{F^{i-1}(\mathbf{f})} F^i(\mathbf{0}) \xrightarrow{F^i(\mathbf{f})} \cdots$$

for $i \in \mathbb{N}$, where $\mathbf{0} = (0_{m_1}, \ldots, 0_{m_n})$ and $\mathbf{f}$ is the tuple of initial homomorphisms. The cocone of $\mathbf{G}$ over the diagram is given by the arrows $F^i(\mathbf{g}): F^i(\mathbf{0}) \to \mathbf{G}$, where $\mathbf{g}: \mathbf{0} \to \mathbf{G}$ is the tuple of initial homomorphisms. The same statement holds with $\mathcal{SG}_m$ instead of $\mathcal{G}_m$.

Definition 6 (system of regular equations, equational hypergraph). *A system of regular equations E is a sequence of equations*

$$x_1 = t_1 \quad \ldots \quad x_k = t_k \qquad (k \in \mathbb{N})$$

with pairwise distinct variables $x_1, \ldots, x_k$, each one having a sort $\rho(x_i)$, and right-hand sides t_i of the form $f(x_{j_1}, \ldots, x_{j_r})$, where the operation f has the correct type $\rho(x_{j_1}) \ldots \rho(x_{j_r}) \rho(x_i)$, and $j_1, \ldots, j_r \in [k]$.

Sorts s and operations are interpreted by ω-complete categories C_s and ω-continuous functors between them, respectively. E induces an ω-continuous functor $E_C: C_{\rho(x_1)} \times \cdots \times C_{\rho(x_k)} \to C_{\rho(x_1)} \times \cdots \times C_{\rho(x_k)}$ from the ω-complete category $C_{\rho(x_1)} \times \cdots \times C_{\rho(x_k)}$ to itself. Therefore, E_C has an initial fixpoint $\mathcal{L}(E, C)$. It is called the least solution of E in the family C. $\mathcal{L}(E, C).x_i$ is its ith component.

A hypergraph is HR-equational iff it is $\mathcal{L}(E,\mathcal{G}).x_1$ for some system E of regular equations involving HR operations. A simple hypergraph is VR-equational iff it is $\mathcal{L}(E,\mathcal{SG}).x_1$ for some system E of regular equations involving VR operations.

3 The Main Theorem

We need a few graph-theoretic notions before we can state the theorem.

Definition 7 (tree-width). *A* tree-decomposition *of a hypergraph $G = (V, E, \text{vert}, \text{lab}, \text{src})$ of sort n consists of a tree T with set of nodes N and a mapping $f \colon N \to \mathcal{P}(V)$ such that:*

1. $V = \bigcup \{f(v) \mid v \in N\}$.
2. *Every hyperedge in E has all its vertices in $f(v)$ for some $v \in N$.*
3. *If $u, v, w \in N$ and if v is on the unique (undirected) shortest path from u to w in T then $f(u) \cap f(w) \subseteq f(v)$.*
4. $\text{src}([n]) \subseteq f(r)$, *where $r \in N$ is the root of T.*

The width *of a tree-decomposition is $\sup\{|f(v)| \mid v \in N\} - 1$ (it may be ∞). The* tree-width *of a hypergraph G is the minimum width of a tree-decomposition of G.*

Definition 8 (clique-graph). *The* clique-graph $\text{cg}(G)$ *derived from a hypergraph G is obtained by the substitution of an undirected clique K_n with vertices $v_1, \ldots, v_n$ for every hyperedge connecting vertices $v_1, \ldots, v_n$, and deleting all labels (and sources).*

Theorem 9. *A VR-equational simple hypergraph G is HR-equational iff $\text{cg}(G)$ has only small (undirected, unlabelled) bipartite graphs $K_{n,n}$ as subgraphs iff G has finite tree-width. This condition is decidable.*

Moreover, a VR-equational simple hypergraph is HR-equational if each of its vertices has finite (even bounded) degree [8, 9]. This, however, is not a condition shared by all HR-equational hypergraphs. It leads to a natural subclass; such hypergraphs are called *context-free* in [21].

The 'only if' parts of the theorem are well-known [14]. Since the bipartite graph $K_{n,n}$ has tree-width n, the 'if' part of the second assertion follows from the first. Finally, it turns out that it actually suffices to exclude the infinite (undirected, unlabelled) bipartite graph $K_{\infty,\infty}$ from $\text{cg}(G)$.

Our proof of the theorem is divided into two parts. The first step is to collect information about the components of the least solution of the defining system of regular equations. We identify subhypergraphs with a finite number of occurrences. This is done by counting substructures satisfying certain logical properties. The technique grew out of the "typing" in [2]. Similar approaches have been used in the literature, in particular [19]. The second step is to get rid of the pqfd operations by transforming the system of equations. If a pqfd operation adds only a finite number of hyperedges, the same effect can be achieved by introducing a sufficient number of sources and using HR operations. Otherwise the pqfd operation is "moved" through the system of equations towards the constants. The creation of hyperedges that span a parallel composition can be delegated to the operands if we supply each of them with a finite number of vertices from the partner. A similar technique was used in [2].

3.1 Counting Substructures

Let E be a system of regular equations and let $(G_1, \ldots, G_k) = \mathcal{L}(E, \mathcal{SG})$ have sorts $(n_1, \ldots, n_k)$. We count the number of occurrences of all subhypergraphs of G_i, $i \in [k]$, up to the size $n_i + \rho(L)$, where $\rho(L) = \max\{\rho(l) \mid l \in L\}$. More formally, we introduce variables $x_1, \ldots, x_{\rho(L)}$ and take Φ_{n_i} as the disjoint union of all $\mathrm{QF}_{n_i}(x_{j_1}, \ldots, x_{j_r})$, $r \in \mathbb{N}$, where $j_1 < \cdots < j_r$ and $j_1, \ldots, j_r \in [\rho(L)]$. Note that Φ_{n_i} is finite. Let M_{n_i} be the set of all multisets over Φ_{n_i}, that is, the set of all mappings from Φ_{n_i} to $\mathbb{N} \cup \{\infty\}$.

Definition 10 (occurrences). *Let $\phi \in \mathrm{QF}_n(x_{j_1}, \ldots, x_{j_r}) \subseteq \Phi_n$ be a qf formula and $G = (V, E, \mathrm{src})$ a simple hypergraph of sort n. We define $\mathrm{sat}_n(G, \phi) = \{(v_1, \ldots, v_r) \in V^r \mid G, v_1, \ldots, v_r \models \phi(x_{j_1}, \ldots, x_{j_r})\}$ and $\mathrm{occ}_n : \mathcal{SG}_n \to M_n$ such that $\mathrm{occ}_n(G)(\phi) = |\mathrm{sat}_n(G, \phi)|$ (this may be ∞).*

We compute $\mathrm{occ}_{n_i}(G_i)$, $i \in [k]$, by solving E a second time. For each $j \in \mathbb{N}$, let us define tuples of hypergraphs $(H_{1,j}, \ldots, H_{k,j}) = E_{\mathcal{SG}}^j(0_{n_1}, \ldots, 0_{n_k})$ and homomorphisms $(h_{1,j}, \ldots, h_{k,j}) = E_{\mathcal{SG}}^j(g_1, \ldots, g_k)$, where $g_i : 0_{n_i} \to G_i$, $i \in [k]$, are the initial homomorphisms. It follows easily from Fact 5 that G_i is the colimit of the diagram $h_{i,0}(H_{i,0}) \rightarrowtail h_{i,1}(H_{i,1}) \rightarrowtail \cdots \rightarrowtail h_{i,j}(H_{i,j}) \rightarrowtail \cdots$. Note that $h_{i,j}(H_{i,j})$ is a subhypergraph of $h_{i,j+1}(H_{i,j+1})$ and of G_i. Moreover, because $E_{\mathcal{SG}}$ is a functor, $(h_{1,j+1}(H_{1,j+1}), \ldots, h_{k,j+1}(H_{k,j+1})) = E_{\mathcal{SG}}(h_{1,j}(H_{1,j}), \ldots, h_{k,j}(H_{k,j}))$. We can compute $\mathrm{occ}_{n_i}(h_{i,j}(H_{i,j}))$ inductively for $j \in \mathbb{N}$. The image of 0_{n_i} in G_i is determined as follows. The QF theory $\mathrm{Th}(G_i) = \{\phi \in \mathrm{QF}_{n_i}() \mid G_i \models \phi\}$ of G_i is decidable (even its MSO theory is). $\mathrm{Th}(G_i)$ induces an equivalence relation $\sim$ on $[n_i]$: $p \sim q$ means that $G_i \models s_p = s_q$. Therefore, $g_i(0_{n_i})$ has $V = [n_i]/\sim$, $E = \emptyset$ and $\mathrm{src}(p) = [p]_\sim$ for $p \in [n_i]$. $\mathrm{sat}_{n_i}(g_i(0_{n_i}), \phi)$, $\phi \in \Phi_{n_i}$, can then be determined explicitly. This gives $(\mathrm{occ}_{n_1}(h_{1,0}(H_{1,0})), \ldots, \mathrm{occ}_{n_k}(h_{k,0}(H_{k,0})))$. It remains to compute $(\mathrm{occ}_{n_1}(h_{1,j+1}(H_{1,j+1})), \ldots, \mathrm{occ}_{n_k}(h_{k,j+1}(H_{k,j+1})))$ from $(\mathrm{occ}_{n_1}(h_{1,j}(H_{1,j})), \ldots, \mathrm{occ}_{n_k}(h_{k,j}(H_{k,j})))$.

Proposition 11. occ_n *can be computed inductively with respect to the operations on simple hypergraphs. (This is a special case of a similar theorem in [19]. The main part was already proved in [15].)*

The pointwise ordering $\leq$ turns M_{n_i} into a complete partial order with least element $(0, \ldots, 0)$. The sequence $\mathrm{occ}_{n_i}(h_{i,j}(H_{i,j}))$, $j \in \mathbb{N}$, is an ascending chain in M_{n_i}. Therefore, it has a least upper bound B_i. It is obvious from the definitions that $B_i = \mathrm{occ}_{n_i}(G_i)$. By analysing the recursions in E, it is possible to compute the fixpoint in a finite number of steps.

3.2 Eliminating VR Operations

Let E be a system of regular equations and let $(G_1,\ldots,G_k) = \mathcal{L}(E,\mathcal{SG})$ have sorts $(n_1,\ldots,n_k)$. We derive a new system of equations for G_1, with new variables and equations for them. The variables have the form $x_i\langle \Gamma, m\rangle$, where Γ is a qf definition scheme, consisting of δ, η_l, $l \in L$, and $\sigma_p \equiv p$ for $p \in [m]$; $m \in \mathbb{N}$, $m \geq n_i$. $x_i\langle \Gamma, m\rangle$ will denote the simple hypergraph $\mathrm{Def}_\Gamma(G_i \parallel_{n_i,m} 0_m)$ with many more sources. The number of additional sources n_i' is determined by the value $\mathrm{occ}_{n_i}(G_i)$: The hypergraph has $r \cdot \mathrm{occ}_{n_i}(G_i)(\phi)$ additional sources for each $\phi \in \mathrm{QF}_{n_i}(x_{j_1},\ldots,x_{j_r})$ such that $\mathrm{occ}_{n_i}(G_i)(\phi) \in \mathbb{N}$. Their order of succession is arbitrary but fixed. Of course, a sequence of r successive sources is used to designate one occurrence of ϕ in G_i. Γ will always delete all those hyperedges from G_i such that the subgraphs induced by their attachment vertices occur only finitely often. In particular, those edges whose attachment vertices are all among the m sources are left out. This trick makes the distinction between parallel composition on simple hypergraphs and hypergraphs in general essentially disappear.

It is possible to write an equation for G_1 in the form $x_1 = \mathrm{ren}_g(F(x_1\langle \Gamma, n_1\rangle))$, where $g\colon [n_1] \to [n_1 + n_1']$ is the inclusion, the expression F adds all missing hyperedges according to $\mathrm{occ}_{n_1}(G_1)$ and Γ is a qf definition scheme as described above. Now the main theorem reduces to the following proposition.

Proposition 12. *It is possible to derive an equation for $x_i\langle \Gamma, m\rangle$ using only HR operations from the equation $x_i = t_i$ in E, provided that the infinite bipartite graph $K_{\infty,\infty}$ is not a subgraph of $\mathrm{cg}(G)$, G being the hypergraph denoted by $x_i\langle \Gamma, m\rangle$. If $\mathrm{cg}(G_1)$ does not contain the infinite bipartite graph $K_{\infty,\infty}$ as a subgraph then the procedure will eventually stop creating new variables. This property is decidable.*

4 Conclusion

Taking together what is known about the relationships between the classes of HR-equational and VR-equational hypergraphs, one realises immediately that they mirror analogous relationships between HR-equational and VR-equational sets of finite hypergraphs. This is no coincidence because every equational hypergraph is the colimit of an equational set of finite hypergraphs equipped with homomorphisms in a natural way. (If $x_i = t_i$, $i \in [k]$, are the defining equations for an equational hypergraph then the equations $x_i = t_i + 0_{n_i}$ define an equational set of finite hypergraphs. Under a different interpretation, these equations define a unique regular tree and the set of its finite approximations, respectively. Evaluating these trees yields the original hypergraphs and induces homomorphisms between them.) However, it does not seem to be easy to exploit this fact directly. As an advantage, the independent approaches we used can be combined to handle equational sets of (possibly infinite) hypergraphs.

Acknowledgements The author likes to thank the referees for their helpful comments.

References

[1] Adámek, J., Koubek, V.: Least Fixed Point of a Functor. J. Comput. System Sci. **19** (1979) 163–178
[2] Barthelmann, K.: How to Construct a Hyperedge Replacement System for a Context-Free Set of Hypergraphs. Tech. Rep. 7, Universität Mainz, Institut für Informatik (1996). Submitted for publication
[3] Barthelmann, K.: On Equational Simple Graphs. Tech. Rep. 9, Universität Mainz, Institut für Informatik (1997). Submitted for publication
[4] Barthelmann, K.: When Can an Equational Simple Graph Be Generated by Hyperedge Replacement?. Tech. Rep. 2, Universität Mainz, Institut für Informatik (1998)
[5] Bauderon, M.: Infinite hypergraphs I. Basic properties. Theoret. Comput. Sci. **82** (1991) 177–214
[6] Bauderon, M.: Infinite hypergraphs II. Systems of recursive equations. Theoret. Comput. Sci. **103** (1992) 165–190
[7] Bauderon, M., Courcelle, B.: Graph Expressions and Graph Rewritings. Math. Systems Theory **20** (1987) 83–127
[8] Caucal, D.: On the regular structure of prefix rewriting. Theoret. Comput. Sci. **106** (1992) 61–86
[9] Caucal, D.: On Infinite Transition Graphs Having a Decidable Monadic Theory. In: auf der Heide, F. M., Monien, B. (eds.): Automata, Languages and Programming (ICALP '96), Lecture Notes in Computer Science, Vol. 1099. Springer (1996) 194–205
[10] Courcelle, B.: Fundamental properties of infinite trees. Theoret. Comput. Sci. **25** (1983) 95–169
[11] Courcelle, B.: The Monadic Second-Order Logic of Graphs, II: Infinite Graphs of Bounded Width. Math. Systems Theory **21** (1989) 187–221
[12] Courcelle, B.: Graph Rewriting: An Algebraic and Logic Approach. In: van Leeuwen [23], Ch. 5, 193–242
[13] Courcelle, B.: The monadic second-order logic of graphs IV: Definability properties of equational graphs. Ann. Pure Appl. Logic **49** (1990) 193–255
[14] Courcelle, B.: The monadic second-order logic of graphs III: Tree-decompositions, minors and complexity issues. RAIRO Informatique théorique et Applications/ Theoretical Informatics and Applications **26**, 3 (1992) 257–286
[15] Courcelle, B.: The monadic second-order logic of graphs VII: Graphs as relational structures. Theoret. Comput. Sci. **101** (1992) 3–33
[16] Courcelle, B.: Structural Properties of Context-Free Sets of Graphs Generated by Vertex Replacement. Inform. and Comput. **116** (1995) 275–293
[17] Courcelle, B.: The Expression of Graph Properties and Graph Transformations in Monadic Second-Order Logic. In: Rozenberg, G. (ed.): Handbook of Graph Grammars and Computing by Graph Transformation, Vol. 1, Foundations. World Scientific (1997) Ch. 5, 313–400
[18] Courcelle, B., Engelfriet, J., Rozenberg, G.: Handle-Rewriting Hypergraph Grammars. J. Comput. System Sci. **46** (1993) 218–270
[19] Courcelle, B., Mosbah, M.: Monadic second-order evaluations on tree-decomposable graphs. Theoret. Comput. Sci. **109** (1993) 49–82
[20] Engelfriet, J.: Context-Free Graph Grammars. In: Rozenberg and Salomaa [22], Ch. 3, 125–213

[21] Muller, D. E., Schupp, P. E.: The theory of ends, pushdown automata, and second-order logic. Theoret. Comput. Sci. **37** (1985) 51–75
[22] Rozenberg, G., Salomaa, A. (eds.): Handbook of Formal Languages, Vol. 3, Beyond Words. Springer (1997)
[23] van Leeuwen, J. (ed.): Handbook of Theoretical Computer Science, Vol. B, Formal Models and Semantics. Elsevier (1990)

Spatial and Temporal Refinement of Typed Graph Transformation Systems *

Martin Große–Rhode[1], Francesco Parisi–Presicce[2], and Marta Simeoni[2]

[1] Dip. di Informatica, Università di Pisa, Corso Italia, 40, I – 56125 Pisa, Italy,
mgr@di.unipi.it
[2] Università di Roma *La Sapienza*, Dip. Scienze dell'Informazione,
Via Salaria 113, I-00198 Rome, Italy,
{parisi,simeoni}@dsi.uniroma1.it

Abstract. Graph transformation systems support the formal modeling of dynamic, concurrent, and distributed systems. States are given by their graphical structure, and transitions are modeled by graph transformation rules. In this paper we investigate two kinds of refinement relations for graph transformation systems in order to support the development of a module concept for graph transformation systems. In a spatial refinement each rule is refined by an amalgamation of rules, in a temporal refinement it is refined by a sequence of rules.

1 Introduction

Graph grammars and graph transformation systems, in their different variations, have become a well accepted approach to the formal modeling of systems. (For a survey see [Roz97].) In this paper we investigate refinement relations between graph transformation systems, a question that has been addressed only few in the literature up to now (see [CH95,HCEL96,Par96,Rib96]). Our main concern are refinement relations that preserve the full behaviour of graph transformation systems, as opposed to [CH95,HCEL96] for instance, whose refinement relation guarantees only the existence of specialised transformations in the refining system, not the whole behaviour. Using typed graph transformation systems ([CEL+96]) refinement also supports the *implementation* of a more abstract system by another more concrete one. Thereby type restriction corresponds to the hiding of implementation details.

A possible application of refinement is the development of a module concept for graph transformation systems. Well investigated in the field of programming languages module concepts have been carried over also to formal specification approaches, as for instance algebraic specification of abstract data types (see e.g. [BEP87,EM90]). Basically, a module is given by an export and an import interface, and a body that *implements* the features offered at the export interface, possibly using the features required at the import interface. A necessary

---

* This research has been supported by the TMR Network GETGRATS, ERB-FMRX-CT960061.

formal means to define such modules for formal specifications are *morphisms* between the specification units for the three parts, that model these relationships appropriately. That means, morphisms are required that model the inclusion of the imported features into the body, and morphisms that model the relation between the exported features and their implementation in the body. Since the latter task is of more general nature there should be an embedding of morphisms of the first kind (inclusions) into morphisms of the second kind (implementations). In [EM90] *horizontal* composition operations have been introduced, such as union and composition via import–export interface matching. The essential requirement on the category of specification units to support these horizontal operations is that pushouts (more generally colimits) of specifications exist. For the special and most important case of import–export interface matching it suffices already, if pushouts of inclusions and implementations exist.

The first kind of morphisms between graph transformation systems, corresponding to inclusions, are mappings between the name sets that are compatible with the associated rules. In a refinement morphism names are mapped to *instructions* that indicate how a rule is refined to a composition of rules of the refining system. In a spatial refinement, several rules of the refining system are glued together in parallel (amalgamated) to obtain the effect of the original rule. That means, the different rules of the refining system must be applied at the same time to different, possibly overlapping parts of the actual graph (state), and their simultaneous application yields the same successor graph as the original rule. In a temporal refinement, a sequential composition of rules refines a given one, i.e. the sequential computation steps are refined.

The paper is organized as follows. In the next two sections graph transformation systems and refinements are introduced for the untyped case. Although this case is not very meaningful for applications, the separated presentation makes the presentation easier. In section 2 basic definitions and facts of graph transformation systems and their behaviour are revisited. In section 3 spatial and temporal refinements are introduced. In section 4 types for graph transformation systems and the extension and restriction constructions associated with type morphisms are revisited. Finally in section 5 the results of the previous sections are put together to obtain the results we consider useful for applications. Full proofs and further examples can be found in the technical reports [GPS97a,GPS97b].

2 Graph Transformation Systems

In this section we briefly review the standard definitions and facts of graph transformation systems. A *graph* $G = (N, E, src, tar)$ is given by a set N of nodes, a set E of edges, and functions $src, tar : E \to N$ that assign source and target nodes to each edge. Thus graphs are unlabeled directed graphs that may have multiple edges and loops. A *graph morphism* $f = (f_N, f_E) : G \to G'$ is given by functions $f_N : N \to N'$ and $f_E : E \to E'$ such that $src' \circ f_E = f_N \circ src$ and

$tar' \circ f_E = f_N \circ tar$. With identities and composition being defined component wise this defines the category **Graph**.

A *graph transformation rule* $p = (L \xleftarrow{l} K \xrightarrow{r} R)$ is given by a left graph L, that is matched to the actual state graph when the rule is applied, a right graph R by which the occurrence of L is replaced, and a span $L \leftarrow K \to R$, given by a gluing graph K and graph morphisms to L and R. The span expresses which items of L are related to which items of R. Intuitively, items related in this way are preserved when the rule is applied, and items in $L - K$ are deleted. A *rule morphism* $mp = (mp_L, mp_K, mp_R) : p \to p'$ is given by graph morphisms $mp_L : L \to L'$, $mp_K : K \to K'$, and $mp_R : R \to R'$, that commute with l and l', and r and r' respectively, i.e. $mp_L \circ l = l' \circ mp_K$ and $mp_R \circ r = r' \circ mp_K$. With component wise identities and composition this defines the category **Rule**. The *amalgamation* of two rules w.r.t. a common subrule is their pushout in **Rule**.

A *graph transformation system* $\mathbf{G} = (P, \pi)$ is given by a set P of names, that is considered as its signature, and a mapping $\pi : P \to |\mathbf{Rule}|$ that assigns to each name a rule, thus specifying the behaviour. A *morphism of graph transformation systems*, $f : \mathbf{G} \to \mathbf{G}'$ is a mapping $f : P \to P'$ between the sets of rule names that is compatible with π and π', i.e. $\pi' \circ f = \pi$. With composition and identity inherited from **Set**, this defines the category **GTS**.

Since **Graph** and **Rule** are (isomorphic to) functor categories to **Set** and **GTS** is a comma category to **Set** all three categories are cocomplete.

Given a graph transformation system $\mathbf{G} = (P, \pi)$ a *direct derivation* $p/m : G \Rightarrow H$ over $\mathbf{G}$ from a graph G via a rule p and a matching morphism $m : L \to G$ is a pair (p, S), where $p \in P$, S is a double pushout diagram

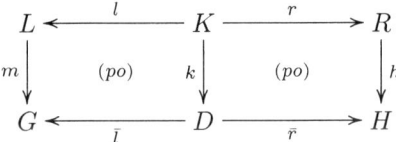

in **Graph**, and $\pi(p) = (L \xleftarrow{l} K \xrightarrow{r} R)$. G is called the *input*, and H the *output* of $p/m : G \Rightarrow H$. A *derivation* $p_1/m_1; \ldots; p_n/m_n : G \Rightarrow H$ over $\mathbf{G}$ from a graph G via rules $p_1, \ldots, p_n$ and matching morphisms $m_1, \ldots, m_n$ is a sequence of direct derivations over $\mathbf{G}$, such that the output of the i'th direct derivation is the input of the $(i+1)$'st direct derivation. The set of all derivations over $\mathbf{G}$ is denoted $Der(\mathbf{G})$. Using amalgamated rules for derivations allows to prescribe synchronized derivations. The expressive power of amalgamated rules is in general higher than sequential composition, see [BFH87]. For a derivation with an amalgamated rule q we use the notation $\bar{q}/n : G \Rightarrow H$. Note that q is a rule here, whereas p in $p/m : G \Rightarrow H$ is a rule name. The set of all derivations over $\mathbf{G}$ with amalgamated rules is denoted $ADer(\mathbf{G})$.

Considering $Der(\mathbf{G})$ as the behaviour of a graph transformation system, morphisms $f : \mathbf{G} \to \mathbf{G}'$ preserve behaviour. I.e., for each derivation $d : G \Rightarrow H$ with $d = (p_1/m_1; \ldots; p_n/m_n)$ in $Der(\mathbf{G})$ there is a derivation $f(d) : G \Rightarrow H$ in $Der(\mathbf{G}')$, where $f(d) = (f(p_1)/m_1; \ldots; f(p_n)/m_n)$. The same holds for $ADer(\mathbf{G})$.

3 Untyped Refinements

As mentioned in the introduction a refinement of a graph transformation system is given by a mapping that associates with each rule name an instruction how to implement the associated rule as a composition of rules of the refining system. In a spatial refinement this composition is an amalgamation, in a temporal refinement a sequence.

Definition 1 (Refinement Instructions). *Let* $\mathbf{G} = (P, \pi)$ *be a graph transformation system. A spatial refinement instruction si on* $\mathbf{G}$ *is defined by:*
$$si = (p_1 \ldots p_k, (\pi(p_i) \xleftarrow{m_{ij}} r_{ij} \xrightarrow{m'_{ij}} \pi(p_j))_{1 \leq i < j \leq k})$$
where $p_1, \ldots, p_k \in P$, $r_{ij} \in |\mathbf{Rule}|$ *and* $\pi(p_i) \xleftarrow{m_{ij}} r_{ij} \xrightarrow{m'_{ij}} \pi(p_j)$ *for* $1 \leq i < j \leq k$ *is a span of morphisms in* **Rule**.

A temporal refinement instruction $ti = p_1 \ldots p_k$ *on* $\mathbf{G}$ *is a string of sort names* $p_1, \ldots, p_k \in P$, *such that* $R_i = L_{i+1}$ *for* $i \in \{1, \ldots, k-1\}$ *if* $\pi(p_j) = (L_j \leftarrow K_j \rightarrow R_j)$.

The sets of spatial refinement instructions and temporal refinement instructions on $\mathbf{G}$ *are denoted* $SRI(\mathbf{G})$ *and* $TRI(\mathbf{G})$ *respectively.*

The rule result(si) of a spatial refinement instruction si is defined as the colimit in **Rule** *of the diagram given by all spans of si with their adjacent rules. The result of a temporal refinement instruction* $ti = p_1 \ldots p_k$ *is given by* $result(ti) = (L_1 \xleftarrow{l} K \xrightarrow{r} R_k)$, *where* K *is the limit of the diagram*
$$L_1 \xleftarrow{l_1} K_1 \xrightarrow{r_1} R_1 = L_2 \xleftarrow{l_2} K_2 \xrightarrow{r_2} \cdots \xleftarrow{l_k} K_k \xrightarrow{r_k} R_k$$
in **Graph**, *and* $l : K \to L_1$ *and* $r : K \to R_k$ *are the corresponding projections.*

Definition 2 (Refinement Morphisms). *Let* $\mathbf{G} = (P, \pi)$ *and* $\mathbf{G}' = (P', \pi')$ *be two graph transformation systems. A spatial (temporal) refinement morphism* $ref : \mathbf{G} \to \mathbf{G}'$ *is a mapping from the set of rule names* P *to the set of spatial (temporal) refinement instructions* $SRI(\mathbf{G}')$ *(resp.* $TRI(\mathbf{G}'))$ *such that, for* $p \in P$, $result(ref(p)) \cong \pi(p)$.

Two refinement morphisms $ref : \mathbf{G} \to \mathbf{G}'$ *and* $ref' : \mathbf{G} \to \mathbf{G}'$ *are equivalent if, for all* $p \in P$, $result(ref(p)) \cong result(ref'(p))$ *in* **Rule**.

From this definition follows immediately that any two refinement morphisms $ref : \mathbf{G} \to \mathbf{G}'$ and $ref' : \mathbf{G} \to \mathbf{G}'$ are equivalent.

Example 1 (Asynchronous Communication). Consider an asynchronous communication of agents P and Q, with writing and reading access to a common channel c. Agent P holds a value a, that it sends to Q. Thus the abstract view of the communication is

   asynch-com :

Refined into intermediate steps this communication would look as follows.

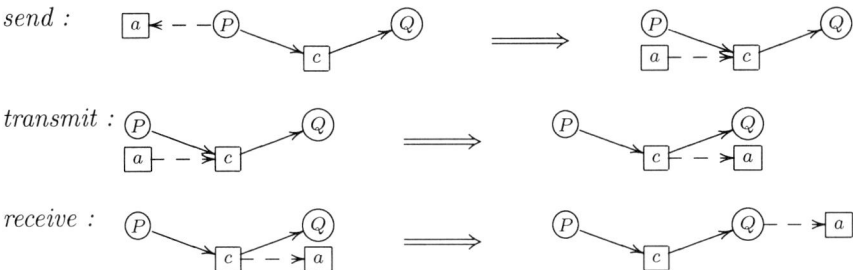

First P sends a to c, then a is transmitted from the input port of c to its output port, where Q can receive it.

The spatial and temporal refinement relations are transitive. Compositions of refinement morphisms are constructed via pullbacks and pushouts respectively ([GPS97b]). Thus it cannot be expected that composition is strictly associative. Moreover, it is impossible to obtain pushouts on this level, that were required for horizontal composition of modules. This is due to the fact that a refinement of a single rule may have arbitrary many component rules, that may be connected in many ways. Thus a common refinement of two given ones, that were required to construct a pushout, does not exist. Therefore we abstract from the concretely given refinement instructions at this point, and proceed with *existence* of spatial refinements alone.

Definition 3 (Categories of Refinements). *The spatial (temporal) refinement category* $\mathbf{SR}_\equiv$ *(resp.* $\mathbf{TR}_\equiv$*) has graph transformation systems as objects and equivalence classes of spatial (temporal) refinements as morphism.*

Since all diagrams in these categories commute, all colimits of refinements exist. They can be constructed as disjoint union (i.e. *coproducts*) of the rule name sets of the components and the induced mappings to the rules. Note that the diagram of morphisms on the underlying sets of names of a colimit diagram in $\mathbf{SR}_\equiv$ or $\mathbf{TR}_\equiv$ need not commute.

Proposition 1 (Colimits of Refinements). *The categories* $\mathbf{SR}_\equiv$ *and* $\mathbf{TR}_\equiv$ *have colimits.*

As mentioned in the introduction pushouts of inclusion morphisms and spatial refinements are required for the horizontal composition of modules. The obvious embedding of morphisms of graph transformation systems into the refinement categories yields a more intuitive way to construct a pushout of (the embedding of) an injective **GTS**–morphism $f : \mathbf{G}_0 \to \mathbf{G}_1$ and a spatial refinement morphism $sr : \mathbf{G}_0 \to \mathbf{G}_2$. In this case the set of names of the pushout can in fact be taken as a pushout in **Set** with the induced mappings to the rules, thus it also commutes.

Since spatial refinements use amalgamations a refining system must be able to use amalgamated rules. This is reflected in the preservation properties for the two kinds of refinements.

Theorem 1 (Preservation of Behaviour). *Let $sr : \mathbf{G} \to \mathbf{G}'$ be a spatial refinement morphism of graph transformation systems. For each derivation $d : G \Rightarrow H$ with $d = (p_1/m_1; \ldots; p_n/m_n)$ in $Der(\mathbf{G})$ there is an amalgamated derivation $\bar{d} : G \Rightarrow H$ in $ADer(\mathbf{G}')$, where $\bar{d} = (q_1/m_1; \ldots; q_n/m_n)$ and $q_i = result(sr(p_i))$.*

Let $\mathbf{G}'$ be a graph transformation system with injective rules, i.e. for each rule name $p' \in P'$ the graph morphisms l_i', r_i' of $\pi'(p_i') = (L_i' \xleftarrow{l_i'} K_i' \xrightarrow{r_i'} R_i')$ are injective, and $tr : \mathbf{G} \to \mathbf{G}'$ be a temporal refinement morphism. Then for each derivation $d : G \Rightarrow H$ with $d = (p_1/m_1; \ldots; p_n/m_n)$ in $Der(\mathbf{G})$ there is a derivation $d' : G \Rightarrow H$ in $Der(\mathbf{G}')$, where $d' = (p_{11}'/m_{11}; \ldots; p_{nk_n}'/m_{nk_n})$ and $tr(p_i) = p_{i1}' \ldots p_{in_i}'$ for $i = 1, \ldots, n$.

4 Typed Graph Transformation Systems

In [CMR96] typed graphs have been introduced as a technical means for the construction of graph processes. This typing, however, may also be considered as a structuring means for graph transformation systems in the usual sense of typing. Morphisms and refinements of graph transformation systems with type graphs as structuring means have been introduced in [CH95,HCEL96,Rib96].

Definition 4 (Categories of Typed Graphs and Typed Rules). *Given a graph TG, a TG-typed graph g is a graph morphism $g : G \to TG$, and a TG-typed graph morphism $k : g \to h$ is a graph morphism with $h \circ k = g$. This defines the category $\mathbf{Graph}_{TG}$.*

The category $\mathbf{Rule}_{TG}$ of TG-typed rules is given by TG-typed graph spans and morphisms, as for untyped rules.

Definition 5 (Retyping, Forgetful Functor and Free Functor). *Let $f : TG \to TG'$ be a graph morphism. f induces a backward retyping functor $f^< : \mathbf{Graph}_{TG'} \to \mathbf{Graph}_{TG}$, $f^<(g') = g^*$ and $f^<(k' : g' \to h') = k^* : g^* \to h^*$ by pullbacks and mediating morphisms as in the following diagram,*

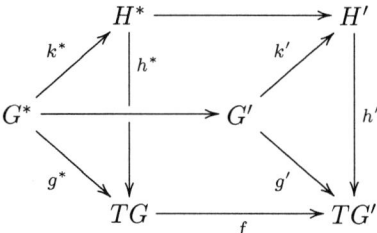

and a forward retyping functor $f^> : \mathbf{Graph}_{TG} \to \mathbf{Graph}_{TG'}$, $f^>(g) = f \circ g$ and $f^>(k : g \to h) = k$ by composition.

Forward and backward retyping functors are left and right adjoints, i.e. for each $f : TG \to TG'$ we have $f^> \dashv f^< : \mathbf{Graph}_{TG} \to \mathbf{Graph}_{TG'}$. Moreover, forward retyping is a functor $\_^> : \mathbf{Graph} \to \mathbf{Cat}$, and backward retyping is a pseudo functor $\_^< :$ **Graph** $\to$ **Cat**^{op}, i.e. $(id_{TG})^< \cong id_{\textbf{Graph}_{TG}}$ and $(e \circ f)^< \cong f^< \circ e^<$.

Typing a graph transformation system means to define a type system as a type graph TG, and have all rules typed w.r.t. TG.

Definition 6 (Typed Graph Transformation System). *A typed graph transformation system* **TG** $= (TG, P, \pi)$ *consists of a type graph TG, a set of rule names P, and a mapping $\pi : P \to |\textbf{Rule}_{TG}|$, associating with each rule name its TG-typed rule.*

A morphism of typed graph transformation systems must first of all relate their type systems, i.e. it must contain a type graph homomorphism. Forward and backward retyping then induce translations to compare the rules of both systems. In general, however, compatibility with the forgetful functor (backward retyping) is too weak to preserve derivations.

Definition 7. *A morphism of typed graph transformation systems, $f = (f_P, f_{TG})$:* **TG** $\to$ **TG**$'$ *is given by a mapping $f_P : P \to P'$ between the sets of rule names and an injective type graph morphism $f_{TG} : TG \to TG'$, such that $f_{TG}^>(\pi(p)) = \pi'(f_P(p))$ for all $p \in P$.*

Typed graph transformation systems and morphisms form a category, called **TGTS**.

Proposition 2 (Colimits). *The category* **TGTS** *has colimits.*

The preservation theorem for morphisms of typed graph transformation systems compares the behaviour of the systems again in both directions. Each derivation of the first system gives rise to a corresponding derivation in the second system, and the forgetful image of the derivation coincides with the original one.

Theorem 2 (Preservation of Behaviour). *Let $f = (f_P, f_{TG})$:* **TG** $\to$ **TG**$'$ *be a TGTS–morphism. For each derivation $d : G \Rightarrow H$ with $d = (p_1/m_1; \ldots; p_n/m_n)$ in Der(**TG**) there is a derivation $f(d) : f_{TG}^>(G) \Rightarrow f_{TG}^>(H)$ in Der(**TG**$'$), where $f(d) = (f(p_1)/f_{TG}^>(m_1); \ldots; f(p_n)/f_{TG}^>(m_n))$. Furthermore $f^<(f(d) : f_{TG}^>(G) \Rightarrow f_{TG}^>(H)) = (d : G \Rightarrow H)$.*

5 Typed Refinements

Having defined refinements and typing we are now in a position to combine the results and obtain the constructions that we consider appropriate for the module concept discussed in the introduction, and other applications. In the following definition the sets $TySRI(\textbf{G}')$ and $TyTRI(\textbf{G}')$ of *typed spatial and temporal refinement instructions* are given by replacing all rules by TG'-typed rules. Their results are the corresponding constructions in $\textbf{Rule}_{TG'}$.

Definition 8 (Typed Refinements). Let $\mathbf{TG} = (TG, P, \pi)$ and $\mathbf{TG}' = (TG', P', \pi')$ be typed graph transformation systems. A typed spatial (temporal) refinement morphism $tr = (r, f_{TG}) : \mathbf{TG} \to \mathbf{TG}'$ is given by a mapping $r : P \to TySRI(\mathbf{G}')$, (resp. $r : P \to TyTRI(\mathbf{G}')$) and an injective type graph morphism $f_{TG} : TG \to TG'$, such that $result(r(p)) \cong f_{TG}^{>}(\pi(p))$ for all $p \in P$.

Theorem 3 (Preservation of Behaviour). Let $ref = (r, f_{TG}) : \mathbf{TG} \to \mathbf{TG}'$ be a typed refinement. For each derivation $d : G \Rightarrow H$ with $d = (p_1/m_1; \ldots; p_n/m_n)$ in $Der(\mathbf{TG})$ there is a derivation $ref(d) : f_{TG}^{>}(G) \Rightarrow f_{TG}^{>}(H)$ over $Der(\mathbf{TG}')$, where $ref(d) = (q_1/f_{TG}^{>}(m_1); \ldots; q_n/f_{TG}^{>}(m_n))$ and $q_i = result(r(p_i))$. Furthermore $f_{TG}^{<}(ref(d) : f_{TG}^{>}(G) \Rightarrow f_{TG}^{>}(H)) = (d : G \Rightarrow H)$.

Example 2. The channel used for the asynchronous communication of agents P and Q in example 1 should be considered as an internal communication infrastructure for the group that contains P and Q, and should not be visible from the outside. With the type graphs for the abstract system (left) and the refined system (right)

(that we implicitly already used in example 1) we obtain as visible (abstract) behaviour the rule

which is given by the backward retyping along the type inclusion of the rule *asynch-com*. Then the rules *send*, *transmit*, and *receive* given in example 1, together with rules to connect and disconnect agents P and Q via channel c, form a typed temporal refinement of this abstract behaviour, using the internal communication channel.

6 Conclusion

In this paper we have started the investigation of refinements of graph transformation systems. For two cases we have shown how these can be defined in a categorical framework. The background has been the use of refinement morphisms as relations between the interfaces and the body of a graph transformation module, similar to algebraic specification modules. The pattern has been the same in both cases, and the investigation of further possibilities of refinement, as e.g. the combination of temporal and spatial refinement as discussed here, would also pursue this pattern. First the shape of the refinement instructions is defined, basing on operations of rules like amalgamation and sequential composition. It must be shown then, that these refinement instructions can be composed vertically in order to obtain transitivity of refinement and to allow for a categorical treatment. The latter proceeds by taking equivalence classes of refinement morphisms, i.e. a pre–order of refinement relationships is considered. This reflects the fact that the vertical composition in most cases is not associative, and pushouts do not exist. However, passing to the abstract level of existence of refinements instead of concrete refinement constructions yields the framework for the investigation and general formulation of further properties, like the existence of pushouts for the general case, and the special case of pushout of a refinement with an inclusion morphism. This particular result is especially important for the application to graph transformation module composition, since it corresponds to the composition of modules via their export and import interfaces.

References

[BEP87]   E.K. Blum, H. Ehrig, and F. Parisi-Presicce. Algebraic specification of modules and their basic interconnections. *JCSS*, 34(2-3):293–339, 1987.

[BFH87]   P. Böhm, H.-R. Fonio, and A. Habel. Amalgamation of graph transformations: a synchronization mechanism. *JCSS*, 34:377–408, 1987.

[CEL+96] A. Corradini, H. Ehrig, M. Löwe, U. Montanari, and J. Padberg. The category of typed graph grammars and its adjunctions with categories of derivations. In *5th Int. Workshop on Graph Grammars and their Application to Computer Science, Williamsburg '94*, Springer LNCS 1073, pages 240–256. 1996.

[CH95]   A. Corradini and R. Heckel. A compositional approach to structuring and refinement of typed graph grammars. *Proc. of SEGRAGRA'95 "Graph Rewriting and Computation", Electronic Notes of TCS*, 2, 1995. http://www.elsevier.nl/locate/entcs/volume2.html .

[CMR96]   A. Corradini, U. Montanari, and F. Rossi. Graph processes. *Fundamenta Informaticae*, 26(3,4):241–266, 1996.

[Ehr79]   H. Ehrig. Introduction to the algebraic theory of graph grammars. In V. Claus, H. Ehrig, and G. Rozenberg, editors, *1st Graph Grammar Workshop* Springer LNCS 73, pages 1–69. 1979.

[EM90]   H. Ehrig and B. Mahr. *Fundamentals of Algebraic Specification 2: Module Specifications and Constraints*, volume 21 of *EATCS Monographs on Theoretical Computer Science*. Springer Verlag, Berlin, 1990.

[GPS97a] M. Große–Rhode, F. Parisi-Presicce, and M. Simeoni. Concrete spatial refinement constructions for graph transformation systems. Technical Report 97-10, Università di Roma *La Sapienza*, 1997.

[GPS97b] M. Große–Rhode, F. Parisi-Presicce, and M. Simeoni. Spatial and temporal refinement of typed graph transformation systems. Technical Report 97-11, Università di Roma *La Sapienza*, 1997.

[HCEL96] R. Heckel, A. Corradini, H. Ehrig, and M. Löwe. Horizontal and vertical structuring of typed graph transformation systems. *MSCS*, pages 1–35, 1996.

[Par96]   F. Parisi-Presicce. Transformation of graph grammars. In *5th Int. Workshop on Graph Grammars and their Application to Computer Science, Williamsburg '94*, Springer LNCS 1073, 1996.

[Rib96]   L. Ribeiro. *Parallel Composition and Unfolding Semantics of Graph Grammars*. PhD thesis, TU Berlin, 1996.

[Roz97]   G. Rozenberg, editor. *Handbook of Graph Grammars and Computing by Graph Transformations, Volume 1: Foundations*. World Scientific Publishing, 1997.

Approximating Maximum Independent Sets in Uniform Hypergraphs*

Thomas Hofmeister and Hanno Lefmann **

LS Informatik 2, Universität Dortmund, D-44221 Dortmund, Germany
{hofmeist,lefmann}@Ls2.cs.uni-dortmund.de

Abstract. We consider the problems of approximating the independence number and the chromatic number of k-uniform hypergraphs on n vertices. For fixed $k \geq 2$, we describe for both problems polynomial time approximation algorithms with approximation ratios $O(n/(\log^{(k-1)} n)^2)$. This extends results of Boppana and Halldórsson [5] who showed for the case of graphs that an approximation ratio of $O(n/(\log n)^2)$ can be achieved in polynomial time. On the other hand, assuming $NP \neq ZPP$, there are no polynomial time algorithms for the independence number and the chromatic number of k-uniform hypergraphs with approximation ratio of $n^{1-\varepsilon}$ for any fixed $\varepsilon > 0$.

1 Introduction

For a graph $G = (V, E)$ with vertex set V and edge set $E \subseteq [V]^2$, a subset $I \subseteq V$ is called *independent*, if G contains no edges on I. The *independence number* $\alpha(G)$ is the size of a largest independent set. Computing an independent set I with $|I| = \alpha(G)$ is an NP-hard problem. Therefore, polynomial time approximation algorithms for this problem have been investigated, cf. [1], [9]. For an optimization instance I, let $OPT(I)$ denote the value of an optimal solution and $A(I)$ the solution found by an algorithm A. The approximation ratio $AR(n)$ of algorithm A is defined by $\max_{I_n} \left\{ \frac{OPT(I_n)}{A(I_n)}, \frac{A(I_n)}{OPT(I_n)} \right\}$, where the maximum is taken over all instances I_n of input size n (e.g., graphs on n vertices).

A greedy strategy for computing independent sets was studied by Halldórsson and Radhakrishnan [11] and approximation ratios in terms of the average degree $d = 2 \cdot |E|/|V|$ and the maximum degree Δ of a graph on $|V| = n$ vertices were derived, namely, $AR(n) \leq (d+2)/2$ and $AR(n) \leq (\Delta + 2)/3$, respectively.

For arbitrary graphs on n vertices an approximation ratio of $O(n/(\log n)^2)$ can be achieved in polynomial time as was shown by Boppana and Halldórsson [5]. They also showed that in graphs G on n vertices with $\alpha(G) \geq n/k+m$, where k is a fixed integer, one can find in polynomial time an independent set of size

---

* This research was supported by the Deutsche Forschungsgemeinschaft as part of the Collaborative Research Center "Computational Intelligence" (SFB 531).
** Part of this work was done during this author's visit at Humboldt-Universität zu Berlin.

at least $\Omega(m^{1/(k-1)})$. This was improved by Alon and Kahale [2] who applied semidefinite programming in conjunction with the Lovász θ-number $\theta(G)$ of a graph G and they showed that if $\theta(G) \geq n/k + m$, then an independent set of size at least $\tilde{\Omega}(m^{3/(k+1)})$ can be found in randomized polynomial time.

As far as inapproximability is concerned, the latest result is due to Håstad [12] who showed that for each fixed $\varepsilon > 0$, there can be no polynomial time algorithm with approximation ratio of $n^{1/2-\varepsilon}$, unless $P=NP$. Under the assumption that $NP \neq ZPP$, the same holds even for an approximation ratio of $n^{1-\varepsilon}$.

The corresponding problem for hypergraphs has been less studied. A *hypergraph* $\mathcal{H} = (V, \mathcal{E})$ is given by a set V of vertices and a set $\mathcal{E}$ of edges where $E \subseteq V$ for every edge $E \in \mathcal{E}$. A hypergraph is called *k-uniform* if $|E| = k$ for every edge $E \in \mathcal{E}$. A subset $I \subseteq V$ is called *independent* if I contains no edges from $\mathcal{H}$, i.e., $E \not\subseteq I$ for each edge $E \in \mathcal{E}$. The size of a largest independent set in $\mathcal{H}$ is called the *independence number* of $\mathcal{H}$ and is denoted by $\alpha(\mathcal{H})$.

Some results are concerned with finding in parallel a maximal independent set in a hypergraph, see e.g. [6], [15]. However, the sizes of maximal independent sets can be much smaller than the size of a maximum independent set. In terms of the average degree $d^{k-1} := k \cdot |\mathcal{E}|/|V|$ of a k-uniform hypergraph $\mathcal{H} = (V, \mathcal{E})$, derandomizing a probabilistic argument of Spencer [19] yields a linear time algorithm with approximation ratio $O(d)$. However, d can be as large as $\Omega(n)$.

We consider here the problem of approximating maximum independent sets in arbitrary k-uniform hypergraphs. It turns out that one can achieve in polynomial time an approximation ratio of $O(n/(\log^{(k-1)} n)^2)$. Here, $\log^{(k)} n$ denotes the k-fold iterated logarithm $\log \cdots \log n$.

By a simple reduction from the case of graphs, inapproximability results of the order $\Omega(n^{1-\varepsilon})$ for each given $\varepsilon > 0$ can be derived.

We also consider the problem of coloring the vertices of a k-uniform hypergraph $\mathcal{H}$ with as few colors as possible. Call a coloring of the vertices of $\mathcal{H}$ *proper* if no edge is monochromatic. If the vertices of $\mathcal{H}$ can be colored properly with l colors, then $\mathcal{H}$ is called *l-colorable*. Let $\chi(\mathcal{H})$ denote the *chromatic number* of a hypergraph $\mathcal{H}$, i.e., the minimum l such that $\mathcal{H}$ is l-colorable.

For graphs, determining the chromatic number is an NP-hard problem. The currently best known polynomial time algorithm for coloring a 3-colorable graph on n vertices is by Blum and Karger [4] and uses $\tilde{O}(n^{3/14})$ colors. For graphs with larger chromatic number, the currently best known result is due to Karger, Motwani and Sudan [14] who showed that graphs on n vertices with maximum degree Δ can be colored in polynomial time using at most
min $\{\tilde{O}(\Delta^{1-2/\chi(G)}), \tilde{O}(n^{1-3/(\chi(G)+1)})\}$ colors.

For arbitrary graphs on n vertices, the first polynomial time approximation algorithm for graph coloring was given by Johnson [13] with an approximation ratio of $O(n/\log n)$. This was improved by Wigderson in [20] to $O(n \cdot (\log \log n/ \log n)^2)$ and further by Boppana and Halldórsson [5] to $O(n/(\log n)^2)$. The latest improvement is by Halldórsson [10] who gave a polynomial time algorithm to achieve an approximation ratio of $O(n \cdot (\log \log n)^2/(\log n)^3)$. Recently,

Feige and Kilian [8] showed that assuming $NP \neq ZPP$, given any fixed $\varepsilon > 0$, there is no polynomial time algorithm which approximates the chromatic number within a factor $O(n^{1-\varepsilon})$.

For the corresponding problem of coloring hypergraphs not much is known. In fact, just recently the first polynomial time algorithm capable of coloring a 2-colorable k-uniform hypergraph on n vertices with a sublinear number of colors was given by Kelsen, Mahajan and Ramesh [16]. Their algorithm uses $O(n^{1-1/k} \cdot (\log n)^{1-1/k})$ colors.

Our algorithm works not only for 2-colorable k-uniform hypergraphs, but for k-uniform hypergraphs without any restrictions on their chromatic number. It achieves an approximation ratio of $O(n/(\log^{(k-1)} n)^2)$ for the chromatic number. Thus, if applied to k–uniform hypergraphs which are l–colorable for a fixed $l > 0$, our algorithm also only uses a sublinear number of colors, though in the case $l = 2$, the bound from [16] is better. Nevertheless, it seems that the technique from [16] can not be applied to hypergraphs with chromatic number at least 3.

By applying semidefinite programming, it was shown in [16] that 2-colorable hypergraphs on n vertices where each edge contains at most 3 vertices can be colored with at most $O(n^{2/9} \cdot (\log n)^{17/8})$ colors in polynomial time. The assumption on 2-colorability was essential in those arguments. The semidefinite programming approach does not seem to be applicable for k-uniform hypergraphs where $k \geq 4$, cf. [16].

Our approximation ratio $O(n/(\log^{(k-1)} n)^2)$ is sublinear, but still rather large. Again, one should note that under the assumption $NP \neq ZPP$, the following holds: For each fixed $\varepsilon > 0$, there is no polynomial time algorithm which approximates the chromatic number of hypergraphs on n vertices with approximation ratio $O(n^{1-\varepsilon})$.

2 An Approximation Algorithm

In addition to the notions given in the introduction, we also use the notion of a *clique* in a k-uniform hypergraph which is a complete subhypergraph of $\mathcal{H}$, i.e., contains, say, l vertices and all $\binom{l}{k}$ edges. Note that an approximation algorithm for computing independent sets can always be applied to the complement of a hypergraph to obtain an approximation algorithm for cliques, with the same approximation ratio.

The case $k=2$, i.e., graphs, has been treated by Boppana and Halldórsson [5], who described the following algorithm 2–Ramsey.

Algorithm 2–Ramsey: Transform the graph into a binary tree, its vertices corresponding to the vertices of the graph, in such a way that for each vertex v the set of its left descendants contains exactly the non-neighbours of v. Given any path from the root to a leaf in this tree, the leaf and the set of vertices on this path which have an outgoing right edge, form a clique C. Likewise, the leaf and the vertices with an outgoing left edge form an independent set I. Compute a path with the largest number of right edges and a path with the largest number of left edges and the corresponding sets C and I. A Ramsey-type argument shows:

Theorem 1. [5] *In graphs on n vertices, algorithm 2–Ramsey returns an independent set I and a clique C such that $|I| \cdot |C| \geq 1/4 \cdot (\log n)^2$.*

Our algorithm uses a transformation which constructs $(k-1)$-uniform hypergraphs from k-uniform hypergraphs:

Definition 1. *Let $2 \leq g < k$, and let $\mathcal{H}_g$ and $\mathcal{H}_k$ be g- and k-uniform hypergraphs, respectively, both with the same totally ordered vertex set V.*

Then, $\mathcal{H}_k$ is the "ordered canonical extension" of $\mathcal{H}_g$ whenever the following holds for all k-element subsets $E = \{v_1, \ldots, v_k\}$ with $v_1 < \ldots < v_k$ of the vertices:
E is an edge in $\mathcal{H}_k$ if and only if $\{v_1, \ldots, v_g\}$ is an edge in $\mathcal{H}_g$.

Moreover, $\mathcal{H}_k$ is the "unordered canonical extension" of $\mathcal{H}_g$ whenever the following holds for all k-element subsets E of the vertices:
E is an edge in $\mathcal{H}_k$ if and only if there is some edge e in $\mathcal{H}_g$ such that $e \subseteq E$.

It is easy to show that if the k-uniform hypergraph $\mathcal{H}_k$ is the (unordered) canonical extension of some g-uniform hypergraph $\mathcal{H}_g$ with $g < k$, then all cliques of size at least k in $\mathcal{H}_g$ are also cliques in $\mathcal{H}_k$ and all independent sets of size at least k in $\mathcal{H}_g$ are also independent sets in $\mathcal{H}_k$.

Define an algorithm k–Ramsey as follows:

Algorithm k–Ramsey: If $k = 2$, apply the procedure by Boppana and Halldórsson. Otherwise, compute some induced subhypergraph of $\mathcal{H}_k$ which is the ordered canonical extension of a $(k-1)$-uniform hypergraph $\mathcal{H}_{k-1}$ and which has $\Omega((\log n)^{1/(k-1)})$ vertices. Apply $(k-1)$–Ramsey to $\mathcal{H}_{k-1}$.
Return the clique C and the independent set I that $(k-1)$–Ramsey returns.

The crucial point is that $\mathcal{H}_{k-1}$ should have a number of vertices which is not too small. The existence of such a hypergraph follows from the next lemma which is due to Erdös and Rado [7], see also [17].

Lemma 1. [7] *Let $k \geq 2$ be fixed. Given a k-uniform hypergraph $\mathcal{H}_k$ on n vertices, one can find in polynomial time a subset S of the vertices with $|S| = \Omega((\log n)^{1/(k-1)})$, and a $(k-1)$-uniform hypergraph $\mathcal{H}_{k-1}$ such that the induced subhypergraph of $\mathcal{H}_k$ on S is the ordered canonical extension of $\mathcal{H}_{k-1}$.*

Proof. Let $A, B \subseteq \{1, \ldots, n\}$ be nonempty sets such that $\max A < \min B$. We say that (A, B) is *good* if the following holds:

For each $(k-1)$–element subset S of A and for all $x, y \in B$ it holds that $S \cup \{x\}$ is an edge in $\mathcal{H}_k$ if and only if $S \cup \{y\}$ is an edge in $\mathcal{H}_k$.

Assume that (A, B) is a good pair. Set $A' := A \cup \{\min B\}$ and let $j := |A|$. We claim that there is a subset $B' \subseteq B$ such that (A', B') is good and such that

$$|B'| \geq (|B| - 1)/2^{\binom{j}{k-2}}. \tag{1}$$

To see this, consider all $(k-1)$-element subsets S' of A'. If S' does not contain $\min B$, then the goodness condition holds for S' since (A, B) is good.

There are $\binom{j}{k-2}$ subsets S_i of A' which have cardinality $(k-1)$ and which contain $\min B$. Mark each $b \in B \setminus \{\min B\}$ with a bitstring of length $\binom{j}{k-2}$ with the meaning that the i-th bit is 1 if $S_i \cup \{b\}$ is an edge in $\mathcal{H}_k$ and 0 otherwise. By the pigeonhole principle, there must be a subset $B' \subseteq B \setminus \{\min B\}$ of cardinality at least $(|B|-1)/2^{\binom{j}{k-2}}$ in which all elements are marked with the same bitstring. Hence, (A', B') is good. It is clear that B' can be found in polynomial time. We apply the following procedure:

$$A = \{1, \ldots, k-2\}; B = \{k-1, \ldots, n\}; j = k-2;$$
while $B \neq \emptyset$ **do**
  compute B' from B as sketched above ;
  $A = A \cup \{\min B\}$; $B = B'$; $j = j+1$;
end;
return A;

The induced subhypergraph of $\mathcal{H}_k$ on the returned vertex set A is the ordered canonical extension of a $(k-1)$-uniform hypergraph on the same vertex set.

It remains to estimate $|A|$. Before each execution of the while-loop, the cardinality of B is $\Omega(n/2^{\binom{j}{k-1}})$: This is clear at the first execution and after that, we have by (1): $|B'| = \Omega(n/(2^{\binom{j}{k-1}} \cdot 2^{\binom{j}{k-2}})) = \Omega(n/2^{\binom{j+1}{k-1}})$. Thus, the while-loop is executed $\Omega((\log n)^{1/(k-1)})$ times which is also a lower bound for $|A|$. □

Theorem 2. *Let $k \geq 2$ be fixed. Given a k-uniform hypergraph $\mathcal{H}_k$ on n vertices, algorithm k-Ramsey returns in polynomial time a clique C and an independent set I such that $|C| \cdot |I| \geq c \cdot (\log^{(k-1)} n)^2$ for some constant $c > 0$.*

Proof. By induction. For $k=2$, see Theorem 1. In the induction step, algorithm $(k-1)$-Ramsey is applied to a hypergraph with $\Omega((\log n)^{1/(k-1)})$ vertices. We then apply the induction hypothesis. □

Algorithm k-Ramsey guarantees that we either find a reasonable large clique or an independent set in a given k-uniform hypergraph. If however the hypergraph contains no large clique, then the size of an independent set should be large. The idea, as in [5], is to remove independent sets.

In step i, algorithm k-Ramsey is applied to a graph on n_i vertices and returns a clique C_i and an independent set I_i. Remove the vertices from I_i and all incident edges from the hypergraph. This is repeated until no vertex is remaining. Let C be the clique C_i of maximum cardinality. Note that the removed independent sets I_i, $i = 1, \ldots, COL$, yield a proper coloring of the original hypergraph.

The following lemma was used in [5] for graphs, it also holds for hypergraphs and an implicit proof of it can be found in [18], Section 1.4.

Lemma 2. *If in k-uniform hypergraphs on i vertices one can find an independent set of size at least $f(i)$, where $f(i)$ is non-decreasing and $f(i) > 0$, then one* can find a proper coloring for every k-uniform hypergraph on n vertices with at most $\sum_{i=1}^{n} 1/f(i)$ colors.

The returned sets C_i and I_i satisfy $|C_i| \cdot |I_i| \geq c \cdot (\log^{(k-1)} n_i)^2$, i.e., $f(i) \geq max\{1, c \cdot (\log^{(k-1)} i)^2 / |C|\}$. By Lemma 2, we have (for $n \geq n_0$ and some constant $D > 0$):

$$COL \leq D + \sum_{i=D}^{n} \frac{1}{c} \cdot \frac{|C|}{(\log^{(k-1)} i)^2} \leq c' \cdot \frac{n \cdot |C|}{(\log^{(k-1)} n)^2}.$$

Let $cl(\mathcal{H})$ be the size of a largest clique in $\mathcal{H}$. Since $cl(\mathcal{H}) \leq (k-1) \cdot \chi(\mathcal{H})$, one obtains

$$\frac{cl(\mathcal{H})}{k-1} \leq \chi(\mathcal{H}) \leq COL \leq c' \cdot \frac{n \cdot |C|}{(\log^{(k-1)} n)^2} \leq c' \cdot \frac{n \cdot cl(\mathcal{H})}{(\log^{(k-1)} n)^2} \leq \frac{c_k \cdot n \cdot \chi(\mathcal{H})}{(\log^{(k-1)} n)^2}.$$

From this chain of inequalities, one obtains (simultaneously) the upper bounds for $cl(\mathcal{H})/|C|$ as well as for $COL/\chi(\mathcal{H})$ expressed in the following theorem:

Theorem 3. *Let $k \geq 2$ be fixed. There is a polynomial time algorithm which approximates the independence number $\alpha(\mathcal{H})$, the clique number $cl(\mathcal{H})$ and the chromatic number $\chi(\mathcal{H})$ of a k-uniform hypergraph $\mathcal{H}$ on n vertices with approximation ratio $O(n/(\log^{(k-1)} n)^2)$.*

One might consider the question how well the algorithm performs on random hypergraphs. The answer is that unfortunately, the approximation ratio which we can guarantee is only slightly better than the trivial algorithms which pick one vertex as the independent set and color the graph with n colors. Consider for example random k-uniform hypergraphs on n vertices where edges are present with probability $1/2$ independently of each other. For such a random hypergraph, one can show that almost always $cl(\mathcal{H}), \alpha(\mathcal{H}) = (1+o(1)) \cdot (k! \cdot \log n)^{\frac{1}{k-1}}$. Moreover, we almost always have $\chi(\mathcal{H}) = (1+o(1)) \cdot n/(k! \cdot \log n)^{\frac{1}{k-1}}$. This can be seen by using similar techniques as in the case of graphs, cf. [3].

Considering the product $\dfrac{cl(\mathcal{H})}{|C|} \cdot \dfrac{COL}{\chi(\mathcal{H})} \leq \dfrac{cl(\mathcal{H})}{|C|} \cdot c' \cdot \dfrac{n \cdot |C|}{(\log^{(k-1)} n)^2} \cdot \dfrac{1}{\chi(\mathcal{H})}$, we see that it is almost always bounded by

$$O\left(\frac{(\log n)^{\frac{2}{k-1}}}{(\log^{(k-1)} n)^2}\right).$$

Note that for $k = 2$, this is a constant, while for $k \geq 3$, it is growing with n.

The trivial strategies on the other hand almost always lead to a product of the approximation ratios of $O((\log n)^{2/(k-1)})$.

3 Excluding l-Cliques

Very often, the approximation ratio can be improved if we know that our input graphs are taken from a certain subset of all graphs only. A particularly interesting subset which has often been considered in the literature is the set of all graphs which for some fixed $l \geq 2$ do not contain any clique on l vertices ("l–clique").

Boppana and Halldórsson observed in [5] that if a graph G on n vertices does not contain any l-clique for some $l \leq 2\log n$, then G contains an independent set of size at least $\Omega(l \cdot n^{1/(l-1)})$, and for fixed $l \geq 2$ such an independent set can be found in polynomial time. They also showed that this implies that in graphs G on n vertices with $\alpha(G) \geq n/l + m$, where $l \geq 3$ is fixed, one can always find in polynomial time an independent set of size at least $\Omega(m^{1/(l-1)})$.

If we are given a hypergraph without l–cliques, then the reduction technique used in the algorithm k–Ramsey also guarantees that the graph to which we apply 2–Ramsey is without l–cliques, hence instead of using 2–Ramsey in the final step of k–Ramsey, any of the procedures mentioned above can also be used. As an example, one obtains

Theorem 4. *In every k–uniform hypergraph on n vertices which does not contain any l–cliques (for some fixed l), one can find in polynomial time an independent set of size at least $\Omega((\log^{(k-2)} n)^{1/(2(l-1))})$.*

4 Negative Results

The negative results follow from a straightforward transformation from the negative results for graphs. We list those results here for completeness. They follow from considering the unordered canonical extensions.

Theorem 5. *Let $k \geq 2$ be a fixed integer and let $\varepsilon > 0$ be fixed. One cannot approximate in polynomial time the size of a maximum independent set (or clique) in k-uniform hypergraphs on n vertices within a factor of $O(n^{1-\varepsilon})$, unless $NP=ZPP$.*

Proof. For $k=2$, the assertion holds by the result of Håstad [12]. Now assume $k > 2$ and that we had an approximation algorithm with ratio $O(n^{1-\varepsilon})$ for k-uniform hypergraphs. Construct an approximation algorithm for graphs with the same ratio as follows: First, we check in polynomial time $O(n^k)$ whether the independence number of a graph G is smaller than k. If this is the case, then we can compute the independence number exactly. Otherwise, let $\mathcal{H}_k$ be the k-uniform hypergraph which is the unordered canonical extension of the graph G. Then, G and $\mathcal{H}_k$ have the same independence number. □

We apply a similar transformation in the case of the chromatic number:

Theorem 6. *Let $k \geq 2$ be a fixed integer and let $\varepsilon > 0$ be fixed. One cannot approximate in polynomial time the chromatic number of a k-uniform hypergraph on n vertices within a factor of $O(n^{1-\varepsilon})$, unless $NP = ZPP$.*

Proof. For $k=2$, the assertion holds by the result of Feige and Kilian [8]. For $k \geq 3$, consider the unordered canonical extension $\mathcal{H}_k$ of a given graph G. Then,

$$\chi(\mathcal{H}_k) \leq \chi(G) \leq (k-1) \cdot \chi(\mathcal{H}_k).$$

The first inequality holds since every proper coloring of G is a proper coloring of $\mathcal{H}_k$ and for the second inequality observe that if we have a coloring of $\mathcal{H}_k$ with color classes $C_1, \ldots, C_l$, then for each class C_i with $|C_i| \geq k$, we can color the corresponding vertices of G with the same color. If, however $|C_i| < k$, then we color the corresponding vertices in G by $|C_i|$ distinct colors. An approximation algorithm for k-uniform hypergraphs can hence be turned into an algorithm for graphs with an approximation ratio which is only worse by at most the constant factor $(k-1)$. □

5 Final Remarks and Questions

Can the approximation algorithm from [10] replace algorithm 2–Ramsey as the underlying procedure in an approximation algorithm for k–uniform hypergraphs?

Is it possible to improve (with respect to polynomial time algorithms) the approximation ratio to $o(n/(\log^{(k-1)} n)^2)$ for the coloring problem or the maximum independent set problem for k-uniform hypergraphs?

It might also be interesting to investigate the approximation ratio with respect to polynomial time algorithms for the maximum independent set problem for l–clique free graphs or hypergraphs on n vertices. The algorithm from [5] can be seen as an approximation algorithm for graphs with approximation ratio of $O(n^{1-\frac{1}{l-1}})$ which shows that the inapproximability results from [12] do not carry over to this case, when l is fixed.

References

1. N. Alon, L. Babai and A. Itai, *A Fast and Simple Randomized Parallel Algorithm for the Maximal Independent Set Problem*, Journal of Algorithms 7, 1986, 567-583.
2. N. Alon and N. Kahale, *Approximating the Independence Number via the θ-Function*, Mathematical Programming, to appear.
3. N. Alon and J. Spencer, *The Probabilistic Method*, Wiley & Sons, NY, 1992.
4. A. Blum and D. Karger, *An $\tilde{O}(n^{3/14})$-coloring Algorithm for 3-colorable Graphs*, Information Processing Letters 61, 1997, 49-53.
5. R. Boppana and M. Halldórsson, *Approximating Maximum Independent Sets by Excluding Subgraphs*, BIT 32, 1992, 180-196, also in: Proc. 2nd Scandin. Workshop on Algorithm Theory (SWAT), Springer, LNCS 447, 1990, 13-25.
6. E. Dahlhaus, M. Karpinski and P. Kelsen, *An Efficient Parallel Algorithm for Computing a Maximal Independent Set in a Hypergraph of Dimension 3*, Information Processing Letters 42, 1992, 309-313.
7. P. Erdös and R. Rado, *Combinatorial Theorems on Classification of Subsets of a Given Set*, Proceedings London Mathematical Society 2, 1952, 417-439.
8. U. Feige and J. Kilian, *Zero Knowledge and the Chromatic Number*, Proc. 11th IEEE Conference on Computational Complexity, 1996, 278-287.
9. M. Goldberg and T. Spencer, *An Efficient Parallel Algorithm that Finds Independent Sets of Guaranteed Size*, SIAM Journal of Discrete Math. 6, 1993, 443-459.

10. M. Halldórsson, *A Still Better Performance Guarantee for Approximate Graph Coloring*, Information Processing Letters 45, 1993, 19-23.
11. M. Halldórsson and J. Radhakrishnan, *Greed is Good: Approximating Independent Sets in Sparse and Bounded-degree Graphs*, Proc. 26th ACM Symposium on the Theory of Computing (STOC), 1994, 439-448.
12. J. Håstad, *Clique is Hard to Approximate Within $n^{1-\varepsilon}$*, Proc. 37th IEEE Symposium on Foundations of Computer Science (FOCS), 1996, 627-636.
13. D. S. Johnson, *Worst Case Behaviour of Graph Coloring Algorithms*, Proc. 5th Southeastern Conference on Combinatorics, Graph Theory and Computing, Congressus Numerantium X, 1974, 513-527.
14. D. Karger, R. Motwani and M. Sudan, *Approximate Graph Coloring by Semidefinite Programming*, Proc. 35th IEEE Symposium on Foundations of Computer Science (FOCS), 1994, 2-13.
15. P. Kelsen, *On the Parallel Complexity of Computing a Maximal Independent Set in a Hypergraph*, Proc. 24th ACM Symposium on the Theory of Computing (STOC), 1992, 339-350.
16. P. Kelsen, S. Mahajan and H. Ramesh, *Approximate Hypergraph Coloring*, Proc. 6th Scandin. Workshop on Algorithm Theory (SWAT), Springer, 1996, 41-52.
17. J. Nešetřil, *Ramsey Theory*, in: Handbook of Combinatorics Vol II, eds. R. L. Graham, M. Grötschel and L. Lovász, North-Holland, 1995, 1331-1403.
18. R. Motwani, P. Raghavan, *Randomized Algorithms*, Cambridge Univ. Press, 1995.
19. J. Spencer, *Turán's Theorem for k-Graphs*, Discrete Math. 2, 1972, 183–186.
20. A. Wigderson, *Improving the Performance Guarantee for Approximate Graph Coloring*, Journal of the ACM 30, 1983, 729-735.

Representing Hyper-Graphs by Regular Languages *

Salvatore La Torre and Margherita Napoli

Dipartimento di Informatica ed Applicazioni
Università degli Studi di Salerno
84081 Baronissi, Italy.
{sallat,napoli}@dia.unisa.it

Abstract. A new compact representation of infinite graphs is investigated. Regular languages are used to represent labelled hyper-graphs which can be also multi-graphs. Our approach is similar to that used by A. Ehrenfeucht et al. for finite graphs since we use a regular prefix-free language as set of vertices, but it differs from that in the representation of the edges. In fact, we use a regular language for the edges instead of a finite loop-free graph. Our approach preserves the finite representation of the edges and of the corresponding labelling mapping and yields to a higher expressive power. As a matter of fact, our graph representation results to be more powerful than the equational graphs introduced by B. Courcelle. Moreover, the use of a regular prefix-free language to represent the vertices allows (fixed the language of the edges) to express a graph by a labelled tree. The advantage to represent graphs by trees is that properties of graphs can be verified by induction on the tree, often leading to efficient algorithms.

1 Introduction

A lot of efforts have been made in the last decade to obtain small specifications of graphs. A well supported idea has been that of representing graphs by expressions or trees [1,6,12]. Recently, A. Ehrenfeucht et al. [9] have introduced a representation of finite graphs by finite prefix-free languages of strings whose alphabets have themselves a graph structure. The strings of the language represent the vertices and there is an edge between two vertices if and only if the pair of the first two symbols, at which the two corresponding strings differ, is an edge in the alphabet.

Another way to specify finite graphs was introduced by M. Bauderon et al. [3]. They define finite hyper-graphs in a compositional way, that is they use graph expressions built from basic ones by applications of simple graph operations corresponding to hyper-edge replacements. Systems of graph equations with this

---
* Partially supported by the M.U.R.S.T. in the framework of "Tecniche formali per la specifica, l'analisi, la verifica, la sintesi e la trasformazione di sistemi software" project.

kind of expressions provide a way to define infinite hyper-graphs, called equational graphs [7]. In the recent paper [2], the class of simple equational hypergraphs is extended by allowing vertex replacement. In this way a characterization is obtained of the class of simple graphs defined in [5] whose monadic second-order theory is decidable and strictly containing the simple graphs among the equational graphs.

In this paper we introduce a new way of specifying infinite hyper-graphs through regular languages. Our approach is similar to that used in [9] for finite graphs since we use a regular prefix-free language as set of vertices, but it differs from that in the representation of the edges. Actually, the finite loop-free graphs turn out to be not sufficient when the goal is the representation of finite graphs. We use a regular language P for the edges instead of a finite loop-free graph with the meaning that the graph has an edge linking the ordered sequence of vertices $x_1, \ldots, x_k$ if their "suffix" belongs to P. Intuitively, by suffix we mean the tuple obtained from $x_1, \ldots, x_k$ by cutting their longest common prefix, if there exists i and j such that $x_i \neq x_j$, and the tuple itself, otherwise. Our approach preserves the finite representation of the (possibly infinite) graphs and allows to specify a meaningful class of infinite hyper-graphs. As a matter of fact, our graph representation results to be more powerful than the equational graphs introduced by B. Courcelle. In a similar way as in [9], the use of regular languages allows us to inherit concepts and ideas from the formal language theory and to use them for graphs. In particular we are interested in the relationships between language operations and graph operations: the relevance of this investigation is due to the possibility of performing graph trasformations by manipulating the regular languages used for the graph representation. Moreover, the use of a regular prefix-free language to represent the vertices allows (fixed the language of the edges) to express a graph by a labelled tree. The advantage to represent graphs by trees is that properties of graphs can be verified by induction on the tree, often leading to efficient algorithms [4,8,10,13].

In section 2 we give some preliminary definitions. In section 3 the graph representation is introduced, some properties of this representation are shown and the relationships between graph substitution and language concatenation is stated. The main result of section 4 is the proof that the graph representation introduced in section 3 is more expressive than the equational graphs defined in [7]. The paper ends with some conclusions in section 5, where we remark the differences between our approach and that in [9] and mention some directions for future works.

Due to the lack of space the proofs are omitted, for a full version of the paper see the URL [14].

2 Preliminaries

In this section we give some basic definitions. We suppose that the reader is familiar with the basic concepts of the formal languages (see for example [11]). We only recall that L is said prefix-free if for every $x, y \in L$ it holds that x is not a prefix of y and remark that in this paper with L^n we denote the Cartesian product $L_1 \times \ldots \times L_n$ when $L_1 = \ldots = L_n = L$. In the following we will use N to denote the set of the positive integers. A multiset ms over a finite set Δ is a mapping from Δ into $N \cup \{0\} \cup \{\infty\}$ and $ms(a)$ is said the multiplicity of a for each $a \in \Delta$. The set of the multisets over a given finite set Δ is denoted by $MS(\Delta)$. We denote with $\mathbf{0}$ the multiset mapping a into 0 for each $a \in \Delta$. Given two multisets ms and ms', we say that $ms \le ms'$ if $ms(a) \le ms'(a)$ for every $a \in \Delta$ and we denote with $ms + ms'$ the multiset which maps each $a \in \Delta$ into $ms(a) + ms'(a)$. Moreover, if $ms'(a) \ne \infty$ for each $a \in \Delta$, then with $ms - ms'$ we denote the multiset which maps a into $ms(a) - ms'(a)$, if $ms(a) - ms'(a) \ge 0$, and 0, otherwise. Obviously, $\infty + k = k + \infty = \infty + \infty = \infty - k = \infty$ holds. Sometimes in the following we denote the multisets which are also sets with the usual set notation.

In this paper we cope with directed labelled hyper-graphs which can be also multi-graphs, that is they can have many hyper-edges linking any ordered tuple of vertices. We consider a directed labelled hyper-edge as given by a sequence of vertices $(v_1, \ldots, v_k)$ and a multiset ms, with the meaning that there are exactly $ms(a)$ directed hyper-edges linking $v_1, \ldots, v_k$ and labelled by $a \in \Delta$. We denote each of them with $((v_1, \ldots, v_k), a)$ and we say that an hyper-edge is incident to each of the vertices $v_1, \ldots, v_k$ that it links. Note that we are not interested in distinguishing among hyper-edges linking the same tuple of vertices and having the same label. Labels are taken from a ranked alphabet, that is a pair (Δ, τ) where Δ is an alphabet and τ is a mapping from Δ into N.

Definition 1. *Let (Δ, τ) be a ranked alphabet. A labelled n-hyper-graph is a tuple $g = (V, E, lab, src)$ where:*
- *V is the set of the vertices;*
- *$lab : \bigcup_{k=1}^{\infty} V^k \to MS(\Delta)$ is a total mapping such that, for every $a \in \Delta$ and $v_1, \ldots, v_k \in V$, $lab(v_1, \ldots, v_k)(a) > 0$ implies $\tau(a) = k$;*
- *$E = \{(v_1, \ldots, v_k) \in \bigcup_{k=1}^{\infty} V^k / lab(v_1, \ldots, v_k) \ne \mathbf{0}\}$.*
- *src is a total mapping from $\{1, \ldots, n\}$ into V.*

The mapping src defines a sequence of n vertices which is called the *sequence of sources* of G and the integer n is the *type* of g. From now on, we will consider only labelled n-hyper-graphs, so we use the word n-graph (or simply graph, when the specification of n is useless) for a labelled n-hyper-graph and its hyper-edges are simply called edges. A subgraph of a graph $g = (V, E, lab, src)$ is a graph $g' = (V', E', lab', src')$ such that $V' \subseteq V$, $lab'(v_1, \ldots, v_k) \le lab(v_1, \ldots, v_k)$ for each $(v_1, \ldots, v_k)$ and $src'(i) = src(i)$. In this case we say that $g' \subseteq g$. Let $g = (V, E, lab, src)$ be a graph and $\approx$ be an equivalence relation on V. We define the quotient graph, denoted by $g/\approx$, the graph $(V/\approx, E/\approx, lab/\approx, src/\approx)$ where: $V/\approx = \{[v]/v \in V\}$, $lab/\approx ([v_1], \ldots, [v_k]) = \sum_{v'_i \in [v_i]} lab(v_1, \ldots, v_k)$ and $src/\approx (i) = [src(i)]$. Let $g_i = (V_i, E_i, lab_i, src_i)$ for $i = 1, 2$ be graphs, an isomorphism between graphs $\phi : V_1 \to V_2$ is a bijective mapping such that: $lab_1(v_1, \ldots, v_k) = lab_2(\phi(v_1), \ldots, \phi(v_k))$ and $src_2(i) = \phi(src_1(i))$. Finally we define the limit of a succession of finite graphs by referring to the intuitive concept of limit. For the sake of simplicity we omit a formal definition and we use the intuitive concept of "arbitrarily close". Then, let $\{g_n\}_{n>0}$ be a succession of finite graphs, we say that a (possibly infinite) graph, denoted by $lim_n g_n$, is the limit for $n \to \infty$ of the succession $\{g_n\}_{n>0}$ if it is always possible to find a graph in the succession which is arbitrarily close to $lim_n g_n$. Given a succession of finite graphs $\{g_n\}_{n>0}$ such that $g_n = (V_n, E_n, lab_n, src)$ and $g_n \subseteq g_{n+1}$, we have that $lim_n g_n = (\bigcup_{n>0} V_n, \bigcup_{n>0} E_n, lim_n lab_n, src)$. Note that $\{lab_n\}_{n>0}$ is a monotonic succession of functions, then its limit always exists. Moreover, it is easy to show the following result.

Lemma 1. *Given two successions of graphs $\{g_n\}_{n>0}$ and $\{g'_n\}_{n>0}$ such that for each $n > 0$ g_n is isomorphic to g'_n, $g_n \subseteq g_{n+1}$ and $g'_n \subseteq g'_{n+1}$, then $lim_n g_n$ is isomorphic to $lim_n g'_n$.*

3 Graph Representation over a Regular Language of Tuples

In this section we introduce a new way of representing graphs which is obtained from the representation introduced in [9]. The new representation is as powerful as the previous one when finite graphs are dealt with. The main difference between them concerns the representation of the edges. In fact, we use a regular language instead of a finite loop-free graph. Moreover, our approach preserves some agreeable features of the previous one and the finite representation of the edges and of the corresponding labelling mapping also for infinite hyper-graphs.

To introduce the new graph representation we define first a notion of regularity for languages of tuples which we call parallel regularity. Let Σ be an alphabet, $\natural \notin \Sigma$ and $x_1, \ldots, x_k \in \Sigma^*$, we denote with $Matrix_\natural(x_1, \ldots, x_k)$ the word $[a_{1\,1} \ldots a_{1\,k}], \ldots, [a_{h\,1}, \ldots, a_{h\,k}]$ over the alphabet $(\Sigma \cup \{\natural\})^k$ where:
- $h = max\{|x_j|/j = 1, \ldots, k\}$;
- for $j = 1, \ldots, k$: $x_j = a_{1\,j} \ldots a_{|x_j|\,j}$ and $a_{r\,j} = \natural$ for $|x_j| < r \leq h$.

That is the i-th symbol $[a_{i\,1} \ldots a_{i\,k}]$ of the word $Matrix_\natural(x_1, \ldots, x_k)$ is the ordered tuple of the i-th symbols of the words $x_1, \ldots, x_k$.

Definition 2. *Let Σ be an alphabet, $\natural \notin \Sigma$ and s be a positive integer. Then, $P \subseteq \bigcup_{i=1}^{s}(\Sigma^*)^i$ is said to be regular in parallel if the language $Matrix_\natural(P) = \{Matrix_\natural(x_1, \ldots, x_k)/(x_1, \ldots, x_k) \in P\}$ is regular.*

Then, we formalize the notion of suffix of a tuple of strings.

Definition 3. *Let Σ be an alphabet and $L \subseteq \Sigma^*$ be a prefix-free language. For each $x_1, \ldots, x_k \in \Sigma^*$ we define the suffix of the k-tuple $(x_1, \ldots, x_k)$, denoted by $suf(x_1, \ldots, x_k)$, as the k-tuple $(a_1 y_1, \ldots, a_k y_k)$, if $x_i = x a_i y_i$, for $a_i \in \Sigma$, and $\exists j, m$ such that $a_j \neq a_m$, and as the k-tuple $(x_1, \ldots, x_k)$, otherwise.*

Intuitively, by suffix we mean the tuple obtained from $x_1, \ldots, x_k$ by cutting their longest common prefix, if there exists i and j such that $x_i \neq x_j$, and the tuple itself, otherwise. The fact that $suf(x_1, \ldots, x_k) = (x_1, \ldots, x_k)$ whenever

$x_1 = \ldots = x_k$ allows us to represent the graphs with loops. In fact, we represent a graph with a prefix-free language (the vertices), a set of tuples P and a labelling function with the meaning that there is an edge with vertices $x_1, \ldots, x_k$ if the suffix of the tuple $(x_1, \ldots, x_k)$ belongs to P and the labelling function maps this suffix in a multiset which differs from $\mathbf{0}$. Formally:

Definition 4. Let Σ be an alphabet, (Δ, τ) be a ranked alphabet, $L \subseteq \Sigma^*$ be a regular prefix-free language, $P \subseteq \bigcup_{j=1}^{\infty}(\Sigma^*)^j$ be such that $P = \bigcup_{i=1}^{m} P_i$ where P_i is regular in parallel for all $i = 1, \ldots, m$, lab be a total mapping from $\bigcup_{j=1}^{\infty}(\Sigma^*)^j$ into $MS(\Delta)$ such that $|lab(P_i)| = 1$ for all $i = 1, \ldots, m$ and src be a total mapping from $\{1, \ldots, n\}$ into L. We denote with $gra(L, P, lab, src)$ the n-graph $g = (V, E, lab', src)$ where:
- $V = L$;
- $lab' : \bigcup_{j=1}^{\infty}(\Sigma^*)^j \to MS(\Delta)$ is the mapping defined as $lab'(x_1, \ldots, x_k) = lab(suf(x_1, \ldots, x_k))$ if $suf(x_1, \ldots, x_k) \in P$ and $lab'(x_1, \ldots, x_k) = \mathbf{0}$, otherwise.

In this case we say that the graph g is representable by regular languages.

Example 1. Let $g = (N, E, lab, src)$ be such that:
1. $E = \{(2i-1, 2i+1)/i \in N\} \cup \{(2i-1, 2i)/i \in N\} \cup \{(2i+2, 2i)/i \in N\}$,
2. $lab(2i-1, 2i+1) = \{a\}$, $lab(2i-1, 2i) = \{b\}$ and $lab(2i+2, 2i) = \{c\}$ and
3. src is the sequence $1, 2$.

We show that g is representable by regular languages. In fact, consider the graph $g' = gra(1^*(v + y + w + z), \{((v,y), (v,w), (w,z), (z,y), (w,1w), (1z,z)\}, lab', src')$ where $lab'(v,y) = lab'(w,z) = \{b\}$, $lab'(v,w) = lab'(w,1w) = \{a\}$, $lab'(z,y) = lab'(1z,z) = \{c\}$ and src' defines the sequence v, y (see Figure 1). Then, it easy to see that g' is isomorphic to g.

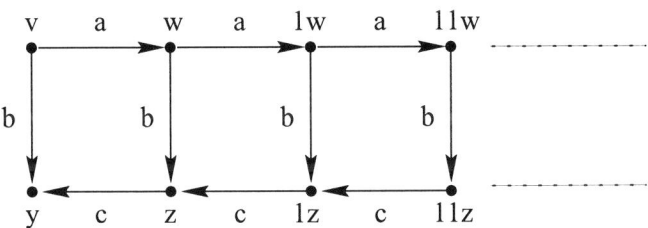

Fig. 1. A graphical representation of g'.

By using this new graph representation, as for the graph representation introduced in [9], the following substitution of graphs corresponds to a language concatenation.

Definition 5. Let $g = (V, E, lab, src)$ be an n-graph and $g_v = (V_v, E_v, lab_v, src_v)$ be an n_v-graph, for $v \in V$. Then, the graph obtained by substituting g_v for v in g, denoted by $g[v \leftarrow g_v]_{v \in V}$, is the graph (V', E', lab', src') where:

- $V' = \bigcup_{v \in V} \{v\} V_v$;
- $lab'(x_1 y_1, \ldots, x_k y_k)$ is equal to $lab_{x_1}(y_1, \ldots, y_k)$, if $x_1 = \ldots = x_k$, and is equal to $lab(x_1, \ldots, x_k)$, otherwise;
- src' defines the sequence $s_1, \ldots, s_n$ where $s_i = x_i src_{x_i}(1), \ldots, x_i src_{x_i}(n_{x_i})$ and $x_1, \ldots, x_n$ is the sequence defined by src.

This graph substitution is said *uniform* if we substitute a unique graph for every vertex v. The following theorem states that uniform graph substitution corresponds to language concatenation in the representation of the graph.

Theorem 1. *Let $g = gra(L, P, lab, src)$ and $g_x = gra(L_x, P, lab, src_x)$, for $x \in L$. It holds that $g[x \leftarrow g_x]_{x \in L} = gra(L', P, lab, src')$, where L' is the language $\bigcup_{x \in L} \{x\} L_x$. Moreover, if $L'' = L_x$ for every $x \in L$ then $L' = LL''$.*

Another interesting property of our notation is related to the continuity of the function gra. In fact, the following result holds.

Theorem 2. *Let $\{gra(L_n, P_n, lab_n, src)\}_{n>0}$ be a succession of graphs such that $L_n \subseteq L_{n+1}$, $P_n \subseteq P_{n+1}$ and $lab_n \leq lab_{n+1}$. Then, $lim_n gra(L_n, P_n, lab_n, src) = gra(\bigcup_{n>0} L_n, \bigcup_{n>0} P_n, lim_n lab_n, src)$.*

4 Equational Graphs vs. Graphs Representable by Regular Languages

In this section we first briefly recall the notion of the equational graphs as defined in [7] and then we compare the expressive power of the equational graphs with our representation.

In the following, some operations on graphs are defined. Let $g = (V, E, lab, src)$ and $g' = (V', E', lab', src')$ be respectively an n-graph and an n'-graph such that $V \cap V' = \emptyset$ and $E \cap E' = \emptyset$. The disjoint union of g and g', denoted by $g \oplus g'$, is defined as the $(n+n')$-graph $(V \cup V', E \cup E', lab \cup lab', src'')$ where src'' is the mapping defining the sequence which is the concatenation of the sequences defined by src and src'. Moreover, let $f : \{1, \ldots, m\} \to \{1, \ldots, n\}$ be a total mapping, the source redefinition map σ_f is defined as $\sigma_f(g) = (V, E, lab, src \circ f)$ where $\circ$ is the usual composition of functions. Let δ be an equivalence relation over $\{1, \ldots, n\}$, the source fusion map θ_δ is defined as $\theta_\delta(g) = g/\approx$, where $g/\approx$ is the quotient graph with respect to the equivalence relation on V defined as:
$$v \approx v' \Leftrightarrow v = v' \text{ or } (v = src(i), v' = src(j) \text{ and } (i,j) \in \delta).$$
Finally, let (Δ, τ) be a ranked alphabet, $(v_1, \ldots, v_{n'})$ be a tuple belonging to E and $a \in \Delta$ be such that $lab(v_1, \ldots, v_{n'})(a) \neq \mathbf{0}$. The graph, which is obtained by substituting g' for an edge $((v_1, \ldots, v_{n'}), a)$ in g and which we denote as $g[((v_1, \ldots, v_{n'}), a) \leftarrow g']$, is the graph $g''/\approx$ where:
- $g'' = \sigma_f(\overline{g} \oplus g')$ with $f : \{1, \ldots, n\} \to \{1, \ldots, n+n'\}$ defined as $f(i) = i$ for $i = 1, \ldots, n$ and $\overline{g} = (V, E, \overline{lab}, src)$ where $\overline{lab}(w_1, \ldots, w_k) = lab(w_1, \ldots, w_k)$ for $(w_1, \ldots, w_k) \neq (v_1, \ldots, v_{n'})$ and $\overline{lab}(v_1, \ldots, v_{n'}) = lab(v_1, \ldots, v_{n'}) - \{a\}$;
- $\approx$ is an equivalence relation on $V \cup V'$ defined as:
$$v \approx v' \Leftrightarrow v = v' \text{ or } (v = v_i \text{ and } v' = src'(i)).$$

Roughly speaking, the substitution of g' for the edge $((v_1, \ldots, v_{n'}), a)$ in g consists of the deletion in g of an edge linking the tuple $(v_1, \ldots, v_{n'})$ and labelled by a and the gluing of g' and g by the fusion of the sources of g' and the vertices in $(v_1, \ldots, v_{n'})$, in the order. Note that the graph $\bar{g}$ is the result of the deletion of one edge among those linking $(v_1, \ldots, v_{n'})$ and labelled by a in g.

The above graph substitution can be generalized by defining a substitution of all the edges which have a given label. Thus, for $i = 1, \ldots, m$ let g_i be an n_i-graph and $a_i \in \Delta$ be such that $\tau(a_i) = n_i$. We denote with $g[a_1 \leftarrow g_1, \ldots, a_m \leftarrow g_m]$ the graph which is obtained by simultaneously substituting g_i for every edge which is labelled by a_i in g.

The next definition gives the notion of graph expression.

Definition 6. *Let (Δ, τ) be a ranked alphabet. The graph expressions are defined as:*

- *a is a graph expression of type $\tau(a)$ for all $a \in \Delta$;*
- *n is a graph expression of type n for all $n \in N \cup \{0\}$;*
- *$e_1 \hat{\oplus} e_2$ is a graph expression of type $n_1 + n_2$ for all the graph expressions e_1 of type n_1 and e_2 of type n_2;*
- *$\hat{\sigma}_f(e)$ is a graph expression of type m for all the graph expressions e of type n and for all the total mappings $f : \{1, \ldots, m\} \to \{1, \ldots, n\}$;*
- *$\hat{\theta}_\delta(e)$ is a graph expression of type n for all the graph expressions e of type n and for all the equivalence relations δ on $\{1, \ldots, n\}$.*

The meaning of the operators $\hat{\oplus}$, $\hat{\theta}_\delta$ and $\hat{\sigma}_f$ is obtained by the meaning of the corresponding operators on graphs. In fact, to each graph expression e of type n it is possible to associate a n-graph, denoted with $val(e)$, which represent the evaluation of the expression e. By considering some unknowns in the set Δ, the graph equations and the systems of graph equations are defined in an obvious way. A solution to such a system is a tuple of graph expressions whose depth (i.e. the nesting of the operators) may be infinite. An evaluation mapping exists associating an unique (up to an isomorphism) graph to an infinite graph expression and we denote it again with val. More details can be found in [7].

Let S be the system of graph equations $\langle u_1 = e_1, \ldots, u_m = e_m \rangle$ in the unknowns $u_1, \ldots, u_m$. If S satisfies the Greibach condition (that none of the e_i is equal to an unknown), then S has a unique solution $(U_1, \ldots, U_m)$ and the elements of the m-tuple $(val(U_1), \ldots, val(U_m))$ are said equational graphs [7]. We can define, for each h, a regular perfix-free language L, a set $P = \bigcup_{i=1}^{p} P_i$ where each P_i is regular in parallel, a labelling mapping lab such that $|lab(P_i)| = 1$ and src, in such a way that $val(U_h)$ is equal to $gra(L, P, lab, src)$. So we have the following theorem.

Theorem 3. *Every equational graph is representable by regular languages.*

The proof of the above theorem gives us a way to obtain a description of every equational graph. For example, if we consider the system $u = \hat{\sigma}_f(\hat{\theta}_\delta(c \hat{\oplus} a \hat{\oplus} b \hat{\oplus} u))$ where δ is $\{(1,3), (2,6), (4,7), (5,8)\}$ and $f : \{1, 2\} \to \{1, 2, 3, 4, 5, 6, 7, 8\}$ is defined as $f(i) = i$ for $i = 1, 2$, then the graph obtained in that way is the one in Figure 1.

The vice-versa of the above theorem is not true. In fact there is a class of graphs which are representable by regular languages but are not equational.

Theorem 4. *Let $g = (V, E, lab, src)$ be a graph with a subgraph $g' = (V', E', lab', src')$ such that: a) V' is infinite and b) for every $v, v' \in V'$ there exists $(v_1, \ldots, v_k) \in E'$ such that $(v_1, \ldots, v_k)$ is incident on v and v'. Then, g is not equational.*

Corollary 1. *The class of the equational graphs is strictly included in the class of the graphs representable by regular languages.*

5 Conclusions

In this paper we have introduced a new way of specifying infinite hyper-graphs through regular languages and have compared the class of representable graphs with the equational graphs considered in [7]. Our approach is similar to that used in [9] where the authors introduce a new representation of finite graphs. They use finite prefix-free languages of strings on alphabets which have themselves a graph structure. The strings of the language represent the vertices of the graph and there is an edge between two vertices if and only if the pair of the first two symbols, at which the two corresponding strings differ, is an edge in the alphabet. One can prove that in the above approach the class of infinite graphs which can be represented through an infinite prefix-free regular language and a finite loop-free graph contains only graphs such that either they have an infinite degree (that is, there is a vertex with infinite edges incident on it) or they are the disjoint union of infinitely many maximal connected subgraphs. So, when the goal is the representation of infinite hyper-graphs, that approach turns out to be strictly less powerful than the one presented in this paper. Two interesting aspects in [9] are the use of prefix-free languages, which can be viewed as trees so that this approch presents the advantages of representing graphs by trees, and the relationships between graph operations and language operations (and then operations on the representation itself). Our aim was that of preserving these advantages also when infinite graphs are dealt with. We have also proved that the class of the equational graphs is strictly contained in the class of graphs that we have considered. For the case of simple graphs, this result could also be derived by comparing the simple graphs in the class considered in our paper with the graphs definedin [5]. However, our direct proof provides a way to obtain the representation by regular languages of a given equational hyper-graph. By introducing some constraints on P and lab (such as, the number of different multisets), it is possible to determine some hierarchies whose investigation could give interesting hints on the use of this representation. Another worthwhile aspect to study concerns the comparisons among the different classes of hyper-graphs defined by different choices of the set P. It would be interesting to check how the relationships among different P's reflect on the corresponding classes.

Acknowledgment We thank the referees for helpful comments and for pointing us to the paper of Caucal.

References

1. S. Arnborg, J. Lagergren and D. Seese, "Easy problems for tree-decomposable graphs", *Journal of Algorithms*, **12** (1991) 308–340.
2. K. Barthelmann, "When Can an Equational Simple Graph Be Generated by Hyperedge Replacement?", this Volume.
3. M. Bauderon and B. Courcelle, "Graph Expressions and Graph Rewritings", *Mathematical System Theory*, **20** (1987) 83–127.
4. H. L. Bodlaender and R. H. Möhring, "The pathwidth and treewidth of cographs", *SIAM Journal on Discrete Mathematics*, **6** (2) (1993) 181–188.
5. D. Caucal, "On infinite transition graphs having a decidable monadic theory", *Proc. of ICALP'96*, (F. M. auf der Heide and B. Monien, Eds.), Lecture Notes in Computer Science, vol. 1099 (1996) 194–205.
6. D. G. Corneil, H. Lerchs and L. Stuart Burlingham, "Complement reducible graphs", *Discrete Applied Mathematics*, **3** (1981) 163–174.
7. B. Courcelle, "The monadic second-order logic of graphs. II. Infinite graphs of bounded width", *Mathematical System Theory*, **21** (1989) 187–121.
8. B. Courcelle, "The monadic second-order logic of graphs. III. Tree-width, forbidden minors and complexity issues", *RAIRO Inform. Théor. Appl.*, **26** (1992) 257–286.
9. A. Ehrenfeucht, J. Engelfriet and G. Rozenberg, "Finite Languages for the Representation of Finite Graphs", *Journal of Computer and System Sciences*, **52** (1996) 170–184.
10. J. Engelfriet, T. Harju, A. Proskurowski and G. Rozenberg, "Characterization and Complexity of Uniformly Non-primitive Labeled 2-Structures", *Theoretical Computer Science*, **154** (1996) 247–282.
11. J. Hopcroft and J. Ullman, "Introduction to Automata Theory, Formal Languages and Computation" *Addison-Wesley Series in Computer Science* (Addison-Wesley Publishing Company) (1979).
12. N. Robertson and P. Seymour, "Graph Minors. II Algorithmic aspects of tree-width", *Journal of Algorithms*, **7** (1986) 309–322.
13. J. Valdez, R. E. Tarjan and E. Lawler, "The recognition of series parallel digraphs", *SIAM Journal of Computing*, **11** (1982) 298–313.
14. www.unisa.it/papers/g.ps.gz

Improved Time and Space Hierarchies of One-Tape Off-Line TMs

Kazuo Iwama[1] and Chuzo Iwamoto[2]

[1] Kyoto University, Kyoto 606-8501, Japan
iwama@kuis.kyoto-u.ac.jp
[2] Hiroshima University, Higashi-Hiroshima 739-8527, Japan
iwamoto@ke.sys.hiroshima-u.ac.jp

Abstract. This paper presents improved time and space hierarchies of one-tape off-line Turing Machines (TMs), which have a single worktape and a two-way input tape: (i) For any time-constructible functions $t_1(n)$ and $t_2(n)$ such that $\inf_{n\to\infty} \frac{t_1(n)\log\log t_1(n)}{t_2(n)} = 0$ and $t_1(n) = n^{O(1)}$, there is a language which can be accepted by a $t_2(n)$-time TM, but not by any $t_1(n)$-time TM. (ii) For any space-constructible function $s(n)$ and positive constant ϵ, there is a language which can be accepted in space $s(n) + \log s(n) + (2+\epsilon)\log\log s(n)$ by a TM with two worktape-symbols, but not in space $s(n)$ by any TM with the same worktape-symbols. The $(\log\log t_1(n))$-gap in (i) substantially improves the Hartmanis and Stearns' $(\log t_1(n))$-gap which survived more than 30 years.

1 Introduction

Although there are many different TM models, most standard ones are one-tape TMs and multitape TMs. As for the one-tape model, it is most simple and its properties are relatively well known. However, it is not an appropriate model for contemporary complexity theory because it is too inefficient. A typical example is that one-tape TMs need $\Omega(n^2)$ time to recognize palindromes. On the other hand, multitape TMs, the most standard model for designing TM-algorithms, are so strong that it appears to be very hard to obtain tight complexity results. For example, no one has succeeded in obtaining non-linear lower-bounds for multitape TMs.

This is the reason why many researchers have been paying attentions to the intermediate model called "one-tape off-line TMs," which have one read-write worktape and a two-way input tape. This model no longer suffers from trivial inefficiency like the one-tape TM. At the same time, it is no longer too powerful to defeat any lower-bound proofs [2]. Actually, there is a huge literature discussing computational complexities on this one-tape off-line TMs [1,2,11,12,13,14,15,16]. These results often provide knowledge for the extension to the multitape case.

It is known that space complexities of one-tape off-line TMs coincide with those of two-worktape models, but time complexities do not. This can be seen in, for example: (i) If time-constructible functions $t_1(n)$ and $t_2(n)$ satisfy $\inf_{n\to\infty} \frac{t_1(n)\log t_1(n)}{t_2(n)} = 0$, then there is a language which is accepted by a $t_2(n)$-time TM, but not by any $t_1(n)$-time TM [8]. (ii) If space-constructible functions $s_1(n)$ and $s_2(n)$ satisfy $\inf_{n\to\infty} \frac{s_1(n)}{s_2(n)} = 0$, then there is a language which is accepted by an $s_2(n)$-space TM, but not by any $s_1(n)$-space TM [7]. The space hierarchy (ii) is tight because of the linear speed-up theorem [7]; on the other hand, the time hierarchy (i) has a large, logarithmic gap between $t_1(n)$ and $t_2(n)$ (no such gap exists for two-tape TMs [4]). This fact might be the reason why some researchers think one-tape off-line TMs are not a proper model to discuss time complexity.

However, is this claim, i.e., no gap for space but a large gap for time, really true? In this paper, we give a partial negative answer, i.e., there does exist a kind of gap for space, and the time gap can be significantly reduced. For the time hierarchy, we show that the above $\log t_1(n)$ gap can be replaced by $\log \log t_1(n)$. We also show that a $\log s(n)$ *additive*-gap exists for the space hierarchy if we fix the number of tape symbols. More precisely, it is shown that for each integer $m \geq 2$ and for any small constant $\epsilon > 0$, there is a language which can be accepted in space $s(n) + \log_m s(n) + (2+\epsilon)\log_m \log_m s(n)$ by a TM with m symbols, but not by any TM in space $s(n)$ with the same m symbols. Note that if we can use a fixed number of symbols, it appears to be hard to save, in general, more than a constant number of cells. Thus the *additive* logarithm function is regarded as a gap. The best previous result is due to [10,19,20,22], where there still exists a constant *multiplicative*-gap.

Both proofs are by standard diagonalization but based on several new ideas: For the time hierarchy, our new idea is to make a novel use of the padding sequence. For the space hierarchy, our result fully depends on the "halting space-bounded computations" by Sipser [21].

The first concrete lower bound argument for one-tape off-line TMs is given by Dietzfelbinger, Maass, and Schnitger [2]; it was shown that transposing an $l \times l$-matrix with elements of bit length p needs $\Omega(n \cdot l/((\log l)/p)^{1/2})$ time. Dietzfelbinger [1] also showed that one-tape off-line TMs can copy a string of length s across d cells in $O(d + sd/\log(\min\{n,d\}))$ time, and that the same model needs $\Omega(sd/\log(\min\{n,d\}))$ time for the same task if $d \geq s \geq \log n$. For TMs with one-way input tape, Maass [15] presented the quadratic lower bound for the simulation of two-tape TMs by one-tape ones. Li, Longpré, and Vitányi [12,13] presented the time lower bounds for the simulation of queue, stacks, and tapes by one worktape with one-way input. Several papers presented simulation results among deterministic, nondeterministic and alternating one-tape off-line TMs (e.g., [11,14]). Among others, Maass and Schorr [16] showed that $(t(n))^{2/3}\log^2 t(n)$-time one-tape off-line TMs with two-alternation can simulate deterministic $t(n)$-time ones if $t(n) \geq n^3$. Geffert [5] showed that for each constant k, nondeterministic $t(n) \geq n^2$ time can be simulated by nondeterministic $t(n)/k$ time even if the number of symbols is two. For multitape TMs, the following separation results are known. Dymond and Tompa [3] showed that DTIME($t(n)$) $\subsetneq$ ATIME($t(n)$), and Paul et al. [17] showed DTIME(n) $\subsetneq$ NTIME(n), but it is unknown whether this separation can be extended to DTIME(n^k) $\subsetneq$? NTIME(n^k) for $k > 1$. Furthermore, Gupta [6] proved DTIME($t(n)$) $\subsetneq \Sigma_2(t(n))$, where $\Sigma_2(t(n))$ is the set of languages accepted by $t(n)$-time multitape TMs with two-alternation.

2   Models, Main Theorems, and Related Results

Our TM model is the so-called *one-tape off-line deterministic TM*, which has one read-only input tape whose both ends are delimited by special end-markers and one semi-infinite read-write worktape whose left end is also delimited by a special end-marker. The worktape symbols of space-bounded (resp. time-bounded) TMs are only 0 and 1 (resp. $0, 1, 2, \cdots$). All space-hierarchy results in this paper can easily be extended to the (fixed) m-symbol case. We call a function $t(n)$ *time-constructible* if there is some $t(n)$-time bounded one-tape off-line TM which given a string of n ones produces the binary representation of $t(n)$ (this definition is based on [18]). Similarly, if there is an $s(n)$-space TM generating $s(n)$ ones, we call $s(n)$ *space-constructible*.

Theorem 1. *Suppose that $t_1(n)$ and $t_2(n)$ are time-constructible functions such that $\inf_{n\to\infty} \frac{t_1(n) \log \log t_1(n)}{t_2(n)} = 0$ and $t_1(n) = n^{O(1)}$. Then, there is a language $L \subseteq \{0,1\}^*$ which can be accepted by a $t_2(n)$-time TM, but not by any $t_1(n)$-time TM.*

If the number of read-write tapes is fixed $k \geq 2$, then it is known [4] that $t_2(n)$-time TMs with k tapes are stronger than $t_1(n)$-time TMs with k tapes for any time-constructible function $t_2(n)$ not bounded by $O(t_1(n))$. For $k = 1$, however, no progress has been made since 1965 [8].

Theorem 2. *Suppose that $s(n)$ is an arbitrary space-constructible function. For any small constant $\epsilon > 0$, there is a language $L \subseteq \{0,1\}^*$ which can be accepted in space $s(n)+\log s(n)+(2+\epsilon)\log\log s(n)$ by a TM T with two worktape-symbols, but not in space $s(n)$ by any TM T_x with the same worktape-symbols, where T always halts but T_x may not halt.*

Since $(s(n) + c)$-space with two symbols can be simulated by $s(n)$-space with the same symbols [9,10], the *additive* term of $\log s(n) + (2 + \epsilon) \log \log s(n)$ is regarded as a gap. The best results previously known [10] is that for any integer $i \geq 1$ and rational constant $1 + \epsilon$, $(1 + \epsilon)cn^i$-space TMs with m symbols are stronger than cn^i-space TMs with the same symbols. It should be noted that Theorem 2 can easily be extended to the m-symbol case, and that cn^i is a space-constructible function. Therefore, the above theorem improves the results of [10] in (i) narrowing the gap from multiplicative $1 + \epsilon$ to additive $\log s(n)$ and (ii) generalizing the space-function $s(n)$. The best results for general space-constructible functions are due to [19,20]; they showed that $(2 + \epsilon)s(n)$-space is stronger than $s(n)$-space. Thus, our result also improves their results. In the case of linear $s(n)$, the optimal result has been known; from the results in [22], $s(n)$-space is stronger than $s_1(n)$-space if $s(n) - s_1(n+1) \neq O(1)$. The model in [22] is the same as [20] except that the worktape has a movable right end-marker.

Remark 1. Our model is the same as [20]; namely, our TM is the "simplest" version of the TM in [20] (i.e., the number of tape symbols is two, and the number of worktape heads is one). On the other hand, the model defined in [10] is slightly different from our model. The TM in [10] has either a two-way infinite worktape or a one-way infinite worktape with no left end-marker. Our space-hierarchy theorem also holds for the same model as [10] if the additive term $\log s(n)$ in Theorem 2 is replaced by $2 \log s(n)$. Hence, our result improves the hierarchy theorem in [10] on this no-end-marker model.

The additive $\log s(n)$ gap in Theorem 2 can be reduced if T may be an alternating TM. This result suggests that alternating TMs are stronger than deterministic TMs if the space is exactly the same.

Theorem 3. *Suppose that $s(n)$ is an arbitrary space-constructible function, and $\psi(n)$ is an arbitrary function such that the string $11\cdots 1$ of length $\psi(n)$ can be generated by some $s(n)$-space TM. Then, for any slowly growing function $\psi(n) \neq O(1)$, there is a language $L \subseteq \{0,1\}^*$ which can be accepted in space $s(n) + \psi(n)$ by an alternating TM with two worktape-symbols, but not in space $s(n)$ by any deterministic TM with two worktape-symbols.* (The proof is a modification of that of Theorem 2. Omitted due to space limitations.)

3 Proof of Theorem 1

All languages in this paper are over $\{0,1\}$. It is known that any TM can be encoded using 0 and 1 by standard encoding, and that it can be checked in linear time whether the given string encodes a proper TM. Let T_x denote the TM whose encoding sequence is x. If x is not a proper encoding sequence, we regard T_x as a TM accepting $\emptyset$ as usual. The language $L(t_1(n))$ is defined as $\{x\#y|\ T_x$ does not accept input x within time $t_1(|x|)\}$, where $\#$ is a boundary string not appearing in x, and $y = 10^{2^b-1}10^{2^b-1}\cdots 10^{2^b-1}$ is a padding sequence. We call each 10^{2^b-1} a *length-2^b portion*. The padding sequence can easily be checked in $O(n)$ time. It is obvious that any $t_1(n)$-time TM cannot accept $L(t_1(n))$. We first construct a TM T accepting $L(t_1(x))$ in time $t_2(n) \neq O(t_1(n)(\log t_1(n))^{1/2})$ in Section 3.1, and then we improve it to $t_2(n) \neq O(t_1(n) \log \log t_1(n))$ in Section 3.2.

3.1 Simulation Using Two Counters

We assume the worktape of TM T is divided into two tracks; track 1 is used for simulating T_x's worktape and track 2 is for holding counters. Track 2 is further divided into sub-tracks. If the worktape head of T_x is placed at the ith cell, then the *small counter* of length $\log d$ is inserted at the position starting at the $(ci+1)$st cell in track 2, where c is a constant depending on T_x. We will fix d and b later. Furthermore, the *big counter* of length $\log t_1(n)$ is inserted at the $(cd+1)$st cell in track 2 from the right end of the small counter. The small and big counters count the numbers of steps from 0 up to d and up to $t_1(n)/d$, respectively.

(Strictly speaking, the big counter has length $\log(t_1(n)/d)$, but the difference between $\log(t_1(n)/d)$ and $\log t_1(n)$ does not influence on the complexity in this case.) The encoding of T_x is included in a sub-track of the small counter.

The big counter is moved (at most) cd cells when the small one is moved d times. If we could use two read-write tapes, we can build a "bucket" on an extra read-write tape; i.e., when moving the big counter, we put some contents of the big counter into this bucket, move the head cd cells, and put the contents in the bucket back into the tape. Thus we need $O(cd(\log t_1(n))/b)$ steps for d-step simulation (i.e., $O(c(\log t_1(n))/b)$ overhead per step), where b is the bucket size. Note that we need $\log t_1(n)$ overhead if we move the big counter step-by-step. If $c/b \ll 1$, we can save a lot. Although our present model has no extra worktape for this bucket, we can still use this "bucket" by exploiting the input tape.

Stage 1: The $t_1(n)$-step simulation is divided into $t_1(n)/d$ *time-segments* of length d. In each time-segment, T makes a d-step simulation of T_x using the small counter. A sub-track of the small counter also contains the position $\pm s$ of the small counter itself from the original position at the beginning of this time-segment. The small counter is always placed at the worktape-head position. The two counters do not collide because cd cells exist between them at the beginning of the time-segment. T can make a d-step simulation in $O(cd \log d)$ steps.

Stage 2: Suppose T has just finished a d-step simulation for one time-segment and that the small counter (stored in the sub-track) moved cs cells to the right during this time-segment (i.e., $c(d-s)$ cells exist between the two counters). Then T moves the big counter cs cells to the right by extending the idea in [1].

(1) T moves its input head to the leftmost 0 within the current length-2^b portion. During this task, T stores the original position (within the length-2^b portion) of the input head into a sub-track in order that T can continue the d-step simulation of the next time-segment. T stores the value s into the length-2^b portion *in unary* as the position of the input head. (Later, one can see $2^b \geq s$.) T moves its worktape head to the leftmost cell of the big counter. Using the (unary) value s, T moves its worktape head to the $(cs+1)$st cell from the left end of the big counter, and T writes a marker $*$ there. The marker $*$ is required for indicating the destination. We need $O(cd)$ steps for this (1) because $s \leq d$.

(2) The big counter is divided into $(\log t_1(n))/b$ blocks of length b. T moves every block cs cells to the right using a length-2^b portion as a unary counter. Note that the contents in each block of length b can be stored in the length-2^b portion in unary. (In [1], the whole input tape is used as a unary counter, and thus the input head must go to the left end of the tape. Our TM T has unary counters which appear repeatedly in the whole area of the padding sequence.) We need $O(2^b)$ steps for storing (loading) a value into (from) a length-2^b portion, and $O(cd)$ steps for moving the worktape head cd cells to the right. Hence we need $O((2^b + cd)(\log t_1(n))/b)$ time for moving the big counter. Finally, T moves its input head back to the original position. Therefore, the time complexity for (1) and (2) is $O\left((2^b + cd)(\log t_1(n))/b\right)$.

T repeats Stages 1 and 2 until the value in the big counter becomes $t_1(n)/d$. Therefore, the time complexity becomes $O(\{cd \log d + (2^b + cd)(\log t_1(n))/b\}$

$\times t_1(n)/d$). During the simulation, if T_x halts with an accepting (rejecting) state, T rejects x (accepts x). If T_x does not halt in the $t_1(n)$ steps, T accepts x. We fix $b = (\log t_1(n))^{1/2}$ and $\log d = (\log t_1(n))^{1/2}$, and thus b and d satisfy $2^b = O(d)$ and $\log d = O((\log t_1(n))/b)$. Therefore, the above complexity can be written as $O(ct_1(n)(\log t_1(n))^{1/2})$, by which $t_2(n)$ is not bounded for any large constant c. Since the input string must be prefixed by some TM, the function b must satisfy that $n - 2^b \neq O(1)$. Note that any polynomial $t_1(n)$ satisfies this condition.

3.2 Extension to the k-Counter Case

The discussion in the previous section allows us to make the following observation. Suppose that we have a small counter of length x/α and a big counter of length x. (i) The big counter has to be moved once in $2^{x/\alpha}$ steps. The amount of moving distance is at most $2^{x/\alpha}$. (ii) With the help of the padding sequence, we can carry out this task in $(2^b + 2^{x/\alpha}) \cdot x/b$ steps. (iii) It appears to be optimal to set $b = x/\alpha$ ($=$ the size of the small counter).

Let $C(x)$ be the cost (overhead per step) of manipulating the big counter to keep the number of steps with the help of the small counter. Then, this $C(x)$ is shown as:

$$C(x) = \min\left\{x, C(x/\alpha) + \frac{1}{2^{x/\alpha}}(2^{x/\alpha} + 2^{x/\alpha}) \cdot \alpha\right\} = \min\{x, C(x/\alpha) + 2\alpha\},$$

where $\frac{1}{2^{x/\alpha}}$ comes from (i), $2^{x/\alpha}$ is the cost for the moving distance and for the manipulation of the padding sequence, and the final α is x/b in (ii) that is equal to α if we set $b = x/\alpha$. Note that if we move the big counter one cell every step, then $C(x) = x$ obviously.

One can verify that if we set $\alpha = \sqrt{x}$, then the solution of $C(x) = \min\{x, C(\sqrt{x}) + 2\sqrt{x}\}$ is obviously $C(x) = O(\sqrt{x})$, which is the result of Section 3.1. Now let us try to solve this equation by setting $\alpha = 2\psi(n)$, where $\psi(n)$ is a slowly growing function not bounded by $O(1)$. Then one can easily see that the solution of $C(x) = \min\{x, C(x/\alpha) + 2\alpha\}$ satisfies $C(x) \leq 2\alpha \log x$, which is $C(x) = 4\psi(n) \log \log t_1(n)$ when the big counter has length $x = \log t_1(n)$. Now we fix $\psi(n) = t_2(n)/(t_1(n) \log \log t_1(n))$ and thus the time complexity for $t_1(n)$-step simulation is not bounded by $O(t_1(n) \log \log t_1(n))$.

Finally, we must not forget one important thing, i.e., the padding sequence. Recall that observation (ii) above assumes that the input head can encounter some unique "mark" within 2^b steps regardless of its initial position. In Section 3.1, it was enough to put 1 at regular intervals of length 2^b as this mark. This time, however, we need different "marks" at regular intervals of length $2^{x/\alpha}, 2^{x/\alpha^2}, 2^{x/\alpha^3}, \cdots$. Here is our solution to this problem. For $j = 1, 2, \cdots$, we regard the $(2^{i-1} + 2^i(j-1))$th cells of the input tape as the ith track. Let 2^l be the shortest interval. Then the above length can be written as $2^l, 2^{\alpha l}, 2^{\alpha^2 l}, \cdots, 2^{x/\alpha}$. The mark appearing at regular intervals of length $2^{\alpha^{i-1}l}$ is stored in the ith track. Thus, the complexity for the manipulation of the padding sequence in the above (ii) is not $2^b = 2^{\alpha^{i-1}l}$ but $2^{\alpha^{i-1}l}2^i$. Furthermore, a counter of length x/α^l

is moved once when the counter of length x/α^{i+1} is moved $2^{x/\alpha^{i+1}}$ times, and thus the amount of moving distance of the big counter is calculated as $2^{x/\alpha} \cdot 2^{x/\alpha^2} \cdots = 2^{x/\alpha + x/\alpha^2 + \cdots} \leq 2^{x/(\alpha-1)}$. Therefore, $2^{x/\alpha}$ in (ii) should be $2^{x/(\alpha-1)}$. These differences do not influence on the time complexity in this case.

T can verify that the padding sequence has the above structure in $O(n)$ time. T first generates a binary counter of length $\log n$ in the worktape. T then moves the input head to the leftmost cell of the padding sequence. T's input head moves to the right, while the value of the counter is increased one by one, doing both simultaneously. If the lowest $i-1$ bits are all 0 and the ith bit is 1, then the input head is placed on a cell in the ith track. Furthermore, if the $(i+1)$st through $(i+\alpha^{i-1}l)$th bits are all 0 and the $(i+\alpha^{i-1}l+1)$st bit is 1, then T verifies that this cell contains the mark. This completes the proof.

Paul [18] also uses many counters, and he succeeded in narrowing the gap to $\log^* t_1(n)$ if the number of worktapes is fixed $k \geq 2$. However, it seems very hard to achieve such a small gap on our one-worktape model. The main disadvantage is that storing value i into the input tape in unary requires i steps, while two-worktape model can store i into an extra worktape in binary in $\log i$ steps.

4 Proof of Theorem 2

The language $L(s(n))$ is defined as $\{x\#y|\ \text{TM}\ T_x \text{ does not accept input } x \text{ within space } s(|x|)\}$, where $y = 00\cdots 0$. Obviously, any $s(n)$-space TM cannot accept $L(s(n))$. We construct a TM T accepting $L(s(n))$ in space $s(n) + (1+\epsilon)\log s(n)$, which is improved to $s(n) + \log s(n) + (2+\epsilon)\log\log s(n)$ later.

Let T'_x be a TM satisfying the following conditions: Let z be an arbitrary string of length n. (i) T'_x accepts z iff T_x accepts z in space $s(n)$. (ii) T'_x uses at most $s(n) + \frac{c}{c-1}\log s(n) + 2c$ cells regardless of the space usage of T_x, where c is some constant. (iii) The accepting configuration of T'_x is unique. We first construct T'_x, from which we will construct T that determines whether T_x accepts x.

4.1 Constructing T'_x

T'_x works as follows: (i) T'_x simulates an $s(n)$-space TM, say, T_s, which generates $1^{s(n)}$, and (ii) T'_x then simulates T_x in space $s(n) + \frac{c}{c-1}\log s(n) + 2c$. T'_x uses a binary counter and the purpose of (i) is to hold the value $s(n)$ in it.

(i) T'_x generates $1^{s(n)}$ by simulating T_s. Then, T'_x constructs a binary counter, say, A, (of constant length initially) having value one at the position starting at the $(s(n)+1)$st cell. T'_x moves A one cell to the left and increases the value in A by one. Repeating this procedure until A reaches the left end, the value in A becomes $s(n)$. We need "delimiters" 1^c at both ends of A, and 0 must be inserted at a regular interval of $c-1$ symbols in A so that 1^c does not appear in it. Thus the length of A is $\frac{c}{c-1}\log s(n) + 2c$.

(ii) T'_x simulates T_x, while T'_x also updates A's value and moves A according to the head position. During the simulation, if the value in A becomes less than 0 (which violates the space limit), then T'_x halts in a rejecting state. If T_x halts in an accepting state (rejecting states), then T'_x moves its worktape head and A to the left end and T'_x halts in the unique accepting state (rejecting states). The number of cells T'_x uses is bounded by $s(n) + \frac{c}{c-1}\log s(n) + 2c$.

4.2 Deciding Whether T_x Accepts x or Not

To decide if T_x accepts x, T simulates T'_x instead of T_x. The simulation of T'_x is similar to Section 4.1. T generates two blocks, called B and C, of length $\psi(n)$ at the left end of the worktape. Here, $\psi(n) = o(\log s(n))$ may be an arbitrary function such that $1^{\psi(n)}$ can be generated in space $s(n)$. As above, 1^c is inserted at either end of each block, and 0 is inserted in B and C at regular intervals of length $c - 1$. We use B for keeping the encoding of T'_x. We will use C later.

Note that T'_x may reject x by looping. If T simulates T'_x from the initial configuration to the final one, T cannot know if T'_x is looping. Here, we implement Sipser's method [21] on our model. Consider a finite directed graph such that each node is a configuration of T'_x and an arc represents transition between two configurations. Since T'_x is deterministic, the unique accepting configuration, say, a, becomes the root of some tree. T'_x accepts x iff T'_x's initial configuration belongs this tree. Thus, T performs a depth-first search of the tree rooted at a. T accepts x iff the initial configuration of T'_x does not exist in the tree.

To traverse all nodes in the tree, T must know (i) T'_x's current configuration, say, c_1, and (ii) the next move function, say, δ_i, by which the previous configuration c_2 has just been changed to the current configuration c_1. (i) T'_x's input-head position, worktape head position, and symbols in the $s(n) + (1 + \epsilon)\log s(n)$ cells are simulated as in Section 4.1. T'_x's state is stored in block C. For (ii), we suppose that the encoding of T'_x (in block B) is the concatenation of the next move functions, and that the encoding of any next move function is prefixed by 11. Then T can replace the first two symbol 11 of the encoding of δ_i by 01.

(a) Suppose the current configuration is c_1. By sweeping the concatenation of T'_x's next move functions in B from left to right, T finds a next move function, say, δ_1, such that there is a configuration c_2 which can be changed to c_1 according to δ_1. If such a δ_1 is found, T changes T'_x's configuration c_1 back to c_2. If c_2 is not the initial configuration, then T again applies procedure (a) to c_2. (b) If such a δ_1 is not found for c_1, T simply simulates the one-step move of T'_x, by which c_1 is changed to, say, c_0. Let δ_2 be the next move function which changes c_1 to c_0. Then, T applies the same procedure as (a) to c_0 except that the sweep of the next move functions starts from δ_2.

4.3 Improving the Space Limitation

We only show the modifications required to improve the space limitation. In Section 4.1, we constructed T'_x working in space $s(n) + \frac{c}{c-1}\log s(n) + 2c$ for an arbitrary $s(n)$-space TM T_x. Here, we construct a T'_x working in space $s(n) + \log s(n) + \frac{2c}{c-1}\log\log s(n) + 3c$. In this section, we use A for the same purpose as in Section 4.1, but 1^c is *not* inserted at either end of A, and 0 is *not* inserted at regular intervals of length $c - 1$. The worktape head works just as in

Section 4.1. To recognize the boundaries of A, two *small counters* are inserted at the worktape-head position. The small counters contain the head positions i and h from the left and right ends of A, respectively. The small counters are moved according to the head position. (Recall that the worktape head must always be inside A or the head must keep within a constant number of cells even if it goes out.) 1^c is inserted at either end of the small counters, and 0 is inserted at regular intervals of length $c-1$ only in the small counters. Thus, the length of A is $\log s(n) + \frac{2c}{c-1} \log \log s(n) + 3c$. Hence T'_x works in space $s(n) + \log s(n) + (2+\epsilon) \log \log s(n)$.

References

1. M. Dietzfelbinger, The speed of copying on one-tape off-line Turing machines, *IPL* **33** (1989) 83–89.
2. M. Dietzfelbinger, W. Maass and G. Schnitger, The complexity of matrix transposition on one-tape off-line Turing machines, *TCS* **82** (1991) 113–129.
3. P.W. Dymond and M. Tompa, Speedups of deterministic machines by synchronous parallel machines, *Proc. IEEE FOCS*, 336–343, 1983.
4. M. Fürer, The tight deterministic time hierarchy, *Proc. ACM STOC*, 8–16, 1982.
5. V. Geffert, A speed-up theorem without tape compression, *TCS* **118** (1993) 49–65.
6. S. Gupta, Alternating time versus deterministic time: a separation, *Proc. IEEE FOCS*, 266–277, 1993.
7. J. Hartmanis, P.M. Lewis and R.E. Stearns, Classification of computations by time and memory requirements, *Proc. IFIP Congress*, 266–277, 1965.
8. J. Hartmanis and R.E. Stearns, On the computational complexity of algorithms, *Trans. Amer. Math. Soc.* **117** (1965) 285–306.
9. O.H. Ibarra, A hierarchy theorem for polynomial-space recognition, *SIAM J. Comput.* **3** 3 (1974) 184–187.
10. O.H. Ibarra and S.K. Sahni, Hierarchies of Turing machines with restricted tape alphabet size, *JCSS* **11** (1975) 56–67.
11. O.H. Ibarra and S. Moran, Some time-space tradeoff results concerning single-tape and offline TM's, *SIAM J. Comput.* **12** 2 (1983) 388–394.
12. M. Li, L. Longpré and P.M.B. Vitányi, The power of the queue, *SIAM J. Comput.* **21** 4 (1992) 697–712.
13. M. Li and P.M.B. Vitányi, Tape versus queue and stacks: the lower bounds, *Inform. and Comput.* **78** (1988) 56–85.
14. M. Liśkiewicz and K. Loryś, Fast simulations of time-bounded one-tape Turing machines by space-bounded ones, *SIAM J. Comput.* **19** 3 (1990) 511–521.
15. W. Maass, Quadratic lower bounds for deterministic and nondeterministic one-tape Turing machines, *Proc. IEEE FOCS*, 401–408, 1984.
16. W. Maass and A. Schorr, Speed-up of Turing machines with one work tape and a two-way input tape, *SIAM J. Comput.* **16** 1 (1987) 195–202.
17. W.J. Paul, N. Pippenger, E. Szemerédi, and W.T. Trotter, On determinism versus non-determinism and related problems, *Proc. IEEE FOCS*, 429–438, 1983.
18. W.J. Paul, On time hierarchies, *JCSS* **19** (1979) 197–202.
19. J.I. Seiferas, Techniques for separating space complexity classes, *JCSS* **14** (1977) 73–99.
20. J.I. Seiferas, Relating refined space complexity classes, *JCSS* **14** (1977) 100–129.
21. M. Sipser, Halting space-bounded computations, *TCS* **10** (1980) 335–338.
22. S. Žák, A Turing machine space hierarchy, *Kybernetika*, **26** 2 (1979) 100–121.

Tarskian Set Constraints Are in NEXPTIME*

Pawel Mielniczuk and Leszek Pacholski

Institute of Computer Science
University of Wroclaw
{mielni,pacholsk}@tcs.uni.wroc.pl

Abstract. In this paper we show that satisfiability of Tarskian set constraints (without recursion) can be decided in exponential time. This closes the gap left open by D.A. McAllester, R. Givan, C. Witty and D. Kozen in [14].

Introduction

Set constraints have a form of inclusions between set expressions built over a set of set-valued variables, constants and function symbols. They have been used in program analysis and type inference algorithms for functional, imperative and logic programming languages [3], [11], [12], [15], [16], [18].

The systems of set constraints used for program analysis were considered as inclusion constraints over the Herbrand universe i.e. a solution consisted of a collection of subsets of the Herbrand universe. To distinguish them from set constraints studied here we shall call them the *Herbrand* set constraints.

The satisfiability problem for Herbrand set constraints have been studied by many authors including N. Heintze and J. Jaffar [10], A. Aiken and E.L. Wimmers [4], R. Gilleron, S. Tison, and M. Tommasi [7], L. Bachmair, H. Ganzinger, and U. Waldmann [5], A. Aiken, D. Kozen, M. Vardi, and E.L. Wimmers [1]. The strongest decidability results were obtained by A. Aiken, D. Kozen, and E. Wimmers [2], R. Gilleron, S. Tison and M. Tommasi [8], K. Stefansson [17], and by W. Charatonik and L. Pacholski [6].

Recently D.A. McAllester, R. Givan, C. Witty and D. Kozen [14] liberalized the notions of set constraints to so-called Tarskian set constraints over arbitrary first-order domain, with a link to modal-logics. Early work on Tarskian set constraints by R. Givan and D. McAllester stemmed from the work on artificial intelligence [9,13].

D. McAllester, R. Givan, C. Witty and D. Kozen [14] gave a complexity analysis of systems of Tarskian set constraints depending on various parameters. They proved that the satisfiability problem for set constraints with *deterministic* operators of arbitrary arity (functions and constants) extended by recursion (μ-operator) is undecidable. They proved that for pure set constraints with deterministic constants but without deterministic function symbols of arity > 0, the satisfiability problem is EXPTIME-complete, and for pure set constraints with

---
* This research was partially supported by a KBN grant 8 T11C 029 13

deterministic function symbols of arity > 0, but without deterministic constants the satisfiability problem is NEXPTIME-complete. Finally they proved that for full (with deterministic constant and function symbols) set constraints without recursion the satisfiability problem is in 2-NEXPTIME and is NEXPTIME-hard, leaving the gap between the upper and the lower bound.

The approach to the upper bound was based on a reduction of the satisfiability problem to a problem of solving systems of so called prequadratic Diophantine inequalities. The result of reduction was of exponential size and the algorithm solving prequadratic Diophantine inequalities presented in [14] worked in nondeterministic exponential time, so the resulting set constraint algorithm worked in 2-NEXPTIME. D.A. McAllester, R. Givan, C. Witty and D. Kozen expressed a hypothesis that the satisfiability problem for systems of prequadratic Diophantine inequalities was in NP, which would give an algorithm of complexity matching the lower bound.

Here we give a NEXPTIME algorithm solving full systems of Tarskian set constraints (without recursion). We do it without a reduction to systems of prequadratic Diophantine inequalities.

1 Basic Definitions

We assume an infinite set of (deterministic and non-deterministic) operator symbols of each arity.

Definition 1. *1. A* set expression *is defined by the grammar:*

$$E ::= F(E_1, \ldots, E_n) \mid E_1 \cup E_2 \mid E_1 \cap E_2 \mid \neg E,$$

where F is an operator symbol of arity n.

2. By $\mathcal{M}$ we denote a first order structure with the universe M. Set expressions are interpreted in $\mathcal{M}$ as follows: $F^{\mathcal{M}}(E_1 \cup E_2) = F^{\mathcal{M}} E_1 \cup F^{\mathcal{M}} E_2$, $F^{\mathcal{M}}(E_1 \cap E_2) = F^{\mathcal{M}} E_1 \cap F^{\mathcal{M}} E_2$, $F^{\mathcal{M}}(\neg E_1) = M \setminus F^{\mathcal{M}} E_1$ and

$$F^{\mathcal{M}}(E_1, \ldots, E_n) = \{x \mid F^{\mathcal{M}}(x_1, \ldots, x_n) = x,\ x_i \in E_i^{\mathcal{M}},\ 1 \leq i \leq n\}.$$

If F is deterministic then, for all $x_1, \ldots, x_n \in \mathcal{M}$, there is exactly one $x \in \mathcal{M}$ such that $F^{\mathcal{M}}(x_1, \ldots, x_n) = x$.

A non-deterministic operator of arity zero can, from the point of view of satisfiability, be considered as a *set variable*. In the following we shall identify non-deterministic operators of arity zero with set variables. For this reason we did not explicitly mention variables in Definition 1. A deterministic operator is called a *function*, if it is of arity zero we call it a *constant*.

Definition 2. *A* constraint *is an expression of the form $E_1 \subseteq E_2$ (*positive constraint*) or $E_1 \not\subseteq E_2$ (*negative constraint*), where E_1, E_2 are set expressions.*

A constraint of the form $E_1 \subseteq E_2$ ($E_1 \not\subseteq E_2$) is satisfied in $\mathcal{M}$ if $E_1^{\mathcal{M}} \subseteq E_2^{\mathcal{M}}$ ($E_1^{\mathcal{M}} \not\subseteq E_2^{\mathcal{M}}$ respectively). We say that a set S of constraints is satisfiable, if there is a structure $\mathcal{M}$ such that every element of S is satisfied in $\mathcal{M}$.

We consider the problem to determine whether a finite set of constraints is satisfiable.

2 Set Constraints as Flat Expressions

We reduce systems of set constraints to one inclusion constraint defined using a flat set expression, where *flat set expression* is a set expression involving terms of depth at most one.

Definition 3. *A flat set expression is defined by the following grammar:*

$$E ::= F(X_1, \ldots, X_n) \mid E_1 \cup E_2 \mid E_1 \cap E_2 \mid \neg E,$$

where $X_1, \ldots, X_n$ are set variables.

The *size* of a system S of set constraints is the number of symbols in S.

Proposition 1. *For every system S of set constraints of size n there is a flat set expression $\mathcal{E}$ of size polynomial in n such that S is equivalent to inclusion constraint $\mathcal{E} \subseteq \emptyset$. Moreover, we can assume that negation is applied only to expressions of the form $F(X_1, \ldots, X_n)$.*

Proof. Set expressions of depth greater than one can be eliminated from S as follows. If $F(E_1, \ldots, E_n)$ appears in S and E_i is not a variable, for some $0 < i \le n$, then S is replaced by $S[X_i/E_i] \cup \{X_i \subseteq E_i, E_i \subseteq X_i\}$, where X_i is a new set variable and $S[X_i/E_i]$ is the result of replacing E_i by X_i everywhere in S.

A negative constraint $E_1 \not\subseteq E_2$ can be replaced by $c \subseteq E_1 \cap \neg E_2$, for a new constant symbol c. Finally, $E_1 \subseteq E_2$ is equivalent to $E_1 \cap \neg E_2 \subseteq \emptyset$, and $E_1 \subseteq \emptyset$, $E_2 \subseteq \emptyset$ is equivalent to $E_1 \cup E_2 \subseteq \emptyset$.

The last sentence follows by de Morgan rules. □

As an exercise the reader can check that a flat version of $f(a) \subseteq a$ is $(f(X) \cap \neg a) \cup (X \cap \neg a) \cup (\neg X \cap a) \subseteq \emptyset$.

3 Description of Structures

Let $\mathcal{E}$ be a flat set expression. Denote by Σ_0 the set of variables occurring in $\mathcal{E}$, and by Σ_1 - the set of set expressions of the form $F(X_1, \ldots, X_n)$, where F occurs in $\mathcal{E}$ and $X_1, \ldots, X_n \in \Sigma_0$. Define $\Sigma = \Sigma_0 \cup \Sigma_1$.

A Σ-type (Σ_0-type, Σ_1-type) is a set expression τ of the form $\bigcap T$ where T is a set of elements of Σ (Σ_0, Σ_1, respectively) and their negations such that for every $E \in \Sigma$ ($E \in \Sigma_0, \Sigma_1$, respectively) exactly one of E and $\neg E$ occurs in τ.

Note that every Σ_0-type (Σ_1-type) can be expressed as a union of Σ-types. Moreover, the number of Σ-types is is at most single exponential in the size of $\mathcal{E}$.

Consider a set expression $F(X_1, \ldots, X_n) \in \Sigma_1$. Let τ be a Σ_1-type, and let $\tau_1, \ldots, \tau_n$ be Σ_0-types such that $\tau \subseteq F(X_1, \ldots, X_n)$ and $\tau_i \subseteq X_i$, for $1 \le i \le n$.

We say that $(\tau_1, \ldots, \tau_n)$ is a *domain* of τ (τ is an *image* of $(\tau_1, \ldots, \tau_n)$) for $F(X_1, \ldots, X_n)$ if

$$(\forall Y_1, \ldots, Y_n \in \Sigma_0)(\tau_i \subseteq Y_i, 1 \leq i \leq n \quad \rightarrow \quad \tau \subseteq F(Y_1, \ldots, Y_n)),$$

which we write as

$$(\tau_1, \ldots, \tau_n) \stackrel{F(X_1,\ldots,X_n)}{\rightarrow} \tau.$$

The expression $(\tau_1, \ldots, \tau_n) \stackrel{F(X_1,\ldots,X_n)}{\rightarrow} \tau$ denotes that it is possible that in a structure $\mathcal{M}$ there exist $x_1, \ldots, x_n$ belonging to $\tau_1^{\mathcal{M}}, \ldots, \tau_n^{\mathcal{M}}$, respectively such that $F^{\mathcal{M}}(x_1, \ldots, x_n) \subseteq \tau^{\mathcal{M}}$. Denote by $\Delta_{F(X_1,\ldots,X_n)}^{(\tau_1,\ldots,\tau_n),\tau}$ the number such n-tuples $x_1, \ldots, x_n$. Formally:

$$\Delta_{F(X_1,\ldots,X_n)}^{(\tau_1,\ldots,\tau_n),\tau} = \#\left\{(x_1, \ldots, x_n) \mid x_i \in \tau_i^{\mathcal{M}}, 1 \leq i \leq n, F^{\mathcal{M}}(x_1, \ldots, x_n) \subseteq \tau^{\mathcal{M}}\right\}.$$

Given a relational structure $\mathcal{M}$, the description of $\mathcal{M}$ is the pair $\langle Card, \Delta \rangle$, where $Card$ is a function giving for a Σ-type τ the number $Card(\tau)$ of elements realizing τ in $\mathcal{M}$ (i.e the number of such x that $x \in \tau^{\mathcal{M}}$), and Δ is a function giving, for each $F(X_1, \ldots, X_n)$ and $(\tau_1, \ldots, \tau_n), \tau$ the cardinality of $\Delta_{F(X_1,\ldots,X_n)}^{(\tau_1,\ldots,\tau_n),\tau}$.

For a given pair $\langle Card, \Delta \rangle$ as above, to determine whether it is a description of a structure it suffices to check the *non-emptiness conditions* and the *cardinality conditions* defined below.

Definition 4. *The* non-emptiness conditions *hold if*

(1) For every $F(X_1, \ldots, X_n) \in \Sigma_1$ and a non-empty Σ_1-type $\tau \subseteq F(X_1, \ldots, X_n)$ there exist non-empty Σ_0-types $\tau_i \subseteq X_i$, for $1 \leq i \leq n$ such that
$(\tau_1, \ldots, \tau_n) \stackrel{F(X_1,\ldots,X_n)}{\rightarrow} \tau.$
(2) For every $f(X_1, \ldots, X_n) \in \Sigma_1$, where f is a function and all non-empty Σ_0-types $\tau_i \subseteq X_i$, $1 \leq i \leq n$ there exists a non-empty Σ_1-type τ such that
$(\tau_1, \ldots, \tau_n) \stackrel{f(X_1,\ldots,X_n)}{\rightarrow} \tau.$

The cardinality conditions *hold if*

(1) $\sum_{\tau \subseteq c} Card(\tau) = 1$, *for every constant symbol c.*
(2) For each function symbol f, each Σ_1-type τ, all variables X_i and Σ_0-types τ_i such that $\tau_i \subseteq X_i$, for $1 \leq i \leq n$,

$$\sum_{\tau} \Delta_{f(X_1,\ldots,X_n)}^{(\tau_1,\ldots,\tau_n),\tau} = \prod_{1 \leq i \leq n} Card(\tau_i).$$

(3) For each function symbol f, each Σ_1-type τ and all variables X_i

$$\sum_{(\tau_1,\ldots,\tau_n)} \Delta_{f(X_1,\ldots,X_n)}^{(\tau_1,\ldots,\tau_n),\tau} \geq Card(\tau).$$

Part (1) of non-emptiness conditions says that if a function has a non-empty image, then it has a non-empty domain and part (2) says the converse. Part (2) of the cardinality conditions says that every member of domain has exactly one image and part (3) that every member of image is the unique image for at least one member of domain.

4 Main Theorem

Consider a system S of set constraints of size n satisfiable in a structure $\mathcal{M}$. In fact, as we have noticed earlier, we can assume that S is without variables, and we do it. So, we assume that $\mathcal{M}$ is a model of S. We shall construct another model $\mathcal{M}'$ of S in which the cardinality of each set defined by a Σ-type is either infinite or at most double exponential in n. A description of $\mathcal{M}'$ can be written in single exponential space.

The idea is to find all Σ-types which always define finite sets and make all other sets infinite. We use the obvious fact that if $X_1, \ldots, X_k$ define finite sets and f is a function then $f(X_1, \ldots, X_k)$ is finite. Note that for non-deterministic operators this is not true.

We first describe a procedure which determines Σ-types which are always finite. This procedure works in steps. To provide some intuitions we give an example.

Let S be the system

$$X \subseteq a \cup c, \quad Y \subseteq f(X, X), \quad W \subseteq f(Y, Z),$$

where a, c are constants, X, Y, W, Z are set variables, and f is a function.

In the preprocessing step we mark as finite some Σ-types which have to be empty. In our example we mark all Σ-types contained in $X \cap \neg(a \cup c)$, $Y \cap \neg f(X, X)$ or $W \cap \neg f(Y, Z)$.

During each step we mark as finite an image of a function for which the domain has been marked earlier as finite and we stop if we can not continue this procedure. In our example a and c will be marked in the first step since the domain of a constant is empty. In fact X (all types containing X) will be marked since $X \subseteq (X \cap \neg(a \cup c)) \cup (X \cap (a \cup c))$ - the first summand being finite by preprocessing and the second being included in $a \cup c$. In the second step we mark Y and the procedure stops after two steps. So, if S is satisfied, the cardinality of X is at most 2, and the cardinality of Y - at most 4. There is no reason, however, to bound the cardinalities of W and Z, so we can assume that they are infinite.

The proof of the main result of this part goes as follows. First we prove that for a system S of set constraints of size n the procedure outlined above stops after n steps and gives a set $\mathcal{F}_n$ of types. Then we use the fact that the cardinality of an image does not exceed the cardinality of the domain to get the doubly exponential bound on the size of sets definable by types in $\mathcal{F}_n$. Finally given a model of S we change it by leaving the types in $\mathcal{F}_n$ unchanged and making all other types infinite. The structure so obtained has a description of exponential size and is a model of S.

Definition 5. *Let $\mathcal{E}$ be a flat constraint.*

1. By $\mathcal{F}$ we denote the operator on the powerset of Σ-types defined as follows:

$$\mathcal{F}(\mathcal{P}) = \{\tau \in \Sigma \mid (\exists f(X_1,\ldots,X_n) \in \Sigma_1)(\exists \tau_1 \in \Sigma_0)\ldots(\exists \tau_n \in \Sigma_0)$$
$$(((\tau_1,\ldots,\tau_n) \xrightarrow{f(X_1,\ldots,X_n)} \tau) \wedge (\forall 1 \leq i \leq n)(\tau_i \subseteq \mathcal{P}))\},$$

where Σ_1-type τ represents all Σ-types contained in τ.

2. We put $\mathcal{F}_0 = \{\tau \mid \tau \subseteq \mathcal{E}\}$, $\mathcal{F}_{n+1} = \mathcal{F}_n \cup \mathcal{F}(\mathcal{F}_n)$. By $\mathcal{F}_\infty$ we denote $\mathcal{F}_n$, for minimal n is such that $\mathcal{F}_{n+1} = \mathcal{F}_n$.

Part 1 of this Definition is quite technical but its meaning is quite clear. The operator $\mathcal{F}$ works as follows. In each step for any function f and a type τ which is the image under f of types which are all in $\mathcal{P}$, τ is included in $\mathcal{F}(\mathcal{P})$. In part 2 we include in $\mathcal{F}_0$ all types which are empty by direct application of given constraints.

Lemma 1. *$\mathcal{F}_\infty = \mathcal{F}_n$, where n is the size of S.*

Proof. Clearly $\mathcal{F}_{n+1} = \mathcal{F}_n$ if each Σ_0 type belonging to $\mathcal{F}_n$ belongs already to $\mathcal{F}_{n-1}$. Moreover, if a Σ_0-type τ_1 is included into $\mathcal{F}$ before Σ_0-type τ_2 then there is a Σ_1-type τ such that $\tau_1 \cap \tau \subseteq \mathcal{E}$ and $\tau_2 \cap \tau \not\subseteq \mathcal{E}$. To prove this assume that in a stage i the operator $\mathcal{F}$ adds a new Σ_0 type τ_1 and that τ_2 is not added neither in this stage nor has been added earlier. Since the only relationship between Σ_0 and Σ_1-types is through $\mathcal{E}$ it means that in the stage i a new Σ_1-type τ has been added and $\tau_1 \cap \neg\tau \subseteq \mathcal{E}$. Moreover, $\tau_2 \cap \neg\tau \not\subseteq \mathcal{E}$ since otherwise τ_2 would have been added in stage i.

Now, we will show that for a flat system $\mathcal{E} \subseteq \emptyset$ of set constraints of size n, if k is an integer for which there exist Σ_0-types $\tau_0,\ldots,\tau_k$ and Σ_1-types $\tau'_1,\ldots,\tau'_k$ such that

$$\tau_{i-1} \cap \tau'_i \subseteq \mathcal{E},\ \tau_i \cap \tau'_i \not\subseteq \mathcal{E},\ \text{for } 1 \leq i \leq k,$$

then $k \leq n$.

We prove it by induction on the size of $\mathcal{E}$. We consider several cases:

Case 1. $\mathcal{E}$ has the form X.

Then for every Σ_0-type τ either τ is contained in $\mathcal{E}$ or τ and $\mathcal{E}$ are disjoint. It is clear that $k \leq 1$.

Case 2. $\mathcal{E}$ has the form $F(X_1,\ldots,X_n)$.

Then for every Σ_1-type τ' either τ' is contained in $\mathcal{E}$ or τ' and $\mathcal{E}$ are disjoint and therefore $k \leq 0$.

Case 3. $\mathcal{E}$ has the form $\mathcal{E} = \mathcal{E}_1 \cup \mathcal{E}_2$.

Then for each Σ_0-type τ and each Σ_1-type τ' if $\tau \cap \tau' \subseteq \mathcal{E}$ then $\tau \cap \tau' \subseteq \mathcal{E}_1$ or $\tau \cap \tau' \subseteq \mathcal{E}_2$. Moreover, for each Σ_0-type τ and each Σ_1-type τ' if $\tau \cap \tau' \not\subseteq \mathcal{E}$ then $\tau \cap \tau' \not\subseteq \mathcal{E}_1$ and $\tau \cap \tau' \not\subseteq \mathcal{E}_2$. From this it follows that $k \leq k_1 + k_2$, where k_1 and k_2 are the maximal values for $\mathcal{E}_1$ and $\mathcal{E}_2$, respectively.

Case 4. $\mathcal{E}$ has the form $\mathcal{E}_1 \cap \mathcal{E}_2$.

Then for each Σ_0-type τ and each Σ_1-type τ' if $\tau \cap \tau' \subseteq \mathcal{E}$ then $\tau \cap \tau' \subseteq \mathcal{E}_1$ and $\tau \cap \tau' \subseteq \mathcal{E}_2$. Similarly for $\tau \cap \tau' \not\subseteq \mathcal{E}$ we have $\tau \cap \tau' \not\subseteq \mathcal{E}_1$ or $\tau \cap \tau' \not\subseteq \mathcal{E}_2$.

From this we deduce that $k \leq k_1 + k_2$, where k_1 and k_2 are the maximal values for $\mathcal{E}_1$ and $\mathcal{E}_2$, respectively.

This completes the proof of Lemma. $\square$

Lemma 2. *In every structure $\mathcal{M}$ each type in $\mathcal{F}_n$ is realized by at most doubly exponential number of elements of M.*

Proof. We use the fact that $Card(f(X_1, \ldots, X_n)) \leq \prod_{1 \leq i \leq n} Card(X_i)$, and the fact that the arity of all function symbols in $\mathcal{E}$ is bounded. $\square$

Theorem 1. *If a system S of set constraints is satisfiable by a structure $\mathcal{M}$ then S is satisfiable by a structure $\mathcal{M}'$ such that the cardinality of every Σ-type interpreted in $\mathcal{M}'$ is infinite or at most double exponential in the size of S.*

Proof. By Proposition 1 we can assume that S is of the form $\mathcal{E} \subseteq \emptyset$ where $\mathcal{E}$ is a flat expression.

To construct a structure $\mathcal{M}'$ from $\mathcal{M}$ we keep all types in $\mathcal{F}_\infty$ unchanged, and make all other types infinite. So be the Lemma 2 of all types in $\mathcal{F}_\infty$ have cardinality at most double exponential.

Moreover, put $\Delta_{F(X_1,\ldots,X_m)}^{(\tau_1,\ldots,\tau_n),\tau} = \infty$ if $\tau_i = \infty$, for some $1 \leq i \leq n$, and leave it unchanged otherwise. Clearly, as in the case of cardinality of types, the Δ is either infinite or at most double exponential. It is easy to check that the non-emptiness and cardinality conditions remain valid after these changes. $\square$

By the Theorem 1 to determine satisfiability of finite set of constraints it suffices to guess an exponential description D_S of a structure to and verify correctness of D_S i.e. the non-emptiness and the cardinality conditions for D_S.

References

1. A. Aiken, D. Kozen, M. Vardi, and E. L. Wimmers. The complexity of set constraints. In *Computer Science Logic'93*, LNCS 832, pages 1–17. Springer-Verlag, 1994.
2. A. Aiken, D. Kozen, and E. L. Wimmers. Decidability of systems of set constraints with negative constraints. Technical Report 93-1362, Computer Science Department, Cornell University, June 1993.
3. A. Aiken and B. Murphy. Static type inference in a dynamically typed language. In *Eighteenth Annual ACM Symposium on Principles of Programming Languages*, pages 279–290, January 1991.
4. A. Aiken and E. L. Wimmers. Solving systems of set constraints (extended abstract). In *Seventh Annual IEEE Symposium on Logic in Computer Science*, pages 329–340, 1992.
5. L. Bachmair, H. Ganzinger, and U. Waldmann. Set constraints are the monadic class. In *Eight Annual IEEE Symposium on Logic in Computer Science*, pages 75–83, 1993.

6. W. Charatonik and L. Pacholski. Set constraints with projections are in nexptime. In *35th Annual IEEE Symposium on Foundations of Computer Science*, pages 642–653, November 1994.
7. R. Gilleron, S. Tison, and M. Tommasi. Solving systems of set constraints using tree automata. In *10th Annual Symposium on Theoretical Aspects of Computer Science*, LNCS 665, pages 505–514. Springer-Verlag, 1993.
8. R. Gilleron, S. Tison, and M. Tommasi. Solving systems of set constraints with negated subset relationships. In *Proceedings of the 34th Symp. on Foundations of Computer Science*, pages 372–380, 1993. A full version *Technical report IT 247, Laboratoire d'Informatique Fondamentale de Lille*.
9. D. Givan, R. McAllester. New results on local inference relations. In M. K. Press, editor, *Principles of Knowledge representation and Reasoning: Proceedings of the Third International Conference*, pages 403–412, 1992.
10. N. Heintze and J. Jaffar. A decision procedure for a class of set constraints (extended abstract). In *Fifth Annual IEEE Symposium on Logic in Computer Science*, pages 42–51, 1990.
11. N. Heintze and J. Jaffar. A finite presentation theorem for approximating logic programs. In *Seventeenth Annual ACM Symposium on Principles of Programming Languages*, pages 197–209, January 1990.
12. N. D. Jones and S. S. Muchnick. Flow analysis and optimization of lisp-like structures. In *Sixth Annual ACM Symposium on Principles of Programming Languages*, pages 244–256, January 1979.
13. D. A. McAllester and R. Givan. Taxonomic syntax for first order inference. *Journal of ACM*, 40:346–283, 1993.
14. D. A. McAllester, R. Givan, C. Witty, and D. Kozen. Tarskian set constraints. In *Proceedings, 11th Annual IEEE Symposium on Logic in Computer Science*, pages 138–147, New Brunswick, New Jersey, July 1996. IEEE Computer Society Press.
15. P. Mishra and U. Reddy. Declaration-free type checking. In *Twelfth Annual ACM Symposium on the Principles of Programming Languages*, pages 7–21, 1985.
16. J. C. Reynolds. Automatic computation of data set definitions. *Information Processing*, 68:456–461, 1969.
17. K. Stefansson. Systems of set constraints with negative constraints are *NEXPTIME*-complete. In *Ninth Annual IEEE Symposium on Logic in Computer Science*, pages 137–141, 1994.
18. J. Young and P. O'Keefe. Experience with a type evaluator. In D. Bjørner, A. P. Ershov, and N. D. Jones, editors, *Partial Evaluation and Mixed Computation*, pages 573–581. North-Holland, 1988.

∀∃*-Equational Theory of Context Unification is Π_1^0-Hard

Sergei Vorobyov

Max-Planck Institut für Informatik, Im Stadtwald, D-66123, Saarbrücken, Germany,
sv@mpi-sb.mpg.de, http://www.mpi-sb.mpg.de/~sv

Abstract. Context unification is a particular case of second-order unification, where all second-order variables are *unary* and only *linear* functions are sought for as solutions. Its decidability is an open problem. We present the simplest (currently known) undecidable quantified fragment of the theory of *context unification* by showing that for every signature containing a ≥ 2-ary symbol one can construct a *context equation* $\mathcal{E}(p, r, \overline{F}, \overline{w})$ with parameter p, first-order variables r, $\overline{w}$, and context variables $\overline{F}$ such that the set of true sentences of the form

$$\forall r\ \exists\ \overline{F}\ \exists\ \overline{w}\ \ \mathcal{E}(p, r, \overline{F}, \overline{w})$$

is Π_1^0-hard (i.e., every co-r.e. set is many-one reducible to it), as p ranges over finite words of a binary alphabet. Moreover, the existential prefix above contains just 5 context and 3 first-order variables.

1 Introduction

The *Context Unification Problem* (CUP for short) is:
– A generalization of the celebrated Markov-Löb's problem of solvability of equations in a free semigroup proved *decidable* by Makanin [7]; CUP coincides with this problem for monadic signatures.
– A specialization of the *Second-Order Unification* (SOU), known to be *undecidable* due to Goldfarb [6,5]. CUP is almost SOU, but with *only unary function variables* allowed and solutions required to be *linear*, i.e., of the form $\lambda x.t(x)$, where $t(x)$ contains *exactly one occurrence* of x.

Context unification is useful in different areas of Computer Science: term rewriting, theorem proving, equational unification, constraint solving, computational linguistics, software engineering [11,9,12]. CUP is stated as follows:
Given a pair of terms t, t' built as usual from symbols of a signature Σ, first-order variables $\overline{w}$, and unary function variables $\overline{F}$, does there exist an assignment θ of terms to $\overline{w}$ and linear second-order functions to $\overline{F}$ such that $\theta(t) = \theta(t')$?

Thus, CUP is a decidability problem for the existentially quantified equations ($\exists^*$-equational theory) of the form

$$\exists\ \overline{F}\ \exists\ \overline{w}\ \ t = t', \tag{1}$$

where the quantified context variables $\overline{F}$ range over *linear* functions.

Currently the decidability of CUP is an open problem [11,9,12]. Most researchers conjecture and hope that CUP is decidable. All the above papers provide some approximations: either prove decidability of particular cases, or settle undecidability of some generalizations, or provide technical results towards decidability of CUP.

Presumably, CUP is very hard to settle, both in decidable and undecidable sense. This is because CUP lies between a technically difficult decidable case of equations in free semigroups (Markov-Löb's problem proved decidable by Makanin [7]), and the undecidable case of SOU settled by Goldfarb [6] and reinforced by Farmer [5]. Farmer's result is also technically quite difficult.

Goldfarb [6] demonstrated that SOU is undecidable for second-order languages containing at least one $\geq$ 2-ary function constant and finitely many *unary* and *ternary* function variables. Later Farmer [5] improved it by showing that SOU remains undecidable in presence of *unary function variables only* (but, unlike CUP, substitutions looked for are *not required to be linear*; in fact, they are not linear in Farmer's proof). It follows from Makanin's result [7] that SOU is decidable when all variable and constant function symbols are unary. Farmer [4] improved this by showing that decidability is preserved if n-ary function variables are allowed in addition to constant function symbols of arity *at most one*.

Thus, CUP represents the only unknown remaining difficult intermediate case between decidable word equations and undecidable SOU (unary variables, n-ary constants, linear solutions). This explains why the progress on CUP has been quite slow. Indeed, decidability of CUP would considerably improve Makanin's result, whereas undecidability would considerably improve Goldfarb-Farmer's undecidability of SOU.

In this paper we show that adding just one outermost universal quantifier to a context equation (1) leads to the Π_1^0-hard class of formulas, where Π_1^0 is the class of all co-recursively enumerable sets.

For comparison, the following undecidability results are known about quantified fragments of context unification. Quine [10] showed that the full first-order theory of free semigroups is undecidable (this corresponds to context unification in unary signatures). Durnev [3] improved it to the undecidability of $\exists\forall\exists^3$-positive (without negation, but with $\wedge$ and $\vee$) theory of free semigroups. Marchenkov [8] improved it to the undecidability of $\forall\exists^4$-positive theory of free semigroups. Durnev [2] improved it to undecidability of $\forall\exists^3$-positive theory of free semigroups. Niehren, Pinkal, and Ruhrberg [9] claimed undecidability of the $\exists^*\forall^*\exists^*$-theory of context unification.

It should be noticed that all known methods to transform a positive formula of the theory of free semigroups into just one equation require a considerable number of *auxiliary existentially quantified variables* (see, e.g., [1]), depending on the number of disjunctions involved. Thus the above undecidability results for $\forall\exists^4$- and $\forall\exists^3$-positive theories of free semigroups yield only undecidability of the $\forall\exists^n$-equational theories of free semigroups with a *very large* number n of existentially quantified variables.

In this paper we show that the situation with context equations is quite different, and *just two* extra existentially quantified context variables suffice to eliminate all disjunctions. This, together with a direct reduction from the halting problem for Turing machines, gives the undecidability of the $\forall\exists^8$-equational theory of context unification, with a reasonably simple quantifier prefix.

The main result of this paper may now be stated as follows.

Main Theorem. *For every signature containing a ≥ 2-ary symbol one can construct a* context equation $\mathcal{E}\,(p, r, C, F, F', G, H, x, y, z)$ *with parameter p, first-order variables r, x, y, z, and context variables C, F, F', G, H such that the set of true sentences of the form*

$$\forall r \; \exists \, C, F, F', G, H \; \exists \, x, y, z \;\; \mathcal{E}\,(p, r, C, F, F', G, H, x, y, z) \tag{2}$$

is Π_1^0-hard (i.e., every co-r.e. set is many-one reducible to it), as p ranges over finite words of a binary alphabet. □

It follows, in particular, that the $\forall\exists^8$-equational theory (without $\land$, $\lor$, $\neg$) of context unification is *undecidable*.

Warning. We would like to stress that in this paper we do not intend to improve the undecidability results of Marchenkov and Durnev [8,2] for $\forall\exists^4$- and $\forall\exists^3$-*positive theories of free semigroups* as to *minimizing* the number of existential quantifiers. However, we do get an improvement over [8,2] (in a different framework of context unification) as to simplicity of the existential prefix in the undecidable $\forall\exists^*$-*equational theory* of context unification. As we mentioned above, all known methods of eliminating disjunctions from positive formulas in free semigroups use a considerable number of auxiliary existentially quantified variables, proportional to the number of disjunctions to be eliminated. We show that in context unification just two extra variables are enough. We do not claim that the quantifier prefix we obtain is minimal. We rather tried to keep proofs intuitive and transparent. We believe that applying the methods of disjunction elimination with a constant number of extra variables (Section 5) directly to the reductions of [8,2] (the proofs absent from [2] are unavailable to the author) will yield the undecidability of the $\forall\exists^5$-equational theory of context unification.

Paper Outline. After Section 2 with preliminaries, in Sections 3, 4 we give the main construction of an $\forall\exists^*$-sentence (expressing non-applicability) from a DTM with a complete Σ_1^0 domain. In Section 5 we eliminate disjunctions.

2 Preliminaries

Context Unification. Let Σ be a fixed finite signature with each symbol assigned a fixed arity, *containing at least one constant*. Let X be an infinite set of first-order variables and $\mathcal{F} = \{F, G, \ldots\}$ be an infinite set of *function variables of arity one*, also called *context variables*.

Definition 1 (Terms). *The set $\mathcal{T}(\Sigma, X)$ of* terms *of signature Σ with variables from X is defined as usual: variables from X are terms and for $f \in \Sigma$ of arity $n \geq 0$ an expression $f(t_1, \ldots, t_n)$ is a term whenever t_i's are terms.* □

We assume all the standard definitions and conventions concerning λ-notation, like β-reduction, normalization, etc.

Definition 2 (Contexts). *A* context *is an expression of the form* $\lambda x.t(x)$, *where* $t(x) \in \mathcal{T}(\Sigma, \{x\})$ *contains exactly* one *occurrence of the variable* x. *A context with β-normal form $\lambda x.x$ is called* empty. □

Remark 1. Note that the '*exactly one*' requirement, absent from the definition of SOU, is a characteristic feature of CUP, distinguishing it from SOU. □

Definition 3 (Context Terms). *Are defined inductively: if $\phi \in \Sigma \cup \mathcal{F}$ is n-ary and $t_1, \ldots, t_n$ are context terms, then $\phi(t_1, \ldots, t_n)$ is a context term.* □

Convention. Writing terms and context terms with unary function symbols and function variables we usually drop parentheses to improve readability.

Definition 4 (CUP, Context Equations). *An instance of CUP, also called a* context equation *(CE for short), is an expression $\tau_1 \stackrel{?}{=} \tau_2$, where $\tau_{1,2}$ are context terms. A* solution *to a CE $\tau_1 \stackrel{?}{=} \tau_2$ is a substitution θ of contexts for context variables such that $\theta(\tau_1) \equiv_\beta \theta(\tau_2)$ (equality modulo the usual β-reduction).* □

Remark 2. The requirement of replacing functional variables with *contexts*, as opposed to arbitrary second-order λ-terms, distinguishes CUP from SOU. When this requirement is dropped, the problem becomes *undecidable* [5]. □

Turing Machine with Complete R.E. Domain. A Σ_1^0-set is a recursively enumerable set. A Π_1^0-set is a complement of a Σ_1^0-set. An Σ_1^0-set (resp., Π_1^0-set) A is called *complete* iff every Σ_1^0-set (resp., Π_1^0-set) B is *many-one reducible* to A, i.e., there exists a total recursive function f such that $x \in B \Leftrightarrow f(x) \in A$.

It is well known that there exists a DTM M whose domain is a complete Σ_1^0-set, and, therefore, the set of elements it does not accept is a complete Π_1^0-set. We may assume, without loss of generality, that the tape alphabet B of M consists of two symbols, 0 and 1, the tape of M is infinite to the right, M never tries to move left from its leftmost tape cell, M may only extend its tape by writing new symbols on the right end, that the states of M are $Q = \{q_0, \ldots, q_f\}$, q_0 is the unique initial state, q_f is the unique final state of M, M always starts by observing its leftmost tape cell, and M immediately stops entering state q_f. Further we assume that M is a fixed such DTM.

The DTM M applies to a nonempty word $b_1 \ldots b_m \in B^+$ iff there exists a finite sequence of IDs (instantaneous descriptions) $id_0, \ldots, id_F$ of M such that $id_0 \equiv q_0 b_1 \ldots b_m$, $id_F \equiv \gamma q_f \delta$ for some $\gamma, \delta \in B^*$, and such that for every pair of IDs id_i, id_{i+1} in the sequence id_{i+1} is obtained from id_i by application of some command of M. We omit the well-known definitions.

We will represent IDs of M by terms constructed from unary function symbols, starting from the constant ε for the empty word. We have a unary function symbol for every symbol in $B \cup Q$, and represent a word $q_0 10101010$ as a term $q_0(1(0(1(0(1(0(1(0(\varepsilon))))))))))$, which for simplicity will always be written as $q_0 10101010$ (i.e., with ( ) and ε omitted), if it does not lead to ambiguity.

We will represent a sequence $id_0, \ldots, id_F$ of IDs as a right-flattened list

$$f(id_0, f(id_1, f \ldots, f(id_{F-1}, f(id_F, \varepsilon)) \ldots)), \tag{3}$$

where id_i's are term representations of words using unary functional symbols as described above. Thus the DTM M applies to an input iff such a term (3) exists.

Signature. Formally, we will use the following signature Σ: 1) ε - constant, for the empty word and list, 2) 0, 1 - unary for the tape alphabet B, 3) $q_0, \ldots, q_f$ - unary for M's states in Q, 4) f - binary list constructor, 5) a - unary, auxiliary.

Convention. If not stated otherwise, everywhere below s denotes a symbol from $B \cup Q$, b denotes a symbol from B, q denotes a symbol from Q.

3 Sentence Expressing Inapplicability

To prove the main claim of this paper we will write a sentence of the following form (with context variables C, F, F', G, H, and ordinary variables r, x, y, z):

$$\forall r \; \exists C, F, F' \; \exists x, y, z \; \exists G, H \;\; \Phi\Big(f(q_0 b_{i_1} \ldots b_{i_p}, r), t_1, \ldots, t_m\Big), \tag{4}$$

which expresses the fact that the DTM M *does not apply* to the input string $\bar{s} \equiv b_{i_1} \ldots b_{i_p}$, because for every r, the term $t \equiv f(q_0 b_{i_1} \ldots b_{i_p}, r)$ *is not* a correct run (3) of M, i.e., $\bar{s} \equiv b_{i_1} \ldots b_{i_p}$ is in the complement of the r.e.-complete domain of M. This implies the main claim of the paper.

Technically, expressing that a term $t \equiv f(q_0 b_{i_1} \ldots b_{i_p}, r)$ *is not* a correct run amounts to saying that 'something goes wrong' in t. For example, t contains 'senseless' subterms like $f(f(\ldots, \ldots), \ldots)$, or $a(\ldots)$, etc. It turns out that this can be expressed by saying 'contains one of the *finitely* many wrong patterns' $t_1, \ldots, t_m$, which can be done by saying $t = Ct_1 \vee \ldots \vee t = Ct_m$, where C is a context variable used to say 'there exists a subterm of t'. All such patterns are expressed by using 2 context variables F, F' and 3 first-order variables x, y, z. In Section 5 we show how to get rid of disjunction by using just two extra existentially quantified context variables (G and H in (4)), independently of the number m of patterns. Thus just one context equation with 5 context and 3 first-order existentially quantified variables as in (2), (4) suffices for undecidability.

4 Forbidden Patterns

We enumerate all forbidden patterns preventing a term to be a correct run.

Forbidden Structural Patterns. A term containing one of the following subterms cannot represent a correct run (for some context F and terms x, y, z):

1. $f(F(a(x)))$. **Reason:** a is an auxiliary symbol and cannot occur in a run.
2. $f(f(x,y),z)$. **Reason:** a run is a right-flattened list of IDs, f cannot occur in the first argument position to itself.

3. $f(x, g(y))$ {for every function symbol $g \in B \cup Q$}. **Reason:** a run is a right-flattened list of IDs.
4. $q(F(q'(x)))$ {for all $q, q' \in Q$}. **Reason:** at most one state symbol per ID.
5. $g(f(x, y))$ {for all $g \in B \cup Q$}. **Reason:** IDs cannot contain f.
6. $f(F(q_f(x)), f(y, z))$. **Reason:** entering q_f DTM stops.
7. $q(s(x))$ {for all $q \in Q \setminus \{q_f\}$ and $s \in B$ such that M has no commands $(q, s \to \ldots)$}.
8. $q(\varepsilon)$ {for all $q \in Q \setminus \{q_f\}$ such that there are no commands to extend the tape on the right end in state q}.

Patterns for Incorrect ID Transitions. Now we enumerate patterns that cannot occur in a correct run of the DTM M for the reason a term matching one of these patterns contains a 'senseless' ID transition. This can be expressed by existence of solutions to a *finite* number of 'local' context equations. As an easy and representative example, note that an ID α cannot be transformed into an ID β, if for some context F, state $q \in Q$, tape symbols $s_{1,2,3,4,5} \in B$, and x, y one has simultaneously $\alpha \equiv Fs_1qs_2x$ and $\beta \equiv Fs_3s_4s_5y$. This is because each TM's transition moves head either left or right, so in a correct transition either s_3 or s_5 should belong to Q (be a state). This illustrates the main idea. By routinely enumerating all possibilities we can assure that a transition is correct iff it *does not match any* of the patterns enlisted.

The reader is invited to check that all these patterns use only three first-order variables x, y, z, and two context variables F, F'. Recall that we use these patterns P_i disjunctively in $t = CP_1 \vee \ldots \vee t = CP_m$ to say that t contains a subterm matching one of the patterns $P_1, \ldots, P_m$ (where the context variable C is used to say 'there exists a subterm'). Since $\exists$ distributes over $\vee$, the variables may be reused and we need just three context C, F, F', and three first-order variables x, y, z. The two extra context variables G, H will be needed to transform a disjunction into an equation in Section 5.

List of the patterns for incorrect transitions.

$f(Fsx, f(Fs'y, z))$ {for $s, s' \in B$, $s \neq s'$}. **Reason:** in two succeeding IDs the leftmost position in which they differ should necessarily contain a state from Q. For example, for a right shift command $(q, s \to q', s', R)$ we have, in a correct ID transition (the first disagreement is $q - s'$):

$$ID_i = s_1 \ldots s_n \; q \; s \; s_{n+1} \ldots s_{n+m}$$
$$ID_{i+1} = s_1 \ldots s_n \; s' \; q' \; s_{n+1} \ldots s_{n+m}$$

For a left shift command $(q, s \to q', s', L)$ we have, in a correct ID transition:

$$ID_i = s_1 \ldots s_{n-1} \; s_n \; q \; s \; s_{n+1} \ldots s_{n+m}$$
$$ID_{i+1} = s_1 \ldots s_{n-1} \; q' \; s_n \; s' \; s_{n+1} \ldots s_{n+m}$$

(the first disagreement is $s_n - q'$). For an 'extension on the right' command $(q, \varepsilon \to q', s')$ we have, in a correct ID transition (first disagreement $q - q'$):

$$ID_i = s_1 \ldots s_n \; q$$
$$ID_{i+1} = s_1 \ldots s_n \; q' \; s'$$

$f(Fsx, f(F(\varepsilon), z))$ {for all $s \in B \cup Q$}. **Reason:** ID_{i+1} cannot be shorter ID_i.

$f(qs_1x, f(s_2s_3y, z))$ {for all $s_3 \in B$}. **Reason:** if the first symbol of the ID_i is a state, then the second symbol in the ID_{i+1} should be a state.

$f(Fs_1qs_2x, f(Fs_3s_4s_5y, z))$ {for all $s_3, s_5 \in B$}. **Reason:** if the k-th symbol in ID_i is a state then either $k+1$-th or $k-1$-th in ID_{i+1} should be a state.

$f(Fqsx, f(Fs_1s_2y, z))$ {for every command $(q, s \to q', s', R)$ and every pair of symbols $s_1, s_2 \in B \cup Q$ such that either $s_1 \neq s'$ or $s_2 \neq q'$}. **Reason:** an ID $Fqsx$ should yield $Fs'q'x$, thus each pair of consecutive IDs $Fqsx, Fs_1s_2y$ is incorrect.

$f(Fqsx, f(Fs_1(\varepsilon), z))$ {for every $s_1 \in B \cup Q$}. **Reason:** ID_{i+1} abruptly ends. Similar patterns must be written for the case of left shift commands below.

$f(Fbqsx, f(Fs_1s_2s_3y, z))$ {for every command $(q, s \to q', s', L)$, every $b \in B$, and every triple of symbols $s_1, s_2, s_3 \in B \cup Q$ such that either $s_1 \neq q'$, or $s_2 \neq b$, or $s_3 \neq s'$}. **Reason:** an ID $Fbqsx$ should yield $Fq'bs'x$, thus each pair of consecutive IDs $Fbqsx, Fs_1s_2s_3y$ is incorrect.

... Similar patterns should be spelled out for the commands extending the tape on the right. We leave it as a straightforward exercise.

$f(FqsF's_1x, f(Fs'q'F's_2y, z))$ {for every command $(q, s \to q', s', R)$ and every pair of different symbols $s_1, s_2 \in B$}. **Reason:** an ID transformation is 'almost' correct, but some symbol to the right of the head is copied erroneously.

... Similar pattern for an 'almost correct' left shift command (exercise).

Let us add a brief explanation of how the above patterns work. In a correct run no ID can contain two states (since we have a pattern $q(F(q'(x)))$). Can a correct run contain and ID with no state symbols at all? Let us show that this is impossible. For, if such a situation was possible, there would exist a pair of neighboring IDs with the least index i such that id_i contains a state symbol and id_{i+1} does not. Then it is easy to see that one of the patterns responsible for the transition correctness would match, thus guaranteeing the run candidate incorrectness.

We thus constructed a finite number of patterns $t_1, \ldots, t_m$ such that the DTM M *does not apply* to the word $\bar{s} = b_{i_1} \ldots b_{i_p}$ if and only if

$$\forall r \; \exists C, F, F' \; \exists x, y, z \; \Big(f(q_0 b_{i_1} \ldots b_{i_p}, r) = Ct_1 \vee \ldots \vee f(q_0 b_{i_1} \ldots b_{i_p}, r) = Ct_m \Big).$$

This already proves that the positive $\forall \exists^6$-theory of context unification is Π_1^0-hard (by the choice of M with a complete r.e.- or Σ_1^0-set). In the next section we proceed to eliminating disjunctions from the sentence above.

5 Trading Disjunctions for Equations

To finish the proof of the main claim we show how to represent a disjunction of context equations of the form $t = t_1 \vee \ldots \vee t = t_m$ by *just one* context equation, using only two auxiliary (existential) context variables, independently of the number m. Conjunctions do not cause any problems, because in presence of function symbols of arity ≥ 2, a conjunction of context equations can be easily expressed by just one equation ($s = s' \wedge t = t'$ iff $f(s,t) = f(s',t')$; similarly for larger arities). When all function symbols are unary, there exists a well-known trick to represent $s = s' \wedge t = t'$ as $satsbt = s'at's'bt'$, where a, b are different unary function symbols. Disjunction elimination is based on the following

Lemma 1. *Let Θ be the substitution $\left\{ Ct_i/x_i \right\}_{i=1}^{m}$, where t_i's are the forbidden patterns enumerated in Section 4. Consider the system of two equations (where $[\,]$ denotes the empty list ε and $[e_0, \ldots] = f(e_0, [\ldots])$):*

$$G(a(H(a))) = \Big[a\Big((\lambda x_1.[x_1,\ldots,x_m])a\Big),$$
$$\ldots$$
$$a\Big((\lambda x_i.[x_1,\ldots,x_m])a\Big), \qquad (5)$$
$$\ldots$$
$$a\Big((\lambda x_m.[x_1,\ldots,x_m])a\Big)\Big] \Theta,$$

$$Hf(q_0\bar{s}, r) = [x_1, \ldots, x_m]\,\Theta. \qquad (6)$$

For every ground term r one has the following equivalence: the system (5), (6) has a solution if and only if ($\Leftrightarrow$) $f(q_0\bar{s}, r)$ is an incorrect run.

Brief Explanation. The term on the right of (5) is (with $a(\varepsilon)$ on the diagonal):

$$\begin{bmatrix} a\,[\ \ a, & Ct_2, & \ldots & Ct_i, & \ldots & Ct_m\ ], \\ a\,[\ Ct_1, & a, & \ldots, & Ct_i, & \ldots, & Ct_m\ ], \\ & & \ldots & & & \\ a\,[\ Ct_1, & Ct_2 & \ldots, & a, & \ldots, & Ct_m\ ], \\ & & \ldots & & & \\ a\,[\ Ct_1, & Ct_2, & \ldots, & Ct_i, & \ldots, & a\ \ ] \end{bmatrix} \qquad (7)$$

Proof. ($\Leftarrow$) is straightforward. Suppose for a ground r the term $f(q_0\bar{s}, r)$ is an incorrect run for the reason it contains one of the forbidden patterns t_i. Let

$$G_i = \lambda u.\,\Big[a\Big((\lambda x_1.[x_1,\ldots,x_m])a\Big),$$
$$\ldots$$
$$a\Big((\lambda x_{i-1}.[x_1,\ldots,x_m])a\Big),$$
$$u, \qquad (8)$$
$$a\Big((\lambda x_{i+1}.[x_1,\ldots,x_m])a\Big),$$
$$\ldots$$
$$a\Big((\lambda x_m.[x_1,\ldots,x_m])a\Big)\Big]\,\Theta,$$

$$H_i \;=\; \Big(\lambda x_i.[x_1,\ldots,x_i,\ldots,x_m]\Big)\,\Theta. \qquad (9)$$

Clearly, substituting such G_i, H_i in (5) yields the identity. Moreover, by substituting H_i in (6) we obtain
$$[Ct_1 \ldots Ct_{i-1}, f(q_0\bar{s}, r), Ct_{i+1} \ldots Ct_m] = [Ct_1 \ldots Ct_{i-1}, Ct_i, Ct_{i+1} \ldots Ct_m],$$
which, obviously, has a solution for C, since $f(q_0\bar{s}, r)$ contains a forbidden pattern t_i by assumption. Thus the system (5), (6) has a solution. □

For the opposite direction ⇒ in Lemma 1 we prove (the contrapositive)

Lemma 2. *Let $f(q_0\bar{s}, r)$ be a correct run. Then the system (5), (6) has no solutions.*

Proof. (5), (6) cannot have solutions of the form (8), (9), because the opposite would mean that $f(q_0\bar{s}, r)$ is incorrect (see the proof of the previous lemma). The other 'possible' solutions split into the following two cases.

Case 1. The outermost a on the left of (5) matches one of the outermost a's on the right, but the innermost a on the left does not match the corresponding a on the right. In other words, for some substitution Θ' either

$$H = \lambda z. [\ldots, \quad a, \quad \ldots, Ct_j\Theta'[z/a(\varepsilon)], \ldots], \text{ or}$$
$$H = \lambda z. [\ldots, Ct_j\Theta'[z/a(\varepsilon)], \ldots, \quad a, \quad \ldots],$$

with a in the i-th ($i \neq j$) place in the list. It is readily seen that none of such 'solutions' can satisfy (6), because $Ct_i\Theta'$ on the right cannot be equal (for any C, Θ') to $a \equiv a(\varepsilon)$ on the left, since all the forbidden patterns t_i enumerated in Section 4 contain function symbols different from a.

Case 2. The whole $a(H(a))$ on the left of (5) matches with some subterm of $Ct_k\Theta'$, $1 \leq k \leq m$, for some substitution Θ' (i.e., neither of a's in $a(H(a))$ on the left matches a visible a on the right of (5); see also (7)).

Since a correct ID $q_0\bar{s}$ cannot match any of Ct_j, in order to satisfy (6), H should be substituted with $\lambda x.[r_1, \ldots, r_l(x), \ldots, r_m]$ for some terms r_i. Consequently, $Ct_k\Theta'$ contains $a[r_1, \ldots, r_l(a), \ldots, r_m]$. Let us show that this cannot yield a solution.

Suppose, $k \neq l$. Then $Ct_k\Theta'$ properly contains r_k, and therefore (6) cannot be satisfied, because it requires $r_k = Ct_k\Theta'$.

Suppose, $k = l$. We have $Ct_k\Theta'$ contains $a[r_1, \ldots, r_k(a), \ldots, r_m]$. Thus $Ct_k\Theta'$ contains $p+2$ occurrences of a. On the other hand, to satisfy (6) we should have $r_k(f(q_0\bar{s}, r)) = Ct_k\Theta'$. Note that $r_k(f(q_0\bar{s}, r))$ contains at most p occurrences of a (compared with $p + 2$ on the right). Hence the above equality cannot hold.

Thus, the system (5), (6) has no solutions. This finishes the proof of Lemmas 1, 2, and the proof of the main claim of his paper. □

6 Conclusions

By reduction from the non-acceptance problem for a deterministic Turing machine with a complete r.e. domain we demonstrated that for a one-parametric set of context equations $\mathcal{E}_p$ the set of true sentences of the form $\forall^1 \exists^8 \mathcal{E}_p$ is co-r.c.-hard as p runs over binary words. It follows that the $\forall^1 \exists^8$-equational theory of context unification is undecidable. This gives the simplest currently known undecidable quantified fragment of context unification. The decidability of the CUP itself remains open. Quite surprisingly, the only known lower bound for the CUP is NP-hardness [12]. Thus, in the absence of the decidability proof it would be very interesting to obtain nontrivial lower complexity bounds for CUP.

Technically, we described an *ad hoc* method to eliminate disjunctions from the positive formulas (of some particular structure) of context unification, which is completely sufficient for the purposes of this paper. It is clear that the method can be generalized so as to apply to *arbitrary* positive formulas. We conjecture that combining techniques presented with ideas of [8,2] will lead to a more tight undecidability results for CUP, with fewer quantifiers.

Thanks to Margus Veanes and anonymous referees for insightful remarks.

References

1. J. R. Büchi and S. Senger. Coding in the existential theory of concatenation. *Archive of Mathematical Logic*, 26:101–106, 1986.
2. V. Durnev. Studying algorithmic problems for free semi-groups and groups. In *Logical; Foundations of Computer Science (LFCS'97)*, volume 1234 of *Lect. Notes Comput. Sci.*, pages 88–101. Springer-Verlag, 1997.
3. V. G. Durnev. Positive theory of a free semigroup. *Soviet Math. Doklady*, 211(4):772–774, 1973.
4. W. Farmer. A unification algorithm for second-order monadic terms. *Annals Pure Appl. Logic*, 39:131–174, 1988.
5. W. Farmer. Simple second-order languages for which unification is undecidable. *Theor. Comput. Sci.*, 87:25–41, 1991.
6. W. D. Goldfarb. The undecidability of the second-order unification problem. *Theor. Comput. Sci.*, 13:225–230, 1981.
7. G. S. Makanin. The problem of solvability of equations in a free semigroup. *Math USSR Sbornik*, 32(2):127–198, 1977.
8. S. S. Marchenkov. Undecidability of the positive $\forall\exists$-theory of a free semigroup. *Siberian Math. Journal*, 23(1):196–198, 1982.
9. J. Niehren, M. Pinkal, and P. Ruhrberg. On equality up-to constraints over finite trees, context unification, and one-step rewriting. In W. McCune, editor, *CADE-14*, volume 1249 of *Lect. Notes Comput. Sci.*, pages 34–48. Springer-Verlag, 1997.
10. W. V. Quine. Concatenation as a basis for arithmetic. *J. Symb. Logic*, 11(4):105–114, 1946.
11. M. Schmidt-Schauß. Unification of stratified second-order terms. Interner Bericht 12/94, University of Frankfurt am Main, 1994.
12. M. Schmidt-Schauß and K. U. Schulz. On the exponent of periodicity of minimal solutions of context equations. In *Rewriting Techniques and Applications'98*, volume 1379 of *Lect. Notes Comput. Sci.*, pages 61–75. Springer-Verlag, 1998.

Speeding–Up Nondeterministic Single–Tape Off–Line Computations by One Alternation
(Extended Abstract)

Jiří Wiedermann[*]

Institute of Computer Science
Academy of Sciences of the Czech Republic
Pod vodárenskou věží 2 , 182 07 Prague 8, Czech Republic
e–mail: wieder@uivt.cas.cz

Abstract. It is shown that any nondeterministic single–tape off–line Turing machine of time complexity $T(n)$ can be speeded–up by one extra alternation by the factor $\log \log T(n)/\sqrt{\log T(n)}$, for any well–behaved function $T(n)$. This leads to the separation $\mathsf{NTIME}_{1+\mathrm{I}}(T(n)) \subset \Sigma_2 - \mathsf{TIME}_{1+\mathrm{I}}(T(n))$ of the respective complexity classes. Analogous result holds also for the complementary classes $\mathsf{co\text{-}NTIME}_{1+\mathrm{I}}(T(n))$ and $\Pi_2\text{-}\mathsf{TIME}_{1+\mathrm{I}}(T(n))$. This is the first occasion where such separation results have been proved for a restricted type of multitape nondeterministic machines. For the general case of multitape nondeterministic machines similar results are not known to hold.

1 Introduction

The separation of complexity classes within the Σ hierarchy $\mathsf{DTIME}(T(n)) \subseteq \mathsf{NTIME}(T(n)) \subseteq \Sigma_2 - \mathsf{TIME}(T(n)) \subseteq \ldots \mathsf{ATIME}(T(n)) \subseteq \mathsf{DSPACE}(T(n))$ is one of the central problems in complexity theory. The first proof that nondeterminism is stronger than determinism comes probably from Hennie [2] who in 1965 proved this result for single tape machines recognizing non–palindroms. For multitape machines the separation[1] $\mathsf{DTIME}(n) \subset \mathsf{NTIME}(n)$ has been proved in 1983 by Paul, Pippenger, Szemerédi, and Trotter [12]. In 1973 Paterson [10] proved that deterministic space is more powerful than deterministic time for single–tape machines. The analogous result for multitape machines $\mathsf{DTIME}(T(n)) \subset \mathsf{DSPACE}(T(n))$ followed in 1977 by Hopcroft, Paul and Valiant [5]. In 1980 Paul, Prauss, and Reischuk [11] proved that unbounded number of alternations can speedup single–tape computations. The separation for multitape machines ensued in 1983 by Dymond and Tompa [3]: $\mathsf{DTIME}(T(n)) \subset \mathsf{ATIME}(T(n))$. Finally, Kannan in 1983 and Maass and Schorr in 1987 proved increasingly better results

---

[*] This research was supported by GA ČR Grant No. 201/98/0717 and by an EU grant INCO–COOP 96–0195 'ALTEC–KIT' jointly with the accompanying grant of the MMT R No. OK–304
[1] The inclusion symbol '⊂' denotes the proper containment.

that bounded number of alternations can speedup deterministic single tape computations (also with a separate input tape). For the case of multitape machines the respective result DTIME$(T(n)) \subset \Sigma_2$-TIME$(T(n))$ proved Gupta [4] in 1988. Thus, merely two alternation were enough to separate the latter complexity classes.

There are two lessons to be taken from the previous short historical excursion in the context of the present paper. The first lesson is that in all cases more general results for multitape machines always followed only after proving similar results for simpler types of machines. Thus, results for restricted machines served as inspiration for looking for similar results on more general models of computing. Second, as a rule, all the respective results separate *deterministic* time from some higher complexity class in the above mentioned hierarchy. This is because it seemed that there were principal reasons that prevented the application of analogous speed–up techniques also in the case of nondeterministic time. Roughly, in some cases it was the impossibility efficiently rerunning a piece of nondeterministic computation twice along the same computational path (cf. [5]), or "unavailability" of nondeterminism without adding a further alternation (cf. [9]).

The problem of efficient speedup of nondeterministic computations has been identified as the major roadblock that prevents any further progress in separating other complexity classes as before (cf. [4]).

Nevertheless, recently a way of solving also this problem has appeared, so far at least for restricted Turing machines. In order to see the key idea of such a solution we have to return to the idea of the best separation result that separates single–tape off–line deterministic time bounded computations from Σ_2 single tape off–line computations [9].

This proof, and also other simulation proofs achieving speedup by two alternations (cf. [4]) share roughly the same strategy. In the first, nondeterministic phase a space efficient description of size $o(T(n))$ of the original deterministic computation is guessed. The correctness of this guess is in turn verified in the second phase by invoking the parallelism offered by co–nondeterminism.

Unfortunately, this natural idea of simulation cannot be straightforwardly transferred to the nondeterministic case. This is because in the above mentioned approach the verification phase requires replaying of some pieces of the original computation. Thus, when the original computation was a nondeterministic one it appeared that there was no way of its efficient verification during one subsequent alternation.

The first solution of the problem of speeding–up single–tape nondeterministic computations by one alternation has been devised in 1996 in [13].

The previously mentioned obstacle was roundabout by reorganizing the above mentioned two phase schema of similar proofs. Namely, the original nondeterministic computation was nondeterministically split into very small pieces in order to achieve that among them many pieces were equal. The equal pieces of computations were eliminated and only the correctness of remaining different ones was checked, still in the first phase. In the second phase the correctness of split and that of the elimination was checked.

This has lead to a speedup by factor $\log T(n)$ by one extra alternation. A separation of the respective time complexity classes followed: $\mathsf{NTIME}_1(T(n)) \subset \Sigma_2\text{-}\mathsf{TIME}_1(T(n))$. In its machine category this has been a much stronger result when compared with all the previous results since it achieves a speed–up by the single alternation and separates two directly neighbouring complexity classes.

The present paper continues attacking the problem of speeding up nondeterministic computations by one alternation for the next more powerful type of restricted Turing machines — viz. Turing machines with one work tape and extra read–only two–way input tape (or shortly: single–tape off–line machines). Note that this type of machines is an intermediate type between single tape machines without an input tape, and two tape off–line machines. Due to the result of Book et al. [1] in nondeterministic case the latter machines are time equivalent to multitape off–line machines.

In Section 2 a speed–up theorem for single–tape off–line nondeterministic machines by one alternation is proved. As compared to previous cases a completely new simulation strategy has been developed here in order to reflect the qualitative change in machine architecture given by the addition of input tape as that of the second tape. This has called for two substantial changes. First, an idea of having a space efficient representation of a *log* of input head movements has been implemented. Second, verification of very short nondeterministic pieces of computations has been moved back to co–nondeterministic phase where their correctness is checked by their *deterministic simulation* in exponential time w.r.t. their length. Efficient realization of both ideas requires a couple of further tricks that make use of a a parallelism offered by single-tape off–line co–nondeterministic computation. Eventually, a speed–up of order $\log \log T(n)/\sqrt{\log T(n)}$ is achieved.

Due to the space limit the proof of the speed–up theorem had to be drastically shortened. For the omitted details of the proof, see the full version of the paper [14].

Finally, in Section 3 it is shown that for single–tape off–line machines nondeterministic time $T(n)$ is strictly contained in Σ_2–time $T(n)$.

The previous results hold also for the case of complementary machines. Thus, for deterministic single–tape off–line machines two alternations lead to more efficient computations than a single alternation.

From methodological point of view a proof technique that is strong enough to capture the efficiency difference between single–tape and (restricted) two tape nondeterministic computations has been devised.

The results presented in this paper have so far no counterpart in the case of nondeterministic multitape machines. It appears that proof techniques that make use of a rectangular representation of nondeterministic single–tape off–line computations have been pushed to their limits in our approach. Our results at least leave open the possibility that similar results hold also for the general case of nondeterministic multitape computations and in fact present a necessary intermediate step towards proving such results.

2 Speed-Up

For the efficient speed–up simulation we are after it is essential that the nondeterministic machine to be simulated works in the small space. Fortunately, it is possible to make this assumption without any loss of generality due to the following theorem [7] which we shall give without a proof:

Theorem 21 *Let $\sqrt{T(n)\log n}$ be constructible in time $T(n)$, with $T(n) \geq n^2/\log n$. Than any $T(n)$–time bounded single–tape off–line nondeterministic Turing machine $\mathcal{M}$ can be simulated in linear time and in space $O(\sqrt{T(n)\log n})$ by a machine of the same type.*

Now we are in a position to formulate and prove our main result. The proof of the respective theorem has been partly inspired by the proof of a similar theorem that was proved in [13] for the case of single–tape nondeterministic TM *without input tape*. Thus, the difference in current proof captures the presence of input tape; this, however, leads to non trivial changes both in simulation strategy, as well as in the respective data representation.

The idea is as follows. The computation of the machine to be simulated is represented with the help of so–called rectangular representation (this techniques goes back to Paterson [10]). The positions of input head at selected times are represented in a compressed way with the help of a so–called *log*, in which in chronological order the differences between any two subsequent positions of input head are recorded. The size of rectangles is selected so as to enable *deterministic* and w.r.t. $T(n)$ also a sublinear time verification of any *nondeterministic* piece of computation as described by the given rectangle, with the help of information from the above mentioned log. Thus, the simulation of the original machine $\mathcal{M}$ by the respective Σ_2–machine $\mathcal{S}$ consists of two phases: in the first, nondeterministic phase the respective rectangular representation and the log of input head movement is guessed and recorded and, in the second, co–nondeterministic (universal) phase the previous guesses are verified in parallel. Moreover, in order to make the verification process efficient enough, prior to the simulation the computation of the original machine is first transformed into an equivalent one, with the help of the previous theorem, and then "cut" into certain time segments that have the property that the contents of working tape is completely rewritten at the end of each time segment. As a result, the history of cell rewritings (that is needed to verify the correctness of rectangular representation) can be completely verified by performing the respective verification in parallel for each time segment.

In order to avoid worries concerning the time–constructibility of $T(n)$ we shall make use of the following definition of time complexity for the alternating machines (cf. [11]). We shall say that an alternating machine $\mathcal{M}$ is of *time complexity* $T(n)$ if for every accepted input of length n the respective computation tree of $\mathcal{M}$ stays accepting if it is pruned at depth $T(n)$.

Theorem 22 *Let $T(n) \geq n^2/\log n$. Than any $T(n)$–time bounded single–tape off–line nondeterministic Turing machine $\mathcal{M}$ can be simulated by a single–tape off–line Σ_2–machine $\mathcal{S}$ in time $O(T(n)\log \log T(n)/\sqrt{\log T(n)})$.*

Proof Outline. According to the statement of theorem 21, w.l.o.g. we can assume that $\mathcal{M}$ is of space complexity $O(\sqrt{T(n)\log n})$.

Split now the computation of $\mathcal{M}$ into $O(T^{1/3}(n))$ *time segments* of length $O(T^{2/3}(n))$ each and introduce a so-called *sweep* at the end of each time segment. A sweep consists of one complete traversal of $\mathcal{M}$'s working head over the entire rewritten part of $\mathcal{M}$'s working tape — i.e., the working head moves to the right end of the rewritten part of working tape, then to the left end and finally returns to its marked original position. Make each sweep a part of the respective time segment. Clearly, this transformation does not influence the time complexity of the resulting machine substantially — the resulting machine still works in time $O(T(n))$.

Now, consider the respective time–space computational diagram (i.e., the sequence of instantaneous descriptions of $\mathcal{M}$'s working tape, written one above the other, starting with the initial and ending with the final accepting instantaneous description), with the recorded trajectory of $\mathcal{M}$'s working head movement during the computation. In the resulting diagram, draw horizontal lines to denote the boundaries between any two subsequent time segments.

In what follows think about each time segment separately.

Split a given time segment by vertical lines into *slots* of equal size $b(n)$ (the last slot can be shorter). The particular choice of $b(n)$ will be determined by the necessity to simulate deterministically the nondeterministic computations of lengths $b^2(n)$ within the so-called second order rectangles (see their definition in the sequel). This deterministic simulation requires then time $O(c^{b^2(n)})$ per square what should be accommodated within the required total asymptotic simulation time. It appears that the choice of $b(n) = \lceil\sqrt{\log(T(n)/\log^3 T(n))}\rceil$ satisfies this condition (for the details, see the full version of the paper).

Consider now the crossing sequence at the boundaries between individual slots (i.e., the sequence of points where the trajectory of input head crosses the above mentioned vertical lines). By shifting all slot boundaries simultaneously along the tape from its origin to the right, while keeping them equidistant, at some position j, with $0 < j < b(n)$, a situation must occur that the sum of lengths of crossing sequences at the current slot boundaries does not exceed $\ell(T^{2/3}(n)) = T^{2/3}(n)/b(n)$, within the time segment at hand. Namely, in the opposite case, if there was not such a position j, then the total sum of lengths of crossing sequences in between every tape cell would exceed $T^{2/3}(n)$, what is impossible in a time segment of the given length.

Still within the time segment at hand, fix the first slot boundary at the position j. This will split the time segment into $\sqrt{T(n)\log n}/b(n) = O(T^{1/2}(n))$ of so-called *first order rectangles*. Further, split horizontally each first order rectangle into *second order rectangles* whose size is maximized, subject to the satisfaction of either of the following two conditions:

- none of the two respective vertical sides is crossed by $\mathcal{M}$'s working head more often than $b(n)$ times;
- the total time spent by $\mathcal{M}$'s working head in a given second order rectangle must not exceed $b^2(n)$.

Clearly, second order rectangles can be created in each first order rectangle, with the possible exception of "too short" slots, or in remainders of slots that are "artificially" cut by the line separating time segments. Call the respective second order rectangles, that could not be created in the "full size", as required by the previous two conditions, as *small rectangles*.

In this way, in each time slot we obtain at most $O(T^{2/3}(n)/b^2(n))$ second order full size rectangles (since the computation within each rectangle "consumes" either $b(n)$ crossing sequence elements, or time $b^2(n)$), plus at most $O(T^{1/2}(n))$ small ones (since the number of small rectangles does not exceed that of first order ones). Thus, in each time segment the total number of second order rectangles is safely bounded by $O(T^{2/3}(n)/b^2(n))$.

In a total, the rectangular representation contains $O(T(n)/b^2(n))$ second order rectangles.

Each second order rectangle will be completely represented by its two horizontal sides of length $b(n)$, giving the contents of the corresponding block on $\mathcal{M}$'s working tape at the respective time steps, and by the description of the history of crossing its two vertical sides by the working head of $\mathcal{M}$. For each side the history of crossing is described by the so–called *crossing sequence* of length ℓ_1 for the left side and ℓ_2 for the right side, respectively, with $\ell_1, \ell_2 \leq b(n)$. Any crossing sequence consists from so–called *crossing sequence elements* that are ordered chronologically in that order in which the head has crossed the respective rectangle side. Each element of a crossing sequence is represented by a pair $\{q, d\}$. Here q denotes the state of $\mathcal{M}$ when crossing the vertical side at hand and $d \in \{left, right\}$ records the direction of the crossing.

Hence, the size of each second order rectangle representation is $\Theta(b(n))$, what in a total gives $O(T^{2/3}(n)/b(n))$ per time segment, or $O(T(n)/b(n))$ for all rectangles.

The rectangular representation of $\mathcal{M}$'s computation pertinent to the given input that is written on $\mathcal{S}$'s input tape will be represented on $\mathcal{S}$'s tape in the following order, from left to right: it is the sequence of individual second order rectangles that is generated for time segment by time segment, and within each time segment, first order rectangle by first order rectangle, and within each first order rectangle, second order rectangle by second order rectangle, in chronological order. Boundaries between individual (first and second order) rectangles, and time segments, respectively, are marked by special symbols on a special track.

To represent the computations of $\mathcal{M}$ in accordance with the idea mentioned before the statement of the theorem 22 we need moreover to record the position of $\mathcal{M}$'s input head that corresponds to each crossing sequence element in each second order rectangle. Consider the sequence of crossing sequence elements ordered chronologically — i.e., in that order in which the boundaries between slots are crossed during $\mathcal{M}$'s computation and consider also the respective associate sequence of corresponding input head positions. This associate sequence has as many elements as is the length $\ell(T(n))$ of our chronologically ordered crossing sequence and the respective elements are integers in the range $< 1..n >$. Now, instead of recording the "absolute" positions of $\mathcal{M}$'s input head on the input tape, record only the *differences* $d_i = p_i - p_{i-1}$ between any two absolute positions p_i and p_{i-1}, respectively, for $i = 0, 1, \ldots, \ell(T(n))$ and $p_0 = 1$.[2] Thus, the differences are integers from the interval $< -n+1..n-1 >$. The size of representation of any $|d_i|$ is at most $\log n$. Superpose now the resulting sequence that, starting from p_0 enables to compute $\mathcal{M}$'s input head positions, with the above mentioned chronologically ordered crossing sequence and call the resulting merged sequence a *log* of $\mathcal{M}$'s head movement. Thus the log elements take the form $\{q, d, d_i\}$ of triples, where q denotes the state of $\mathcal{M}$, d the direction of $\mathcal{M}$'s working head movement, and d_i the difference between current and previous position of $\mathcal{M}$'s input head, with all values within the i–th triple pertinent to the moment when $\mathcal{M}$'s working

---

[2] The idea of encoding head movement by differences in its positions at predetermined times has been previously used by [8].

head is crossing for the i–th time a boundary between any two slots. Ignoring items of constant size, for the log elements it holds that $\sum_{i=1}^{\ell(T(n))} |d_i| \leq T(n)$. Therefore, the length of the corresponding representation can be bounded by $\sum_{i=1}^{\ell(T(n))} \lceil \log |d_i| \rceil =$

$$= O(\log \prod_{i=1}^{\ell(T(n))} |d_i|) \leq O\left(\ell(T(n)) \log \frac{\sum_{i=1}^{\ell(T(n))} |d_i|}{\ell(T(n))}\right) = O\left(T(n) \frac{\log \log T(n)}{\sqrt{\log T(n)}}\right)$$

The log representation will be also represented on $\mathcal{S}$'s working tape in a natural way from left to right, with special separators in between the time segments that split the log into segments that correspond to the computations of $\mathcal{M}$ of length $T^{2/3}(n)$. It is not difficult to show by a similar argument as above that the length of representation of a part of the log corresponding to any time segment is $O(T^{2/3}(n) \log \log T(n) / \sqrt{\log T(n)})$.

In order to have the possibility to work extra, but in parallel with each time segment we will insist that both rectangular representation and the log representation will be written one above the other on two parallel tracks on $\mathcal{S}$'s working tape. Moreover, we shall require that the respective time segment separators both in the rectangular representation and in the log will find themselves at the same positions.

It is clear that this can be done while preserving the tape length proportional to the previously estimated total length of log. Namely, comparing the previous estimates of the length of a time segment in the log and in the rectangular representation we see that the former representation is always greater. Thus, by padding the representation of a time segment in the rectangular representation with suitable symbols we can always achieve that both representations will be of equal length, not exceeding $O(T^{2/3}(n) \log \log n / \sqrt{\log T(n)})$.

This estimate will be important for the complexity estimation of actions performed within one time segment. Nevertheless, it is obvious that the total length of the joint representation of the log and of all second order rectangles is of order $O(T(n) \log \log T(n) / \sqrt{\log T(n)})$, i.e., is bounded by the length of the log.

Now, the idea of simulation is first to guess and record the above rectangular representation simultaneously with the log of $\mathcal{M}$'s computations and then to verify the correctness of the above guesses. The verification process consists of two main phases. First, we have to verify whether the guess of rectangular representation was correct — i.e., whether all rectangles 'fit' together and whether the size ant the format of rectangular representation and of the log have been guessed correctly (a so–called *global correctness*). Second, we have to attest whether each rectangle represents a valid piece of M's computation. Such a computation starts in "partial" configuration as described by the upper horizontal side of the rectangle at hand and ends in a configuration as described by the lower side of the rectangle. Moreover, in such a computation $\mathcal{M}$'s working head must leave and re–enter rectangles in accordance both with the respective two crossing sequence at both rectangle's vertical sides and with the symbol read by $\mathcal{M}$'s input head at that time (so–called *local correctness*).

This leads to the design of the simulation scheme in which $\mathcal{M}$ is simulated by a single–tape off–line Σ_2-machine $\mathcal{S}$ in two main phases: in the first, nondeterministic (existential) one, all guesses will be performed, whereas in the second, universal one, the verification of all previous guesses will be done.

The efficient realization of this idea requires a number of auxiliary data structures. In these structures, e.g. the number of second order rectangles, and of crossing sequence elements, within each time segment, is kept. Addresses of each second order rectangle within the rectangular representation are needed as well. Moreover, there must be information about the position of input head within each log segment, plus the index of the corresponding slot in which the working head finds itself at that time, etc.

Making use of all these structure the simulation machine is able to orient itself in the rectangular representation and log, separated into individual time segments, as written on the working tape.

The simulation ends successfully when all the verifications performed terminate successfully and in the universal subphase an accepting rectangle has been discovered.

The time complexity of simulation is dominated by the time needed to generate the log in the existential phase. Therefore its length $O(T(n) \log \log T(n)/\sqrt{\log T(n)})$ is at the same time the bound on the time complexity of the entire simulation.

□

3 Separation Result

In terms of complexity classes the previous theorem says that for a suitable $T(n)$ the class of nondeterministic $T(n)$ time bounded single-tape off-line computations is a subset of the $T(n) \log \log T(n)/\sqrt{\log T(n)}$-time bounded computations on a Σ_2 machine of the same type:

$$\mathsf{NTIME}_{1+\mathrm{I}}(T(n)) \subseteq \Sigma_2\text{-}\mathsf{TIME}_{1+\mathrm{I}}(T(n) \log \log T(n)/\sqrt{\log T(n)})$$

To prove a separation result related to the above mentioned complexity classes that are bounded by the same function we shall need the following hierarchy theorem for nondeterministic single–tape of–line machines by Loryś and Liśkiewicz [7]:

Theorem 31 *Let $T_2(n)$ be a fully time constructible function, with $n \log n \in o(T_2(n))$ and such that there exists a deterministic off–line Turing machine which for each input of length n writes on the work tape the binary representation of $T_2(n)$ in time $T_2(n)$. Let $T_1(n+1) \in o(T_2(n))$. Then*

$$\mathsf{NTIME}_{1+\mathrm{I}}(T_1(n)) \subset \mathsf{NTIME}_{1+\mathrm{I}}(T_2(n))$$

Now we are in a position to prove the separation result we are after:

Theorem 32 *Let $T_1(n) = T(n)$ and $T_2(n) = T(n) \log \log n$ fulfill the assumptions of Theorem 31 and let $T_2(n)$ fulfills the assumptions of Theorem 22. Then*

$$\mathsf{NTIME}_{1+\mathrm{I}}(T(n)) \subset \Sigma_2\text{-}\mathsf{TIME}_{1+\mathrm{I}}(T(n))$$

Proof Outline: From Theorem 31 we know the proper inclusion
$$\mathsf{NTIME}_{1+\mathrm{I}}(T(n)) \subset \mathsf{NTIME}_{1+\mathrm{I}}(T(n) \log \log T(n))$$
According to the Theorem 22 the latter class is contained in
$$\Sigma_2\text{-}\mathsf{TIME}_{1+\mathrm{I}}\left(T(n) \log \log T(n) \frac{\log \log(T(n) \log \log T(n))}{\sqrt{\log(T(n) \log \log T(n))}}\right) \subseteq \Sigma_2\text{-}\mathsf{TIME}_{1+\mathrm{I}}(T(n))$$
□

It appears that all the previous theorems can be reworked to hold also for complementary classes co-$\mathsf{NTIME}_{1+\mathrm{I}}(T(n))$ and Π_2-$\mathsf{TIME}_{1+\mathrm{I}}(T(n))$. However, from space reasons we will abstain from the formulation of the respective proofs (cf. [13] for a similar procedure for case of single tape machines without an input tape).

4 Conclusions

It has been shown that for a large class of identically time bounded computations, adding of one more alternation to a nondeterministic or co-nondeterministic Turing machine with one work tape and extra input tape leads to strictly larger complexity classes. Thus, for such machines the first and the second level of the respective alternating hierarchy of complexity classes do not collapse. This seems to be the first occasion where such a general result has been achieved for a restricted type of multitape nondeterministic Turing machines.

References

1. Book, R.V. — Greibach, S.A. — Wegbreit, B.: Time– and Tape–Bounded Turing Acceptors and AFLs. *JCSS 4*, Vol. 6, Dec. 1970, pp. 602–621
2. Hennie, F.C.: One–Tape, Off–line Turing Machine Computations. *Information and Control*, Vol. 8, 1965, pp. 553–578
3. Dymond, P.W. — Tompa, M.: Speedups of Deterministic Machines by Synchronous Parallel Machines. In *Proc. 24th Annual IEEE Symposium on Foundations of Computer Science*, pp. 336–364, 1983
4. Gupta, S.: Alternating Time Versus Deterministic Time: A Separation. *Proc. of Structure in Complexity*, San Diego, 1993
5. Hopcroft, J. — Paul, W. — Valiant L.: On time versus space and related problems. *Proc. IEEE FOCS* **16**, 1975, pp. 57–64
6. Kannan, R.: Alternation and the power of nondeterminism (Extended abstract). *Proc. 15–th STOC*, 1983, pp. 344–346
7. Loryś, K. — Liśkiewicz, M.: Two Applications of Führers Counter to One–Tape Nondeterministic TMs. *Proceedings of the MFCS'88*, LNCS Vol. 324, Springer Verlag, 1988, pp. 445–453
8. Maass, W.: Combinatorial Lower Bound Arguments for Deterministic and Nondeterministic Turing Machines.*Trans. Am. Math. Soc.* 292, 1985, pp. 675–693
9. Maass, W. — Schorr, A.: Speed-up of Turing Machines with One Work Tape and Two–way Input Tape. *SIAM J. Comput.*, Vol. **16**, no. 1, 1987, pp. 195–202
10. Paterson, M.: Tape Bounds for Time–Bounded Turing Machines, *JCSS*, Vol. 6, 1972, pp. 116–124
11. Paul, W. — Prauss, E. J. — Reischuk, R.: On Alternation. *Acta Informatica*, **14**, 1980, pp. 243–255
12. Paul, W.J. — Pippenger, N. — Szemerédi, E. — Trotter, W.T.: On determinism versus nondeterminism and related problems. In *Proc. 24th Annual IEEE Symposium on Foundations of Computer Science*, pp. 429–438, 1983
13. Wiedermann, J.: Speeding–up Single–Tape Nondeterministic Computations by Single Alternation, with Separation Results. *Proceedings of the 23–rd International Colloquium on Automata, Languages, and Programming*, ICALP'96, LNCS Vol. 1099, Springer Verlag, Berlin, 1996
14. Wiedermann, J.: Accelerating Nondeterministic Single–Tape Off–Line Computations by One Alternation. Technical Report V-725–97, Institute of Computer Science, Prague, 1998

Facial Circuits of Planar Graphs and Context-Free Languages [*]

Bruno Courcelle[1] and Denis Lapoire[2]

[1] LaBRI (UMR 5800, CNRS), Université Bordeaux
351, cours de la Libération 33405 Talence Cedex (France)
`courcell@LaBRI.U-Bordeaux.fr`
[2] University of Bremen, P.O. Box 33 04 40, D-28334 Bremen (Germany)
`ldenis@informatik.uni-bremen.de`

Abstract. It is known that a language is context-free iff it is the set of borders of the trees of recognizable set, where the border of a (labelled) tree is the word consisting of its leaf labels read from left to right.
We give a generalization of this result in terms of planar graphs of bounded tree-width. Here the border of a planar graph is the word of edge labels of a path which borders a face for some planar embedding. We prove that a language is context-free iff it is the set of borders of the graphs of a set of (labelled) planar graphs of bounded tree-width which is definable by a formula of monadic second-order logic.

Thatcher and Wright [12] (see also Doner [5]) characterize context-free languages as the images of the recognizable sets of finite trees under a mapping *border* that produces for each given tree the sequence of symbols labeling its leaves, read from left to right.

Our aim is to extend such a characterization to Monadic Second Order definable sets of graphs. Here, the border mapping concerns *special graphs*, that have a unique path of labeled edges. Such a mapping, whose study is investigated by Engelfriet and Heyker [6], associates with every special graph the word read on the path. We know that the language generated by a MS-definable set of graphs is in general not context-free (see Example 1).

Example 1. The noncontext-free language $\{a^n b^n c^n \mid n \geq 1\}$ is defined by the MS-definable set of special graphs $\{G_n \mid n \geq 1\}$ described in Fig. 1.

Thus, we impose two restrictions. First, we consider *special planar* graphs, that admit a planar embedding in which the labeled path borders a face (see Example 2). The path is actually made from a circuit by means of a special edge marked #. Secondly, we consider sets of graphs of uniformly bounded *tree-width*. The tree-width is a measure of complexity for graphs introduced by Robertson and Seymour [11].

Example 2. The context-free language $\{a^n b^n \mid n \geq 1\}$ is defined by the MS-definable set of special-planar graphs $\{H_n \mid n \geq 1\}$ described in Fig. 2.

---

[*] Research partly supported by the EC TMR Network GETGRATS (General Theory of Graph Transformation Systems).

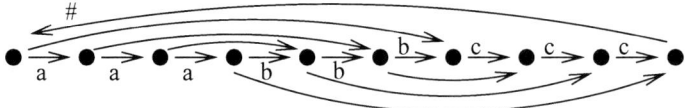

Fig. 1. A special graph G_3 that defines the word $a^3b^3c^3$.

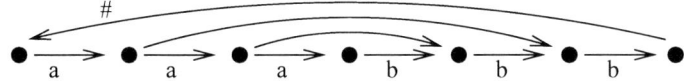

Fig. 2. A special-planar graph H_3 that defines the word a^3b^3.

Our main result is to characterize the context-free languages as the sets of borders of MS-definable sets of special-planar graphs of bounded tree-width.

1 Special Graphs

Definition 3. A *hypergraph* G is a tuple (V, E, f), denoted by $(\mathbf{V}_G, \mathbf{E}_G, \text{vert}_G)$, where V is the finite set of *vertices*, where E is the finite set of *hyperedges* which is supposed disjoint with V and where f associates with every $e \in E$ the sequence of its *extremities* $f(e)$, supposed pairwise distinct (there is no loop) and denoted by $(f(e,1), \ldots, f(e,n))$. The *initial* (resp. *terminal*) extremity of e is $f(e,1)$ (resp. $f(e,n)$). A *graph* is a hypergraph such that each hyperedge is an *edge*, that has two extremities. A vertex x and an edge e are *incident* if x is an extremity of e.

A *path* of some graph G is a sequence $p = (o_1, \ldots, o_m) \in (\mathbf{V}_G \cup \mathbf{E}_G)^+$ for some $m \geq 1$ where o_i and o_{i+1} are incident for every $i \in [m-1]$ and where o_1 (resp. o_m) is a vertex, its *initial* (resp. *terminal*) *extremity*. All vertices and edges of a path, except eventually its extremities, are supposed pairwise distinct. A path is a *cycle* if $o_1 = o_m$. It is said *oriented* if for every arc o_i with $i \in [2, m-1]$, we have $o_{i-1} = \text{vert}_G(o_i, 1)$ and $o_{i+1} = \text{vert}_G(o_i, 2)$. A *circular path* is an oriented path having identical extremities. To simplify, an oriented path is represented by a sequence of edges. The graph G is *connected* if all two vertices are the extremities of some path of G.

Let H and K be two hypergraphs. H is a *subhypergraph* of K if $\mathbf{V}_H \subseteq \mathbf{V}_K$, $\mathbf{E}_H \subseteq \mathbf{E}_K$ and $\text{vert}_H \subseteq \text{vert}_K$. If $\text{vert}_H(d) = \text{vert}_K(d)$ for each $d \in \mathbf{E}_H \cap \mathbf{E}_K$, the *union* of H and K, denoted by $H \cup K$, is the hypergraph $(\mathbf{V}_H \cup \mathbf{V}_K, \mathbf{E}_H \cup \mathbf{E}_K, \text{vert}_H \cup \text{vert}_K)$.

Definition 4. A *tree* is a connected graph that contains no cycle. A vertex (resp. edge) of a tree T is called a *node* (resp. *arc*). Their set is denoted by $\mathbf{N}_T$ (resp. $\mathbf{A}_T$). If T is a tree, if $f \in \mathbf{A}_T$ and s is an extremity of f, we let $T(f,s)$ denote the subtree of T consisting of s and all paths in T containing s but not f. If a tree T is given with a root r, we direct its arcs from the root towards the leaves. Every node except the root has then a unique father and zero, one or several sons. A leaf has no son. If $x \in \mathbf{N}_T$, we let $T/x = T$ if $x = r$, $T/x = T(f, x)$ if f is the edge between x and its father. The root of T/x is x.

In this article, $\#$ denotes a special label.

Definition 5 (Special graphs). Let X be an alphabet that does not contain $\#$. A *special* graph over X is a graph G with a unique edge labeled $\#$, denoted by $\#_G$, and with a family $(E_a)_{a \in X}$ where $E_a \subseteq \mathbf{E}_G$, the sets E_a are pairwise disjoint and $E_X = \bigcup \{E_a \mid a \in X\} \cup \{\#_G\}$ is the set of edges of a unique circular path $\gamma(G)$ of G having as last edge $\#_G$; moreover, we suppose that no two distinct edges having same sequence of extremities and same label (resp. no label). We shall say that $e \in \mathbf{E}_G$ has *label* a iff $e \in E_a$. Some edges of G may have no label.

For each special graph G, we let $\underline{bd}(G) \in X^+$ be the nonempty word $a_1 \ldots a_n$ where $(e_1, \ldots, e_n, \#_G)$ is the circular path $\gamma(G)$ and where for each $i \in [n]$, e_i is labeled a_i. We call it the *border* of G. If $\pi = (e_i, e_{i+1}, \ldots, e_j)$ is a path in $\gamma(G)$, we denote by $\underline{bd}(\pi)$ the word $a_i a_{i+1}, \ldots, a_j$ which is a factor of $\gamma(G)$.

Such a graph will be represented by the *relational structure* $|G|_2 = < \mathbf{V}_G \cup \mathbf{E}_G, \mathrm{inc}_G, (lab_{aG})_{a \in X \cup \{\#\}} >$ where $lab_{aG}(e)$ holds iff $e \in \mathbf{E}_a$ (or $e = \#_G$) and $\mathrm{inc}_G = \{(e, x, y) \mid e \in \mathbf{E}_G, e \text{ links } x \text{ to } y\}$. Hence, sets of graphs can be *defined* by formulas in Monadic Second Order logic or in *Counting* MS logic, a refinement of MS logic using special predicate expressing cardinality of sets modulo fixed integers; see [2,3,4]. Such sets are said *MS-definable* (resp. *CMS-definable*), for short. MS logic is the extension of First-Order logic with set variables. For words and binary trees, MS-definability equals recognizability.

Definition 6 (Tree-width). A *tree-decomposition* is a pair (T, g) where T is a tree and where g associates with every node t of T a graph $g(t)$ such that:

- $\mathbf{E}_{g(s)} \cap \mathbf{E}_{g(t)} = \emptyset$, for all distinct nodes s, t of T.
- for all nodes s, u of T and every node t of the path of T from s to u, we have: $\mathbf{V}_{g(s)} \cap \mathbf{V}_{g(u)} \subseteq \mathbf{V}_{g(t)}$.

The *width* of (T, g) is the maximum of $\mathbf{card}(\mathbf{V}_{g(t)}) - 1$ taken over all $t \in \mathbf{N}_T$. The *tree-width* of a hypergraph G, denoted by $\underline{twd}(G)$, is the minimum width of all tree-decompositions (T, g) of G (such that $\bigcup_{t \in \mathbf{N}_T} g(t) = G$).

Definition 7. Let (T, g) be a tree-decomposition of a graph G. If T' is a subtree of T, we denote by $G[T', g]$ the subgraph of G defined as $\bigcup \{g(s) \mid s \in \mathbf{N}_{T'}\}$. Thus $(T', g/T')$ is a tree-decomposition of $G[T', g]$. Let now G be special. A tree-decomposition (T, g) of G is *special* if:

1. T has a root r which verifies $\#_G \in \mathbf{E}_{g(r)}$.
2. for every $s \in \mathbf{N}_T$ and incident edge f, the set of edges $E_X \cap \bigcup \{\mathbf{E}_{g(u)} \mid u \in T(f, s)\}$ is either empty, or is E_X or forms a noncircular path, that we shall denote by $\gamma(G, f, s)$.

We say that such a tree-decomposition is *compact* if for all f and s, $\gamma(G, f, s)$ is a (nonempty) noncircular path. Let t be a node of some compact tree-decomposition (T, g), d the eventual edge linking t to its parent. Its sons $t_1, \ldots, t_n$ are linked to t by $d_1, \ldots, d_n$. Clearly, if t is (resp. is not) the root, $\gamma(G)$ (resp. $\gamma(G, d, t)$) admits a unique expression of the form $w_0 \gamma(G, d_{\pi(1)}, t_{\pi(1)}) w_1 \ldots \gamma(G, d_{\pi(n)}, t_{\pi(n)}) w_n$ for some permutation π on $[n]$. We denote by $\hat{t}$ the sequence of words $(\underline{bd}(w_0), \ldots, \underline{bd}(w_n))$.

Here is some intuition: a path in a graph given with a tree-decomposition yields a traversal of a portion of the tree of the tree-decomposition. Special means that the distinguished path yields a single traversal of each subtree. Compact means that every box of the tree-decompositions contains one edge of the distinguished path.

Example 8. Figures 3 and 4 represents two tree-decompositions (T, g) and (T, h) of a same graph G (in the left part of Fig. 4) and having a same tree T (in the left part of Fig. 3). The node 1 is the root of T. (T, g) is not special: the graph $\gamma(G) \cap G[T/5, g]$ consists of two paths with corresponding words bb and dd.

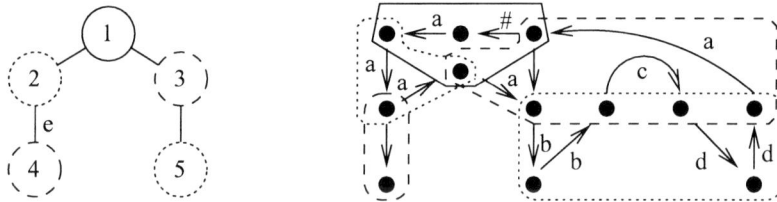

Fig. 3. A non special tree-decomposition.

(T, h) is special: for example, $\gamma(G) \cap G[T/5, h]$ consists of a path with corresponding word $bbcdd$. (T, h) is not compact, because $\gamma(G, e, 4)$ is empty.

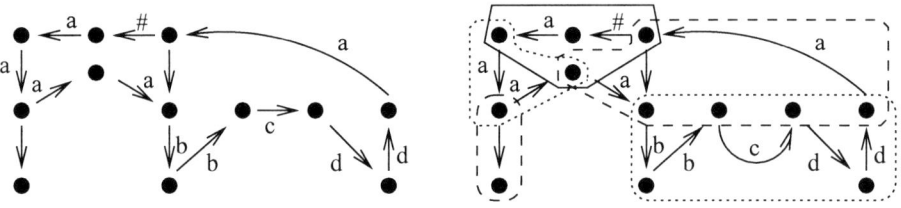

Fig. 4. A special but non compact tree-decomposition.

Definition 9. Let F be a finite signature and $\mathbf{T}(F)$ be the set of finite terms built over F. Let X be a finite alphabet. For every $f \in F$ of arity k, let be given a sequence of words $\hat{f} = (u_0, \ldots, u_k)$, $u_0, \ldots, u_k \in X^*$. We define a mapping $\alpha : \mathbf{T}(F) \to X^*$ by letting: $\alpha(f(t_1, \ldots, t_k)) = u_0 \alpha(t_1) u_1 \alpha(t_2) \ldots \alpha(t_k) u_k$ where $(u_0, \ldots, u_k) = \hat{f}$, $f \in F$, $t_1, \ldots, t_k \in \mathbf{T}(F)$. If $k = 0$ then $\hat{f} = (u_0)$ and we let $\alpha(f) = u_0$. Such a mapping α is called a *homomorphism*: $\mathbf{T}(F) \to X^*$.

Proposition 10. *If $K \subseteq \mathbf{T}(F)$ is a recognizable set of terms and $\alpha : \mathbf{T}(F) \to X^*$ is a homomorphism, then $\alpha(K)$ is a context-free language.*

Proof. One constructs a context-free grammar the derivation trees of which are the terms in K. The result is well known (see [5,12]). □

2 Tree-Decompositions as Algebraic Terms

In this section, k denotes a fixed integer and X a finite alphabet.

Definition 11 (Hyperedge-substitution). We denote by $\mathcal{G}$ the set of all hypergraphs G labeled over $X \cup \{\#\}$ equipped with a sequence of pairwise distinct vertices, called the sequence of *sources* and denoted by $\underline{src}_G$. Let $H, K \in \mathcal{G}$ with $\mathbf{E}_H \cap \mathbf{E}_K = \emptyset$ and let e be an edge of H such that $\mathbf{vert}_G(e)$ equals $\underline{src}_K$ and contains every vertex of $\mathbf{V}_H \cap \mathbf{V}_K$, we denote by $H\{K/e\}$ the hypergraph obtained from $H \cup K$ by forgetting the edge e and having as sequence of sources $\underline{src}_H$. It results from the *substitution* in H of K for e.

Notation 12 We denote by $\mathcal{T}_k$ the set of all compact tree-decompositions (T, g) of width at most $k - 1$ such that every $s \in \mathbf{N}_T$ has maximum degree k. Let $<$ be a linear ordering on $\mathbf{V}_G \cup \mathbf{E}_G$ such that $e_1 < e_2 < e_3 < \ldots < e_p$ where $(e_1, e_2, \ldots, e_p, \#_G) = \gamma(G)$. For every s in $\mathbf{N}_T$, we order its set of sons as a sequence $(t_1, \ldots, t_n)$, in such a way that the $\leq$-smallest edge in $G[T/t_i, g]$ is strictly smaller with respect to $\leq$ than the $\leq$-smallest one in $G[T/t_{i+1}, g]$ and we make $g(s)$ into a hypergraph with sources, denoted by $H(s)$, as follows:

a) its sequence of sources is $\underline{src}_G$ if $s = r$, and, otherwise, consists of $\mathbf{V}_{g(s)} \cap \mathbf{V}_{g(t_0)}$ in increasing order for $\leq$ with t_0 the father of s.
b) for each $i = 1, \ldots, n$, we add to $g(s)$ a hyperedge h_i with sequence of vertices $\mathbf{V}_{g(s)} \cap \mathbf{V}_{g(t_i)}$ in increasing order for $\leq$.

For each isomorphism class of hypergraphs of the form $H(s)$ with n hyperedges and for each $n + 1$-tuple of the form $\hat{s}$ (see Definition 7), we get a function f for which $\hat{f} = \hat{s}$ and that mapps an n-tuple $(G_1, \ldots, G_n)$ of graphs with sources into a graph with or without sources (If $n = 0$ then this function takes no argument and is a constant denoting a graph with sources). We denote by F_k such a signature and by α the homomorphism $\mathbf{T}(F_k) \to X^*$ induced by $\{(f, \hat{f}) \mid f \in F_k\}$.

It follows that a tree-decomposition (T, g) of a graph as above can be represented by a term $t \in \mathbf{T}(F_k)$. We shall denote by $\mathbf{val} : \mathbf{T}(F_k) \to \{graphs\}$ the mapping from terms to isomorphism classes of graphs.

With these notations, we have:

Claim 13 *For every node s of some $(T, g) \in \mathcal{T}_k$, we have $G[T/s, g] = (H(s)\{G[T/t_1, g]/h_1\} \ldots)\{G[T/t_n, g]/h_n\}$, where for each $i = 1, \ldots, n$, $G[T/t_i, g]$ is equipped with a sequence of sources consisting of $\mathbf{V}_{g(t_i)} \cap \mathbf{V}_{g(s)}$ ordered in increasing way w.r.t $\leq$.*

Proof. Obvious from Definition 11 and the above construction. □

Since X is finite, since multiple edges of any graph have distinct labels, and by the limitations given by the bound k, there are finitely many isomorphism classes of hypergraphs $H(s)$. Hence, F_k is finite.

The following result will allow us to represent compact tree-decompositions by terms in $\mathbf{T}(F_k)$.

Lemma 14. *The degree of a compact tree-decomposition of width at most $k - 1$ is at most k.*

Proof. Easy. □

Proposition 15. *Let L be a CMS-definable set of special graphs, let L' be the set of graphs in L having a compact tree-decomposition of width at most $k - 1$. Then $\underline{bd}(L')$ is a context-free language.*

Proof. Let (T, g) be a compact tree-decomposition of some graph G and let $\leq$ be the total order on $\mathbf{V}_G \cup \mathbf{E}_G$ defined in Notation 12. A term $t \in \mathbf{T}(F_k)$ *represents* (T, g) if it is constructed from (T, g) as in Notation 12 and if for every occurrence t of a symbol $f \in F_k$ associated with a node u of T, we have: $\hat{f} = \hat{u}$. For every arc e from s to u in T, we have $\alpha(t/u) = \underline{bd}(\gamma(G, e, u))$. This is straightforward by induction on the depth of t/u from the definitions of $\hat{f}$ and $\hat{u}$. It follows that $\alpha(t) = \underline{bd}(\gamma(G)) = \underline{bd}(G)$.

Let L be a CMS-definable set of special graphs, L' be the subset of those having a compact tree-decomposition of width at most $k - 1$. Let $K \subseteq \mathbf{T}(F_k)$ be the set of terms representing all these tree-decompositions. Thus, every term $t \in \mathbf{T}(F_k)$ belongs to K iff

1. **val**$(t) \in L$.
2. t actually specifies a compact tree-decomposition of width at most $k - 1$ (some terms may specify tree-decompositions that are even not special or define graphs that are not special).

Condition (1) is CMS-definable (because $\mathbf{val}^{-1}$ preserves CMS-definability, see Courcelle [3]); condition (2) is easy to express in MS-logic. It follows that K is CMS-definable, hence, recognizable by the theorem of Doner et al. [5,2]. As a consequence of Lemma 14, each compact (T, g) is represented by some term $t \in \mathbf{T}(F_k)$ which verifies $\alpha(t) = \underline{bd}(\gamma(G))$. It follows that $\underline{bd}(L') = \underline{bd}(\mathbf{val}(t)) = \alpha(K)$, hence is a context-free language by Proposition 10. □

Our aim is now to extend Proposition 15 to sets of graphs having special tree-decompositions of bounded width (as opposed to compact ones). We need some preliminary results from Courcelle (see [2] or the survey [4]).

Notation 16 (Parallel-composition) Let $n \in \mathbb{N}$, let H and K be two edge-disjoint graphs having the same sequence of n sources. If every common vertex of H and K is a source, then we denote by $H\Box_n K$ the graph with no source obtained from the union $H \cup K$ by forgetting the sources. We denote by $\sim$ the isomorphism of graphs with and without sources (the isomorphism must preserve the sequence of sources in an obvious way).

Let L be a CMS-definable set of graphs. It is recognizable (see [2]), hence we have the following:

Fact 17 *For every n, there exists a finite set $\{K_1, \ldots, K_p\}$ of graphs with a sequence of n sources such that for every graph K with n sources, there exists i such that for every graph G if $G = H\Box_n K$ then:*

$$G \in L \Leftrightarrow H\Box_n K'_i \in L \text{ for some } K'_i \sim K_i$$

(This means that in a graph G, a "factor" K can be replaced by one of bounded size, which is "syntactically" equivalent w.r.t. L). For later reference, we let $\kappa(L, n)$ denote $\max\{\mathbf{card}(\mathbf{V}_{K_i}) - n \mid i = 1, \ldots, p\}$.

Proposition 18. *Let $m \in \mathbb{N}$ and L be a CMS-definable set of special graphs, each of them having a special tree-decomposition of width at most m. There exists an integer k and a subset L' of L such that $\underline{bd}(L') = \underline{bd}(L)$ and every graph in L' has a compact tree-decomposition of width at most k.*

Proof. Let $G \in L$ and (T, g) be a special tree-decomposition of G of width at most m which is not compact. We let $N_0 \subseteq \mathbf{N}_T$ be the set of nodes s of T such that $g(s)$ contains at least one edge of $\gamma(G)$. We let T' be the subgraph of T consisting of the (undirected) paths in T linking two nodes of N_0. It is clear by the properties of tree-decompositions that T' is connected hence is a tree. Note that $r \in T'$. We shall define $G' \in L$ such that $\underline{bd}(G') = \underline{bd}(G)$ and having a compact tree-decomposition of width at most k, where k is large enough, but fixed and defined from m and the CMS-formula φ which characterizes L.

Let $u \in \mathbf{N}_{T'}$ which has neighbors in T which do not belong to T'; we call them $s_1, \ldots, s_l$ (since $r \in \mathbf{N}_{T'}$, the father of u in T is also in T'). We let $K = G[T/s_1, g] \cup \ldots \cup G[T/s_l, g]$ and H be the unique subgraph of G such that $G = H\Box K$. Note that $S = \mathbf{V}_H \cap \mathbf{V}_K \subseteq \mathbf{V}_{g(u)}$ hence has cardinality n at most $m + 1$. The circular path $\gamma(G)$ is included in H.

Let K' be the graph with n sources "syntactically equivalent" to K w.r.t. L obtained by Fact 17. Consider now $G' = H\Box_n K'$; it belongs to L, hence is a special graph; the circular path $\gamma(G)$ being in H is also in G'. We obtain a tree-decomposition for G' from (T, g) by deleting $T/s_1, \ldots, T/s_l$ and replacing

$g(u)$ by $g(u)\square_n K'$. We repeat this step for each node u of T' having neighbors in T that are not in $\mathbf{N}_{T'}$, and we eliminate these neighbors. Doing this, we reduce the size of G, we restrict T to T' but we increase by the constant additive factor $\max\{\kappa(L,n) \mid n \leq m+1\}$ (see Fact 17). Hence, we get at the end a graph $G' \in L$ such that $\underline{bd}(G') = \underline{bd}(G)$, G' has a compact tree-decomposition of width at most $k = m + \max\{\kappa(L,n) \mid n \leq m+1\}$. □

Theorem 19. *Let $m \in \mathbb{N}$ and L be a CMS-definable set of special graphs, each of them having a special tree-decomposition of width at most m. Then, $\underline{bd}(L)$ is a context-free language.*

Proof. Immediate consequence of Propositions 15 and 18. □

3 Special-Planar Graphs

A special graph G is *special-planar* if it is planar and if $\gamma(G)$ borders a face in some planar embedding of G. Such graphs verify the following property, which is a consequence of a more general result of [8] or [7]. We have not enough space to include the proof.

Proposition 20. *Every special-planar graph G admits a special tree-decomposition of width $\underline{twd}(G)$.*

Proposition 21. *Every context free language $L \subseteq X^+$ is $\underline{bd}(K)$ for some MS-definable set K of special-planar graphs of $\underline{twd} \leq 2$.*

Proof. The proof is available in a more complete version at URL http://dept-info.labri.u-bordeaux.fr/~courcell/ActSci.html.

Now, we can establish our main theorem.

Theorem 22. *A language $L \subseteq X^+$ is context-free if and only if it is $\underline{bd}(K)$ for some CMS-definable set K of special-planar graphs of bounded tree-width.*

Proof. As a consequence of Propositions 20, 21 and Theorem 19. □

As an immediate consequence of Theorem 22, a very simple proof of a well-known result on context-free languages (see [1]). Two words v, w are said to be *conjugates* if there exist words t, u such that $v = t \cdot u$ and $w = u \cdot t$. If L is a language, its conjugacy closure, denoted by $L^\sim$, is the set L augmented with all conjugates of all its words.

Corollary 23. *The class of context-free languages is closed under conjugacy closure.*

Proof. Let L be a context-free language. By Theorem 22, there is a CMS-definable set K of special-planar graphs of bounded tree-width such that $L = \underline{bd}(K)$. Two special graphs G and H are said *conjugate* if there is $p \in [n]$ such that H is obtained from G by relabelling every e_i with the label of $e_{f(i)}$ with $\gamma(G) = (e_1, \ldots, e_n)$ and where f is defined by $f(i) = i$ (resp. $= n, = i-1$) for each $i \in [p-1]$ (resp. $i = p$, $i \in [p+1, n]$). Let $K^\sim$ be the conjugate closure of K. Clearly, $K^\sim$ is CMS-definable, hence, verifies the conditions of Theorem 22 and $\underline{bd}(K^\sim) = L^\sim$. Then, $L^\sim$ is context-free. □

Discussion:

In Theorems 19 and 22, one can replace "CMS-definable" by "recognizable" since it is proved in [9] that CMS-definability equals recognizability for sets of graphs of bounded tree-width (see [2] or [3] or [4] for recognizable sets of graphs).

The limitation to "bounded tree-width" cannot be eliminated since it is proved in [10] that context-sensitive languages can be defined similarly as borders of grids.

With the limitation of $\gamma(G)$ to border a face, one can obtain noncontext-free languages as shown in Example 1.

One may ask about a similar characterization of linear languages in terms of special-planar graphs of bounded parh-with.

References

1. J.-M. Autebert. *Langages algébriques*. Etudes et recherches en informatique. Masson, 1987.
2. B. Courcelle. The monadic second-order logic of graphs I: Recognizable sets of finite graphs. *Information and Computation*, 85:12–75, 1990.
3. B. Courcelle. The monadic second-order logic of graphs V: On closing the gap between definability and recognizability. *Theor. Comput. Sc.*, 80:153–202, 1991.
4. B. Courcelle. The expression of graph properties and graph transformations in monadic second-order logic. In *Handbook of Graph Grammars and Computing by Graph Transformations. Vol. I: Foundations*, chapter 5, pages 313–400. World Scientific, 1997.
5. J. Doner. Tree acceptors and some of their applications. *J. Comp. Syst. Sc.*, 4:406–451, 1970.
6. J. Engelfriet and L. Heyker. The string generating power of context-free hypergraph grammars. *J. Comp. Syst. Sc.*, 43:328–360, 1991.
7. D. Lapoire. Treewidth and duality for planar hypergraphs. submitted.
8. D. Lapoire. *Structuration des graphes planaires*. PhD thesis, Université Bordeaux I, Novembre 1996.
9. D. Lapoire. Recognizability equals monadic second-order definability for sets of graphs of bounded tree-width. In *Proc. STACS'98*, volume 1373 of *LNCS*, pages 618–628. Springer Verlag, 1998.
10. M. Latteux and D. Simplot. Context-sensitive string languages and recognizable picture languages. *Information and Computation*, 138,2:160–169, 1997.
11. N. Robertson and P. D. Seymour. Graph minors. III. Planar tree-width. *J. Comb. Theory Ser. B*, 36:49–64, 1984.
12. J.-W. Thatcher and J. Wright. Generalized finite automata theory with an application to a decision problem in second-order logic. *Math. Systems Theory*, 3:57–81, 1968.

Optimizing OBDDs Is Still Intractable for Monotone Functions *

Kazuo Iwama[1], Mitsushi Nozoe[1], and Shuzo Yajima[2]

[1] Kyoto University, Kyoto 606, Japan
[2] Kansai University, Takatsuki, Osaka 569, Japan
nouzoe@kuis.kyoto-u.ac.jp

Abstract. Optimizing the size of Ordered Binary Decision Diagrams is shown to be NP-complete for monotone Boolean functions. The same result for general Boolean functions was obtained by Bollig and Wegener recently.
Keywords. Ordered Binary Decision Diagrams, NP-completeness, Monotone Functions

1 Introduction

It is well known that optimization problems are generally hard. However, the degree of this hardness is exceptionally high for the optimization of Boolean circuits. Let $size^*(C)$ denote the smallest size of the circuit that is equivalent to the given circuit C. Then (i) even the problem asking whether $size^*(C) = 0$ (i.e., whether C's output is always 0 or always 1) is co-NP-complete, which is so-called the tautology problem. Furthermore (ii) any concrete family of circuits C of n variables such that $size^*(C) > 4n$ has never been known[12] although its average for the whole circuits is easily calculated to be exponential[9]. These facts suggest that Boolean circuits, which appear to be the only way of realizing Boolean functions, might not be a good way of *representing* Boolean functions.

Another approach that might be better for this purpose is to use Ordered Binary Decision Diagrams(OBDDs), first introduced by Bryant[3]. There are lots of merits including: (i) Several problems such as the equivalence problem of two OBDDs can be solved in polynomial time. (Hence the question $size^*(\mathcal{B}) = 0$? for a given OBDD $\mathcal{B}$ is now easy.) (ii) We can prove exponential lower bounds of $size^*(\mathcal{B})$ for several kind of OBDDs[e.g.,5,6]. More importantly, it is also known that $size^*(\mathcal{B})$ is polynomial for many practical functions such as the binary addition.

Not surprisingly, therefore, there is a huge literature dealing with the optimization problem of OBDDs[e.g.,7]. Unfortunately, however, those papers mostly depend on heuristic approaches, and we had few concrete results on its computational complexity for long time. The first breakthrough was in 1993. Tani, Hamaguchi and Yajima proved that optimizing SBDDs is intractable[10], where

---
* Supported in part by Scientific Research Grant, Ministry of Education, Japan

SBDDs are an extension of OBDDs that can express multiple functions. There appeared to be a somewhat large gap between proving the intractability for SBDDs and OBDDs. However, the second breakthrough was given by Bollig and Wegener[2] who succeeded in filling this gap, i.e., proving the intractability of optimizing OBDDs.

In this paper, we show that the OBDD optimization problem is still intractable for monotone functions. It should be noted that there again appear to be a gap between our result and that of Bollig and Wegener. To see this, for example, one should remember that exponential lower bounds are known for monotone circuits[1] against only $4n$ for general circuits as mentioned above. Namely, we can use specific proof techniques for monotone circuits that are not applicable in general. In the case of OBDDs, we can use so-called non-crossing OBDDs to represent monotone functions and our result holds for this subclass of OBDDs.

Another contribution of this paper, we believe, is that although it follows the basic structure of [2] (and [10]), our intractability proof is simpler than [2]. As is well known, NP-hardness proofs must involve some kind of lower-bound proofs, which are sometimes very hard to read. However, this difficulty can be reduced by paying much care to the design of the reduction and of course by inventing nice lemmas characterizing the target system which is now OBDDs. Needless to say, both contribute a lot to better understanding of OBDDs, which is even as important as the improved intractability result itself.

It is easily seen that the problem (both for OBDDs and for SBDDs) is in P for majority functions. Henceforce, the next important goal is to (dis)prove that minimizing OBDDs is NP-hard for threshold functions.

2 Monotone Functions and OBDDs

A *literal* is a (Boolean) variable, x, or its negation $\bar{x}$. x is sometimes called a *positive literal* and $\bar{x}$ a *negative literal*. A Boolean function $f(x_1, x_2, ..., x_n)$ is said to be *monotone* if f can be given as a formula that includes only positive literals[8]. A typical example of a monotone function is a majority function which becomes 1 iff more than one half of the whole variables are assigned 1.

See Fig.6. An *ordered binary decision diagram*(OBDD) is a directed acyclic graph $\mathcal{B}(f)$, which represents a Boolean function f. For its formal definition, see, e.g., [10]. The OBDD in Fig.6 represents function $f = x_1 + x_2 x_3$. An OBDD is given with a *variable ordering* $\pi = (\pi[1], \pi[2], ..., \pi[|X|])$, like $\mathcal{B}(f, \pi)$, where $\pi[i] = j$ means variable x_j appears at level i. Thus, in each path of $\mathcal{B}$, the node for variable x_j (if any) must appear at level i *only once* if $\pi[i] = j$. In the OBDD in Fig.6, its variable ordering is $\pi = (3, 2, 1)$. Note that each node v in $\mathcal{B}$ represents the Boolean function which is obtained from f by assigning 0 or 1 to the variables which already appeared above that node. If two nodes v_1 and v_2 represent the same function, then v_1 and v_2 are said to be *equivalent*. Two or more equivalent nodes can be merged.

Suppose that both 1-edge and 0-edge from a node v go to the same node. Then v is said to be *redundant* and can be removed. Thus some variables may not appear (i.e., no node exists) in each path of $\mathcal{B}$ (see Fig.6). If $\mathcal{B}$ does not include redundant or equivalent nodes, then G is said to be *reduced*. The size of $\mathcal{B}$, $|\mathcal{B}|$, is defined as the number of total nodes including the two constant nodes. Now here is the most fundamental property of OBDDs.

Proposition 1[3]: Reduced $\mathcal{B}(f, \pi)$ is unique if f and π are fixed.

In this paper, OBDDs mean reduced OBDDs unless otherwise stated. Thus the size of an OBDD $\mathcal{B}$ is also fixed if the variable ordering is fixed.

3 Main Results

Our problem in this paper is optimizing OBDDs, which is equivalent to finding the best variable ordering by Proposition 1. More precisely, the OBDD Optimization Problem(OPT-OBDD) gives us a (not necessary reduced) OBDD $\mathcal{B}$ and an integer k, and asks whether there is a variable ordering under which $|\mathcal{B}|$ is at most k. If this given $\mathcal{B}$ is of size exponential in the number of variables, then the problem is not interesting, and that is why only polynomial-size OBDDs appear in this paper. Also recall that we shall prove that OPT-OBDD is NP-complete for monotone functions. It is not hard to see that monotone functions can be represented by *non-crossing* OBDDs which can be drown in such a way that there is no crossing between any two 0-edges or between any two 1-edges. Conversely, non-crossing OBDDs always represent monotone functions. All OBDDs from now on are non-crossing, i.e., our results holds for this subclass of OBDDs.

As in [2], our NP-hardness proof uses a reduction from the Optimal Linear Arrangement Problem (OLA)[4]. An instance of OLA is a graph $G = (V, E)$ and a positive integer K. Its question is whether there is a one-to-one function (specifying an order of vertices) $\psi : V \rightarrow \{1, 2, ..., |V|\}$ such that the cost of G given by $\sum_{(u,v) \in E} |\psi(u) - \psi(v)|$, is at most K. See the example shown in Fig.6(a), where the best order of the vertices is $(1, 2, 3, 4)$, whose cost is 5.

Main Theorem: OPT-OBDD is NP-complete for monotone functions.

Proof: The problem is obviously in NP since the general OPT-OBDD is in NP[10]. Its NP-hardness is proved using the remaining three sections: In Sec.4, we describe how OLA is reduced to OPT-OBDD. For a given graph $G = (V, E)$, we introduce two monotone functions called *edge functions* and a *penalty function*. These functions are merged into a single function f using extra variables and then the OBDD for f is computed. There are several important points here: Our penalty function needs much larger width than an edge function in their OBDDs, merging into f also includes new ideas, and so on. Those will greatly contribute to make our later arguments simpler and more readable. In Sec.5, we will summarize several useful properties on the transformed OBDDs. Sec.6 is the main section where we prove that our reduction is correct, i.e., it keeps a tight relation between the size of the optimal vertex order in OLA and the size of the optimal OBDD. □

4 Polynomial-Time Reduction

Two major building blocks in our reduction are edge functions and a penalty function. An edge function is denoted by $EF(x_i, x_j)$ which is the following monotone function of p variables: $EF(x_i, x_j) = (x_i + x_j) \prod_{k \in \{1,2,...,p\} \setminus \{i,j\}} x_k$. See Fig.6(b) for an example of OBDDs for edge functions, where the names of the variables are a little modified for some purpose later given. Note that $|\mathcal{B}(EF(i,j))|$ depends on the positions of variables x_i and x_j.

Lemma 4.1: Suppose that $\pi[l_1] = x_i$ and $\pi[l_2] = x_j$ for some $1 \le l_1, l_2 \le p$. Then $|\mathcal{B}(EF(i,j), \pi)| = |l_1 - l_2| + p + 1$.

Proof: See Fig.6(b) again. Note that the OBDD in this figure satisfies the above condition for its size. One can see easily that two nodes on the same level are not equivalent. Thus the lemma holds. □

The penalty function is simply the majority function of n majority functions, which is denoted by MAJMAJ in this paper:

$$\text{MAJMAJ} = \text{MAJ}\Big(\text{MAJ}(x_{11}, x_{12}, ..., x_{1m}, y_{11}, y_{12}, ..., y_{1m}),$$
$$\text{MAJ}(x_{21}, x_{22}, ..., x_{2m}, y_{21}, y_{22}, ..., y_{2m}), \cdots,$$
$$\text{MAJ}(x_{n1}, x_{n2}, ..., x_{nm}, y_{n1}, y_{n2}, ..., y_{nm})\Big).$$

Now we shall give the reduction from OLA to OPT-BDD. Let $G(V, E)$ be a given graph as an instance of OLA, $V = \{v_1, v_2, ..., v_n\}$ and $|E| = m$. Then we first introduce $T = 2nm + 4m$ variables, $x_{11}, ..., x_{1m}, x_{21}, ..., x_{2m}, ..., x_{n1}, ..., x_{nm}$, $y_{11}, ..., y_{1m}, y_{21}, ..., y_{2m}, ..., y_{n1}, ..., y_{nm}, a_1, ..., a_{2m}$ and $b_1, ..., b_{2m}$. Then our goal is to construct a monotone function, $\mathcal{H}^*$, in the following procedure:

(i) For each edge (v_i, v_j) in E, we construct two edge functions $f_k = EF(x_{ik}, x_{jk})$ and $g_k = EF(y_{ik}, y_{jk})$. Thus we have $2m$ edge functions, $f_1, f_2, ..., f_m$ and $g_1, g_2, ..., g_m$, in total (see Fig.6). The reason for introducing $2m$ functions instead of m ones is just technical, i.e., for easy calculation.

(ii) Construct one penalty function of $2mn$ variables, which is exactly the same as MAJMAJ given above.

(iii) Those $2m + 1$ functions are combined into a single monotone function as follows (see Fig.6):

$$\mathcal{H}_1 = a_1 f_1 + b_1 \text{MAJMAJ} + a_1 b_1 \quad \mathcal{H}_2 = a_2 g_1 + b_2 \mathcal{H}_1 + a_2 b_2 \cdots$$
$$\mathcal{H}_{2i-1} = a_{2i-1} f_i + b_{2i-1} \mathcal{H}_{2i-2} + a_{2i-1} b_{2i-1} \quad \mathcal{H}_{2i} = a_{2i} g_i + b_{2i} \mathcal{H}_{2i-1} + a_{2i} b_{2i} \cdots$$
$$\mathcal{H}_{2m-1} = a_{2m-1} f_m + b_{2m-1} \mathcal{H}_{2m-2} + a_{2m-1} b_{2m-1}$$
$$\mathcal{H}_{2m} = \mathcal{H}^* = a_{2m} g_m + b_{2m} \mathcal{H}_{2m-1} + a_{2m} b_{2m}.$$

The remaining job is to construct an OBDD which represents this $\mathcal{H}^*$. For this purpose, we adopt what we call "the standard variable order"(or SVO). The resulting OBDD is denoted by $\mathcal{B}^*(G)$. SVO is given as $(\pi[1], \pi[2], ..., \pi[2m], \pi[2m+1], \pi[2m+2], ..., \pi[2mn+2m], \pi[2mn+2m+1], \pi[2mn+2m+2], ..., \pi[T]) = (b_{2m}, b_{2m-1}, ..., b_1, y_{11}, y_{12}, ..., x_{mn}, a_1, a_2, ..., a_{2m})$. Recall that $\mathcal{B}^*(G)$ is not necessarily reduced. Construction of $\mathcal{B}^*(G)$ is straightforward and therefore we shall use an example to explain it. Let $m - n = 4$: (i) $\mathcal{B}^*(\text{MAJ}(x_1, x_2, ..., x_8))$ is constructed as in Fig.6. $\mathcal{B}^*(\text{MAJMAJ})$ is its simple extension. $\mathcal{B}^*(f_j)$ was already given in Fig.6. To combine all those OBDDs, we use a_i and b_i. To see this, let us consider a function $h = af + bg + ab$. One can see that if f and g are monotone, then the whole h is also monotone. Combining $\mathcal{B}^*(\text{MAJMAJ}), \mathcal{B}^*(f_1), \mathcal{B}^*(g_1)$, ..., $\mathcal{B}^*(f_m), \mathcal{B}^*(g_m)$ is again a simple extension of this construction (Fig.6).

Lemma 4.2: $|\mathcal{B}^*(G)|$ can be written as $2K^* + \Phi(G)$, where K^* is the cost of the graph G in OLA when the vertex order is $v_1, v_2, ..., v_n$, and $\Phi(G)$ is given as the following formula,

$$\Phi(G) = 2(n-1)m + (m^2 + m) \times (\lfloor n/2 \rfloor + 1) \times \lceil n/2 \rceil + 4m + 2.$$

Proof: Just by a simple counting. □

Now the graph G and Integer K is transformed into the OBDD $\mathcal{B}^*(G)$ with the target cost of $2K + \Phi(G)$. One can see that this reduction can be done in polynomial time.

Thus the building blocks are different, but the underlying idea is the same as [2,10]: (i) As shown in Fig.6, OBDDs for edge functions provide a high cost if the distance of the two nodes for a single edge is large. (ii) No two edge-function OBDDs share the same variables, i.e., there are different nodes for the same vertex of G. Hence, if we have only edge function OBDDs then there exists a trivial ordering which minimizes the size. (iii) To prevent it, the penalty function is introduced. It gives us a "penalty size" if we separates the nodes for the same vertex of G.

5 Useful Properties

This section is for introducing more definitions needed in later discussion and summarizing lemmas being used to justify the reduction given in the previous section. First of all, the OBDD for $\text{MAJ}(x_1, x_2, ..., x_8)$ in Fig.6 is optimal.

Lemma 5.1: For any variable ordering, the size of the OBDD for $\text{MAJ}(x_1, x_2, ..., x_p)$ is given as follows (Proof is omitted):

$$|\mathcal{B}(\text{MAJ}(x_1, x_2, ..., x_p))| = \begin{cases} (p/2 + 1) \times p/2 + 2 & \text{(if } p \text{ is even)} \\ \lceil p/2 \rceil^2 + 2 & \text{(if } p \text{ is odd)}. \end{cases}$$

Let $B_i = \{x_{ij} | 1 \leq j \leq m\} \cup \{y_{ij} | 1 \leq j \leq m\}$, which is called the ith *block*. If the variables in each B_i are arranged consecutively in a variable ordering, then it is called *grouped*, otherwise *ungrouped*. If a variable ordering is grouped and looks like $(B_{i_1}, B_{i_2}, ..., B_{i_n})$, then we say that its *high-level ordering*, denoted by $(\varphi(1), \varphi(2), ..., \varphi(n))$, is $(i_1, i_2, ..., i_n)$. If $(\pi[i], \pi[2nm + 4m - i + 1]) = (b_{2m-i+1}, a_{2m-i+1})$ for $1 \leq i \leq 2m$, i.e., if all a-variables are placed on the top and all b-variables on the bottom, then this ordering is said to be *splitted*, otherwise *unsplitted*.

Lemma 5.2: For any grouped variable ordering π,

$$|\mathcal{B}(\text{MAJMAJ}, \pi)| = S(2m)S(n) + 2,$$

where $S(i)$ is the size of an OBDD for i-variable majority functions except two constant nodes (Proof is omitted).

For a p-variable function $f(x_1, x_2, ..., x_p)$, $f|_{x_1=b_1, x_2=b_2, ..., x_q=b_q}$, $b_i = 0$ or 1, denotes the partially assigned(PA) function. A *properly partially assigned(PPA) function* means $1 \leq q \leq p-1$, i.e., there is at least one assigned and unassigned variables. If $q = p$, then we call it a *totally assigned(TA) function* and if $q = 0$, a *zero assigned(ZA) function*. Recall that MAJMAJ is the majority function of n majority functions $\text{MAJ}_i(x_{i1}, x_{i2}, ..., x_{im}, y_{i1}, y_{i2}, ..., y_{im}), 1 \leq i \leq n$. $\text{MAJ}_i^{(j_1, j_2)}$ denotes a PA MAJ_i such that j_1 variables are assigned 0 or 1 out of which j_2 ones are assigned 1.

Note that the value of $\text{MAJ}_i^{(j_1, j_2)}$ may be fixed to 0 or 1 even if it is not TA. Such $\text{MAJ}_i^{(j_1, j_2)}$ is said to be *fixed*. $\text{MAJMAJ}^{(j_1, j_2)}$ denotes a PA MAJMAJ such that j_1 MAJ_i's are fixed out of which j_2 are fixed to 1. Then, the following lemma holds (Proof is omitted). Intuitively speaking, MAJMAJ has a large width.

Lemma 5.3: For PA MAJMAJ the following property holds:

(i) If the number of TA MAJ_is is k_1, then the number of different (i.e., pairwise inequivalent) PA MAJMAJs is $k_1 + 1$ if $k_1 \leq \lceil n/2 \rceil - 1$, and $n - k_1 + 2$ if $k_1 \geq \lceil n/2 \rceil$,
(ii) If the number of PPA MAJ_is is k_2 and $k_2 \leq \lfloor n/2 \rfloor$, then the number of different PA MAJMAJs is more than 2^{k_2}.

6 Justification of the Reduction

In this paper, we assume $m \geq 4n$, $m \geq 13$ and $n^2 m^2 \leq 2^{n/6}$. The last inequality automatically holds when $n \geq 281$ in a given graph in OLA. OLA is NP-hard under this assumption.

Lemma 6.1: An arbitrary unsplitted $\mathcal{B}(\mathcal{H}^*)$ can be transformed into a functionally equivalent splitted $\mathcal{B}(\mathcal{H}^*)$ with less nodes.

Proof: Only the idea is given bliefly: See Fig.6 where the OBDD changes from G to G' by moving some b-variable to the bottom and then changes to G'' by moving some b-variable to the bottom. Let us look at what happens when we move $b = b_{2m}$ from level k ($2 \leq k \leq T - 1$) to level 1. Recall that G consists of

OBDDs $\mathcal{H}^*$ for the penalty function and g's for edge functions. We now assume that the level k is intermediate for both $\mathcal{H}$ and g's (and we also have to consider other cases).

We define $\mathcal{F}_i$ as the set of different functions which can be obtained by assigning 0 or 1 to each variable of level $\pi[i], \pi[i+1], ..., \pi[T]$ of $\mathcal{H}^*$. Then $\mathcal{F}_{k+1}$ is expressed as follows:

$$\mathcal{F}_{k+1} = \{b\mathcal{H}^1, b\mathcal{H}^2, ..., b\mathcal{H}^{k_1}, g_m^1 + b, g_m^2 + b, ..., g_m^{k_2} + b\},$$

where $\mathcal{H}^i$ and g_m^i are the functions which can be obtained by assigning 0 or 1 to each variable of $\pi[k+1], \pi[k+2]..., \pi[T]$ of $\mathcal{H}$ and g_m, respectively. By the above assumption, there are obviously at least two functions $\mathcal{H}^{k'}$ and $g_m^{k''}$, which are not constant. Both $b\mathcal{H}^{k'}$ and $g_m^{k''} + b$ depend on b and one can prove that the two functions are not equivalent. Therefore, the number of nodes labeled b in G is at least two, which is larger than that in G'. The number of nodes at level $(\pi[k+1], ..., \pi[T])$ of G is the same as that of G' because the variable ordering there does not change. Also, it is not hard to prove that the number of nodes at level $(\pi[1], ..., \pi[k-1])$ of G is equal to that of G'. Thus we can move b to the bottom without increasing the size. □

Now we can assume that an optimal $\mathcal{B}(\mathcal{H}^*)$ is splitted. It is furthermore shown below that there is an optimal $\mathcal{B}(\mathcal{H}^*)$ which is both splitted and grouped.

Lemma 6.2: An arbitrary splitted and ungrouped $\mathcal{B}(\mathcal{H}^*)$ can be transformed into a functionally equivalent splitted and grouped $\mathcal{B}(\mathcal{H}^*)$ with less nodes.

Before proving this lemma, let us finish the proof of our Main Theorem.

Lemma 6.3: Let $\tilde{\mathcal{B}}$ be a splitted and grouped $\mathcal{B}(\mathcal{H}^*)$ which is constructed in the same way as $\mathcal{B}^*(G)$ (see Sec.4). Then $\tilde{\mathcal{B}}$ is reduced and $|\tilde{\mathcal{B}}| = 2\tilde{K} + \varphi(G)$ where $\tilde{K}$ is the cost of the OLA graph G for the variable ordering that matches the high-level ordering of $\tilde{\mathcal{B}}$(Proof is omitted).

Now we are almost done with the justification of our reduction: Suppose that G can be drawn in cost K. Then there is a vertex ordering ψ that realizes this cost K. This implies that we can change the high-level variable ordering of $\mathcal{B}^*(G)$ so as to match ψ. By Lemma 6.3, the size of this OBDD is bounded by the target cost. Suppose conversely that $\mathcal{B}^*(G)$ can be transformed into some equivalent OBDD $\mathcal{B}$ of cost $2K + \Phi(G)$. Then $\mathcal{B}$ can be transformed into a splitted and grouped OBDD of cost $2K + \Phi(G)$ by Lemma 6.2. Then we can derive the vertex order ψ for G from its high-level ordering and the cost of ψ is at most K by Lemma 6.3.

(**Proof of Lemma 6.2**) A splitted and not grouped $\mathcal{B}(\mathcal{H}^*)$ G is transformed into an equivalent splitted and grouped G' with less nodes by the following procedure: Since G is splitted the top $2m$ variables are a-variables and the bottom $2m$ ones b-variables. So we first take the lowest non-ab variable (at level $2m+1$), say, x. Then we "gather" all the variables that are in the same group as x, say B_{u_1}. Then we move them to the bottom, i.e., at levels $\pi[2m+1], \pi[2m+2], ..., \pi[4m]$ without changing their relative order. This is what we call stage (1a). Now G changes into G_{1a}.

In stage (1b), we gather all the variables in the same group, say B_{u_2}, as the non-ab variable existing at the top of G_{1a}. They are moved to the top similarly as before. In stage (2a), another group of variables are moved the bottom and so on. The key idea of this approach is that when moving a variable x, downward at some stage, x does not come from too high positions since those high positions were already grouped. One should observe how this is important to our argument below.

In the following, we prove that the number of nodes does not increase at each stage. To evaluate the variation in the nodes of the splitted and ungrouped $\mathcal{B}(\mathcal{H}^*)$, we observe the f_j-parts and g_j-parts separately from MAJMAJ-part.

(*Case1*:(wa) stage ($2 \leq w \leq \lfloor n/2 \rfloor$)): Suppose that after the (($w-1$)b) stage, the variable at $\pi[2mw+1]$ is in the block B_u and that there are q variables which are not in B_u among $\pi[2mw+1], \pi[2mw+2], ..., \pi[\max\{p|\pi[p] \in B_u\}]$. If $q = 0$, all the variables between $\pi[2mw+1]$ and $\pi[\max\{p|\pi[p] \in B_u\}] = \pi[2m(w+1)]$, are already all in B_u after the (($w-1$)b) stage. Thus at the (wa) stage the variable ordering does not change. We assume $q > 0$ in the following of this proof. We denote $\min\{p|\pi[p] \in B_u\}$ and $\max\{p|\pi[p] \in B_u\}$ by $p_{\min}$ and $p_{\max}$ respectively, and the decrease in the number of the nodes of MAJMAJ-part at the (wa) stage by $-\Delta$MAJMAJ.

First, we observe the variation in the nodes of $\mathcal{B}(f_j)$ and $\mathcal{B}(g_j)$ ($j = 1, 2, ..., m$) at the (wa) stage. We denote the increase in the number of nodes of each $\mathcal{B}(f_j)$ and each $\mathcal{B}(g_j)$ at the (wa) stage by Δf_j and Δg_j, respectively. For a similar reason as [10], it holds that $\sum_{j=1}^m (\alpha_j + \beta_j) \leq q$.

Secondly, we compute ΔMAJMAJ. Let the OBDD G_1 turn into the OBDD G_2 at the (wa) stage. We divide the whole $2nm$ variables (if we say variables in this proof, they are non-ab variables) into three parts, which are called an *A-part*, a *B-part* and a *C-part*, respectively. The variables which do not move at the (wa) stage belong to the A-part. The variables which move and are in B_u belong to the B-part. The variables which move and are not in B_u, i.e., those that are "passed by" by B-part variables, belong to the C-part.

As the variable ordering in the A-part does not change, the number of nodes in the A-part does not change either. The number of nodes in the B-part does not increase by Lemma 5.2, because there are ($w - 1$) consecutive variable blocks at the top. For each level in the C-part, the number of nodes in the level decreases by at least one for the following reason: Assume $\pi[k_1]$ is in the C-part in G_1 and $\pi[k_1]$ moves to $\pi[k_2] (k_1 < k_2)$ at the (wa) stage, i.e., $\pi[k_1]$ in G_1 and $\pi[k_2]$ in G_2 have the same label. Let the number of nodes in level k_2 is n_2 in G_2, then the functions representing the nodes in level k_2 is expressed as follows: $\mathcal{F}_{k_2+1}^2 = \{\text{MAJMAJ}_2^1, ..., \text{MAJMAJ}_2^{n_2}\}$, where MAJMAJ$_2^i$ is the function which can be obtained by assigning 0 or 1 to each variable of $\pi[k_2+1], ..., \pi[2nm]$ in G_2. In G_2 all the variables of B_u are lower than level k_2, hence the function MAJ$_u$ in MAJMAJ$_2^i$ is ZA, i.e., MAJMAJ$_2^i$ = MAJ(MAJ$_1^i$, ..., MAJ$_u$, ..., MAJ$_n^i$)(MAJ$_j^i$ is a PA MAJ$_j$). In G_1, the function MAJ$_u$ in the functions representing the nodes in level k_1 is PA by the definition of the C-part. Then we can get n_2 functions in level k_1 of G_1, MAJMAJ$_1^1$, ..., MAJMAJ$_1^{n_2}$, where MAJMAJ$_1^i$ = MAJ(MAJ$_1^i$, ..., MAJ$'_u$, ..., MAJ$_n^i$)

and MAJ'_u is a PA MAJ_u which is not constant. We will show these n_2 functions are different. Because MAJMAJ^i_2 and MAJMAJ^j_2 are inequivalent, so are $\text{MAJ}(\text{MAJ}^i_1, ..., a_u, ..., \text{MAJ}^i_n)$ and $\text{MAJ}(\text{MAJ}^j_1, ..., a_u, ..., \text{MAJ}^j_n)$, where a_u is a TA MAJ_u, i.e., 0 or 1. By a proper assignment, we can make $\text{MAJ}'_u = a_u$. Therefore, MAJMAJ^i_1 ($1 \leq i \leq n_2$) are different. Moreover we choose an assignment for G_1 such that $\text{MAJMAJ}^{1a}_1 = \text{MAJ}(\text{MAJ}^1_1, ..., \text{MAJ}''_u, ..., \text{MAJ}^1_n)$, where MAJ''_u is another PA MAJ_u. Then MAJMAJ^{1a}_1 is not equivalent to MAJMAJ^1_1 because by a proper assignment, we can make $(\text{MAJ}'_u, \text{MAJ}''_u) = (0,1)$ or $(1,0)$. MAJMAJ^{1a}_1 is not equivalent to MAJMAJ^i_1 ($2 \leq i \leq n_2$) because by a proper assignment, we can make $\text{MAJ}'_u = \text{MAJ}''_u = a_u$. Hence, the number of nodes in level k_1 is at least $n_2 + 1$. As there are q levels in the C-part, $-\Delta\text{MAJMAJ} \geq q$.

Thus we obtain: $\sum_{j=1}^{m}(\Delta f_j + \Delta g_j) + \Delta\text{MAJMAJ} \leq 0$. Therefore one can now see that at the (wa) stage, the number of nodes of the $\mathcal{B}(\mathcal{H}^*)$ does not increase. For the $(1a)$ stage, our proof goes in exactly the same way.

(*Case2*:(wb) stage ($1 \leq w \leq \lfloor n/2 \rfloor$)): Suppose that after the (wa) stage, $\pi[2m(n+2-w)]$ is in B_u and that there are q variables which are not in B_u among $\pi[\min\{p|\pi[p] \in B_u\}], ..., \pi[2m(n+2-w)-1], \pi[2m(n+2-w)]$. If $q = 0$, all the variables between $\pi[\min\{p|\pi[p] \in B_u\}](= \pi[2m(n+1-w)])$ and $\pi[2m(n+2-w)]$ are already all in B_u after the (wa) stage. Thus at the (wb) stage the variable ordering does not change. We assume $q > 0$ in the following of this proof. We denote the amount of decrease in the number of the nodes of MAJMAJ-part at the (wb) stage by $-\Delta\text{MAJMAJ}$. In the same way as Case1, the increase in the number of nodes of Edge functions is less than q.

We compute ΔMAJMAJ. Let the OBDD G_2 turn into the OBDD G_3 at the (wb) stage. We divide all $2nm$ variables into three parts, exactly as before.

Claim 6.1: Let z_k be a variable in the B-part where k is its position among B_u (which has not changed). Let $a(k)$ be the number of z_k-nodes in an OBDD for MAJ_u with the variable ordering $(z_1, ..., z_{2m})$. Then there are at least $(w+1) \times a(k)$ z_k-nodes in G_2. Moreover, there are exactly $w \times a(k)$ z_k-nodes in G_3. The proof of this claim is omitted.

Let $r \geq 2$ be the number such that the top variable in the C-part in G_2 is in level $2m(n+2-w)-r+1$. This implies that the number of variables in the B-part is $2m-r+1$. We can apply claim 6.1 to these variables.

If $r \leq m+1$, then $2m-r+1 \geq m$. Hence, the number of nodes in the B-part in G_3 is less than that in G_2 by at least $(|\mathcal{B}(\text{MAJ}(x_1, ..., x_{2m}))|-2)/2-1 \geq m(m+1)/2 \geq 2nm \geq r$. Here, we use the assumption that $m \geq 4n$.

If $r \geq m+2$, for each level in the C-part, the number of nodes in the level decreases by at least one. As there are q levels in the C-part, $-\Delta\text{MAJMAJ} \geq q$. Thus we obtain $\sum_{j=1}^{m}(\Delta f_j + \Delta g_j) + \Delta\text{MAJMAJ} \leq 0$. Thus we have shown that at the (wb) stage, the number of nodes of the $\mathcal{B}(\mathcal{H}^*)$ does not increase. □

Acknowledgments

Thanks are due to Y.Okabe and K.Yasuoka for valuable discussions.

References

1. N.Alon and R.B.Boppana: The monotone circuit complexity of Boolean functions, Combinatorica 7(1)a, pp.1-22(1987).
2. B.Bollig and I.Wegener: Improving the variable ordering of OBDDs Is NP-complete, IEEE Trans. Comput. Vol.45,No.9,pp.993-1002(1996).
3. R.E.Bryant: Graph-based algorithms for Boolean function manipulation, IEEE Trans. Comput. Vol.C35,N0.8,pp.677-691(1986).
4. M.R.Garey, D.S.Johnson and L.Stockmeyer: Some simplified NP-complete graph problems, Theoretical Computer Science, Vol.1, pp.237-267(1976).
5. K.Hosaka, Y.Takenaga, T.Kaneda and S.Yajima: Size of ordered binary decision diagrams representing threshold functions, Theoret. Comput. Sci. 180,pp.47-60(1997).
6. S.Jukuna, A.Razborov, P.Savicky and I.Wegener: On P versus NP∩co-NP for decision trees and read-once branching problems, Proc.27nd MFCS, pp.319-326(1997).
7. M.R.Mercer, R.Kapur and D.E.Ross: Functional approached to generating orderings for efficient symbolic representation, Proc. 29th ACM/IEEE Design Automation Conference, pp.614-619(1992).
8. S.Muroga: Threshold logic and its applications, John Wiley & Sons(1971).
9. C.E.Shannon: The synthesis of two-terminal switching circuits, Bell Systems Tech.J.28(1), pp.59-98(1949).
10. S.Tani,K.Hamaguchi and S.Yajima: The complexity of the optimal variable ordering problems of a shared binary decision diagram, IEICE Trans.Inf.& Syst., Vol.E79-D, No.4, pp.271-281(1996).Also, in Proc.ISAAC93.
11. E.Tardos: The gap between monotone and non-monotone circuit complexity is exponential, Combinatorica, Vol.8,No.1,pp.141-142(1988).
12. U.Zwick: A $4n$ lower bound on the combinational complexity of certain symmetric Boolean functions over the basis of unate dyadic Boolean functions, SIAM J. COMPUT., Vol.20, No.3, pp.499-505(1991).

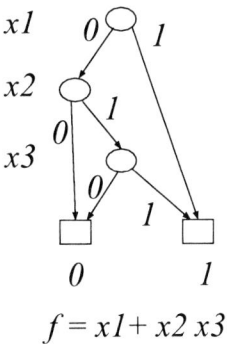

Fig. 1. an example of an OBDD

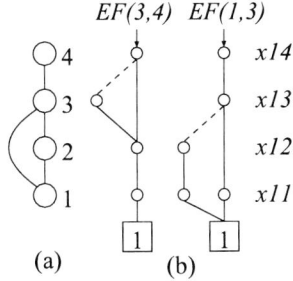

Fig. 2. a graph of OLA and OBDDs representing edge functions

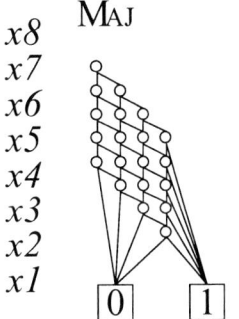

Fig. 3. an OBDD for $\mathrm{MAJ}(x_1, x_2, ..., x_8)$

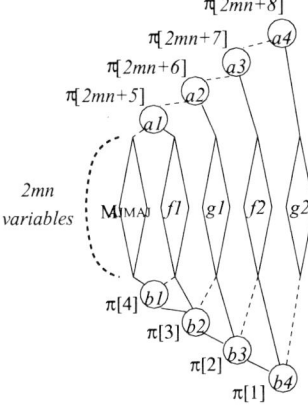

Fig. 4. an OBDD for $\mathcal{H}^*$ where $m = 2$

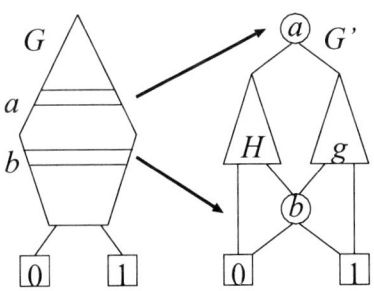

Fig. 5. splitted MOLAD G'(left) and unsplitted MOLAD G(right)

Blockwise Variable Orderings for Shared BDDs
Extended Abstract

Harry Preuß[1] and Anand Srivastav[2]

[1] Institut für Informatik, Humboldt-Universität zu Berlin,
Unter den Linden 6, 10099 Berlin, Germany
(preuss@informatik.hu-berlin.de)

[2] Mathematisches Seminar, Universität zu Kiel,
Ludewig-Meyn-Str. 4, 24098 Kiel, Germany
(asr@numerik.uni-kiel.de)

Abstract. A state-of-the-art data structure for the representation of Boolean functions are ordered binary decision diagrams (OBDDs). The size of an OBDD representing a Boolean function depends on the variable ordering. Finding a variable ordering with optimal (i.e. minimum) OBDD size is a central, but NP-hard problem. Thus it is of great interest to characterize optimal variable orderings from the structure of the given function. In this paper we investigate the problem of characterizing optimal variable orderings for shared OBDDs of two Boolean functions $f_i = g_i \otimes_i h_i$, $i = 1, 2$, where $\otimes_i$ is an operator from the base B_2^*, (the full binary basis consisting of all ten binary operations depending essentially on both inputs) and g_i (resp. h_i) depends only on x-variables (resp. y-variables). Tree-like circuits provide an example for such functions. In the special case $f_1 = \overline{f_2}$, Sauerhoff, Wegener and Werchner [6] proved that there is some optimal ordering where all x-variables are tested before all y-variables or vice versa (blockwise variable ordering). We show that this is also true for arbitrary f_1, f_2 provided that $\otimes_1 = \wedge$ and $\otimes_2 = \vee$, and for shared OBDDs with complemented edges and arbitrary f_1, f_2 provided that $\otimes_1 = \wedge$ and $\otimes_2 = \oplus$. For all other combinations of $\otimes_1$ und $\otimes_2$ we give counterexamples.

1 Introduction

Shared binary decision diagrams (briefly denoted by SBDD) introduced by Minato, Ishiura and Yajima [5] are a generalization of ordered binary decision diagrams (OBDD) (introduced by Bryant [2]).

For a Boolean function $f : \{0,1\}^n \mapsto \{0,1\}$ with variables $x_1, \ldots, x_n$ a variable ordering is a permuted sequence $x_{\pi(1)}, \ldots, x_{\pi(n)}$ where π is a permutation on $\{1, \ldots, n\}$. An OBDD for a Boolean function f and a variable ordering π for f is a rooted directed acyclic graph with one source and two sinks, one labeled by 0 (called 0-sink) and the other labeled by 1 (called 1-sink). The non-sink nodes are labeled by the Boolean variables $x_1, \ldots, x_n$ and have two outgoing edges, one labeled by 0 (called 0-edge) and the other by 1 (called 1-edge) . If an edge leads from a node labeled by x_i to a node labeled by x_j, then x_i has to precede x_j in the variable ordering π. Furhermore, the OBDD must represent f: each input $a \in \{0,1\}^n$ corresponds to an unique path in the OBDD starting in the root and ending in the $f(a)$-sink. A SBDD is roughly speaking a data structure representing multiple OBDDs for a variable ordering π in one diagram by sharing nodes which represent the same subfunction (so isomorphic subgraphs do not co-exist in the SBDD; for a technical definition we refer to Minato et al. [5]) The OBDD and resp. SBDD for given functions and a given variable ordering is unique (up to isomorphisms).

As for OBDDs, the size of SBDDs heavily depends on the variable ordering. Finding an optimal variable ordering, i.e. an ordering with minimum OBDD (resp. SBDD size), has been shown to be a NP-hard problem [1] resp. [9]. This is one motivation to characterize optimal variable orderings for interesting classes of Boolean functions.

For OBDDs of Boolean functions represented by tree-like circuits the problem has been resolved by Sauerhoff, Wegener and Werchner [6], who proved that there is some depth-first-search traversal leading to an optimal variable ordering; moreover an optimal variable ordering and the resulting OBDD can be computed in linear time. This result is a consequence of a more general result for Boolean functions which are a $\otimes$-product of two functions not sharing variables, where $\otimes$ is an operator from the full binary basis B_2^* of all ten binary operations depending essentially on both inputs $x, y \in \{0, 1\}$, namely $x \wedge y$ (AND), $x \vee y$ (OR), $\overline{x \wedge y}$ (NAND), $\overline{x \vee y}$ (NOR), $\overline{x} \wedge y$, $x \wedge \overline{y}$, $\overline{x} \vee y$, $x \vee \overline{y}$, $x \oplus y$ (EXOR), $\overline{x \oplus y}$ (NEXOR). The result of Sauerhoff, Wegener and Werchner can be stated as follows: For $i = 1, 2$ let g_i and h_i be Boolean functions, where g_i (resp. h_i) depends only on some x-variables (resp. y-variables), so $g_i = g_i(x_1, \ldots, x_k)$ and $h_i = h_i(y_1, \ldots, y_m)$, and consider the product functions $f_i = g_i \otimes_i h_i$, with $\otimes_i \in B_2^*$. Then OBDD(f_i) admits an optimal variable ordering in which *all* x-variables are tested before all y-variables or vice versa. Such orderings are called *blockwise* variable orderings. The authors also prove that for $f_1 = \overline{f}_2$ the SBDD(f_1, f_2) always has an optimal variable ordering which is blockwise. In this paper we ask whether this result extends to arbitrary f_1, f_2 (decomposable as above). We prove that there is an optimal blockwise variable ordering for the SBDD (SBDD with complemented edges respectively) of arbitrary f_1, f_2 whenever $\otimes_1 = \wedge$ and $\otimes_2 = \vee$ ($\otimes_1 = \wedge$ and $\otimes_2 = \oplus$ respectively). Moreover for SBDDs this is also true, if $\otimes_1 = \otimes_2$ and $h_1 = h_2$. For all other combinations of $\otimes_1, \otimes_2$ there are examples for which all blockwise variable orderings are not optimal.

2 Basic Notations and Known Results

In this section we briefly recall some basic notations and definitions. For a Boolean function f, $f_{|x_i=b_i}$ denotes the Boolean function resulting from f by replacing the variable x_i by the constant $b_i \in \{0, 1\}$. We say f *depends essentially* on a variable x_i, iff $f_{|x_i=1} \neq f_{|x_i=0}$. In this notation B_2^* can be regarded as the set of functions from B_2 depending essentially on both variables. An OBDD with respect to a variable ordering π representing the Boolean function f is denoted by π-OBDD(f). The size of a given OBDD G is defined as the number of its inner nodes and is denoted by $|G|$. For a given variable z and a given OBDD G we denote the number of inner nodes of G with label z by $|G|_z$.

The following theorem of Sieling and Wegener [8] gives a relation between the number of nodes labeled by z and the number of functions that are represented at a node of an OBDD.

Theorem 1 (Sieling and Wegener [8]). *If f is a Boolean function defined on $x_1, \ldots, x_n$ and π is the variable ordering $(x_1, \ldots, x_n)$, then $|\pi\text{-}OBDD(f)|_{x_i}$ is equal to the number of different subfunctions $f_{|x_1=a_1,\ldots,x_{i-1}=a_{i-1}}$ for $a_1, \ldots, a_{i-1} \in B$ which depend essentially on x_i.*

An extension of OBDDs and SBDDs are *OBDDs with complemented edges* ($OBDD_{ce}$) and *SBDDs with complemented edges* ($SBDD_{ce}$), also introduced by Minato et al. [5]. In OBDDs and SBDDs with complemented edges a subfunction and its complement are represented by the same node by complementing some 1-edges. In the best case this can reduce the size by a factor of 2.

For a Boolean function arising from a tree-like-circuits Sauerhoff et al. gave a linear-time algorithm to compute an optimal variable ordering. The underlying structure theorem is the following one:

Theorem 2 (Sauerhoff, Wegener, Werchner [7]). *Let*

$$f(x_1, \ldots, x_k, y_1, \ldots, y_m) := g(x_1, \ldots, x_k) \otimes h(y_1, \ldots, y_m)$$

for $\otimes \in B_2^*$. Then there is an optimal variable ordering for f where all x-variables are tested before all y-variables or vice versa. The same holds for $SBDD(f, \overline{f})$.

If a Boolean function f is defined as in theorem 2, we call a variable ordering for f *blockwise*, if all x-variables are tested before all y-variables or vice versa.

3 SBDDs Without Complemented Edges

Let $g_1(x_1, \ldots, x_k), g_2(x_1, \ldots, x_k), h_1(y_1, \ldots, y_m), h_2(y_1, \ldots, y_m)$ be Boolean functions, where every function depends essentially on all its variables. Further let

$$f_1(x_1, \ldots, x_k, y_1, \ldots, y_m) = g_1(x_1, \ldots, x_k) \otimes_1 h_1(y_1, \ldots, y_m) \quad (1)$$
$$f_2(x_1, \ldots, x_k, y_1, \ldots, y_m) = g_2(x_1, \ldots, x_k) \otimes_2 h_2(y_1, \ldots, y_m)$$

for arbitrary $\otimes_1, \otimes_2 \in B_2^*$.

3.1 Affirmative Answer for $\otimes_1 = \wedge$ and $\otimes_2 = \vee$

The main positive result of this section is:

Theorem 3. *For the $SBDD(f_1, f_2)$ with f_1, f_2 defined by (1) and $\otimes_1 = \wedge$ and $\otimes_2 = \vee$ there is always an optimal variable ordering which is blockwise.*

Proof. Let π be an arbitrary variable ordering and w.l.o.g. let the last variable in π be a y-variable. Let π' be the variable ordering, where first all x-variables are tested in the same order as prescribed by π and then all y-variables are tested in the same order as prescribed by π. Therefore we can assume that π' is given by $(x_1, \ldots, x_k, y_1, \ldots, y_m)$. Let G (resp. G') be the π-SBDD(f_1, f_2) (resp. the π'-SBDD(f_1, f_2)).

We show that for every variable z, $|G'|_z \leq |G|_z$. So if π is optimal for the SBDD(f_1, f_2), π' is optimal, too. We distinguish two cases: z is an x-variable or a y-variable.

Case 1 (Variable x_i): From theorem 1 it follows that $|\pi'\text{-OBDD}(f_1)|_{x_i}$ is equal to the number of different functions

$$g_1\,_{|x_1=a_1,\ldots,x_{i-1}=a_{i-1}} \wedge h_1$$

which depend essentially on x_i.

We assume that j y-variables are tested in π before x_i. So $|\pi\text{-OBDD}(f_1)|_{x_i}$ is equal to the number of different subfunctions

$$g_1\,_{|x_1=a_1,\ldots,x_{i-1}=a_{i-1}} \wedge h_1\,_{|y_1=b_1,\ldots,y_j=b_j}$$

depending essentially on x_i.

Because h_1 depends essentially on all its variables and because $j < m$ (we remember: the last variable in π is a y-variable), we can choose fixed constants $b'_1, \ldots, b'_j$ such that $h_1\,_{|y_1=b'_1,\ldots,y_j=b'_j}$ is not constant and obtain

$$|\pi'\text{-OBDD}(f_1)|_{x_i} = |\pi\text{-OBDD}(f_1\,_{|y_1=b'_1,\ldots,y_j=b'_j})|_{x_i} \leq |\pi\text{-OBDD}(f_1)|_{x_i}.$$

In the same manner we choose fixed constants $b''_1, \ldots, b''_j$ such that $h_2\,_{|y_1=b''_1,\ldots,y_j=b''_j}$ is not constant and

$$|\pi'\text{-OBDD}(f_2)|_{x_i} = |\pi\text{-OBDD}(f_2\,_{|y_1=b''_1,\ldots,y_j=b''_j})|_{x_i} \leq |\pi\text{-OBDD}(f_2)|_{x_i}.$$

Now we show that no x_i-node of the π-OBDD$(f_1\,_{|y_1=b'_1,\ldots,y_j=b'_j})$ can be merged with an x_i-node of the π-OBDD$(f_2\,_{|y_1=b''_1,\ldots,y_j=b''_j})$, if we construct the π-SBDD$(f_1\,_{|y_1=b'_1,\ldots,y_j=b'_j}, f_2\,_{|y_1=b''_1,\ldots,y_j=b''_j})$ by merging the two OBDDs. Let us assume that there are x_i-nodes in the π-OBDD$(f_1\,_{|y_1=b'_1,\ldots,y_j=b'_j})$ which are merged with x_i-nodes in the π-OBDD$(f_2\,_{|y_1=b''_1,\ldots,y_j=b''_j})$.

From this assumption it follows that there must exist fixed constants $a'_1, \ldots, a'_{i-1}$ and $a''_1, \ldots, a''_{i-1} \in B$ such that $g_1\,_{|x_1=a'_1,\ldots,x_{i-1}=a'_{i-1}}$ and $g_2\,_{|x_1=a''_1,\ldots,x_{i-1}=a''_{i-1}}$ depend essentially on x_i and

$$g_1\,_{|x_1=a'_1,\ldots,x_{i-1}=a'_{i-1}} \wedge h_1\,_{|y_1=b'_1,\ldots,y_j=b'_j} = g_2\,_{|x_1=a''_1,\ldots,x_{i-1}=a''_{i-1}} \vee h_2\,_{|y_1=b''_1,\ldots,y_j=b''_j}. \tag{2}$$

Since $h_1\,_{|y_1=b'_1,\ldots,y_j=b'_j}$ is not constant, there are constants $b'_{j+1}, \ldots, b'_m$ such that

$$h_1\,_{|y_1=b'_1,\ldots,y_m=b'_m} = 0$$

and therefore
$$g_1 |_{x_1=a'_1,\ldots,x_{i-1}=a'_{i-1}} \wedge h_1 |_{y_1=b'_1,\ldots,y_m=b'_m} = 0.$$

Now
$$h_2 |_{y_1=b''_1,\ldots,y_j=b''_j, y_{j+1}=b'_{j+1},\ldots,y_m=b'_m} = 1$$
is a contradiction to (2). But
$$h_2 |_{y_1=b''_1,\ldots,y_j=b''_j, y_{j+1}=b'_{j+1},\ldots,y_m=b'_m} = 0$$
also leads to a contradiction to (2), because
$$g_2 |_{x_1=a''_1,\ldots,x_{i-1}=a''_{i-1}} \vee h_2 |_{y_1=b''_1,\ldots,y_j=b''_j, y_{j+1}=b'_{j+1},\ldots,y_m=b'_m}$$
depends essentially on x_i, while
$$g_1 |_{x_1=a'_1,\ldots,x_{i-1}=a'_{i-1}} \wedge h_1 |_{y_1=b'_1,\ldots,y_m=b'_m}$$
does not. So no x_i-node of the π-OBDD$(f_1 |_{y_1=b'_1,\ldots,y_j=b'_j})$ can be merged with an x_i-node of the π-OBDD$(f_2 |_{y_1=b''_1,\ldots,y_j=b''_j})$.

Hence
$$|G'|_{x_i} \leq |\pi\text{-SBDD}(f_1 |_{y_1=b'_1,\ldots,y_j=b'_j}, f_2 |_{y_1=b''_1,\ldots,y_j=b''_j})|_{x_i}.$$

Because each x_i-node of the π-SBDD$(f_1 |_{y_1=b'_1,\ldots,y_j=b'_j}, f_2 |_{y_1=b''_1,\ldots,y_j=b''_j})$ represents a function which must also be represented by an x_i-node in G, we have
$$|\pi\text{-SBDD}(f_1 |_{y_1=b'_1,\ldots,y_j=b'_j}, f_2 |_{y_1=b''_1,\ldots,y_j=b''_j})|_{x_i} \leq |G|_{x_i}.$$

All together we get $|G'|_{x_i} \leq |G|_{x_i}$.

Case 2 (Variable y_j): The proof is analogous to the proof in case 1.

3.2 Counter Examples

If we choose other operators from B_2^* for $\otimes_1$ and $\otimes_2$, there are always functions f_1 und f_2 for which all blockwise variable orderings are not optimal. Let f_1 and f_2 be defined as in (1).

- If $\otimes_1 = \wedge$, $\otimes_2 = \oplus$ and
$$\begin{aligned} g_1(x_1, x_2, x_3, x_4) &= x_1 \wedge x_2 \wedge (x_3 \oplus x_4) \\ g_2(x_1, x_2, x_3, x_4) &= x_1 \vee x_2 \vee (x_3 \oplus x_4) \\ h_1(y_1) = h_2(y_1) &= y_1, \end{aligned} \quad (3)$$
the SBDD(f_1, f_2) for variable orderings, where the y-variable is tested first or last, contains at least 12 nodes, while the SBDD with the optimal variable ordering $(x_1, x_2, y_1, x_3, x_4)$ contains 11 nodes.

- If $\otimes_1 = \otimes_2 = \wedge$ and

$$g_1(x_1, x_2) = x_1 \wedge x_2$$
$$g_2(x_1, x_2) = \overline{x_1} \wedge x_2$$
$$h_1(y_1, y_2) = y_1 \wedge y_2$$
$$h_2(y_1, y_2) = \overline{y_1} \wedge y_2,$$

the SBDD(f_1, f_2) for the optimal variable ordering (x_1, y_1, x_2, y_2) contains 6 nodes, while it contains at least 7 nodes for every blockwise variable ordering.
- If $\otimes_1 = \otimes_2 = \oplus$ and

$$g_1(x_1, x_2, x_3) = (x_1 \vee x_2) \oplus x_3$$
$$g_2(x_1, x_2, x_3) = (x_1 \wedge x_2) \oplus x_3 \qquad (4)$$
$$h_1(y_1, y_2, y_3) = y_1 \vee (y_2 \oplus y_3)$$
$$h_2(y_1, y_2, y_3) = y_1 \wedge (y_2 \oplus y_3),$$

the SBDD(f_1, f_2) for every blockwise ordering contains at least 16 nodes, while the SBDD with the variable ordering $(x_1, x_2, y_1, y_2, y_3, x_3)$ contains 14 nodes and is optimal.

With theorem 3 and these counterexamples we have covered all cases of $\otimes_1$ and $\otimes_2$ from B_2^*. The results are summarized in the following table.

		$f_2 =$		
		$g_2 \wedge h_2$	$g_2 \vee h_2$	$g_2 \oplus h_2$
	$g_1 \wedge h_1$	-	+	-
$f_1 =$	$g_1 \vee h_1$	+	-	-
	$g_1 \oplus h_1$	-	-	-

"+" in line $f_1 = g_1 \otimes_1 h_1$ and column $f_2 = g_2 \otimes_2 h_2$ means: there is an optimal variable ordering for the SBDD(f_1, f_2) which is blockwise, and "-" means, there are functions f_1, f_2 for which no optimal variable ordering for SBDD(f_1, f_2) is blockwise.

3.3 Special Cases

Now we consider the special case of (1), where $h_1 = h_2$. Due to this restriction we must also handle the NAND-Funktion $\overline{\wedge} \in B_2^*$ in order to cover all cases.

- If $\otimes_1 = \wedge$ and $\otimes_2 = \vee$, or $\otimes_1 = \wedge$ and $\otimes_2 = \overline{\wedge}$, we conclude from Theorem 3, that there is always an optimal variable ordering which is blockwise.
- If $\otimes_1 = \wedge$ and $\otimes_2 = \oplus$, example (3) serves as a counterexample.

- If $\otimes_1 = \vee$, $\otimes_2 = \overline{\wedge}$,

$$g_1(x_1, x_2) = x_1 \vee x_2$$
$$g_2(x_1, x_2) = x_1 \wedge \overline{x_2}$$
$$h_1(y_1, y_2) = h_2(y_1, y_2) = y_1 \oplus y_2$$

and f_1 and f_2 are defined by (1), the SBDD for the best blockwise variable ordering contains 8 nodes, while it contains 7 nodes for the optimal variable ordering (x_1, y_1, y_2, x_2).

So only the case $\otimes_1 = \otimes_2$ is open. Here we have the following nice characterization even for the SBDD of more than two functions.

Let $p \in \mathbf{N}$, $p \geq 2$, and let $g_1(x_1, \ldots, x_k), \ldots, g_p(x_1, \ldots, x_k)$ be p pairwise differrent functions, each depending essentially on all its variables. Furthermore let $h(y_1, \ldots, y_m)$ be another Boolean function depending essentially on all its variables, $\otimes$ an operator from B_2^* and

$$f_1(x_1, \ldots, x_k, y_1, \ldots, y_m) = g_1(x_1, \ldots, x_k) \otimes h(y_1, \ldots, y_m)$$
$$\vdots$$
$$f_p(x_1, \ldots, x_k, y_1, \ldots, y_m) = g_p(x_1, \ldots, x_k) \otimes h(y_1, \ldots, y_m).$$

Theorem 4. *There is always an optimal variable ordering for the SBDD(f_1, ..., f_p), where all x-variables are tested before any y-variable is tested.*

Proof. We introduce $p-1$ additional variables $z_1, \ldots, z_{p-1}$. Apply theorem 2 to OBDD(f) where f is the function

$$f(z_1, \ldots, z_{p-1}, x_1, \ldots, x_k, y_1, \ldots, y_m)$$
$$= z_1 \cdot f_1 \vee \overline{z_1} \cdot z_2 \cdot f_2 \vee \ldots \vee \overline{z_1} \cdots \overline{z_{p-2}} \cdot z_{p-1} \cdot f_{p-1} \vee \overline{z_1} \cdots \overline{z_{p-1}} \cdot f_p.$$

The results for the special case $h_1 = h_2$ are summarized in the following table.

		$f_2 =$			
		$g_2 \wedge h$	$g_2 \vee h$	$g_2 \oplus h$	$\neg(g_2 \wedge h)$
$f_1 =$	$g_1 \wedge h$	+	+	-	+
	$g_1 \vee h$	+	+	-	-
	$g_1 \oplus h$	-	-	+	-
	$\neg(g_1 \wedge h)$	+	-	-	+

4 SBDDs with Complemented Edges

4.1 Affirmative Answer for $\otimes_1 = \wedge$ and $\otimes_2 = \oplus$

The starting point for the results in this section is the following lemma which generalizes a theorem of Sauerhoff, Wegener, and Werchner [7].

Lemma 1. *For arbitrary Boolean functions $f_1, \ldots, f_p, p \in \mathbf{N}$ and every variable ordering π for $f_1, \ldots, f_p$ it holds*

$$|\pi\text{-}SBDD(f_1, \ldots, f_p, \overline{f_1}, \ldots, \overline{f_p})| = 2|\pi\text{-}SBDD_{ce}(f_1, \ldots, f_p)|.$$

Our main positive results for SBDDs with complemented edges is:

Theorem 5. *For the $SBDD_{ce}(f_1, f_2)$ with f_1, f_2 defined by (1) and $\otimes_1 = \wedge$ and $\otimes_2 = \oplus$ there is always an optimal variable ordering which is blockwise.*

Sketch of Proof: Let π be an arbitrary variable ordering for f_1, f_2, w.l.o.g. let the last variable in π be a y-variable. Let π' be the variable ordering, where first all x-variables are tested in the same order as prescribed by π and after that all y-variables are tested in the same order as prescribed by π. Thus we can assume w.l.o.g. that π' is the variable ordering $(x_1, \ldots, x_k, y_1, \ldots, y_m)$.

Let G be π-SBDD$(f_1, f_2, \overline{f_1}, \overline{f_2})$ and let G' be π'-SBDD$(f_1, f_2, \overline{f_1}, \overline{f_2})$. We show that for every variable z, $|G'|_z \leq |G|_z$. Then by lemma 1 it follows that

$$|\pi'\text{-SBDD}_{ce}(f_1, f_2)| \leq |\pi\text{-SBDD}_{ce}(f_1, f_2)|.$$

So if π is optimal for the SBDD$_{ce}(f_1, f_2)$, π' is optimal, too. Again we have to distinguish the cases whether a node is labeled with an x-variable or a y-variable. For the technical proof we must refer to the full paper.

4.2 Counter Examples

If $\otimes_1$ and $\otimes_2$ are other operators from B_2^*, we can find counterexamples. Let f_1, f_2 be defined by (1). If $\otimes_1 = \otimes_2 = \wedge$ and

$$g_1(x_1, x_2, x_3) = x_1 \wedge (x_2 \oplus x_3)$$
$$g_2(x_1, x_2, x_3) = \overline{x_1} \wedge (x_2 \oplus x_3)$$
$$h_1(y_1) = h_2(y_1) = y_1,$$

the SBDD$_{ce}(f_1, f_2)$ contains 5 nodes for the optimal variable ordering (x_1, y_1, x_2, x_3), while it contains at least 6 nodes for every blockwise variable ordering. If $\otimes_1 = \otimes_2 = \oplus$, (4) serves as a counterexample. The SBDD$_{ce}(f_1, f_2)$ contains 9 nodes for the optimal ordering $(x_1, x_2, y_1, y_2, y_3, x_3)$, while it contains at least 10 nodes for every blockwise variable ordering. If we complement f_1 and/or f_2, the size of the SBDD$(f_1, f_2, \overline{f_1}, \overline{f_2})$ does not change. Thus for every $\otimes_1$ and $\otimes_2$ we can decide now, whether for every f_1, f_2 satisfying (1) the SBDD$_{ce}(f_1, f_2)$ has an optimal variable ordering which is blockwise.

4.3 Special Cases

For the special case of (1) with $h_1 = h_2$ similar results as in section 3.3 can be proved.

5 Concluding Remarks

Finding an optimal variable ordering for OBDDs or SBDDs is a NP-hard problem. Therefore it is promising to look for special classes of Boolean functions for which the problem can be solved efficiently. Tree-like-circuits are a special class of Boolean functions with an optimal blockwise variable ordering, and for which such an ordering can be computed in linear time by a depth-first-search traversal [6]. It is a natural and relevant question asking whether this is also true for the SBDD of functions of tree-like-circuits.

Our results classifies the situations when an optimal blockwise variable ordering for the SBDD of two functions each represented by a tree-like-circuit does exist and when all blockwise variable orderings are non optimal.

We leave open the question of finding an optimal blockwise variable ordering for the SBDD of two tree-like circuits in polynomial time, whenever such an ordering exists. Also it would be interesting to derive upper bounds for the size of the optimal SBDD of two functions given by tree-like circuits. A solution for this problem in case of OBDDs is given in [6].

References

1. B. Bollig, I. Wegener: *Improving the variable ordering of OBDDs is NP-complete.* IEEE Transactions on Computers 45, pages 993-1002, 1996
2. R. E. Bryant: *Graph-based algorithms for Boolean function manipulation.* IEEE Transactions on Computers, vol. C-35(8), pages 677-691, 1986.
3. R. E. Bryant: *Symbolic Boolean manipulation with ordered binary decision diagrams.* ACM Computing Surveys, vol. 24, pages 293-318, 1992.
4. S. J. Friedman, K. J. Supowit: *Finding the optimal variable ordering for binary decision diagrams.* IEEE Trans. on Computers 39, pages 710-713, 1990.
5. S. Minato, N. Ishiura, S. Yajima: *Shared binary decision diagrams with attributed edges for efficient Boolean function manipulation.* Design Automation Conference, pages 52-57, 1990.
6. M. Sauerhoff, I. Wegener, R. Werchner: *Optimal ordered binary decision diagrams for fanout-free circuits.* Proc. of SASIMI 1996.
7. M. Sauerhoff, I. Wegener, R. Werchner: *Optimal ordered binary decision diagrams for tree-like circuits.* Forschungsbericht Nr. 613, Universität Dortmund 1996.
8. D. Sieling, I. Wegener: *NC-algorithms for operations on binary decision diagrams.* Parallel Processing Letters, vol. 3, pages 3-12, 1993
9. S. Tani, K. Hamaguchi, S. Yajima: *The complexity of the optimal variable ordering problems of a shared binary decision diagram.* Proc. 4th Int. Symp. on Algorithms and Computation ISAAC, Lecture Notes in Computer Science 762, pages 389-398, 1993.

On the Composition Problem for OBDDs with Multiple Variable Orders

Anna Slobodová *

Institute of Telematics, Trier, Germany
slobodova@ti.fhg.de

Abstract. Ordered Binary Decision Diagram (OBDD) is a favorite data structure used for representation Boolean functions in computer-aided synthesis and verification of digital systems. The secret of its success is the efficiency of the algorithms for Boolean operations, satisfiability and equivalence check. However, the algorithms work well under condition only that the variable order of considered OBDDs is the same.
In this paper, we discuss the problem of Boolean operations on OBDDs with multiple variable orders, which naturally appears, e.g., in the connection with minimization techniques based on dynamic variable reordering. Our goal is to place the problem with respect to its complexity and to point out the difficulties in finding an acceptable solution.

1 Some Motivation Remarks

The main application of Ordered Binary Decision Diagrams (OBDDs) up to now is in computer-aided synthesis, verification and testing of digital systems. The operations like equivalence test, variable quantification and binary Boolean operations, that are crucial for CAD can be performed efficiently, if the considered Boolean functions are represented by OBDDs that are built with respect to the same variable order [2]. The variable order plays an important role in minimization of OBDDs, since the size of an OBDD representation for a function may vary exponentially with different orders. The problem of finding an optimal variable order is known to be NP-complete [11, 4]. Hence, a big effort is focused to development of heuristics that find an acceptable order. Heuristics that are based on a variable reordering are of a special interest, as they allow to optimize OBDDs dynamically whenever their size grows too much during computations. Many tasks cannot be completed without dynamic minimization at all. The trick is that although variable order varies during a computation, at any time point, it is the same order for all stored OBDDs, and this assures an efficient performance of all operations. In some applications, like traversal of large finite state machines that model sequential circuits, the growing size of OBDDs forces to partition them, keep the OBDDs that are necessary in the next step of the computation and to store the rest into secondary memory. The OBDDs are loaded back at the point when they are required for the computations. As

---
* This work has been supported by German Research Society project Me 1077/12-1

a consequence, we are faced with the *problem of different variable orders* in the current and loaded OBDDs.

Another example of manipulation of OBDDs with multiple variable orders appears in combinatorial verification of a circuit. It is one of the main CAD tasks that can be formulated as equivalence test between outputs of the circuit an its specification that may be another circuit. The OBDD representation of circuit outputs is obtained by symbolic simulation of the circuit – a proccess that usually requires a dynamic reordering. The final variable order of the resulting OBDDs depends on many factors, e.g., used dynamic heuristic and structure of the circuit, and is unpredictable. The usual way of testing equivalence of two circuits is the application of XOR operation on OBDD representation of the corresponding outputs (which is more informative than a simple equivalence check). It is not known how to perform an operation on OBDDs with different variable orders (output) efficiently. This is the central problem in our paper. We investigate its complexity and try to estimate our chances to find an efficient solution at least for somehow restricted input. The obtained results have served as the first step towards development of heuristics for a feasible solution of the described problem [5].

2 Preliminaries

In order to make the paper self-contained, we present brief definitions of the notions used. For formal definition and the basic properties of OBDDs, we refer to [2, 3]. An *Ordered Binary Decision Diagram P* over a set of Boolean variables X_n is a single-rooted DAG with two sink-nodes labelled by 0 and 1. Each internal node is labelled by a variable from X_n and has two ordered successors called *low* and *high* son. On any root-to-sink path, any variable occurs at most once, and the occurrence of variables satisfies a fixed order. Let $X_n^k \subset X_n$ be the set of the k top-most variables with respect to the variable order in P. An assignment $\alpha^{(k)} : X_n^k \mapsto \{0,1\}^k$ naturally defines a computational path with initial point in the root: if the path contains a node v labelled by x_i, and $\alpha^{(k)}(x_i) = 0$ ($\alpha^{(k)}(x_i) = 1$), then the path contains the low (respectively, high) son of v. The size of OBDD P is measured by the number of its internal nodes and is denoted by $|P|$. An OBDD P represents a Boolean function $f(x_1, \ldots, x_n)$, $f : \{0,1\}^n \mapsto \{0,1\}$, if for each assignment $\alpha^{(n)}$, the corresponding path terminates in the sink labelled by $f(\alpha^{(n)}(x_1), \ldots, \alpha^{(n)}(x_n))$. An OBDD for f is denoted by OBDD(f), or by πOBDD(f), where π is an order of variables in the OBDD. Figure 1 depicts an OBDD for a function r_i defined in Section 3. Multiple functions are represented by OBDDs that share isomorphic subgraphs. Any node v of an OBDD(f) represents a subfunction of f, namely a cofactor of f with respect to an assignment α that corresponds to some root-to-v path. This cofactor is denoted by $f_{|\alpha}$. An OBDD is called *reduced*, if all its nodes represent different functions. Any OBDD can be reduced in linear time [9]. Throughout the paper, we will work with reduced OBDDs without explicitly pointing it out. A *dependence* of f

on a variable x means that the cofactors of f with respect to x differ. A *support* of f ($sup(f)$) is the set of all variables variables f depends on.

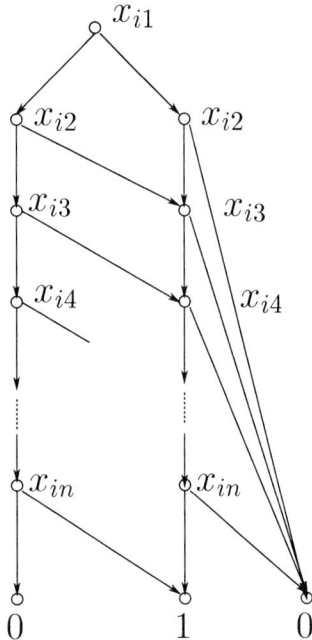

Fig. 1. $\sigma_1 \text{OBDD}(r_i)$ (the low son of a node appears left from its high son).

We represent variable orders by strings, e.g., $\pi = [x_0, \ldots, x_{k-1}]$ have meaning of $x_i <_\pi x_j$, for $i < j$. Let π be an order over X_n and $A \subset X_n$. Then $\pi - A$ denotes the order obtained from π by removing all variables in A. For two orders π_1 and π_2, $\pi_1 \cap \pi_2$ is the set of variables that appear in both strings. If $\pi_1 \cap \pi_2 = \emptyset$, $\pi_1 \cdot \pi_2$ is defined as an order that corresponds to concatenation of the strings. $\mathcal{NP}$ denotes the set of all decision problems that are solvable in nondeterministic polynomial time [6]. A problem is *output efficient*, if its time complexity is bounded polynomially with respect to the size of input and output. $\mathcal{P}_{\pi\text{OBDD}}$ denotes the set of Boolean functions with a polynomially bounded πOBDD representations (with respect to the number of variables). $\mathcal{P}_{\text{OBDD}} = \bigcup_\pi \mathcal{P}_{\pi\text{OBDD}}$.

2.1 Problems

We start by defining the problems considered in the paper, either as objects of our investigation or as references in the proofs. The problem of optimal variable order for OBDDs can be formulated as decision and/or construction problem.

OPTORDER

input   : $\pi OBDD(f)$, an integer s
question : Is there any variable order σ such that $|\sigma OBDD(f)| \leq s$?

OPTORDER*

input   : $\pi OBDD(f)$, an integer s
output   : a variable order σ such that $|\sigma OBDD(f)| \leq s$

The central problem of this paper is an optimal variable order for an OBDD representation of the function resulting from a binary Boolean operation. The standard way of implementing binary Boolean operations in the OBDD packages used for CAD (e.g., [1]) is via an *ite*-operator defined by:

$$ite(f, g, h) = fg + \bar{f}h$$

The advantage of a single operator is in the use of a common computed table for all Boolean functions, that avoids recomputations. For the sake of generality, we will formulate the considered problem for this operator.

OPTORDER_ITE

input   : $\pi_1 OBDD(f_1)$, $\pi_2 OBDD(f_2)$, $\pi_3 OBDD(f_3)$, an integer s
question : Is there any variable order π such that $|\pi OBDD(ite(f_1, f_2, f_3))| \leq s$?

It is easy to see that the problem OPTORDER* is polynomially reducible to that of finding an optimal variable order, and vice versa. Similarly, we can formulate the problem of finding an optimal variable order for a result of the *ite*-operation. We will show that the problem is hard even for the instances with optimal input OBDDs.

OPTORDER_ITE*

input   : $\pi_1 OBDD(f_1)$, $\pi_2 OBDD(f_2)$, $\pi_3 OBDD(f_3)$, such that
           π_1, π_2, π_3 are optimal for f_1, f_2, f_3, respectively
output   : an optimal variable order π for $ite(f_1, f_2, f_3)$

3 Results

In this section we investigate OPTORDER_ITE and OPTORDER_ITE*. Since the problems are of high practical interest, the lower bound estimation of their complexity (in our case it is $\mathcal{NP}$-hardness) is the first step in understanding the limits of our abilities to develop efficient algorithms for their solution. Being equipped by this knowledge, we stress our attention on finding at least a solution for restricted subsets of all possible inputs. The results could be considered as a starting point for development of heuristics.

Theorem 1. *[4, 11] OPTORDER is $\mathcal{NP}$-complete*

Theorem 2. *OPTORDER_ITE is $\mathcal{NP}$-hard*

Proof. The problem OPTORDER is reducible to OPTORDER_ITE by the fact that any Boolean function f can be expressed by means of $f = ite(f, 1, 0)$, where 1 and 0 are the constant functions.

The next question is, whether OPTORDER_ITE is in $\mathcal{NP}$. In the case of OPTORDER problem, we can guess an optimal order, construct an optimal OBDD and check whether it is smaller than s. Since there are known output efficient algorithms for rebuilding of an OBDD to an equivalent OBDD with respect to a given order of variables [8, 10] that can be interrupted whenever the output OBDD turns to be larger than the input, OPTORDER is in $\mathcal{NP}$. In the case of OPTORDER_ITE , we cannot assure that the function resulting from the *ite*-operation is in $\mathcal{P}_{OBDD}$. Even if it would be the case, it is not known, how to construct the corresponding OBDD efficiently (even if the target order would be known). An efficient procedure for checking the size of OBDDs for a function without its construction is not known, too. These facts make the problem OPTORDER_ITE hard. We try to find solutions for special cases (when some additional restrictions are put onto its instances) and investigate how far they cover all possible cases.

The first case is less interesting, but it is the basis to which some other cases can be easily reduced.

Lemma 1. *OPTORDER_ITE restricted to the instances with the same variable order is in $\mathcal{NP}$.*

Proof. We use two facts:

1. Let $P_i = \pi OBDD(f_i)$, for $i \in \{1, 2, 3\}$ are given. $\pi OBDD(ite(f_1, f_2, f_3))$ can be constructed within time and space in $\mathcal{O}(|P_1| \cdot |P_2| \cdot |P_3|)$, see e.g.,[1].
2. There is an output-efficient procedure that for any given $\pi OBDD(f)$ and σ constructs $\sigma OBDD(f)$, see e.g., [8].

The $\mathcal{NP}$-algorithm for the problem applies the *ite*-operation, guesses a target order and rebuilds the result with respect to this order.

Let us try to extend this result.

Definition [good order] Let π be a variable order over a set of variables X_n, $A = \{\pi_1 OBDD(f_1), \ldots, \pi_k OBDD(f_k)\}$ be a set of OBDDs over X_n. Let m be the size of the maximal OBDD in A. π is a *good order for A*, if there is a polynom p such that $\forall i \quad : \quad |\pi OBDD(f_i)| \leq |p(m)|$.

Theorem 3. *All instances of OPTORDER_ITE with the property that there is a good variable order for the input OBDDs, are solvable in nondeterministic polynomial time.*

Proof. In the considered case, all input OBDDs can be rebuilt with respect to a common variable order in nondeterministic polynomial time. Afterwards, Lemma 1 is applicable.

If A contains an OBDD of exponential size, i.e., m is exponential with respect to n, then any variable order is good for A. Such instances are consequently "easy".

Corollary 1 *Instance of OPTORDER_ITE with an exponentially large OBDD (wrt the number of variables), are solvable in nondeterministic polynomial time.*

Difficult Examples The last corollary covers all cases when one of the considered functions has exponentially large OBDD representations only. This fact induces the question, whether it is possible that all considered functions have polynomial OBDD representations but there is no good order for them. We explicitly show that this case may appear by proving a stronger result.

Lemma 2.

$$\exists f, g \, \exists \sigma_1 \sigma_2 : \quad (f \in \mathcal{P}_{\sigma_1} OBDD) \,\&\, (g \in \mathcal{P}_{\sigma_2} OBDD) \,\&\, (\forall \sigma : f \wedge g \notin \mathcal{P}_\sigma OBDD)$$

Proof. Let M be an $n \times n$ matrix over $\{0, 1\}$. We define

$$f(M) = \begin{cases} 1 & : \quad \text{each row of M has exactly one 1} \\ 0 & : \quad \text{otherwise} \end{cases}$$

$$g(M) = \begin{cases} 1 & : \quad \text{each column of M has exactly one 1} \\ 0 & : \quad \text{otherwise} \end{cases}$$

We define a function $h(M)$ that equal 1 iff M is a permutation matrix, i.e., $h(M) = (f \wedge g)(M)$. Let $M = (x_{i,j})_{i,j=1}^n$. Let σ_1 and σ_2 order variables according to rows, respectively to columns, i.e.,

$$\sigma_1 = [\underbrace{x_{11}, x_{12}, \ldots, x_{1n}}_{x_1^r}, \underbrace{x_{21}, x_{22}, \ldots, x_{2n}}_{x_2^r}, \quad \ldots \quad , \underbrace{x_{n1}, x_{n2}, \ldots, x_{nn}}_{x_n^r}]$$

$$\sigma_2 = [\underbrace{x_{11}, x_{21}, \ldots, x_{n1}}_{x_1^c}, \underbrace{x_{12}, x_{22}, \ldots, x_{n2}}_{x_2^c}, \quad \ldots \quad , \underbrace{x_{1n}, x_{2n}, \ldots, x_{nn}}_{x_n^c}]$$

f can be decomposed to functions $r_1, \ldots, r_n$ defined by

$$r_i(x_i^r) = \begin{cases} 1 & : \quad \text{there is exactly one 1 in } x_i^r \\ 0 & : \quad \text{otherwise} \end{cases}$$

For any i, the reduced $\sigma_1 OBDD(r_i)$ consists of $2n - 1$ nodes (Fig. 1). Since $f = \bigwedge_{i=1}^n r_i$ and $\bigcap_{i=1}^n sup(r_i) = \emptyset$, $\sigma_1 OBDD(f)$ can be obtained by identifying the 1-sink of r_i with the root of r_{i+1}, for $1 \leq i \leq n-1$. Therefore

$$|\sigma_1 OBDD(f)| = \sum_{i=1}^n |\sigma_1 OBDD(r_i)| = n(2n-1)$$

By a similar construction, $|\sigma_2 OBDD(g)| = n(2n-1)$. According to the result by Krause, Meinel, Waack [7], $h \notin \mathcal{P}_{OBDD}$.

Theorem 4. *There are functions f_1, f_2, f_3, and variable orders π_1, π_2, and π_3 s.t. no variable order is good for $\{\pi_1 OBDD(f_1), \pi_2 OBDD(f_2), \pi_3 OBDD(f_3)\}$.*

Proof. Let $f, g, h, \sigma_1, \sigma_2$ have the same meaning as in the previous lemma. Let us define $f_1 := f$, $\quad f_2 := g$, $\quad f_3 = 0$, $\quad \pi_1 := \sigma_1$, $\quad \pi_2 := \sigma_2$, π_3 as any order. Let us assume σ be a good order for $\{\pi_1 OBDD(f_1), \pi_2 OBDD(f_2), \pi_3 OBDD(f_3)\}$. Then $\sigma OBDD(h) = \sigma OBDD(ite(f_1, f_2, f_3))$ is polynomially bounded. This contradicts Lemma 2.

The result of the *ite*-operation from Theorem 4, was a function that has an exponential OBDD representation with respect to any variable order. From the practical point of view, the construction of the function is uninteresting in this case, even if the optimal variable order would have been known. Is there any example where f_1, f_2, f_3 and $ite(f_1, f_2, f_3)$ have polynomial OBDD representations but there is no good order for those OBDDs? If such functions exist, then it makes sense to look for an algorithm that constructs the *ite*-result directly from the given OBDDs, without rebuilding them to a common order (that is suitable for the result but not for all argument functions), and avoids an exponential size explosion of the intermediate results.

Theorem 5. *There are functions F and G, and variable orders τ_1, τ_2, and τ_3 that fulfil all following properties simultaneously:*

1. $F \in \mathcal{P}_{\tau_1} OBDD$
2. $G \in \mathcal{P}_{\tau_2} OBDD$
3. $(F \wedge G) \in \mathcal{P}_{\tau_3} OBDD$
4. $\forall \tau : \quad F \in \mathcal{P}_\tau OBDD \Rightarrow G \notin \mathcal{P}_\tau OBDD$

Proof. Let f, g, σ_1 and σ_2 be functions, respectively, variable orders defined in the proof of Lemma 2. Let x be a variable that does not belong to $sup(f) \cup sup(g)$. For $F := x \wedge f$, $G := x \wedge g$, the conjunction $F \wedge G$ equals constant zero function. We define the variable orders for OBDD representations of F and G by appending the missing variable x to the top of σ_1, respectively, σ_2: $\tau_1 := x\sigma_1$, $\tau_2 := x\sigma_2$. τ_3 is any variable order. According to the size estimations of $\sigma_1 OBDD(f)$ and $\sigma_2 OBDD(g)$ and the definitions of F, G, τ_1, τ_2 and τ_3, the first three conditions are fulfilled.
Assume that there is τ such that $(F \in \mathcal{P}_\tau OBDD) \& (G \in \mathcal{P}_\tau OBDD)$. f and g are cofactors of F, respectively G. This implies that $f \in \mathcal{P}_{\tau - \{x\}} OBDD$ and $g \in \mathcal{P}_{\tau - \{x\}} OBDD$. Consequently, $(f \wedge g) \in \mathcal{P}_{\tau - \{x\}} OBDD$, too. This contradicts the exponential lower bound for $f \wedge g$ in [7].

Corollary 2 *There are functions f_1, f_2, f_3, $f = ite(f_1, f_2, f_3)$ with polynomial OBDD representations with respect to variable orders π_1, π_2, π_3 and π, respectively, such that there is no good variable order for the set of OBDDs $\{\pi_1 OBDD(f_1), \pi_2 OBDD(f_2), \pi_3 OBDD(f_3), \pi OBDD(ite(f_1, f_2, f_3))\}$.*

Close Variable Orders We have seen that even if all argument functions and also the result of the *ite*-operation have polynomial OBDD representations there need not be a common variable order that is suitable for all functions involved. Our hope is, that such variable order exists if the suitable variable orders for the argument functions are close to each other. We define a metric on the space of variable orders over the same variable set as a minimal number of variables that must be removed for obtaining the same order. It will be shown that if orders are close to each other wrt this metric, then a new order can be found (deterministically) that allows an efficience performance of the *ite*-operation.

Definition [witness of difference] Let π_1, π_2 be variable orders over X_n. Let W be a minimal subset of X_n (minimal with respect to the number of elements) such that $\pi_1 - W = \pi_2 - W$. W is called a *witness of difference* and its size $\rho(\pi_1, \pi_2) = |W|$ a *difference*.

The notions of witness and difference defined above can be easily extended to any finite set of variable orders.

Lemma 3. *Let π_1, π_2, and π_3 be variable orders over X_n. Let $P_1 = \pi_1 OBDD(f_1)$, $P_2 = \pi_2 OBDD(f_2)$, $P_3 = \pi_3 OBDD(f_3)$, and $\rho(\pi_1, \pi_2, \pi_3) = d$. Then there is a variable order π such that the reduced $\pi OBDD(ite(f_1, f_2, f_3))$ can be constructed from P_1, P_2 and P_3 in time and space $\mathcal{O}(2^d |P_1||P_2||P_3|)$*

Proof. Let $W = \{x_{i_1}, \ldots, x_{i_k}\}$ be a witness of the difference for π_1, π_2 and π_3. We will construct an $OBDD(ite(f_1, f_2, f_3))$ P with respect to variable order $\pi = [x_{i_1}, \ldots, x_{i_k}] \cdot [\pi_1 - W]$. The top part of the OBDD is a complete tree of depth d. The nodes on the j-th level are labelled by the variable x_{i_j}. A leaf of this tree reachable via the path associated with an assignment α represents the function $ite((f_1)_{|\alpha}, (f_2)_{|\alpha}, (f_3)_{|\alpha})$. Since α is an assignment to all variables in W, the cofactors of f_1, f_2, and f_3 with respect to α have the same support $\pi_1 - W$ and their OBDD representations $(P_1)_{|\alpha}, (P_2)_{|\alpha}$, and $(P_3)_{|\alpha}$ are built with respect to the same variable order $\pi_1 - W$. The time and space complexity of these particular *ite*-operations is bounded by $|(P_1)_{|\alpha}| \cdot |(P_2)_{|\alpha}| \cdot |(P_3)_{|\alpha}| \leq |P_1||P_2||P_3|$.

Corollary 3 *Let $\pi_1 OBDD(f_1), \pi_2 OBDD(f_2), \pi_3 OBDD(f_3)$, are OBDDs of an OPTORDER_ITE instance of size N. Let $\rho(\pi_1, \pi_2, \pi_3) \in \mathcal{O}(\log N)$. Then OPTORDER_ITE for this instance is solvable in nondeterministic polynomial time.*

Lemma 4. *For any constant number of variable orders over X_n, a witness of their difference can be found in time and space $\mathcal{O}(n^2)$.*

Proof. The problem of finding a witness of difference for two variable orders is a special case of finding a longest common substring (not necessarily a continuous one) of two strings. For this problem, we design an efficient algorithm that is easily extendable for an application to *several* orders and to finding *all* longest suborders. The basic idea is to assign a graph to given orders such that there is one-to-one correspondence between longest paths in the graph and longest common suborders.

Let π_1 and π_2 be two variable orders over X_n. We can associate a graph $G = (V, E)$ with them, s.t. $V = X_n$, and $(x_i, x_j) \in E$ iff $(x_i <_{\pi_1} x_j) \& (x_i <_{\pi_2} x_j)$. $|V| = n$ and $|E| \leq n^2$. It holds:

1. G can be constructed in time and space $\mathcal{O}(n^2)$.
2. For any edge $(x_{\pi_1(i)}, x_{\pi_1(j)}) \in E$, it holds: $i < j$. Consequently, G is acyclic.

$LongestPath(\pi_1, \pi_2)$
```
/* Let π₁ = [v₁,...,vₙ].*/
length(vₙ) = 1;
next(vₙ) =NIL;
for (i = n − 1, to 1, step -1) {
    Find a successor u of vᵢ with maximal value of length
    length(vᵢ) = length(u) + 1;
    next(vᵢ) = u;
}
Find node v with maximal value of length;
while (v ≠NIL) {
    print(v);
    v = next(v);
}
```

Lemma 5 (Correctness). *Let G be a graph that corresponds to variable orders π_1 and π_2 as described above. $LongestPath(\pi_1, \pi_2)$ computes a largest suborder consistent with π_1 and π_2.*

Proof. It is sufficient to prove, that the procedure correctly computes the longest path in G. It follows from the following facts:

1. Before the labels of v_i are computed, the labels of $\{v_{i+1}, v_{i+2}, \ldots, v_n\}$ are known.
2. If the labels of v are computed, then $length(v)$ equals the length of the longest path that starts in v and goes via $next(v)$.

The second statement can be proved by induction on $i = n, \ldots, 1$, for $v = v_i$.

Lemma 6 (Complexity). *The time complexity (under unit cost) of the algorithm LongestPath is $\mathcal{O}(|E| + n)$, where E are the edges of the graph G that correspond to the considered variable orders over X_n.*

Theorem 6. *Let $N := |\pi_1 OBDD(f_1)| + |\pi_2 OBDD(f_2)| + |\pi_3 OBDD(f_3)|$ and $\rho(\pi_i, \pi_j) \in \mathcal{O}(\log N)$, for $i, j \in \{1, 2, 3\}$. Time complexity for a deterministic construction of an OBDD representation for $ite(f_1, f_2, f_3)$ is bounded by a polynomial of N.*

Theorem 7. *OPTORDER_ITE* is $\mathcal{NP}$-hard.*

Proof. OPTORDER* is many-one reducible to OPTORDER_ITE* by means of the procedure *OptOrder*. Let $\pi OBDD(f)$ P_f be an input. If *OptOrderIte* solves OPTORDER_ITE* , then $OptOrder(Root(P_f))$ returns an optimal variable order for f. The use of computed table assures that there are $2|P_f|$ recursive calls of *OptOrder* invoked by the top-most call. Since rebuilding of an OBDD wrt an optimal order can be performed in polynomial time, OPTORDER is polynomially reducible to OPTORDER* . This proves that OPTORDER* and consequently OPTORDER_ITE* are $\mathcal{NP}$-hard.

OptOrder(v)
  if *IsConstant(v)* return [];
  if *IsComputed(f)* return result;
  /* any node is represented by a variable, low son and high son */
  $\sigma_0 = OptOrder(v.low)$;
  $\sigma_1 = OptOrder(v.high)$;
  $\sigma = OptOrderIte(v.var), Rebuild(v.high, \sigma_1), Rebuild(v.low, \sigma_0))$;
  StoreInComputed(v, σ)
  return σ

Theorem 8. *All instances of OPTORDER_ITE* with a good variable order for the set of input OBDDs are solvable nondeterministically in polynomial time. In particular, all instances for which the difference of the variable orders is bounded by $\mathcal{O}(\log N)$, where N is the size of the instance, are solvable nondeterministicaly in polynomial time.*

References

[1] K.S. Brace, R.L. Rudel, and R.E. Bryant. Efficient Implementation of a BDD Package. *ACM/IEEE Proc. Design Automation Conference*, pages 40–45, 1990.

[2] R.E. Bryant. Graph Based Algorithms for Boolean Function Manipulation. *IEEE Transaction on Computers*, C-35:677–691, 1986.

[3] R.E. Bryant. Symbolic Boolean Manipulation With Ordered Binary Decision Diagrams. *Comp. Surveys*, 24:293–318, 1992.

[4] B. Bollig and I. Wegener. Improving the Variable Ordering of OBDDs is NP–complete. *IEEE Transaction on Computers*, 45(9):993–1002, 1996.

[5] G. Cabodi and S. Quer and Ch. Meinel and H. Sack and A. Slobodová and Ch. Stangier. Binary Decision Diagrams and Multiple Variable Order Problem *International Workshop on Logic Synthesis'98, Lake Tahoe, CA*

[6] M.R. Garey and D.S. Johnson. *Computers And Intractability - A guide to NP-Completness*. Freeman, 1979.

[7] M. Krause, Ch. Meinel, and S. Waack. Separating the Eraser Turing Machine Classes L_e, NL_e and P_e. *TCS*, 86:267–275, 1991.

[8] Ch. Meinel and A. Slobodová. On the Complexity of Constructing Optimal Ordered Binary Decision Diagrams. *Proc. of MFCS*, LNCS 841:515–525, 1994.

[9] D. Sieling and I. Wegener. Reduction of bdds in linear time. *Information Processing Letters*, 48(3):139–144, 1993.

[10] P. Savický and I. Wegener. Efficient algorithms for the transformation between different types of binary decision diagrams. *Acta Informatica*, 34:245–256, 1997.
[11] S. Tani, K. Hamaguchi, and S. Yajima. The Complexity of the Optimal Variable Ordering Problem of Shared Binary Decision Diagrams. In *Proc. ISAAC*, LNCS 762, pages 389–398. Springer, 1993.

Equations in Transfinite Strings

Christian Choffrut[1] and Sandor Horvath[2]

[1] LIAFA, Université Paris 7
Tour 55–56, 2 Place Jussieu, 75251 Paris Cedex 05
cc@liafa.jussieu.fr
[2] Dept. of Computer Science, Eötvös Loránd University,
Budapest, Múzeum körut 6–8., II. em. 2–3, H–1088, Hungary
horvath@cs.elte.hu

Abstract. We address the question of extending the theory of equations in finite strings to transfinite strings. We give a full description of the solutions of the equations in two unknowns as well as those of the equations of the form $x^m y^p = z^q$ for $m, p, q \geq 2$. By so doing we introduce some new notions that we believe are interesting for their own sake.

1 Introduction

Transfinite strings have long been introduced. Logicians were the first to study them by extending to transfinite strings Büchi's result showing that with every sentence of the second order logic of one successor one can associate a rational subset of infinite strings, [2] and [4]. In the area of theoretical computer science such objects can be viewed as modelling the behaviour of sequential processes consisting of infinitely many successive actions whose times define a convergent sequence, also known in the literature as Zeno strings. The vivid area of timed automata which takes the actual duration of transitions of finite automata into account makes explicit use of this notion, [1].

Combinatorics in strings is an old and wide area whose origin is usally traced back to Axel Thue, but that matured in the sixties with the works of Marcel-Paul Schützenberger and his followers. Numerous publications have already given account of the state of art (e. g. [9] or the chapter [3]) and they testify for the richness and profoundness of the topic. One of the main aspects of this theory is concerned with solving equations with strings as unknowns which gave rise to difficult issues. Let us mention for example the effective possibility of determining whether or not an equation with constants has a solution established by Makanin's or the possibility of expressing all solutions of an equation with at most three variables with "parameters" due to Hmelevskii, see [10] and [6]. The picture for transfinite strings is quite different since apart from isolated results (see, e. g., [7]) little is developped yet. We believe though that some elementary combinatorial properties on transfinite strings might help the study of subsets of transfinite strings as they have helped the study of subsets of finite strings.

Among the variety of issues on transfinite strings, we focus as said on that of solving equations, i. e., on describing the set of their solutions. Our main contribution is a thorough description of the solutions of equations in two unknowns as well as those of the form $x^m y^p = z^q$ for $m, p, q \geq 2$. Observe that not much more is known for finite strings. As a by-product, we discover new notions that are slight modifications of standard ones. Indeed, instead to the property for two strings x and y to be powers of a third one or equivalently to commute, here we need a more general property that says that xy is a prefix of yx or vice versa. We also tackle the standard notions of conjugacy, of primitivity and of periodicity. In particular the famous result attributed to Fine and Wilf concerning the periods of a (finite) string can be extended to transfinite strings with little damage.

For lack of space the proofs of the results are omitted.

2 Preliminaries

We refer the reader to [11] for a comprehensive exposition of the theory on ordinals. In the present work, Ord denotes the class of all ordinals α satisfying $\alpha < \omega^\omega$. Every ordinal α admits a unique polynomial representation $\omega^n a_n + \omega^{n-1} a_{n-1} + \ldots + \omega a_1 + a_0$ where $n, a_n, \ldots, a_1, a_0 < \omega$. The integer n is the *degree* of the ordinal also denoted ∂_α.

The sum of two ordinals satisfies for all $a, b < \omega$

$$\omega^n a + \omega^p b = \begin{cases} \omega^p b & \text{if } p > n \\ \omega^n (a+b) & \text{if } p = n \end{cases}$$

Addition is not commutative but characterizing the pairs of ordinals that commute is easy.

Proposition 1. *(e.g., [11], Thm. 1., p. 346) For two ordinals α, β the following conditions are equivalent*
 i) $\alpha + \beta = \beta + \alpha$
 ii) there exist an ordinal γ and two integers $q, p \geq 0$ such that $\alpha = \gamma q$ and $\beta = \gamma p$.
 iii) either $\alpha\beta = 0$ or else for some $0 \leq p, q, n < \omega$ and $\mu < \omega^n$ we have $\alpha = \omega^n p + \mu$ and $\beta = \omega^n q + \mu$

Given a finite alphabet Σ, a *string* is a mapping u of $\alpha < \omega^\omega$ into Σ. Equivalently, u is a sequence of Σ indexed by the ordinal α and we write indifferently u_i or $u(i)$ for all $i < \alpha$. The set of all strings is denoted by Σ^{Ord}. The ordinal α is the *length* of u, denoted by $|u|$. By extension, the *degree* of a string x is the degree of its length and it is denoted by ∂_x. For $a \in \Sigma$, $|u|_a$ denotes the *length in the letter* a of the string u, i. e., the ordinal of the subsequence consisting of all the positions $i < \alpha$ for which $u_i = a$.

Given a string u, we denote by u^ω the string of length $|u|\omega$ defined by

$$u^\omega_{\ell k + i} = u_i$$

where $\ell = |u|$, $k < \omega$ and $i \in \ell$.

The *concatenation* of two strings u and v of length α and β respectively is the mapping

$$(uv)_i = \begin{cases} u_i & \text{if } i < \alpha \\ v_j & \text{if } i = \alpha + j, j < \beta \end{cases}$$

Observe that $|uv| = |u| + |v|$ for all $u, v \in \Sigma^{Ord}$, but the condition $uv = w$, $|u| > 0$ does not imply $|v| < |w|$ (consider $u = a, v = w = a^\omega$), which is a main departure from usual finite strings.

The notions of *prefix*, *suffix* extends naturally from finite to infinite strings. We may also say that two transfinite strings $x, y \in \Sigma^{Ord}$ are *comparable* if x is a prefix of y or y is a prefix of x. Levi's Lemma trivially holds. We recall it here.

Proposition 2. *Let* $x, y, z, t \in \Sigma^{Ord}$ *satisfy the equality* $xy = zt$. *Then there exists* $u \in \Sigma^{Ord}$ *such that either* $x = zu$ *and* $t = uy$ *or* $z = xu$ *and* $y = ut$ *holds*.

3 Equations

First we extend the notions of equations on finite strings to infinite strings.

Given a finite subset $\Xi = \{x, y, \ldots\}$ of *unknowns*, an *equation* is a pair of strings of the free monoid Ξ^* written

$$L(x, y, \ldots) = R(x, y, \ldots) \tag{1}$$

An equation is *trivial* if the strings $L(x, y, \ldots)$ and $R(x, y, \ldots)$ are the same (as elements of Ξ^*). A *solution* of equation (1) is an assignment of a string of Σ^{Ord} to each unknown $x \in \Xi$ such that the two handsides of the equations denote the same transfinite string on the alphabet Σ. It is *homogeneous* if the lengths of the variables have the same degree.

Actually, we will conform to the tradition that indulges in using the same symbol x for the unknown ($x \in \Xi$) and its assignment in Σ^{Ord}. E. g., the equation $xy = y$ admits the solution $x = a$ and $y = a^\omega a$. We say that the pair of strings $x = u, y = v$, with $u, v \in \Sigma^{Ord}$ *is a solution in length of the equation* $L(x, y) = R(x, y)$ if the equation (in Σ^{Ord}) $|L(u, v)| = |R(u, v)|$ holds. E. g., $x = a$ and $y = b^\omega$, $a \neq b \in \Sigma$ is a solution in length of the equation $xy = y$ but is not a solution of $xy = y$.

We concentrate on the equations in two unknowns. We start with two technical Lemmas whose proofs are left to the reader.

Lemma 1. *Let* x, y *be a solution of a non trivial equation in two unknowns* $L(x, y) = R(x, y)$. *If* $\partial_x < \partial_y$, *then the equation is of the form*

$$L_1(x, y) y x^p = R_1(x, y) y x^p \text{ where } |L_1(x, y)|_y = |R_1(x, y)|_y \text{ with } p \geq 0 \tag{2}$$

and there exists $z \in \Sigma^{Ord}$ *such that* $y = x^\omega z$. *Conversely every pair* x, y *with* $y = x^\omega z$ *is a solution of each equation satisfying condition (2)*.

Lemma 2. Let x, y be a solution of a non trivial equation in two unknowns $L(x,y) = R(x,y)$. Then there exist $z, t \in \Sigma^{Ord}$ and four integers $n, m, p, q \geq 0$ such that

$$x = (z^\omega t)^n z^p \text{ and } y = (z^\omega t)^m z^q \tag{3}$$

The next results extend the classical property on equations with two unknowns on finite strings to infinite strings. They characterize the set of solutions of an arbitrary equation relative to the suffixes of the two hand sides. The proofs are quite straightforward.

Proposition 3. Let $L(x,y) = R(x,y)$ be a non trivial equation where $L(x,y)$ and $R(x,y)$ end with the same letter x or y. Then x, y is a solution if and only if the following two conditions hold
   i) x, y is a solution in length of $L(x,y) = R(x,y)$
   ii) there exist $z, t \in \Sigma^{Ord}$ and four integers $n, m, p, q \geq 0$ such that $x = (z^\omega t)^n z^p$ and $y = (z^\omega t)^m z^q$.

Proposition 4. Let $L(x,y) = R(x,y)$ be a non trivial equation where $L(x,y)$ and $R(x,y)$ end with different letters, say x and y respectively. Then x, y is a solution if and only if the following two conditions hold
   i) x, y is a solution in length of $L(x,y) = R(x,y)$
   ii) there exist $z \in \Sigma^{Ord}$, and two integers $n, m \geq 0$ such that $x = z^n$ and $y = z^m$.

4 Periodicity

4.1 Fine and Wilf Revisited

As a result of Proposition 4, two strings commute if and only if they are powers of a third string. To draw further the analogy with the finite strings, we observe that under the hypothesis $|xy| = |yx|$, a direct analog of Fine and Wilf theorem holds. Observe that because of Proposition 1, $|x|$ and $|y|$ have a maximal common left divisor $|x| \wedge |y|$ such that $|x| = (|x| \wedge |y|)p$ and $|y| = (|x| \wedge |y|)q$ holds for some integers $0 < p, q < \omega$ and $p \wedge q = 1$.

Proposition 5. Let $x, y \in \Sigma^{Ord}$ satisfy $|xy| = |yx|$. Then the following conditions are equivalent
   i) $xy = yx$
   ii) there exist $r, s < \omega$ and z such that $x = z^r$ and $y = z^s$
   iii) there exist $r, s < \omega$ such that $x^r = y^s$
   iv) x^ω and y^ω have a common prefix of length $|x| + |y| - (|x| \wedge |y|)$.

Observe that the condition is sharp for any degree. E. g., with $\Sigma = \{a,b,c\}$, $x = a^\omega aba^\omega ac$ and $y = a^\omega aba^\omega aca^\omega ab$ the strings x^ω and y^ω have a common prefix of length equal to the predecesor of $|x| + |y| - (|x| \wedge |y|)$.

A similar result holds when considering suffixes instead of prefixes. However, it needs a special treatment as the "mirror image" of a transfinite string is not a transfinite string.

Proposition 6. *Let $x, y \in \Sigma^{Ord}$. If for some $k, l < \omega$, x^k and y^l have a common suffix of length $|xy|$ then $xy = yx$.*

Observe that the conditions are sharp (consider $x = ba^\omega b^2 a^\omega b^2$ and $y = a^\omega b^3 a^\omega b^2 a^\omega b^2$).

4.2 Roots – Primitivity

The previous section yields, as in the case of finite strings, a few interesting consequences. In particular the notions of roots and primitive strings make sense in this more general setting. Indeed, given $1 \neq x \in \Sigma^{Ord}$ there exists a unique (length-) minimal element $z \in \Sigma^{Ord}$ and a unique integer n such that $x = z^n$. Then z is the *root* of x and n its *exponent*. By convention, the root of the empty string is the empty string itself and its exponent is 0. A string x is *primitive* if it can not be written as $x = z^n$ where $n > 1$. E. g., $a^\omega a^2$ is primitive.

5 Conjugacy

We recall that two elements x, y of a monoid are *conjugate* if there exist two elements z, t such that $x = zt$ and $y = tz$. The proof of the next property connects the notions of primitivity and conjugacy. Its proof follows the same line as in the case of finite strings and it is left to the reader.

Proposition 7. *Two conjugate strings have conjugate roots and equal exponents.*

As in the case of free monoids there is an alternative definition of the notion of conjugacy. The following result is a direct extension of [8].

Proposition 8. *The strings $x, y, z \in \Sigma^{Ord}$, $y \neq 1$ satisfy the condition $xz = zy$ if and only if there exist $u, v \in \Sigma^{Ord}$ and $n < \omega$ such that $x = uv$, $y = vu$ and $z = (uv)^n u$*

Observe that the equality $xx^\omega = x^\omega 1$ holds. Therefore the condition $y \neq 1$ is necessary.

5.1 Strong Conjugacy

The following is a refinement of the notion of conjugacy.

Definition 1. *Two strings $x, y \in \Sigma^{Ord}$ are strongly conjugate if and only if there exist $z, t \in \Sigma^{Ord}$ and two integers m, n such that*

$$x = z^\omega t z^n \text{ and } y = z^\omega t z^m \tag{4}$$

Strong conjugacy has nice properties which we proceed to state.

Proposition 9. *Let $x, y \in \Sigma^{Ord}$. The following conditions are equivalent*
i) x, y are strongly conjugate
ii) $xy = yy$ (resp. $yx = xx$)
iii) x, y are conjugate and x is a prefix of y or y is a prefix of x
iv) there exist z, t such that $x = zt, y = tz$ and either $zt = t$ or $tz = z$
Furthermore, the relation "being strongly conjugate" is an equivalence relation

As a consequence we have

Corollary 1. *If two strings x, y are strongly conjugate then so are their roots.*

5.2 Quasi-Cyclicity

From the section 3, it should be clear that the following concept is useful.

Definition 2. *Two strings $x, y \in \Sigma^{Ord}$ are quasi-cyclic if there exist two strings u, v such that $x, y \in (u^\omega v)^* u^*$*

Proposition 10. *In the previous definition, u and $u^\omega v$ may be assumed primitive.*

The following are the main results on quasi-cyclicity that are used in the proof of Theorem 1.

Proposition 11. *Two strings $x, y \in \Sigma^{Ord}$ are quasi-cyclic if and only if so are their roots.*
Furthermore, the relation "x and y are quasi-cyclic" reduced to the strings of equal degree is an equivalence relation.

Observe that reducing the relation to the strings having the same degree is necessary. Indeed, with $x = a^\omega a$, $y = a$, $z = (a^\omega b)^\omega a = a^\omega (ba^\omega)^\omega a$ we have that x, y are quasi-cyclic and so are y, z but x, z are not.

We may reformulate the notion of quasi-cyclicity by refining it.

Proposition 12. *Given two strings $x, y \in \Sigma^{Ord}$, the following conditions are equivalent.*

i) x, y are quasi-cyclic

ii) there exist $u, v \in \Sigma^{Ord}$ such that u and $u^\omega v$ are primitive, and $x, y \in (u^\omega v)^* u^*$.

iii) xy and yx are comparable.

As an application we may state a weaker version than that of Proposition 5 in the case of $|xy| \neq |yx|$.

Proposition 13. *Let $x, y \in \Sigma^{Ord}$. Then the following conditions are equivalent*

i) $x^\omega = y^\omega$

ii) x and y are quasi-cyclic and $\partial_x = \partial_y$

iii) x^ω and y^ω have a common prefix of length $\min\{|xy|, |yx|\}$

6 The Equation $x^m y^p = z^q$

We extend the notion of quasi-cyclicity to arbitrary subsets X of strings by imposing that $X \subseteq (u^\omega v)^* u^*$ holds for some strings $u, v \in \Sigma^{Ord}$. We first state a technical result (for finite strings this would mean "if two of the three unknowns are powers of a third string then so are the three unknowns"). The proof makes use of Propositions 13, 6 and assertion 12, ii).

Proposition 14. *Let x, y, z be a solution of the equation $x^m y^p = z^q$ where $2 \leq m, p, q \leq \omega$. If two of the three unknowns are quasi-cyclic, then so are the three unknowns.*

The following is the main result of our paper.

Theorem 1. *All solutions of the equation $x^m y^p = z^q$ with $m, p, q \geq 2$ are quasi-cyclic.*

For lack of space, we can not reproduce the proof. It first consists in showing that by applying Propositions 5 and 6, the cases $q \geq 4$ and $m, p \geq 3, q = 3$ may be ruled out. The rest of the proof, which is more technical, proceeds by examining the remaining cases and by repeatedly applying the results of the previous sections.

In order to give the reader a flavour of the techniques used, we reproduce the verification of the case $p = m = 2, q = 3$. We are able to prove that there exist some $u, v \in \Sigma^{Ord}$, and $r \geq 0$ (the case $r = 0$ can be handled directly) such that

$$x = (uv)^{r+1}u, \quad y^2 = vu(uv)^{r+2}uuv \text{ and } z = (uv)^{r+1}uuv \qquad (5)$$

Subcase 1) $r + 1 = 2r'$ for some $r' > 0$. Observe first that $\partial_u \geq \partial_v$ holds else we have $vu(uv)^{r'} = u(vu)^{r'}uv$ meaning that u and v are quasi-cyclic. By standard considerations on the length, equation (5) implies

$$y = vu(uv)^{r'}u_1 = u_2(vu)^{r'}uv \tag{6}$$

with $u_1u_2 = u$. If $\partial_u > \partial_v$ then by considering the lengths of both handsides we have $|u_1| = |u_2v|$. Furthermore, the two handsides have the same prefix of length $|u_1| = |vu_1| = |u_2v|$ thus $vu_1 = u_2v$ implying $u_1 = vu_1$.

We are left with $\partial_u = \partial_v$. By considering the lengths of both handsides we have $|vu_1| = |u_2v|$ and then by considering both prefixes we get $vu_1 = u_2v$. By cancelling out this common prefix, we observe that u_1u_2 and u_2u_1 are comparable, i. e., for some k, i, j, we get $u_1 = (a^\omega b)^r a^i$ and $u_2 = (a^\omega b)^r a^j$ by Property 12. Then equality $vu_1 = u_2v$ yields $v = (a^\omega b)^s a^i$ for some $s \geq 0$ and thus $uv = (a^\omega b)^{r+s} a^i = (a^\omega b a^i)^{r+s}$. Finally, $y^2, z \in (a^\omega b a^i)^*$ which closes this subcase.

Subcase 2) $r + 1 = 2r' + 1$ for some $r' \geq 0$. Observe first that $\partial_u \leq \partial_v$ holds else we have $vu(uv)^{r'+1} = (uv)^{r'}uuv$ meaning that u and v are quasi-cyclic.
Using considerations on the lengths we have

$$y = vuu(vu)^{r'}v_1 = v_2u(vu)^{r'}uv \tag{7}$$

By considering the lengths of both handsides, we have $|v_1| = |v_2|$. Since v_1 and v_2 are prefixes of y we get $v_1 = v_2$. Thus equation (7) is an equation in the two unknowns, and the verification follows from Proposition 3.

References

1. B. Bérard and C. Picaronny. Accepting Zeno words without stopping time. In *Proceedings of the 22th International Symposium MFCS'97*, number 1295 in Lecture Notes in Computer Science, pages 148–158. Springer, 1997.
2. J. Büchi. The monadic theory of ω_1. In *Decidable Theories II*, number 328 in Lecture Notes in Mathematics. Springer, 1973.
3. C. Choffrut and J. Karhumäki. Combinatorics of words. In G. Rozenberg and A. Salomaa, editors, *Handbook of Formal Languages*, volume 1, pages 329–438. World Scientific, 1997.
4. S. C. Choueka. Finite automata, definable sets and regular expressions over ω^n-tapes. *J. Comb. Sys. Sci.*, 17:81–97, 1978.
5. N. J. Fine and H. S. Wilf. Uniqueness theorems for periodic functions. *Proc. Am. Math. Soc.*, 3(2):109–114, 1965.
6. Y. I. Hmelevskii. *Equations in free semigroups*, volume 107 of *Am. Math. Soc. Transl.* Proc. Steklov and Insti. Mat, 1976.
7. S. Horvath. Continuum-long squarefree words on three letters. 1996.
8. A. Lentin and M. P. Schützenberger. A combinatorial problem in the theory of free monoids. In R. C. Bose and T. E. Bowlings, editors, *Combinatorial Mathematics*, pages 112–144. North Carolina Press, Chapel Hill, N. C., 1967.
9. Lothaire. *Combinatorics on Words*, volume 17 of *Encyclopedia of Mathematics and its Applications*. Addison-Wesley, 1983.

10. G. S. Makanin. On the rank of equations in four unknowns in a free semigroup. *Proc. Am. Math. Soc.*, 100:285–311, 1976. (English trans. in *Math. USSR Sb., 29* 257-280).
11. W. Sierpinski. *Cardinal and Ordinal Numbers*. Warsaw: PWN, 1958.

Minimal Forbidden Words and Factor Automata

M. Crochemore[*,1], F. Mignosi[2], and A. Restivo[2]

[1] Institut Gaspard-Monge mac@univ-mlv.fr
[2] Università di Palermo [mignosi,restivo]@altair.math.unipa.it

Abstract. Let $L(M)$ be the (factorial) language avoiding a given antifactorial language M. We design an automaton accepting $L(M)$ and built from the language M. The construction is effective if M is finite.
If M is the set of minimal forbidden words of a single word v, the automaton turns out to be the factor automaton of v (the minimal automaton accepting the set of factors of v).
We also give an algorithm that builds the trie of M from the factor automaton of a single word. It yields a non-trivial upper bound on the number of minimal forbidden words of a word.

Keywords: factorial language, anti-factorial language, factor code, factor automaton, forbidden word, avoiding a word, failure function.

1 Introduction

Let $L \subseteq A^*$ be a *factorial* language, *i.e.*, a language containing all factors of its words. A word $w \in A^*$ is called a *minimal forbidden word* for L if $w \notin L$ and all proper factors of w belong to L. We denote by $MF(L)$ the language of minimal forbidden words for L.

The study of combinatorial properties of $MF(L)$ helps investigate the structure of the language L or of the system it describes. For instance, locally testable factorial languages (cf [8]) are characterized by the fact that the corresponding languages of minimal forbidden words are finite. In the context of Symbolic Dynamics they correspond to systems of finite type.

Another example is given by a language L that is the set of factors of an infinite word: in this case, as shown in [2], the elements of $MF(L)$ are closely related to the *bispecial* factors (cf. [6], [7] and [3]) of the infinite word.

A measure of complexity of the language L is introduced in [2] based on the function F_L, that counts, for any n, the number of words of length n in $MF(L)$. Authors prove that the growth of $F_L(n)$ as well as the topological entropy of $MF(L)$ are topological invariants of the dynamical system defined by L. This result provides a usefull tool to show that some systems are not isomorphic, which comes in addition to other notion like the ordinary notion of entropy and the zeta function, for example.

Finally, [5] considers properties of languages defined by finite forbidden sets of words. Authors define the Möbius function for these languages.

---
[*] Work by this author is supported in part by Programme "Génomes" of C.N.R.S.

In this paper we focus on the transformations between L and $MF(L)$. We first design an automaton accepting $L(M)$ and that is built from the language M. When M is a finite set the transformation is effective. Moreover, if M is given by its digital tree, that is, its tree-like deterministic automaton, the algorithm is very similar to the algorithm of Aho and Corasick that builds a pattern-matching machine for a finite set of words [1].

In a second part we consider the particular situation of a language that is the set of factors of a single word v. The construction of its *factor automaton*, the minimal deterministic automaton accepting the factors of v (see [4]) is known to be rather intricate. It is remarkable that the preceding transformation yields exactly the factor automaton of v when the input if the set M of minimal forbidden words of v. We also give an algorithm that realizes the converse transformation, building the trie of M from the factor automaton of v. A corollary of the algorithm is a non-trivial upper bound on the number of minimal forbidden words of a word.

The complexities of algorithms described in this paper are all linear in the size of their input or output. Therefore, the design of possible faster algorithms relies on different representations of objects, which is not the aim of the paper.

2 Avoiding an Anti-Factorial Language

Let A be a finite alphabet and A^* be the set of finite words drawn from the alphabet A, the empty word ϵ included. Let $L \subseteq A^*$ be a *factorial language*, i.e. a language satisfying: $\forall u, v \in A^*$ $uv \in L \implies u, v \in L$. The complement language $L^c = A^* \setminus L$ is a (two-sided) ideal of A^*. Denote by $MF(L)$ the base of this ideal, we have $L^c = A^* MF(L) A^*$.

The set $MF(L)$ is called the set of *minimal forbidden words* for L. A word $v \in A^*$ is forbidden for the factorial language L if $v \notin L$, which is equivalent to say that v occurs in no word of L. In addition, v is minimal if it has no proper factor that is forbidden.

One can note that the set $MF(L)$ uniquely characterizes L, just because

$$L = A^* \setminus A^* MF(L) A^*. \tag{1}$$

The following simple observation provides a basic characterization of minimal forbidden words.

Remark 1 *A word $v = a_1 a_2 \cdots a_n$ belongs to $MF(L)$ iff the two conditions hold:*

- *v is forbidden, (i.e., $v \notin L$),*
- *both $a_1 a_2 \cdots a_{n-1} \in L$ and $a_2 a_3 \cdots a_n \in L$ (the prefix and the suffix of v of length $n-1$ belong to L).*

The remark translates into the equality:

$$MF(L) = AL \cap LA \cap (A^* \setminus L). \tag{2}$$

As a consequence of both equalities (1) and (2) we get the following proposition.

Proposition 1 *For a factorial language L, languages L and $MF(L)$ are simultaneously rational, that is, $L \in Rat(A^*)$ iff $MF(L) \in Rat(A^*)$.*

The set $MF(L)$ is an *anti-factorial language* or a *factor code*, which means that it satisfies: $\forall u, v \in MF(L)$ $u \neq v \Longrightarrow u$ is not a factor of v, property that comes from the minimality of words of $MF(L)$.

We introduce a few more definitions.

Definition 1 *A word $v \in A^*$ avoids the set M, $M \subseteq A^*$, if no word of M is a factor of v, (i.e., if $v \notin A^*MA^*$). A language L avoids M if every words of L avoid M.*

From the definition of $MF(L)$, it readily comes that L is the largest (according to the subset relation) factorial language that avoids $MF(L)$. This shows that for any anti-factorial language M there exists a unique factorial language $L(M)$ for which $M = MF(L)$. The next remark summarizes the relation between factorial and anti-factorial languages.

Remark 2 *There is a one-to-one correspondence between factorial and anti-factorial languages. If L and M are factorial and anti-factorial languages respectively, both equalities hold: $MF(L(M)) = M$ and $L(MF(L)) = L$.*

We also refer to the next definition that is to be considered in the context of dynamical systems (see [9] for example).

Definition 2 *The factorial language L is said to be* of finite type *when $MF(L)$ is finite.*

Finally, with an anti-factorial finite language M we associate the finite automaton $\mathcal{A}(M)$ as described below. The automaton is deterministic and complete, and, as shown at the end of the section by Theorem 3, the automaton accepts the language $L(M)$.

The automaton $\mathcal{A}(M)$ is the tuple (Q, A, i, T, F) where

- the set Q of states is $\{w \mid w$ is a prefix of a word in $M\}$,
- A is the current alphabet,
- the initial state i is the empty word ϵ,
- the set T of terminal states is $Q \setminus M$.

States of $\mathcal{A}(M)$ that are words of M are sink states. The set F of transitions is partitioned into the three (pairwise disjoint) sets F_1, F_2, and F_3 defined by:

- $F_1 = \{(u, a, ua) \mid ua \in Q, a \in A\}$ (forward edges or tree edges),
- $F_2 = \{(u, a, v) \mid u \in Q \setminus M, a \in A, ua \notin Q, v$ longest suffix of ua in $Q\}$ (backward edges),
- $F_3 = \{(u, a, u) \mid u \in M, a \in A\}$ (loops on sink states).

The transition function defined by the set F of arcs of $\mathcal{A}(M)$ is noted δ.

Remark 3 *One can easily prove from definitions that*
1. *if $q \in Q \setminus (M \cup \{\epsilon\})$, all transitions arriving on state q are labeled by the same letter $a \in A$,*
2. *from any state $q \in Q$ we can reach a sink state, i.e., q can be extended to a word of M.*

Definition 3 *For any $v \in A^*$, q_v denotes the state $\delta(\epsilon, v)$, target of the unique path in $\mathcal{A}(M)$ starting at the initial state and labeled by v.*

Since $\mathcal{A}(M)$ is a complete automaton, q_v is always defined. In the automaton $\mathcal{A}(M)$ states are words, but to avoid misunderstandings we sometimes write "the word corresponding to q_v" instead of just "the word q_v".

Remark 4 *Note that if v is a state of $\mathcal{A}(M)$ we have $q_v = v$.*

We are now ready to state the next lemma (which proof is by induction on v) that is used in the proof of Theorem 3, the main result of the section.

Lemma 2 *Let M be an anti-factorial language and consider $\mathcal{A}(M)$. Let $v \in A^*$ be such that, for any proper prefix u of v, q_u is not a sink state ($q_u \notin M$). Then,*
1. *the word q_v is a suffix of v,*
2. *q_v is the longest suffix of v that is also a state of $\mathcal{A}(M)$ (or $\forall q \in Q$ q suffix of $v \Longrightarrow q$ suffix of q_v).*

Proof. By induction on $|v|$. ◻

Denoting by $Lang(\mathcal{A})$ the language accepted by an automaton $\mathcal{A}$, we get the main theorem of the section.

Theorem 3 *For any anti-factorial language M, $Lang(\mathcal{A}(M)) = L(M)$.*

Proof. We first prove $L(M) \subseteq Lang(\mathcal{A}(M))$. We have to show that if v is a word that avoids M then $v \in Lang(\mathcal{A}(M))$. Assume *ab absurdo* that $v \notin Lang(\mathcal{A}(M))$; therefore q_v is a sink state. Let u be the shortest prefix of v for which q_u is a sink state (note that $q_u = q_v$). By lemma 2 statement 1, q_u is a suffix of u, but q_v is by definition an element of M, and so v does not avoid M, a contradiction.

We then prove $Lang(\mathcal{A}(M)) \subseteq L(M)$. Let $v \in Lang(\mathcal{A}(M))$. Let us suppose *ab absurdo* that v does not avoid M, i.e., $v = uwz$ for some $w \in M, u, z \in A^*$. We choose uw as the shortest prefix of v that belongs to A^*M. Since $w \in M$ it is by definition a state of $\mathcal{A}(M)$; since w is a state that is a suffix of uw, by Lemma 2 statement 2, w is a suffix of q_{uw}. But q_{uw}, which is by definition a state of $\mathcal{A}(M)$, is a prefix of an element w' of M. Since w is a suffix of a prefix of w', w is a factor of w', a contradiction because M is anti-factorial. ◻

The above definition of $\mathcal{A}(M)$ turns into the algorithm below, called L-AUTOMATON, that builds the automaton from a finite anti-factorial set of words. The input is the trie $\mathcal{T}$ that represents M. It is a tree-like automaton accepting the set M and, as such, it is noted (Q, A, i, T, δ'). The procedure can be adapted to test whether $\mathcal{T}$ represents an anti-factorial set, or even to generate the trie of the anti-factorial language associated with a set of words.

In view of Equality 1, the design of the algorithm remains to adapt the construction of a pattern matching machine (see [1] or [4]) The algorithm uses a function f called a *failure function* and defined on states of $\mathcal{T}$ as follows. States of the trie $\mathcal{T}$ are identified with the prefixes of words in M. For a state au ($a \in A$, $u \in A^*$), $f(au)$ is $\delta'(i,u)$, quantity that may happen to be u itself. Note that $f(i)$ is undefined, which justifies a specific treatment of the initial state in the algorithm.

```
L-AUTOMATON (trie T = (Q, A, i, T, δ'))
 1. for each a ∈ A
 2.     if δ'(i, a) defined
 3.         set δ(i, a) = δ'(i, a);
 4.         set f(δ(i, a)) = i;
 5.     else
 6.         set δ(i, a) = i;
 7. for each state p ∈ Q \ {i} in width-first search and each a ∈ A
 8.     if δ'(p, a) defined
 9.         set δ(p, a) = δ'(p, a);
10.         set f(δ(p, a)) = δ(f(p), a);
11.     else if p ∉ T
12.         set δ(p, a) = δ(f(p), a);
13.     else
14.         set δ(p, a) = p;
15. return (Q, A, i, Q \ T, δ);
```

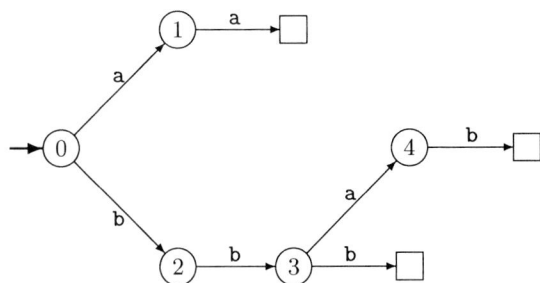

Fig. 1. Trie of the factor code $\{aa, bbaa, bbb\}$ on the alphabet $\{a, b\}$. Squares represent terminal states.

Example. Figure 1 displays the trie that accepts $M = \{aa, bbaa, bbb\}$. It is an anti-factorial language. The automaton produced from the trie by algorithm L-AUTOMATON is shown in Figure 2. It accepts the prefixes of $(ab \cup b)(ab)^*ba$ that are all the words avoiding M.

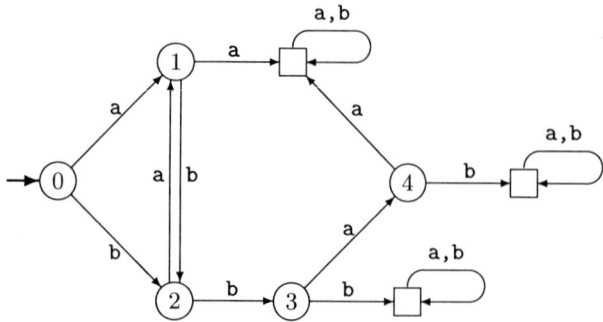

Fig. 2. Automaton accepting the words that avoid the set $\{aa, bbaa, bbb\}$. Squares represent non-terminal states (sink states).

Theorem 4 *Let T be the trie of an anti-factorial language M. Algorithm L-AUTOMATON builds a complete deterministic automaton accepting $L(M)$.*

Proof. The automaton produced by the algorithm has the same set of states as the input trie. It is clear that the automaton is deterministic and complete.

Let $u \in A^+$ and $p = \delta(i, u)$. A simple induction on $|u|$ shows that the word corresponding to $f(p)$ is the longest proper suffix of u that is a prefix of some word in M. This notion comes up in the definition of the set of transitions F_2 in the automaton $\mathcal{A}(M)$. Therefore, the rest of the proof just remains to check that instructions implement the definition of $\mathcal{A}(M)$. ⋈

Theorem 5 *Algorithm L-AUTOMATON runs in time $O(|Q| \times |A|)$ on input $T = (Q, A, i, T, \delta')$ if transition functions are implemented by transition matrices.*

3 Factor Automaton of a Single Word

In this section we specialize the previous results to the language of factors of a single word. It is proved below that the contruction of Section 2 yields the factor automaton (minimal dterministic automaton accepting the factors) of the word (see Theorem 7). The minimality of the automaton seems to be exceptional because, for example, the same construction applied to the set $\{aa, ab\}$ does not provide a minimal automaton.

The reverse construction that produces the trie of minimal forbidden words from the factor automaton is described in the next section.

We consider a fixed word $v \in A^*$ and denote by $\mathcal{F}(v)$ be the language of factors of v.

Proposition 6 *The language $\mathcal{F}(v)$ is of finite type.*

Proof. Indeed, factors of v, of lengths less than $|v| + 1$, avoid all words of length exactly $|v| + 1$. Therefore, every minimal forbidden word of $\mathcal{F}(v)$ has length at most $|v| + 1$. ⋈

The result of the previous proposition is made more precise in the next section, but an immediate consequence of it and of the definition of the automaton $\mathcal{A}(M)$ for an anti-factorial language M, the automaton $\mathcal{A}(MF(\mathcal{F}(v)))$ has a finite number of states. The next statement gives a complete characterization of the automaton as the factor automaton of v.

Theorem 7 *For any $v \in A^*$, the automaton obtained from $\mathcal{A}(MF(\mathcal{F}(v)))$ by removing its sink states is the minimal deterministic finite automaton accepting the language $\mathcal{F}(v)$ of factors of v.*

Proof. The automaton $\mathcal{A}(MF(\mathcal{F}(v)))$ is already a deterministic finite automaton that accepts the language $\mathcal{F}(v)$ by Theorem 3. We only have to prove that it is minimal after removing the sink states.

Suppose *ab absurdo* that there exist two equivalent non-sink states p, q in Q. By the standard equivalence relation of undistinginshability and by construction $p, q \in \mathcal{F}(v)$. Hence, $v = xpy$ and $v = x'qy'$ and we can choose x and x' of minimal length. We consider two cases:

(i) $|xp| \neq |x'q|$,
(ii) $|xp| = |x'q|$.

Case (i). We can suppose for example that $|xp| < |x'q|$ (the case $|xp| > |x'q|$ is handled symmetrically). Then, $xpy \in \mathcal{F}(v)$ implies that $\delta(p, y)$ is not a sink state, hence, by the equivalence $\delta(q, y)$ is not a sink state, that is, $qy \in \mathcal{F}(v)$ by Remark 4. Therefore, $v = x"qyz$ where $|x"| \geq |x'|$ by the choice of x' (of minimal length). Hence, $|v| \geq |x'| + |q| + |y| + |z| > |xp| + |y| = |v|$, a contradiction.
Case (ii). The equality $|xp| = |x'q|$ implies either that p is a suffix of q or the converse. Let us suppose for example that $p = sq$ for some word $s \neq \epsilon$. By Remark 3 statement 2, there exists $w = pz$ that belongs to $MF(\mathcal{F}(v))$. By the equivalence, qz is also a sink state and, again by the equivalence, for no proper prefix u of qz, q_u is a sink state. Hence, by Lemma 2.1, q_{qz} is an element of $MF(\mathcal{F}(v))$, that is, a suffix of qz. Since $p = sq, s \neq \epsilon$, q_{qz} is a proper suffix of pz against the anti-factorial property of $MF(\mathcal{F}(v))$. A contradiction again.

After cases (i) and (ii) it appears that there cannot exist two different non-sink states p, q in Q that are equivalent. Therefore the automaton without sink states is minimal, which ends the proof. ⋈

4 Minimal Forbidden Words of a Word

We end the article by an algorithm that builds the trie accepting the language $MF(\mathcal{F}(v))$ of minimal words avoided by v. This is an implementation of the inverse of the transformation described in Section 2. Its design follows Equality 2. A corollary of the transformation gives a bound on the number of minimal forbidden words of a single word, which improves on the bound coming readily from Proposition 6.

> MF-TRIE (factor automaton $\mathcal{A} = (Q, A, i, T, \delta)$ and its suffix function s)
> 1. **for** each state $p \in Q$ in width-first search from i **and** each $a \in A$
> 2.   **if** $\delta(p, a)$ undefined **and** ($p = i$ **or** $\delta(s(p), a)$ defined
> 3.   set $\delta'(p, a) = $ new sink;
> 4.   **else if** $\delta(p, a) = q$ **and** q not already treated
> 5.   set $\delta'(p, a) = q$;
> 6. **return** $(Q, A, i, \{sinks\}, \delta')$;

The input of algorithm MF-TRIE is the factor automaton of word v. It includes the failure function defined on the states of the automaton and called s. This function is a by-product of efficient algorithms that build the factor automaton (see [4]). It is defined as follows. Let $u \in A^+$ and $p = \delta(i, u)$. Then, $s(p) = \delta(i, u')$ where u' is the longest suffix of u for which $\delta(i, u) \neq \delta(i, u')$. It can be shown that the definition of $s(p)$ does not depend on the choice of u.

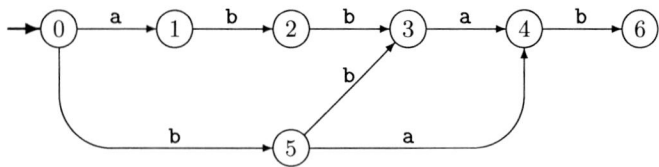

Fig. 3. Factor automaton of abbab; all states are terminal.

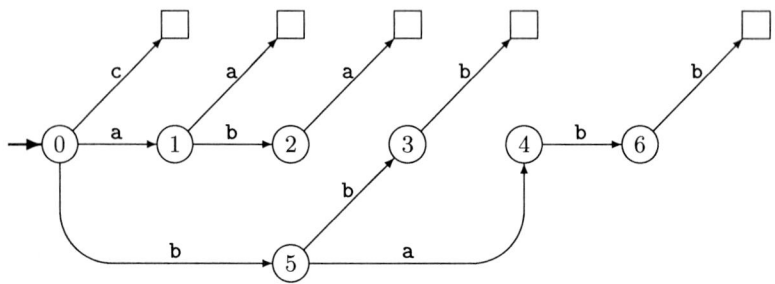

Fig. 4. Trie of minimal forbidden words of $\mathcal{F}(\text{abbab})$ on the alphabet $\{a, b, c\}$. Squares represent terminal states.

Example Consider the word $v = $ abbab on the alphabet $\{a, b, c\}$. Its factor automaton is displayed in Figure 3. The failure function s defined on states has values: $s(1) = s(5) = 0$, $s(2) = s(3) = 5$, $s(4) = 1$, $s(6) = 2$. Algorithm MF-TRIE produces the trie of Figure 4 that represents the set of five words $\{\text{aa, aba, babb, bbb, c}\}$.

Theorem 8 *Let $\mathcal{A}$ be the factor automaton of a word $v \in A^*$. (It accepts the language $\mathcal{F}(v)$.) Algorithm* MF-TRIE *builds the tree-like deterministic automaton accepting $MF(\mathcal{F}(v))$ the set of minimal forbidden words of $\mathcal{F}(v)$.*

Corollary 9 *A word $v \in A^*$ has no more than $2(|v|-2)(|A_v|-1)+|A|$ minimal forbidden words if $|v| \geq 3$, where A_v is the set of letters occurring in v. The bound becomes $|A|+1$ if $|v| < 3$.*

Proof. The number of words in $MF(\mathcal{F}(v))$ is the number of sink states created during the execution of algorithm MF-TRIE. These states have exactly one ingoing arc originated at a state of the factor automaton $\mathcal{A}$ of v. So, we have to count these arcs.

From the initial state of $\mathcal{A}$ there is exactly $|A| - |A_v|$ such arcs. From the (unique) state of $\mathcal{A}$ without outgoing arc, there are at most $|A_v|$ such arcs. From other states there are at most $|A_v| - 1$ such arcs.

For $|v| \geq 3$, it is known that $\mathcal{A}$ has at most $2|v|-2$ states (see [4]). Therefore, $|MF(\mathcal{F}(v))| \leq (|A|-|A_v|)+|A_v|+(2|v|-4)(|A_v|-1) = 2(|v|-2)(|A_v|-1)+|A|$. When $|v| < 3$, it can be checked directly that $|MF(\mathcal{F}(v))| \leq |A|+1$. ⋈

Theorem 10 *Algorithm* MF-TRIE *runs in time $O(|v| \times |A|)$ on input word v if transition functions are implemented by transition matrices.*

References

1. A. V. Aho and M. J. Corasick. Efficient string matching: an aid to bibliographic search, *Comm. ACM* **18:6** (1975) 333–340.
2. M.-P. Béal, F. Mignosi, and A. Restivo. Minimal Forbidden Words and Symbolic Dynamics, in (C. Puech and R. Reischuk, eds., LNCS 1046, Springer, 1996) 555–566.
3. J. Cassaigne. Complexité et Facteurs Spéciaux, *Bull. Belg. Math. Soc.* **4** (1997) 67–88.
4. M. Crochemore, C. Hancart. Automata for matching patterns, in *Handbook of Formal Languages*, G. Rozenberg, A. Salomaa, eds.", Springer-Verlag", 1997, Volume 2, *Linear Modeling: Background and Application*, Chapter 9, 399–462.
5. V. Diekert, Y. Kobayashi. Some identities related to automata, determinants, and Möbius functions, Report 1997/05, Fakultät Informatik, Universität Stuttgart, 1997.
6. A. de Luca, F. Mignosi. Some Combinatorial Properties of Sturmian Words, *Theor. Comp. Sci.* **136** (1994) 361–385.
7. A. de Luca, L. Mione. On Bispecial Factors of the Thue-Morse Word, *Inf. Proc. Lett.* **49** (1994) 179–183.
8. R. McNaughton, S. Papert. *Counter-Free Automata*, M.I.T. Press, MA 1970.
9. D. Perrin. Symbolic Dynamics and Finite Automata, invited lecture in *Proc. MFCS'95*, LNCS **969**, Springer, Berlin 1995.

On Defect Effect of Bi-Infinite Words*

Juhani Karhumäki, Ján Maňuch, and Wojciech Plandowski

Department of Mathematics and Turku Centre for Computer Science,
University of Turku, SF-20014 Turku, Finland
karhumak@cs.utu.fi, manuch@cs.utu.fi, wojtekpl@mimuw.edu.pl

Abstract. We prove the following two variants of the defect theorem. Let X be a finite set of words over a finite alphabet. Then if a nonperiodic bi-infinite word w has two X-factorizations, then the combinatorial rank of X is at most $\mathrm{card}(X) - 1$, i.e. there exists a set F such that $X \subseteq F^+$ with $\mathrm{card}(F) < \mathrm{card}(X)$. Further, if $\mathrm{card}(X) = 2$ and a bi-infinite word possesses two X-factorizations which are not shift-equivalent, then the primitive roots of the words in X are conjugates. Moreover, in the case $\mathrm{card}(F) = \mathrm{card}(X)$, the number of periodic bi-infinite words which have two different X-factorizations is finite and in the two-element case there is at most one such bi-infinite word.

1 Introduction

Defect theorem is one of the fundamental results on words, cf [Lo]. Intuitively it states that if n words satisfy a non-trivial relation then these words can be expressed as products of at most $n-1$ words. Actually, as discussed in [CK], for example, there does not exist just one defect theorem but several ones depending on restrictions put on the required $n-1$ words.

It is also well-known that the nontrivial relation above can be replaced by a weaker condition, namely by the nontrivial one-way infinite relation. The goal of this note is to look for defect theorems for bi-infinite words. In a strict sense such results do not exist: the set $X = \{ab, ba\}$ of words satisfies a bi-infinite nontrivial relation since $(ab)^{\mathbb{Z}} = (ba)^{\mathbb{Z}}$, but there exists no word ϱ such that $X \subseteq \varrho^+$. However, we are going to prove two results which can be viewed as defect theorems for bi-infinite words.

In terms of factorizations of words defect theorem can be stated as follows: Let $X \subseteq \Sigma^+$ be a finite set of words. If there exists a word $w \in \Sigma^+$ having two different X-factorizations, then the rank of X is at most $\mathrm{card}(X) - 1$. Here the rank of X can be defined in different ways, cf again [CK]. For example, it can be defined as a combinatorial rank $r_c(X)$ denoting the smallest number k such that $X \subseteq Y^+$ with $\mathrm{card}(Y) = k$.

To describe our results let w be a bi-infinite word, i.e. an element of $\Sigma^{\mathbb{Z}}$, and X a finite subset of Σ^+. We say that w has an X-factorization if $w \in X^{\mathbb{Z}}$, and that w has two different X-factorizations, if it has two X-factorizations such

---

* Supported by Academy of Finland under Grant No. 14047.

that they do not match at least in one point of w. We are going to prove the following two results:

i) If a nonperiodic bi-infinite word w has two different X-factorizations, then the combinatorial rank $r_c(X)$ of X is at most $\text{card}(X) - 1$. Moreover, if $r_c(X) = \text{card}(X)$ then the number of bi-infinite words with two different X-factorizations is finite.

ii) Let $\text{card}(X) = r_c(X) = 2$, so that X is a code. If a bi-infinite word w has two different X-factorizations which are not shift-equivalent, then the primitive roots of words in X are conjugates. Moreover, there is at most one bi-infinite word possessing two different X-factorizations.

Note that case ii) is a strict sharpening of case i) for two-element sets. We also want to emphasize that a restriction to nonperiodic words in case i) is necessary, and even more that this theorem requires to consider the combinatorial rank. This seems to be the first result where the defect effect is realized only by the combinatorial rank, and not by the other types of ranks, cf again [CK].

The first part of the result in ii) is related to the main result of [lRlR], and, we believe, deducible from considerations of that paper. However, our proof is self-contained and essentially shorter, and moreover formulated directly to yield a defect-type of theorem.

Our paper is organized as follows. In Section 2 we fix our terminology and present the auxiliary results needed for our proofs. In Section 3 we prove our general defect theorem for bi-infinite words, i.e. i) above. In Section 4 we prove, as our main result, a defect theorem for binary sets X satisfying a nontrivial bi-infinite relation, i.e. above ii). In Section 5 we prove the second part of ii), i.e. the uniqueness of the bi-infinite word in the two-element case. The last section contains conclusions and open problems.

The full version of this paper with complete proofs appears in [KMP].

2 Preliminaries

In this section we fix our terminology and recall a few lemmas on combinatorics of words needed for our proofs. For undefined notions we refer to [Lo] or [CK].

Let Σ be a finite alphabet and X a finite subset of Σ^+. The set of all finite, infinite and bi-infinite words are denoted by Σ^*, $\Sigma^{\mathbb{N}}$ and $\Sigma^{\mathbb{Z}}$, respectively. Hence, formally a bi-infinite word is a mapping $f_w : \mathbb{Z} \to \Sigma$, usually written as

$$w = \ldots a_{-1}a_0a_1\ldots \qquad \text{with } a_i = f_w(i).$$

An X-factorization of w is any sequence of words from X yielding w as their products. Formally, an X-factorization of $w \in \Sigma^{\mathbb{Z}}$ is a mapping $F : \mathbb{Z} \to X \times \mathbb{Z}$ such that for each $k \in \mathbb{Z}$ if $F(k) = (\alpha, i)$ and $F(k+1) = (\beta, j)$, then $a_i a_{i+1} \ldots a_{j-1} = \alpha$, i.e. the position i is a starting position of the factor α in w. We say that two X-factorizations F_1 and F_2 of a bi-infinite word are

- *different*, whenever there is a $k_0 \in \mathbb{Z}$ such that for each $k \in \mathbb{Z}$, $F_1(k_0) \neq F_2(k)$,
- *disjoint*, whenever the starting positions of all factors in F_1 are distinct from the ones in F_2,
- *shift-equivalent*, if there is a k_0 such that whenever $F_1(k) = (\alpha, i)$ and $F_2(k_0 + k) = (\beta, j)$, then $\alpha = \beta$.

Example 1. Let $X = \{a, bab, baab\}$. The word $(baa)^{\mathbb{Z}}$ has two different X-factorizations, namely the ones depicted as

$$\ldots baabaab \ldots \quad .$$

They are clearly shift-equivalent. On the other hand the word

$$w = \ldots bababaabaab \cdots = {}^{\mathbb{Z}}(ba)b(aab)^{\mathbb{Z}}$$

also has two different X-factorizations which, however, are not shift-equivalent

$$\ldots bababaabaa \ldots \quad .$$

Clearly, in both of the above cases the two factorizations are disjoint.

We define the *combinatorial rank* of $X \subseteq \Sigma^+$ by the formula

$$r_{\mathrm{c}}(X) = \min\{\mathrm{card}(Y) \mid X \subseteq Y^+\}.$$

For the sake of completeness we remind that

$$r_{\mathrm{c}}(X) \leq r_{\mathrm{f}}(X) \leq \mathrm{card}(X),$$

where $r_{\mathrm{f}}(X)$ denotes the *free rank* (or simply the *rank*) of X defined as the cardinality of the base of the smallest free semigroup containing X, cf [CK].

Example 1 (continued). Clearly, $r_{\mathrm{c}}(X) = 2$, since $X \subseteq \{a, b\}^+$, but for no word ϱ the inclusion $X \subseteq \varrho^+$ holds. On the other hand, since X is a code we conclude that $r_{\mathrm{f}}(X) = 3$.

Example 2. Let $X = \{ab, bc, ca\}$. Then we have $r_{\mathrm{c}}(X) = r_{\mathrm{f}}(X) = \mathrm{card}(X)$. Note also that the word $(abc)^{\mathbb{Z}}$ has two disjoint, but shift-equivalent, X-factorizations:

Next we recall a few basic results on words that we shall need in our later considerations, for their proofs the reader is referred to [Lo] or [CK].

Lemma 1. *Let $u, v \in \Sigma^+$. If the words $u^{\mathbb{N}}$ and $v^{\mathbb{N}}$ have a common prefix of length at least $|u| + |v| - \gcd(|u|, |v|)$, then u and v commute.*

Lemma 2. *No primitive word ϱ satisfies a relation $\varrho\varrho = s\varrho p$ with $s, p \neq 1$.*

Lemma 3. *If two words u and v satisfy the relation $ut = tv$ for some $u, v, t \in \Sigma^+$, i.e. if they are conjugates, then there exist words p and q such that pq is primitive and*

$$u = (pq)^i, \quad v = (qp)^i \quad \text{and} \quad t \in p(qp)^* \quad \text{for some } i \geq 1.$$

In Section 5 we shall need also the following claim, cf [LyS].

Lemma 4. *Consider nonempty words x, y, z satisfying equation $x^m = y^n z^p$, where $m, n, p \geq 2$. Then all words x, y, z are powers of a common word.*

In order to formulate our fifth, and the most crucial lemma, we need some terminology, cf [CK] or [HK]. We associate a finite set $X \subseteq \Sigma^+$ with a graph $\mathcal{G}_X = (V_X, E_X)$, called *the dependency graph* of X, as follows: the set V_X of vertices of $\mathcal{G}_X$ equals to X, and the set E_X of edges of $\mathcal{G}_X$ is defined by the condition

$$(x, y) \in E_X \quad \text{iff} \quad xX^{\mathbb{N}} \cap yX^{\mathbb{N}} \neq \emptyset.$$

Then we have

Lemma 5. *For each finite set $X \subseteq \Sigma^+$, the combinatorial rank of X is at most the number of connected components of $\mathcal{G}_X$.*

As we shall see Lemma 5 is particularly suitable for our subsequent considerations. Indeed, in that lemma it is crucial that words in X are nonempty, and that indeed is satisfied in the proofs of our Theorems 1 and 2.

3 The General Case

In this section we prove our first defect theorem for bi-infinite words. To make ideas of our proofs clearer we shall frequently use pictures as illustrations. Let us fix some notations used in pictures. A horizontal line expresses a bi-infinite word with two X-factorizations F_1, F_2. The sequences of words in the factorization F_1 are depicted above the line by consecutive arcs, similarly the sequences of words in F_2 are depicted by arcs, which are below the line. For example, in Figure 2 we consider the words $f_1, f_1', f_2, f_2' \in X^*$, such that f_1, f_1' are parts of the factorization F_1 and f_2, f_2' are parts of F_2.

Theorem 1. *Consider a set $X = \{\alpha_1, \ldots, \alpha_n\} \subseteq \Sigma^+$. Let w be a bi-infinite word over Σ and F_1, F_2 two different X-factorizations of w. Then the combinatorial rank of X is at most $n - 1$, or both the word w and the X-factorizations F_1, F_2 are periodic. Moreover, if the rank of X is n, then the number of periodic bi-infinite words with two different X-factorizations is finite.*

Proof. If F_1 and F_2 are not disjoint the result follows from Lemma 5. Now we study all words $t \in \Sigma^+$ as depicted in Figure 1 with $x, y \in X$. More precisely, we take the beginning of any $x \in X$ in the lower factorization F_2, i.e. the point A, and we find the closest end of $y \in X$ in the upper factorization F_1 to the right from the point A, i.e. the point B. Now the *X-difference t* is defined as the word between points A and B. Note that there are infinitely many occurrences of X-differences, but only finitely many of different values of t's, since all t's are proper suffixes of words in X. By the pigeon hole principle there exists a t such that the X-difference t occurs infinitely many times in the bi-infinite word w.

Consider now two occurrences of an X-difference t. Let the part of the factorization F_1 between the end of the first t and the end of the second t be $f_1 \in X^+$ and the part of F_2 between the beginnings of the t's be $f_2 \in X^+$. We shall call the pair (f_1, f_2) the *t-pair*. Notice that for any t-pair (f_1, f_2) it holds $tf_1 = f_2 t$. Further, we shall call a t-pair (f_1, f_2) *minimal*, if there is no other occurrence of the X-difference t inside the f_1, f_2.

Let us look at 3 consecutive occurrences of an X-difference t, which has infinitely many occurrences in w, see Figure 2. Clearly, the pairs (f_1, f_2) and (f_1', f_2') are minimal t-pairs.

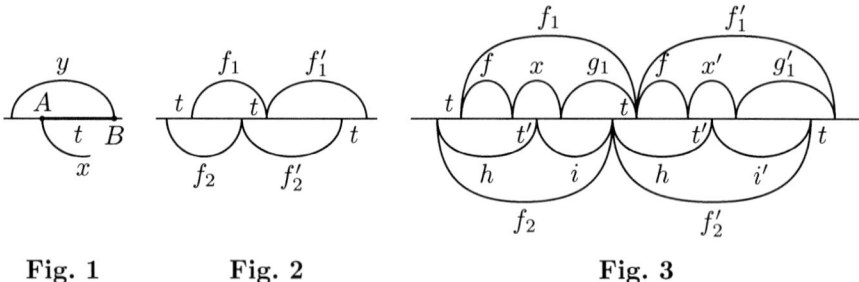

Fig. 1 **Fig. 2** **Fig. 3**

In the case that for any such 3 occurrences we have

$$f_1 = f_1', \qquad f_2 = f_2', \tag{1}$$

the word w and both of the X-factorizations are periodic with the length of a period equal to $|f_1|$. So assume that we found such 3 occurrences for which at least one of the equalities (1) does not hold. Take the pair f_i, f_i' in which the factorizations differ earlier. Without loss of generality we can assume that it is the pair $f_1 \neq f_1'$. Let $f \in X^*$ be the longest common prefix of f_1, f_1' and $g, g' \in X^*$ their differently starting suffixes. Similarly define $h, i, i' \in X^*$ for f_2, f_2'. Now either one of the words g, g' is empty or they start with different words $x, x' \in X$.

If, for example, g is empty, then by the choice of the pair f_1, f_1', the word i must be also empty. But then we immediately have an intermediate occurrence of the X-difference t in f_1', f_2', i.e. the pair (f_1', f_2') is not minimal. So assume that g, g' are both nonempty and their factorizations start with different words $x, x' \in X$, i.e. that $g = xg_1$, $g' = x'g_1'$, where $g_1, g_1' \in X^*$. By the choice of f_1

and f_1' the common prefix h ends after the beginning of x and x'. In Figure 3 it is shown the case when it ends inside x and x', but the following considerations work also in the other cases. We have 3 equations over $X \cup \{t, t'\}$:

$$t'it = xg_1, \qquad t'i't = x'g_1', \qquad tf_1 = f_2 t,$$

where $x \neq x'$. So by Lemma 5 we obtain $r_c(X) \leq n-1$.

On the other hand, this implies that in the case $r_c(X) = \text{card}(X)$, for each value of t there is at most one minimal t-pair. By periodicity this minimal t-pair specifies the whole bi-infinite word w, and so for each value of t there is also at most one bi-infinite word containing X-difference t. But any t must be a proper suffix of a word from X, hence we have only finitely many bi-infinite words with two different X-factorizations. □

In the two-element case we have the following consequence.

Corollary 1. *Consider set $X = \{\alpha, \beta\} \subseteq \Sigma^+$. Let w be a bi-infinite word over Σ and F_1, F_2 two different X-factorizations of w. Then the words α, β commute or both the word w and the X-factorizations F_1, F_2 are periodic.*

Theorem 1 deserves a few comments. First, the possibility that the two factorizations are both periodic cannot be ruled out. This follows from Example 2. Second, the combinatorial rank cannot be replaced by the free rank, for example. This indeed follows from Example 1. This latter remark is quite interesting since in all previous defect theorems, see [CK], either of our notions of the rank, or even some others, can be used to witness the defect effect.

4 The Two-Element Case

In this section we generalize the result of the previous section in the case of two-element sets. First we recall that in a strict sense we cannot have a defect theorem for bi-infinite words even in this simple case.

Example 3. The set $X = \{ab, ba\}$ is of combinatorial rank 2 although the word $(ab)^{\mathbb{Z}}$ has two disjoint, and even non-shift-equivalent, X-factorizations.

As a main result of this paper we, however, show that the above example, and its natural variants, are the only exceptions which may occur. And even in these cases the roots of words in X are conjugates, i.e. they are cyclic permutations of powers of a common word.

Theorem 2. *Consider set $X = \{\alpha, \beta\}$ with $\alpha, \beta \in \Sigma^+$. Let w be a bi-infinite word over Σ and F_1, F_2 two different X-factorizations of w containing together both elements of X. Then one of the following possibilities holds:*

i) α and β commute, or
ii) the roots of α and β are conjugates and $F_1 \in \alpha^{\mathbb{Z}}$, $F_2 \in \beta^{\mathbb{Z}}$, or vice versa, or
iii) the two X-factorizations F_1, F_2 are shift-equivalent and there exists an $n \geq 1$ such that $F_1, F_2 \in (\alpha \beta^n)^{\mathbb{Z}}$ and α is primitive or $F_1, F_2 \in (\beta \alpha^n)^{\mathbb{Z}}$ and β is primitive.

The proof of this is rather long and can be found in [KMP]. It uses essentially Lemma 5.

Theorem 2 deserves a few comments. The number of different X-factorizations of the bi-infinite word w having an X-factorization is very different in cases *i)–iii)*. In case *i)* there exist nondenumerably many such X-factorizations, in case *ii)* there are finitely many different X-factorizations and if we consider shift-equivalent X-factorization as the same, then there are exactly two of them. Finally, in case *iii)* there are also finitely many different X-factorizations, which are all shift-equivalent. This actually means that in case *iii)* no bi-infinite word can be expressed in two different ways as a product of words from X. Hence, indeed, Theorem 2 shows a defect effect of a two-element set for bi-infinite factorizations.

In Theorem 2 we showed that if the words of X do not commute and are not conjugates then only the case *iii)* is possible. But if they do not commute and are conjugates Theorem 2 allows either case *ii)* or *iii)*. It can be proved that only case *ii)* is possible. We can formulate the following lemma and its corollaries, cf [KMP].

Lemma 6. *If pq is primitive, then $p(qp)^n$ and $q(pq)^n$ are not conjugates for any $n \geq 1$.*

Corollary 2. *If α and β are different conjugates, then $\alpha\beta$ must be primitive.*

In fact Corollary 2 is a special case of the claim in [LeS] which states under the additional assumption that α, β are primitive, that $\alpha\beta^m$ is primitive for all natural numbers m.

Corollary 3. *Consider set $X = \{\alpha, \beta\}$ with $\alpha, \beta \in \Sigma^+$. Let w be a bi-infinite word over Σ and F_1, F_2 two different X-factorizations of w containing together both elements of X. If the roots of α, β are non-commuting conjugates, then $F_1 \in \alpha^{\mathbb{Z}}$, $F_2 \in \beta^{\mathbb{Z}}$, or vice versa.*

5 The Uniqueness of the Bi-Infinite Word

In Section 3 we proved that if the rank of the set X equals to card(X), then the number of bi-infinite words possessing two X-factorizations is finite. In this section we shall prove that in the two-element case there is at most one such bi-infinite word. This holds also in the case when $r_c(X) = 1$, since then both elements of X are powers of a common word t and the only possible bi-infinite word is $t^{\mathbb{Z}}$. The situation is also trivial in the case when roots of elements of $X = \{\alpha, \beta\}$ are conjugates, by Corollary 3 the only possible bi-infinite word is $w = \alpha^{\mathbb{Z}} = \beta^{\mathbb{Z}}$. So we need to consider only the case when the roots of α and β are not conjugates.

In this case, by Theorem 2, we know that a bi-infinite word possessing two X-factorizations must be of the form $(\alpha\beta^n)^{\mathbb{Z}}$ or $(\alpha^n\beta)^{\mathbb{Z}}$. Moreover, since w has two X-factorizations, the word $\alpha\beta^n$ or the word $\alpha^n\beta$ cannot be primitive, by Lemma 2. As we stated in the previous section, if α and β are conjugates, then the words $\alpha\beta^n$ and $\alpha^n\beta$ are primitive for all natural n. We have showed a similar result for α,β being non-conjugates, i.e. we have showed that at most one word of $\alpha\beta^n$, $\alpha^n\beta$ is not primitive, cf [KMP]. By this result, we have that also in the last case there is at most one bi-infinite word possessing two different X-factorizations.

Theorem 3. *Consider set $X = \{\alpha,\beta\}$, $\alpha,\beta \in \Sigma^+$. There is at most one bi-infinite word over Σ possessing two different X-factorizations.*

6 Conclusions and Open Problems

Our Theorem 2 is closely related to the main result of [IRIR], where it is characterized when a finite word can have two disjoint X-interpretations for a binary set X. Our result could be concluded, with some effort, from the considerations in this paper. However, our proof is simpler, due to the use of the graph lemma (5), and moreover directly formulated to obtain a defect type of theorems.

We pose an open problem asking whether Theorem 2 can be extended to arbitrary sets.

Open Problem 1. Let $X \subseteq \Sigma^+$ be a finite set such that $r_c(X) = \text{card}(X)$. Does there exist a bi-infinite word w having two X-factorizations F_1 and F_2 satisfying:

i) both F_1 and F_2 contain all elements of X, and
ii) F_1 and F_2 are not shift-equivalent.

Observe here that even in the case of a two-element set X without the assumption that all elements of X occur in both factorizations the answer to this problem is trivially negative.

The answer is also negative if the condition $r_c(X) = \text{card}(X)$ is replaced by a weaker one involving free rank: $r_f(X) = \text{card}(X)$. This is verified by the following example.

Example 4. Let $X = \{\alpha_1, \alpha_2, \alpha_3\}$ where $\alpha_1 = babab$, $\alpha_2 = abbabba$ and $\alpha_3 = babbab$. Then $(\alpha_1\alpha_2\alpha_3)^{\mathbb{Z}} = (\alpha_2\alpha_1\alpha_3)^{\mathbb{Z}}$ and we have $r_f(X) = 3 = \text{card}(X)$, since X is a code. Clearly, $r_c(X) = 2 < \text{card}(X)$.

Another open problem asks whether Corollary 3 can be generalized for an arbitrary finite X.

Open Problem 2. Let $X \subseteq \Sigma^+$ be a finite set satisfying $r_c(X) = \text{card}(X)$. Suppose that primitive roots of all elements of X are conjugates and that a bi-infinite word w has at least two different X-factorizations. Are all X-factorizations of w of the form $\alpha^{\mathbb{Z}}$, where $\alpha \in X$?

Example 5. The answer to the above question is negative if we omit the assumption $r_c(X) = \text{card}(X)$. Indeed, let $X = \{\alpha_1, \alpha_2, \alpha_3\}$, where $\alpha_1 = baa$, $\alpha_2 = aba$, $\alpha_3 = aab$. Then clearly $\alpha_1 \sim \alpha_2 \sim \alpha_3$ and the word $(abaaab)^{\mathbb{Z}}$ has two different X-factorizations: $(\alpha_1\alpha_2)^{\mathbb{Z}}$ and $(\alpha_2\alpha_3)^{\mathbb{Z}}$.

In Section 5 we proved that in the two-element case there is at most one bi-infinite word possessing two different X-factorizations. In Section 3 we showed that if $r_c(X) = \text{card}(X)$, then there are finitely many such bi-infinite words, or more precisely, at most $s(X) - \text{card}(X)$ such words, where $s(X)$ is the sum of lengths of words in X. We conclude with our last open problem.

Open Problem 3. Let $X \subseteq \Sigma^+$ be a finite set satisfying $r_c(X) = \text{card}(X)$. Find a good upper bound for the number of bi-infinite words possessing two different X-factorizations. Does there exist an upper bound which depends only on $\text{card}(X)$?

Acknowledgement

The authors are grateful to Dr Tero Harju for useful discussions.

References

[CK] Choffrut, C., Karhumäki, J., Combinatorics of words, in G. Rozenberg and A. Salomaa (eds), Handbook of Formal Languages, Springer, 1997.
[lRlR] Le Rest, E., Le Rest, M. Sur la combinatoire des codes a deux mots, Theoretical Computer Science 41, 61–80, 1985.
[Lo] Lothaire, M., Combinatorics on words, Addison-Wesley, 1983.
[HK] Harju, T., Karhumäki, J., On the defect theorem and simplifiability, Semigroup Forum 33, 199–217, 1986.
[LeS] Lentin, A., Schützenberger, M.P., A combinatorial problem in the theory of free monoids, Proc. University of North Carolina, 128–144, 1967.
[LyS] Lyndon, R.C., Schützenberger, M.P., The equation $a^m = b^n c^p$ in a free group, Michigan Mathematical Journal 9, 289–298, 1962.
[KMP] Karhumäki, J., Mañuch, J., Plandowski, W., On defect effect of bi-infinite words, TUCS Report 181, 1998.

On Repetition-Free Binary Words of Minimal Density

(Extended Abstract)

Roman Kolpakov[1] *, Gregory Kucherov[1], and Yuri Tarannikov[3]

[1] INRIA-Lorraine/LORIA, 615, rue du Jardin Botanique, B.P. 101, 54602 Villers-lès-Nancy, France, e-mail: {roman,kucherov}@loria.fr.
[2] Faculty of Discrete Mathematics, Department of Mechanics and Mathematics, Moscow State University, 119899 Moscow, Russia, e-mail: yutaran@nw.math.msu.su

1 Introduction

In this paper we continue with the study initiated in [12]. The general problem behind this study can be described as follows. Assume we have specified a set of "prohibited" words $P \subseteq A^*$ and we are interested in the set $F \subseteq A^*$ of words that don't contain words from P as subwords. Words of F are said to *avoid* P. If the set F is infinite, the set P is called *avoidable*, otherwise it is called *unavoidable*. One might specify, for example, a finite number of prohibited subwords P. Properties of unavoidable finite sets of words were studied in [13]. The set P of prohibited subwords can be infinite, in which case it may be specified by one or several *patterns*, i.e. words composed with variables and possibly with alphabet letters. Pattern avoidability has been subject of many works, and we refer to [7,6] for an introduction to this area and a survey of known results. One might think of other ways of specifying the set of subwords to be avoided, e.g. as a language specified by a grammar. Note that for any set P of prohibited subwords, the set F of avoiding words is closed under taking subwords, and vice versa, any set F closed under subwords is the set of avoiding words for some P (just take $P = A^* \setminus F$). Therefore, being closed under subwords can be considered as a characterization for the sets of words that can be specified by means of prohibited subwords.

The case when the prohibited subwords are those of the form u^n, for some $n \geq 2$, has been extensively studied. Such subwords are called n-repetitions or n-powers, and words that don't contain such subwords are called n-th power-free. Back in the beginning of the century, Thue proved that there exist infinite 2-nd power-free (square-free) words over the three-letter alphabet, and 3-rd power-free (cube-free) words over the two-letter alphabet [15,16] (see also [4]). To recast it in terms of pattern avoidability, Thue showed that pattern xx is avoidable on the three-letter alphabet, and pattern xxx is avoidable on the two-letter alphabet. Note that xx is unavoidable on the 2-letter alphabet, and this illustrates the

---

* On leave from French-Russian Institute for Informatics and Applied Mathematics, Moscow University, 119899 Moscow, Russia

fact that a set of subwords (or patterns) can be avoidable on some alphabet and unavoidable on a smaller alphabet. Paper [2] contains an example of a pattern which is avoidable on four letters but not on three letters. The question whether a given pattern is avoidable on the k-letter alphabet (for a given k) is not known to be decidable (see [9]) even if patterns are composed only of variables. In contrast, the question whether a given pattern is unavoidable on *any* alphabet has been shown decidable in [3,1].

This paper is motivated by the following general question: If a pattern p is unavoidable on k letters but avoidable on $(k+1)$ letters, what is the minimal proportion (density) of a letter in words over $(k+1)$ letters avoiding p? In other terms, what is the minimal contribution (in terms of relative number of occurrences) of the $(k+1)$-st letter that allows to create words of unbounded length avoiding p? Note that the "minimal proportion" is understood here as the limit minimal proportion as the length of words goes to infinity. An answer to this question would establish a relationship between two properties of different kind: avoidance of a certain pattern (regularity) and proportion of occurrences of a letter.

To the best of our knowledge, minimal density has been first studied in a related paper [12]. However, some work had been done on counting limit densities of subwords in words defined by DOL-systems (cf e.g. [11]). In [12], this study was undertaken for the case of n-th power-free words on the 2-letter alphabet, and some first results were obtained. Here we continue with this analysis, and considerably extend the results of [12]. First, we analyse the very notion of minimal limit proportion (density) of a letter by comparing different possible definitions. In particular, we prove that two natural definitions, through finite and infinite words, lead actually to the same quantity. This confirms the significance of this notion and the interest of studying it. We then analyse the minimal proportion $\rho(n)$ of one letter in n-th power-free binary words. In [12] it has been shown that $\rho(n) = \frac{1}{n} + \mathcal{O}(\frac{1}{n^2})$. Here we obtain a much more precise estimate by computing the first four terms of the asymptotic expansion of $\rho(n)$. Specifically, we show that $\rho(n) = \frac{1}{n} + \frac{1}{n^3} + \frac{1}{n^4} + \mathcal{O}(\frac{1}{n^5})$. Then we turn to the analysis of the generalized minimal density $\rho(x)$, defined for all real $x > 2$. This generalization, based on the notion of period of a word, was introduced in [12]. It was shown, in particular, that $\rho(x)$, considered as a real function, is discontinuous, as it admits a jump at $x = \frac{7}{3}$. Here we prove much more, namely that $\rho(x)$ has actually an infinity of discontinuity points, as those are all integer points $n \geq 3$. Futhermore, we give an estimate for $\rho(n+0)$ – the right limit of $\rho(x)$ at integer points $n \geq 3$ – and prove that $\rho(n+0) = \frac{1}{n} - \frac{1}{n^2} + \frac{2}{n^3} - \frac{2}{n^4} + \mathcal{O}(\frac{1}{n^5})$.

As usual, A^* denotes the free monoid over an alphabet A. $u \in A^*$ is a *subword* of $w \in A^*$ if w can be written as $u_1 u u_2$ for some $u_1 u_2 \in A^*$. $|u|$ stands for the length of $u \in A^*$. A^ω stands for the set of *one-way infinite* words, often called ω-words, over A, that are defined as mappings $\mathbb{N} \to A$. For $n \in \mathbb{N}$, the word w obtained by concatenating n copies of a word v is called the *n-th power* of v and denoted by v^n. A word v is a period of w iff w is a subword of v^n for some $n \in \mathbb{N}$.

2   Minimal Density: General Definition and Properties

Assume we have an infinite set $F \subseteq A^*$ which is *closed under subwords*, that is if a word w is in F, then any subword of w belongs to F too. As noted in Introduction, the property of being closed under subwords characterizes the class of languages that can be specified by a set of prohibited subwords. As F is infinite and closed under subwords, there exist an infinite word from A^ω such that its every finite subword belongs to F. With interpretation of subword avoidance, this allows to speak about infinite words avoiding the set of subwords. We denote by F^ω the set of infinite words of A^ω with every finite subword belonging to F.

Let $a \in A$ be a distinguished letter, and we are interested in the minimal limit proportion of a's in words of F of unbounded length. For $w \in F$, define $c_a(w)$ to be the number of occurrences of a in w and $\rho_a(w) = \frac{c_a(w)}{|w|}$. Denote $F(l) = \{w \in F \mid |w| = l\}$.

Definition 1. *For every $l \in \mathbb{N}$, let $\rho_a(F, l) = \frac{1}{l} \min_{w \in F(l)} c_a(w)$ and $\rho_a(F) = \underline{\lim}_{l \to \infty} \rho_a(F, l)$. $\rho_a(F)$ is called the* minimal (limit) density *of a in F.*

Note that the type of argument of ρ_a will always make it clear if the density of an individual word, or the minimal density is meant.

Obviously, all numbers $\rho_a(F, l)$ belong to $[0, 1]$ and therefore $\rho_a(F)$ belongs to $[0, 1]$ too. The following two Lemmas clarify the behaviour of the sequence $\{\rho_a(F, l)\}_{l=1}^\infty$ with respect to $\rho_a(F)$. They are direct generalizations of Propositions 1,2 from [12] and are given without proof.

Lemma 1. *For every $l \in \mathbb{N}$, $\rho_a(F, l) \leq \rho_a(F)$.*

Lemma 2. $\rho_a(F) = \lim_{l \to \infty} \rho_a(F, l) = \sup_{l \geq 1} \rho_a(F, l)$.

By Lemma 2, the lower limit in Definition 1 can be replaced by the simple limit. Thus, the definition $\rho_a(F) = \lim_{l \to \infty} \min_{w \in F(l)} \rho_a(w)$ is correct and seems to capture in a right way the notion of the minimal density. However, there is another natural way to define the minimal limit density directly in terms of infinite words F^ω, and one may ask if this can lead to a different density value.

For a word $w \in F \cup F^\omega$, let $w[1:j]$ denotes the prefix of w of length j. The density of letter a in an infinite word $v \in F^\omega$ is naturally defined as the limit $\lim_{j \to \infty} \rho_a(v[1:j])$. Obviously, this limit may not exist. However, below we show that among all words for which this limit exists, there is one that realizes the minimum of these limits, which is equal to $\rho_a(F)$. This confirms that $\rho_a(F)$ is the right quantity caracterizing the limit density.

We define an auxiliary measure $\sigma_a(F, l) = \min_{w \in F(l)} \max_{1 \leq j \leq l} \rho_a(w[1:j])$. The following lemma gives a key argument.

Lemma 3. *For every $l \in \mathbb{N}$, $\rho_a(F, l) \leq \sigma_a(F, l) \leq \rho_a(F)$.*

Proof. It is easily seen that $\sigma_a(F, l) \geq \rho_a(F, l)$. Let us prove that $\sigma_a(F, l) \leq \rho_a(F)$ for all $l \in \mathbb{N}$. Assume that $\sigma_a(F, L) > \rho_a(F)$ for some $L \in \mathbb{N}$. This means that every word $v \in F$ of length at least L has a prefix $v[1 : j]$ with $\rho_a(v[1 : j]) > \rho_a(F)$. Let $\varepsilon = \min\{\rho_a(v[1 : j]) - \rho_a(F)\}$ where minimum is taken over all such prefixes. Take any word $w \in F(N)$ with $N > \frac{2L}{\varepsilon}(\rho_a(F) + \varepsilon)$. Find a decomposition $w = w_1 w_2 \ldots w_m$ such that $|w_j| \leq L$ and $\rho_a(w_j) > \rho_a(F)$ for every j, $1 \leq j \leq m - 1$, and $|w_m| < L$. Then $c_a(w) \geq (\rho_a(F) + \varepsilon)(|w| - L)$ and $\rho_a(w) \geq \rho_a(F) + \varepsilon - L(\frac{\rho_a(F)+\varepsilon}{|w|}) \geq \rho_a(F) + \frac{\varepsilon}{2}$. Since w was chosen arbitrarily, this contradicts to $\rho_a(F, N) \leq \rho_a(F)$ (Lemma 1).

Corollary 1. *The limit $\lim_{l \to \infty} \sigma_a(F, l)$ exists and is equal to $\rho_a(F)$.*

Lemma 4. *There exists a word $v \in F^\omega$ such that $\lim_{j \to \infty} \rho_a(v[1 : j])$ exists and is equal to $\rho_a(F)$.*

Proof. From Lemma 3 it follows that for every $l \in \mathbb{N}$, there exists a word $w \in F(l)$ with $\rho(w) = \sigma_a(F, l) \leq \rho_a(F)$, that is $\max_{1 \leq j \leq |w|} \rho_a(w[1 : j]) \leq \rho_a(F)$. Moreover, every prefix of w verifies the same inequality. Therefore, the set of words w verifying the inequality forms an infinite tree with respect to the prefix relation such that the parent of a word w in the tree is its immediate prefix, obtained by removing the rightmost letter. Since the alphabet A is finite, the tree is finitely branching. By König's Lemma, there exists an infinite path in this tree which defines the infinite word v with $\rho_a(v[1 : j]) \leq \rho_a(F)$ for all $j \in \mathbb{N}$. Since $\rho_a(F, j) \leq \rho(v[1 : j]) \leq \rho_a(F)$, the result follows from Lemma 2.

Lemma 5. $\min_{v \in F^\omega} \lim_{j \to \infty} \rho_a(v[1 : j]) = \rho_a(F)$, *where minimum is taken over all $v \in F^\omega$ for which the limit exists.*

Proof. By Lemma 4, there exists a word $v \in F^\omega$ such that $\lim_{j \to \infty} \rho_a(v[1 : j]) = \rho_a(F)$. Therefore, $\inf_{v \in F^\omega} \lim_{j \to \infty} \rho_a(v[1 : j]) \leq \rho_a(F)$. On the other hand, since $v[1 : j] \in F(j)$, then $\rho_a(v[1 : j]) \geq \rho_a(F, j)$, then $\lim_{j \to \infty} \rho_a(v[1 : j]) \geq \lim_{j \to \infty} \rho_a(F, j) = \rho_a(F)$ and $\inf_{v \in F^\omega} \lim_{j \to \infty} \rho_a(v[1 : j]) \geq \rho_a(F)$. The lemma follows.

Lemmas 4 and 5 imply that there exists a word $v \in F^\omega$ that realizes the minimal limit $\lim_{j \to \infty} \rho_a(v[1 : j])$ among all words of F^ω for which the limit exists. Moreover, this minimum is equal $\rho_a(F)$. To avoid the problem of existence of the limit, we could replace it by the lower limit and define the quantity $\inf_{v \in F^\omega} \underline{\lim}_{j \to \infty} \rho_a(v[1 : j])$ where the infimum is taken over *all* words $v \in F^\omega$. The proof of Lemma 5 shows that this value is also equal to $\rho_a(F)$, and the infimum is reached on some word $v \in F^\omega$.

The equivalence of different definitions gives a strong evidence that $\rho_a(F)$ is an interesting quantity to study. In this paper, we undertake this study for a particular family of sets F – the sets of n-th power-free binary words.

3 Minimal Letter Density in n-th Power-Free Binary Words

Consider an alphabet A. For a natural $n \geq 2$, a word $w \in A^*$ is called *n-th power-free* iff it does not contain a subword which is the n-th power of some non-empty word. We denote $PF(n) \subseteq A^*$ the set of n-th power-free finite words. Words from $PF(2)$ are called *square-free*, and words from $PF(3)$ are called *cube-free*. If $w \in A^*$ does not contain a subword uua, where u is a non-empty word and a is the first letter of u, then w is called *strongly cube-free*. An equivalent property (see [14]) is overlap-freeness – w is *overlap-free* if it does not contain two overlapping occurrences of a non-empty word u. Well known Thue's results [15,16] state that there exist square-free words of unbounded length on the 3-letter alphabet, and strongly cube-free words of unbounded length on the 2-letter alphabet. Note that the existence of infinite strongly cube-free words on the 2-letter alphabet implies that for that alphabet the set $PF(n)$ is infinite for every $n \geq 3$.

From now on we fix on the binary alphabet $A = \{0, 1\}$. Our goal is to compute, for all $n > 2$, the value $\rho_1(PF(n))$ – minimal density of 1 in the words $PF(n)$. Note that by symmetry, $\rho_1(PF(n)) = \rho_0(PF(n))$, and to simplify the notation, we denote $\rho_1(PF(n))$ (respectively $\rho_1(PF(n), l)$) by $\rho(n)$ (respectively $\rho(n, l)$) in the sequel. Similarly, we will drop the index in $c_1(w)$ and $\rho_1(w)$, and will write $c(w)$ and $\rho(w)$ instead.

In [12] it has been proved that $\rho(n) = \frac{1}{n} + \mathcal{O}(\frac{1}{n^2})$. Here, using a different method, we prove the following more precise estimation, that corresponds to the first four terms in the asymptotic expansion of $\rho(n)$.

Theorem 1. $\rho(n) = \frac{1}{n} + \frac{1}{n^3} + \frac{1}{n^4} + \mathcal{O}(\frac{1}{n^5})$.

We first establish the upper bound

$$\rho(n) \leq \frac{1}{n} + \frac{1}{n^3} + \frac{1}{n^4} + \mathcal{O}(\frac{1}{n^5}). \quad (1)$$

The proof is based on the following lemma. Denote by α_i the word $0^i 1$.

Lemma 6. *Let $k \geq 3$. For i, j, $0 \leq i, j \leq k$ and $i \neq j$, consider a morphism $h : \{0, 1\}^* \to \{0, 1\}^*$ defined by $h(0) = \alpha_i$, $h(1) = \alpha_j$. For a word $w \in \{0, 1\}^*$, if $w \in PF(k)$ then $h(w) \in PF(k + 1)$.*

Proof. First observe that $\{h(0), h(1)\}$ is a prefix code, i.e. the inverse image w of any word $h(w)$ is unique. Furthermore, for any $u \in \{0, 1\}^*$, the occurrences of 1 in $h(u)$ delimit the images of individual letters of w. This means that any subword of $h(w)$ which ends with 1 and is preceeded by 1 (or starts at the beginning of $h(w)$) is the image of some subword of w.

To prove the lemma, assume by contradiction that for some $w \in PF(k)$, $h(w)$ contains a subword v^{k+1}. Proceed by case analysis on the number of 1's in v. If v contains no 1's, then v^{k+1} contains at least $k + 1$ consecutive 0's which is impossible as $h(w)$ is a concatenation of words α_i, α_j. If v contains one 1, then $v = 0^l 10^m$, and $v^{k+1} = 0^l 1(0^{l+m} 1)^k 0^m$. Since $h(w) \in \{\alpha_i, \alpha_j\}^*$, we conclude that $l+m \in \{i,j\}$ and w must contain k consecutive occurrences of the letter $h^{-1}(0^{l+m}1)$. Finally, if v contains s 1's, then $v = 0^l 1 \alpha_{i_1} \ldots \alpha_{i_{s-1}} 0^m$, and $v^{k+1} = 0^l 1 (\alpha_{i_1} \ldots \alpha_{i_{s-1}} 0^{l+m} 1)^k 0^m$. Again, $l+m \in \{i,j\}$ and w contains the k-th power of the inverse image $h^{-1}(\alpha_{i_1} \ldots \alpha_{i_{s-1}} 0^{l+m} 1)$.

Lemma 7. *For every $n \geq 4$,*

$$\rho(n) \leq \frac{1}{n - \rho(n-1)} \tag{2}$$

Proof. For $l \in \mathbb{N}$, take a word $w \in PF(n-1)$ with $|w| = l$ and $\rho(w) = \rho(n-1, l)$. Denote by h the morphism defined by $h(0) = \alpha_{n-1}$, $h(1) = \alpha_{n-2}$. Let $u = h(w)$. By Lemma 6, $u \in PF(n)$. Since $c(u) = |w|$, and $|u| = (n-1)c(w) + n(|w|-c(w)) = n|w| - c(w)$, we have $\rho(n, |u|) \leq \rho(u) = \frac{c(u)}{|u|} = \frac{1}{n - \rho(w)} = \frac{1}{n - \rho(n-1, l)}$. Taking the limit for $l \to \infty$, and then $|u| \to \infty$, we have $\rho(n) \leq \frac{1}{n - \rho(n-1)}$.

Upper bound (1) is now proved as follows. (2) trivially implies $\rho(n) \leq \frac{1}{n - 1/2} = \frac{1}{n} + \mathcal{O}(\frac{1}{n^2})$. Then, from (2) again, $\rho(n) \leq \frac{1}{n - \frac{1}{n-1} + \mathcal{O}(\frac{1}{n^2})} = \frac{1}{n} + \frac{1}{n^3} + \mathcal{O}(\frac{1}{n^4})$. Substituting this into (2) again, we have $\rho(n) \leq \frac{1}{n - (\frac{1}{n-1} + \frac{1}{(n-1)^3} + \mathcal{O}(\frac{1}{n^4}))} = \frac{1}{n} + \frac{1}{n^3} + \frac{1}{n^4} + \frac{3}{n^5} + \mathcal{O}(\frac{1}{n^6})$. This subsumes (1).

Now we turn to bounding $\rho(n)$ from below, and prove the following lower bound.

$$\rho(n) \geq \frac{n-1}{n^2 - n - 1} \tag{3}$$

for all $n \geq 3$.

Consider an arbitrary finite n-th power-free word w. First, group its letters into blocks $\alpha_i = 0^i 1$, $0 \leq i \leq n-1$. For a technical reason we assume that w does not start with α_{n-1}. If it does, we temporarily remove the first symbol 0. w is uniquely decomposed into a concatenation of α_i's and a suffix of at most $n-1$ 0's. Then, we group occurrences of α_i's into larger blocks $\beta(m,k) = (\alpha_{n-1})^m \alpha_k$, $0 \leq m \leq n-1$, $0 \leq k \leq n-2$. Informally, blocks β are delimited by occurrences of α_i with $i \leq n-2$. Again, w is uniquely decomposed into blocks β and the remaining suffix Q of length at most $n^2 - 1$ ($n-1$ occurrences of α_{n-1} followed by $n-1$ 0's). We proceed by grouping blocks β into yet more large blocks. Let

$$\gamma(l, k_0, k_1, \ldots, k_s) = \beta(l, k_0) \beta(n-1, k_1) \ldots \beta(n-1, k_s) = (\alpha_{n-1})^l \alpha_{k_0} (\alpha_{n-1})^{n-1} \alpha_{k_1} \ldots (\alpha_{n-1})^{n-1} \alpha_{k_s},$$

where $0 \leq l \leq n-2$, $s \geq 0$, $0 \leq k_0, k_1, \ldots, k_s \leq n-2$. Blocks γ are delimited by each occurrence of $\beta(l, k)$ with $l \leq n-2$. Note that since w starts with α_k, $k \leq n-2$, it starts with $\beta(0, k)$ and therefore the first block γ starts at the beginning of w. Thus, the decomposition of w is uniquely defined with a possibly remaining suffix Q of length up to $n^2 - 1$. Taking into account the first possibly removed 0, we have $w = Pw'Q$, where $|P| \leq 1$, $|Q| \leq n^2 - 1$, and w' is uniquely decomposed into blocks γ.

Let us now compute the minimal possible ratio of 1's in blocks γ. Consider a block $\gamma(l, k_0, k_1, \ldots, k_s)$. We distingish two cases:

Case $s \geq 1$: We claim that $k_j + k_{j+1} \leq n-2$ for every j, $0 \leq j \leq s-1$. Indeed, consider the subword $\alpha_{k_j}(\alpha_{n-1})^{n-1}\alpha_{k_{j+1}}$ of $\gamma(l, k_0, k_1, \ldots, k_s)$. If $k_j + k_{j+1} \geq n-1$ then it has the prefix $(0^{k_j}10^{n-1-k_j})^n$ which contradicts to the n-th power-freeness of w.

Using this observation, we can bound $\sum_{j=0}^{s} |\alpha_{k_j}|$ by $\frac{s+1}{2}n$ when s is odd, and by $\frac{s}{2}n + (n-1)$ when s is even. Then $|\gamma(l, k_0, k_1, \ldots, k_s)| \leq \frac{s}{2}n + (n-1) + sn(n-1) + ln = s(n^2 - \frac{n}{2}) + nl + n - 1$. Since the number of 1's in $\gamma(l, k_0, k_1, \ldots, k_s)$ is $ns + l + 1$, we have $\rho(\gamma(l, k_0, k_1, \ldots, k_s)) \geq \frac{ns+l+1}{s(n^2-\frac{n}{2})+nl+n-1}$. The right-hand side minimizes when l is maximal ($l = n-2$) and s is minimal ($s = 1$). We then obtain $\rho(\gamma(l, k_0, k_1, \ldots, k_s)) \geq \frac{2n-1}{2n^2-\frac{3n}{2}-1}$.

Case $s = 0$: In this case $\gamma(l, k_0) = \beta(l, k_0)$, $|\gamma(l, k_0)| = ln + k_0 + 1$, and $\rho(\gamma(l, k_0)) = \frac{l+1}{ln+k_0+1}$. The right-hand mininizes when both l and k_0 are maximal ($l = k_0 = n-2$), which gives $\rho(\gamma(l, k_0)) \geq \frac{n-1}{n^2-n-1}$.

The second case gives a smaller bound for all $n \geq 3$ and we conclude that $\rho(\gamma(l, k_0, \ldots, k_s)) \geq \frac{n-1}{n^2-n-1}$. Since w' is a concatenation of blocks γ, this implies $\rho(w') \geq \frac{n-1}{n^2-n-1}$. Returning to w, we have $c(w) \geq c(w') \geq \frac{n-1}{n^2-n-1}|w'| \geq \frac{n-1}{n^2-n-1}(|w| - n^2)$, and then $\rho(w) = \frac{c(w)}{|w|} \geq \frac{n-1}{n^2-n-1}(1 - \frac{n^2}{|w|})$. As w is an arbitrary n-th power-free word, we have $\rho(n, l) \geq \frac{n-1}{n^2-n-1}(1 - \frac{n^2}{l})$ for all l. Taking the limit for l going to infinity, we obtain $\rho(n) \geq \frac{n-1}{n^2-n-1}$. This implies in particular that

$$\rho(n) \geq \frac{1}{n} + \frac{1}{n^3} + \frac{1}{n^4} + \frac{2}{n^5} \tag{4}$$

Lower bound (4) together with upper bound (1) implies Theorem 1.

4 Generalized Minimal Density Function

Following [12], we consider in this Section a natural generalization of function $\rho(n)$ to real arguments. Recall that the exponent of a word w is the ratio $\frac{|w|}{\min |v|}$, where the minimum is taken over all periods v of w. The exponent is a useful notion often used in word combinatorics (see [10,5,8]), that generalizes the notion of n-th power. For example, Dejean proved that on the 3-letter alphabet, there exist infinite words that don't contain any subword of exponent more than $\frac{7}{4}$. This strengthens Thue's result on the existence of square-free words (i.e. words without subwords of exponent 2) over the 3-letter alphabet.

Using periods, function $\rho(n)$ can be defined on real numbers in the following way. For a real number x, define $PF(x)$ (resp. $PF(x+\varepsilon)$) to be the set of binary words that do not contain a subword of exponent greater than or equal to (resp. strictly greater than) x.

Note that $PF(2+\varepsilon)$ is precisely the class of strongly cube-free words. For the binary alphabet, the existence of infinite cube-free words implies that $PF(x)$ (resp. $PF(x+\varepsilon)$) is infinite for $x > 2$ (resp. for $x \geq 2$). Using the results of

Section 2, values $\rho_1(PF(x))$ and $\rho_1(PF(x+\varepsilon))$ are well-defined for $x > 2$ and $x \geq 2$ respectively. Similar to the previous section, we denote them respectively by $\rho(x)$ and $\rho(x + \varepsilon)$. Notation $\rho(x, l)$ and $\rho(x + \varepsilon, l)$ is defined accordingly. Observe that for natural values of $x > 2$, $\rho(x)$ coincides with $\rho(n)$ studied in the previous section.

Functions $\rho(x), \rho(x+\varepsilon)$ are non-increasing with values from $[0, \frac{1}{2}]$. This implies the existence, for every $x > 2$, of the right limit $\rho(x + 0)$, that verifies $\rho(x + 0) = \sup_{y>x} \rho(y)$. The following lemma is from [12].

Lemma 8. *For every $x > 2$, $\rho(x + 0) = \rho(x + \varepsilon)$.*

In [12], it has been shown that $\rho(x) = \frac{1}{2}$ for $x \in (2, \frac{7}{3}]$, and then proved that the right limit of $\rho(x)$ at $x = \frac{7}{3}$ is strictly smaller than $\frac{1}{2}$, implying that $\rho(x)$ has a jump to the right of $x = \frac{7}{3}$. Here we complement this result by proving that $\rho(x)$ has an infinite number of discontinuity points. We show that, besides $x = 7/3$, the function $\rho(x)$ is discontinuous to the right at all integer points $x \geq 3$. The following lemma is somewhat similar to Lemma 6. Recall that $\alpha_i = 0^i 1$.

Lemma 9. *Let $A = \{a_1, \ldots, a_k\}$ and $n \geq 3$. Let $h : A \to \{0, 1\}$ be a morphism such that $h(a_i) = \alpha_{m_i}$, where $m_i \leq n$ for all $1 \leq i \leq k$, and $m_i \neq m_j$ for all $i \neq j$. Then for every $(n-1)$-th power-free word $w \in A^*$, $h(w)$ is $(n + \varepsilon)$-th power-free.*

We skip the proof which goes along the same lines as that of Lemma 6.

Lemma 10. *For every $n \geq 4$,*

$$\rho(n + \varepsilon) \leq \frac{1}{n + 1 - \rho(n-1)} \qquad (5)$$

Proof. Denote by $h_n : \{0,1\}^* \to \{0,1\}^*$ the morphism defined by $h_n(0) = \alpha_n$, $h_n(1) = \alpha_{n-1}$. Let w_l be an $(n-1)$-th power-free word of length l with minimal number of 1's ($\rho(w_l) = \rho(n-1, l)$). Clearly, $|h_n(w_l)| = (n+1)(l - c(w_l)) + nc(w_l) = (n+1)l - c(w_l)$, and $c(h_n(w_l)) = l$. By Lemma 9, $h_n(w_l)$ is $(n+\varepsilon)$-th power-free, and we have

$$\rho(n + \varepsilon, |h_n(w_l)|) \leq \rho(h_n(w_l)) = \frac{l}{(n+1)l - c(w_l)} = \frac{1}{n + 1 - \rho(n-1, l)}$$

By taking the limit for $l \to \infty$ (see Lemma 2), inequality (5) follows.

Inequality (5) together with the trivial inequality $\rho(n-1) \leq 1/2$ gives $\rho(n + \varepsilon) \leq \frac{1}{n-1/2} < \frac{1}{n}$ for $n \geq 4$. On the other hand, from lower bound (3) it follows that $\rho(n) \geq \frac{n-1}{n^2-n-1} > \frac{1}{n}$. This implies that $\rho(n + 0) = \rho(n + \varepsilon) < \rho(n)$, that is $\rho(x)$ has a jump to the right of all integer points $n \geq 4$.

For $n = 3$, inequality (5) does not make sense ($\rho(2)$ is not defined). Therefore, the case $n = 3$ should be analysed separately.

Lemma 11. $\rho(3+\varepsilon) \leq \frac{1}{3}$.

Proof. Take a 3-letter alphabet $A = \{1, 2, 3\}$. For $w \in A^*$, let $c_i(w)$ ($i = 1, 2, 3$) denote the number of occurrences of i in w. For any $l \in \mathbb{N}$, choose a square-free word $w_l \in A^*$ of length l such that $c_1(w) \leq c_2(w) \leq c_3(w)$. Note that for all $l \in \mathbb{N}$, w_l is well-defined, which follows from the existence of infinite square-free words on the 3-letter alphabet. Consider the morphism $h : A^* \to \{0,1\}^*$ defined by $h(1) = 01$, $h(2) = 001$, $h(3) = 0001$. Then $|h(w_l)| = 2c_1(w_l) + 3c_2(w_l) + 4c_3(w_l) = 3l + (c_3(w_l) - c_1(w_l)) \geq 3l$, and $\rho(h(w_l)) \leq \frac{l}{3l} = \frac{1}{3}$. By Lemma 9, word w_l is $(3+\varepsilon)$-th power-free, and then $\rho(3+\varepsilon, |h(w_l)|) \leq \frac{1}{3}$. Taking the limit for $l \to \infty$ and using Lemma 2, we get $\rho(3+\varepsilon) \leq \frac{1}{3}$.

On the other hand, from lower bound (3) it follows that $\rho(3) \geq \frac{2}{5}$. Therefore, $\rho(x)$ has a jump to the right of $x = 3$. Putting all together, we obtain

Theorem 2. $\rho(x)$ *is discontinuous to the right of* $x = \frac{7}{3}$ *as well as to the right of all natural points* $n \geq 3$.

Finally, we compute first four terms in the asymptotic expansion of $\rho(n + \varepsilon)$. Together with Theorem 1, this will give an estimate to the size of jumps postulated by Theorem 2. Recall that $\rho(n + \varepsilon) = \rho(n + 0)$ by Lemma 8.

The following lower bound holds for all $n \geq 3$.

$$\rho(n+\varepsilon) \geq \frac{n-1}{n^2-2} \qquad (6)$$

We omit the proof which follows closely the proof of lower bound (3).

Expanding the right-hand side of (6), we have

$$\rho(n+\varepsilon) \geq \frac{1}{n} - \frac{1}{n^2} + \frac{2}{n^3} - \frac{2}{n^4} + \mathcal{O}(\frac{1}{n^5}). \qquad (7)$$

To obtain an upper bound to $\rho(n+\varepsilon)$ that matches lower bound (7), it suffices to substitute into inequality (5) the upper bound $\rho(n-1) \leq \frac{1}{n-1} + \mathcal{O}(\frac{1}{n^3})$ implied by (1) (instead of trivial upper bound $\rho(n-1) \leq \frac{1}{2}$ as above). We then get

$$\rho(n+\varepsilon) \leq \frac{1}{n+1-\frac{1}{n-1}+\mathcal{O}(\frac{1}{n^3})} = \frac{1}{n} - \frac{1}{n^2} + \frac{2}{n^3} - \frac{2}{n^4} + \mathcal{O}(\frac{1}{n^5})$$

Together with (7), this gives

Theorem 3. $\rho(n+\varepsilon) = \frac{1}{n} - \frac{1}{n^2} + \frac{2}{n^3} - \frac{2}{n^4} + \mathcal{O}(\frac{1}{n^5})$.

5 Concluding Remarks

In this paper we have continued with the study of minimal density function $\rho(x)$, introduced in [12]. We analysed this notion in a general framework, and proved that different possible definitions are actually equivalent. Then, for the case of repetition-free binary words, we have extended several results of [12].

Specifically, we have given a more precise estimation for the values of $\rho(n)$, and we proved that $\rho(x)$, considered as a function on real argument, is discontinuous to the right of all integer values of x. Finally, we gave an estimate of values $\rho(n)$ and $\rho(n+\varepsilon)$.

Many questions about minimal density function $\rho(x)$ remain open. Does it have other discontinuities? What are they? Is this function piece-wise constant? All these questions are still to be answered.

Acknowledgements This work was supported by the French-Russian A.M.Liapunov Institut of Applied Mathematics and Informatics. The first and third authors were supported by the Russian Foundation of Fundamental Research (grant 96-01-01068). We are grateful to Alexander Ugolnikov and Vladimir Grebinski for their help, and to the anonymous referees for their useful remarks.

References

1. А.И. Зимин. Блокирующие множества термов. *Математический Сборник*, 119(3):363–375, 1982. English Translation: A.I.Zimin, Blocking sets of terms, Math. USSR Sbornik 47 (1984), 353-364.
2. K. Baker, G. McNulty, and W. Taylor. Growth problems for avoidable words. *Theoret. Comp. Sci.*, 69:319–345, 1989.
3. D. Bean, A. Ehrenfeucht, and G. McNulty. Avoidable patterns in strings of symbols. *Pacific J. Math.*, 85(2):261–294, 1979.
4. J. Berstel. Axel thue's work on repetitions in words. Invited Lecture at the 4th Conference on Formal Power Series and Algebraic Combinatorics, Montreal, 1992, June 1992. accessible at http://www-litp.ibp.fr:80/berstel/.
5. J. Berstel and D. Perrin. *Theory of codes*. Academic Press, 1985.
6. J. Cassaigne. *Motifs évitables et régularités dans les mots*. Thèse de doctorat, Université Paris VI, 1994.
7. C. Choffrut and J. Karhumäki. Combinatorics of words. In G. Rozenberg and A. Salomaa, editors, *Handbook on Formal Languages*, volume I. Springer, Berlin-Heidelberg-New York, 1996.
8. M. Crochemore and P. Goralcik. Mutually avoiding ternary words of small exponent. *International Journal of Algebra and Computation*, 1(4):407–410, 1991.
9. J. Currie. Open problems in pattern avoidance. *American Mathematical Monthly*, 100:790–793, 1993.
10. F. Dejean. Sur un théorème de Thue. *J. Combinatorial Th. (A)*, 13:90–99, 1972.
11. M. Dekking. On the Thue-Morse measure. *Acta Univ. Carolin. Math. Phis*, 33(2):35–40, 1992.
12. R. Kolpakov and G. Kucherov. Minimal letter frequency in n-power-free binary words. In *Proceedings of the 22nd International Symposium on Mathematical Foundations of Computer Science (MFCS), Bratislava (Slovakia)*, volume 1295 of *Lecture Notes in Computer Science*, pages 347–357. Springer Verlag, 1997.
13. L. Rosaz. Making the inventory of unavoidable sets of words of fixed cardinality. *Theoretical Computer Science*, 1998. to appear.
14. A. Salomaa. *Jewels of formal language theory*. Computer Science Press, 1986.
15. A. Thue. Über unendliche Zeichenreihen. *Norske Vid. Selsk. Skr. I. Mat. Nat. Kl. Christiania*, 7:1–22, 1906.
16. A. Thue. Über die gegenseitige Lage gleicher Teile gewisser Zeichenreihen. *Norske Vid. Selsk. Skr. I. Mat. Nat. Kl. Christiania*, 10:1–67, 1912.

Embedding of Hypercubes into Grids *

S.L. Bezrukov[1], J.D. Chavez[2], L.H. Harper[3], M. Röttger[1], and U.-P. Schroeder[1]

[1] Department of Math. and Computer Science, University of Paderborn,
D-33102 Paderborn, Germany
[2] Department of Math., California State University, San Bernandino, CA 92407, USA
[3] Department of Math., California State University, Riverside, CA 92521, USA.

Abstract. We consider one-to-one embeddings of the n-dimensional hypercube into grids with 2^n vertices and present lower and upper bounds and asymptotic estimates for minimal dilation, edge-congestion, and their mean values. We also introduce and study two new cost-measures for these embeddings, namely the sum over $i = 1, ..., n$ of dilations and the sum of edge-congestions caused by the hypercube edges of the ith dimension. It is shown that, in the simulation via the embedding approach, such measures are much more suitable for evaluating the slowdown of uniaxial hypercube algorithms then the traditional cost measures.

1 Introduction

The power of message-passing multiprocessor systems strictly depends on the structure of their interconnection network. Various types of network topologies have gained favor and are in use today. Among these, grids are emerging as one of the most popular network architectures. An obvious reason for the popularity of grids lies in their simple structure, which provides easy construction and scaling-up of such systems. On the other hand, a large number of algorithms and programming techniques (e.g. the ascend-descend algorithms) are developed for the hypercube. Its popularity, efficiency, and versatility as a programming network model is mostly due to its recursive structure that is well suited for many algorithms. In order to run these efficient hypercube algorithms on a grid-based parallel computer, one has to simulate their communication requirements. A natural approach to such simulation would be via graph embedding.

We denote by $V(G)$ and $E(G)$ the vertex and edge set of a graph G. An embedding $f = (\phi_V, \phi_E)$ of a graph G into a graph H is an injective mapping $\phi_V : V(G) \mapsto V(H)$ together with a routing scheme $\phi_E : E(G) \mapsto 2^{E(H)}$, which assigns for each edge $e = \{u, v\} \in E(G)$ some path $\phi_E(e)$ from $\phi_V(u)$ to $\phi_V(v)$ in H. The quality of an embedding f may be expressed in terms of *dilation* and *edge-congestion*, in order to satisfy the demands of process locality and communication efficiency. For an edge $e \in E(G)$ its dilation $\mathbf{dil}_f(e)$ in the

---

* This work was supported by the DFG-Sonderforschungsbereich 376 "Massive Parallelität: Algorithmen, Entwurfsmethoden, Anwendungen" and by the EC ESPRIT Long Term Research Project 20244 "ALCOM-IT".

embedding f is defined as the length of the path $\phi_E(e)$ in H. Now, for an edge $e' \in E(H)$ we put its congestion $\mathbf{con}_f(e')$ in the embedding f to be equal to $|\{e \in E(G) \mid e' \in \phi_E(e)\}|$. In this paper we consider embeddings of hypercubes into grids. As usual by the *hypercube* of dimension $n \in \mathbb{N}$ (denotation Q^n), we mean the graph with $V(Q^n) = \{0,1\}^n$ and $E(Q^n) = \{\{\mathbf{x},\mathbf{y}\} \mid \mathbf{x},\mathbf{y} \in V(Q^n), \rho(\mathbf{x},\mathbf{y}) = 1\}$, where $\rho(\mathbf{x},\mathbf{y})$ is the Hamming distance (i.e. the number of entries where $\mathbf{x},\mathbf{y}$ differ). The d-dimensional *grid* G^d is defined as the Cartesian product of d linear arrays with p vertices each. We assume throughout the paper that the grid G^d has 2^n vertices, i.e. d divides n, and so $p = 2^{\frac{n}{d}}$. For the class $\mathcal{F}$ of all considered embeddings of Q^n into G^d, denote

$$\mathbf{dil}(n,d) = \min_{f \in \mathcal{F}} \max_{e \in E(Q^n)} \mathbf{dil}_f(e), \quad \overline{\mathbf{dil}}(n,d) = \min_{f \in \mathcal{F}} \frac{1}{|E(Q^n)|} \sum_{e \in E(Q^n)} \mathbf{dil}_f(e),$$

$$\mathbf{con}(n,d) = \min_{f \in \mathcal{F}} \max_{e' \in E(G^d)} \mathbf{con}_f(e'), \quad \overline{\mathbf{con}}(n,d) = \min_{f \in \mathcal{F}} \frac{1}{|E(G^d)|} \sum_{e' \in E(G^d)} \mathbf{con}_f(e').$$

In the case $d = 1$ the numbers $\mathbf{dil}(n,1)$ and $\mathbf{con}(n,1)$ are called *bandwidth* and *cutwidth* and are well studied. For $t \in \mathbb{N}$ and for $A \subseteq V(Q^n)$, we denote $\Gamma_t A = \{v \in V(Q^n) \setminus A \mid \min_{w \in A} \rho(v,w) \le t\}$, $\partial A = \{\{u,v\} \in E(Q^n) \mid u \in A, v \in V(Q^n) \setminus A\}$. It is known (see [6] and [2,5,9], respectively) that

$$\mathbf{dil}(n,1) = \max_m \min_{|A|=m} |\Gamma_1 A| = \sum_{i=0}^{n-1} \binom{i}{\lfloor i/2 \rfloor}, \tag{1}$$

$$\mathbf{con}(n,1) = \max_m \min_{|A|=m} |\partial A| = \frac{1}{3}\left(2^{n+1} - 2 + (n \bmod 2)\right). \tag{2}$$

We denote by f_{ban} and f_{lex} the embeddings of Q^n into the linear array G^1, which have the corresponding parameters as in (1) and (2), respectively. To define these embeddings we first number the vertices of the linear array by $1, 2, ..., 2^n$ from one end to the other, then introduce the Bandwidth order $\mathcal{B}$ and the Lexicographic order $\mathcal{L}$ on the vertices of Q^n and finally map the ith vertex of Q^n in the order $\mathcal{B}$ (resp. $\mathcal{L}$) to the vertex of the linear array numbered by i for $i = 1, ..., 2^n$. Moreover, we assume that the routing scheme is given by using the shortest paths.

Let $\mathbf{x}, \mathbf{y} \in V(Q^n)$ with $\mathbf{x} = (x_1, ..., x_n)$, $\mathbf{y} = (y_1, ..., y_n)$. We say that $\mathbf{x}$ is greater than $\mathbf{y}$ in order $\mathcal{L}$ (denotation $\mathbf{x} >_{\mathcal{L}} \mathbf{y}$) iff $\sum_{i=1}^n x_i \cdot 2^{n-i} > \sum_{i=1}^n y_i \cdot 2^{n-i}$. Similarly, replacing n with d and 2 with p in this formula, we define the Lexicographic order of the vertices of G^d. Concerning the order $\mathcal{B}$ for Q^n, we write $\mathbf{x} >_{\mathcal{B}} \mathbf{y}$ iff $\sum_{i=1}^n x_i > \sum_{i=1}^n y_i$, or $\sum_{i=1}^n x_i = \sum_{i=1}^n y_i$ and $\mathbf{x} <_{\mathcal{L}} \mathbf{y}$.

In [5,6] it is shown that for each $m \in [1, 2^n]$, the minima in (1) and (2) are attained on the subsets of $V(Q^n)$ represented as the collections of the first m elements in the orders $\mathcal{B}$ and $\mathcal{L}$, respectively.

The case $d = 2$ was studied by Zienicke [10] (see also [1]), who proved that $\mathbf{dil}(n,2) = O(\sqrt{2^n/n})$ and $\mathbf{con}(n,2) = O(\sqrt{n\, 2^n})$. The result concerning the dilation were sharpened by Lai and Sprague [8] (cf. also [3]), who showed that

$0.89 \leq \lim_{n\to\infty} \frac{\mathbf{dil}(n,2)}{\sqrt{2^n/n}} \leq \frac{2}{\sqrt{\pi}} \approx 1.128$. In [2] we solved the congestion problem and proved $\mathbf{con}(n,d) = \mathbf{con}(n/d,1) = \frac{1}{3}\left(2^{\frac{n}{d}+1} - 2 + \left(\frac{n}{d} \bmod 2\right)\right)$.

In this paper we extend the approach of [6], which allows us to get a good lower bound for $\mathbf{dil}(n,d)$ for higher-dimensional grids. The constructions for all the upper bounds which are presented in the next sections are based on embeddings of hypercubes into linear arrays and are of the following type. Taking into account that d divides n, we represent Q^n as the Cartesian product of d copies of $Q^{n/d}$. Now, embed each $Q^{n/d}$ into a linear array P_i with $2^{n/d}$ vertices in accordance with some embedding f, the same for each $i = 1, ..., d$. Since $P_1 \times \cdots \times P_d$ is isomorphic to G^d, we obtain an embedding of the whole Q^n into G^d, which we denote by $f^\times$. We show (cf. [3,8] for $d=2$)

Theorem 1 $\qquad \sqrt{\frac{1}{3}} \leq \lim_{\substack{n,d\to\infty \\ d=o(n)}} \frac{\mathbf{dil}(n,d)}{\sqrt{\frac{d}{n}2^{\frac{n}{d}}}} \leq \sqrt{\frac{2}{\pi}}.$

Moreover, concerning the minimal average values of the dilation and edge-congestion, we present the following results:

Theorem 2 $\qquad$ Let $n \to \infty$. Then $\overline{\mathbf{dil}}(n,d) \sim \frac{d}{n} 2^{\frac{n}{d}}$, $\overline{\mathbf{con}}(n,d) \sim \frac{1}{2} 2^{\frac{n}{d}}$.

In the analysis of interconnection networks it is usually assumed that two neighboring nodes of the network can communicate only in discrete time units. Thus, the standard parameters of an embedding, such as the dilation and the edge-congestion, are appropriate for the evaluation of its quality only on the assumption that in each time unit any edge of Q^n may be used for the communication equiprobably. However, for a wide class of hypercube algorithms, namely for so-called *uniaxial algorithms*, it is signified that at each time unit only the edges of the same dimension are used. We say that an edge of Q^n is of the ith dimension, $i = 1, ..., n$, if the endpoints of this edge differ in the ith entry. We denote by E_i the set of all edges of Q^n of the ith dimension. The slowdown of the simulation of one time unit of an uniaxial algorithm that uses the edges of E_i can still be estimated by dilations d_i and edge-congestions c_i only caused by the edges of E_i. Hence, an effective simulation of the whole uniaxial algorithm corresponds to the minimization not of d_i or c_i, but rather of their sums. Furthermore, let $w_1, ..., w_n$ with $w_1 \geq w_2 \geq \cdots \geq w_n$ be some non-negative weights which correspond to the frequency of the communication links of E_i during the run-time of an uniaxial algorithm. In accordance with this, we introduce and study two new quality measures of an embedding f. Now, instead of $\mathbf{dil}(n,d)$ and $\mathbf{con}(n,d)$, consider the functions

$$\mathbf{Sdil}(n,d) = \min_{f \in \mathcal{F}} \sum_{i=1}^{n} w_i \cdot \max_{e \in E_i} \mathbf{dil}_f(e),$$

$$\mathbf{Scon}(n,d) = \min_{f \in \mathcal{F}} \sum_{i=1}^{n} w_i \cdot \max_{e' \in G^d} \{\mathbf{con}_f(e') \mid e' \in \phi_E(e),\ e \in E_i\}.$$

These functions precisely describe the total slowdown of the run-time of an uniaxial algorithm by its simulation on the d-dimensional grid. We are able to compute the value $\mathbf{Sdil}(n,d)$ exactly only for $d = 1$. For larger values of d, an exact formula is difficult to obtain without special knowledge about the weights. This concerns mostly the upper bound, so we assume that $w_1 = \cdots = w_n = 1$ for $d > 1$. However, our arguments provide a lower bound for $\mathbf{Sdil}(n,d)$ for arbitrary weights as well.

Theorem 3 $\mathbf{Sdil}(n,1) = \sum_{i=1}^{n} w_i \cdot 2^{i-1}$,

$\mathbf{Sdil}(n,d) \sim d\, 2^{\frac{n}{d}}$ for $d > 1, w_1 = \cdots = w_n = 1$ as $n \to \infty$.

2 Standard Measures

2.1 Bounds for the Dilation

Proof of Theorem 1.
Considering the embedding $f_{\mathrm{ban}}^{\times}$, we immediately get $\mathbf{dil}(n,d) \leq \mathbf{dil}(n/d, 1)$. This inequality in combination with (1) and $\mathbf{dil}(n,1) \sim \binom{n}{\lfloor n/2 \rfloor}$ as $n \to \infty$ (see [3] for a complete proof) provides the upper bound in Theorem 1.

Let us turn to the lower bound. For $\mathbf{x} = (x_1, ..., x_d) \in V(G^d)$ denote $\|\mathbf{x}\| = \sum_{i=1}^{d} x_i$. We introduce the levels L_i of the grid G^d defined by $L_i = \{\mathbf{x} \in V(G^d) \mid \|\mathbf{x}\| = i\}$ for $i = 0, ..., d(p-1)$ and consider a set $D = \bigcup_{i=0}^{k} L_i \cup D'$ for some k and $D' \subset L_{k+1}$. As it follows from the vertex isoperimetric problem for the grid [4], for any fixed h such a set D (with an appropriate choice of the subset D') has a minimal number of vertices at distance at most h in the grid (outside of the set itself) among all subsets of the grid of the same cardinality. Denote $|D| = m$.

Let A be the collection of vertices of Q^n mapped onto D in an embedding f and consider the set F of vertices of the grid, which are images of the vertices of $\Gamma_t A$ in the embedding. Let $y \in F$ be a vertex with maximal value of $\|\mathbf{y}\|$ and denote $q = \|\mathbf{y}\|$. Now $q - k$ is the width of a band in the grid located between its levels L_k and L_q, which is required to contain the vertices of $\Gamma_t A$ for $|A| = |D| = m$. We denote $W_t(m) := q - k - 1$.

Furthermore, let u be a vertex of $\Gamma_t A$, whose image in the embedding is $\mathbf{y}$ and let $f(v) = \mathbf{x} \in D$ for some $v \in A$. Considering a shortest path P connecting the vertices u and v in Q^n, we conclude that there exists an edge $e \in P$, so that $\mathbf{dil}_f(e) \geq (\|\mathbf{y}\| - \|\mathbf{x}\| - 1)/\rho(u,v) \geq W_t(m)/t$. This leads to the lower bound

$$\mathbf{dil}(n,d) \geq \max_{m} \max_{t} \frac{W_t(m)}{t}.$$

Let us choose m and t of the form $m = \sum_{i=0}^{(n-s\sqrt{n})/2} \binom{n}{i}$ and $t = s\sqrt{n}$, with some positive constant s, which will be an optimization parameter (here and below we omit the integer parts for brevity). Let H be the ball of radius $n - s\sqrt{n}$ (in the Hamming metric) centered in $(0, ..., 0) \in V(Q^n)$. Now, $|H| = m$ and from the vertex isoperimetric problem for Q^n [7] it is known that

$$g_t(m) := \min_{|A|=m} |\Gamma_t A| = |\Gamma_t H| = \sum_{i=(n-s\sqrt{n})/2}^{(n+s\sqrt{n})/2} \binom{n}{i}. \tag{3}$$

This shows that $|V(Q^n) \setminus (A \cup \Gamma_t(A))| \leq m$, which implies $|V(G^d) \setminus (D \cup F)| \leq m$. Since the grid G^d has $d(p-1)+1$ levels, we get $W_t(m) \geq z := d(p-1) - 1 - 2k$. In other words, z is defined in such a way, that the sum of the z greatest numbers $|L_j|$ ($0 \leq j \leq d(p-1)$) asymptotically equals $g_t(m)$ (cf. (3)). Since the largest levels of G^d are located symmetrically around its middle level (with $j = d(p-1)/2$), we have $z = 2x$ with x determined by the equation

$$\sum_{j=d(p-1)/2-x}^{d(p-1)/2+x} L_j \sim g_t(m). \tag{4}$$

For further applications we have to estimate the sums in (3) and (4) asymptotically. Therefore, we use some facts taken from the probability theory. Let ξ be a continuous random variable and $F(x)$ be its distribution function (i.e. $F(x) = P(\xi \leq x)$). We say that ξ is *normally distributed* in $(-\infty, \infty)$ if $F(x) = \Phi(x) := \frac{1}{\sqrt{2\pi}} \int_{-\infty}^{x} e^{-z^2/2} dz$. Furthermore, consider a sequence of discrete random variables ξ_n, taking on a finite number of values x_n^i with corresponding probabilities p_n^i. Let $F_n(x) = P(\xi_n \leq x) = \sum_{x_n^i \leq x} p_n^i$ be the distribution function of ξ_n, and μ_n and σ_n^2 be its mean and variance, respectively. We say that ξ_n is *asymptotically normal* with mean μ_n and variance σ_n^2 if $\lim_{n \to \infty} \sum_{x_n^i \leq \mu_n + x\sigma_n} p_n^i = \Phi(x)$, for every $x \in (-\infty, \infty)$. The random variables $\xi_1, ..., \xi_n$ are called *independent* if $P(\xi_1 < x_1, ..., \xi_n < x_n) = P(\xi_1 < x_1) \cdots P(\xi_n < x_n)$.

Theorem 4 (Central Limit Theorem) *If $\xi_1, ..., \xi_n$ are independent random variables with the common distribution function F, mean μ, and variance σ^2, then the random variable $\zeta_n = \xi_1 + \cdots + \xi_n$ is asymptotically normal with mean $n\mu$ and variance $n\sigma^2$.*

We apply Theorem 4 to the independent discrete random variables $\xi_1, ..., \xi_n$, which take on the values $0, ..., l-1$ each with probability $1/l$. Then, the random variable $\zeta_n = \xi_1 + \cdots + \xi_n$ is asymptotically normal with mean and variance respectively $\mu_{n,l} = n\frac{l-1}{2}$, $\sigma_{n,l}^2 = n\frac{l^2-1}{12}$. Denote by L_j the number of vertices of the $l \times \cdots \times l$ grid G^n on distance j from the vertex $(0, ..., 0)$, i.e. the size of its jth level. Clearly, ζ_n takes on the values $0, ..., n(l-1)$ each with probability $1/l^n$, so the distribution function $F_n(x)$ of ζ_n is $F_n(x) = \frac{1}{l^n} \sum_{j \leq x} L_j$. Therefore, $\lim_{n \to \infty} \sum_{x_n^i \leq \mu_{n,l} + x\sigma_{n,l}} p_n^i = \lim_{n \to \infty} \frac{1}{l^n} \sum_{j=0}^{\mu_{n,l} + x\sigma_{n,l}} L_j = \Phi(x)$. This implies $\sum_{j=\mu_{n,l} - x\sigma_{n,l}}^{\mu_{n,l} + x\sigma_{n,l}} L_j \sim \theta(x) l^n$ as $n \to \infty$, where $\theta(x) = \Phi(x) - \Phi(-x) = \frac{1}{\sqrt{2\pi}} \int_{-x}^{x} e^{-z^2/2} dz$.

Now turning back to the estimation of the sum in (4), let us consider x represented in the form $x = y\,\sigma_{d,p}$. Applying Theorem 4 with $n = d$, $l = p = 2^{n/d}$, $\mu_{n,l} = \mu_{d,p} = d(p-1)/2$, $\sigma_{n,l} = \sigma_{d,p}$, we get

$$g_t(m) \sim \sum_{j=d(p-1)/2-x}^{d(p-1)/2+x} L_j = \sum_{j=dp/2-y\,\sigma_{d,p}}^{dp/2+y\,\sigma_{d,p}} L_j \sim \theta(y)\,2^n. \tag{5}$$

Similarly, for the estimation of the sum in (3) we apply Theorem 4 to Q^n with $l = 2$ and $\mu_{n,2} = n/2$. Rewriting t in the form $t = 2s\,\sigma_{n,2}$, we get

$$g_t(m) = \sum_{j=n/2-s\,\sigma_{n,2}}^{n/2+s\,\sigma_{n,2}} \binom{n}{j} \sim \theta(s)\,2^n. \tag{6}$$

From (5) and (6) it follows that $y \sim s$. Thus, $W_t(m) \geq z \sim 2s\,\sigma_{d,p}$. Finally, taking into account that $t = 2s\,\sigma_{n,2} = s\sqrt{n}$ and $\sigma_{d,p} \sim p\sqrt{d/12}$, one has

$$\frac{W_t(m)}{t} \geq \frac{2s\,\sigma_{d,p}}{2s\,\sigma_{n,2}} \sim \frac{2p\sqrt{d/12}}{\sqrt{n}} = \sqrt{\frac{d}{3}}\,\frac{2^{n/d}}{\sqrt{n}}.$$

and the lower bound in Theorem 1 is proved. □

2.2 Average Case Analysis

Note that for any embedding $f \in \mathcal{F}$ of a graph G into H it holds

$$\sum_{e \in E(G)} \mathrm{dil}_f(e) = \sum_{e' \in E(H)} \mathrm{con}_f(e'). \tag{7}$$

Proof of Theorem 2.
Consider an embedding f of Q^n into G^d and let us compute the sum on the right hand side of (7), which we denote by Σ_f. Let $A \subseteq V(Q^n)$ and D be its image in f in the grid G^d. Consider the edge-cut ∇D separating D from its complement in G^d. Then,

$$\sum_{e \in \nabla D} \mathrm{con}_f(e) \geq |\partial A|. \tag{8}$$

Indeed, the image of each edge of ∂A passes through the edges of ∇D. However, maybe there are some other paths connecting two vertices of D (or of its complement), which are also cut by ∇D. This could make the sum in (8) greater than $|\partial A|$. Now, let D_m be the collection of the first m vectors of G^d in the Lexicographic order and $A_m \subseteq V(Q^n)$ be the vertices mapped into D_m in the embedding f. Then $|A_m| = m$ and (8) implies

$$\sum_{m=1}^{2^n} \sum_{e \in \nabla D_m} \mathrm{con}_f(e) \geq \sum_{m=1}^{2^n} |\partial A_m|. \tag{9}$$

In [5] it is shown
$$\sum_{m=1}^{2^n} |\partial A_m| \geq 2^{n-1}(2^n - 1), \tag{10}$$
where the equality takes place if each A_m is the collection of the first m vertices of Q^n taken in the Lexicographic order. Let us estimate the double sum in (9). We say that an edge $e = \{u,v\} \in E(G^d)$ is an i-edge, if the vectors u,v differ in the ith entry. Denote by E^i the set of all i-edges of G^d. It is easily shown by induction on d that, due to the choice of the Lexicographic order, $\mathbf{con}_f(e)$ for each i-edge e appears in the double sum at most p^{i-1} times. Therefore, denoting $c_i = \sum_{e \in E^i} \mathbf{con}_f(e)$ and taking into account (9) and (10), one has $\Sigma_f = \sum_{i=1}^{d} c_i$ and $\sum_{i=1}^{d} c_i p^{i-1} \geq \sum_{m=1}^{2^n} \sum_{e \in \nabla D_m} \mathbf{con}_f(e) \geq 2^{n-1}(2^n - 1)$.

Thus, we are able to estimate the sum of $c_i p^{i-1}$. To get a bound for Σ_f, however, we need to estimate the sum of c_i. In order to do this, let us define a *squashed* Lexicographic order $\mathcal{L}_j$. Consider the permutation

$$\pi_j = \begin{pmatrix} 1 & \cdots & j-1 & j & j+1 & \cdots & d \\ d-j+2 & \cdots & d & 1 & 2 & \cdots & d-j+1 \end{pmatrix},$$

where the bottom row is a cyclic shift of the top row on j entries. Now, for $\mathbf{x} = (x_1, ..., x_d)$, $\mathbf{y} = (y_1, ..., y_d) \in V(G^d)$ we say that $\mathbf{x}$ is greater $\mathbf{y}$ in order $\mathcal{L}_j$ if $\sum_{i=1}^{d} x_i \cdot p^{\pi_j(i)-1} \geq \sum_{i=1}^{d} y_i \cdot p^{\pi_j(i)-1}$. Clearly, the order $\mathcal{L}_j$ is isomorphic to the Lexicographic order up to rotations of the grid G^d. Therefore, considering instead of the set D_m, the collections of the first m vertices of G^d in order $\mathcal{L}_j$ and applying the arguments mentioned above, for any $j \in \{1, ..., d\}$ one has $\sum_{i=1}^{d} c_i p^{\pi_j(i)-1} \geq 2^{n-1}(2^n - 1)$. Summarizing these inequalities for $j = 1, ..., d$, we finally get $\sum_{i=1}^{d} c_i \sum_{j=1}^{d} p^{j-1} \geq d\, 2^{n-1}(2^n - 1)$, which, with $p = 2^{n/d}$, implies as $n \to \infty$

$$\Sigma_f = \sum_{i=1}^{d} c_i \geq \frac{d\, 2^{n-1}(2^n - 1)}{\frac{p^d - 1}{p - 1}} = \frac{d}{2} 2^{n(d+1)/d}(1 - o(1)). \tag{11}$$

To get an upper bound for $\min_{f \in \mathcal{F}} \Sigma_f$, consider the embedding $f_{\text{lex}}^{\times}$. We call a set of p vertices of G^d a *column*, if all these vertices agree in some $p-1$ entries. Denote by C_d the number of columns in G^d. Since the image of each edge of Q^n in the embedding $f_{\text{lex}}^{\times}$ belongs to some column, denoting by A_m the collection of the first m vertices of $Q^{n/d}$ in the Lexicographic order, we get (cf. (10), [5])

$$\Sigma_{f_{\text{lex}}^{\times}} \leq C_d \cdot \sum_{e \in E(G^1)} \mathbf{con}_{f_{\text{lex}}^{\times}}(e) = \sum_{m=1}^{2^{n/d}} |\partial A_m| = C_d \cdot 2^{n/d-1}(2^{n/d} - 1). \tag{12}$$

Simple arguments show that C_d satisfies the recursion $C_d = pC_{d-1} + p^{d-1}$, which with $C_1 = 1$ gives $C_d = d p^{d-1}$. Therefore, (11) and (12) imply as $n \to \infty$ $\min_{f \in \mathcal{F}} \Sigma_f \sim \frac{d}{2} 2^{n(d+1)/d}$. Finally, combining this fact with (7), one has

$\overline{\mathrm{dil}}(n,d) = \min_{f \in \mathcal{F}} \Sigma_f / |E(G^d)|$ and $\overline{\mathrm{con}}(n,d) = \min_{f \in \mathcal{F}} \Sigma_f / |E(Q^n)|$, which, after taking into account $|E(G^d)| = (p-1)C_d \sim d\, 2^n$ and $|E(Q^n)| = n\, 2^{n-1}$, completes the proof of Theorem 2. □

3 Simulation of the Uniaxial Algorithms

Proof of Theorem 3.
Consider first the case $d = 1$. We apply induction on n. For an embedding f and $i = 1, ..., n$ denote $d_i = \frac{1}{|E_i|}\sum_{e \in E_i} \mathrm{dil}_f(e)$. Note that $\max_{e \in E_i} \mathrm{dil}_f(e) \geq d_i$ and $\sum_{i=1}^n d_i \geq 2^n - 1$ (cf. (11) and (10)). Now assuming $w_1 \geq w_2 \geq \cdots \geq w_n$ and applying the inductive hypothesis with $w'_i = w_i - w_n$, $i = 1, ..., n-1$, one has

$$\mathbf{Sdil}(n,1) \geq \sum_{i=1}^{n} w_i \cdot d_i = \sum_{i=1}^{n-1} w'_i \cdot d_i + w_n \sum_{i=1}^{n} d_i \geq \sum_{i=1}^{n-1} w'_i \cdot 2^{i-1} + w_n(2^n - 1) = \sum_{i=1}^{n} w_i \cdot 2^{i-1}.$$

For $d > 1$ using the same method and (11) again, we get

$$\mathbf{Sdil}(n,d) \geq \sum_{i=1}^{n} d_i = \frac{2}{2^n} \sum_{i=1}^{n} \sum_{e \in E_i} \mathrm{dil}_f(e) = \frac{2}{2^n} \sum_{e \in E(Q^n)} \mathrm{dil}_f(e) \geq d\, 2^{\frac{n}{d}}(1 - o(1)).$$

Clearly, for the embedding f_{lex} it holds $d_i = 2^{i-1}$ for $i = 1, ..., n$ and, thus, $\mathbf{Sdil}(n,1) \leq \sum_{i=1}^{n} w_i \cdot 2^{i-1}$, which matches the lower bound. Furthermore, the embedding $f_{\mathrm{lex}}^{\times}$ with $w_1 = \cdots = w_n = 1$ provides $\mathbf{Sdil}(n,d) \leq d \cdot \mathbf{Sdil}(n/d, 1) \leq d\, 2^{\frac{n}{d}}$, which completes the proof of Theorem 3. □

Note that in the case of arbitrary weights w_i and $d > 1$ our approach provides a lower bound for $\mathbf{Sdil}$ of the form $2^{\frac{1-d}{d}n} \sum_{i=1}^{n} w_i\, 2^{i-1}(1 - o(1))$. Concerning the quality measure $\mathbf{Scon}(n,d)$ let us mention some results without giving the proofs because of space limitation. In the case $w_1 = \cdots = w_n = 1$ we obtain the estimation $\frac{1}{3} \cdot 2^{\frac{n}{d}} \leq \mathbf{Scon}(n,d) \leq d \cdot \mathbf{Scon}(n/d, 1) \leq d\, 2^{\frac{n}{d}}$, where the upper bound is given by the embedding $f_{\mathrm{lex}}^{\times}$. In the case of arbitrary weights we can prove that the above given lower bound for $\mathbf{Sdil}$ also gives a lower bound for $\mathbf{Scon}(n,d)$ and in the case $d = 1$ this bound is asymptotically equal to the upper bound given by the embedding $f_{\mathrm{lex}}^{\times}$.

4 Applications

It is interesting to note that concerning the ordinary parameters (dilation and edge-congestion) an embedding optimal for one of them is not optimal for the other, while for our new cost-measures the embedding $f_{\mathrm{lex}}^{\times}$ seems to be optimal for both of them. This makes it highly probable that this embedding provides good practical results for the simulation. Below, we present some experimental results concerning data transfer in the Parsytec GCel 1024 parallel computer, consisting of 1024 transputers T-800 configured in the 32×32 grid. The message-routing in this computer is supported by a software implementation of the wormhole routing. We simulated the communication behavior of the uniaxial algorithms with $w_1 = \cdots = w_n$ for $n = 1, ..., 10$, by sending an amount of 6.4 Mbyte of data split into small packets of size 120 bytes through each edge of E_i in Q^n for $i = 1, ..., n$. For each mentioned value of n, we embedded Q^n into a subgrid of the 32×32 grid and for each $i = 1, ..., n$ we measured the maximal time t_i taken for a message to reach its destination in the grid along the paths corresponding to the images of the edges of E_i in the embedding. Let $\bar{t} = (t_1 + \cdots + t_n)/n$, then the average performance $P(f)$ of the communication in the embedding f equals $P(f) = 6.4/\bar{t}$ Mbyte/s. In

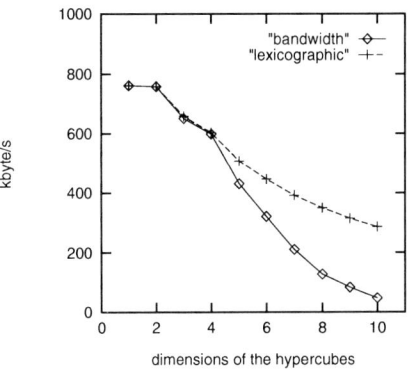

Fig. 1. Emulation of uniaxial hypercube algorithms.

Fig. 1 we show the results for $P(f_{\text{lex}}^\times)$ and $P(f_{\text{ban}}^\times)$. As one can see from the figure, the performance of each embedding is almost the same for $n \leq 4$, which is explained by the similarity of the orders $\mathcal{B}$ and $\mathcal{L}$ for small n. However, for larger n, the difference increases and reaches the factor of approximately 6 for $n = 10$, i.e. the whole communication in uniaxial algorithms is done 6 times faster in the embedding $f_{\text{lex}}^\times$ compared to $f_{\text{ban}}^\times$. Therefore, the embedding $f_{\text{lex}}^\times$ is not only asymptotically optimal, as we showed above, but also provides very good practical results for concrete instances.

References

1. Annexstein, F.: Embedding hypercubes and related networks into mesh-connected processor arrays. *J. Parall. Distr. Comput.*, **23** (1994), 72–79.
2. Bezrukov, S.L., Chavez, J.D., Harper, L.H., Röttger M., Schroeder U.-P.: The congestion of n-cube layout on a rectangular grid, to appear in *Disc. Mathematics*.
3. Bezrukov, S.L., Röttger, M., Schroeder, U.-P.: Embedding of hypercubes into grids. *Technical Report* tr-sfb-95-1, University of Paderborn, (1995).
4. Bollobás, B., Leader, I.: Compressions and isoperimetric inequalities. *J. Comb. Th.*, **A-56** (1991), 47–62.
5. Harper, L.H.: Optimal assignment of numbers to vertices. *J. Sos. Ind. Appl. Math*, **12** (1964), 131–135.
6. Harper, L.H.: Optimal numberings and isoperimetric problems on graphs. *J. Comb. Theory*, **1** (1966), No.3, 385–393.
7. Katona, G.O.H.: A theorem on finite sets. In: *Theory of Graphs*, Akademia Kiado, Budapest, (1968), 187–207.
8. Lai, T.-H., Sprague, A.P.: Placement of the processors of a hypercube. *IEEE Trans. Comp.*, **40** (1991), No.6, 714–722.
9. Nakano, K.: Linear layout of generalized hypercubes. In: *Proc. Graph-Theoretic Concepts in Computer Science*, LNCS **790**, Springer Verlag, (1994), 364–375.
10. Zienicke, P.: Embedding hypercubes in 2-dimensional meshes. *Humboldt-Universität zu Berlin*, (manuscript).

Tree Decompositions of Small Diameter*

Hans L. Bodlaender[1] and Torben Hagerup[2]

[1] Department of Computer Science, Utrecht University, P.O. Box 80.089,
3508 TB Utrecht, the Netherlands hansb@cs.uu.nl
[2] Fachbereich Informatik, Johann Wolfgang Goethe-Universität Frankfurt,
Robert-Mayer-Str. 11-15, D-60054 Frankfurt am Main, Germany
hagerup@informatik.uni-frankfurt.de

Abstract. Motivated by applications in parallel and dynamic graph algorithms, we investigate the tradeoff between width and diameter of tree decompositions. For all integers n, k and K with $1 \leq k \leq K \leq n-1$, denote by $D(n, k, K)$ the maximum, over all n-vertex graphs G of treewidth k, of the smallest diameter of a tree decomposition of G of width K. We determine $D(n, k, K)$, up to a constant factor, for all values of n, k and K. When K is bounded by a constant (the case of greatest practical relevance), $D(n, k, K)$ is $\Theta(n)$ for $K \leq 2k-1$, $\Theta(\sqrt{n})$ for $2k \leq K \leq 3k-2$, and $\Theta(\log n)$ for $K \geq 3k - 1$. We provide much more accurate bounds for the case $K \leq 2k - 1$.

1 Introduction

A *tree decomposition* of a graph G is a tree that, informally, imparts a tree structure to G (precise definitions are provided in the next section). A tree decomposition is characterized by a nonnegative integer called its *width*, and the *treewidth* of a graph G is the smallest width of any tree decomposition of G. The treewidth of a graph G, informally, is a measure of how closely G resembles a tree—the smaller the treewidth, the more tree-like G is.

It has been observed that a tree decomposition of small width of a graph G is a powerful aid in the solution of any of a large number of computational problems on G. E.g., if the width is bounded by a constant, many NP-complete graph problems can be solved in linear time on G by using the tree decomposition to guide the computation (a recent overview of such results can be found in [2]). In the traditional setting of graph algorithms, any tree decomposition of G of sufficiently small width will do. For the newer models of parallel and dynamic computation, however, a tree decomposition of small width of an input graph is as useful as in the traditional setting, but a second parameter of the tree decomposition acquires importance, namely its *diameter*, the maximal distance between two nodes in the tree decomposition. This is because, after rooting the

---

* This research was partially supported by ESPRIT Long Term Research Project 20244 (project ALCOM-IT: *Algorithms and Complexity in Information Technology*). The work was carried out while the first author was with the Max-Planck-Institut für Informatik in Saarbrücken, Germany.

tree decomposition at an arbitrary node, a parallel algorithm typically processes the nodes in the tree decomposition one level at a time, while a dynamic algorithm answers a query or executes an update by traversing a root-to-leaf path. In both cases, the resulting time bound is (at least) proportional to the height of the tree, which in turn is at least half the diameter. Algorithms of this kind were described, e.g., in [1,3,4,5,6,7,8].

It was shown by Bodlaender [1] that every n-vertex graph of treewidth k has a tree decomposition of width at most $3k + 2$ and diameter $O(\log n)$. By allowing a width slightly larger than the minimum width, we can thus reduce the diameter to a very low value. However, enlarging the width ever so slightly is highly undesirable, because the running times of the algorithms that employ a tree decomposition typically increase dramatically—at least exponentially—with the width of the decomposition. It is therefore natural to ask whether it is really necessary to go from treewidth k to treewidth $3k + 2$ in order to ensure a logarithmic diameter. We answer this question completely. The answer turns out to be "Almost, but not quite". More precisely, we show that a logarithmic diameter can be preserved while the width is reduced to $3k - 1$, whereas it is impossible in general to achieve a logarithmic diameter and a width of $3k - 2$ simultaneously.

More generally, we investigate the complete tradeoff between width and diameter of tree decompositions. For all integers n, k and K with $1 \leq k \leq K \leq n-1$, denote by $D(n, k, K)$ the maximum, over all n-vertex graphs G of treewidth k, of the smallest diameter of a tree decomposition of G of width K. We determine the value of $D(n, k, K)$, up to a constant factor, for all combinations of n, k and K. Our findings are summarized in the theorem below. Note that we intend $E_1 = \Theta(E_2)$, where E_1 and E_2 are nonnegative expressions, to mean that there are constants $c, c' > 0$ such that $cE_2 \leq E_1 \leq c'E_2$ for *all* values of the parameters occurring in E_1 and E_2. We allow E_2 to be zero, in which case E_1 is zero as well.

Theorem 1. *Let n, k and K be integers with $1 \leq k \leq K \leq n - 1$. Then $D(n, k, K)$ has the following value:*

$$\begin{cases} \Theta\left(\left\lfloor \dfrac{n-k-1}{K-k+1} \right\rfloor\right), & \text{if} \quad k \leq K \leq 2k-1; \\ \Theta\left(\left\lceil \dfrac{\sqrt{(K-2k+1)n+k^2}-K+k-1}{K-2k+1} \right\rceil\right), & \text{if} \quad 2k \leq K \leq 3k-2; \\ \Theta\left(\left\lceil \log_2\left(\dfrac{n-k}{K-k+1}\right) \bigg/ \log_2(K/k) \right\rceil\right), & \text{if} \quad 3k-1 \leq K. \end{cases}$$

In rough approximation and for small values of k and K relative to n, we can say that $D(n, k, K)$ is linear (in n) if K is below $2k$, "square-rootic" if K is between $2k$ and $3k$, and logarithmic if K is above $3k$.

In an algorithmic context, it is not sufficient to know that tree decompositions of small width and diameter exist; we must also be able to construct them.

When K is bounded by a constant—the most interesting case—our upper bounds can be realized by efficient sequential and parallel algorithms. We omit further discussion of the algorithmic aspect.

2 Definitions

A *tree decomposition* of an undirected graph $G = (V, E)$ is a tree $T = (X, F)$, each of whose nodes $x \in X$ is labeled with a subset U_x of V, called the *bag* of x, such that

- $\bigcup_{x \in X} U_x = V$ (every vertex in G occurs in some bag);
- for all $\{u, v\} \in E$, there exists an $x \in X$ such that $\{u, v\} \subseteq U_x$ (every edge in G is "internal" to some bag);
- for all $x, y, z \in X$, if y is on the path from x to z in T, then $U_x \cap U_z \subseteq U_y$ (every vertex in G occurs in the bags in a connected part of T, i.e., in a subtree).

The *width* of the tree decomposition T is $\max_{x \in X} |U_x| - 1$, i.e., one less than the size of a largest bag. We will refer to the third requirement imposed on tree decompositions above as the "connectedness property". A *path decomposition* is a tree decomposition that is a path.

It is obvious that every n-vertex graph has a tree decomposition of width $n-1$ and diameter 0 (put all vertices in a single bag), and the case of treewidth 0 (no edges) is trivial and without interest. Thus we lose nothing essential through the condition $1 \leq k \leq K \leq n-1$ imposed throughout the paper.

3 Upper Bounds

For all integers n, k and K with $1 \leq k \leq K \leq n - 1$, let $f(n, k, K) = \lfloor (n - k - 1)/(K - k + 1) \rfloor$. We prove that $f(n, k, K) \leq D(n, k, K) \leq f(n, k, K) + 1$ for $K \leq 2k - 1$. Since $D(n, k, K) = 0 \Leftrightarrow K = n - 1 \Leftrightarrow f(n, k, K) = 0$, this implies that $D(n, k, K) = \Theta(f(n, k, K))$ for this range of K.

Theorem 2. *(the linear upper bound) Let n, k and K be integers with $1 \leq k \leq K \leq n-1$. Then every n-vertex graph of treewidth k has a tree decomposition of width K with at most $f(n, k, K)$ nodes of degree greater than 1. In particular, $D(n, k, K) \leq f(n, k, K) + 1$.*

Proof. Let $G = (V, E)$ be an n-vertex graph of treewidth at most k and let $T = (X, F)$ be a tree decomposition of G in which every bag contains exactly $k + 1$ vertices—it is easy to obtain such a tree decomposition from an arbitrary tree decomposition of G of treewidth at most k by adding vertices to bags with too few vertices. We root T at an arbitrary node $r \in X$ and process the nodes in T in postorder, i.e., process all children of a node $x \in X$ before processing x itself. The processing of a node $x \in X$ consists in computing the union U of the bags of x and its children and, if and only if $|U| \leq K + 1$, contracting all edges between x and its children (in any order) and assigning U as the bag of the new node resulting from the contractions. It is easy to see that this yields a tree decomposition T' of G of width at most K. It now suffices to show that the size of the set Z of internal nodes in T' is bounded by $f(n, k, K)$. To this end, define the *high point* of each vertex $v \in V$ as the node of minimal depth in T' whose bag contains v; it follows from the connectedness property of tree decompositions that there is a unique such node. During the construction of T' described above, whenever a node $x \in X$ is processed but the edges between x and its children are not contracted, the bags of x and of its children together contain at least $K + 2$ vertices. At most $k + 1$ of these occur in the bag of x, and the high points of all the remaining at least $K + 2 - (k + 1) = K - k + 1$ vertices are children of x. This argument assigns at least $K - k + 1$ vertices in V uniquely to each node in Z. Moreover, the $k + 1$ vertices belonging to the bag of r are not assigned to any node. It follows that $|Z| \le \lfloor (n - (k + 1))/(K - k + 1) \rfloor = f(n, k, K)$. □

Lemma 3. *For all integers $n, k \ge 1$, every graph that has a path decomposition of width k with n nodes has a tree decomposition of width at most $2k$ with diameter at most $2\sqrt{n}$.*

Proof. Suppose that we are given a graph G together with a path decomposition of G of width k and with n nodes. We begin by preprocessing the given path decomposition as follows: First, we contract every edge whose endpoints have identical bags. Second, we place a new node "on" every (remaining) edge and give each such node x a bag that is the intersection of the bags of the neighbors of x. It is easy to see that these steps produce a new path decomposition of G whose number N of nodes is bounded by $2n$ and with the property that every other node has a bag of size at most k. For $i = 1, \ldots, N$, let U_i be the bag of the ith node in this path decomposition, counted from one fixed end.

We use a tree decomposition of a form illustrated in Fig. 1. In general, for all integers $m \ge 1$, there is a tree T_m of diameter $2m - 2$ with the following properties:

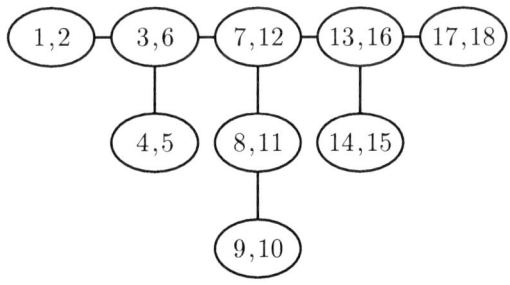

Fig. 1. The tree T_m for $m = 3$

1. Every node in the tree is labeled by two integers in the range $\{1,\ldots,2m^2\}$, and every integer in this range labels exactly one tree node.
2. For $i = 1,\ldots,2m^2-1$, the integers i and $i+1$ label the same node or adjacent nodes.
3. Every node is labeled by an odd integer and an even integer.

We choose m minimal so that a node in T_m has the label N, i.e., so that $2m^2 \geq N$, and associate each node x in T_m with a bag that is the union of U_i over the at most two integers i in $\{1,\ldots,N\}$ labeling x. We must show that this yields a tree decomposition of G. By Condition (1) above, the first two properties of a tree decomposition are clearly inherited from the path decomposition, and the connectedness property is implied by Condition (2). By Condition (3) and the fact that every other node in the modified path decomposition has a bag of size at most k, all bag sizes in the tree decomposition are bounded by $2k + 1$; i.e., the width of the tree decomposition is at most $2k$. What remains is to show that the diameter $2m - 2$ of T_m is bounded by $2\sqrt{n}$. But $m = \lceil\sqrt{N/2}\rceil$ and hence $2m - 2 \leq 2\sqrt{N/2} \leq 2\sqrt{n}$. □

Lemma 4. *Let l be a positive integer and let $a_1,\ldots,a_l$ be nonnegative real numbers such that, for $i = 2,\ldots,l$, $a_i \geq a_{i-1} + a_{i-2} + \cdots + a_1$. Then $\sum_{i=1}^{l}\sqrt{a_i} \leq (1+\sqrt{2})\sqrt{\sum_{i=1}^{l} a_i}$.*

Proof. Omitted. □

By a *clique* in a graph $G = (V, E)$ we mean a subset of V that induces a complete subgraph of G. We say that a clique C in G is *covered* by a node x in a tree decomposition of G if C is a subset of the bag of x. It is well-known that for every clique C in a graph G and for every tree decomposition (X, F) of G, C is covered by some node in X; we will refer to this as the *clique-covering property*.

Theorem 5. *For all integers $m, k \geq 1$, every graph that has an m-node tree decomposition of width k has a tree decomposition of width at most $2k$ with diameter at most $\lfloor 4(1+\sqrt{2})\sqrt{m}\rfloor + 2\lfloor\log_2 m\rfloor \leq 12\sqrt{m}$.*

Proof. Let T be a rooted m-node tree decomposition of width k of the given graph G. Define the *weight* of a node in T as the number of its descendants. For each internal node x in T, fix a child y of x of maximum weight and call the edge from x to y *heavy*; all other edges in T are *light*. This partitions the nodes in T into *heavy paths* (some of which may consist of a single node).

Independently for each heavy path P, consider for each bag U of a node on P the complete graph on the vertex set U, and let H_P be the union of these complete graphs. We process H_P according to Lemma 3, of course using P as the path decomposition of H_P. For each heavy path P, this yields a tree decomposition T_P of H_P that we will call the *comb* of P. We combine the combs into a single graph T' by, for each light edge $\{x,y\}$ that joins heavy paths P_x

and P_y, adding an edge between nodes in the combs of P_x and P_y whose bags are supersets of U_x and U_y, respectively—such nodes exist, by the clique-covering property. It is clear that T' is a tree and that it inherits the two first properties of a tree decomposition from the combs. The way in which the combs are stitched together guarantees that the connectedness property is also satisfied; i.e., T' is a tree decomposition of G.

What remains is to bound the diameter of T'. We root T' at an arbitrary node of the comb of the heavy path containing the root of T. The diameter of T' is at most twice the number of edges on a longest leaf-to-root path in T'. Consider such a path P and assume that the combs that it touches are $T_{P_1}, \ldots, T_{P_l}$, in that order. For $i = 1, \ldots, l$, define a_i to be the number of nodes in T that are descendants of a node on P_i, but not, if $i > 1$, of a node on P_{i-1}. For $i = 2, \ldots, l$, the edge that joins P_{i-1} to P_i is light, which implies that $a_i \geq a_{i-1} + \cdots + a_1$ and that $l \leq \lfloor \log_2 m \rfloor + 1$. Moreover, $\sum_{i=1}^{l} a_i = m$. By Lemma 4, $\sum_{i=1}^{l} \sqrt{a_i} \leq (1 + \sqrt{2})\sqrt{m}$. Since a_i is an upper bound on the number of nodes on P_i, for $i = 1, \ldots, l$, Lemma 3 can now be seen to imply that the length of P is at most $2(1 + \sqrt{2})\sqrt{m} + l - 1$, from which the theorem follows. □

Corollary 6. (*the square-rootic upper bound*) *For all integers n, k and K with $1 \leq k < 2k \leq K \leq n - 1$, $D(n, k, K) \leq 12\sqrt{f(n, k, \lfloor K/2 \rfloor)} + 2$.*

Proof. Given an n-vertex graph G of treewidth k, we first use the algorithm of Theorem 2 to obtain a tree decomposition T of G of treewidth $\lfloor K/2 \rfloor$ with at most $m = f(n, k, \lfloor K/2 \rfloor)$ nodes of degree greater than 1. For each bag U of a node in T of degree greater than 1, consider the complete graph on the vertex set U, and let H be the union of these complete graphs. We now use the algorithm of Theorem 5 to obtain a tree decomposition of width K with diameter at most $12\sqrt{m}$ of H. By the clique-covering property, we can obtain a tree decomposition of G of width at most K by sticking back the original nodes of degree 1, which increases the diameter by at most 2. □

We omit the proof of the upper bound of Theorem 1 for the case $K \geq 3k - 1$, but provide the following brief outline of some of the ideas involved: In a natural way, the proof of Theorem 2 can be viewed as partitioning a tree decomposition of width k, now assumed without loss of generality to be binary, of an n-vertex graph into at most $2f(n, k, K) + 1$ vertex-induced subtrees, such that the union of the bags of the nodes in each subtree contains at most $K + 1$ vertices. We refine this partition to ensure several additional properties, while keeping the number of subtrees $O(f(n, k, K) + 1)$. In particular, each subtree is required to have at most two *boundary nodes*, nodes adjacent to nodes in other subtrees. We then remove all nonboundary nodes, introducing an edge between each pair of boundary nodes in the same subtree, after which we apply a more refined variant of the tree-contraction algorithm of [1]. A detailed analysis shows that this yields a binary tree of diameter $O(\log(f(n, k, K) + 1))$ in which each node has a bag of size at most $3k$. Subsequently we reduce the diameter to $O(\log(f(n, k, K) + 1)/\log(K/k))$ while allowing the maximum bag size to grow to $K + 1$ by contracting subtrees of height $\Theta(\log(K/k))$ to single nodes. The resulting tree is not yet a tree decomposition of the input graph (in particular, many vertices may not be represented in bags at all), but we can turn it into one by adding appropriate nodes of degree 1, which increases the diameter by at most 2.

4 Lower Bounds

Throughout this section, n, k and K are fixed integers with $1 \leq k \leq K \leq n-1$. We show the constructions of the previous section to be essentially optimal in the case of the specific graph $G_{n,k} = (\{1,\ldots,n\}, \{\{i,j\} : 1 \leq i < j \leq n \text{ and } |i-j| \leq k\})$ (see Fig. 2). Using the clique-covering property, one can easily show the treewidth of $G_{n,k}$ to be exactly k.

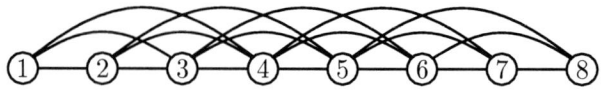

Fig. 2. $G_{8,3}$

Let $T = (X, F)$ be a tree decomposition of width at most K of $G_{n,k}$ and denote the bag of x by U_x, for all $x \in X$. We can assume without loss of generality that the bag of every node in T of degree 1 contains at least one vertex that does not occur in any other bag—otherwise the node is superfluous and can be removed without increasing the diameter. For each node x in T of degree 1, fix such a vertex and call it the *unique vertex* of x. Our proofs consist of two components: On the one hand, we show that T must be a path if $K \leq 2k-1$, a caterpillar (suitably defined later) if $K \leq 3k - 2$, and a tree whose degree is bounded by a function of K/k in general. On the other hand, we prove lower bounds for the number of nodes in T by exploiting the fact that every element of the family $\mathcal{C} = \{\{i,\ldots,i+k\} : 1 \leq i \leq n-k\}$ of $n-k$ cliques of size $k+1$ in $G_{n,k}$ is covered by some node in X.

Define an *interval* as a finite set of consecutive integers and the *interval-covering number* of a finite set U of integers as the number of maximal intervals of size at least k contained in U.

Lemma 7. *Let $x \in X$ and suppose that U_x has interval-covering number m. Then x covers at most $K + 1 - mk$ cliques in $\mathcal{C}$, and x covers no cliques in $\mathcal{C}$ if $m = 0$.*

Proof. Let $I_1, \ldots, I_m$ be the m maximal intervals of size at least k contained in U_x and note that each clique in $\mathcal{C}$ covered by x, itself being an interval of size $k+1$, must be contained in I_j for some $j \in \{1,\ldots,m\}$. By considering the cliques in $\mathcal{C}$ in the natural order given by their smallest vertices, it is easy to see for arbitrary $l \geq 0$ and for $j = 1, \ldots, m$ that if I_j contains the union of l distinct cliques in $\mathcal{C}$, then $|I_j| \geq k + l$, i.e., there is an "overhead" of k per interval. The lemma now follows by recalling that $|U_x| \leq K + 1$. □

Theorem 8. (*the linear lower bound*) *For all integers n, k and K with $1 \leq k \leq K \leq 2k - 1$, $D(n, k, K) \geq f(n, k, K) = \lfloor (n - k - 1)/(K - k + 1) \rfloor$.*

Proof. Since all neighbors of the unique vertex of a node of degree 1 in T occur in the bag of a single node, all unique vertices must be drawn from the set $\{1, \ldots, k\} \cup \{n - k + 1, \ldots, n\}$ of vertices in $G_{n,k}$ with fewer than $2k$ neighbors. Moreover, since all vertices in $\{1, \ldots, k\}$ are neighbors, at most one node of degree 1 can have a vertex from this set as its unique vertex, and the same is true of the set $\{n - k + 1, \ldots, n\}$. Therefore T has at most two nodes of degree 1: T is a path.

By Lemma 7, no node in T can cover more than $K - k + 1$ cliques in $\mathcal{C}$. It follows that T contains at least $\lceil |\mathcal{C}|/(K - k + 1) \rceil = f(n, k, K) + 1$ nodes and, since T is a path, that the diameter of T is at least $f(n, k, K)$. □

Theorems 2 and 8 together show the validity of the first bound of Theorem 1.

Let us say that a node $y \in X$ *separates* integers i_x and i_z if $i_x, i_z \notin U_y$, but there are nodes $x, z \in X$ with $i_x \in U_x$ and $i_z \in U_z$ such that y lies on the path in T between x and z.

Lemma 9. *Suppose that $y \in X$ separates integers i_x and i_z with $i_x < i_z$. Then U_y contains at least k elements of $\{i_x + 1, \ldots, i_z - 1\}$.*

Proof. Let $x, z \in X$ be as in the definition above; i.e., $i_x \in U_x$, $i_z \in U_z$, and y lies on the path in T between x and z. Let T_x be the subtree of T induced by the nodes that can be reached from x without touching y, and let j be the smallest integer larger than i_x that does not belong to the bag of a node in T_x. It is easy to see that such an integer indeed exists, and that $j \leq i_z$. Since i_x does not belong to U_y and each of $i_x + 1, \ldots, \min\{i_x + k, n\}$ is a neighbor of i_x, we have $j > i_x + k$. By definition of j, thus, each element of the k-element set $W = \{j - k, \ldots, j - 1\}$ belongs to the bag of a node in T_x. On the other hand, each element of W, being a neighbor of j, also belongs to the bag of a node not in T_x, and thus to U_y. A final observation is that $W \subseteq \{i_x + 1, \ldots, i_z - 1\}$. □

We define a *caterpillar* as a tree in which no node is of degree more than 3 and all nodes of degree 3 lie on a single path. A *spine* of a caterpillar J is a maximal path in J that contains all nodes of degree 3. For every node x on a fixed spine S of a caterpillar J, we define the *leg attached at x* as the subgraph of J induced by the nodes in J closer to x than to any other node on S.

Lemma 10. *For every integer $d \geq 0$, a caterpillar of diameter d contains at most $\frac{1}{4}(d + 2)^2$ nodes.*

Proof. Let J be a caterpillar of diameter d with a maximal number of nodes and fix a particular spine S of J. A little thought shows that the number of nodes on two legs of J attached at neighboring nodes of S cannot differ by more than one. The two legs attached at the endpoints of S by definition consist of just one node (which belongs to S), and thus the total number of nodes in J cannot exceed $1 + 2 + 3 + \cdots + 3 + 2 + 1$, where the sum has exactly $d + 1$ terms. The value of this sum is bounded by $\frac{1}{4}(d+2)^2$. □

Theorem 11. *(the square-rootic lower bound) Let n, k, and K be integers with $1 \leq k < 2k \leq K \leq 3k - 2$. Then $D(n, k, K) \geq \lceil g(n, k, K) \rceil$, where*

$$g(n, k, K) = 2 \frac{\sqrt{(K - 2k + 1)n + k^2} - K + k - 1}{K - 2k + 1}.$$

Proof. Let z and z' be nodes in T whose bags contain the cliques $\{1, \ldots, k+1\}$ and $\{n-k, \ldots, n\}$, respectively. We assume that 1 occurs in no bag other than that of z and that n occurs in no bag other than that of z'—this condition can be enforced simply by removing all offending occurrences of 1 and n.

We begin by showing that T is a caterpillar whose spine can be chosen to contain z and z'. If this is not the case, T contains a node y that either is of degree at least 4 or is of degree 3 and does not belong to the path from z to z'. In either case, removing y splits T into subtrees, at least two of which, say, T_1 and T_2, contain neither z nor z'. Choose i_1 and i_2 as unique vertices of nodes of degree 1 in T that belong to T_1 and T_2, respectively, and assume without loss of generality that $i_1 < i_2$. Now y separates i_1 and i_2 and therefore, by Lemma 9, U_y contains at least k elements of $\{i_1 + 1, \ldots, i_2 - 1\}$. Similarly, $y = z$ or y separates 1 and i_1, so U_y contains at least k elements of $\{1, \ldots, i_1 - 1\}$. Finally, U_y contains at least k elements of $\{i_2 + 1, \ldots, n\}$, contradicting the fact that $|U_y| \leq K + 1 < 3k$.

Thus T is a caterpillar, and we can fix a spine S in T that contains z and z'. An argument similar to the one used above shows that if y is a node in T of degree at least 2 whose removal disconnects some node of degree 1 in T from both z and z', then U_y has interval-covering number at least 2. On each leg L, at most one node is not of this kind, namely the node on L farthest from S. Thus each leg contributes at most one node whose bag has interval-covering number 1.

Since all $n - k$ cliques in $\mathcal{C}$ are covered by some node in X, Lemma 7 implies that $|X|(K - 2k + 1) + |S|k \geq n - k$. Let d be the diameter of T. Then $|S| \leq d + 1$ and, by Lemma 10, $|X| \leq \frac{1}{4}(d+2)^2$. It follows that $p(d) \geq 0$, where p is the quadratic polynomial with $p(m) = \frac{1}{4}(m+2)^2(K - 2k + 1) + (m+2)k - n$. Since $p(-2) = -n < 0$, p has two roots, the smaller of which is negative. A standard analysis shows that the larger of the two roots is precisely $g(n, k, K)$. It is now easy to see that $D(n, k, K) \geq \lceil g(n, k, K) \rceil$. □

It can be shown that the smaller of the bounds of Theorem 2 and Corollary 6 is within a constant factor of the bound of Theorem 11 for $K \geq 2k$, which proves the validity of the second bound of Theorem 1.

Theorem 12. (*the logarithmic lower bound*) *Let n, k and K be integers with $1 \leq k \leq K \leq n - 1$. Then*

$$D(n,k,K) \geq \left\lceil \log_2 \left(\frac{n-k}{K-k+1} \right) \bigg/ \log_2 \left\lfloor \frac{K+k+1}{k} \right\rfloor \right\rceil.$$

Proof. If a node y in T is of degree $r \geq 2$, removing y splits T into r subtrees, and we can pick unique vertices $i_1, \ldots, i_r$, one from each subtree, such that $i_1 < i_2 < \cdots < i_r$. Then, for $j = 2, \ldots, r$, y separates i_{j-1} and i_j and hence, by Lemma 9, U_y contains at least k elements of $\{i_{j-1} + 1, \ldots, i_j - 1\}$. Thus $(r-1)k \leq |U_y| \leq K + 1$ and therefore $r \leq \lfloor (K+k+1)/k \rfloor$.

It is easy to see that a tree of maximum degree at most $r \geq 2$ and with diameter d contains at most r^d nodes. As already observed in the proof of Theorem 8, Lemma 7 implies that T contains at least $(n-k)/(K-k+1)$ nodes. Theorem 12 follows by combining this with the bound on the maximum degree of T established above. □

Acknowledgment

We thank the organizers of ICALP'95 and the city of Szeged for providing the hospitable environment in which this research was begun.

References

1. H. L. Bodlaender, NC-algorithms for graphs with small treewidth, in Proc. 14th International Workshop on Graph-Theoretic Concepts in Computer Science, J. van Leeuwen, ed., Lecture Notes in Computer Science, Springer-Verlag, Berlin, vol. 344, 1989, pp. 1–10.
2. H. L. Bodlaender, Treewidth: Algorithmic techniques and results, in Proc. 22nd International Symposium on Mathematical Foundations of Computer Science, I. Prívara and P. Ružička, eds., Lecture Notes in Computer Science, Springer-Verlag, Berlin, vol. 1295, 1997, pp. 19–36.
3. H. L. Bodlaender and T. Hagerup, Parallel algorithms with optimal speedup for bounded treewidth, in Proc. 22nd International Colloquium on Automata, Languages and Programming, Z. Fülöp and F. Gécseg, eds., Lecture Notes in Computer Science, Springer-Verlag, Berlin, vol. 944, 1995, pp. 268–279. To appear in SIAM J. Computing, 1998.
4. N. Chandrasekharan, *Fast Parallel Algorithms and Enumeration Techniques for Partial k-Trees*, Ph.D. thesis, Clemson University, 1989.
5. N. Chandrasekharan and S. T. Hedetniemi, Fast parallel algorithms for tree decomposing and parsing partial k-trees, in Proc. 26th Annual Allerton Conference on Communication, Control, and Computing, Urbana-Champaign, Illinois, 1988, pp. 283–292.
6. S. Chaudhuri and C. D. Zaroliagis, Optimal parallel shortest paths in small treewidth digraphs, in Proc. 3rd Annual European Symposium on Algorithms, P. Spirakis, ed., Lecture Notes in Computer Science, Springer-Verlag, Berlin, vol. 979, 1995, pp. 31–45. To appear in Theoretical Computer Science, 1998.

7. T. Hagerup, Dynamic algorithms for graphs of bounded treewidth, in Proc. 24th International Colloquium on Automata, Languages and Programming, P. Degano, R. Gorrieri, and A. Marchetti-Spaccamela, eds., Lecture Notes in Computer Science, Springer-Verlag, Berlin, vol. 1256, 1997, pp. 292–302.
8. T. Hagerup, J. Katajainen, N. Nishimura, and P. Ragde, Characterizations of k-terminal flow networks and computing network flows in partial k-trees, in Proc. 6th Annual ACM-SIAM Symposium on Discrete Algorithms, 1995, pp. 641–649.

Degree-Preserving Forests

Hajo Broersma[1], Andreas Huck[2], Ton Kloks[3], Otto Koppius[1], Dieter Kratsch[4], Haiko Müller[4], and Hilde Tuinstra[1]

[1] University of Twente
Faculty of Applied Mathematics
P.O. Box 217
7500 AE Enschede, the Netherlands
{broersma,tuinstra}@math.utwente.nl

[2] University of Hannover
Institute of Mathematics
Welfengarten 1
30167 Hannover, Germany
huck@math.uni-hannover.de

[3] Department of Applied Mathematics
Charles University
Malostranské nám. 25
11800 Praha 1, Czech Republic
kloks@kam.ms.mff.cuni.cz

[4] Friedrich-Schiller-Universität Jena
Fakultät für Mathematik und Informatik
07740 Jena, Germany
{kratsch,hm}@minet.uni-jena.de

Abstract. We consider the degree-preserving spanning tree (DPST) problem: given a connected graph G, find a spanning tree T of G such that as many vertices of T as possible have the same degree in T as in G. This problem is a graph-theoretical translation of a problem arising in the system-theoretical context of identifiability in networks, a concept which has applications in e.g., water distribution networks and electrical networks. We show that the DPST problem is NP-complete, even when restricted to split graphs or bipartite planar graphs. We present linear time approximation algorithms for planar graphs of worst case performance ratio $1 - \epsilon$ for every constant $\epsilon > 0$. Furthermore we give exact algorithms for interval graphs (linear time), graphs of bounded treewidth (linear time), cocomparability graphs ($O(n^4)$), and graphs of bounded asteroidal number.

1 Description of the Problem and Its Practical Niche

Analysis of communication or distribution networks is often concerned with finding spanning trees (or forests) of those networks fulfilling certain criteria. Also in other contexts spanning trees show up as important tools in modeling and analyzing problems. Therefore, a myriad of problems on spanning trees have been studied in literature (see [5,6,8,9]). This paper deals with a virtually unexplored problem concerning spanning trees which we call the degree-preserving spanning tree (DPST) problem: given a connected graph G, find a spanning tree T of G with a maximum number of degree-preserving vertices, i.e., with a maximum number of vertices having the same degree in T as in G.

Some closely related questions were studied before by Lewinter et al. [1,12,13] from a purely theoretical point of view. They published a number of short notes on the subject. For example, Lewinter [12] introduced the concept of degree-preserving spanning trees and he proved that the number of degree-preserving vertices interpolates on the set of spanning trees of a given connected graph G. In other words: if spanning trees exist with k and l degree-preserving vertices respectively and $k < l$, then there exists a spanning tree with exactly m degree-preserving vertices for every m with $k < m < l$.

Our attention was initially turned to this problem through a practical application in water distribution networks (see [14]), which makes the DPST problem a nice example of theory and practice going hand-in-hand.

Suppose that we have to determine (or control) all flows in such a network by installing and using a small number of flow meters and/or pressure gauges. The network can be regarded as an undirected connected graph G and the flow through each edge of G is described by an orientation of that edge and a non-negative flow value. Since the sum of all flow values of edges entering a vertex is always the same as the sum of all flow values of edges leaving that vertex, except for possible sources and sinks, it is not difficult to derive all flows in the network from the flows through all edges of a cotree C of G (i.e., C is obtained from G by removing the edges of a spanning tree). Hence it would suffice to install flow meters at the edges of C. However the costs of installing a flow meter is much higher than those of installing a water pressure gauge at some vertex. Alternatively, we can derive the flow through an edge from the water pressure drop between the two incident vertices. If we only use pressure gauges, and want to minimize the costs, the problem becomes that of finding a cotree whose edges are incident with a minimum number of vertices (in order to minimize the number of pressure gauges that have to be installed) or, equivalently, of finding a spanning tree T whose complement in G has as many isolated vertices as possible, i.e., T has a maximum number of degree-preserving vertices. Rahal [15] independently discovered the cotree approach in his investigation of a steady state formulation for water distribution networks.

Our problem of determining all flows in the network with minimal costs of measuring (installing pressure gauges) is a so-called identifiability problem (see Walter [16]). The concrete water distribution network that we considered has 80 vertices and 98 edges, making it a very sparse network. Our network is planar and it has outerplanarity 2. Especially this latter fact enables us to solve the DPST problem in our case by a linear time algorithm.

2 Preliminaries

Throughout let $G = (V, E)$ be a graph and let $n = |V|$ and $m = |E|$.

For a subset $S \subseteq V$ we use $G[S]$ to denote the subgraph of G induced by the vertices of S. For a subset $S \subseteq V$ we also write $G - S$ for $G[V \setminus S]$, and for a vertex x of G we write $G - x$ instead of $G - \{x\}$.

For a vertex x of G we use $N_G(x)$ to denote the set of neighbors of x in G, and we write $N_G[x] = \{x\} \cup N_G(x)$ for the closed neighborhood of x in G; the degree of x in G is $d_G(x) = |N_G(x)|$. A pendant vertex of G is a vertex with degree one in G. We omit the subscript G from the above expressions if it is clear which graph G we consider.

Definition 1. *A subset $S \subseteq V$ is* realizable *if there exists a spanning forest T of G such that the degree of every vertex $x \in S$ is preserved in T (i.e., if $d_T(x) = d_G(x)$ for every vertex $x \in S$). If T is such a spanning forest, then we call T an S-preserving* forest. *If, moreover, T is chosen in such a way that $|S|$ is maximum, then we call T a* maximum degree-preserving *forest, and $|S|$ the* degree-preserving number *(of T or G). The DPST problem is the problem to find for a given graph G a maximum degree-preserving spanning forest.*

As an example, the degree-preserving number of a tree, a unicyclic graph, and a complete graph ($\neq K_2$) on n vertices are respectively n, $n - 2$, and 1.

Notice that to solve the DPST problem, it is sufficient to compute a maximum (cardinality) realizable set S since, given S, an S-preserving spanning forest is then easy to find. By $p(G)$ we denote the cardinality of a maximum realizable set in G. Clearly $p(G)$ is the sum of $p(C)$ taken over all 2-edge-connected components C of G. Therefore we can restrict to 2-edge-connected graphs.

Let W be a set of vertices of a graph G. By $G[\![W]\!]$ we denote the graph with vertex set $N[W]$ containing all edges of G incident with a vertex in W.

Lemma 1. *Let S be a nonempty set of vertices of a graph $G = (V, E)$. Then S is a realizable set of G if and only if $G[\![S]\!]$ is a forest.*

3 Hardness Results

A graph $G = (V, E)$ is called a *split graph* (*bipartite graph*) if V can be partitioned into an independent set I and a clique C (into two independent sets X and Y) of G. Such a graph is also denoted by $G = (I, C, E)$ ($G = (X, Y, E)$).

Let $G = (V, E)$ be a graph. We define a split graph H with independent set V and clique $E \times \{1, 2\}$ as follows. A pair $\{v, (e, i)\}$ is an edge of H if and only if $v \in V$, $e \in E$, $i \in \{1, 2\}$ and $v \in e$. It is easy to see that a set $W \subseteq V$ is an independent set of G if and only if W is a realizable set in H. Moreover, if G has no isolated vertices (i.e., vertices with degree zero), then for every realizable set W of H with $|W| > 1$ we have $W \subseteq V$. These simple observations lead to the following theorem showing that the DPST problem restricted to split graphs is NP-complete.

Theorem 1. *For a given split graph H and a given integer k it is NP-complete to decide whether H contains a realizable set of cardinality k.*

Proof. The reduction is from the NP-complete graph problem INDEPENDENT SET. As seen before a graph G has an independent set of cardinality k if and only if the corresponding split graph H has a realizable set of cardinality k. □

Next we apply the same idea to bipartite graphs. Let $G = (V, E)$ be a graph. We define a bipartite graph $B = (V \cup (E \times \{2, 4, 6, 8\}), E \times \{1, 3, 5, 7\}, F_1 \cup F_2)$, where

$$F_1 = \{\{v, (e, i)\} : v \in V, e \in E, i \in \{1, 5\}, v \in e\}$$
$$F_2 = \{\{(e, 1), (e, 2)\}, \{(e, 2), (e, 3)\}, \{(e, 3), (e, 4)\}, \{(e, 4), (e, 1)\},$$
$$\{(e, 5), (e, 6)\}, \{(e, 6), (e, 7)\}, \{(e, 7), (e, 8)\}, \{(e, 8), (e, 5)\} : e \in E\}.$$

Note that for the maximum degrees $\Delta(B)$ and $\Delta(G)$ of B and G we get $\Delta(B) = \max\{4, 2 \cdot \Delta(G)\}$. Moreover, B is planar if and only if G is planar.

We observe that for every edge $e \in E$ and every realizable set S of B, $|S \cap (\{e\} \times \{1, 2, 3, 4\})| \leq 2$. In what follows we may assume $S \subseteq V \cup (E \times \{2, 3, 6, 7\})$ for all realizable sets S of B, since for every other realizable set T the set $T' = (T \cap V) \cup (E \times \{2, 3, 6, 7\})$ is also realizable and fulfills $|T| \leq |T'|$.

Next observe that $W \subseteq V$ is an independent set of G if and only if W is a realizable set of B. This leads to the following theorem showing that the DPST problem restricted to bipartite planar graphs is NP-complete.

Theorem 2. *For a given bipartite planar graph B of maximum degree six and a given integer k, it is NP-complete to decide whether B contains a realizable set of cardinality k.*

Proof. The reduction is from the INDEPENDENT SET problem restricted to cubic (i.e., 3-regular) planar graphs [9]. Let (G, k) be an instance of this NP-complete problem where $G = (V, E)$ with $|E| = m$. As seen before a planar graph G has an independent set of cardinality k if and only if the corresponding bipartite planar graph B has a realizable set of cardinality $k + 4m$. □

Our problem remains NP-complete even when restricted to bipartite planar graphs of maximum degree three [7].

The INDEPENDENT SET problem is not only NP-complete, it is also hard to approximate. More precisely for every $\epsilon > 0$, there is no polynomial time approximation algorithm for the MAXIMUM INDEPENDENT SET problem with worst case ratio $n^{1/4-\epsilon}$ unless P=NP [4], and there is no polynomial time approximation algorithm with worst case ratio $n^{1-\epsilon}$ unless co-NP=NP [11]. By the reduction used in the proof of Theorem 1, approximating an optimal solution to the DPST problem is as hard as approximating MAXIMUM INDEPENDENT SET, even when DPST is restricted to split graphs.

Theorem 3. *For every $\epsilon > 0$, there is no polynomial time algorithm to approximate a maximum realizable set of a given split graph with worst case ratio $n^{1/4-\epsilon}$ unless P=NP (respectively with worst case ratio $n^{1-\epsilon}$ unless co-NP=NP).*

4   Approximation for Planar Graphs

In this section we apply an idea of Baker [3] to establish linear time approximation algorithms for the DPST problem when restricted to planar graphs. We will prove the following theorem.

Theorem 4. *For every $\epsilon > 0$ there is a linear time approximation algorithm of worst case performance ratio $1 - \epsilon$ for the DPST problem restricted to planar graphs.*

Let $W \subseteq V$ be a set of forbidden vertices. A realizable set R of G is called *maximum W-avoiding realizable set* if $R \cap W = \emptyset$ and $|R| \geq |R'|$ for every realizable set R' of G with $R' \cap W = \emptyset$.

Let $G = (V, E)$ be a planar graph given with a fixed embedding. We partition V into levels $L_1, L_2, \ldots, L_d$. The level L_1 contains all vertices on the outer face of G. For $i > 1$, the level L_i contains all vertices on the outer face of $G - \bigcup_{j=1}^{i-1} L_j$. Let d be the largest index such that $L_d \neq \emptyset$. For technical reason set $L_i = \emptyset$ for $i > d$ or $i < 1$. A planar graph is *k-outerplanar* if and only if it has an embedding defining at most k nonempty levels.

We decompose the planar graph G into k-outerplanar graphs. Each k-outerplanar graph consists of k consecutive levels of G. More precisely, let k and r be integers with $1 \leq r \leq k$. For $i = 0, 1, \ldots, q$ with $q = \lceil \frac{d-r}{k} \rceil$ we define

$$G_{k,r,i} = G\bigl[\bigcup_{j=(i-1)k+r+1}^{ik+r} L_j\bigr] \quad \text{and} \quad W_{k,r,i} = L_{(i-1)k+r+1} \cup L_{ik+r}.$$

Note that $W_{k,r,i}$ contains all vertices in the outer and inner level of $G_{k,r,i}$.

Lemma 2. *For $i = 0, 1, \ldots, q$ let $R_{k,r,i}$ be a $W_{k,r,i}$-avoiding realizable set of $G_{k,r,i}$. Then $\bigcup_{i=0}^{q} R_{k,r,i}$ is a realizable set of G.*

Proof. For all i the set $W_{k,r,i}$ contains the vertices on the outer and the inner level of the k-outerplanar graph $G_{k,r,i}$. Hence the endpoints of an arbitrary edge of $G[\![R_{k,r}]\!]$ belong to the same k-outerplanar graph. □

Lemma 3. *For every $k \geq 1$ there is an index $r(k)$ with $1 \leq r(k) \leq k$ such that*

$$|R \setminus \bigcup_{i=0}^{q} W_{k,r(k),i}| \geq \tfrac{k-2}{k} p(G).$$

Proof. Let R be a maximum realizable set of G and let $W_{k,r} = \bigcup_{i=0}^{q} W_{k,r,i}$. For every level L_j, $j = 1, 2, \ldots, d$, of G there exist at most two $r \in \{1, 2, \ldots, k\}$ with $L_j \subset W_{k,r}$. Hence $\sum_{r=1}^{k} |R \cap W_{k,r}| \leq 2|R|$, which implies that there is an $r = r(k)$ such that $|R \cap W_{k,r(k)}| \leq \tfrac{2}{k}|R|$. □

Let $k \geq 1$. For every $r = 1, 2, \ldots, k$ and every $i = 1, 2, \ldots, q$ let $R_{k,r,i}$ be a maximum $W_{k,r,i}$-avoiding realizable set of $G_{k,r,i}$. By Lemma 2, $R_{k,r} = \bigcup_{i=0}^{q} R_{k,r,i}$ is a realizable set of G. Consequently,

$$\max\{|R_{k,r}| : 1 \leq r \leq k\} \geq \tfrac{k-2}{k} p(G).$$

For every k we develop an exact linear time algorithm computing a maximum W-avoiding realizable set for k-outerplanar graphs. Using standard techniques for graphs of bounded treewidth, it can be shown that a linear time algorithm, exists [2]. Notice that the treewidth of a k-outerplanar graph is at most $3k - 1$. Consequently, for every fixed k we obtain a linear time approximation algorithm of worst case performance ratio $\frac{k-2}{k}$.

5 Interval Graphs

Definition 2. *A graph is* chordal *if it contains no induced cycle of length more than three.*

Notice that for chordal graphs, the problem of finding a maximum realizable set is NP-complete, since the class of split graphs is a proper subclass of the class of chordal graphs. However, for the class of interval graphs, which is another important subclass of the class of chordal graphs, we can give a fast algorithm. For an introduction into these graph class we refer to [10].

Our first result shows that for chordal graphs we can restrict our search for realizable sets to independent sets. Remember that we may restrict to 2-edge connected graphs.

Theorem 5. *If G is a 2-edge connected chordal graph, then any realizable set S of G is an independent set of G.*

Proof. Let $G = (V, E)$ be a 2-edge connected chordal graph and assume $\{x, y\} \in E$ for two distinct vertices $x, y \in S$. Since G is 2-edge connected, $\{x, y\}$ is contained in a cycle of G, and, since G is chordal this implies $\{x, y\}$ is contained in some triangle of G. This contradicts Lemma 1. □

If a graph G is disconnected, then a maximum realizable set of G is simply the union of maximum realizable sets of all components of G. If a connected graph G (or a component) has a bridge e, then to compute a maximum realizable set of G delete e and compute maximum realizable sets S_1 and S_2 for both components. Let T_1 be an S_1-preserving forest and T_2 be an S_2-preserving forest. Adding e as an edge between T_1 and T_2 gives a forest T which is $S_1 \cup S_2$-preserving, and $S_1 \cup S_2$ is a maximum realizable set in G. We will use the above observations and the following properties of 2-edge connected interval graphs.

Definition 3. *An* interval graph *is a graph for which one can associate with each vertex an interval on the real line such that two vertices are adjacent if and only if their corresponding intervals have a nonempty intersection.*

Interval graphs can be recognized in linear time, and, given an interval graph, an interval model for it can be found in linear time [10]. In the following we assume that an interval model of the graph is given, and we identify the vertices of the graph with the corresponding intervals. Without loss of generality we may assume that no two intervals have an endpoint in common.

Definition 4. *An interval and its corresponding vertex are called* minimal *if it is minimal with respect to inclusion, i.e., if it does not contain any other interval.*

Lemma 4. *Let G be a 2-edge connected interval graph. Then there exists a maximum realizable set S of G such that for every vertex $p \in S$ the corresponding interval is minimal.*

Proof. Let S be a maximum realizable set containing a vertex x which is not minimal. Then there exists an interval y contained in the interval x. By Theorem 5 we know that a realizable set can contain only one of x and y and hence $y \notin S$. Now $N(y) \subseteq N[x]$, and hence, there exists a maximum realizable set $S' = \{y\} \cup S \setminus \{x\}$. Repeating the arguments we can prove the assertion of the theorem. □

Consider the ordering of the minimal intervals defined by the left endpoints.

Lemma 5. *Let G be a 2-edge connected interval graph with corresponding interval model and let x be the first minimal interval (i.e., with the leftmost left endpoint). There exists a maximum realizable set S of G with $x \in S$.*

Proof. Consider a maximum realizable set S of G containing only minimal intervals. If $x \in S$ there is nothing to prove. Otherwise, let y be the first interval in S. The other intervals of S lie totally to the right of y because S is an independent set by Theorem 5. The right endpoint of y must be to the right of the right endpoint of x since the interval x is minimal. It follows that $S' = \{y\} \cup S \setminus \{x\}$ is also realizable, since x lies totally left of $S \setminus \{y\}$ and $N(z) \cap N(x) \subseteq N(z) \cap N(y)$ for all $z \in S \setminus \{y\}$. □

Theorem 6. *There is a linear time algorithm to compute a maximum realizable set S for given interval graph G.*

Proof. Locate the set of bridges B in G and compute maximum cardinality realizable sets for each component of $G - B$. This can be done as follows.

Consider an interval model for a 2-edge connected component. First mark the minimal intervals. Take the minimal interval with the leftmost left endpoint as the first element of S. Consider the endpoints one by one, from left to right. We keep track of the last minimal interval in S which is totally left of the current position. We also keep a counter for the number of intervals that have one endpoint to the left of the current position and that overlap with the last interval in S. If we encounter a left endpoint of a minimal interval which starts to the right of the last interval in S so far, and if there is at most one interval overlapping the current position and the last interval of S, then we put this new minimal interval in S.

Let S' be a maximum realizable set such that $S \neq S'$. By the previous lemmas we may assume that S' contains minimal intervals only and that S and S' have a common first interval. Suppose y is the first interval of S' which is not in S, and that $x_1, x_2, \ldots, x_p$ are common intervals of S and S' and $x_{p+1} \neq y$ is the next interval of S chosen by the above procedure. We complete the proof by showing that y in S' can be replaced by x_{p+1}. This follows by the same arguments as in the proof of Lemma 5 and the following observations. By the choice of $x_1, x_2, \ldots, x_p$, for all $i, j \in \{1, \ldots, p\}$ with $i \neq j$, x_i and x_j have at most one common neighbor and $N(x_{p+1}) \cap N(x_i) \subseteq N(x_{p+1}) \cap N(x_{i+1})$ ($i = 1, \ldots, p-1$). If the addition of x_{p+1} to $\{x_1, \ldots, x_p\}$ would cause a cycle in $G[\![\{x_1, x_2, \ldots, x_p, x_{p+1}\}]\!]$, then such a cycle would already exist in $G[\![\{x_1, \ldots, x_p\}]\!]$, a contradiction to the choice of $x_1, x_2, \ldots, x_p$. □

6 Other Classes of Graphs

In this section we list further results proven in the full version.

Theorem 7. *The degree preserving spanning tree problem is solvable in linear time for graphs of bounded treewidth.*

Definition 5. *A graph $G = (V, E)$ is a* cocomparability graph *if and only if there is an ordering $v_1, v_2, \ldots, v_n$ of V such that $i < j < k$ and $\{v_i, v_k\} \in E$ implies either $\{v_i, v_j\} \in E$ or $\{v_j, v_k\} \in E$. Hence $N(v_j) \cap \{v_i, v_k\} \neq \emptyset$ for all j with $i < j < k$. Such an ordering is called* cocomparability ordering.

Theorem 8. *There is an algorithm to compute a maximum degree-preserving forest of a cocomparability graph in time $O(n^4)$.*

Definition 6. *An independent set A is called an* asteroidal set *if for every vertex $a \in A$, the set $A \setminus \{a\}$ is contained in a component of $G - N[a]$. The* asteroidal number *of a graph G, is the maximum cardinality of an asteroidal set in G.*

Theorem 9. *There is an algorithm to solve the degree preserving spanning tree problem for any graph G in time $O(2^{k^3} n^{k+3} \log n)$, where k is the asteroidal number of G.*

References

1. Aaron, M. and M. Lewinter, 0-deficient vertices of spanning trees, *NY Acad. Sci. Graph Theory Notes* **XXVII**, (1994), pp. 31–32.
2. Arnborg S., J. Lagergren and D. Seese, Easy problems for tree-decomposable graphs, *Journal of Algorithms* **12**, (1991), pp. 308–340.
3. Baker, B. S., Approximation algorithms for NP-complete problems on planar graphs, *J. ACM* **41**, (1994), pp. 153–180.
4. Bellare, M., O. Goldreich and M. Sudan, Free bits, PCPs and non-approximability - towards tight results, *SIAM J. Comput.* **27** (1998), pp. 804–915.

5. Camerini, P. M., G. Galbiati and F. Maffioli, Complexity of spanning tree problems: Part I, *Eur. J. Oper. Res.* **5**, (1980), pp. 346–352.
6. Camerini, P. M., G. Galbiati and F. Maffioli, The complexity of weighted multi-constrained spanning tree problems, *Colloq. Math. Soc. Janos Bolyai* **44**, (1984), pp. 53–101.
7. Damaschke, P., Degree-preserving spanning trees and coloring bounded degree graphs, Manuscript 1997.
8. Dell'Amico, M., M. Labbé and F. Maffioli, Complexity of spanning tree problems with leaf-dependent objectives, *Networks* **27**, (1996), pp. 175–181.
9. Garey, M. R. and D.S. Johnson, *Computers and Intractability: A guide to the Theory of NP-completeness*, Freeman, New York, 1979.
10. Golumbic, M. C., *Algorithmic Graph Theory and Perfect Graphs*, Academic Press, New York, 1980.
11. Håstad, J., Clique is hard to approximate within $n^{1-\epsilon}$, Proc. 37th Ann. IEEE Symp. on Foundations of Comput. Sci., (1996), IEEE Computer Society, pp. 627–636.
12. Lewinter, M., Interpolation theorem for the number of degree-preserving vertices of spanning trees, *IEEE Trans. Circ. Syst.* **CAS-34**, (1987), 205.
13. Lewinter, M. and M. Migdail-Smith, Degree-preserving vertices of spanning trees of the hypercube, *NY Acad. Sci. Graph Theory Notes* **XIII**, (1987), 26–27.
14. Pothof, I. W. M. and J. Schut, *Graph-theoretic approach to identifiability in a water distribution network*, Memorandum 1283, Faculty of Applied Mathematics, University of Twente, Enschede, the Netherlands, (1995).
15. Rahal, A co-tree flows formulation for steady state in water distribution networks, *Adv. Eng. Softw.* **22**, (1995), pp. 169–178.
16. Walter, E., *Identifiability of state space models with applications to transformation systems*, Springer-Verlag, New York NY, USA, 1982.

A Parallelization of Dijkstra's Shortest Path Algorithm

A. Crauser, K. Mehlhorn, U. Meyer, and P. Sanders

Max-Planck-Institut für Informatik,
Im Stadtwald, 66123 Saarbrücken, Germany.
{crauser,mehlhorn,umeyer,sanders}@mpi-sb.mpg.de
http://www.mpi-sb.mpg.de/{~crauser,~mehlhorn,~umeyer,~sanders}

Abstract. The single source shortest path (SSSP) problem lacks parallel solutions which are fast and simultaneously work-efficient. We propose simple criteria which divide Dijkstra's sequential SSSP algorithm into a number of phases, such that the operations within a phase can be done in parallel. We give a PRAM algorithm based on these criteria and analyze its performance on random digraphs with random edge weights uniformly distributed in $[0,1]$. We use the $\mathcal{G}(n, d/n)$ model: the graph consists of n nodes and each edge is chosen with probability d/n. Our PRAM algorithm needs $\mathcal{O}(n^{1/3} \log n)$ time and $\mathcal{O}(n \log n + dn)$ work with high probability (whp). We also give extensions to external memory computation. Simulations show the applicability of our approach even on non-random graphs.

1 Introduction

Computing shortest paths is an important combinatorial optimization problem with numerous applications. Let $G = (V, E)$ be a directed graph, $|E| = m$, $|V| = n$, let s be a distinguished vertex of the graph, and c be a function assigning a non-negative real-valued *weight* to each edge of G. The *single source shortest path problem* (SSSP) is that of computing, for each vertex v reachable from s, the weight dist(v) of a minimum-weight path from s to v; the weight of a path is the sum of the weights of its edges.

The theoretically most efficient sequential algorithm on digraphs with non-negative edge weights is Dijkstra's algorithm [8]. Using Fibonacci heaps its running time is $\mathcal{O}(n \log n + m)$[1]. Dijkstra's algorithm maintains a partition of V into *settled*, *queued* and *unreached* nodes and for each node v a *tentative distance* tent(v); tent(v) is always the weight of some path from s to v and hence an upper bound on dist(v). For unreached nodes, tent(v) = ∞. Initially, s is queued, tent(s) = 0, and all other nodes are unreached. In each iteration, the queued node v with smallest tentative distance is selected and declared settled and all edges (v, w) are relaxed, i.e., tent(w) is set to min{tent(w), tent(v) + $c(v, w)$}.

---
[1] There is also an $\mathcal{O}(n + m)$ time algorithm for undirected graphs [20], but it requires the RAM model instead of the comparison model which is used in this work.

If w was unreached, it is now queued. It is well known that $\text{tent}(v) = \text{dist}(v)$, when v is selected from the queue.

The queue may contain more than one node v with $\text{tent}(v) = \text{dist}(v)$. All such nodes could be removed simultaneously, the problem is to identify them. In Sect. 2 we give simple sufficient criteria for a queued node v to satisfy $\text{tent}(v) = \text{dist}(v)$. We remove all nodes satisfying the criteria simultaneously.

Although there exist worst-case inputs needing $\Theta(n)$ phases, our approach yields considerable parallelism on random directed graphs: We use the random graph model $\mathcal{G}(n, d/n)$, i.e., there are n nodes and each theoretically possible edge is included into the graph with probability d/n. Furthermore, we assume random edge weights uniformly distributed in $[0, 1]$: In Sect. 3 we show that the number of phases is $\mathcal{O}(\sqrt{n})$ using a simple criterion, and $\mathcal{O}(n^{1/3})$ for a more refined criterion with high probability (whp)[2].

Sect. 4 presents an adaption of the phase driven approach to the CRCW PRAM model which allows p processors (PUs) concurrent read/write access to a shared memory in unit cost (e.g. [13]). We propose an algorithm for random graphs with random edge weights that runs in $\mathcal{O}(n^{1/3} \log n)$ time whp. The work, i.e., the product of its running time and the number of processors, is bounded by $\mathcal{O}(n \log n + dn)$ whp.

In Sect. 5 we adapt the basic idea to external memory computation (I/O model [22]) where one assumes large data structures to reside on D disks. In each I/O operation, D blocks from distinct disks, each of size B, can be accessed in parallel. We derive an algorithm which needs $\mathcal{O}(\frac{n}{D} + \frac{dn}{DB} \log_{S/B} \frac{dn}{DB})$ I/Os on random graphs whp and can use up to $D = \mathcal{O}(\min\{n^{2/3}/\log n, \frac{S}{B}\})$ independent disks. S denotes the size of the internal memory.

In Sect. 6 we report on simulations concerning the number of phases needed for both random graphs and real world data. Finally, Sect. 7 summarizes the results and sketches some open problems and future improvements.

Previous Work

PRAM algorithms: There is no parallel $\mathcal{O}(n \log n + m)$ work PRAM algorithm with sublinear running time for general digraphs with non-negative edge weights. The best $\mathcal{O}(n \log n + m)$ work solution [9] has running time $\mathcal{O}(n \log n)$. All known algorithms with polylogarithmic execution time are work-inefficient. ($\mathcal{O}(\log^2 n)$ time and $\mathcal{O}(n^3 (\log \log n / \log n)^{1/3})$ work for the algorithm in [11].) An $\mathcal{O}(n)$ time algorithm requiring $\mathcal{O}((n + m) \log n)$ work was presented in [3].

For special classes of graphs, like planar digraphs [21] or graphs with separator decomposition [6], more efficient algorithms are known. Randomization was used in order to find approximate solutions [5]. Random graphs with *unit* weight edges are considered in [4]. The solution is restricted to dense graphs ($d = \Theta(n)$) or edge probability $d = \Theta(\log^k n/n)$ ($k > 1$). In the latter case $\mathcal{O}(n \log^{k+1} n)$ work is needed. Properties of shortest paths in complete graphs ($d = n$) with

---
[2] Throughout this paper "whp" stands for "with high probability" in the sense that the probability for some event is at least $1 - n^{-\beta}$ for a constant $\beta > 0$.

random edge weights are investigated in [10, 12]. In contrast to all previous work on random graphs, we are most interested in the case of small, even constant d.
External Memory: The best result on SSSP was published in [16]. This algorithm requires $\mathcal{O}(n + \frac{m}{DB} \log_2 \frac{m}{B})$ I/Os. The solution is only suitable for small n because it needs $\Theta(n)$ I/Os.

2 Running Dijkstra's Algorithm in Phases

We give several criteria for dividing the execution of Dijkstra's algorithm into phases. In the first variant (OUT-version) we compute a threshold defined via the weights of the *outgoing* edges: let $L = \min\{\text{tent}(u) + c(u,z) : u \text{ is queued and } (u,z) \in E\}$ and remove all nodes v from the queue which satisfy $\text{tent}(v) \leq L$. Note that when v is removed from the queue then $\text{dist}(v) = \text{tent}(v)$. The threshold for the OUT-criterion can either be computed via a second priority queue for $o(v) = \text{tent}(v) + \min\{c(v,u) : (v,u) \in E\}$ or even on the fly while removing nodes.

The second variant, the IN-version, is defined via the *incoming* edges: let $M = \min\{\text{tent}(u) : u \text{ is queued}\}$ and $i(v) = \text{tent}(v) - \min\{c(u,v) : (u,v) \in E\}$ for any queued vertex v. Then v can be safely removed from the queue if $i(v) \leq M$. Removable nodes of the IN-type can be found efficiently by using an additional priority queue for $i(\cdot)$.

Finally, the INOUT-version applies both criteria in conjunction.

3 The Number of Phases for Random Graphs

In this section we first investigate the number of delete-phases for the OUT-variant of Dijkstra's algorithm on random graphs. Then we sketch how to extend the analysis to the INOUT-approach. We start with mapping the OUT-approach to the analysis of the reachability problem as provided in [14] and [1, Sect. 10.5] and give lower bounds on the probability that many nodes can be removed from the queue during a phase.

Theorem 1. OUT-approach. *Given a random graph from $\mathcal{G}(n, d/n)$ with edge labels uniformly distributed in $[0,1]$, the SSSP problem can be solved using $r = \mathcal{O}(\sqrt{n})$ delete-phases with high probability.*

We review some facts of the reachability problem using the notation of [1].

The following procedure determines all nodes reachable from a given node s in a random graph G from $\mathcal{G}(n, d/n)$. Nodes will be neutral, active, or dead. Initially, s is active and all other nodes are neutral, let time $t = 0$, and $Y_0 = 1$ the number of active nodes. In every time unit we select an arbitrary active node v and check all theoretically possible edges (v, w), w neutral, for membership in G. If $(v, w) \in E$, w is made active, otherwise it stays neutral. After having treated all neutral w in that way, we declare v dead, and let Y_t equal the new number of active nodes. The process terminates when there are no active nodes.

The connection with the OUT-variant of Dijkstra's algorithm is easy: The distance labels determine the order in which queued vertices are considered and declared dead, and time is partitioned into intervals (=phases): If a phase of the OUT-variant removes k nodes this means that the time t increases by k.

Let Z_t be the number of nodes w that are reached for the first time at time t. Then $Y_0 = 1$, $Y_t = Y_{t-1} + Z_t - 1$ and $Z_t \sim B[n - (t-1) - Y_{t-1}, d/n]$ where $B[n, q]$ denotes the binomial distribution for n trials and success probability q.

Let T be the least t for which $Y_t = 0$. Then T is the number of nodes that are reachable from s. The recursive definition of Y_t is continued for all t, $0 \leq t \leq n$. We have $Y_t \sim B[n-1, 1-(1-d/n)^t] + 1 - t$.

It is shown in [1] that the number of nodes reachable from s is either very small (less than $\mathcal{O}(\log n)$) or concentrates around $T_0 = \alpha_0 n$, where $0 < \alpha_0 < 1$, and $\alpha_0 = 1 - e^{-d\alpha_0}$. Only the case $T \approx T_0$ requires analysis; if $T = \mathcal{O}(\log n)$ the number of phases is certainly small. Chernoff bounds yield:

Lemma 1. *Except for small t ($t \leq \sqrt{n}$) and large t ($t \geq T_0 - n^{1/2+\epsilon}$) Y_t is $(1 \pm o(1/n^2))\mathbf{E}[Y_t]$ with high probability.*

The *yield* of a phase in the OUT-variant is the number of nodes that are removed in a phase. We call a phase starting at time t *profitable* if its yield is $\Omega(\sqrt{Y_t/d})$ and *highly profitable* if its yield is $\Omega(\sqrt{(Y_{t/2} - t/2)t/n})$ and show:

Lemma 2. *A phase is profitable with probability at least $1/8$. A phase starting at time t with $\frac{n \ln d}{d} \leq t \leq \alpha_0 n - n/d$ is highly profitable with probability at least $1/8$.*

Theorem 1 follows fairly easily from lemmas 1 and 2: We call a phase with starting time t *early extreme* if $t \leq \sqrt{n}$, *early intermediate* if $\sqrt{n} < t \leq (n \ln d)/d$, *early central* if $(n \ln d)/d < t \leq n/2$, *late central* if $n/2 < t \leq \alpha_0 n - n/d$, *late intermediate* if $\alpha_0 n - n/d < t \leq \alpha_0 n - n^{1/2+\epsilon}$, and *late extreme* if $\alpha_0 n - n^{1/2+\epsilon} < t$, and show that there are only $\mathcal{O}(\sqrt{n})$ phases of each kind with high probability. Consider, for example, the late intermediate phases. A profitable late intermediate phase starting at time t has yield $\Omega(\sqrt{Y_t/d}) = \Omega(\sqrt{\mathbf{E}[Y_t]/d}) = \Omega(\sqrt{(\alpha_0 n - t)/d})$, where the first equality holds with high probability by Lemma 1. Let $t' := \alpha_0 n - t$. The number of profitable phases with $2^i \leq t' < 2^{i+1}$ is therefore $\mathcal{O}(\sqrt{2^i d})$ and the number of profitable phases with $\alpha_0 n - n/d \leq t = \alpha_0 n - t'$ is therefore $\sum_{i \leq \log(n/d)} \mathcal{O}(\sqrt{2^i d}) = \mathcal{O}(\sqrt{n})$. Since a phase is profitable with probability at least $1/8$, the number of phases is also $\mathcal{O}(\sqrt{n})$ with high probability. The number of early extreme phases is $\mathcal{O}(\sqrt{n})$ trivially. For the number of late extreme phases we argue as follows. We first show that $T \leq \alpha_0 n + n^{1/2+\epsilon}$ with high probability and then consider the first time t_1, $t_1 \geq \alpha_0 n - n^{1/2+\epsilon}$, with $Y_{t_1} \leq n^{1/4}$. Lemma 1 implies that the number of late extreme phases starting before t_1 is $\mathcal{O}(\sqrt{n})$. If the number of phases starting after t_1 is $\sqrt{n}$ or more, then $Z_{t_1} + Z_{t_1+1} + \cdots + Z_{t_1+\sqrt{n}} \geq \sqrt{n} - n^{1/4} \geq \sqrt{n}/2$. The probability of this event is bounded by $\mathbf{P}\left[B[n^{1/2}(n - (n - n^{1/2+\epsilon})), d/n] \geq \sqrt{n}/2\right]$, which is exponentially small.

The idea for the proof of Lemma 2 is as follows. Let $v_1, v_2, \ldots, v_q$, $q = Y_t$, be the queued nodes in order of increasing tentative distances, and let L' be the value of L in the previous phase. The distance labels $\text{tent}(v_i)$ are random variables in $[L', L'+1]$. We show that their values are independent and their distributions are biased towards smaller values (since $\text{tent}(v_i) = \min\{\text{dist}(v) + c(v, v_i), v \text{ settled and } (v, v_i) \in E\}$, $\text{dist}(v) \leq L'$, $c(v, v_i)$ uniform in $[0, 1]$. The value of $\text{tent}(v_r)$ is therefore less than r/q with constant probability for arbitrary r, $1 \leq r \leq q$. The number of edges out of $v_1, \ldots, v_r$ is $r(d/n)n = rd$ on the average and not much more with constant probability. The shortest of these edges has length about $\frac{1}{rd}$. We remove $v_1, \ldots, v_r$ from the queue if $\text{tent}(v_r)$ is smaller than the length of the shortest edge out of $v_1, \ldots, v_r$. This is the case (with constant probability) if $r/q \leq \frac{1}{rd}$ or $r \leq \sqrt{q/d}$.

For the phases starting at time t with $(n \ln d)/d \leq t \leq \alpha_0 n - n/d$ we refine the argument as follows. We call a node queued at time t old if it was already queued before time $t/2$ and show that the number of old queued nodes at time t is at least $Y_{t/2} - t/2$. Each old queued node has an expected indegree from settled nodes of at least $\frac{t}{2}\frac{d}{n}$. We use this fact to deduce that $\text{tent}(v_r)$ is less than $r/(\frac{td}{2n}(Y_{t/2} - t/2))$ with constant probability and then proceed as above.

INOUT Approach. If both IN- and OUT-criterion are applied together, the tentative distance labels of queued nodes may spread over a range as large as $[L', L' + 2)$, while the edge weights are only in $[0, 1]$. In order to reuse the analysis of the OUT-part we analyze a slightly slower version which alternates the two criteria in the following way:

I-**Step:** Let q be the current queue size. Apply the IN-criterion to the $g(q)$ nodes with smallest tentative distances where g is a function we are free to choose[3]. Let L be the largest distance of any removed node. Switch to *O*-Step.

O-**Step:** Repeatedly apply the OUT-criterion until no tentative distance is smaller than L. Then switch back to *I*-Step.

The function $g()$ is chosen in such a way that there is both a constant probability for a large yield in an *I*-Step and the expected number of subsequent *O*-Steps is constant. The function $g()$ is chosen dependent of the current phase type. For example, during late intermediate phases we take $g(q) = cq^{2/3}/d^{1/3}$ for some constant c. A super-phase consisting of an *I*-Step and series of *O*-Steps is now profitable if at most a constant number of *O*-Steps is needed and if its total yield is $\Omega(Y_t^{2/3}/d^{1/3})$, highly profitable if its yield is $\Omega((Y_{t/2} - t/2)^{2/3}/(n/t)^{1/3})$. Then one has to show again that a super-phase is (highly) profitable with constant probability.

Theorem 2. INOUT-approach. *Given a random graph from $\mathcal{G}(n, d/n)$ with edge labels uniformly distributed in $[0, 1]$, the SSSP problem can be solved using $r = \mathcal{O}(n^{1/3})$ delete-phases with high probability.*

---

[3] Note that the implementation does not need to know this function since it uses the faster combined criterion.

4 Parallelization

We now show how the sequential OUT-variant of Sect. 2 can be efficiently implemented on an arbitrary-write CRCW PRAM for random graphs from $\mathcal{G}(n, d/n)$ and random edge weights. The actual number of edges is $m = \Theta(dn)$ whp.

The algorithm keeps a global array tent($\cdot$) for all tentative distance values. Each processor P_i, $0 \leq i < p$ is responsible for two sequential priority queues: Q_i and Q_i^*. Each pair (Q_i, Q_i^*) only deals with a subset of nodes, the distribution is made randomly and stored in a global array ind(). Furthermore, each PU maintains a buffer array for incoming requests.

The queues Q_i handle tentative node distances for the nodes they are responsible for, the key of a node $v \in Q_i^*$ is given by tent$(v) + \delta_o(v)$ where $\delta_o(v) := \min\{c(v,w) : (v,w) \in E\}$; $\delta_o(v)$ is precomputed once and for all upon initialization. The Q_i^* queues are used to efficiently derive the criterion of the OUT-version indicating whether a node can be deleted in a phase. The queues are implemented as relaxed heaps [9] because they provide worst-case running times: `findMin`, `insert` and `decreaseKey` are performed in $\mathcal{O}(1)$ time and `delete`/`deleteMin` in $\mathcal{O}(\log q)$ time where q denotes the local queue size.

Let r be the number of delete-phases which are needed, e.g. for the OUT-variant $r = \mathcal{O}(\sqrt{n})$ whp. For the analysis we fix the number of processors as $p = \max\{\frac{n}{r \log n}, \frac{dn}{r \log^2 n}\}$; so from now on a time bound T implies a work bound pT.

The algorithm works similar to Dijkstra's algorithm: The queues start with only s in $Q_{\text{ind}(s)}$ and $Q_{\text{ind}(s)}^*$ and all other local queues empty. This and the initialization of other arrays and buffers (ind(), outgoing edges, ...) can be done in time $\mathcal{O}((n+m)/p) = \mathcal{O}(r \log^2 n)$ whp, even if the input uses an adjacency-list representation.

While any queue is nonempty the algorithm performs a phase consisting of five steps. These steps are now further explicated together with the most interesting part of their analysis, namely for the case that at most n/r nodes are deleted in this phase.

Step 1 finds the global minimum L of all elements in all Q_i^* and can clearly be performed in $\mathcal{O}(\log p) \leq \mathcal{O}(\log n)$ time.

In **Step 2** each PU i removes the nodes with tent$(v) \leq L$ from Q_i and Q_i^*. Let $\check{R}$ denote the union of all these sets of deleted nodes. Our index distribution ensures that no PU has to deal with more than $\mathcal{O}(\log p + |\check{R}|/p)$ `deleteMins` whp. A single `deleteMin` or `delete` operation takes $\mathcal{O}(\log n)$ time, thus due to $|\check{R}| \leq n/r$ and $p = \max\{\frac{n}{r \log n}, \frac{dn}{r \log^2 n}\}$ Step 2 can be performed in $\mathcal{O}(\log^2 n)$ time whp.

In **Step 3** all PUs cooperate to generate a set Req $:= \{(w, \text{tent}(v) + c((v,w))) : v \in \check{R} \text{ and } (v,w) \in E\}$ of *requests*. By compacting $\check{R}$ and using prefix sums to schedule the PUs this task can be perfectly load balanced. Since |Req| $= \mathcal{O}\left(d|\check{R}| + \log n\right)$ whp for $|\check{R}| \leq n/r$, this step can be performed in time $\mathcal{O}(m/(rp) + \log n) = \mathcal{O}(\log^2 n)$ whp.

Step 4 permutes the requests such that (w,x) is put into a buffer array $B_{\text{ind}(w)}$. Altogether there are at most $\mathcal{O}(d|\check{R}|)$ requests whp that are spread over p buffers, thus, because of the random node distribution, each buffer gets $\mathcal{O}(\log n + d|\check{R}|/p) = \mathcal{O}(\log^2 n)$ requests whp (Chernoff bounds, $|\check{R}| \leq n/r$, $p = \max\{\frac{n}{r \log n}, \frac{dn}{r \log^2 n}\}$). The requests are placed by "randomized dart throwing" [18]. If each processor is responsible for the placement of a group of $\mathcal{O}(\log^2 n)$ requests (which may go to different buffers) Step 4 takes $\mathcal{O}(\log^2 n)$ time whp. The dart throwing progress is regularly monitored. In the unlikely case of stagnation (buffers are chosen too small), the buffer sizes are adapted.

Finally, in **Step** 5 PU i scans buffer i and for each request (w,x) with $x < \text{tent}(w)$ it updates $\text{tent}(w)$ to x and calls $\texttt{decreaseKey}(Q_i, w, x)$, $\texttt{decreaseKey}(Q_i^*, w, x + \delta_o(w))$ (respectively $\texttt{insert}$ for new nodes). Each operation can be executed in $\mathcal{O}(1)$ time, so for $|\check{R}| \leq n/r$ Step 5 needs time $\mathcal{O}(\log^2 n)$ whp.

Phases with $|\check{R}| > n/r$ show whp at least as balanced queue access patterns as those phases deleting less elements, thus time and work of a phase increase at most linearly. Let k_i denote the number of nodes removed in phase i. Then $\sum_{i \leq r} k_i \leq n$. The total time over all phases is $T = \mathcal{O}(\sum_{i \leq r} \lceil k_i r/n \rceil \log^2 n) = \mathcal{O}(r \log^2 n + (nr/n) \log^2 n) = \mathcal{O}(r \log^2 n)$ whp.

For $d > r \log^2 n$ more than n PUs can be used by dropping explicit queues: n global bits denote whether an element is "queued" or not and p/n PUs take care of each buffer area in order to cope with the increased number of requests. Alternatively, one can apply an initial filtering step because all but the $c \log n$ smallest edges per node, c some constant, can be ignored whp without changing the shortest paths [10, 12].

The INOUT-version is supported by p additional priority queues. Initialization of $\delta_i(v) := \min\{c(w,v) : (w,v) \in E\}$ involves collecting the weights of edges that are potentially distributed over $\Omega(d)$ adjacency-lists. For random graphs, the number of incoming edges of $k = \Omega(\log n)$ randomly selected nodes is $\mathcal{O}(dk)$ whp. Thus, we can use the randomized dart throwing to perform the initialization using $\mathcal{O}(dn)$ work whp.

Theorem 3. *If the number of delete-phases is bounded by r then the SSSP can be solved in $\mathcal{O}(r \log^2 n)$ time and $\mathcal{O}(n \log n + m)$ work whp. using $\max\{\frac{n}{r \log n}, \frac{m}{r \log^2 n}\}$ processors on a CRCW PRAM.*

The running time can be improved by a factor of $\mathcal{O}(\log n)$ if we choose an alternative implementation for the queues based on the parallel priority queue data structure from [19] which supports $\texttt{insert}$ and $\texttt{deleteMin}$ for $\mathcal{O}(p)$ elements in time $\mathcal{O}(\log n)$ using p PUs whp. In [7] we show how to augment this data structure so that $\texttt{decreaseKey}$ and $\texttt{delete}$ are also supported.

A queue is represented by three relaxed heaps: A main heap Q_1, a buffer Q_0 for newly inserted elements plus the $\mathcal{O}(\log n)$ smallest ones and Q_d for elements whose key drops below a bound L' due to a $\texttt{decreaseKey}$. Deleted elements in Q_1 are only marked as deleted. More generally, $\texttt{delete}$ and $\texttt{deleteMin}$ are most of the time only performed on Q_0 and Q_d and only every $\mathcal{O}(\log n)$ phases a function `cleanUp` is called which guarantees that Q_0 and Q_d do not grow too large. For an analysis we refer to [19, 7].

Corollary 1. *SSSP on random graphs with random edge weights uniformly distributed in $[0,1]$ can be solved on a CRCW PRAM in $\mathcal{O}(n^{1/3} \log n)$ time and $\mathcal{O}(n \log n + m)$ work whp.*

The approach is relatively easy to adapt to distributed memory machines. The ind-array can be replaced by a hash-function and randomized dart throwing by routing. For random graphs, the PU scheduling for generating requests is unnecessary, if the number of PUs is decreased by a logarithmic factor.

The algorithm can also be adapted to a $\mathcal{O}(n^{1/3+\epsilon})$ time and $\mathcal{O}(n \log n + m)$ work EREW PRAM for an arbitrary small constant $\epsilon > 0$. Concurrent write accesses only occur during the randomized dart throwing. It can be replaced by $1/\epsilon$ reordering phases (essentially radix sorting), such that phase i groups all request for a subset of $p^{1-\epsilon i}$ queue pairs. Processors are rescheduled after each phase. After the last phase all requests to a certain queue pair are grouped together and can be handled sequentially.

5 Adaption to External Memory

The best previous external memory SSSP algorithm is due to [16]. It requires at least n I/Os and hence is unsuitable for large n. For our improved algorithm we use D to denote the number of disks and B to denote the block size. Let r be the number of delete-phases and assume for simplicity that each phase removes n/r elements from the queue.

Furthermore, we assume that $D \log D \leq n/r$ and that the internal memory, S, is large enough to hold *one* bit per node. It is indicated in [7] how to proceed if this reasonable assumption does not hold. We partition the adjacency-lists into blocks of size B and distribute the blocks randomly over the disks. All requests to adjacency-lists of a single phase are first collected in D buffers, in large phases they are possibly written to disk temporarily. At the end of a phase the requests are performed in parallel. If $D \log D \leq n/r$, the n/r adjacency-lists to be considered in a phase will distribute almost evenly over the disks whp, and hence the time spent in reading adjacency-lists is $\mathcal{O}(n/D + m/(DB))$ whp. We use a priority queue without `decreaseKey` operation (e.g. buffer trees [2]) and insert a node as often as it has incoming edges (each edge may give a different tentative distance). When a node is removed for the first time its bit is set. Later values for that node are ignored.

The total I/O complexity for this approach is given by $\mathcal{O}(\frac{n}{D} + \frac{m}{DB} \log_{S/B} \frac{m}{B})$ I/Os whp. The number of disks is restricted by $D = \mathcal{O}(\min\{\frac{n}{r \log n}, \frac{S}{B}\})$.

We note that it is useful to slightly modify the representation of the graph (provide each edge (v,w) with $\delta_o(w)$, the minimum weight of any edge out of w). This allows us to compute the L-value while deleting elements from the queue without the auxiliary queue Q^*. This online computing is possible because the nodes are deleted with increasing distances and the L-value initialized with findMin() + 1 can only decrease. The preprocessing to adapt the graph takes $\mathcal{O}(\frac{n+m}{DB} \log_{S/B} \frac{m}{B})$ I/Os.

Theorem 4. *SSSP with r delete-phases can be solved in external memory using $\mathcal{O}(\frac{n}{D} + \frac{m}{DB} \log_{S/B} \frac{m}{B})$ I/Os whp if the number of disks is $D = \mathcal{O}(\min\{\frac{n}{r \log n}, \frac{S}{B}\})$ and S is large enough to hold one bit per node.*

6 Simulations

Simulations of the algorithm have greatly helped to identify the theoretical bounds to be proven. Furthermore, they give information about the involved constant factors.

For the OUT-variant on random graphs with random edge weights we found an average value of $2.5\sqrt{n}$ phases. The refined INOUT-variant needs about $6.0\, n^{1/3}$ phases on the average. A modification of the INOUT-approach which switches between the criteria as described in Sect. 2 takes about $8.5\, n^{1/3}$ phases.

We also ran tests on planar graphs taken from [15, GB_PLANE] where the nodes have coordinates uniformly distributed in a two-dimensional square and edge weights denote the Euclidean distance between respective nodes. The OUT-version finished in about $1.2\, n^{2/3}$ phases; taking random edge weights instead, about $1.7\, n^{2/3}$ phases sufficed on the average. The performance of the INOUT-version is less stable on these graphs; it seems to give only a constant factor improvement over the simpler OUT-variant.

Motivated from the promising results on planar graphs we tested our approach on real-world data: starting with a road-map of a town ($n = 10,000$) the tested graphs successively grew up to a large road-map of Southern Germany ($n = 157,457$). While repeatedly doubling the number of nodes, the average number of phases (for different starting points) only increased by a factor of about $1.63 \approx 2^{0.7}$; for $n = 157,457$ the simulation needed $6,647$ phases.

7 Conclusions

We have shown how to subdivide Dijkstra's algorithm into delete phases and gave a simple CRCW PRAM algorithm for SSSP on random graphs with random edge weights which has sublinear running time and performs $\mathcal{O}(n \log n + m)$ work whp. Although the bounds only hold with high probability for random graphs, the approach shows good behavior on practically important real-world graph instances.

Future work can tackle the design and performance of more refined criteria for safe node deletions, in particular concerning non-random inputs.

Another promising approach is to relax the requirement of $\text{tent}(v) = \text{dist}(v)$ for deleted nodes. In [7, 17] we also analyze an algorithm which allows these two values to differ by an amount of Δ. While this approach yields more parallelism for random graphs, the safe criteria do not need tuning parameters and can better adapt to inhomogeneous distributions of edge weights over the graph.

Acknowledgements

We would like to thank Volker Priebe for fruitful discussions and suggestions.

References

[1] N. Alon, J. H. Spencer, and P. Erdős. *The Probabilistic Method*. Wiley, 1992.
[2] L. Arge. *Efficient external-memory data structures and applications*. PhD thesis, University of Aarhus, BRICS-DS-96-3, 1996.
[3] G. S. Brodal, J. L. Träff, and C. D. Zaroliagis. A parallel priority queue with constant time operation. In *11th IPPS*, pages 689–693. IEEE, 1997.
[4] A. Clementi, L. Kučera, and J. D. P. Rolim. A randomized parallel search strategy. In A. Ferreira and J. D. P. Rolim, editors, *Parallel Algorithms for Irregular Problems: State of the Art*, pages 213–227. Kluwer, 1994.
[5] E. Cohen. Polylog-time and near-linear work approximation scheme for undirected shortest paths. In *26th STOC*, pages 16–26. ACM, 1994.
[6] E. Cohen. Efficient parallel shortest-paths in digraphs with a separator decomposition. *Journal of Algorithms*, 21(2):331–357, 1996.
[7] A. Crauser, K. Mehlhorn, U. Meyer, and P. Sanders. Parallelizing Dijkstra's shortest path algorithm. Technical report, MPI-Informatik, 1998. in preparation.
[8] E. Dijkstra. A note on two problems in connexion with graphs. *Num. Math.*, 1:269–271, 1959.
[9] J. R. Driscoll, H. N. Gabow, R. Shrairman, and R. E. Tarjan. Relaxed heaps: An alternative to Fibonacci heaps with applications to parallel computation. *Communications of the ACM*, 31(11):1343–1354, 1988.
[10] A. Frieze and G. Grimmett. The shortest-path problem for graphs with random arc-lengths. *Discrete Appl. Math.*, 10:57–77, 1985.
[11] Y. Han, V. Pan, and J. Reif. Efficient parallel algorithms for computing all pairs shortest paths in directed graphs. In *4th SPAA*, pages 353–362. ACM, 1992.
[12] R. Hassin and E. Zemel. On shortest paths in graphs with random weights. *Math. Oper. Res.*, 10(4):557–564, 1985.
[13] J. Jájá. *An Introduction to Parallel Algorithms*. Addison-Wesley, 1992.
[14] R. M. Karp. The transitive closure of a random digraph. *Rand. Struct. Alg.*, 1, 1990.
[15] D. E. Knuth. *The Stanford GraphBase : a platform for combinatorial computing*. Addison-Wesley, New York, NY, 1993.
[16] V. Kumar and E. J. Schwabe. Improved algorithms and data structures for solving graph problems in external memory. In *8th SPDP*, pages 169–177. IEEE, 1996.
[17] U. Meyer and P. Sanders. Δ-stepping: A parallel shortest path algorithm. In *6th ESA*, LNCS. Springer, 1998.
[18] G. L. Miller and J. H. Reif. Parallel tree contraction and its application. In *26th Symposium on Foundations of Computer Science*, pages 478–489. IEEE, 1985.
[19] P. Sanders. Randomized priority queues for fast parallel access. *Journal Parallel and Distributed Computing*, 49:86–97, 1998.
[20] M. Thorup. Undirected single source shortest paths in linear time. In *38th Annual Symposium on Foundations of Computer Science*, pages 12–21. IEEE, 1997.
[21] J. L. Träff and C. D. Zaroliagis. A simple parallel algorithm for the single-source shortest path problem on planar digraphs. In *Irregular' 96*, volume 1117 of *LNCS*, pages 183–194. Springer, 1996.
[22] J. S. Vitter and E. A. M. Shriver. Algorithms for parallel memory I: Two-level memories. Technical Report CS-90-21, Brown University, 1990.

Comparison Between the Complexity of a Function and the Complexity of Its Graph

Bruno Durand[1] and Sylvain Porrot[2]

[1] LIP, ENS-Lyon CNRS, 46 Allée d'Italie, 69634 Lyon CEDEX 07, France;
Bruno.Durand@ens-lyon.fr
[2] LAIL and LIFL, Bât. P2, Université des Sciences et Technologies de Lille,
59655 Villeneuve d'Ascq CEDEX, France;
porrot@lifl.fr

Abstract. This paper investigates in terms of *Kolmogorov complexity* the differences between the information necessary to compute a recursive function and the information contained in its graph. Our first result is that the complexity of the initial parts of the graph of a recursive function, although bounded, has almost never a limit. The second result is that the complexity of these initial parts approximate the complexity of the function itself in most cases (and in the average) but not always.

Introduction

The goal of this paper is to compare the information contained in the graph of a function to the information needed to compute the function. Our approach is based on *Kolmogorov complexity* (also known as *Algorithmic Information Theory*). In this framework we compare the Kolmogorov complexity of a recursive function f, *i.e.* the size of a smallest program that computes f, with the conditional Kolmogorov complexities of initial parts of the graph of f. As far as we know, the only result in this field is a theorem of Meyer (see Theorem 2 in this paper), reported in the well-known article of Loveland [4]. A proof of this theorem is also given in the fundamental article of Zvonkin and Levin [7]. However, the point of view of these papers is different from ours: they are mainly interested in non-recursive sequences and in randomness. They also investigate varieties of Kolmogorov complexity (see on this topic the paper of Uspensky and Shen [6]). We focus on recursive sequences (or functions).

Our study is also motivated by the analysis of data flows (see also [5]). Imagine a flow that, step by step, produces integer numbers. The information contained in the flow up to time t can be understood as the conditional Kolmogorov complexity of the outputs obtained before time t, knowing t. Our goal is to analyze the variations of this information when t varies.

Our results are rather surprising: the first one is that this information is bounded when the function is recursive, but has no limit, except for a finite number of functions (Theorem 1). Our second result is that the complexities of the initial parts of a graph do not always constitute an approximation of the complexity of the function (Theorem 3). In the case of data flows it means that, if we consider any recursive family of systems producing data flows, the amount of information issued by some of the systems is much lower than the information contained in the systems themselves. But we prove in Theorem 4 and in Corollary 1 that this strange behaviour appears rather rarely in the family, and that, in the average, this approximation is justified.

A more theoretical field of investigation is to compare the maximum of the complexity of the graph, its lim sup, the Kolmogorov complexity of the function and some other varieties of definitions of its Kolmogorov complexity relativised to oracles (e.g. the standard oracle set $\mathbb{K}$, also called $0'$ in recursion theory), see [2].

1 Preliminaries

1.1 Kolmogorov Complexity

Theory of *Kolmogorov complexity* [3], also called *Algorithmic Information Theory* [1], gives rigorous mathematical foundations to the notion of "information content" of an object x (represented by a word over the binary alphabet $\{0,1\}$). This quantity $K(x)$ is the length of a smallest program that halts and outputs x on an empty input. The programming language must satisfy an important technical property called *additive optimality* which is true in all "natural" programming languages:

$$\forall K_1, K_2 \ \exists C \ \forall x \ |K_1(x) - K_2(x)| < C,$$

where $K_1(x)$ and $K_2(x)$ are Kolmogorov complexities defined for two different additively optimal programming languages.

In order to talk about the complexity of integers, we use the following one-to-one mapping between words and integers: we associate each word with its index in the ordering, first by length, then lexicographically.

1.2 Definitions of Models

We study recursive functions and their graphs $\mathcal{G}_f = \{\langle x,y \rangle, \ y = f(x)\}$. We denote by $\mathcal{G}_f^n$ the initial part of the graph $\mathcal{G}_f$ i.e. $\mathcal{G}_f^n = \{\langle x,y \rangle, \ x \leq n, y = f(x)\}$. As it is defined here, $\mathcal{G}_f^n$ is a set. So there is a choice of different representations for this set. We choose to identify $\mathcal{G}_f^n$ with $\langle f(0), f(1), \ldots, f(n) \rangle_n$ where $\langle . \rangle_n$ denotes the standard encoding of $\mathbb{N}^n$ in $\mathbb{N}$. Any other recursively equivalent definition could have been chosen.

Note that the special case where the images of the functions are restricted to the pair $\{0,1\}$ is equivalent to the study of recursive infinite sequences. All results in the sequel remain valid if recursive functions from $\mathbb{N}$ to $\mathbb{N}$ are replaced by recursive sequences over the alphabet $\{0,1\}$.

Definition 1. *A program P is a* weak model *of a function f over a domain D if D is an infinite subset of $\mathbb{N}$ and $\forall n \in D \ P(n)$ halts and outputs $\mathcal{G}_f^n$.*

Note that if $n \notin D$, either $P(n)$ does not halt, or $P(n)$ halts and its output can be different from $\mathcal{G}_f^n$.

Definition 2. *The* weak complexity *of a function f is defined by*

$$K_w(f) = \limsup_{n \to \infty} K(\mathcal{G}_f^n | n).$$

Note that any limit point of $K(\mathcal{G}_f^n | n)$ (reached over an infinite subset D of $\mathbb{N}$) corresponds to the size of a smallest weak model of f over D.

Definition 3. *A* strong model *of a function f is a program P accepting one input n and that, for all n, halts and outputs $\mathcal{G}_f^n$, i.e. $\forall n \in \mathbb{N} \ P(n) = \mathcal{G}_f^n$.*

We could present a different definition of a strong model e.g. $\forall n \in \mathbb{N} \ P(n) = f(n)$ instead of $\forall n \in \mathbb{N} \ P(n) = \mathcal{G}_f^n$. It gives, up to an additive constant, the same notion of strong complexity (see below).

Definition 4. *The* strong complexity $K_s(f)$ *of a function f is the length of a smallest strong model of f if it exists, the infinity otherwise.*

Remark that f is recursive if and only if $K_s(f)$ is finite.

2 A Study of Weak Models

A strong model of a function is also a weak model of this function. Thus we get the following straightforward proposition.

Proposition 1. *If f is a recursive function, then there exists a strong model of f and we have $K_w(f) \leq K_s(f)$.*

If other (equivalent) definitions of K_w and K_s are given, then an additive constant is added to the previous proposition: $K_w(f) \leq K_s(f) + C$.

We now present some results concerning the series $K(\mathcal{G}_f^n | n)$ in order to justify the choice of the upper limit in the definition of the weak complexity. Indeed,

1. the limit of $K(\mathcal{G}_f^n | n)$ exists at most for a finite number of recursive functions f (Theorem 1);
2. the lower limit of $K(\mathcal{G}_f^n | n)$ is bounded by a constant not depending on f (Lemma 1), due to the fact that the whole information describing f can be infinitely often found encoded in the parameter n;
3. the upper limit of $K(\mathcal{G}_f^n | n)$ is the size of a smallest weak model for which the input n provides no information on f; thus a counting argument proves that it cannot be uniformly bounded (Lemma 2).

Lemma 1. *There exists a universal weak model for recursive functions: there exists a program $P_u(n)$ such that for all recursive function f, P_u is a weak model of f.*

Proof. Let $P_u(n)$ be the following program: if there exist k and x such that $n = 1^k 0x$, P_u simulates program x on n, otherwise P_u loops indefinitely.

Let f be a recursive function of strong model P_g. P_u is a weak model of f since for all n in $\{1^k 0 P_g | k \in \mathbb{N}\}$, $P_u(n)$ computes $\mathcal{G}_f^n$.

Lemma 2. *Given $A > 0$, the set of functions having a weak complexity bounded by A is finite.*

This lemma is obtained by a standard counting argument (we skip the proof). Lemmas 1 and 2 clearly imply the following theorem:

Theorem 1. *The set of recursive functions such that $\lim_{n \to \infty} K(\mathcal{G}_f^n | n)$ exists is finite.*

In this theorem the finite number of functions such that a limit exists depends on the programming system. Let us first present a system in which this number is zero. Consider a standard enumeration of partial recursive functions ϕ_i and let us define the following programming system: $\psi_0 = f_u$, f_u being the partial recursive function computed by P_u, $\psi_1, \ldots, \psi_{1998}$ are functions of which indexes are programs that always diverge, $\psi_i = \phi_i$ for all $i > 1998$. In this system $\liminf K(\mathcal{G}_f^n | n) = 0$ and clearly $\limsup K(\mathcal{G}_f^n | n) \geq \text{length}(1998)$.

Now let us present a programming system in which the limit exists for some functions : $\gamma_0, \ldots, \gamma_{1998}$ are distinct total recursive functions, $\gamma_{1999} = f_u$ and $\gamma_i = \phi_i$ for all $i > 1999$.

3 Comparison Between Strong and Weak Models

3.1 Existence of a Strong Model

A well known theorem (here Theorem 2) due to Meyer and reported in [4, 7] states that if $K(\mathcal{G}_f^n | n)$ is bounded over an infinite recursively enumerable domain D, then f is recursive. A weaker version of this theorem states that if the weak complexity of f is finite then f is recursive. In other terms the hypothesis is that there is a finite number of weak models computing $\mathcal{G}_f^n$ for all n. This result is not obvious (and is rather strong) since we do not know *a priori* which one of all weak models computes $\mathcal{G}_f^n$ for a given n in D.

Theorem 2 (Meyer). *A function f is recursive if and only if there exists an infinite recursively enumerable set $D \subset \mathbb{N}$ where $K(\mathcal{G}_f^n | n)$ is bounded.*

3.2 Comparing Weak and Strong Complexities

We have just seen that a finite weak complexity implies the existence of a strong model. Does weak complexity approximate strong complexity? The proof of Theorem 2 does not provide any answer to this question, because it is not constructive. Indeed, no proof of this theorem can be constructive as shown below in Theorem 3. As Kolmogorov complexity is defined up an additive constant we need a family of functions to express this fact.

Theorem 3. *Let $\mathcal{F} = \{f_i\}_{i \in \mathbb{N}}$ be any recursive family of distinct recursive functions. Then $\forall C \; \exists f_i \in \mathcal{F} \; K_s(f_i) - K_w(f_i) > C$.*

More precisely, a recursive family of recursive functions is a family such that $\exists \mathcal{P} \; \forall i, x \; \mathcal{P}(i,x)$ halts and outputs $f_i(x)$. In the sequel, we prove a stronger result: K_s is infinitely often of order k when K_w is of order $\log(k)$.

Example: The family $\mathcal{F}$ defined by $\forall n \leq i \; f_i(n) = 1$ and $\forall n > i \; f_i(n) = 0$ satisfies the hypothesis of Theorem 3, and therefore we cannot bound by a constant the difference between the weak and the strong complexities of functions of this family.

In general, according to Lemma 1 and Theorem 3, the behaviour of the series $K(\mathcal{G}_f^n|n)$ as a function of n is illustrated by Figure 1.

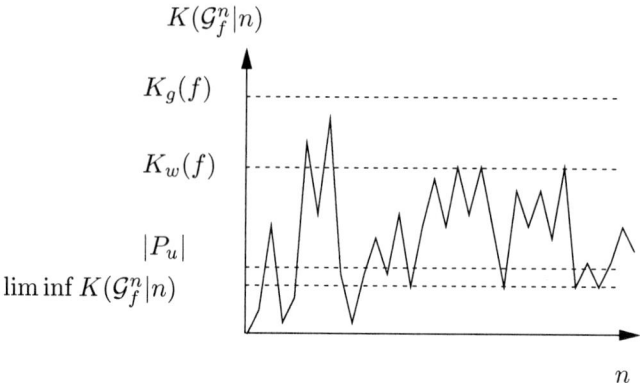

Fig. 1. Graph of $K(\mathcal{G}_f^n|n)$

In order to prove Theorem 3 we need some preliminary results. In the following, the standard notation $\text{Step}(P, x, t)$ denotes the program that simulates t steps of program P with x as input, gives as result $P(x) + 1$ if convergence is observed, else gives as result 0. In the proofs, the notation $\mathcal{O}_k(1)$ (resp. $\mathcal{O}_{k,n}(1)$) denotes a function of k (resp. of k and n) that is bounded by a constant.

Definition 5. *Let P be a program and f a function. We define the extension properties $P \subset f$ and for all $n \; P \subset_n f$ by*

$$(P \subset f) \Leftrightarrow (\text{if } P(x) \text{ converges, then } P(x) = f(x)),$$
$$(P \subset_n f) \Leftrightarrow (\forall x \leq n \; \text{Step}(P, x, n) \neq 0 \Rightarrow \text{Step}(P, x, n) = f(x) + 1).$$

Definition 6. *(goodness) We say that i is good for k, $i \leq 2^k$, if:*

$$\exists P, |P| < k, \; (P \subset f_i \text{ and } \forall i' \leq 2^k, i' \neq i, P \not\subset f_{i'});$$

We say that i is n-good for k, $i \leq 2^k$, if:

$$\exists P, |P| < k, \ (P \subset_n f_i \text{ and } \forall i' \leq 2^k, i' \neq i, P \not\subset_n f_{i'}).$$

We say that i is *bad* for k (resp. *n-bad*) if it is not good (resp. n-bad) i.e. respectively

$$\forall P, |P| < k, \ (P \subset f_i) \Rightarrow (\exists i' \leq 2^k, i' \neq i, P \subset f_{i'}),$$
$$\forall P, |P| < k, \ (P \subset_n f_i) \Rightarrow (\exists i' \leq 2^k, i' \neq i, P \subset_n f_{i'}).$$

Lemma 3. *For all n, k there exists $i \leq 2^k$ such that i is n-bad for k.*

Proof. There is 1 more function $\{f_i\}_{i \leq 2^k}$ than programs of length $\leq k$.

The relation $\subset_n$ is an approximation of $\subset$ as proved in the following lemma, but beware that the convergence speed of $\subset_n$ to $\subset$ is not computable.

Lemma 4. *Given k, for all large enough n, $i(n,k)$ is stationary and equal to $i(k)$.*

We skip the proof. The idea is that only finitely many n can change the n-goodness of the finitely many (i,k).

Proof (of Theorem 3, see acknowledgements). It is sufficient to prove the following inequalities:

1. $\forall k \ K_s(f_{i(k)}) \geq k$,
2. $\forall k \ \limsup_{n \to \infty} K(\mathcal{G}^n_{f_{i(k)}} | n) \leq \log(k) + \mathcal{O}_k(1)$.

Step 1: Suppose there exists k such that $K_s(f_{i(k)}) < k$. Then there exists a program $P(n)$, $|P| < k$, computing $f_{i(k)}(n)$ for all n. Since $i(k)$ is bad for k, there exists $i' \neq i(k)$ such that $P \subset f_{i'}$. Since P always converges, it means that for all n, $P(n)$ halts and outputs $f_{i'}(n)$. Thus $f_{i(k)} = f_{i'}$ which is false according to the hypothesis on the family $\mathcal{F}$.

Step 2: Since $\mathcal{F}$ is a recursive family of recursive functions there exists a program $\mathcal{P}$ such that for all i, x, $\mathcal{P}(i,x)$ halts and outputs $f_i(x)$. Consider the following program $P(n,k)$ accepting two inputs n and k and computing $i(n,k)$:

```
Program P(n,k)
    Simulate n steps of all programs of length less than k with
    all inputs less or equal to n.
    Compute the 2^k first functions f_i on [0,n] using program P.
    Compare the results and output the smallest i that is bad
    for k.
End
```

Lemma 3 indicates that $P(n,k)$ computes $i(n,k)$ for all n and all k.

Because of the recursivity hypothesis on family $\mathcal{F}$, we can compute $\mathcal{G}_{f_i}^n$ from input i and n using $\mathcal{P}$, thus $\forall i\; K(\mathcal{G}_{f_i}^n|n) \leq K(i|n) + \mathcal{O}_n(1)$. Moreover, since for all n,k, $P(n,k)$ computes $i(n,k)$, we have

$$\forall n,k\; K(\mathcal{G}_{f_{i(n,k)}}^n|n) \leq |P(.,k)| + \mathcal{O}_{n,k}(1)$$
$$\leq \log(k) + \mathcal{O}_{n,k}(1).$$

With Lemma 4, given k, there exists n_k such that $\forall n \geq n_k\; i(n,k) = i(k)$. Therefore we get $\forall k\; \exists n_k\; \forall n \geq n_k\; K(\mathcal{G}_{f_{i(k)}}^n|n) \leq \log(k) + \mathcal{O}_{n,k}(1)$ and finally we obtain $\limsup K(\mathcal{G}_{f_{i(k)}}^n|n) \leq \log(k) + \mathcal{O}_k(1)$.

We have just seen that even for reasonable family of functions, the complexities of the initial parts of a graph does not always constitute an approximation of the complexity of the function itself, since some functions appear to be pathological. However, in most cases, both complexities are close to each other, as shown in the following theorem.

Theorem 4. *Let $\mathcal{F} = \{f_i\}_{i \in \mathbb{N}}$ be any recursive family of distinct recursive functions. For all ν and all k, there are at most a proportion of $\frac{A}{2^\nu}$ functions, among the k first functions f_i, for which the difference between strong and weak complexities is greater than ν, i.e.*

$$\exists A\; \forall \nu\; \forall k\; \frac{Card\{i \leq k | K_s(f_i) - K_w(f_i) \geq \nu\}}{k} \leq \frac{A}{2^\nu}$$

Proof. Since the family $\mathcal{F}$ is a recursive family of recursive functions, there exists a program computing f_i for all i, thus we have

$$\exists A\; \forall i\; K_s(f_i) \leq log(i) + A. \tag{1}$$

Since $K_w(f_i) = \limsup\limits_{n \to \infty} K(\mathcal{G}_{f_i}^n|n)$ we get

$$\forall i\; \exists n'_i\; \forall n \geq n'_i\; K_w(f_i) \geq K(\mathcal{G}_{f_i}^n|n) \tag{2}$$

and from equations 1 and 2 we obtain

$$\exists A\; \forall i\; \exists n'_i\; \forall n \geq n'_i\; K_s(f_i) - K_w(f_i) \leq \log(i) + A - K(\mathcal{G}_{f_i}^n|n). \tag{3}$$

Let k and ν be fixed. Since all functions of family $\mathcal{F}$ are distinct, there exists n_k such that, for all $n \geq n_k$, all truncated graphs $(\mathcal{G}_{f_i}^n)_{i<k}$ are distinct. Let $S(\nu,k)$ denote the set $\{i \leq k, K_s(f_i) - K_w(f_i) \geq \nu\}$. Given $n \geq \max\{n_k, (n'_i)_{i \leq k}\}$ and according to Equation 3 we have

$$S(\nu,k) \subset \{i \leq k, \log(i) + A - K(\mathcal{G}_{f_i}^n|n) \geq \nu\}$$
$$= \{i \leq k, K(\mathcal{G}_{f_i}^n|n) \leq \log(i) + A - \nu\}$$
$$\subset \{i \leq k, K(\mathcal{G}_{f_i}^n|n) \leq \log(k) + A - \nu\}.$$

Since $n \geq n_k$ all truncated graphs $(\mathcal{G}_{f_i}^n)_{i \leq k}$ are distinct. Thus all smallest programs computing these truncated graphs are distinct too and we can conclude that

$$Card(S(\nu,k)) \leq 2^{\log(k)+A-\nu}.$$

By a combinatorial proof we obtain the following corollary (we skip the proof).

Corollary 1. *Let $\mathcal{F} = \{f_i\}_{i \in \mathbb{N}}$ be any recursive family of distinct recursive functions. The difference between strong and weak complexities is in average bounded by a constant:*

$$\exists C \ \forall k \ \frac{\sum_{i \leq k}(K_s(f_i) - K_w(f_i))}{k} \leq C.$$

4 Open Problems

In this paper we have studied graphs of recursive functions. These graphs are relations on $\mathbb{N} \times \mathbb{Z}$ with some additional properties: not only the relation is recursive but also all its projections to given abscissas are uniformly recursive. If we consider a data flow, we may represent it by a relation R such that for all x there exists a finite number of y such that $R(x,y)$. It means that at each time step a finite (but not bounded) number of outputs is issued. It may be interesting to extend this study to these relations or, more generally, to any recursive relation.

Acknowledgements

The authors are indebted to, chronologically, Pr. Alexander Shen and Pr. Nikolaï Vereshchagin for their help in proving Theorem 3 and earlier versions (see also [2]). We thank Pr. Max Dauchet for his constructive remarks, comments and encouragements.

References

[1] C. Calude. *Information and Randomness, an Algorithmic Perspective*. Springer-Verlag, 1994.
[2] B. Durand, Alexander Shen, and Nikolai Vereshchagin. Descriptive complexity of computable sequences. *Draft*, 1998.
[3] M. Li and P. Vitányi. *An Introduction to Kolmogorov Complexity and its Applications*. Springer-Verlag, 2 edition, 1997.
[4] D. W. Loveland. A variant of the Kolmogorov concept of complexity. *Information and Control*, 15:510–526, 1969.
[5] S. Porrot, M. Dauchet, B. Durand, and N. Vereshchagin. Deterministic rational transducers and random sequences. In *FOSSACS'98*, volume 1378 of *Lecture Notes in Computer Science*, pages 258–272. Springer-Verlag, march 1998.
[6] V. A. Uspensky and A. Shen. Relations between varieties of Kolmogorov complexities. *Math. Syst. Theory*, 29(3):270–291, 1996.
[7] A.K. Zvonkin and L.A. Levin. The complexity of finite objects and the development of the concepts of information and randomness by means of theory of algorithms. *Russian Math. Surveys*, 25(6):83–124, 1970.

IFS and Control Languages

Henning Fernau[1] and Ludwig Staiger[2]

[1] Wilhelm-Schickhard-Institut für Informatik, Universität Tübingen,
Sand 13, D-72076 Tübingen, Germany
email: `fernau@informatik.uni-tuebingen.de`

[2] Institut für Informatik, Martin-Luther-Universität Halle-Wittenberg,
Kurt-Mothes-Str. 1, D-06120 Halle, Germany
email: `staiger@informatik.uni-halle.de`

Abstract. Valuations — morphisms from $(\Sigma^*, \cdot, e)$ to $((0, \infty), \cdot, 1)$ — are a generalization of Bernoulli morphisms introduced in [7]. Here, we show how to generalize the notion of entropy (of a language) in order to obtain new formulae to determine the Hausdorff dimension of fractal sets (also in Euclidean spaces) especially defined via regular ω-languages. In this way, we can sharpen and generalize earlier results [1,10,11,20,29].

1 Introduction

Several approaches exist using formal languages to describe pictures (fractals). Probably the most prominent examples are L systems [22,23] and finite automata [3,4,5]. Further links provide collage grammars [14,34] and cellular automata [33,15]. Another approach to fractals is via *Iterated Function Systems* (IFS) (cf. [2,6,8]). An IFS is composed of a metric space $\mathcal{X}$ and a set $\{\mathcal{F}(1), \ldots, \mathcal{F}(n)\}$ of (contracting) mappings on $\mathcal{X}$. A IFS $\mathcal{F}$ defines a fractal as the smallest nonempty and (topologically) closed solution of $K = \bigcup_{i=1}^{n} \mathcal{F}(i)(K)$. Likewise, K can be defined in the following way (cf. [2]):

1. For every mapping $\xi : \mathbb{N} \to \{1, \ldots, n\}$ the sequence
$$((\mathcal{F}(\xi(1)) \circ \cdots \circ \mathcal{F}(\xi(j)))(a))_{j=1}^{\infty} \qquad (1)$$
converges independently of the starting point $a \in \mathcal{X}$ to a limit point $a_\xi \in K$ depending only on ξ.
2. Thus to every mapping ξ can be seen as address of $\phi_\mathcal{F}(\xi) := a_\xi$.
3. The above-mentioned solution K equals $\{\phi_\mathcal{F}(\xi) : \xi : \mathbb{N} \to \{1, \ldots, n\}\}$.

Mappings $\xi : \mathbb{N} \to \Sigma$ where Σ is a finite set (alphabet) are known in the theory of formal languages as ω-words (cf. [31]).

It makes also sense to consider IFS where not all mappings $\mathcal{F}(i)$ are contractive (cf. [1,4,20]). In that case, Item 1 from above is not fulfilled for all ξ. So, one has to single out those ξ for which the sequence (1) converges independently of the start $a \in \mathcal{X}$. In practice, this condition is very intriguing and depends heavily on the mappings $\mathcal{F}(i)$. In the above-mentioned papers finite graphs (automata) were used as control structures to guarantee the convergence of (1).

Evidently, finite automata are simple control structures, thus leaving aside, in general, a great part of ω-words ξ for which convergence can be guaranteed (even if we take into account—as it was done in [1,4,20]—only the contraction coefficients $\beta_{\mathcal{F}}(i)$ of the mappings $\mathcal{F}(i)$). Thus it is natural to ask what happens if we allow for more general control structures. Here, we will not deal with a particular type of automata, instead we will consider languages (or ω-languages) as devices controlling the process of convergence of (1).

One of the main problems encountered in fractal geometry is the determination of the Hausdorff dimension of fractals. We have shown earlier that, for certain types of fractals described by formal languages, this calculation may be simplified considerably by using results on unambiguous regular expressions and unambiguous contextfree grammars. We will show how this task can be accomplished for more general types of languages.

To this end, we generalize here the notion of entropy.[1] Moreover, we investigate special metrizations (induced by valuations) of spaces of ω-words; Hausdorff measure and dimension within these spaces are directly related to those entities within Euclidean spaces. Further material is contained in other works of the authors [9,10,11,21,27,28,29,30]. Proofs of the results of this paper can be found in [13].

Conventions: $\Sigma_n = \{1,\ldots,n\} \subset \mathbb{N}$ denotes our standard alphabet. Σ (without subscript) denotes some at most countable alphabet. A language L is a subset of the word monoid Σ^* generated by the alphabet Σ, where e is the neutral element of the monoid, called empty word. Mostly, the monoid operation called catenation is just denoted by juxtaposition, sometimes made explicit using $\cdot$ between the words. The monoid generated by the language $L \subseteq \Sigma^*$ is denoted by L^*, and the semigroup generated by L is denoted by L^+.

We also consider ω-languages F over the alphabet Σ, i.e., sets of one-sided infinite words, $F \subseteq \Sigma^\omega$ for short. If $L \subseteq \Sigma^*$, $L^\omega = \{v_0 \cdot v_1 \cdot v_2 \cdots : \forall i \in \mathbb{N}(v_i \in L\setminus\{e\})\}$. Further notions and denotations are introduced throughout the paper.

2 Valuations

We call a monoid morphism β mapping from $(\Sigma_n^*, \cdot, e)$ to $((0,\infty), \cdot, 1)$ a *valuation*. Any valuation can be extended to languages $L \subseteq \Sigma_n^*$ defining $\beta(L) = \sum_{w \in L} \beta(w)$. As an example, consider the valuation β_n defined by $\beta_n(a) = 1/n$ for every $a \in \Sigma_n$. Basic facts on valuations can be found in [10,11].

We call $\beta^s(L) := \sum_{w \in L} \beta(w)^s$ the *s-dimensional valuation* of the language $L \subseteq \Sigma_n^*$. We allow here valuations β having $\beta(w) \geq 1$ for some words, which we did not consider in detail before. For a fixed $L \subseteq \Sigma_n^*$ we consider the s-dimensional valuation as a function $\beta^{\cdot}(L) : [0,\infty) \to [0,\infty]$. We summarize some properties:

Properties 1 *Let $L \subseteq \Sigma_n^*$ and $\beta: \Sigma_n^* \to (0,\infty)$ be a valuation, and let $\beta^s(L) < \infty$ for some $s \in [0,\infty)$. Then there is an $\alpha \in [0,\infty)$ such that $\beta^s(L) = \infty$ for*

---
[1] This entity corresponds to the Besicovitch-Taylor index defined in connection with IFS [35].

$s < \alpha$, $\beta^s(L) < \infty$ for $s > \alpha$, and the function $\beta^{\cdot}(L)$ is continuous on (α, ∞) and satisfies $\lim_{s\downarrow\alpha} \beta^s(L) = \beta^\alpha(L)$.[2] If, moreover, $\beta(w) < 1$ for all $w \in L$, then the function $\beta^{\cdot}(L)$ is strictly decreasing and $\lim_{s\to\infty} \beta^s(L) = 0$.

The *β-entropy* of the language $L \subseteq \Sigma_n^*$, written H_L^β, is defined as the point α defined above, which is a "change-over-point" of the function $\beta^{\cdot}(L)$, i.e., $H_L^\beta := \inf\{s : s \geq 0 \wedge \beta^s(L) < \infty\}$,[3] so that $H_L^\beta < \infty$ iff $\exists s(s \in (0, \infty) \wedge \beta^s(L) < \infty)$.

Remark 1. One can construct valuations β and languages L such that $\beta(w) < 1$ for $w \in L$ and nevertheless $\beta^s(L) = \infty$ for all $s \in [0, \infty)$. In the sequel, however, we are not interested in such pathological cases. If $\beta(a) < 1$ for every $a \in \Sigma_n$, then there is a finite change-over-point α of the function $\beta^s(L)$ for any $L \subseteq \Sigma_n^*$.

It was shown in [21,28,29] that the entropy of languages introduced by Chomsky and Miller (cf. [16]) is a useful tool for the calculation of the Hausdorff dimension of certain subsets of the Cantor space Σ_n^ω or of the Euclidean space $\mathbb{R}^d$. Here, we will see the usefulness of the generalized notion of β-entropy, especially leading to similar calculation formulae for the Hausdorff dimension of subsets of $(\Sigma_n^\omega, \rho_\beta)$ and of $\mathbb{R}^d$, thereby generalizing results of [1,20]. The main tool is to generalize properties of the entropy of languages (see [26], [29, Section 2]). So, we find:

$$H^\beta_{W \cup V} = H^\beta_{W \cdot V} = \max\{H^\beta_W, H^\beta_V\} \quad \text{if } W \cdot V \neq \emptyset, \text{ and} \qquad (2)$$
$$H^\beta_L = 0 \quad \text{if } L \text{ is finite.} \qquad (3)$$

3 β-Entropy of Languages

In this section we show for two classes of languages that their β-entropy can be computed. The first class is the class of regular languages. Here we rely on results of [1,20]. Moreover, we show the close relationship between the β-entropy of a regular language and the β-entropy of its language of subwords.
The second subsection deals with the approximation of the β-entropy of a star-language by the β-entropy of their finitely generated sublanguages.[4] It is interesting to note that this approximation is valid for arbitrary star-languages.

3.1 The β-Entropy of Regular Languages

We can characterize the *regular* languages with finite β-entropy. The *state* of $M \subseteq \Sigma_n^* \cup \Sigma_n^\omega$ derived from $w \in \Sigma_n^*$ is defined as: $M/w := \{p : p \in \Sigma_n^* \cup \Sigma_n^\omega \wedge w \cdot p \in M\}$. $M \subseteq \Sigma_n^* \cup \Sigma_n^\omega$ is called *finite-state* if it has a finite number of distinct states. It is well-known that $L \subseteq \Sigma_n^*$ is finite-state iff it is regular,

---
[2] permitting the value ∞ for $\beta^\alpha(L)$
[3] Here we follow the convention $\inf \emptyset = \infty$.
[4] In general, the β-entropy is in no way continuous, that is, $\lim_{i\to\infty} L_i = L$ does not necessarily imply that $H^\beta_{L_i}$ tends to H^β_L.

whereas every regular ω-language [5] is finite-state but the converse does not hold (see e.g. [25]). $w \in \Sigma_n^*$ is called *prefix* of a string $p \in \Sigma_n^* \cup \Sigma_n^\omega$ provided $p = w \cdot p'$ for some $p' \in \Sigma_n^* \cup \Sigma_n^\omega$ (abbreviated by $w \sqsubseteq p$). For $M \subseteq \Sigma_n^* \cup \Sigma_n^\omega$ its set of finite prefixes is denoted by $\mathbf{A}(M)$ and its set of subwords (infixes) is $\mathbf{T}(M) := \{v : v \in \Sigma_n^* \wedge \exists p \exists w (w \in \Sigma_n^* \wedge w \cdot v \cdot p \in M)\}$.

Properties 2 *If $L \subseteq \Sigma_n^*$ is a regular language and $\beta : \Sigma_n^* \to (0, \infty)$ is a valuation, then the following conditions are equivalent.*

1. *There is an $s \geq 0$ such that $\beta^s(L) < \infty$.*
2. *$\forall w, v(v \neq e \wedge L/w = L/w \cdot v \neq \emptyset \to \beta(v) < 1)$.* [6]
3. *There are an $\ell \in \mathbb{N}$ and a positive constant $c < 1$ such that for all $u \in \mathbf{T}(L)$ with $|u| \geq \ell$ it holds $\beta(u) \leq c^{|u|}$.*

We obtain the following relations between the β-entropies of L, $\mathbf{A}(L)$ and $\mathbf{T}(L)$:

Properties 3 *Let L be regular. Then, $\beta^s(L) < \infty$ iff $\beta^s(\mathbf{A}(L)) < \infty$ iff $\beta^s(\mathbf{T}(L)) < \infty$. Further, $H_L^\beta = H_{\mathbf{A}(L)}^\beta = H_{\mathbf{T}(L)}^\beta$.*

Next, we show how to compute the β-entropy for a regular set $L \neq \emptyset$. Let $\{L_1 = L, L_2, \ldots, L_k\}$ be its set of nonempty states. Define $\mathcal{A}_L^{\beta,s} = (a_{s;i,j})_{1 \leq i,j \leq k}$ by $a_{s;i,j} := \sum_{L_i/x = L_j} (\beta(x))^s$. Then, $\beta^s(\mathbf{A}(L) \cap \Sigma_n^\ell) = (1, 0, \ldots, 0) \cdot (\mathcal{A}_L^{\beta,s})^\ell \cdot \mathbb{1}$ where $\mathbb{1}$ is the all ones column vector. Let $\Phi_L(s) := \lim_{\ell \to \infty} \sqrt[\ell]{\|(\mathcal{A}_L^{\beta,s})^\ell\|}$ be the spectral radius of $\mathcal{A}_L^{\beta,s}$. By [20, Thm. 2], Φ_L is strictly decreasing,[7] $\Phi_L(0) \geq 1$, and $\lim_{s \to \infty} \Phi_L(s) = 0$. Thus, $\beta^s(\mathbf{A}(L)) = (1, 0, \ldots, 0) \cdot \sum_{\ell \in \mathbb{N}} (\mathcal{A}_L^{\beta,s})^\ell \cdot \mathbb{1}$ converges iff $\Phi_L(s) < 1$. So, $H_L^\beta = H_{\mathbf{A}(L)}^\beta = H_{\mathbf{T}(L)}^\beta = \alpha$ iff $\Phi_L(\alpha) = 1$. Hence, we have:

Corollary 1. *$\beta^\alpha(L) = \infty$ for $\alpha = H_L^\beta$ if L is an infinite regular set.*

3.2 The β-Entropy of Star-Languages

As $H_L^\beta \leq H_{L^*}^\beta$ and $H_{L^*}^\beta = \infty$ if $\beta(w) \geq 1$ for some $w \in L \setminus \{e\}$, only $H_L^\beta < \infty$ and $\beta(w) < 1$ for $w \in L \setminus \{e\}$ is of interest, so that $\beta(w) < 1$ if $w \in L^* \setminus \{e\}$.

Proposition 1. *Let $e \notin L$, $\alpha = H_L^\beta < \infty$ and $\beta(w) < 1$ for all $w \in L$. Then*

1. *$H_{L^*}^\beta \leq \inf\{s : \beta^s(L) \leq 1\}$ and,*
2. *if L is a code and $\beta^\alpha(L) \geq 1$, then $H_{L^*}^\beta$ is the unique solution of $\beta^s(L) = 1$.*

Such, for codes $C \subseteq \Sigma_n^*$ we get the formula $H_{C^*}^\beta = \inf\{s : \beta^s(C) \leq 1\}$. We obtain a condition sufficient for the inequality $H_{L^*}^\beta > H_L^\beta$:

---
[5] Regular ω-languages are defined as finite unions of sets of the form $W \cdot V^\omega$ where W, V are regular languages.
[6] This is just another formulation of the "contracting cycles property" of [20] and [1].
[7] More precisely, there is a c, $0 < c < 1$, such that for all $\varepsilon > 0$ the inequality $\Phi_L(s + \varepsilon) \leq c^\varepsilon \cdot \Phi(s)$ holds.

Lemma 1. *If L is a finite union of k codes which satisfies $\beta^\alpha(L) > k$ for $\alpha = H_L^\beta < \infty$, then $H_{L^*}^\beta > H_L^\beta$.*

Corollary 2. *If $L \subseteq \Sigma_n^*$ is regular and a finite union of codes and $H_L^\beta < \infty$, then $H_{L^*}^\beta > H_L^\beta$. (See Cor. 1.)*

Next we consider the approximation of $H_{L^*}^\beta$ via $H_{U^*}^\beta$ where U is a finite subset of L. We derive an analogue to the theorem of [26] stated in Thm. 5. In [26] we used the real numbers λ_m defined as the smallest (positive) roots of the equation $1 = \lambda_m + (\lambda_m)^m$.[8] In the sequel we assume there is a positive constant $c < 1$ such that every word $w \in L$ (and, hence also every $w \in L^*$) satisfies $\beta(w) \leq c^{|w|}$, i.e., $L^* \subseteq V_{\beta,c}$ where $V_{\beta,c} := \{w : w \in \Sigma_n^* \wedge \beta(w) \leq c^{|w|}\}$. Note that $V_{\beta,c}^* \subseteq V_{\beta,c}$.

Theorem 4. *Let $L \subseteq V_{\beta,c}$, $L \neq \emptyset$. Then for $m \leq \min\{|w| : w \in L \setminus \{e\}\}$ and $\varepsilon_m := \log_c \lambda_m$ we have $\beta^s(L^*) \leq \sum_{i \in \mathbb{N}} (\beta^s(L))^i \leq \beta^{s-\varepsilon_m}(L^*)$ for all $s \geq \varepsilon_m$.*

If $\beta^\theta(L) = 1$, then on the one hand $H_{L^*}^\beta \leq \theta$ and on the other hand, according to Thm. 4, $\beta^{\theta-\varepsilon_m}(L^*) = \infty$, i.e., $H_{L^*}^\beta \geq \theta - \varepsilon_m$. So, we get:

Corollary 3. *Let $L \subseteq V_{\beta,c}$ for some $c < 1$, $e \notin L$ and $\min\{|w| : w \in L\} \geq m > 0$. Then $0 \leq \theta - H_{L^*}^\beta \leq \varepsilon_m$ whenever $\beta^\theta(L) = 1$.*

Theorem 5. *Let $L \subseteq V_{\beta,c}$ for some $c < 1$. Then for every $\varepsilon > 0$ there is a finite subset $U \subseteq L$ such that $H_{L^*}^\beta - H_{U^*}^\beta < \varepsilon$.*

We derive an upper bound to the β-entropy of $V_{\beta,c} \subseteq \Sigma_n^*$ where $c < 1$: $H_{V_{\beta,c}}^\beta \leq -\log_c n$, for $\beta^s(V_{\beta,c}) \leq \sum_{i \in \mathbb{N}} n^i \cdot c^{s \cdot i} = \sum_{i \in \mathbb{N}} (n \cdot c^s)^i < \infty$ if only $n \cdot c^s < 1$.

4 ω-Languages and Hausdorff Dimension

Now, we apply our results on valuations of languages to the calculation of the Hausdorff dimension in the spaces $(\Sigma_n^\omega, \rho_\beta)$ where the metric ρ_β is derived from the valuation $\beta : \Sigma_n \to (0, \infty)$ by $\rho_\beta(\xi, \eta) = \inf\{\beta(w) : w \in \mathbf{A}(\xi) \cap \mathbf{A}(\eta)\}$. The case when $\beta(a) = \beta_n(a) = \frac{1}{n}$ for $a \in \Sigma_n$ was investigated in [29]; here, we generalize those results. Particular results for arbitrary valuations (with finite automata as control structure) were obtained in [1,20].

First we list some properties of the metric ρ_β. It turns out that there is a crucial distinction between the behaviour of the metrics derived from various valuations β, mainly depending on the fact whether $\beta(a) < 1$ for all $a \in \Sigma_n$ [9] or not.

---

[8] It is well-known that $\lambda_m^{-\ell}$ upperbounds the number of compositions (ordered partitions) of the number ℓ into parts not smaller than m, and it holds $0 < \lambda_m < \lambda_{m+1} < 1$ and $\lim_{m \to \infty} \lambda_m = 1$ (cf. [26]).

[9] Valuations having this property will be called *contractive*. In that case, $(\Sigma_n^\omega, \rho_\beta)$ is compact and balls are exactly the sets of the form $w \cdot \Sigma_n^\omega$.

1. ρ_β satisfies the *ultrametric inequality*.
2. Sets of the form $w \cdot \Sigma_n^\omega$ are always open. Therefore, ω-languages $E \subseteq \Sigma_n^\omega$ satisfying $E = \{\xi : \mathbf{A}(\xi) \subseteq \mathbf{A}(E)\} = \Sigma_n^\omega \setminus (\Sigma_n^* \setminus \mathbf{A}(E)) \cdot \Sigma_n^\omega$ are closed in every space $(\Sigma_n^\omega, \rho_\beta)$; we call them *strongly closed*.
3. If β is noncontractive, $(\Sigma_n^\omega, \rho_\beta)$ need not be *compact*, and there may be *isolated points*, i.e., $\xi \in \Sigma_n^\omega$ with $\rho_\beta(\xi, \eta) > \epsilon$ for some $\epsilon > 0$ and all $\eta \neq \xi$.
4. Balls $\overline{\mathbb{B}}_\epsilon(\xi) = \{\eta : \rho_\beta(\xi, \eta) \leq \epsilon\}$ with center ξ and radius $\epsilon > 0$ (they are simultaneously open and closed) are described by:

$$\overline{\mathbb{B}}_\epsilon(\xi) = \begin{cases} \{\xi\} &, \text{if } \forall \eta(\xi \neq \eta \to \rho_\beta(\xi, \eta) > \epsilon), \text{ and} \\ w_\beta(\xi, \epsilon) \cdot \Sigma_n^\omega &, \text{otherwise } (w_\beta(\xi, \epsilon) \text{ exists!}); \end{cases} \quad (4)$$

where $w_\beta(\xi, \epsilon)$ is the shortest prefix (provided it exists) $w \sqsubset \xi$ with $\beta(w) \leq \epsilon$. Any $\eta \in \overline{\mathbb{B}}_\epsilon(\xi)$ can be chosen to be the center, which shows that its radius is an upper bound of its *diameter*. In contrast to the contractive case, neither all balls are subsets of the form $w \cdot \Sigma_n^\omega$ nor all subsets of the form $w \cdot \Sigma_n^\omega$ are balls; e.g., take $12 \cdot \Sigma_2^\omega$ in $(\Sigma_2^\omega, \rho_\beta)$ where $\beta(1) < 1$ and $\beta(2) \geq 1$.
5. Let $\mathbb{I}_\beta := \{\xi : \inf\{\beta(w) : w \sqsubset \xi\} > 0\}$ be the set of all isolated points. $\mathbb{I}_\beta$ is open. $\mathbb{I}_\beta = \emptyset$ iff β is contractive. For noncontractive β we have $\mathbb{I}_\beta = \Sigma_n^* \cdot a^\omega$ iff $\beta(a) = 1$ and $\beta(b) < 1$ for $b \in \Sigma_n \setminus \{a\}$, and else $\mathbb{I}_\beta$ is uncountable.
6. We call $\mathbb{F}_\beta := \Sigma_n^\omega \setminus \mathbb{I}_\beta$ β-*fundamental*. $\mathbb{F}_\beta$ is closed. If $\mathbb{F}_\beta \neq \emptyset$, it is not strongly closed unless β is contractive, since $\mathbb{F}_\beta = \mathbb{F}_\beta / w$ for all $w \in \Sigma_n^*$ and only two strongly closed one-state ω-languages exist in Σ_n^ω: Σ_n^ω and $\emptyset$.

In order to introduce the Hausdorff dimension of subsets of $(\Sigma_n^\omega, \rho_\beta)$ we define the α-*dimensional outer Hausdorff measure* induced by ρ_β:

$$\nu_\beta^\alpha(F) := \liminf_{\epsilon \to 0} \{\sum_{i \in \mathbb{N}} (\text{diam } F_i)^\alpha : F \subseteq \bigcup_{i \in \mathbb{N}} F_i \wedge \text{diam } F_i < \epsilon\}.[10] \quad (5)$$

Then the *Hausdorff dimension* (HD) of $F \subseteq \Sigma_n^\omega$ in $(\Sigma_n^\omega, \rho_\beta)$ is defined as $\dim^{(\beta)} F := \inf\{\alpha : \nu_\beta^\alpha(F) = 0\} = \sup\{\alpha : \alpha = 0 \vee \nu_\beta^\alpha(F) = \infty\}$. Here we mention that the HD is countably stable. If $F \subseteq \mathbb{F}_\beta$, we could show the following characterization of ν_β^α:

$$\nu_\beta^\alpha(F) = \liminf_{\epsilon \to 0} \{\beta^\alpha(L) : F \subseteq L \cdot \Sigma_n^\omega \wedge \forall w(w \in L \to \beta(w) \leq \epsilon)\} \quad (6)$$

Next, we derive some relations between the β-entropy of languages and the HD of ω-languages in $(\Sigma_n^\omega, \rho_\beta)$. First we get analogues to [29, L. 3.8 and 3.10]. We introduce the δ-*limit* $V^\delta := \{\xi : \xi \in \Sigma_n^\omega \wedge \mathbf{A}(\xi) \cap V \text{ is infinite}\}$ of $V \subseteq \Sigma_n^*$.

Lemma 2. *If $\beta^\alpha(V) < \infty$, then $\nu_\beta^\alpha(V^\delta) = 0$.*

Lemma 3. *Let $F \subseteq \mathbb{F}_\beta$. Then $\nu_\beta^\alpha(F) = 0$ iff there is a language $L \subseteq \Sigma_n^*$ such that $F \subseteq L^\delta$ and $\beta^\alpha(L) < \infty$.*

---
[10] If F contains uncountably many isolated points, we have always $\nu_\beta^\alpha(F) = \infty$.

As consequences of the HD definition, we get the following relations:

$$\dim{}^{(\beta)} V^\delta \leq H_V^\beta \quad , \text{and} \tag{7}$$

$$\dim{}^{(\beta)} F = \inf\{\dim{}^{(\beta)} W^\delta : F \subseteq W^\delta\} \quad , \text{if } F \subseteq \mathbb{F}_\beta . \tag{8}$$

Utilizing the results of [1,20] and Prop. 3 we can relate HD and measure of strongly closed finite-state $F \subseteq \Sigma_n^\omega$ and the β-entropy of $\mathbf{A}(F)$.

Lemma 4. 1. If $c < 1$ then $V_{\beta,c}^\delta \subseteq \mathbb{F}_\beta$, and this inclusion is proper if $\mathbb{F}_\beta \neq \emptyset$ and β is not contractive. ($V_{\beta,c}$ was defined in Section 3.2).
2. For every finite-state and strongly closed ω-language $E \subseteq \mathbb{F}_\beta$ there are $c \in (0,1)$ and $\ell \in \mathbb{N}$ such that $E \subseteq \{w : |w| = \ell \wedge \beta(w) \leq c^\ell\}^\omega$.

In [1,20] it is shown that $\alpha = \dim{}^{(\beta)}(F)$ is the solution of $\Phi_{\mathbf{A}(F)}(s) = 1$, and that $\nu_\beta^\alpha(F) > 0$. Together with our ideas how to compute $\beta^\alpha(\mathbf{A}(L))$, this yields:

Theorem 6. If $\emptyset \neq F \subseteq \mathbb{F}_\beta$ finite-state and strongly closed, then $H_{\mathbf{A}(F)}^\beta = \dim{}^{(\beta)}(F)$; further, if $\alpha = \dim{}^{(\beta)}(F)$, then $\nu_\beta^\alpha(F) > 0$.

Since U^ω is finite-state and strongly closed if only U is finite, in view of $\mathbf{A}(U^\omega) = \mathbf{A}(U^*)$ and Prop. 3 the HD of any $U^\omega \subseteq \mathbb{F}_\beta$ is obtained as $\dim{}^{(\beta)} U^\omega = H_{U^*}^\beta$. By an approximation as in Thm. 5, we got a general formula for the HD of L^ω:

Lemma 5. If $c \in (0,1)$ and $L \subseteq V_{\beta,c}$, then $\dim{}^{(\beta)} L^\omega = \dim{}^{(\beta)}(L^*)^\delta = H_{L^*}^\beta$.

Lemma 6. If $c \in (0,1)$ and $L \subseteq V_{\beta,c}$, then $\nu_\beta^\alpha((L^*)^\delta) \leq 1$ for $\alpha = H_{L^*}^\beta$.

Define the *strong closure* of an ω-language E as $\mathrm{cl}(E) := (\mathbf{A}(E))^\delta$; it is the smallest strongly closed ω-language containing E, thus independently of β it contains the smallest ρ_β-closed set $F \subseteq \Sigma_n^\omega$ with $E \subseteq F$. By Lemma 5 and Prop. 3, we obtain:

Corollary 4. If $c \in (0,1)$ and $L \subseteq V_{\beta,c}$ is regular, $\dim{}^{(\beta)} L^\omega = \dim{}^{(\beta)} \mathrm{cl}(L^\omega)$.

Corollary 5. If $c \in (0,1)$ and $L \subseteq V_{\beta,c}$ is regular and a finite union of codes and $\alpha = H_{L^*}^\beta$, then $0 < \nu_\beta^\alpha(L^\omega) = \nu_\beta^\alpha(\mathrm{cl}(L^\omega)) \leq 1$.

Remark 2. More involved calculations as in [21, Thm. 6] show $0 < \nu_\beta^\alpha((L^*)^\delta) = \nu_\beta^\alpha(\mathrm{cl}(L^\omega)) \leq 1$ for arbitrary regular $L \subseteq V_{\beta,c}$, but in the case of nonregular W one might even have $\dim{}^{(\beta_n)} W^\omega < \dim{}^{(\beta_n)} \mathrm{cl}(W^\omega)$ (cf. [29, Ex. 6.3 and 6.5]).

5 IFS and Fractal Geometry

One of the most popular ways to describe fractals is IFS [2]. We restrict ourselves in the following to *Euclidean spaces* $\mathcal{X} \subseteq \mathbb{R}^m$ equipped with the Euclidean distance ρ_E. Denoting the set of contractive similitudes $f : \mathcal{X} \to \mathcal{X}$ by $\mathcal{S}(\mathcal{X})$, we can describe an IFS $\mathcal{F}$ as a map $\mathcal{F} : \Sigma_n \to \mathcal{S}(\mathcal{X})$. We sketch some well-known properties of IFS in the following: An IFS $\mathcal{F}$ gives a contractive valuation $\beta_\mathcal{F} : \Sigma_n^* \to (0, \infty)$, where $\beta_\mathcal{F}(i)$ (for $i \in \Sigma_n$) denotes the similarity factor of $\mathcal{F}(i)$. So, $w \in \Sigma_n^+$ can be seen as a similitude $\phi_\mathcal{F}(w) \in \mathcal{S}(\mathcal{X})$, where $\phi_\mathcal{F}$ is a semigroup morphism $(\Sigma_n^+, \cdot) \to (\mathcal{S}(\mathcal{X}), \circ)$. Recall the notion of *address* derived from Eq. (1) in the introduction. Further, the map $\phi_\mathcal{F} : (\Sigma_n^\omega, \rho_{\beta_\mathcal{F}}) \to (\mathcal{X}, \rho_E)$ is Lipschitz. Call $A_\mathcal{F} = \phi_\mathcal{F}(\Sigma_n^\omega)$ *limit set* of $\mathcal{F}$.

Given an IFS $\mathcal{F} : \Sigma_n \to \mathcal{S}(\mathcal{X})$, we interpret a finite (m-element) language $L = \{w_1, \ldots, w_m\} \subset \Sigma_n^+$ as an IFS $\mathcal{F}_L : \Sigma_m \to \mathcal{S}(\mathcal{X}), i \mapsto \phi_\mathcal{F}(w_i)$. We have $A_{\mathcal{F}_L} = \phi_\mathcal{F}(L^\omega)$. Similarly, infinite L lead to infinite IFS (IIFS) [36,9,19] whose theory is more involved but analogous to IFS theory. We can still define a set described by an IIFS $\mathcal{F}_L$ (based on the IFS $\mathcal{F}$ and the language L), the *limit set* $\phi_\mathcal{F}(L^\omega)$.[11] In $\mathcal{X} = (\Sigma_n^\omega, \rho_{\beta_n})$, we interpret any $L \subseteq \Sigma_n^+$ as an (I)IFS by $w : \Sigma_n^\omega \to \Sigma_n^\omega, x \mapsto w \cdot x$ with limit set L^ω. For $L \subset \Sigma_n^+$, call $\mathrm{vd}_\beta(L) = \inf\{s : \beta^s(L) \leq 1\}$ *valuation dimension* (VD). Prop. 1 shows the close relation of $\mathrm{vd}_\beta(L)$ and $H_{L^*}^\beta$. VD corresponds to the similarity dimension in IFS theory.

We denote the s-dimensional outer Hausdorff measure on $(\mathcal{X}, \rho_E)$ by $\mathcal{H}^s$, and the corresponding HD by $\dim_H$. For IFS, *Moran's open set condition (OSC)* is well-known as an assumption alleviating the determination of the HD of $A_\mathcal{F}$ [2,6,8]: Provided there is an open bounded non-empty test set $M \subseteq \mathcal{X}$ such that $\mathcal{F}(i)(M) \subseteq M$ for any $i \in \Sigma_n$, and that, furthermore, for any $i, j \in \Sigma_n$, $i \neq j$, $\mathcal{F}(i)(M) \cap \mathcal{F}(j)(M) = \emptyset$, then, for $\alpha = \mathrm{vd}_{\beta_\mathcal{F}}(\Sigma_n)$, $0 < \mathcal{H}^s(A_\mathcal{F}) < \infty$, and $\alpha = \dim_H(A_\mathcal{F})$. Generally, it is not trivial to find a test set for some $\mathcal{F}$. But, if we knew that $\mathcal{F}$ fulfills OSC, (when) could we say something about $\mathcal{F}_L$? Here, we need two further notions [30]. Call $V \subseteq \Sigma_n^*$ *OSC-code* iff there is a $\emptyset \neq W \subseteq \Sigma_n^*$ (OSC-*witness*) verifying:

$$\forall v(v \in V \to v \cdot W \cdot \Sigma_n^\omega \subseteq W \cdot \Sigma_n^\omega), \text{ and} \tag{9}$$

$$\forall v, v'(v, v' \in V \land v \neq v' \to v \cdot W \cdot \Sigma_n^\omega \cap v' \cdot W \cdot \Sigma_n^\omega = \emptyset). \tag{10}$$

Any OSC-code is a code, and any prefixcode is an OSC-code. Moreover, any regular code is an OSC-code [30]. Note further the correspondence with the Euclidean case: Interpreting V as an (I)IFS in Σ_n^ω, V satisfies the OSC with open test set $W \cdot \Sigma_n^\omega$ iff V is an OSC-code with OSC-witness W.

Theorem 7. *Let $\mathcal{F} = (\varphi_1, \ldots, \varphi_n)$ where $\varphi_i : \mathbb{R}^d \to \mathbb{R}^d$ be an IFS satisfying OSC, and let $C \subseteq \Sigma_n^*$ be an OSC-code. Then (I)IFS $\mathcal{F}_C$ satisfies OSC, too.*

Togetner with [9, Thm. 3.11], we obtain $\dim_H(\phi_\mathcal{F}(L^\omega)) = \mathrm{vd}_{\beta_\mathcal{F}}(L)$ if the IFS $\mathcal{F} : \Sigma_n \to \mathcal{S}(\mathcal{X})$ satisfies the OSC and L is an OSC-code. The previous sections

---

[11] When restricting one's attention to compact sets, take its closure instead.

together with Thm. 3 of [1], however, allow to strengthen the mentioned result and to generalize it to not necessarily contractive valuations.

In [1,20], IFS have been generalized to systems $\mathcal{F}$ with arbitrary similitudes. To guarantee the convergence of (1) one has to restrict the set of admissible ω-words ξ. In [1, Thm. 3], it is shown that $\phi_{\mathcal{F}} : (E, \rho_{\beta_{\mathcal{F}}}) \to (\mathcal{X}, \rho_E)$ is Lipschitz whenever E is a strongly closed finite-state subset of $\mathbb{F}_{\beta_{\mathcal{F}}}$. Related to this, the following generalization of the OSC for pairs $(\mathcal{F}, E)$ satisfying the above-mentioned property is introduced. Let $\mathcal{M}$ be a finite set of open subsets of $(\mathcal{X}, \rho_E)$. To every $w \in \Sigma_n^*$ we assign a set $M_w \in \mathcal{M}$. We say that the assignment is *compatible* with E iff $M_w = \emptyset \to w \notin \mathbf{A}(E)$, $\bigcup_{i=1}^n \varphi_i(M_{w \cdot i}) \subseteq M_w$, and $\varphi_i(M_{w \cdot i}) \cap \varphi_j(M_{w \cdot j}) = \emptyset$, for $i \neq j$. We say that a pair $(\mathcal{F}, E)$ satisfies the *Generalized Open Set Condition (GOSC)* iff E is a finite-state strongly closed subset of $\mathbb{F}_{\beta_{\mathcal{F}}}$ and there are an $\mathcal{M}$ and an assignment $w \mapsto M_w \in \mathcal{M}$ compatible with E. By the first condition, for every finite-state strongly closed subset $F \subseteq E$ the pair $(\mathcal{F}, F)$ satisfies GOSC if $(\mathcal{F}, E)$ satisfies GOSC. Thm. 3 of [1] gives:

Theorem 8. *Let E be a finite-state strongly closed subset of $\mathbb{F}_{\beta_{\mathcal{F}}}$ so that $(\mathcal{F}, E)$ satisfies GOSC. Then, we have: $\dim_H \phi_{\mathcal{F}}(E) = H_{\mathbf{A}(E)}^{\beta_{\mathcal{F}}} = \dim^{(\beta_{\mathcal{F}})} E =: \alpha$ with $\mathcal{H}^\alpha(\phi_{\mathcal{F}}(E)) > 0$.*

We proceed with the strengthening of $\dim_H(\phi_{\mathcal{F}}(L^\omega)) = \mathrm{vd}_{\beta_{\mathcal{F}}}(L)$.

Theorem 9. *Let $(\mathcal{X}, \rho_E)$ be a Euclidean space, $\mathcal{F} : \Sigma_n \to \mathcal{S}(\mathcal{X})$, E be a finite-state and strongly closed subset of $\mathbb{F}_\beta$, and let $L \subseteq \Sigma_n^*$ such that $L^\omega \subseteq E$. Assume the pair $(\mathcal{F}, E)$ satisfies the GOSC. Then $\dim_H(\phi_{\mathcal{F}}(L^\omega)) = \dim^{(\beta_{\mathcal{F}})} L^\omega$, and provided L is a code, we have $\dim_H(\phi_{\mathcal{F}}(L^\omega)) = \mathrm{vd}_{\beta_{\mathcal{F}}}(L)$.*

Remark 3. An analogue for IIFS satisfying the OSC (using the notion of topological pressure function) is given in [19, Thm. 3.15]. Confer also [12, Thm. 10].

Theorem 10. *Let $(\mathcal{X}, \rho_E)$ be a Euclidean space, $\mathcal{F} : \Sigma_n \to \mathcal{S}(\mathcal{X})$, and let $L \subseteq \Sigma_n^*$ be a regular language such that $\beta_{\mathcal{F}}(w) < 1$ for all $w \in L \setminus \{e\}$. Then $\mathrm{cl}(L^\omega) \subseteq \mathbb{F}_{\beta_{\mathcal{F}}}$ and $\dim_H(\phi_{\mathcal{F}}(\mathrm{cl}(L^\omega))) = \dim_H(\phi_{\mathcal{F}}(L^\omega)) = \dim^{(\beta_{\mathcal{F}})} L^\omega$. If, moreover, L is a finite union of codes then $\mathcal{H}^s(\phi_{\mathcal{F}}(L^\omega)) = \mathcal{H}^s(\phi_{\mathcal{F}}(\mathrm{cl}\, L^\omega))$ for $s \in [0, \infty)$.*

Note 1. In [9, Remark 3.12], the question was raised whether requiring an OSC for each IFS-part $\mathcal{F}_n = (\mathcal{F}(1), \ldots, \mathcal{F}(n))$ of a given IIFS $\mathcal{F}$ is weaker than requiring an OSC for $\mathcal{F}$ itself. We can show the following here: If all $\mathcal{F}_n$ fulfill an OSC, then $\mathcal{F}$ itself does not necessarily satisfy an OSC. Namely, take as basic IFS $\mathcal{F} : \Sigma_2 \to \mathcal{S}(([0,1], \rho_E))$ defined by $\mathcal{F}(1)(x) = x/2$ and $\mathcal{F}(2)(x) = x/2 + 1/2$. Clearly, $A_{\mathcal{F}} = [0,1]$. Consider the suffixcode $L = \{w12^{|w|} : w \in \Sigma_2^*\}$ (which is no OSC-code) from [30, Example 1]. The IIFS $\mathcal{F}_L$ doesn't satisfy an OSC.

Acknowledgment: The first author has been supported by DFG La 618/3-2.

References

1. C. Bandt. Self-similar sets 3. *Monatsh. Math.*, 108:89–102, 1989.
2. M. F. Barnsley. *Fractals Everywhere*. Acad. Press, 1988.
3. J. Berstel and M. Morcrette, Compact representations of patterns by finite automata, in: Proc. Pixim '89, Hermes, Paris, 1989, pp. 387 – 402.
4. K. Čulik II and S. Dube, Affine automata and related techniques for generation of complex images, *TCS*, 116:373–398, 1993.
5. K. Čulik II and J. Kari, in: [24], pp. 599–616.
6. G. A. Edgar. *Measure, Topology, and Fractal Geometry*. Springer, 1990.
7. S. Eilenberg. *Automata, Languages, and Machines, A*. Acad. Press, 1974.
8. K.J. Falconer, *Fractal Geometry*. Wiley, 1990.
9. H. Fernau. Infinite IFS. *Mathem. Nachr.*, 169:79–91, 1994.
10. H. Fernau. Valuations of languages, with applications to fractal geometry. *TCS*, 137(2):177–217, 1995.
11. H. Fernau. Valuations, regular expressions, and fractal geometry. *AAECC*, 7(1):59–75, 1996.
12. H. Fernau and L. Staiger. Valuations and unambiguity of languages, with applications to fractal geometry. In S. Abiteboul and E. Shamir, eds., *ICALP'94*, vol. 820 of *LNCS*, pp. 11–22, 1994.
13. H. Fernau and L. Staiger. Valuations and unambiguity of languages, with applications to fractal geometry. TR No. 94-22, RWTH Aachen, 1994.
14. A. Habel, H.-J. Kreowski, and S. Taubenberger. Collages and patterns generated by hyperedge replacement. *Lang. of Design*, 1:125–145, 1993.
15. F. Haeseler, H.-O. Peitgen, and G. Skordev. Cellular automata, matrix substitutions and fractals. *Ann. Math. and Artif. Intell.*, 8:345–362, 1993.
16. W. Kuich. On the entropy of context-free languages. *IC*, 16:173–200, 1970.
17. R. Lindner and L. Staiger. *Algebraische Codierungstheorie; Theorie der sequentiellen Codierungen*. Akademie-Verlag, 1977.
18. B. Mandelbrot. *The Fractal Geometry of Nature*. Freeman, 1977.
19. R. D. Mauldin and M. Urbański. Dimensions and measures in infinite iterated function systems. *Proc. Lond. Math. Soc.*, III. Ser. 73, No.1, 105–154, 1996.
20. R. D. Mauldin and S. C. Williams. Hausdorff dimension in graph directed constructions. *Trans. AMS*, 309(2):811–829, Oct. 1988.
21. W. Merzenich and L. Staiger. Fractals, dimension, and formal languages. *RAIRO Inf. théor. Appl.*, 28(3–4):361–386, 1994.
22. P. Prusinkiewicz and A. Lindenmayer. *The Algorithmic Beauty of Plants*. Springer, 1990.
23. P. Prusinkiewicz et al. in: [24], pp. 535–597.
24. G. Rozenberg and A. Salomaa (eds.) *Handbook of Formal Languages, Vol. 3*, Springer, 1997.
25. L. Staiger. Finite-state ω-languages. *JCSS*, 27:434–448, 1983.
26. L. Staiger. Ein Satz über die Entropie von Untermonoiden. *TCS*, 61:279–282, 1988.
27. L. Staiger. Quadtrees and the Hausdorff dimension of pictures. In *"Geobild'89"*, pp. 173–178, Akademie-Verlag, 1989.
28. L. Staiger. Hausdorff dimension of constructively specified sets and applications to image processing. In *Topology, Measures, and Fractals*, pp. 109–120. Akademie-Verlag, 1992.

29. L. Staiger. Kolmogorov complexity and Hausdorff dimension. *IC*, 103:159–194, 1993.
30. L. Staiger. Codes, simplifying words, and open set condition. *IPL*, 58:297–301, 1996.
31. L. Staiger, in: [24], pp. 339–387.
32. L. Staiger. Rich ω-words and monadic second-order arithmetic, to appear in: Proc. CSL'97, *LNCS*, 1998.
33. S. Takahashi. Self-similarity of linear cellular automata. *JCSS*, 44:114–140, 1992.
34. S. Taubenberger. Correct translations of generalized iterated function systems to collage grammars. TR 7/94, Universität Bremen, Fachbereich Mathematik und Informatik, 1994.
35. C. Tricot. Douze définitions de la densité logarithmique. *CR de l' Académie des Sciences (Paris), série I*, 293:549–552, Nov. 1981.
36. K. Wicks. *Fractals and Hyperspaces*, vol. 1492 of *LNM*. Springer, 1991.

One Quantifier Will Do in Existential Monadic Second-Order Logic over Pictures

Oliver Matz

Institut für Informatik und Praktische Mathematik
Christian-Albrechts-Universität Kiel, 24098 Kiel, Germany
oma@informatik.uni-kiel.de

Abstract. We show that every formula of the existential fragment of monadic second-order logic over picture models (i.e., finite, two-dimensional, coloured grids) is equivalent to one with only one existential monadic quantifier.
The corresponding claim is true for the class of word models ([Tho82]) but not for the class of graphs ([Ott95]).
The class of picture models is of particular interest because it has been used to show the strictness of the different (and more popular) hierarchy of quantifier alternation.

1 Introduction

We study monadic second-order logic (MSO) over finite structures. For a given class of structures, one can consider the following two hierarchies:

1. the quantifier alternation hierarchy, where properties are classified wrt. the number of monadic quantifier alternations required for an MSO-sentence;
2. the existential-quantifier depth hierarchy, where properties in the existential fragment of monadic second-order logic (i.e., on the lowest level of the alternation hierarchy) are classified wrt. the number of existential quantifiers required.

Both hierarchies are strict for graphs, i.e., on each level there are graph properties that are not on the previous. The proof of the strictness of the first ([MT97]) goes via another domain, namely the class of (finite, two-dimensional) *pictures*, i.e., arrays over a finite alphabet. The proof of the strictness of the second ([Ott95]) also uses grid-like structures, but different ones.

In contrast to that, for the class of finite strings, both hierarchies collapse, i.e., every MSO-sentence over strings is equivalent to a sentence whose monadic quantifier prefix consists of only one existential quantifier. The proof for the collapse of the second hierarchy can be found in [Tho82].

In the present paper we show (by an adaption of this proof) that the second hierarchy also collapses for the class of pictures, i.e., in a formula of the existential fragment of monadic second-order logic over picture models, the length of the quantifier prefix can be reduced to one.

2 Definitions

For $1 \leq j \leq n$, we denote $\{j, \ldots, n\}$ by $[j, n]$ and $[1, n]$ by $[n]$.

A *picture of size* (m, n) over a given finite alphabet Γ is an $(m \times n)$-matrix over Γ, i.e. a mapping $[m] \times [n] \to \Gamma$. If P is a picture of size (m, n), we call m and n the *height* and *width* of P, denoted by $\overline{P}$ and $|P|$, respectively. If $i \leq i' \leq \overline{P}$ and $j \leq j' \leq |P|$, then the subblock of P on the rectangle $[i, i'] \times [j, j']$ is denoted by

$$P([i, i'] \times [j, j']) = \begin{pmatrix} P(i,j) & \ldots & P(i,j') \\ \vdots & \ddots & \vdots \\ P(i',j) & \ldots & P(i',j') \end{pmatrix}.$$

By *picture languages* we refer to sets of pictures. The language of all pictures (all picture of size (m, n), or of height m, or of width n, respectively) over Γ is denoted by $\Gamma^{+,+}$ (or $\Gamma^{m,n}$, or $\Gamma^{m,+}$, or $\Gamma^{+,n}$, respectively).

If $P \in \Gamma^{+,+}$, $Q \in \Omega^{+,+}$ are pictures of the same size over two alphabets Γ and Ω, then we denote by $P \otimes Q$ the picture over $\Gamma \times \Omega$ for which $(P \otimes Q)(x) = (P(x), Q(x))$ for every $x \in \mathrm{dom} P$.

2.1 Monadic Second-Order Logic over Picture and Word Models

In the case of pictures, formulas are in the signature $\sigma_\Gamma := \{S_1, S_2, (Q_a)_{a \in \Gamma}\}$, where the Q_a are unary and S_1, S_2 are binary relation symbols. To every $m \times n$-picture P over Γ we associate the *picture model* $\underline{P} := ([m] \times [n], S_1^P, S_2^P, (Q_a^P)_a)$, where $S_1^P = \{((i,j), (i+1,j)) \mid (i,j) \in [m-1] \times [n]\}$ and $S_2^P = \{((i,j), (i,j+1)) \mid (i,j) \in [m] \times [n-1]\}$, and $Q_a^P = P^{-1}(a)$ is the set of all positions that carry the letter $a \in \Gamma$. When the picture P is displayed the usual way, the relations S_1^P and S_2^P are the vertical (respectively horizontal) successor relations.

We use $x, y, z, \ldots$ as first-order variables (ranging over elements of the universe) and $X, Y, \ldots$ as monadic second-order variables (ranging over subsets of the universe).

Consequently, atomic formulas over picture models are of the form $X(x)$ (saying that position x is in the set X), $x = y$ (saying that positions x and y are equal), $Q_a(x)$ (saying that position x carries letter a), $S_1(x, y)$ (saying that x is a vertical successor of y), or $S_2(x, y)$ (saying that x is a horizontal successor of y).

Formulas of monadic second-order logic (*MSO-formulas*) over picture models are built inductively from atomic ones by using (1) boolean connectives $\vee, \neg$, (2) first-order quantifications of the form $\exists x \varphi$, and (3) second-order quantifications of the form $\exists X \varphi$.

First-order formulas are MSO-formulas in which no second-order quantifier occurs. The existential fragment *EMSO* of MSO consists of formulas of the form $\exists X_1 \ldots \exists X_t \varphi$, where φ is first-order. We write $\varphi(X_1, \ldots, X_t)$ if φ is a formula with free second-order variables among $X_1, \ldots, X_t$. If $X_1, \ldots, X_t \subseteq \mathrm{dom} M$ for a structure M such that φ holds in M under the assignment mapping $\boldsymbol{X}_i$ to X_i, we write $M \models \varphi[X_1, \ldots, X_t]$.

The *picture language* (over alphabet Γ) *defined by* a sentence φ of signature σ_Γ) is the set of pictures whose associated picture models make φ true. If a picture language is defined by some *EMSO*-sentence over σ_Γ then it is called *EMSO-definable*.

In the case of words, formulas are in the signature $\varrho_\Gamma := \{S, (Q_a)_{a\in\Gamma}\}$ with the binary relation symbol S. The word model associated to a nonempty word $w \in \Gamma^+$ is the ϱ_Γ-structure $\underline{w} := (\mathrm{dom}\underline{w}, S^w, (Q_a^w)_{a\in\Gamma})$, where $\mathrm{dom}\underline{w} = \{1, \ldots, |w|\}$ is the set of positions of w, and S^w is the successor relation on $\{1, \ldots, |w|\}$, and Q_a^w is the set of those positions of w that carry the letter a. Atomic formulas over word models are of the form $\boldsymbol{X}(\boldsymbol{x})$, $\boldsymbol{x} = \boldsymbol{y}$, $Q_a(\boldsymbol{x})$, or $S(\boldsymbol{x}, \boldsymbol{y})$, the latter saying that y is the succesor of x.

3 Compression of Existential Quantifier Block

[Tho82] exploits the connection of EMSO-definable word languages to local languages. This connection is presented in the next subsection. Afterwards we recall how locality is transferred to pictures, here using the notion of *domino-local* picture languages as in [LS94,Mat95], and then use this notion to prove Theorem 4.

3.1 EMSO vs. Locality

A word language $L \subseteq \Gamma^+$ is *local* iff there are sets $A, B \subseteq \Gamma$ and $C \subseteq \Gamma^2$ such that $L = (A\Gamma^* \cap \Gamma^*B) \setminus (\Gamma^*C\Gamma^*)$. The following remark about regular word languages is folklore.

Remark 1. Every word language definable in existential monadic second-order logic is a projection of a local word language.

The above remark also holds if the word "existential" is removed. The following is shown in [Tho82].

Theorem 2. *Let L be a a projection of a local word language. Then there exists a first-order formula $\varphi(\boldsymbol{X})$ in the signature ϱ_Γ such that $L = \mathrm{Mod}(\exists \boldsymbol{X} \varphi(\boldsymbol{X}))$.*

The proof idea is a follows: Let M be a local word language over alphabet Γ and $L = \pi(M) \subseteq \Sigma^*$ for an alphabet projection $\pi : \Gamma \to \Sigma$. A word $u \in M$ is called a *run* on $\pi(u)$, and letters from Γ are called *states*. A word w over Σ is partioned into sufficiently large sequences such that a $\{0,1\}$-colouring of such a sequence can encode the first state of the corresponding substring of a run on w. Now the existence of a run on w can be checked by a formula of the required form: φ checks that $\boldsymbol{X}$ corresponds to a $\{0,1\}$-colouring that encodes the first states of all sequences of a run on w.

The above two results give the following "compression corollary" that says that the number of existential quantifiers in *EMSO*-formulas can be reduced to one.

Corollary 3. *Every sentence of existential monadic second-order logic over words is equivalent to a sentence of the form $\exists \boldsymbol{X} \varphi(\boldsymbol{X})$, where φ is first-order.*

The contribution of this paper is to transfer the above proof and result to pictures languages, i.e., we will show the following analogue of Corollary 3:

Theorem 4. *Every sentence of existential monadic second-order logic over picture models is equivalent to a sentence of the form $\exists \boldsymbol{X} \varphi(\boldsymbol{X})$, where φ is first-order.*

This theorem is proved in the following three subsections.

Before that, let us note that we can conclude the following corollary, which states that the "compression" of an existential quantifier block also works in the presence of free variables.

Corollary 5. *Every formula of existential monadic second-order logic over picture models is equivalent to a formula of the form $\exists \boldsymbol{X} \varphi(\boldsymbol{X})$, where φ is first-order.*

The proof is straightforward. The essential idea is to pass to a sentence over a larger alphabet that encodes the free variables, and then to apply Theorem 4.

3.2 Domino-Local Picture Languages

The first step is to transfer the notion of locality from words to pictures.

Definition 6. *Let P be a picture, and Δ_1 (and Δ_2) be a set of pictures of size $(2,1)$ (or $(1,2)$, respectively) over the same alphabet. Δ_1 (or Δ_2) tiles P iff all subblocks of P of size $(2,1)$ (or $(1,2)$) are in Δ_1 (or Δ_2, respectively)*

Let P be a picture over Γ. The picture that results from P by surrounding it with the fresh boundary symbol $\#$ is denoted by $\hat{P}$.

A picture language L over Γ is domino-local *iff there exist sets $\Delta_1 \subseteq (\Gamma \cup \{\#\})^{2,1}$ and $\Delta_2 \subseteq (\Gamma \cup \{\#\})^{1,2}$ such that for every picture $P \in L$, both Δ_1 and Δ_2 tile $\hat{P}$. In that case, (Δ_1, Δ_2) is called a* domino tiling system *that recognizes L.*

(The notion of "locality" has been introduced in [GRST96] in another way, using (2×2)-"tiles" instead of (2×1)- and (1×2)-tiles. The slightly different and more convenient notion presented here has been studied in [LS94,Mat95]. See [GR96] for a comprehensive survey.)

By *projection* we refer to a mapping from one alphabet to another. A projection is lifted to pictures, words, picture languages, and word languages the obvious way.

Then we have indeed the analogue to Remark 1.

Theorem 7. *([GR96,LS94,Mat95]) Every EMSO-definable picture language is a projection of some domino-local picture language.*

Thus it suffices to show the following in order to obtain Theorem 4, which will be done in the next two subsections

Theorem 8. *Let L be a projection of a domino-local picture language. Then there exists some first-order formula $\varphi(\boldsymbol{X})$ such that $L = \mathrm{Mod}(\exists \boldsymbol{X} \varphi(\boldsymbol{X}))$*

3.3 Pictures of Bounded Height

Firstly, we consider a domino-local picture language restricted to pictures of a fixed height. In this case, the compression of existential quantifier prefixes of EMSO-formulas over picture models can easily be reduced to the case of word models.

Theorem 9. *Let L be a projection of a domino-local picture language over Ω and $m \geq 1$ a fixed height. Then there exists an first-order formula $\varphi(\boldsymbol{X})$ (in the signature σ_Ω) such that $\mathrm{Mod}(\exists \boldsymbol{X} \varphi(\boldsymbol{X}))$ is the set of pictures in L that have height m.*

Proof (Sketch). Let L_0 be the word language over alphabet $\Omega^{m,1}$ that contains all words in $(\Omega^{m,1})^*$ that are (as a picture of size height m over Ω) in L.

Application of Theorem 2 to the local word language L_0 yields a first-order formula in the signature $\varrho_{\Omega^{m,1}}$, which can be translated to a first-order formula φ in the signature σ_Ω in a straightforward way.

3.4 Pictures of Unbounded Height

In this subsection, we consider projections of domino-local picture languages without the restriction of a fixed height. We will sketch the main construction.

The aim is to construct a formula of the existential fragment of monadic second-order logic whose models are exactly the pictures of some projection of a given domino-local picture language $M \subseteq \Omega^{+,+}$ under an alphabet projection π. Let (Δ_1, Δ_2) be a domino tiling system recognizing M.

Without the additional limitation to one single monadic quantifier, such a formula may, informally speaking, work as follows: (1) Guess an Ω-colouring of the input picture, and then (2) check that the local restrictions are fulfilled, i.e., the picture obtained this way is tiled by Δ_1 and Δ_2.

Since we are restricted to one single monadic quantifier, the formula is not able to guess the Ω-colouring in step (1) but only a $\{0,1\}$-colouring. However, if we partition the picture into sufficiently large blocks, then it is possible to guess the Ω-colouring *of the border* of each block and store this in a $\{0,1\}$-colouring of the complete block. Then a finite disjunction may check if there really is a Ω-colouring of the block with that border such that Δ_1 and Δ_2 tile the inside of it. What remains to be done is to check whether the colouring of the borders of neighboured blocks fit to each other in the sense that the 2×1 or 1×2-subblocks (of the Ω-coloured picture) along the edges of blocks are in Δ_1 or, respectively, Δ_2. This can be checked (in the actual $\{0,1\}$-coloured picture) by a first-order

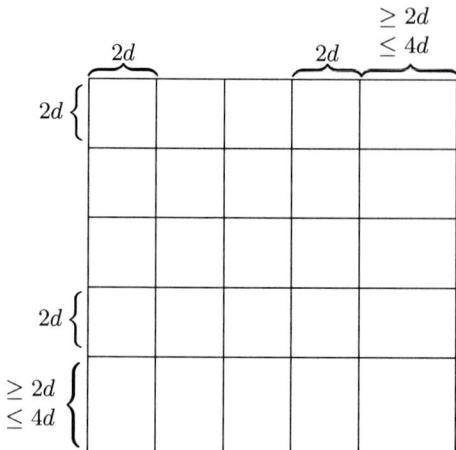

Fig. 1. Partition into Blocks

formula because the information about the Ω-colouring of the border of the block is coded in the $\{0,1\}$-colouring of the inside.

We prepare the proof with some definitions.

Definition 10. Let $P \in \Gamma^{+,+}$ be a picture of size (m, n). Then

$$\text{left}(P) = (P(1,1) \ldots P(m,1))^\top$$
$$\text{right}(P) = (P(1,n) \ldots P(m,n))^\top$$
$$\text{top}(P) = (P(1,1) \ldots P(1,n))$$
$$\text{bottom}(P) = (P(m,1) \ldots P(m,n))$$
$$\text{border}(P) = (\text{left}(P), \text{right}(P), \text{top}(P), \text{bottom}(P))$$

The next definition will help to define the partition of a picture into blocks.

Definition 11. Let $m \geq d \geq 1$. Choose $n \geq 0$ and $r < d$ in such a way that $m = (n+1)d + r$. The tuple $(1, d+1, 2d+1, \ldots, nd+1, m+1)$ is called the d-step sequence in m.

Note that if $(i_0, \ldots, i_{n+1})$ is the d-step sequence in m, then $d \leq i_{n+1} - i_n < 2d$.

Let P be some picture of size (m, n) (with $m, n \geq d$). Let i, i' (respectively j, j') be consecutive components in the d-step sequence in m (respectively n). The d-block of P at position (i, j) is the subblock $\text{Block}(P, d, (i, j)) = P([i, i'-1] \times [j, j'-1])$ of P.

If i'' (respectively j'') are components following i' (respectively j') in these sequences, then $P[i', i''-1] \times [j, j'-1]$ (respectively $P[i, i'-1] \times [j', j''-1]$) will be called a horizontally (respectively vertically) following d-block of P.

Figure 1 illustrates how a picture of size $\geq (2d, 2d)$ is split into $2d$-blocks.

We will make use of the fact that every picture P over Ω whose width and height are $\geq 2d$ can be split into $2d$-blocks, and there are only singly exponentially many essentially different (wrt. tilability by (Δ_1, Δ_2)) types of $2d$-blocks.

Proof. (of Theorem 8.) Let $L \subseteq \Gamma^{+,+}$ be a projection of a domino-local picture language M. W.l.o.g. we may assume that M is over alphabet $\Omega = \Theta \times \Gamma$ and that L is the image of M under the alphabet projection $\pi : \Omega \to \Gamma$, $(a, b) \mapsto b$. Let (Δ_1, Δ_2) be a domino tiling system that recognizes M.

Choose d such that there exists an injective mapping

$$f : \bigcup_{2d \leq m,n < 4d} (\Omega^m \times \Omega^m \times \Omega^n \times \Omega^n) \to \{0,1\}^{d \times d}.$$

We will only show that there exists a first-order formula $\psi(\boldsymbol{X})$ such that for every picture P over Γ we have $P \in L$ iff $P \models \exists \boldsymbol{X} \psi(\boldsymbol{X})$ and $\overline{P}, |P| \geq 2d$.

The claim will then follow from Theorem 9 because for every $m < 2d$, there are formulas $\varrho_m(\boldsymbol{X})$ and $\gamma_m(\boldsymbol{X})$ such that $\mathrm{Mod}(\exists \boldsymbol{X} \varrho(\boldsymbol{X})) = L \cap \Gamma^{m,+}$ and $\mathrm{Mod}(\exists \boldsymbol{X} \gamma(\boldsymbol{X})) = L \cap \Gamma^{+,m}$, and hence for $\varphi = \psi \vee \bigvee_{m<2d}(\varrho_m \vee \gamma_m)$ we have

$$L = (L \cap \Gamma^{\geq 2d, \geq 2d}) \cup \bigcup_{m<2d} (L \cap \Gamma^{m,+}) \cup (L \cap \Gamma^{+,m})$$

$$= \mathrm{Mod}(\exists \boldsymbol{X} \psi \vee \bigvee_{m<2d} (\exists \boldsymbol{X} \varrho_m \vee \exists \boldsymbol{X} \gamma_m)$$

$$= \mathrm{Mod}(\exists \boldsymbol{X} \varphi).$$

We proceed with the construction of ψ. For a block B over Ω whose width and height are $\geq 2d$ and $< 4d$, let $\mathrm{ess}(B)$ be the picture of same size as B such that

$$\mathrm{ess}(B)(i,j) = \begin{cases} 1 & \text{if } i = 0 \vee j = 0 \\ f(\mathrm{border}(B))(\frac{1}{2}(i+1), \frac{1}{2}(j+1)) & \text{if } i \equiv_2 j \equiv_2 1 \wedge i,j < 2d \\ 0 & \text{else.} \end{cases}$$

Intuitively, $\mathrm{ess}(B)$ carries all the essential information of a block B.

The following are equivalent for every picture P over Γ:

1. $P \in L$
2. There exists a picture Q over Θ of the same size such that Δ_1 and Δ_2 tile $\widehat{(Q \otimes P)}$.[1]
3. There exists a picture Q over Θ of the same size such that for every $2d$-block B of $(Q \otimes P)$ we have:
   (a) for the horizontally next $2d$-block B_1 of $(Q \otimes P)$ (if present), Δ_2 tiles $(\mathrm{right}(B) \ominus \mathrm{left}(B_1))^\top$;
   (b) for the vertically next $2d$-block B_1 of $(Q \otimes P)$ (if present), Δ_1 tiles $\mathrm{bottom}(B) \ominus \mathrm{top}(B_1)$;

---
[1] Remember that $\hat{P}$ means P surrounded with the boundary symbol $\#$.

(c) if B is a top-most 2d-block, then Δ_1 tiles $\#^{|B|} \ominus top(B)$;
   (d) if B is a bottom-most 2d-block, then Δ_1 tiles $bottom(B) \ominus \#^{|B|}$;
   (e) if B is a leftmost 2d-block, then Δ_2 tiles $(\#^{\overline{B}} \ominus left(B))^\top$;
   (f) if B is a rightmost 2d-block, then Δ_2 tiles $(right(B) \ominus \#^{\overline{B}})^\top$;
   (g) Δ_1 and Δ_2 tile B.
4. There exists a picture Q' over $\{0,1\}$ of the same size such that: for every i from the 2d-step sequence of $\overline{P}$ and every j from the 2d-step sequence of $|P|$, if $B' = Block(Q', 2d, (i,j))$, then
   (a) there is some $B \in ess^{-1}(B') \cap \pi^{-1}(Block(P, 2d, (i,j)))$ that is tiled by Δ_1 and Δ_2,
   (b) some (and hence any) 2d-block $B \in ess^{-1}(B')$ satisfies 3a to 3f.

The equivalence of 3 an 4 is due to the fact that properties 3a to 3f only depend on $border(B)$ and hence only on $ess(B)$.

The properties 4b and 4a can be checked by a first-order formula $\psi(\boldsymbol{X})$, where $\boldsymbol{X}$ encodes the $\{0,1\}$-picture Q'. To see the latter we observe that every picture Q' all of whose 2d-blocks have property 4a has the property that two horizontally (respectively vertically) consecutive 1-positions mark a row (respectively column) whose index is in the 2d-step sequence of the height (respectively width) of a picture. And for a picture that has this property, it is possible to determine consecutive 2d-blocks by first-order formulas and hence to check 4b by finite disjunctions ranging over all possible 2d-blocks over $\{0,1\}$.

This completes the proof of Theorem 8.

From Theorems 7 and 8 we can conclude our main result Theorem 4, i.e., that the existential quantifier prefix of an EMSO-sentence over picture models can be compressed to length one.

4 Concluding Remarks

The proof that the number of quantifiers in formulas of existential monadic second-order logic over words can be limited to one has been transfered to pictures, which are the 2-dimensional analogue to words. It is quite obvious that this proof can be transfered to arbitrary finite dimensions by induction.[2]

For the four classes of models mentioned in the introduction and the two monadic hierarchies of quantifier alternation and existential quantifier depth, the situation looks as follows. Here, "collapse" always means "collapses to the existential fragment EMSO".

	Quantifier Alternation	Existential Quantifier Depth
Word Models	collapse	collapse (see [Tho82])
Picture Models	strict (see [MT97])	collapse (this paper)
Otto-Grids	(open)	strict (see [Ott95])
Graphs	strict	strict

---
[2] The crucial point is to reprove Theorem 7 for higher dimensions, which has not been done yet though it is straightforward.

The results of the bottom row are infered by the results of the second and third row, respectively, by encoding techniques. I conjecture that for the class of "Ottogrids" (as defined in [Ott95]), the quantifier alternation hierarchy collapses.

The class of pictures is of special interest because it has a strict quantifier alternation hierarchy. Regarding this hierarchy, the following question is natural: Can the number of quantifiers in each block of a formula of minimal quantifier alternation be limited to one?

Acknowledgments. I thank Wolfgang Thomas and Thomas Wilke for their explanations of Corollary 3, which was the starting point for this paper. Furthermore, I thank Ina Schiering for valid criticism concerning the presentation.

References

[GR96]   D. Giammarresi and A. Restivo. Two-dimensional languages. In G. Rozenberg and A. Salomaa, editors, *Handbook of Formal Language Theory*, volume III. Springer-Verlag, New York, 1996.

[GRST96] D. Giammarresi, A. Restivo, S. Seibert, and W. Thomas. Monadic second-order logic and recognizability by tiling systems. *Information and Computation*, 125:32–45, 1996.

[LS94]   M. Latteux and D. Simplot. Recognizable picture languages and domino tiling. Internal Report IT-94-264, Laboratoire d'Informatique Fondamentale de Lille, Université de Lille, France, 1994.

[Mat95]   O. Matz. *Klassifizierung von Bildsprachen mit rationalen Ausdrücken, Grammatiken und Logik-Formeln*. Diploma thesis, Christian-Albrechts-Universität Kiel, 1995. (German).

[MT97]   O. Matz and W. Thomas. The monadic quantifier alternation hierarchy over graphs is infinite. In *Twelfth Annual IEEE Symposium on Logic in Computer Science*, pages 236–244, Warsaw, Poland, 1997. IEEE.

[Ott95]   M. Otto. Note on the number of monadic quantifiers in monadic Σ_1^1. *Information Processing Letters*, 53:337–339, 1995.

[Tho82]   W. Thomas. Classifying regular events in symbolic logic. *Journal of Computer and System Sciences*, 25:360–376, 1982.

On Some Recognizable Picture-Languages [*]

Klaus Reinhardt

Wilhelm-Schickhard Institut für Informatik, Universität Tübingen
Sand 13, D-72076 Tübingen, Germany
reinhard@informatik.uni-tuebingen.de

Abstract. We show that the language of pictures over $\{a,b\}$, where all occurring b's are connected is recognizable, which solves an open problem in [Mat98]. We generalize the used construction to show that monocausal deterministically recognizable picture languages are recognizable, which is surprisingly nontrivial. Furthermore we show that the language of pictures over $\{a,b\}$, where the number of a's is equal to the number of b's is nonuniformly recognizable.

1 Introduction

In [GRST94] pictures are defined as two-dimensional rectangular arrays of symbols of a given alphabet. A set (language) of pictures is called recognizable if it is recognized by a finite tiling system. It was shown in [GRST94] that a picture language is recognizable iff it is definable in existential monadic second-order logic. In [Wil97] it was shown that star-free picture expressions are strictly weaker than first-order logic. A comparison to other regular and context-free formalisms to describe picture languages can be found in [Mat97, Mat98]. Characterizations of the recognizable picture languages by automata can be found in [IN77] and [GR96], where also the subclasses, which are defined by a restriction from nondeterminism to determinism or unambiguity are considered.

We show in chapter 2 that the language of pictures over $\{a,b\}$, where all occurring b's are connected is recognizable, which solves an open problem in [Mat98]. (Connectedness is not recognizable in general [FSV95].) The technique which is used here is generalized in the following chapter to show that monocausal deterministically recognizable languages are recognizable. The notion of deterministic recognizability, which we use is stronger than the determinism in [GR96], has more closure properties (for example rotation), which promises practical relevance.

Furthermore we show in the last chapter that the language of pictures over $\{a,b\}$, where the number of a's is equal to the number of b's is nonuniformly recognizable. Hereby we use counters similar to those used in [Für82].

Definition 1. *[GRST94] A picture over Σ is a two-dimensional array of elements of Σ. The set of pictures of size (m,n) is denoted by $\Sigma^{m,n}$. A picture*

---

[*] This research has been supported by the DFG Project La 618/3-1 KOMET.

language is a subset of $\Sigma^{*,*} := \bigcup_{m,n \geq 0} \Sigma^{m,n}$.
For a $p \in \Sigma^{m,n}$, we have $\hat{p} \in \Sigma^{m+2,n+2}$
adding a frame of symbols $\# \notin \Sigma$.

$$\hat{p} := \begin{array}{|c|c|c|c|c|c|} \hline \# & \# & \# & \# & \# & \# \\ \hline \# & & & & & \# \\ \hline \# & & & p & & \# \\ \hline \# & & & & & \# \\ \hline \# & & & & & \# \\ \hline \# & \# & \# & \# & \# & \# \\ \hline \end{array}$$

Let $T_{2,2}(p)$ be the set of all sub-pictures of p with size $(2,2)$.

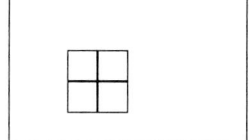

A picture language $L \subseteq \Gamma^{*,*}$ is called local if there is a Δ with $L = \{p \in \Gamma^{*,*} | T_{2,2}(\hat{p}) \subset \Delta\}$. A picture language $L \subseteq \Sigma^{*,*}$ is called recognizable if there is a mapping $\pi : \Gamma \to \Sigma$ and a local language $L' \subset \Gamma^{*,*}$ with $L = \pi(L')$.

A necessary condition for recognizability is the following:

Lemma 1. *[Mat98] Let $L \subseteq \Gamma^{*,*}$ be recognizable and $(M_n \subseteq \Gamma^{n,*} \times \Gamma^{n,*})$ be pairs with $\forall n, \forall (l, r) \in M_n \; lr \in L$ and $\forall (l, r) \neq (l', r') \in M_n \; lr' \notin L$ or $l'r \notin L$, then $|M_n| \in 2^{O(n)}$.*

$$lr = \begin{array}{|c|c|} \hline l & r \\ \hline \end{array} \Big\} n$$

Considering pictures where the width is in $2^{\omega(n)}$ for the height n, we can find $2^{\omega(n)}$ pairs (l, r) such that the number of a's in lr is equal to the number of b's in lr but all the l's have a different differences of numbers of a's and b's. By contradiction we get the following:

Corollary 1. *The language of pictures over $\{a,b\}$, where the number of a's is equal to the number of b's (and where the size (n,m) is only restricted to $f^{-1}(n) \leq m \leq f(n)$ for a function $f \in 2^{\omega(n)}$) is not recognizable.*

In Section 4 we will see that we can not make such a conclusion, if we restrict the width of the pictures to be at most exponential to the height (and vice versa), by showing nonuniform recognizability in this case.

2 Connectedness in a Planar Grid

An interesting question for picture recognition is, whether an object is 'in one piece', which means the subgraph of the grid having a special color respectively letter is connected.

Theorem 1. *The language of pictures over $\{a,b\}$, where all occurring b's are connected is recognizable.*

Proof. Since the recognizable languages are closed under concatenation according to [GRST94], it suffices to show that the language of pictures p_r over $\{a,b\}$, where all occurring b's are connected and one of them touches the left side is recognizable. Then we can concatenate with the language of pictures p_l consisting only of a's. The result is the language of pictures $p_l p_r$ of the theorem.

The idea is that the b's are connected if and only if they can be organized as a tree being rooted at the lowest b on the left side. (All a's under this b are distinguished from the other a's.) Hereby π and Δ are constructed in a way such that every g in Γ with $\pi(g) = b$ encodes the direction to the parent node.

To achieve this Δ has to contain tiles having for example the form

but contain **no** tiles of the form

which would build a cycle or not connect to parent nodes. But the picture to the right side shows that it is not so easy: A problem of this naive approach is that cycles could exist independently from the root.

To solve this problem we additionally encode *tentacles* into each cell, where they can occur at the lower and the right side and must occur at the lower right corner.

(This means we interpret one cell as a two-dimensional structure of four cells like embedding a grid into a grid with double resolution.) These tentacles also have to build trees, which can have their roots at any $\#$ or a_r. Furthermore we do not allow a tentacle crossing a connection of the tree of b's. Each lower right side of a cell must be a tentacle part, which needs a way to a $\#$ or a_r, therefore the b's can not have a cycle around such a spot and thus no cycle at all. Analogously we have to avoid cycles in a tentacle tree. Therefore we also organize the a's in trees which are rooted to the tree of b's. This means the tree of b's and a's and the tentacle trees completely intrude the spaces between the other tree and hereby avoid any cycle.

The alphabet Γ contains for example

, 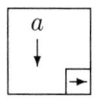, 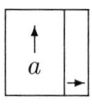, , , , ,...

The first kind allows 2 possible parent directions for the a (the other 2 direction would cross the tentacle) and 4 possible parent directions for the tentacle, which lead to 8 possible combinations; the second kind allows 4 possible parent directions for the a and 2 possible parent directions for the tentacle, which again leads to 8 possible combinations; the third and fourth kind allow 3 possible parent directions for the a and 3 possible parent directions for the tentacle, which leads to 9 possible combinations. This means 34 elements for a and the same number for b thus together with a_r we have $|\Gamma| = 69$.

Our tiling Δ allows neighboring cells if tentacles have a parent direction pointing to a tentacle, # or a_r, if furthermore b's have a parent direction pointing to a b or downward to an a_r and a's have a parent direction pointing to an a or a b like for example (only half of a tile is shown):

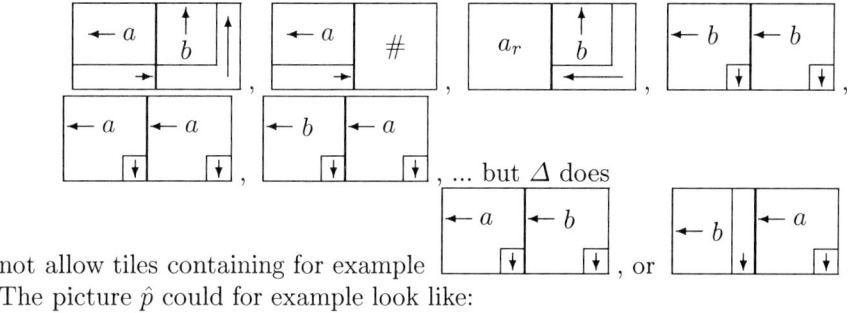

not allow tiles containing for example , or .
The picture $\hat{p}$ could for example look like:

3 The Recognition of a Picture as a Deterministic Process

Recognizing a given picture p can be viewed as a process finding a picture p' over Γ in the local language with $\pi(p') = p$. One major feature for recognizable languages is that this process is nondeterministic.

For practical applications however, we would like to have an appropriate deterministic process starting with a given picture p over Σ and ending with the local p' over Γ. The intermediate configurations are pictures over $\Sigma \cup \Gamma$. One step is a replacement of an $s \in \Sigma$ by a $g \in \Gamma$ with $s = \pi(g)$, which can be performed only if it is locally the only possible choice. This means formally:

Definition 2. *Let $\Sigma \cap \Gamma = \emptyset$, $\pi : \Gamma \to \Sigma$ and $\Delta \subseteq (\Gamma \cup \{\#\})^{1,2} \cup (\Gamma \cup \{\#\})^{2,1}$, which means we consider two kinds of tiles (We conjecture that using 2×2-tiles would make non recognizable picture languages deterministically recognizable):*

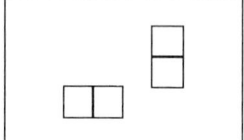

Extend Δ to $\Delta' = \Delta \cup \{\begin{array}{|c|}\hline s\\\hline r\\\hline\end{array}, \begin{array}{|c|}\hline s\\\hline f\\\hline\end{array}, \begin{array}{|c|}\hline g\\\hline r\\\hline\end{array}, \begin{array}{|c|c|}\hline q & o\\\hline\end{array}, \begin{array}{|c|c|}\hline q & d\\\hline\end{array}, \begin{array}{|c|c|}\hline e & o\\\hline\end{array}, \begin{array}{|c|}\hline g\\\hline f\\\hline\end{array}, \begin{array}{|c|c|}\hline e & d\\\hline\end{array} \in \Delta, s = \pi(g), r = \pi(f), q = \pi(e), o = \pi(d),\}$ *by also allowing the image symbols in the tiling.*

For two intermediate configurations $p, p' \in (\Sigma \cup \Gamma \cup \{\#\})^{m,n}$, $n, m > 0$ we allow a replacement step $p \underset{(\Delta,\pi)}{\Rightarrow} p'$ if for all $i \leq m, j \leq n$ we have either $p(i,j) = p'(i,j)$ or $p(i,j) = \pi(p'(i,j))$ and all of the 4 tiles containing this $p'(i,j)$ namely $\begin{array}{|c|}\hline p(i,j\text{-}1)\\\hline p'(i,j)\\\hline\end{array}$, $\begin{array}{|c|}\hline p'(i,j)\\\hline p(i,j+1)\\\hline\end{array}$, $\begin{array}{|c|c|}\hline p(i\text{-}1,j) & p'(i,j)\\\hline\end{array}$ *and* $\begin{array}{|c|c|}\hline p'(i,j) & p(i+1,j)\\\hline\end{array}$ *are in Δ' and if the choice of $p'(i,j)$ was 'forced', that means there is no other $g \neq p'(i,j)$ in Γ with $p(i,j) = \pi(g)$ such that replacing $p(i,j)$ in p by g would result in each of the 4 tiles containing this g is in Δ'. If the choice of $p'(i,j)$ was forced even if 3 of the neighbors where in Σ (or regarded as their image of π), then the replacement step $p \underset{m(\Delta,\pi)}{\Rightarrow} p'$ is called* monocausal.

The accepted language is $\mathcal{L}_d(\Delta, \pi) := \{p \in \Sigma^{*,*} | \hat{p} \underset{(\Delta,\pi)}{\overset{*}{\Rightarrow}} p' \in (\Gamma \cup \{\#\})^{*,*}\}$.

and analogously $\mathcal{L}_{md}(\Delta, \pi) := \{p \in \Sigma^{*,*} | \hat{p} \underset{m(\Delta,\pi)}{\overset{*}{\Rightarrow}} p' \in (\Gamma \cup \{\#\})^{*,*}\}$. A picture language $L \subseteq \Sigma^{*,*}$ is called deterministically recognizable *if there are Δ, π with $L = \mathcal{L}_d(\Delta, \pi)$ and analogously* monocausal deterministically recognizable *if $L = \mathcal{L}_{md}(\Delta, \pi)$.*

Clearly $\mathcal{L}_{md}(\Delta, \pi) \subseteq \mathcal{L}_d(\Delta, \pi) \subseteq \mathcal{L}(\Delta, \pi)$. Furthermore it is easy to see that $\underset{(\Delta,\pi)}{\Rightarrow}$ is confluent on pictures (and their intermediate configurations), which are in $\mathcal{L}_d(\Delta, \pi)$, if we regard possible replacements as voluntarily. (But even if the generated picture $p' \in \Gamma$ is unambiguous, this does not mean that that a deterministically recognizable language is unambiguously recognizable in the sense of [GR96], since the simulation of order of replacements might be ambiguous.) This gives us a simple algorithm to simulate the process by adding those neighbors of a cell, which has just been replaced, to a queue if they are now forced to be replaced and not already in the queue.

Corollary 2. *Deterministically recognizable picture languages can be accepted in linear time.*

As an exercise for the following Theorem 2 we show:

Lemma 2. *The language of pictures over $\{a,b\}$, where all occurring b's are connected to the bottom line is monocausal deterministically recognizable.*

Proof. The language is $\mathcal{L}_d(\Delta, \pi)$ for $\pi(x_i) = x$ and $\Delta = \{$

[picture rules diagram] $|i \in \{c,u\}\}$. Clearly a's can only be replaced by a_c. The b's could possibly be b_c or b_u. In the first step only the b's at the bottom line can be replaced by b_c since b_u can not occur there. Then in the following steps b's, which are neighbors of an b_c can be replaced by b_c since a b_u can not occur beside a b_c. In this way all connected b's are replaced by b_c.

Theorem 2. *The language of pictures over $\{a,b\}$, where all occurring b's are connected is (monocausal) deterministically recognizable.*

Proof. The language is $\mathcal{L}_d(\Delta, \pi)$ for $\pi(x_i) = x$ and $\Delta = \{$

[picture rules diagram] $|i \in \{l,c,r\}\}$.

The deterministic process starts in the lower left corner. If there is an a then a_{sr} is the only possible choice here since a_l can not occur over $\#$ and a_c and a_r can not occur right of $\#$. Then the right neighbor can only be a_{sr} since no other a_i can be right of an a_{sr}. This continues along the bottom line. Then the process proceeds on the right lower corner. If there is an a then a_{sl} is the only possible choice here since a_r can not occur over a_{sr} and a_c and a_l can not occur left of #. Analogously the second line becomes a_{sl}. This continues in snakelike manner producing a_{sr} on the way right and a_{sl} on the way left until the first b is found, which is then forced to become a b_c since neither b_l nor b_r can be left of a_{sl} or right of a_{sr}. Then all connected b's must become b_c and all remaining a's become a_l, a_r or a_c depending on their position.

#	#	#	#	#	#	#	#	#
#	a_l	a_c	a_c	b_c	b_c	b_c	a_r	#
#	b_c	b_c	b_c	b_c	a_c	b_c	a_r	#
#	a_l	b_c	a_c	b_c	b_c	b_c	a_r	#
#	a_l	a_c	a_c	b_c	a_c	a_c	a_r	#
#	a_{sr}	a_{sr}	b_c	b_c	a_c	a_c	a_r	#
#	a_{sl}	a_{sl}	a_{sl}	a_{sl}	a_{sl}	a_{sl}	a_{sl}	#
#	a_{sr}	a_{sr}	a_{sr}	a_{sr}	a_{sr}	a_{sr}	a_{sr}	#
#	#	#	#	#	#	#	#	#

#	#	#	#	#	#	#	#	#
#	a_l	a_c	a_c	b_c	a_c	b	a_r	#
#	b	a_c	b_c	b_c	a_c	b	a_r	#
#	a_l	b_c	a_c	b_c	a_c	b	a_r	#
#	a_l	a_c	a_c	b_c	a_c	a_c	a_r	#
#	a_{sr}	a_{sr}	b_c	b_c	a_c	a_c	a_r	#
#	a_{sl}	a_{sl}	a_{sl}	a_{sl}	a_{sl}	a_{sl}	a_{sl}	#
#	a_{sr}	a_{sr}	a_{sr}	a_{sr}	a_{sr}	a_{sr}	a_{sr}	#
#	#	#	#	#	#	#	#	#

But if the b's are not connected, then some of them can not be determined to b_c, b_l or b_r and the process stops, as shown in the right picture. Note that the tiling is not monocausal since an a left of a b_c can only become a a_c if the a under that a became an a_c or a_{sr} (and not an a_{sl}); but it could be made monocausal by introducing two more b_i-symbols for the first b.

The fact that $\mathcal{L}_{md}(\Delta,\pi)$ and $\mathcal{L}(\Delta,\pi)$ might be different makes the following non trivial:

Theorem 3. *Every monocausal deterministically recognizable language is recognizable.*

Proof. (Sketch) The idea is a generalization of the tentacle method in the proof of Theorem 1. The tree which was used there corresponds to the order of the replacements in Theorem 2: A b was replaced by b_c if the 'parent' b had been replaced by b_c before. Every cell contains encoded *tentacles* like in Theorem 2 and additional the images of the 4 neighbors (which is checked by the tiling) and one third pointer. These third pointers use the same ways (but not the same direction) as the causal pointers and connect all cells to a forest rooted to #'s, which together with the tentacle forest guarantees the cycle freeness. The tilings simulate the monocausal replacement.

Conjecture Every deterministically recognizable language is recognizable.

What changes in the general case is that several (up to 4) neighbors together can force one cell to be replaced, which means instead of a tree we need a planar directed acyclic graph to simulate the order of replacements in the deterministic process.

Open problem Are the deterministically recognizable languages closed under complement?

4 Nonuniform Counting

Nonuniformity is a widely used principle in theoretical computer science. It says that we do not need one algorithm, Turing machine, grammar or whatever to recognize a language but we may use a hole family of them, where each is used only for words of one special size. Connections of nonuniformity and counting can for example be found in [RA97] and [AR98]. A common characterization of nonuniformity is by advice strings. One major observation is that most lower bounds of problems or statements saying that a problem does not belong to a certain class also hold for the nonuniform version of the measure or class. This also holds for Lemma 1. It is easy to see that, as long as we keep the size of the alphabet constant, the lemma does not make use of uniformity. It is an open problem, whether the language of pictures over $\{a,b\}$, where the number of a's is equal to the number of b's and having a size (n,m) with $\log n \leq m \leq 2^n$ is recognizable. The following result shows that it is not possible showing its non recognizability using Lemma 1.

Definition 3. *For 2 pictures $p \in \Sigma^{*,*}$ and $q \in \Gamma^{*,*}$ of size (m,n) the product $p \times q \in (\Sigma \times \Gamma)^{*,*}$ is defined by $(p \times q)(i,j) = (p(i,j), q(i,j))$.*

A picture language $L \subseteq \Sigma^{*,*}$ is called nonuniformly local *if there is an infinite 2-dimensional array of advice-pictures $(a_{m,n} \in \Gamma^{*,*})$ and a local picture language $L' \subseteq (\Sigma \times \Gamma)^{*,*}$ with*
$$p \in L \Leftrightarrow p \times a_{m,n} \in L'$$
for every picture p of size (m,n).

Theorem 4. *The language of pictures over $\{a,b\}$, where the number of a's is equal to the number of b's and having a size (n,m) with $\log n \leq m \leq 2^n$ is nonuniformly recognizable.*

For the proof we need the following closure property:

Definition 4. *A* horizontal (or vertical) folding *is a function $f : \Sigma^{*,*} \mapsto (\Sigma^2)^{*,*}$, which maps a picture $p \in \Sigma^{*,*}$ of size $(m, 2n)$ (or $(2m,n)$) to $f(p)$ of size (m,n) with $(f(p))(i,j) = (p(i,j), p(i, 2n-1-j))$ (or $(f(p))(i,j) = (p(i,j), p(2m-1-i,j))$). For a picture language L we define the folding $f(L) = \{f(p) | p \in L\}$.*

Lemma 3. *The (nonuniformly) recognizable languages are closed under folding.*

Proof. (Sketch of Theorem 4) Because of the last lemma it suffices to restrict to those cases, where we only have to count the difference of a's and b's in the upper left quadrant (we may for example assume the rest to be filled with alternating stripes of a's and b's); by a finite number of foldings and projecting to the upper left quadrant we get the original language. Furthermore w.l.o.g. we assume the width to be greater than the height.

The essential idea of the proof is that the counter is constructed from small constant size counters which have different order. The orders are powers of 2. The number of occurrences of a counter with order 2^{2i} is exponentially decreasing with i similar to the counter used in [Für82]. Since the order can not be known on the local level (the alphabet is finite but not the order), the advice is needed to tell, when counters can be combined.

The constant size counters of a column represent a counter state, which holds the difference of a's and b's left of this column. We simulate a process moving the counter from left to right performing an increment for each a and a decrement for each b. Let every cell c have 3 counters c_1, c_2, c_3, with $-2 \leq c_i \leq 2$. The order of the first counter is always 1, the order of the second counter depends only on the row:

- In the upper half the order is 2^{2i} in all rows $2^{i-1} + j2^i$ for every i, j.
- In the lower half the order is 2^{2i} in the row i for every i.

The order of the third counter is 2^{2i-1} if column + n - row= $2^{i-1} + j2^i$, where n is the height of the picture. A cell c has an advice $c_a \in \{0, s, g, 2, 4\}$ and an effect $c_e \in \{-1, +1\}$ which is $c_e = 1$ if $\pi(c) = a$ and $c_e = -1$ if $\pi(c) = b$, this means an a increments the counter and a b decrements the counter. An example for the order of the counters is the following:

1	4	2	1	4	512	1	4	2
1	16	8	1	16	2	1	16	512
1	4	2	1	4	8	1	4	2
1	64	32	1	64	2	1	64	8
1	4	2	1	4	32	1	4	2
1	16	8	1	16	2	1	16	32
1	4	2	1	4	8	1	4	2
1	256	128	1	256	2	1	256	8
1	4	2	1	4	128	1	4	2
1	16	8	1	16	2	1	16	128
1	64	2	1	64	8	1	64	2
1	256	32	1	256	2	1	256	8
1	1024	2	1	1024	32	1	1024	2
1	4096	8	1	4096	2	1	4096	32
1	16384	2	1	16384	8	1	16384	2
1	65536	512	1	65536	2	1	65536	8

with the advice

4	0	4
g	2	0
4	s	4
g	2	0
4	0	4
g	2	s
4	s	4
g	2	0
4	0	4
g	0	0
0	0	0
0	0	0
0	0	0
0	0	0
0	0	0
0	0	0

We allow tiles of the form $\begin{array}{|c|c|}\hline c & d \\\hline e & f \\\hline\end{array}$ in the following 5 cases:

- $f_a = 0$, $e_1 + f_e = f_1$, $e_2 = f_2$, $c_3 = f_3$, which means the first counter has to take the effect, the second counter is just moved to the next column and the third counter is moved to the next column simultaneously changing the row
- $f_a = s$, $e_1 + f_e = f_1$, $e_2 + 2c_3 = f_2 + 2f_3$, which means additionally the second counter has the double order of the third counter, which allows transfer between them
- $f_a = g$, $e_1 + f_e = f_1$, $2e_2 + c_3 = 2f_2 + f_3$, which means the second counter has the half order of the third counter

- $f_a = 2$, $e_1 + f_e + 2c_3 = f_1 + 2f_3$, $e_2 = f_2$, which means the first counter has the half order of the third counter
- $f_a = 4$, $e_1 + f_e + 2c_3 + 4e_2 = f_1 + 2f_3 + 4f_2$, which means the first counter has the half order of the third counter and the forth order of the second counter.

A counter of a certain order will meet two times a counter of half order and take their load until it meets a counter of double order, where it can get rid of its load. If it has for example a 1, then it can nondeterministically decide to keep it or to get -1 and increment the counter with double order by 1. Third counters are only allowed leaving the picture at the bottom if they are zero. A picture is in the language iff the tiling system can simulate a process of a counter starting at the leftmost column with zero and ending at the rightmost column with zero.

Acknowledgment: We thank V. Diekert, H. Fernau, K.-J.Lange, P. McKencie, O. Matz, W. Thomas and T. Wilke for helpful discussions.

References

[AR98]   E. Allender and K. Reinhardt. Isolation matching and counting. to appear in Proc. of 13th Computational Complexity, 1998.

[FSV95]   Ronald Fagin, Larry J. Stockmeyer, and Moshe Y. Vardi. On monadic NP vs. monadic co-NP. *Information and Computation*, 120(1):78–92, July 1995.

[Für82]   Martin Fürer. The tight deterministic time hierarchy. In *Proceedings of the Fourteenth Annual ACM Symposium on Theory of Computing*, pages 8–16, San Francisco, California, 5–7 May 1982.

[GR96]   D. Giammarresi and A. Restivo. Two-dimensional languages. In G. Rozenberg and A. Salomaa, editors, *Handbook of Formal Language Theory*, volume III. Springer-Verlag, New York, 1996.

[GRST94]   Dora Giammarresi, Antonio Restivo, Sebastian Seibert, and Wolfgang Thomas. Monadic second-order logic over pictures and recognizability by tiling systems. In P. Enjalbert, E.W. Mayr, and K.W. Wagner, editors, *Proceedings of the 11th Annual Symposium on Theoretical Aspects of Computer Science, STACS 94 (Caen, France, February 1994)*, LNCS 775, pages 365–375, Berlin-Heidelberg-New York-London-Paris-Tokyo-Hong Kong-Barcelona-Budapest, 1994. Springer-Verlag.

[IN77]   K. Inoue and A. Nakamura. Some properties of two-dimensional on-line tessellation acceptors. *Information Sciences*, 13:95–121, 1977.

[Mat97]   Oliver Matz. Regular expressions and context-free grammars for picture languages. In *14th Annual Symposium on Theoretical Aspects of Computer Science*, volume 1200 of *lncs*, pages 283–294, Lübeck, Germany, 27 February– March 1 1997. Springer.

[Mat98]   Oliver Matz. On piecewise testable, starfree, and recognizable picture languages. In Maurice Nivat, editor, *Foundations of Software Science and Computation Structures*, volume 1378 of *Lecture Notes in Computer Science*, pages 203–210. Springer, 1998.

[RA97]   K. Reinhardt and E. Allender. Making nondeterminism unambiguous. In *38 th IEEE Symposium on Foundations of Computer Science (FOCS)*, pages 244–253, 1997.

[Wil97]   Thomas Wilke. Star-free picture expressions are strictly weaker than first-order logic. In Pierpaolo Degano, Roberto Gorrieri, and Alberto Marchetti-Spaccamela, editors, *Automata, Languages and Programming*, volume 1256 of *Lect. Notes Comput. Sci.*, pages 347–357, Bologna, Italy, 1997. Springer.

On the Complexity of Wavelength Converters*

Vincenzo Auletta[1], Ioannis Caragiannis[2], Christos Kaklamanis[2], and Pino Persiano[1]

[1] Dipartimento di Informatica ed Appl.
Universitá di Salerno, 84081 Baronissi, Italy
{auletta,giuper}@dia.unisa.it
[2] Computer Technology Institute
Dept. of Computer Engineering and Informatics
University of Patras, 26500 Rio, Greece
caragian@ceid.upatras.gr, kakl@cti.gr.

Abstract. In this paper we present a greedy wavelength routing algorithm that allocates a total bandwidth of $w(l)$ wavelengths to any set of requests of load l (where load is defined as the maximum number of requests that go through any directed fiber link) and we give sufficient conditions for correct operation of the algorithm when applied to binary tree networks. We exploit properties of Ramanujan graphs to show that (for the case of binary tree networks) our algorithm increases the bandwidth utilized compared to the algorithm presented in [3]. Furthermore, we use another class of graphs called dispersers, to implement wavelength converters of asymptotically optimal complexity with respect to their size (the number of all possible conversions). We prove that their use leads to optimal and nearly–optimal bandwidth allocation even in a greedy manner.

1 Introduction

Optical fiber is rapidly becoming the standard transmission medium for networks. Networks using optical transmission and maintaining optical data paths through the nodes are called all–optical networks. Wavelength division multiplexing (WDM) technology establishes connectivity by finding transmitter–receiver paths and assigning a wavelength to each path such that no two paths going through the same link use the same wavelength. Optical bandwidth is the number of available wavelengths.

Current techniques for optical bandwidth allocation cannot guarantee high bandwidth utilization under the worst conditions. A promising solution for efficient use of bandwidth is wavelength conversion. Devices called wavelength converters are located at the nodes of the network and they can change the wavelength assigned to a transmitter–receiver path up to a node and allocate a different wavelength at the rest of the path.

---

* This work has been partially supported by Progetto MURST 40% Algoritmi, Modelli di Calcolo e Strutture Informative, EU Esprit Project 20244 ALCOM–IT, and Project 3943 of the Greek General Secretariat of Research and Technology.

Related Work. Several authors have already addressed the case of no wavelength conversion in tree networks. Raghavan and Upfal [10] showed that routing requests of maximum load l per link of undirected trees can be satisfied using $3l/2$ optical wavelengths and their arguments extend to give a $2l$ bound for the directed case. Mihail et al. [8] address the directed case. Their main result is a $15l/8$ upper bound which was improved to $7l/4$ in [4] and independently in [6]. Kaklamanis et al. [5] present a greedy algorithm that routes a set of requests of maximum load l using at most $5l/3$ and prove that no greedy algorithm can go below $5l/3$ in general.

Models for wavelength routing with converters in trees have been studied in [1,2,3]. The authors present in [1] how to obtain optimal routing in binary tree networks that support l wavelengths using converters of degree $2\sqrt{l}-1$. The model used is actually the one addressed throughout the current paper. A model with many wavelength converters in each node of the network is studied in [2]. In that work it is shown how to obtain nearly optimal and optimal bandwidth allocation using converters of constant degree. This result refers to binary trees as well. A wavelength routing algorithm of any pattern of requests of load l in arbitrary tree networks with $3l/2 + o(l)$ wavelengths using converters of polylogarithmic degree is also presented in [2]. Gargano in [3] presents an algorithm that guarantees efficient wavelength routing in arbitrary trees under a different network model of limited wavelength conversion. She also extends the optimal result of [2] in quasi–binary trees.

Network Model. We model the underlying fiber network as a directed graph. Connectivity requests are ordered pairs of nodes, to be thought of as transmitter–receiver paths. For networks with unique transmitter–receiver paths (such as trees), the load l of a directed fiber link is the number of paths going through the link. Each directed fiber link can support $w(l)$ wavelengths $\omega_1, \omega_2, ..., \omega_{w(l)}$, with distinct optical frequencies.

Current approaches to the wavelength assignment problem in trees use greedy algorithms [4,5,6,8]. Intuitively we can think of wavelengths as colors and the procedure of wavelength assignment as coloring. A greedy algorithm visits the network in a top to bottom manner and at each vertex v colors all requests that touch vertex v and are still uncolored. Moreover, once a request has been colored it is never recolored again. Although greedy algorithms are important as they are very simple and amenable of being implemented in a distributed setting, they cannot guarantee a bandwidth utilization higher than 60% [5]. Furthermore, no algorithm can guarantee bandwidth utilization better than 80% [6] if wavelength conversion is not supported.

A wavelength converter is represented by a bipartite graph $G(U, V, E)$. For each wavelength ω_i, there exist two vertices $u_i \in U$ and $v_i \in V$ in the bipartite graph ($|U| = |V| = w(l)$). The set of edges E is defined as follows: $(u_i, v_j) \in E \Leftrightarrow$ *the wavelength ω_i can be converted to the wavelength ω_j.*

Previous work on wavelength conversion consider that the cost of the converters depends on their *wavelength degree*, i.e. the maximum degree of any node $u \in U$ of the corresponding bipartite graph. Another factor that is expected to influence the cost of a converter is its *size*, i.e. the number of edges in its bipartite graph (the number of all possible conversions).

We can define the class of greedy algorithms for networks that support wavelength conversion. Such algorithms visit the network in a DFS manner but the functionality in a node u that supports wavelength conversion is different. In this case, the greedy algorithm colors the segments of a request that touch u. Thus, a greedy algorithm may assign a different color for a request that has been colored in a previous step, so that the wavelength corresponding to the old color can be converted to the new one by the wavelength converter located at node u which is responsible for the conversion of that request.

Converters are placed at network nodes as follows. The tree network is rooted at a predefined node. At each non–leaf node u of degree $d+1, d > 1$ with a parent f and children $v_1, ..., v_d$, there are $2d$ converters $C_1, C_2, ..., C_{2d-1}, C_{2d}$. Converter $C_{2i-1}, 1 \leq i \leq d$ is responsible for the conversion of the wavelengths assigned to the set of requests R_{2i-1}, which comes from the parent node f and goes to the child v_i. Converter $C_{2i}, 1 \leq i \leq d$ is responsible for the conversion of the wavelengths assigned to the set of requests R_{2i}, which comes from the child v_i and goes to the parent node f.

Summary of results. We start with some preliminary definitions and lemmas in section 2. In section 3 we present a greedy wavelength routing algorithm for binary tree networks with converters. As a first application, we use Ramanujan graphs as converters and prove that our algorithm utilizes the two thirds of the bandwidth wasted by the algorithm presented in [3] (when a Ramanujan graph of a given degree is used as converter in both cases). Next we exploit properties of dispersers to present upper bounds on the size of the converters in order to greedily achieve optimal and almost optimal ($w(l) = l + o(l)$) bandwidth allocation. These results are presented in section 4. In section 5 we prove that our upper bounds are asymptotically tight for greedy algorithms. Also, we prove that the $2\sqrt{l} - 1$ upper bound on the degree of the converters presented in [1] is asymptotically tight as well.

2 Preliminaries

As in [2,3] we will exploit expansion properties of k–regular bipartite graphs to build wavelength converters. The following lemma states Tanner's inequality which relates the expansion of a graph with the value of its second eigenvalue.

Lemma 1. *Let $G(U, V, E)$ with $|U| = |V| = n$ be a k–regular bipartite graph. For any $X \subseteq U$,*

$$\frac{|N(X)|}{|X|} \geq \frac{k^2}{\lambda^2 + (k^2 - \lambda^2)|X|/n}$$

where λ is the second largest eigenvalue of the adjacency matrix of G in absolute value.

Ramanujan graphs have the property that their second eigenvalue is upper bounded by $2\sqrt{k-1}$. Furthermore, these graphs have been explicitly constructed in [7]. In this work we also exploit properties of another class of bipartite graphs called dispersers.

Definition 1. *[11] A bipartite graph $G = (U, V, E)$ is an (K, ε) disperser if for each subset $A \subseteq U$ of size K there are at least $(1 - \varepsilon)|V|$ vertices of V that are adjacent to A.*

Sipser showed in [11] that such graphs exist. An almost optimal explicit construction of dispersers is reported in [12].

Lemma 2. *[11] There exist a (K, ϵ)-disperser $G(U, V, E)$ with $|U| = N$ and $|V| = M$, such that each node $v \in U$ has degree $\max\{\frac{2M}{K}(1 + \ln 1/\epsilon), \frac{2}{\epsilon}(1 + \ln N/K\}$.*

The properties of dispersers have been used for the construction of asymptotically optimal depth–two superconcentrators.

Definition 2. *A N-superconcentrator is a directed graph with N distinguished vertices called inputs, and N other distinguished vertices called outputs, such that for any $1 \leq k \leq N$, any set X of k inputs and any set Y of k outputs, there exist k vertex–disjoint paths from X to Y. The size of a superconcentrator G is the number of edges in it, and the depth of G is the number of edges in the longest path from an input to an output.*

Lemma 3. *[9] Depth–two N-superconcentrators have size $\Theta\left(N\frac{\log^2 N}{\log\log N}\right)$.*

The construction of [9] produces a depth–two superconcentrator $H(U, V, W, E)$ with $|U| = |W| = N$ and $|V| = 2N$. We slightly extend their arguments to obtain the following lemma.

Lemma 4. *There exist a depth two superconcentrator $G(U, V, W, E)$ of size $O\left(N\frac{\log^2 N}{\log\log N}\right)$ with $U = \{u_i | 1 \leq i \leq N\}$, $V = \{v_i | 1 \leq i \leq N\}$, $W = \{w_i | 1 \leq i \leq N\}$, such that for any $1 \leq i, j \leq N$, $(u_i, v_j) \in E \Leftrightarrow (v_i, w_j) \in E$.*

3 The Wavelength Routing Algorithm

In this section we describe a greedy wavelength routing algorithm that allocates optical bandwidth of $w(l)$ available wavelengths to any set of communication requests of load l on a binary tree network. Four wavelength converters C_1, C_2, C_3, C_4 are placed at each node as described. We denote by S the set of available wavelengths (colors).

Starting from a node, the algorithm computes a DFS numbering of the nodes of the tree. The algorithm proceeds in phases, one per each node u of the tree.

The nodes are considered following their depth–first numbering. The phase associated with node u assumes that we already have a partial proper coloring where the segments of requests (paths) that touch nodes with numbers strictly smaller than u's have been colored and no other segments have been colored.

Consider a phase of the algorithm associated with a node u. Let f be the parent node of u and v,w its children. Let A_1 be the set of colors assigned to the set R_1 of the requests from f to v, A_2 the set of colors assigned to the set R_2 of the requests from v to f, A_3 the set of colors assigned to the set R_3 of requests from f to w, and A_4 the set of colors assigned to the set R_4 of requests from w to f. These colors are used only in segment (f, u). Also let R_5 be the set of requests from v to w and R_6 the set of requests from w to v. We ignore requests than start or end at u since they can be colored easily. The algorithm performs two independent steps:

Step 1: Converters C_1 and C_4 are set in such way that:

- C_1 converts the color assigned to the segment (f, u) of each request $r \in R_1$ to a color that is assigned to the segment (u, v) of r.
- C_4 converts a color that is assigned to the segment (w, u) of each request $r \in R_4$ to the color assigned to the segment (u, f) of r.

Let A'_1 and A'_4 the set of colors assigned to the segments (u, v) and (w, u) of the requests R_1 and R_4, respectively. The algorithm maintains the following invariants:

1. Segments (u, v) of the requests R_1 are assigned different colors ($|A'_1| = |A_1|$) and segments (w, u) of the requests R_4 are assigned different colors ($|A'_4| = |A_4|$).
2. $|A'_1 \cap A'_4| \geq \min\{|R_1|, |R_4|\} - w(l) + l$.

Requests R_6 are assigned colors from $S \setminus (A'_1 \cup A'_4)$.

Step 2: This step is symmetric to step 1. Converters C_2 and C_3 are set and the uncolored segments of requests R_2, R_3, and R_5 are colored in a similar way.

The following lemma gives sufficient conditions for the correctness of our algorithm.

Lemma 5. *Let $H(A, B, E(H))$ be the bipartite graph that corresponds to the wavelength converter ($A = \{a_i | 1 \leq i \leq w(l)\}$ and $B = \{b_i | 1 \leq i \leq w(l)\}$). Consider the three–level graph $G(U, V, W, E(G))$ such that $U = \{u_i | 1 \leq i \leq w(l)\}$, $V = \{v_i | 1 \leq i \leq w(l)\}$, $W = \{w_i | 1 \leq i \leq w(l)\}$, and*

$$E(G) = \{(u_i, v_j) | (a_i, b_j) \in E(H)\} \cup \{(v_j, w_i) | (a_i, b_j) \in E(H)\}.$$

The wavelength routing algorithm correctly assigns $w(l)$ colors to any set of communication requests of load l on a binary tree network with wavelength converters H if for any sets $\Gamma_1 \subseteq U$ and $\Gamma_2 \subseteq W$ of cardinalities $|\Gamma_1|, |\Gamma_2| \leq l$, the following conditions hold:

1. There exist sets $X \subseteq \Gamma_1$ and $Z \subseteq \Gamma_2$ of cardinality $k = |X| = |Z| \geq \min\{|\Gamma_1|,|\Gamma_2|\} - w(l) + l$ such that there exist k vertex disjoint paths from X to Z.
2. Let $Y \subseteq V$ the set of vertices of V that belongs to the k disjoint paths. The set $\Gamma_1 \setminus X$ has a matching of cardinality $|\Gamma_1 \setminus X|$ with vertices of $V \setminus Y$, and the set $\Gamma_2 \setminus Z$ has a matching of cardinality $|\Gamma_2 \setminus Z|$ with vertices of $V \setminus Y$.

Proof. We concentrate on step 1 of the algorithm at a node u. The proof is identical for step 2. Assume that at the current step the algorithm has colored the segments (f, u) of the requests R_1 and R_4. Consider the color assigned to the segment (f, u) of a request $r_1 \in R_1$ as a vertex of U and a color assigned to the segment (u, f) of a request $r_4 \in R_4$ as a vertex of W. The mate vertex of V that is connected with an edge (implied by the two conditions) to a vertex of U is the color assigned to the segment (u, v) of r_1, while the mate vertex of V that is connected with an edge to a vertex of W is the color that will be assigned to the segment (w, u) of r_4.

Both conditions maintain that requests of R_1 are assigned different colors in segment (u, v) (similarly for the requests of R_4 in segment (w, u)). Furthermore, condition 1 guarantees that the invariant 2 of the algorithm is satisfied. Thus the algorithm can assign the remaining colors (corresponding to vertices of V that have no mates to U or V) to requests of R_6. □

The algorithm is correct even if sets Γ_1 and Γ_2 have the same cardinality. The second condition can be eliminated if the bipartite graph H of the wavelength converter has a perfect matching (which always holds). Formally

Lemma 6. *Let H and G be the graphs defined in lemma 5. H has a perfect matching and for any sets $\Gamma_1 \subseteq U$, $\Gamma_2 \subseteq W$ of the same cardinality g, there exist sets $X \subseteq \Gamma_1$ and $Z \subseteq \Gamma_2$ of cardinality $k = |X| = |Z| \geq g - w(l) + l$ such that there exist k vertex disjoint paths from X to Z. Let $Y \subseteq V$ the set of vertices of V that belongs to the k disjoint paths. Then the set $\Gamma_1 \setminus X$ has a matching of maximum cardinality $g - k$ with vertices of $V \setminus Y$, and the set $\Gamma_2 \setminus Z$ has a matching of maximum cardinality $g - k$ with vertices of $V \setminus Y$.*

As a corollary, when the cardinality of R_1 and R_4 is small, no vertex disjoint paths need to be found. In particular,

Corollary 1. *Consider a node u of a binary tree network of $w(l) > l$ available wavelengths and wavelength converters H, and a pattern of requests of maximum load l. If $|R_1| = |R_4| \leq w(l) - l$, then the uncolored segments of R_1, R_4, and R_6 have a proper wavelength assignment with $w(l)$ wavelengths.*

The following lemma gives a condition for the existence of vertex disjoint paths when the cardinality of R_1 and R_4 is large.

Lemma 7. *Let $G(U, V, W, E)$ be a three level graph with $|U| = |W| = w(l)$. If for any sets $\Gamma_1 \subseteq U$ and $\Gamma_2 \subseteq W$ of cardinality k with $w(l) - l < k \leq l$ there exist $k - w(l) + l$ common neighbors in V, then for any sets $A \subseteq U, B \subseteq W$ of cardinality $k \leq l$, there exist subsets $X \subseteq A$ and $Z \subseteq B$ such that there exist $k - w(l) + l$ vertex disjoint paths from X to Z.*

Proof. By Menger's theorem proving that the minimum cut has size at least $k - w(l) + l$. □

4 Upper Bounds

Theorem 1. *Let T be a binary tree network and $w(l)$ be the available number of wavelengths on each link. Using (explicitly constructible) converters of degree k, it is possible to greedily assign wavelengths to any set of requests of load $l \leq \left(1 - \frac{4(k-1)}{3(k-2)^2}\right) w(l)$.*

Proof. We use a k–regular Ramanujan graph H as converter with $w(l)$ wavelengths. We construct the three level graph G. Let $X \subseteq U$, such that $|X| > w(l) - l$. It can be verified that

$$|N(X)| \geq \frac{k^2 |X|}{4(k-1) + (k-2)^2 |X|/w(l)} \geq \frac{l + |X|}{2}$$

where $N(X)$ is the neighborhood of X in V. Thus, for any sets $\Gamma_1 \subseteq U$ and $\Gamma_2 \subseteq W$ of cardinality k with $w(l) - l < k \leq l$, there exist $k - w(l) + l$ common neighbors in V. H has a perfect matching, thus, by lemmas 7 and 6 the conditions of lemma 5 hold. The theorem follows. □

The result of [3] and theorem 1 imply the following.

Theorem 2. *Let $1 < f(l) = o(l)$. There exist converters of size $O(lf(l))$ that allow routing of requests of load l using at most $l + \frac{l}{f(l)}$ wavelengths.*

Next we show better tradeoffs between the unutilized bandwidth and the size of the converters under our network model.

Lemma 8. *Let $f(l) = o(l)$. There exists a three level graph $G(U,V,W,E)$ with $|U| = |V| = |W| = l + \frac{l}{f(l)}$ with size $O\left(l \frac{\log^2 f(l)}{\log \log f(l)}\right)$ such that for any sets $X \subseteq U$ and $Y \subseteq W$ with cardinality k with $\frac{l}{f(l)} < k \leq l$ there exist $l - \frac{l}{f(l)}$ common neighbors.*

Proof. The proof is based on [9]. Let $w(l) = l + \frac{l}{f(l)}$. We build a three level graph $(A = [w(l)], C = [w(l)], B = [w(l)], E)$. Let $C = C_{i_s}$ where C_{i_s} is defined as follows. Let $i_s = \frac{\log \frac{w(l)}{3}}{\log \log f(l)} - 1$, $i_0 = \frac{\log l - \log f(l)}{\log \log f(l)}$, and $C_i = [3 \log^{i+1} f(l)]$, $i = i_0, \ldots i_s$, such that $C_i \subseteq C_{i+1}$, for $i_0 \leq i \leq i_s - 1$. For every i put a $\left(K = \log^i f(l), \epsilon = \frac{1}{3}\right)$-disperser $D_i = (A, C_i, E_i)$ and another (K, ϵ)-disperser between B and C_i. Also, put a copy of the edges between A, C between C, A and C, B and (symmetric) 15–regular ramanujan graphs $H_1 = (A, C, E(H_1))$ and $H_2 = (B, C, E(H_2))$ (in order graphs (A, C) and (C, B) can correspond to identical wavelength converters).

For $\log^i f(l) \leq K \leq \log^{i+1} f(l)$, for every set $X \subseteq A$ of cardinality K, its neighborhood $N(X)$ has size at least $\frac{2|C_i|}{3}$. Similarly for each set $Y \subseteq B$ of cardinality K, $|N(Y)| \geq \frac{2|C_i|}{3}$. Thus the number of the common neighbors of X and Y is at least $\frac{|C_i|}{3} \geq K \geq K - \frac{l}{f(l)}$. This holds for any K with $\frac{l}{f(l)} = \log^{i_0} f(l) \leq K \leq \log^{i_s+1} f\left(\frac{w(l)}{3}\right) = \frac{w(l)}{3}$.

Applying the Tanner's inequality to the ramanujan graph H_1, we have that for any set $X \subseteq A$ of cardinality $K > \frac{w(l)}{3}$, there are $\frac{w(l)+K}{2}$ neighbors in C. The same argument holds for H_2 and thus any two sets $X \subseteq A$ and $Y \subseteq B$ of cardinality $K > \frac{w(l)}{3}$ have at least $K > K - \frac{l}{f(l)}$ common neighbors.

It can be verified that the total size of the converters used in the above construction is $O\left(l \frac{\log^2 f(l)}{\log \log f(l)}\right)$. □

Theorem 3. *Let $f(l) = o(l)$. There exist converters of size $O\left(l \frac{\log^2 f(l)}{\log \log f(l)}\right)$ that allow greedy routing of requests of load l using at most $l + \frac{l}{f(l)}$ wavelengths.*

Proof. We use the converter implied by lemma 8. The theorem follows by lemmas 7, 6, and 5. □

Using as a converter the half part of the depth–two superconcentrator of lemma 4, we obtain the following.

Theorem 4. *There exist converters of size $O\left(l \frac{\log^2 l}{\log \log l}\right)$ that allow greedy routing of requests of load l using at most l wavelengths.*

5 Lower Bounds

In [1] it is shown that if the binary tree network has wavelength converters with degree $2\sqrt{l}-1$, under this model it is possible to route all possible sets of requests of load l with l wavelengths. Next we show that this result is asymptotically tight.

Theorem 5. *Let T be a binary tree network supporting l wavelengths. Then, wavelength converters of degree $\Omega(\sqrt{l})$ are necessary to guarantee that all sets of requests of load l can be routed on T by a greedy deterministic algorithm.*

Proof. Consider a node v of the tree. Assume that there is one request from the parent p of v to the left child u colored with a color c_1, one request from the right child w to the parent colored with c_2 and $l-1$ requests from w to u. Since no conversion is supported for requests from w to u, the color c_1 must be converted to a color that can be converted to c_2. We can create a pattern on a sufficiently large tree T such that a color c_1 must be converted to colors that can be converted to all colors $c_1, ...c_l$. Assume that the converter translates c_1 to k colors. Then there must be a color c_i that can be converted to at least l/k colors. Thus the degree of the converter must be at least $\min\{k, l/k\} = \sqrt{l}$. □

The following theorems state that the upper bounds of theorems 3 and 4 are asymptotically tight as well. The proofs are omitted.

Theorem 6. *Let T be a tree network supporting l wavelengths. Then, wavelength converters of size $\Omega\left(l\frac{\log^2 l}{\log\log l}\right)$ are necessary to guarantee that all sets of requests of load l can be routed on T by a greedy deterministic algorithm.*

Theorem 7. *Let $f(l) = o(l)$ and T be a tree network supporting $w(l) = l + \frac{l}{f(l)}$ wavelengths. Then, wavelength converters of size $\Omega\left(l\frac{\log^2 f(l)}{\log\log f(l)}\right)$ are necessary to guarantee that all sets of requests of load l can be routed on T by a greedy deterministic algorithm.*

References

1. V. Auletta, I. Caragiannis, C. Kaklamanis, P. Persiano, "Bandwidth Allocation Algorithms on Tree–Shaped All–Optical Networks with Wavelength Converters". In *Proc. of SIROCCO '97*, 1997.
2. V. Auletta, I. Caragiannis, C. Kaklamanis, P. Persiano, "Efficient Wavelength Routing in Trees with Low–Degree Converters". In *Proc. of the DIMACS Workshop on Optical Networks*, 1998, to appear.
3. L. Gargano, "Limited Wavelength Conversion in All–Optical Tree Networks". In *Proc. of ICALP '98*, 1998, to appear.
4. C. Kaklamanis, P. Persiano, "Efficient Wavelength Routing on Directed Fiber Trees". In *Proc. of the 4th European Symposium on Algorithms (ESA '96)*, LNCS, Springer Verlag, 1996, pp. 460–470.
5. C. Kaklamanis, P. Persiano, T. Erlebach, K. Jansen, "Constrained Bipartite Edge Coloring with Applications to Wavelength Routing". In *Proc. of ICALP '97*, LNCS 1256, Springer Verlag, 1997, pp. 493–504.
6. V. Kumar, E. Schwabe, "Improved Access to Optical Bandwidth in Trees". In *Proc. of the 8th Annual ACM–SIAM Symposium on Discrete Algorithms*, 1997.
7. A. Lubotsky, R. Philipps, P. Sarnak, "Ramanujan Graphs". *Combinatorica*, vol. 8, pp. 261–278, 1988.
8. M. Mihail, C. Kaklamanis, S. Rao, "Efficient Access to Optical Bandwidth". In *Proc. of the 36th Annual Symposium on Foundations of Computer Science*, pp. 548–557, 1995.
9. J. Radhakrishnan, A. Ta-Shma, "Tight Bounds for Depth-Two Superconcentrators". In *Proc. of the 38th Annual Symposium on Foundations of Computer Science*, 1997.
10. P. Raghavan, E. Upfal, "Efficient Routing in All-Optical Networks". In *Proc. of the 26th Annual ACM Symposium on the Theory of Computing*, 1994, pp. 133–143.
11. M. Sipser, "Expanders, randomness, or time versus space". Journal of Computer and System Sciences, 36:379–383, 1988.
12. A. Ta–Shma, "Almost Optimal Dispersers". In *Proc. of STOC '98*, 1998, to appear.

On Boolean vs. Modular Arithmetic for Circuits and Communication Protocols

Carsten Damm

University of Trier, Department of Computer Science
D-54296 Trier, Germany
damm@uni-trier.de
http://www.informatik.uni-trier.de/~damm/

Abstract. We compare two computational models that appeared in the literature in a Boolean setting and in an analog setting based on modular arithmetic. We prove that in both cases the arithmetic version can to some extend simulate the Boolean version. Although the models are very different, the proofs rely on the same idea based on the Schwartz-Zippel-Theorem.
In the first part we prove that depth d semi-unbounded Boolean circuits can be simulated by depth $2d + O(\log d + \log n)$ semi-unbounded arithmetic circuits, regardless of the size. This is an improvement on a similar construction in [3] that achieves depth $3d + O(\log s + \log n)$, where s is the size of the original circuit. Our construction is simpler and uses fewer random bits. In the second part we prove, that two-party parity communication protocols can approximate nondeterministic communication protocols. A strict simulation of one by the other is impossible as was shown in [2].

1 Introduction

We present two randomized simulations of Boolean computational models by their corresponding "parity" models. The first results of this kind are due to [4,3], who prove $NL/poly \subseteq \oplus L/poly$ and $SAC^1/poly \subseteq \oplus SAC^1/poly$, where the latter are the classes of Boolean functions computable by polynomial size, logarithmic depth semi-unbounded circuits with unbounded OR- and $PARITY$-gates, respectively (definitions to follow).

Our results are inspired by a recent paper by Beimel and Gál [1], who give an improved simulation for $NL/poly \subseteq \oplus L/poly$. The original proof for this (as well as for the result $SAC^1/poly \subseteq \oplus SAC^1/poly$) in [3] was shown on base of the Isolation Lemma of Mulmuley et al. [7]. The Isolation Lemma is a means to isolate with high probability a unique minimal certificate from a non-empty set of certificates by adding random weights. Hence, an odd number of certificates is isolated, which can be used to translate an existential computation into a parity computation. Actually, uniqueness is more than is needed to do this translation. It was shown in [8] that this aspect can be exploited further. The proof of [1] relies on testing of polynomial identities. We apply the technique to semi-unbounded circuits. We obtain an alternative and simpler proof for the containment $SAC^1/poly \subseteq \oplus SAC^1/poly$ first shown in [3]. Further this gives better depth bounds and the depth of our simulation is independent from the size of the circuits.

In the second part we consider a similar question, namely to what extend can parity communication protocols simulate nondeterministic protocols. A simulation as in the circuit case is not possible, since there are functions whose parity communication complexity is exponentially larger than their nondeterministic communication complexity. So the best we can hope for is an approximation result. We show basically along the same lines as in the first part, that parity communication protocols can approximate nondeterministic protocols with only small increase in length.

There is nothing special about considering parity, i.e., arithmetic modulo 2. When working modulo a fixed prime p, everything remains true[1]. The same holds for the case of a composite modulus: e.g., acceptance modulo 6 can simulate acceptance modulo 2 *and* acceptance modulo 3, as has been formally proved for a lot of other computational models. We will not go further into these issues.

Throughout log stands for $\log_2$.

2 Definitions

2.1 Circuits

A *Boolean semi-unbounded circuit* on $X_n = \{x_1, \ldots, x_n, \neg x_1, \ldots, \neg x_n\}$ is a pair $C = (G, \lambda)$, where $G = (V, E)$ is a directed acyclic graph with set of vertices V, set of edges E, and vertex labeling $\lambda : V \to X_n \cup \{\vee, \wedge\}$. The vertices of G are called *the gates* of C. If $(g', g) \in E$ then g' is called an *input to* g. We denote the set of inputs to a specific gate g by $I(g)$. The size of $I(g)$ is the *fan-in* of g. According to their label we distinguish the gates into input gates, $\vee$-gates and $\wedge$-gates. Further we require that input gates have fan-in 0. The other gates have fan-in at least 1 and the fan-in of $\wedge$-gates is at most 2. The size of a circuit is it's number of gates, the depth is the length (number of edges) of the longest path from an input to an output gate (these are the gates that are not input to any gate). Finally we assume the circuit has one distinguished output gate g^*.

In the usual way an input $\alpha \in \{0,1\}^n$ assigns to each gate g of C a Boolean value $val(g, \alpha)$ which is used to define the Boolean function computed at this gate. Formally:

1) If $I(g) = \emptyset$ then $\lambda(g)$ is x_i or $\neg x_i$ for some i. In this case $val(g, \alpha) = \alpha_i$ or $val(g, \alpha) = 1 - \alpha_i$, respectively.
2) If $I(g) \neq \emptyset$ then $val(g, \alpha)$ is the logical OR or the AND of the $val(g', \alpha)$ with $g' \in I(g)$ according to whether $\lambda(g) = \vee$ or $\lambda(g) = \wedge$.

---

[1] An appropriate notion of "modulo p acceptance" is: accept, if the number of certificates is not divisible by p.

The function computed by C is defined by $C(\alpha) = val(g^*, \alpha)$.

Analogously we define *arithmetic semi-unbounded circuits over the field* $GF(2) = (\{0,1\}, \oplus, \cdot)$ by replacing in the definition above $\vee$ by $\oplus$, $\wedge$ by $\cdot$ (multiplication), and logical OR by the sum modulo 2. If no confusion can arise or if statements hold for both models we speak just of *semi-unbounded circuits*.

2.2 Communication Protocols

Let X and Y be disjoint sets. We consider two-party communication games on $X \times Y$: the players share an input $(x, y) \in X \times Y$, where access is restricted to x for one player and to y for the other. They exchange bits subject to a given *communication protocol*. In the deterministic version the communicated bit is determined by the accessible part of the input and the communication history. In a *nondeterministic* protocol (which is basic for our considerations) there may additionally be situations, in which the player in turn may nondeterministically choose the bit to be sent. In either case, at each step the bit to be communicated is chosen from a nonempty set of bits determined by the accessible part of the input and the communication history. The protocol specifies a prefix-free set of complete communication strings (we call them simply *communications*). Hence players can tell from the communication string whether the game is over. A subset of the communications is distinguished as *accepting*. The function $f_P : X \times Y \to \{0,1\}$ computed by P depends on the number $a_P(x,y)$ of accepting computations of P on (x,y): If P is interpreted *nondeterministically*, then $f_P(x,y) = 1$ iff $a_P(x,y) > 0$, if P is interpreted as *parity protocol*, then $f_P(x,y) = 1$ iff $a_P(x,y)$ is odd.

Now we give a formal definition: a communication protocol on $X \times Y$ is a directed full binary tree $P = (V, E)$ with specific edge and node labeling. The edges leaving an inner node are labeled 0 and 1, respectively. The inner nodes are partitioned into X- and Y-nodes. Let Z be any of the types X and Y. Each Z-node v is labeled by some mapping $b_v : Z \to \{\{0\}, \{1\}, \{0,1\}\}$, where we write $b_v(x,y)$ to uniquely denote $b_v(x)$ or $b_v(y)$. Each node v is a state in the communication process and the type of the actual node determines the player in turn. $b_v(x,y)$ is the set of bits from which this player chooses the next message upon seeing her input and the exchanged bits v. Further the leaves of the tree are labeled by 0 or 1.

Adopting this formalism, a communication of P on input $(x,y) \in X \times Y$ is a non-extendible path $p = p_1 p_2 \cdots p_t$ of the tree such that for $1 \leq i \leq t$ holds $p_i \in b_{p_0 p_1 \cdots p_{i-1}}(x,y)$, where we identify nodes with the string of edge labels along the path from the root to the node. Accepting communications are those paths that reach a leaf labeled by 1. The length of a protocol is the maximum over all inputs (x,y) of the maximal lengths of communications on this inputs. Consider a function $f : X \times Y \to \{0,1\}$. The *nondeterministic communication complexity of f* is the minimum length of a nondeterministic communication protocol computing f. The *parity communication complexity* is the minimum length of a parity communication protocol computing f. We denote the nondeterministic and parity communication complexities of f by $c_n(f)$ and $c_\oplus(f)$, respectively.

3 Simulating Boolean Circuits by Arithmetic Circuits

We are going to prove the following result:

Theorem 1 *Any Boolean semi-unbounded circuit C of size s and depth $d = d(n)$ can be simulated by a size $s^{O(1)}$ arithmetic semi-unbounded circuit over $GF(2)$ with depth $2d + \log(d+2) + \log n + 2$.*

To prove this we need some technical preparations.

Observation 2 *Any semi-unbounded circuit can be transformed into strictly alternating form, i.e., into a circuit whose gates are organized in levels, such that each gate has inputs only on the preceding level and the type of gates alternates from one level to another. Further we require, that the first level above the input gates is a $\wedge$-level. Translation of an arbitrary circuit into this form requires doubling of depth and only polynomial size blow-up.*

Let k be a field. We identify the Booleans 0 and 1 with the field elements $0, 1$. Let E be the set of edges in the graph G underlying C. Consider a sequence $z = (z_e)_{e \in E}$ of indeterminates. Each $\alpha \in \{0,1\}^n$ defines polynomials $p_{g,\alpha} \in k[z]$ in the following way:

1) If g is an input gate, then $p_{g,\alpha} = val(g, \alpha) \in k[z]$.
2) If g is an inner gate with inputs $g_1, \ldots, g_r$ then if $\lambda(g) = \wedge$ put $p_{g,\alpha} = \prod_{i=1}^{r} p_{g_i,\alpha} \cdot z_{e_i}$ and if $\lambda(g) = \vee$ put $p_{g,\alpha} = \sum_{i=1}^{r} p_{g_i,\alpha}$, where $e_i = (g_i, g) \in E$ for $i = 1, \ldots, r$.

We denote the polynomials $p_{g^*,\alpha}$ assigned to the output gate by p_α.

Observation 3 *If the depth of C is d, then $\deg p_\alpha < 2^{d+1}$. More accurately, if on each path from an input gate of the circuit to the output gate there are at most d $\wedge$-gates, then $\deg p_\alpha < 2^{d+1}$.*

Lemma 4 $p_\alpha = \mathbf{0}$ *(the zero polynomial) if and only if $C(\alpha) = 0$.*

Proof: Let a *certificate* for $C(\alpha) = 1$ be a subgraph $G' = (V', E') \subseteq G$ such that $g^* \in V'$ and for each gate g in G' holds: 1) $val(g, \alpha) = 1$ and 2) if g is a $\wedge$-gate, then each input g' to g belongs to G', and if g is a $\vee$-gate, then exactly one input g' to g belongs to G'. By doubling of edges a certificate G' for $g(\alpha) = 1$ can be expanded into a tree $T_{G'}$. For each certificate G' for $g(\alpha) = 1$ consider the monomial $M_{G'} = \prod_{e \in E'} z_e^{m(e)}$, where $m(e)$ is the multiplicity of e in $T(G')$. Now it is easy to see, that $p_\alpha = \sum_{G'} M_{G'}$, where the sum runs over all certificates for $C(\alpha) = 1$. Since different certificates lead to different monomials, the monomials cannout cancel out each other, regardless of the field. Clearly, $C(\alpha) = 1$, if and only if there is a certificate for $C(\alpha) = 1$. Hence, $p_\alpha = \mathbf{0}$ if and only if $val(g^*, \alpha) = 0$. □

Lemma 5 *If $|k| \geq 2^{d+2}$ then there exist assignments $w_1, \ldots, w_n : E \to k$ such that for all $\alpha \in \{0,1\}^n$ holds $C(\alpha) = 0$ if and only if $p_\alpha(w_i) = 0$ for all $i = 1, \ldots, n$.*

For the proof we need the following result (see [6]):

Lemma 6 (Schwartz-Zippel Theorem) *Let $p \in k[z_1, \ldots, z_t]$ be a polynomial of total degree D and let $S \subseteq k$ be a finite set. If $r_1, \ldots, r_t$ are chosen independently and uniformly at random from S, then*

$$\Pr[p(r_1, \ldots, r_t) = 0 | p \neq \boldsymbol{0}] \leq \frac{D}{|S|}.$$

Proof of Lemma 5. By Observation 3 and Lemmas 4 and 6 for uniformly and randomly chosen $w : E \to k$ holds $\Pr[p_\alpha(w) = 0] < 1/2$ if $C(\alpha) \neq 0$. Taking n independent random trials $w_1, \ldots, w_n$, we obtain $\Pr[p_\alpha(w_1) = \ldots = p_\alpha(w_n) = 0] < 1/2^n$ if $C(\alpha) \neq 0$. This means for less than a $1/2^n$-fraction of all tuples $(w_1, \ldots, w_n)$ there is joint disagreement. By a standard counting argument there is one tuple $(w_1, \ldots, w_n)$ for which there is joint disagreement on less than a $1/2^n$-fraction of all input assignments α. Since there are only 2^n input assignments, for this choice holds: $C(\alpha) \neq 0 \Rightarrow \exists i : p_\alpha(w_i) \neq 0$. On the other hand, if $C(\alpha) = 0$ then clearly $p_\alpha(w_i) = 0$ for all i. This proves the claim. □

Now we introduce weighted arithmetic circuits similar to the arithmetic branching programs introduced in [1]. A *weighted arithmetic circuit over a field* k is a circuit C together with a weight function $w : E \to k$ (where $G = (V, E)$ is the underlying graph). The gates perform field arithmetics over the weighted inputs.

We restrict consideration to weight functions that assign weight 1 to edges feeding into ∨-gates

Weighted arithmetic circuits define *Boolean functions* in the sense that the computed value is 1 if and only if the value (a field element) propagated to the output gate is different from $0 \in k$. By fixing a weight function w we can to each Boolean circuit C assign an arithmetic circuit (C, w) just by interpreting Boolean gates as corresponding arithmetic gates (∧ for product and ∨ for sum).

More formally: Let $C = ((V, E), \lambda)$ and $w : E \to k$. In the following way an input $\alpha \in \{0,1\}^n \subseteq k$ assigns to each gate g of C a value $val_w(g, \alpha) \in k$:

1) If g is an input gate, then $p_{g,\alpha} = val(g, \alpha) \in \{0,1\} \subseteq k$.
2) If g is an inner gate with inputs $g_1, \ldots, g_r$ then if $\lambda(g) = \wedge$ put $val_w(g, \alpha) = \prod_{i=1}^{r} val_w(g_i, \alpha) \cdot w(e_i)$ and if $\lambda(g) = \vee$ put $val_w(g, \alpha) = \sum_{i=1}^{r} val_w(g_i, \alpha)$, where $e_i = (g_i, g) \in E$ for $i = 1, \ldots, r$.

The Boolean function computed by (C, w) is defined by $C_w(\alpha) = 0$ if $val_w(g^*, \alpha) = 0$ and $C_w(\alpha) = 1$ otherwise.

Observe that for all $w : E \to k$ as above and $\alpha \in \{0,1\}^n$ holds $val_w(g^*, \alpha) = p_\alpha(w)$. Thus by Lemma 5 we obtain immediately:

Corollary 7 Let C be a Boolean semi-unbounded circuit on n variables. If $|k| \geq 2^{d+2}$ then there exist weighted arithmetic circuits $(C, w_1), \ldots, (C, w_n)$ over k such that for $\alpha \in \{0,1\}^n$ holds $C(\alpha) = \bigvee_{i=1}^{n} C_{w_i}(\alpha)$.

□

In the sequel we consider only finite fields k. Throughout let $T = |k|$.

Lemma 8 Let (C, w) be a depth d semi-unbounded weighted arithmetic circuit over k. There is an semi-unbounded weighted arithmetic circuit (C', w') of depth $d + \lceil \log T \rceil + 1$ with $C'_{w'} = \neg C_w$.

Proof: Let $a \in k, a \neq 0$. Since $k^* = k \setminus \{0\}$ is a group of order $T - 1$ under multiplication, we have $a^{T-1} = 1$ (see, e.g., [5]). This exponentiation can be modeled by augmenting a binary tree of product-gates on top of a weighted arithmetic circuit. In the resulting circuit each $\alpha \in \{0,1\}^n$ assigns only values $v = 1$ or $v = 0$ to the output gate. The negation of v is $1 - v$, which can be accomplished by a final $\vee$-gate with appropriate weights at incoming edges. For the weighted arithmetic circuit (C', w') thus obtained holds $C'_{w'} = \neg C_w$. □

Lemma 9 Let C be a Boolean semi-unbounded circuit of depth d on n variables. If $|k| \geq 2^{d+2}$ then there exists a weighted arithmetic circuit (C', w) over k such that for $\alpha \in \{0,1\}^n$ holds $C(\alpha) = C'_w(\alpha)$ and the depth of C' is at most $d + \lceil \log |k| \rceil + \lceil \log n \rceil + 2$.

Proof: Observe that by the corollary $C(\alpha) = \neg(\bigwedge_{i=1}^{n} \neg C_{w_i})$ for appropriate $w_1, \ldots, w_n$. By Lemma 8 the inner negations increase the depth by $\lceil \log |k| \rceil + 1$ and the product increases the depth by $\lceil \log n \rceil$. The resulting circuit is "0-1-valued", hence the final negation needs only one additional gate on top. □

Note that in both constructions the increase in size is a polynomial in the size of the original circuit and the size of the field.

Now we are ready to prove the simulation: small depth Boolean semi-unbounded circuits can be simulated by small depth arithmetic semi-unbounded circuits.

Proof of Theorem 1: First we describe a simulation that achieves depth $3d + 2\lceil \log n \rceil + 4$. We describe a further improvement afterwards.

Translate C into strictly alternating normal form as in Observation 2. This increases the depth to at most $2d$, however the degree of the p_α is still bounded by 2^{d+1} by Observation 3. By Lemma 9 there exists a polynomial size weighted arithmetic circuit C' over the field $k = GF(2^{d+2})$ with depth $3d + \lceil \log n \rceil + 4$ computing the same function. This circuit can be simulated by a semi-unbounded circuit G'' over $GF(2)$, similar to the simulation of arithmetic branching programs over $GF(q^d)$ by those over $GF(q)$ in [1]. The simulation relies on the fact, that each field element of k is an univariate polynomial over $GF(2)$ reduced modulo an irreducible polynomial of degree $d + 2$. Hence, field elements $a \in k$ can naturally be represented as vectors v_a in $(GF(2))^{d+2}$, where addition in k translates to component-wise addition in $GF(2)$, and multiplication in k translates into a sum of products in $GF(2)$ (details omitted). Thus, levels of sum-gates in C' are simulated by a single sum-level in C'' and levels of product gates in C' are simulated by a sum-level followed by a product level. Since the first $2d$

levels of C' are strictly alternating, those can be simulated by $2d$ levels in C''. The tree of product-gates in C' requires $2\log n$ depth in C'', hence altogether we have depth $3d + 2\lceil \log n \rceil + 4$.

The improvement to depth $2d + \log(d+2) + \log n + 2$, as promised, can be achieved by the following strategy (instead of the above): 1) translate into alternating form, 2) simulate each (C, w_i) over $GF(2)$ to 0-1-output, and 3) compute the OR of the results. For step 2) $\log(d+2)$ levels are sufficient (to test whether at least one of $d+3$ components is different from 0). □

Remark 10 *The simulation in [3], achieves depth of order $3d + \log n + \log s + O(1)$, where s is the size of the original circuit. Observe that in our simulation the depth of the simulating circuit does not depend on the size of the original circuit. Further their construction uses at least $\Omega(n(m \log m + d))$ random bits, where m is the number of edges in the Boolean circuit. Our construction uses only $O(n \cdot m' \cdot d)$ random bits, where m' is the number of input edges to $\wedge$-gates of the Boolean circuit, which is much less in case of large ciruit size.*

4 Approximating Nondeterministic Protocols by Modular Protocols

It is known, that nondeterministic and parity communication complexity are incomparable: there are functions whose nondeterministic communication complexity is exponentially larger than it's parity communication complexity and vice versa [2]. However, we prove that the *approximative* parity communication complexity of a function is not much greater than it's nondeterministic communication complexity. To give an exact formulation, we denote for given $\varepsilon > 0$ by $c_\oplus^{1,\varepsilon}(f)$ the minimal length of a parity communication protocol computing f correctly on all inputs with exception of at most an ε-fraction of the inputs in $f^{-1}(1)$, i.e., for the function $g : X \times Y \to \{0, 1\}$ computed by the parity protocol holds $\mathbf{Pr}[g(x,y) = 0|f(x,y) = 0] = 1$ and $\mathbf{Pr}[g(x,y) = 0|f(x,y) = 1] \leq \varepsilon$ if $(x,y) \in X \times Y$ is chosen at random.

Theorem 11 *Let $f : X \times Y \to \{0, 1\}$ and $\varepsilon > 0$. Then*

$$c_\oplus^{1,\varepsilon}(f) = O(c_\mathrm{n}(f) \cdot \log c_\mathrm{n}(f)).$$

For the proof we introduce the following model. Let k be a field. An *arithmetic communication protocol* over k with respect to the input space $X \times Y$ is a communication protocol $P = (V, E)$ on $X \times Y$ together with a weight function $w : E \to k$. The notions "communication", "accepting comunication", and "length" carry over from the base model. The function $f : X \times Y \to \{0, 1\}$ computed by (P, w) is defined as follows: for a communication $p = p_1 p_2 \ldots p_t$ of P let $val_w(p) = \prod_{e \in p} w(e)$, where $e \in p$ means, that e is an edge in the path p. We define $f(x, y) = 1$ iff $\sum_{p \in A_P(x,y)} val_w(p) \neq 0$, where $A_P(x,y)$ denotes the set of accepting communications of P on (x, y).

The *arithmetic communication complexity of f over k* is the minimum length of an arithmetic communication protocol over k that computes f. We denote this complexity by $c_k(f)$. Finally, we denote by $c_k^{1,\varepsilon}(f)$ the minimal length of an arithmetic communication protocol over k that computes a function $g : X \times Y \to \{0, 1\}$ such that $\mathbf{Pr}[g(x, y) = 0 | f(x, y) = 0] = 1$ and $\mathbf{Pr}[g(x, y) = 0 | f(x, y) = 1] \leq \varepsilon$ if $(x, y) \in X \times Y$ is chosen at random.

Lemma 12 *Let $\varepsilon > 0$ and $f : X \times Y \to \{0, 1\}$. If k is a field with $|k| \geq c_n(f)/\varepsilon$, then*
$$c_k^{1,\varepsilon}(f) \leq c_n(f).$$

Proof. Let $P = (V, E)$ be a nondeterministic communication protocol of length $t = c_n(f)$ that computes f. For each edge $e \in E$ consider an indeterminate z_e. Obviously the polynomial $q_{(x,y)}((z_e)_{e \in E}) := \sum_{p \in A_P(x,y)} \prod_{e \in p} z_e$ vanishes if and only if there is no accepting computation of P on (x, y) if and only if $f(x, y) = 0$. Observe that the degree of $q_{(x,y)}$ is t. Let $S \subseteq k$ be a set of size at least $c_n(f)/\varepsilon$. By Lemma 6 we know that, if $q_{(x,y)} \neq 0$, then for randomly chosen $w = (w_e)_{e \in E} \in S^{|E|}$ holds
$$\mathbf{Pr}[q_{(x,y)}(w) = 0] \leq t/|S| \leq \varepsilon.$$

By a standard double counting argument there is at least one vector $w^* \in S^{|E|}$ such that for at most a fraction ε of all inputs $(x, y) \in f^{-1}(1)$ holds $q_{(x,y)}(w^*) = 0$. On the other hand, clearly $q_{(x,y)} = 0$ if $f(x, y) = 0$. Hence, the arithmetic communication protocol (P, w^*) computes f in the above mentioned approximative sense.

Now we consider the special case of finite fields k of characteristic 2.

Lemma 13 *Let d be a positive integer. Then for $f : X \times Y \to \{0, 1\}$ holds*
$$c_\oplus(f) \leq 1 + d \cdot (c_{GF(2^d)}(f) + 1).$$

Proof. Recall that elements in $GF(2^d)$ can be naturally represented as length d binary vectors. We denote the j-th component of this representation of a field element a by $[a]_j$.

Let (P, w) be an arithmetic communication protocol over $GF(2^d)$ that computes f. We consider modified copies $P_1, \ldots, P_d$ of P. The modification concerns the labels of the leaves, while the labeling of the edges and of the inner nodes is preserved in each copy. Leaf p will be labeled 1 in P_j if and only if $[val_w(p)]_j = 1$ (remember the correspondence between communications = maximal paths and leaves). Let f_j denote the function computed by P_j as parity communication protocol. By definition

$$f(x, y) = 0 \text{ iff } \sum_{p \in A_P(x,y)} val_w(p) = 0$$
$$\text{iff } \forall j : f_j(x, y) = 0$$
$$\text{iff } \forall j : \neg f_j(x, y) = 1.$$

Denote by P'_j the parity communication protocol obtained from P_j by "complementation", i.e., by augmenting a new root with one accepting branch and one branch leading to the root of P_j. Clearly P'_j computes $\neg f_j$. From copies of these protocols we construct a new parity communication protocol P' in d stages: stage 1 consists of P'_1. In stage $i < d$ we identify each accepting leaf of the present construction with a copy of P'_{i+1}. It is easy to check, that the number of accepting communications of P' on (x, y) is odd if and only if the number of accepting communications of P'_j on (x, y) is odd for each j. Hence, P' computes $\neg f$. A final complementation step on P' gives a parity communication protocol that computes f as desired.

Proof of Theorem 11: Let $k = GF(2^d)$ with $d = \lceil \log(c_n(f)/\varepsilon) \rceil$. Then by Lemma 12 there is an arithmetic communication protocol over k that computes a function $g : X \times Y \to \{0, 1\}$ that agrees with f on all of $f^{-1}(0)$ and disagrees with f on at most a fraction of ε of $f^{-1}(1)$. Simulating this protocol as in Lemma 13 yields a parity communication protocol of length
$\lceil \log c_n(f) - \log \varepsilon \rceil \cdot (c_n(f) + 1) + 1$. □

References

1. Amos Beimel and Anna Gál. On arithmetic branching programs. *Proc. Conf. on Computational Complexity, 1998* (Preliminary version: Technical Report 97-81, Center for Discrete Mathematics and Theoretical Computer Science 1997)
2. Carsten Damm, Matthias Krause, Christoph Meinel, and Stephan Waack. On Relations Between Counting Communication Complexity Classes. *Journal on Computer System Sciences* (to appear).
3. Anna Gál and Avi Wigderson. Boolean complexity classes vs. their arithmetic analogs. *Random Structures and Algorithms*. John Wiley & Sons, Inc., 1996. *see also:* Electronic Colloquium on Computational Complexity, Report 95-49, http://www.eccc.uni-trier.de/eccc/, 1995.
4. Avi Wigderson. $NL/poly \subseteq \oplus L/poly$. Proc. of the 9th Conference on Structure in Complexity Theory, pp. 59–62, 1994.
5. Rudolf Lidl and Harald Niederreiter. *Introduction to finite fields and their applications*. Cambridge University Press, 1986.
6. Rajeev Motwani and Prabhakar Raghavan. *Randomized Algorithms*. Cambridge University Press, 1995.
7. K. Mulmuley, U. Vazirani, and V. Vazirani. Matching is as easy as matrix inversion. In *Proceedings of the 19th STOC*, pages 345–354, 1987.
8. Klaus Reinhardt and Eric Allender. Making nondeterminism unambigous. Electronic Colloquium on Computational Complexity, Report 97-14, http://www.eccc.uni-trier.de/eccc/, 1997.

Communication Complexity and Lower Bounds on Multilective Computations*
(Extended Abstract)

Juraj Hromkovič

Dept. of Computer Science I, RWTH Aachen
Ahornstr. 55, 52074 Aachen, Germany
jh@i1.informatik.RWTH-Aachen.de

Abstract. Communication complexity of two-party (multiparty) protocols has established itself as a successful method for proving lower bounds on the complexity of concrete problems for numerous computing models. While the relations between communication complexity and oblivious, semilective computations are usually transparent and the main difficulty is reduced to proving nontrivial lower bounds on the communication complexity of given computing problems, the situation essentially changes, if one considers non-oblivious or multilective computations. The known lower bound proofs for such computations are far from being transparent and the crucial ideas of these proofs are often hidden behind some nontrivial combinatorial analysis. The aim of this paper is to create a general framework for the use of two-party communication protocols for lower bound proofs on multilective computations. The result of this creation is not only a transparent presentation of some known lower bounds on the complexity of multilective computations on distinct computing models, but also the derivation of new nontrivial lower bounds on multilective VLSI circuits.

1 Introduction

The communication complexity of two-party protocols has been introduced by Abelson [Ab78] and Yao [Ya79]. The initial goal was to develop a method for proving lower bounds on the complexity of distributed and parallel computations.

Informally let $f : \{0,1\}^n \to \{0,1\}$, $n \in \mathbb{N}$, be a Boolean function over a set X of n Boolean variables, and let $\pi = (X_1, X_2)$ be a partition of X. A **two-party (communication) protocol** D **computing** f **according to** π consists of two computers C_I and C_{II} with unbounded computational power. At the beginning C_I obtains an input $x : X_1 \to \{0,1\}$ and C_{II} obtains an input $y : X_2 \to \{0,1\}$. Then C_I and C_{II} communicate according to the protocol by exchanging binary messages until one of them knows the result $f(x,y)$. The complexity of the protocol computation on the input (x, y) is the sum of the

---

* This work has been supported by the DFG Project HR 14/3-1.

lengths of messages exchanged. The complexity of the protocol D, $cc(D)$, is the maximum of the complexities over all inputs from $z : X \to \{0,1\}$ [1]. The **communication complexity of f according to** π, $cc(f, \pi)$, is the complexity of the best protocol computing f according to π.

There are several ways how to define the communication complexity of a Boolean function f. The choice depends on the application considered. The most common definition used in many applications is to consider the **communication complexity of f**, $cc(f)$, as the minimum of $cc(f, \pi)$ over all "almost balanced"[2] partitions of input variables.

In the almost 20 years of its existence communication complexity has established itself as a method for proving lower bounds on several fundamental complexity measures of sequential and parallel computations. Its success in the applications is comparable with that of Kolmogorov complexity in computability and complexity theory.

The simplest standard application of communication complexity is based on the division of the hardware of the computing model considered (circuit, input tape, etc.) into two parts, in such a way, that each part contains approximately half the inputs. Obviously, such a cut corresponds to an almost balanced partition of the set of input variables. So $cc(f)$ gives a lower bound on the amount of information that must be exchanged between these two parts of hardware. Lower bounds on the size of the hardware or on some tradeoffs between hardware size and time follow. Another standard possibility is to cut time in some discrete moment t in such way that the number of input bits read before t is approximately the same as the number of input values read after t. Again this corresponds to an almost balanced partition of the set of input variables. Then $cc(f)$ is a lower bound on the amount of information transfered between two time units of the computation. To realize this transfer the size of the hardware (memory, circuit) has to be large enough. The standard applications mentioned above are elegant and transparent and the main technical difficulty lies in proving nontrivial lower bounds on $cc(f)$ of a function f of interest. But these cuts with required properties can be found only if the computing model is oblivious[3] and semilective[4]. If the model is 2-multilective (each variable may enter at most twice) the ideas for the above standard applications do not work anymore, because there are no cuts corresponding partitions of the set of input variables as defined above. This should be not surprising because usually multilective computing models are much more powerful than their

---

[1] We consider an input as an assignment of values to the input variables.

[2] Usually almost balanced means that at least one third of the input is assigned to each of the two computers of a protocol.

[3] Obliviousness means that the position (time), where (when) an input enter our computing device is fixed for every input variable, i.e. independent of the values of specific inputs.

[4] Semilectivity means that each variable enters the computing device exactly once, i.e. is read exactly once.

semilective counterparts[5]. So the known lower bounds proofs for multilective computations are far from beeing obvious and transparent (see, for instance, [BRS93,DG93,Gr91,HKMW92,KMW89,Ok93,Sau97,Sa84,Tu89,Ya81]) and the crucial ideas of these proofs are often hidden behind some nontrivial combinatorial analysis.

The main aim of this paper is to create a general framework enabling to present some lower bound proofs on multilectivity as a well-structured transparent method based on two-party communication protocols. Another consequence of our effort are some new applications for proving lower bounds for multilective VLSI circuits and multilective planar Boolean circuits. Moreover, using our approach one can get lower bounds for numerous concrete functions and not only for one (or a few) function as it is usual for lower bound proofs.

The paper is organized according to three steps in which we consecutively present our method for proving lower bounds on multilective computations. Section 2 gives the definition of a so called "overlapping" communication complexity introduced in [Hr97] in a slightly different form. This captures the fact, that for multilective devices we are unable to cut them in such way, that the cut corresponds to a partition of the set of input variables X into two disjoint subsets. But what is possible to find, is a cut corresponding to a partition of X into X_1 and X_2 in such a way that $X_1 - X_2$ and $X_2 - X_1$ are "large enough". Because we have methods to prove nontrivial lower bounds on communication complexity according such "overlapping" partitions this concept has good chances to be applied.

Section 3 is devoted to the problem how to search for a cut of a multilective device (computation). The cuts can not be found so easily as in the semilective case. Usually we need to partition the device (computation) considered into small pieces and then to build the cut by "sticking" some small pieces together. The method explaining how to partition and how to stick together is based on a combinatorial lemma. We present this lemma having broad applications in Section 3 and explain there its relation to overlapping communication complexity introduced in Section 2.

In Section 4 we illustrate the applications of the concept developed in Sections 2 and 3 for proving lower bounds on multilective VLSI circuits and some versions of k-time-only branching programs. [6]

2 Overlapping Communication Complexity

For the reasons explained in the introduction we give the formal definition of overlapping communication complexity here.

Definition 1. *Let $f : \{0,1\}^n \to \{0,1\}$ be a Boolean function defined over a set of variables $X = \{x_1, x_2, \ldots, x_n\}$, $n \in \mathbb{N}$. Let $U_0 \subseteq X$, $V_0 \subseteq$*

---

[5] Consider for instance branching programs, VLSI circuits, space bounded Turing machines, finite automata, etc.
[6] Following our aim we rather prefer to present ideas instead of technical proofs in this extended abstract.

X, $|U_0| = |V_0|$ be two disjoint subsets of X. Let k be a positive integer. A pair $\pi = (\pi_L, \pi_R)$ is called a **(U_0, V_0, k)-overlapping partition of** X, if:
1. $\pi_L \cup \pi_R = X$, and
2. there exist $U \subseteq U_0 \cap \pi_L$ and $V \subseteq V_0 \cap \pi_R$ such that $U \cap \pi_R = V \cap \pi_L = \emptyset$ and $|U| \geq |U_0|/32^k$, $|V| \geq |V_0|/32^k$.

$Par(X, U_0, V_0, k)$ denotes the set of all (U_0, V_0, k)-overlapping partitions of X.

The reason to consider such partitions is the following one. May be, one knows that if C_I knows values of variables from U_0 but no variable from V_0 and C_{II} knows all from V_0 but none of U_0 then the communication complexity must be large. But one is unable[7] to find a cut separating U_0 from V_0. The idea is to find a cut where at least some parts of input variables $U \subseteq U_0$ and $V \subseteq V_0$ are separated. To have a chance to prove the necessity of a long communication the sizes of U and V may not be too small compared with U_0 and V_0 respectively

In what follows we say that a communication has k-rounds if exactly k messages between C_I and C_{II} have been exchanged for any $k \in \mathbb{N}$. A protocol is called k-**rounds**[8] if for every input the communication of the protocol consists of at most k rounds.

Definition 2. *Let k be a positive integer. Let f, X, U_0, V_0 have the same meaning as in Definition 1. For every $\pi \in Par(X, U_0, V_0, k)$ we define the* **overlapping $2k$-rounds communication complexity of f according to π**, $occ_{2k}(f, \pi)$, *as the complexity of the best $2k$-rounds protocol computing f according to π.*

For all disjoint subsets $U_0, V_0 \subseteq X$ we define the **overlapping $2k$-rounds communication complexity of f according to U_0 and V_0** *as*

$$occ_{2k}(f, U_0, V_0) := \min\{occ_{2k}(f, \pi) \mid \pi \in Par(X, U_0, V_0, k)\}. \quad (1)$$

Finally, the **overlapping $2k$-rounds communication complexity of f** *is*

$$occ_{2k}(f) := \max\{occ_{2k}(f, U_0, V_0) \mid U_0 \subseteq X, V_0 \subseteq X,$$
$$|U_0| = |V_0| \geq |X|/8, U_0 \cap V_0 = \emptyset\}.$$

In what follows we also want to apply a new version of overlapping communication complexity. Let $\overline{occ}_{2k}(f, \pi)$ and $\overline{occ}_{2k}(f)$ be defined in the same way as $occ_{2k}(f, \pi)$ and $occ_{2k}(f)$, resp. with the only difference that we give no bounds on the number of rounds (i.e. k is related only to k-overlapping partitions). Obviously $\overline{occ}_{2k}(f, \pi) \leq occ_{2k}(f, \pi)$ for every f and π.

Overlapping $2k$-rounds communication complexity has been introduced in order to be applied for lower bounds on k-multilective computations[9]. We see

---
[7] Usually, such cut even does not exists.
[8] For formal definition and the study of k-rounds protocol see [DGS84].
[9] In a k-multilective computation each variable can be read at most k-times.

that k is strongly related to the size of the input variable subsets U and V (see Definition 1, (2)) separated by a cut. With the growth of the multilectivity the sizes of subsets, one is able to separate, decreases. Why the speed-up of the decrease of $|U|$ is related to $|U_0|/3^{2k}$ we shall see in the next section. The reason to consider $2k$-rounds protocols is that one is able to find such cuts of k-multilective devices (computations) that the information flow crossing this cuts can be described by the exchange of $2k$-binary messages between the two parts given by the cuts.

Before using $occ_{2k}(f)$ and $\overline{occ}_{2k}(f)$ to get lower bounds on k-multilective computations we should mention, that one is able to prove high lower bounds on $\overline{occ}_{2k}(f)$. This seems to be hard because following Definition 2, $\overline{occ}_{2k}(f)$ is the minimum over all $\pi \in Par(X, U_0, V_0, k)$ and over the communication complexities of all protocols computing f according to π. Despite of this we have standard methods (in communication complexity theory) that can be used to prove nontrivial (even linear[10]) lower bounds on the communication complexity of concrete computing problems. On the other hand a detailed, technical presentation of a lower bound proof on $\overline{occ}_{2k}(f)$ for a specific function f would be to long for this extended abstract. Because of this we prefer to explain one of the possible ideas only.

Let f be defined over a set of input variables $X = X_1 \cup X_2 \cup X_3$, where $X_i \cap X_j = \emptyset$ for $i \neq j$ and $|X_1| \geq |X|/4$, $|X_2| \geq |X|/4$, $|X| = n$. Let the values of variables in X_3 determine which pairs $(u, v) \in X_1 \times X_2$ are in some relation (for instance have to have the same value), and so they must be somehow compared. To prove $\overline{occ}_{2k}(f) \geq n/(4 \cdot 3^{2k})$ one may choose $U_0 = X_1$ and $V_0 = X_2$ [11]. Now we have to prove $occ_{2k}(f, \pi) \geq n/(4 \cdot 3^{2k})$ for every $\pi \in Par(X, X_1, X_2, k)$. Let $\pi = (\pi_L, \pi_R)$ be an arbitrary (X_1, X_2, k)-overlapping partition of X. Then, there exist $U \subseteq X_1 \cap \pi_L$ and $V \subseteq X_2 \cap \pi_R$ such that $U \cap \pi_R = V \cap \pi_L = \emptyset$, and $|U|$ and $|V|$ are at least $n/(4 \cdot 3^{2k})$. [12] Now, one can choose the set of input assignments by fixing the values of variables in X_3 in such a way that $n/(4 \cdot 3^{2k})$ different pairs form $U \times V$ have to be compared. The standard methods like the fooling set method and the rank method (see for instance [DHS96,AUY83,Hr97,KN97]) are able to establish $\overline{occ}_{2k}(f, \pi) \geq n/(4 \cdot 3^{2k})$. Since k is a constant independent of n, we have $\overline{occ}_{2k}(f) = \Omega(n)$. [13]

3 A Combinatorial Lemma

In what follows we present a lemma giving a very general concept for searching for cuts of multilective computations. This lemma has been proved in several

---
[10] Note that the communication complexity is at most linear.
[11] Note that $\overline{occ}_{2k}(f)$ is defined as the maximum over the choices of U_0 and V_0.
[12] For the explanation see the next section.
[13] Note that the idea described above means that one can get numerous linear lower bounds on overlapping communication complexity. In fact, for every function f and every balanced partition π one can construct a function F_f such that $\overline{occ}_{2k}(F_f) \geq cc(f, \pi)$.

versions in the literature (see for instance [DG93,Hr97]) and so we omit to present its proof.

Lemma 1. *Let m, n and k be positive integers, $m \leq n/3^{2k}$, $k < \frac{1}{2}\log_2 n$. Let U_0, V_0 be two disjoint subsets of a set X, $|U_0| \geq n$, $|V_0| \geq n$. Let $W = W_0, W_1, \ldots, W_d$ be a sequence of subsets of X with the properties $|W_i| \leq m$ for every $i = 1, \ldots, d$ and for every $x \in X$, x belongs to at most k sets of W. Then there exist $U \subset U_0$ and $V \subset V_0$ and integers $t_0 = -1, t_1, \ldots, t_b$, $b \in \mathbb{N}$, such that the following five conditions hold:*
1. $|U| \geq n/3^{2k}$, $|V| \geq n/3^{2k}$
2. $b \leq 2k$, $t_a \in \{1, \ldots, d\}$ for $a = 1, 2, \ldots, b$ and $t_0 < t_1 < \cdots < t_b$
3. *if* $U \cap (\bigcup_{j=t_i+1}^{t_{i+1}} W_j) \neq \emptyset$ *for some* $i = \{0, \ldots, b-1\}$ *then*
$$V \cap (\bigcup_{j=t_i+1}^{t_{i+1}} W_j) = \emptyset \text{ and}$$
4. *if* $V \cap (\bigcup_{j=t_i+1}^{t_{i+1}} W_j) \neq \emptyset$ *for some* $i = \{0, \ldots, b-1\}$ *then*
$$U \cap (\bigcup_{j=t_i+1}^{t_{i+1}} W_j) = \emptyset,$$
5. $(U \cup V) \cap (\bigcup_{j=t_b+1}^{d} W_j) = \emptyset.$

Now let us explain the relation of Lemma 1 to our lower bound proof concept by describing the interpretation of symbols (objects) appearing in Lemma 1. As before X denotes the set of input variables of a computing problem, that has to be solved in a k-multilective computation of a computing device. The sets U_0 and V_0 have the same meaning as in Definition 1 of an overlapping partition of X. These two subsets of X one may choose arbitrarily[14].

The idea to apply Lemma 1 in the search for a cut of the hardware or of the computation of a k-multilective computing device corresponding to an overlapping partition from $Par(X, U_0, V_0, k)$ is as follows. Partition the hardware (or the computation) into d "very small" pieces (or time intervals), where d may be arbitrarily large. The pieces have to be so small, that the number of variables entering one piece (the number of variables read in one interval) is bounded by $m \leq n/3^{2k}$. Obviously each variable enters (is read in) at most k different pieces (intervals). Then Lemma 1 says that one can stick the pieces corresponding to $W_0, W_1, \ldots, W_d$ together into at most $b+1 \leq 2k+1$ larger pieces[15]

$$\overline{W_i} = \bigcup_{j=t_i+1}^{t_{i+1}} W_j, \quad \overline{W} = \bigcup_{j=t_b+1}^{d} W_j \qquad (2)$$

---

[14] Obviously, the quality of the resulting lower bound essentially depends on the appropriate choice of U_0 and V_0. The fact that one may choose U_0 and V_0 is the consequence of the fact that $occ_{2k}(f)$ is defined as the maximum over all choices of U_0 and V_0.
[15] See (2) of Lemma 1.

for $i = 0, 1, \ldots, b-1$ in such a way that there exist $U \subseteq U_0$ and $V \subseteq V_0$ with the properties $\overline{W_i} \cap U = \emptyset$ or $\overline{W_i} \cap V = \emptyset$ for all $i = 0, 1, \ldots, b-1$ [16] and $(U \cup V) \cap \overline{W} = \emptyset$ [17].

The final product is a cut of the hardware (time) into only two parts L and R, where L is the union of all parts of corresponding W_i's containing no variable from V, and R is the union of the rest (i.e. the complement). The cut (L, R) corresponds to a partition $\pi = (\pi_L, \pi_R) \in Par(X, U_0, V_0, k)$. Thus we have the lower bound striven for, because $occ_{2k}(f, \pi) \geq occ_{2k}(f, U_0, V_0)$ bits must flow via the boundary between the parts L and R.

So Lemma 1 provides a general strategy for proving lower bounds on multilective computations by communication complexity. But there are a few free parameters in this strategy and these parameters decide about the success of this method. The first free parameter is the choice of U_0 and V_0. A deep analysis of the inner structure of the computing problem is necessary to find the best possibility. The second free parameter is the manner in which the hardware (time) is partitioned into small pieces correspoding to $W_0, W_1, \ldots, W_d$. The partition should be done in such a way that after sticking the pieces together [18] the resulting border between L and R is as small as possible. [19]

4 Applications

We start to illustrate the applications on VLSI circuits. [20] First we give a transparent presentation of the method introduced by Ďuriš and Galil for proving lower bounds on the area of multilective VLSI circuits. Let for a given circuit S, $A(S)$ denote the area complexity of S, $T(S)$ denote the time complexity of S, and $P(S)$ denote the number of processors of S. Obviously $P(S) \leq A(S)$ for every S. The idea of the proof of the following theorem is to consider the sets W_i as the sets of inputs read in the time t by the circuit S.

Theorem 1. *Let f be a Boolean function of n' variables, for a positive integer n'. Let $k < \frac{1}{2}\log_2 n' - 2$ be a positive integer. Then, for every k-multilective VLSI circuit computing f,*

$$A(S) \geq P(S) \geq occ_{2k}(f)/2k. \tag{3}$$

Now, we present a new result by showing that overlapping communication complexity may be even used to prove lower bounds on AT^2-tradeoff of multilective VLSI-circuits. This generalizes a similar result [Th79] for the relation between communication complexity and (semilective) VLSI circuits. The idea of the proof is to consider W_i as the set of inputs read by the i-th processor of the circuit.

---

[16] See (3) and (4) of Lemma 1.
[17] See (5) of Lemma 1.
[18] First to $\overline{W_i}$'s and $\overline{W}$ and then to L and R.
[19] The second free parameter does not depend on the computing problem considered, but on the multilective computing model.
[20] The formal definition of k-multilective VLSI circuits may be found in [Hr97,Sa84].

Theorem 2. Let k and n' be positive integers, $k < \frac{1}{2}\log_2 n' - 2$. Let f be a Boolean function depending on all its n' variables. Then, for every k-multilective VLSI circuit S computing f,

$$A(S) \cdot (T(S))^2 \geq (\overline{occ}_{2k}(f)/4k)^2. \tag{4}$$

Now, we consider branching programs [HKMW92,PZ83,We88]. In [KMW89] an exponential lower bound on the size of k-times-only oblivious branching programs has been proved. The proof does not explicitely use the method based on communication complexity. We show that by using overlapping communication complexity we will not only get a transparent proof of this fact, but even a more powerful lower bound. The k-time-only oblivious branching program consists of levels. All nodes of every level read the same variable and every variable is read at at most k levels. Using overlapping communication complexity we can remove obliviousness and allow several variables to be read in one level.

Theorem 3. Let f be a Boolean function of n variables, $n \in \mathbb{N}$. Let k, m be positive integers such that $k < \frac{1}{2}\log_2 n - 2$ and $m \leq n/8 \cdot 3^{2k}$. The size of every branching program reading at most m variables on every level, asking for every variable on at most k distinct levels, and computing f is at least:

$$2^{occ_{2k}(f)/2k}. \tag{5}$$

So Theorem 3 enables to prove $2^{\Omega(n)}$ lower bounds on the size of k-time-only branching programs (with the above restriction) computing specific functions. Note that there are already known exponential lower bounds on syntactic k-times-only branching programs [BRS93, Ok93, Sue97] that are a more powerful model of branching programs than those considered in Theorem 4.3. The next assertion shows how overlapping communication complexity can be used to get lower bounds on syntactic k-times-only branching programs. Unfortunately, this application is not so transparent and easy to use as the previous ones.

Theorem 4. Let f be a Boolean function of n variables, $n \in \mathbb{N}$. Let k be a positive integer such that $k < \frac{1}{2}\log_2 n - 2$. Let D be a syntactic k-times-only branching program computing f.

Let the width of D be bounded by r. Then there exists a partition of f into $c = r^{8k \cdot 3^{2k}}$ partial Boolean functions $f_1, f_2, ..., f_c$ such that

$$occ_{2k}(f_i) \leq 2k \cdot log_2(8k \cdot 3^{2k} \cdot r), \tag{6}$$

for every $i = 1, ..., c$.

The last application, we can mention only very shortly in this extended abstract, is to prove lower bounds on multilective planar Boolean circuits. Using the Planar separator theorem of Lipton and Tarjan [LT79] and our general concept of overlapping communication complexity we can obtain transparent proofs of essentially stronger lower bounds than the lower bounds presented in [Gr91,Tu89], where multiparty communication complexity has been applied.

References

[Ab78] Ableson, H.: Lower bounds on information transfer in distributed computations, *Proc. 19th IEEE FOCS*, IEEE 1978, pp. 151-158.

[AUY83] Aho, A.V., Ullman, J.D., Yanakakis, M.: On notions of informations transfer in VLSI circuits, *Proc. 15th ACM STOC*, ACM 1983, pp. 133-139.

[BRS93] Borodin,A., Razborov,A., Smolensky,R.: On lower bounds for read-k-times branching programs. *Computational Complexity* 3 (1993), 1-18.

[DG93] Ďuriš, P., Galil, Z.: On the power of multiple read in chip, *Information and Computation* 104 (1993), pp. 277-287.

[DGS84] Ďuriš, P., Galil, Z., Schnitger, G.: Lower bounds on communication complexity, *Proc. 16th ACM STOC*, ACM 1984, pp. 81-91.

[DHS96] Dietzfelbinger, M., Hromkovič, J., Schnitger, G.: A comparison of two lower bounds methods for communication complexity, *Theoretical Computer Science* 168 (1996), pp. 39-51.

[Gr91] Gröger, H.D.: A new partition lemma for planar graphs and its application to circuit complexity, In: *Proc. FCT'91, Lecture Notes in Computer Science* 529, Springer-Verlag 1991, pp. 220-229.

[HKMW92] Hromkovič, J., Krause, M., Meinel, Ch., Waack, S.: Branching programs provide lower bounds on the area of multilective deterministic and nondeterministic VLSI circuits, *Information and Computation* 95 (1992), pp. 117-128.

[Hr97] Hromkovič, J.: *Communication Complexity and Parallel Computing*, EATCS Series, Springer 1997, 336p.

[KMW89] Krause, M., Meinel, Ch., Waack, S.: Separating complexity classes related to certain input oblivious logarithmic space-bounded Turing machines, In: *Proc. Structure in Complexity Theory 1989*, pp. 240-249.

[KN97] Kushilevitz, E., Nisan, N.: *Communication Complexity*, Cambridge University Press 1997.

[LT79] Lipton, R.J.,Tarjan, R.E.: A separator theorem for planar graphs, *SIAM J. Applied Mathematics* 36 (1979), pp. 177-189.

[PZ83] Pudlák, P., Žák, S.: Space complexity of computations, Tech. Report, Prague 1983.

[Ok93] Okolnishkova,E.A.: On lower bounds for branching programs. *Siberian Advances in Mathematics* 3 (1993), 152-166.

[Sau97] Sauerhoff,M.: Lower bounds for randomized read-k-times branching programs. In: *Proc. STACS'98, Lecture Notes in Computer Science* 1373, pp. 105-115.

[Sa84] Savage, J.E.: Multilective VLSI algorithms, *JCSS* (1984), pp. 243-273.

[Th79] Thompson, C.D.: Area-time complexity for VLSI, *Proc. 11th ACM STOC*, ACM 1979, pp. 81-88.

[Tu89] Turán, Gy.: On restricted Boolean circuits, In: *Proc. FCT'89, Lecture Notes in Computer Science* 380, Springer-Verlag 1989, pp. 460-469.

[We88] Wegener, I.: On the complexity of branching programs and decision trees for clique funcion, *J. of ACM* 35 (1988), pp. 461-471.

[Ya79] Yao, A.C.: Some complexity questions related to distributive computing, *Proc. 11th ACM STOC*, ACM 1979, pp. 209-213.

[Ya81] Yao, A.C.: The entropic limitations on VLSI computations, *Proc. 11th ACM STOC*, ACM 1979, pp. 209-213.

A Finite Hierarchy of the Recursively Enumerable Real Numbers

Klaus Weihrauch and Xizhong Zheng[*]

Theoretische Informatik I,
FernUniversität Hagen,
58084 Hagen, Germany

Abstract. For any set A of natural numbers, denote by x_A the corresponding real number such that A is just the set of "1" positions in its binary expansion. In this paper we characterize the number x_A for some classes of recursively enumerable sets A. Applying finite injury priority methods we show that there is a d-r.e. set A such that x_A is not a semi-computable real number (which corresponds to the limit of computable monotonic sequence of rational numbers) and that there is an ω-r.e. set A such that x_A can't be represented as a sum of two semi-computable real numbers.

1 Introduction

A real number x is *computable*, if there is a computable Cauchy sequence $(r_n)_{n \in \mathbb{N}}$ of rational numbers which converges effectively to x (see e.g. [3,5,6].) Where a sequence $(r_n)_{n \in \mathbb{N}}$ of rational numbers is *computable* means that there are recursive functions $a, b, c : \mathbb{N} \to \mathbb{N}$ such that $r_n = (a(n) - b(n))/(c(n) + 1)$ for all $n \in \mathbb{N}$, and the sequence $(r_n)_{n \in \mathbb{N}}$ *converges effectively* means that $|r_{n+m} - r_n| < 2^{-n}$ holds for all $m, n \in \mathbb{N}$ ($\mathbb{N}$ is the set of all natural numbers.) We denote the class of all computable real numbers by $\mathbf{C}_0$. Here the effectivity of the convergence is crucial, because there are computable sequences of rational numbers which converge (non-effectively, of course) to non-computable real numbers (see [5,11]). A standard example is the real number $x_A := \sum_{n \in A} 2^{-n}$ for a nonrecursive r.e. set $A \subseteq \mathbb{N}$. Let $a : \mathbb{N} \to \mathbb{N}$ be an 1-1 recursive enumeration function of A, i.e., rang$(a) = A$, then the increasing computable sequence $(x_n)_{n \in \mathbb{N}}$ defined by $x_n := \sum_{i=0}^{n} 2^{-a(i)}$ converges noneffectively to the noncomputable real number x_A (see [4]). In fact, it is easy to see that x_A is a computable real number iff A is a recursive set.

Although the real number x_A for a nonrecursive r.e. set A is not computable, it is still quite "effective" in the sense that we can approximate it effectively from below. We call a real number x *left computable* (*right computable*), if there is an increasing (decreasing) computable sequence of rational numbers which converges to x. A real number is called *semi-computable*, if it is either left computable or right computable. The class of all semi-computable real numbers

---

[*] Contact author. Email address: xizhong.zheng@fernuni-hagen.de

is denoted by $\mathbf{C}_1$. The above example shows that $\mathbf{C}_0 \subsetneq \mathbf{C}_1$. Obviously, x_A is (right) left computable, if A is (co-)r.e. On the other hand, C. G. Jockusch (see [9]) has observed that if B is a non-recursive r.e. set, then the non-r.e. set $B \oplus \overline{B} := \{2n : n \in B\} \bigcup \{2n+1 : n \notin B\}$ still corresponds to a left computable real number $x_{B \oplus \overline{B}}$. Note that the set $B \oplus \overline{B}$ above is in fact a d-r.e. set (difference of r.e. sets.) We can extend Jockusch's result to show that there are a real $(k+1)$-r.e set (which is not k-r.e.) and a real ω-r.e. set (which is not k-r.e for any k), respectively, such that their corresponding real numbers are still left computable.

Because $x_A = x_{B \cup C} - x_C$ if $A = B \setminus C$ (set difference,) the above results show that there are many semi-computable real numbers whose differences are still semi-computable. Let $\mathbf{C}_2$ denote the closure of $\mathbf{C}_1$ under the arithmetic operations of "+" and "−". Then it is natural to ask whether $\mathbf{C}_1 = \mathbf{C}_2$ holds. The answer is no. We can construct by finite injury priority method a d-r.e. set A such that x_A is neither left computable nor right computable, thus $\mathbf{C}_1 \subsetneq \mathbf{C}_2$ holds. $\mathbf{C}_2$ is an interesting class of real numbers. It forms in fact a field, i.e. it is closed under the arithmetical operations "+", "−", "×" and "÷". We will give another characterization of the elements of $\mathbf{C}_2$ by the limits of the "weakly effectively convergent sequences", namely, $x \in \mathbf{C}_2$ iff there is a computable sequence $(x_n)_{n \in \mathbb{N}}$ of rational numbers such that $\sum_{n=0}^{\infty} |x_{n+1} - x_n|$ is bounded and converges to x. Compairing with the effectively convergent sequence $(y_n)_{n \in \mathbb{N}}$ which satisfies that $|y_{n+m} - y_n| \leq 2^{-n}$, for every $n, m \in \mathbb{N}$, hence $\sum_{n=0}^{\infty} |y_{n+m} - y_n|$ is bounded for every $m \in \mathbb{N}$, we can say that the sequence $(x_n)_{n \in \mathbb{N}}$ above converges to x *weakly effectively*. So the real numbers in $\mathbf{C}_2$ can be naturally called *weakly computable*.

Our last result shows that not every convergent computable sequence converges weakly effectively. Again by a priority injure construction we show that there is an ω-r.e. set A such that x_A is not in $\mathbf{C}_2$, i.e. there is no computable sequence of rational numbers which converges to x_A weakly effectively. We call a real number *recursively enumerable*, if there is a computable sequence of rational numbers which converges to it and denote by $\mathbf{C}_3$ the class of all such real numbers. Our results show that $(\mathbf{C}_i : i \leq 3)$ forms a noncollapsed hierarchy.

2 Computable and Semi-Computable Real Numbers

This section discusses the computable and semi-computable real numbers. By definition, it is easy to see that x is left computable iff $-x$ is right computable. Left and right computabilities are incomparable and x is computable iff it is both left and right computable. If $A \subseteq \mathbb{N}$ is a r.e. (co-r.e.) set, then x_A is a left (right) computable real number. We will show that the inverse is not true.

Definition 1 ((cf. [10])). A r.e. set $A \subseteq \mathbb{N}$ is called 1-*r.e.* For any $k \geq 1$, set $A \subseteq \mathbb{N}$ is called $(k+1)$-*r.e*, if there are r.e. set $B \subseteq \mathbb{N}$ and k-r.e. set $C \subseteq \mathbb{N}$ such that $A = B \setminus C$. 2-r.e. set is usually called *d-r.e.*

For any finite set $E \subset \mathbb{N}$, we define its *canonical index* i by $i := \sum_{j \in E} 2^j$. A finite set with canonical index i is denoted by D_i. A sequence $(E_n)_{n \in \mathbb{N}}$ of finite subsets of $\mathbb{N}$ is called *computable*, iff there is a recursive function $f : \mathbb{N} \to \mathbb{N}$ such that $E_n = D_{f(n)}$ for all $n \in \mathbb{N}$.

Proposition 1. *A set $A \subseteq \mathbb{N}$ is k-r.e iff there is a computable sequence $(A_s)_{s \in \mathbb{N}}$ of finite subsets of $\mathbb{N}$ such that*

1. $A = \lim_{n \to \infty} A_n := \bigcup_{n=0}^{\infty} \bigcap_{s=n}^{\infty} A_s$, *and*
2. $|\{s \in \mathbb{N} : n \in A_{s+1} \triangle A_s\}| \leq k$, *for all $n \in \mathbb{N}$.*

where $A \triangle B := (A \backslash B) \bigcup (B \backslash A)$ is the symmetric difference of sets A and B, and $|B|$ denotes the cardinality of set B. The sequence $(A_s)_{s \in \mathbb{N}}$ above is usually called an effective k-enumeration *of A.*

Jockusch's example shows that there is a d-r.e. set A such that x_A is left computable. R.I. Soare [9] extended this result in several directions. He has shown, e.g., that there is a dominant d-r.e. set A and a cohesive set C such that x_A and x_C are left computable. Where A is dominant iff the principal function of A dominates every recursive function, and C is cohesive iff C is infinite and there is no r.e. set W such that $W \cap C$ and $\overline{W} \cap C$ are both infinite (see [9] for exact definitions.) The next theorems give another extension of Jockusch's observation and induce to an infinite hierarchy of left computable real numbers.

Theorem 1. *For any $k \geq 2$, there are k-r.e. sets A, B which are not $(k-1)$-r.e. such that x_A and x_B are left and right computable, respectively.*

The concept of k-r.e. set can be generalized to ω-r.e. by the Proposition 1.

Definition 2. *Set $A \subseteq \mathbb{N}$ is called ω-r.e., iff there is a computable sequence $(A_s)_{s \in \mathbb{N}}$ of finite subsets of $\mathbb{N}$ such that*

1. $A = \bigcup_{n=0}^{\infty} \bigcap_{s=n}^{\infty} A_s$, *and*
2. $\forall n \in \mathbb{N} \exists k \in \mathbb{N}(|\{s \in \mathbb{N} : n \in A_{s+1} \triangle A_s\}| \leq k)$.

The sequence $(A_s)_{s \in \mathbb{N}}$ is called an effective ω-enumeration *of A.*

Theorem 2. *There is an ω-r.e. sets A which is not k-r.e., for any $k \in \mathbb{N}$, such that x_A is left computable. The same claim holds for right computability as well.*

3   Weakly Computable Real Numbers

From the last section we know that there are many noncomputable semi computable real numbers whose sums and differences are still semi-computable. We will show in this section that this is not always the case, i.e., there are left computable real numbers y and z such that their difference $x := y - z$ is neither left computable nor right computable. Hence $\mathbf{C}_1$ is not closed under the arithmetical operations "+" and "−". We introduce a new notion at first.

Definition 3. A real number x is called *weakly computable*, if there are two left computable real numbers y, z such that $x = y - z$. The set of all weakly computable real numbers is denoted by $\mathbf{C}_2$.

Equivalently, x is weakly computable iff there are left computable real y and right computable real z such that $x = y + z$. If A is a d-r.e. set, then x_A is weakly computable. More generally, we can show by an easy induction on $k \geq 1$, that, if A is a k-r.e set, then x_A is a weakly computable real number. Now we give another characterization of weakly computable real numbers by the "weak convergence" of sequences.

Definition 4. A sequence $(x_n)_{n\in\mathbb{N}}$ of real numbers is *weakly effectively convergent* (w.e. convergent for short) if , if $\sum_{n=0}^{\infty} |x_{n+1} - x_n|$ is bounded.

If a sequence $(x_n)_{n\in\mathbb{N}}$ converges weakly effectively, i.e., $\sum_{n=0}^{\infty} |x_{n+1} - x_n| < \infty$, then "big" jumps may occur in the sequence and then may occur very late. If, however, the sequence is effectively convergent, then the big jumps must occur early. So, weakly effective convergence is a kind of weak version of effective convergence. Any effectively convergent sequence of rational numbers converges to computable real numbers. For w.e. convergent sequences, we have

Theorem 3. *A real number x is weakly computable, iff there is a computable sequence $(x_n)_{n\in\mathbb{N}}$ of rational numbers which converges to x weakly effectively.*

Proof. "$\Rightarrow$". Let x be a weakly computable real number. Then there are two computable increasing sequences $(y_n)_{n\in\mathbb{N}}$ and $(z_n)_{n\in\mathbb{N}}$ of rational numbers such that $y := \lim_{n\to\infty} y_n$ and $z := \lim_{n\to\infty} z_n$ exist and $x = y - z$. Let $x_n := y_n - z_n$. Then $(x_n)_{n\in\mathbb{N}}$ is a computable sequence of rational numbers which satisfies

$$\sum_{n=0}^{\infty} |x_{n+1} - x_n| \leq \sum_{n=0}^{\infty} (y_{n+1} - y_n) + \sum_{n=0}^{\infty} (z_{n+1} - z_n) = y - y_0 + z - z_0$$

So $(x_n)_{n\in\mathbb{N}}$ converges to x weakly effectively.

"$\Leftarrow$". Let $(x_n)_{n\in\mathbb{N}}$ be a computable sequence of rational numbers which converges to x weakly effectively. Define computable sequences $(y_n)_{n\in\mathbb{N}}$ and $(z_n)_{n\in\mathbb{N}}$ of rational numbers by

$$y_n := x_0 + \sum_{i=0}^{n} (x_{i+1} \dotminus x_i) \quad \text{and} \quad z_n := \sum_{i=0}^{n} (x_i \dotminus x_{i+1})$$

Obviously, they are both non-decreasing and bounded. Hence $y := \lim_{n\to\infty} y_n$ and $z := \lim_{n\to\infty} z_n$ exist and they are the left computable real numbers which satisfy

$$y - z = \lim_{n\to\infty}(y_n - z_n) = \lim_{n\to\infty}\left(x_0 + \sum_{i=0}^{n}(x_{i+1} \dotminus x_i) - \sum_{i=0}^{n}(x_i \dotminus x_{i+1})\right)$$

$$= \lim_{n\to\infty}\left(x_0 + \sum_{i=0}^{n}(x_{i+1} - x_i)\right) = \lim_{n\to\infty} x_n = x.$$

That is, x is weakly computable.

Corollary 1. *A real number x is weakly computable iff there is a computable sequence $(x_n)_{n \in \mathbb{N}}$ of rational numbers such that $\sum_{n=0}^{\infty} |x_{n+i} - x_n|$ is bounded for every $i \in \mathbb{N}$ and $\lim_{n \to \infty} x_n = x$.*

Proposition 2. *If $(x_n)_{n \in \mathbb{N}}$ is a computable sequence of computable real numbers which converges w.e. to x, then x is weakly computable.*

Theorem 4. *The class $\mathbf{C}_2$ is a field generated by $\mathbf{C}_1$.*

Now we will show that $\mathbf{C}_2$ extends $\mathbf{C}_1$ properly.

Theorem 5. *There is a d-r.e. set A such that x_A is neither left computable, nor right computable.*

Proof. We construct the set A effectively in stages so that, for any $s \in \mathbb{N}$, the number $2s$ is always enumerated into A at the beginning. They can be removed from A at a later stage and then will never enter A again. The number $2s+1$ can only be enumerated into A and can not be removed from A. This makes sure that A is a d-r.e. set. We use A_s to denote the set constructed at the end of stage s. Our proof is a finite injury priority construction. The detail explanations about such kind of constructions can be found in [10].

To guarantee that x_A is neither left computable nor right computable, we let x_A diagonalize all semi-computable real numbers. Let $(M_n)_{n \in \mathbb{N}}$ be the standard effective enumeration of all Turing machines, $\varphi_n :\subseteq \mathbb{N} \to \mathbb{N}$ the function computed by M_n. Function $\varphi_{n,s} :\subseteq \mathbb{N} \to \mathbb{N}$ is defined by $\varphi_{n,s}(x) := y$, if $M_n(x)$ halts and outputs y in s steps and $\varphi_{n,s}(x)$ undefined otherwise. Let

$$a_n := \sup(\{\nu_Q(i) \leq 2 : i \in \mathrm{dom}(\varphi_n)\} \cup \{0\})$$

$$b_n := \inf(\{\nu_Q(i) \geq 0 : i \in \mathrm{dom}(\varphi_n)\} \cup \{2\})$$

$$a_{n,s} := \max(\{\nu_Q(i) \leq 2 : i \in \mathrm{dom}(\varphi_{n,s})\} \cup \{0\})$$

$$b_{n,s} := \min(\{\nu_Q(i) \geq 0 : i \in \mathrm{dom}(\varphi_{n,s})\} \cup \{2\})$$

where $\nu_Q : \mathbb{N} \to \mathbb{Q}$ is an effective enumeration of rational numbers. For any increasing computable sequence $(x_s)_{s \in \mathbb{N}}$ of rational numbers from $[0;2]$, there is an n such that $\forall s \in \mathbb{N}(x_s = a_{n,s})$. And $a_n = \lim_{s \to \infty} a_{n,s}$ for every $n \in \mathbb{N}$. Then $L := \{a_n : n \in \mathbb{N}\}$ consists of all left computable real numbers of interval $[0;2]$. Similarly, $R := \{b_n : n \in \mathbb{N}\}$ consists of all right computable real numbers of interval $[0;2]$.

It suffices now to make sure that the set A satisfies, for all $n \in \mathbb{N}$, the following requirements:

$$R_{2n} \ : \ x_A \neq a_n,$$
$$R_{2n+1} \ : \ x_A \neq b_n.$$

The strategy to satisfy a single requirement R_{2n} is simple: we put at the beginning all even numbers into A_0. If, at some stage $s+1$, there is some i such that $x_{A_s} - 2^{-2i} < a_{n,s}$, then reduce the value of x_{A_s} by removing the number $2i$ from A_s. That is, define $A_{s+1} := A_s \setminus \{2i\}$ (we call R_{2n} is attacked at this stage.) Furthermore, we define a restraint $r(2n, s+1) := \mu j(x_{A_s} - 2^{-2i} + 2^{-j} < a_{n,s})$. At any stage $t+1 > s+1$, all elements $x \le r(2n, s+1)$ are not allowed to be put into or taken out from A_t. Therefore, the set $A := \lim_{s \to \infty} A_s$ satisfies that $x_A \le x_{A_{s+1}} + 2^{-r(2n,s+1)} = x_{A_s} - 2^{-2i} + 2^{-r(2n,s+1)} < a_{n,s} \le a_n$. Hence R_{2n} is satisfied. If no such stage $s+1$ exists, then $a_{n,s} \le x_{A_s} - 2^{-2i}$ holds for all s and i, hence, say, $a_n \le x_A - 2^{-2} < x_A$. That is, R_{2n} is satisfied too. The strategy for R_{2n+1} is similar.

To accommodate all requirements R_n simultaneously, we use finite injury priority method. We give all requirements the priority in order $R_0, R_1, R_2, \ldots$, i.e., R_m has higher priority than R_n iff $m < n$. At any stage $s+1$, if we want to enumerate an element k into A_s while attacking the requirement R_{2n+1}, we do not need to take care for lower priority requirements R_m (with $m > 2n+1$) but we have to do that for all higher priority requirements R_m (with $m < 2n+1$.) In this case, the element k can be put into A_s at this stage only, if this does not injury any higher priority requirement, i.e., k must be bigger than all restraint $r(m, s)$ for all $m \le 2n+1$. Note that, for any requirement R_m, one attack suffices to satisfy it, if it is not injured any more. Then any requirement R_m can be injured only finitely often (at most $2^m - 1$ times for R_m.) Eventually, every requirement can be satisfied by this strategy.

The construction of $(A_s)_{s \in \mathbb{N}}$ is as following:

Stage 0: Define $A_0 := \{2i : i \in \mathbb{N}\}$ and $r(n, 0) := 0$ for all $n \in \mathbb{N}$. All requirements are set to the state of *unsatisfied*.

Stage $s+1$: Given A_s and $r(n, s)$ for all n. The requirement R_{2n} *requires attention*, if there is an $i \le s$ such that (i) $r(m, s) < 2i$ for all $m \le 2n$; (ii) $a_{n,s} > x_{A_s} - 2^{-2i}$ and (iii) R_{2n} is in the state of unsatisfied. The requirement R_{2n+1} requires attention, if there is an $i \le s$ such that (i)' $r(m, s) < 2i+1$ for all $m \le 2n+1$; (ii)' $b_{n,s} < x_{A_s} + 2^{-(2i+1)}$ and (iii)' R_{2n+1} is in the state of unsatisfied.

Choose $m \le s$ minimal such that R_m requires attention. If $m = 2n$, then define $A_{s+1} := A_s \setminus \{2i_0\}$; $r(e, s+1) := r(e, s)$, if $e \ne m$ and $r(m, s+1) := \mu i(a_{n,s} > x_{A_s} - 2^{-2i_0} + 2^{-i})$, where i_0 is the least i which satisfies the condition (i) – (iii).

If $m = 2n+1$, then define $A_{s+1} := A_s \bigcup\{2i_0 + 1\}$; $r(e, s+1) := r(e, s)$, if $e \ne m$ and $r(m, s+1) := \mu i(b_{n,s} < x_{A_s} + 2^{-2i_0+1} - 2^{-i})$, where i_0 is the least i which satisfies the condition (i)' – (iii)'.

In both cases, we say that R_m *receives attention* and set it now to the state of *satisfied*. Furthermore, all requirements $R_{m'}$ with $m' > m$ are set to the state of *unsatisfied*. If $R_{m'}$, $m' > m$, was satisfied at stage s, then it is *injured* at this stage by R_m.

If there is no $m \le s$ such that R_m requires attention, then define simply $A_{s+1} := A_s$ and $r(n, s+1) := r(n, s)$ for all $n \in \mathbb{N}$.

This ends the construction. It is easy to see that the construction succeeds by the following claims whose proofs are omitted here.

Claim 1 $A := \lim_{n\to\infty} A_n$ is a d-r.e. set.

Claim 2 For any $n \in \mathbb{N}$, R_n receives attentions at most finitely many times.

Claim 3 For any $n \in \mathbb{N}$, R_n is eventually satisfied.

It follows immediately from above theorem the following corollary.

Corollary 2. *There are non-semi-computable weakly computable real numbers, i.e. $\mathbf{C}_1 \subsetneq \mathbf{C}_2$.*

4 Recursively Enumerable Real Numbers

We have shown that x_A is weakly computable, if A is k-r.e. for any $k \geq 1$ and there are also ω-r.e. set A such that x_A is semi-computable (Theorem 2), hence weakly computable. It comes the question: is every x_A weakly computable, if A is ω-r.e.? Or more generally, is there any convergent computable sequence of rational numbers which converges not weakly effectively? This section will answer these question positively.

Theorem 6. *There is a r.e. real number which is not weakly computable. Thus, $\mathbf{C}_2 \subsetneq \mathbf{C}_3$ holds.*

Proof. We construct a computable sequence $(x_n)_{n\in\mathbb{N}}$ of rational numbers in stages such that the limit $x := \lim_{n\to\infty} x_n$ diagonalizes all differences of left computable real numbers. Let a_n and $a_{n,s}$ be same as in the proof of Theorem 5 and define, for all $n, m \in \mathbb{N}$, that

$$d_{\langle n,m\rangle} := a_n - a_m \quad \text{and} \quad d_{\langle n,m\rangle,s} := a_{n,s} - a_{m,s}. \tag{1}$$

where $\langle \cdot, \cdot \rangle : \mathbb{N}^2 \to \mathbb{N}$ is the Cantor's pairing function defined by $\langle n, m \rangle := (n+m)(n+m+1)/2 + m$. Then $D := \{d_n : n \in \mathbb{N}\}$ is the set of all weakly computable real numbers by the Definition 3. It suffices now to make sure that x is a r.e. real number which satisfies all the following requirements:

$$R_n: \quad x \neq d_n.$$

We will construct x in such a way that $x = \sum_{i=0}^{\infty} w_i \, 8^{-(i+1)}$ and x is a limit of some computable sequence of rational numbers, where $w_i \in \{0, 4\}$ for all $i \in \mathbb{N}$. To satisfy the requirement R_n, we change the value of w_n from 0 to 4 or from 4 to 0, if it is necessary, so that $|x - d_n| \geq 8^{-(n+1)}$ holds. For single R_n, the strategy is simple. Let $r_n := \sum_{i=0}^{n-1} w_i \, 8^{-(i+1)}$ (for any given rational numbers w_i with $i < n$) and consider the interval $[r_n; r_n + 8^{-n}]$. At the beginning, let $w_n := 0$. We redefine $w_n := 4$, if there is an s such that $d_{n,s} \leq r_n + 2 \cdot 8^{-(n+1)}$, set w_n back to 0 later, if some $t > s$ appears such that $d_{n,t} \geq r_n + 3 \cdot 8^{-(n+1)}$ and let $w_n = 4$ again, if there is another $s' > t$ such that $d_{n,s'} \leq r_n + 2 \cdot 8^{-(n+1)}$, and so on. Every change of the value of w_n is called an *attack* to R_n. Since $(d_{n,s})_{s\in\mathbb{N}}$ is not necessarily monotonic, w_n may be changed many times. We denote by w_n^s the value of w_n at the end of stage s. Then w_n^s can't be changed infinitely often because $(d_{n,s})_{s\in\mathbb{N}}$ converges, hence $w_n := \lim_{s\to\infty} w_n^s$ exists. We define simply $x := r_n + w_n 8^{-(n+1)}$. By the definition, $|d_{n,s} - (r_n + w_n^s 8^{-(n+1)})| \geq 8^{-(n+1)}$ holds for all $s \in \mathbb{N}$. Therefore $|d_n - (r_n + w_n 8^{-(n+1)})| = |d_n - x| \geq 8^{-(n+1)}$ holds. That is, x satisfies R_n.

To treat all requirements simultaneously, we apply priority injury method again. Here is the construction of $(x_n)_{n\in\mathbb{N}}$:

Stage $s = 0$. Set $x_0 := w_0^0 := 0$.

Stage $s + 1$. Given $x_s := \sum_{i=0}^s w_i^s 8^{-(i+1)}$ with $w_i^s \in \{0, 4\}$ for all $i \leq s$. The requirement R_n, $n \leq s$, *requires attention*, if either

$$w_n^s = 0 \ \& \ d_{n,s} \leq \sum_{i=0}^{n-1} w_i^s 8^{-(i+1)} + 2 \cdot 8^{-(n+1)}, \quad \text{or} \tag{2}$$

$$w_n^s = 4 \ \& \ d_{n,s} \geq \sum_{i=0}^{n-1} w_i^s 8^{-(i+1)} + 3 \cdot 8^{-(n+1)} \tag{3}$$

holds. Choose the least $n \leq s$ such that R_n requires attention. If (2) holds, then define

$$w_m^{s+1} := \begin{cases} w_m^s & \text{if } m < n; \\ 4 & \text{if } m = n; \\ 0 & \text{if } n < m \leq s+1. \end{cases} \tag{4}$$

If (3) holds, then define

$$w_m^{s+1} := \begin{cases} w_m^s & \text{if } m < n; \\ 0 & \text{if } n \leq m \leq s+1. \end{cases} \tag{5}$$

In both cases we say then R_n *receives attention*.

If no requirement R_n, $n \leq s$, requires attention, then define, for all $m \leq s$, simply

$$w_m^{s+1} := w_m^s \quad \text{and} \quad w_{s+1}^{s+1} := 0. \tag{6}$$

At last, set $x_{s+1} := \sum_{i=0}^{s+1} w_i^{s+1} 8^{-(i+1)}$. This ends the construction. We can show our construction succeeds by the flowing Claims whose proofs are omitted here.

Claim 1 For every $n \in \mathbb{N}$, the requirement R_n receives attentions at most finitely often.

Claim 2 For every $n \in \mathbb{N}$, the limit $w_n := \lim_{s\to\infty} w_n^s$ exists.

Claim 3 Let $x := \sum_{i=0}^\infty w_i 8^{-(i+1)}$. Then $x = \lim_{n\to\infty} x_n$ holds.

Claim 4 The real number x of Lemma 3 satisfies all requirements R_n for $n \in \mathbb{N}$.

Corollary 3. *There is an ω-r.e. set A such that x_A is a r.e. real number which is not weakly computable.*

Proof. Let w_i^s $(s \in \mathbb{N}, i \leq s)$ be same as in the proof of the Theorem 6. Note that $4 \cdot 8^{-(n+1)} = 2^{-(3n+1)}$. Define a computable sequence $(A_s : s \in \mathbb{N})$ of finite sunsets of $\mathbb{N}$ by $A_s := \{3i + 1 \in \mathbb{N} : i \leq s \ \& \ w_i^s = 4\}$. Since there are only finite many s such that $w_n^s \neq w_n^{s+1}$, there are only finite many s such that $A_{s+1} \triangle A_s$ contains n, for every $n \in \mathbb{N}$. That is $(A_s : s \in \mathbb{N})$ is an effective ω-enumeration of the set $A := \lim_{s \to \infty} A_s$, hence A is an ω-r.e. set. On the other hand, it is easy to see that $A = \{3n + 1 \in \mathbb{N} : w_n = 4\}$. Then $x_A = \sum\{2^{-(3n+1)} : w_n = 4\} = \sum_{n=0}^{\infty} w_n \, 8^{-(n+1)} = x$. By the proof of Theorem 6, x_A is a r.e. real number which is not weakly computable.

Corollary 4. $(\mathbf{C}_i : i \leq 3)$ *forms a noncollapsed hierarchy.*

References

1. Ker-I Ko   On the definitions of some complexity classes of real numbers, *Math. Systems Theory* 16(1983) 95–100.
2. A. H. Lachlan   Recursive real numbers. *J. Symbolic Logic* 28(1963), 1–16.
3. J. Myhill   Criteria of constructivity for real numbers, *J. Symbolic Logic* 18(1953), 7–10.
4. M. Pour-El & J. Richards   *Computability in Analysis and Physics.* Springer-Verlag, Berlin, Heidelberg, 1989.
5. H. G. Rice   Recursive real numbers, *Proc. Amer. Math. Soc.* 5(1954), 784–791.
6. R. M. Robinson   Review of "R. Peter: ' Rekursive Funktionen', Akad. Kiado. Budapest, 1951", *J. Symb. Logic* 16(1951), 280.
7. H. Jr. Rogers   *Theory of Recursive Functions and Effective Computability* McGraw-Hill, Inc. New York, 1967.
8. R. Soare   Recursion theory and Dedekind cuts,   *Trans, Amer. Math. Soc.* 140(1969), 271–294.
9. R. Soare   Cohensive sets and recursive enumerable Dedekind cuts *Pacific J. of Math.* 31(1969), no.1, 215–231.
10. R. Soare   *Recursively Enumerable Sets and Degrees,* Springer-Verlag, Berlin, Heidelberg, 1987.
11. E. Specter   Nicht konstruktive beweisbare Sätze der Analysis, *J. Symbolic Logic* 14(1949), 145–158
12. K. Weihrauch   *Computability.* EATCS Monographs on Theoretical Computer Science Vol. 9, Springer-Verlag, Berlin, Heidelberg, 1987.

One Guess One-Way Cellular Arrays

Thomas Buchholz, Andreas Klein, and Martin Kutrib

Institute of Informatics, University of Giessen
Arndtstr. 2, D-35392 Giessen, Germany
{buchholz,kutrib}@informatik.uni-giessen.de

Abstract. One-way cellular automata with restricted nondeterminism are investigated. The number of allowed nondeterministic state transitions is limited to a constant. It is shown that a limit to exactly one step does not decrease the language accepting capabilities. We prove a speed-up result that allows any linear-time computation to be sped-up to real-time. Some relationships to deterministic arrays are considered. Finally we prove several closure properties of the real-time languages.

1 Introduction

Linear arrays of finite automata can be regarded as models for massively parallel computers. Mainly they differ in how the automata are interconnected and in how the input is supplied. Various types have been studied for a long time. Here we are investigating arrays with a parallel input mode and a very simple interconnection pattern. Each node is connected to its right immediate neighbor only. They are usually called one-way cellular automata (OCA).

Although deterministic and nondeterministic finite automata have the same computing capability, nondeterminism can strengthen the power of OCAs under some time resource bounds.

Nondeterministic OCAs have been investigated e.g. in [7] where $\mathscr{L}(\text{NOCA}) = \mathscr{L}(\text{NCA})$ was proved, and in [10] where it was shown in terms of homogeneous trellis automata that $\mathscr{L}_{rt}(\text{NOCA})$ contains the ε-free context-free languages as well as a NP-complete language, and is an AFL closed under intersection.

Here we consider arrays with restricted nondeterminism. We limit the number of allowed nondeterministic transitions. Moreover, all nondeterministic transitions have to appear before the deterministic ones. The main object of the present paper is to investigate arrays that are limited to exactly one nondeterministic transition step.

2 Basic Notions

We denote the integers by $\mathbb{Z}$, the positive natural numbers $\{1, 2, \ldots\}$ by $\mathbb{N}$, the set $\mathbb{N} \cup \{0\}$ by $\mathbb{N}_0$ and the powerset of a set S by 2^S.

A nondeterministic one-way cellular automaton is a linear array of nondeterministic finite automata, sometimes called cells, each of them is connected to its nearest neighbor to the right. For our convenience we identify the cells by natural numbers. The state transition depends on the actual state of each cell and the actual state of its neighbor. The transition function is applied to all cells synchronously at discrete time steps. More formally:

Definition 1. *A* nondeterministic one-way cellular automaton *(NOCA) is a system* $(S, \delta, \#)$, *where*

1. S *is the finite, nonempty set of* states,
2. $\# \in S$ *is the* boundary state,
3. $\delta : S^2 \to 2^S$ *is the* local transition function *satisfying* $\forall s_1, s_2 \in S : (\delta(s_1, s_2) \neq \emptyset)$ *and* $(\delta(s_1, s_2) = \{\#\} \iff s_1 = \#)$.

Let $\mathcal{M}$ be an NOCA with n cells. A configuration of $\mathcal{M}$ at some time $i \geq 0$ is a description of its global state, which is actually a mapping $c_i : [1, \ldots, n] \to S$. During its course of computation an NOCA steps nondeterministically through a sequence of configurations. The configuration c_0 at time 0 is defined by the initial sequence of states in an NOCA, while subsequent configurations are chosen according to the global transition Δ:

Let $n \in \mathbb{N}$ be an arbitrary natural number and c resp. c' be defined by $s_1, \ldots, s_n \in S$ resp. $s'_1, \ldots, s'_n \in S$.

$$c' \in \Delta(c) \iff s'_1 \in \delta(s_1, s_2), s'_2 \in \delta(s_2, s_3), \ldots, s'_n \in \delta(s_n, \#)$$

The i-fold composition of Δ is defined as follows:

$$\Delta^0(c) := c, \qquad \Delta^{i+1}(c) := \bigcup_{c' \in \Delta^i(c)} \Delta(c')$$

$\pi_i(s_1 \cdots s_n) := s_i$ selects the ith component of $s_1, \ldots, s_n$. If the state set is a Cartesian product of some smaller sets $S = S_0 \times S_1 \times \cdots \times S_r$, we will use the notion 'register' for the single parts of a state.

If the flow of information is extended to two-way, the resulting device is a *nondeterministic two-way cellular automaton* (NCA). I.e. the next state of each cell depends on the state of the cell itself and the states of its both immediate neighbors (to the left and to the right).

An NOCA (NCA) is deterministic if $\delta(s_1, s_2)$ ($\delta(s_1, s_2, s_3)$) is a singleton for all states $s_1, s_2, s_3 \in S$. Deterministic cellular arrays are denoted by OCA resp. CA.

Definition 2. *Let A be an Alphabet and $\mathcal{M} = (S, \delta, \#)$ be an NOCA with $A \subseteq S$.*

1. *A word $w \in A^+$ is* accepted by $\mathcal{M}$ *with* final states $F \subseteq S$ *in t time steps if there exists a $t_0 \leq t$ such that there exists a configuration $c_{t_0} \in \Delta^{t_0}(c_0)$ where $\pi_1(c_{t_0}) \in F$.*
2. $L(\mathcal{M}) = \{w \in A^+ \mid w \text{ is accepted by } \mathcal{M}\}$ *is the* language accepted by $\mathcal{M}$.
3. *Let $t : \mathbb{N} \to \mathbb{N}$, $t(n) \geq n$, be a mapping. If all $w \in L(\mathcal{M})$ are accepted within $t(|w|)$ time steps, then L is said to be of* time complexity t.

The family of all languages which can be accepted by an NOCA with time complexity t is denoted by $\mathscr{L}_{t(n)}$(NOCA). If t equals the *identity function* $id(n) := n$ acceptance is said to be in real-time and we write $\mathscr{L}_{rt}$(NOCA).

There is a natural way to restrict the nondeterminism of the arrays. One can limit the number of allowed nondeterministic state transitions of the cells.

For the following let us suppose the local transition consists of a deterministic and nondeterministic part δ_d and δ_{nd}, where $\delta_d(s_1, s_2) \in \delta_{nd}(s_1, s_2)$. At a whole $\delta = \delta_{nd}$ remains nondeterministic, but with this distinction the restriction is easily defined. Δ_{nd} denotes the global transition based on δ_{nd} and Δ_d the deterministic one based on δ_d.

Let $g : \mathbb{N} \to \mathbb{N}$ be a mapping for which $g(n) \leq t(n)$ holds. g gives the number of allowed nondeterministic transitions. An NOCA of length n for which the i-fold global transition Δ^i is defined as

$$\Delta^i := \begin{cases} \Delta_{nd}^i & \text{if } i \leq g(n) \\ \Delta_d^{i-g(n)} \left(\Delta_{nd}^{g(n)} \right) & \text{otherwise} \end{cases}$$

is denoted by gG-OCA (g guess OCA). Observe that all nondeterministic transitions have to be applied before the deterministic ones. In the sequel we are mainly interested in NOCAs allowed to guess constant times (i.e. $g(n) = k$, $k \in \mathbb{N}_0$).

3 Speed-Up and Guess Reduction

It is known [4] that deterministic OCAs can be sped-up by a constant amount of time as long as the remaining time complexity does not fall below real-time. For constructions it is sometimes convenient to have a corresponding result for 1G-OCAs: For example, after the first nondeterministic step a deterministic $t(n)$-time OCA can be simulated and subsequently the resulting $(t(n)+1)$-time 1G-OCA can be sped-up to a $t(n)$-time 1G-OCA again. Observe that in case of real-time it is not possible to speed-up the deterministic OCA by 1 time step before its simulation.

Lemma 3. *Let $t : \mathbb{N} \to \mathbb{N}$, $t(n) \geq n$, be a mapping and $k \in \mathbb{N}_0$ be a constant number. Then $\mathscr{L}_{t(n)+k}$(1G-OCA) = $\mathscr{L}_{t(n)}$(1G-OCA) holds.*

Proof. Let $\mathcal{M}$ be an 1G-OCA with time complexity $t(n)+k$ which passes through the configurations c_0 to $c_{t(n)+k}$. An 1G-OCA $\mathcal{M}'$ which simulates $\mathcal{M}$ in time $t(n)$ works as follows. Each cell consists of $k+1$ registers each may contain a state from S, thus $S' := S^{k+1}$. k of the registers are initially empty.

The rightmost cell 'knows' its input in advance (i.e. the border state). Therefore it can compute the states $c_1(n), \ldots, c_{k+1}(n)$ in the first transition and store them in its registers. Subsequently it simulates one step of cell n of $\mathcal{M}$ in every time step. At the second time step cell $n-1$ observes the filled registers of its neighbor and can compute the states $c_2(n-1), \ldots, c_{k+2}(n-1)$ in one time step. Again, subsequently it simulates one transition per time step. The behavior of cells $n-1$ to 1 is identical. Thus at time n the first cell computes the states $c_n(1),\ldots,c_{n+k}(1)$, and at time $t(n)$ the state $c_{t(n)+k}(1)$. □

Deterministic OCAs can be sped-up from $(n+t(n))$-time to $(n+\frac{t(n)}{k})$-time [1, 11]. Thus linear-time (i.e. k times real-time, $k \geq 1$) is close by real-time. By the way, it is not the same, since the inclusion $\mathscr{L}_{rt}(\text{OCA}) \subset \mathscr{L}_{(1+\varepsilon)\cdot id}(\text{OCA}) = \mathscr{L}_{rt}(\text{CA})$ is a proper one [17, 4]. For 1G-OCAs we have the following stronger result, from which follows that real-time is as powerful as linear-time.

Theorem 4. *Let* $t : \mathbb{N} \to \mathbb{N}$, $t(n) \geq n$, *be a mapping and* $k \in \mathbb{N}$ *be a constant number. Then* $\mathscr{L}_{k\cdot t(n)}(\text{1G-OCA}) = \mathscr{L}_{t(n)}(\text{1G-OCA})$ *holds.*

Proof. The inclusion $\mathscr{L}_{t(n)}(\text{1G-OCA}) \subseteq \mathscr{L}_{k\cdot t(n)}(\text{1G-OCA})$ follows from the definition. In order to prove $\mathscr{L}_{k\cdot t(n)}(\text{1G-OCA}) \subseteq \mathscr{L}_{t(n)}(\text{1G-OCA})$ let L be a language belonging to $\mathscr{L}_{k\cdot t(n)}(\text{1G-OCA})$ and let $\mathcal{M}$ be an 1G-OCA that accepts L with time complexity $k \cdot t(n)$. We construct an 1G-OCA $\mathcal{M}'$ that simulates $\mathcal{M}$ in time $t(n)$.

The idea is as follows: on an input of length n each cell i with $1 \leq i \leq n/k$, of $\mathcal{M}'$, guesses the initial states of the cells $k(i-1)+1, k(i-1)+2, \ldots, ki$ and additionally what each cell of $\mathcal{M}$ might have guessed with respect to these initial states. (For simplicity here we assume that n is a multiple of k. The other cases are omitted since their handling is only a technical challenge.) Based on this compressed representation $\mathcal{M}'$ can simulate k time steps of $\mathcal{M}$ per time step.

In parallel $\mathcal{M}'$ has to check whether the guesses of the initial states were correct. Therefore each cell $(n/k)j + i$ with $1 \leq i \leq n/k$ and $0 \leq j \leq k-1$ guesses the initial states of the cells $(i-1)n/k+1, (i-1)n/k+2, \ldots in/k$, too. So we are concerned with an interim configuration of the form $x_1 x_2 \ldots x_{n/k}$ where $|x_i| = k$ and each x_i might contain the compressed initial input. Now $\mathcal{M}'$ subsequently verifies – the first checking task – that the initial states guessed in the cells corresponding to x_j and x_{j+1}, are the same, $1 \leq j < n/k$. Additionally it checks – the second one – whether the guessed initial states $x_{n/k}$ are really the packed initial states of all cells. So in total it can be ensured that the simulation of $\mathcal{M}$ is based on the correct data.

To complete the proof we have to show how the two checking tasks can be realized. For the first task w.l.o.g. we may assume that $k = 2$. Further we may assume that the first $n/2$ cells and the last $n/2$ cells are distinguishable which can be provided by guessing and a simple verification task.

The first checking task is then performed as follows. The last $n/2$ cells shift there guessed initial states in some register with maximum speed to the left. Two initially empty registers of each of the first $n/2$ cells work as a queue in a first-in first-out manner through which the arriving symbol stream is successively piped (cf. Fig. 1). Additionally in the rightmost cell a signal is generated in the first time step which moves leftward with maximum speed. If it enters one of the first $n/2$ cells it checks whether the cell's guessed initial states which were stored in some register are equal to the initial states that are currently in the position to leave the queue next. If they are not equal the signal prohibits the cell and the cells left from this cell to become final.

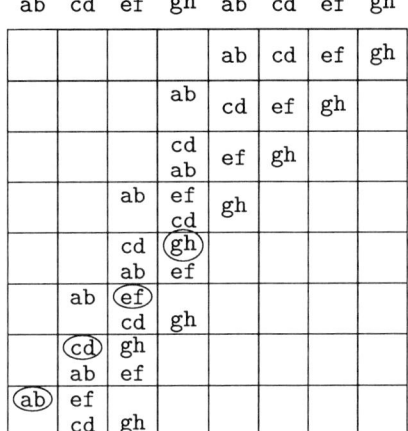

Fig. 1. Example to the proof of Theorem 4 with $k = 2$

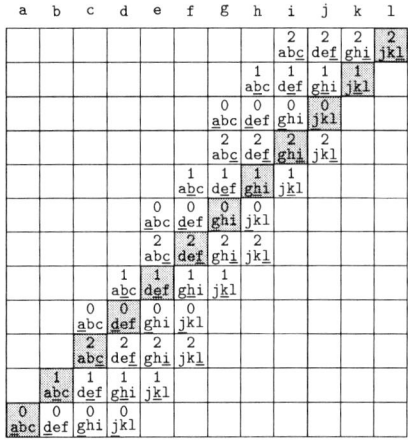

Fig. 2. Example to the proof of Theorem 4 with $k = 3$

To perform the second checking task each cell is equipped with a counter modulo k which is initialized to $k-1$ and decremented by one at each time step. Further at each time step a cell takes over the guessed packed input symbols of its right neighbour if its counter differs from $k-1$ such that they are shifted through the cells. Otherwise a cell holds the packed input symbols which it actually contains (cf. Fig. 2). Again in the rightmost cell a signal is generated in the first time step which moves leftward with maximum speed. If it enters a cell it checks whether the symbol in the packed representation at position $r+1$ equals to the (real) initial state of that cell where r denotes the value of its counter. Similarly if there exists some cell where the equality check fails this signal prohibits this cell and the cells left from this cell to become final. □

The next result shows that $k+1$ guesses per cell are not better than k guesses.

Theorem 5. *Let* $g : \mathbb{N} \to \mathbb{N}$, $g(n) \leq t(n)$, *be a mapping and* $k \in \mathbb{N}_0$ *be a constant number. Then* $\mathscr{L}_{t(n)}((g+k)\text{G-OCA}) = \mathscr{L}_{t(n)}(g\text{G-OCA})$ *holds for all mappings* $t : \mathbb{N} \to \mathbb{N}$, $t(n) \geq n$.

Proof. It suffices to show the theorem for $k = 1$.

Let $\mathcal{M}$ be a $(g+1)$G-OCA. We construct a gG-OCA $\mathcal{M}'$ which simulates $\mathcal{M}$ without any loss of time.

In its first (nondeterministic) step $\mathcal{M}'$ simulates the first step of $\mathcal{M}$ and, additionally, another nondeterministic step of $\mathcal{M}$ for all possible pairs of states of $\mathcal{M}$. The second result is stored in an additional register. It is a table $S^2 \times S$, which contains one row for every $(s_1, s_2) \in S^2$. After $g(n)$ time steps the first deterministic step of $\mathcal{M}'$ is as follows. Every cell takes the actual state of itself and its neighbor and selects the corresponding row in the table. The next state is the third component of that row. Since the third components were nondeterministically chosen a nondeterministic transition is simulated deterministically. From time $g(n) + 2$ to $t(n)$ $\mathcal{M}'$ simulates $\mathcal{M}$ directly. □

4 Comparisons with Deterministic Cellular Arrays

In order to compare the real-time computing power of 1G-OCAs to the well-investigated deterministic devices we prove the following theorem which states that the computing power of real-time OCAs is strictly increased by adding one nondeterministic step to that device.

Theorem 6. $\mathscr{L}_{rt}(\text{OCA}) \subset \mathscr{L}_{rt}(\text{1G-OCA})$

Proof. Obviously, we have an inclusion between the families since the nondeterministic part of the state transition can be designed to be deterministic. In [3] $L = \{w^{|w|} \mid w \in A^+\} \in \mathscr{L}_{rt}(\text{1G-OCA})$ has been shown for an arbitrary alphabet A.

In [16] it was shown that it does not belong to $\mathscr{L}_{rt}(\text{OCA})$: L intersected with the regular language $\{a\}^+$ is the unary language $\{a^{n^2} \mid n \in \mathbb{N}\}$ which is not regular and thus not a real-time OCA language. Thus the inclusion is a proper one. □

It is known that $\mathscr{L}_{rt}(\text{OCA})$ is closed under inverse homomorphism [14], injective length multiplying homomorphism [5, 6] and inverse deterministic gsm mappings [10] but is not closed under ε-free homomorphism [14]. There is another relation between $\mathscr{L}_{rt}(\text{OCA})$ and $\mathscr{L}_{rt}(\text{1G-OCA})$. If we build the closure under ε-free homomorphisms of $\mathscr{L}_{rt}(\text{OCA})$ we obtain exactly the family $\mathscr{L}_{rt}(\text{1G-OCA})$. To prove the assertion we need the result from [3] that $\mathscr{L}_{rt}(\text{OCA})$ is closed under another weak kind of homomorphism.

Definition 7. *Let* $h : A^* \to B^*$ *be an ε-free homomorphism. h is* structure preserving *if for every two* $a, a' \in A$ *with* $h(a) = b_1 \cdots b_m$ *and* $h(a') = b'_1 \cdots b'_n$ *the sets* $\{b_1, \ldots, b_m\}$ *and* $\{b'_1, \ldots, b'_n\}$ *are disjoint if* $a \neq a'$.

Lemma 8. $\mathscr{L}_{rt}(\text{OCA})$ *is closed under structure preserving homomorphism.*

Theorem 9. *1. Let L be a language belonging to $\mathscr{L}_{rt}(\text{OCA})$ and h be an ε-free homomorphism. Then $h(L)$ belongs to $\mathscr{L}_{rt}(\text{1G-OCA})$.*
  2. Let L be a language belonging to $\mathscr{L}_{rt}(\text{1G-OCA})$. Then there exist an ε-free homomorphism and a language $L' \in \mathscr{L}_{rt}(\text{OCA})$ such that $h(L') = L$ holds.

Proof. a) Let L be a language over the alphabet $A = \{a_1, \ldots, a_m\}$ and a homomorphism $h : A^* \to B^*$ be defined according to
$$h(a_1) = b_{1,1} \cdots b_{1,n_1}, \ldots, h(a_m) = b_{m,1} \cdots b_{m,n_m},$$
where $b_{i_j} \in B$.

We introduce an alphabet $\bar{B} := \{\bar{b}_{1,1}, \ldots, \bar{b}_{1,m_1}, \bar{b}_{2,1}, \ldots, \bar{b}_{m,n_m}\}$ of different symbols and a structure preserving homomorphism $h' : A^* \to \bar{B}^*$:
$$h'(a_1) = \bar{b}_{1,1} \cdots \bar{b}_{1,n_1}, \ldots, h'(a_m) = \bar{b}_{m,1} \cdots \bar{b}_{m,n_m}.$$
Since $\mathscr{L}_{rt}(\text{OCA})$ is closed under structure preserving homomorphism $h'(L)$ is a real-time OCA language. Define an ε-free length preserving homomorphism $h'' : \bar{B}^* \to B^*$: $h''(\bar{b}_{1,1}) = b_{1,1}, \ldots, h''(\bar{b}_{m,n_m}) = b_{m,n_m}$.

Obviously, we have $h(L) = h''(h'(L))$.

An 1G-OCA $\mathcal{M}'$ that accepts $h(L)$ in $n + 1$ time steps works as follows. Since h'' is length preserving, in the first time step every cell can guess the inverse image of its initial state under h''. During the next n time steps $\mathcal{M}'$ simulates a real-time OCA $\mathcal{M}$ that accepts $h'(L)$. As shown in Lemma 3 we can speed-up $\mathcal{M}$ by one time-step.

b) Let $\mathcal{M}$ be an 1G-OCA accepting L in real-time. $\mathcal{M}$ can be simulated by another 1G-OCA $\mathcal{M}'$ which works in $(n+1)$-time as follows. In the first step $\mathcal{M}'$ guesses the state of each cell of $\mathcal{M}$ at time 2. In the second step $\mathcal{M}'$ verifies its guess. During the steps 3 to $n+1$ $\mathcal{M}'$ works exactly as $\mathcal{M}$ during the steps 2 to n. Now let $\mathcal{M}''$ be an OCA which simulates the computation of $\mathcal{M}'$ during the time steps 2 to $n+1$ and h be the ε-free homomorphism that maps a pair of states of $\mathcal{M}$ to the first component $h(s_1, s_2) = s_1$. Thus $h(L(\mathcal{M}'')) = L$. □

Adding two-way communication to deterministic cellular arrays yields more powerful real-time devices. It is well known that $\mathscr{L}_{rt}(\text{CA})$ is a proper superset of $\mathscr{L}_{rt}(\text{OCA})$ [14]. The following two theorems relate both augmentations.

Theorem 10. $\mathscr{L}_{rt}(\text{CA}) \subseteq \mathscr{L}_{rt}(\text{1G-OCA})$

Proof. In [4, 17] it has been shown that $L \in \mathscr{L}_{rt}(\text{CA})$ if and only if $L^R \in \mathscr{L}_{2id}(\text{OCA})$, where L^R denotes the reversal of L. From Theorem 4 we know $\mathscr{L}_{2id}(\text{1G-OCA}) = \mathscr{L}_{rt}(\text{1G-OCA})$. The inclusion $\mathscr{L}_{2id}(\text{OCA}) \subseteq \mathscr{L}_{2id}(\text{1G-OCA})$ holds due to structural reasons. In [3] the closure of $\mathscr{L}_{rt}(\text{1G-OCA})$ under reversal was shown what proves the assertion. □

Theorem 11. $\mathscr{L}_{rt}(\text{1G-OCA}) \subseteq \mathscr{L}(\text{CA})$

Proof. The idea is to construct a brute-force CA which tries all possible choices of the $\mathscr{L}_{rt}(\text{1G-OCA})$. In order to realize such a behavior we need two mechanisms. One has to select successively all possible choices. The other mechanism is the simulation of the $\mathscr{L}_{rt}(\text{1G-OCA})$ on the actual choice.

To control the mechanisms we use synchronization by a modified fssp. If the synchronization is started at both border cells simultaneously it can be done in exactly n time steps, where n is the length of the array. This process can be repeated such that the array fires every n time steps.

Let S denote the state set of a real-time 1G-OCA. To generate all of its possible time step 1 configurations it suffices to set up a $|S|$-ary counter. At most every possible number on the counter corresponds to one configuration. To increment the counter a signal is send from the border cell containing the least significant digit to the opposite. It needs n time steps.

Subsequently n time steps of the $\mathscr{L}_{rt}$(1G-OCA) are simulated in a straightforward manner. The input is accepted if one of the choices leads to an accepting simulation. Otherwise it is rejected when the border cell containing the most significant digit generates a carry-over. □

5 Closure Properties

The family $\mathscr{L}_{rt}$(1G-OCA) has strong closure properties.

Lemma 12. *$\mathscr{L}_{rt}$(1G-OCA) is closed under union, intersection and set difference.*

Proof. Using the same two channel technique of [15] and [7] the assertion is easily seen. Each cell consists of two registers in which acceptors for both languages are simulated in parallel. □

Theorem 13. *$\mathscr{L}_{rt}$(1G-OCA) is an AFL (i.e. is closed under intersection with regular sets, inverse homomorphism, ε-free homomorphism, union, concatenation and positive closure).*

Proof. Closure under intersection with regular sets and union have been shown in Lemma 12.

Assume there is a language $L' \in \mathscr{L}_{rt}$(1G-OCA) and an ε-free homomorphism h' such that $L'' := h'(L') \notin \mathscr{L}_{rt}$(1G-OCA). From Theorem 9 follows that there exists a language $L \in \mathscr{L}_{rt}$(OCA) and an ε-free homomorphism h such that $h(L) = L'$. Therefore we have $L'' = h'(h(L))$. Since $h' \circ h$ is an ε-free homomorphism too, Theorem 9 is contradicted. The closure under ε-free homomorphism follows.

Let $L \in \mathscr{L}_{rt}$(1G-OCA) be a language over A and $h : B^* \to A^*$ be a homomorphism. From Theorem 9 we obtain a real-time OCA language L' over A' and a length preserving homomorphism $h' : A'^* \to A$ with $h'(L') = L$. Let h_1 be the homomorphism with $h_1((x, x')) = x'$ for every $x \in B$, $x' \in A'$ with $h(x) = h'(x')$. Further let p_1 be an ε-free homomorphism with $p_1((x, x')) = x$ for $x \in B$, $x' \in A'$. Then $p_1(h_1^{-1}(L')) = h^{-1}(h'(L')) = h^{-1}(L)$. Since $\mathscr{L}_{rt}$(OCA) is closed under inverse homomorphism [14], $h_1^{-1}(L')$ belongs to $\mathscr{L}_{rt}$(OCA). Now Theorem 9 implies $h^{-1}(L) \in \mathscr{L}_{rt}$(1G-OCA) which proves the closure under inverse homomorphism.

Now let $L_1, L_2 \in \mathscr{L}_{rt}$(1G-OCA) and $\mathcal{M}_1, \mathcal{M}_2$ be acceptors for L_1 and L_2. We construct a 2G-OCA $\mathcal{M}'$ that accepts the concatenation $L_1 L_2$ in $n+1$ time steps. To accept an input $w_1 w_2$, $w_1 \in L_1$, $w_2 \in L_2$, $\mathcal{M}'$ guesses in its first time step the cell in which the first symbol of w_2 occurs. In the remaining time steps 2 to $n+1$ $\mathcal{M}'$ simulates $\mathcal{M}_1$ in the left part on w_1 and $\mathcal{M}_2$ in the right part of the array on w_2. Due to Theorem 5 we can construct an 1G-OCA that accepts $L_1 L_2$ in time $n+1$ which according to Lemma 3 can be sped-up to work in real-time.

The closure under positive closure follows analogously. □

Corollary 14. $\mathscr{L}_{rt}$(1G-OCA) *is closed under ε-free substitution.*

Proof. In [8] it has been shown that an AFL that is closed under intersection is also closed under ε-free substitution. Thus the assertion follows from Lemma 12 and Theorem 13. □

In [3] it has been shown that $\mathscr{L}_{rt}$(1G-OCA) is closed under reversal and that theorem 11 proves the inclusion $\mathscr{L}_{rt}$(1G-OCA) $\subseteq \mathscr{L}$(CA). Up to now it is not known whether the family is closed under complement. A negative answer would imply that there exists a CA language which is not a real-time CA language.

References

[1] Bucher, W., Čulik II, K.: On real time and linear time cellular automata. RAIRO Inform. Théor. **18** (1984) 307–325
[2] Buchholz, Th., Kutrib, M.: On time computability of functions in one-way cellular automata. Acta Inf. **35** (1998) 329–352
[3] Buchholz, Th., Klein, A., Kutrib, M.: One guess one-way cellular arrays. Research Report 9801, Institute of Informatics, University of Giessen, Gießen, 1998.
[4] Choffrut, C., Čulik II, K.: On real-time cellular automata and trellis automata. Acta Inf. **21** (1984) 393–407
[5] Čulik II, K., Gruska, J., Salomaa, A.: Systolic trellis automata I. Internat. J. Comput. Math. **15** (1984) 195–212
[6] Čulik II, K., Gruska, J., Salomaa, A.: Systolic trellis automata II. Internat. J. Comput. Math. **16** (1984) 3–22
[7] Dyer, C. R.: One-way bounded cellular automata. Inform. Control **44** (1980) 261–281
[8] Ginsburg, S., Hopcroft, J. E.: Two-way balloon automata and AFL. J. Assoc. Comput. Mach. **17** (1970) 3–13
[9] Ibarra, O. H., Jiang, T.: On one-way cellular arrays. SIAM J. Comput. **16** (1987) 1135–1154
[10] Ibarra, O. H., Kim, S. M.: Characterizations and computational complexity of systolic trellis automata. Theoret. Comput. Sci. **29** (1984) 123–153
[11] Ibarra, O. H., Palis, M. A.: Some results concerning linear iterative (systolic) arrays. J. Parallel and Distributed Comput. **2** (1985) 182–218
[12] Kutrib, M.: Pushdown cellular automata. Theoret. Comput. Sci. (1998)
[13] Kutrib, M., Richstein, J.: Real-time one-way pushdown cellular automata languages. In: Developments in Language Theory II. At the Crossroads of Mathematics, Computer Science and Biology, World Scientific, Singapore, (1996) 420–429
[14] Seidel, S. R.: Language recognition and the synchronization of cellular automata. Technical Report 79-02, Department of Computer Science, University of Iowa, 1979
[15] Smith III, A. R.: Real-time language recognition by one-dimensional cellular automata. J. Comput. System Sci. **6** (1972) 233–253
[16] Terrier, V.: Language recognizable in real time by cellular automata. Complex Systems **8** (1994) 325–336
[17] Umeo, H., Morita, K., Sugata, K.: Deterministic one-way simulation of two-way real-time cellular automata and its related problems. Inform. Process. Lett. **14** (1982) 158–161
[18] Vollmar, R.: Algorithmen in Zellularautomaten. Teubner, Stuttgart, 1979

Topological Definitions of Chaos Applied to Cellular Automata Dynamics

Gianpiero Cattaneo[1] and Luciano Margara[2]

[1] Dipartimento di Scienze dell'Informazione, Università di Milano,
Via Comelico 39, 20135 Milano, Italy.
cattang@dsi.unimi.it

[2] Dipartimento di Scienze dell'Informazione, Università di Bologna,
Mura Anteo Zamboni 7, 40127 Bologna, Italy.
margara@cs.unibo.it

Abstract. We apply the two different definitions of chaos given by Devaney and by Knudsen for general discrete time dynamical systems (DTDS) to the case of 1-dimensional cellular automata. A DTDS is chaotic according to the Devaney's definition of chaos iff it is topologically transitive, has dense periodic orbits, and it is sensitive to initial conditions. A DTDS is chaotic according to the Knudsen's definition of chaos iff it has a dense orbit and it is sensitive to initial conditions. We continue the work initiated in [3], [4], [5], and [14] by proving that an easy-to-check property of local rules on which cellular automata are defined–introduced by Hedlund in [11] and called *permutivity*–is a sufficient condition for chaotic behavior. Permutivity turns out to be also a necessary condition for chaos in the case of elementary cellular automata while this is not true for general 1-dimensional cellular automata. The main technical contribution of this paper is the proof that permutivity of the local rule either in the leftmost or in the rightmost variable forces the cellular automaton to have dense periodic orbits.

1 Introduction

The notion of chaos is very appealing, and it has intrigued many scientists (see [1,2,9,13,16] for some works on the properties that characterize a chaotic process). There are simple deterministic dynamical systems that exhibit unpredictable behavior. Though counterintuitive, this fact has a very clear explanation. The lack of *infinite precision* in the description of the state of the system causes a loss of *information* which is dramatic for some processes which quickly loose their deterministic nature to assume a non deterministic (unpredictable) one. A chaotic phenomenon can indeed be viewed as a deterministic one, in the presence of infinite precision, and as a nondeterministic one, in the presence of finite precision constraints. Thus one should look at chaotic processes as at processes merged into time, space, and precision bounds, which are the key resources in the science of computing. A nice way in which one can analyze this finite/infinite dichotomy is by using cellular automata (CA) models. CA are dynamical systems consisting of a regular lattice of variables which can take a finite number of discrete values. The global state of the CA, specified by the values of all the variables at a given time, evolves in synchronous discrete time steps according to a given *local rule* which acts on the value of each single variable.

Consider the 1-dimensional CA $\langle X, \sigma \rangle$, where $X = \{0,1\}^{\mathbf{Z}}$ and σ is the left-shift map on X associating to any configuration $c \in \{0,1\}^{\mathbf{Z}}$ the next time step configuration $\sigma(c) \in \{0,1\}^{\mathbf{Z}}$ defined by $[\sigma(c)](i) = c(i+1)$, $i \in \mathbf{Z}$. In order to completely describe the elements of X, we need to operate on two-sided sequences of binary digits of infinite length. Assume for a moment that this is possible. Then the shift map is completely predictable, i.e., one can completely describe $\sigma^n(x)$, for any $x \in X$ and for any integer n. In practice, only finite objects can be computationally manipulated. Let $x \in X$. Assume we know a portion of x of length n. One can easily verify that $\sigma^n(x)$ completely depends on the unknown portion of x. In other words, if we have finite precision, the shift map becomes unpredictable, as a consequence of the combination of the finite precision representation of x and the *sensitivity* of σ.

In the case of discrete time dynamical systems $\langle X, F \rangle$, many definitions of chaos are based on the notion of sensitivity to initial conditions (see for example [9,12]). Here, we assume that the *phase space* X is equipped with a distance d and that the *next state map* $F : X \mapsto X$ is continuous on X according to the topology induced by the metric d.

Definition 1 (Sensitivity). *A DTDS $\langle X, F \rangle$ is sensitive to initial conditions iff*

$$\exists \delta > 0 \ \forall x \in X \ \forall \epsilon > 0 \ \exists y \in X \ \exists n \in \mathbf{N}: \quad (d(x,y) < \epsilon \text{ and } d(F^n(x), F^n(y)) \geq \delta.)$$

Constant δ is called the sensitivity constant.

Intuitively, a map is sensitive to initial conditions, or simply sensitive, if there exist points arbitrarily close to x which eventually separate from x by at least δ under iteration of F. We emphasize that not all points near x need eventually separate from x, but there must be at least one such point in every neighborhood of x.

In the case of continuous dynamical systems defined on a metric space, there are many possible definitions of chaos, ranging from measure theoretic notions of randomness in ergodic theory to the topological approach we will adopt here. We now recall some other properties which are central to topological chaos theory namely, having a *dense orbit*, *topological transitivity*, and *denseness of periodic points*.

Definition 2 (Dense orbit). *A dynamical system $\langle X, F \rangle$ has a dense orbit iff*

$$\exists x \in X \ \forall y \in X \ \forall \epsilon > 0 \ \exists n \in \mathbf{N}: \quad d(F^n(x), y) < \epsilon.$$

The existence of a dense orbit implies topological transitivity.

Definition 3 (Transitivity). *A dynamical system $\langle X, F \rangle$ is topologically transitive iff for all nonempty open subsets U and V of X*

$$\exists n \in \mathbf{N}: \ F^n(U) \cap V \neq \emptyset.$$

Intuitively, a topologically transitive map has points which eventually move under iteration from one arbitrarily small neighborhood to any other. As a consequence, the dynamical system cannot be decomposed into two disjoint closed sets which are invariant under the map (*undecomposability* condition).

Definition 4 (Denseness of periodic points). *A dynamical system $\langle X, F \rangle$ has dense periodic points iff the set of all the periodic points of F defined by*

$$Per(F) = \left\{ x \in X \mid \exists k \in \mathbf{N}: \ F^k(x) = x \right\},$$

is a dense subset of X, i.e., $\forall x \in X \ \forall \epsilon > 0 \ \exists p \in Per(F): \ d(x, p) < \epsilon$.

Denseness of periodic points is often referred to as the *element of regularity* a chaotic dynamical system should exhibit.

The popular book by Devaney [9] isolates three components as being the essential features of topological chaos. They are formulated for a continuous map $F: X \mapsto X$, on some metric space (X, d).

Definition 5 (D-chaos). *Let $F: X \mapsto X$, be a continuous map on a metric space (X, d). Then the dynamical system $\langle X, F \rangle$ is chaotic according to the Devaney's definition of chaos (D-chaotic) iff*
(**D_1**): *F is topologically transitive,*
(**D_2**): *F has dense periodic points (topological regularity), and*
(**D_3**): *F is sensitive to initial conditions.*

It has been proved in [2] that for a generic DTDS, transitivity and denseness of periodic points imply sensitivity to initial condition. A stronger result has been proved in [6] by one of the authors in the case of CA dynamical systems: topological transitivity alone implies sensitivity to initial conditions. As a consequence of these results, in order to prove that a DTDS $\langle X, F \rangle$ is chaotic in the sense of Devaney, one has only to prove properties D_1 and D_2.

Knudsen in [13] proved that in the case of a dynamical system which is chaotic according to Devaney's definition, the restriction of the dynamics to the set of periodic points (which is clearly invariant) is Devaney's chaotic too. Due to the lack of nonperiodicity this is not the kind of system most people would consider labeling chaotic. In view of these considerations, Knudsen proposed the following definition of chaos which excludes chaos without non-periodicity [13].

Definition 6 (K-chaos). *Let $F: X \mapsto X$, be a continuous map on a metric space (X, d). Then the dynamical system $\langle X, F \rangle$ is chaotic according to the Knudsen's definition of chaos (K-chaotic) iff*
(**K_1**): *F has a dense orbit, and*
(**K_2**): *F is sensitive to initial conditions.*

The two-sided shift dynamical system $\langle \mathcal{A}^{\mathbf{Z}}, \sigma \rangle$ on a finite alphabet $\mathcal{A}$ is a paradigmatic example of both Devaney's and Knudsen's chaotic system. In the case of perfect compact DTDS, i.e., DTDS whose phase space is a perfect and compact metric space, we have that topological transitivity is equivalent to have a dense orbit. In addition, in this case the next state map is surjective. As we will see later, the phase space of DTDS induced by CA local rules is perfect and compact. As a consequence, the following immediately follows.
1- If a compact DTDS $\langle X, F \rangle$ is D-chaotic then it is K-chaotic.
2- In the case of a DTDS $\langle \mathcal{A}^{\mathbf{Z}}, F_f \rangle$ induced by a CA local rule f, the dynamical system is K-chaotic iff it is topologically transitive.
In the case of 1-dimensional CA, there have been many attempts of classification according to their asymptotic behavior (see for example [5,7,10,15,17]), but none of them completely captures the notion of chaos. As an example, Wolfram divides 1-dimensional CA in four classes according to the outcome of a large number of experiments. Wolfram's classification scheme, which does not rely on a precise mathematical definition, has been formalized by Culik and Yu [8] who split CA in three classes of increasing complexity. Unfortunately membership in each of these classes is shown to be undecidable.
In this paper we complete the work initiated in [3], [4], [5], and [14], where the authors for the first time apply the definition of chaos given by Devaney and by Knudsen to CA. More precisely:
- In [5] the authors make a detailed analysis of the behavior of the elementary CA based on a particular non-additive rule (rule 180) and prove its chaoticity according to the Devaney's definition of chaos.
- In [14] the authors completely classify 1-dimensional additive CA defined over any alphabet of prime cardinality according to the Devaney's definition of chaos.
- In [4] the authors completely characterize topological transitivity for every D-dimensional additive CA over $\mathbf{Z}_m$ ($m \geq 2$, and $D \geq 1$) and denseness of periodic points for any 1-dimensional additive CA over $\mathbf{Z}_m$ ($m \geq 2$).
- In [3] the authors classify a number of non-additive elementary CA (ECA), i.e., binary 1-dimensional CA with radius 1, according to the Devaney definition of chaos and leave open the problem of the classification of all ECA.
In this paper we apply both the Devaney's and the Knudsen's definitions of chaos to the class of ECA. To this extent, we introduce the notion of *permutivity* of a map in a certain variable. A boolean map f is permutive in the variable x_i if $f(\ldots, x_i, \ldots) = 1 - f(\ldots, 1 - x_i, \ldots)$. In other words, f is permutive in the variable x_i if any change of the value of x_i causes a change of the output produced by f, independently of the values assumed by the other variables. The main results of this paper can be summarized as follows.

- Every 1-dimensional CA based on a local rule f which is permutive either in the first (leftmost) or in the last (rightmost) variable is Devaney, and then Knudsen, chaotic.
- An ECA based on a local rule f is Devaney chaotic if and only if f is permutive either in the first (leftmost) or in the last (rightmost) variable (in this case Devaney and Knudsen chaoticity are equivalent).

– There exist chaotic CA based on local rules that are not permutive in any variable.

We wish to emphasize that in this paper we propose the first complete classification of the ECA rule space based on a widely accepted rigorous mathematical definition of chaos.

2 Notations and Definitions

For $m \geq 2$, let $\mathcal{A} = \{0, 1, \ldots, m-1\}$ denote the ring of integers modulo m with the usual operations of addition and multiplication modulo m. We call $\mathcal{A}$ the *alphabet* of the CA. Let $f : \mathcal{A}^{2k+1} \to \mathcal{A}$, be any map depending on the $2k + 1$ variables $x_{-k}, \ldots, x_k$. We say that k is the *radius* of f. A 1-dimensional CA based on the *local rule* f is the pair $\langle \mathcal{A}^{\mathbf{Z}}, F \rangle$, where

$$\mathcal{A}^{\mathbf{Z}} = \{c : \mathbf{Z} \mapsto \mathcal{A}, \ i \to c(i)\}$$

is the *space of configurations* and $F : \mathcal{A}^{\mathbf{Z}} \mapsto \mathcal{A}^{\mathbf{Z}}$ is the global *next state map*, defined as follows. For any configuration $c \in \mathcal{A}^{\mathbf{Z}}$ and for any $i \in \mathbf{Z}$

$$[F(c)](i) = f(c(i-k), \ldots, c(i+k)).$$

Throughout the paper, $F(c) \in \mathcal{A}^{\mathbf{Z}}$ will denote the result of the application of the map F to the configuration $c \in \mathcal{A}^{\mathbf{Z}}$ and $c(i) \in \mathcal{A}$ will denote the ith element of the configuration c. We recursively define $F^n(c)$ by $F^n(c) = F(F^{n-1}(c))$, where $F^0(c) = c$.

In order to specialize the notions of sensitivity to the case of D-dimensional CA, we introduce the following distance (known as *Tychonoff distance*) over the space of configurations. For every $a, b \in \mathcal{A}^{\mathbf{Z}}$

$$d(a, b) = \sum_{i=-\infty}^{+\infty} \frac{1}{m^{|i|}} |a(i) - b(i)|,$$

where m is the cardinality of $\mathcal{A}$. It is easy to verify that d is a metric on $\mathcal{A}^{\mathbf{Z}}$ and that the metric topology induced by d coincides with the product topology induced by the discrete topology of $\mathcal{A}$. With this topology, $\mathcal{A}^{\mathbf{Z}}$ is a complete, compact and totally disconnected space and F is a (uniformly) continuous map, whatever be the CA local rule inducing this global next state map. We now give the definition of permutive local rule and that of leftmost [resp., rightmost] permutive local rule, respectively.

Definition 7. *[11]* f *is permutive in* x_i, $-k \leq i \leq k$, *iff for any given sequence*

$$\overline{x}_{-k}, \ldots, \overline{x}_{i-1}, \overline{x}_{i+1}, \ldots, \overline{x}_k \in \mathcal{A}^{2k}$$

we have

$$\{f(\overline{x}_{-k}, \ldots, \overline{x}_{i-1}, x_i, \overline{x}_{i+1}, \ldots, \overline{x}_k) : \ x_i \in \mathcal{A}\} = \mathcal{A}.$$

Definition 8. *The CA local rule f is said to be* leftmost *[resp.,* rightmost*] permutive iff there exists an integer $i : -k \leq i \leq 0$ [resp., $i : 0 \leq i \leq k$] such that*
1. $i \neq 0$,
2. f *is permutive in the ith variable, and*
3. f *does not depend on x_j, $j < i$, [resp., $j > i$].*

Definition 9. *A 1-dimensional CA based on a local rule $f : \mathcal{A}^{2k+1} \mapsto \mathcal{A}$, is an* elementary *CA (ECA) iff $k = 1$ and $\mathcal{A} = \{0, 1\}$.*

We enumerate the $2^{2^3} = 256$ different ECA as follows. The ECA based on the local rule f is associated with the natural number n_f, where

$$n_f = f(0,0,0) \cdot 2^0 + f(0,0,1) \cdot 2^1 + \cdots + f(1,1,0) \cdot 2^6 + f(1,1,1) \cdot 2^7.$$

In the case of ECA, a rule $f : \{0,1\}^3 \mapsto \{0,1\}$ is leftmost permutive iff

$$\forall x_0, x_1 : \quad f(0, x_0, x_1) \neq f(1, x_0, x_1).$$

Similarly, it is rightmost permutive iff

$$\forall x_{-1}, x_0 : \quad f(x_{-1}, x_0, 0) \neq f(x_{-1}, x_0, 1).$$

3 Chaotic Cellular Automata

In this section we analyze global dynamics of 1-dimensional CA according to both Knudsen's and Devaney's Definitions of Chaos.

Leftmost and/or Rightmost Permutive CA: D-Chaos.
We recall the following result due to one of the authors.

Theorem 1. *[14] Let $\langle \mathcal{A}^{\mathbf{Z}}, F \rangle$ be any leftmost and/or rightmost permutive 1-dimensional CA defined on a finite alphabet $\mathcal{A}$. Then $\langle \mathcal{A}^{\mathbf{Z}}, F \rangle$ is topologically transitive (K-chaotic).*

We now prove now that Leftmost [Rightmost] Permutive 1-dimensional CA have dense periodic points. To this extent we need some preliminary definitions and lemmas. We say that a configuration $x \in \mathcal{A}^{\mathbf{Z}}$ is spatially periodic iff there exists $s \in \mathbf{N}$ such that $\sigma^s(x) = x$.

Lemma 1. *Let $\langle \mathcal{A}^{\mathbf{Z}}, F \rangle$ be a surjective CA. Every predecessor according to F of a spatially periodic configuration is spatially periodic*

Proof. Let $x, y \in \mathcal{A}^{\mathbf{Z}}$ be such that $F(x) = y$ and $\sigma^s(y) = y$ for some $s \in \mathbf{N}$. For every $i \in \mathbf{Z}$ we have

$$F(\sigma^{is}(x)) = \sigma^{is}(F(x)) = \sigma^{is}(y) = y.$$

Assume that x is not spatially periodic. Then there exist infinitely many predecessors of y according to F namely, $\sigma^{is}(x)$, $i \in \mathbf{Z}$. Since every 1-dimensional surjective CA have a finite number of predecessors (see [11]), we have a contradiction.

We now give the definition of *Right [Left]* CA.

Definition 10. Let $\langle \mathcal{A}^\mathbf{Z}, F \rangle$ be a CA based on the local rule $f(x_{-r}, \ldots, x_r)$. F is a Right [Left] CA iff f does not depend on $x_{-r}, \ldots, x_0$ [$x_0, \ldots, x_r$].

We have the following Lemma.

Lemma 2. Let $\langle \mathcal{A}^\mathbf{Z}, F \rangle$ be a Right [Left] CA. Then $G = I - F$ is surjective, where I denotes the identity map.

Proof. Since F is a Right [Left] CA, we have that $I - F$ is Leftmost [Rightmost] permutive and then surjective.

In the next theorem we prove that for surjective Right [Left] CA periodic configurations are also spatially periodic.

Theorem 2. Let $\langle \mathcal{A}^\mathbf{Z}, F \rangle$ be a Right [Left] CA (non necessarily surjective). Then for every $x \in \mathcal{A}^\mathbf{Z}$ we have

$$\left(\exists t \in \mathbf{N}:\ F^t(x) = x\right) \Rightarrow \left(\exists s \in \mathbf{N}:\ \sigma^s(x) = x\right)$$

Proof. If x is periodic for F, i.e., $F^n(x) = x$, then x is a predecessor of the all-zero configuration $(\ldots, 0, 0, 0, \ldots)$ according to $G = I - F^n$. Since F is a Right [Left] CA then F^n is again a Right [Left] CA and then, from Lemma 2, we have that $G = I - F^n$ is surjective for every $n \in \mathbf{N}$. From Lemma 1 we conclude that x is spatially periodic.

Corollary 1. Let $\langle \mathcal{A}^\mathbf{Z}, F \rangle$ be a (non necessarily surjective) CA. Let $n \in \mathbf{Z}$ be such that $G = \sigma^n F$ is a Right [Left] CA global map. Every periodic configuration for G is periodic also for F, i.e.,

$$\left(\exists t \in \mathbf{N}:\ G^t(x) = x\right) \Rightarrow \left(\exists t' \in \mathbf{N}:\ F^{t'}(x) = x\right)$$

Proof. Let x be such that $G^t(x) = x$. From Theorem 2 we have that there exists $s \in \mathbf{N}$ such that $\sigma^s(x) = x$. We have

$$x = G^t(x) = G^{ts}(x) = (\sigma^n F)^{ts}(x)$$
$$= \sigma^{nts} F^{ts}(x) = F^{ts} \sigma^{nts}(x) = F^{ts}(x).$$

We are now ready to prove the main result of this section.

Theorem 3. *Rightmost [Leftmost] permutive 1-dimensional CA have dense periodic points.*

Proof. Assume without loss of generality that F is Rightmost permutive. Let $s \in \mathbf{Z}$ be such that $G_s = \sigma^s F$ is a Right CA. We now prove that G_s has dense periodic orbits. Let

$$w = (w_{-k} \cdots w_0 \cdots w_k) \in \mathcal{A}_m^{2k+1}$$

be any finite configuration of length $2k+1$. Let $V_0 \in \mathcal{A}_m^{\mathbf{Z}}$ be the following configuration
$$V_0 = \cdots \alpha_2 \alpha_1 w_{-k} \cdots w_0 \cdots w_k \beta_1 \beta_2 \cdots,$$
where w is centered at the origin of the lattice, i.e., $V_0(0) = w_0$. Since G_s is a rightmost permutive CA then, in view of Theorem 1, it is topologically transitive and there exist $n \in \mathbf{N}$ and
$$W_0 = \cdots \alpha'_2 \alpha'_1 w_{-k} \cdots w_0 \cdots w_k \beta'_1 \beta'_2 \cdots$$
such that
$$G^n(V_0) = W_0.$$
Let
$$W_1 = \cdots \alpha'_2 \alpha'_1 w_{-k} \cdots w_0 \cdots w_k \beta_1 \beta'_2 \cdots.$$
Since G_s is a Right and Rightmost permutive CA one can find suitable β''_i, $i \geq 2$, such that
$$G_s^n(\cdots \alpha_3 \alpha_2 \alpha_1 w_{-k} \cdots w_0 \cdots w_k \beta_1 \beta''_2 \beta''_3 \cdots) = W_1.$$
Let
$$V_1 = \cdots \alpha_3 \alpha_2 \alpha'_1 w_{-k} \cdots w_0 \cdots w_k \beta_1 \beta''_2 \beta''_3 \cdots.$$
Since G_s is a Right CA, we have
$$G_s^n(V_1) = \cdots \alpha'''_3 \alpha''_2 \alpha'_1 w_{-k} \cdots w_0 \cdots w_k \beta_1 \beta'_2 \beta'_3 \cdots,$$
for some α'''_i, $i \geq 2$.
By repeating the above procedure we are able to construct a sequence of pairs of configurations (V_i, W_i) such that $G^n(V_i) = W_i$ and $V_i(j) = W_i(j)$ for $j = -i-k, \ldots, k+i$ and $i = 1, 2, \ldots$. Since $\mathcal{A}^{\mathbf{Z}}$ is a complete space we have
$$\lim_{i \to \infty} W_i = \lim_{i \to \infty} V_i = W \quad \text{and} \quad G^n(W) = W.$$
Since w can be arbitrarily chosen, we conclude that G has dense periodic orbits. Finally, from Corollary 1 we conclude that F has dense periodic points.

Leftmost and/or Rightmost Permutive ECA: Topological Chaos.
Since boolean CA are based on the alphabet $\{0,1\}$ which has prime cardinality and from Theorem 1, we have the following corollary.

Corollary 2. *All the leftmost and/or rightmost permutive ECA are D-chaotic.*

Next result (whose proof will be given in the full paper) shows that if an ECA is neither leftmost nor rightmost permutive, then it is not surjective and then not topologically transitive.

Theorem 4. *Let $\langle \{0,1\}^{\mathbf{Z}}, F \rangle$ be a non trivial ECA based on the local rule $f : \{0,1\}^3 \mapsto \{0,1\}$. If F is surjective, then f is either leftmost or rightmost permutive.*

We summarize the results of this paper in the following corollary.

Corollary 3. *Let* $\langle \{0,1\}^{\mathbf{Z}}, F \rangle$ *be an ECA based on the local rule* f. *Then, the following statements are equivalent.*
1. *f is either leftmost or rightmost permutive (or both).*
2. *F is Devaney-chaotic.*
3. *F is Knudsen-chaotic.*
4. *F is surjective and non-trivial.*

References

1. D. Assaf, IV and W. A. Coppel, Definition of Chaos. *The American Mathematical Monthly* **99**, 865, 1992.
2. J. Banks, J. Brooks, G. Cairns, G. Davis, and P. Stacey, On the Devaney's Definition of Chaos. *The American Mathematical Monthly* **99**, 332–334, 1992.
3. G. Cattaneo, M. Finelli, L. Margara, Topological Chaos for Elementary Cellular Automata. *Italian Conference on Algorithm and Complexity (CIAC'97), LNCS n. 1203.*
4. G. Cattaneo, E. Formenti, G. Manzini, and L. Margara, Ergodicity, Transitivity, and Regularity for Linear Cellular Automata Over $\mathbf{Z}_m$. *Theoretical Computer Science*, to appear. A preliminary version of this paper has been presented to the *Symposium of Theoretical Computer Science (STACS'97), LNCS n. 1200.*
5. G. Cattaneo, L. Margara, Generalized Sub-shifts in Elementary Cellular Automata. The "Strange Case" of Chaotic Rule 180. *Theoretical Computer Science,* to appear.
6. B. Codenotti and L. Margara, Transitive Cellular Automata are Sensitive. *The American Mathematical Monthly* **103**, *58–62, 1996.*
7. K. Culik, L. P. Hurd, and S. Yu, Computation Theoretic Aspects of Cellular Automata. *Physica D* **45**, *357–378, 1990.*
8. K.Culik and S. Yu, Undecidability of Cellular Automata Classification Schemes. *Complex Systems* **2**(2), *177–190, 1988.*
9. R. L. Devaney, An Introduction to Chaotic Dynamical Systems. *Addison Wesley,* 1989.
10. H. A. Gutowitz, A Hierarchical Classification of Cellular Automata. *Physica D* **45**, 136–156, 1990.
11. G. A. Hedlund, Endomorphism and Automorphism of the Shift Dynamical System. *Mathematical System Theory* **3**(4), *320–375, 1970.*
12. C. Knudsen, Aspects of Noninvertible Dynamics and Chaos, *Ph.D. Thesis,* 1994.
13. C. Knudsen, Chaos Without Nonperiodicity, *The American Mathematical Monthly* **101**, *563–565, 1994.*
14. P. Favati, G. Lotti and L. Margara, Additive One Dimensional Cellular Automata are Chaotic According to Devaney's Definition of Chaos. *Theoretical Computer Science* **174**(1-2), *157–170, 1997.*
15. K. Sutner, Classifying Circular Cellular Automata. *Physica D* **45**, *386–395, 1990.*
16. M. Vellekoop and R. Berglund, On Intervals, Transitivity = Chaos. *The American Mathematical Monthly* **101**, *353–355, 1994.*
17. S. Wolfram, Theory and Application of Cellular Automata. *Word Scientific Publishing Co., Singapore,* 1986.

Characterization of Sensitive Linear Cellular Automata with Respect to the Counting Distance

Giovanni Manzini[1,2]

[1] Dipartimento di Scienze e Tecnologie Avanzate, Università Piemonte Orientale, 15100 Alessandria, Italy.
[2] Istituto di Matematica Computazionale, CNR, 56126 Pisa, Italy.

Abstract. In this paper we give sufficient and necessary conditions for a linear 1-dimensional cellular automaton F to be sensitive with respect to the counting distance defined by Cattaneo et al. in [MFCS '97, pagg. 179–188]. We prove an easy-to-check characterization in terms of the coefficients of the local rule, and an alternative characterization based on the properties of the iterated map F^n.

1 Introduction

One-dimensional Cellular Automata (CA) are dynamical systems consisting of a bi-infinite array of cells which can take a finite number of discrete values. The *global state* of the CA, specified by the values of all the cells at a given time, evolves in synchronous discrete time steps according to a given *local rule* which acts on the value of each single cell. CA have been widely studied in a number of disciplines (e.g., computer science, physics, mathematics, biology, chemistry) with different purposes (e.g., simulation of natural phenomena, pseudo-random number generation, image processing, analysis of universal model of computations, cryptography).

Since their introduction, CA have been analyzed using tools derived from the theory of discrete time dynamical systems. That is, the concepts of sensitivity, expansivity, transitivity, ergodicity, *etc.* have been considered for CA, and many efforts have been made to determine when a CA satisfies one of these properties. This study has provided many insights into the relationship between the local rule governing the evolution of the CA and its long term behavior (see for example [1,2,5,6,8,12,10,11,13]).

The study of some of the dynamical properties of CA requires the introduction of a distance over the space of configurations. The most commonly used distance is the so-called *Tychonoff distance*. According to this distance, two configurations are close if they differ only in cells which are far away from the center of the bi-infinite array. As a result, the study of, say sensitivity or transitivity, with respect to this metric provides useful information on the extent to which differences far away from the center of the array can influence the central region.

If we study the dynamical properties with respect to a different distance, we will likely get quite different results. Obviously, there is nothing like a "true" distance for CA: every distance deserves to be investigated if it provides useful information on the behavior of CA. In [3], the authors introduce a new distance for CA which has many attractive features. This distance, that we call *counting distance*, weights all cells equally. The distance between two configurations x, y is based on the asymptotic ratio between the number of cells in which x and y differ and the number of cells in which they coincide. Loosely speaking, when we use this metric the dynamical properties measure the effects of modifying a small fraction of the cells, regardless of where they are located.

In this paper we characterize which linear CA over the ring $\mathbf{Z}_m$ are sensitive with respect to the counting distance. We provide two characterizations. The first one (Corollary 1) is based on gcd computations involving the coefficients of the local rule. The second one (Corollary 2) is based on the number of nonzero cells generated by the CA starting with a configuration containing a single nonzero cell and executing n iterations. The proofs of these results required techniques which are different, and in general more complex, from those employed in [10] for the characterization of sensitive CA with respect to the Tychonoff distance. Comparing the results of this paper with those in [10], we get that linear CA which are sensitive with respect to the counting distance are sensitive with respect to the Tychonoff distance, whereas the vice versa does not hold.

Sensitivity with respect to the counting distance has been studied also in [3]. There the authors prove that the linear CA based on the elementary rules 60 and 90 are sensitive, and conjecture that sensitivity is related to the Hausdorff dimension of the CA limit set. We believe that the results obtained in this paper can be useful to prove or disprove this conjecture.

Due to the limited space we omit the proof of some of the technical lemmas. Full details are given in [9].

2 Cellular Automata

For $m \geq 2$, let $\mathbf{Z}_m = \{0, 1, \ldots, m-1\}$ denote the ring of integers modulo m. We consider the *space of configurations* $\mathcal{C}_m = \{c \mid c\colon \mathbf{Z} \to \mathbf{Z}_m\}$, which consists of all functions from $\mathbf{Z}$ into $\mathbf{Z}_m$. Each element of $\mathcal{C}_m$ can be visualized as a bi-infinite array in which each cell contains an element of $\mathbf{Z}_m$.

Let $r \geq 0$. A (1-dimensional) CA of *radius* r is a map $F\colon \mathcal{C}_m \to \mathcal{C}_m$ defined as follows

$$[F(c)](i) = f(c(i-r), c(i-r+1), \ldots, c(i+r-1), c(i+r)), \qquad c \in \mathcal{C}_m,\ i \in \mathbf{Z}.$$

In other words, the content of cell i in the configuration $F(c)$ is a function of the contents of cells $i-r, \ldots, i+r$ in the configuration c. Note that the same *local rule* f determines the new value $[F(c)](i)$ for all $i \in \mathbf{Z}$. In this paper we consider *linear* CA, that is, CA which have a local rule of the form $f(x_{-r}, \ldots, x_r) =$

$\sum_{i=-r}^{r} a_i x_i \mod m$, where at least one of a_{-r} and a_r is nonzero. Using this notation, the global map F becomes

$$[F(c)](i) = \sum_{j=-r}^{r} a_j c(i+j) \mod m, \quad c \in \mathcal{C}_m, \ i \in \mathbf{Z}.$$

Throughout the paper, $F(c)$ will denote the result of the application of the map F to the configuration c and $c(i)$ will denote the value assumed by c in i. For $n \geq 0$, we recursively define $F^n(c)$ by $F^n(c) = F(F^{n-1}(c))$, where $F^0(c) = c$. Given two configurations $a, b \in \mathcal{C}_m$ we define their sum $a+b$ by the rule $(a+b)(i) = a(i) + b(i) \mod m$. If F is linear we have $F(a+b) = F(a) + F(b)$. A special configuration is the *null* configuration $\mathbf{0}$ which has the property that $\mathbf{0}(i) = 0$ for all $i \in \mathbf{Z}$. In the following we use $\rho(F)$ to denote the radius of a CA F. Note that $\rho(F^n) \leq n\rho(F)$. The equality always holds when m is prime, but does not hold in general. For example, if $m = 4$ and $f(x_0, x_1) = x_0 + 2x_1$ we have $\rho(F) = 1$ and $\rho(F^2) = 0$.

Example 1. An important CA is the *right shift map* $(\mathcal{C}_m, \sigma)$ defined by $[\sigma(c)](i) = c(i-1)$. The map σ corresponds to the local rule $f(x_{-1}, x_0, x_1) = x_{-1}$. The inverse of σ is the *left shift map* defined by $[\sigma^{-1}(c)](i) = c(i+1)$ which corresponds to the local rule $f(x_{-1}, x_0, x_1) = x_1$. For $j \geq 0$ the iterated map σ^j is such that

$$[\sigma^j(c)](i) = c(i-j), \quad c \in \mathcal{C}_m, \ i \in \mathbf{Z}. \tag{1}$$

In the following we use σ^j with $j < 0$ to denote the map σ^{-1} iterated $|j|$ times. Note that, using this notation, (1) holds for any $j \in \mathbf{Z}$. A fundamental property of the shift map is that it commutes with any other CA $(\mathcal{C}_m, F)$. That is, for any $i \in \mathbf{Z}$, we have $\sigma^i \circ F = F \circ \sigma^i$. □

A convenient notation for the study of linear CA, is the *formal power series* (fps) representation of the configuration space $\mathcal{C}_m$ (see [7, Sec. 3] for details) in which to each configuration $c \in \mathcal{C}_m$ we associate the fps $P_c(X) = \sum_{i \in \mathbf{Z}} c(i) X^i$. The advantage of this representation is that the computation of a linear map is equivalent to power series multiplication. Let $F \colon \mathcal{C}_m \to \mathcal{C}_m$ be a linear map with local rule $f(x_{-r}, \ldots, x_r) = \sum_{i=-r}^{r} a_i x_i$. We associate to F the finite fps $A(X) = \sum_{i=-r}^{r} a_i X^{-i}$. Then, for any $c \in \mathcal{C}_m$ we have $P_{F(c)}(X) = P_c(X) A(X) \mod m$.

3 Sensitivity and the Counting Distance over $\mathcal{C}_m$

Sensitivity is a central notion in the study of the qualitative behavior of discrete time dynamical systems (see for example [4]). The general definition of sensitivity to initial conditions is given for a metric space (X, d) and a map $F \colon X \to X$ continuous on X according to the topology induced by $d \colon X \times X \to \mathbf{R}_+$.

Definition 1 (Sensitivity). *A dynamical system (X, F) is sensitive to initial conditions if and only if there exists $\delta > 0$ such that for any $y \in X$ and for any $\epsilon > 0$, there exist $z \in X$ and $n > 0$, such that $d(y, z) < \epsilon$ and $d(F^n(y), F^n(z)) > \delta$. The value δ is called the sensitivity constant.* □

For CA, sensitivity, as well as other metric properties such as expansivity and transitivity, have been studied with respect to the *Tychonoff distance* d_T (as defined for example in [1]). A CA is sensitive with respect to d_T if we can modify cells which are arbitrarily far away from the center of the array in such a way that the iteration of the map F will eventually "move" this "perturbation" close to the center. In other words, a map is sensitive if there exist "perturbations" which "propagate" for an arbitrarily large distance (the speed of these perturbations have been analyzed in [5]). Linear CA which are sensitive with respect to d_T have been characterized in [10] where it is shown that a linear CA $(\mathcal{C}_m, F)$, with local rule $f(x_{-r}, \ldots, x_r) = \sum_{i=-r}^{r} a_i x_i$, is sensitive if and only if there exists a prime p such that $p \mid m$ and $p \nmid \gcd(a_{-r}, \ldots, a_{-1}, a_1, \ldots, a_r)$.

In this paper we characterize CA which are sensitive with respect to the counting distance over $\mathcal{C}_m$ introduced in [3]. This distance is defined as follows. For any pair of configurations $x, y \in \mathcal{C}_m$ and $n > 0$ let

$$\Delta_{[n]}(x, y) = \#\{i \in [-n, n] \mid x(i) \neq y(i)\}.$$

The value $\Delta_{[n]}(x, y)$ gives the number of cells with an index i, $|i| \leq n$, in which the configurations x and y differ. The counting distance $d(x, y)$ is defined by

$$d(x, y) = \limsup_{n \to \infty} \frac{\Delta_{[n]}(x, y)}{2n + 1}, \qquad (2)$$

and measures the asymptotic ratio between the number of differences and the number of cells. In [3] the authors have shown that d is a pseudo-metric which makes $\mathcal{C}_m$ a non-compact perfect space. In addition, CA are uniformly continuous with respect to d. One of the main reason for introducing the counting distance is that d "weights" all cells equally, whereas this is not true for the Tychonoff distance d_T. One can easily prove that, for all $i \in \mathbf{Z}$, $d(x, y) = d(\sigma^i(x), \sigma^i(y))$, where σ is the shift map defined in Example 1. We say that d is *shift invariant*. Note that d is also *translation invariant*, that is, for all $a, b, c \in \mathcal{C}_m$ we have $d(a, b) = d(a + c, b + c)$.

Loosely speaking, a CA F is sensitive with respect to the counting distance if we can modify each configuration in an arbitrarily small fraction of the cells in such a way that the repeated iteration of F will increase the number of differences up to a constant fraction of the total number of cells (this fraction being determined by the sensitivity constant δ).

Let $m = pq$, with $\gcd(p, q) = 1$. It is well known that the ring $\mathbf{Z}_m$ is isomorphic to the direct product $\mathbf{Z}_p \otimes \mathbf{Z}_q$. That is, each element $a \in \mathbf{Z}_m$ can be replaced by the pair $\langle a \bmod p, a \bmod q \rangle$ with sums and products done componentwise. We can extend this isomorphism to $\mathcal{C}_m$ which can be seen as the direct product $\mathcal{C}_p \otimes \mathcal{C}_q$. To each configuration $x \in \mathcal{C}_m$ we associate the pair $\langle x_p, x_q \rangle$ such that, for all $i \in \mathbf{Z}$, $x_p(i) = x(i) \bmod p$, and $x_q(i) = x(i) \bmod q$. Define $F_p: \mathcal{C}_p \to \mathcal{C}_p$ (resp. $F_q: \mathcal{C}_q \to \mathcal{C}_q$) by $F_p(x) = F(x) \bmod p$ (resp. $F_q(x) = F(x) \bmod q$). The above discussion suggests that the properties of $(\mathcal{C}_m, F)$ can be inferred by the properties of the two CA $(\mathcal{C}_p, F_p)$, and $(\mathcal{C}_q, F_q)$. The following lemma shows that this is indeed the case for sensitivity.

Lemma 1. Let $(\mathcal{C}_m, F)$ denote a linear CA. Let $m = pq$ with $\gcd(p,q) = 1$. Then, $(\mathcal{C}_m, F)$ is sensitive if and only if at least one of $(\mathcal{C}_p, F_p)$ and $(\mathcal{C}_q, F_q)$ is sensitive. □

Note that by the above lemma, in order to characterize sensitive linear CA it suffices to consider the case in which the alphabet size is a prime power.

We conclude this section with a characterization of sensitive linear CA that will be used to establish our main results.

Lemma 2. Let $(\mathcal{C}_m, F)$ denote a linear CA. The map F is sensitive to initial conditions if and only if there exists a constant c and a sequence $\{x_i, n_i\}_{i \in \mathbb{N}}$, $x_i \in \mathcal{C}_m$, $n_i \in \mathbb{N}$ such that

$$\lim_{i \to \infty} d(x_i, \mathbf{0}) = 0, \quad \text{and} \quad d(F^{n_i}(x_i), \mathbf{0}) \geq c. \tag{3}$$

□

4 Characterization of Sensitive 1-Dimensional Linear CA

Let $z_0 \in \mathcal{C}_m$ denote the configuration such that $z_0(0) = 1$, and $z_0(i) = 0$ for $i \neq 0$. For any $x \in \mathcal{C}_m$ we denote by $\|x\|_\#$ the number of nonzero cells in the configuration x (assuming this number is finite). That is, $\|x\|_\# = \lim_{n \to \infty} \Delta_{[n]}(x, \mathbf{0})$. With a little abuse of notation, in the following we use $\|A(X)\|_\#$ to denote the number of nonzero coefficients in the finite fps $A(X)$.

The following lemma is equivalent to Theorem 6 in [3].

Lemma 3. Let $(\mathcal{C}_m, F)$ denote a linear CA. If

$$\lim_{n \to \infty} \rho(F^n) = \infty, \quad \text{and} \quad \limsup_{n \to \infty} \frac{\|F^n(z_0)\|_\#}{\rho(F^n)} > 0$$

then F is sensitive to initial conditions. □

The next lemma is the main technical result of this section. It essentially establishes a sufficient condition for sensitivity for linear CA over $\mathbf{Z}_p$ with p prime.

Lemma 4. Let $(\mathcal{C}_p, F)$ denote a linear CA over $\mathbf{Z}_p$, p prime, with radius $r > 0$. Let $A(X) = \sum_{i=-r}^{r} a_i X^{-i}$ denote the finite fps associated with F. If $\|A(X)\|_\# \geq 2$ then

$$\limsup_{n \to \infty} \frac{\|F^n(z_0)\|_\#}{\rho(F^n)} > 0.$$

Proof. Note that, since the finite fps associated with F^n is $A^n(X)$, we have $\|F^n(z_0)\|_\# = \|A^n(X)\|_\#$. In addition, since p is prime, we have $\rho(F^n) = nr$. Assuming $\|A(X)\|_\# \geq 2$, we write $A(X)$ as $A(X) = X^h B(X)$ where $B(X) = \sum_{i=0}^{m} b_i X^i$ with $b_0 \neq 0$ and $b_m \neq 0$. Obviously, $\|A^n(X)\|_\# = \|B^n(X)\|_\#$ for all $n \geq 1$.

We prove the lemma considering the sequence $n_i = p^i - 1$. If $\|B(X)\|_\# = 2$, that is, $b_1 = b_2 = \cdots = b_{m-1} = 0$, we have

$$B^{n_i}(X) = (b_0 + b_m X^m)^{n_i} = \sum_{j=0}^{n_i} \binom{n_i}{j} b_0^{n_i - j} b_m^j X^{jm}.$$

We claim that, for $j = 0, \ldots, n_i$, $p \nmid \binom{n_i}{j}$. We have

$$\binom{n_i}{j} = \binom{p^i - 1}{j} = \frac{(p^i - 1)(p^i - 2) \cdots (p^i - j)}{1 \cdot 2 \cdots j},$$

hence, $p^t | (p^i - h)$ implies $p^t | h$ (for $t < i$) and every factor p in the numerator appears also in the denominator and $p \nmid \binom{n_i}{j}$ as claimed. Thus, $\binom{n_i}{j} b_0^{n_i - j} b_m^j \neq 0$ (mod p). This yields

$$\frac{\|F^{n_i}(z_0)\|_\#}{\rho(F^{n_i})} = \frac{\|A^{n_i}(X)\|_\#}{n_i r} = \frac{\|B^{n_i}(X)\|_\#}{n_i r} = \frac{n_i + 1}{n_i r}, \tag{4}$$

and the lemma follows. If $\|A(X)\|_\# > 2$ we can prove a result analogous to (4) using a slightly more complex argument. Being $\|B(X)\|_\# > 2$ there exists an index i, with $0 < i < m$, such that $b_i \neq 0$. Let b_d be such that $b_d \neq 0$ and $b_1 = \cdots b_{d-1} = 0$. It is well known that, working modulo p, for all $a, b \in \mathbf{Z}_p$, we have

$$(a + b)^{p^k} = \sum_{i=0}^{p^k} \binom{p^k}{i} a^i b^{p^k - i} = a^{p^k} + b^{p^k} = a + b.$$

This yields

$$\begin{aligned} B(X) B^{n_i}(X) &= (b_0 + b_d X^d + \cdots + b_m X^m)^{p^i} \\ &= (b_0 + b_d X^{dp^i} + \cdots + b_m X^{mp^i}). \end{aligned} \tag{5}$$

In other words, $\|B^{n_i + 1}(X)\|_\# = \|B(X)\|_\#$. In addition, in $B^{n_i + 1}(X)$ all powers X^j with $0 < j < dp^i$ have a zero coefficient. We show that this implies $\|B^{n_i}(X)\|_\# \geq (d/m) n_i - 1$ for all i such that $p^i > m$. More precisely, we claim that among the first $dp^i - m$ coefficients of $B^{n_i}(X)$ we cannot have a sequence of m consecutive zero coefficients. Let $B^{n_i}(X) = \sum_{j=0}^{mn_i} c_j X^j$. Assume by contradiction there exist j, k, with

$$0 \leq j < dp^i - m, \quad \text{and} \quad k > j + m,$$

such that

$$c_j \neq 0, \quad c_{j+1} = \cdots = c_{k-1} = 0. \tag{6}$$

The coefficient of X^{j+m} in $B^{n_i + 1}(X)$ is given by $\sum_{h=0}^{m} b_h c_{j+m-h}$, which, by (6), is equal to $b_m c_j$ and is therefore nonzero. This is impossible by (5) since $j + m < dp^i$. Since $c_0 = b_0^{n_i} \neq 0$, we conclude that $B^{n_i}(X)$ must have the form

$$B^{n_i}(X) = c_0 + c_{i_1} X^{i_1} + c_{i_2} X^{i_2} + \cdots + c_{i_k} X^{i_k} + \cdots + c_{mn_i} X^{mn_i}.$$

where

$$i_1 \leq m, \qquad i_j \leq i_{j-1} + m, \quad \text{for } j = 2, \ldots, k, \qquad i_k \geq dp^i - m.$$

Hence, $\|B^{n_i}(X)\|_\# \geq (dp^i - m)/m = (d/m)p^i - 1 \geq (d/m)n_i - 1$. Reasoning as in (4) we get

$$\frac{\|F^{n_i}(z_0)\|_\#}{\rho(F^{n_i})} = \frac{\|B^{n_i}(X)\|_\#}{n_i r} \geq \frac{(d/m)n_i - 1}{n_i r}$$

and the lemma follows. □

The following lemma establishes a necessary condition for sensitivity in terms of the radius of the map F^n.

Lemma 5. *Let $(\mathcal{C}_m, F)$ denote a linear CA. If there exists a constant M such that $\rho(F^n) \leq M$ for all $n > 0$, the map F is not sensitive.*

Proof. We use the characterization of Lemma 2. Assume by contradiction that there exist a constant c and a sequence $\{x_i, n_i\}_{i \in \mathbf{N}}$ for which (3) holds. Let $y_i = F^{n_i}(x_i)$, $c' = c/2$. Since $d(y_i, \mathbf{0}) \geq c$, there exists a sequence m_k such that

$$\lim_{k \to \infty} m_k = \infty, \qquad \text{and} \qquad \Delta_{[m_k]}(y_i, \mathbf{0}) \geq c'(2m_k + 1).$$

Since F^{n_i} is linear and $\rho(F^{n_i}) \leq M$, it follows that $y_i(j) \neq 0$ implies there exists j' with $|j' - j| \leq M$ such that $x_i(j') \neq 0$. Hence,

$$\Delta_{[m_k]}(y_i, \mathbf{0}) \geq c'(2m_k + 1) \implies \Delta_{[m_k + M]}(x_i, \mathbf{0}) \geq \frac{c'(2m_k + 1)}{2M + 1}.$$

By (2) we have

$$d(x_i, \mathbf{0}) \geq \lim_{k \to \infty} \frac{\Delta_{[m_k + M]}(x_i, \mathbf{0})}{2(m_k + M) + 1}$$

$$\geq \lim_{k \to \infty} \frac{c'}{2M + 1} \left(\frac{2m_k + 1}{2(m_k + M) + 1} \right)$$

$$= \frac{c'}{2M + 1}.$$

Hence $d(x_i, \mathbf{0})$ does not converge to zero, and F is not sensitive. □

We are now able to prove a necessary and sufficient condition for sensitivity in terms of the coefficients of the local rule when m is a prime power.

Theorem 1. *Let $(\mathcal{C}_{p^k}, F)$ denote a linear CA over $\mathbf{Z}_{p^k}$, p prime, with local rule $f(x_{-r}, \ldots, x_r) = \sum_{i=-r}^{r} a_i x_i$. The map F is sensitive if and only if there exist two coefficients a_i, a_j such that $\gcd(a_i, p) = \gcd(a_j, p) = 1$.*

Proof. We first prove the if part. Let $(\mathcal{C}_p, F_p)$ denote the linear CA over $\mathbf{Z}_p$ associated to the local rule $f'(x_{-r}, \ldots, x_r) = \sum_{i=-r}^{r} a'_i x_i$ where $a'_i = a_i \bmod p$. By hypothesis at least two coefficients a'_i, a'_j are nonzero. Hence, $\rho(F_p) > 0$, and by Lemma 4 we have

$$\limsup_{n \to \infty} \frac{\|F_p^n(z_0)\|_{\#}}{\rho(F_p^n)} > 0. \tag{7}$$

We show that F is sensitive using the characterization of Lemma 3. The finite fps associated with F^n (resp. F_p^n), is $[A(X)]^n = (\sum_{i=-r}^{r} a_i X^{-i})^n$ (resp. $[A'(X)]^n = (\sum_{i=-r}^{r} a'_i X^{-i})^n$). Since $[A'(X)]^n = [A(X)]^n \bmod p$ we have

$$n\rho(F) \geq \rho(F^n) \geq \rho(F_p^n) = n\rho(F_p), \tag{8}$$

where the last equality holds since p is prime. Since $F_p^n(z_0) = F^n(z_0) \bmod p$, we have

$$\|F^n(z_0)\|_{\#} \geq \|F_p^n(z_0)\|_{\#}. \tag{9}$$

From (8) it follows that $\lim_{n \to \infty} \rho(F^n) = \infty$. Combining (8), and (9) we get

$$\frac{\|F^n(z_0)\|_{\#}}{\rho(F^n)} \geq \frac{\|F_p^n(z_0)\|_{\#}}{n\rho(F)} \geq \frac{\|F_p^n(z_0)\|_{\#}}{\rho(F_p^n)} \frac{\rho(F_p)}{\rho(F)}.$$

By (7), $\limsup_{n \to \infty}(\|F^n(z_0)\|_{\#})/(\rho(F^n)) > 0$ and the map F is sensitive by Lemma 3.

To prove the only if part we use Lemma 5. If $\gcd(a_i, p) > 1$ for $i = -r, \ldots, r$ then, for all $x \in \mathcal{C}_{p^k}$ we have $F^k(x) = \mathbf{0}$. Hence, for all $n > 0$, $\rho(F^n) \leq k\rho(F)$ and F is not sensitive.

Assume now there exists a unique coefficient a_i such that $\gcd(a_i, p) = 1$, and assume by contradiction that F is sensitive. Since d is shift invariant, using for example Lemma 2 it is straightforward to verify that $F' = \sigma^i F$ should be sensitive as well. By construction, the finite fps associated to F' is $B(X) = X^i A(X)$. Hence, $B(X)$ has the form $B(X) = a_i + pB'(X)$. A simple proof by induction shows that for $j \geq 1$ we have $(a + pb)^{p^j} \equiv a^{p^j} \pmod{p^{j+1}}$. Hence,

$$[B(X)]^{p^{k-1}} \equiv a_i^{p^{k-1}} \pmod{p^k},$$

which implies

$$\sup_{n \geq 1} \rho((\sigma^i F)^n) \leq \max_{1 \leq n \leq p^{k-1}} \rho((\sigma^i F)^n),$$

hence, $\rho((\sigma^i F)^n)$ is bounded by a constant. By Lemma 5, $\sigma^i F$, and therefore F, is not sensitive.

This completes the proof. □

Combining the above result with Lemma 1 we get the following characterization of linear sensitive CA for an arbitrary alphabet size.

Corollary 1. *Let $(\mathcal{C}_m, F)$ denote the linear CA with local rule $f(x_{-r}, \ldots, x_r) = \sum_{i=-r}^{r} a_i x_i$. The map F is sensitive if and only if there exist a prime p and two coefficients a_i, a_j such that $p|m$, and $\gcd(a_i, p) = \gcd(a_j, p) = 1$.* □

Finally, we are able to prove that the sufficient condition for sensitivity given by Lemma 3 is also a necessary condition.

Corollary 2. *Let $(\mathcal{C}_m, F)$ denote a linear CA. The map F is sensitive if and only if*

$$\lim_{n \to \infty} \rho(F^n) = \infty, \quad \text{and} \quad \limsup_{n \to \infty} \frac{\|F^n(z_0)\|_\#}{\rho(F^n)} > 0$$

Proof. The if part is precisely Lemma 3. If F is sensitive, by Corollary 1 we know there exists p and two coefficients a_i, a_j of the local rule f such that $p|m$ and $\gcd(a_i, p) = \gcd(a_j, p) = 1$. The thesis follows repeating verbatim the proof of the if part of Theorem 1. □

References

1. F. Blanchard, P. Kůrka, and A. Maass. Topological and measure-theoretic properties of one-dimensional cellular automata. *Physica D*, 103:86–99, 1997.
2. G. Cattaneo, E. Formenti, G. Manzini, and L. Margara. Ergodicity, transitivity, and regularity for additive cellular automata over Z_m. *Theoretical Computer Science*. To appear.
3. G. Cattaneo, E. Formenti, L. Margara, and J. Mazoyer. A shift-invariant metric on S^Z inducing a non-trivial topology. Technical Report 97-21, LIP — Ecole Normale Supérieure de Lyon, 1997. A preliminary version appeared in Proc. MFCS '97, LNCS n. 1295, Springer Verlag.
4. R. L. Devaney. *An Introduction to Chaotic Dynamical Systems*. Addison-Wesley, Reading, MA, USA, second edition, 1989.
5. M. Finelli, G. Manzini, and L. Margara. Lyapunov exponents vs expansivity and sensitivity in cellular automata. *Journal of Complexity*. To appear.
6. M. Hurley. Attractors in cellular automata. *Ergodic Theory and Dynamical Systems*, 10:131–140, 1990.
7. M. Ito, N. Osato, and M. Nasu. Linear cellular automata over Z_m. *Journal of Computer and System Sciences*, 27:125–140, 1983.
8. P. Kůrka. Languages, equicontinuity and attractors in cellular automata. *Ergodic theory and dynamical systems*, 17:417–433, 1997.
9. G. Manzini. Characterization of sensitive linear cellular automata with respect to the counting distance. Technical Report B4-98-05, Istituto di Matematica Computazionale, CNR, Pisa, Italy, 1998.
10. G. Manzini and L. Margara. A complete and efficiently computable topological classification of D-dimensional linear cellular automata over Z_m. In *24th International Colloquium on Automata Languages and Programming (ICALP '97)*. LNCS n. 1256, Springer Verlag, 1997.
11. G. Manzini and L. Margara. Attractors of D-dimensional linear cellular automata. In *15th Annual Symposium on Theoretical Aspects of Computer Science (STACS '98)*, pages 128–138. LNCS n. 1373, Springer Verlag, 1998.
12. G. Manzini and L. Margara. Invertible linear cellular automata over Z_m: Algorithmic and dynamical aspects. *Journal of Computer and System Sciences*, 56:60–67, 1998.
13. T. Sato. Ergodicity of linear cellular automata over Z_m. *Information Processing Letters*, 61(3):169–172, 1997.

Additive Cellular Automata over $\mathbb{Z}_p$ and the Bottom of (CA,$\leq$)

Jacques Mazoyer and Ivan Rapaport

LIP-École Normale Supérieure de Lyon
46 Allée d'Italie, 69364 Lyon Cedex 07, France
{Jacques.Mazoyer,Ivan.Rapaport}@ens-lyon.fr

Abstract. In a previous work we began to study the question of "how to compare" cellular automata (CA). In that context it was introduced a preorder (CA,$\leq$) admitting a global minimum and it was shown that all the CA satisfying very simple dynamical properties as nilpotency or periodicity are located "on the bottom of (CA,$\leq$)". Here we prove that also the (algebraically amenable) additive CA over $\mathbb{Z}_p$ are located on the bottom of (CA,$\leq$). This result encourages our conjecture that says that the "distance" from the minimum could represent a measure of "complexity" on CA. We also prove that the additive CA over $\mathbb{Z}_p$ with p prime are pairwise incomparable. This fact improves our understanding of (CA,$\leq$) because it means that the minimum, even in the canonical order compatible with $\leq$, has infinite outdegree.

1 Introduction

One-dimensional cellular automata with radius 1, or simply CA, are infinite arrays of finite-state machines called cells and indexed by $\mathbb{Z}$. These identical cells evolve synchronously at discrete time steps following a local rule by which the state of a cell is determined as a function of its own state together with the states of its two neighbors. These devices, despite their simplicity, may exhibit very complex behavior.

In order to "understand" these CA one should find some criteria capable of structuring them into natural classes or hierarchies. In this direction, the classification of S. Wolfram [11], though heuristic and coarse, corresponds to the best-known attempt. Wolfram, by "observing" the long-term behavior of "arbitrary" periodic configurations, distinguishes four CA classes. Some efforts have been made in order to formalize this classification [6] or, typically by dynamical systems arguments, to introduce new classification schemes [5, 4]. Unfortunately, this last approach yields to some paradoxes: the shift CA, for instance, appears to be chaotic.

CA may also be seen as computational devices. In fact, it is easy to exhibit a CA that simulates any Turing machine [7]. In other words, the CA model is Turing-universal. The question whether the CA model is intrinsic-universal or, in other words, whether there exists a CA capable of simulating any other, remained open for some years. Notice that the CA can not be simulated by a Turing machine because the latter have a unique head which obviously will never visit the whole tape. J. Albert and K. Čulik II exhibited in [1] an intrinsic-universal CA. The "intrinsic-reducibility" notion induces a preorder on the set of CA. Unfortunetly, the study of this preorder structure is very difficult: it is based on the evolution of all the possible configurations (which are uncountable) and it does not take explicitly into account the CA transition tables. In addition, the "simulation" notion is so broad that a pair of CA with extremely different dynamics could appear to be "equivalent".

Another approach is to consider CA as algebraic objects. In this context, with the purpose of endowing the set of CA with an order relation, it would be sufficient to say that A is a subautomaton of B if the transition table of A is contained (after a suitable relabeling of the states) in the transition table of B. This notion is extremely restrictive. In fact, if A is a subautomaton of B then the space-time diagrams of A are "cell by cell equivalent" to the corresponding space-time diagrams of B (space-time diagrams are representations of a CA from a particular initial configuration in $\mathbb{Z}^2$). In other words, A and B may not be associated by the subautomaton relation even with their respective space-time diagrams being identical after suitable "changes of scale".

It seems therefore very natural to try to replace the subautomaton relation by a new one which could take into account potential changes of scale. This can be done by defining the powers of a CA. More precisely, let us denote by X^i the CA that generates the i-scaled space-time diagram of X and which is simply obtained by grouping i cells (or states) into blocks and by considering as transitions the interactions of neighbor blocks. Let us also note $A \leq B$ when some power of A is a subautomaton of some power of B or, equivalently, when the space-time diagrams of A are "block by block equivalent" to the corresponding space-time diagrams of B.

In [9] it was shown that (CA,$\leq$) is a preorder with no maximum. It was also proved that (CA,$\leq$) admits a global minimum and that all the CA satisfying very simple dynamical properties (nilpotents, periodics, shift-like) are located "on the bottom of (CA,$\leq$)". In addition, the fact that an algorithmically non-trivial "synchronization CA" was separated from the minimum by an infinite chain led us to conjecture that the "distance" from the minimum could represent a measure of "complexity" on CA.

In this paper we give more "evidence" supporting the intuitive expectation that says that the "simplest" CA should be located on the bottom of (CA,$\leq$) or, more precisely, that the "simplest" CA should be located immediately above the global minimum. By following an algebraic criterion of simplicity we decide to study the class of additive CA over $\mathbb{Z}_p$. In fact, these CA have been extensively studied because of their amenability to algebraic analysis [8, 2]. We prove that, for p prime, the additive CA over $\mathbb{Z}_p$ are located on the bottom of (CA,$\leq$). We also show that the additive CA over $\mathbb{Z}_p$ with p prime are pairwise incomparable. Therefore, if we note by CA* the set of CA modulo the canonical equivalence relation induced by $\leq$, then the minimum of the order (CA*,$\leq$) has infinite outdegree. Until now, we had no examples of unbounded outdegrees in (CA*,$\leq$).

2 Preliminaries

In this section we formally introduce the preorder (CA,$\leq$) and we recall some already known results. First, a CA is defined by a couple (Q, δ) where Q is a finite set of states and $\delta : Q^3 \to Q$ is a transition function. We say that (Q_1, δ_1) is a subautomaton of (Q_2, δ_2), and we note $(Q_1, \delta_1) \subseteq (Q_2, \delta_2)$, if there exists an injection $\varphi : Q_1 \to Q_2$ such that for all $x, y, z \in Q_1$:

$$\varphi(\delta_1(x,y,z)) = \delta_2(\varphi(x), \varphi(y), \varphi(z)).$$

When the function φ is a bijection we say that (Q_1, δ_1) and (Q_2, δ_2) are isomorphic and we note $(Q_1, \delta_1) \cong (Q_2, \delta_2)$.

Let $\mathbb{N}^* = \mathbb{N} - \{0\}$. For any CA (Q, δ) the evolution of a finite block of states looks like a light-cone (see Figure 1-i). This basic fact inspires the notion of the n-block evolution function $\delta^n : Q^{2n+1} \to Q$, which is recursively defined for all $n \in \mathbb{N}^*$ as follows:

$$\delta^1(w_{-1}, w_0, w_1) = \delta(w_{-1}, w_0, w_1),$$
$$\delta^n(w_{-n}, \cdots, w_0, \cdots, w_n) = \delta^{n-1}(\delta(w_{-n}, w_{-n+1}, w_{-n+2}) \cdots \delta(w_{n-2}, w_{n-1}, w_n)).$$

By grouping several states into blocks and by letting interact triplets of blocks as schematically appears in Figure 1-ii, we generate CA with (exponentially) more states. Formally, the n-power of a CA (Q, δ) is the CA $(Q, \delta)^n = (Q^n, \delta^n_{\mathcal{G}})$, where $\vec{q} \in Q^n$ is denoted by $(q_1, \cdots, q_n)$ and for all $\vec{x}, \vec{y}, \vec{z} \in Q^n$:

$$(\delta^n_{\mathcal{G}}(\vec{x}, \vec{y}, \vec{z}))_i = \delta^n(x_i, \cdots, x_n, y_1, \cdots, y_i, \cdots, y_n, z_1, \cdots, z_i).$$

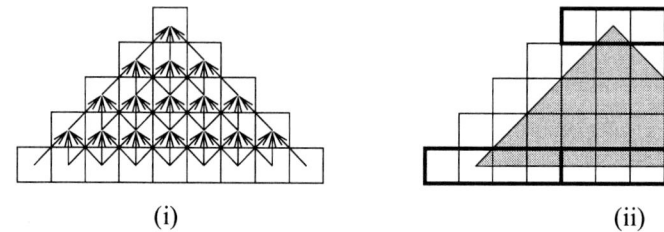

(i)   (ii)

Fig. 1. (i) Dependencies diagram representing a block of states evolution as a light-cone. (ii) Interaction of three blocks.

We relate two CA by $\leq$ when some power of the first is a subautomaton of some power of the second. More precisely, for (Q_1, δ_1) and (Q_2, δ_2):

$$(Q_1, \delta_1) \leq (Q_2, \delta_2) \iff \exists n, m \in \mathbb{N}^* : (Q_1, \delta_1)^n \subseteq (Q_2, \delta_2)^m.$$

In [9] it was shown that (CA,$\leq$) is a preorder. In addition, it was proved that (CA,$\leq$) admits a global minimum corresponding to the (set of isomorphic)

CA having a single state. This fact allows us to define the "bottom of (CA,≤)". More precisely, a CA is said to belong to the bottom of (CA,≤) if there is no other CA located strictly between the minimum and itself. In other words, (Q^*, δ^*) belongs to the bottom of (CA,≤) if for any other non-singleton CA (Q, δ), $(Q, \delta) \leq (Q^*, \delta^*) \Longrightarrow (Q^*, \delta^*) \leq (Q, \delta)$. Finally, the following lemma is going to be used later:

Lemma 1. *If* $(Q, \delta) \leq (\tilde{Q}, \tilde{\delta})$ *then there exist* $i_0, j_0 \in \mathbb{N}^*$ *and* $\overrightarrow{q_0} \in Q^{i_0}$ *such that* $(Q, \delta)^{i_0} \subseteq (\tilde{Q}, \tilde{\delta})^{j_0}$ *and* $\delta_{\mathcal{G}}^{i_0}(\overrightarrow{q_0}, \overrightarrow{q_0}, \overrightarrow{q_0}) = \overrightarrow{q_0}$.

Proof. Let $(Q, \delta) \leq (\tilde{Q}, \tilde{\delta})$. By definition, there exist $i, j \in \mathbb{N}^*$ such that $(Q, \delta)^i \subseteq (\tilde{Q}, \tilde{\delta})^j$. By the finiteness of Q there exist $q \in Q$ and $k \in \mathbb{N}^*$ such that $\delta^k(q \cdots q) = q$. Considering the fact that $(Q, \delta)^{ik} \subseteq (\tilde{Q}, \tilde{\delta})^{jk}$ [9], the lemma is concluded for $i_0 = ik, j_0 = jk$ and $\overrightarrow{q_0} = (q \cdots q)$. □

3 Permutive CA

The notion of permutive CA has been extensively used (see for instance [10]). A given CA (Q, δ) is said to be right permutive if for all $a, b \in Q$ the function $\delta(a, b, \cdot) : Q \to Q$ is bijective. A CA (Q, δ) is said to be left permutive if for all $a, b \in Q$ the function $\delta(\cdot, a, b) : Q \to Q$ is bijective. A CA (Q, δ) is said to be permutive if it is right and left permutive.

Here we prove that all the subautomata and all the powers of a given permutive CA are permutive.

Lemma 2. *Let* (Q_2, δ_2) *be a permutive CA. If* (Q_1, δ_1) *is such that* $(Q_1, \delta_1) \subseteq (Q_2, \delta_2)$ *then* (Q_1, δ_1) *is also permutive.*

Proof. Direct. □

Lemma 3. *Let* (Q, δ) *be a CA and let* $n \in \mathbb{N}^*$. (Q, δ) *is permutive if and only if* $(Q, \delta)^n$ *is permutive.*

Proof. We prove the equivalence for the right permutivity. For the left permutivity the proof is identical. Let us therefore assume (Q, δ) to be right permutive. It is easy to prove by induction that for all $m \in \mathbb{N}^*$, for all $\overrightarrow{a} \in Q^{2m}$, and for all $x, y \in Q$: $\delta^m(\overrightarrow{a} x) = \delta^m(\overrightarrow{a} y) \Rightarrow x = y$. Let $\overrightarrow{a}, \overrightarrow{b}, \overrightarrow{x}, \overrightarrow{y} \in Q^n$. If $\overrightarrow{x} \neq \overrightarrow{y}$ then there is an index $i \in \{1, \cdots, n\}$ such that $i = \min\{j \in \{1, \cdots, n\} : x_j \neq y_j\}$. It follows that $\delta^n(a_i \cdots a_n b_1 \cdots b_n x_1 \cdots x_i) \neq \delta^n(a_i \cdots a_n b_1 \cdots b_n x_1 \cdots y_i)$, and therefore $(\delta_{\mathcal{G}}^n(\overrightarrow{a}, \overrightarrow{b}, \overrightarrow{x}))_i \neq (\delta_{\mathcal{G}}^n(\overrightarrow{a}, \overrightarrow{b}, \overrightarrow{y}))_i$.

Let us assume now that $(Q, \delta)^n$ is right permutive (notice that the non-trivial case is when $n > 1$). Let $a, b, x, y \in Q$. If $\delta(a, b, x) = \delta(a, b, y)$ then:

$$\delta_{\mathcal{G}}^n(a \cdots a, a \cdots a, \underbrace{a \cdots a}_{n-2} bx) = \delta_{\mathcal{G}}^n(a \cdots a, a \cdots a, \underbrace{a \cdots a}_{n-2} by),$$

and therefore $(\underbrace{a \cdots a}_{n-2} bx) = (\underbrace{a \cdots a}_{n-2} by)$, which implies that $x = y$. □

4 Additive Cellular Automata over $\mathbb{Z}_p$

This section is the core of the present work. Here we prove that, for p prime, the additive CA over $\mathbb{Z}_p$ are pairwise incomparable (Corollary 1) and that they are all located on the bottom of (CA,$\leq$) (Corollary 2).

Let us start by denoting, for each $p \in \mathbb{N}^*, p > 1$, the additive abelian group of integers modulo p by $(\mathbb{Z}_p, +)$. For each $n \in \mathbb{N}^*$ we denote the canonical product group by $(\mathbb{Z}_p^n, +)$. More precisely, for all $(x_1, \cdots, x_n), (y_1, \cdots, y_n)$:

$$(x_1, \cdots, x_n) + (y_1, \cdots, y_n) = (x_1 + y_1, \cdots, x_n + z_n).$$

Stated for arbitrary groups of finite order, the following proposition appears in any introductory textbook of Algebra (see for instance [3]).

Proposition 1. *Let $p, n \in \mathbb{N}^*, p > 1$, and let $\mathcal{X} \subseteq \mathbb{Z}_p^n$ be a nonempty set. If $(\mathcal{X}, +)$ is such that for all $\vec{x}, \vec{y} \in \mathcal{X} : \vec{x} + \vec{y} \in \mathcal{X}$, then $(\mathcal{X}, +)$ is a subgroup of $(\mathbb{Z}_p^n, +)$ and $|\mathcal{X}| \mid p^n$. Moreover, if p is prime then:*

$$\mathcal{X} = \prod_{k=1}^n \mathcal{X}_k, \text{ with } \mathcal{X}_k = \mathbb{Z}_p \text{ or } \mathcal{X}_k = \{0\} \text{ for all } k \in \{1, \cdots, n\}.$$

To the abelian group $(\mathbb{Z}_p, +)$ we associate in the canonical way the CA $(\mathbb{Z}_p, \oplus)$ such that for all $x, y, z \in \mathbb{Z}_p : \oplus(x, y, z) = x + y + z$. Similarly, to the product group $(\mathbb{Z}_p^n, +)$ we associate the CA $(\mathbb{Z}_p^n, \oplus)$ in such a way that for all $\vec{x}, \vec{y}, \vec{z} \in \mathbb{Z}_p^n : \oplus(\vec{x}, \vec{y}, \vec{z}) = \vec{x} + \vec{y} + \vec{z}$. Finally, the n-power of the CA $(\mathbb{Z}_p, \oplus)$ corresponds, by definition, to $(\mathbb{Z}_p, \oplus)^n = (\mathbb{Z}_p^n, \oplus_\mathcal{G}^n)$.

Remark 1. Notice that in $\mathbb{Z}_p^n$ the operations $\oplus$ and $\oplus_\mathcal{G}^n$ are not the same. For instance, if we consider the set $\mathbb{Z}_3^4$ (see Figure 2):

$$\oplus((2,2,1,2),(1,1,0,2),(0,1,2,1)) = (0,1,0,2) \text{ and}$$
$$\oplus_\mathcal{G}^4((2,2,1,2),(1,1,0,2),(0,1,2,1)) = (2,1,1,1).$$

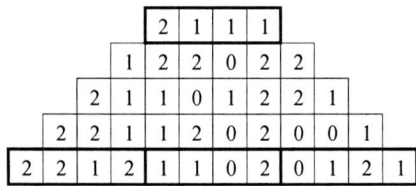

Fig. 2. $\oplus_\mathcal{G}^4((2,2,1,2),(1,1,0,2),(0,1,2,1)) = (2,1,1,1)$.

Lemma 4. *For all $p, n \in \mathbb{N}^*, p > 1$, the CA $(\mathbb{Z}_p, \oplus)^n$ is permutive.*

Proof. By Lemma 3 together with the permutivity of $(\mathbb{Z}_p, \oplus)$. □

The following (and easy to prove) "superposition principle" [8] reflects why additive CA are so amenable to algebraic analysis.

Proposition 2 (The superposition principle). *Let $p, n \in \mathbb{N}^*, p > 1$. For all $\vec{x_1}, \vec{x_2}, \vec{y_1}, \vec{y_2}, \vec{z_1}, \vec{z_2} \in \mathbb{Z}_p^n$ it holds the following:*

$$\oplus_{\mathcal{G}}^n(\vec{x_1} + \vec{x_2}, \vec{y_1} + \vec{y_2}, \vec{z_1} + \vec{z_2}) = \oplus_{\mathcal{G}}^n(\vec{x_1}, \vec{y_1}, \vec{z_1}) + \oplus_{\mathcal{G}}^n(\vec{x_2}, \vec{y_2}, \vec{z_2}).$$

Proposition 3. *Let $p, n \in \mathbb{N}^*, p > 1$, and let $\mathcal{X} \subseteq \mathbb{Z}_p^n$ with $\vec{0} = (0, \cdots, 0) \in \mathcal{X}$. It holds that if $(\mathcal{X}, \oplus_{\mathcal{G}}^n) \leq (\mathbb{Z}_p^n, \oplus_{\mathcal{G}}^n)$ then $(\mathcal{X}, +)$ is a subgroup of $(\mathbb{Z}_p^n, +)$.*

Proof. Let $(\mathcal{X}, \oplus_{\mathcal{G}}^n) \subseteq (\mathbb{Z}_p^n, \oplus_{\mathcal{G}}^n)$ and let $\vec{x_0}, \vec{y_0} \in \mathcal{X}$. Considering Lemma 2 together with Lemma 4, it follows that $(\mathcal{X}, \oplus_{\mathcal{G}}^n)$ is permutive and therefore there exists $\vec{\alpha_0} \in \mathcal{X}$ such that $\oplus_{\mathcal{G}}^n(\vec{\alpha_0}, \vec{x_0}, \vec{x_0}) = \vec{0}$. By the superposition principle:

$$\oplus_{\mathcal{G}}^n(\vec{\alpha_0}, \vec{x_0}, \vec{x_0} + \vec{y_0}) = \oplus_{\mathcal{G}}^n(\vec{\alpha_0}, \vec{x_0}, \vec{x_0}) + \oplus_{\mathcal{G}}^n(\vec{0}, \vec{0}, \vec{y_0})$$
$$= \oplus_{\mathcal{G}}^n(\vec{0}, \vec{0}, \vec{y_0}) = \vec{y_*} \in \mathcal{X}.$$

On the other hand, again by permutivity of $(\mathcal{X}, \oplus_{\mathcal{G}}^n)$, there exists $\vec{x_*} \in \mathcal{X}$ such that $\oplus_{\mathcal{G}}^n(\vec{\alpha_0}, \vec{x_0}, \vec{x_*}) = \vec{y_*}$. Finally, now by permutivity of $(\mathbb{Z}_p^n, \oplus_{\mathcal{G}}^n)$, $\vec{x_*} = (\vec{x_0} + \vec{y_0}) \in \mathcal{X}$. □

Proposition 4. *Let (Q, δ) be a CA and let $p \in \mathbb{N}^*, p > 1$. If $(Q, \delta) \leq (\mathbb{Z}_p, \oplus)$ then there exist $i, j \in \mathbb{N}^*$ and an injection $\psi : Q^i \to \mathbb{Z}_p^j$ such that:*

$$(Q, \delta)^i \cong (\psi(Q^i), \oplus_{\mathcal{G}}^j) \subseteq (\mathbb{Z}_p^j, \oplus_{\mathcal{G}}^j),$$

with $(\psi(Q^i), +)$ being a subgroup of $(\mathbb{Z}_p^j, +)$.

Proof. Let us suppose that $(Q, \delta) \leq (\mathbb{Z}_p, \oplus)$. By definition, there exist $i, j \in \mathbb{N}^*$ such that $(Q, \delta)^i \subseteq (\mathbb{Z}_p, \oplus)^j$ by some injection φ. Moreover, by Lemma 1, we can assume that there exists $\vec{q_0} \in Q^i$ such that $\delta_{\mathcal{G}}^i(\vec{q_0}, \vec{q_0}, \vec{q_0}) = \vec{q_0}$ and therefore $\oplus_{\mathcal{G}}^j(\varphi(\vec{q_0}), \varphi(\vec{q_0}), \varphi(\vec{q_0})) = \varphi(\vec{q_0})$. Let us define the injection $\psi : Q^i \to \mathbb{Z}_p^j$ in such a way that, for all $\vec{q} \in Q^i : \psi(\vec{q}) = \varphi(\vec{q}) - \varphi(\vec{q_0})$. Let us denote $\mathcal{X} = \psi(Q^i)$. Notice that $\vec{0} = (0, \cdots, 0) \in \mathcal{X}$ because $\vec{0} = \varphi(\vec{q_0}) - \varphi(\vec{q_0})$.

In order to prove that $(Q, \delta)^i \cong (\mathcal{X}, \oplus_{\mathcal{G}}^j) \subseteq (\mathbb{Z}_p^j, \oplus_{\mathcal{G}}^j)$ it suffices to prove that $(\varphi(Q^i), \oplus_{\mathcal{G}}^j) \cong (\mathcal{X}, \oplus_{\mathcal{G}}^j)$ because $(Q, \delta)^i \cong (\varphi(Q^i), \oplus_{\mathcal{G}}^j)$. Let $\eta : \varphi(Q^i) \to \mathcal{X}$ be such that $\eta(\vec{x}) = \vec{x} - \varphi(\vec{q_0})$. The function η is obviously a bijection and, in addition, for all $\vec{x}, \vec{y}, \vec{z} \in \varphi(Q^i)$:

$$\eta(\oplus_{\mathcal{G}}^j(\vec{x}, \vec{y}, \vec{z})) = \oplus_{\mathcal{G}}^j(\vec{x}, \vec{y}, \vec{z}) - \oplus_{\mathcal{G}}^j(\varphi(\vec{q_0}), \varphi(\vec{q_0}), \varphi(\vec{q_0}))$$
$$= \oplus_{\mathcal{G}}^j(\eta(\vec{x}), \eta(\vec{y}), \eta(\vec{z})).$$

From Proposition 3, it follows that $(\psi(Q^i), +)$ is a subgroup of $(\mathbb{Z}_p^j, +)$. □

Remark 2. The process of searching additive CA $(\mathbb{Z}_p, \oplus)$ located on the bottom of (CA,$\leq$) may be restricted to p prime because if $a, b \in \mathbb{N}^*$ are such that $a|b$ then $(\mathbb{Z}_a, \oplus) \leq (\mathbb{Z}_b, \oplus)$. In fact, it suffices to notice that $(\mathbb{Z}_a, \oplus) \subseteq (\mathbb{Z}_b, \oplus)$ by the injection $\varphi : \mathbb{Z}_a \to \mathbb{Z}_b$ which assigns to each $x \in \mathbb{Z}_a$ the value $\varphi(x) = \frac{bx}{a} \in \mathbb{Z}_b$.

Corollary 1. *Let $p, q > 1$ be prime numbers. If $p \neq q$ then $(\mathbb{Z}_p, \oplus) \not\leq (\mathbb{Z}_q, \oplus)$.*

Proof. Let us suppose that $(\mathbb{Z}_p, \oplus) \leq (\mathbb{Z}_q, \oplus)$. Then, by Proposition 4, there exist $i, j \in \mathbb{N}^*$ and an injection $\psi : \mathbb{Z}_p^i \to \mathbb{Z}_q^j$ such that $(\psi(\mathbb{Z}_p^i), +)$ is a subgroup of $(\mathbb{Z}_q^j, +)$. By Proposition 1, $p^i | q^j$. This is a contradiction. □

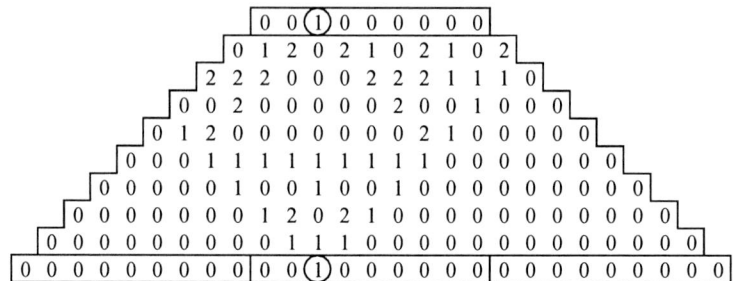

Fig. 3. $\oplus_{\mathcal{G}}^{3^2}(\vec{0}, \vec{e_3}, \vec{0}) = \oplus_{\mathcal{G}}^{3^2}(\vec{0}, \vec{0}, \vec{e_3}) = \vec{e_3}$.

Proposition 5. *For all $\alpha \in \mathbb{N}^*$ and $p > 1$ prime it holds that:*

$$(\mathbb{Z}_p^{p^\alpha}, \oplus) \cong (\mathbb{Z}_p, \oplus)^{p^\alpha}.$$

On the other hand, for every $\mathcal{X} \subseteq \mathbb{Z}_p^{p^\alpha}$ such that $|\mathcal{X}| > 1$ and $(\mathcal{X}, \oplus_{\mathcal{G}}^{p^\alpha}) \subseteq (\mathbb{Z}_p^{p^\alpha}, \oplus_{\mathcal{G}}^{p^\alpha})$ with $(\mathcal{X}, +)$ being a subgroup of $(\mathbb{Z}_p^{p^\alpha}, +)$, it holds that $(\mathbb{Z}_p, \oplus) \subseteq (\mathcal{X}, \oplus_{\mathcal{G}}^{p^\alpha})$.

Proof. By using the well-known dipolynomial representation of additive CA that appears in [8], together with the fact that a dipolynomial over $\mathbb{Z}_p$ of the form $\sum x^i$ satisfies that $(\sum x^i)^{p^\alpha} = \sum x^{ip^\alpha}$, it can be proved that (see Figure 3): $\oplus_{\mathcal{G}}^{p^\alpha}(\vec{e_k}, \vec{0}, \vec{0}) = \oplus_{\mathcal{G}}^{p^\alpha}(\vec{0}, \vec{e_k}, \vec{0}) = \oplus_{\mathcal{G}}^{p^\alpha}(\vec{0}, \vec{0}, \vec{e_k}) = \vec{e_k}$, with $\vec{e_k} = (\underbrace{0, \cdots, 0, 1}_{k \in \{1, \cdots, p^\alpha\}}, 0, \cdots, 0) \in \mathbb{Z}_p^{p^\alpha}$ and $\vec{0} = (0, \cdots, 0)$.

The previous result, together with the superposition principle, allows us to conclude, as it appears in the example of Figure 4, that $(\mathbb{Z}_p^{p^\alpha}, \oplus) \cong (\mathbb{Z}_p, \oplus)^{p^\alpha}$. Finally, let $\mathcal{X} \subseteq \mathbb{Z}_p^{p^\alpha}$ be such that $|\mathcal{X}| > 1$ and $(\mathcal{X}, \oplus_{\mathcal{G}}^{p^\alpha}) \subseteq (\mathbb{Z}_p^{p^\alpha}, \oplus_{\mathcal{G}}^{p^\alpha})$ with $(\mathcal{X}, +)$ being a subgroup of $(\mathbb{Z}_p^{p^\alpha}, +)$. By considering Proposition 1 we conclude the existence of an index $k_0 \in \{1, \cdots, p^\alpha\}$ such that $\mathcal{X}_{k_0} = \mathbb{Z}_p$. It follows that $(\mathbb{Z}_p, \oplus) \subseteq (\mathcal{X}, \oplus_{\mathcal{G}}^{p^\alpha})$ by the injection $\varphi : \mathbb{Z}_p \to \mathcal{X}$ that assigns, to each $x \in \mathbb{Z}_p$, the image $\varphi(x) = (\underbrace{0, \cdots, 0, x}_{k_0}, \cdots, 0)$. □

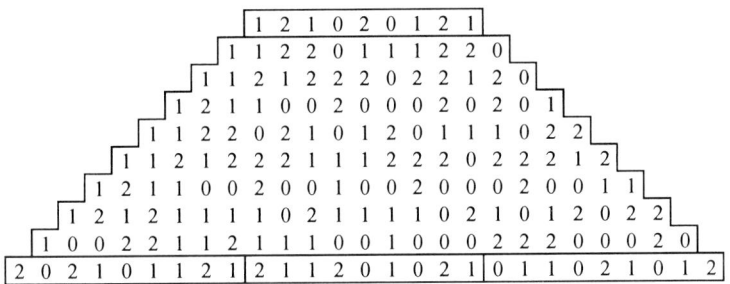

Fig. 4. $(\mathbb{Z}_3^{3^2}, \oplus) \cong (\mathbb{Z}_3, \oplus)^{3^2}$.

Proposition 6. *Let $n \in \mathbb{N}^*$ and let $p > 1$ be prime. If $\mathcal{X} \subseteq \mathbb{Z}_p^n$ is such that $|\mathcal{X}| > 1$ and $(\mathcal{X}, \oplus_{\mathcal{G}}^n) \subseteq (\mathbb{Z}_p^n, \oplus_{\mathcal{G}}^n)$ with $(\mathcal{X}, +)$ being a subgroup of $(\mathbb{Z}_p^n, +)$, then there exists $m \in \mathbb{N}^*$ satisfying $(\mathbb{Z}_p^m, \oplus_{\mathcal{G}}^m) \subseteq (\mathcal{X}, \oplus_{\mathcal{G}}^n)$.*

Proof. Let us consider the decomposition $n = \lambda p^\alpha$ with $\alpha \in \mathbb{N}$ and $\lambda \in \mathbb{N}^*$ such that $p \nmid \lambda$. Let us assume first that $\lambda = 1$. If $\alpha = 0$ then, considering that $(\mathcal{X}, +)$ is a subgroup of $(\mathbb{Z}_p, +)$ together with the fact that $|\mathcal{X}| > 1$, it can be concluded by Proposition 1 that $\mathcal{X} = \mathbb{Z}_p$ and therefore $(\mathcal{X}, \oplus) \cong (\mathbb{Z}_p, \oplus)$. On the other hand, if $\alpha > 0$ then, by considering Proposition 5, we conclude that $(\mathbb{Z}_p, \oplus) \subseteq (\mathcal{X}, \oplus_{\mathcal{G}}^{p^\alpha})$. Let us assume now that $\lambda > 1$. By Proposition 1:

$$\mathcal{X} = \prod_{k=1}^{\lambda p^\alpha} \mathcal{X}_k, \text{ with } \mathcal{X}_k = \mathbb{Z}_p \text{ or } \mathcal{X}_k = \{0\} \text{ for all } k \in \{1, \cdots, \lambda p^\alpha\}.$$

Let us suppose that there exist $k_1, k_2 \in \{1, \cdots, \lambda p^\alpha\}$ with $k_2 - k_1 = p^\alpha$ such that $\mathcal{X}_{k_1} = \{0\}$ and $\mathcal{X}_{k_2} = \mathbb{Z}_p$. Let $\overrightarrow{e_{k_2}} = (\underbrace{0, \cdots, 0, 1}_{k_2}, 0, \cdots, 0)$. It follows, as it is schematically shown in Figure 5, that $(\oplus_\mathcal{G}^{\lambda p^\alpha}(\overrightarrow{e_{k_2}}, \overrightarrow{0}, \overrightarrow{0}))_{k_1} = \lambda \mod p \neq 0$, and then $(\oplus_\mathcal{G}^{\lambda p^\alpha}(\overrightarrow{e_{k_2}}, \overrightarrow{0}, \overrightarrow{0}))_{k_1} \notin \mathcal{X}_{k_1}$, which is a contradiction. We can therefore assume that $\mathcal{X}$ is such that for any pair of indexes $k_1, k_2 \in \{1, \cdots, \lambda p^\alpha\}$ satisfying that $k_2 - k_1 = p^\alpha$ it holds that $\mathcal{X}_{k_1} = \mathcal{X}_{k_2}$. Let $k \in \{1, \cdots, p^\alpha\}$ be such that $\mathcal{X}_k = \mathbb{Z}_p$. This k does exist because $|\mathcal{X}| > 1$. It follows, as it is shown in the example of Figure 6, that $(\mathbb{Z}_p, \oplus)^\lambda \subseteq (\mathcal{X}, \oplus_\mathcal{G}^{\lambda p^\alpha})$. In fact, it suffices to consider the injection $\varphi : \mathbb{Z}_p^\lambda \to \mathcal{X}$ such that for all $\overrightarrow{x} = (x_1, \cdots, x_\lambda) \in \mathbb{Z}_p^\lambda$ and for all $i \in \{1, \cdots, \lambda p^\alpha\}$:

$$(\varphi(\overrightarrow{x}))_i = \begin{cases} (\overrightarrow{x})_{\frac{i-k}{p^\alpha}+1} & \text{if } i = k \mod p^\alpha \\ 0 & \text{otherwise.} \end{cases}$$

□

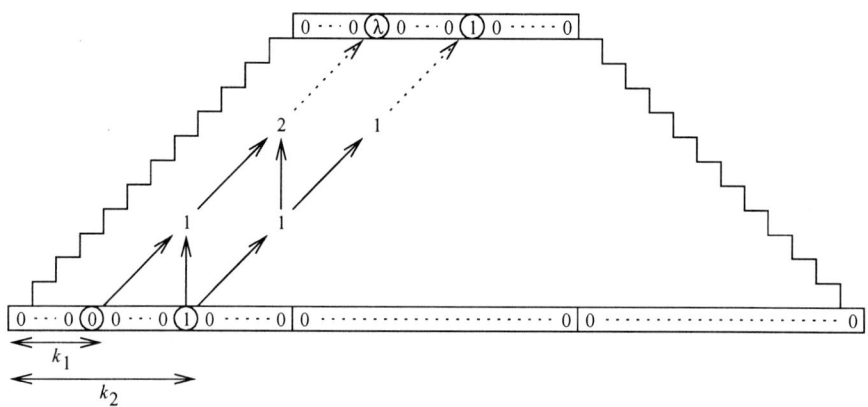

Fig. 5. $(\oplus_\mathcal{G}^n(\overrightarrow{e_{k_2}}, \overrightarrow{0}, \overrightarrow{0}))_{k_1} = \lambda \mod p$.

Corollary 2. *Let (Q, δ) be a CA with $|Q| > 1$ and let $p > 1$ be prime. It holds that if $(Q, \delta) \leq (\mathbb{Z}_p, \oplus)$ then $(\mathbb{Z}_p, \oplus) \leq (Q, \delta)$.*

Proof. Let us suppose that $(Q, \delta) \leq (\mathbb{Z}_p, \oplus)$. From Proposition 4 there exist $i, j \in \mathbb{N}^*$ and an injection $\psi : Q^i \to \mathbb{Z}_p^j$ such that $(Q, \delta)^i \cong (\psi(Q^i), \oplus_\mathcal{G}^j) \subseteq (\mathbb{Z}_p^j, \oplus_\mathcal{G}^j)$, with $(\psi(Q^i), +)$ being a subgroup of $(\mathbb{Z}_p^j, +)$. Then, by Proposition 6, there exists $m \in \mathbb{N}^*$ such that $(\mathbb{Z}_p, \oplus)^m \subseteq (\psi(Q^i), \oplus_\mathcal{G}^j) \cong (Q, \delta)^i$. □

Remark 3. If we denote by $\sim$ the canonical equivalence relation induced by $\leq$, then the minimum of the canonical order $(CA/\sim, \leq)$ has infinite outdegree.

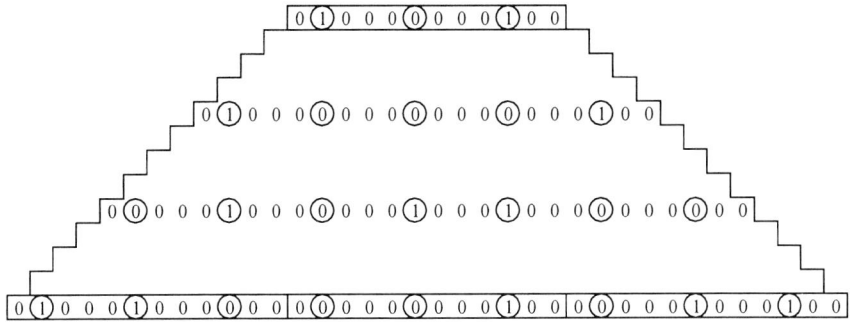

Fig. 6. $(\mathbb{Z}_2, \oplus)^3 \subseteq (\mathcal{X}, \oplus_{\mathcal{G}}^{3\times 2^2})$ with the index $k = 2 \in \{1, \cdots, 2^2\}$.

5 Concluding Remarks

We have shown that, according to (CA,≤), all the additive CA over $\mathbb{Z}_p$ with p prime have the same "complexity": they are all incomparable and located immediately above the minimum. Moreover, for finding the position of any additive CA in (CA,≤) one should follow simple number divisibility considerations. Nevertheless, one question remains open: given $x, y, i, j \in \mathbb{N}^*$ with $x^i | y^j$, is it true or false that $(\mathbb{Z}_x, \oplus) \leq (\mathbb{Z}_y, \oplus)$?

References

[1] J. Albert and K. Čulik II. A simple universal cellular automaton and its one-way and totalistic version. *Complex Systems*, 1:1–16, 1987.

[2] J.-P. Allouche, F. von Haeseler, H.-O. Peitgen, and G. Skordev. Linear cellular automata, finite automata and Pascal's triangle. *Discrete Applied Mathematics*, 66:1–22, 1996.

[3] G. Birkhoff and S. Mac Lane. *A survey of modern algebra*, chapter VI. MacMillan, New York, 1953. (Theorem 15, Theorem 18).

[4] F. Blanchard, A.Maass, and P.Kurka. Topological and measure-theoretic properties of one-dimensional cellular automata. *Physica D*, 103:86–99, 1997.

[5] G. Braga, G. Cattaneo, P. Flocchini, and C. Quaranta Vogliotti. Pattern growth in elementary cellular automata. *Theoretical Computer Science*, 45:1–26, 1995.

[6] K. Čulik II and S. Yu. Undecidability of CA classification schemes. *Complex Systems*, 2:177–190, 1988.

[7] A.R. Smith III. Simple computation-universal cellular spaces. *Journal ACM*, 18:339–353, 1971.

[8] O. Martin, O. Odlyzko, and S. Wolfram. Algebraic properties of cellular automata. *Communications in Mathematical Physics*, 93:219–258, 1984.

[9] J. Mazoyer and I. Rapaport. Inducing an order on cellular automata by a grouping operation. In *STACS'98*, volume 1373 of *Lecture Notes in Computer Science*, pages 116–127, 1998.

[10] C. Moore. Predicting nonlinear cellular automata quickly by decomposing them into linear ones. *Physica D*, 111:27–41, 1998.

[11] S. Wolfram. Universality and complexity in cellular automata. *Physica D*, 10:1–35, 1984.

Author Index

Aldaz M., 167
Alekhnovich M., 176
Amano K., 399
Ambainis A., 409
Ambos-Spies K., 465
Auletta V., 771
Ausiello G., 1

Baaz M., 203
Barrington D.M., 409
Barthe G., 316
Barthelmann K., 543
Benke M., 326
Bentzien L., 474
Bezrukov S.L., 693
Bierman G.M., 336
Bodlaender H.L., 702
Bonner R., 213
Börger E., 17
Broersma H., 713
Buchholz Th., 807
Buss S., 176

Caragiannis I., 771
Cattaneo G., 816
Chavez J.D., 693
Choffrut Ch., 656
Chrobak M., 185
Chrząszcz J., 346
Ciabattoni A., 203
Clerbout M., 533
Courcelle B., 616
Crauser A., 722
Crochemore M., 665

Damm C., 780
Di Pierro A., 446
Durand B., 732
Dürr Ch., 185

Emerson E.A., 427

Fermüller Ch., 203
Fernau H., 740
Ferreira M.C.F., 239
Freivalds R., 213
Fülöp F., 248

Gastin P., 356
Giacobazzi R., 366
Goldsmith J., 483
Grohe M., 437
Große-Rhode M, 553

Hagenah Ch., 277
Hagerup T., 702
Harel D., 36
Harju T., 503
Harper L.H., 693
Heintz J., 167
Hemaspaandra L.A., 418
Hermann M., 257
Hofmeister Th., 562
Horvath S., 656
Hromkovič., 789
Huck A., 713
Hune Th., 378

Italiano G.F., 1
Iwama K., 580, 625
Iwamoto Ch., 580

Jurvanen E., 248

Kaklamanis Ch., 771
Karhumäki J., 674
Kesner D., 239
Kesten Y., 54
Klein A., 807
Kloks T., 713
Köbler J., 493
Kolpakov R., 683
Koppius O., 713
Kratsch D., 713
Kucherov G., 683
Kuich W., 512
Kutrib M., 807
Kuzjurin N., 194

Lapiņš J., 213
Lapoire D., 616
La Torre S., 571
Latteux M., 286

Lefmann H., 562
Lempp S., 465
LêThanh H., 409
Lopez L.-M., 522
Lukjanska A., 213

Maass W., 72
Mainhardt G., 465
Maňuch J., 674
Manzini G., 825
Margara L., 816
Maruoka A., 399
Mateescu A., 503
Matera G., 167
Matz O., 751
Mazoyer J., 834
Mehlhorn K., 84, 722
Meyer R., 356
Meyer U., 722
Micali S., 94
Mielniczuk P., 589
Mignosi F., 665
Montaña J.L., 167
Moran S., 176
Müller H., 713
Muscholl A., 277

Näher S., 84
Nanni U., 1
Napoli M., 571
Narbel P., 522
Nielsen M., 117, 378
Nielson F., 220
Nielson H.R., 220
Nozoe M., 625

Ogihara M., 483

Pacholski L., 589
Pardo L.M., 167
Parisi–Presicce F., 553
Persiano P., 771
Petersen H., 296
Petit A., 356
Pitassi T., 176
Plandowski W., 674
Pnueli A., 54
Porrot S., 732
Preuß H., 636

Pudlák P., 129
Puel L., 239

Rabinovich A., 229
Ranzato F., 366
Rapaport I., 834
Reinhardt K., 760
Restivo A., 665
Roos Y., 533
Rothe J., 418, 483
Röttger M., 693
Ryl I., 533

Salomaa A., 503
Salzer G., 257
Sanders P., 722
Schroeder U.-P., 693
Schuler R., 493
Schulte W., 17
Schwentick T., 437
Scozzari F., 366
Sénizergues G., 305
Simeoni M., 553
Simplot D., 286
Simpson A.K., 456
Slobodová A., 645
Srba J., 388
Srivastav A., 636
Staiger L., 740
Steinby M., 248
Stirling C., 142

Tarannikov Y., 683
Terlutte A., 286
Touzet H., 267
Trefler R.J., 427
Tuinstra H., 713

Vágvölgyi S., 248
Veith H., 203
Vorobyov S., 597

Weihrauch K., 798
Wiedermann J., 152, 607
Wiklicky H., 446

Yajima S., 625

Zheng X., 798

Springer and the environment

At Springer we firmly believe that an international science publisher has a special obligation to the environment, and our corporate policies consistently reflect this conviction.
We also expect our business partners – paper mills, printers, packaging manufacturers, etc. – to commit themselves to using materials and production processes that do not harm the environment. The paper in this book is made from low- or no-chlorine pulp and is acid free, in conformance with international standards for paper permanency.

Lecture Notes in Computer Science

For information about Vols. 1–1376

please contact your bookseller or Springer-Verlag

Vol. 1377: H.-J. Schek, F. Saltor, I. Ramos, G. Alonso (Eds.), Advances in Database Technology – EDBT'98. Proceedings, 1998. XII, 515 pages. 1998.

Vol. 1378: M. Nivat (Ed.), Foundations of Software Science and Computation Structures. Proceedings, 1998. X, 289 pages. 1998.

Vol. 1379: T. Nipkow (Ed.), Rewriting Techniques and Applications. Proceedings, 1998. X, 343 pages. 1998.

Vol. 1380: C.L. Lucchesi, A.V. Moura (Eds.), LATIN'98: Theoretical Informatics. Proceedings, 1998. XI, 391 pages. 1998.

Vol. 1381: C. Hankin (Ed.), Programming Languages and Systems. Proceedings, 1998. X, 283 pages. 1998.

Vol. 1382: E. Astesiano (Ed.), Fundamental Approaches to Software Engineering. Proceedings, 1998. XII, 331 pages. 1998.

Vol. 1383: K. Koskimies (Ed.), Compiler Construction. Proceedings, 1998. X, 309 pages. 1998.

Vol. 1384: B. Steffen (Ed.), Tools and Algorithms for the Construction and Analysis of Systems. Proceedings, 1998. XIII, 457 pages. 1998.

Vol. 1385: T. Margaria, B. Steffen, R. Rückert, J. Posegga (Eds.), Services and Visualization. Proceedings, 1997/1998. XII, 323 pages. 1998.

Vol. 1386: T.A. Henzinger, S. Sastry (Eds.), Hybrid Systems: Computation and Control. Proceedings, 1998. VIII, 417 pages. 1998.

Vol. 1387: C. Lee Giles, M. Gori (Eds.), Adaptive Processing of Sequences and Data Structures. Proceedings, 1997. XII, 434 pages. 1998. (Subseries LNAI).

Vol. 1388: J. Rolim (Ed.), Parallel and Distributed Processing. Proceedings, 1998. XVII, 1168 pages. 1998.

Vol. 1389: K. Tombre, A.K. Chhabra (Eds.), Graphics Recognition. Proceedings, 1997. XII, 421 pages. 1998.

Vol. 1390: C. Scheideler, Universal Routing Strategies for Interconnection Networks. XVII, 234 pages. 1998.

Vol. 1391: W. Banzhaf, R. Poli, M. Schoenauer, T.C. Fogarty (Eds.), Genetic Programming. Proceedings, 1998. X, 232 pages. 1998.

Vol. 1392: A. Barth, M. Breu, A. Endres, A. de Kemp (Eds.), Digital Libraries in Computer Science: The MeDoc Approach. VIII, 239 pages. 1998.

Vol. 1393: D. Bert (Ed.), B'98: Recent Advances in the Development and Use of the B Method. Proceedings, 1998. VIII, 313 pages. 1998.

Vol. 1394: X. Wu. R. Kotagiri, K.B. Korb (Eds.), Research and Development in Knowledge Discovery and Data Mining. Proceedings, 1998. XVI, 424 pages. 1998. (Subseries LNAI).

Vol. 1395: H. Kitano (Ed.), RoboCup-97: Robot Soccer World Cup I. XIV, 520 pages. 1998. (Subseries LNAI).

Vol. 1396: E. Okamoto, G. Davida, M. Mambo (Eds.), Information Security. Proceedings, 1997. XII, 357 pages. 1998.

Vol. 1397: H. de Swart (Ed.), Automated Reasoning with Analytic Tableaux and Related Methods. Proceedings, 1998. X, 325 pages. 1998. (Subseries LNAI).

Vol. 1398: C. Nédellec, C. Rouveirol (Eds.), Machine Learning: ECML-98. Proceedings, 1998. XII, 420 pages. 1998. (Subseries LNAI).

Vol. 1399: O. Etzion, S. Jajodia, S. Sripada (Eds.), Temporal Databases: Research and Practice. X, 429 pages. 1998.

Vol. 1400: M. Lenz, B. Bartsch-Spörl, H.-D. Burkhard, S. Wess (Eds.), Case-Based Reasoning Technology. XVIII, 405 pages. 1998. (Subseries LNAI).

Vol. 1401: P. Sloot, M. Bubak, B. Hertzberger (Eds.), High-Performance Computing and Networking. Proceedings, 1998. XX, 1309 pages. 1998.

Vol. 1402: W. Lamersdorf, M. Merz (Eds.), Trends in Distributed Systems for Electronic Commerce. Proceedings, 1998. XII, 255 pages. 1998.

Vol. 1403: K. Nyberg (Ed.), Advances in Cryptology – EUROCRYPT '98. Proceedings, 1998. X, 607 pages. 1998.

Vol. 1404: C. Freksa, C. Habel. K.F. Wender (Eds.), Spatial Cognition. VIII, 491 pages. 1998. (Subseries LNAI).

Vol. 1405: S.M. Embury, N.J. Fiddian, W.A. Gray, A.C. Jones (Eds.), Advances in Databases. Proceedings, 1998. XII, 183 pages. 1998.

Vol. 1406: H. Burkhardt, B. Neumann (Eds.), Computer Vision – ECCV'98. Vol. I. Proceedings, 1998. XVI, 927 pages. 1998.

Vol. 1407: H. Burkhardt, B. Neumann (Eds.), Computer Vision – ECCV'98. Vol. II. Proceedings, 1998. XVI, 881 pages. 1998.

Vol. 1409: T. Schaub, The Automation of Reasoning with Incomplete Information. XI, 159 pages. 1998. (Subseries LNAI).

Vol. 1411: L. Asplund (Ed.), Reliable Software Technologies – Ada-Europe. Proceedings, 1998. XI, 297 pages. 1998.

Vol. 1412: R.E. Bixby, E.A. Boyd, R.Z. Ríos-Mercado (Eds.), Integer Programming and Combinatorial Optimization. Proceedings, 1998. IX, 437 pages. 1998.

Vol. 1413: B. Pernici, C. Thanos (Eds.), Advanced Information Systems Engineering. Proceedings, 1998. X, 423 pages. 1998.

Vol. 1414: M. Nielsen, W. Thomas (Eds.), Computer Science Logic. Selected Papers, 1997. VIII, 511 pages. 1998.

Vol. 1415: J. Mira, A.P. del Pobil, M.Ali (Eds.), Methodology and Tools in Knowledge-Based Systems. Vol. I. Proceedings, 1998. XXIV, 887 pages. 1998. (Subseries LNAI).

Vol. 1416: A.P. del Pobil, J. Mira, M.Ali (Eds.), Tasks and Methods in Applied Artificial Intelligence. Vol.II. Proceedings, 1998. XXIII, 943 pages. 1998. (Subseries LNAI).

Vol. 1417: S. Yalamanchili, J. Duato (Eds.), Parallel Computer Routing and Communication. Proceedings, 1997. XII, 309 pages. 1998.

Vol. 1418: R. Mercer, E. Neufeld (Eds.), Advances in Artificial Intelligence. Proceedings, 1998. XII, 467 pages. 1998. (Subseries LNAI).

Vol. 1419: G. Vigna (Ed.), Mobile Agents and Security. XII, 257 pages. 1998.

Vol. 1420: J. Desel, M. Silva (Eds.), Application and Theory of Petri Nets 1998. Proceedings, 1998. VIII, 385 pages. 1998.

Vol. 1421: C. Kirchner, H. Kirchner (Eds.), Automated Deduction – CADE-15. Proceedings, 1998. XIV, 443 pages. 1998. (Subseries LNAI).

Vol. 1422: J. Jeuring (Ed.), Mathematics of Program Construction. Proceedings, 1998. X, 383 pages. 1998.

Vol. 1423: J.P. Buhler (Ed.), Algorithmic Number Theory. Proceedings, 1998. X, 640 pages. 1998.

Vol. 1424: L. Polkowski, A. Skowron (Eds.), Rough Sets and Current Trends in Computing. Proceedings, 1998. XIII, 626 pages. 1998. (Subseries LNAI).

Vol. 1425: D. Hutchison, R. Schäfer (Eds.), Multimedia Applications, Services and Techniques – ECMAST'98. Proceedings, 1998. XVI, 532 pages. 1998.

Vol. 1427: A.J. Hu, M.Y. Vardi (Eds.), Computer Aided Verification. Proceedings, 1998. IX, 552 pages. 1998.

Vol. 1430: S. Trigila, A. Mullery, M. Campolargo, H. Vanderstraeten, M. Mampaey (Eds.), Intelligence in Services and Networks: Technology for Ubiquitous Telecom Services. Proceedings, 1998. XII, 550 pages. 1998.

Vol. 1431: H. Imai, Y. Zheng (Eds.), Public Key Cryptography. Proceedings, 1998. XI, 263 pages. 1998.

Vol. 1432: S. Arnborg, L. Ivansson (Eds.), Algorithm Theory – SWAT '98. Proceedings, 1998. IX, 347 pages. 1998.

Vol. 1433: V. Honavar, G. Slutzki (Eds.), Grammatical Inference. Proceedings, 1998. X, 271 pages. 1998. (Subseries LNAI).

Vol. 1434: J.-C. Heudin (Ed.), Virtual Worlds. Proceedings, 1998. XII, 412 pages. 1998. (Subseries LNAI).

Vol. 1435: M. Klusch, G. Weiß (Eds.), Cooperative Information Agents II. Proceedings, 1998. IX, 307 pages. 1998. (Subseries LNAI).

Vol. 1436: D. Wood, S. Yu (Eds.), Automata Implementation. Proceedings, 1997. VIII, 253 pages. 1998.

Vol. 1437: S. Albayrak, F.J. Garijo (Eds.), Intelligent Agents for Telecommunication Applications. Proceedings, 1998. XII, 251 pages. 1998. (Subseries LNAI).

Vol. 1438: C. Boyd, E. Dawson (Eds.), Information Security and Privacy. Proceedings, 1998. XI, 423 pages. 1998.

Vol. 1439: B. Magnusson (Ed.), System Configuration Management. Proceedings, 1998. X, 207 pages. 1998.

Vol. 1441: W. Wobcke, M. Pagnucco, C. Zhang (Eds.), Agents and Multi-Agent Systems. Proceedings, 1997. XII, 241 pages. 1998. (Subseries LNAI).

Vol. 1443: K.G. Larsen, S. Skyum, G. Winskel (Eds.), Automata, Languages and Programming. Proceedings, 1998. XVI, 932 pages. 1998.

Vol. 1444: K. Jansen, J. Rolim (Eds.), Approximation Algorithms for Combinatorial Optimization. Proceedings, 1998. VIII, 201 pages. 1998.

Vol. 1445: E. Jul (Ed.), ECOOP'98 – Object-Oriented Programming. Proceedings, 1998. XII, 635 pages. 1998.

Vol. 1446: D. Page (Ed.), Inductive Logic Programming. Proceedings, 1998. VIII, 301 pages. 1998. (Subseries LNAI).

Vol. 1447: V.W. Porto, N. Saravanan, D. Waagen, A.E. Eiben (Eds.), Evolutionary Programming VII. Proceedings, 1998. XVI, 840 pages. 1998.

Vol. 1448: M. Farach-Colton (Ed.), Combinatorial Pattern Matching. Proceedings, 1998. VIII, 251 pages. 1998.

Vol. 1449: W.-L. Hsu, M.-Y. Kao (Eds.), Computing and Combinatorics. Proceedings, 1998. XII, 372 pages. 1998.

Vol. 1450: L. Brim, F. Gruska, J. Zlatuška (Eds.), Mathematical Foundations of Computer Science 1998. Proceedings, 1998. XVII, 846 pages. 1998.

Vol. 1451: A. Amin, D. Dori, P. Pudil, H. Freeman (Eds.), Advances in Pattern Recognition. Proceedings, 1998. XXI, 1048 pages. 1998.

Vol. 1452: B.P. Goettl, H.M. Halff, C.L. Redfield, V.J. Shute (Eds.), Intelligent Tutoring Systems. Proceedings, 1998. XIX, 629 pages. 1998.

Vol. 1453: M.-L. Mugnier, M. Chein (Eds.), Conceptual Structures: Theory, Tools and Applications. Proceedings, 1998. XIII, 439 pages. (Subseries LNAI).

Vol. 1454: I. Smith (Ed.), Artificial Intelligence in Structural Engineering. XI, 497 pages. 1998. (Subseries LNAI).

Vol. 1456: A. Drogoul, M. Tambe, T. Fukuda (Eds.), Collective Robotics. Proceedings, 1998. VII, 161 pages. 1998. (Subseries LNAI).

Vol. 1457: A. Ferreira, J. Rolim, H. Simon, S.-H. Teng (Eds.), Solving Irregularly Structured Problems in Prallel. Proceedings, 1998. X, 408 pages. 1998.

Vol. 1458: V.O. Mittal, H.A. Yanco, J. Aronis, R-. Simpson (Eds.), Assistive Technology in Artificial Intelligence. X, 273 pages. 1998. (Subseries LNAI).

Vol. 1459: D.G. Feitelson, L. Rudolph (Eds.), Job Scheduling Strategies for Parallel Processing. Proceedings, 1998. VII, 257 pages. 1998.

Vol. 1461: G. Bilardi, G.F. Italiano, A. Pietracaprina, G. Pucci (Eds.), Algorithms – ESA'98. Proceedings, 1998. XII, 516 pages. 1998.

Vol. 1464: H.H.S. Ip, A.W.M. Smeulders (Eds.), Multimedia Information Analysis and Retrieval. Proceedings, 1998. VIII, 264 pages. 1998.